AF774050

Small Particles and Inorganic Clusters

Small Particles and Inorganic Clusters

Proceedings of the Fourth International Meeting on Small Particles and Inorganic Clusters
University Aix-Marseille III
Aix-en-Provence, France, 5–9 July 1988

Editors: C. Chapon, M.F. Gillet, and C.R. Henry

With 441 Figures

Springer-Verlag Berlin Heidelberg NewYork
London Paris Tokyo Hong Kong

Dr. Claude Chapon
CRMC2-CNRS, Campus de Luminy, Case 913
F-13288 Marseille Cedex 9, France

Professor Dr. Marcel F. Gillet
Laboratoire de Microscopie et Diffraction Electroniques,
Faculté des Sciences et Techniques, rue Henri Poincaré
F-13387 Marseille Cedex 13, France

Dr. Claude R. Henry
CRMC2-CNRS, Campus de Luminy, Case 913
F-13288 Marseille Cedex 9, France

This book originally appeared as the journal
Zeitschrift für Physik D – Atoms, Molecules and Clusters, Volume 12, Number 1–4
(ISSN 0178-7683) © Springer-Verlag Berlin, Heidelberg 1989

International Advisory Board

J.M. Basset, Lyon
K.H. Bennemann, Berlin
R.S. Berry, Chicago
J.M. Cowley, Tempe
D.E. Ellis, Evanston
O.F. Hagena, Karlsruhe
M. Ito, Sendai
J. Jortner, Tel Aviv
K. Kern, Marseille
W.D. Knight, Berkeley
J. Koutecky, Berlin
G. Mühlschlegel, Köln
M. Ozin, Toronto
E. Recknagel, Konstanz
R.E. Smalley, Houston
A.M. Stoneham, Oxford
S. Sugano, Tokyo

Scientific Organizing Committee

Chairman: J. Friedel, Orsay
Co-chairman: M.F. Gillet, Marseille

F. Bassani, Pise
C. Bréchignac, Orsay
A. Chatelain, Lausanne
P. Gallezot, Villeurbanne
H. Haberland, Freiburg
J.P. Malrieu, Toulouse
A. Masson, Paris
A. Renou, Marseille
A. Tramer, Orsay

ISBN-13: 978-3-642-74915-5 e-ISBN-13: 978-3-642-74913-1
DOI: 10.1007/ 978-3-642-74913-1

This work is subject to copyright. All rights are reserved, whether the whole or part of the material is concerned, specifically the rights of translation, reprinting, reuse of illustrations, recitation, broadcasting, reproduction on microfilms or in other ways, and storage in data banks. Duplication of this publication or parts thereof is only permitted under the provisions of the German Copyright Law of September 9, 1965, in its version of June 24, 1985, and a copyright fee must always be paid. Violations fall under the prosecution act of the German Copyright Law.

© Springer-Verlag Berlin Heidelberg 1989
Softcover reprint of the hardcover 1st edition 1989

The use of registered names, trademarks, etc. in this publication does not imply, even in the absence of a specific statement, that such names are exempt from the relevant protective laws and regulations and therefore free for general use.

Typesetting: Universitätsdruckerei H. Stürtz AG, D-8700 Würzburg
2156/3150-543210 – Printed on acid-free paper

Contents

Dynamics

Stability and Fragmentation

Electronic Structure and Properties

Optical Properties

Adsorption and Chemical Reactions

Preface

After the Symposia held in London in 1976, in Lausanne in 1980 and in Berlin in 1984, the Fourth International Symposium on Small Particles and Inorganic Clusters took place in Aix-en-Provence during July 5–9, 1988. It attracted about 500 participants. The Symposium programme covered all physical and chemical properties of free and supported clusters or small particles – static, dynamical, electronic, magnetic and optical – adsorption, and chemical phenomena.

The contributions were presented in three different forms: invited lectures, consisting of general reviews; progress reports, suggested or chosen from among the submitted papers; and posters. Eight invited lectures were scheduled; 24 progress reports, and all the relevant poster contributions, were accepted.

Like the preceding meetings, this Symposium demonstrated the great vitality and interdisciplinary character of cluster research as well as the usefulness of synthesizing its many parts. It contributed to the development of a common language and a better mutual understanding for all the researchers involved in the various aspects of this endeavour.

The papers presented at the Symposium appear in these proceedings. The editors wish to express their gratitude to the many referees who reviewed, and when necessary revised, all the contributions during the conference.

March 1989 C. Chapon, M.F. Gillet, C.R. Henry

Z. Phys. D – Atoms, Molecules and Clusters 12, 3–6 (1989)

Atoms, Molecules
Zeitschrift für Physik D and Clusters
© Springer-Verlag 1989

The smoke ion source: a device for the generation of cluster ions via inert gas condensation

K.M. McHugh[1], H.W. Sarkas[1], J.G. Eaton[1], C.R. Westgate[2], and K.H. Bowen[1]

[1] Department of Chemistry, Johns Hopkins University, Baltimore, MD, 21218 USA
[2] Department of Electrical Engineering, Johns Hopkins University, Baltimore, MD 21218, USA

Received 20 July 1988

We report the development of an ion source for generating intense, continuous beams of both positive and negative cluster ions. This device is the result of the marriage of the inert gas condensation method with techniques for injecting electrons directly into expanding jets. In the preliminary studies described here, we have observed cluster ion size distributions ranging from $n=1$–400 for Pb_n^+ and Pb_n^-, and from $n=12$–5700 for Li_n^-.

PACS: 36.40, 34.80G

Introduction

In both science and high technology there is a need for high intensity sources of large cluster ions comprised of relatively high temperature materials. In recent years, both spectroscopic and kinetic studies of cluster ions have begun to explore the world of aggregated phenomena lying in the size regime between single atoms and molecules and the solid state [1, 2]. At the same time, interest has continued to grow in the potential applications of cluster ions to thin film formation, size-specific catalyst preparation, ion beam sputtering, and ion lithography [3, 4].

Here, we report the development of a source for generating intense, continuous beams of large cluster ions made up of lead or lithium atoms. This device is the result of combining the inert gas condensation method with techniques for injecting electrons directly into expanding jets. Inert gas condensation is a proven approach for generating strong beams of large neutral clusters comprised of relatively high temperature materials [5–10]. In inert gas condensation cells an oven evaporates the material of interest into a bath of cool inert gas. In this environment the evaporated material condenses to form a dilute smoke composed of ultra-small particles and clusters. An orifice allows the inert gas, along with its entrained smoke, to exit the cell into a high vacuum region where a beam is formed. The injection of low energy electrons directly into the high density portion of supersonic expansions has been shown over the past few years to be a highly efficient method for generating both positive and negative cluster ions [11–13]. The electrons can be provided either by an electron gun or by a biased hot filament. In the past, usually in the course of mass spectrometric characterization of their inert gas condensation sources, several investigators have generated positive cluster ions from inert gas condensation cells by subjecting the neutral cluster beams to electron bombardment or laser ionization well downstream of their cells' exit apertures [7–10, 14, 15]. In the present work, electrons from a biased filament are injected, in a close-coupled manner, directly into the weak jet expansion of smoke-containing inert gas as it leaves the condensation cell to generate intense beams of large positive and negative cluster ions composed of lead or lithium. We refer to the unique union of these two techniques as the Smoke Ion Source. We anticipate that large cluster ions of much higher temperature materials can also be generated with this source. Below, we describe the smoke ion source and its associated apparatus along with our preliminary experiments with the test materials, lead and lithium.

Experimental

A schematic diagram of the Smokc Ion Source is presented in Fig. 1. The material of interest is evaporated from a heat shielded oven (O) by direct resistive heating. The oven assembly is separated from the inert gas condensation cell (CC) by a water cooled copper box (CB) which serves to thermally isolate the cool inert gas environment from the high temperature region. Vapor effusing from the oven then enters the cool inert gas environment. The condensation cell typically contains from 0.5 to 10 torr of helium which can be maintained at constant temperatures between 77 K and 285 K by a coolant reservoir (CR). The cool inert gas thermally quenches the vapor causing supersaturation with subsequent condensation and cluster growth. The condensation cell is coupled to high vacuum by a small (1.0–1.5 mm diameter) aperture (A). The flow of the helium entrains the clusters and transports them into the high vacuum region via a weak jet expansion. The smoke flux is rather high and reasonably directional as evidenced by the rapid formation of opaque deposits on glass targets placed about 1 cm in front of the aperture.

Electrons are injected into the smoke-containing helium flow by a negatively biased hot filament (I) immediately as it leaves the aperture. Axial magnetic fields in this region were found to greatly enhance cluster ion production. The entire source is biased at either ±500 V or ±1 kV with respect to ground potential. The generation of both positive and negative cluster ions utilizes the same electron injection configuration. When switching from negative to positive ions, necessary voltage changes involve reversing appropriate electrode polarities and using a higher filament bias voltage.

The resulting beam of cluster ions and accompanying neutrals is skimmed before entering the rest of the apparatus. Briefly, this apparatus consists of an ion optical beam line, an **E** × **B** mass separator (Wien filter), and a Faraday cup for ion detection. The Wien filter can be operated at a high electrostatic field where it achieves normal mass resolution over a limited mass range, or at a low electrostatic field where it exhibits poor resolution but over a much larger mass range. The latter condition is especially useful for detecting very large cluster ions. At Wien filter electrostatic fields of a few volts and with a beam voltage of 1 kV, our available mass range extends up to 80,000 amu. Also, given the high ion currents observed, we can use a Faraday cup for ion detection. Since Faraday cups measure only impinging charges, they are able to detect both high and low mass ions with equal effici-

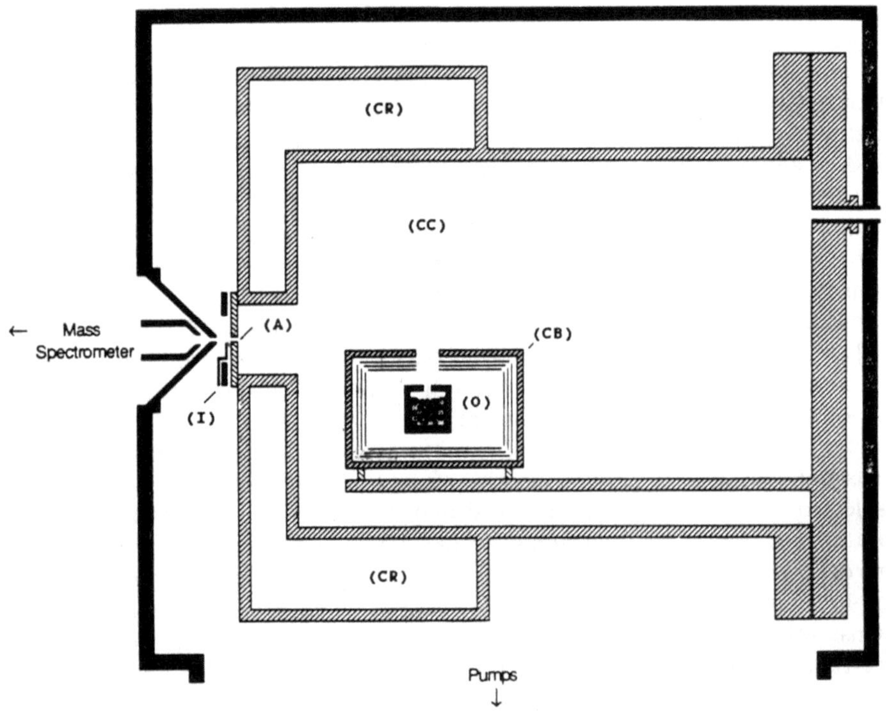

Fig. 1. Schematic diagram of the Smoke Ion Source

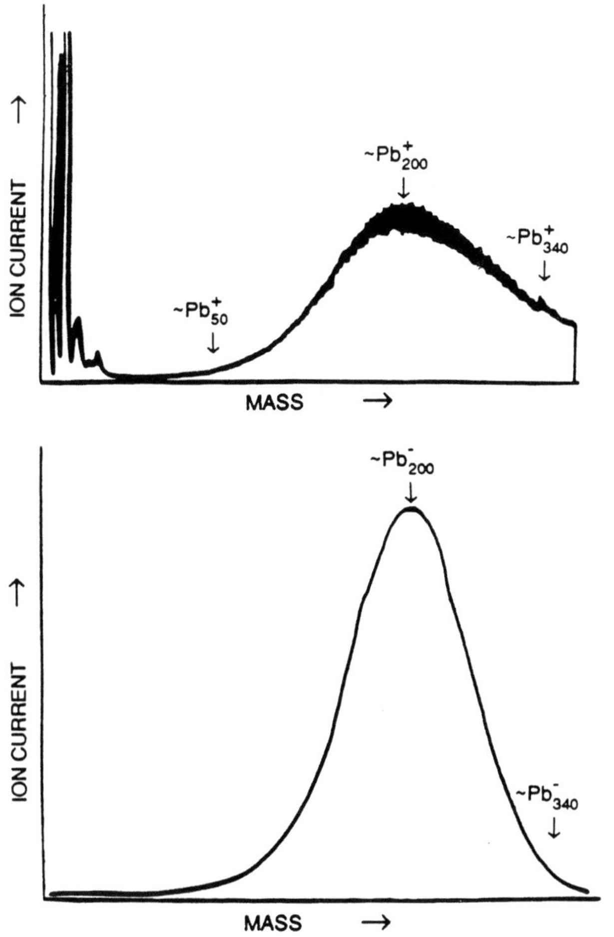

Fig. 2. Mass spectra of large positive and negative cluster ions of lead generated using the Smoke Ion Source

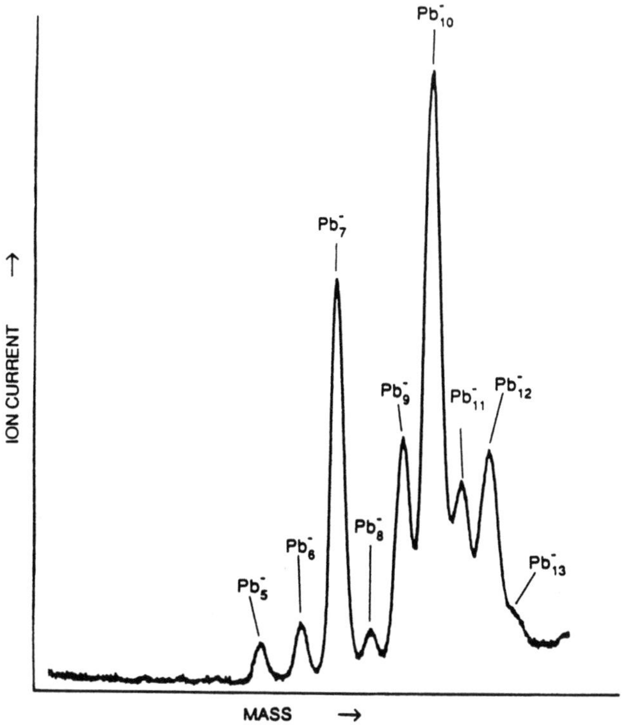

Fig. 3. Mass spectrum of small lead cluster anions generated using the Smoke Ion Source

ency, bypassing the difficulties associated with detecting large cluster ions with particle multipliers.

Source performance: generation of cluster ions

Here, we describe the results of our preliminary experiments with the test materials, lead and lithium. Lead cluster ions were generated under two different sets of source conditions. The first set utilized a source aperture diameter of 1.0 mm, a helium pressure of 6.0 torr maintained at 195 K, and an oven temperature of 1460 K. Figure 2 presents mass spectra for both positive and negative lead cluster ions recorded under these conditions. In order to obtain these spectra, the Wien filter was operated in its high mass range mode. Both spectra exhibit a progression of unresolved cluster ion peaks ranging from approximately 40 to 400 atoms per cluster ion. For both polarities, the maximum in the size distribution corresponds to about 200 atoms per cluster ion. In the anion spectrum, an ion current of 600 pA was observed at this maximum. If average currents of cluster anions are compared, this is about five orders of magnitude more intense than those available via laser vaporization techniques [16]. In the cation spectrum, we observe a series of low mass peaks due to Pb^{++} and $Pb^+_{n=1-3}$ in addition to its high mass distribution. Interestingly, these low mass peaks are absent in the anion spectrum. This difference suggests that the small lead cluster cations may result from fragmentation. The rough similarity between the high mass distributions of these spectra, on the other hand, may indicate that they reflect the neutral cluster distribution.

In an effort to explore the source's ability to generate smaller cluster ions, a second set of source conditions was selected. This set utilized a source aperture diameter of 1.5 mm, a helium pressure of 1.6 torr at 273 K, and an oven temperature of 1460 K. Figure 3 shows the resultant lead cluster anion mass spectrum with the Wien filter operating in its low mass range mode. A variety of small lead cluster anions were observed demonstrating the feasibility of shifting the cluster ion size distribution by manipulating source conditions. We note that the low mass size distribution of our lead cluster anion spectra is reminiscent of Sattler's lead cluster cation distribution [8].

Experiments with lithium were performed using a 1.5 mm diameter source aperture, 1.0 torr of He at 273 K, and an oven temperature of 1150 K. This set of source conditions produced a lithium cluster anion distribution that ranged from 12 to 5700 atoms per cluster anion. In addition, we have observed total anion currents as high as 45 nA for this system.

We are grateful to A. Yokozeki, K. Sattler, T.P. Martin, W. Schulze, J.L. Gole, E. Recknagel, S.J. Riley, and the late G.D. Stein for helpful discussions. The development of this source was supported by the Semiconductor Research Corporation under contract #83-01-034 and by the National Science Foundation under grant #CHE 8511320.

References

1. Pettiette, C.L., Yang, S.H., Craycraft, M.J., Conceicao, J., Laaksonen, R.T., Cheshnovsky, O., Smalley, R.E.: J. Chem. Phys. **88**, 5377 (1988)
2. Jarrold, M.F., Bower, J.E.: Physics and Chemistry of Small Clusters, p. 801–806. New York, Plenum Press 1986
3. Yamada, I., Takagi, T.: Thin Solid Films **80**, 105 (1981)
4. Johar, S.S., Thompson, D.A.: Surf. Sci. **90**, 319 (1979)
5. Yokozeki, A., Stein, G.D.: J. Appl. Phys. **49**, 2224 (1978)
6. Mann, D.M., Broida, H.P.: J. Appl. Phys. **44**, 4950 (1973)
7. Martin, T.P.: J. Chem. Phys. **81**, 4426 (1984)
8. Mühlbach, J., Pfau, P., Sattler, K., Recknagel, E.: Z. Phys. B – Condensed Matter **47**, 233 (1982)
9. Frank, F., Schulze, W., Tesche, B., Urban, J., Winter, B.: Surf. Sci. **156**, 90 (1985)
10. Riley, S.J., Parks, E.K., Mao, C.R., Pobo, L.G., Wexler, S.: J. Phys. Chem. **86**, 3911 (1982)
11. Haberland, H., Schindler, H.G., Worsnop, D.R.: Ber. Bunsenges. Phys. Chem. **88**, 270 (1984)
12. Coe, J.V., Snodgrass, J.T., Freidhoff, C.B., McHugh, K.M., Bowen, K.H.: J. Chem. Phys. **84**, 618 (1986)
13. Alexander, M.L., Johnson, M.A., Levinger, N.E., Lineberger, W.C.: Phys. Rev. Lett. **57**, 976 (1986)
14. Kuiper, A.E.T., Thomas, G.E., Schouten, W.J.: J. Cryst. Growth **51**, 17 (1981)
15. Farley, R.W., Ziemann, P.J., Keesee, R.G., Funasaka, H., Castleman, A.W.: Z. Phys. D – Atoms, Molecules and Clusters **12** (1989) (this issue)
16. Cheshnovsky, O., Pettiette, C.L., Smalley, R.E.: Ion and cluster ion spectroscopy and structure. Amsterdam: Elsevier Science 1989

Z. Phys. D – Atoms, Molecules and Clusters 12, 7–9 (1989)

Atoms, Molecules
Zeitschrift für Physik D and Clusters
© Springer-Verlag 1989

Electron diffraction studies of pulsed cluster beams

L.S. Bartell and R.J. French
Department of Chemistry, University of Michigan, Ann Arbor, MI 48109, USA

Received 5 July 1988

A pulsed nozzle source has been installed in the electron diffraction unit at the University of Michigan. The pulsed mode of operation is found to offer several important advantages in the investigation of clusters generated in nozzle flow. These advantages, including the feasibility of operating without a skimmer, are discussed. Design features and characteristic results are briefly outlined.

PACS: 36.40; 47.55.E; 61.14.F

1. Introduction

Modifications of the electron diffraction unit at the University of Michigan have greatly facilitated research on molecular clusters generated by homogeneous nucleation in nozzle flow. Included are a pulsed nozzle and auxiliary equipment to characterize the supersonic jets produced. Advantages of pulsed operation are several. Briefly stated, sample pressures can be increased by an order of magnitude without alteration of the pumping system, and substantially lower temperatures can be attained. It is now possible to produce cluster beams of such volatile substances as Ne (at 10 °K) that previously resisted condensation in the Michigan apparatus. The increased range of expansion conditions markedly augments the control that can be exercised over the internal structure of clusters formed in nozzle flow. Moreover, the improved quality of the supersonic jets makes it possible under some conditions to obtain serviceable electron diffraction patterns of clusters without skimming the jets and thereby interfering with their flow characteristics.

Although spectroscopists have long taken advantage of the attractive features of pulsed jets, diffractionists have been slow to adopt such a system because of perceived problems associated with its use. It turns out that the problems need not be severe. A description of the pulsed system is given in the next section together with a discussion of measures taken to overcome the interferences encountered when a pulser operates in the vicinity of an electron beam.

2. Description of apparatus

The pulsed nozzle system consists of a commercial electromagnetic solenoid valve, designed for spectroscopic research, controlling the flow through a miniature Laval nozzle. A high-speed shutter actuating the 40 kev electron beam can be timed to examine any desired interval of the gas pulses. A beam chopper and flight tube monitored by a fast ionization gauge can be introduced into the system to display the distribution of molecular and cluster velocities in a jet. In other aspects the Michigan electron diffraction unit is unchanged from its configuration described elsewhere [1]. Under some operating conditions, however, diffraction patterns are recorded with the skimmer removed. Such a mode of operation will be discussed later.

Of greatest concern in initial planning was the possibility that electromagnetic pulses associated with the triggering of the pulsed valve might severely disturb the electron beam and spoil diffraction patterns. This potential source of trouble has been overcome, partly by ensheathing the solenoid with three layers of magnetic shielding to reduce the deflection of the electron beam to a few hundredths of an *s*-unit. In addition, the unwanted magnetic pulse decays to a small fraction of its maximum value in the nearly 1 ms time it takes the jet to reach the electron beam after the initial surge of current through the solenoid. During the brief, reproducibly timed period that the electron beam is switched on to probe the jet, any deflection is so small and static that no degradation of the diffraction pattern has been detectable.

Contained in a unit less than 2.5 cm in diameter and 5 cm in length, the pulser valve and electromagnetic solenoid driving it are so compact that they could be introduced into the apparatus with only minor alterations. Materials to which gas samples are exposed besides the nozzle are teflon, viton, and stainless steel. The valve is rated for operating pressures up to 86 bars and is designed to produce pulse widths from 250 μs FWHM to continuous with a 160 μs opening time, 350 μs closing time, and a repetition rate exceeding 250 Hz for short pulses. A 300 V pulse, 150 μs FWHM opens the valve followed by a sustained application of about 10 V to hold the valve open.

In the commonest configuration of the pulsed source a miniature glass Laval nozzle, throat diameter ≈0.012 cm, is joined to the valve, being cushioned at both ends by teflon gaskets. The primary pressure seal is made by an O-ring around the nozzle near its throat. Between the nozzle and the valve is a dead space of about 10^{-3} cm^3 whose half-life for emptying after the valve is closed is less than 1 ms. At the repetition rate normally used in experiments the dead space is exhausted long before a new pulse begins.

Diffracted electrons are filtered, as usual, by a rotating sector (r^1 shape for small-angle studies and r^2 for intermediate angles) before being recorded on photographic plates. During individual pulses the motion of the rotating sector is nearly frozen by the brief 0.2 to 1 ms electron shutter time and the individual contributions to the accumulated diffraction patterns are far from symmetric, axially. Integrated patterns, typically consisting of 10^3 exposures registered on a plate, are satisfactorily averaged and not significantly degraded because the reciprocity law is well obeyed for exposures by fast electrons.

Pulsed operation makes it possible to monitor the spatial distribution of the effluent supersonic jet (skimmer removed) directly in the diffraction chamber with a fast ion gauge without interference by the substantial background gas that accumulates during operation in the continuous flow mode. Characteristic times for a gas pulse to pass a probing electron beam or ion gauge, rebound from the wall, and return to the probe are ≈0.6 ms. Measurements of pulse densities are easily made before this interference can occur. Therefore, neither cryopumping nor fast diffusion pumps are required to keep effects of background gas minimal.

3. Characteristic results

In the absence of a skimmer, the spatial distribution of gas flowing from our standard Laval nozzle can be represented by a second-order Witch of Agnesi [2]. Profiles of cluster beams are nearly Gaussian and much narrower. Of special interest is the great fractionation between clusters and carrier gas implied by these profiles. Narrow cluster beams may be enriched, on axis, relative to initial concentrations in the carrier by factors of several hundred. It is this characteristic of nozzle flow that enables useful diffraction patterns to be taken under some conditions even when no skimmer is present.

As the initial pressure and/or molecular weight of monoatomic carrier gas are increased, the gas jet narrows as might be expected from considerations of inertial versus viscous effects. As the gas jet narrows, however, the cluster beam broadens by virtue of the increased scattering of clusters by collisions with gas molecules. The natural fractionation in micronozzles, then, is a strong function of pressure and carrier composition, and it depends on nozzle diameter and shape, as well.

Although the gas jets are broad compared with the cluster beams, they are narrow in comparison with those formed in free jet expansion. Time of flight measurements show that such concentrated jets as were commonly studied at Michigan in the past suffer a severe skimmer interaction. Clusters generated in flow through Laval nozzles tend to be so large, however, that they are able to plow through the nearly stagnant gas in the vicinity of the skimmer throat with little loss of velocity and, usually, only minor consequences. It is now possible to reduce such interactions drastically.

In prior research [3] we discovered that we can often alter the internal structures of clusters produced by varying flow conditions. In some cases clusters can be generated either in liquid or in solid form. In others, clusters can be induced to grow in two or even three different solid phases in pure form or mixed. Underlying principles are poorly understood. Our present experiments also show that high carrier pressures can produce clusters that are appreciably colder than clusters condensed from neat gas and cooled solely by evaporation.

Overall, the new pulsed system shows great promise in the study of clusters by electron diffraction. A much wider range of expansion conditions can be explored, and experiments are freer of deleterious interactions that complicate analyses of results. Results are expected to shed light on nucleation and mechanisms of nonequilibrium growth.

References

1. Bartell, L.S., Heenan, R.K., Nagashima, M.: J. Chem. Phys. **78**, 234 (1983).
2. Bartell, L.S., Kacner, M.A., Goates, S.R.: J. Chem. Phys. **75**, 2730 (1981)
3. Bartell, L.S., Harsanyi, L., Valente, E.J.: NATO ASI Ser., B. **158**, 31 (1987)

Z. Phys. D – Atoms, Molecules and Clusters 12, 11–13 (1989)

Atoms, Molecules
Zeitschrift für Physik D and Clusters
© Springer-Verlag 1989

Equilibrium intermetallic particles in the gas phase – A Knudsen cell mass spectrometric study

A. Hartmann* and K.G. Weil

Institut für Physikalische Chemie der Technischen Hochschule Darmstadt, Petersenstrasse 20, D-6100 Darmstadt, Federal Republic of Germany

Received 5 July 1988

Using the Knudsen cell effusion technique it is possible to generate a molecular beam containing intermetallic molecules. Mass spectrometric studies were carried out with the systems: cesium/antimony, cesium/antimony/bismuth, cesium/antimony/indium, cesium/antimony/tin and cesium/antimony/bismuth/tin. In all these systems intermetallic particles with extraordinary stoichiometries could be detected. A brief discussion of the appearance energies and the bonding principle of these molecules or clusters are given.

PACS: 07.75; 36.40; 34.80.G

Introduction

In the phase diagrams of Na/Sb, K/Sb and Cs/Sb, the existence of congruently and of incongruently melting compounds is documented. Examples for compounds with especially high stability are for instance CsSb and Cs_3Sb [1]. Knudsen cell mass spectrometric studies of these systems indicate that in the gas phase intermetallic molecules of different stoichiometries are present [2, 3]. For example over Cs/Sb mixtures the existence of the species Cs_2Sb, $CsSb_2$, Cs_2Sb_2 and Cs_2Sb_4 as parent molecules was reported [2]. In the vapor over liquid NaSb and KSb the large particles Na_6Sb_4 and K_4Sb_4 were found, in addition to other compounds [3].

Table 1 shows the appearance energies of Na_6Sb_4, Na_2Sb_4, K_4Sb_4, $K_2Sb_4^{\cdot}$, K_2Sb_2, Cs_2Sb_4, Cs_2Sb_2 together with the ionization energies of the pure alkali metals.

Na_6Sb_4 and K_4Sb_4 show unusual low ionization energies. For Na_6Sb_4 the value is close or even equal to that of the sodium atom. In contrast, the appearance energies of Na_2Sb_4, K_2Sb_4, K_2Sb_2, Cs_2Sb_4 and Cs_2Sb_2 are relatively high. We assume that this different behavior indicates different bond types, being more metallic in the large particles and predominantly ionic in the smaller molecules [4]. In the present paper we will demonstrate that large cluster-like molecules with low ionization energies can also be detected in other systems. Furthermore we will show that the Zintl-Klemm-Busmann principle [5], which up to now was only applied to condensed phases, also seems to be applicable in some intermetallic molecules.

Experimental

We used a single focussing magnetic sector field mass spectrometer MM 30 K from VG Instruments. Ion intensities were measured with a photomultiplier and monitored with a fast counter. In order to avoid mass discrimination effects a Daly converter was used. The magnetic flux was determined with a Hall probe. The Knudsen cell temperature could either be kept at a constant value or could be changed with a constant rate. Mass spectra were automatically monitored and recorded with the help of A/D and D/A converters and a personal computer (Olivetti M24). All the necessary software was developed in our laboratory [6].

* Part of the Dissertation of A. Hartmann, Darmstadt (D17)

Table 1. Appearance energies of selected intermetallics and pure alkali metals (Lit. values)

Particle	Appearance energy or ionization energy/eV
Na_2Sb_4	7.3 [4]
Na_6Sb_4	5.3 [4]
Na	5.14 [11]
K_2Sb_2	6.7 [4]
K_2Sb_4	5.3 [4]
K_4Sb_4	5.2 [4]
K	4.34 [11]
Cs_2Sb_2	6.0 [2]
Cs_2Sb_4	4.4 [2]
Cs	3.89 [11]

The Knudsen cells were machined from stainless steel with an orifice diameter of 0.8 mm. Cesium, antimony, bismuth, indium and tin all were of puriss. quality. Desired fractions were mixed in a glove box under argon and stored in quartz ampoules. The mixtures were annealed for 24 h at temperatures above the melting point of the highest melting compound in the respective system. The Knudsen cells were filled in the glove box and then inserted into the mass spectrometer.

Results

In all studied intermetallic mixtures the predominant species in the gas phase are atoms and homonuclear particles like Cs, Sb_2, Sb_4, Bi, Bi_2 and Bi_4. If bismuth and antimony are simultaneously present in the cell then Sb_3Bi, Sb_2Bi_2, $SbBi_3$ [7] dominate in addition to the monomers, dimers, and tetramers of Sb and Bi.

In this section we will describe some new and unusual intermetallic particles, which are found in addition to the above mentioned molecules. In all cases they are identified by their isotopic pattern and in some cases characterized by their appearance energies.

The system cesium/antimony

This system was studied previously [2]. A reexamination with increased sensitivity and resolution of the computerized system confirmed the older results. In addition, the new particle Cs_6Sb_4 could be detected and unanimously identified by its isotope pattern, see Fig. 1. The appearance potential of this molecule was determined by linear extrapolation of the ionization efficiency curve. It is 3.9 ± 0.3 eV. As in the case of

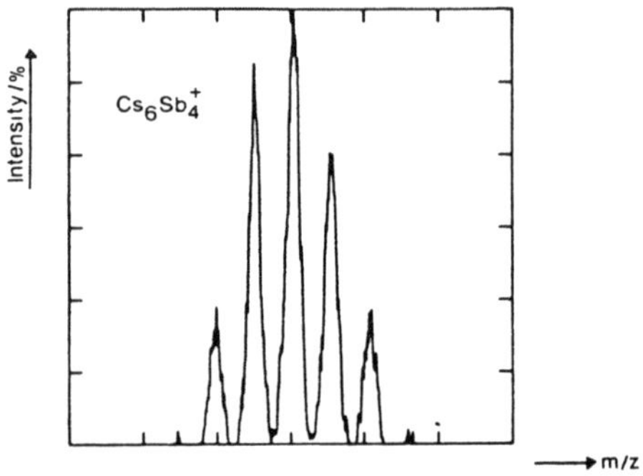

Fig. 1. High resolution mass spectrum, Cs_6Sb_4

Na_6Sb_4 and K_4Sb_4 this value is within the error limits equal or even smaller than that of the respective alkali atom (E_A(Cs) = 3.89 eV).

The system cesium/antimony/bismuth

In this system, we find a small molecule $CsBi_2$. More interesting within the scope of this paper are the large species $Cs_2(BiSb_3)$, $Cs_2(Bi_2Sb_2)$, $Cs_2(Bi_3Sb)$ and Cs_2Bi_4. This series demonstrates that antimony atoms in the Sb_4 unit can successively be replaced by bismuth. The same phenomenon can be found for the sixfold coordinated Sb_4 unit. The newly detected species are: $Cs_6(BiSb_3)$ and $Cs_6(Bi_2Sb_2)$. Because of the more metallic behavior of bismuth compared to antimony the tendency to form Bi_4 units is much smaller. It is therefore plausible that particles with the compositions $Cs_6(Bi_3Sb)$ and Cs_6Bi_4 could not be detected.

The system cesium/antimony/indium

In the binary systems antimony/indium and bismuth/indium two intermetallic particles of different stoichiometry are known:

DeMaria et al. [8] report on InSb and $InSb_2$, while Rieckert et al. [9] described InBi and $InBi_2$. In the presence of cesium, we found a third stoichiometry, namely $Cs_2(InSb_3)$ [10].

The systems cesium/antimony/tin and cesium/antimony/bismuth/tin

In the gas phase above these systems we found $Cs(SnSb_3)$ and $Cs_2(Sn_2Sb_2)$ as stable compounds. As an example we show in Fig. 2 the mass signal, attributed to $Cs(SnSb_3)$.

When we add bismuth to the ternary mixture then in addition to $Cs(SnSb_3)$ we detect also $Cs(SnBiSb_2)$ and $Cs(SnBi_2Sb)$. Table 2 shows the appearance energies of these new gaseous compounds [10].

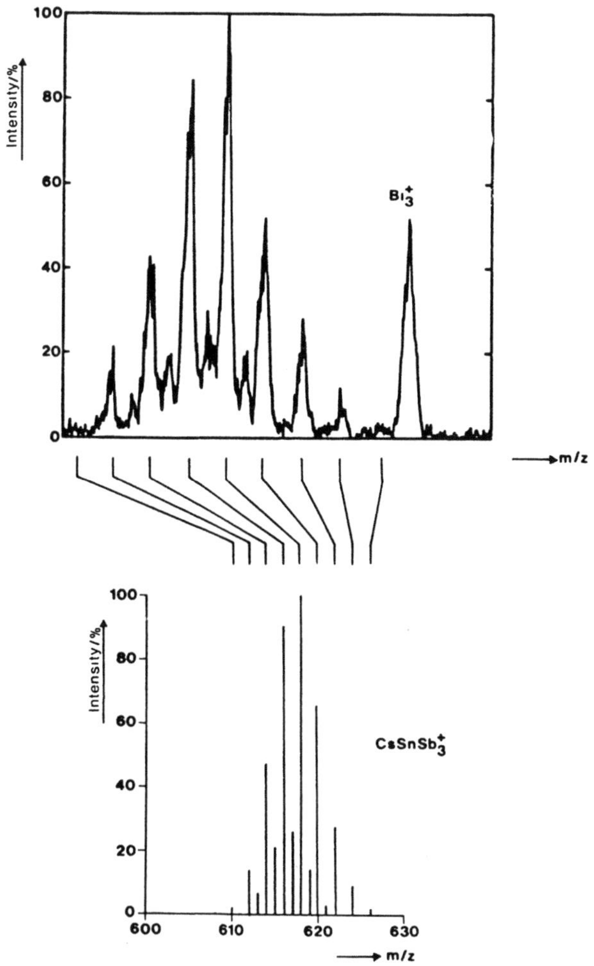

Fig. 2. Measured and calculated mass spectra, $CsSnSb_3$

Table 2. Appearance energies of intermetallic gaseous Zintl molecules

Particle	Appearance energy/eV
Cs_2InSb_3	(*a*)
$CsSnSb_3$	5.6 ± 0.2
$CsSnBiSb_2$	5.3 ± 0.2
$CsSnBi_2Sb$	5.0 ± 0.3
$Cs_2Sn_2Sb_2$	(*a*)

[a] Not yet determined

Discussion

From the relatively high appearance potentials of the molecules Cs_2Sb_2 and $Cs_2(Sb_xBi_{4-x})$ $(0 \leqslant x \leqslant 4)$ we conclude, that these molecules are highly ionic. The electronic structure of double negative Sb_2 dumbbells is equal to that of the very stable oxygen molecule. We therefore propose the structure $Cs^+Sb_2^{2-}Cs^+$. When charge is transferred to the Sb_4-tetrahedron or to one of the distorted $Sb_{4-x}Bi_x$-tetrahedra, Jahn–Teller distortion will occur, leading to a flat structure for the doubly charged species. As a limiting case, one can discuss the structure $Cs^+(Sb_{4-x}Bi_x)^{2-}Cs^+$.

A completely different bond type seems to be realized in the species containing group III or group IV elements. In these cases, $(InSb_3)^{2-}$, $(SnSb_3)^-$, and $(Sn_2Sb_2)^{2-}$ are isoelectronic with Sb_4, which is known to exist as a stable tetrahedron in the gas phase. That means that one or two cesium atoms must be attached to the tetrahedral centres in order to supply the necessary charge. This bonding principle, the Zintl–Klemm–Busmann concept [5] is well known in condensed phase chemistry. To our knowledge, the above mentioned cases are the first ones, in which this principle can be applied to gas phase molecules.

A third bond mechanism seems to be realized in the clusters with six alkali atoms. The clusters Cs_6Sb_4 and Na_6Sb_4 both have extremely low ionization energies. Furthermore, we could show in a previous paper [3], that the bond energy per alkali atom is larger in these clusters than in the smaller molecules. These findings support the conclusion, that in these particles the interaction between alkali atoms contributes to the stabilization of the clusters. Close packing considerations point to a structure, in which six alkali atoms are arranged octahedrically around the tetrahedral Sb_4 nucleus. One may speculate that clusters of the type Me_6Sb_4 and the similar ones containing six cesium atoms together with antimony and bismuth constitute examples which represent extremely small metallic clusters.

The Deutsche Forschungsgemeinschaft generously sponsored this research, especially within the framework of the Schwerpunkt: Physik anorganischer Cluster. Financial assistance from the Fonds der Chemischen Industrie is greatly acknowledged.

References

1. Elliot, R.P.: Constitution of binary alloys, p. 370. New York: McGraw-Hill 1958
2. Busse, B., Weil, K.G.: Ber. Bunsenges. Phys. Chem. **86**, 93 (1982)
3. Scheuring, T., Weil, K.G.: Ber. Bunsenges. Phys. Chem. **89**, 811 (1985)
4. Scheuring, T., Weil, K.G.: Surf. Sci. **156**, 457 (1985)
5. Schäfer, H., Eisenmann, B.: Rev. Inorg. Chem. **3**, 29 (1981)
6. A. Hartmann, Diplomarbeit, Darmstadt 1986
7. Kohl, F.J., Carlson, K.D.: J. Am. Chem. Soc. **90**, 4814 (1968)
8. DeMaria, G., Drowart, J., Inghram, M.G.: J. Chem. Phys. **31**, 1076 (1959)
9. Riekert, G., Harbach, G.R., Lamparter, P., Steeb, S.: Z. Metallkd. **72**, 406 (1981)
10. Hartmann, A., Weil, K.G.: Angew. Chem. Int. Ed. Engl. **27**, 1091 (1988)
11. Weast, R.C. (ed.): Handbook of chemistry and physics, 55th Edn. Cleveland, OH: CRC Press 1974

Z. Phys. D – Atoms, Molecules and Clusters 12, 15–18 (1989)

Atoms, Molecules
Zeitschrift für Physik D and Clusters
© Springer-Verlag 1989

Neutralization and matrix deposition experiments on mass-selected silver atoms and clusters

D.M. Lindsay*, F. Meyer, and W. Harbich

Institut de Physique Expérimentale, Ecole Polytechnique Fédérale de Lausanne, CH-1015 Lausanne, Switzerland

Received 5 July 1988

A new apparatus for depositing intense beams of neutral, mass selected metal clusters in rare gas matrices or on a solid surface has been constructed. Metal cluster cations, produced by Ar^+ sputtering, may be energy analyzed and then mass selected in a quadrupole mass filter. The clusters are neutralized by low energy electrons and then co-deposited with a rare gas on a 4 K sapphire window. Matrix samples were interrogated in situ by UV-optical absorption spectroscopy. Good quality neutral Ag spectra were obtained following a 30 min. Kr deposition of a ca. 2.5 nA cm^{-2} beam of Ag^+. Line strength measurements imply matrices containing about 10^{13} atoms cm^{-2}, corresponding to a neutralization efficiency of about 50%. Preliminary experiments on Ag_2^+ yielded only Ag spectra, but with a lesser yield, perhaps indicative of incomplete fragmentation.

PACS: 33.20.N; 36.40

1. Introduction

While the two extrema, the free atom and the bulk metal, are quite well understood, much less is known about the transition region in between. Particularly noticeable is the lack of detailed, structural (geometrical and electronic) information for metal clusters other than the dimers. While broad conclusions can be drawn from photoelectron [1] and photofragmentation [2] studies, well resolved gas phase spectra have only been obtained for Cu_3 [3], Na_3 [4] and Al_3 [5]. The majority of reported spectroscopic measurements have employed matrix isolation methods [6], most often in combination with optical absorption, Raman or ESR spectroscopy. With few exceptions [7], however, measurements on larger clusters are severely hampered by ambiguities in assigning the spectral carrier.

The assignment problem can be overcome by preparing matrices which contain known, single sized clusters. In principle, this may be accomplished by mass selecting ion clusters in the gas phase and neutralizing these either before or during deposition. This requires an intense source of cluster ions and (more challenging) that these clusters be efficiently "soft landed" (i.e. collected without fragmentation) in the matrix. Here, we report on the construction of such an apparatus, the first (to our knowledge) specifically designed for soft landing experiments in rare gas matrices. Initial experiments on monomers demonstrate that good quality optical absorption spectra may be obtained when mass selected Ag^+ is deposited with electrons in a Kr matrix.

2. Experimental

Figure 1 shows a schematic of the newly constructed cluster deposition source, which is in concept similar to the ion cluster apparatus developed by Fayet and Wöste [8,9]. Silver cluster ions were formed by sputtering from a water-cooled, rotatable metal target (diameter 25 mm). An intense, high energy (typically

* Permanent address: Department of Chemistry, City University of New York, The City College, New York, NY 10031, USA

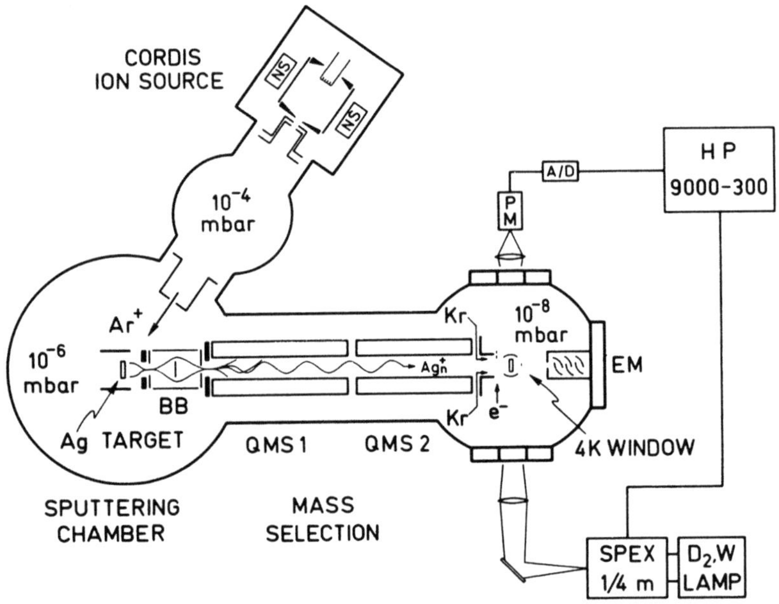

Fig. 1. Schematic of the cluster deposition source

2 mA at 20 keV) primary beam of Ar^+ ions was produced by a Cold Reflex Discharge Ion Source ("CORDIS") designed and manufactured by Rokion Ionenstrahl-Technologie (Darmstadt, W. Germany). The primary beam was collimated by a 10 mm aperture which also served to separate the source and sputtering chambers, each of which was differentially pumped with a 500 1/s turbomolecular pump. The 50° inclination between the primary and secondary beam lines corresponds closely to the optimum ejection angle for material sputtered from a metal surface [10].

Silver cluster cations were extracted and focused into a "Bessel Box" type (BB) energy filter [11] which serves to reduce the approximately 10 eV energy spread of the sputtered secondary ions [8]. The Bessel Box also acts as a beam stop for both high energy (primary and secondary) ions and for the intense flux of neutrals which would otherwise be the dominant species in the secondary beam line. The (unblocked) neutral atom intensity is estimated (see below) to be 10^{11}–$10^{12}\,s^{-1}\,cm^{-2}$ at the matrix target. The cluster ion of interest was mass selected by a commercial (Extrel 7–162–8) quadrupole mass filter (QMS1, length 20 cm) after which it enters a second, specially built quadrupole (QMS2, length 17.5 cm) which is normally operated without mass discrimination i.e. as an RF guide. Both quadrupoles have the same rod diameter (about 9.5 mm) and were driven by a common (1.2 MHz) RF frequency source. The relative phases of the RF on the two rod systems were adjusted to give maximum transmission for a particular ion. Aside from permitting a certain flexibility in overall length, the tandem arrangement is advantageous in that the RF guide acts as a "delayed DC ramp" which reduces the effects of fringing fields [12, 13]. Accordingly, the ions exiting QMS2 are relatively more focused than would be the case for a single quadrupole arrangement.

As shown in more detail in Fig. 2, mass selected silver ions were mixed with a matrix gas (here krypton, injected co-axially from a ring of (16) 0.5 mm holes) and directed (with the aid of the cylindrical lens, L2) onto a liquid helium cooled sapphire window (Insaco,

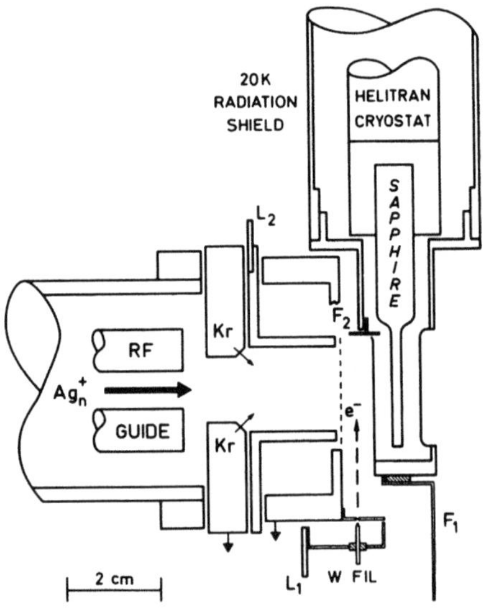

Fig. 2. Detail of the deposition arrangement

Quakertown PA). Cluster cations could be neutralized with electrons produced by a heated tungsten filament and introduced about 1 cm from the matrix surface. Electron currents (measured using a Faraday plate, F2 in Fig. 2) and energies were controlled by a repeller voltage L1 and/or by biasing the filament or the Faraday plate. The neutralization system is somewhat analogous to that employed by Rivoal et al. [14]. The sapphire window (an 8 mm rod with two polished 25 mm × 8 mm surfaces, designed to fit into an EPR cavity) was soldered (using Woods metal) into an OFHC copper support and screwed to an extended length Heli-tran cryostat (APD Cryogenics, LT-3B-110A). Except for two 10 mm × 25 mm apertures, the deposition window was enclosed by a grounded radiation shield screwed to the (nominally 20K) collar of the Heli-tran unit. The deposition chamber was separately pumped by a 500 1/s turbomolecular pump. The complete cryostat could be either rotated (as when taking optical spectra) or raised (as much as 40 cm) by means of a screw-driven metal bellows (welded diaphragm fabricated by Reinhold SA, Geneva Switzerland). This latter option allows the cryostat plus matrix sample to be withdrawn into a sealed chamber for transport to fixed spectroscopic instruments, as for example an ESR spectrometer. With the cryostat raised, cluster currents could be measured using either an electron multiplier (EM in Fig. 1), or more directly on a Faraday plate (F1 in Fig. 2).

For the experiments discussed here, matrix samples were interrograted in-situ using UV-optical absorption spectroscopy. The single beam spectrometer consists of a Spex minimate-2 monochromator (focal length 227 mm, dispersion 3.7 nm/mm) plus a broad band radiation source (Spex 1682BF) incorporating both a deuterium and tungsten-halogen lamp. As illustrated in Fig. 1, the dispersed light was focused (through a sapphire window) onto the matrix sample and then detected by a Hamamatsu R928 Photomultiplier tube (PM) operated in a pulse counting mode. The signal from the PM was fed via a Spex DM302 amplifier discriminator (A/D) to a specially built 20 MHz counter interfaced to a Hewlett-Packard computer which was also interfaced to the monochromator drive unit. In order to avoid saturating the counting electronics, the slit size of the monochromator was normally kept well below 100 μ.

3. Results and discussion

Figure 3 shows the characteristic [15] triplet absorption band of atomic silver in a krypton matrix. The figure represents the difference of two (single beam) spectra (five 40s scans each) corresponding to successively deposited krypton matrices: one with ions and electrons and one without. The former involved a 30 min co-deposit of mass selected Ag^+ (2.5 nA cm^{-2} as measured [16] on F1), electrons and Kr whose deposition rate is estimated [17] to be about 3×10^{15} atoms s^{-1} (thickness: 2 nm s^{-1}). In general, we found that absorption spectra were most intense when the current as monitored on F2 was maintained near zero during a deposition. Typically, the total electron emission current was −5 to −10 nA and the electrons had an energy of 2–3 eV. Deposition without all ions gave a spectrum approximately 20 times weaker (when scaled for equivalent conditions) than that shown in Fig. 3. Without either ions or electrons, no atomic absorption features were observed thus implying that few neutrals reach the matrix target.

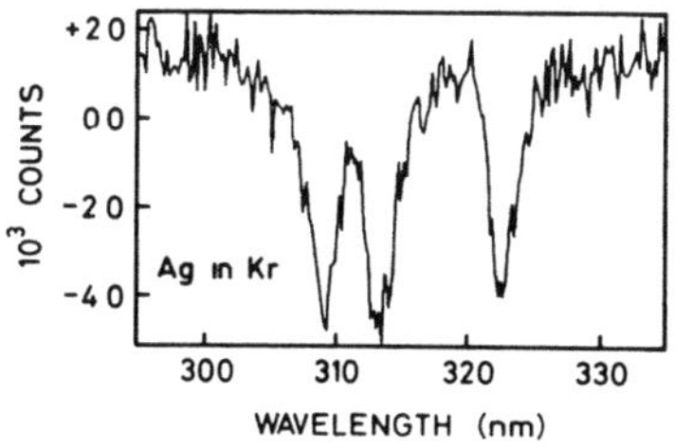

Fig. 3. Absorption Spectrum (including background subtraction) of Ag in a Kr matrix. The peak at 322.6 nm has a maximum absorbance of about 4×10^{-3}

The measured positions of the three (approximately equally intense) peaks are 322.6, 313.5 and 309.3 nm in good accord with those obtained previously [15]. The density of neutrals in the matrix (N cm^{-2}) may be estimated from the gas-phase expression [18],

$$N = 2.6 \times 10^{19} \frac{1}{f\lambda^2} \int A(\lambda) d\lambda \tag{1}$$

where $A(\lambda)$ is the measured absorbance, f is the (absorption) oscillator strength for the transition and λ is the wavelength in nm. Equation (3) was applied to the 322.6 nm peak, whose shape was assumed to be triangular with area equal to the product of the full width at half maximum (FWHM = 3.3 nm) and the maximum absorbance (A_{max}). From Fig. 3 and the incident photon flux of 5.8×10^5 c/s, we calculate $A_{max} = 3.7 \times 10^{-3}$. Using the gas-phase oscillator strength for the $^2S_{1/2}$–$^2P_{1/2}$ transition of atomic silver, $f = 0.22$ [19], we find $N = 1.4 \times 10^{13}$ atoms cm^{-2} corresponding to 7.7 $\times 10^9$ atoms cm^{-2} s^{-1} (i.e. 1.2 nA cm^{-2}). Since the incident Ag^+ flux was 2.5 nA cm^{-2}, Fig. 3 pertains to an atomic neutralization/collection efficiency (E_1) of about 50%.

Neutralization experiments using Ar as the matrix were markedly less successful than in the case of Kr.

The maximum value of E_1 found for Ar was about 5%. With the neutral beam unblocked, intense atomic absorption bands (plus weaker features arising from higher aggregates) were obtained for both Ar and Kr matrices. For example, a 32 min. deposition using Kr gave $A_{max}=0.28$, corresponding to an incident flux of 5×10^{11} atoms $cm^{-2}s^{-1}$.

Preliminary experiments in which Ag_2^+ was co-deposited with Kr and electrons produced no spectra other than that of the atom. However, under nearly identical deposition conditions to those described above, the atomic intensities yielded an E_1 value (counting 2 atoms/cluster) of 25%, which may indicate that not all the (0.7 nA cm^{-2}) dimer ions were fragmented, even though deposited at an energy of about 90 eV.

In conclusion, we have demonstrated that mass-selected atoms may be efficiently neutralized by electrons and deposited in rare gas matrices with sufficient numbers to be observed spectroscopically. An optimization and better understanding of the processes involved awaits the outcome of further, more carefully controlled experiments which are currently in progress. The central problem, a "soft landing" of neutral, unfragmented cluster species, remains unresolved at present and will only be decided by the definitive observation of spectral carriers other than atoms.

The authors wish to express their thanks to P. Fayet, W. A. Saunders and J. C. Rivoal for their expert advice and aid during the course of these experiments. One of us (DML) is grateful to the EPFL for generous financial and other support during his sabbatical year in Lausanne. This experiment is part of the project "Catalysis" of the EPFL, the financia' support of which is acknowledged. We wish to thank M. Claude Blanc and M. François Jeanquartier for their skillful fabrication of the cluster deposition apparatus.

References

1. For example: Leopold, D.G, Ho, J., Lineberger, W.C.: J. Chem. Phys. **86**, 1715 (1987); Pettiette, C.L., Yang, S.H., Craycraft, M.J., Conceicao, J., Laaksonen, R.T., Chesnovsky, O., Smalley, R.E.: J. Chem. Phys. **88**, 5377 (1988)
2. de Heer, W.A., Selby, K., Kresin, V., Masui, J., Vollmer, M., Châtelain, A., Knight, W.D.: Phys. Rev. Lett. **59**, 1805 (1987)
3. For example: Morse, M.D., Hopkins, J.B., Langridge-Smith, P.R.R., Smalley, R.E.: J. Chem. Phys. **79**, 5316 (1983); Rohlfing, E.A., Valentini, J.J.: Chem. Phys. Lett. **126**, 113 (1986)
4. For example: Delacrétaz G., Grant, E.R., Whetten, R.L., Wöste, L., Zwanziger, J.W.: Phys. Rev. Lett. **56**, 2598 (1986)
5. Fu, Z., Lemire, G.W., Hamrick, Y.M., Taylor, S., Shui, J-C., Morse, M.: J. Chem. Phys. **88**, 3524 (1988)
6. For a recent review see: Moskovits, M.: Metal clusters, Chap. 7. New York: Wiley 1986
7. See, for example: Garland D.A., Lindsay, D.M.: J. Chem. Phys. **80**, 4761 (1984)
8. Fayet, P., Wöste, L.: Z. Phys. D – Atoms, Molecules and Clusters **3**, 177 (1986)
9. Fayet, P.: PhD Thesis No. 684, Ecole Polytechnique Fédérale de Lausanne, 1987
10. Besocke, K., Berger, S., Hofer, W.O., Littmark, U.: Radiat. Eff. **66**, 35 (1982)
11. Allen, J.D., Wolfe, J.P., Schweitzer, G.K.: Int. J. Mass Spectrom. Ion Phys. **8**, 81 (1972)
12. Brubaker, W.M.: Adv. Mass Spectrom. **4**, 293 (1968)
13. Dawson, P.H.: Int. J. Mass Spectrom. Ion Phys. **6**, 33 (1971)
14. Rivoal, J.C., Grisolia, C., Lignieres, J., Kreisle, D., Fayet, P., Wöste, L.: Z. Phys. D – Atoms, Molecules and Clusters 12 (1989) (this issue)
15. Schultze, W., Becker, H.U., Abe, H.: Chem. Phys. **35**, 177 (1978)
16. By moving F1 vertically, we were able to measure the ion beam profile and estimate that it's area at the sapphire window was about 1.5 cm^2
17. From gas dynamic considerations and by observation of interference fringes
18. This expression may be derived from results given in (e.g.): Steinfeld, J.I.: Molecules and radiation. Cambridge MA, London: MIT press 1978
19. Radzig, A.A., Smirnov B.M.: Reference data on atoms, molecules and ions, Chap. 7. Berlin, Heidelberg, New York: Springer 1985

Z. Phys. D – Atoms, Molecules and Clusters 12, 19–22 (1989)

Atoms, Molecules and Clusters
Zeitschrift für Physik D
© Springer-Verlag 1989

Palladium particles growth on various aluminas

P. Légaré[1], F. Finck[1], R. Roche[2], and G. Maire[1]

[1] Laboratoire de Catalyse et Chimie des surfaces. Université Louis Pasteur, 4 rue Blaise Pascal, F-67070 Strasbourg Cedex, France

[2] P. S. A. Etudes et Recherches, 18 rue des Fauvelles, F-92250 La Garenne, France

Received 5 July 1988; final version 30 September 1988

Pd was deposited from a filament onto an α-Al_2O_3 single-crystal. The Pd $3dP^{5/2}$ binding energy change versus deposition time was used as a calibration of the Pd clusters size. The experiment was then reproduced on $\gamma-Al_2O_3$. The Pd $3d^{5/2}$ binding energy showed that the clusters growth mechanism was now different. Thermal treatments produced a dispersion of the metal or a clusters size increase depending on the temperature or the nature of the support. Oxygen exposure dispersed Pd in every case. A mechanism is proposed to interpret this phenomenon.

PACS: 79.60

1. Introduction

The influence of the support on the mechanism of metal particles growth and behaviour is a general problem for supported catalysts [1]. The geometric shape and consequently the electronic and chemical properties of supported particles are largely determined by the nature of the support as well as the thermal and chemical treatments [2, 3, 4].

On another hand, the growth of supported metal aggregates and the development of their electronic system, from the atomic to the solid state, has been the field of a high activity in the ten past years [5, 6]. It was shown that small metal clusters deposited on a poorly conducting or insulating support suffer in XPS a high binding energy shift of the metal core levels and photoemission threshold for decreasing particle size. This effect has been attributed [6] to the electrostatic effect of the positive hole, unneutralized during the photoemission time. This means that this effect would be independent on the nature of the metal atoms in agreement with the experimental observations. The use of poorly conducting supports, as α-Al_2O_3 [7] and amorphous carbon [8] showed that Pd exhibited qualitatively this behaviour. However the basic attribution of this phenomenon to final or initial state effects [8, 9] is still debated.

Our aim is to take profit of this high binding energy shift to get some insight on the influence of the support on the growth mechanism of small Pd particles, and the effects of thermal and chemical treatments. In this purpose, we used first the (0001) surface of an α-Al_2O_3 single-crystal for Pd clusters growth. Then the deposition was reproduced on a γ-Al_2O_3 substrate. In the two cases, the effects of thermal treatments and oxygen exposure were observed.

2. Experimental

The experiments were performed in a VG ESCA III system already described [10]. It is composed by two interconnected vacuum chambers. The first one devoted to sample preparation, was equipped with an Ar-ion bombardment gun and a metal evaporation source using an electrically heated palladium wire. The thermal treatments and oxygen exposures were also performed in this part of the apparatus. The second chamber was used for analysis with an Al-Kα X-Ray source and a hemispherical analyzer. The total resolution was around 0.9 eV. All the binding energies were measured with respect to the Pd Fermi level of the most important Pd deposit corresponding to an evaporation time of 20 mn, with the assumption that the Al

2s core level peak remained at a constant energy during all the experiments, in order to correct for electrical charge building.

The (0001) α-Al_2O_3 single-crystal was cleaned by Ar ion bombardment until no surface contaminant could be detected by photoemission. This procedure implies that the surface was atomically rough. The other sample was obtained by a colloïdal alumina coating of a steel foil. X-Ray analysis showed that alumina was of the γ phase. This type of support has been elaborated for an industrial aim. Therefore we wanted to use it in its initial state, without decontamination of its surface. So this support was not submitted to ion bombardment: its main contaminants were *C* and *S*.

3. Results and discussion

3.1. Pd deposition and size calibration

Pd was progressively condensed on the clean α-Al_2O_3 support for exposure times up to 20 mn. Figure 1 shows the binding energy changes of the Pd $3d^{5/2}$ peak and the photoemission threshold versus the deposition time. The quasi-parallel movement of these two features during the clusters deposition is in good qualitative agreement with the results already reported [6, 7, 11].

In order to use these curves as a calibration to follow changes of the aggregate sizes in subsequent treatments, it was interesting to have an approximation of the mean cluster size versus the time of deposition. This was obtained using a model of hemispherical particles deposited on a flat surface. The changes of core level intensities for the support and Pd could be calculated versus the escape depth, the radius of the particles and their number per unit area.

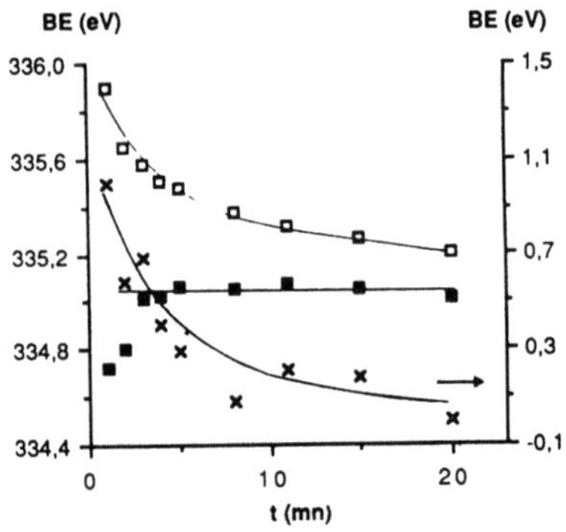

Fig. 1. Binding energy change of the Pd $3d^{5/2}$ (□) peak and photoemission threshold (*x*) versus deposition time on α-alumina. ■: Pd $3d^{5/2}$ binding energy change after subtraction of a A/R term (see text)

It was then possible to calculate the deposited quantity, which showed a good linearity with time (this was not postulated by the model), and the mean radius of the particles. In the limits of our model, the radius *R* of the aggregates was estimated to vary from 2.5 Å to 11 Å along our experiments. The results of Fig. 1 were then analysed considering that the binding energy change versus the radius of the aggregates was of the A/R form [6]. The B.E. (Pd $3d^{5/2}$) $=f(t)$ curve in Fig. 1 was then corrected, choosing the parameter A in order to get a corrected binding energy as flat as possible. This was obtained with $A=2$ (*R* in Å). With this value, the Pd 3*d* binding energy remains almost constant during all the deposition process, except for the two first steps where the correction was too high, as can be seen on Fig. 1. As indicated in [6], for spherical particles the B.E. change would follow an $e^2/2R$ law, which would correspond to $A=7.2$. This discrepancy can be due to several factors. First of all the particles are certainly not well simulated by spheres, and anyway *R* was obtained assuming hemispherical particles. The result confirms however that the proposal of DiCenzo and Wertheim is reasonable as they checked themselves in [8]. It remains that the shape of the curves shown in Fig. 1 would characterize the growth mechanism of the Pd clusters on α-Al_2O_3 as well-dispersed aggregates. The other point to outline is the constant value of the Pd $3d^{5/2}$ binding energy with respect to the actual Fermi level during all the depositions (335.15 ± 0.05 eV). This is slightly, but constantly lower than the value recorded during the same work on a Pd foil (335.35 ± 0.05 eV). We ascribe this discrepancy to the dominant contribution of low-coordinated atoms in the aggregates.

These deposits have been reproduced with exactly the same procedure (power and time) on γ-Al_2O_3. Figure 2 shows the evolution of the Pd $3d^{5/2}$ binding energy during these depositions. Clearly this evolution corresponds to a very small change when compared to the change obtained on α-Al_2O_3 (0.9 eV). Actually, during the first 8 min., the binding energy remains practically constant (335.4 eV) and decreases only to 335.25 eV after 25 min. First this could be attributed to a growth of constant sized particles during all the depositions on γ-Al_2O_3. This would correspond to a linear increase of the number of particles versus the deposited quantity. Such a nucleation mode does not seem to be reasonable. So we have to understand why the A/R binding energy shift which plays satisfactorily for α-Al_2O_3 does not hold with γ-Al_2O_3. First we have to outline that this law can be valid for quasi-spherical clusters only. If the aggregates growth as rafts, their charging energy would certainly be lower. Such a suggestion is conforted by the observation by electron microscopy of an epitaxial growth of Pd on γ-Al_2O_3

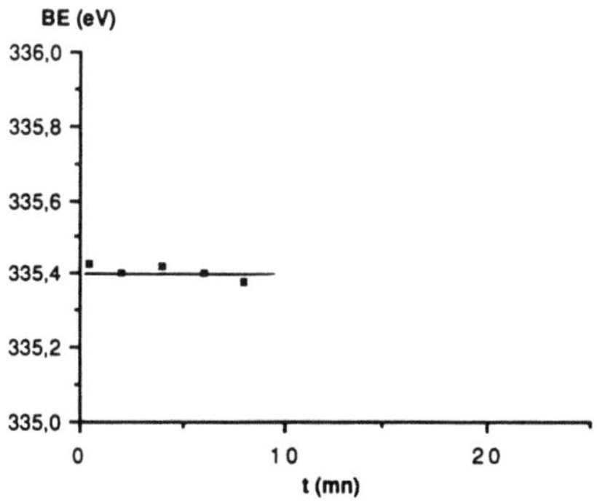

Fig. 2. Binding energy of the Pd $3d^{5/2}$ peak during deposition on γ-alumina

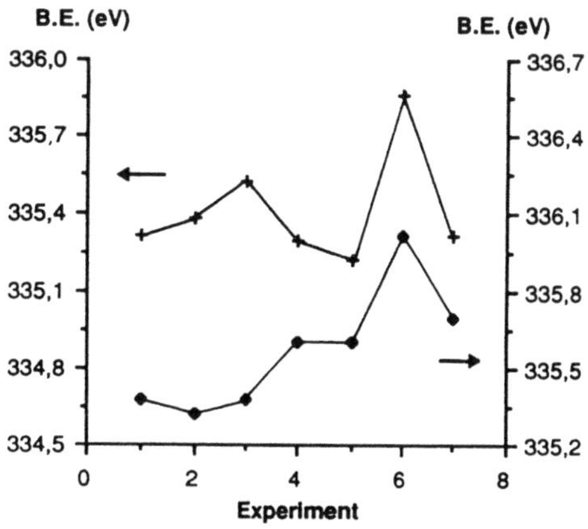

Fig. 3. Binding energy change of the Pd $3d^{5/2}$ peak on α-alumina (*left arrow*) and γ-alumina (*right arrow*) after Pd deposition (1), after heatings at 100°C (2), 200°C (3), 350°C (4), 500°C (5), oxygen exposure and pumping to UHV (6), subsequent heating at 500°C (7)

[3]. We have also to take into account the surface OH groups on γ-Al_2O_3. Recent calculations showed that the polarisability of the support can more or less compensate the core level shift of the clusters [12] and the OH dipoles could induce a serious difference between the two kinds of alumina. Moreover it has been shown that the core level lineshapes in the case of metal aggregates was largely due to a broad size distribution [6, 13]. This is true only if the deviation around the mean size results in a strong deviation of the electrostatic effect. This is the case on the α-Al_2O_3 support where the full width at half maximum (FWHM) for the Pd $3d^{5/2}$ peak is about 2 eV, but not on γ-Al_2O_3 where the FWHM was the same that the value recorded on the Pd foil (1.5 eV). An interesting consequence is the possibility to observe comfortably the electronic changes due to size reduction on γ-Al_2O_3.

3.2. *Thermal and chemical treatments*

Two deposits on α-Al_2O_3 and γ-Al_2O_3 corresponding to dosing times of 8 mn were then submitted to various treatments: heatings under UHV (100, 200, 350, 500°C), exposure to oxygen at atmospheric pressure and room temperature during one hour followed by pumping to UHV for 3 h, then heating again under UHV at 500°C. The subsequent evolution of the Pd $3d^{5/2}$ binding energy is reported on Fig. 3. We can note that it increases slightly at 100°C and 200°C for deposits on α-Al_2O_3, and remains constant on γ-Al_2O_3. At 350 and 500°C, it decreases on α-Al_2O_3 and increases on γ-Al_2O_3. On the two supports, oxygen exposure (and pumping to UHV) resulted in an increase of the Pd 3*d* B.E., followed by a decrease after heating under UHV.

If we interprete the B.E. increase (or decrease) as a consequence of aggregates size decrease (increase) the above results can be analyzed as follow. The heating at 100 and 200°C dispersed the metal on α-Al_2O_3 which aggregated again on heating at 350 and 500°C. On the contrary the Pd clusters appear to be anchored on γ-Al_2O_3 as no B.E. change appeared at 100 and 200°C, but the atoms dispersed at 350 and 500°C. The high binding energy shift after oxygen exposure cannot be ascribed to the formation of PdO as the Pd $3d^{5/2}$ peak would appear at 336.95 eV, as checked separately on an oxidized Pd foil. So this shift corresponds to a high dispersion of the metal. To explain this effect we can refer to the literature. Using TEM, Schmidt [14] has observed similar phenomenon with Rh particles on SiO_2. He showed that Rh particles oxidized at 500°C under O_2 then redispersed after reduction (H_2, 500°C). Moreover, Burwell et al [15] have shown with EXAFS study that Pd small particles can be oxidised at room temperature. We can propose the following evolution in our experiments. It consists in the oxidation of the Pd clusters (or chemisorption of O_2 on Pd) followed by a reduction of the particles during the pumping to UHV, resulting in a dispersion of the metal. This proposal can be confirmed by our observation that a quasi-2*d* form of PdO resulted in a spontaneous decomposition under vacuum to metal particles [16]. The subsequent heating at 500°C results in the formation of bigger aggregates by coalescence.

In conclusion we have used the size dependent binding energy of the Pd core levels to follow the evolution of metal aggregates. On α-Al_2O_3, the growth mechanism results in the formation of three dimensional aggregates. On γ-Al_2O_3, the growth is clearly different and could correspond to mostly bidimensional islands or rafts. The effect of temperature on the metal dispersion has been observed. This leads to the idea that the Pd atoms mobility is lower on γ-Al_2O_3

than on α-Al_2O_3. The exposure of the Pd deposits to oxygen at room temperature results in a dispersion of the metal.

References

1. Bond, G.C.: Metal support and metal additive effects in Catalysis. Stud. Surface Sci. Catal., Imelik, B et al. (eds.) Vol. 11. Amsterdam, New York: Elsevier 1982
2. Fuentes, S., Vàsquez, A., Pérez, J.G., Yacaman, M.J.: J. Catal. **99**, 492 (1986)
3. Dexpert, A., Freund, E., Lesage, E., Lynch, J.P.: Metal support and metal additive effects in catalysis. Stud. Surface Sci. Imelik, B. et al. (eds.), Vol. 11. Amsterdam, New York: Elsevier 1982
4. Beck, D.D., Carr, C.J.: J. Catal. **110**, 285 (1988)
5. Mason, M.G.: Phys. Rev. B **27**, 748 (1983)
6. Dicenzo, S.B., Wertheim, G.K.: Comm. Solid State Phys. **11**, 203 (1985)
7. Kohiki, S.: Appl. Surf. Sci. **25**, 81 (1986)
8. Wertheim, G.K., DiCenzo, S.B.: Phys. Rev. **B 37**, 844 (1988)
9. Bagus, P.S., Nelin, C.J.: Z. Phys. D – Atoms, Molecules and Clusters **12** (1989) (this issue)
10. Hilaire, L., Diaz Guerrero, G., Légaré, P., Maire, G., Krill, G.: Surf. Sci. **146**, 569 (1984)
11. Wertheim, G.K., Dicenzo, S.B.: Phys. Rev. B **33**, 5384 (1986)
12. Castellani, N.J., Leroy, D.B.: Phys. Stat. Solidi B **148**, 121 (1988)
13. Fritsch, A., Légaré, P.: Surf. Sci. **162**, 742 (1985)
14. Lee, C., Schmidt, L.D.: J. Catal. **101**, 123 (1986)
15. Nandi, R.K., Georgopoulos, P., Cohen, J.B., Butt, J.B., Burwell, R.L.: J. Catal. **77**, 421 (1982)
16. Légaré, P., Finck, F., Roche, R., Maire, G.: Surf. Sci (in press)

Z. Phys. D – Atoms, Molecules and Clusters 12, 23–26 (1989)

Atoms, Molecules
Zeitschrift für Physik D and Clusters
© Springer-Verlag 1989

Comparison between the growth of gold particles onto MgO and amorphous carbon substrates by in situ electrical conductance measurements

G. Desrousseaux[1], B. Robrieux[1], A. Renou[1], A. Carlan[2], H. Schaffar[2], and M. Gillet[1]

[1] Laboratoire de Microscopie et de Diffractions Electroniques U.A. 797 Faculté des Sciences de Saint-Jérôme, F-13397 Marseille Cedex 13, France

[2] Laboratoire des Propriétés Physiques des Couches Minces Faculté des Sciences de Saint-Jérôme, F-13397 Marseille Cedex 13, France

Received 9 July 1988; final version 30 September 1988

The sheet conductance of gold condensates on MgO and *a–C* is expressed in terms of granulometric quantities. This expression enables us to follow the evolution of metal repartition on its substrate during the gold evaporation. The difference between the growth modes of these gold layers on MgO and on *a–C* is explained by the predominance of respectively nucleation and coalescence phenomena

PACS: 81.15.G

1. Introduction

We have previously shown [1–4] how it is possible, from the simultaneous recordings of the sheet conductance and mass thickness of a discontinuous metallic condensate, to follow the metal repartition on amorphous carbon (*a–C*) substrate during the evaporation.

This method of granulometric study will be used to compare the respective evolutions of the granulometries of gold condensates on MgO and on *a–C* during the metal evaporation.

2. Experimental

The sheet conductance of the metal condensate is deduced from the planar current intensity at dc constant voltage. The measurement device has been fully previously described [1–4]. The intensity I_s of the current flowing in the substrate is measured before the gold condensation. Then the increase I of the intensity is measured during the metal evaporation. The metal repartition onto each substrate is computed from this intensity I and the mass thickness e_m. Two electron microscopy grids, previously covered with carbon, are placed in the plane of the electrodes. Each grid thus receives the same deposits as the glass supports. A piezoelectric quartz monitor enables us to measure the mass thickness of substrate and the metal condensation rate. A thermocouple indicates the temperature of the substrate.

The MgO substrates are prepared at room temperature by evaporation from 4*N*-pure monocrystals by means of an electron gun at a rate about 1 nm/s. The carbon substrates are deposited by flash evaporations from spectroscopically pure rods. The substrate thicknesses are respectively about 20 nm for MgO and 10 nm for *a–C*. The TEM studies show (Fig. 1) that the MgO substrate looks like a mosaic of flat desoriented polygonal monocrystals with a size ranging between 10 and 15 nm. In contrast the carbon substrates have an amorphous structure.

Gold is evaporated from a Knudsen cell source with an atom flux calibrated in the range 0.01–0.04 nm/min. The sheet conductance measurements show that the gold repartition on MgO remains

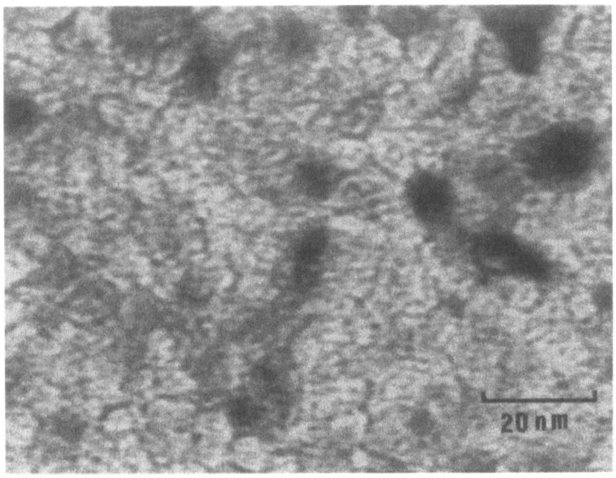

Fig. 1. Micrograph of the bare MgO substrate

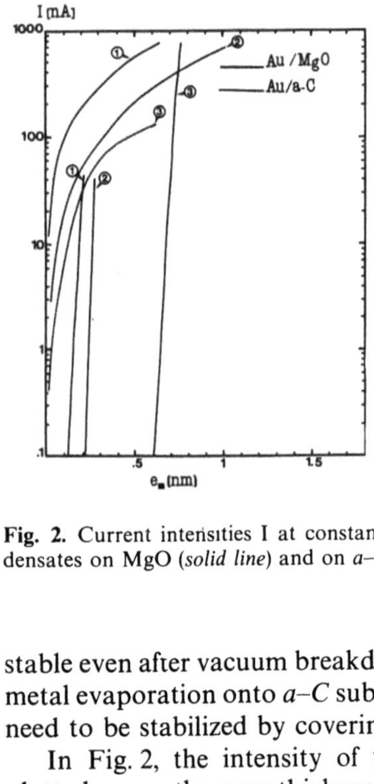

Fig. 2. Current intensities I at constant voltage versus gold condensates on MgO (*solid line*) and on *a–C* (*dashed line*)

stable even after vacuum breakdown. In contrast, after metal evaporation onto *a–C* substrate the condensates need to be stabilized by covering with a carbon film.

In Fig. 2, the intensity of the planar current is plotted versus the mass thickness of the gold deposit at a given evaporation rate on *a–C* (dashed line) and on MgO (solid line) for the three experiments labelled 1, 2, 3 on the graphs. It should be noted that the slopes of the curves are higher for Au/MgO than for Au/*a–C*. In addition in this log-normal plotting scale, Au/MgO curves have a constant slope, whereas it decrease for Au/*a–C* due to the prevalence of coalescence on nucleation. At the end of the gold condensation, the samples deposited on grids are observed in TEM in order to compare the results of the granulometric computation from I-measurements and statistical values of the dimensional analysis performed on micrograph. Such a micrograph enlargement is shown in Fig. 3a for Au on MgO and in Fig. 3b for Au on *a–C*.

3. Results and discussion

Above the threshold of the intensity measurement (10 pA), the calculations show that in a weak electric field and at room temperature, the tunnelling between aggregates via their substrate is the prevalent conduction process. The metal grains are considered to be oblate spheroids with an axis perpendicular to the substrate; their eccentricity is labelled η. The supplement α of the contact angle define their troncature in the substrate plane. α and η are graphically determined by minimizing the difference between the calculated granulometric quantities and the results of the dimensional analysis. A computation program finds by iteration the diameter a of the grains and their edge to edge distance s which, for a given mass thickness e_m, leads to the intensity I. At the end of the iteration, the program moreover calculates two other important granulometric quantities: the density n_g and the fractional area c of the substrate covered by the metal. By using the theory of Hill [5] in its new formulation of Barr and Finney [6], the iteration computes a and s.

The barrier height and the tunnel electron mass values are taken from the bibliographic data [7, 8]. Then we take into account the possible creation in the gap of "trap" levels which can reduce these tunnelling parameters [9, 10].

The three computed granulometric quantities are plotted versus the atomic coverage n_a, (number of atoms condensed on substrate area unit) in Fig. 4 for n_g and in Fig. 5 for a and c. Within the studied atomic coverage range, for the growth of Au on MgO, the nucleation prevails over the coalescence. This gives rise to a steep increase of the surface density of grains (Fig. 4) and correspondingly to a diminution of their mean size (Fig. 5). Such a behaviour corresponds to the displacement of the peak of histogram towards lower sizes due to the appearance of many small particles. The slope of n_g and a curves can be explained by the appearance of backnucleation phenomena which takes place on sites which are released by the coalescence of clusters. In contrast, the growth of Au onto *a–C* quickly reaches the stage where the coalescence compensates the nucleation before exceeding it.

The dimensional analyses on micrographs give statistical values for n_g and a which are close to the computed granulometric quantities at the curve extremity. The relative error between statistical and computed values is less than 30%. This margin is less than that observed between the granulometries obtained on two meshs of the same grid.

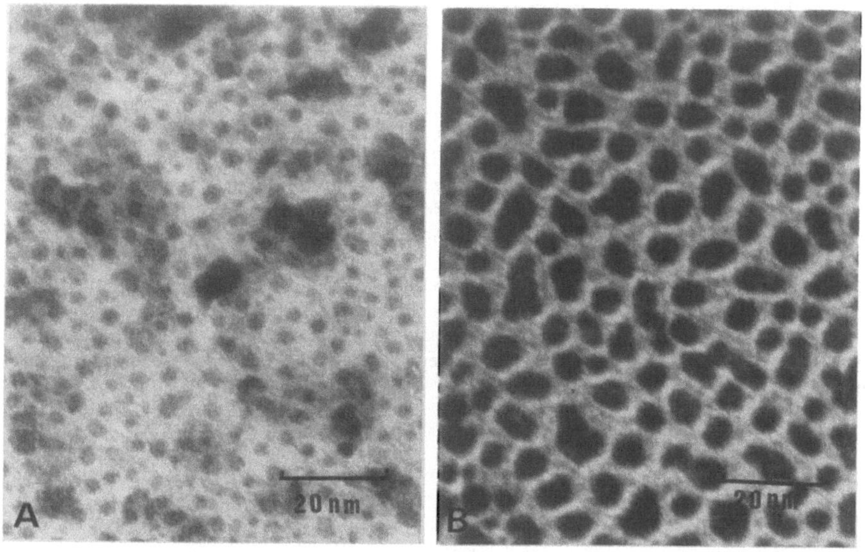

Fig. 3a and b. Micrograph of gold condensates on MgO (**a**) and on *a–C* (**b**). They give the metal repartition at the end of condensation for 2nd run

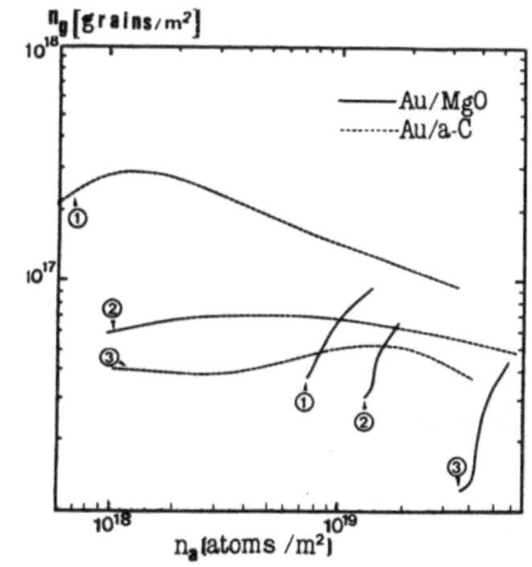

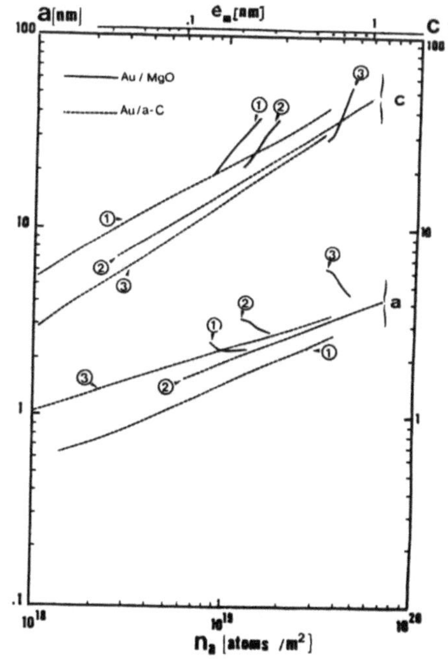

Fig. 5. Diameter *a* of grains and substrate fractional area *c* versus coverage n_a

4. Conclusion

Measurements of tunnelling current intensity performed versus mass thickness of gold deposits onto MgO and *a–C* have shown an interesting difference. In the case of Au on *a–C*, the regular increase of *a* and *c* is accompanied by a low increase then a fast decrease of n_g. For Au/MgO, the covered fractional area *c* is reached more quickly by means of a fast increase of the n_g density concomitant with a strong decrease of *a*.

In order to explain this difference between the evolution of the distribution of gold on *a–C* and on MgO during the metal condensation, we suggest the following chain of events. In the course of the growth of gold deposit on *a–C*, the coalescence very soon prevails over the nucleation. The metal is quickly distributed in great grains the capture area of which cover the whole surface of the substrate. Consequently the overlap of these area prevents some secondary nucleation on the sites released by the grains which have coalesced.

In contrast, for the growth of gold on MgO, the capture radius and diffusion distance are shorter. Therefore some active sites occupied at the time of the primary nucleation and released after coalescence of grains, can again become available for a backnucleation. This secondary nucleation [11] lowers the mean diameter a the more so as the shortening lifetime of these new grains is reducing their growth.

References

1. Desrousseaux, G., Carlan, A., Robrieux, B., Schaffar, H., Trompette, J., Dussaulcy, J.P., Faure, R.: J. Phys. Chem. Solids **46**, 929 (1985)
2. Desrousseaux, G., Robrieux, B., Carlan, A., Schaffar, H., Faure, R., Dussaulcy, J.P.: J. Phys. C. **19**, 2217 (1986)
3. Robrieux, B., Desrousseaux, G., Gillet, M.: J. Microsc. Spectrosc. Electron. **12**, 423 (1987)
4. Schaffar, H., Desrousseaux, G., Carlan, A., Robrieux, B.: Rev. Phys. Appl. **22**, 1139 (1987)
5. Hill, R.M.: Trans. R. Soc. London Ser. A. **309**, 377 (1969)
6. Barr, A., Finney, R.D.: Thin Solid Films **24**, S11 (1974)
7. Desrousseaux, G., Trompette, J., Faure, R., Schaffar, H., Dussaulcy, J.P.: Thin Solid Films **98**, 139 (1982)
8. Tirodkar, P.P., Pratap, R.: Ind. J. Pure Appl. Phys. **16**, 582 (1978)
9. Duke, C.B.: Tunnelling in Solids. New-York: Academic Press 1969
10. Tunnelling phenomena in solids. Burstein, E., Lundqvist, S. (eds.), New York: Plenum Press 1969
11. Lewis, B., Anderson, J.C.: Nucleation and growth of thin films, pp. 332–335. New York, San Francisco, London: Academic Press 1978

Z. Phys. D – Atoms, Molecules and Clusters 12, 27–29 (1989)

Atoms, Molecules and Clusters
Zeitschrift für Physik D
© Springer-Verlag 1989

Morphology determination of small particles by electron microscopy and electrical conduction measurements

B. Robrieux, G. Desrousseaux, A. Renou, and M. Gillet

Laboratoire de Microscopie et de Diffractions Electroniques, U.A. 797 Faculté des Sciences et Techniques de Saint-Jérôme, F-13397 Marseille Cedex 13, France

Received 5 July 1988

In this paper, we show that it is possible to deduce the actual morphology of small particle condensed onto an insulator by combining the granularity analysis from electron micrographs and the electrical sheet conductance of the deposit on its substrate. Assuming the particles are truncated ellipsoids, we determine the excentricity and the contact angle with the substrate for Au on amorphous carbon and MgO substrates.

PACS: 68.55.Jk; 61.16.Di; 72.15.v

Introduction

Many properties of small particles are related to their morphology. The knowledge of the density of particles on their support as well as their size and shape is a problem which is still of interest.

Whereas it is generally accepted that the density of these particles is easily determined from electron micrographs, their actual shape is often difficult to determine by transmission microscopy, with which in most cases only a projection of the particle can be seen. On the other hand, these observations generally ask for the particles to be removed from the vacuum and exposed to the air after their growth. In addition, in the case of thick support, it is necessary to make a carbon replica for particle observation.

We have developed a method to determine the granulometry and shape of particles. This method enables us to follow the particle changes occurring in situ during the growth of metallic particles on an insulating support. In this article, we describe the method which allows particle morphology determination.

Experimental

Gold particles onto amorphous carbon (*a–C*) or MgO substrates are deposited by thermal evaporation under UHV, with Knudsen cell. During the deposition, substrates are at room temperature. In our experiments, condensation rate ranges from 1 to 8×10^{16} atoms $m^{-2} s^{-1}$, and the mass thickness from 0.2 to 3 nm.

The method used to determine the particle shape is based on the measurement of the electric current I passing among the particles by tunnelling effect through the substrate. We showed [1] that, for a given

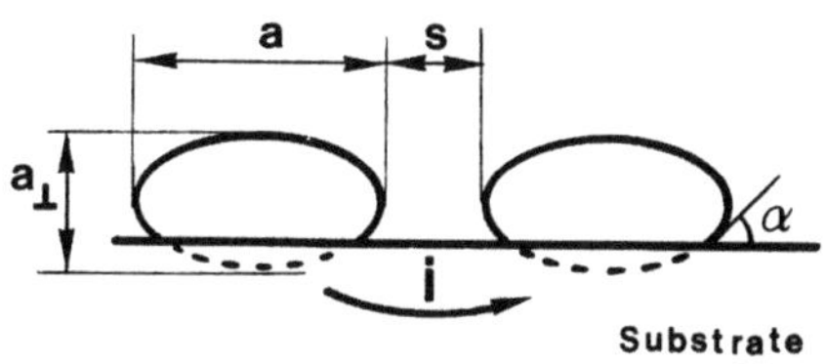

Fig. 1. Schematic representation of Au particles onto their substrate

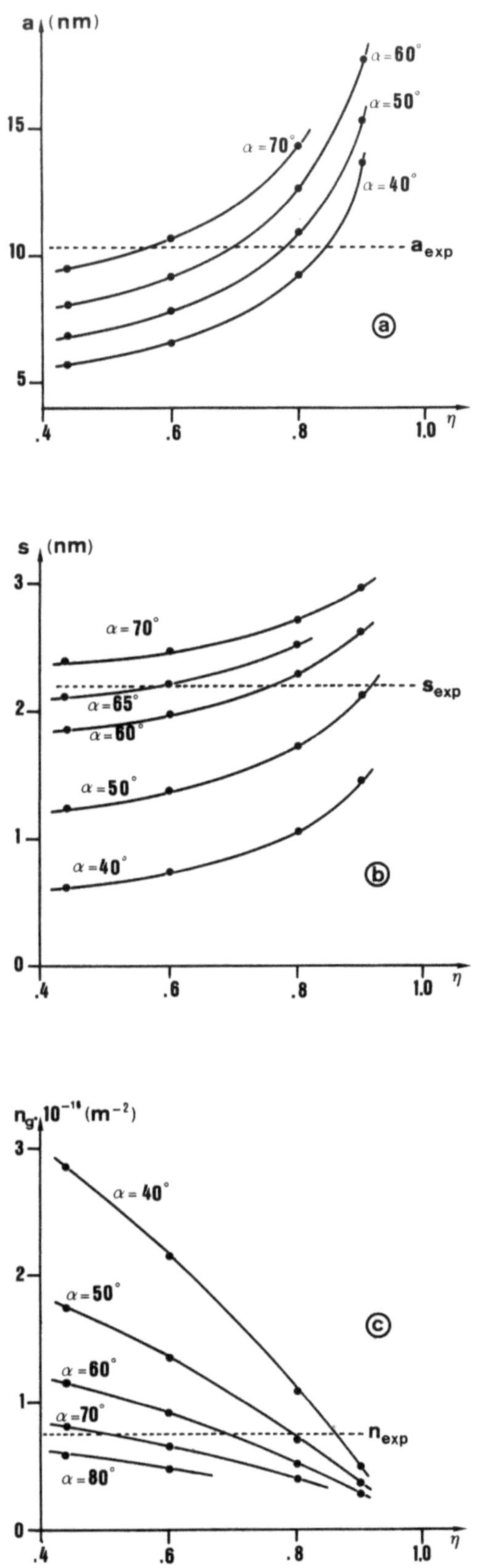

Fig. 2a–c. Computed values, corresponding to the experimental current I, for various α and η: **a)** particle size $a=f(\eta)$; **b)** edge to edge distance $s=f(\eta)$; **c)** particle density $n_g=f(\eta)$

mean deposit thickness e, this current I depends on the particle size characterized by a, their edge to edge distance s and shape. The particle shape is considered as a truncated ellipsoid with $a_{\perp}/a=\gamma$ and an angle of contact with the support α (Fig. 1). We showed [2] that, for a given particle shape (η, α), simultaneous measurement of I and the mean thickness e allows to follow the evolution of particle size (a), edge to edge distance (s) and density (n_g) during their growth. Conversely, particle shape (η, α) can be determined if the values a, s and n_g are known. Excentricity η is given by: $\eta=(1-\gamma^2)^{\frac{1}{2}}$

Results

By measuring I, corresponding to a deposit of mean thickness e, and using a micrograph of the deposit at a given stage of growth, we determined the particle shape (η, α) for gold growth on an amorphous carbon

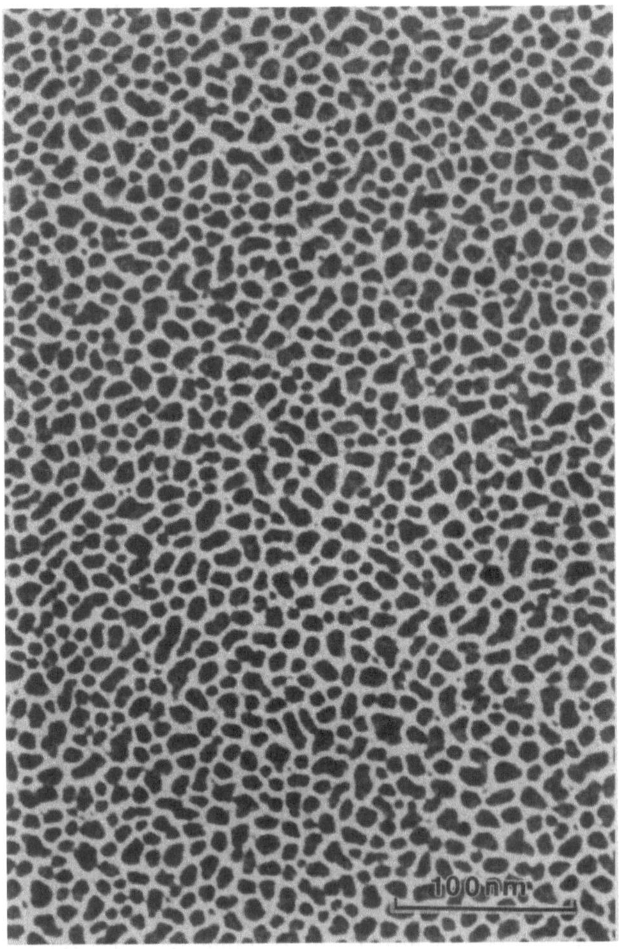

Fig. 3. Electron micrograph of Au particles on a–C substrate (e = 2.4 nm)

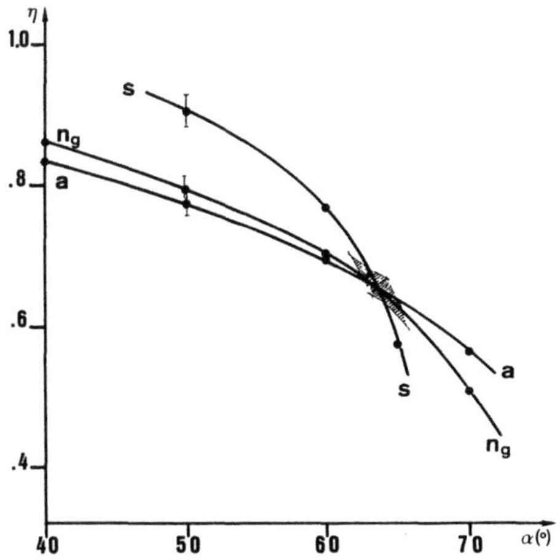

Fig. 4. Couples (α, η) giving experimental values respectively of *a*, *s* or n_g

support and a crystalline MgO support. We computed the values of *a*, *s* and n_g corresponding to an experimental value *I* for various values of η and α. As an example, the corresponding curves are given in Fig. 2 in the case of Au on carbon. From micrographs (Fig. 3 for Au on carbon) we deduced $a = a_{exp}$, $s = s_{exp}$, $n_g = n_{exp}$. By considering the intersection of the obtained curves with the straight lines $a = a_{exp}$, $s = s_{exp}$, $n_g = n_{exp}$ (Fig. 2) we obtain relations $\eta = f(\alpha)$: Fig. 2a gives couples (η, α) which account for the experimental value of *a*, Fig. 2b for the value of *s*, and Fig. 2c for n_g. They are plotted in the curves of Fig. 4. Taking into account the uncertainties on a_{exp}, s_{exp}, n_{exp}, the values of η and α can be found in the hachured area. For gold on carbon, the values thus determined from various experiments are $0.55 < \eta < 0.65$, $57° < \alpha < 63°$; for gold on MgO, the values thus determined are $0.56 < \eta < 0.62$, $80° < \alpha < 84°$.

References

1. Desrousseaux, G., Carlan, A., Schaffar, H., Robrieux, B. Dussaulcy, J.P.. Thin Solid Films **109**, 313 (1983)
2. Desrousseaux G., Carlan, A., Robrieux, B., Tompette, J., Schaffar, H., Faure, R., Dussaulcy, J.P.: J. Non-Cryst. Solids **59 & 60**, 529 (1983).

Z. Phys. D – Atoms, Molecules and Clusters 12, 31–35 (1989)

Atoms, Molecules
Zeitschrift für Physik D and Clusters
© Springer-Verlag 1989

Size-dependent thermodynamic properties of silver aggregates. Simulation of the photographic development process

M. Mostafavi, J.L. Marignier, J. Amblard, and J. Belloni
Laboratoire de Physico-Chimie des Rayonnements, UA 75 du CNRS, Université Paris-Sud, Bât. 350, F-91405 Orsay Cedex, France

Received 5 July 1988; final version 30 September 1988

Pulse radiolysis and time-resolved optical spectroscopy were used to observe the earliest steps of silver aggregation in the presence of an electron donor whose electrochemical potential is close to those of usual photographic developers. A critical size has to be reached by the silver aggregate before it is thermodynamically able to accept electrons from the donor, then to grow by an autocatalytic process with alternate adsorption of Ag^+ ions and electron scavenging. From a computer simulation the critical aggregation number n_c is found to be 4 and hence the electrochemical potential of the critical aggregate is ≈ -0.4 V/NHE. The size-dependence of the electrochemical potential in an aqueous solution is discussed and compared with that of the ionization potential in the gas phase. A tentative extrapolation of the present interpretation to the mechanism of the photographic development is proposed.

PACS: 82.50; 36.40; 35.20.V

1. Introduction

Quantization effects exert a crucial influence upon the structure and fate of metal aggregates. In the gas phase an important shift of their properties with the aggregation number n has been emphasized [1]. In solution, more and more data support the concept of a size-dependent electrochemical potential [2, 3]. The same size-dependence governs the reactivity towards the medium of the earliest aggregates and hence their nucleation-growth dynamics [4, 5]. These earliest aggregates are short-lived and can be studied only by pulse techniques [3]. As done classically for transient species, the electrochemical potential of a given aggregate in an aqueous solution may be derived from its reactivity towards known redox systems.

Silver appears to be the best candidate for this approach since many data are now available [3, 6] and since silver nucleation is the basis of photography [7–11]. As a matter of fact, photography may be regarded as a two-step process: *i)* a physical step involving the reduction of silver ions into atoms by photons and leading to the so-called latent image; *ii)* a chemical step of further reduction by an electron donor of those among the AgBr crystallites which contain silver atoms. Various arguments support the hypothesis that a minimum number of atoms (depending on the developer) is required to make an aggregate of the latent image (speck) developable. This was also recently demonstrated by experiments in which aggregates of known size were grafted onto the crystallites of an emulsion and then developed [10, 11]. But why such a critical size is required still remains a puzzling question [7].

The aim of this work is to investigate by pulse radiolysis techniques how, when and why a population of silver aggregates, in evolution from isolated atoms produced by one electron pulse, is able to exchange electrons with a known redox couple S/S^- playing the role of a photographic developer.

The possible reactions of S/S^- with the various oligomers Ag_n depend on their electrochemical poten-

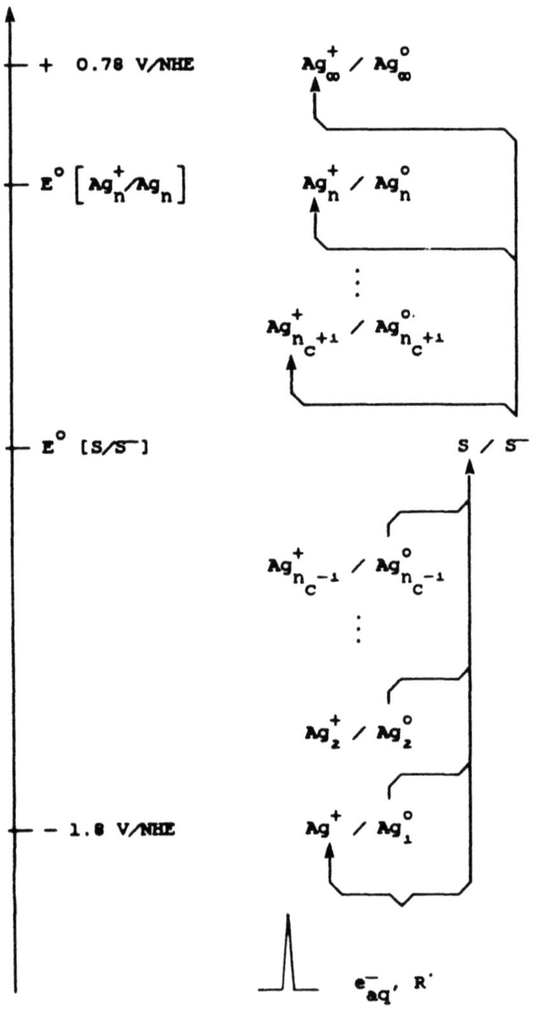

Fig. 1. Possible electron transfers between a redox couple S/S^- and silver aggregates. Silver atoms and S^- are formed by the same radiation pulse. Postulating an increasing variation of E^0 $[Ag_n^+/Ag_n]$ with n, the electron transfer may occur from Ag_n toward S at small n and is reversed from S^- (or a developer) to Ag_n^+ beyond a critical value n_c

tial, which is related to n. As shown in Fig. 1, the aggregates may be either electron donors, for n small and for E^0 $[Ag_n^+/Ag_n] < E^0$ $[S/S^-]$, or electron acceptors, for higher values of n, when E^0 $[Ag_n^+/Ag_n]$ is higher than E^0 $[S/S^-]$.

Both transfers may compete successively with the silver growth process involving the mere coalescence of atoms Ag_1^0. The critical aggregate Ag_{n_c} corresponds to the thermodynamic limit beyond which the electron transfer is reversed and to which a potential higher than E^0 $[S/S^-]$ may be assigned. The zwitterion sulfonato-propylviologen (SPV) was chosen as S [12, 13]: its reduced form absorbs at 396 and 603 nm. The standard potential is E^0 $[SPV/SPV^-] = -0.41$ V/NHE.

II. Experimental

The 0.6 MeV, 3 ns pulse radiolysis facilities and the quartz irradiation flow cell have been described elsewhere [14]. The dose per pulse ranges from 1 to 4.10^{18} eV·ml^{-1}. Typically, the aqueous solutions to be irradiated contain 0.2 M isopropyl alcohol to scavenge OH· and H· radicals, silver sulfate (from 2.10^{-4} to 3.6×10^{-3} M) and SPV (from 3 to 9.10^{-4} M). The solvated electrons and alcohol radicals partly reduce Ag^+ into Ag_1^0 and SPV into SPV^-. The silver solutions were prepared at room temperature and stored in the dark before use. The oxidized form SPV was synthesized by the reaction between bipyridyl and 1,3-propane sultone [13].

III. Results and discussion

The spontaneous evolution of silver atoms is well known [3, 6], but has been revisited to ascertain a better comparison with the study in presence of SPV [12]. When $[Ag^+] = 2.10^{-4}$ M, the optical spectrum of Ag_n shows an absorption peak at $\lambda_{max} = 400$ nm beyond ≈ 100 ms [12]. The aggregation number can be derived from the time of the observation (for a given initial concentration of atoms Ag_1^0) provided that the coalescence rate constant is known.

III.1. Time delay before electron transfer

Figure 2 shows the typical profile of the variation with time of the SPV^- absorbance at $\lambda = 603$ nm and of the $Ag_{n>2}$ absorbance (with a minor SPV^- component) at $\lambda = 430$ nm. The radical anion SPV^- is stable for ≈ 1 ms before starting to decay. Correspondingly, the 430 nm absorbance is first constant during the same

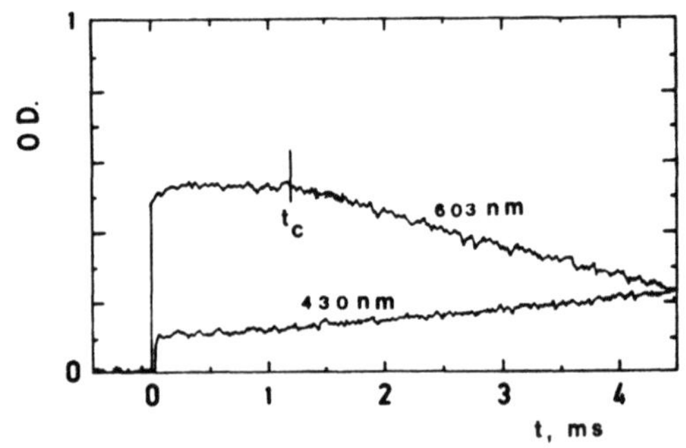

Fig. 2. Pulse radiolysis of a deaerated solution containing 2.10^{-4} Ag^+, 8.10^{-4} SPV and 0.2 M $(CH_3)_2CHOH$. Then $[Ag_1^0]_{t=0} = 5.10^{-6}$ and $[SPV^-]_{t=0} = 4.4 \times 10^{-5}$ M. Decay of SPV^- at 603 nm, absorbance of both SPV^- and Ag_n at 430 nm

period but then increases, due to silver growth. At longer time, SPV^- has disappeared, and the yield of the silver colloid Ag_{coll} is now the same as without SPV^-, as if a reaction had totally transferred the electrons from the relay SPV^- to the acceptor Ag_{n+x}^{x+} [12]. Conversely, no electron transfer has been observed within 1 ms from the earliest silver species toward SPV as acceptor.

It is clear that the time delay t_c corresponds to the critical minimum number of silver aggregation n_c required to allow the transfer. This time-delay strongly depends on the relative concentrations of Ag^+ and SPV and on the dose per pulse, which together decide the initial amount of Ag_1^0. The SPV^- decay depends strictly on the initial concentrations of Ag_1^0 and SPV^-, and not on excess Ag^+ or SPV (Fig. 3). If the bimolecular processes of coalescence are supposed up to $n=n_c$ identical to those without SPV^-, it follows that the smaller $[Ag_1^0]_{t=0}$, the longer t_c to reach the critical size Ag_{nc} (Fig. 3).

The computer simulation below indicates that the product $t_c \times [Ag_1^0]_{t=0}$ must be constant, and the plot of experimental values of $1/t_c$ vs. $[Ag_1^0]_{t=0}$ is indeed a linear function. The linearity factor depends on the bimolecular rate constant k_d of the cascade of aggregation reactions ($Ag_{n-p} + Ag_p \rightarrow Ag_n$, where n and p are the numbers of reduced atoms). If k_d is taken equal to $4.10^8\ M^{-1} \cdot s^{-1}$ [6], then n_c equals 4. The smallest charged aggregate acting as an acceptor is therefore Ag_{4+1}^+.

III.2. Kinetics of SPV^- decay

After t_c, the SPV^- decay lasts several milliseconds. It presents at least two successive pseudo first-order components with a marked change of rate (Fig. 3). Since $[Ag_n] \ll [SPV^-]$, the decay of SPV^- could not obey pseudo-first order kinetics unless the resulting species is able itself to accept another electron, so that the acceptor concentration remains constant. This feature is characteristic of an autocatalytic process. It is known that the adsorption of Ag^+ ions (rate constant k_a) is a fast process, as compared to the electron transfer (k_t). Both processes keep on competing with the aggregation processes (k_d). When $k_t \times [SPV^-]$ becomes smaller than $k_d \times [Ag_n]$, the competition favors the aggregation, which rapidly reduces $[Ag_n]$ by a factor of ≈ 2, and again the pseudo-first order autocatalytic electron transfer becomes preponderant with a concentration of Ag_{2n} two times smaller. A third component may sometimes be observed (Fig. 3, curves 1 and 5).

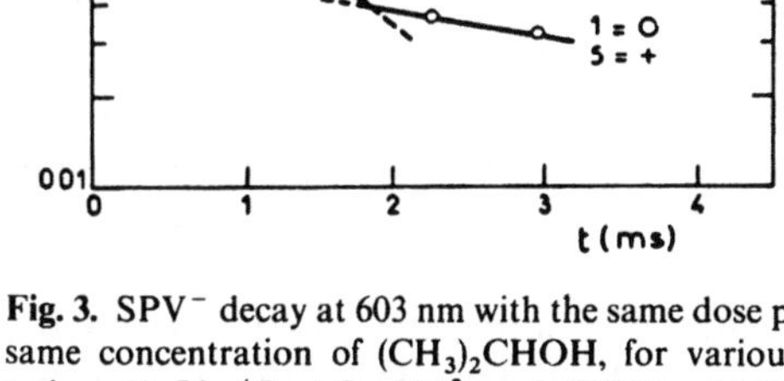

Fig. 3. SPV^- decay at 603 nm with the same dose per pulse and the same concentration of $(CH_3)_2CHOH$, for various $[Ag^+]/[SPV]$ ratios. 1) $[Ag^+] = 1.2 \times 10^{-3}$ and $[SPV] = 3.10^{-4}$ M. 2) 8 and 3.10^{-4} M. 3) 6 and 3.10^{-4} M. 4) 2 and 8.10^{-4} M. 5) 3.6×10^{-3} and 9.10^{-4} M

III.3. Computer simulation

The profiles of the time-depending evolution of $[SPV^-]$ and $[Ag_n]$ found experimentally have been compared with those given by a computer simulation where the initial concentrations of SPV^- and Ag_1^0 are the same as those yielded in the pulse radiolysis study. The reactions considered are

i) the coalescence of two aggregates disregarding their size (or their charge due to Ag^+ adsorption) with the rate constant k_d,

ii) the electron transfer from SPV^- to Ag_{n+1}^+ (restricted to $n \geqslant n_c$) with the rate constant k_t.

For $n < n_c$, the evolution of Ag_n is the same with or without SPV^-. For k_d, $[SPV^-]_{t=0}$ and k_t given, the change of $[Ag_1^0]_{t=0}$ causes a parallel shift of the curves of Fig. 4 giving the Ag_n distribution and the SPV^- decay. The latter occurs after a critical time t_c, proportional to the reciprocal of $[Ag_1^0]_{t=0}$. The adjustment of the calculated and the observed critical time indicates that $n_c = 4$. Then the best fitting between experimental and calculated $[SPV^-]$ decay yields $k_t = 7.10^8\ M^{-1} \cdot s^{-1}$.

The influence of SPV^- on the Ag_n evolution is, as expected, drastic, especially when the $[SPV^-]_{t=0}/[Ag_1^0]_{t=0}$ ratio is high. While $Ag_{n \leqslant 3}$ is unchanged, the following term Ag_4 is rapidly transformed into Ag_5, then Ag_6, and so on (Fig. 4a and b). The corresponding fractions of atoms incorporated in Ag_n also increase as fast as SPV^- decays, as in an autocatalytic electron transfer. The rapid storage of Ag atoms on these growing centers due to the pseudo-first order discharge of SPV^- accelerates the growth of a restricted number of aggregates (depending on $[Ag_1^0]_{t=0}$) where the atoms are concentrated.

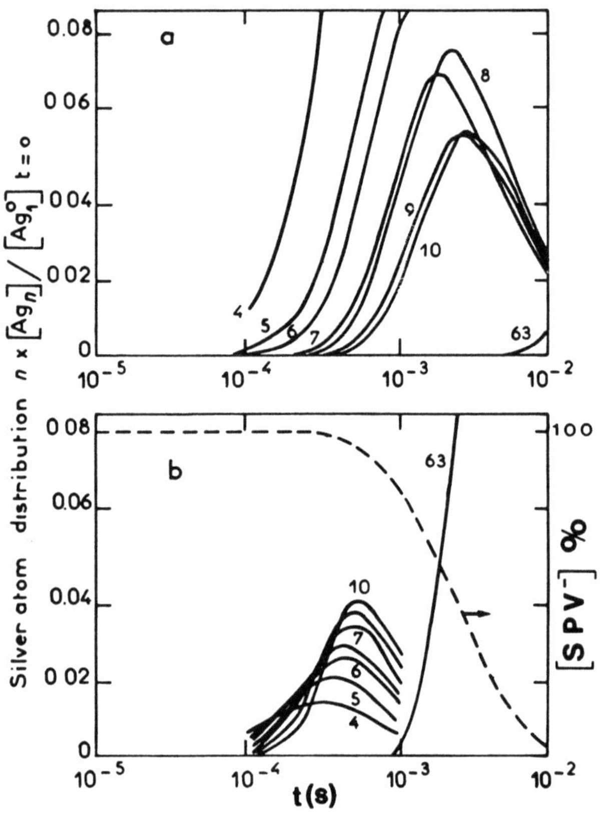

Fig. 4a and b. Time evolution of the Ag_1^0 distribution in the various $Ag_{n \geq 4}$ transients, under the conditions of Fig. 2. Computer simulation. **a**) $k_d = 4.10^8\ M^{-1} \cdot s^{-1}$, no SPV. **b**) $[SPV^-]_{t=0} = 4.7 \times 10^{-5}$ M, same other conditions. SPV^- decay (*dashed line*) starts with the autocatalytic growth of Ag_4

III.4. Size dependence of E^0 $[Ag_n^+/Ag_n]$

The computer simulation with the sole adjustment of k_t accounts for the experimental results and yields n_c. Since the potential of Ag_{n_c} has a value close to that of E^0 $[SPV/SPV^-]$, it follows: E^0 $[Ag_5^+/Ag_5] \approx -0.4$ V.

Figure 5 compares this value with a few published data [3] about the size-dependent potential of Ag_n. A monotonous increase is observed from E^0 $[Ag_{n=1}]$ to E^0 $[Ag_{n \to \infty}]$ of bulk metal in the condensed phase. It is worthwhile to compare this variation on a scale of energies E′(eV) with respect to the vacuum: E′ corresponds to the work function for extracting one electron from Ag_n to yield Ag_n^+. Figure 5 gives also the known values of the corresponding energy in the gas phase (i.e. the ionization potential IP) for a few Ag_n [1, 15]. Generally [1], this energy increases with the aggregation number *n*, although a few oscillations may exist at low *n* values, as was found for copper [15].

The increasing stability of Ag_n with *n* (with respect to the loss of e^-) in solution is certainly the major factor which enables the growth of an aggregate by catalytic reduction above a critical limit. A reverse variation (as in the gas phase) would rapidly restrict the electron transfer to the smallest sizes and exclude any growth or any photographic development [7], which is contrary to the observations.

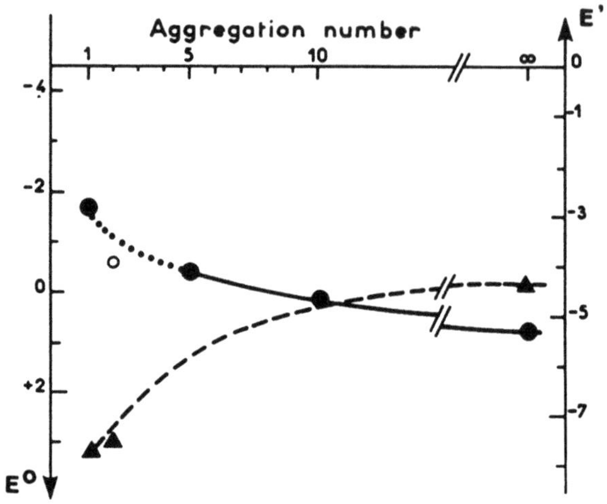

Fig. 5. Size-dependence of E^0 $[Ag_n^+/Ag_n]$ relative to NHE in an aqueous solution (● = experimental and ○ = calculated value), and of the ionization potential of Ag_n in the gas phase (▲). In solution, the E'_{sol} (eV) scale is derived from the electrochemical potential E^0_{sol} (V) according to $E'_{sol} = -|e| \times E^0_{sol} - 4.5$ eV

III.5. Comparison with photographic development

Insofar as quite similar situations are encountered in the case of a nascent aggregate surrounded by excess Ag^+ ions in an electron donor solution, and in the case of a silver speck of the latent image embedded near the surface of an AgBr crystallite in contact with the developer solution, the above conclusions are of close interest for the understanding of the photographic development.

An obvious difference lies in the fact that the formation, during exposure to light, of the specks able to initiate the development, is quite distinct from the development itself, whereas both steps coexist in the present study. Another difference is that in the liquid phase silver atoms benefit from a free diffusion, while, in photography, atoms are irregularly produced by light in the AgBr pellets, and their diffusion (or the diffusion of e^-) is strictly confined within the boundaries of the pellet, which hence constitutes the elementary unit of the image information. At the stage of the development, the initiating specks are isolated in a large excess of the developer, and will catalyze, by progressive accretion of reduced silver atoms, the reduction of the whole pellet.

In both cases the aggregates are surrounded by the developer solution, then a parallel increasing variation of the potential vs. n is to be expected, resulting in the autocatalytic reduction of the whole crystal above a critical size. The latter must obviously depend on the potential of the developer as observed in photography [7–11]. The critical number n_c can be slightly changed in AgBr pellets relatively to sulfate solutions, as it is changed by adding doping ions or sensitizers to AgBr. Whatever the phase (solution or AgBr), as long as $n < n_c$, the aggregate is unable to accept any electron. Beyond n_c, the electron transfer becomes thermodynamically possible. From step to step the process (i.e. electron transfer and Ag^+ adsorption) is more and more favoured, so that it cannot stop unless Ag^+ ions are exhausted.

In the same way, the developing threshold, its change with a preliminary exposure creating randomly-distributed pre-existent atoms, the fogging phenomena, become now more easily understandable. The present concept which considers the specks as non-noble micro-electrodes with a n-dependent potential, clearly explains that below n_c they cannot initiate any efficient development and even are fragile towards oxidation, which would have been misunderstood in regarding silver aggregates as mere particles of a noble metal [9].

IV. Conclusion

A time-resolved observation of the fast reactions between an electron donor (acting as a developer) and silver aggregates has shown that a critical size of the aggregate was required to make it capable of accepting electrons and therefore to start an autocatalytic growth. From the comparison with other data, it appears that the electrochemical potential of these microelectrodes does roughly increase with their aggregation number.

The role of the solvent on the thermodynamic properties of Ag_n is obviously expressed in the opposite sign of their variation vs. n in the gas phase and in the solution. While some ab initio calculations have been published concerning the energy levels of oligomers of a few metal atoms [1, 16, 17], no model has yet attempted to predict the structure and properties of aggregates surrounded by solvent molecules and particularly the energy ($-E'$) required for their ionization. No doubt that such models would afford decisive information on nucleation dynamics in the condensed phase, e.g. in the photographic development process, in the ordered growth of bimetallic aggregates [18], or in electroless processes involving pre-existent growth centers.

References

1. Morse, M.D.: Chem. Rev. **86**, 1049 (1986)
2. Delcourt, M.O., Belloni, J.: Radiochem. Radioanal. Lett. **13**, 329 (1973)
3. Henglein, A.: Current Chem. **143**, 113 (1987)
4. Belloni, J., Delcourt, M.O, Marignier, J.L., Amblard, J.: In: Radiation chemistry. Hedvig, P., Nyikos, L., Schiller, R. (eds.), p. 89. Budapest: Akadémiai Kiadó 1987
5. Marignier, J.L., Belloni, J.: J. Chim. Phys. **85**, 21 (1988)
6. Henglein, A., Tausch-Treml, R.: J. Coll. Interf. Sci. **80**, 84 (1981)
7. Hamilton, J.F.: In: Growth and Properties of Metal Clusters. Bourdon, J. (ed), p. 289. Amsterdam: Elsevier 1980
8. Cyrot-Lackmann, F.: Entre l'atome et le cristal· les agrégats, (ed.) Les Ulis Editions de Physique 1981
9. Moisar, E.: In: Contribution of clusters physics to materials science and technology. Davenas, J., Rabette, P.M. (eds.) NATO ASI Series, Series E: Applied Sciences – No. 104, p. 311. Dordrecht: Nijhoff 1986
10. Fayet, P., Granzer, F., Hegenbart, G., Moisar, E., Pischel, B., Wöste, L.: Phys. Rev. Lett. **55**, 3002 (1985)
11. Fayet, P., Granzer, F., Hegenbart, G., Moisar, E., Pischel, B., Wöste, L.: Z. Phys. D – Atoms, Molecules and Clusters **3**, 299 (1986)
12. Mostafavi, M., Marignier, J.L., Amblard, J., Belloni, J.: Radiat. Phys. Chem. (in press)
13. Delaire, J.A., Sanquer-Barrie, M., Webber, S.E.: J. Phys. Chem. **92**, 1252 (1988)
14. Belloni, J., Billiau, F., Cordier, P., Delaire, J.A., Delcourt, M.O.: J. Phys. Chem. **82**, 532 (1978)
15. Zheng, L.S., Karner, C.M., Brucat, P.J., Yang, S.H., Pettiette, C.L., Craycraft, M.J., Smalley, R.E.: J. Chem. Phys. **85**, 1681 (1986)
16. Baetzold, R.C.: In: Contributions of clusters physics to materials science and technology. Davenas, J. Rabette, P.M. (eds.) NATO ASI Series, Series E: Applied Sciences – No.104, p. 195. Dordrecht: Nijhoff 1986
17. Ozin, G.A., McIntosh, D.F.: J. Phys. Chem. **90**, 5756 (1986)
18. Marignier, J.L., Belloni, J., Delcourt, M.O., Chevalier, J.P.: Nature **317**, 344 (1985)

Z. Phys. D – Atoms, Molecules and Clusters 12, 37–40 (1989)

Atoms, Molecules
Zeitschrift für Physik D and Clusters
© Springer-Verlag 1989

Sodium clusters in zeolites

P.J. Grobet, L.R.M. Martens*, W.J.M. Vermeiren, D.R.C. Huybrechts, and P.A. Jacobs
Laboratorium voor Oppervlaktechemie, K.U. Leuven, Kardinaal Mercierlaan 92, B-3030 Leuven, Belgium

Received 5 July 1988

The method of loading sodium clusters in zeolites, consisting of the controlled thermal decomposition of physisorbed sodium azide, is discussed. The influence of the azide loading, the azide decomposition rate and the sintering process on the amount of ionic and metallic sodium clusters in zeolite Y was followed by ESR. The method is compared to other metal deposition techniques.

PACS: 36.40; 33.30; 33.20.E

1. Introduction

Zeolites are promising supports for the preparation of very small metal clusters [1]. Due to the porous structure of zeolites, consisting of cavities and channels of well defined aperture (ranging from 0.5 to 1.2 nm) [2], the size of the formed metal particles is well controlled and high dispersion can be achieved. The restricted size distribution of the particles gives an unique opportunity to study the magnetic properties of small sodium clusters.

The synthesis of sodium clusters in zeolites is mainly performed by exposing the zeolite material to an external sodium vapour source [3–8]. Using this method only a small part of the zeolite structure is impregnated and most of the sodium is deposed on the outer surface of the zeolite powder particles and on the wall of the reaction vessel. To avoid these drawbacks of this vapour deposition technique we developed a new impregnation method [9] consisting of the decomposition of physisorbed sodium azide. By loading sodium metal on zeolites, these materials with inherent acidic catalytic properties are transformed in new heterogeneous basic catalysts [10–12].

In the present work we discuss the influence of the azide loading, the azide decomposition rate and sintering process on the formation of ionic and metallic clusters in zeolite Y. Our impregation method is compared to the vapour deposition technique and to a liquid phase deposition method, developed very recently [13].

* Present address: Exxon Chemical International Inc., Machelen, Belgium

2. Experimental

The starting material was NaY zeolite obtained from Union Carbide Corp., Linde Division; the ideal chemical composition of the unit cell was $Na_{56}[(AlO_2)_{56}(SiO_2)_{136}]$. The zeolite samples were repeatedly exchanged with Na^+, calcined at 723 K and finally impregnated with a slurry of NaN_3 in methanol. The loadings are given in Table 1. These zeolite-azide adducts were dried at 523 K under inert atmosphere and activated in situ in the high temperature ESR cavity. During the standard activation procedure the temperature was raised at a rate of 3 K per minute up to 653 K, followed by cooling to the analysis temperature, which was normally room temperature.

The ESR spectra were recorded on a Bruker ER 200-SRC operating in the X-band. The incident microwave power was 200 μW. Spin densities are determined using a DPPH reference sample.

3. Results and discussion

The ESR spectra of the different NaN_3-NaY samples after standard activation are shown in Fig. 1. With

Table 1. The original amount of NaN_3 loading and spin densities in the activated samples

Symbol	NaN_3 loading (mol./u.c.)	Spin densities (10^{-2} spins/u.c.) Na_4^{3+}	Na_x^0
Y1	10	0.014	—
Y2	15	4.8	1.6
Y3	20	10.4	8.0
Y4	30	9.7	32
Y5	40	9.6	40
Y6	50	9.6	45
Y7	60	9.6	32

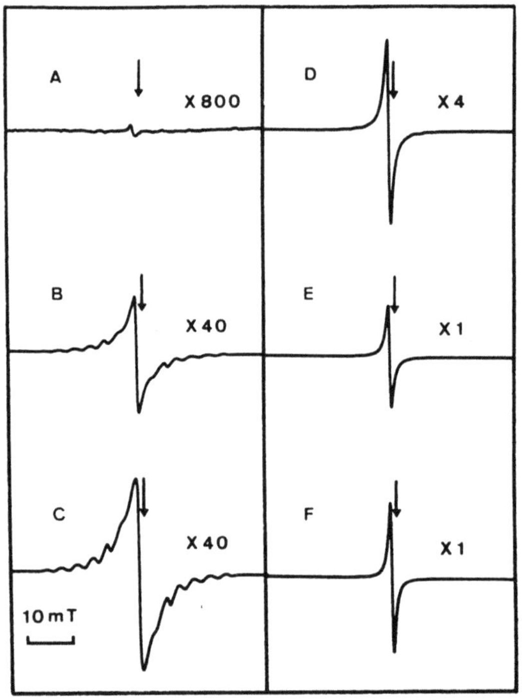

Fig. 1. Room temperature ESR spectra of NaY loaded with *A*, 10, *B*, 15, *C*, 20, *D*, 30, *E*, 40 and *F*, 50 NaN_3 mol/u.c. The marker corresponds to $g = 2.0002$

increasing NaN_3 concentrations two superimposed signals appear: a sharp line ($g = 2.0002$, linewidth of 0.82 ± 0.01 mT) and a 13 line hyperfine signal ($g = 2.0002$, hyperfine coupling constant $a = 3.32 \pm 0.01$ mT), due respectively to sodium metal particles Na_x^0 in the supercages (or α-cages) and to ionic Na_4^{3+} clusters formed in the sodalite cages (or β-cages) of the Y zeolite [3–13]. The evolution of both signal intensities, expressed in spin density per unit cell (Table 1), is seen in Fig. 2. The spin density of the Na_4^{3+} increases from $1.4\,10^{-4}$ (Y1) to a maximum of $1.04\,10^{-1}$ spins/u.c. (Y3) at a NaN_3 loading of 20

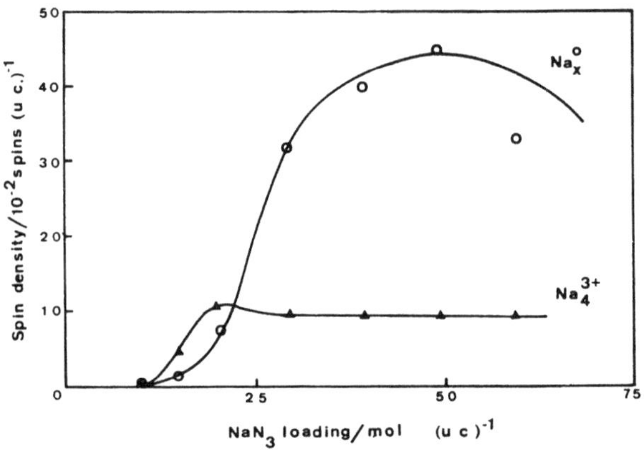

Fig. 2. Spin density of the metallic Na_x^0 and ionic Na_4^{3+} cluster signals at 293 K

molecules per unit cell. The CESR signal increases in a sigmoidal way with loading and a maximum of $4.5\,10^{-1}$ spins/u.c. is found at a loading of 50 molec./u.c. (Y6). At higher loadings the signal intensity tends to decrease (Y7).

Compared to the classical vapour deposition technique the cluster loadings obtained with the azide decomposition method are at least 50 times larger [11]. Recently Yoon et al. [13] employed solutions of *n*-butyl-lithium as reductant to form sodium clusters in zeolites. From a quenching reaction with e.g. water they estimated the concentration of Na_4^{3+} to be eight per unit cell; the metallic cluster concentration was not determined.

The following reactions account for the formation of the different particles during the thermal decomposition of NaN_3:

$$NaN_3 \xrightarrow{623\,K} Na^0 + 3/2\,N_2 \qquad (1)$$

$$Na^0 + 4Na^+ \longrightarrow Na_4^{3+} + Na^+ \qquad (2)$$

$$x\,Na^0 \longrightarrow Na_x^0 \qquad (3)$$

Reaction 1 is the azide decomposition; reaction 2 describes the formation of ionic clusters inside the sodalite cages, via an electron transfer process from a sodium atom to the sodalite cage, thus associating 4 sodalite cations. The formation of neutral clusters inside the faujasite supercages is described by reaction 3. From the ESR results shown in Figs. 1 and 2 we can derive that at low azide loadings (10–20 NaN_3 mol./u.c.) reaction 2 is favoured and that mainly Na_4^{3+} clusters are formed in the sodalite cages; at higher loadings (>25 NaN_3 mol./u.c.) reaction 3 is strongly favoured.

The hyperfine coupling constant *a* of the ionic cluster signal is found to be constant for all NaN_3 loadings. This indicates that the nature of the charged cluster is independent of the loading. However, the *a*-value depends on temperature, as seen in Fig. 3; this temperature dependence can be ascribed to a change of the free electron spin density on the Na_4^{3+} cluster, caused by a change of the bond strength between the cluster and the sodalite lattice or between the cluster cations mutually.

The susceptibility of the ionic *and* the metallic clusters, which is proportional to the intensity of the respective ESR signals, follows the Curie law, as shown in Fig. 4 for the metallic clusters of Sample Y6. Such a Curie-like temperature dependence of the magnetic susceptibility, instead of a constant Pauli-like susceptibility, is evidence that the clusters are so small that quantum size effects are important [6]. This is also borne out the symmetry of the CESR signal. Due to the impregnation and decomposition technique of NaN_3 inside the zeolite, most of the metal particles reside in the supercages of the faujasite structure. From the dimension of the supercage an upper value of 1.14 nm can be estimated for the size of the metallic particles.

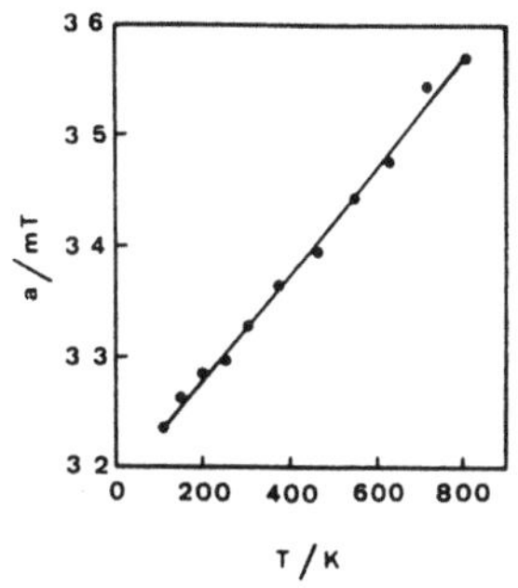

Fig. 3. Temperature dependency of the hyperfine coupling constant, *a*, of the Na_4^{3+} cluster

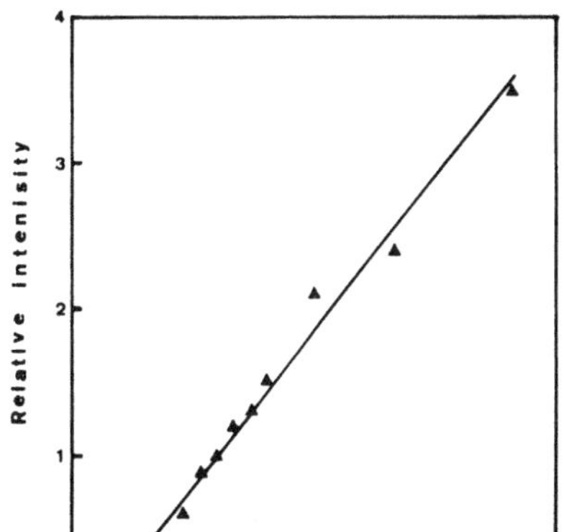

Fig. 4. Temperature dependency of the Na_x^0 signal intensity

The amount of metallic clusters in the supercages is not only depending on the loading but also on the decomposition rate of the NaN_3. The ESR spectra obtained after slow heating (3K/min) and fast heating (30K/min) of samples with a loading of 50 mol./u.c. are given in Fig. 5. Slow heating results in the formation of mainly metallic clusters; a small underlying ionic signal is only detectable at higher amplifications. Fast azide decomposition results in the formation of mainly ionic clusters, characterized by the 13 lines spectrum; the broad underlying resonance stems from iron impurities in the zeolite and can not be assigned to metallic clusters on the outer zeolite surface, as stated in former work [9–11].

Figure 6 displays the room temperature ESR spectra recorded after 20 min. sintering at 653 K, 723 K and 823 K respectively of the same loaded samples.

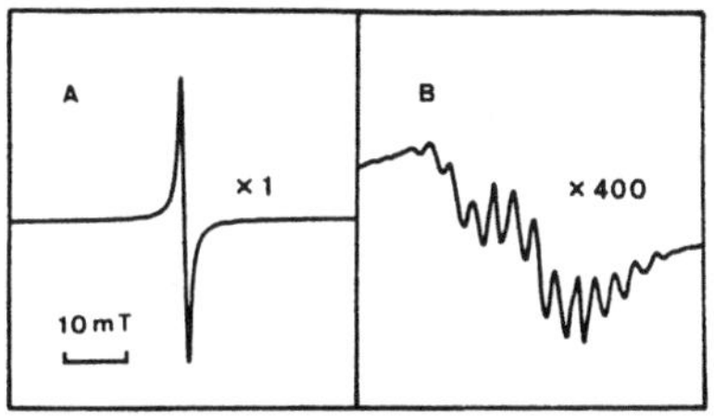

Fig. 5. ESR spectra of sodium species obtained after *A*, slow decomposition of NaN_3 on NaY and *B*, fast decomposition; the azide loading is 50 mol./u.c

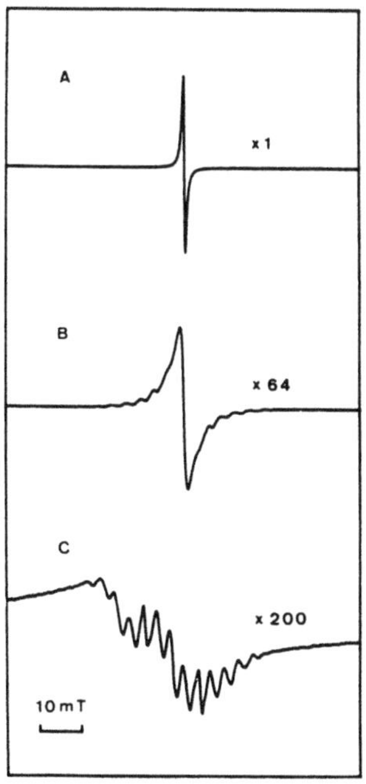

Fig. 6. Room temperature ESR spectra of a Sample Y6, recorded after a 20 minutes thermal sintering period at 653 K (*A*), 723 K (*B*) and 823 K (C)

The spectra show that thermal sintering results in the decrease of the intensity and a broadening of the Na_x^0 signal, while the ionic cluster seems to be resistant to the heat treatment (the broad underlying signal at $g=2.078$ in Fig. 6 *C* is again assigned to iron impurities). Probably during sintering there is a migration of metal sodium from the zeolite supercage towards the external surface where it agglomerates to larger particles; the resulting broadening of the CESR linewidth is observed.

4. Conclusion

The azide decomposition technique is a convenient method to produce sodium clusters in zeolites. By ESR spectroscopy the parameters influencing the formation of ionic and metallic sodium clusters in zeolite Y were determined. A maximum loading of matallic sodium in the supercages, parallel to a maximum catalytic basic activity [11], is obtained at 50 NaN_3 molecules per unit cell.

D.R.C. Huybrechts, P.J. Grobet and P.A. Jacobs, acknowledge the Belgian National Fund of Scientific Research for a grant as Research Assistant, Senior Research Associate and Research Director, respectively.

References

1. Jacobs, P.A.: Stud. Surf. Sci. Catal. **12**, 71 (1982)
2. Breck, D.W.: Zeolite molecular sieves. New York: John Wiley & Sons 1974
3. Rabo, J.A., Angell, C.L., Kasai, P.H., Schomaker, V.: Discuss. Faraday Soc. **41**, 328 (1966)
4. Rabo, J.A., Kasai, P.H.: Prog. Solid State Chem. **9**, 1 (1975)
5. Edwards, P.P., Harrison, M.R., Klinowski, J., Ramdas, S., Thomas, J.M., Johnson, D.C., Page, C.J.: J. Chem. Soc. Chem. Commun. 982 (1984)
6. Harrison, M.R., Edwards, P.P., Klinowski, J., Thomas, J.M., Johnson, D.C., Page, C.J.: J. Solid State Chem. **54**, 330 (1984)
7. Westphal, U., Geismar, G.: Z. Anorg. Allg. Chem. **508**, 165 (1984)
8. Smeulders, J.B.A.F., Hefni, M.A., Klaassen, A.A.K., de Boer, E., Westphal, U., Geismar, G.: Zeolites **7**, 347 (1987)
9. Martens, L.R.M., Grobet, P.J., Jacobs, P.A.: Nature **315**, 568 (1985)
10. Martens, L.R.M., Grobet, P.J., Vermeiren, W.J.M., Jacobs, P.A.: Stud. Surf. Sci. Catal. **28**, 935 (1986)
11. Martens, L.R.M., Vermeiren, W.J.M., Grobet, P.J., Jacobs, P.A.: Stud. Surf. Sci. Catal. **31**, 531 (1987)
12. Martens, L.R.M., Vermeiren, W.J.M., Huybrechts, D.R., Grobet, P.J., Jacobs, P.A.: Proc. 9th Int. Congress on Catalysis, Calgary, (1988) (accepted for publication)
13. Yoon, K.B., Kochi, J.K.: J. Chem. Soc. Chem. Commun. 510 (1988)

Z. Phys. D – Atoms, Molecules and Clusters 12, 41–44 (1989)

Atoms, Molecules
Zeitschrift für Physik D and Clusters
© Springer-Verlag 1989

Minimal spanning tree analysis of regular, random and cluster structures

C. Dussert[1], G. Rasigni[1], M. Rasigni[1], J. Palmari[1] and A. Llebaria[2]

[1] Département de Physique des Interactions Photons-Matière, Faculté des Sciences de Saint-Jérôme, F-13397 Marseille Cedex 13, France
[2] Laboratoire d' Astronomie Spatiale, F-13012 Marseille, France

Received 5 July 1988; final version 30 September 1988

It is shown that a graph called the minimal spanning tree (MST) is capable of detecting several kinds of cluster structure in arbitrary points set. From a MST it is possible to deduce the average edge length m and the standard deviation σ, which are characteristics for the corresponding distribution. The use of a diagram involving both m and σ makes it possible to determine the degree of order by taking a simple reading in the (m, σ) plane. It is also possible, through angular diagrams retrieved from the MST, to bring out directional properties. The method is applied to the study of the growth of granular thin films.

PACS: 82.70

1. Introduction

The need for knowledge of physical and chemical properties of small particles and inorganic clusters arises in a variety of contexts ranging from purely theoretical problems (electronic states in small particles, surface-enhanced Raman scattering ...) to practical applications (adsorption, catalysis, microlithography ...). The field of nucleation of thin films is a good example where cluster detection is useful. At the initial stage of its formation a thin deposit may be granular, made up of small particles having various shapes and sizes. Let us consider for instance a thin gold film deposited on a NaCl substrate previously covered with a carbon film, by using sputtering technique. To fix the condensed gold particles an additional carbon film was evaporated on top. Figure 1 shows, at the same magnification, the gold deposit sandwiched between the two carbon films at two stages of growth: A and B. Are the gold particles arranged randomly, or aggregated, or clustered, or regularly spaced? Is the degree of order related to sample A different from the one related to sample B? To answer these questions requires the use of mathematical approaches which make possible a more accurate detection than the simple visual one.

The power of graph theory methods in the context of detecting and describing inherent cluster structure in arbitrary point sets with a distance function is well known [1]. It is shown in this paper that a graph called the Minimal Spanning Tree (MST) may be successfully implemented, through the concept of order, in the study of regular, random and cluster structures, and can lead to a better understanding of the fundamental processes involved in the very early stages of the formation of thin granular deposits.

2. The minimal spanning tree (MST)

Basic definitions related to the graph theory have previously been given [2]. Let us recall that an edge-weighted linear graph is composed of a set of points called nodes and a set of node pairs called edges with a number called weight (the Euclidean distance in this paper) assigned to each edge. A MST is a tree which contains all the nodes and where the sum of the edges weight is minimal.

A microdensitometer analysis [3] of the micrograph related to sample A makes possible the determination of the position of particles considered as point objects (Fig. 2a). For convenience only a fraction of particles have been represented whereas full computations were based on $N = 5602$. Figure 2b shows the MST of Fig. 2a computed using Rohlf algorithm [4]. Depending on the starting point there may be more than one MST for a given set of points, but all the MSTs have the same edge-length histogram (Fig. 2c). It follows that statistical information deduced from this histogram such as the average edge-length m and the standard deviation σ may be used as characteristics for the corresponding distribution. In order to be able to compare different distributions regardless the particles density and the sampling window, the values of m and σ must be normalized. For this purpose the process described by Hoffman and Jain [5] has been used.

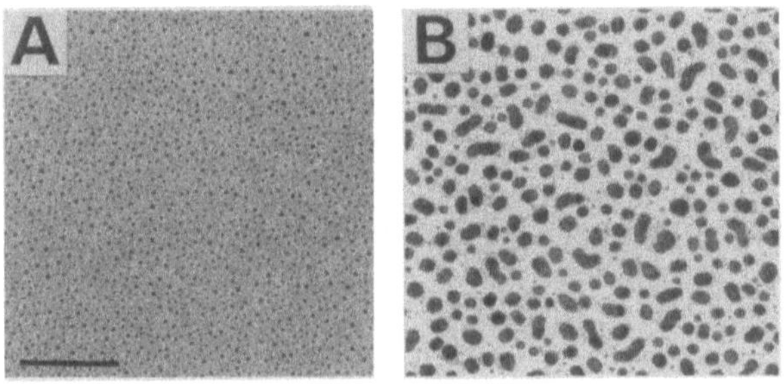

Fig. 1A and B. Thin gold film deposited on NaCl substrate previously covered with a carbon film, by sputtering technique, at two stages of its formation: **A** and **B**, same magnification. The line represents 500 Å

3. Characterization of a particles distribution

The use of a diagram involving both the values of m and σ enables us to determine the degree of order of any distribution by taking a simple reading in the (m, σ) plane on which well characterized distributions have previously been located (Fig. 3). First, perfectly orderly distributions are obviously characterized by $\sigma = 0$. There exist eleven possibilities permitting the arrangement of a set of points on a plane along regular or semi-regular mosaics [6]. The normalized m values corresponding to each model are reported in the Fig. 3. Each of those arrangements may be randomized by giving each point a new position deduced from its previous position using a Gaussian distribution with a standard deviation of ω. Then some trajectories in the

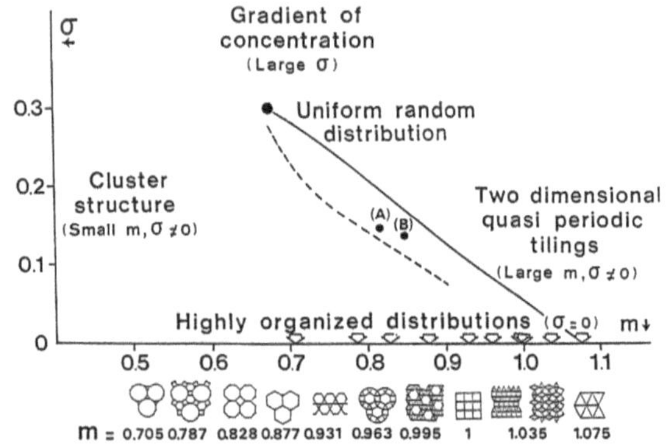

Fig. 3. Quantization of the (m, σ) plane. The eleven possibilities of arrangement of a set of points on a plane along regular or semi regular mosaics ($\sigma = 0$) and the m value corresponding to each model are reported on the figure. *Full line*: trajectory related to the triangular lattice. *Dashed line*: hardcore model

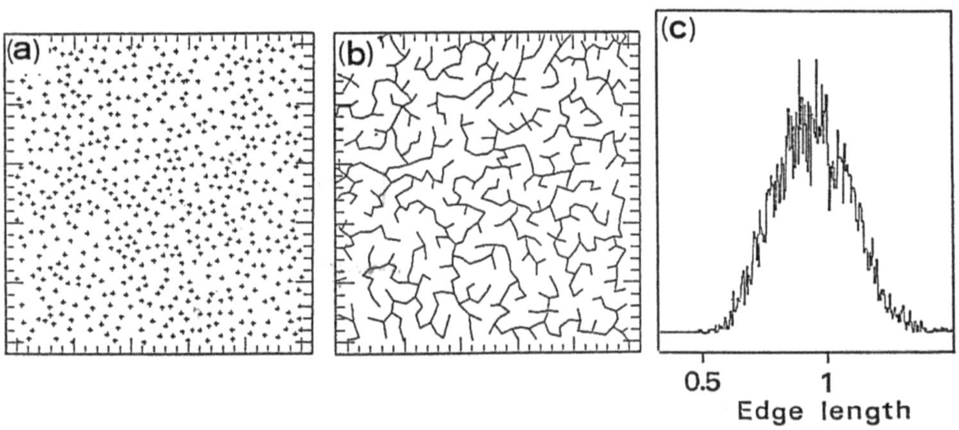

Fig. 2. a Positions of gold particles determined from microdensitometer analysis for sample A. **b** Minimal spanning tree concerning the distribution of **a. c** Histogram of edge lengths for the MST of the whole sample

(m, σ) plane are obtained, each point of the trajectories corresponding to a certain degree of order [2]. For a certain ω value the uniform random distribution is reached. The corresponding values $m=0.650$ and $\sigma=0.300$ – reported in the Fig. 3 – have been corroborated by Monte Carlo simulations of random distributions generated using the linear congruential method [7]. On the (m, σ) plane are also marked in different areas: cluster structures (small m, $\sigma \neq 0$), gradient of concentration (large σ), two dimensional quasi periodic tilings (large m, $\sigma \neq 0$).

Thanks to the normalization process any actual distribution can be plotted in the (m, σ) plane and easily compared to the above mentioned distributions. The values $m=0.816$ and $\sigma=0.151$ related to the gold particles of sample A are reported in Fig. 3. It appears that those particles are not randomly distributed. This result, which has been found on other metallic granular deposits [2] takes a special interest in the framework of theories about nucleation processes.

Let us now compare the particles distribution related to samples A and B. Clearly the values of m and σ reported on Fig. 3. show that the degree of order is higher for B than for A. Moreover those two distributions may be compared to a more realistic model, the hardcore one [9]. The hardcore model (also called inhibition model) is based on randomly distributed points with a constraint consisting of a minimal interpoint distance d which is related to the diameter of hard disks representing the particles. The hardcore curve which is related to disks having various diameters is reported on Fig. 3. From this comparison it is possible to deduce diameter d of the disks whose distribution has a degree of order which fits at best the distributions of A and B. It is found $d_A=0.60$ and $d_B=0.65$ in normalized values. The d_B value compares favourably to the mean diameter of the actual particles of B, namely $D_B=0.64$. Thus the geometrical behavior of B may be explained and means that this deposit approaches the continuous stage. On the other hand d_A is quite different from the mean diameter of the particles of A, namely $D_A=0.34$. This means that the particles of A are more ordered than which could be expected from the simple geometry and could be accounted for by a medium range interaction between the particles or by a quite regular distribution of nucleation germs on the substrate. The foregoing shows that MST analysis not only makes possible the determination of the degree of order for small particles scattered on a plane but also turns out to be a powerful tool to study the growth of small particles and thin films.

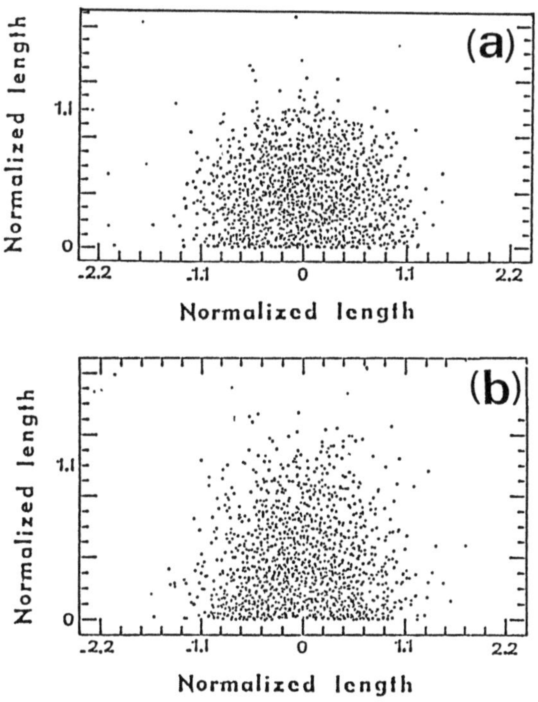

Fig. 4a and b. Angular diagrams obtained from the MST's of an isotropic distribution (**a**) and an anisotropic one (**b**). Due to graphical problems, the set of points in Fig. 4(b) is slightly anamorphosed

4. Angular diagrams

The MST may also be implemented to tackle the problem of the possible anisotropy for a distribution of particles located on a surface. Such a problem arises when dealing with the oriented growth of thin deposit upon a crystalline substrate [10]. For this purpose it is convenient to use an angular diagram (AD) which may be built in the following way [11]. The lower node of each edge is replotted to the centre O of the plane, and the upper one is put in the upper half of the plane. The set of points located in that upper half of the plane is the AD related to the initial distribution. For an isotropic distribution of points, the MST is made up of edges scattered isotropically. Therefore the corresponding AD is a set of points isotropically distributed on a half disk (Fig. 4a). When the distribution is no longer isotropic, the AD is altered. From this alteration it is possible to deduce information about the anisotropy of the distribution. In the Fig. 4b for example, the symmetry of the AD is broken, and a large part of points are grouped below the normalized length 0.2 along the y-axis, although the others are more sparsely distributed. This indicates a preferential orientation of particles along the x-axis, in the original distribution.

5. Conclusion

The above presented method makes possible the detection and the quantization of different kinds of cluster structure in arbitrary points set. It offers several advantages when compared to conventional statistical approaches. Combined with electron microscope information, microdensitometer analysis and reflection and transmission electron diffraction measurements, (m, σ) diagrams as well as AD's should lead, through the concept of order, to a better understanding of fundamental processes involved in the growth of clusters and the formation of thin deposits.

References

1. Zahn, C.T.: IEEE Trans. Comput. C **20**, 68 (1971)
2. Dussert, C., Rasigni, G., Rasigni, M., Palmari, J., Llebaria, A.: Phys. Rev. B **34**, 3528 (1986)
3. Rasigni, M., Rasigni, G., Palmari, J.P., Llebaria, A.: J. Opt. Soc. Am. **71**, 1124 (1981)
4. Rohlf, F.J.: Computer. J. **16**, 93 (1973)
5. Hoffman, R., Jain, A.K.: Pattern Recognit. Lett. **1**, 175 (1983)
6. Stevens, P.J.: Les formes dans la nature. Paris: Editions du Seuil 1978
7. Knuth, D.E.: The art of computer programming. New York: Addison Wesley 1969
8. Venables, J.A., Spiller, G.D.T., Nanücken, M.: Rep. Prog. Phys. **47**, 399 (1984)
9. Diggle, P.J.: Statistical analysis of spatial points patterns. London: Academic Press 1983
10. Bassett, G.A., Menker, J.W., Pashley, D.W.: In: Structure and properties of thin films. Neugebauer, C.A., New-Kirk, J.B., Vermilyea, D.A., (eds.), p. 11. New York: Wiley 1959
11. Dussert, C., Rasigni, G., Llebaria, A.: J. Theor. Biol. **135**, 295 (1988)

Z. Phys. D – Atoms, Molecules and Clusters 12, 45–51 (1989)

Atoms, Molecules and Clusters
Zeitschrift für Physik D
© Springer-Verlag 1989

On the structure and stability of small metal particles: high-resolution UHV electron microscope study

M. Mitome[1], Y. Tanishiro[2], and K. Takayanagi[1]

[1] Department of Materials Science and Engineering, Tokyo Institute of Technology, Nagatsuta, Midori-ku, Yokohama, Kanagawa 227, Japan
[2] Department of Physics, Tokyo Institute of Technology, Oh-okayama, Meguro-ku, Tokyo 152, Japan

Received 6 July 1988

Structures of gold clusters consisting of several to several tens of atoms and of small particles of 20–30 Å are seen by high-resolution UHV electron microscopy. The small particles have multiply twinned structures and show orientational fluctuation on a graphitized carbon substrate at room temperature, but no structural change from twinned to the fcc structure occurred during the observation with electron beam of 10–15 Amp/cm^2 on the specimen. Their structure changes at substrate temperature of 500°C. Small clusters show configurational change at room temperature from 5-fold arrangement to fcc one. These fluctuating motions are concluded to be due to thermal effect by estimating the temperature rise of the particles due to the electron beam to be less than 20–50°C from the melting temperature of lead particles.

PACS: 61.16.Di, 61.50.F

1. Introduction

Atomic clusters attract much interest because of their structures and physical properties. The stability of sodium [1], gold, silver and copper [2, 3] clusters has been investigated, and "magic numbers" for stable cluster sizes have been explained as the shell-closing number of the electron shell [1, 4]. The superiority of complete-shell has been predicted for free xenon clusters from these "magic numbers" [5].

Structures of small metal particles have been observed by electron microscopy, and multiply twinned particles have found to be stable structure form [6–8]. Recently structural anomaly was observed in gold particles of less than 50 Å in diameter [9–11]: The gold particles exhibit rapid configurational change from the face centered cubic (fcc), to multiply twinned (or twinned) structures such as icosahedron and decahedron under very strong electron beam irradiation (1.3×10^5 electrons/Å$^2 \cdot$ s $= 200$ Amp/cm^2 on specimen). Iijima [11] attributed the rapid structure change as due to charging effects of the particles through the emission of secondary electron rather than specimen temperature rise through the electron irradiation. Although a phenomenon suggesting charging effect [12] was reported previously, most of the epitaxial particles such as Au, Ag grown on MoS_2, MgO were caused only a little change in their orientation by electron irradiation [12]. The origin of the configurational change [11] (structural and orientational changes) and mobility of small particles [13–15] (clusters) are still a matter of debate.

In the present investigation, we used a high-resolution UHV electron microscope to observe structures and configurational changes of small metal particles (clusters) (Au, Pb) on atomic scale which were deposited on a graphitized carbon surface in situ in the electron microscope [16, 17]. This is to keep particle surfaces free from any contaminations, while in previous observations the surface cleanness is not certain because of poor vacuum condition inside electron microscopes [6] and/or of observations for post-mortem specimens (specimens were transferred into a electron microscope through atmospheric circumstance).

The present paper outlines structures of small gold particles and clusters having clean surfaces and structural changes of them by heating and by electron irradiation. The origin of the configurational changes is concluded to be due not to charging up through the imaging electrons [10, 11], but due chiefly to thermal effect of the substrate temperature; heating of small particles by electron irradiation is within 20–50 °C.

2. Experimental

The high-resolution UHV electron microscope, used has a 200 kV accelerating voltage and a point-to-point resolution of 0.22 nm. The vacuum at the specimen position is 10^{-8} Pa level and surfaces of Au and Si are kept clean; the reconstructed structures of Au(110) 2×1, Au(001) 5×1 and Si (111)· 7×7 of the clean surfaces are obtained in the microscope [16]. Gold (or lead) was deposited on a graphitized carbon film kept at room temperature and the structure of small particles which nucleate and grow on the lateral surface of the film (Fig. 1) were observed by the profile-view imaging on atomic scale. Structural changes of them during deposition and heating procedure were recorded continuously on VTR frames at every 1/30 s. The temperatures of the substrate was measured by a Pt-Pt 10% Rh thermo-couple, which was calibrated by a pyrometer within an accuracy of 30°C. The electron beam intensity on the specimen, measured by a Faraday cup at the imaging plane (the fluorescence screen), was about 10–15 Amp/cm^2. The intensity was intentionally increased to see effects caused by electron beam irradiation. To know the temperature increase by the electron beam irradiation, lead particles were deposited on a graphitized carbon surface in the same manner as gold particles and their melting temperatures were observed.

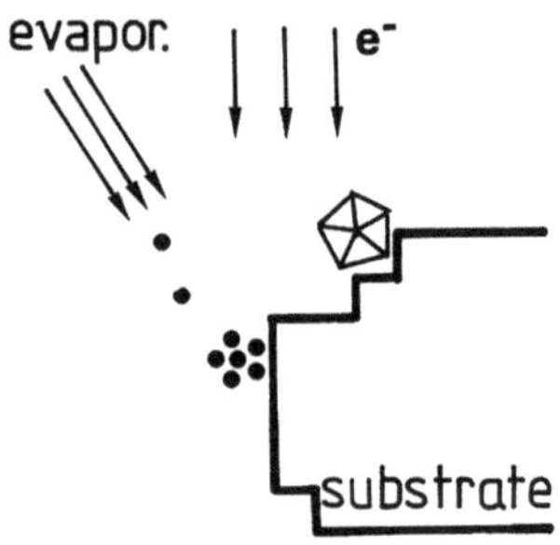

Fig. 1. Geometry of in-situ deposition in a high-resolution UHV electron microscope. Metal particles grown on the lateral surface of a graphitized carbon film are observed by profile imaging on atomic scale

3. Results

3.1. Growth of gold clusters

Figure 2 reproduces a growth process of a gold cluster on a graphitized carbon film at room temperature from the same VTR frames reported previously [16]. Fringes appearing in the lower half of each figure are the image of the basal lattice planes (*c*-plane) of the graphitized carbon (the *c*-planes are nomal to the plane of the figures). Each dot (single dot and agglomerates of dots) appearing in the upper side of the fringe images is a gold atom or an atomic row of gold atoms aligned in the direction parallel to the incident electron beam (normal to the plane of the figures). The contrasts of the dots indicate the number of gold atoms in the atomic rows. At the beginning of the deposition, dark dots were noticed to hop around the surface. It was noticed that not only single atoms but also clusters consisting of a few atoms hopped on the surface, since the number and arrangement of dots in a cluster were seen to vary for every hopping, as well as the contrast of the dots. A stable cluster (an agglomerate of dots on the right hand side of each figure) grew to form an icosahedral particle which has the multiply twinned structure [16] (hereafter called as MTP [6]). Atomic configuration of a small cluster changes as the cluster grows as seen in Fig. 2, and it is quite difficult to say that the growing gold clusters have a definite stable structure. It is seen that dark dots tend to arrange locally into tetrahedral configurations. The image in Fig. 2(h) is a profile image of an icosahedral particle (cf. Fig. 3(a)) along a 2-fold axis. The atomic rows in the surface layers are seen in the image: the contrast of the atomic rows decreases from one at the apex toward the other two apexes, indicating the decrease of atom numbers (4, 3, 2 and 1 atoms) in each atomic row.

3.2. Orientational fluctuation at room temperature

Large gold particles, having multiply twinned structure, were seen to have orientational fluctuation without impinging gold atoms as shown in Fig 3(a)–(c), where the MTP orientation changes from 2-fold to 3-fold and 5-fold. Small clusters also showed configuration change during observations without the impinging gold atoms. Although structures of such particles could not be determined well, a 5-fold configuration and a configuration indicating the fcc structure were often observed during structural variation. An example is shown in Fig. 4 for a small cluster (composed of about 30 atoms). Configurations of 5-fold symmetry in

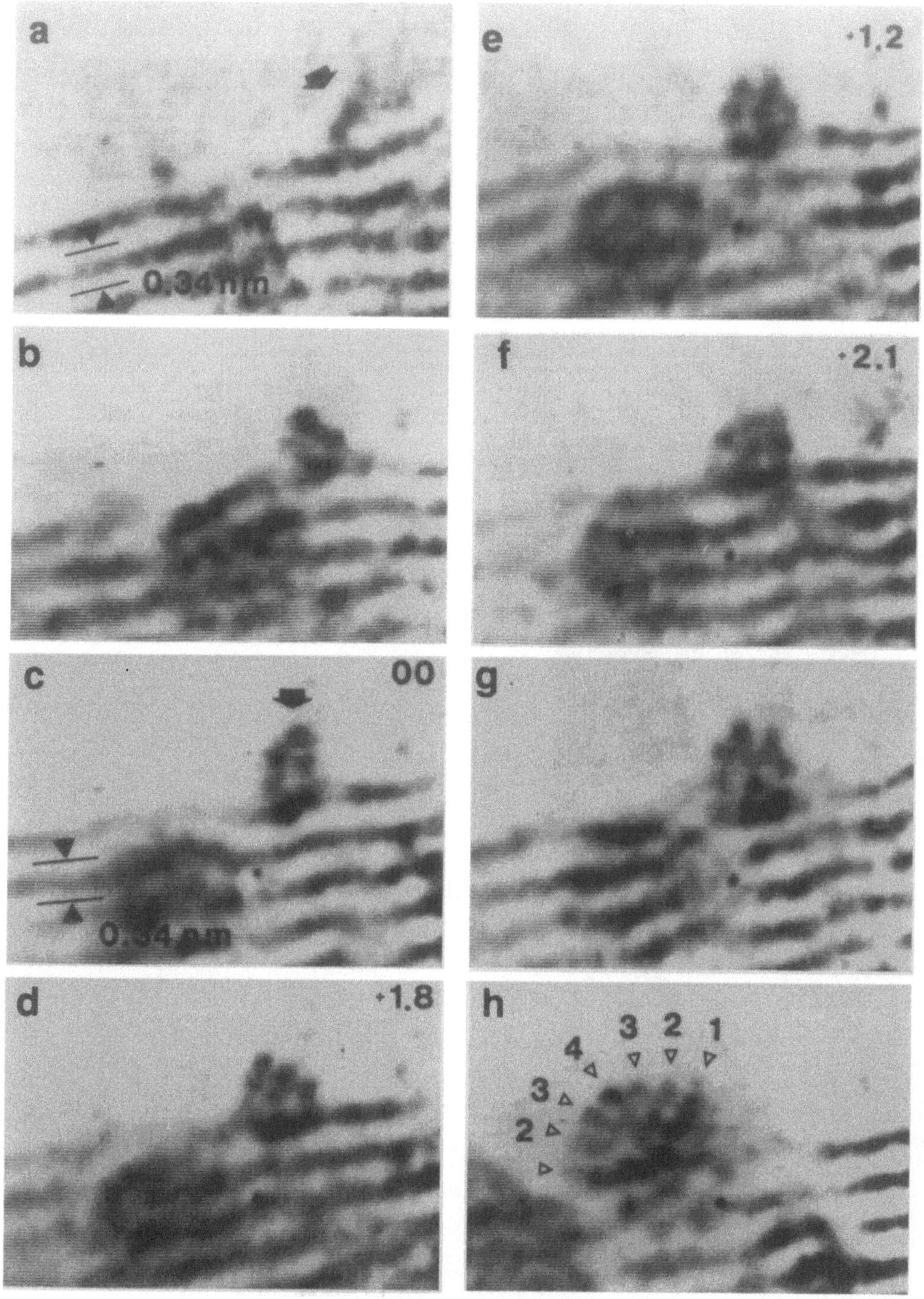

Fig. 2a-h. Growth process of a gold cluster on graphitized carbon. A dark dot in (**a**) and clusters of dots are images of gold clusters. Fringes in the lower half of each figure are lattice fringes of the graphitized carbon. A cluster of two dark dots in (**a**) indicated by an arrow is stable one and grow further in (**b**)–(**h**) to form a multiply twinned particle. Time differences between frames from (**b**) to (**f**) are indicated. Note changes in atomic configurations of clusters

Fig. 4(b), (c) and (d) and another configuration of fcc structure are noticed locally in each image.

In summary, gold particles of 20–30 Å in diameter on room temperature substrate were found to have multiply twinned configurations as stable structures. Their orientations fluctuated, but no structural change to fcc structure occurred during observations with electron beam flux of 10–15Amp/cm^2. The structural change was not observed even for stronger electron beam irradiation (about 80 Amp/cm^2). Small clusters

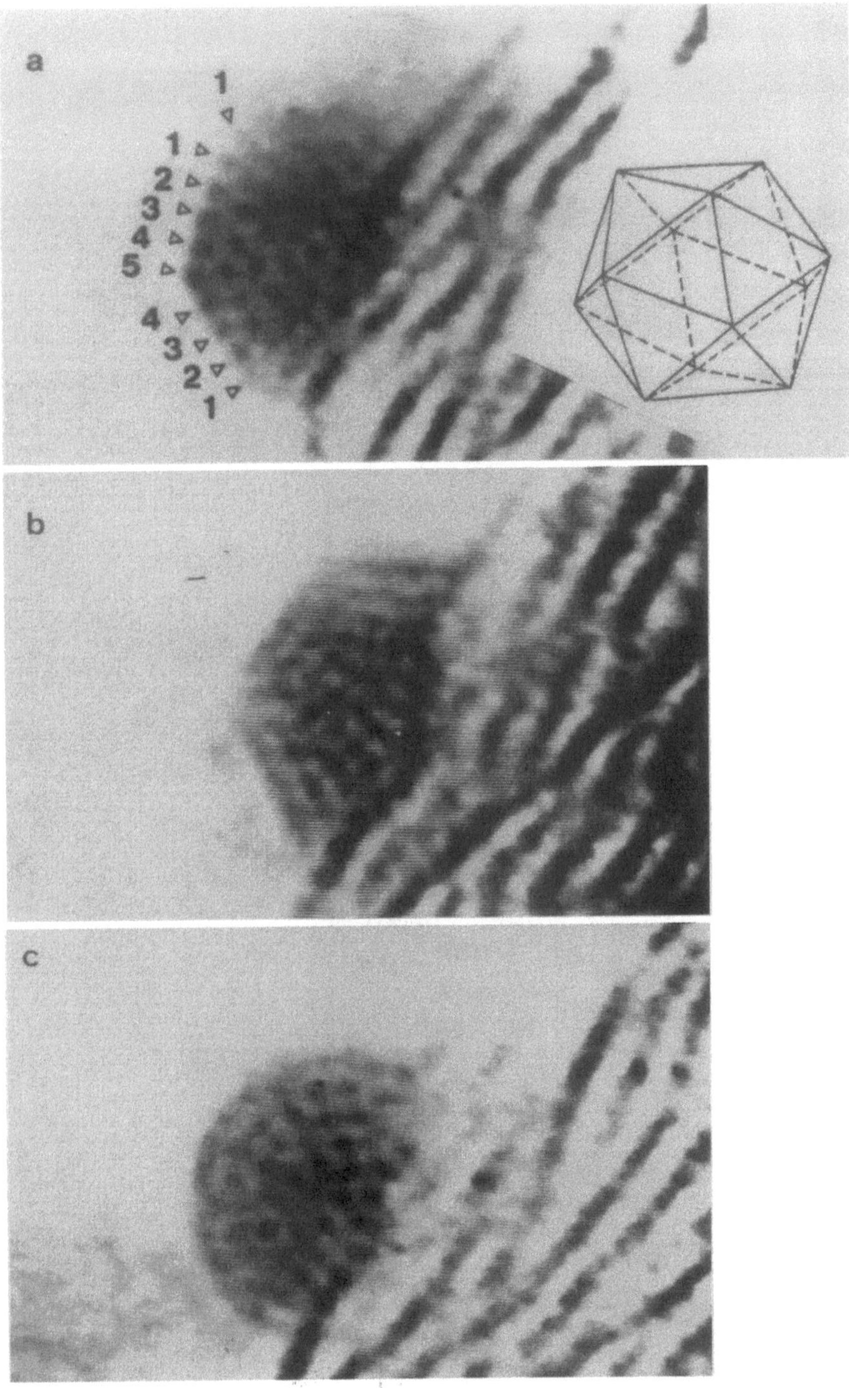

Fig. 3a–c. Orientational fluctuation of a gold multiply twinned icosahedron (not complete icosahedron). The images in (a)–(c) have 2-fold, 3-fold and 5-fold symmetry, respectively. Gold deposition has been stopped

consisting of several to several tens of atoms changed their configurations. Orientational and structural fluctuations became rapidly as the electron beam intensity was increased.

3.3. Structural changes at elevated temperature

Changes in the structure and the orientation of MTP's were observed at elevated temperatures (a range up to

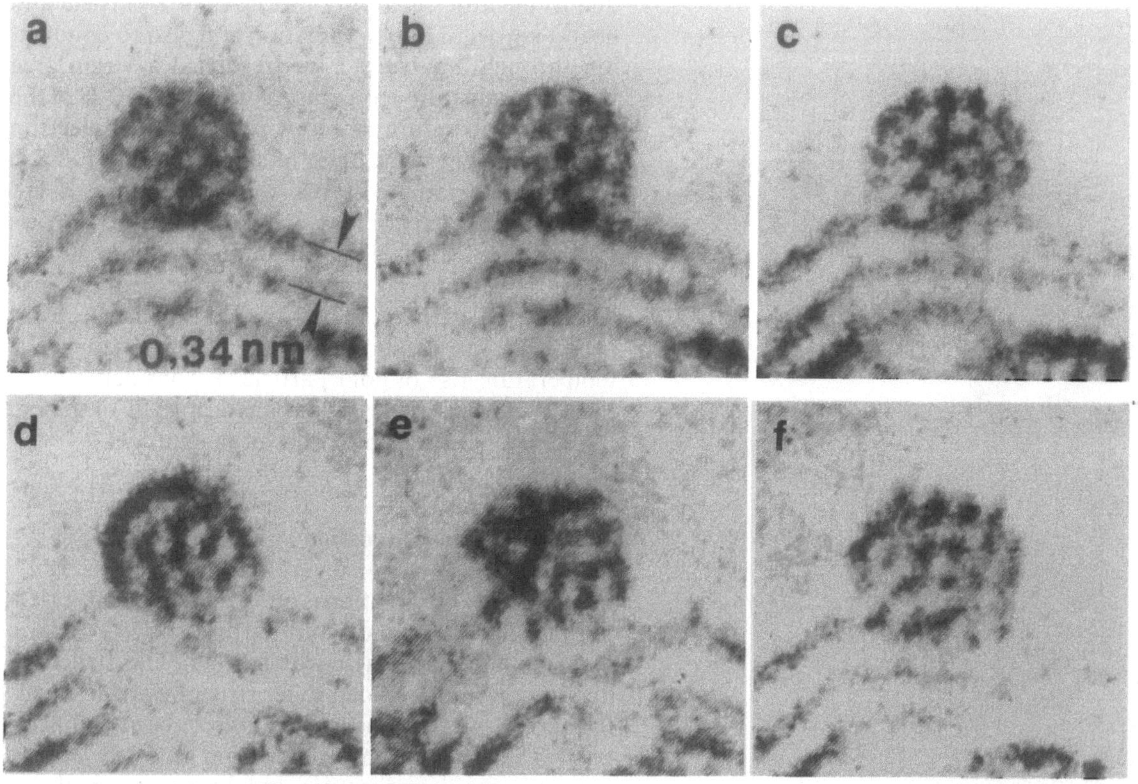

Fig. 4a–f. Configurational change of gold cluster composed with about 30 atoms. Note the changes in arrangements of *dark dots* of the clusters from **a**–**f**. The substrate is room temperature

840°C). As shown in Fig. 5(a) and (b), multiply twinned particles of about 30 Å in diameter were transformed to fcc structure (by elimination of twins) and re-transformed to multiply twinned structure during the observation. In some occasion lattice images of the particle were disappeared as shown in (c). This disappearance of the fringe image was surmised by Iijima as due to "melting". However, the particle is still in solid phase since it has a crystalline habit and the lattice fringes were recognized in VTR frames soon after. Such structural fluctuations of the particle were noticed to occur only above about 500°C (larger particles showed the structural fluctuation at higher temperatures). It was noticed further that no such structural fluctuations of the particle occurred below 500°C even when the incident beam intensity was raised. This observation indicates that thermal energy activates the structural transformation from twinned to the fcc structure and vice versa.

The higher was the substrate temperature, the faster became the motion of the dark dots (lattice fringes) in clusters. The smaller were the particles, the lower became the melting temperature [18]. And the smaller was the particle, the lower became the temperature which the structural fluctuation from twinned to the fcc structure start at. Therefore, it is quite certain that the rapid configurational change, structural and orientational fluctuation, is caused by thermal effect.

3.4. Temperature rise by the incident beam

It has been discussed for a long time whether the incident beam raise the substrate temperature or not. Although the amount of the temperature rise is quite difficult to be measured, it depends on the substrate material, the supporting of the substrate and the kind of metal. We estimated the amount by melting of lead particles. Since small particles have lower melting point than the bulk ($t_m = 327$°C) (super cooling effect [18]), we choose lead particles of 100–150 Å in diameter and determined their melting temperature by changing the substrate temperature under the incident electron beam of 10–15 Amp/cm^2. The melting temperature of lead particles of 130 Å (60 Å) in diameter was observed to be about 310°C (180°C): Lead particles, after melting, did not show lattice fringes during a prolonged observation time, while particles before melting show regular lattice fringes' image and structure-less images occasionally. Thus, the melting

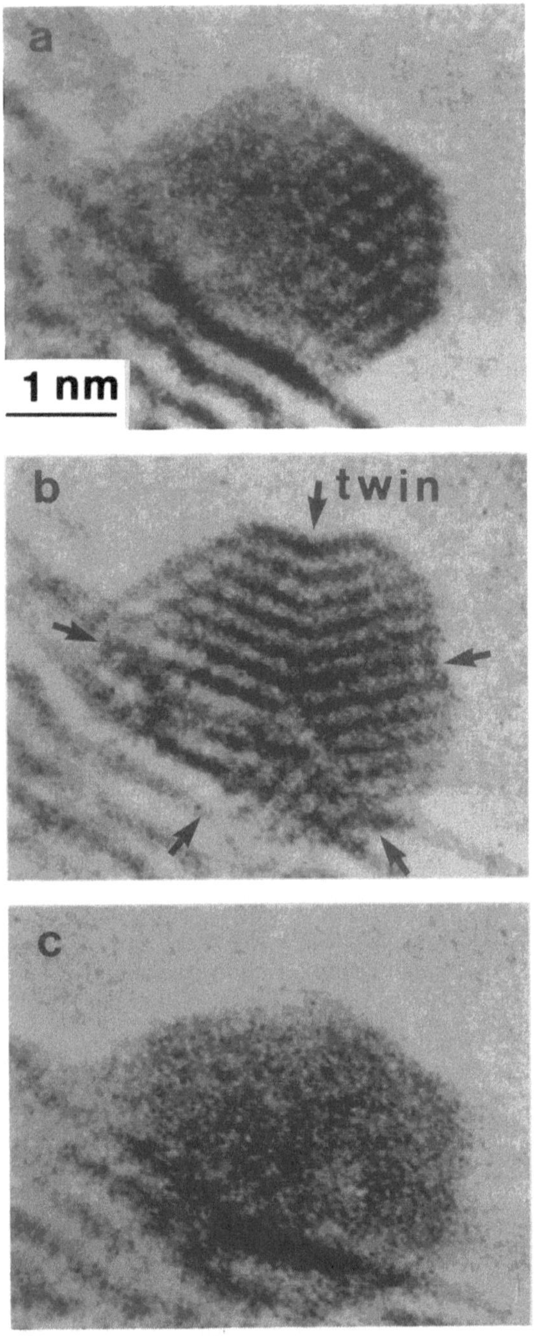

Fig. 5. Structural fluctuation of a gold cluster of 30 Å in diameter at 570°C. The structure in **a** has no twin (fcc structure), while in **b** has twins (decahedral MTP)

experiment suggests that temperature rise by electron irradiation of 10–15 Amp/cm^2 is within 20°C (do not exceed 50°C by taking into account of the inaccuracy of the calibration). To estimate the energy (heat) transferred to a particle from the irradiated electron, which is determined by a balance between heat gain due to inelastic scattering processes in a particle and heat dissipation due to radiation and also to conduction through a substrate, we irradiated a virgin area while the substrate temperature was kept 310°C of the melting temperature of 130 Å particle under electron irradiation. Then, particles of similar sizes (170 Å) in the freshly irradiated area were melt after a few minutes. This experiment confirms that the particle temperature increases but the amount is about 20°C at most. Take into account the specific heat of lead (0.126 J/g·K at 0°C) the heat brought by the electrons is estimated to be about 3×10^{-5} eV/s per atom. The temperature rise of gold particles is considered to be the same order since the specific heat (0.127 J/g·K at 20 − 100°C) is the same order of magnitude.

4. Summary and discussion

The present observations on structural and orientational fluctuations of gold particles of 20 − 30 Å in diameter are summarized as follows. The structural fluctuation from twinned to fcc structure is due chiefly to thermal effect and is excited above about 500°C for gold. The orientational fluctuation is easily excited even at room temperature and becomes rapidly by strong irradiation (about 80 Amp/cm^2). The temperature rise for such irradiation, however, was not enough to excite the structural fluctuation observed at 500°C. The structural change from twinned to fcc structure observed by Iijima under strong electron beam irradiation which is more than twenty times stronger than ours is surmised due to heating of the specimen above 500°C, although Iijima discussed charging effect.

Structures of small clusters less than 10 Å consisting of 30–50 atoms were observed for the first time, and they were found to fluctuate at room temperature. Configurations of the 5-fold and/or fcc structure exist locally in the particles during the fluctuation. Since atoms (surface atoms) in small clusters, which have the lower melting temperatures, are more mobile than those in large particles, the fluctuating motions of small clusters at room temperature is the intrinsic nature. The temperature rise due to the electron beam irradiation, of the order of 20°C, does not affect the mobility much. Examinations at low temperatures are interesting for determining the structure of small clusters, because phonon excitations (corresponding to the observed rapid configurational change) are reduced much. Similar experiments on semiconductors having a different physical nature from metals are also interesting.

Present authors express their thanks to Mr. K. Kobayashi for his skillful assistance in experiments.

References

1. Knight, W.D., Clemenger, K., de Heer, W.A., Sanders, W.A., Chou, M.Y., Cohen, M.L.: Phys. Rev. Lett. **52**, 2141 (1981)
2. Katakuse, I., Ichihara, T., Fujita, Y., Matsumoto, T., Sakurai, T., Matsuda, H.: Int. J. Mass Spectrom Ion Proc. **67**, 229 (1985)
3. Katakuse, I., Ichihara, T., Fujita, Y., Matsuo, T., Sakurai, T., Matsuda, H.: Int. J. Mass Spectrom. Ion Proc. **74**, 33 (1986)
4. Ishii, Y., Ohnishi, S., Sugano, S.: Phys. Rev. B **33**, 527 (1986)
5. Echt, O., Sattler, K., Reckenagel, E.: Phys. Rev. Lett. **47**, 1121 (1981)
6. Ogawa, S., Ino, S., Kato, T., Ota, H.: J. Phys. Soc. Jpn **21**, 1963 (1966)
7. Ino, S.: J. Phys. Soc. Jpn **27**, 941 (1967)
8. Marks, L.D., Howie, A.: Philos. Mag. A **49**, 81 (1984)
9. Iijima, S., Ichihashi, T.: Jpn. J. Appl. Phys. **24**, L125 (1985)
10. Iijima, S., Ichihashi, T.: Phys. Rev. B **56**, 616 (1986)
11. Iijima, S.: In: Microclusters, Sugano, S., Nishina, Y., Ohnishi, S., (eds.), p. 186. Berlin, Heidelberg, New York: Springer 1987
12. Some gold particles (dendrite crystal) grown on a graphite substrate moved on the surface by oscillatory motion which became faster as increase of the incident electron beam intensity. [13]
13. Honjo, G., Takayanagi, K., Kobayashi, K., Yagi, K.: Phys. Status Solidi **A255**, 353 (1979)
14. Kern, R., Masson, A., Metois, J.J.: Surf. Sci. **27**, 483 (1971)
15. Metois, J.J., Heinemann, K., Poppa, H.: Philos. Mag. **35**, 1413 (1977)
16. Takayanagi, K., Tanishiro, Y., Kobayashi, K., Akiyama, K., Yagi, K.: Jpn. J. Appl. Phys. **26**, L957 (1987)
17. Takayanagi, K.: In: Proceedings of the Asia Pacific Symposium on Surface Physics, Xie Xide (ed.), p. 23 Singapore: World Scientific 1987
18. Buffat, Ph., Borel, J.-P.: Phys. Rev. A **13**, 2287 (1976)

Z. Phys. D – Atoms, Molecules and Clusters 12, 53–59 (1989)

Atoms, Molecules and Clusters
Zeitschrift für Physik D
© Springer-Verlag 1989

A group theoretical paradigm for describing the skeletal molecular orbitals of cluster compounds. Part 3*

D.M.P. Mingos and Lin Zhenyang
Inorganic Chemistry Laboratory, University of Oxford, South Parks Road, Oxford OX1 3QR, UK

Received 5 July 1988

The group theoretical consequences of the Tensor Surface Harmonic Theory have been developed for $[ML_2]_n$, $[ML_4]_n$ and $[ML_5]_n$ clusters where either the xz and yz or x^2-y^2 and xy components to L_d^π and L_d^δ do not contribute equally to the bonding. The closed shell requirements for such clusters are defined and the orbital symmetry constraints pertaining to the interconversion of conformers of these clusters are described.

PACS: 31.20N; 36.40

Introduction

The Polyhedral Skeletal Electron Pair Theory has proved to be useful for accounting for the structures of cluster compounds of transition metals and main group atoms [1, 2]. The theoretical basis for this approach has been provided by the Tensor Surface Harmonic Theory developed initially by Stone [3, 4] for deltahedral clusters and subsequently extended to three and four connected, condensed and high nuclearity clusters [5–8]. The group theoretical consequences of the Tensor Surface Harmonic Theory have been defined by Ceulemans [9], Fowler [10, 11], Quinn [12] and Johnston and Mingos [13–16]. These conclusions have been based primarily on the parity relationships which connect L^π and $\bar{L}^\pi$ molecular orbitals of cluster compounds. The parity relationship is defined as the local rotation of each p_π component of a molecular orbital by a rotation of 90° in the same clockwise or counterclockwise sense. The effect of such a rotation is to convert a bonding molecular orbital (L^π) into an equally antibonding molecular orbital ($\bar{L}^\pi$). Group theoretically the parity operation is the one dimensional pseudo-scalar representation, Γ_u^0, which is symmetric to all proper operations and antisymmetric to all improper operations. For example, for a tetrahedral cluster the $P^\pi(t_2)$ and $\bar{P}^\pi(t_1)$ surface molecular orbitals are related by the parity operator, $\hat{P}$, which transforms as a_2 in the tetrahedral point group.

$$t_2 \xrightarrow[a_2]{\hat{P}} t_1$$

Or, generally Γ_u^0 in a point group, i.e.

$$\Gamma_{L^\pi} \otimes \Gamma_u^0 = \Gamma_{\bar{L}^\pi}$$

If the above direct product generates the original representation then the representation is described as self-conjugate:

$$\Gamma_A \otimes \Gamma_u^0 = \Gamma_A$$

The occurrence of an odd number of self-conjugate representations for a cluster is indicative of non-bonding molecular orbitals and alternative electron counts. For example, in the tetrahedron the e molecular orbitals are self-conjugate:

$$e \otimes \Gamma_u^0 = e \otimes a_2 = e$$

Such a self-conjugate representation must be equally well described as L^π and $\bar{L}^\pi$ and is consequently non-bonding, i.e. e in tetrahedron is $D^\pi/\bar{D}^\pi$. These two orbitals remain unoccupied in B_4Cl_4, but is filled in P_4 and C_4R_4.

* Parts 1 and 2 are Refs. 15 and 16

In a large number of clusters of the transition metals the ligand arrangement around a specific metal atom does not have axial symmetry and therefore the p_π orbitals do not comprise a complete set of basis functions. In this paper, a group theoretical analysis for clusters based on ML_2 and ML_4 fragments with only one p_π component and ML_5 fragments with only one d_δ component is presented and used to account for their closed shell requirements.

ML_2 and ML_4C_{2v} fragments

Evans and Mingos [17, 18] have completed semi-empirical molecular orbital calculations on clusters based on ML_2 and ML_4 fragments and these results have been used to confirm the group theoretical arguments developed in this paper. In $[ML_2]_n$ clusters the absence of axial symmetry for the ML_2 fragment leads to the possibility of alternative high symmetry conformations. The alternative conformations of this type for tetrahedral and triangular clusters are illustrated in (1) and (3) and (4) and (6). In these instances

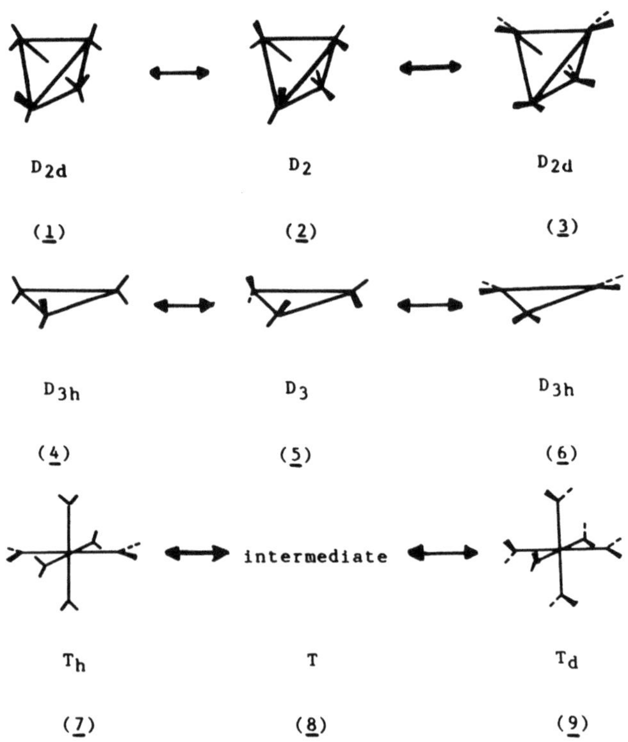

although the conformations belong to the same point group they represent very different bonding situations. They can be interconverted by simultaneous rotations of the ML_2 fragments by 90° involving intermediates belonging to the pure rotation groups D_2 (2) and D_3 (5) respectively. In contrast the $[ML_2]_6$ clusters illustrated in (7) (T_h point group) and (9) (T_d point group) are converted into each other by simultaneous rotations of the ML_2 fragments by 45° and involve the intermediate (8) belonging to the T pure rotation group, i.e. $T_h \rightarrow T \rightarrow T_d \rightarrow T \rightarrow T_h$. A rotation of 90° leads to an identical geometry, i.e. $T_h \rightarrow T_h$ and $T_d \rightarrow T_d$.

For a tetrahedral cluster derived from conical *B-H* or ML_3 fragments the tangential skeletal molecular orbitals transform as $t_2(P^\pi)$ bonding, $e(D^\pi/\bar{D}^\pi)$ non-bonding and $t_1(\bar{P}^\pi)$ antibonding. Table 1 summarises the relevant orbital correlations for the descent of symmetry from T_d to D_{2d} to D_2 and also the symmetry properties of the pseudo-scalar parity operator Γ_u^0. In the pure rotation groups C_n, D_n, 0 and T the parity operator must by definition belong to the totally symmetric representation. Therefore, L^π and $\bar{L}^\pi$ molecular orbitals must occur in pairs belonging to the same irreducible representations, such that:

$$\Gamma_{L^\pi} \otimes A = \Gamma_{\bar{L}^\pi} = \Gamma_{L^\pi}$$

For example, in Table 1 for D_2 L^π and $\bar{L}^\pi$ transform as $2a + 2b_1 + 2b_2 + 2b_3$. Furthermore, for clusters derived from angular fragments one subset is associated with $L_x^\pi/\bar{L}_x^\pi$ (or $L_{xz}^\pi/\bar{L}_{xz}^\pi$ for a transition metal) and the other with $L_y^\pi/\bar{L}_y^\pi$ (or $L_{yz}^\pi/\bar{L}_{yz}^\pi$). The x and xz orbitals are defined locally to lie in the angular ML_2 plane. An alternative proof of this generalisation is based on Quinn's relationship [12]:

$$\Gamma_{\pi x} + \Gamma_{\pi y} = \Gamma_\sigma \otimes \Gamma_z + \Gamma_\sigma \otimes \Gamma_{Rz}$$

Since both z and R_z transform in an identical fashion in the rotational groups pairs of matching L_x^π and L_y^π representations are generated.

If the fragments retained axial symmetry the resultant L_x^π and L_y^π molecular orbitals would mix extensively and regenerate molecular orbitals which closely resembled those in the parent tetrahedron. In angular ML_2 and ML_4 fragments the xz and yz components not only have different symmetry characteristics, but also very different bonding capabilities [17, 18]. Molecular orbital calculations have shown that the spectrum of tangential skeletal molecular orbitals is dominated by the xz component. The yz components give

Table 1. Correlation for descent in symmetry for $T_d \rightarrow D_{2d} \rightarrow D_2$

	T_d	D_{2d}	D_2
$(\bar{P}^\pi)$	t_1	a_2+e	$b_1+b_2+b_3$
$(D^\pi/\bar{D}^\pi)$	e	a_1+b_1	$2a$
(P^π)	t_2	b_2+e	$b_1+b_2+b_3$
Γ_u^0	a_2	b_1	a

rise to a narrow spectrum of bonding, non-bonding and antibonding molecular orbitals which are completely filled by $2n$ electrons in a $[ML_2]_n$ cluster and have a neutral effect on the bonding requirements of the cluster.

For the D_{2d} geometries illustrated in (1) and (3) the tangential skeletal molecular orbitals derived from xz give rise to the following components:

T_d	(1)	(3)
t_1	$e(\bar{L}^{\pi}_{xz})$	$a_2(\bar{L}^{\pi}_{xz})$
e	$a_1(L^{\pi}_{xz}/\bar{L}^{\pi}_{xz})$	$b_1(L^{\pi}_{xz}/\bar{L}^{\pi}_{xz})$
t_2	$b_2(L^{\pi}_{xz})$	$e(L^{\pi}_{xz})$

In D_{2d} Γ^0_u transforms as b_1 and leads to the following relationships:

$$e \otimes \Gamma^0_u = e$$
$$b_2 \otimes \Gamma^0_u = a_2$$
$$a_1 \otimes \Gamma^0_u = b_1$$

Therefore, the parity operation has the effect of transforming a bonding skeletal molecular orbital in (1) into an antibonding skeletal molecular orbital in (3). Each conformation has one component of the original non-bonding e molecular orbitals and they are also related by the parity transformation. The resultant spectra of molecular orbitals are summarised in Fig. 1a. The results are in complete agreement with the molecular orbital calculations reported previously [17]. The conformation (3) is clearly the more stable and is associated with $3(e+b_1)$ bonding and non-bonding L^{π}_{xz} skeletal bonding molecular orbitals. Since the tetrahedron generally has 5 L^{π} bonding and non-bonding skeletal molecular orbitals it follows that $[ML_2]_4$ clusters will be associated with 56 rather than 60 valence electrons. $[Pt_4(\mu_3-H)_4(PPr^i_2Ph)_4]$ with 56 valence electrons provides an example of such a cluster compound [17].

It is also apparent from Fig. 1 that the interconversion (1)→(3) through (2) is an orbitally forbidden process [19] when the spectrum of L^{π}_{xz} molecular orbitals is filled with 2 to 6 valence electrons. Furthermore, the a, b_1, b_2 and b_3 molecular orbitals in the intermediate D_2 geometry must be approximately non-bonding.

Similar arguments can be developed for the interconversion of the two D_{3h} conformations of a triangular cluster (4) and (6) through the D_3 intermediate (5). The relevant orbital correlation diagram is illustrated in Fig. 1b, and the orbital correlations are summarised in Table 2. The more stable conformer is (6) since this is associated with 2 bonding tangential skeletal molecular orbitals of e' symmetry. This conformation is observed in 48 electron $M_3(CO)_{12}$ (M = Ru or Os) [20] and 42 electron platinum cluster, e.g. $[Pt_3(\mu_3$-

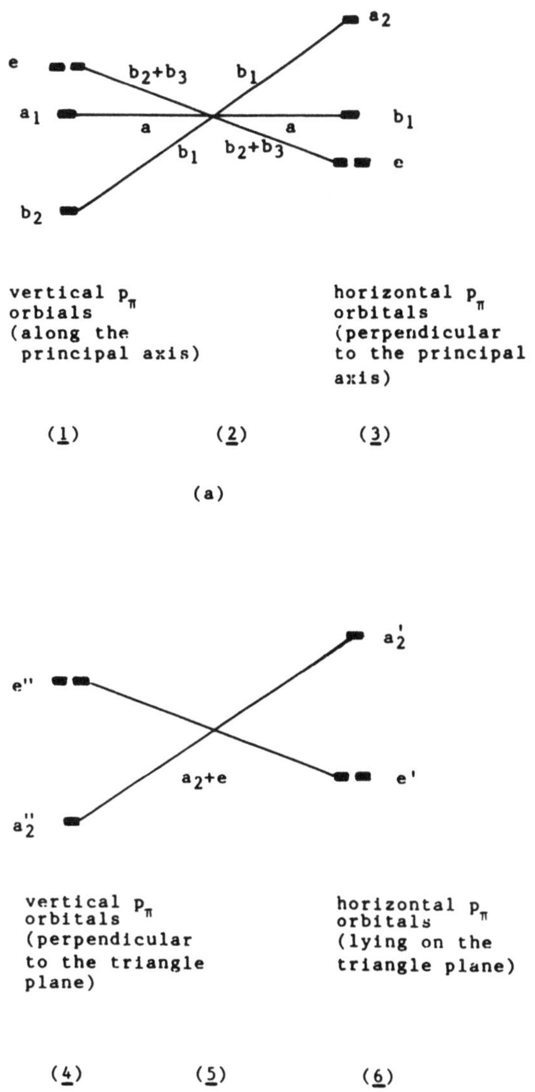

Fig. 1. The correlation diagrams of L^{π}'s (**a**) for (1), (2) and (3); and (**b**) for (4), (5) and (6)

Table 2. Correlation for $D_{3h} \rightarrow D_3$

	D_{3h}	D_3
$(\bar{P}^{\pi})$	$e''+a'_2$	$e+a_2$
(P^{π})	a''_2+e'	a_2+e
Γ^0_u	a''_1	a_1

$H)_3H_3(PPr^i_2Ph)_3]$ [21]. The interconversion of (4) to (6) through a D_3 intermediate is once again a symmetry forbidden process.

The following general points emerge from the analysis discussed above.

1. The bonding, non-bonding and antibonding components for the two conformations of $[ML_2]_n$, which are related by 90° rotations of the ML_2 fragments, correlate with the L^{π} and $\bar{L}^{\pi}$ skeletal molecular orbitals of a $[ML_3]_n$ cluster only when they are taken together.

2. The molecular orbitals of one conformer are one subset of the $L^{\pi}/\bar{L}^{\pi}$ spectrum and are not related by the parity operator Γ_u^0.

3. The parity operator does relate the molecular orbitals of the conformers A and B in the following manner:

$$\Gamma_{L^{\pi}}^{A\,(\mathrm{bonding})} \otimes \Gamma_u^0 \rightarrow \Gamma_{\bar{L}^{\pi}}^{B\,(\mathrm{antibonding})}$$
$$\Gamma_{L^{\pi}/\bar{L}^{\pi}}^{A\,(\mathrm{non-bonding})} \otimes \Gamma_u^0 \rightarrow \Gamma_{L^{\pi}/\bar{L}^{\pi}}^{B\,(\mathrm{non-bonding})}$$

For example, for (1) and (3)

$$b_2^{(1)} \otimes \Gamma_u^0 = a_2^{(3)}$$
$$a_1^{(1)} \otimes \Gamma_u^0 = b_1^{(3)}$$
$$e^{(1)} \otimes \Gamma_u^0 = e^{(3)}$$

4. Since the parity operator does not relate bonding and antibonding molecular orbitals within the same conformer these clusters do not have a symmetrical spread of molecular orbitals, and the more stable conformer is that which has the larger number of bonding molecular orbitals, i.e. (3) for $[ML_2]_4$ and (6) for $[ML_2]_3$.

5. The interconversion of the conformers by a simultaneous rotation of the angular fragments is a symmetry forbidden process.

In contrast the octahedral clusters (7) and (9) are converted into themself by simultaneous rotations of the ML_2 fragments by 90°. The parent octahedral cluster is characterised by $t_{2g}(D^{\pi})$ and $t_{1u}(P^{\pi})$ bonding and $t_{1g}(\bar{P}^{\pi})$ and $t_{2u}(\bar{D}^{\pi})$ antibonding skeletal molecular orbitals. The relevant orbital correlations in O_h, T, T_h and T_d are given in Table 3. The two equivalent conformations of (7) are characterised by the following components of the octahedral molecular orbitals.

O_h	(7)	(7)	(9)	(9)
t_{2u}				
t_{1g}	t_g	t_g	t_1	t_1
t_{1u}	t_u	t_u	t_2	t_2
t_{2g}				

Since the conformations are identical the parity operator does now relate the bonding and antibonding skeletal orbitals in one conformer:

Table 3. Correlation for descent in symmetry $O_h \rightarrow T_h \rightarrow T_d \rightarrow T$

	O_h	T_h	T_d	T
$(\bar{D}^{\pi})$	t_{2u}	t_u	t_1	t
$(\bar{P}^{\pi})$	t_{1g}	t_g	t_1	t
(P^{π})	t_{1u}	t_u	t_2	t
(D^{π})	t_{2g}	t_g	t_2	t
Γ_u^0	a_{1u}	a_u	a_2	a_1

$$t_u^{(7)} \otimes \Gamma_u^0 = t_g^{(7)}$$
$$t_2^{(9)} \otimes \Gamma_u^0 = t_1^{(9)}$$

Furthermore the intermolecular rearrangement of (7) into (9) and then into itself is now a symmetry allowed process. It follows also from the above analysis that deltahedral clusters based on ML_2 fragments (except the tetrahedron) will be characterised by $n/2+1$ bonding skeletal molecular orbitals if they have two identical conformers related by 90° rotations.

L^{δ} and $\bar{L}^{\delta}$ inequivalences in ML_5 fragments

The halide and sulfide clusters of niobium, tantalum, molybdenum and tungsten in their lower oxidation states [22], e.g. $[M_6(\mu_3\text{-Cl})_8L_6]$ (10) and $[M_6(\mu_2\text{-Cl})_{12}L_6]$ (11), provide examples of clusters based on the ML_5 fragments. The bonding in these clusters has been studied by Ceulemans and Fowler [23], and Johnston and Mingos [24] independently within the framework of Stone's tensor surface harmonic theory (TSH). Ceulemans and Fowler's analysis is based on ML_5 fragments with four effective valence orbitals. They also noted the parity relationship between the d_{δ} orbital pattern of face-bridged cluster (10) and that of the edge-bridged cluster (11). In this section, the group theoretical analyses for the octahedral clusters (10) and (11) are extended to their *nido* and *arachno* derivatives. Figure 2 illustrates the structures of *closo*, *nido* and *arachno* chloride-bridged clusters based on (10) and (11).

The molecular orbital diagram for an ML_5 fragment is shown in Figure 3. The frontier fragment orbitals are $hy(\sigma)$ and the "t_{2g}" set. In carbonyl clusters, the "t_{2g}" orbitals usually are core-like because of the strong stabilisation by CO π^* vacant orbitals. In the π-donor clusters, e.g. sulfide- and chloride-bridged clusters, the "t_{2g}" orbitals are more important for the metal-metal interactions. The three "t_{2g}" orbitals are

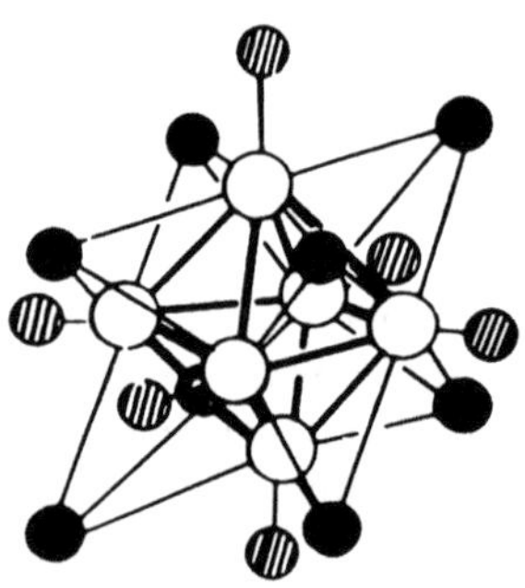

$[M_6Cl_8L_6]^{4+}$ (M=Mo or W)

(10)

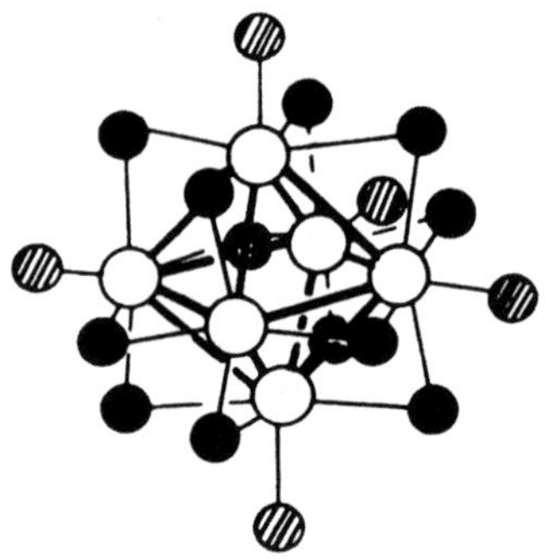

$[M_6Cl_{12}L_6]^{2+}$ (M=Nb or Ta)

(11)

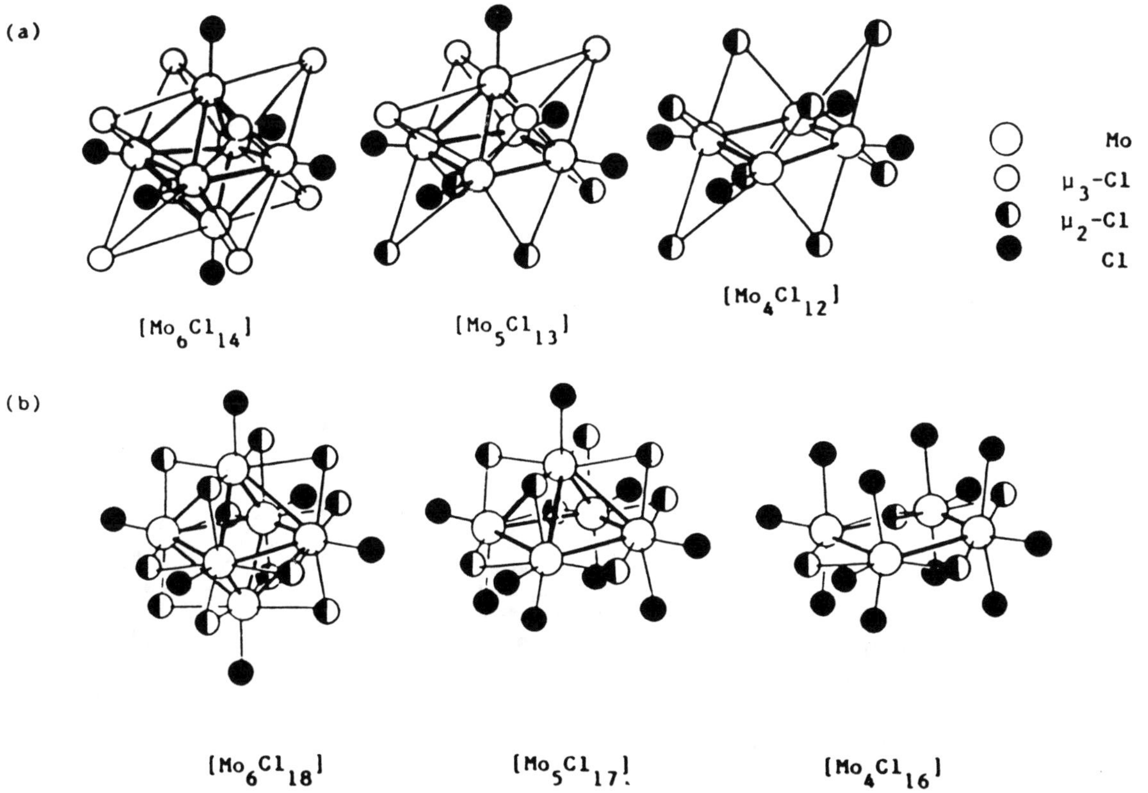

Fig. 2. *Closo, nido* and *arachno* chloride-bridged clusters based on (**a**) $[Mo_6Cl_8L_6]$ (10) and (**b**) $[Mo_6Cl_{12}L_6]$ (11)

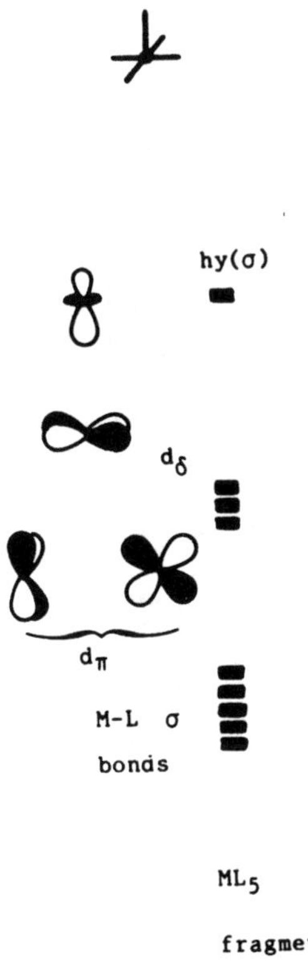

Fig. 3. The MO level diagram of ML_5 fragment

divided into a d_π pair and one d_δ component. The molecular orbitals derived from the interactions of $hy(\sigma)$ and d_π pair of the ML_5 fragments are entirely equivalent to those derived from the interactions of the three frontier orbitals of $M(CO)_3$ fragments, and is independent of the arrangement of ligands. In contrast, the d_δ–d_δ interactions depend greatly on the arrangement of ligands.

In a local coordinate frame, the ML_5 fragment in (10) appears to be rotated through 45° about the radial vector relative to the ML_5 fragment in (11). The single d_δ component is defined as $\delta(x^2-y^2)$ for ML_5 the fragments in (10) and $\delta(xy)$ for those in (11). In the TSH theory L_d^δ (bonding) is related to $\bar{L}_d^\delta$ (antibonding) by a local rotation of 45° at each vertex about the radial vector.

$$\hat{P} \otimes L^\delta = \bar{L}^\delta$$

Therefore, the d_δ orbital pattern in (10) can be related by the parity operation to that in (11). In a similar manner to that discussed above for ML_2 fragments, i.e.

$$\hat{P} \otimes L_d^\delta \text{ (face-bridged cluster)} = \bar{L}_d^\delta$$

(edge-bridged cluster)

$$\text{Bonding} \overset{\hat{P}}{\leftrightarrow} \text{Antibonding}$$

An intermediate structure between the face- and edge-bridged clusters belongs to the pure rotation point group O, where the parity operator transforms according to the totally symmetric representation. The linear combinations of the d_δ components are therefore $2a_2+2e+2t_2$. The following correlation is thereby established for the octahedral clusters:

O_h	O	O_h
$(a_{2g}+e_g+t_{2u})$ ⟵	(a_2+e+t_2)	⟶ $(a_{2u}+e_u+t_{2g})$
face-bridged	single d_δ component	edge-bridged
(10) $d_{x^2-y^2}$	$d_{x^2-y^2}$ or d_{xy}	(11) d_{xy}

In the octahedral point group the parity operator transforms as a_{1u} and relates the L_d^δ and $\bar{L}_d^\delta$ orbitals in the following fashion:

$$a_{2g}\otimes\Gamma_u^0\rightarrow a_{2u}$$
$$e_g\otimes\Gamma_u^0\rightarrow e_u$$
$$t_{2u}\otimes\Gamma_u^0\rightarrow t_{2g}$$

The results of the molecular orbital calculations of Johnston and Mingos [24] which are reproduced in Figs. 4, 5 and 6 confirm the argument developed above. In these figures, the MO levels in the left side show the results for (10) and its *nido*, *arachno* derivatives and those in the right side are the results of (11) and its corresponding derivatives (see Fig. 2). It also can be seen that the L_d^π and $\bar{L}_d^\pi$ pattern is entirely equivalent to the L_{pd}^π and $\bar{L}_{pd}^\pi$ pattern derived for $M(CO)_3$ fragments.

Ceulemans and Fowler [23] have previously noted that L_d^π/L_d^δ and $\bar{L}_d^\pi/\bar{L}_d^\delta$ mixings have to be taken into account to rationalise the relative energies of the $t_{2u}(\bar{D}_d^\delta)$ and $t_{2g}(D_d^\delta)$ components illustrated in the Figure (see ref. [23]). They classified the symmetries of L_d^π, $\bar{L}_d^\pi$ and those derived from the single d_δ component in the O_h finite point group and then took into account

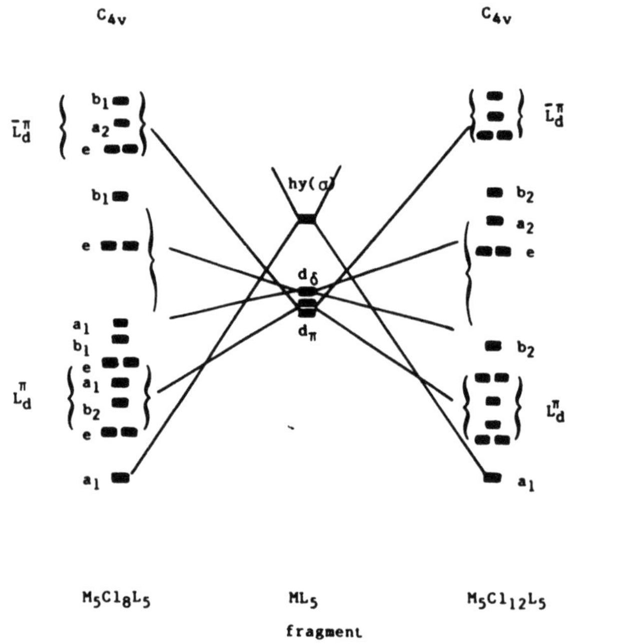

Fig. 5. The correlation MO diagram of the *nido* chloride-bridged clusters derived from (10) and (11)

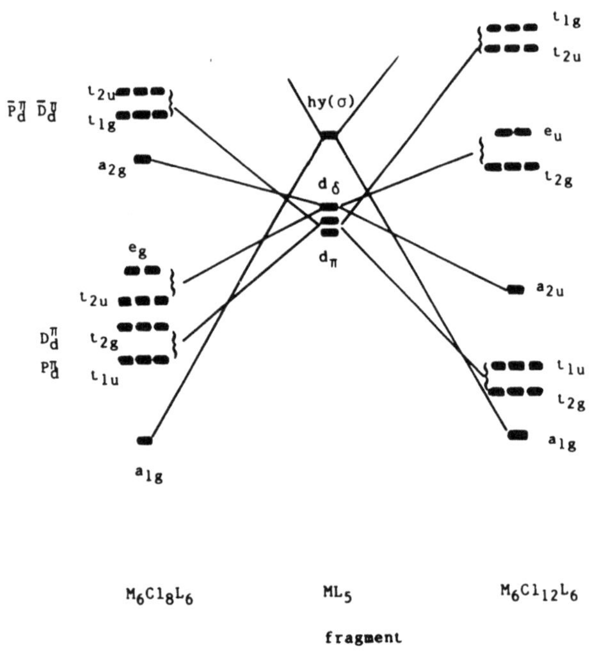

Fig. 4. The correlation MO diagram of (10) and (11)

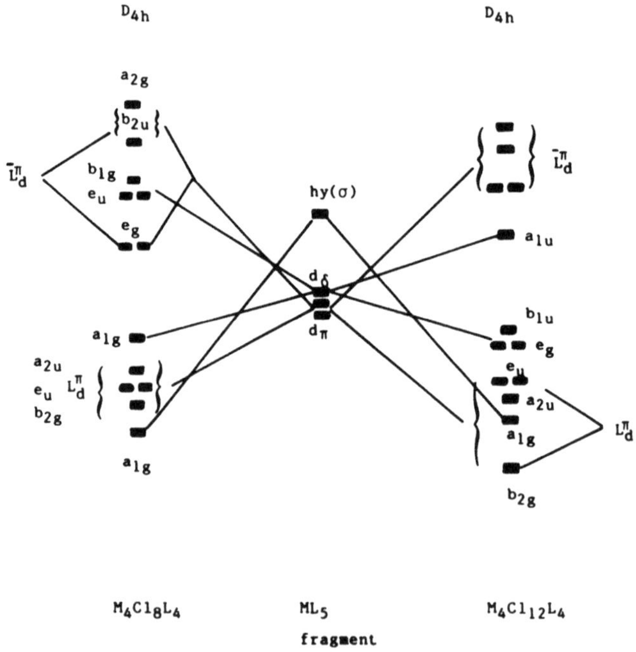

Fig. 6. The correlation MO diagram of the *arachno* chloride-bridged clusters derived from (10) and (11)

the interactions between equisymmetric orbitals. The ideas of Ceulemans and Fowler are extended to the *nido* and *acrachno* structures in Fig. 2. Tables 4 and 5 list the symmetries of molecular orbitals derived from the d_π pair, $\delta(x^2-y^2)$ and $\delta(xy)$ basis set for *nido* (square pyramid) and *arachno* (square) derivatives of ($\underline{10}$) and ($\underline{11}$).

The *nido* structure $M_5Cl_8L_5$ has been derived from the face-bridged cluster ($\underline{10}$) (see Table 4). The d_δ component, available for cluster bonding, is $\delta(x^2-y^2)$. The non-bonding $D^\pi/\bar{D}^\pi(e)$ set becomes strongly bonding as a result of interaction with $D^\delta(e)$ derived from the $\delta(x^2-y^2)$ basis. The other two e (i.e. P_d^π and $\bar{P}_d^\pi$) similarly repel each other but this does not affect the bonding pattern. As a net result 9 bonding orbitals are obtained, $S^\sigma(a_1)$, $P_d^\pi(a_1+e)$, $D_{2s}^\pi(b_2)$, $D_d^\pi/\bar{D}_d^\pi(e)$ and $D_d^\delta(a_1)$, $D_d^\delta(b_1)$ (see the left hand side of Fig. 5). For the *nido* structure $M_5Cl_{12}L_5$ derived from the edge-bridged cluster ($\underline{11}$) (see Table 4), the δ component is $\delta(xy)$. The strong mixing between $D^\pi/\bar{D}^\pi(e)$ and $D_d^\delta(e)$ leads to one e bonding and the other e antibonding (see the right hand side of Fig. 5). Therefore, 8 bonding orbitals are obtained, $S^\sigma(a_1)$, $P_d^\pi(a_1+e)$, $D_{2s}^\pi(b_2)$, $D_d^\pi/\bar{D}_d^\pi(e)$ and $F_d^\delta(b_2)$.

In the square clusters $[M_4Cl_8L_4]$ and $[M_4Cl_{12}L_4]$ (see Table 5), the mixing between the equisymmetric d_π and d_δ orbitals does not affect the bonding pattern (see Fig. 6). Both structures have one S^σ and four L_d^π bonding molecular orbitals and the molecular orbitals derived from the single d_δ components. The parity operation relates the bonding components in one conformer to the antibonding components in the second conformer.

The following points can be noted from the above discussion.

Table 4. Symmetries of skeletal molecular orbitals in *nido* (C_{4v}) structures derived from ($\underline{10}$) and ($\underline{11}$) (see Fig. 2). The $\delta(x^2-y^2)$ bases are needed in face-bridged cluster $M_5Cl_8L_5$. The $\delta(xy)$ bases are needed in edge-bridged cluster $M_5Cl_{12}L_5$

	$\pi_{xy,yz}$	$\delta(x^2-y^2)$	$\delta(xy)$
bonding	$P^\pi(a_1+e)$, $D_{2s}^\pi(b_2)$	$D_0^\delta(a_1)$, $D^\delta(b_1)$	$D^\delta(e)$, $F^\delta(b_2)$
non-bonding	$D^\pi/\bar{D}^\pi(e)$		
antibonding	$\bar{P}^\pi(a_2+e)$, $\bar{D}_{2s}^\pi(b_1)$	$\bar{D}^\delta(e)$, $\bar{F}^\delta(b_1)$	$\bar{D}_0^\delta(a_2)$, $\bar{D}^\delta(b_2)$

Table 5. Symmetries of skeletal molecular orbitals in square structures derived from ($\underline{10}$) and ($\underline{11}$) (see Fig. 2). The $\delta(x^2-y^2)$ bases are needed in $M_4Cl_8L_4$. The $\delta(xy)$ bases are needed in $M_4Cl_{12}L_4$

	$\pi_{xz,yz}$	$\delta(x^2-y^2)$	$\delta(xy)$
bonding	$P^\pi(a_{2u}+e_u)$, $D_{2s}^\pi(b_{2g})$	$D_0^\delta(a_{1g})$	$D^\delta(e_g)$, $D^\delta(b_{1u})$
antibonding	$\bar{P}^\pi(a_{2g}+e_g)$, $\bar{D}_{2s}^\pi(b_{2u})$	$\bar{D}^\delta(e_u)$, $\bar{D}^\delta(b_{1g})$	$\bar{D}_0^\delta(a_{1u})$

1. The skeletal molecular orbitals for clusters based on ML_5 fragments are derived from the interactions of one $hy(\sigma)$, a pair of d_π and one d_δ component.

2. The skeletal molecular orbitals derived from $hy(\sigma)$ and d_π pair are entirely equivalent to those derived from the three frontier orbitals of $M(CO)_3$ and *B-H* fragments.

3. The molecular orbitals derived from the single d_δ components are one subset of $L_d^\delta/\bar{L}_d^\delta$ spectrum and are not related by the parity operator Γ_u^0. The parity operator does relate the $L_d^\delta/\bar{L}_d^\delta$ orbitals of face-bridged and edge-bridged clusters which are interconverted by a local rotation of 45° at each ML_5 fragment.

4. L_d^π/L_d^δ and $\bar{L}_d^\pi/\bar{L}_d^\delta$ mixings have to be taken into account in order to rationalise the energies of the $L_d^\delta/\bar{L}_d^\delta$ molecular orbitals. The mixings do not affect the conclusions 2 and 3 because the L_d^π/L_d^δ and $\bar{L}_d^\pi/\bar{L}_d^\delta$ mixings occur simultaneously.

The S.E.R.C. and the Chinese Academy of Sciences are thanked for their financial support and Dr R.L. Johnston for some useful discussions.

References

1. Wade, K.: Adv. Inorg. Chem. Radiochem. **18**, 1 (1976)
2. Mingos, D.M.P.: Nature (London) **236**, 99 (1972)
3. Stone, A.J.: Mol. Phys. **41**, 1339 (1980)
4. Stone, A.J.: Inorg. Chem. **20**, 563 (1981)
5. Johnston, R.L., Mingos, D.M.P.: J. Organomet. Chem. **280**, 407 (1985)
6. Johnston, R.L., Mingos, D.M.P.: J. Organomet. Chem. **280**, 419 (1985)
7. Mingos, D.M.P.: J. Chem. Soc. Chem. Commun. 706 (1983)
8. (a) Mingos, D.M.P.: J. Chem. Soc. Chem. Commun. 1353 (1985); (b) Mingos, D.M.P., Lin Zhenyang: J. Chem. Soc. Dalton Trans. 1657 (1988)
9. Ceulemans, A.: Mol. Phys **54**, 161 (1985)
10. Fowler, P.W.: Polyhedron **4**, 2051 (1985)
11. Fowler, P.W., Quinn, C.M.: Theor. Chim. Acta. **70**, 333 (1986)
12. Redmond, D.B., Quinn, C.M., McKiernan, J.G.R: J. Chem. Soc. Faraday Trans. 2 **79**, 1791 (1983)
13. Johnston, R.L., Mingos, D.M.P.: Polyhedron **5**, 2059 (1986)
14. Johnston, R.L., Mingos, D.M.P.: Inorg. Chem. **25**, 3321 (1986)
15. Johnston, R.L., Mingos, D.M.P.: J. Chem. Soc. Dalton Trans. 647 (1987)
16. Johnston, R.L., Mingos, D.M.P.: J. Chem. Soc. Dalton Trans. 1445 (1987)
17. Evans, D.G., Mingos, D.M.P.: J. Organomet. Chem. **240**, 321 (1982)
18. Evans, D.G., Mingos, D.M.P.: Organometallics **2**, 435 (1983)
19. Woodward, R.B., Hoffmann, R.: The conservation of orbital symmetry. New York: Academic Press 1970
20. Churchill, M.R., DeBoer, B.G.: Inorg. Chem. **16**, 878 (1977)
21. Gregson, D., Howard, J.A.K., Murray, M., Spencer, J.L.: J. Chem. Soc. Chem. Commun. 716 (1981)
22. (a) McCarley, R.E.: Philos. Trans. R. Soc. London A **308**, 141 (1982), and references therein; (b) Chevrel, R., Hirrien, M., Sergent, M.: Polyhedron. **5**, 87 (1986)
23. Ceulemans, A., Fowler, P.W.: Inorg. Chim. Acta. **105**, 75 (1985)
24. Johnston, R.L., Mingos, D.M.P.: Inorg. Chem. **25**, 1661 (1986)

Z. Phys. D – Atoms, Molecules and Clusters 12, 61–64 (1989)

Atoms, Molecules
Zeitschrift für Physik D and Clusters
© Springer-Verlag 1989

Theoretical studies on carbon and silicon clusters: comparison of the structures and stabilities of neutral and ionic forms

Krishnan Raghavachari
AT&T Bell Laboratories, Murray Hill, NJ 07974, USA

Received 5 July 1988

Optimized molecular geometries and electronic structures are determined for neutral, positively charged, and negatively charged carbon and silicon clusters containing up to ten atoms. The effects of polarization functions and electron correlation are included in these claculations. Carbon clusters have linear or monocyclic ground state geometries whereas silicon clusters containing five or more atoms have three-dimensional ground state structures. Neutral C_4, C_6 and C_8 all have linear and monocyclic isomers of comparable stability whereas the ionic forms appear to be generally more stable as linear geometrical arrangements. In the case of neutral and positively charged carbon clusters, the odd-numbered clusters are significantly more stable than the adjacent even-numbered clusters whereas the opposite order of stability occurs for the negative ions. This is due to the large values of the electron affinities of the linear forms of even-numbered clusters such as C_4 and C_6. The relative stabilities of silicon clusters does not change with the charge state of the clusters.

PACS: 31.20; 36.40

Introduction

The field of atomic and molecular clusters [1] has recently witnessed an upsurge of experimental [2–5] and theoretical [6–9] work focussed on small clusters of carbon and silicon. Interesting questions such as the relative stabilities of the clusters ("magic numbers") and the nature of the fragmentation processes have been addressed [7, 9]. However, much of the theoretical discussion involving these clusters has been restricted to the neutral molecules while experimentally all the clusters have been detected as ionic species. The correlation between the structures and properties of the neutral and ionic clusters is thus an extremely important question which has to be addressed carefully.

In previous works, we have carried out systematic studies of the structures and stabilities of neutral carbon [7] and silicon [9] clusters (C_n, Si_n, $n=2$–10). The calculated ground state structures are represented in Figs. 1 and 2. We have now carried out a similar study on both positively and negatively charged clusters for $n=2$–6. In this work, we summarize some of our results comparing the neutral clusters with their corresponding ionic forms.

Theoretical methods

The computational techniques employed are similar to the methods used in our previous studies on neutral clusters [10]. The Hartree-Fock (HF) method with the 6–31G* basis set (double-zeta *sp*- plus a set of *d*-type polarization functions on each atom) was used in the determination of all molecular geometries. For open-shell systems, the spin-unrestricted version of HF

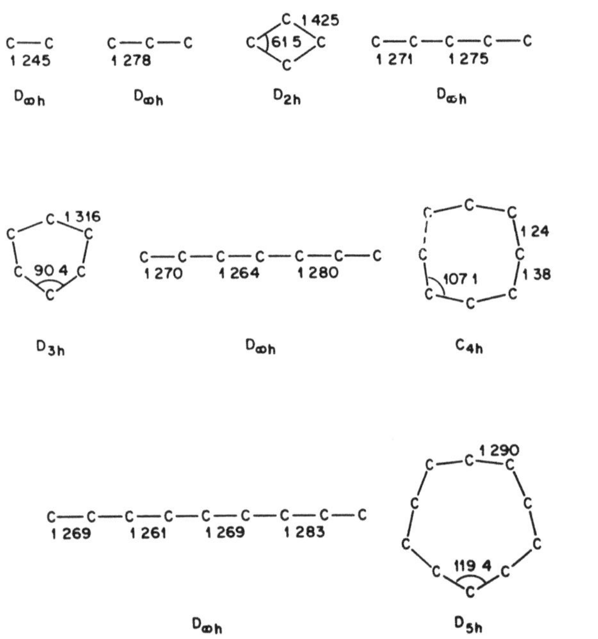

Fig. 1. Calculated ground state structures of neutral carbon clusters. Note that there are also low-lying linear structures which are close in energy to the cyclic forms in the case of C_4, C_6 and C_8

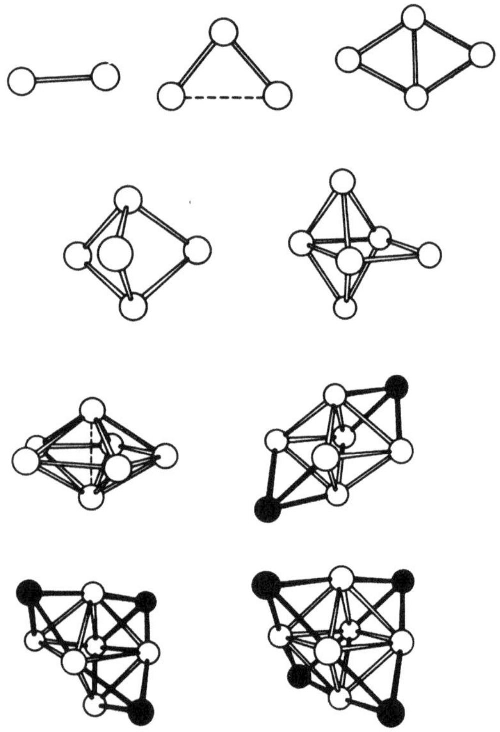

Fig. 2. Calculated ground state structures of neutral silicon clusters. Note that there is another low-lying isomer (tetracapped trigonal prism) for Si_{10} which is not shown

theory (UHF) was used. Electron correlation effects were included by means of fourth order Møller-Plesset perturbation (MP4) theory or by means of augmented coupled cluster methods in cases where the convergence of perturbation theory was poor.

Results and discussion

As seen in Fig. 1, carbon clusters adopt linear or monocyclic geometries [7] indicating the stability of π-bonding for carbon. There is also a significant odd-even oscillation in the nature of the ground state geometries. The odd-numbered clusters up to C_9 all have clear cut linear geometries. Both monocyclic and linear structures have been found for C_4, C_6 and C_8 with the former being very slightly more stable [7]. Silicon clusters containing five or more atoms [9], however, have three-dimensional ground state geometries (Fig. 2) in strong contrast to carbon clusters. In the following discussion, we point out some of the interesting aspects of individual ionic clusters which we have calculated and then point out some of the general conclusions resulting from our study.

Two atom cluster ions

Neutral C_2 has a $^1\Sigma_g^+$ ground state [7] with four π-electrons and a low-lying $^3\Pi_u$ state (three π-electrons) which is only about 0.1 eV higher in energy. The ground state of C_2^+ is $^4\Sigma_g^-$ (two π-electrons) which is best considered as obtained by removing an electron from the $^3\Pi_u$ state. Since this state has only two π-electrons, the optimized bond length (1.38 Å) is significantly larger than that of neutral C_2 (1.24 Å). The $^2\Pi_u$ state of C_2^+ containing three π-electrons is about 0.6 eV higher in energy at the MP4/6-31G* level of theory. C_2^- has a $^2\Sigma_g^+$ ground state with an optimized bond length of 1.25 Å which is similar to that of neutral C_2.

Neutral Si_2 has a $^3\Sigma_g^-$ ground state containing two π-electrons with the $^3\Pi_u$ state (three π-electrons) being only slightly higher in energy. Si_2^+ again has a $^4\Sigma_g^-$ state which is similar to that of C_2^+. However, in the case of Si_2^-, the $^2\Sigma_g^+$ state was found to be almost isoenergetic with a $^2\Pi_u$ state (three π-electrons) using larger basis sets, in agreement with recent experimental studies [11] which show the presence of two low-lying states for Si_2^-.

Three atom cluster ions

Neutral C_3 has a linear geometry [6, 7] and a $^1\Sigma_g^+$ electronic configuration with the orbital occupation

.... $(\sigma_g)^2(\sigma_u)^2(\pi_u)^4$. Ionization from these three orbitals gives rise to $^2\Sigma_g^+$, $^2\Sigma_u^+$ and $^2\Pi_u$ states for C_3^+. At the HF/6–31 G* level, the $^2\Pi_u$ lies lowest in energy, consistent with the π_u being the highest occupied molecular orbital for neutral C_3. However, electron correlation effects favor the $^2\Sigma_u^+$ considerably so that at the highest level of theory it is 1.3 eV more stable than the $^2\Pi_u$ state. There are several such instances reported in this paper where electron correlation effects are extremely important in determining the ground state structures, demonstrating the inadequacy of simplistic orbital type ideas in such cases.

Vibrational analysis reveals that while the $^2\Pi_u$ state is a local minimum at the HF/6–31G* level, the $^2\Sigma_u^+$ state has a pair of degenerate imaginary frequencies indicating that bending the molecule causes energy lowering. Optimization of this state eventually leads to the formation of a C_{2v} structure ($^2B_2$ state) with a bond angle of 71°. Single point calculations using these geometries at higher levels of theory reveal that the C_{2v} form is slightly more stable than the linear form. The cyclic nature of the C_{2v} form is roughly consistent with the structure of C_3^+ as determined in the recent experiments by Faibis et al. [12].

The lowest unoccupied orbital of C_3 is a π_g orbital. Thus, the negative ion C_3^- has a $^2\Pi_g$ ground state with the bond length stretching only slightly by $\simeq$0.002 Å on going to the anion.

Neutral Si_3 has two low-lying electronic states [8, 9], a $^1A_1$ state (C_{2v}) with a bond angle of $\simeq$78° and a D_{3h} $^3A_2'$ state. On ionization to give Si_3^+, there are two C_{2v} doublets ($^2B_2$ and $^2A_1$) and a D_{3h} quartet ($^4A_2'$) which are all within 0.1–0.2 eV of each other. The ground state appears to be $^2B_2$ with a bond angle of $\simeq$71° which is analogous to the bent geometry of C_3^+ mentioned earlier.

Unlike C_3^- which is linear, Si_3^- has a cyclic structure with a pair of states ($^2A_1$ and $^2B_2$) which are related by means of a Jahn-Teller pseudorotation being very close in energy (<0.1 eV of each other).

Four atom cluster ions

Neutral C_4 has two low-lying isomers [6, 7] which are almost isoenergetic, a rhombus singlet structure (^1A_g with two π-electrons) and a linear triplet structure ($^3\Sigma_g^-$ with six pi-electrons). Thus it is of interest to see if the ionic states have linear or cyclic geometries.

The linear form of C_4^+ has a $^2\Pi_g$ ground state (five π-electrons). The rhombus has two low-lying states, ^2A_g and $^2B_{1u}$, corresponding to ionization from the lone-pair type orbitals. The ^2A_g is lower in energy at the HF level whereas the $^2B_{1u}$ state is more stable at higher levels of theory, similar to the behavior of C_3^+ mentioned earlier. It is interesting to note that the linear $^2\Pi_g$ state and the rhombic $^2B_{1u}$ state are very close in energy. Thus C_4^+ also has two low-lying isomers similar to the case of neutral C_4.

The negative ion C_4^- is very different. In this case, the linear $^2\Pi_g$ form (seven π-electrons) is found to be more stable than the rhombic $^2B_{2g}$ state by$\simeq$1.2 eV. Thus the negative ion has a clear cut linear ground state in agreement with the interpretation of the photoelectron spectrum of C_4^- by Yang et al. [3]. The electron affinity of C_4 is found to be $\simeq$2 eV greater than that of C_3 and is consistent with the triplet ground state of linear C_4 involving partially occupied orbitals.

Neutral Si_4 has a clear cut rhombus ground state [9] and thus it is not surprising that both Si_4^+ and Si_4^- are derived from this structure. For Si_4^+, the ground state is $^2B_{1u}$ and this is similar to the corresponding form of C_4^+. For Si_4^-, the ground state is $^2B_{2g}$ with two π-electrons, again analogous to the rhombic form of C_4^-.

Five atom cluster ions

Neutral C_5 has a linear ground state [6, 7] analogous to that of C_3. The positive and negative ion states are also correspondingly similar. For C_5^+, the $^2\Sigma_u^+$ state is again found to be more stable than the $^2\Pi_g$ state. However, the energy difference of 0.4 eV is significantly smaller than the corresponding energy difference of 1.3 eV for C_3^+. Thus it appears likely by extrapolation that in the case of the next analogue, linear C_7^+, the $^2\Pi_u$ state will be more stable than the $^2\Sigma_u^+$ state.

C_5^- has a $^2\Pi_u$ ground state with nine π-electrons, as expected. The calculated electron affinity of C_5 is significantly smaller than that of C_4, consistent with the experimental measurements of Yang et al. [3].

Unlike the ground state of C_5, Si_5 has a three-dimensional trigonal bipyramidal geometry [9] with the orbital occupation... $(a_1')^2(e')^4$. This results in two low-lying electronic states for Si_5^+, a $^2A_1'$ (D_{3h}) and a Jahn-Teller distorted $^2B_2$ with C_{2v} symmetry. The $^2A_1'$ is more stable at the HF level whereas the $^2B_2$ is lower in energy at correlated levels of theory, a situation similar to that of C_3^+, C_4^+ and Si_4^+.

Si_5^- has a $^2A_2''$ ground state consistent with the occupation of the lowest unoccupied orbital of Si_5 having a_2'' symmetry. Unlike Si_5 which has a compressed trigonal bipyramidal geometry, Si_5^- has a more traditional capped tetrahedral type structure where there is significant additional bonding between the equatorial atoms. The electron affinity of Si_5 is thus significantly larger than that of Si_4, in agreement with the measurements of Cheshnovsky et al. [5].

Six atom cluster ions

Neutral C_6 has been calculated to have a cyclic ground state [7] with the linear isomer being about 0.4 eV higher in energy. Our calculations show that both C_6^+ and C_6^- probably have linear ground states. For linear C_6^+, the ground state is $^2\Pi_u$ with nine π-electrons. The ion resulting from the cyclic isomer appears to be higher in energy though we do not have a reliable estimate of this energy difference due to spin contamination in our UHF wave functions.

The situation for C_6^- is very similar to that of C_4^-. The $^2\Pi_u$ state of linear C_6^- is more stable than the $^2A_1'$ cyclic state (planar D_{3h} symmetry) by $\simeq 1.4$ eV. A similar situation is likely for C_8 also. Thus while C_4, C_6, and C_8 have low-energy linear and cyclic structures, the negative ions in all three cases have clear cut linear geometries. The linear geometry of C_6^- is again consistent with the interpretations of Yang et al. [3].

Neutral Si_6 has been calculated to have a C_{2v} geometry [9] ($^1A_1$) which is best characterized as an edge-capped trigonal bipyramid. This may also be considered as a distorted form of a compressed tetragonal bipyramid (D_{4h}). On ionization to give Si_6^+, the ground state is $^2B_1$, consistent with an electron being removed from the highest occupied orbital. Surprisingly, the negative ion Si_6^-, on optimization, yields the more symmetric D_{4h} geometry. Otherwise, the situation is somewhat analogous to that of Si_5^- with the added electron being in an orbital (a_{2u} for Si_6^-) which is antibonding with respect to the apex atoms thus resulting in a geometry which is near octahedral.

General aspects and conclusions

The largest changes on ionization occur in the case of carbon clusters. Neutral C_3 is linear whereas C_3^+ appears to be bent. Neutral C_4, C_6 and C_8 all have linear and monocyclic isomers of comparable stability whereas the corresponding ionic forms (with the possible exception of C_4^+) appear to have linear ground states. This is particularly dramatic for the negative ions where the even-numbered clusters have electron affinities which are substantially larger than that of the adjacent odd-numbered clusters. In fact, the difference in the electron affinities is so large that the relative stabilities of the clusters are reversed completely. These can be calculated by considering the energy required for simple fragmentation reactions of these cluster ions, i.e. the energy required for the reactions $C_n^\pm \rightarrow C_{n-1}^\pm + C$. In the case of the neutral and positively charged carbon clusters, the odd-numbered clusters are significantly more stable (i.e., have higher fragmentation energies) than the adjacent even-numbered clusters. For C_n^-, however, even-numbered clusters are more stable than the adjacent odd-numbered clusters. This is consistent with the prominence of the odd-numbered clusters for C_n^+ and the even-numbered clusters for C_n^- as seen in the cluster distribution patterns [2].

The silicon clusters do not show any such dramatic behavior. Though there are some variations in the ionization potentials and electron affinities with cluster size (e.g., larger electron affinity of Si_5 as mentioned earlier) the variations are not large enough to cause changes in the relative stabilities with the charge state of the clusters. In a previous study [9], we had calculated that Si_4, Si_6, Si_7 and Si_{10} are the stable clusters for neutral silicon clusters. From this study it appears likely that the same clusters are likely to be the "magic numbers" for the ionic states also.

References

1. Brown, W.L., Freeman, R.R., Raghavachari, K., Schlüter, M.: Science **235**, 860 (1987)
2. Rohlfing, E.A., Cox, D.M., Kaldor, A.: J. Chem. Phys. **81**, 3322 (1984); Bloomfield, L.A., Geusic, M.E., Freeman, R.R., Brown, W.L.: Chem. Phys. Lett. **121**, 33 (1985); Kroto, H.W., Heath, J.R., O'Brien, S.C., Curl, R.F., Smalley, R.E.: Nature (London) **318**, 162 (1985)
3. Yang, S., Taylor, K.J., Craycraft, M.J., Conceicao, J., Pettiette, C.L., Cheshnovsky, O., Smalley, R.E.: Chem. Phys. Lett. **144**, 431 (1988)
4. Bloomfield, L.A., Freeman, R.R., Brown, W.L.: Phys. Rev. Lett. **54**, 2246 (1985); Martin, T.P., Schaber, H.: J. Chem. Phys. **83**, 855 (1985); Liu, Y., Zhang, Q.-L., Tittle, F.K., Curl, R.F., Smalley, R.E.: J. Chem. Phys. **85**, 7434 (1986)
5. Cheshnovsky, O., Yang, S.H., Pettiette, C.L., Craycraft, M.J., Liu, Y., Smalley, R.E.: Chem. Phys. Lett. **138**, 119 (1987)
6. Whiteside, R.A., Krishnan, R., DeFrees, D.J., Pople, J.A., v.R. Schleyer, P.v.R.: Chem. Phys. Lett. **78**, 538 (1981); Ewing, D.W., Pfeiffer, G.V.: Chem. Phys. Lett. **134**, 413 (1987); Pacchioni, G., Koutecky, J.: J. Chem. Phys. **88**, 1066 (1988)
7. Raghavachari, K., Whiteside, R.A., Pople, J.A.: J. Chem. Phys. **85**, 6623 (1986); Raghavachari, K., Binkley, J.S.: J. Chem. Phys. **87**, 2191 (1987)
8. Balasubramanian, K.: Chem. Phys. Lett. **135**, 283 (1987); Pacchioni, G., Koutecky, J.: J. Chem. Phys. **84**, 3301 (1986); Tománek, D., Schlüter, M.: Phys. Rev. Lett. **56**, 1055 (1986); Ballone, P., Andreoni, W., Car, R., Parrinello, M.: ibid **60**, 271 (1988); Phillips, J.C.: Chem. Rev. **86**, 619 (1986)
9. Raghavachari, K., Logovinsky, V.: Phys. Rev. Lett. **55**, 2853 (1985); Raghavachari, K.: J. Chem. Phys. **83**, 3520 (1985); Raghavachari, K.: ibid, **84**, 5672 (1986); Raghavachari, K., Rohlfing, C.M.: Chem. Phys. Lett. **143**, 428 (1988)
10. For a discussion of the methods and basis sets used in this paper, see: Hehre, W.J., Radom, L., Schleyer, P.v.R., Pople, J.A.: Ab Initio Molecular Orbital Theory. New York: John Wiley 1986
11. Nimlos, M.R., Harding, L.B., Ellison, G.B.: J. Chem. Phys. **87**, 5116 (1987)
12. Faibis, A., Kanter, E.P., Tack, L.M., Bakke, E., Zabransky, B.J.: J. Phys. Chem. **91**, 6445 (1987)

Z. Phys. D – Atoms, Molecules and Clusters 12, 65–68 (1989)

Atoms, Molecules
Zeitschrift für Physik D and Clusters
© Springer-Verlag 1989

Quantum mechanics and the structure of noble-gas clusters

G. Franke and J. Schulte
Fachbereich Physik, Universität Oldenburg D-2900 Oldenburg, Federal Republic of Germany

Received 4 July 1988; final version 30 September 1988

The importance of quantum mechanics for the thermal behaviour of Argon clusters with special respect to phase transitions is investigated applying a Path-Integral Monte-Carlo (PIMC) method. Results for the small-angle neutron scattering function (SANS) for various cluster sizes and temperatures are presented.

PACS: 36.40

Introduction

In the investigation of molecules and clusters, the quantum effects arising from the motion of the atoms as quantum constituents of the system are expected to be very small and thus generally neglected. The wave functions have little overlap and the uncertainty in space is smaller than the interatomic distances. Because of the comparatively weak and soft binding of noble-gas atoms, for noble-gas clusters one should expect more extended atomic wave-functions and a higher importance of quantum contributions such as the groundstate energy. The quantum effects should be most important at a temperature where the spatial uncertainty is comparable to the statistical motion and the number of thermally accessible states is not too high. Here we want to study these quantum effects and their temperature dependence specifically for Argon-clusters and compare the results to classical ones.

So far, noble-gas clusters have been investigated by best-single-cluster approximations, Molecular Dynamics (MD) [1, 2], and classical Monte Carlo (MC) [3, 4, 2] simulations. Special attention has been directed to phase transitions of those clusters with the idea that the structure of the cluster is also important for its stability. Generally, the transitions are less sharp than in infinite systems but smeared out over some temperature range. Here we will specifically investigate an isomerization transition of Ar_6 because here it is easy to survey what happens on the microscopic scale.

The method

The multiparticle quantum statistics is reformulated in a pseudoclassical configuration space, assigning to every particle a "path" through this space along a closed imaginary "time"-axis. Statistical averages are calculated by integrating over the paths with a weight function $e^{-S/\hbar}$, S being the classical action of the path:

$$Z = \int Dy(\tau)\exp\left\{-\int_0^{\hbar\beta} d\tau\left(V(y(\tau)) + \tfrac{1}{2}m\dot{y}(\tau)^2\right)\right\} \qquad (1)$$

where $Dy(\tau)$ is running over all imaginable closed paths in configuration space of length $\hbar\beta$. The idea in applying this formalism is that for a system, for which quantum mechanics is not essential, the path space can be approximated by a few "time slices" and the remaining finite-dimensional integral can be calculated by a Metropolis-MC-algorithm [6]. (Note that in the limit of only one "time slice" the kinetic term of the action integral disappears and the classical configuration integral remains.) For the realization, a modification is necessary because the direct method of varying the path step-by-step causes extremely poor convergence. The configurational and kinetic energy terms must be decoupled by Fourier-transformation of the paths in the time-direction to allow the variation of the Fourier-components with different step sizes, reflecting their relative importance.

With this Path-Integral MC-algorithm we investigated the thermal behavior of Ar-clusters applying free boundary conditions (no constraining potential was used). For the atomic interaction we took a standard Lennard-Jones potential with parameters $\sigma = 3.405$ Å, $\varepsilon = 10.3$ meV (120 K) and a mass of 40 amu. For investigations of the Ar_6 isomerization, we used around 10^6 iterations; when the cluster is dominated by only one structure, many fewer cycles are necessary to obtain reproducible results.

As information about the structure of the clusters and as connection to experimental work, we calculate the small angle elastic neutron scattering (SANS) function. Structure determination with neutron scattering gives directly the regions in which the nuclei move due to their thermal motion and not their convolution with the density distribution of core electrons as in the X-ray case. In addition, in experiments like those done by Oberthür et al. [8] at the D11 of the ILL Grenoble, cluster growth may be observed directly. For neutral noble-gas clusters the scattering function $I(q) = 4\pi \sin\theta/\lambda$, calculated here from the pair correlation function $g(r)$ out of the PIMC simulation, is

$$I(q) = 4\pi \int g(r) \frac{\sin(qr)}{qr} r^2 \mathrm{d}r. \tag{2}$$

Results of the calculations

In the low temperature region, the quantum effects are expected to be most important because the zero-point energy is a large fraction of the total binding energy and prevents the cluster-atoms from taking full advantage of potential minima in configuration space. In this sense, it is somewhat in analogy with thermal energy at higher temperature. For the clusters investigated (Ar_6 and Ar_{13}), the zero point energy lowers the binding energy by 10–15% and corresponds thus to thermal energies at ≈ 15 K. On the other hand, with increasing temperature its influence decreases; above 30 K quantum effects cannot be observed in binding energies or in scattering functions.

The scattering function of an ensemble of Ar_6 clusters calculated by PIMC shows a shifting of maxima and amplitudes of about 10% at temperatures between 8.5 K and 12.5 K (Fig. 1). For clusters of this size, the shifting causes significant changes in both their internal structure and their shape. This indicates that there is some transition of the Ar_6-cluster between these temperatures. The distinct structure in the region of high q dermonstrates that the cluster geometry is dominated by few configurations. As it is seen from Fig. 2, there is a correction by quantum mechanics (**Q**) to the classical calculation (**C**) in the exact atomic position, but not in the overall shape (low momenta).

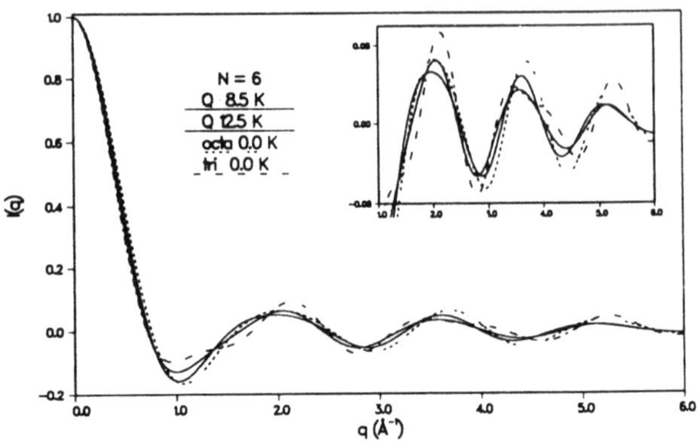

Fig. 1. Scattering function of Ar_6 calculated by PIMC

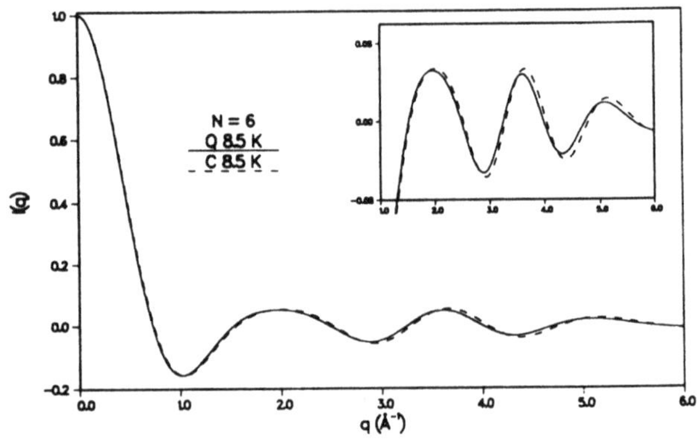

Fig. 2. Scattering function of Ar_6 from classical (**C**) and PIMC (**Q**) calculations

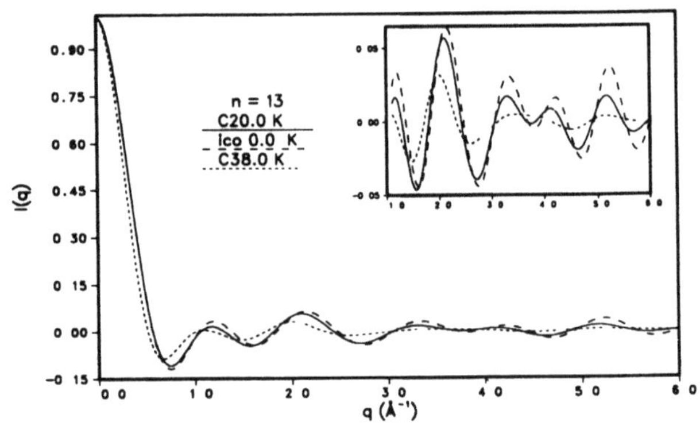

Fig. 3. Scattering function of Ar_{13} from classical calculations

The ensemble of the Ar_{13}-clusters has a very well defined icosahedral structure below its phase transition (Fig. 3) while it looks amorphous above. The two maxima between 3 Å^{-1} and 4 Å^{-1} merge to one broadened one. Similar "strong" and amorphous icosahedral structures have been found by Cargill [9] in X-ray scattering experiments with the amorphous metals

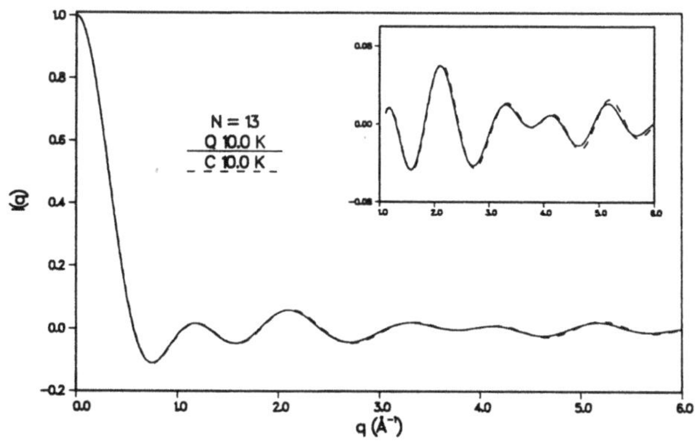

Fig. 4. Scattering function of Ar_{13} from classical (**C**) and PIMC (**Q**) calculations

Pd-Si and Au-Si. Figure 4 shows that quantum mechanical calculation (PIMC) yields observable corrections to the internal cluster structure (as we have proven up to temperatures of some 30 K), but leaves the cluster-shapes unchanged.

The two transitions observed in the calculations have different characteristics: while the Ar_{13} transition behaves very solid/liquid-like (in accordance to recent investigations [3, 2]) in the sense that above the transition temperature the cluster looks amorphous, for Ar_6 at the transition only one configuration (built up of three tetrahedra) becomes important in addition to the octahedral ground-state structure. The SANS-functions for these structures are plotted in Fig. 1; their importance is confirmed by investigations of the pair correlation function and viewing single configurations produced in the PIMC-process [7].

In the temperature range of the Ar_{13}-"melting" at ≈ 35 K, quantum mechanical corrections are not relevant, but also for Ar_6 we were not able to recognize an influence of quantum mechanics on the transition temperature with the precision of our calculations and a resolution of 0.5 K.

Discussion

In this section, we want to discuss the different characters of transitions occurring for Ar_6 and Ar_{13}. In the first case, in the transition region only two distinct configurations are relevant: the octahedron and the tri-tetrahedron. If we consider only these two (meta)stable states assuming that the volume of reachable phase space around them is equal, we can roughly estimate the transition temperature T_t. At T_t, the probabilities of finding the cluster in one of the two states are equal:

$$I_{oct} e^{-\frac{E_{oct}}{kT_t}} = I_{tt} e^{-\frac{E_{oct}+\Delta E}{kT_t}} \Rightarrow kT_t = \frac{\Delta E}{\log I_{tt}/I_{oct}} \tag{3}$$

Where $\Delta E = 0.41\varepsilon = k \times 47$ K is the potential energy difference between the structures and I_{oct}, I_{tt} are the numbers of possible isomers of octahedron, and tri-tetrahedron, respectively. To determine them, we first "divide" by permutations and (continuous) rotations because they are common symmetries for all kinds of structures. Rotations that are equivalent to permutations — i.e. rotations which map the structure onto itself — have then been considered doubly. Although both the octahedron and tri-tetrahedron are unique structures, the ratio I_{tt}/I_{oct} equals $\frac{24}{2} = 12$ because the order of rotational symmetry groups of the bodies are 24 (rotational subgroup of the octahedron-group) for the first and 2 (one 180°-axis perpendicular to the central pair-axis) for the second one. This yields a transition temperature of $T_t \approx 19$ K. Compared to the simulation results, the temperature is overestimated by this simple model a factor of ≈ 1.8. Probably, this is due to the higher mobility of the atoms of the tri-tetrahedron; for example the outer ones may move around the central pair.

Although this model allows only rough estimations of the transition characteristics, it should be useful for comparison of the behaviour of different clusters. Applying it to Ar_{13}, it is less clear from the simulation data which metastable configurations are important. Good candidates are only structures which completely break the icosahedral symmetry of order 60. Therefore we calculated the transition temperatures for a set of clusters obtained from the icosahedron by picking one or two atoms from the surface and placing them on top of it elsewhere. Distinguishing whether the moved particles are neighbouring or not, the model yields 3 very similar transition temperatures in the range of 62–67 K.

Compared to the experimental data cited above, we also have an overestimation of the temperatures of about 1.8. But besides this factor, we recognize the following fact: in contrast to Ar_6, where just one isomerization takes place at T_t, here we have at least three overlapping transitions in a range of $\frac{5\text{ K}}{1.8} = 3$ K. (Probably also more complex metastable states are thinkable but there it becomes difficult to survey what happens.) From the viewpoint of isomerizations, this fact is the reason for the sudden arising of liquid-like behaviour of Ar_{13} which led Berry et al. to the introduction of a phenomenological melting-model for clusters [5]. Our interpretation implies that the melting-like behaviour of Ar_{13} is accidental; having more distinct transition temperatures for the most easily achieved states would lead to a more complex temperature dependence.

References

1. Briant, C.L., Burton, J.J.: J. Chem. Phys. **63**, 2045 (1975)
2. Davis, H.L., Jellinek, J., Berry, S.: J. Chem. Phys. **86**, 6456 (1987)
3. Etters, R.D., Kaelberer, J.: J. Chem. Phys. **66**, 5112 (1977)
4. Quirke, N., Sheng, P.: Chem. Phys. Lett. **110**, 63 (1984)
5. Berry, R.S., Jellinek, J., Natanson, G.: Chem. Phys. Lett. **107**, 227 (1984)
6. Doll, J.D., Freeman, D.L.: J. Chem. Phys. **82**, 465 (1985)
7. Franke, G., Hilf, E.R., Polley, L.: Z. Phys. D. – Atoms, Molecules and Clusters **9**, 343–349 (1988)
8. Oberthür, R.C.: Rev. Phys. Appl. **19**, 663 (1984); Oberthür, R.C., Schulte, J., Hilf, E.R.: ILL Internal Report, Grenoble, 1987
9. Cargill, G.S.: J. Appl. Phys. **41**, 12 (1970)

Z. Phys. D – Atoms, Molecules and Clusters 12, 69–71 (1989)

Atoms, Molecules
Zeitschrift für Physik D and Clusters
© Springer-Verlag 1989

Binding energy of large icosahedral and cuboctahedral Lennard-Jones clusters

J.A. Northby[1], J. Xie[1], David L. Freeman[2], and J.D. Doll[3]

[1] Department of Physics, University of Rhode Island, Kingston, RI 02881, USA
[2] Department of Chemistry, University of Rhode Island, Kingston, RI 02881, USA
[3] Los Alamos National Laboratory, Los Alamos, NM 87545, USA

Received 5 July 1988; final version 30 September 1988

It is widely believed that the lowest energy configurations for small rare gas clusters have icosahedral symmetry. This contrasts with the bulk crystal structures which have cuboctahedral fcc symmetry. It is of interest to understand the transition between this finite and bulk behavior. To model this transition in rare gas clusters we have undertaken optimization studies within the Lennard-Jones pair potential model. Using a combination of Monte Carlo and Partan Search optimization methods, the lowest energy relaxed structures of Lennard-Jones clusters having icosahedral and cuboctahedral symmetry were found. Studies were performed for complete shell clusters ranging in size from one shell having 13 atoms to 14 shells having 10,179 atoms. It was found that the icosahedral structures are lower in energy than the cuboctahedral structures for cluster sizes having 13 shells or fewer. Additional studies were performed using the more accurate Aziz-Chen [HFD-C] pair potential parameterized for argon. The conclusions appear to be relatively insensitive to the form of the potential.

PACS: 36.40; 61.55

One of the more intriguing theoretical issues concerning the study of clusters is the properties of the minimal energy structures of finite systems interacting via short range isotropic pair potentials. In this paper we investigate the energetics of clusters within a strictly classical model. We pose the mathematical question of the location of the absolute minimum of a $3N$-dimensional potential energy surface as a function of the number of atoms in the system, N. This study is motivated by the related question of the structure of free rare gas clusters. It is more or less agreed [1] that small clusters are characterized by an icosahedrally derived non-crystalline symmetry. On the other hand, it is known that bulk rare gas crystals have a face centered cubic (fcc) crystalline symmetry [2]. It is natural to ask at what size the fcc structure becomes lower in energy than the icosahedral (ic) structure. Since closed shell clusters of icosahedral and cuboctahedral fcc symmetry contain the same number of atoms we will consider the more restricted but mathematically more precise question of which closed shell isomer is more tightly bound.

In the current work, we have used the standard Lennard-Jones (LJ) model for the interatomic forces and have tested the sensitivity of the results to the form of the pair potential by using an improved potential introduced by Aziz and Chen [3]. Our procedure has been to begin with atomic coordinates placed on a regular Mackay icosahedral lattice [4] and also on a regular closed-shell fcc lattice. We then performed computer calculations to relax the structures until the force on each atom became insignificant. We verified that the relaxed structures obtained in this fashion maintained the original point group symmetry. Our

Table 1. Coefficients of the fitting function (1) for the icosahedral [ic] and cuboctahedral fcc [fcc] isomers for Lennard-Jones [LJ] and HFD-C pair potentials

	A	B	C	D	Points FIT
ic/LJ	+9.8248958	+1.5534957	−14.217539	+8.5326356	$n=2-14$
fcc/LJ	+4.4050797	+5.8929195	−16.046618	+8.6106745	$n=2-14$
ic/HFD-C	+4.9594809	+1.1613538	−11.592934	+7.6721252	$n=1\rightarrow 7, 10$
fcc/HFD-C	+3.0731707	+3.6507413	−13.154471	+7.7466403	$n=1\rightarrow 7, 10$

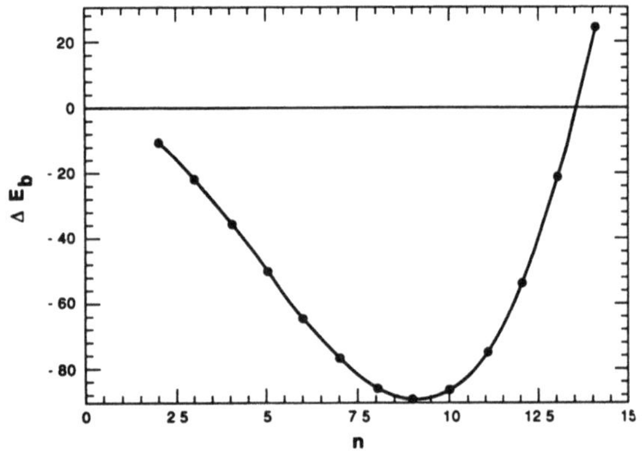

Fig. 1. The difference in binding energy, $\Delta E_b = E_b$ (fcc) $- E_b$ (ic), between the cuboctahedral and icosahedral Lennard-Jones clusters as a function of shell number, n. The energy unit is the Lennard-Jones well depth, the points are the calculated values, and the solid line is obtained by fitting the results to Eq. (1)

relaxation procedure has involved both Monte Carlo [5] techniques and the gradient based relaxation method (continued partan [6]) which has been described previously [7]. Further details of the computational techniques and the geometric features of the relaxed structures so obtained, will appear separately [8]. As indicated, we have defined a structure to be completely relaxed when the forces on each atom vanish. In practice we terminated the relaxations when the maximum force on any of the atoms was four orders of magnitude less than the maximum force after an initial relaxation step. Using this criterion, we found that the energy was stable to better than one part in 10^6 for the largest structures and even more stable for the smaller structures.

The results of the calculations using the LJ model are given in Fig. 1. In the figure $\Delta E_b = E_b$(fcc)-E_b(ic) is the difference in binding energy between the fcc and ic structures in units of the LJ well-depth and n is the shell number. The maximum difference in energy between the two structures occurs at shell number 9, and we find fcc is lower in energy for the first time at shell number 14 ($N=10, 179$). In the figure, the solid line is obtained by fitting the results to a function of the form

$$E_b = A + B\,N^{1/3} + C\,N^{2/3} + D\,N \qquad (1)$$

where E_b is the binding energy of each cluster. The coefficients in units of the LJ well depth are given in Table 1. The function reproduces our calculated values to 0.1 energy units or better. The fitting function predicts a crossing at $N=9184$. The exact value of 9184 has no real significance, however, since the calculations have been limited to closed shell structures. The favored structures for non-closed shell values of N are completely unknown.

To test the sensitivity of the conclusions to the precise form of the potential, we have performed the same calculations using the potential (HFD-C) developed by Aziz and Chen [3] and parameterized for argon. This potential has a larger force constant in the vicinity of the minimum than the LJ potential, and has somewhat larger well-depth (143.2K vs. 119.8K for the LJ potential). It is believed that the HFD-C potential is a significantly more accurate representation of the argon-argon interaction than the LJ model. The HFD-C potential is a more complicated function than the LJ potential and its evaluation consumed considerably more computer time. Since we found that (1) fit the LJ results to high accuracy, we felt confident that a similar fit to fewer data points would be adequate to predict the crossing in the HFD-C case. Consequently we have relaxed structures within the HFD-C model only for shell numbers $n=1\rightarrow 7$ and 10. The fitting coefficients are given in Table 1, and the energy differences are shown in Fig. 2. The energy unit in this case is the HFD-C well depth. The solid line in the figure is obtained from the fitting functions. The minimum occurs at approximately eight shells and the fcc becomes lower in energy at the thirteenth shell. This is in qualitative, although not quantitative, agreement with the LJ results. This agreement is a measure of the insensitivity of the conclusion to the exact form of the potential. While the crossing point is determined by extrapolation in this case, a fit to the same cluster sizes in the LJ model using (1) predicted the same crossing shell number as the fit to the set of LJ data.

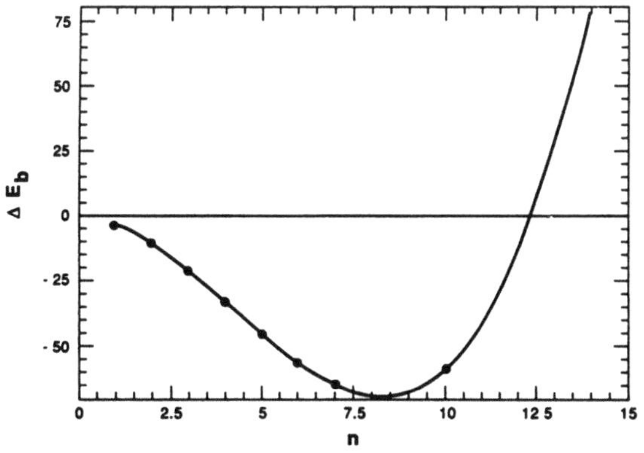

Fig. 2. The difference in binding energy, $\Delta E_b = E_b$ (fcc) $- E_b$ (ic), between the cuboctahedral and icosahedral HFD-C clusters as a function of shell number, *n*. The energy unit is the HFD-C well depth, the points are the calculated values, and the *solid line* is obtained by fitting the results to (1)

Consequently, we are reasonably confident that a full calculation would give the same crossing point.

David L. Freeman acknowledges the donors of the Petroleum Research Fund, administered by the American Chemical Society and Research Corporation for partial support of this work. JAN acknowledges the National Science Foundation (DMR 8405190) for partial support of this work. Acknowledgement is also made to the University of Rhode Island Academic Computer Center for additional support.

References

1. Hoare, M.R.: Adv. Chem. Phys. **40**, 49 (1979)
2. Ashcroft, N.W., Memin, N.D.: Solid state physics. Philadelphia: Saunders College Publishing 1976
3. Aziz, R.A., Chen, H.H.: J. Chem. Phys. **67**, 5719 (1977)
4. Mackay, A.L.: Acta Crystallogr. **15**, 916 (1962)
5. Metropolis, N., Rosenbluth, A.W., Rosenbluth, M.N., Teller, A.H., Teller, E.: J. Chem. Phys. **21**, 1087 (1953)
6. Pierre, D.A.: Optimization theory with applications. New York: John Wiley 1969
7. Northby, J.A.: J. Chem. Phys. **87**, 6166 (1987)
8. Xie, J., Northby, J.A., Freeman, D.L., Doll, J.D.: J. Chem. Phys. (in press)

Z. Phys. D – Atoms, Molecules and Clusters 12, 73–75 (1989)

Atoms, Molecules
Zeitschrift für Physik D and Clusters
© Springer-Verlag 1989

Energy barriers and structural transitions of small Al clusters

T. Halicioglu
Department of Materials Science and Engineering, Stanford University, Stanford, CA 94305, USA

Received 5 July 1988

Energy barriers between low-lying configurational states were calculated for several small Al clusters. Calculations were carried out considering a model potential comprising two- and three-body interactions. Parameters used in this potential energy function were evaluated from fits to high level ab initio calculations for Al clusters. Energy barriers separating low-lying configurations were found to be in varying heights. Results indicate that isomeric transformations between several low-lying high-symmetry forms of small Al clusters are quite likely and may take place at relatively low temperatures.

PACS: 36.40; 31.20

Recent experimental and theoretical investigations on small clusters indicate that certain microclusters exhibit isomeric transformations in the gas phase [1–7]. In order to thoroughly understand and analyze those isomerization processes, all the stationary points (at least for the low-lying configurations) of the potential energy surface associated with the cluster, must be estimated [8]. Stationary points include not only the minima of the surface, but also saddle point configurations [9]. At the present time, highly accurate calculations based on ab initio methods, are possible only for geometry optimizations of relatively smaller clusters [10]. A high level calculation for the lowest saddle point of paths connecting two minima, however, is a formidable job even with today's supercomputers. In this work, using a model potential energy function, which takes into account two- and three-body interactions, we calculated saddle point structures and corresponding energies between several low-lying equilibrium configurations of Al_4, Al_5 and Al_6. The model potential was parametrized by fitting it to highly accurate ab initio results and it has been shown that it is able to reproduce many equilibrium structures of small Al clusters (up to 6 atoms) with an acceptable accuracy [10]. The functional form of the potential and values of parameters are given in reference 10. Configurations corresponding to energy minima were calculated in this work, by an energy gradient method [9]. Low-lying configurations were selected from a large number of results calculated with different initial geometries. Saddle points like minima were also treated as stationary points on the energy surface. Several different paths between two configurational minima were taken into consideration and the saddle point structure corresponding to the path with the lowest barrier is reported in this study. Calculated results are summarized in Table 1.

For the Al_4 case, the two lowest minimum energy structures corresponding to three-dimensional (3*d*) and two-dimensional (2*d*) configurations, are the regular tetrahedron $(T)_4$ and the rhombus $(R)_4$ structures, respectively. The saddle point structure for the $(T)_4 \rightarrow (R)_4$ transformation is a highly distorted tetrahedron with one elongated side. The barrier for this $3d \rightarrow 2d$ transition was estimated as 0.0299 eV/atom.

Energetically low-lying 3*d* configurations for the Al_5 case are the triangular bipyramid $(TBP)_5$ which is the most compact structure and the square pyramid $(SP)_5$. The energetically most favorable Al_5 structure is the 2*d* close-packed $(CP)_5$ configuration. Two saddle point structures were calculated, in this case. For the transition between two 3*d* structures, the saddle point configuration was found to be a distorted triangular bipyramid. Its two apex atoms are shifted toward to one side of the triangular base. Calculated energy for this saddle point configuration corresponds to a

Table 1. Calculated energies and symmetry groups for equilibrated and transition structures of Al clusters

Cluster	Configuration	Energy (eV/atom)	Symmetry
Al_4	Tetrahedral $(T)_4$	−1.1069	T_d
Al_4	$[(T)_4 \rightarrow (R)_4]^{\ddagger}$	−1.0770	C_{2v}
Al_4	Rhombus $(R)_4$	−1.1255	D_{2h}
Al_5	Triangular bipyramid $(TBP)_5$	−1.2469	D_{3h}
Al_5	$[(TBP)_5 \rightarrow (SP)_5]^{\ddagger}$	−1.2414	C_{2v}
Al_5	Square pyramid $(SP)_5$	−1.2431	C_{4v}
Al_5	$[(SP)_5 \rightarrow (CP)_5]^{\ddagger}$	−1.2172	C_1
Al_5	Close-packed $(CP)_5$	−1.2744	C_{2v}
Al_6	Square bipyramid $(SBP)_6$	−1.3424	O_h
Al_6	$[(SBP)_6 \rightarrow (TP)_6]^{\ddagger}$	−1.3307	C_{2v}
Al_6	Tripyramid $(TP)_6$	−1.3480	C_{2v}
Al_6	$[(TP)_6 \rightarrow (PP)_6]^{\ddagger}$	−1.3402	C_s
Al_6	Pentagonal pyramid $(PP)_6$	−1.3709	C_{5v}
Al_6	$[(PP)_6 \rightarrow (CP)_6]^{\ddagger}$	−1.3353	C_s
Al_6	Close-packed $(CP)_6$	−1.3801	C_{2v}
Al_6	Triangular close-packed $(TCP)_6$	−1.3926	D_{3h}

0.0055 eV/atom barrier height for the $(TBP)_5 \rightarrow (SP)_5$ transformation. For the $3d \rightarrow 2d$ transition, among many possible alternatives, the path with the lowest saddle point energy was found for the $(SP)_5 \rightarrow (CP)_5$ transformation. This saddle point structure resembles very much the $(CP)_5$ but it is not planar. The barrier for this $(SP)_5 \rightarrow (CP)_5$ transition was calculated to be 0.0259 eV/atom.

For Al_6, we considered three low-lying $3d$ configurations. The square bipyramid $(SBP)_6$ structure is the most compact configuration. The tripyramid $(TP)_6$ is formed by three slightly distorted tetrahedra. Structurally it is similar to $(TBP)_5$ with an additional Al atom positioned above one of its faces. The pentagonal pyramid $(PP)_6$ is the energetically most stable $3d$ configuration. The low-lying $2d$ configurations for Al_6 include the close-packed $(CP)_6$ and the triangular close-packed $(TCP)_6$ structures. Three transition structures were investigated in this case. The saddle point configuration for the $(SBP)_6 \rightarrow (TP)_6$ transition, resembles the $(TP)_6$ structure with three distorted tetrahedra. The commonly shared edge of three tetrahedra, in this configuration, was found to be considerably elongated, however, the C_{2v} symmetry was still maintained. The energy of this transition structure yields a barrier of 0.0117 eV/atom for the $(SBP)_6 \rightarrow (TP)_6$ conversion reaction. The other $3d \rightarrow 3d$ transformation is the $(TP)_6 \rightarrow (PP)_6$ transition. The three tetrahedra (which were the building blocks of $(TP)_6$) in this saddle point structure, were highly distorted. This is the only second order saddle point [9] found in this study. Among many paths connecting equilibrium $(TP)_6$ and $(PP)_6$ configurations no first order saddle point was found with a lower energy. The calculated energy value for this saddle point configuration produces a barrier of 0.0078 eV/atom for the $(TP)_6 \rightarrow (PP)_6$ transformation. The other saddle point structure, considered in this work, belongs to the $(PP)_6 \rightarrow (CP)_6$ transition which is a $3d \rightarrow 2d$ transformation. This saddle point structure is a highly distorted pentagonal pyramid. The calculated barrier height for the $(PP)_6 \rightarrow (CP)_6$ transition was found to be 0.0356 eV/atom.

For $(TBP)_5 \rightarrow (SP)_5$ and $(SBP)_6 \rightarrow (TP)_6$ transformations, barrier heights (in both forward and reverse directions) were found to be quite low. This implies that isomerization reactions can easily take place at even low temperatures, and isomers may coexist in the gas phase for a certain length of time. Energy barriers for $3d \rightarrow 2d$ transformations, for the cases of Al_5 and Al_6, were found to be generally higher than $3d \rightarrow 3d$ transitions, even though $2d$ structures are energetically more stable than $3d$ structures. This indicates that $3d \rightarrow 2d$ transformation reactions for Al_5 and Al_6 would proceed much slower.

In gas-phase experiments, in general, the time necessary for a complete equilibration (including all the fragmentation and structural transformation processes) is not a simple quantity to obtain. In a cluster generation process there are many factors (such as the temperature, the type of the carrier gas, dimensions of the expansion chamber, etc.) affecting the initial formation and structural characteristics of metal atom clusters. Unfortunately, present condensation theories are generally inadequate to provide a definitive prediction about the clustering conditions of metal vapors in free jet experiments [11]. Based on purely statistical considerations, however, in the very early stages of the clustering process, the probability of a $3d$ cluster formation is higher than that of the formation of a $2d$ cluster. After the equilibration period, at relatively lower temperatures, the probability of finding a cluster in its lowest energy configuration is, of course, the highest. For the Al_5 case, for example, let us compare $(SP)_5 \rightarrow (CP)_5$ and $(SP)_5 \rightarrow (TBP)_5$ transformations which are $3d \rightarrow 2d$ and $3d \rightarrow 3d$ processes, respectively. If preexponential factors of these processes are assumed to be approximately equal, then, based on the absolute rate theory (say, at $T \approx 300°K$) the rate for the $(SP)_5 \rightarrow (CP)_5$ reaction is more than 100 times slower than the rate of the $(SP)_5 \rightarrow (TBP)_5$ transformation. (For transition energies see Table 1.) This indicates that even though the low energy structure of Al_5 is planar, the newly formed Al_5 may remain $3d$, fluctuating back and forth between $(SP)_5$ and $(TBP)_5$ configurations a great many times before the complete equilibration. It has already been shown that kinetics may play a major role in many of the processes occurring in the gas-phase [12]. In particular, if the lowest energy structure of a cluster is separated by relatively high barriers from other low-lying configurations (like Al_5 and Al_6 cases examined in this paper), then, the equilibration time may be long and it may become an important consideration in cluster beam experiments. Barrier heights between isomers may also play an important role in the kinetics of fragmentation which is an important process affecting the gas-phase composition. Among many isomeric configurations of a cluster perhaps one (or, a few) isomeric structure(s) may undergo to a fragmentation process. Accordingly, the fragmentation depends on the concentration of a particular isomer in the beam. The rate of formation of that particular isomer (which serves as a window for the fragmentation) then, becomes an important step, and the energy barriers separating this isomer from other low-lying configurations, may control the rate of the fragmentation process in the beam.

The model potential energy function employed in this study can satisfactorily reproduce highly accurate ab initio results for a variety of minimum energy configurations of small Al clusters [10]. However, care should be exercized in a more quantitative interpretation of the saddle point results obtained here. This is because, at the present time, no accurate calculations based on first principles are available for the barrier heights corresponding to isomeric transformation of small Al clusters for comparison.

The author wishes to thank Drs. C.W. Bauschlicher, Jr and P. Taylor for many fruitful discussions. Furthermore, the support from NASA, Ames Research Center (via Grant NCC 2-297) is gratefully acknowledged.

References

1. Zakin, M.R., Brickman, R.O., Cox, D.M., Kaldor, A.: J. Chem. Phys. **88**, 3555 (1988)
2. Hamrick, Y., Taylor, S., Lemire, G.N., Fu, Z.-W., Shui, J.-C., Morse, M.D.: J. Chem. Phys. **88**, 4095 (1988)
3. Honeycutt, J.D., Andersen, H.C.: J. Phys. Chem. **91**, 4950 (1987)
4. Berry, R.S., Beck, T.L., Davis, H.L., Jellinek, J.: In: Physics and chemistry of small clusters. Jena, P., Rao, B.K., Khanna, S.N. (eds.) NATO ASI Series, pp. 185. New York: Plenum 1987
5. Davis, H.L., Jellinek, J., Berry, R.S.: J. Chem. Phys. **86**, 6456 (1987)
6. Kondow, T.: J. Phys. Chem. **91**, 1307 (1986)
7. Gspann, J.: Z. Phys. D – Atoms, Molecules and Clusters **3**, 143 (1986)
8. Halicioglu, T.: Surf. Sci. **197**, L233 (1988)
9. Schlegel, H.B.: Adv. Chem. Phys. **67**, 249 (1987)
10. Pettersson, L.G.M., Bauschlicher, Jr., C.W., Halicioglu, T.: J. Chem. Phys. **87**, 2205 (1987)
11. Hagena, O.F.: Z. Phys. D – Atoms, Molecules and Clusters **4**, 291 (1987)
12. Phillips, J.C.: J. Chem. Phys. **83**, 3330 (1985)

Z. Phys. D – Atoms, Molecules and Clusters 12, 77–79 (1989)

Atoms, Molecules and Clusters
Zeitschrift für Physik D
© Springer-Verlag 1989

Model potential for beryllium clusters

E. Blaisten-Barojas* and S.N. Khanna

Department of Physics, Virginia Commonwealth University, 1020 West Main Street, Richmond, VA 23284-2000, USA

Received 5 July 1988; final version 30 September 1988

A many-body potential is developed for beryllium based on the electronic information extracted from the total energy surface of clusters with up to 5 atoms. The cluster sequence of growth generated with the potential is in excellent agreement with the quantum mechanical calculations in the literature. The hcp lattice of bulk beryllium is correctly found to be more stable than fcc and bcc lattices.

PACS: 34.20.Gj; 31.20.Sy; 36.40.+d; 61.45.+s

1. Introduction

The last few years have seen research in cluster physics and chemistry develop into an intensively studied area of science [1]. The common approach in simulation studies of cluster structure has been to use potentials developed to fit bulk properties [2, 3, 4] with the hope of bridging between cluster and condensed matter theories. However, recent molecular dynamics studies of Si clusters [3] using bulk potentials have predicted structures in disagreement with ab-initio calculations [5]. A different scheme was proposed by Car and Parrinello [6] where the need to solve the electronic and nuclear dynamics separately is bypassed. In this work we propose another approach. It too, like the Car and Parrinello technique, relies on the local density description (LD) of the electronic structure of the system. We develop a new class of potentials containing two- and three-body functions built simultaneously on the LD energy surface data of the dimer, trimer, tetramer and pentamer.

The binding in Be clusters increases dramatically with size due to the increasing *s-p* hybridization [7]. Namely, the dimer and trimer are weakly bonded [8, 9] whereas beryllium is metallic in the solid phase [7]. Ab-initio CI calculations predict Be_4 to be a stable tetrahedron [9, 10, 11], Be_5 to have a trigonal bipyramid structure [10], Be_6 to be a square bipyramid and Be_7 to acquire the pentagonal bipyramid structure. The fcc and hcp structures of Be_{13} have been analyzed [10, 11, 12] with the prediction that the fcc polyhedron is the most stable. Beryllium bulk calculations yield the hcp lattice [13] as the most stable in agreement with experiment.

2. Many-body potential for beryllium

In order to keep the problem computationally tractable we divide the total potential energy of a cluster into pair potentials and three-body, potentials. Next we construct a model V_2+V_3 so as to give a good simultaneous fit to values of the total energy surface of Be_2 up to Be_5 calculated within the local spin density (LSD) approximation. We have considered 23 values for the dimer, 76 points for the trimer in various angular conformations, 19 points for tetramer corresponding to isotropic expansions of the tetrahedron and one point for the pentamer trigonal bipyramid conformation. By so doing 4- and 5-body contributions are partially accounted for in the construction of a scaled 3-body term.

We write the binding energy of a system with N atoms as the sum over pairs plus the sum over the

---

* On sabbatical leave from Instituto de Física, Universidad Nacional Autónoma de México, Apartado Postal 20-364, 01000 México DF, México

orientational 3-body potentials [14] as:

$$E_N = \sum_{i<j}^{N} (A\exp(-\alpha r_{ij}) - Cf_B(r_{ij})/r_{ij}^5) + \sum_{i<j<k}^{N} \left(-D(N)\exp(-\beta(r_{ij}+r_{ik}+r_{jk})) + \frac{C'}{(r_{ij}r_{ik}r_{jk})^3} \right) \times h(r_{ij}, r_{ik}, r_{jk}) f_{B'}(r_{ij}) f_{B'}(r_{ik}) f_{B'}(r_{jk}) \quad (1)$$

where $f_B(r) = \exp[-B(r_0/r - 1)^5]$ if $r < r_0$ and $f_B(r) = 1$ otherwise. The function $h(r_{ij}, r_{ik}, r_{jk}) = 1 + 3\cos\theta_{ijk} \cos\theta_{kij}\cos\theta_{jki}$ depends upon the angles θ_{ijk} subtended by any triplet of atoms *ikj*. The dependence of the *s-p* binding on cluster size is modelled through the scaled parameter $D(N) = D - G(0.25 - 1/N)$ if $N > 4$ and $D(N) = D$ otherwise. The parameters values are in atomic units: $A = 77.27716$, $\alpha = 1.71169$, $B = 0.6961$, $C = 87.39774$, $D = 9.65426$, $\beta = 0.485767$, $G = 35.945$, $C' = 673.4099$, $B' = 0.333333$, $r_0 = 4.04$. The quality of the fit was excellent leading to a sum of squares less than one thousand.

3. Results and discussion

With this potential we have generated clusters consisting of up to 13 beryllium atoms by a Monte Carlo simulation [15]. The resulting most stable structures of the dimer up to the pentamer are the same as those given by the LSD calculation in what concerns both binding energies and geometries. The structures obtained for the most stable Be_5, Be_6 and Be_7 are in remarkably good agreement with ab-initio CI results [10]—bond distances differing by only 2–3%. However, the binding energies obtained with the potential for these clusters are lower than those obtained up to now with CI calculations [15]. This was expected since for Be the LSD approximation is known to overestimate the binding [7]. The most stable structures obtained with the potential for clusters with 8 up to 12 atoms are plotted in Fig. 1. A careful inspection shows that the square bipyramid and the pentagonal pyramid are the two building blocks contributing to the aggregation of the larger clusters. The 11-atom cluster is made up of two trigonal bipyramids sharing one atom and rotated 45° with respect to one another. This cluster is the first in the series of growth to have one internal atom (not drawn in the figure) with 10-fold coordination. The 12-atom cluster is an icosahedron with one missing atom. The potential predicts Be_{13} to have the icosahedral structure as the most stable, the hcp and fcc polyhedra following in binding energy. This seems to contradict the CI results stating the fcc structure to be more stable than the hcp structure [11, 12]. However these results do not emphasize the need of geometry optimization at the CI level. Besides, no quantum mechanical calculation exists for icosahedral Be. If the binding energy of the hcp and fcc structures are calculated with the potential in exactly the geometries given in [12], the energy order is reversed, i.e., −0.897 a.u. (fcc) and −0.889 a.u. (hcp). Thus, geometry optimization is as important as the search of the most significant electronic configurations in the CI calculations.

In Fig. 2a we show that in this series the cluster binding energy (full line and left scale) tends to the bulk value as the cluster increases its size. The dotted line and right scale shows the point derivative of the binding energy. It puts in evidence that the 4-, 6- and 13-atom clusters are the most stable in the series. In Fig. 2b we have plotted the cluster average bond distance as a function of size. It is clearly shown how dramatic a change occurs near 4–5 beryllium atoms. This fact points exactly to the cluster size where *s-p* hybridization has acquired almost the bulk values. For larger clusters the bond distances change mostly as a result of the increasing number of atoms and evolves with oscillations towards the bulk lattice constants.

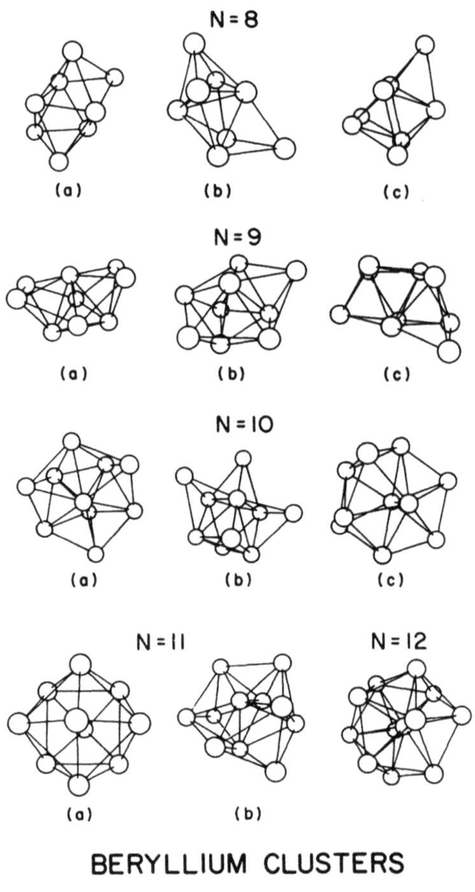

Fig. 1. Cluster configurations

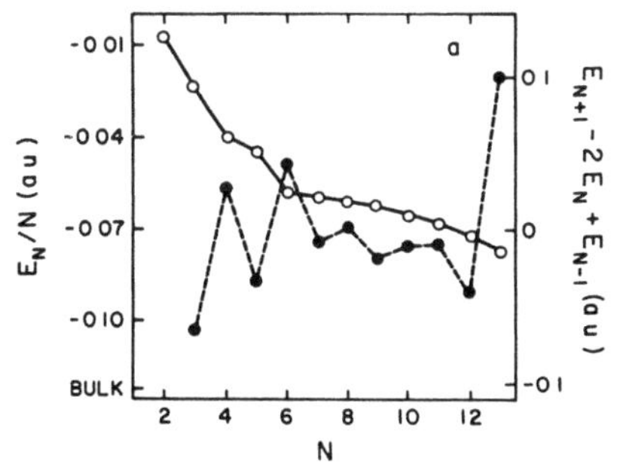

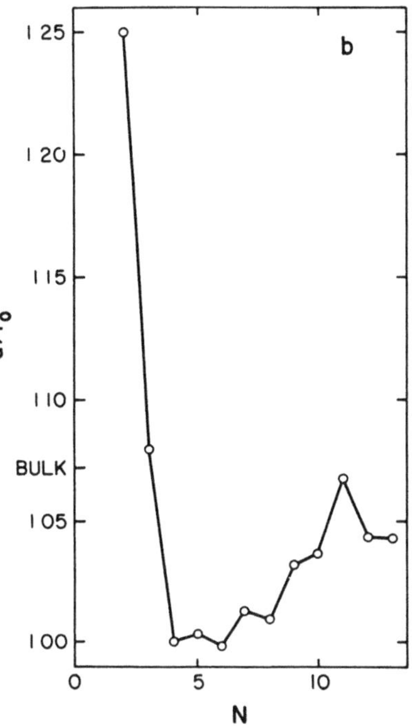

Fig. 2. a Potential energy per atom as a function of cluster size (*full line, left scale*) and its point derivative (*dashed line, right scale*); **b** Average bond distance as a function of cluster size

We have also calculated the cohesive energies E of the hcp, fcc and bcc lattices as a function of density [15]. The hcp lattice is the most stable with $E = -3.597$ eV/atom, lattice constants $a = 2.254$ Å and $c = 3.683$ Å and compressibility $K = 1.001\ 10^{-12}$ cm/dyn. These values are in excellent agreement with experiment ($E = 3.32$ eV/atom, $a = 2.285$ Å, $c = 3.585$ Å) as well as with bulk ab-initio calculations [13]. Besides the potential predicts the fcc lattice to be the next in stability at the room temperature density, but at low densities it predicts the bcc structure to become more stable than even the hcp lattice. This again is in agreement with experiment since beryllium bcc beryllium is stable at high temperatures.

The calculations we have reported describe a very successful strategy that makes possible to incorporate the information contained in the electronic structure of small clusters to the development a classical force field.

EBB thanks support from the international agreement between CONACyT (Mexico) and the NSF (USA) and from the Army Research Office (DAAG 29-85-K-0244). SNK acknowledges support from the Department of Energy (DE-FG05-87ER45316).

References

1. See for example references in: Benedek, G., Martin, T.P., Pacchioni, G. (eds.) Elemental and molecular clusters. Springer Series in Materials Science, Vol. 6. Berlin, Heidelberg, New York: Springer 1988
2. Honeycutt, D., Andersen, H.C.: J. Phys. Chem. **91**, 4950 (1987); Blaisten-Barojas, E., Garzon, I. L., Avalos-Borja, M.: Phys. Rev. B **36**, 8447 (1987)
3. Blaisten-Barojas, E., Levesque, D.: Phys. Rev. B **34**, 3910 (1986); Feuston, B.P., Kalia, R.K., Vashishta, P.: Phys. Rev. B **35**, 6222 (1987)
4. Martin, T.P.: Phys. Rep. **95**, 167 (1983); Landman, U., Sharf, D. Jortner, J.: Phys. Rev. Lett. **54**, 1860 (1985)
5. Ragavachari, K.: J. Chem. Phys. **84**, 5672 (1986); Pacchioni, G., Koutecky, J.: J. Chem. Phys. **84**, 3301 (1986); Tomanek, D., Schluter, H.: Phys. Rev. Lett. **56**, 1055 (1986)
6. Car, R., Parrinello, M.: Phys. Rev. Lett. **55**, 2471 (1985); Ballone, P., Andreoni, W., Car, R., Parrinello, M.: Phys. Rev. Lett. **60**, 271 (1988)
7. Jones, A.O.: In.: Ab-initio methods in quantum mechanics-I. Lawley, K.P. (ed.), p. 413. New York: John Wiley 1987
8. Lengsfield, III, B., McLean, A.D. Yoshimine, M., Liu, B.: J. Chem. Phys. **79**, 1891 (1983)
9. Harrison, R.J., Handy, N.C.: Chem. Phys. Lett. **123**, 321 (1986)
10. Ermler, W.C., Kern, C.W., Pitzer, R.M., Winter, N.W.: J. Chem. Phys. **84**, 3937 (1986); Marino, M., Ermler, W.C.: J. Chem. Phys. **86**, 6283 (1987)
11. Rohlfing, C.M., Binkley, J.S.: Chem. Phys. Lett. **134**, 110 (1987)
12. Bauschlicher, Jr., C.W., Pettersson, L.G.M.: J. Chem. Phys. **84**, 2226 (1986)
13. Chu, M.Y., Lam, P.K., Chen, M.L.: Phys. Rev. B **28**, 4179 (1983)
14. Bruch, L.W., Blaisten-Barojas, E., Novaro, O.: J. Chem. Phys. **67**, 4701 (1977); Blaisten-Barojas, E., Novaro, O., Bruch, L.W.: Mol. Phys. **37**, 599 (1979)
15. Blaisten-Barojas, E., Khanna, S.N.: Phys. Rev. Lett. **61**, 1477 (1988)

Z. Phys. D – Atoms, Molecules and Clusters 12, 81–83 (1989)

Atoms, Molecules
Zeitschrift für Physik D and Clusters
© Springer-Verlag 1989

Structure and dynamics of Lennard-Jones clusters with impurities

I.L. Garzón[1], X.P. Long[2], R. Kawai[2], and J.H. Weare[2]

[1] Instituto de Física, Universidad Nacional Autonoma de México, Apartado Postal 2681, 22800 Ensenada, Baja California, México, and Department of Chemistry, B-040, University of California, San Diego, La Jolla, CA 92093, USA

[2] Department of Chemistry, B-040, University of California, San Diego, La Jolla, CA 92093, USA

Received 5 July 1988

Molecular dynamics simulations and Lennard-Jones potentials have been used to study binary mixed clusters. The low temperature structures, impurity solvation and the melting and freezing transitions for different values of the relative atomic size and interaction energy have been studied for $A_{13}B$, $A_{12}B$, $A_{55}B$ and $A_{13}B_{13}$ clusters. For $A_{13}B$ large impurities do not solvate even for high interaction energy. For $A_{55}B$ larger impurities remain on the surface for low interaction energies but solvate as the energy of interaction increases. The presence of the impurity very strongly affects the solid-liquid transition. Icosahedral structures remain as the minimum energy configurations for $A_{13}B_{13}$ clusters with atoms of the same size and different interaction energies. For $A_{13}B_{13}$ clusters with atoms of the same interaction energy and different size, smaller atoms go inside surrounded by the bigger surface atoms.

PACS: 36.40.+d.

Molecular Dynamics (MD) methods have been used to simulate structural and dynamical properties of pure isolated [1] and adsorbed [2] van der Waals clusters whose interaction are well described by Lennard-Jones (LJ) potentials. The simulation of impure and compound van der Waals clusters is also important but has received less attention. A novel spectroscopic technique developed by Scoles et al. [3] allows the detection and measurement of the structural and dynamical properties of argon clusters containing active infrared molecules. Experimental evidence suggests that structure, temperature, phase, diffusion, solvation and even reactivity of the molecule-cluster system can be probed through the IR photodepletion spectroscopy [3].

In this communication we report the low temperature structures of impure ($A_{13}B$, $A_{55}B$) and compound ($A_{13}B_{13}$) clusters obtained using constant energy MD method and LJ potentials with different sets of parameters. The caloric curve for $A_{12}B$ also is studied. Our objective is to investigate the following questions:

a) What is the structural change in a very stable cluster containing a magic number of atoms (13 or 55 for LJ clusters) when it combines with an impurity that differs in size and interaction energy with respect to the host cluster atoms?

b) What is the effect of the impurity size and interaction energy in the melting and freezing transitions of the host cluster?

c) What are the structure and rearrangements of a compound cluster grown from the combination of two smaller clusters with atoms of different sizes and well depth potential parameters?

Systematic annealing processes were produced on the impure ($A_{12}B$, $A_{13}B$ and $A_{55}B$) and compound ($A_{13}B_{13}$) clusters to obtain the low temperature structures and caloric curves. The clusters were heated and cooled by using constant energy MD method and LJ potentials with the Verlet's algorithm [4].

Tables 1 and 2 show the low temperature structures obtained for $A_{13}B$ and $A_{55}B$ for different values of $\sigma_i = \sigma_b/\sigma_a$ and $\varepsilon_i = \varepsilon_b/\varepsilon_a$ where σ and ε are LJ potential parameters. The low temperature structure of the $A_{13}B$ clusters is strongly dependent on the σ_i values. Three different structures have been observed for $\sigma_i = 0.7$, 1.0, and 1.5. For $\sigma_i = 0.7$ the B atom is always inside the host cluster independently of the ε_i value. When $\sigma_i = 1.0$ and $\varepsilon_i = 1.5$, 2.0, 3.0 the B atom accommodates at the center of the cluster surrounded by the A atoms in an icosahedral configuration. For $\sigma_i = 1.0$ and $\varepsilon_i = 1.0$, 0.7 the B atom sits in the surface and adsorbs on it respectively, still maintaining the icosahedral symmetry. The five-fold symmetry is destroyed when the B atom size increase up to $\sigma_i = 1.5$. In this case the B atom adsorbs on the host cluster producing a big hole on its surface forming a bowl-like structure (see Fig. 1).

The low temperature structure of the $A_{55}B$ are similar to those obtained for $A_{13}B$ except for the fact that in the $A_{55}B$ cluster two icosahedral shells are present. For $\sigma_i = 0.7$ the B atom is always inside the cluster between their first and second shells independent of the ε_i value. For $\sigma_i = 1.0$ and $\varepsilon_i = 0.7$ the impurity adsorbs on the cluster surface but when the ε_i value increases it is solvated by the cluster going from the surface ($\varepsilon_i = 1.0$ and 1.5) to the center ($\varepsilon_i = 2.0$ and 3.0). For bigger impurities ($\sigma_i = 1.5$) we obtain low temperature configurations in which the impurity is adsorbed on the surface of a distorted icosahedral cluster ($\varepsilon_i = 0.7$, 1.0 and 1.5) or solvated inside a disordered cluster ($\varepsilon_i = 2.0$ and 3.0).

Two types of 26-atom compound clusters were studied, in the first one we combined atoms of the same atomic size and distinct interaction energy ($\sigma_A = \sigma_B$ and $\varepsilon_A = 1.5\,\varepsilon_B$). In the second case atoms of different size but equal interaction energy ($\sigma_A = 1.5\sigma_B$ and $\varepsilon_A = \varepsilon_B$) were mixed. Figure 2a shows the low temperature structure for the first case where stronger interacting particles are inside the compound cluster forming an icosahedral shape cluster core while the weaker interacting particles are outside forming a partial cluster surface following the five-fold icosahedral symmetry. Figure 2b shows that the low temperature configuration is a cluster in which the smaller atoms form an internal disordered subcluster surrounded by the bigger surface atoms. This result indicate that in a compound cluster the smaller atoms are bulk-like while the cluster surface is done from the bigger

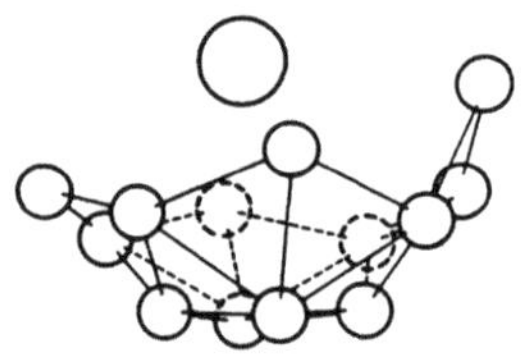

Fig. 1. Bowl-like structure for the $A_{13}B$ cluster with B atom adsorbed on the surface

Table 1. Low temperature structure of the $A_{13}B$ clusters

ε_i	$\sigma_i = 0.7$	$\sigma_i = 1.0$	$\sigma_i = 1.5$
0.7	Icosahedron with B atom inside	Icosahedron with B atom adsorbed	Icosahedron with B atom adsorbed
1.0	Icosahedron with B atom inside	Icosahedron with B atom in the surface	Distorted icosahedron with B atom on the surface
1.5	Icosahedron with B atom inside	Icosahedron with B atom at the center	Very distorted icosahedron with B atom on the surface.
2.0	Icosahedron with B atom inside	Icosahedron with B atom at the center	Bowl-like structure* with B atom on the surface
3.0	Icosahedron with B atom inside	Icosahedron with B atom at the center	Bowl-like structure* with B atom on the surface

* See Fig. 1

Table 2. Low temperature structure of the $A_{55}B$ clusters

ε_i	$\sigma_i = 0.7$	$\sigma_i = 1.0$	$\sigma_i = 1.5$
0.7	Icosahedron with B atom inside	Icosahedron with B atom adsorbed	Icosahedron with B atom adsorbed
1.0	Icosahedron with B atom inside	Icosahedron with B atom in the surface	Icosahedron with B atom adsorbed surface
1.5	Icosahedron with B atom inside	Icosahedron with B atom in the surface	Distorted icosahedron with B atom on the surface
2.0	Icosahedron with B atom inside	Icosahedron with B atom at the center	Disordered cluster with B atom inside
3.0	Icosahedron with B atom inside	Icosahedron with B atom at the center	Disordered cluster with B atom inside

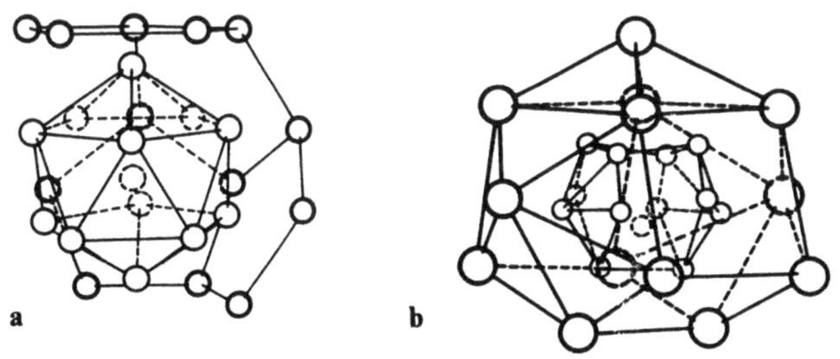

Fig. 2a. Low temperature structure for the $A_{13}B_{13}$ cluster with $\sigma_i = 1.0$ and $\varepsilon_A = 1.5\varepsilon_B$; **b** same as in a) but $\sigma_A = 1.5\sigma_B$ and $\varepsilon_A = \varepsilon_B$

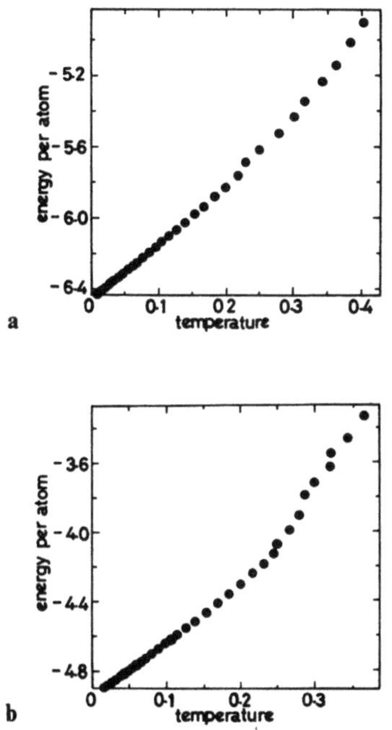

Fig. 3. Caloric curves for the $A_{13}B_{13}$, **a** $\sigma_i = 1.0$ and $\varepsilon_A = 1.5\varepsilon_B$; **b** $\sigma_A = 1.5\sigma_B$ and $\varepsilon_A = \varepsilon_B$

particles. These compound clusters have different dynamical properties as the caloric curves of Fig. 3 show.

The $A_{12}B$ cluster was simulated in order to study the effect of the impurity interaction energy in the melting transition. Figure 4 shows the total energy of the system as a function of the temperature for $\sigma_{AB}/\sigma_{AA} = 1.0$ and $\varepsilon_{AB}/\varepsilon_{AA} = 0.5$, 1.0, 1.5, 2.0, and 2.5. From the "caloric curves" we found that when weak interacting particle is present, evaporation of the impurity take place before the melting of the host cluster. When strong interacting impurity is present, it couples with larger energy to the host atoms hindering the melting processes when the total energy was increased.

New structural properties were found for compound LJ clusters. The size and interaction energy of one impurity atom can modify the well known low

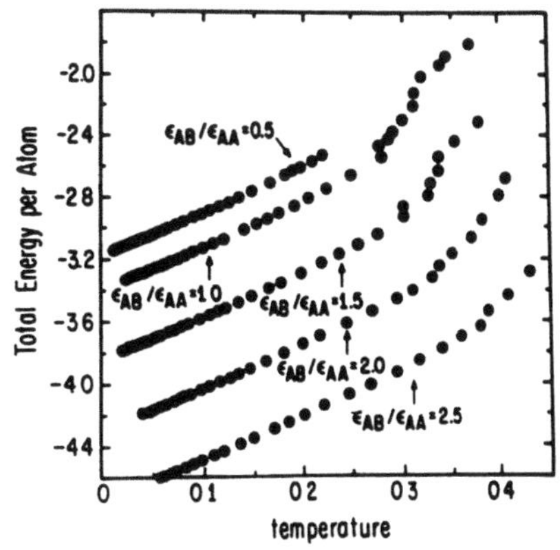

Fig. 4. Caloric curves for the $A_{12}B$ cluster. All curves were obtained with $\sigma_{AB}/\sigma_{AA} = 1.0$. The $\varepsilon_{AB}/\varepsilon_{AA}$ value is 0.5, 1.0, 1.5, 2.0, 2.5 from the top to the bottom

temperature structures and the melting mechanisms found for pure clusters. The MD simulations of the $A_{13}B_{13}$ clusters also indicate that the relative atomic size and the well depth potential parameters are determinant in the cluster structures at low temperatures. We believe that the low temperature structures obtained in these simulations correspond to the minimum energy configurations because a careful annealing using a slow rate in the cooling processes was used. The present simulations can be useful in the understanding of solvation mechanisms in finite systems.

We wish to acknowledge the support of the SDIO/TA/USAF (ONR-N00014-87-K-0675) for this project. We thank the San Diego Supercomputing Center for the use of their computing facilities. One of us (I.L.G.) wishes to thank the Consejo Nacional de Ciencia y Tecnologia (México) for the partial support received during the realization of this work.

References

1. McGinty, D.J.: J. Chem. Phys. **58**, 4733 (1973); Briant, C.L., Burton, J.J.: J. Chem. Phys. **63**, 2045 (1975); Jellinek, J., Beck, T.L., Berry, R.S.: J. Chem. Phys. **84**; 2783 (1986); Berry, R.S., Beck, T.L., Davis, H.L., Jellinek, J.: Adv. Chem. Phys. Vol. LXX (1988)
2. Garzón, I.L., Avalos, M., Blaisten-Barojas, E.: In: Proceedings of the International Symposium on the Physics and Chemistry of Small Clusters, NATO ASI series, ed. Jena, P., Rao, B.K., Khanna, S.N. (eds.) p. 193, New York: Plenum Press 1987; Blaisten-Barojas, E., Garzón, I.L., Avalos-Borja, M.: Phys. Rev. B **36**, 8447 (1987)
3. Gough, T.E., Knight, D.G., Scoles, G.: Chem. Phys. Lett. **97**, 155 (1983); Gough, T.E., Mengel, M., Rowntree, P.A., Scoles, G.: J. Chem. Phys. **83**, 4958 (1985)
4. Verlet, L.: Phys. Rev. **159**, 98 (1964)

Z. Phys. D – Atoms, Molecules and Clusters 12, 85–87 (1989)

Atoms, Molecules
Zeitschrift für Physik D and Clusters
© Springer-Verlag 1989

Stability of relaxed Lennard-Jones models made of 500 to 6000 atoms

B. Raoult, J. Farges, M.F. De Feraudy, and G. Torchet
Laboratoire de Physique des Solides, LA2-Bâtiment 510, Université de Paris-Sud, F-91405 Orsay, France

Received 5 July 1988

We present a study of the stability of clusters models made of a number N of atoms in the range 500 to 6000 atoms, freely interacting through the Lennard-Jones potential. The potential energy per atom, calculated for relaxed models, shows that stable models belong to an icosahedral sequence when $N < 1600$ and to a decahedral sequence beyond. A coexistence size range of both structures is discussed in connection with experimental results on argon clusters in free jet expansions.

PACS: 36.40; 61.55

Introduction

Since the basic paper of Ino [1] on stability of multiply-twinned particles (MTPs), the energetics of MPTs has been widely investigated. Recently Marks [2] made a major contribution by developing a modified Wulff construction in which twin boundaries are included: the total surface energy of the decahedral MTP is then sandwiched between that of an icosahedral MTP and that of a single crystal in an intermediate size range. But because neither type of MTP is completely space-filling, it is necessary to incorporate some form of internal distorsion. Unfortunately the strain energy in MTPs is difficult to estimate precisely and, from the overall energy balance between single crystals and MTPs, one cannot predict any true stability for decahedral particle.

On the other hand, in the case of small clusters, the overall energy calculation can be carried out by using a more rigorous method based on models made of atoms freely interacting through an assumed interatomic potential, i.e. the Lennard-Jones (LJ) potential for argon clusters. In this paper we investigate relaxed LJ models made of 500 to 6000 atoms. Attention is focussed on the crystalline, icosahedral and decahedral sequences of models of minimum potential energy. The relative stability of the models is discussed in connection with experimental observations on argon clusters in cluster beams.

Models

In their initial configuration models have an homogeneous structure: a crystalline model has a fcc structure, a decahedral or an icosahedral model can be described respectively as a collection of five or twenty distorted fcc tetrahedra twin-related on their adjoining faces, with body centered orthorhombic or rhombohedral crystal symmetry. All models are relaxed freely: each atom can move under the action of the resultant force created by its $N-1$ neighbours in the model until the resultant value is practically zero [3]. The symmetry of the model is introduced in the routine to save calculation time. The reduced potential energy U^* is then calculated from the reduced interatomic distances $r_{ij}^* = r_{ij}/\sigma$ by using the formula

$$U^* = U/\varepsilon = (1/N) \sum_{i=1}^{N-1} \sum_{j=i+1}^{N} 4(r_{ij}^{*\,-12} - r_{ij}^{*\,-6})$$

according to the classical expression of the LJ potential

$$V(r_{ij}) = 4\varepsilon \{(r_{ij}/\sigma)^{-12} - (r_{ij}/\sigma)^{-6}\}$$

In a given structure, models keeping the same external shape, as they increase their size, define a particular sequence. This sequence is characterized by discrete values of the number N of atoms, often called magic numbers. For instance, one gets the sequence of full icosahedra: $N = 13$, 55, 147, 309, 561 . . . In each sequence magic numbers are different by construction according to the shapes or structures. Thus the relative stability of models is studied, through a set of lines $U^* = f(N^{-1/3})$ drawn in each sequence, instead of a set of constructed isomers.

Results

In the size range $500 < N < 6000$ the sequence which contains models of minimum energy U^* has been found for the three structures investigated [4]: the icosahedral sequence is made up of icosahedra without their 12 vertex atoms, i.e. $N = 549$, 911, 1403, 2045, etc. . . Both crystal line and decahedral sequences consist of polyhedra the surface of which is composed of (111), (100) and (110) faces: crystalline polyhedra are truncated octahedra with 26 faces and decahedral polyhedra are truncated decahedra with 45 faces as shown in Fig. 1.

These three sequences are compared on Fig. 2 showing the related energy U^* over a limited but significant size range. In the whole investigated size range $500 < N < 6000$, the energy line of the crystalline models (circles) is above the dashed energy line of the decahedral sequence and remains as such probably until $N \sim 10^5$. Stable models belong to the icosahedral sequence (triangles) up to $N \sim 1600$ and to the decahedral one (stars) when $N > 1600$.

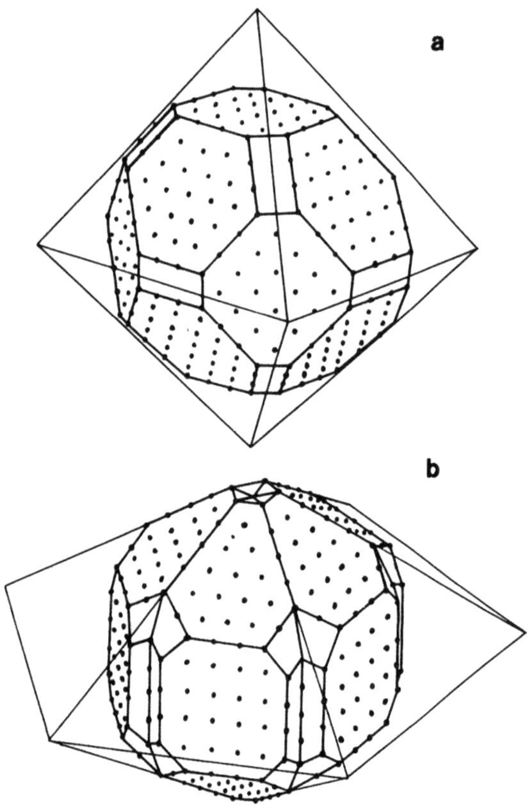

Fig. 1. a Truncated octahedron made of 1071 atoms. The surface of the model is composed of eight (111) faces, six (100) faces and twelve (110) faces. **b** truncated decahedron made of 1103 atoms. The surface of the model is composed of ten (111) faces, five (100) faces twenty (110) faces and ten (111) reentrant faces in the neighbourhood of the twins, the two vertex atoms are also removed

Discussion

The preceding results deal with well defined N values in the size range 500 to 6000 atoms. Therefore it is not sure that stable models are icosahedra for any N value lower than 1600 and decahedra for any $N > 1600$. For instance when $N = 1403$, corresponding to an icosahedral magic number, it is definite that the stable model is an icosahedron because the energy of its expected isomers in the decahedral or crystalline sequence cannot be under the dotted or dashed line. The great stability of magic icosahedra initiates from their quasi spherical surface consisting exclusively of (III) faces, owing to their five fold symmetry. But this energetical advantage tends to diminish when their external shell is not fulfilled, that is between two icosahedral magic numbers N.

At this point it is necessary to recognize the small energy difference between the energy lines of the icosahedral and decahedral sequences in the crossing region of Fig. 2: the stable models could be decahedral for particular N values lower than 1600, in spite of the advantage of the icosahedral line, if icosahedral models, far from magic numbers, cannot avoid a small but sufficient excess in surface energy.

In these conditions we have to consider a more or less extended coexistence N range around 1600 and especially below, where stable models could be alternatively icosahedra or decahedra when N is near the respective magic numbers. It cannot even be excluded a priori that the decahedron $N = 1071$ could be the stable model.

Finally concerning the structure of argon clusters produced in a free jet expansion, the main experimental results can be summarized: the diffraction pattern of Argon clusters show icosahedral features from 50 to about 750 atoms [5] and beyond this size several anomalous features giving to the patterns a twin fcc character [6].

The present paper strongly suggests that these anomalous features could be due to the coexistence of decahedral and icosahedral clusters within the size distribution in the argon cluster beam. The twin fcc character can indeed be accounted for by a decahedral diffraction function.

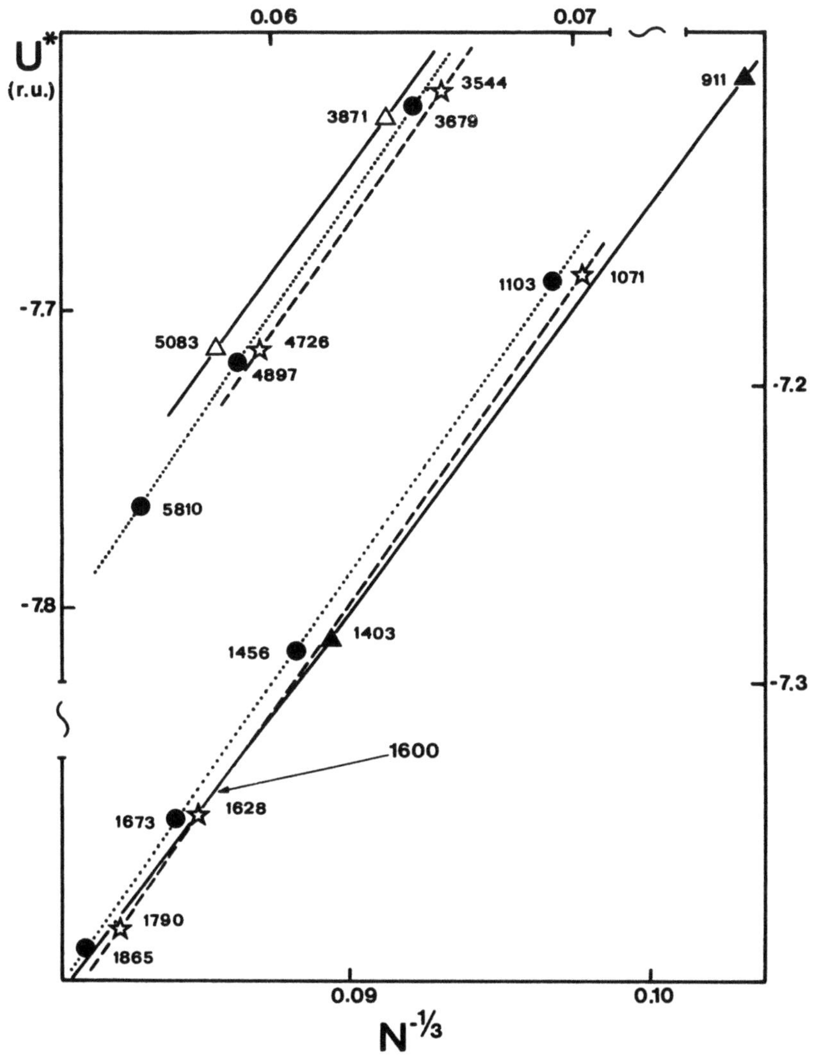

Fig. 2. Reduced potential energy per atom U^* of relaxed LJ models vs $N^{-1/3}$, scaling with the inverse model diameter. The number N of atoms is indicated for each model: truncated octahedron (*circle*) truncated decahedron (*star*), full icosahedron (*white triangle*), icosahedron without vertex atoms (*black triangle*)

Nevertheless, according to the exceptional precision of diffraction patterns and theoretical diffraction functions, some details in the experimental patterns could not be explained only by the coexistence of icosahedral and decahedral clusters. We have then undertaken recently the construction and relaxation of mono twin fcc models: first results indicate that their energy U^* is slightly lower than that of their crystalline isomers. Therefore such models could constitute new candidates in the coexistence region in addition to the icosahedral and decahedral models.

References

1. Ino, S.: J. Phys. Soc. Jpn. **27**, 941 (1969)
2. Marks, L.D.: J. Cryst. Growth **61**, 556 (1983); Marks, L.D.: Philos. Mag. A **49**, 81 (1984)
3. Farges, J., de Feraudy, M.F., Raoult, B., Torchet, G.: Acta Cryst. allogr. Sect. A **38**, 656 (1982)
4. Raoult, B., Farges, J., de Feraudy, M.F., Torchet, G.: Philos. Mag., (in press)
5. Farges, J., de Feraudy, M.F., Raoult, B., Torchet, G.: J. Chem. Phys. **84**, 3491 (1986)
6. Farges, J., Raoult, B., Torchet, G.: J. Chem. Phys. **59**, 3454 (1973)

Z. Phys. D – Atoms, Molecules and Clusters 12, 89–92 (1989)

Atoms, Molecules
Zeitschrift für Physik D and Clusters
© Springer-Verlag 1989

Description of metallic and covalent clusters with icosahedral symmetry: the polytope model

R. Mosseri[1] and J.F. Sadoc[2]

[1] Laboratoire de Physique des Solides, CNRS, F-92195 Meudon Cedex, France
[2] Laboratoire de Physique des Solides, Universite Paris-Sud, F-91405 Orsay Cedex, France

Received 6 July 1988

Compact and tetravalent clusters with icosahedral local or global symmetries are generated by mapping from ideal structures in curved space onto a tangent euclidean 3D space. The observed elastic energy of the clusters can thus be interpreted as an intrinsic curvature associated to a frustrated local order. It is then proposed a kind of classification of the very rich family of possible clusters using a limited set of parameters.

PACS: 36.40; 61.55

1. Introduction

The physics of clusters constitutes a very interesting field in between the classical domains of atomic physics on one hand and infinite solids in the other. Rare gas and metal clusters have been studied for long and the relation between their geometry and their physical properties is better understood. In the rare gas case the nature of the interatomic forces is fairly well known [1]; they are additive and central to a good approximation. A Lennard-Jones 6–12 potential can be confidently used as a model [2] and the clusters total energy minimized with respect to their geometry. Close-packed arrangements are favored with a special mention to clusters with noncrystalline symmetries like the icosahedral ones [3]. In metals the situation is more complicated, nevertheless there is some analogies with rare gases and the compact structures are found more stable. Covalent clusters (mostly column IV like Si or Ge atoms) have been intensively studied in the past few years. The relative importance of the radial and the angular part of the interatomic potential is less well known. A previous study [4] suggests that in small silicon clusters an approximate rule is to maximize the number of bonds provided that the bond angles stay greater than about 90°. But this already allows some degeneracies increasing with cluster size. Here again clusters with noncrystalline symmetries can be favored.

Looking to the available literature in the field, it appears urgent to define some rules for the classification of clusters. One goal of this paper is to present a general method which allows for such taxonomy. Moreover the simultaneous presence of topological and of metrical parameters gives insights to the nature of the energy costs during the clusters growth. The basic idea (to use curved spaces) goes back to the theoretical study of amorphous solids [5]. Ideal structures, the polytopes, with prescribed local order (incompatible with perfect propagation in the euclidean space) are defined on a 3D curved space (the hypersphere $S3$) of appropriate curvature. Finite clusters in $R3$ with slightly distorted local order and symmetry are then generated by mapping finite parts of the ideal polytope onto a tangent $R3$.

II. The polytope model for clusters

II.1. The curved space approach

We shall not describe in detail this approach. More informations can be found in [6] [7]. We look to the simplest case, namely the close packing problem.

Let us try to pack spheres as densely as possible in 3 Dimensions (this is a first approximation to amorphous metals as well as rare gas and metallic clusters). Four spheres build a regular tetrahedron, but a perfect tetrahedral packing is impossible because the tetrahedron dihedral angle is not a submultiple of 2π (Fig. 1a). It is possible to arrange 5 tetrahedra around a common edge but a small unfilled space remains. This is called a *frustration effect*. This is analogous to the problem of packing pentagons in $R2$ (Fig. 1b). In that case a perfect pentagon tiling (a dodecahedron) is possible provided the underlying space is curved and transformed into a sphere $S2$ of appropriate radius. Similarly a perfect tetrahedral packing becomes possible if the 3D space is curved. 600 regular tetrahedra fill the (hyper) surface of a hypersphere $S3$ and form a polytope (the analogue of a polyhedron in higher dimension). $S3$ can be embedded in the 4D euclidean space but it is a 3D (curved) manifold. The polytope, called $\{3,3,5\}$, has 120 vertices, each of which having 12 first neighbours forming a perfect icosahedron.

The initial geometrical frustration has been suppressed and is now encoded into curvature. Fixing to unity the first neighbour distance, the underlying space $S3$ inherits a curvature which is characteristic of the frustrated local order. To other frustrated local orders it may correspond other values for this curvature. For example in tetravalent structures, if the local rule is to pack 3 dodecahedra around each edge, this local rule leads to frustration. An ideal template, the polytope $\{5,3,3\}$ (dual to the $\{3,3,5\}$) satisfy the rule but the corresponding curvature is lower.

II.2. Cluster indexation

We want to index finite clusters which can be derived from a given polytope P. A finite portion of P is mapped onto a tangent hyperplane $R3$. We need to specify the tangent point T and the polar angle α which limits the region to be mapped. This is shown on Fig. 2. The metrical properties depends on the precise location of the projection point Q which is defined by its distance d from T. Note that this position may not be unique for all the values of β ($\alpha = \max(\beta)$). The general notation for the cluster is

$$\mathscr{C} = [P, T, \alpha, d]$$

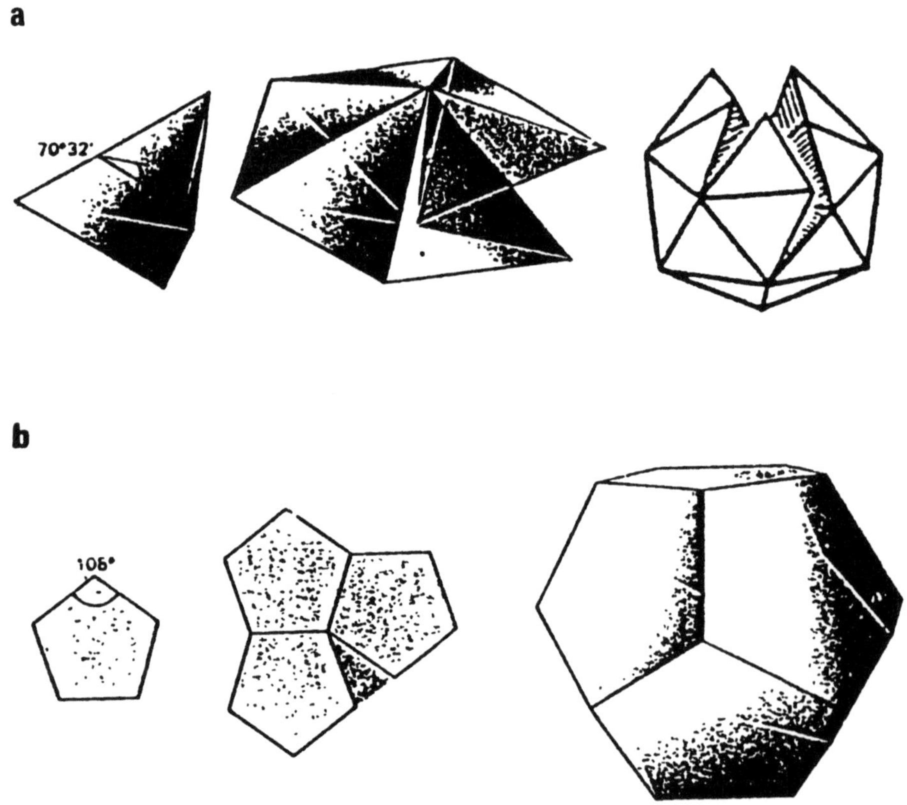

Fig. 1a and b. Geometrical frustration **a** Regular tetrahedra do not tile the euclidean space R3. **b** A similar problem in 2D with pentagons and its solution on S2: a dodecahedron

The first three quantities completely define the cluster from a topological point of view, while last one is of metrical nature.

III. Examples

III.1. Compact clusters

The polytope {3,3,5} and icosahedral local order in R3. Clusters mapped from the {3,3,5} should be the closest to perfect icosahedral order. The price to pay is that these clusters are fairly small (recall that the polytope has 120 vertices). Table 1 describes these clusters as well as their energy per site calculated within the simplest L-J potential ($V(r)=1/r^{12}-2/r^6$).

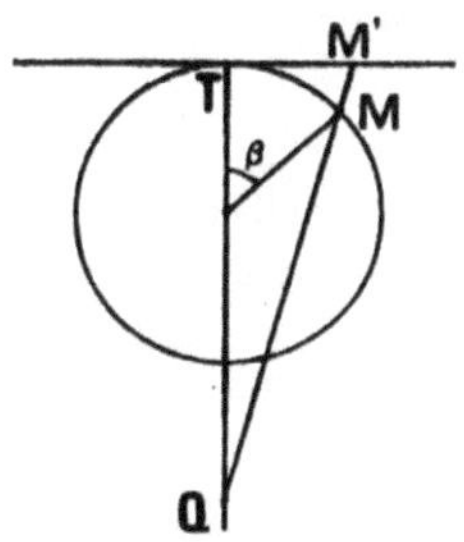

Fig. 2. Partial mapping. The polytope is embedded on S3. M (with polar ang β) is mapped onto M'. T is the tangency point; Q is the projection point and d is the distance TQ

Table 1. Clusters mapped from the {3,3,5} polytope and their energy per site

$\alpha(rd)$	N	$E1$	$E2$
$\pi/5$	13	−3.41	−3.41
$\pi/3$	33	−4.28	−4.32
$2\pi/5$	45	−4.15	−4.5
$\pi/2$	75	−4.16	−4.54
$2\pi/3$	87	−3.06	−4.22
$3\pi/5$	107	−3	−4.3

$E1$ correspond to a unique d optimized for each cluster as a whole, while for $E2$ d is shell-dependent. The cases $\alpha>2\pi/3$ are not considered because unrealistic. The first 4 clusters, shown on Fig. 3, are very familiar in this field and where recently used to model the growth of quasicrystals [8]. Note that by rotating $S3$ we could obtain clusters with the same local order but with tetrahedral global symmetry. This is true for any of the clusters described below.

Hypergeodesic domes. Regular triangulations of $S2$ based on the icosahedron have been popularized by B. Fuller and are called geodesic domes. Each of the 20 triangles of an icosahedron is iteratively dissected into portions of a triangular lattice, which are then mapped onto the sphere $S2$. It is possible to generalize this idea on $S3$ [9]. Upon iteration the original {3, 3, 5} cells are filled by larger and larger portions of fcc lattice. We call G_i these successive hypergeodesic polytopes (G_0 being the {3, 3, 5}). Table 2 describes the mapped clusters and their corresponding energy.

III.2. Tetravalent clusters

The polytope {5,3,3} and dodecahedral arrangements in R3. Tilton [10] described dodecahedral packings of increasing size. He noted that, with one central dodecahedron, the more symmetrical clusters contain 13

Table 2. Clusters mapped from the G1 polytope and their energy per site

$\alpha(rd)$	N	$E1$
$\pi/10$	13	−3.37
0.56	43	−4.45
$\pi/5$	55	−4.83
0.82	115	−5.36
$3\pi/10$	127	−5.19
1.02	157	−5.26
$\pi/3$	177	−5.32
1.14	237	−5.1

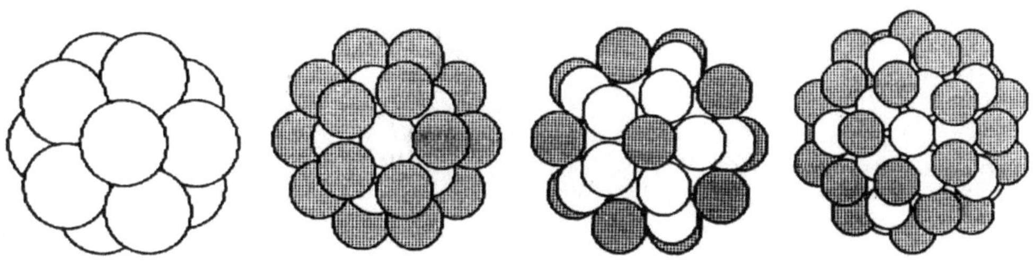

a b c d

Fig.3a–d. The first 4 clusters ($N=13$, 33, 45, 75) obtained by mapping the {3, 3, 5}

Table 3. Comparison between Coleman-Thomas cluster and the $\{5,3,3\}$ polytope

Number of neighbour	4	12	24	12	4	24	24	28
Coleman-Thomas	2.35	3.83	5.45	6.3	6.67	7.53	8.6	9.12
Polytope (on S3)	2.35	3.82	5.45	6.27	6.62	7.43	8.48	9.08
Mapped polytope	2.35	3.83	5.47	6.3	6.66	7.49	8.56	9.17

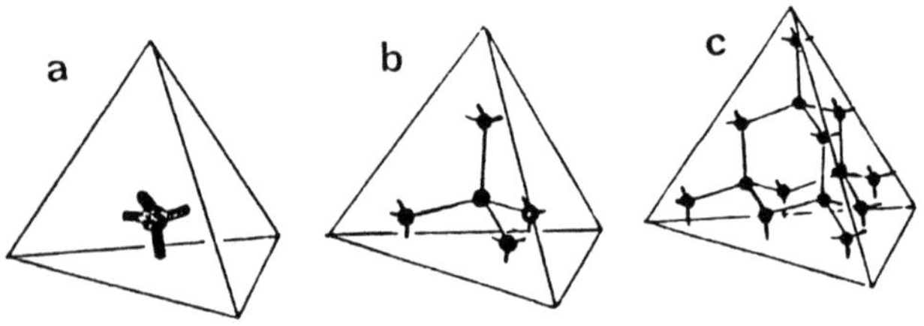

Fig. 4a–c. First stages of decoration of the $\{3,3,5\}$ tetrahedral cells **a** 1 vertex, which generates the polytope $\{5,3,3\}$ **b** 5 vertices. **c** 14 vertices

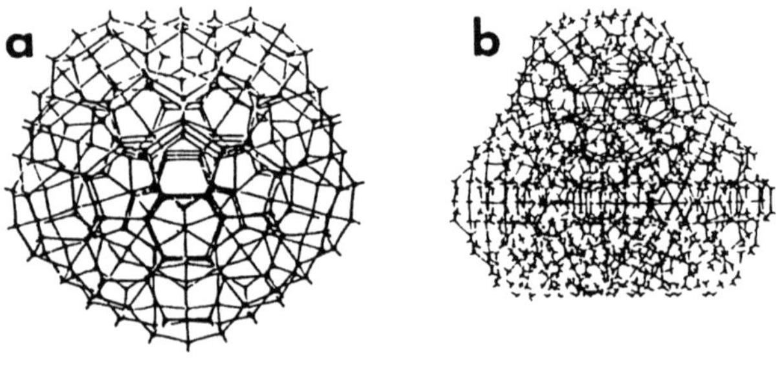

Fig. 5a and b. Tetravalent clusters described by Gaskell (Ref. 13) **a** the 280-vertex model with icosahedral symmetry **b** the 798-vertex model with tetrahedral symmetry

and 45 units. If we call C the centre of $\{5,3,3\}$ cell they correspond to:

$[\{5,3,3\}, C, 0.388\ rd, d]$: one dodecahedron
$[\{5,3,3\}, C, 0.962\ rd, d]$: 13 dodecahedra
$[\{5,3,3\}, C, 1.435\ rd, d]$: 45 dodecahedra

d is not known, which means that our mapped clusters are only topologically equivalent to those built by Tilton. He also described a cluster centred on a vertex with 54 dodecahedra. It is equivalent to $[\{5,3,3\}, V, 1.475\ rd, d]$ where V is a vertex of the $\{5,3,3\}$. Note that none of the clusters have $\alpha > \pi/2$. Tilton noted that larger clusters would have too much strain which is evident from the polytope point of view since the next shells preimage would belong to the $S3$ south hemi-(hyper)-sphere. A more metrical comparison can be done with the work of Coleman and Thomas [11] who gives not only the number of sites but also the distances from a central vertex. Their dodecahedral cluster corresponds to $[\{5,3,3\}, V, \pi/3, d]$. Table 3 compares their data with those calculated for the polytope (with geodesic distance and after mapping with $d = 2.75*R$). The lengths have been scaled to a first neighbour distance of 2.35 (appropriate to silicon atoms). The small differences between the hand-built and polytope mapped models simply means that the former cannot be considered as being mapped from a single point but rather from a collection of neighbouring points.

Decorated hypergeodesic domes. It is possible to decorate the above fcc regions in the G_i polytopes in such a way as to give diamond lattice regions. We call T_i the resultant tetracoordinated polytopes. The matching around the edges of the original $\{3,3,5\}$ polytope involves pentagonal rings. The first two stages of the decoration are shown on Fig. 4. Now it is possible to identify the large cluster built by Dandoloff et al. [12] as $[T_1, T, \alpha, d]$ and the ones studied by Gaskell [13] as $[T_2, T, \alpha, d]$. More precisely in the latter case the 280 vertex model with icosahedral symmetry is $[T2, C, \pi/5, d]$, while the 798 vertex model with tetrahedral symmetry is very close to $[T2, V, 0.776\ rd., d]$ (Fig. 5).

References

1. Friedel, J.: J. Phys. **38**, C2-1 (1977)
2. Hoare, M.R., Pal, P.: Adv. Phys. **20**, 161 (1971)
3. Farges, J., de Feraudy, M.F., Raoult, B., Torchet, G.: J. Phys. **36**, 62 (1975)
4. Mosseri, R., Gaspard, J.P.: J. Phys. **42**, C4-245 (1981)
5. Sadoc, J.F.: J. Phys. **41**, C8-326 (1980); J. Non-Cryst. Solids **44**, 1 (1981)
6. Sadoc, J.F, Mosseri, R.: Philos. Mag. **45**, 467 (1982)
7. Sadoc, J.F., Mosseri, R.: In: Proc. of the 1982 Tanigushi Symposium on the nature of topological disorder. Berlin, Heidelberg, New York: Springer 1982
8. Romeu, D.: Preprint (1987)
9. Sadoc, J.F., Mosseri, R.: J. Phys. **46**, 1809 (1985)
10. Tilton, L: J. Res. Natl. Bur. Standard **59**, 139 (1957)
11. Coleman, M., Thomas, D.: Phys. Stat. Sol. **24**, K111 (1969)
12. Dandoloff, R., Dohler, G., Bilz, H.: J. Non-Cryst. Solids **35–36**, 537 (1980)
13. Gaskell, P.: Philos. Mag. **B32**, 211 (1975)

Z Phys. D – Atoms, Molecules and Clusters 12, 93–96 (1989)

Atoms, Molecules
Zeitschrift für Physik D and Clusters
© Springer-Verlag 1989

Structural transition in SF_6 clusters

G. Torchet, J. Farges, M.F. de Feraudy, and B. Raoult
Laboratoire de Physique des Solides, LA 2, Bâtiment 510, Université de Paris Sud, F-91405 Orsay, France

Received 6 July 1988

SF_6 clusters are produced during the free jet expansion of a Ne-SF_6 mixture and studied by electron diffraction methods. Present experiments have been performed at constant nozzle diameter, stagnation pressure and stagnation temperature, adjustable parameters being the SF_6 mole fraction X and the nozzle-to-skimmer distance x_s/d. It is possible to warm up SF_6 clusters by increasing either X, and thus the cluster size, or x_s/d, which makes them collide with background molecules downstream of the Mach disk. In both cases, cold clusters made of several hundreds of molecules experience a structural transition, similar to that observed in bulk material, from the triclinic to the bcc plastic structure. A molecular dynamics simulation accounts correctly for intermediate stages of the transition which are visible in experimental patterns. Contrary to bulk results, cluster results show that the structural transition occurs over some temperature range.

PACS: 61.14; 36.40

1. Introduction

The structure of SF_6 solid phases has been the subject of numerous studies during the last ten years. The high temperature plastic phase of bulk SF_6 was found in warm clusters produced either in Ar-SF_6 nozzle expansions [1, 2] or in pure SF_6 free jet expansions [3]. The low temperature phase has been elucidated recently as being triclinic from neutron experiments on bulk material [4]. Moreover the phase transition has been simulated via molecular dynamics calculations in both bulk SF_6 [5] and 512 molecule clusters [6]. However, triclinic clusters needed some efforts to be detected as long as a correct structure was not available. In Ar-SF_6 nozzle expansions, the electron diffraction patterns recorded with low SF_6 mole fraction exhibit mainly the bcc lines of the plastic structure with a few weak triclinic lines, in addition to lines due to Ar condensation. SF_6 lines are attributed to the coexistence of cubic and triclinic clusters in the cluster beam [7]. On the other hand, free jet expansion experiments were carried on in Orsay, using Ne instead of Ar in order to avoid any condensation of the carrier gas [3]. Several strong variations were visible in the line intensities according to the SF_6 mole fraction but, in the absence of any new line, it was concluded that no phase transition had been detected. This conclusion was revised after systematic experimental studies [8] and thanks to recent models, it is now established that cold clusters made of several hundreds of molecules are purely triclinic. Furthermore, a molecular dynamics calculation reproduces correctly the progressive variation of experimental patterns, leading to a description of the transition.

2. Experiment

SF_6 clusters are produced during the free jet expansion of a Ne + X SF_6 mixture. Initial conditions are the following: nozzle diameter $d = 0.1$ mm, stagnation pressure $p_0 = 20$ bar and stagnation temperature $T_0 = 300$ K. Clusters scatter a 50 keV electron beam and diffraction patterns are recorded in the form of Debye-Sherrer patterns with cylindrical symmetry around the electron beam axis. It is possible to vary the SF_6 mole fraction X and the reduced nozzle-to-skimmer distance x_s/d. For large x_s/d values, the skimmer is located dowstream of the Mach disk and thus samples clusters which have experienced collisions from the background gas at room temperature. According to the evolution of the total scattered intensity with x_s/d [8], clusters are then progressively swept off the beam axis but surviving ones have the possibility to reach higher temperatures.

Figure 1 shows several diffraction patterns recorded for different X and x_s/d values. Similar transformations are produced when increasing independently these parameters. At constant X=0.03 (left side), the pattern $x_s/d=135$ transforms progressively into the bcc pattern $x_s/d=165$. Above $x_s/d\approx 145$, clusters having crossed the Mach disk are warmer, as visible in the strong damping of several lines at large scattering angles ($s\geqslant 6$ Å^{-1}), and have experienced some structural evolution as indicated by striking changes in line heights in the range $s=3$ to 5 Å^{-1}. During the transformation, clusters are expected to keep the same mean size $\bar{D}\approx 42$ Å, corresponding to a mean number of molecules $\bar{N}\approx 400$, estimated from the 200 line width in the bcc pattern. Starting from the lower pattern and keeping $x_s/d=135$, an increase in X from 0.03 to 0.15 leads to the curves shown in the right side of Fig. 1. The progressive large angle damping and several line inversions must be attributed to warmer clusters. At the same time, the thinning of the lines proves that clusters are getting larger, the mean size being $\bar{D}\approx 65$ Å ($\bar{N}\approx 1400$) at X=0.07 and $\bar{D}\approx 80$ Å ($\bar{N}\approx 3000$) at X=0.15. A higher SF_6 mole fraction results in a higher condensation rate, that is in larger clusters, and consequently in a progressive heating of the carrier gas, that is in warmer clusters. Lines in the pattern X=0.15 can be indexed according to the bcc structure.

3. Structural transition

Such evolutions have been observed in a wide set of experimental patterns on which two characteristic quantities have been measured, the 200 line width β and the ratio of two line intensities I_{310}/I_{431}. This ratio and β^{-1}, which scales with the mean cluster size in the case of a single line, are plotted in Fig. 2

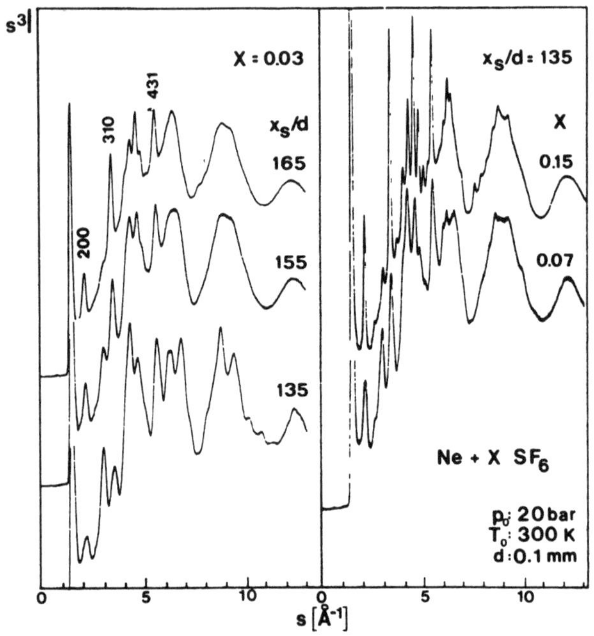

Fig. 1. Electron diffraction patterns recorded for different nozzle-to-skimmer distances (left side) or for different SF_6 mole fractions (right side). $s=(4\pi/\lambda)\sin\theta/2$ is the scattering parameter with λ, the electron wavelength, and θ, the diffraction angle. Measurements on lines 200, 310, 431, labeled according to the bcc structure, are used in Fig. 2

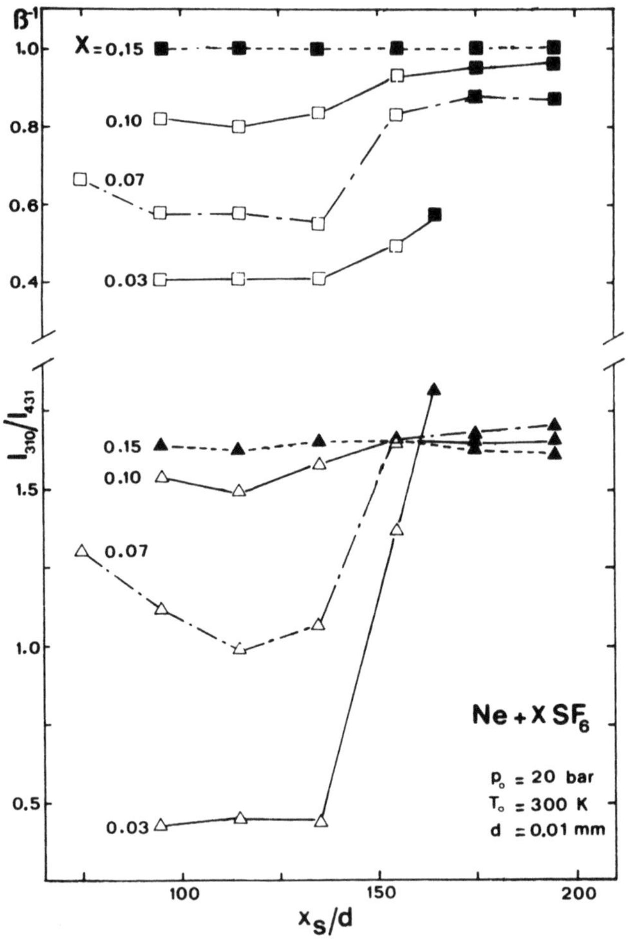

Fig. 2. Plots of the inverse 200 line width (β^{-1}) and of the ratio of 310 and 431 line heights, as measured on experimental patterns recorded for different SF_6 mole fractions and different nozzle-to-skimmer distances. Black symbols refer to high temperature bcc patterns

according to x_s/d for various SF_6 mole fractions. Black symbols refer to bcc patterns. The increase of both quantities downstream of the Mach disk, above $x_s/d \approx 145$, shows that clusters experience a structural change when they are warmed up. At low temperature and for small clusters, the 200 line is broadened due to an overlapping of several lines which correspond to a structure with low symmetry, and at high temperature it turns into the narrow effective 200 line of the highly symmetric bcc structure. When X = 0.15, bcc clusters are produced in the expansion and the crossing of the Mach disk does not affect them significantly, the constant β^{-1} value corresponding to no appreciable evaporation. Two more comments can be made about Fig. 2. In the bcc high temperature structure, the intensity ratio may be related to the temperature through the Debye-Waller factor. Small differences visible in the intensity plot at $x_s/d > 150$ are in fact due to a size effect since larger intensity ratios correspond to smaller clusters in the β^{-1} plot. On the other hand, as the plotted quantities are found to scale with the cluster temperature, one can see that when relatively warm clusters are produced in the free jet, for X = 0.07, they can be cooled down as long as Ne expansion is going on.

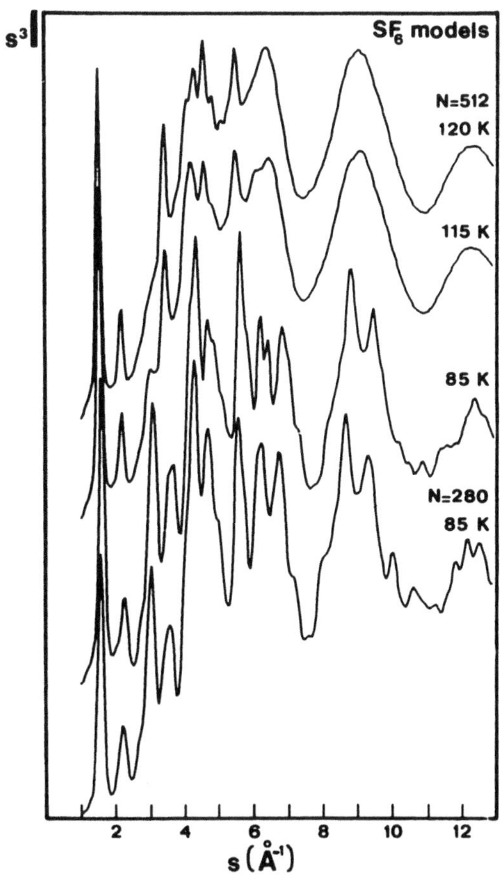

Fig. 3. Lower curve: calculated diffraction function of an homogeneous model $N = 280$ with the triclinic structure of bulk material reported in [4]. Upper curves: diffraction functions of a model $N = 512$ calculated via a molecular dynamics simulation at three different temperatures. These curves reproduce the experimental patterns of Fig. 1 (left side)

4. Cluster structure

4.1. Cubic bcc structure

Upper patterns in Fig. 1 show the bcc structure of the high temperature phase of bulk SF_6. An homogeneous, that is unrelaxed, model made of 339 molecules with bcc structure has been used to calculate a diffraction function which is in good agreement with the experimental pattern X = 0.03, $x_s/d = 165$ [3]. From lattice parameter measurements, the temperature of cubic clusters, whatever the cluster size, is estimated to be in the range 110–120 K.

4.2. Triclinic structure

An homogeneous model made of 280 molecules showing the triclinic structure has been constructed according to the data provided by neutron diffraction experiments on bulk solid at 85 K [4]. The diffraction function, which has been calculated by introducing a few mean dynamical parameters, is reproduced in Fig. 3 (lower curve). A very good agreement is found between this function and the experimental pattern X = 0.03, $x_s/d = 135$ of Fig. 1. Thus it can be asserted that pure triclinic clusters are produced in free jet expansions.

4.3. Molecular dynamics model

In order to account for the experimental patterns showing characteristic features intermediate between those of triclinic and cubic structures, a molecular dynamics calculation has been undertaken with a 512-molecule model. The intermolecular potential is modelled by using a Lennard-Jones atom-atom potential for the interaction between any pair of fluorine atoms [6]. The model, initially at 100 K with the bcc structure is cooled down to an equilibrium temperature of 50 K and then progressively warmed up to 120 K. A set of configurations corresponding to several intermediate temperatures is used to calculate diffraction functions [9]. Functions at three temperatures are reproduced in Fig. 3. There is a remarkable agreement between these curves and the experimental ones in Fig. 1 (left side). The cubic and triclinic patterns are well reproduced, the triclinic pattern showing a few weak discrepancies, with respect to the previous calculation $N=280$, which are possibly due to the assumed intermolecular potential. The most striking result is that the intermediate pattern $x_s/d=155$ is well accounted for by the calculated pattern at 115 K. This means that intermediate patterns, which could be thought as the superposition of cubic and triclinic patterns, must be attributed to single clusters experiencing the structural transition.

5. Discussion

Looking at the model at increasing temperatures makes it possible to give a description of the structural evolution [9]. At low temperatures, $T<90$ K, the triclinic structure is similar to that of bulk material [4]. It can be roughly described as a distorted cubic structure in which one third of the SF_6 molecules has been tilted from the cubic orientation. Observing the model, it is found that increasing the temperature makes the tilted molecules recover progressively, one after the other, the cubic orientation while the triclinic lattice transforms progressively into the cubic one. Thus to each intermediate temperature corresponds some proportion of tilted molecules which decreases down to zero at 120 K. This differs from the proposed two-stage process suggested recently for the transition [4] since neither an intermediate hexagonal structure nor any collective motion of the tilted molecules are observed in the present model. Now one question is still to be considered. The bulk phase transition is known to occur at about 95 K. Even if SF_6 cluster temperatures are not precisely estimated, both experimental and molecular dynamics studies give evidence of some temperature range for the transition. How this temperature range should be related to a size effect would need further experiments on clusters of various sizes, but large cold clusters are almost impossible to be produced in expansions, and even on microcrystals around the transition temperature.

References

1. de Boer, B.G., Kim, S.S., Stein, G.D.: 11th International Symposium on Rarefied Gas Dynamics, Vol. 2, pp. 1151–1160. Paris: CEA 1979
2. Valente, E.J., Bartell, L.S.: J. Chem. Phys. **79**, 2683 (1983)
3. Farges, J., de Feraudy, M.F., Raoult, B., Torchet, G.: Surf. Sci. **156**, 444 (1985)
4. Powell, B.M., Dove, M.T., Pawley, G.S., Bartell, L.S.. Molec. Phys. **62**, 1127 (1987)
5. Dove, M.T., Pawley, G.S.: J. Phys. C **16**, 5969 (1983)
6. Fuchs, A.H., Pawley, G.S.: J. Phys. France **49**, 41 (1988)
7. Bartell, L.S., Valente, E.J., Caillat, J.C.: J. Phys. Chem. **91**, 2498 (1987)
8. Farges, J., de Feraudy, M.F., Raoult, B., Torchet, G.: 15th International Symposium on Rarefied Gas Dynamics, Vol. 2, pp. 208–216. Stuttgart: B.G. Teubner 1986
9. Work in collaboration with Edinburgh (to be published)

Note added in proof

Neutron data reported in [4] have been reanalysed recently in terms of a structure with higher symmetry than triclinic, namely the monoclinic structure which then turns to be that of the coldest clusters. Dove, M.T. et al.: Mol. Phys. **65**, 353 (1988)

Z. Phys. D – Atoms, Molecules and Clusters 12, 97–101 (1989)

Atoms, Molecules and Clusters
Zeitschrift für Physik D
© Springer-Verlag 1989

Electron diffraction on unsupported ultrafine silver particles

B.D. Hall, M. Flüeli, R. Monot, and J.-P. Borel
Institut de Physique Expérimentale, EPFL, CH-1015 Lausanne, Switzerland

Received 5 July 1988

Electron diffraction has been performed on small unsupported Ag particles with mean sizes between 1 nm and 4 nm. Particles are produced by inert gas nucleation and are examined, in situ, in a molecular beam. The results clearly show a change from fcc structure at larger sizes to MTP structure below 3 nm. Icosahedral and decahedral structures co-exist in the beam in roughly equal proportions.

PACS: 36.40.d; 61.14.f; 61.50.j

Introduction

Considerable interest has been shown in the structure and formation of ultrafine particles (diameter $\leqslant 1000$ Å) since the late sixties when it was discovered that certain fcc materials tend to form as small multiply twinned particles (MTPs) [1, 2, 3]. Although preparation techniques vary widely, samples are inevitably collected on a support before measurement and the influence of this support on the particles has to be considered.

With this in mind an inert gas nucleation source has been incorporated with an electron diffractometer. The need for a support is thus eliminated as diffraction information can be obtained directly from freely moving small particles in a supersonic molecular beam of inert carrier gas.

In the paragraphs that follow we describe the experiment and present results concerning small particles of silver (10–40 Å diameter). These results are then interpreted by comparing the diffraction patterns obtained with calculated diffraction patterns for standard MTP and crystal structures.

Experimental procedure

The small particle source was designed in collaboration with the late Prof. G.D. Stein and is based on an earlier experiment of his [4]. It consists of (see Fig. 1.) three co-axial, differentially pumped chambers. Nucleation and growth occur in the innermost chamber only. The two outer chambers serve as collimating and pumping stages producing a supersonic molecular beam of carrier gas in which the small particles are transported to the diffraction chamber.

Silver vapour is produced by heating molten silver in a boron nitride crucible. The hot metallic vapour cools in contact with an inert carrier gas (He), introduced at room temperature to the chamber near the crucible. Homogeneous nucleation occurs in a supersaturated region above the crucible and growth continues by accreation from the vapour and coalescence of the nucleated clusters [4].

The final particle size depends in a complicated way on the conditions in the nucleation chamber. The choice of carrier gas [5] and the rate of flow in the nucleation region [6] are important. The former influences the extent of supersaturation region and the latter determines the time growing particles will spend in the vapour rich region near the crucible. During an experiment the carrier gas pressure and crucible temperature can be varied. The product of these parameters is indicative of the particle size [4, 6], so for example raising or lowering the temperature at constant pressure should produce larger or smaller particles respectively. Problems with convection are however important and efficient source operation occurs only in certain configurations. These have to be

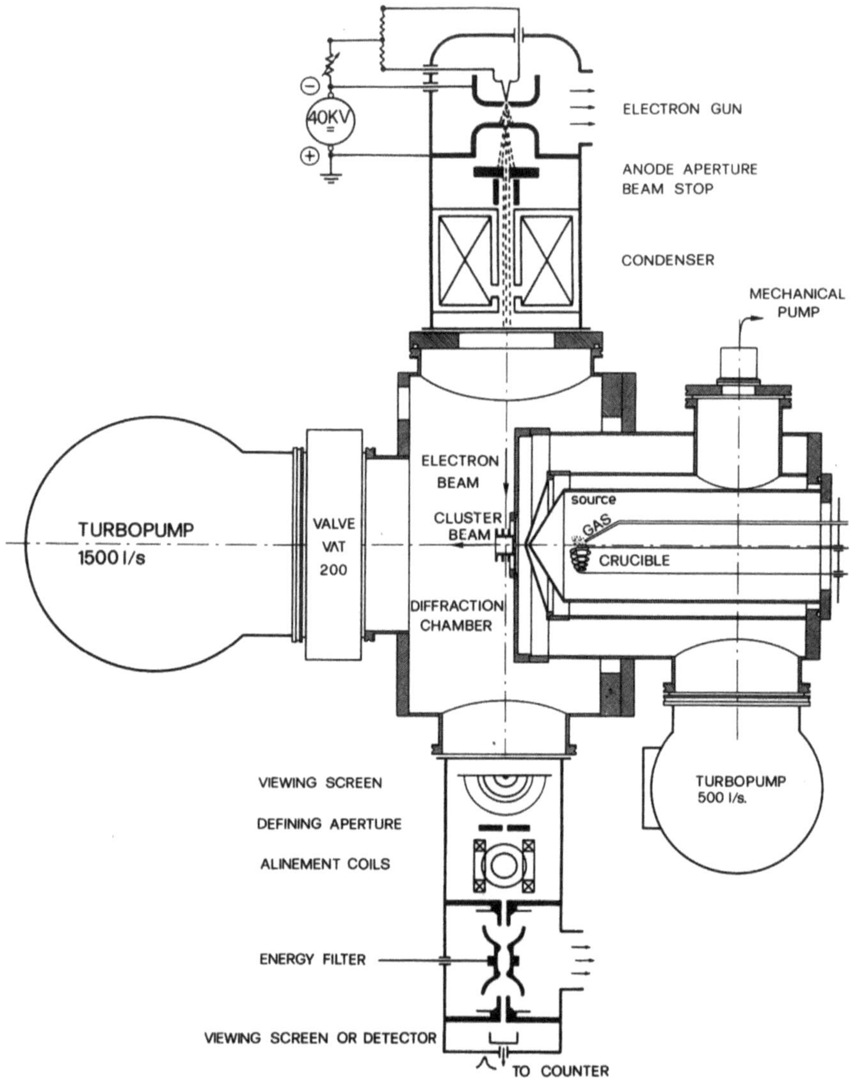

Fig. 1. Experimental apparatus

found by moving the gas inlet position during an experiment.

The diffraction apparatus has been described in detail by Solliard [7]. A 40 keV electron beam crosses the molecular beam as it leaves the source. A pair of Helmholtz coils, spanning the beam crossing, is used to sweep across a diameter of the diffraction pattern. Electrons diffracted through a given angle are magnetically deflected back onto the vertical axis of the experiment and energy filtered (40 keV − 2 eV) before detection and counting.

A sample of the particles in the beam is taken by exposing an amorphous carbon coated electron microscope grid very briefly ($\simeq 10$ ms) to the beam. These samples are examined later under an electron microscope to estimate the distribution of particle sizes.

Results and discussion

The very small size of particles under study here introduces important geometrical effects in their diffraction patterns. Furthermore diffraction from MTPs cannot be derived by considering boundary conditions applied to the bulk as is the case for small crystals.

We have chosen to calculate diffraction patterns for specific fcc and MTP particles and compare these simulations with our data.

Three structures have been used: a regular icosahedron [8, 9], an fcc cubooctahedron with square (100) and triangular (111) facets, and a decahedron [10, 9] truncated perpendicular to the plane of mirror symmetry by five (100) planes. The structures were generated using the bulk lattice parameter for silver at room temperature and no attempt has been made to relax them. By adding atoms to form complete outer layers, particles containing from tens to several thousands of atoms have been created for each structure.

For each particle the diffraction pattern is obtained from the Debye equation:

$$I(S)=I_0 \sum_{i,j} f_i(S) f_j(S) \frac{\sin(2\pi r_{ij} S)}{2\pi r_{ij} S}$$

where I_0 is the incident electron intensity, $f_i(S)$ and $f_j(S)$ the electron scattering factors for atoms i and j, r_{ij} the distance separating atoms i and j, and $S=2\sin(\theta_B)/\lambda$ is the scattering parameter. This relation assumes that kinematical diffraction conditions apply, an assumption which will be discussed below. In an attempt to account for the spread in particle sizes, diffraction patterns have been combined according to the measured size distributions. The procedure is illustrated in Figs. 2–4. In Fig. 2 the size distribution for a particular measurement is shown and has been fitted to a log normal distribution function [11]. In Fig. 3, for each structure, the single particle diffraction patterns have been combined according to the proportions observed. Each curve in Fig. 3 therefore simulates diffraction from a size dispersed pure structure in the beam. In Fig. 4 these 'pure' beams have been combined in proportions chosen to match as well as possible the experimental data.

Experimental diffraction patterns are presented in Figs. 4–6 with experimental conditions summarized in Table 1. The data have been smoothed, corrected for experimental geometry and an estimate of the background due to carrier gas diffraction – a structureless monotonically decreasing function of scattering angle – has been subtracted.

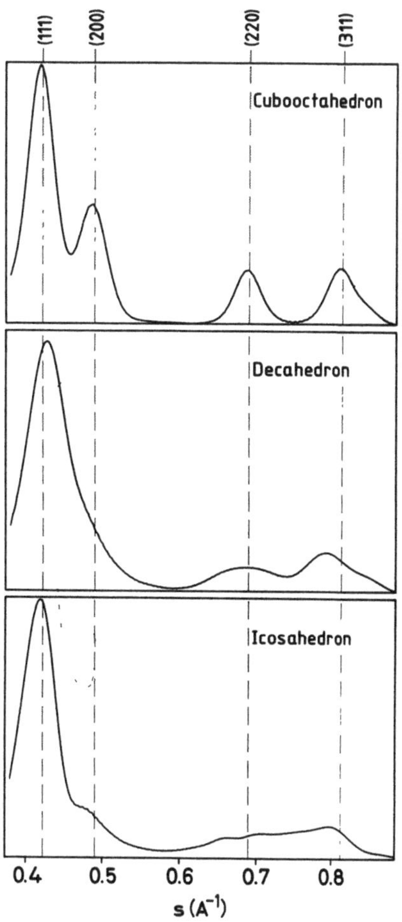

Fig. 3. Pure structure simulations: each diffraction pattern corresponds to a linear combination of single particle diffraction patterns from the same structure. The weighting of different sizes is determined by the size distribution of Fig. 2. Dotted lines mark the position for the bulk diffraction peaks of the fcc structure

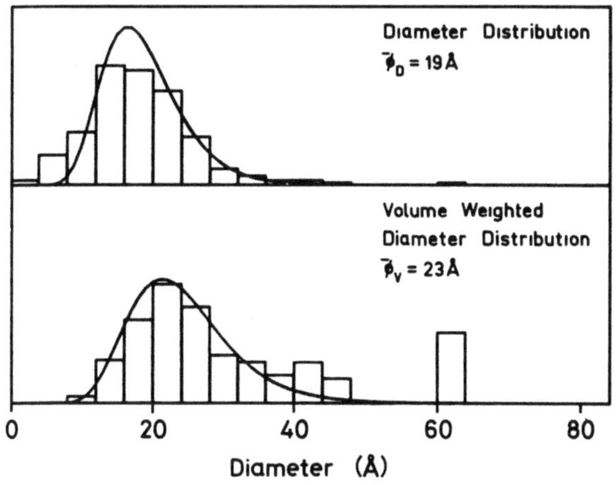

Fig. 2. Particle size distributions corresponding to the data in Fig. 4. The diameter distribution (upper histogram) has been fitted to a log normal distribution function [11] and then transformed to obtain the volume weighted distribution (lower histogram) which is characteristic of the intensity diffracted by the particles. $\bar{\phi}_V = \Sigma_i n_i \frac{\pi}{6}\phi_i^3 / \Sigma_i \frac{\pi}{6}\phi_i^3$, $\quad \bar{\phi}_D = \Sigma_i n_i \phi_i / \Sigma_i n_i$.

For the smallest particles (Fig. 4) the experimental data are in sharp contrast to the diffraction expected from fcc particles. The absence of the (200) peak and an abnormal intensity in the region between the (220) and (311) peaks are inexplicable in terms of size broadening of the fcc structure (see Fig. 3). Even for the largest particles (Fig. 6) where it dominates the beam, an fcc structure is unable to reproduce the experimental results. Only when contributions from MTP structures are included is a reasonable match obtained.

In Fig. 4 a shift is apparent in the experimental (111) peak. A smaller shift occurs also in Fig. 5. A shift towards larger scattering angles implies a contraction of the interplanar distance between planes giving rise to the reflection. Lattice contraction as a function of particle size has been observed and studied in detail

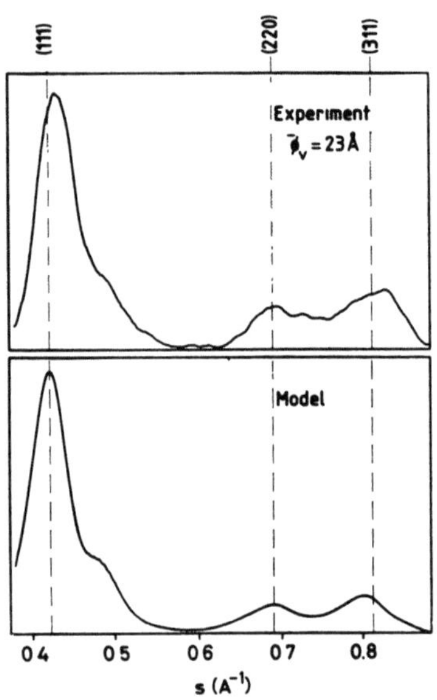

Fig. 4. Experimental and calculated diffraction profiles. The three structures shown in Fig. 3. have been combined in proportions 2:2:1 (icosahedra:decahedra:cubooctahedra). Dotted lines mark the position for the bulk diffraction peaks of the fcc structure

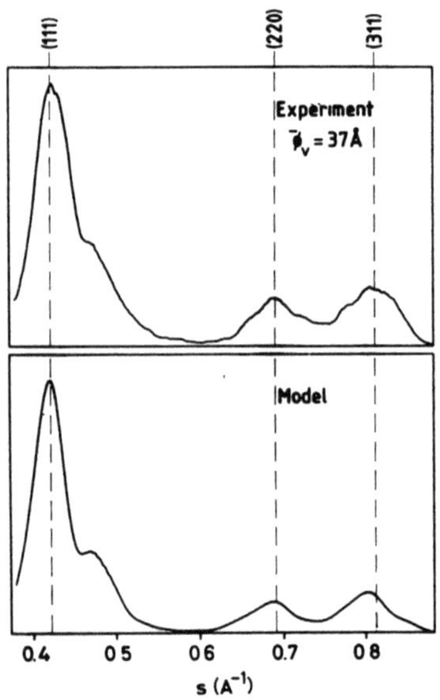

Fig. 5. Experimental and calculated diffraction profiles. Combined in proportions: 6:6:1 (icosahedra:decahedra:cubooctahedra). Dotted lines mark the position for the bulk, diffraction peaks of the fcc structure

[12]. Our results, although in qualitative agreement, cannot be directly compared with this study as the temperature of the particles in the beam is not known.

The arbitrary combination of structures in our diffraction simulation falls short of a quantitative structure analysis of the beam. The proportions used in the simulations are not unique, however in all cases

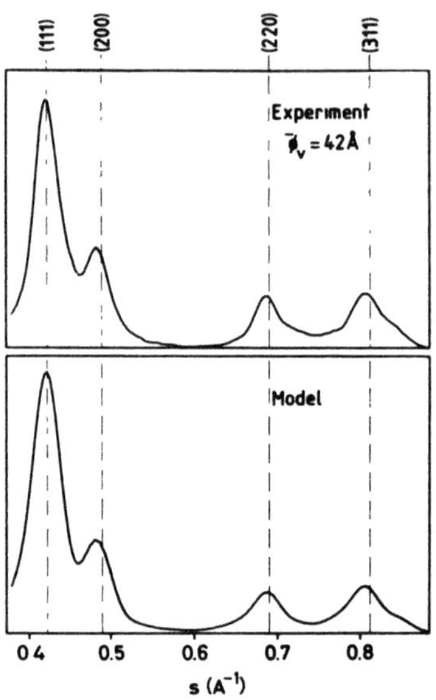

Fig. 6. Experimental and calculated diffraction profiles. Combined in proportions: 2:4:5 (icosahedra:decahedra:cubooctahedra). Dotted lines mark the position for the bulk diffraction peaks of the fcc structure

Table 1. Experimental conditions

Figure	$\bar{\phi}_V$ (Å)	$\bar{\phi}_D$ (Å)	P_0(mbar)	T_0(K)	P_{Dif}(Torr)
Fig. 4	23	19	3.3	1457	8×10^{-6}
Fig. 5	37	35	2.6	1437	2.0×10^{-6}
Fig. 6	42	28	1.5	1528	2.0×10^{-6}

$\bar{\phi}_V = \Sigma_i n_i \frac{\pi}{6}\phi_i^3 / \Sigma_i \frac{\pi}{6}\phi_i^3$
$\bar{\phi}_D = \Sigma_i n_i \phi_i / \Sigma_i n_i$
P_0 = carrier gas pressure in nucleation chamber
T_0 = crucible temperature
P_{Dif} = background pressure in diffraction chamber

a combination of the three structures is required for a reasonable match with experiment. The presence of a large proportion of MTP particles, both icosahedra and decahedra, is confirmed by high resolution electron microscopy (HREM) on the silver aggregates collected from the beam.

Also apparent in each of Figs. 4–6 is a discrepancy in the relative intensity at high scattering angles. The stronger diffraction observed in the experimental data is an indication of possible dynamic diffraction effects [13, 14].

Conclusion

Electron diffraction has been used to study unsupported silver particles, formed by inert gas evaporation and transferred in a molecular beam to a diffraction chamber.

The results clearly show a predominance of fcc structure when the mean particle size exceeds 40 Å. A change to MTP dominance occurs when the mean size is inferior to 30 Å. Comparison with calculated diffraction profiles for both decahedral and icosahedral particles indicate that these structures co-exist in the beam and are in roughly equal proportions.

We thank A. Hoareau for his aid and encouragement during the experimental phase of this work. This project receives financial support from the Swiss Nation Research Foundation, grant: FN 2.454-0.87.

References

1. Ino, S.: J. Phys. Soc. Jpn. **21**, 346 (1966)
2. Ino, S., Ogawa, S.: J. Phys. Soc. Jpn. **27**, 941 (1969)
3. Allpress, J.G., Sanders, J.V.: Aus. J. Phys. **23**, 23 (1970)
4. Yokozeki, A., Stein, G.D.: J. Appl. Phys. **49**, 2224 (1978)
5. de Boer, B.G., Stein, G.D.: Surf. Sci. **106**, 84 (1981)
6. Abe, H., Schulze, W., Tesche, B.: Chem. Phys. **47**, 95 (1980)
7. Solliard, C.: Rev. Sci. Instrum. **52**, 1321 (1981)
8. Mackay, A.L.: Acta Cryst. **15**, 916 (1962)
9. Yang, C.Y.: J. Cryst. Growth **47**, 274 (1979)
10. Bagley, B.G.: Nature **208**, 674 (1965)
11. Granquist, C.G., Buhrman, R.A.: J. Appl. Phys. **47**, 2200 (1976)
12. Solliard, C., Flüeli, M.: Surf. Sci. **156**, 487 (1985)
13. Bartell, L.S., Raoult, B., Torchet, G.: J. Chem. Phys. **66**, 5387 (1977)
14. Bryan Monosmith, W., Cowley, J.M.: Ultramicroscopy **12**, 177 (1984)

Z. Phys. D - Atoms, Molecules and Clusters 12, 103–105 (1989)

Atoms, Molecules and Clusters
Zeitschrift für Physik D
© Springer-Verlag 1989

Structure of Ag, Fe and Ge microclusters

P.A. Montano[1,2], J. Zhao[2], M. Ramanathan[2,3], G.K. Shenoy[3] and W. Schulze[4]

[1] Department of Physics, Brooklyn College of CUNY, Brooklyn, NY 11212, USA
[2] Department of Physics, West Virginia University, Morgantown, WV 26506, USA
[3] Argonne National Laboratory, Argonne, IL 60439, USA
[4] Fritz Haber Institute der Max-Planck-Gesellschaft, 1000 Berlin 33, Germany

Received 5 July 1988; final version 30 September 1988

The structures of Ag, Fe and Ge microclusters were determined using EXAFS. The measurements were performed over a wide range of clusters sizes. The clusters were prepared using the gas aggregation technique and isolated in solid argon at 4.2 K. The measurements were performed at the National Synchrotron Light Source (NSLS) at beam line X-18B. A strong contraction of the interatomic distances was observed for Ag dimers and multimers. Silver clusters larger than 12 A mean diameter show a small contraction of the *nn* distance and a structure consistent with an fcc lattice. By contrast clusters smaller than 12 A show the presence of a small expansion and a strong reduction or absence of *nnn* in the EXAFS signal. This points towards a different crystallographic structure for Ag microclusters with diameter less than 12 A. In iron clusters we observe a gradual reduction of the *nn* distance as the cluster size decreases. The interatomic distance for iron dimers was determined to be 1.94 A, in good agreement with earlier measurements. The iron microclusters show a bcc structure down to a mean diameter of 9 A. Iron clusters with 9 A mean diameter show a structure inconsistent with a bcc lattice. The new structure is consistent with an fcc or hcp lattice. The measurements on Ge clusters show the presence of only nearest neighbors. There was clear evidence of temporal annealing as determined by variations in the near edge structure of the K-absorption edge. Absorption edge measurements were also performed for free Ge clusters travelling perpendicular to the direction of the synchrotron radiation beam. The measurements performed on the free clusters were consistent with those obtained for matrix isolated clusters.

PACS: 36.40, 78.70

Introduction

The properties of microclusters lie between the well established domains of atomic and molecular physics and that of condensed matter. From a fundamental point of view, it is crucial to understand how the electronic and structural properties change as atoms come together to form dimers, trimers, clusters and so on until finally a large unit emerges that has the full characteristics of the bulk. In microclusters the surface atoms represent a large fraction of the material, and they are expected to play a fundamental role in determining the crystallographic and electronic properties of the cluster. From a technological point of view, it is important to understand the properties of microclusters and their relation to catalysis, information storage, photographic and electronic imaging, microcircuitry, and the production of better quality thin films, to name a few applications.

The development of new methods to produce well defined microclusters of metals and semiconductors has facilitated the study of the physical properties of such systems. Finding the crystallographic structure of microclusters is a crucial and challenging problem.

The advent of synchrotron radiation sources has opened new opportunities to perform structural studies of microclusters. In previous years we have employed such facilities to study metal clusters isolated in rare gas solids [1, 2, 3]. We employed extended X-ray absorption fine structure (EXAFS) as a microstructural probe. In the present work we report a detailed study of the structure of silver microclusters as a function of size. We also report the variation in crystallographic structure of Fe as a function of particle size, we observed that the structural evolution of the microclusters does not follow the same crystallographic trends as in the solid. We also performed a structural study of matrix isolated microclusters of Ge. Such clusters show a strong dependence on the method of preparation. We were able to observe the EXAFS signal for free Ge clusters. In the following paragraphs we describe the experimental procedures and summarize the results for Ag, Fe, and Ge microclusters.

Experimental

The microclusters were prepared using the gas-aggregation technique. Narrow size distribution of metal and semiconductor clusters can be obtained using this method [4]. In this technique, the metal and semiconductor atoms are evaporated from a Knudsen cell in an argon atmosphere [2, 3, 4]. The clusters are then transported by the gas stream through an aperture into a liquid-helium cryopump, where most of the gas is condensed. A collimated molecular beam passes through a second aperture on the axis of the cryopump and it is used for gas phase studies or deposited on a substrate. The clusters were isolated in argon at 4.2 K, and the metal concentration was around 0.1 atomic%. The structure and size distribution of such clusters have been the subject of careful electron microscopy, EXAFS, and mass spectrometric studies. The free cluster measurements were performed using a cross beam method with the cluster beam perpendicular to the photon beam. The fluorescence data were collected using a proportional counter at 90 degrees with respect to the photon and clusters beams. The X-ray absorption and fluorescence spectra were recorded at beam line X-18B at NSLS. An intrinsic Ge solid state detector was used for the Ag and Fe fluorescence measurements. Energy calibration was performed using ultrahigh purity samples of Ag, Fe, and Ge. The standards were cooled to liquid nitrogen temperature and were analyzed to obtain the EXAFS phase shifts and amplitudes. For each sample several measurements were performed until a good signal-to-noise ratio was obtained from the added data. Details of the method of analysis can be found in reference [2]. For the silver samples measurements were also performed at the Stanford Synchrotron Radiation Laboratory, this was done to test the reproducibility of the results, in particular below 17 A mean diameter.

Results

Ag: Silver metal has an fcc lattice with a *nn* distance of 2.87 A at 78 K. The EXAFS spectrum of silver at 78 K allows measurements up to the 5-th shell (6.44 A). For silver particles smaller than 14 A (mean diameter) we detect only the presence of *nn* and no higher shell are observable by EXAFS. Two factors can contribute to the absence of *nnn* peaks, one is the presence of a disordered structure, and second a strong reduction in the coordination number due to the presence of a large number of surface atoms. We detected the presence of Ag dimers and multimers (trimers and quadrumers) for samples containing the smallest particles. The interatomic distance for the dimer was found to be 2.47 ± 0.02 A. Multimers of Ag also show a strong contraction of the interatomic distance (see Fig. 1). No significant contraction of the interatomic distances is observed for particles with a mean diameter less than 17 A. For particles larger than 25 A a significant contraction of the *nn* distance is observed [2], but as the particle size decreases, a small expansion is observed instead of further contraction. This might be an indication of a structural change in silver microclusters. The average size of the microclusters was determined from the average coordination number obtained from EXAFS and independently corroborated by mass spectroscopic methods [5]. Figure 1 shows the *nn* distance as a function of particle size.

Fe: Four microcluster samples were prepared with mean diameter from 9 A to 15 A, the uncertainty in the

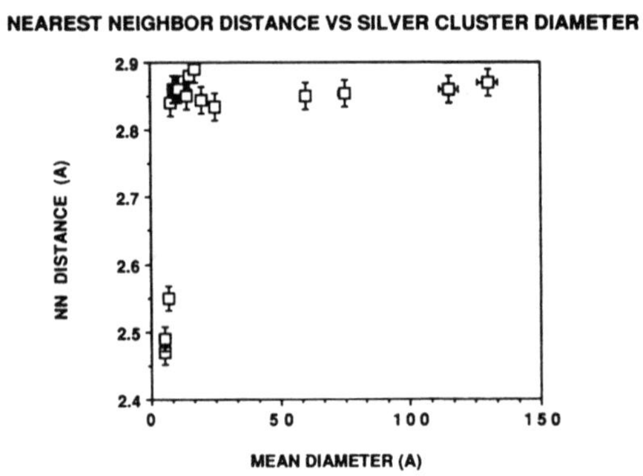

Fig. 1. Nearest neighbor distance vs cluster diameter for Ag. Dark dots from ref 2

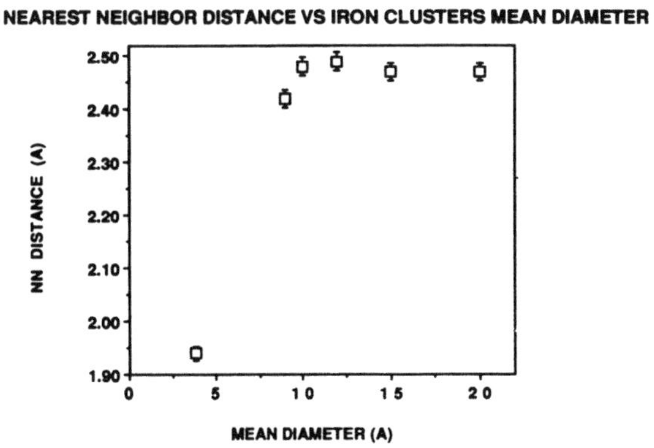

Fig. 2. Nearest neighbor distance vs cluster size for Fe

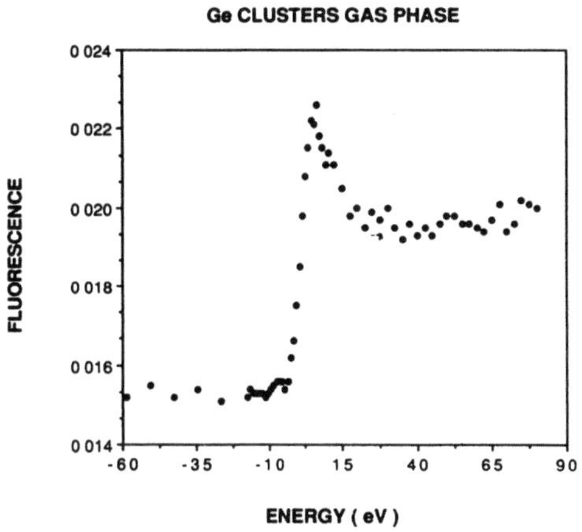

Fig. 3. Near edge structure of free clusters of Ge produced by the gas aggregation technique

diameter determination was between 1 to 2 A. Alpha iron has a bcc lattice, EXAFS measurements at 78 K allow the detection of up to five shells (4.94 A). We used iron metal as our standard. The EXAFS measurements of the iron microclusters show the presence of 3–4 shells. The *nn* vs particle size are plotted in Fig. 2, a small contraction of about 1% in reference to the bulk is observed with the exception of the smallest particles. A large contraction is observed for the smallest particles of about 3%. This sample also shows the presence of an iron distance at 3.52 A such a distance does not exist in bcc iron. The ratio of the first and second shell distances of the 9 A microclusters is 0.682 and compares well with the ratio for fcc or hcp lattices, 0.707. The structural change in the 9 A is also observable in the near edge structure. We consider that the EXAFS results are clear evidence of a crystallographic transformation around a 9 A diameter.

Ge: We observed a great difference in the absorption edge structure for Ge clusters deposited very fast on the matrix (10 A/s). There was a clear evolution of the near edge structure with a significant increase in the number of empty *p*-states. Samples prepared by slow deposition (0.5 to 1 A/s) do not show any time dependence of the near edge structure. The EXAFS analysis of the clusters (mean sizes 7–8 A and 10–15 A) does not show the presence of *nnn*; only *nn* were observed. The *nn* distances measured were 2.42 A (diameter 7–8 A) and 2.45 A (diameters 10–15 A); these values are equal to bulk Ge (2.44 A) within the experimental error of EXAFS. We were also able to measure the near edge structure for free Ge microclusters travelling perpendicular to the photon beam. Figure 3 shows such a spectrum. The results for the free particles were not much different than for matrix isolated species. However, they show the great possibilities this technique offers in the characterization of small clusters.

The authors acknowledge the support of the U.S. DOE and the DFG (SFB 337)

References

1. Montano, P.A., Shenoy, G.K.: Solid State Commun. **35**, 53 (1980)
2. Montano, P.A., Schulze, W., Tesche, B., Shenoy, G.K., Morrison, T.I.: Phys. Rev. B **30**, 672 (1984)
3. Montano, P.A., Shenoy, G.K., Alp, E.E., Schulze, W., Urban, J.: Phys. Rev. Lett. **56**, 2076 (1986)
4. Abe, H., Schulze, W., Tesche, B.: Chem. Phys. **47**, 95 (1980)
5. Mass spectrometric measurements performed at the Fritz-Haber Institute

Z. Phys. D – Atoms, Molecules and Clusters 12, 107–111 (1989)

Atoms, Molecules and Clusters
Zeitschrift für Physik D
© Springer-Verlag 1989

Structure investigation of multiply – twinned Ni particles by electron investigation

M.F. Gillet[1] **and M. Brieu**[2]

[1] Laboratoire de Microscopie et Diffractions Electroniques – UA 797 Faculté des Sciences de Saint Jérôme, F-13397 Marseille Cedex 13, France

[2] Laboratoire de Microscopie et Structure des Matériaux – LA 40759 Université Paul Sabatier, F-31062 Toulouse Cedex, France

Received 15 July 1988; final version 30 September 1988

Nickel particles are prepared by reducing $NiBr_2$ in organic solvent. These Ni particles are investigated by electron diffraction. Some of them are icosahedral or decahedral. We have compared the experimental results and the theoretical patterns obtained with models composed of units with a fcc structure or with either rhombohedral or orthorhombic structure for icosahedral or decahedral particles respectively. The results show that the particles are composed of units with fcc structure.

PACS: 61.14

Introduction

Particles with decahedral or icosahedral shape have often been observed in thin films [1–8]. If these particles have frequently been observed in thin gold films obtained by vapor condensation in vacuum, few examples are available for other metallic particles obtained by other methods.

The atomic structure of these particles is another point which is still of interest. These particles can be considered either as composed of fcc tetrahedra with a gap and atomic defects [9–12] or with a new structure – rhombohedral structure for the icosahedral particles, orthorhombic for the decahedral ones [13, 14].

In this work we study by electron microscopy and electron diffraction icosahedral and decahedral particles prepared by reduction of nickel halides in organic medium [15]. High resolution electron diffraction enabled us to obtain diffraction patterns on individual particles and analyze their atomic structure by comparing the experimental results and theoretical patterns deduced from models of different structures.

Nickel particles preparation

Nickel particles were prepared by reduction of nickel bromide with magnesium in ethanol at a temperature of 78°C. These nickel particles were examined with the JEOL 200 CX electron microscope equipped with a sample-holder allowing rotation and tilt.

The micrograph in Fig. 1 reveals that the particles display various profiles, often with well-defined geometrical outlines (square, triangular, hexagonal shape). Their contrast is in some cases, characteristic of their structure. Icosahedral particles have an hexagonal outline (Fig, 2a) and decahedral particles exhibit a pentagonal shape (Fig. 4b).

Microdiffraction study of icosahedral particles

An example of diffraction pattern is given in Fig. 2b, c, which was obtained from the particle of Fig. 2a. This particle has an "edge" orientation, the beam being parallel to the twofold axis.

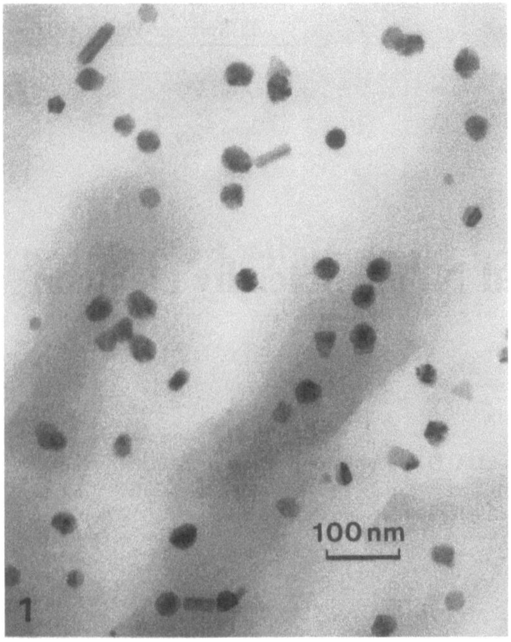

Fig. 1. Electron micrograph of nickel particles

The interpretation of this pattern was performed considering that the icosahedral particle is composed of 20 tetrahedra of rhombohedral or of fcc structure. In the latter case, the gap is uniformly distributed between the tetrahedron interfaces.

Figure 3a is the theoretical pattern obtained from a model with a rhombohedral structure and Fig. 3b is the theoretical pattern in the case of a model fcc structure.

We compare (Tables 1 and 2) the theoretical patterns and the experimental one. In Table 1 we report the distances D_C between the diffraction points determined for the atomic planes (h_R, k_R, l_R) or (h_C, k_C, l_C), for the model with rhombohedral structure and the fcc model respectively. We compared D_C with the distances D_e, measured for the (hkl) planes. Linewise, Table 2 illustrates the comparison of angles between certain atomic planes in the considered models and angles measured from diffraction patterns.

It appears from these comparisons that the more likely structure would be this one corresponding to tetrahedra in twinned position with the gap uniformly distributed between the different interfaces.

Microdiffraction study of decahedral particles

Generally decahedral particles exhibit a pentagonal outline, but in the case studied here the particles often have an elongated shape (Fig. 4a). Figures 4b, c, are electron micrographs for two positions ($\alpha=0°$, $\alpha=90°$)

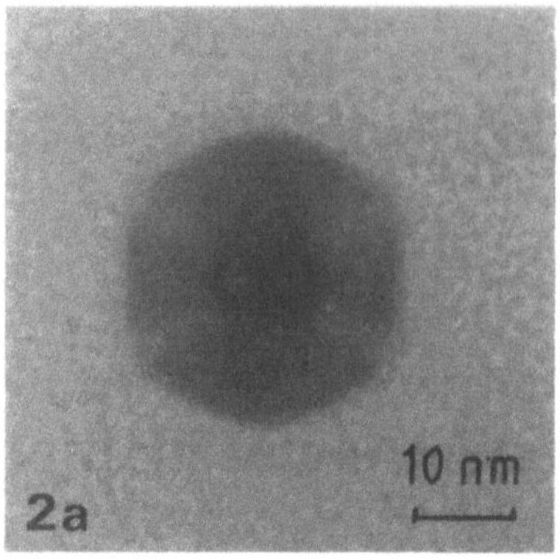

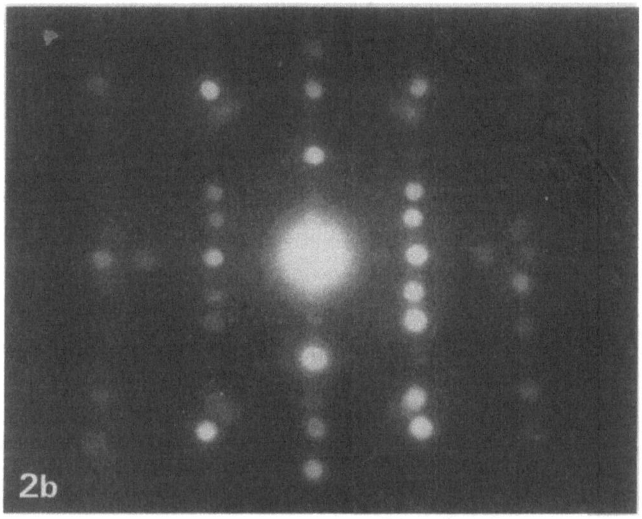

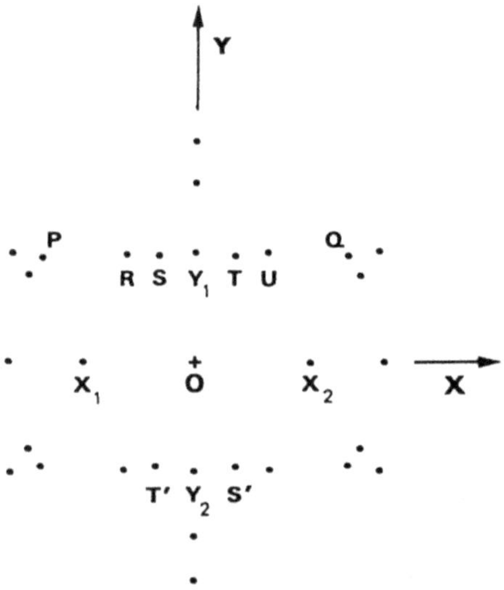

2c

Fig. 2a–c. Icosahedral particles with a edge orientation. a) Electron micrograph of an icosahedral particle. b) Electron diffraction pattern. c) Schematic representation of the diffraction pattern

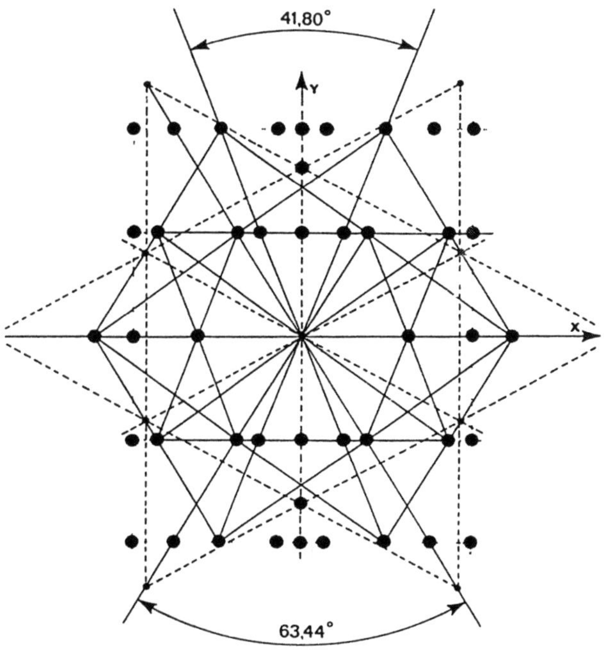

3a

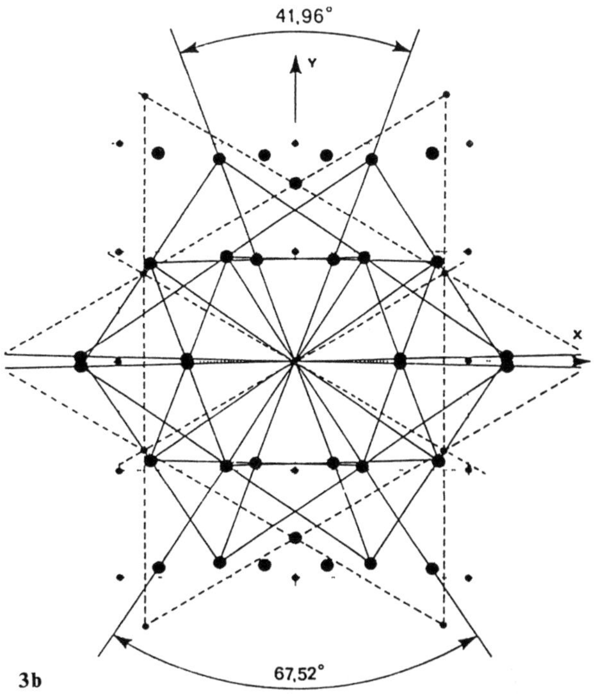

3b

Fig. 3a and b. Computed electron diffraction pattern. **a)** from a model with a rhombohedral structure. **b)** from a model with a fcc structure

Table 1. Comparison theoretical patterns and experimental values for icosahedral particles

Theoretical pattern					Experimental pattern
Rhombohedral structure			fcc structure		(see Fig. 2c)
$h_R k_R l_R$	$h_C k_C l_C$	D_C (mm)	h k l	D_C (mm)	D_e (mm)
0 0 1	1 1 $\bar{1}$		1 1 $\bar{1}$		
0 1 0	1 $\bar{1}$ 1	26.4	1 $\bar{1}$ 1		$X_1X_2 = 27.5$
1 0 0	$\bar{1}$ 1 1		$\bar{1}$ 1 1	27.5	
1 1 1	1 1 1	28.3	1 1 1		$SS' = 28$ $TT' = 27.5$

obtained by tilting the particle around an axis perpendicular to the fivefold axis.

The morphology of these particles has been studied in detail [16]. We present here some structural results deduced from the study of high resolution microdiffraction patterns. Figure 5a is the diffraction pattern corresponding to the particle of Fig. 4c and Fig. 5b the theoretical pattern from a model with an orthorhombic structure. As for icosahedral particles, we compare theoretical and experimental patterns. Table 3 gives the calculated distances D_C between the diffraction spots from two $(hkl)_o$ and $(hkl)_c$ planes in an orthorhombic structure or a fcc structure and the measured distances D_e on the experimental pattern. Similarly,

Table 2. Comparison theoretical patterns and experimental values for icosahedral particles

Angles (°) between planes (hkl)	Values calculated with the rhombohedral struc.	Values calculated with fcc structure	Measure values
(200) and (111)	52.62	54.74	OR, OT = 55.50 OU, OS = 55.75
(200) and (022)	85.83	90	OR, OQ = 90.25 OU, OP = 89.75

Table 3. Comparisons between the calculated value and the experimental ones D_e for decahedral particles

(hkl)	D_c orthorhombic structure	D_c fcc structure	D_e
(220)	61,57	64,74	64,00
($\bar{2}\bar{2}$0)	64,74	64,74	64,50
(420)	97,86	102,36	101,75
($\bar{4}\bar{2}$0)	101,87	102,36	103,00
(042)	100,96	102,36	102,50
(131)	73,42	75,90	75,50
(400)	89,34	91,55	91,00

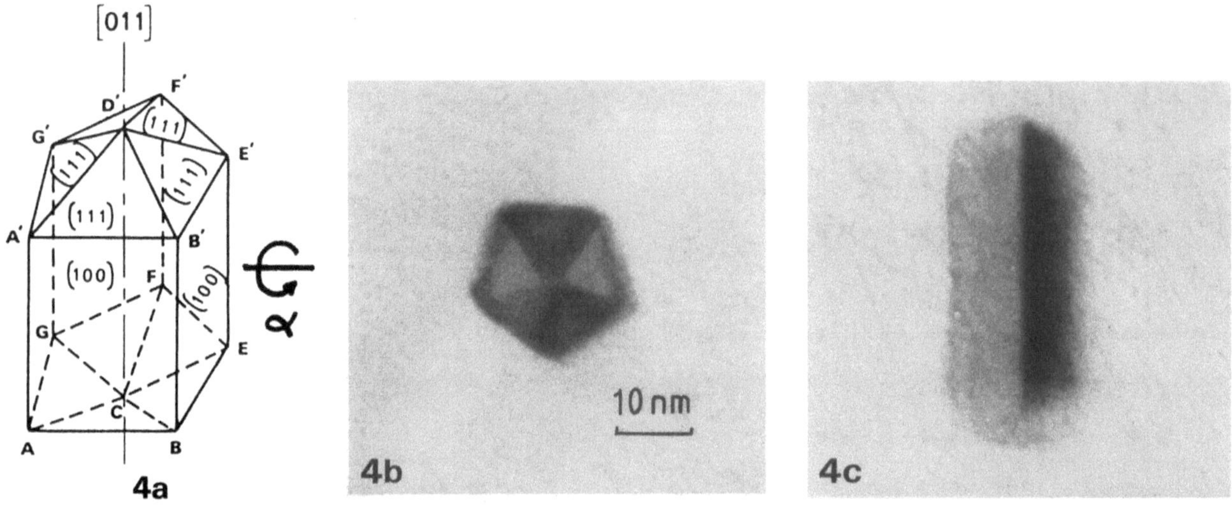

Fig. 4. a Model for an "elongated decahedral" particle. **b** and **c** electron micrographs of an elongated decahedral particle. $\alpha=0°$ and $\alpha=90°$

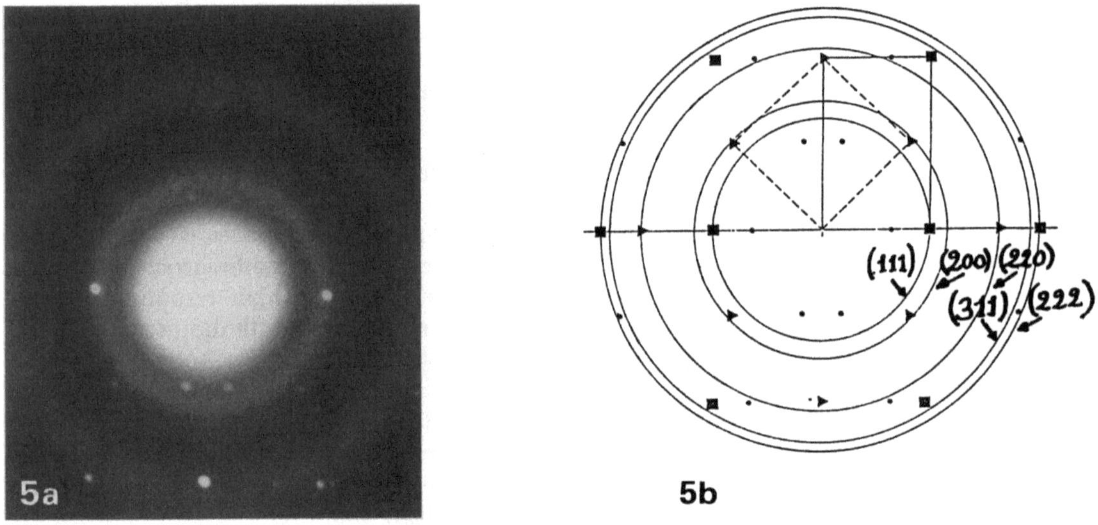

Fig. 5. Electron diffraction pattern for an "elongated decahedral" particle ($\alpha=90°$). **a)** experimental, **b)** theoretical pattern for a model with an orthorhombic structure or fcc structure respectively

Table 4 gives examples for the comparison between the angles calculated from the above mentioned models and the angles deduced from the experimental diffraction patterns. In the case of elongated decahedral particles, the structure deduced from electron diffraction patterns is fcc as for icosahedral particles.

Conclusion

The electron diffraction results obtained for icosahedral or decahedral nickel particles are consistent with models composed of units with fcc structure.

Table 4. Comparisons between the calculated value and the experimental ones D_e for decahedral particles

Angles (°) between planes (*hkl*)	Values calculated with an orthorhombic structure	Values calculated with a fcc structure	Measured values
(200) and (020)	92, 87	90, 00	90, 00
(131) and (311)	66, 04	62, 96	63, 50

However it must be noted that the particles studied here have relatively large sizes (over 10 nm) and different structures may be considered for smaller sized particles (below 10 nm).

References

1. Ino, S.: J. Phys. Soc. Jpn. **21**, 346 (1966)
2. Allpress, J.G., Sanders, J.V.: Surf. Sci. **7**, 1 (1967)
3. Komoda, T.: Jpn. J. Appl. Phys. **7**, 27 (1968)
4. Gillet, E., Gillet, M.: Thin Solid Film **4**, 171 (1969)
5. Yagi, K., Takayanagi, K., Kobayashi, K., Honjo, G.: J. Cryst. Growth **28**, 117 (1975)
6. Yang, C.Y., Heineman, K., Yacaman, M.J., Poppa, H.: Thin Solid Film **58**, 163 (1979)
7. Marks, L.D., Smith, D.J.: J. Microsc. **130**, 249 (1983)
8. Hofmeister, H.. Thin Solid Films **116**, 151 (1984)
9. Ino, S.: J. Phys. Soc. Jpn. **27**, 941 (1969)
10. de Wit, R.: J. Phys. C **5**, 529 (1972)
11. Saito, Y., Yatsuya, S., Mihama, K., Uyeda, R.: Jpn. J. Appl. Phys. **17**, 1149 (1978)
12. Marks, L.D., Smith, D.J.: J. Cryst. Growth **54**, 425 (1981)
13. Bagley, B.G: Nature **208**, 674 (1965)
14. Yang, C.Y.: J. Cryst. Growth **47**, 274 (1979)
15. Mauret, P., Alphonse, P.: J. Org. Chem. **47**, 3322 (1982)
16. Brieu, M.: Thèse de Doctorat d'Etat, Toulouse (1986)

Z. Phys. D – Atoms, Molecules and Clusters 12, 113–114 (1989)

Atoms, Molecules and Clusters
Zeitschrift für Physik D
© Springer-Verlag 1989

Structure of an icosahedral palladium particle

J.M. Penisson[1] and A. Renou[2]

[1] Service de Physique Département de Recherche Fondamentale CEN Grenoble 85X, F-38041 Grenoble Cedex, France

[2] Laboratoire de Microscopie et Diffraction Electroniques Faculté des Sciences et Techniques St. Jérome, F-13397 Marseille Cedex 13, France

Received 7 July 1988; final version 30 September 1988

Small icosahedral palladium particles have been studied using high resolution electron microscopy. In these small particles ($d \leqslant$ 15 nm) no dislocations have been found so that internal strains are necessary to maintain their coherency. The structural deformations induced by these strains have been investigated and compared to existing models.

PACS: 61.70; 68.55

Introduction

Multiply twinned particles exist in a wide range of materials. Among them, decahedral and icosahedral particles are very interesting objects because of the presence of one (decahedra) or several (icosahedra) five fold symmetry axes. The presence of these non crystallographical symmetry elements induces internal strains. These strains can be relieved in large particles by dislocations or stacking faults. However, in small particles (diameter $d \leqslant$ 15 nm) no dislocations are present and these internal strains induce structural deformations. In this paper, icosahedral palladium particles have been observed using very high resolution electron microscopy and the deformations have been analysed.

Experimental procedure and results

Specimen preparation has been published in detail by Renou et al. [1] together with preliminary results [2]. High resolution electron microscopy has been done using a JEOL 4000 EX microscope. This microscope has a very good resolution power and atomic spacings down to 0.12 nm can be imaged [3]. Icosahedral particles are relatively easy to observe and two different crystallographic orientations are present: ⟨112⟩ edge orientation and ⟨111⟩ face orientation. Only the first one has been investigated in detail. The mean diameter of the particles is in between 10 and 15 nm.

Figure 1 shows the high resolution image of an icosahedral particle in the ⟨112⟩ edge orientation.

Several different parts can be distinguished.

– Two large areas A and B on Fig. 1 in which the orientation is ⟨011⟩. These two areas are symmetrical with respect to an horizontal mirror plane; a twin is present in both areas. (arrowed on Fig. 1)

Areas C and D: 3 sets of 60° atomic fringes can be seen. Their spacing can be measured either on the micrographs or on their optical diffraction pattern. By comparison with the 111 planes (0.22 nm) present in A and B areas, these planes are identified as 220 (0.137 nm) planes so that the orientation of C and D is ⟨111⟩.

– The central area E shows one set of 111 planes (horizontal) and one set of 220 planes vertical leading to ⟨112⟩ orientation.

A detailed analysis of the contrast in areas A and B shows that in fact the observed structure differs from a perfect fcc crystal. The angular relationship between the twinned variants is also different from a perfect fcc twinned crystal. In the ⟨011⟩ orientation two different sets of 111 planes are imaged. The first one is parallel to the external surface and the second is parallel to the twin plane. Their spacing is different. If d_s is the spacing of 111 planes parallel to the surface and d_T the spacing

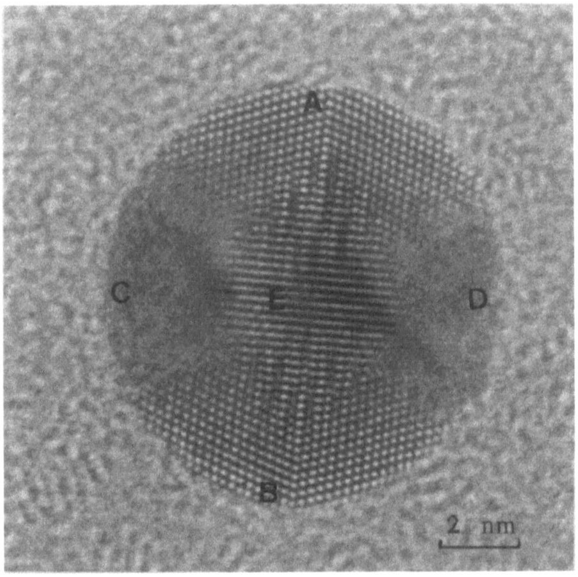

Fig. 1. High resolution image of an icosahedral palladium particle. The orientation is ⟨112⟩ edge. *A* and *B* areas have a ⟨011⟩ orientation while *C* and *D* are seen along ⟨111⟩ and *E* along ⟨112⟩

of the planes parallel to the twin plane.

$$\frac{d_s}{d_T} = 0.97$$

The measured angle between these two sets of 111 planes is $69 \pm 0.5°$ instead of $70.53°$ in the perfect fcc structure.

The trace of the twin plane is a common ⟨112⟩ direction. The angle between this particular direction and the normal to the 111 planes parallel to the outer surface is $20.5°$ instead of $19.47°$ in the perfect fcc structure.

Interpretation of the results

The icosahedral particle have often been described as the stacking of 20 tetrahedrons which are slightly deformed to completely fill the space. The observation of ⟨111⟩ and ⟨112⟩ oriented areas is in good agreement with this description of the particle observed in the ⟨112⟩ edge orientation. This particular orientation corresponds to the stereographic plot of the non crystallographic icosahedral group ($m\bar{3}\bar{5}$) (Fig. 2).

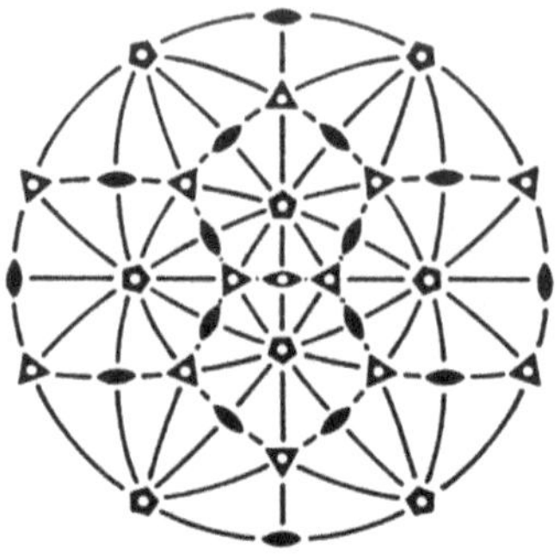

Fig. 2. Stereographic plot of the non crystallographical group ($m\bar{3}\bar{5}$) showing that the particle has the same overall symmetry

Some of the symmetry elements can be located on the images.

– a two fold axis is parallel to the incident electron beam.

– two perpendicular mirror planes are parallel to the incident beam. One of them is parallel to the twin plane.

– four three fold axes lie in the image plane at $20.9°$ of the twin plane (mirror plane). This angle corresponds exactly to the experimental angle measured between ⟨112⟩ and ⟨111⟩. It should be remarked that according to the homogeneous deformation model given by INO [4] the normal to the external face of each building tetrahedron preserves its 3 fold symmetry while the internal faces loose it.

From this correspondence between the experimental and theoretical angle it can be concluded that the particle has the overall symmetry of the icosahedral group. In each building tetrahedra, the crystal lattice is deformed. The elastic model described by Ino [4] indicates that a 5% deformation of each tetrahedron is necessary to obtain an icosahedral particle without internal gaps. The experimentally measured deformations and angular relationships are in good agreement with this model.

References

1. Renou, A., Rudra, A: Surf. Sci. **69**, 1156 (1985)
2. Renou, A., Pénisson, J-M.. J. Cryst. Growth **78**, 357 (1986)
3. Bourret, A., Pénisson, J. M.: J. Microsc. Spectrosc. Electron. **12**, 18a (1987)
4. Ino, S.. J. Phys. Soc. Jpn. **27**, 941 (1969)

Z. Phys. D – Atoms, Molecules and Clusters 12, 115–118 (1989)

Atoms, Molecules
Zeitschrift für Physik D and Clusters
© Springer-Verlag 1989

Structural investigations of metal clusters by HREM

S. Giorgio[1] and J. Urban[2]

[1] CRMC2-CNRS, Campus de Luminy Case 913, F-13288 Marseille Cédex 9, France
[2] Fritz-Haber-Institut der MPG, Elektronenmikroskopie, Faraday Weg 4–6, D-1000 Berlin 33, Germany

Received 9 July 1988; final version 30 September 1988

Silver clusters grown by the inert gas aggregation technique have been investigated by HREM. Undistorted cuboctahedra and icosahedra together with twinned particles have been observed. Three types of nucleation and growth mechanisms are proposed which can explain the observed particle structures. The untwinned large particles are created by the coalescence of liquid- like small clusters. The twins are produced by the coalescence of solid subunits.

PACS: 36.40.+d; 61.16.Di

1. Introduction

The technique of inert gas aggregation (GAT) of metal clusters allows the preparation of particles with well-defined size distributions. Together with HREM, it is possible to resolve the atomic structure of the clusters provided they are oriented properly with respect to their support. Besides the study of the structure of the particles, our aim was to investigate the nucleation and growth of the particles from the TEM observations.

2. Preparation and observation

Silver atoms and dimers were evaporated from a Knudsen cell into a liquid nitrogen cooled argon atmosphere as shown in Fig. 1 and deposited on amorphous carbon film [1]. The TEM observations were performed with the DEEKO 100 microscope of the Fritz-Habber-Institut (100 kV, Cs = 0.5 mm) [2].

Figure 2a shows the micrograph together with the histogram of the size distribution for a deposition time smaller than 1 s, and a deposition rate of 0.4 Å/s. cm^2. Figure 2b represents the same preparation with a deposition time larger than 1 s. The peak corresponding to the small particles in the binary distribution of Fig. 2a disappears while the peak centered at larger particles increases as can be seen in Fig. 2b for larger deposition times. Furthermore, for the latter case the amount of multiple twins increases with the cluster size while in the first case, most of the particles are single crystals.

3. Microstructure and conclusion

The classifications of single crystals and twins in histograms according to the deposition times have been described above. We believe that these findings can be interpreted by three types of growth mechanisms:

1) Nucleation and growth by adding single atoms to the nucleus which is reflected by the first peak at 1.5 nm in Fig. 2a.

2) Coalescence of small clusters in the argon atmosphere of the nucleation cell, i.e. the particles in the second peak at 6.5 nm.

3) Coalescence of the particles on the carbon substrate at larger deposition times leading to the existence of multiple twins. The negligiable amount of twins in Fig. 2a favours the assumption that the smaller particles which have coalesced already in the nucleation cell represent a liquid-like structure. This permits

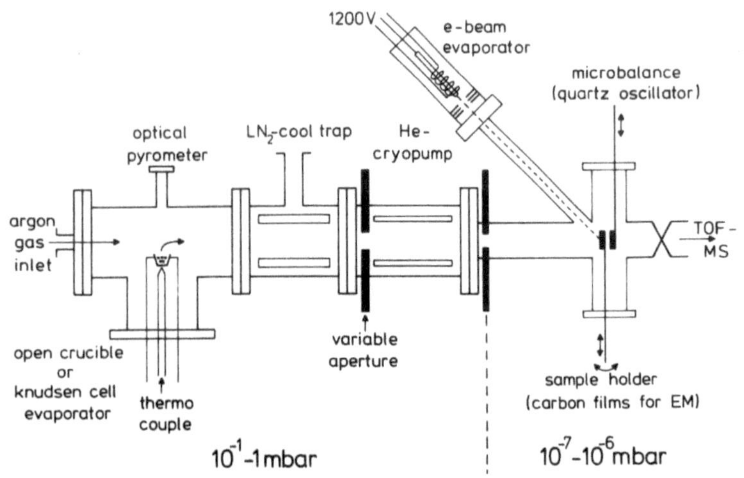

Fig. 1. Nucleation cell for the production of clusters by the inert gas aggregation technique

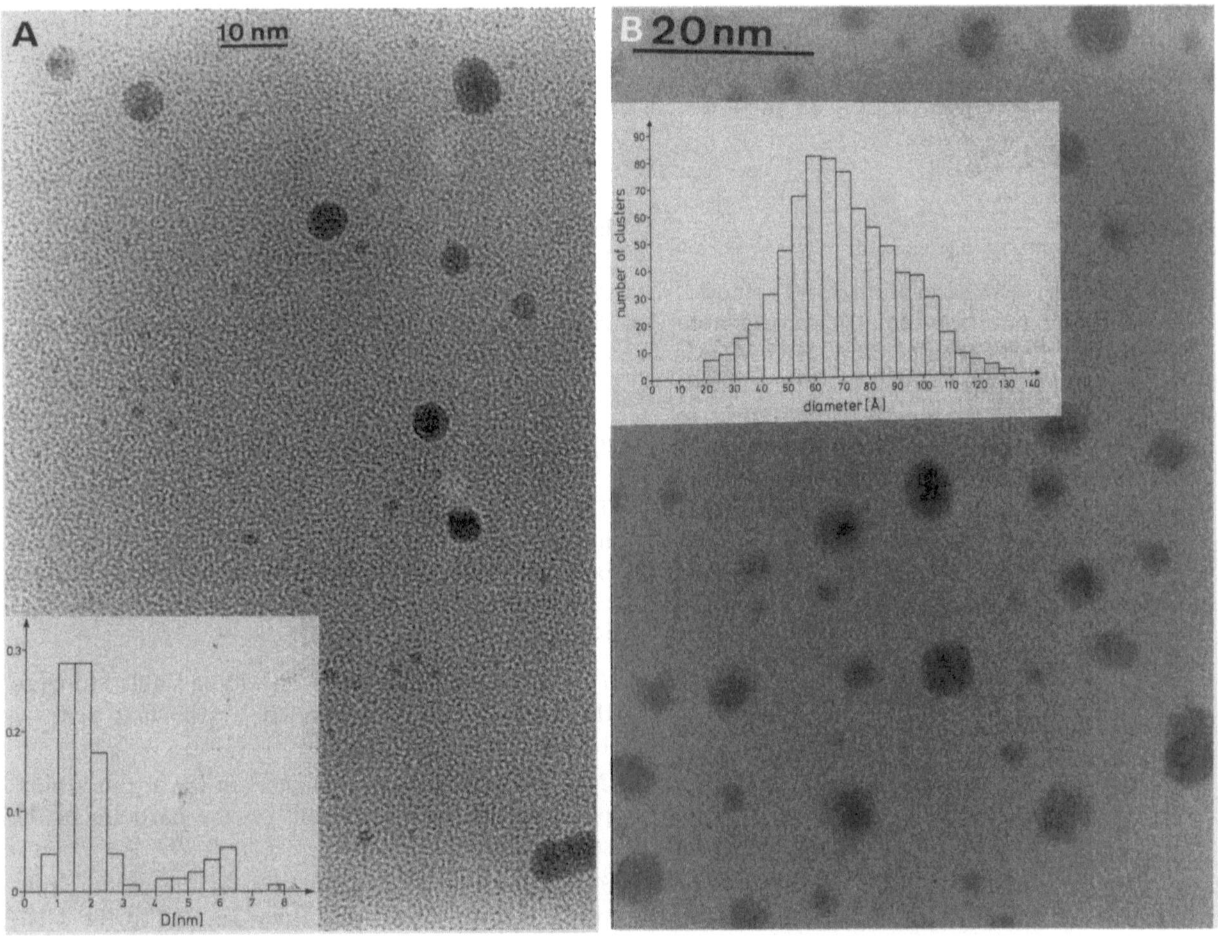

Fig. 2. a Electron micrograph of silver clusters, deposition time $t < 1$ s. The inset represents the histogram for this size distribution. **b** Electron micrograph of silver clusters, deposition time $t > 1$ s, with the size distribution shown in the inset

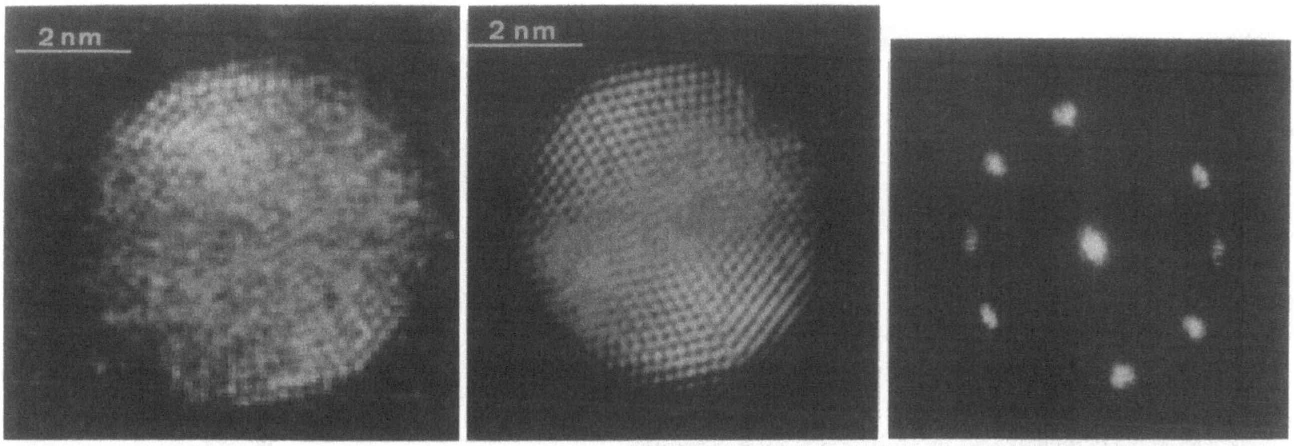

Fig. 3. Perfect icosahedron along the ⟨112⟩ direction, i.e. 5-fold axis of symmetry parallel to the support, (original image, calculated power spectrum, filtered image)

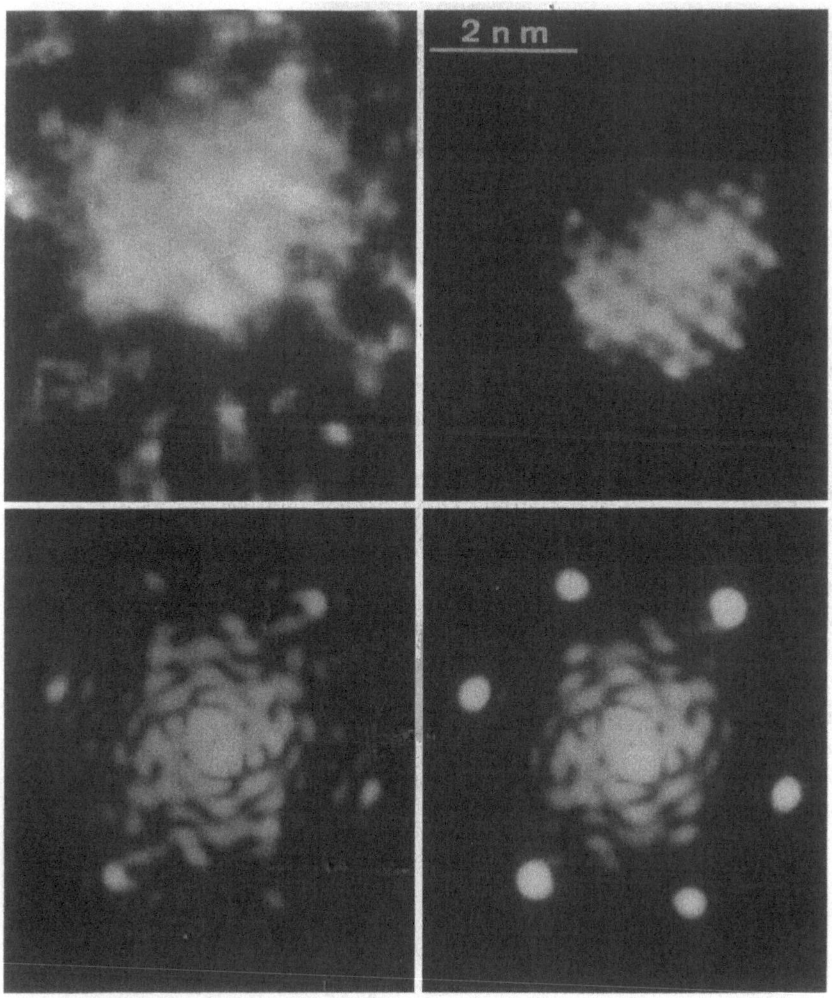

Fig. 4. Perfect cuboctahedron in the 110 orientation

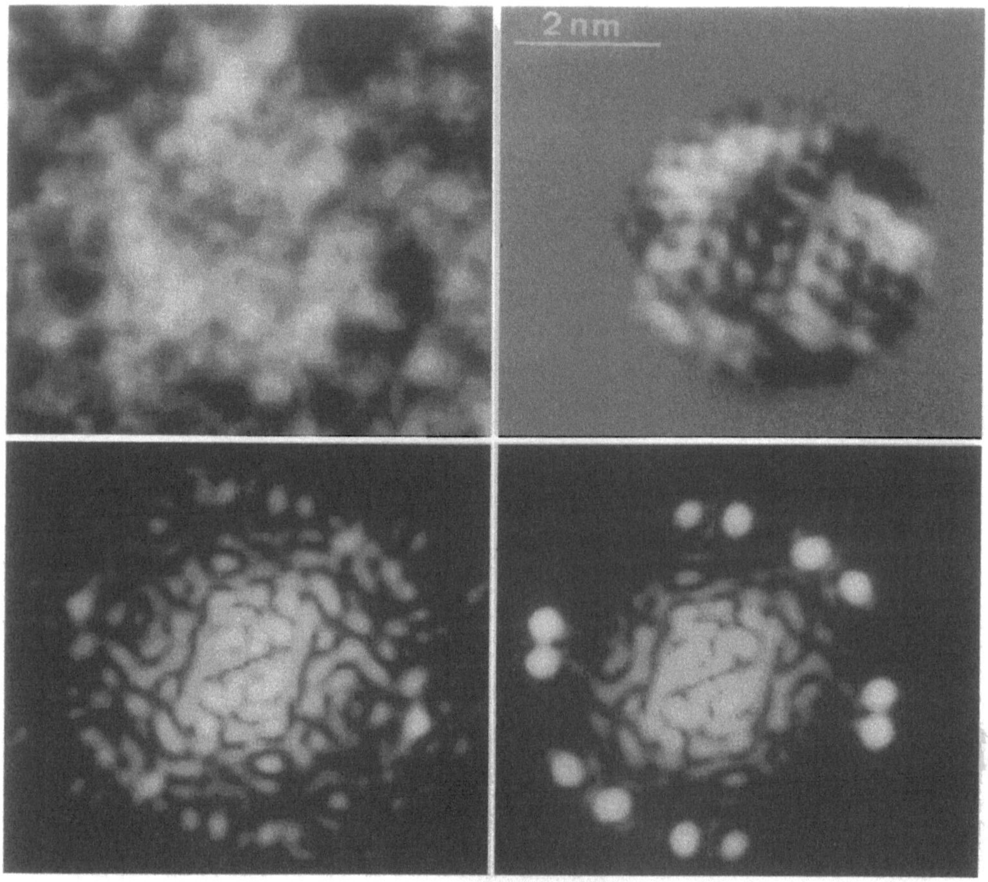

Fig. 5. Twinned cuboctahedron in the 110 orientation

the growth of larger edifices in the shape of either icosahedra or cuboctahedra as will be shown below.

For energetic reasons, undistorted large icosahedra cannot be formed by the coalescence of solid-like small particles, i.e. a total rearrangement of all atoms in the solid would be necessary. Therefore, twin boundaries are created as is reflected in the distribution of Fig. 2b.

Three examples of high resolution images showing icosahedral and cuboctahedral structures are given in Figs. 3, 4 and 5 respectively. In this case we present cuboctahedra either single crystalline or twinned (cf. Figs. 4, 5). All the pictures were digitalized and the noise has been reduced by image treatment. The particles have been extracted from the background by multiplying the image by a Gaussian shape function:

$S(x) = 1$ for $x < xo$

$S(x) = \exp(-\alpha(x - xo)^2)$ for $x > xo$,

xo representing the outline of the particles.

Then, their Fourier Transform has been calculated and multiplied by the same shape function. The inverse Fourier Transform leads to the filtered image of the particle [3]. These filtered images are shown on the right side with their power spectrum.

The cluster project was supported by the Deutsche Forschungsgemeinschaft, Sfb 337. For his interest in the present work, we thank Prof. Dr. E. Zeitler.

References

1. Franck, F., Schultze, W., Tesche, B., Urban, J., Winter, B.: Surf. Sci. **156**, 90 (1985)
2. Kunath, W., Zemlin, F., Weiss, K.: Optik **76**, (4) 122–31 (1987)
3. Giorgio, S., Urban, J.: Appl. Phys. Lett. **52**, (18) 1467–8 (1988)

Z. Phys. D – Atoms, Molecules and Clusters 12, 119–121 (1989)

Atoms, Molecules and Clusters
Zeitschrift für Physik D
© Springer-Verlag 1989

The structure of small penta-twinned gold particles

Pei-Yu Gao[1], W. Kunath[2], H. Gleiter[1], and K. Weiss[2]

[1] Universität des Saarland, D-6600 Saarbrücken, Federal Republic of Germany
[2] Fritz-Haber-Institut der MPG, D-1000 Berlin 33, Germany

Received 6 July 1988

The structural feathers of penta-twinned gold particles (size between 2 and 6 nm) generated by gas evaporation have been investigated by high resolution TEM. The structural characteristic of penta-twinned particles is different from that of quasi-crystals that the five coherent or incoherent twin boundaries separating the twin oriented segments do not join up along a common edge. The lattice parameter is reduced by 4–5% in comparison to that of bulk gold. The formation of the penta-twinned particles is proposed to occur by particle collision. The particles were observed to be crystalline at ambient temperature.

PACS: 61.14; 81.15.L

The structure of small, metallic particles is of interest for the industrial application (e.g. catalysts etc.) as well as basic research (e.g. interatomic forces etc.). The properties of small particles are known to be different from those of bulk materials when the sizes of the particles approaches atomic dimensions. For larger particles (e.g. 100 nm in diameter) the surface morphology may be interpreted in terms of the Wulff construction [1, 2]. More complex particle structures, particularly multiply-twinned particles [3–7], can be understood in terms of a modified Wulff construction [6, 8]. The present paper reports observations by high resolution electron microscopy on small (about 2 to 6 nm in diameter) gold particles containing five twin boundaries arranged in a radial fashion intersecting along a common [110] direction.

The purpose of the study is to investigate the atomic structure and conceivable processes of formation of penta-twinned gold particles supported on amorphous carbon (4 nm thick).

The micrographs were taken with the TEM DEEKO 100 and have been subsequently processed by computer to increase the signal to noise ratio. The specimen preparation and the performance of the DEEKO-100 TEM has been summarized in refs. [9, 10]. Several micrographs of penta-twinned particles and their structural models are shown in Fig. 1–3. The particle shown in Fig. 1 shows the smallest penta-twinned particle observed. It has a diameter of about 20 Å and consists of about 200 atoms. The particle contains two coherent and three incoherent twin boundaries. Figure 2 shows a particle consisting of twin-related segments of different sizes. The particle contains one coherent twin boundary and consists of about 370 atoms. A larger penta-twinned particle (about 600 atoms) may be seen in Fig. 3. The characteristic features emerging from the evolution of the structure of the penta-twinned particles may be summarized as follows:

1. The five radial twin boundaries do not join in a common edge contrary to earlier observations [11].
2. The twin boundaries may be of the coherent and/or incoherent type.
3. The particles are faceted on {111} or {100} planes.
4. The lattice parameter measured from the diffraction patterns of its high resolution micrographs is reduced by about 4–5% in comparison to bulk gold.
5. The radial angles of five segments in penta-twinned particles vary from 65° to 76°.

The characteristic features of the penta-twinned gold particles suggest that these particles are formed by collision and sintering of small crystals in the gas

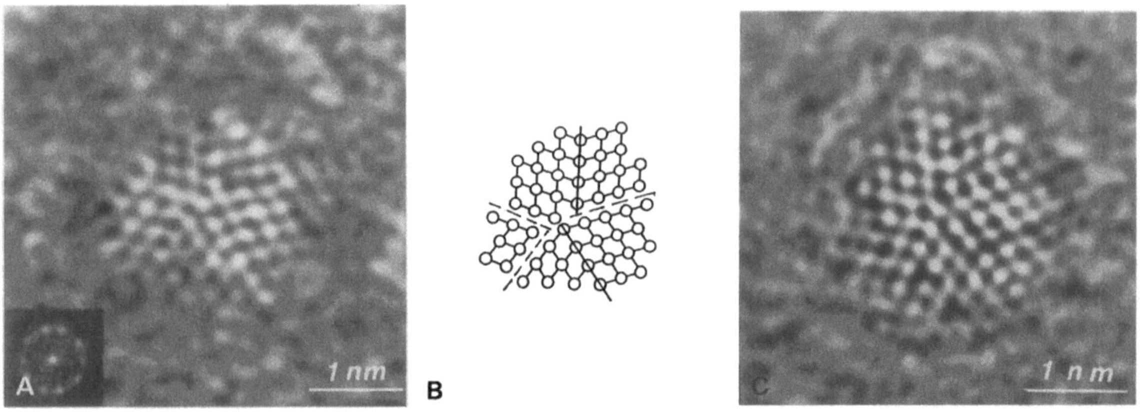

Fig. 1. **a** Electron micrograph showing a penta-twinned gold particle viewed along the [110] direction. The particle is faceted on {111} and {100} planes. **b** Model of atomic arrangement of a. The thick lines which go through the circle represent coherent twin boundaries on {111}. The dashed lines represent incoherent twin boundaries. **c** A simulated image from model in b. The particle contains about 200 atoms

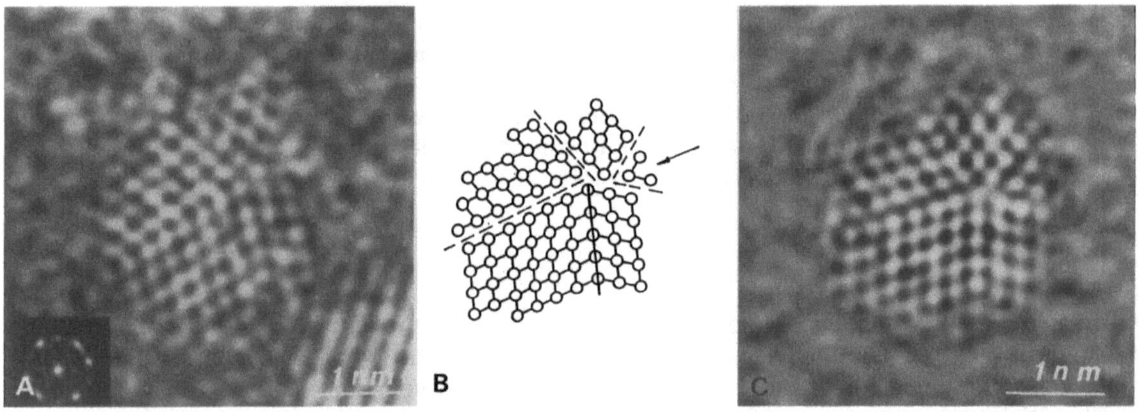

Fig. 2. **a** Electron micrograph showing a penta-twinned gold particle viewed along the [110] orientation. The particle is faceted on {111} and {100} planes. **b** Model for atomic arrangement of the particle shown in a. One of the twin boundaries is a coherent twin boundary. One of the segments indicated by an arrow consists of only a few columns of atoms. **c** A simulated image from model in b. The particle contains about 370 atoms

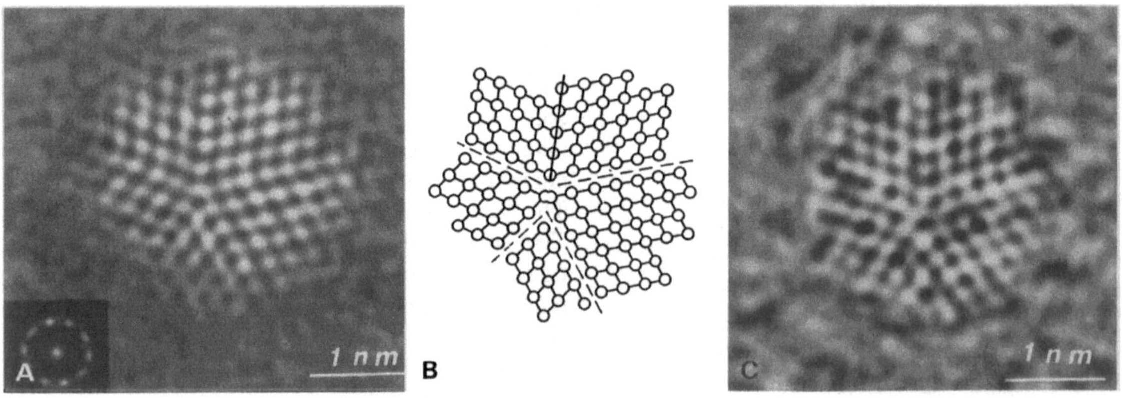

Fig. 3. **a** Electron micrograph showing a penta-twinned gold particle viewed along the [110] orientation. The particle is faceted on {111} planes. **b** Model for the atomic arrangement of a. There is one coherent twin boundary and two segments which are in contact in the center. **c** A simulated image from model in b. The particle contains about 600 atoms

atmosphere. If two small particles collide in the gas phase, the grain boundary formed between them may reduce its free energy by the diffusion induced coherent rotation of both particles with respect to one another into a twin orientation relationship. The existence of a coherent particle rotation has been demonstrated for metals [12, 13] as well as ionic crystals [14]. This model is consistent with lower frequency of formation of penta-twinned particles than of particles with a smaller number of twin boundaries [9]. The non-common edge among the twinned segments may be understood by minimizing the elastic strain energy of the particles [15]. The five twin related segments do not fill space without elastic distortion. The different radial angles of the segments in penta-twinned particles are believed by different relaxation of each segment before collision. The hypothesis of a liquid structure of small gold particles at ambient temperature is not supported by the observations.

The authors acknowledge the continual encouragement of this study by Prof. Dr. E. Zeitler. The financial support by the BMFT (contract no. 03M 00234) is appreciated.

References

1. Herring, C. Phys. Rev **82**, 87 (1951)
2. Lindford, R.G. Solid state surface science, p. 1. New York· Dekker 1973
3 Ino, S J. Phys. Soc Jpn **21**, 346 (1966)
4. Ino, S., Ogawan, T.: J. Phys. Soc Jpn **22**, 1369 (1967)
5. Marks, L.D., Smith, D J.· J. Microsc. **130**, 249 (1982)
6. Marks, L D.: Philos Mag **49**, 81 (1984)
7. Howie, A., Marks, L.D.: Philos. Mag. **49**, 95 (1984)
8. Marks, L.D : J. Cryst. Growth **61**, 556 (1983)
9. Gao, P., Gleiter, H.: Acta Metallo **35**, 1571 (1987)
10. Kunath, W., Zemlin, F., Weiss, K.: Optik **76**, 122 (1987)
11. Iijima, S., Ichihashi, T.· Phys. Rev. Letter **56**, 616 (1986)
12. Hermann, H., Gleiter, H., Baro, G.: Acta Metall. **24**, 353 (1976)
13. Shewman, P.G.: Energy and structure of grain boundaries, p. 165. Metal Park, Cleveland: ASM Publ. 1966
14 Chandhan, P., Matthews, J.W.: Appl. Phy Letters **17**, 115 (1970)
15. Mindlin, R.D, Salvador, M.G.: Handbook of experimental stress analysis. Heteny, M. (ed.), pp. 775. New York: Wiley 1950

Z. Phys. D – Atoms, Molecules and Clusters 12, 123–125 (1989)

Atoms, Molecules and Clusters
Zeitschrift für Physik D

© Springer-Verlag 1989

Electron microscope studies of the regular twinning structure and the grain boundary structure in a system of cobalt particles

A.R. Thölén

Laboratory of Applied Physics, Technical University of Denmark, DK-2800 Lyngby, Denmark

Received 5 July 1988; final version 30 September 1988

Martensitic transformation in small gas evaporated cobalt particles (< 100 nm) has earlier been discussed using conventional electron microscopy. The results can shortly be described as an extremely regular banded twinning structure within the particles. In this paper it will be presented new results on the same transformation and also on the particle formation and interaction but now using high resolution electron microscope methods.

PACS: 61.70; 68.00

Introduction

Gas-evaporated particles form a characteristic structure with branching networks, or in the case of magnetic particles, one dimensional chains. The particle size-distribution is log-normal. Kinetic theories of particle nucleation and coalescence have been used to elucidate the history of such particles and fractal theories have more recently been applied on network formation [1, 2] but no detailed knowledge of the processes exists. In order to learn more about particle formation, interaction and also the rapid martensitic transformation within the particles we have started to investigate cobalt particles with high resolving electron microscopical methods. The ultimate goal is a precise knowledge of the above mentioned processes.

Experimental

Cobalt was evaporated from a tungsten basket in an argon atmosphere and the so formed particles were collected on a holey carbon film. Most of the evaporations were made at a pressure of 20 torr and the strings of cobalt particles were investigated in a Philips 430 electron microscope using the twin lens.

Results

A regular banding (twinning) has earlier been reported in cobalt particles [3, 4] as well as a finer twinning. A particle with dense faulting on two different planes is shown in Fig. 1. The whole particle is divided into two twin related sections and it is noticable that within each (twin) region the planar faults are parallel, i.e. there is no crossing of fine twins.

A transformed particle is observed along the (111)-planes in Fig. 2. The distance between the closest dark lines is 0.4 nm, which corresponds to the thickness of two (111)-planes in fcc cobalt or the length of the c-axis in hcp cobalt.

One particle with three thin twin bands showing lattice resolution is seen in Fig. 3. According to earlier investigations [3, 4], there is often a marked regularity in the banding resulting in a constant bandwidth. This is also observed here with the three "bands" spaced about 10 nm apart. Further, the transition zone to the next particle is twin related to the main particle, a very common observation. Twinning can here be considered as a way to break down a high angle grain boundary into a low energy configuration (two twin boundaries and possibly some additional rotation).

Are grain boundaries between small particles similar to those in bulk material? Figure 4a and b are

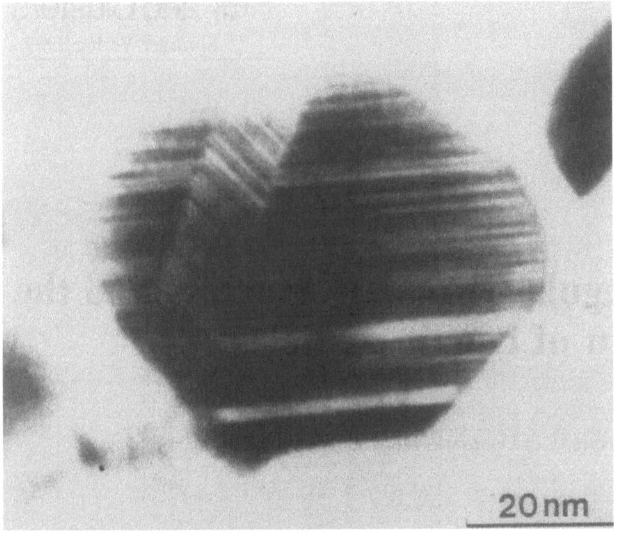

Fig. 1. A transformed cobalt particle. Dense faulting on two sets of planes is visible

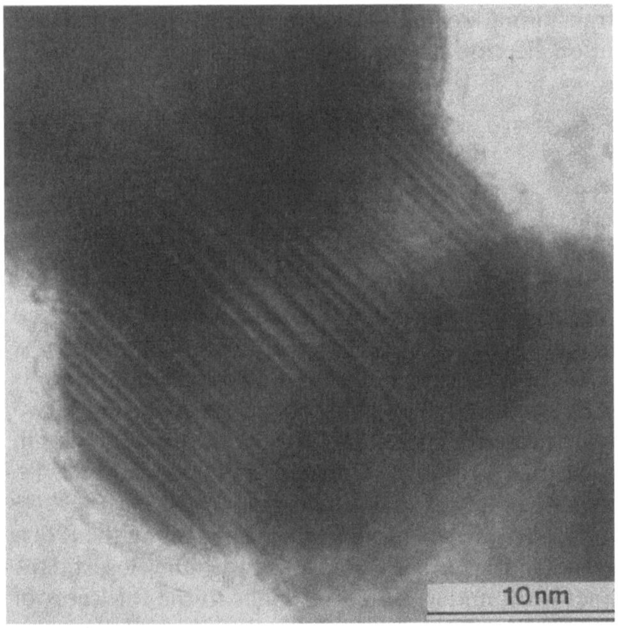

Fig. 2. A transformed cobalt particle. The (111)-planes are seen end on

obtained from the same boundary but with different diffraction conditions. Localized strain fields at the boundary indicate grain boundary dislocations. In spite the proximity to the surface these strain centres have been conserved and it looks very much like a bulk boundary.

Figure 5 shows a twinned particle where transformation has occurred half way along the twin

Fig. 3. Particle with three bands of fine twins showing lattice resolution

boundary which is heavily distorted. But the circumference of the particle is surprisingly smooth indicating a rapid diffusion after the deformation has occurred.

A moiré pattern develops when there is a small rotation between the two crystals or when there is a slight difference in lattice constant. It gives a kind of magnified image of the boundary. A moiré pattern with wavy lines in two directions is seen in Fig. 6. Dislocations at the boundary will distort this picture.

The diffraction pattern from a group of gas-evaporated particles mainly indicates a fcc structure. The traces of hcp are faint, in spite, the bulk structure of cobalt is hcp.

A direct method to determine orientation relationships across particle (grain) boundaries is to take diffraction patterns from each side. Accordingly diffraction patterns were obtained from each single particle along some chains. It is obvious both from these diffraction patterns and from the contrast of the particles that there generally is no simple orientation-relationship between neighbouring grains and there is no long range order. This is also in accordance with the diffraction pattern from an agglomerate of particles showing no preferred orientations.

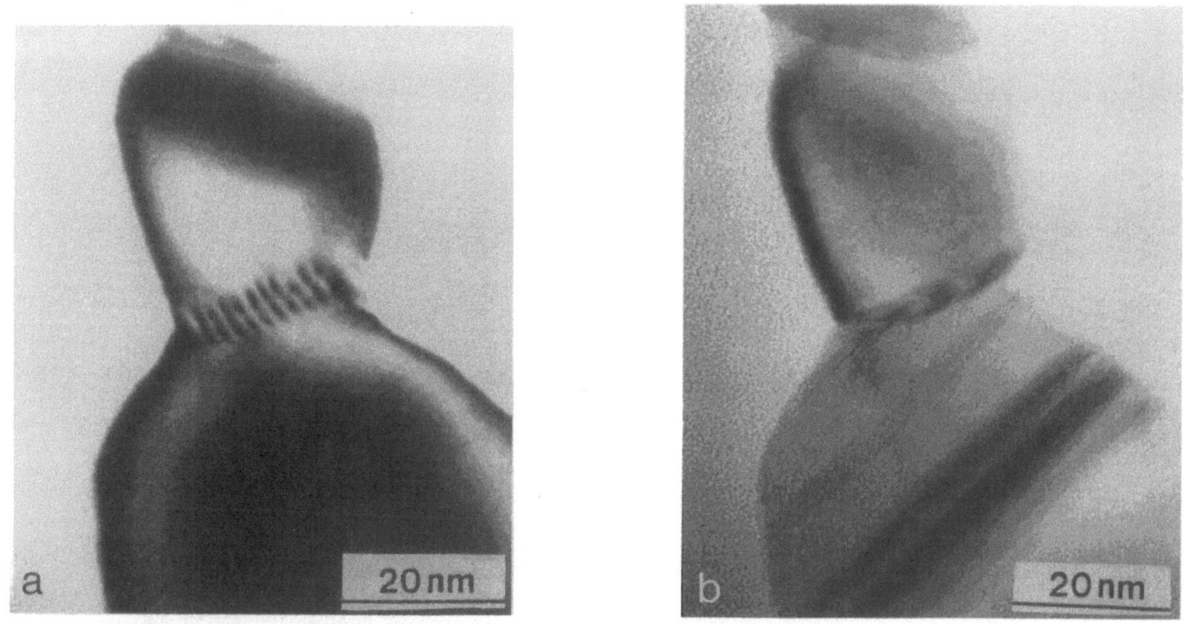

Figs. 4a and b. Grain boundary between two particles taken with different diffraction conditions Dislocations are visible at the boundary

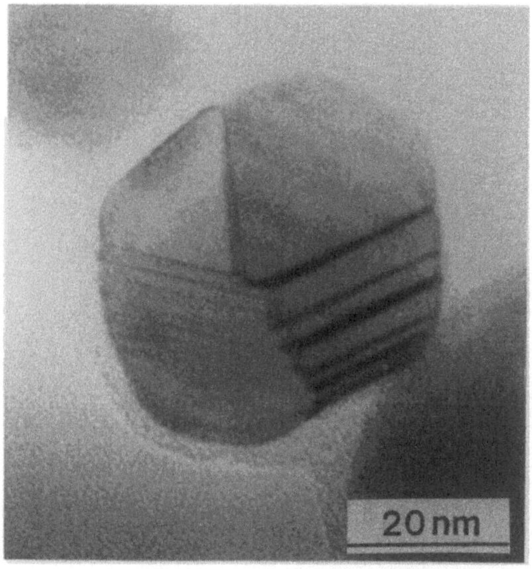

Fig. 5. Twinned particle, where one half is heavily deformed

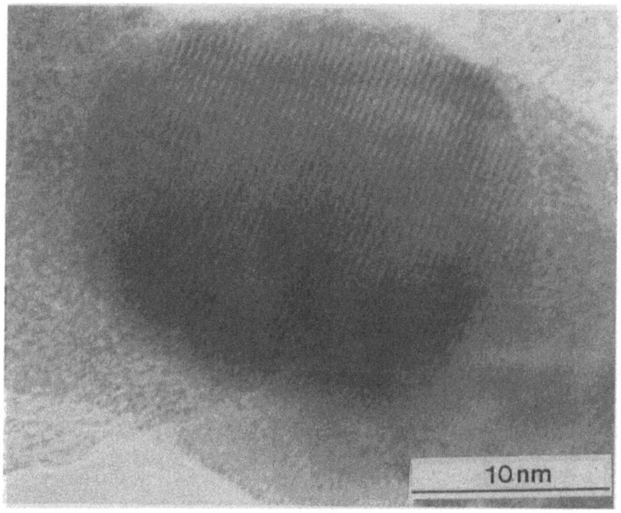

Fig. 6. Moiré pattern from overlapping particles

Conclusion

High resolution microscopy is an efficient way to learn more about particle formation and the martensitic transformation, which is extremely complicated even on a microscale. The further work along this direction should also include an investigation of the precise direction relationships between the various constituencies.

References

1. Granqvist, C.G , Buhrman, R.A.: J. Appl Phys. **47**, 2200 (1976)
2. Niklasson, G.A., Torebring, A., Larsson, C., Granqvist, C.G., Farestam, T.: Phys. Rev. Lett. (submitted)
3. Thölén, A.R.: Surf. Sci. **106**, 70 (1981)
4. Thölén, A.R.: Philos. Mag. A, **53**, 259 (1986)

Z. Phys. D – Atoms, Molecules and Clusters 12, 127–129 (1989)

Atoms, Molecules and Clusters
Zeitschrift für Physik D
© Springer-Verlag 1989

Fine particles of icosahedral quasicrystal formed by gas evaporation technique

Y. Saito[1], T. Maki[1], T. Okazaki[2], and K. Mihama[1]

[1] Department of Applied Physics, Nagoya University, Nagoya 464-01, Japan
[2] Department of Physics, Meijo University, Nagoya 468, Japan

Received 5 July 1988

Al-T (T = Mn, V, Cr) particles in the icosahedral phase have been prepared by evaporation and condensation of metals in an inactive gas. The particles were spherical in shape, with diameters ranging from 50 to 300 nm, and showed speckled contrast in bright and dark field images. Each particle consisted of fine grains of several nm with small misorientations (less than a few degrees). Patterson functions of the icosahedral Al-Mn were also calculated from electron diffraction data. Among several structure models examined in the present study, a model with an array of two-shell Mackay icosahedra gave the best fit with the Patterson functions.

PACS: 81.20; 61.14

1. Introduction

Since the discovery of Al-Mn quasicrystals (QC's) in melt-spun ribbon samples [1], various methods such as ion-beam-assisted thermal processes [2], annealing of amorphous alloys [3], electro-hydro-dynamic atomization [4], and vacuum deposition [5] have been found to form QC's. The advent of new technique for producing QC's sheds light on growth mechanism and stability of this phase. Another principal problem in the study of the QC is the determination of its structure, a prescription for the location of atoms. In this paper we describe the production and characterization of the icosahedral Al-T (T = Mn, V, Cr) QC's in a form of fine particles, which are prepared by gas evaporation technique [6]. For structure analysis of the icosahedral QC, we derived Patterson functions from electron diffraction data.

2. Preparation method

Al-based alloy particles were produced by evaporation of a piece of Al alloy from a single source or by simultaneous evaporation of two metals from two sources in an atmosphere of low pressure Ar gas (4–27 kPa). The latter method of evaporation (double source) is superior to the former (single source) in respect of the control of vapor pressure of the metals. Tungsten boats were used as evaporation sources. For the double source method, evaporation temperatures were as follows; 1800–2200 °C for Al, 1340–1700 °C for Mn, around 2500 °C for V, and around 1840 °C for Cr. Detailed condition of evaporation has already been described elsewhere [7].

The vapor of metals flows outward from their sources by diffusion through the inactive gas. Since the heating of the sources brings about a convective flow of the gas, the metal vapor is exposed to this gas flow and cooled rapidly. A supersaturation sufficient for nucleation is attained and nuclei of the particles are formed. From temperature gradient and flow velocity around the source, the cooling rate of the particles is estimated to be approximately 10^5 °C/s at the solidification temperature of the alloy particles [7].

3. Diagnostics by electron microscopy

Figure 1 shows electron micrographs of an Al-Mn particle in the icosahedral phase taken along 2-fold

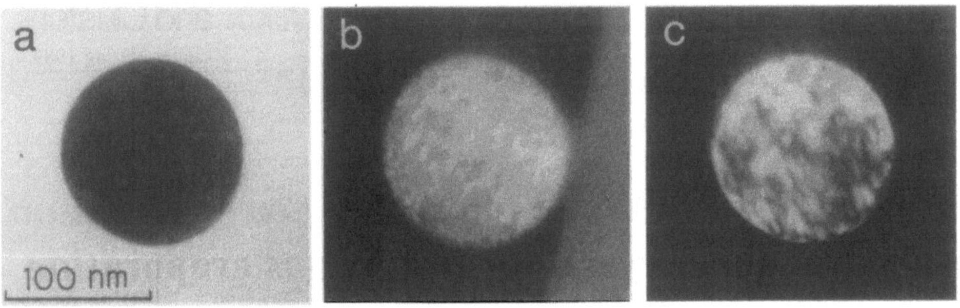

Fig. 1a–c. Electron micrographs of an icosahedral Al-Mn particle taken along 2-fold axis. (**a**) Bright field, (**b**) and (**c**) dark field images taken with reflections B and C shown in Fig. 2(a), respectively

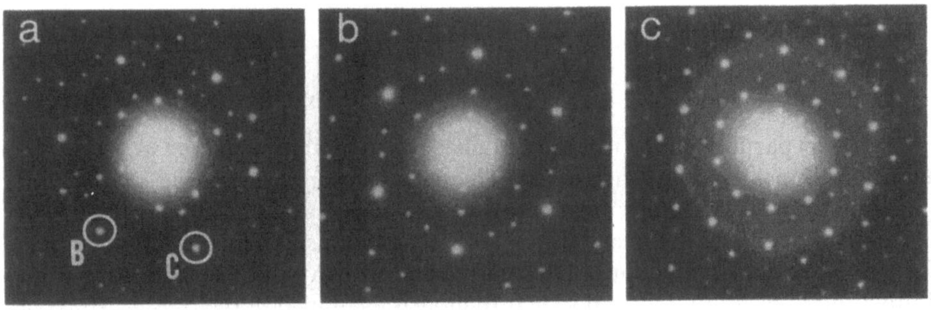

Fig. 2a–c. Electron diffraction patterns taken from the icosahedral Al-Mn particle shown in (**a**) 2-fold, (**b**) 3-fold, and (**c**) 5-fold axes

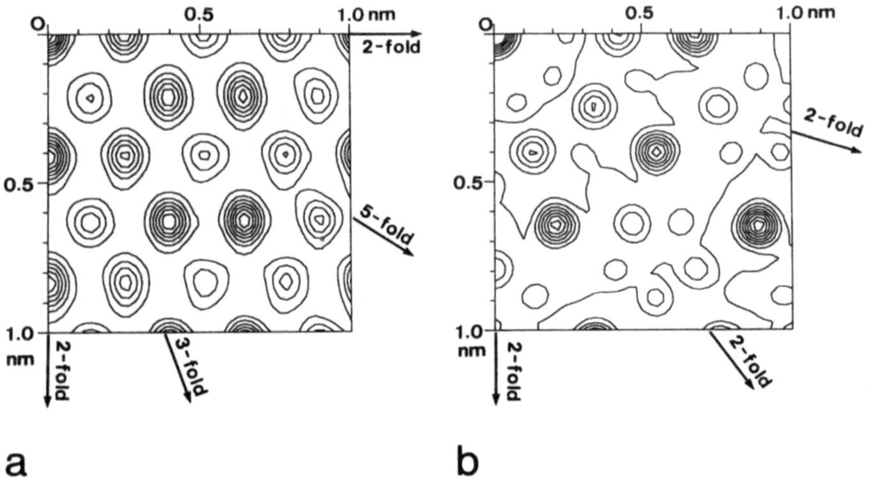

Fig. 3a and b. Patterson functions projected along (**a**) 2-fold and (**b**) 5-fold axes

symmetry axis. Electron diffraction patterns perpendicular to 2-fold, 3-fold and 5-fold axes taken from this particle are shown in Figs. 2(a), (b) and (c), respectively. Dark field images in Figs. 1(b) and (c) were respectively taken with reflections, B and C indicated in Fig. 2(a). The dark field images do not show uniform contrast but granular one, showing that this particle consists of fine grains with sizes of several nm. Some QC particles showed splitting of diffraction spots, corresponding to misorientation of a few degrees [8].

Morphology and texture of the QC particles mentioned above are commonly found among Al-Mn, Al-V and Al-Cr systems. However, diffraction patterns from the three alloy systems showed slightly different features with each other. The details of this matter will be described elsewhere [8].

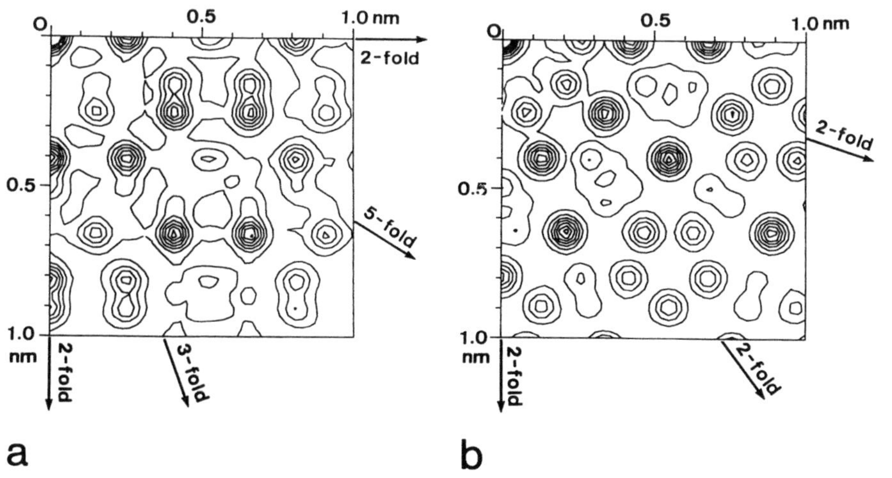

Fig 4a and b. Interatomic vector functions of a structure model in which the Mackay icosahedra are located on the icosahedral sites (see text). Projected along (**a**) 2-fold and (**b**) 5-fold axes

4. Patterson synthesis

The QC particles have a character of "mosaic": A single particle consists of fine grains with a size of several nm, and each grains are misoriented each other within a few degrees. Intensity of diffracted beams from such a specimen is dominantly governed by kinematical diffraction. Therefore, we postulated that the intensity of reflections is proportional to the square of structure factors. Patterson functions were obtained by Fourier transform of the squared structure factors which were derived from electron diffraction intensity.

Figure 3a and b show Patterson functions projected along 2-fold and 5-fold axes, respectively. In the calculation of the functions, reflections up to $Q = 6.1$ Å^{-1} were included. It has already been shown that atoms with strong scattering power, i.e., Mn, are located at vertices of the 3-dimensional Penrose lattice [9]. Concerning positions of Al atoms, following models were considered: *(1)* edge center, *(2)* face center, *(3)* body center, *(4)* 2-site on long-face-diagonal [10], and *(5)* Mackay icosahedra [11]. Among these models, the last one, in which Mackay icosahedra consisting of 12 Mn and 42 Al atoms are located at lattice points with local icosahedral symmetry and between them are "glue" atoms, reproduced best the Patterson functions, though the quantitative agreement was poor (Fig. 4). Recently, this model was shown to explain X-ray and neutron diffraction intensity, chemical composition, and density of icosahedral Al-Mn [12].

References

1. Schechtman, D., Blech, I., Gratias, D., Cahn, J.W.: Phys. Rev. Lett. **53**, 1951 (1984)
2. Lilienfeld, D.A., Nastasi, M., Johnson, H.H., Ast, D.G., Mayer, J.W.: Phys. Rev. Lett. **55**, 1587 (1985)
3. Poon, S.J., Drehman, A.J., Lawless, K.R.: Phys. Rev. Lett. **55**, 2324 (1985)
4. Bendersky, L.A., Ridder, S.D.: J. Mater. Res. **1**, 405 (1986)
5. Follstaedt, D., Knapp, J.A.: Phys. Rev. Lett. **56**, 1827 (1986)
6. Uyeda, R.: J. Cryst. Growth **45**, 485 (1978)
7. Saito, Y., Mihama, K.: In: Quasicrystals. Kuo, K.H. (ed.), p. 601. Aedermannsdorft, Switzerland: Trans. Tech. Publications 1987
8. Okazaki, T., Maki, T., Saito, Y., Mihama, K.: Jpn. J. Appl. Phys. **28**, 195 (1989)
9. Saito, Y.: Jpn. J. Appl. Phys. **25**, L693 (1986)
10. Hiraga, K, Hirabayashi, M.: In: Proc. 1st Intern. Symp. for Science on Form, Tsukuba, 1986, p. 621.
11. Elser, V., Henley, C.L.: Phys. Rev. Lett. **55**, 2883 (1985)
12. Yamamoto, A., Hiraga, K.: Phys. Rev. B **37**, 6207 (1988)

Z. Phys. D – Atoms, Molecules and Clusters 12, 131–134 (1989)

Atoms, Molecules
Zeitschrift für Physik D and Clusters
© Springer-Verlag 1989

Lattice imaging of bismuth clusters

M. Treilleux, G. Fuchs, F. Santos Aires, P. Melinon, A. Hoareau, and B. Cabaud
Département de Physique des Matériaux (U.A. CNRS 172), Université Claude Bernard-Lyon I,
Boulevard du 11 Novembre 1918, F-69622 Villeurbanne Cédex, France

Received 5 July 1988; final version 30 September 1988

Small bismuth clusters with narrow size distribution ($2 \rightarrow 10$ atoms) were generated in an inert gas condensation source and vacuum deposited, at room temperature, on amorphous carbon. The deposit has been characterized by Transmission Electron Microscopy (TEM) which has shown a growth process involving the formation of large particles: diameter from 1.5 to 20 nm. Axial illumination lattice imaging with a high resolution electron microscope has been performed. Particles of about 20 nm diameter have been formed with a single crystalline core edged by an amorphous shell. Cluster images in the $[\bar{2}20\bar{1}]$ direction are presented in detail and compared with simulated images of bismuth single crystal.

PACS: 81.15.G; 61.00

1. Introduction

The early stages of thin film growth on amorphous carbon or single crystal (NaCl, KBr) has been extensively studied during these last years. Most of results deal with the atomic deposition of noble metals (Ag, Au, Pt or Pd) [1–4] for which the knowledge of the formation (nucleation and growth) of the deposit has been specified. In order to study the nucleation and growth kinetics, the substrates are used to be heated after, or during the evaporation. The cluster distribution deposited on the substrate has been analysed as a function of many parameters such as the temperature of the substrate, the deposition rate and the waiting time after evaporation.

The internal and surface structure of these small metal particles have also been characterized in order to precise their role not only in nucleation and thin metal film growth but also in catalytic processes [5–8]. The recent technical developments of cluster deposition have induced a revival of interest in this field of study. More, there is a technological interest to study the cluster beam deposition since some experimental results have shown that such a technology should allow the deposition of high quality films on a wide variety of substrate surface [9].

The purpose of this paper is to present the first results obtained by Transmission Electron Microscopy (TEM) on bismuth cluster deposition on amorphous carbon films. The production of clusters by the inert gas condensation technique as well as their deposition will be discussed in detail in another paper presented at this conference [10]. The present work is the first part of the study of deposited bismuth clusters using High Resolution Transmission Electron Microscopy (HRTEM). The analysis of the deposited metal aggregates at an atomic resolution shows that the lattice images of the bismuth particles present the same contrast as bismuth single crystal ones. Simulated HRTEM images confirm this result. Among the numerous directions in which HRTEM images have been recorded, the $[\bar{2}20\bar{1}]$ orientation is presented in detail.

2. Experimental procedure

Small bismuth clusters were generated in an inert gas condensation source (helium pressure $\simeq 450$ Pa) and deposited at room temperature, on 5 nm thickness amorphous carbon films hold on copper microscopy grids. The pressure during the evaporation was 1×10^{-4} Pa. The size distribution of clusters has been analysed by time of flight mass spectrometry. The deposition rate was $0.1\,nm \cdot s^{-1}$ and the equivalent thickness of deposit measured on a crystal quartz located near the grids was 1 nm. After a transfer into air, the deposit was observed at 200 kV with a JEOL 200 CX high resolution electron microscope equipped with a LaB_6 single crystal filament, a high resolution pole piece and a top entry stage. The specifications given by the manufacturer are: spherical aberration coefficient $C_S = 0.8$ mm, chromatic aberration coefficient $C_C = 1.4$ mm. The focus spread due to chromatic aberration was $\Delta f = 10$ nm. The beam divergence defined by half critical angle θ_C previously measured from diffraction patterns was found about 1 mrad. Values of defocus were determined both from a precalibration of the objective lens potentiometers and from optical diffractograms of micrographs [11]. The spherical aberration coefficient was measured from a precalibration of the objective lens excitation.

For bismuth aggregates oriented in the $[\bar{2}20\bar{1}]$ direction parallel to the incident beam, experimental micrographs performed with all diffracted beams have been compared with HRTEM images calculated using the SHRLI multislice program [12]. For this zone axis the maximum value in reciprocal space (DSMAX) was taken as $50\,nm^{-1}$. Consequently, the number of beams (NBEAM) used in the multislice iteration was 250. It was checked that intensity and extinction distance of the main beams were not modified for higher value of DSMAX and NBEAM. Absorption effects were neglected in our calculations and Debye Waller factors were fixed at $0.01\,nm^2$. Simulated HRTEM images have been calculated on an IBM-PS 8580-071 allowing a sampling of the slices of up to 128 × 128 points.

3. TEM and HRTEM of bismuth clusters

3.1. T.E.M. results

The micrograph and electron diffraction pattern of the room temperature deposit of bismuth clusters on the amorphous carbon grid is presented on Fig. 1. The diffraction pattern exhibits rings which have been indexed with the usual rhombohedral bismuth structure. A diffuse ring mainly due to the amorphous carbon film can also be noticed. The size and the shape of observed aggregates vary on a large range. The diameter of the deposited particles ranges from 1.5 to 20 nm. Some aggregates (arrowed on Fig. 1) show a dark core surrounded by an uniform lighter contrast.

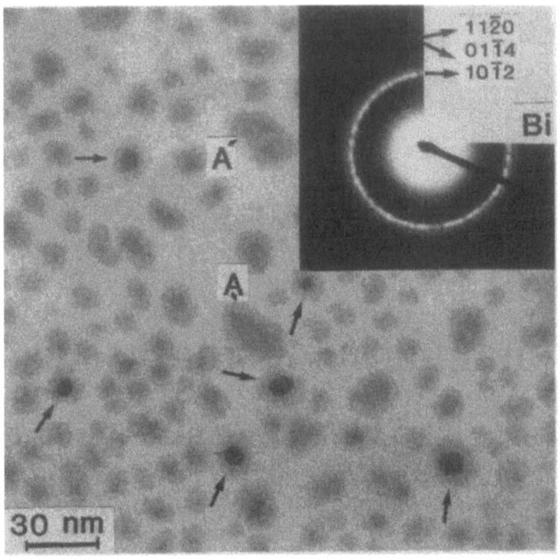

Fig. 1. 1 nm mean thickness bismuth deposit produced by cluster beam deposition on amorphous carbon film and corresponding electron diffraction. Arrowed particles show a single crystalline core surrounded by an amorphous material. Elongated particles (A) probably are the result of coalescence of two particles

The dark contrast corresponds to a well crystallized particle in diffraction condition meanwhile the surrounded zone seems to be amorphous.

As it was previously reported on other systems [13, 14] transmission electron microscopy observations reveal that the size of clusters in the beam (during evaporation) and on the grid (after deposition) is quite different. The size of clusters analysed by time of flight mass spectrometry ranges from 2 to 10 atoms. It has to be compared to the size of deposited particles: 1.5 to 20 nm corresponding to 50 and 10^5 atoms respectively. Nevertheless it has to be underlined that the coalescence of particles on the carbon surface can also be observed for larger ones. During observation, the increase of temperature induced by electron beam has made coalesce some quite spherical aggregates into elongated ones as the one labelled (A) on Fig. 1.

3.2. H.R.T.E.M. observations

Among the bismuth particles presenting a dark core, those which was in favorable orientation conditions have been selected in order to perform lattice images. Supported bismuth aggregates have been imaged with atomic resolution in various zone axis ($[\bar{2}20\bar{1}]$, [0001], $[\bar{4}401]$, $[01\bar{1}0]$). However the most frequent observed orientation was the $[\bar{2}20\bar{1}]$ one. Typical images of aggregates oriented in this direction are presented on Fig. 2a, b, c, for various magnification and defocus values. The two sets of planes ($(01\bar{1}2)$ and $(10\bar{1}2)$) corresponding to 0.328 nm lattice spacing can be clearly seen (Fig. 2b, c). The lattice spacing as well as

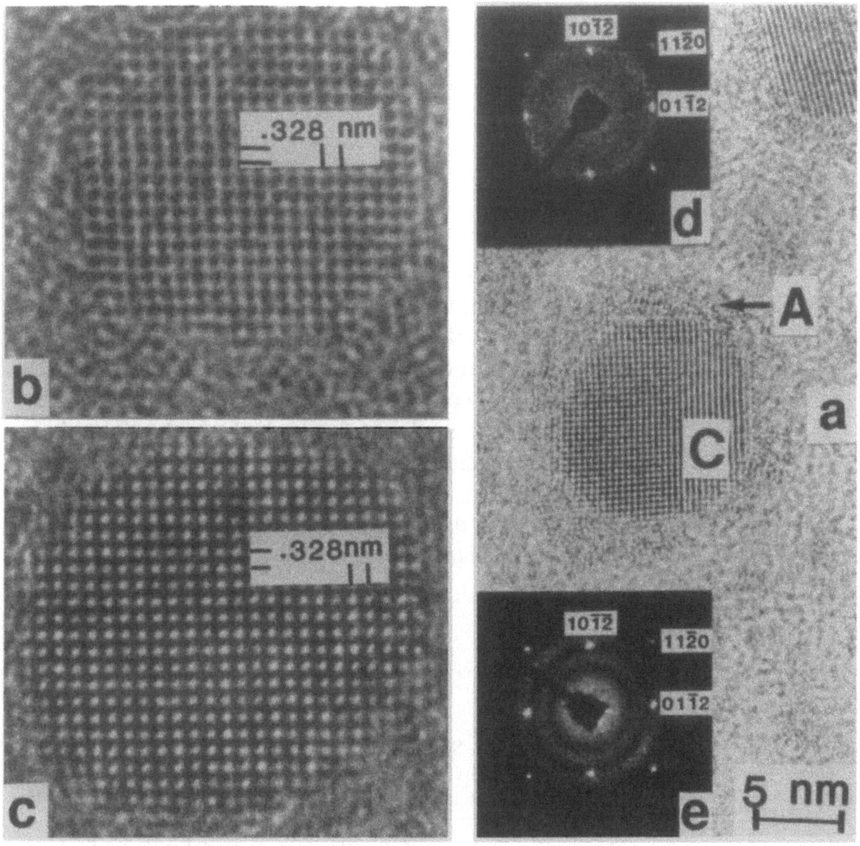

Fig. 2a–e. HRTEM images of bismuth particles imaged along the [$\bar{2}20\bar{1}$] axis: **a)** single crystalline core *C* and amorphous shell *A*: **b)** Bi particle imaged at $\Delta f = -45$ nm; **c)** Bi particle imaged at $\Delta f = -115$ nm. **d)** and **e)** respectively, optical microdiffraction corresponding to b) and c) images

the angle between these planes (87°6) are in agreement with the bismuth rhombohedral structure. These particles of about 7 nm diameter are edged by (01$\bar{1}$2), (10$\bar{1}\bar{2}$), (1$\bar{1}$0$\bar{4}$) and (11$\bar{2}$0) planes. In contrast with the well-known multiply twinned particles of silver or gold (MTPs) [6, 7], the contrast of deposited bismuth particles presented in Fig. 2 is quite constant from one edge of the crystallized core to the other. The perfect set of ordered dark spots (Fig. 2b) or white spots (Fig. 2c) has been compared to simulated images of bismuth single crystals in order to precise the range of the cluster thickness.

In order to simulate the apparent contrast inversion between micrographs of Fig. 2b. and c., for the two values of defocus measured from the optical diffractograms, series of simulated images were performed for various thicknesses with a stretched contrast (Fig. 3a.). It is shown that for thicknesses ranged from 1 to 9.9 nm the black dots observed on Fig. 3a. reproduce both symmetry and location of the bismuth atomic columns (Fig. 3b.). These simulations lead to notice that the set of ordered dark spots (Fig. 2b.) seems to correspond to simulated images in the thickness range between 3 and 5 nm. Then the dark spots (Fig. 2b.) reproduce the location of the bismuth atomic columns. In the case of the image recorded at $\Delta f = -115$ nm (Fig. 2c.), the best match is obtained for a comparable thickness range (from 2 to 5 nm). Although the evolution of the aggregate thickness cannot be determined from one edge of the cluster to the other, the mean thickness has been estimated to 4 nm, value which roughly correspond to the half diameter one.

As expected from TEM images, HRTEM observations of particles show a single crystalline core, with atomic resolution, surrounded by a disordered, probably amorphous, region (Fig. 2a.). It is supposed that the shell surrounding the Bi core is the result of the chemical reaction of bismuth with oxygen during or after the deposition. Taking into account the thickness of the amorphous shell (2 to 5 nm), it is expected that very small bismuth particles could be entirely oxidized which can explain the impossibility to observe these small aggregates with atomic resolution. More, on larger particles, single crystalline core of diameter

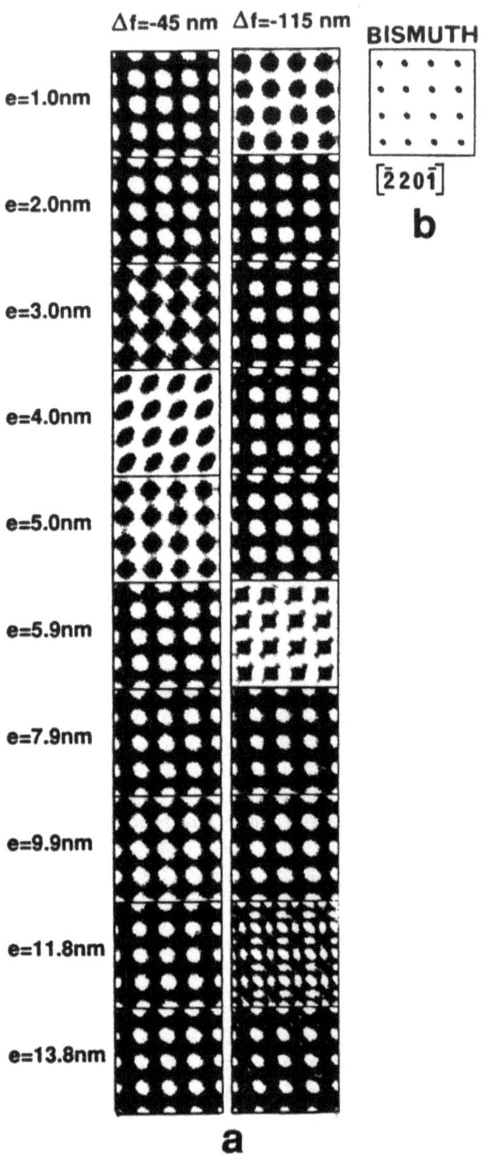

Fig. 3a and b. Simulated images of bismuth single crystal viewed along the [$\bar{2}20\bar{1}$] direction, as a function of the thickness for the two experimental defocus values of figure 2b, c. Computed images were calculated with a stretched contrast **(a)**: **(b)** projected potential of the structure

lower than 5 nm are never observed. It is probably this amorphous shell and the carbon film which prevent this observation.

4. Conclusion

The main feature of the present observations is the manner in which very small Bi clusters deposited on amorphous carbon agglomerate to form larger particles. TEM and HRTEM have revealed that these particles are composed of a single crystalline core surrounded by an amorphous shell which probably is the result of chemical reactions. HRTEM images of particles have been suitably fitted with simulated images of bismuth single crystal confirming the usual structure of bismuth aggregates.

References

1. Takeuchi, K., Shinohara, G., Ishii, K., Kinosita, K.: Thin Solid Films **32**, 191 (1976)
2. Henry, C.R., Chapon, C., Mutaftschiev, B.: Thin Solid Films **46**, 157 (1977)
3. Chapon, C., Henry, C.R., Chemam, A.: Surf. Sci. **162**, 747 (1985)
4. Gillet, M.: J. Microsc. Spectrosc. Electron. **11**, 449 (1986)
5. Smith, D.J., Marks, L.D.: Philos. Mag. A, **44-3**, 735 (1981)
6. Smith, D.J., Marks, L.D.: J. Cryst. Growth **54**, 433 (1981)
7. Marks, L.D.: Ultramicroscopy **18**, 445 (1985)
8. Wallenberg, L.R., Smith, D.J.: J. Mater. Sci. Lett. **5**, 1301 (1986)
9. Takagi, T.: Vacuum **36**, (1–3) 27 (1986)
10. Santos Aires, F., Treilleux, M., Fuchs, G., Hoareau, A., Melinon, P., Cabaud, B.: Z. Phys. D–Atoms, Molecules and Clusters **12** (1989) (this issue)
11. Spence, J.C.J.: Experimental high resolution electron microscopy. New York: Oxford 1980
12. O'Keeff, M.A., Buseck, P.R.: Trans. Ann. Crystallogr. Assoc. **15**, 27 (1979)
13. Urban, J., Tesche, B.: Thin Solid Films **137**, 325 (1986)
14. Sattler, K.: Physics and chemistry of small clusters Jena, P., Rao, B.K., Khanna, (eds.), Vol. 713, New York· Plenum Press 1987

Z. Phys. D – Atoms, Molecules and Clusters 12, 135–138 (1989)

Atoms, Molecules
Zeitschrift für Physik D and Clusters
© Springer-Verlag 1989

Modes of deformation under varied tensile directions in laminar, colloidal gold

S. Maruyama
Osaka College, Hirao, Mihara, Minamikawachi, Osaka 587, Japan

Received 7 July 1988; final version 30 September 1988

In perfect crystals of laminar colloidal gold, free from dislocations and stacking faults, the modes of deformation have been studied systematically, as a function of resolved shear stress or normal stress for each mode, using electron microscope and diffraction. The main modes of plastic deformation are cleavage crack, single glide and mechanical twinning. The cleavage crack is normally found under the tensile direction which gives the maximum normal stress to the cleavage surface. There is, however, another case of cleavage crack caused as a result of the glide deformation non-parallel to the crystal surface. This is the mode of cleavage characteristic of thin crystals. The single glide along the direction parallel to the surface, and the mechanical twinning obey to the resolved shear stress criterion. The latter is in contrast with the mechanical twinning in bulk crystals where in general the criterion does not hold. The twinning shear demonstrated at the edge of the crystal is much larger than that estimated. High resolution lattice images reveal the consecutive occurrence of twinning and detwinning or simple glide in atomic scale. This is another example exclusively found in thin crystals.

PACS: 61.16.Di; 81.40.Lm

Introduction

Laminar micro-crystals of gold prepared under acidic condition have seldom grown-in dislocations [1, 2], nor stacking faults.

Their mechanical behaviour is quite different from that of bulk crystals. They deform elastically as far as 4% [3, 4], which is counted as the theoretical strength. When they exceed it over, plastic deformations occur; cleavage crack, single glide and twinning, creating probably new total or partial dislocations at the edges of the crystals [4].

It is the matter of interest whether the modes of deformation depend on the tensile direction with respect to the plane of deformation, and whether the resolved shear stress criterion holds in dislocation free crystals. It is also the matter of interest to survey in an atomic scale by high resolution electron microscopy, what happens in the mechanical twins which give anomalous twinning shears at the edges of the crystals [3, 4].

Experimental

The micro-crystals, about 10 nm in thickness and a few micron in lateral dimension, have been prepared from about 50 ml of 0.06% aqueous solution of auric chloride by reduction with 2 ml of saturated aqueous solution of salicylic acid after Suito and Uyeda [1, 2]. About one week after, a drop of the mixture solution is put on a Mylar sheet stretched by a tensile device. After having been dried, the device is immersed in water for one day to remove the precipitates of salicylic acid. After having been completely dried, the Mylar sheet is stretched by 20%, resulting in the deformation of the crystals on it through adhesion.

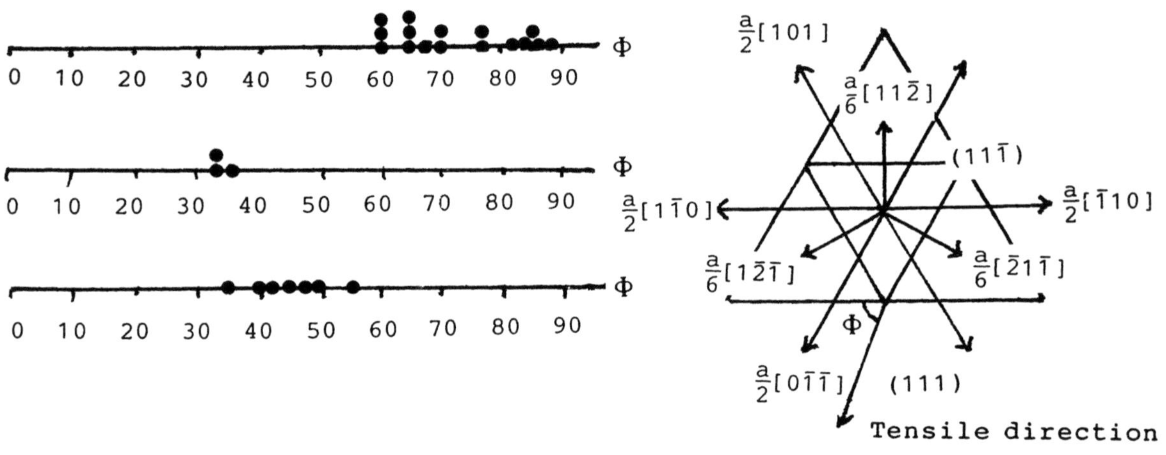

Fig. 1. Tensile direction and modes of deformation. The abscissa denotes the angle between the tensile direction and the trace of the deformation plane (11$\bar{1}$) on the crystal surface (111)

To know the tensile direction, platinum shadowing is sometimes applied along the direction of stretching. Then a drop of polyacrylic acid is put on the specimen. After solidification, the polyacrylic acid is stripped off from the Mylar sheet, together with the gold crystals. Then the surface is backed with carbon. The polyacrylic acid is dissolved in water, and the specimens are put on a grid covered with Formvar film, or on a micro-grid for high resolution images.

Result

The correlation of tensile direction with the mode of deformation is found as below. In Fig. 1, the angle between the tensile direction and the trace of each deformation plane on the crystal surface is plotted in each case. In Fig. 2, the resolved shear or normal stress in each deformation is plotted against the value ϕ.

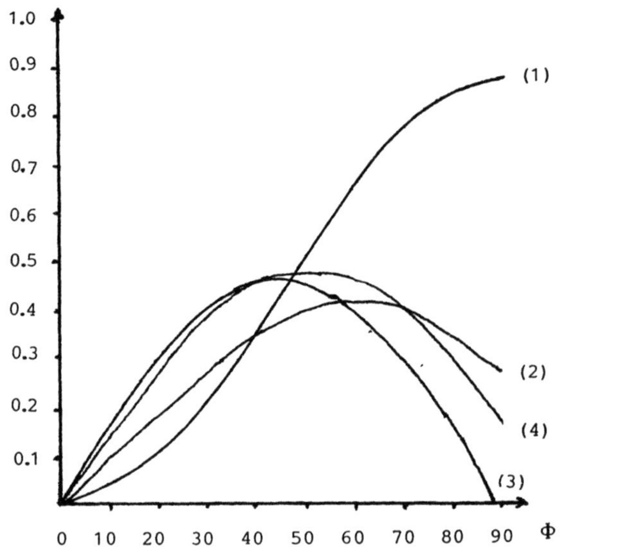

Fig. 2. Resolved normal stress (r.n.s.) or resolved shear stress (r.s.s.) on the deformation plane against ϕ. (1) R.n.s. on the cleavage surface. (2) R.s.s. on the glide plane in the glide direction $(a/2)$ $[0\bar{1}\bar{1}]$ non-parallel to the crystal surface. (3) R.s.s. on the glide plane in the glide direction $(a/2)$ $[1\bar{1}0]$ parallel to the crystal surface. (4) R.s.s. on the twin plane in the twinning shear $(a/6)$ $[1\bar{2}\bar{1}]$

1) Cleavage crack: There are two peaks; $\phi = 85°$ and $\phi = 60° \sim 70°$. The former is near the maximum normal stress on the cleavage surface, as is shown by the curve (1) in Fig. 2. This is the typical case as shown in Fig. 3. The latter ϕ gives the maximum glide shear stress in the direction $(a/2)$ $[0\bar{1}\bar{1}]$, shown by (2) in Fig. 2, which is non-parallel to the crystal surface. The cleavage crack occurs as a result of the glide as is seen by the displacements at the crystal edges (Fig. 4). This is the characteristic mode of deformation in thin films.

2) Single glide: Occasionally, single glide occurs up to 200 nm parallel to the crystal surface, as shown in Fig. 5. This is exclusively found around $\phi = 35°$ which is near the maximum value of resolved shear stress along $(a/2)$ $[1\bar{1}0]$, (3 in Fig. 2), parallel to the crystal surface. As the internal stress is enhanced enough [3, 4], the introduction of a glide dislocation at the crystal edge and the reflection at the opposite edge, like that of a longitudinal wave at a free edge, would operate instead of multiplication of the glide dislocation.

3) Mechanical twinning: The twins as seen in Fig. 6 are mostly found between $40° \sim 50°$, which include the

Fig. 3. Cleavage crack produced under normal stress on the cleavage plane. $\phi = 90°$. Tensile stress is along the shadowing

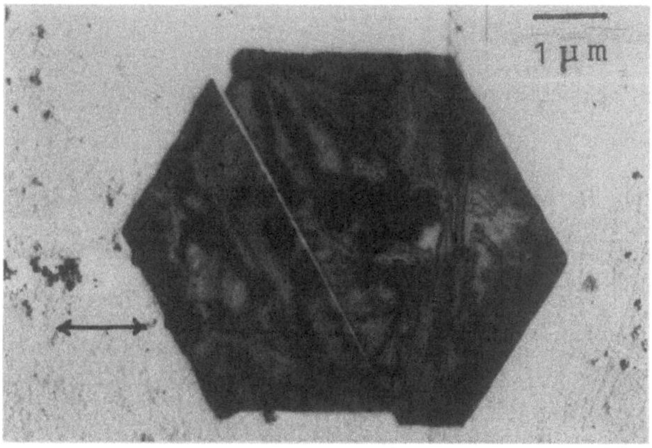

Fig. 4. Cleavage crack as a result of the glide non-parallel to the crystal surface. $\phi = 60°$. Note the displacements at the crystal edges

maximum resolved twinning shear $(a/6)$ $[1\bar{2}\bar{1}]$, (4 in Fig. 2). This is in contrast with bulk crystals, where generally the resolved shear stress criterion does not hold [5] except in Ag–Au alloys [6]. In bulk crystals, the theoretical mechanical twinning stress is much higher than the flow stress. Some stress concentration is necessary for the nucleation of mechanical twins. In the dislocation free crystals, however, the applied stress itself is enhanced up to the theoretical elastic limit 0.03 G, which is comparable to the theoretical twinning stress $G\gamma/30$ where $\gamma = 0.7$.

The twinning shear demonstrated at the edge of the crystal is much larger than that estimated. High resolution lattice images reveal the consecutive occurrence of twinning and detwinning caused by another twinning $(a/6)$ $[2\bar{1}1]$ in the original twin, in atomic scale

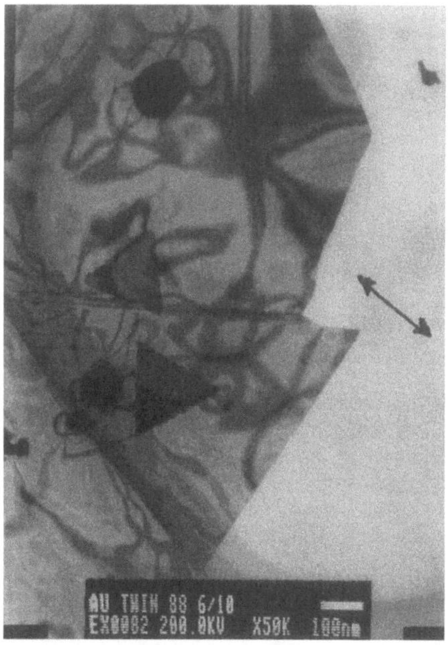

Fig. 5. Single glide parallel to the crystal surface. $\phi = 35°$. After that mechanical twinning occurred along the glide plane

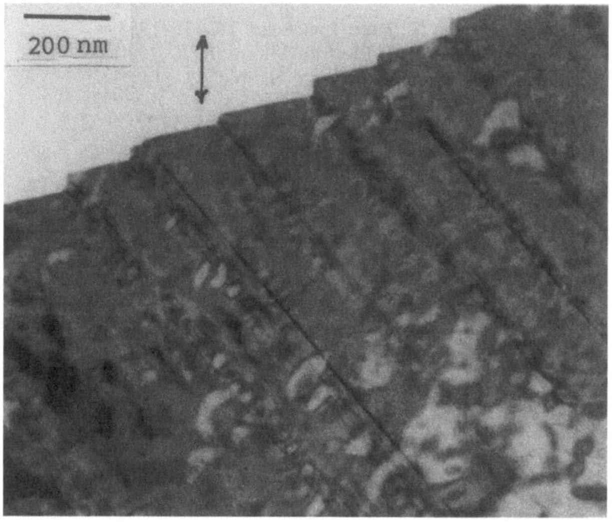

Fig. 6. Mechanical twinning at $\phi = 42°$. Note the large shear at the edge of the crystal

(Fig. 7). This is another example exclusively found in thin crystals.

Some part of this study was done at Osaka Women's University when the author was there. He expresses his cordial thanks to Prof. K. Yamamoto for kindly make use of JEM 100CX, and to the members of EM Application Lab. in JEOL for kindly make use of JEM 2000EX top entry for high resolution lattice images.

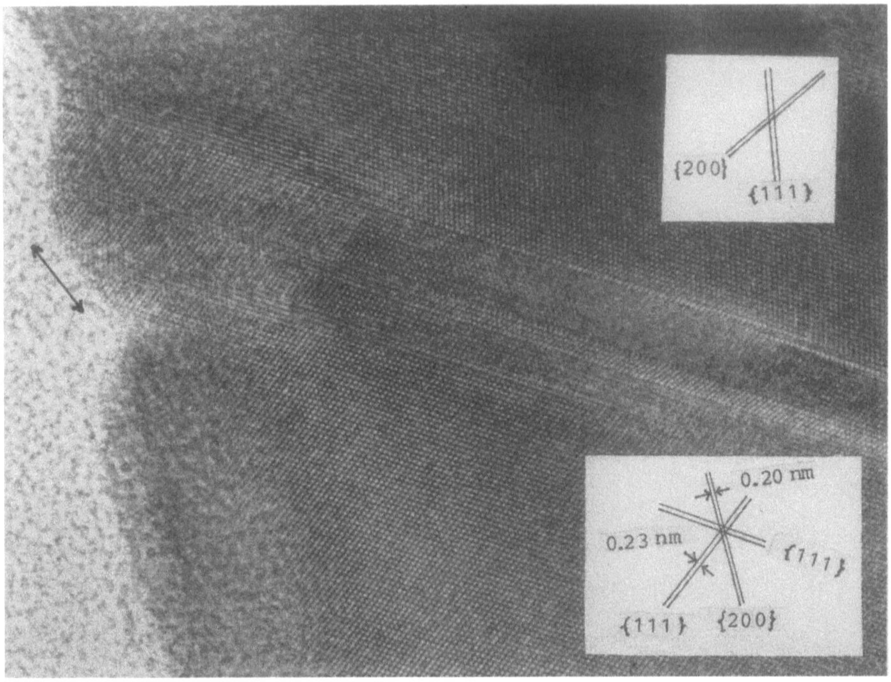

Fig. 7. High resolution lattice images of the twins and detwins caused probably by double twins in the original twins

References

1. Suito, E., Uyeda, N.: Proc. Jpn. Acad. **29**, 331 (1953)
2. Suito, E., Uyeda, N.: J. Electron Micros. **8**, 25 (1960)
3. Kiho, H., Maruyama, S., Uyeda, N.: ibid. **17**, 113 (1968)
4. Kiho, H., Maruyama, S., Uyeda, N.: Proc. 7th Int. Congr. Electronmicros., Grenoble, **2**, 567 (1970)
5. Cahn, R.W.: Deformation Twinning, pp. 10–13. New York, London, Paris: Gordon and Breach Science Publishers 1963
6. Suzuki, H., Barrett, C.S.: Acta Metall. **6**, 156 (1958)

Z. Phys. D – Atoms, Molecules and Clusters 12, 139–144 (1989)

Atoms, Molecules and Clusters
Zeitschrift für Physik D
© Springer-Verlag 1989

Atomic structural rearrangements induced by electron beam heating of coalesced palladium particles

A. Renou[1], J.M. Penisson[2], and M.F. Gillet[1]

[1] Laboratoire de Microscopie et Diffractions Electroniques, Faculté des Sciences et Techniques de Saint-Jérôme UA 797, F-13397 Marseille Cedex 13, France

[2] Service de Physique, Département de Recherche Fondamentale CEN, F-85X-38041 Grenoble Cedex, France

Received 9 July 1988; final version 30 September 1988

Change in the atomic surface structure of small coalesced palladium particles is shown by High Resolution Electron Microscopy imaging; these atomic surface restructuration events during observation are induced by the electron beam irradiation. They include atom evaporation from specific crystallite facets, surface diffusion of atoms which consecutively affects the particle morphology especially in the case of multiple-twinned particles and the growth of crystallite parts and resulting strains.

PACS: 61.16.Di; 61.70.Ng

Introduction

During the growth of small metallic particles onto an insulating substrate, several post nucleation phenomena take place; among them dynamic coalescence prevails, that is to say the migration of crystallites followed by a recrystallisation of these particles; this phenomenon generally leads either to monocrystalline particles with defects or elongated composite particles.

These particles are used as model catalyst when their size is in the range 1–4 nm. It is well known that catalytic properties of metallic particles chiefly depend on morphology and atomic surface structure, especially steps, corners and facet orientation [1]. This morphology can sensibly evolve at high temperature during catalytic phenomena [2] altering the nature of surface sites of particles.

Our aim is on the one hand to show by TEM that atomic rearrangements can occur on the surface of particles or at the interface between them induced by electron beam heating and on the other hand to investigate how to explain them.

Experimental procedure

The palladium particles were condensed inside an ultra high vacuum chamber at a pressure of less than 6×10^{-6} Pa onto insitu epitaxial MgO thin films evaporated from an electron beam source; the MgO films were grown by a double epitaxy phenomenon such as MgO/LiF/NaCl; further details on MgO substrate preparation are given elsewhere [3]. These samples look like flat monocrystals with rectangular outlines oriented $\langle 001 \rangle$, their size ranging within 25–250 nm. Palladium deposits were immediately evaporated from a Knudsen cell with a condensation rate of 1 Å/mn. The substrate temperature was in the range 293–693 K. For TEM studies, the samples were stripped from MgO substrate in water after backing them with carbon. Particle sizes were ranging between 1.5 and 30 nm. A JEOL-4000 EX HREM in the top entry configuration was equipped with an imaging amplification system so that the lattice structure could be directly viewed on a TV screen with a magnification ranging between 1,000,000 and 22,000,000. Electron beam current density was typically at the samples in the range 20–50 A/cm^2.

Particle morphology and structure

On a preceding paper [3] the structure and morphology of Pd particles condensed onto epitaxial thin films MgO substrates was reported in details. It was shown that the structure was chiefly depending on substrate temperature (Ts). When 573 < Ts < 673 K, among (001) monocrystalline fcc particles, a lot of particles with pentagonal and hexagonal outlines were observed, which have already been interpreted as arranged from fcc multiply-twinned tetrahedra with fivefold symmetry axis. The HREM 200 kV observation [4] has suggested that pentagonal particles have a decahedral structure with ⟨011⟩ or ⟨001⟩ orientation, whereas hexagonal particles have an icosahedral structure with ⟨112⟩ orientation and no local lattice strain but only small distortions in the fcc lattice of tetrahedra distributed homogeneously in the whole particle. More recently [5–6] a fine image analysis of these hexagonal particles observed in HREM 400 kV has indicated that deformations inside each tetrahedron are inhomogeneous and suggested they could be described as the presence of screw disclinations. In particular conditions, namely at high temperature, these fcc and multiply-twinned particles can form more or less long chains (Fig. 1). Their growth can be explained either by twinning onto one or several facets (labelled A and B) of an icosahedral particle such as viewed in Fig. 2, or by dynamic coalescence. In these composite particles, fivefold symmetry particles are imbedded and generally keep their own internal structure [3/7].

Surface atomic structures: direct observation of rearrangements of coalesced particles

HREM observations have been made on images of particles sequentially recorded within a period of thirty seconds. Typical example of atomic surface restructuration can be viewed in comparing Fig. 3a and 3b. Figure 3a shows a Pd aggregate with a size of 10 nm. The resolution of (111) and (200) lattice planes enables to distinguish two parts with different crystallographic structure. Part (1) is bi-lamellar twinned whereas part (2), observed elsewhere (4), shows a multiply twinned decahedral particle viewed along the ⟨011⟩ zone axis with (111) and (200) facets and twin planes accentuated by notches labelled A, B, C. Straight chains of black dots on the top surface represent the projection of rows of Pd atoms such as on (111) surfaces labelled D and E and (200) surfaces labelled F and G. Along the edges, the chains are often interrupted indicating step surfaces labelled D, E, F, G, for example.

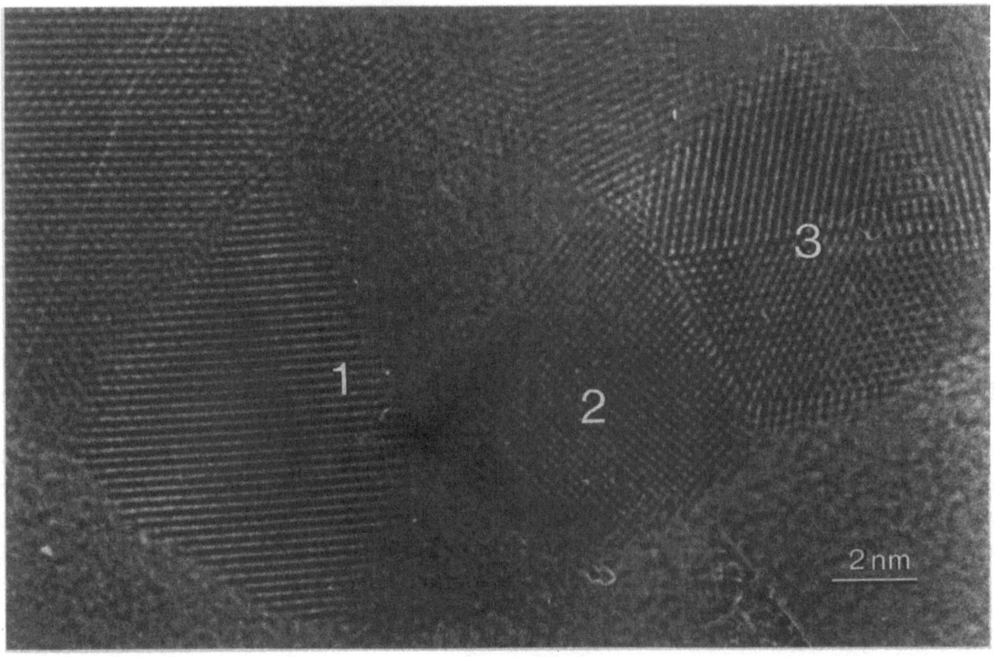

Fig. 1. Chain of coalesced Pd particles; the decahedra (1) and (3) are sharing the ⟨001⟩ oriented monocrystal (2).

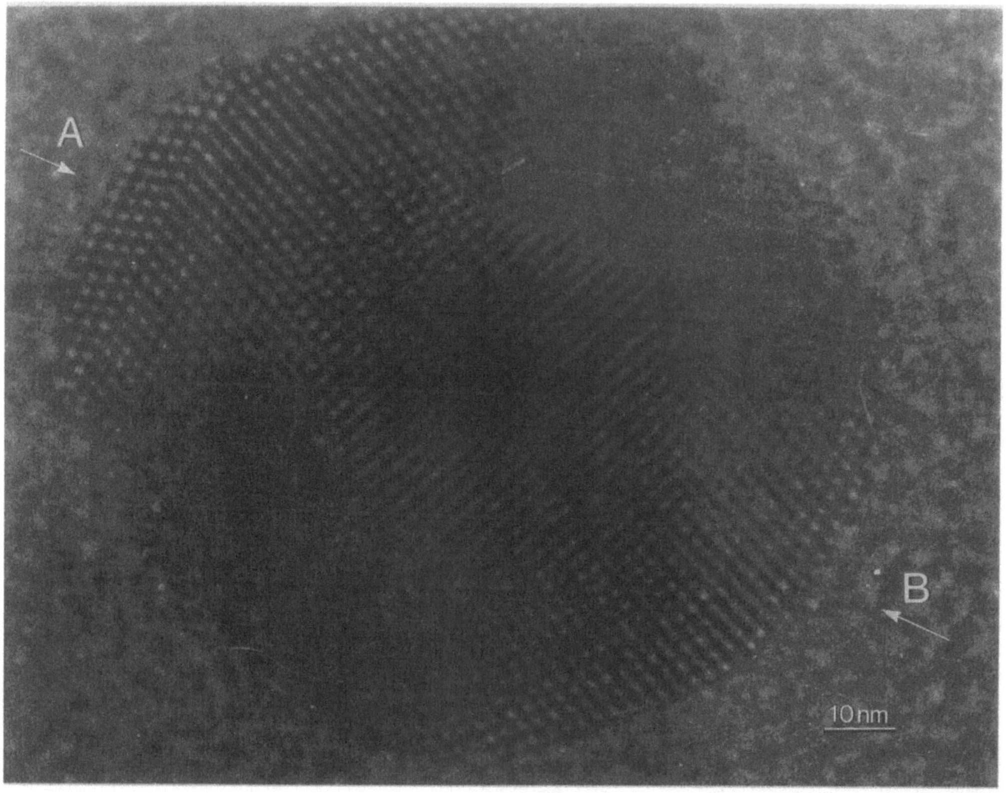

Fig. 2. Twinning on (111) facets A and B (*arrowed*) of ⟨112⟩ oriented icosahedral particle is commonly observed and appears to be part of the particle growth mechanism of Fig. 1

Within thirty seconds, a change in step structure produced by the electron beam current is revealed in Fig. 3b; for example on D and E (111) facets or F (200) facets, or in the notch B the first row of atoms has disappeared, whereas a growth of atomic columns occurs in the F facet and in the notch C. It should be noted that during observation internal structures of that coalesced particle has not been modified.

Interface displacement during observation of coalesced particles

A typical example of displacement of an interface (arrowed) between coalesced particles induced by heat irradiation is shown on time series Fig. 4. This figure shows a chain of coalesced particles. From their shape, and (111), (200), (220) lattice plane resolution, three parts can be distinguished: part (1) shows a Pd monocrystalline ⟨001⟩ cuboctahedral structure without internal defects whereas parts (2) and (3) correspond to multiply twinned decahedra in ⟨011⟩ orientation. Along the edges, steps can easily be noticed either on A, B (111) facets or on C, D (200) facets and near the notch E. The most important feature to be noticed in this chain is the edge-on displacement interface (arrowed) between parts (1) and (2). It should be noted that within 30 s of observation the thickness of crystalline tetrahedral unit of part (2) labelled FG decreases while the interface moves from (1) to (2) in a perpendicular direction, meaning that the (001) monocrystalline particle influences the growth and resulting reorientation of the decahedron part 2. In Fig. 5, which represents an enlarged Fig. 4d, it should be noticed the presence of dislocations and steps (arrowed) at the interface between 1 and 2 parts; they can be explained by the lattice misfit interface (15%) between (002) planes and (111) planes and resulting strains.

Discussion – conclusion

The recording of high resolution surface and interface profile images of Pd coalesced multiply twinned particles projected along ⟨011⟩ direction has revealed several interesting features: under the electron beam irradiation required to imaging *(i)* rearrangement of atom surface along the edges occurs; the disappear-

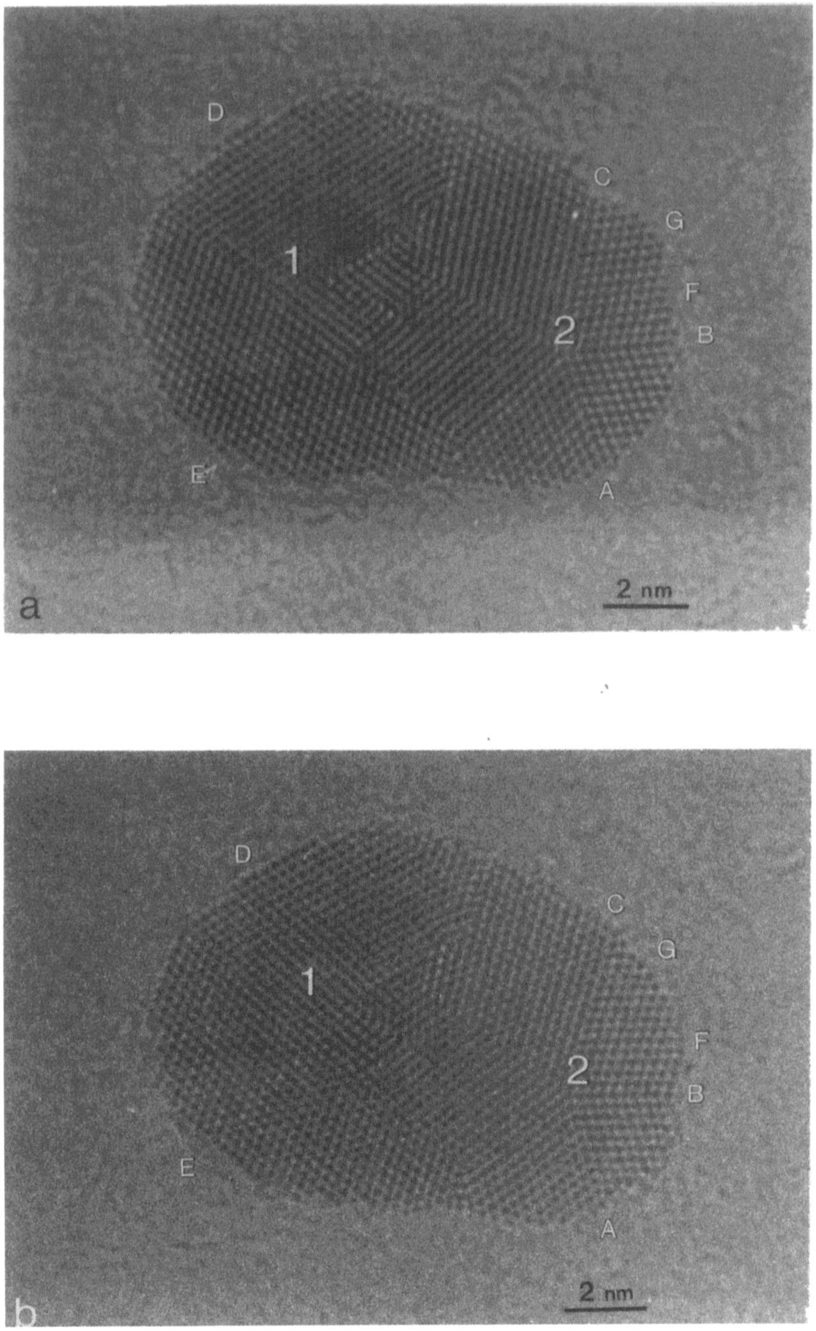

Fig. 3. a HREM image of a decahedral (part 2) and bilamellar twinned Pd particle (part 1). At the edges, steps are labelled D, E, F and notches A, B, C. **b** Same particle 30s later; changes in shape are affected by the e-beam irradiation; one row of atoms is missing on (111) facets D, E and notch B while additive rows appear on (001) facet F and notch C.

ance of Pd atomic columns chiefly from (111) edges (see in D, E Fig. 3b) with no change on the nearest facets suggests that vaporisation is taking place; however, it seems that atom columns can move from notches (see in B) and diffuse at the (111) surface to positions where they are accreted, for example in steps (see in F) or in the notch C. *(ii)* atoms columns can move at the interface (001)/(011) between two particles and dislocation and steps develop consequence of the lattice misfit between (001) and (111) lattice planes stimulated by the electron beam. The displacement of this interface during observation reveals how the coalescence growth mechanism proceeds; this growth is followed by a restructuration of the particle and strain formation. From these surface processes we can imagine why and how small particles as "model" catalyst continuously change their morphology and edge nature during catalysis experiments.

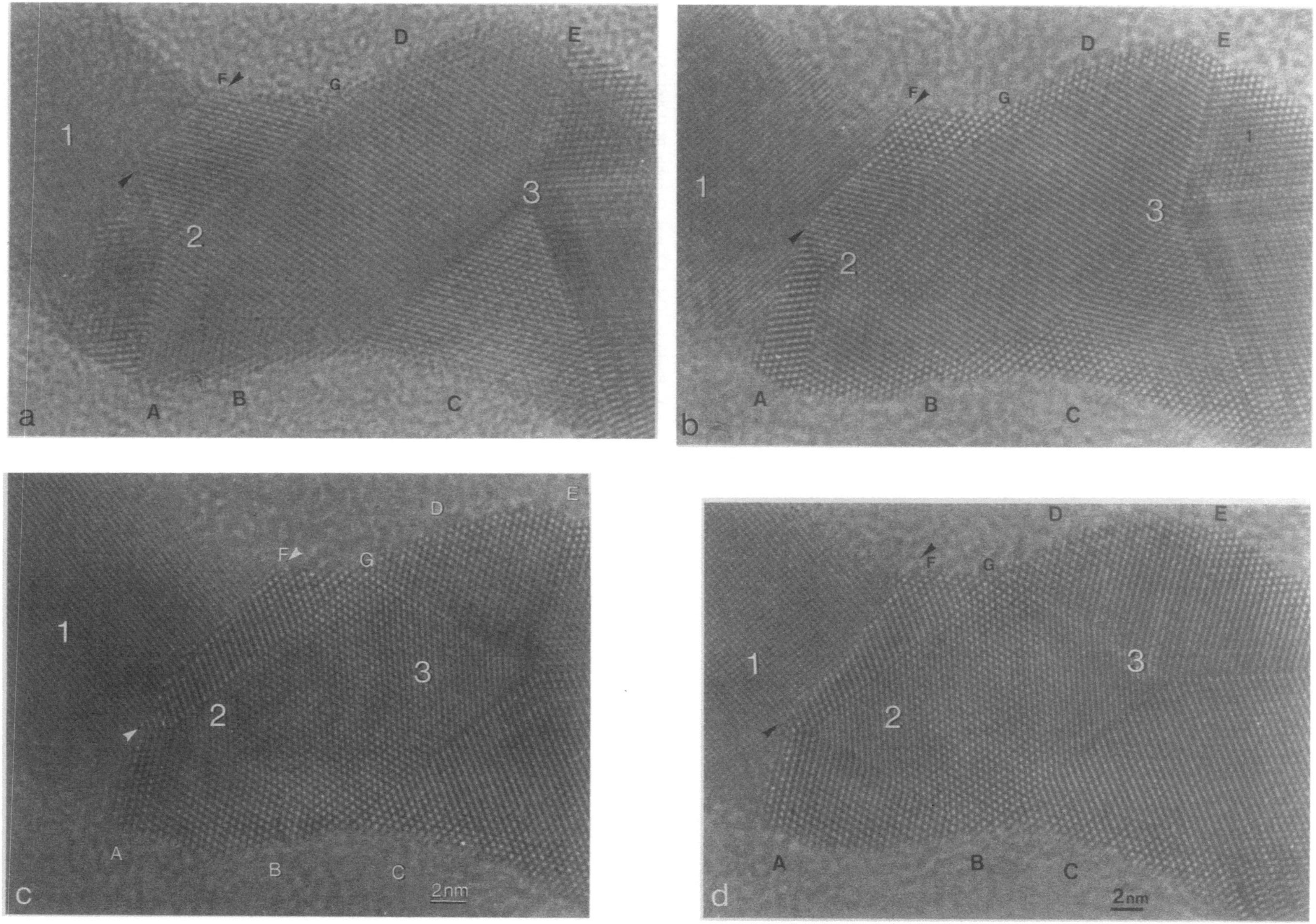

Fig.4a–d. Time series showing interface (001)/(011) displacement (see **a**, **b**, **c** and **d**) which occurs during particle growth induced by heat irradiation of the electron beam. On the sample note the thickness decreasing of the crystalline part labelled FG. Elapsed time of 30s between each photograph.

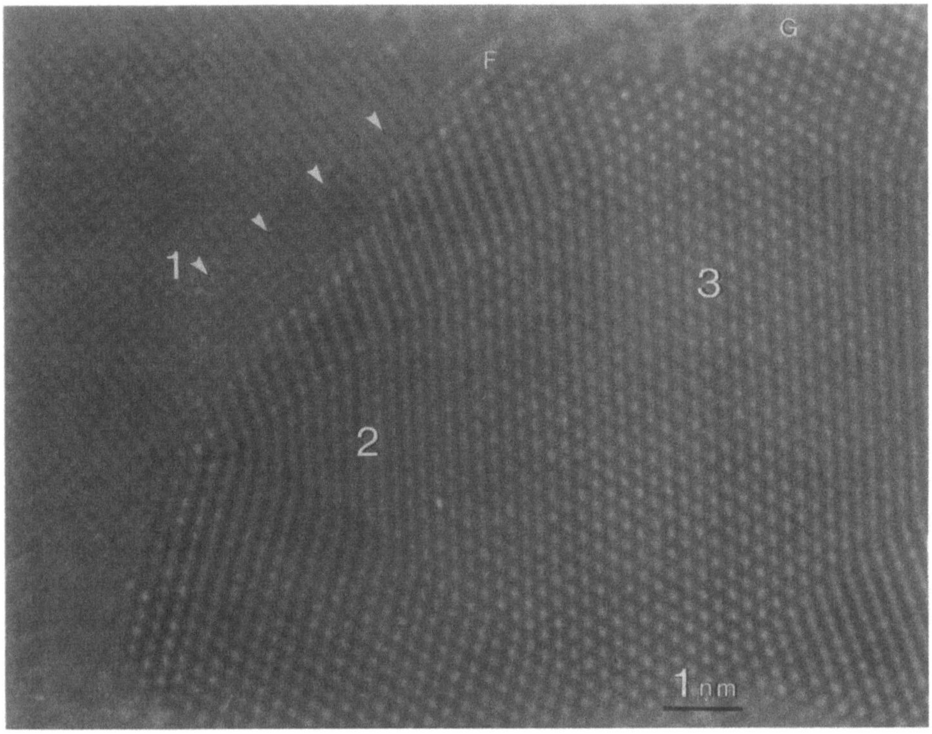

Fig. 5. Enlarged HREM image of the edge-on (001)/(011) interface of Fig. 4d showing misfit dislocations (arrowed) and steps

References

1. Gillet, E., Channakhone, S., Matolin, V.: Part II. J. Catal. **97**, 437 (1986)
2. Gillet, M.F., Channakhone, S.: Part I. J. Catal. **97**, 427 (1986)
3. Renou, A., Rudra, A.: Surf. Sci. **156**, 69 (1985)
4. Renou, A., Penisson, J.M.: J. Cryst. Growth **78**, (1986) 357
5. Penisson, J.M., Renou, A.: Z. Phys. D – Atoms, Molecules and Clusters 12 (1989) (this issue)
6. Penisson, J.M., Renou, A.: Inst. Phys. Conf. Ser. No. 93, Vol. 2, Chap. 8
7. Smith, D.T., Marks, L.D.: J. Cryst. Growth **54**, 433 (1981)

Z. Phys. D – Atoms, Molecules and Clusters 12, 145–148 (1989)

Atoms, Molecules
Zeitschrift für Physik D and Clusters
© Springer-Verlag 1989

Structure of small palladium particles studied by HRTEM

C.R. Henry[1], C. Chapon[1], J.M. Penisson[2], and G. Nihoul[3]

[1] CRMC2-CNRS, Campus de Luminy, case 913, F-13288 Marseille Cedex 9, France
[2] CENG-DRF 85X, F-38041 Grenoble Cedex, France
[3] Université de Toulon, Avenue de l'Université, F-83130 Lagarde, France

Received 7 July 1988

The structure of small palladium particles is investigated by HRTEM in order to know if structural variations, as a function of particles size, could explain some of the size effects encountered in heterogeneous catalysis. Small Pd particles are obtained by vacuum deposition on epitaxial MgO thin films. Imaging of the atomic planes down to a particle diameter of 1.8 nm shows that the particles are (100) (mainly) or (110) oriented on the substrate. Smallest particles (<2.5 nm) are perfectly fcc while some multiply-twinned particles are observed only for larger sizes.

PACS: 61.16.Di; 61.14.Hg

1. Introduction

Small catalyst particles present size effects in heterogeneous catalysis [1]. These effects have been often explained by a structural variation with the particle size. Especially a fivefold symmetry is advanced [2] because of the discovery of icosahedral (or decahedral) particles in rare gas [3] and supported metallic [4] clusters. It is noteworthy that only rare gas clusters have a true icosahedral structure, metallic particles are multi-twinned particles (MTP) composed of fcc tetrahedra. However the origin of the 5-fold MTP is believed to be a perfect 13-atomic icosahedral nucleus, as theoretical calculations have shown that this structure is more stable than the fcc one [5]. In this work the structure of Pd particles supported on MgO is investigated by HRTEM in the size range: 1.5–3.5 nm.

2. Experimental

MgO substrates are thin epitaxial films prepared in UHV following the method of Reichelt [6]. It consists in depositing on a freshly cleaved NaCl crystal kept at 300°C, 300 Å of LiF followed by 600 Å of MgO with a resistively heated carbon crucible and an electron beam evaporator respectively. The Pd particles are in-situ deposited with a calibrated Knudsen cell at a rate of 1.10^{13} atoms $cm^{-2}\,s^{-1}$ at substrate temperatures of 300 and 400°C. For HRTEM observation a thin (~200 Å) carbon layer is deposited in-situ on the particles. After removing the sample from the vacuum chamber the layered substrate: NaCl, LiF, MgO is dissolved in distilled water. Five hours, at least, are necessary to remove all the MgO. Finally the carbon layer containing the particles is mounted on TEM copper grids and observed in a JEOL 200 CX HRTEM.

3. Results

Figure 1 shows a diffraction pattern from a deposit composed of the LiF and MgO layers and the Pd particles. It is clear that all the 3 materials are (100) oriented like the NaCl substrate. Thus it is possible to obtain the lattice parameter of the Pd particles taking bulk values for the substrates. These measurements are indicative of an expansion of the Pd lattice as the size of the particles decreases. We obtain 1.4, 1.6 and 2.3% of dilatation for mean diameters of 5, 4 and 3 nm respectively. This result has already been pointed out

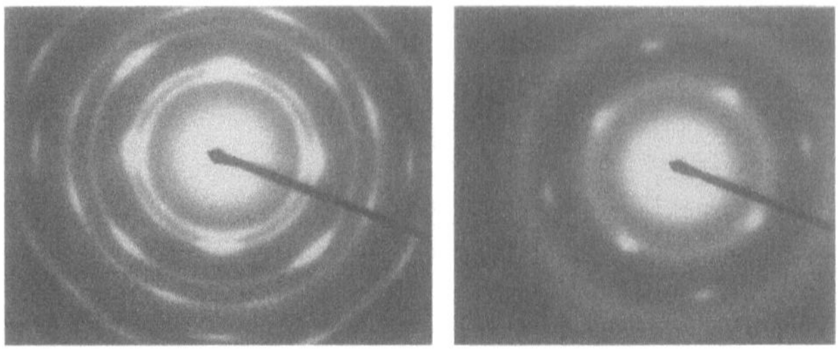

Fig. 1. Diffraction pattern of the Pd particles on the substrate Pd particles, MgO and LiF layers are in (100) epitaxy

Fig. 2. Diffraction patterns of Pd particles, without the substrate, deposited at 300°C (on the left) and at 400°C (on the right)

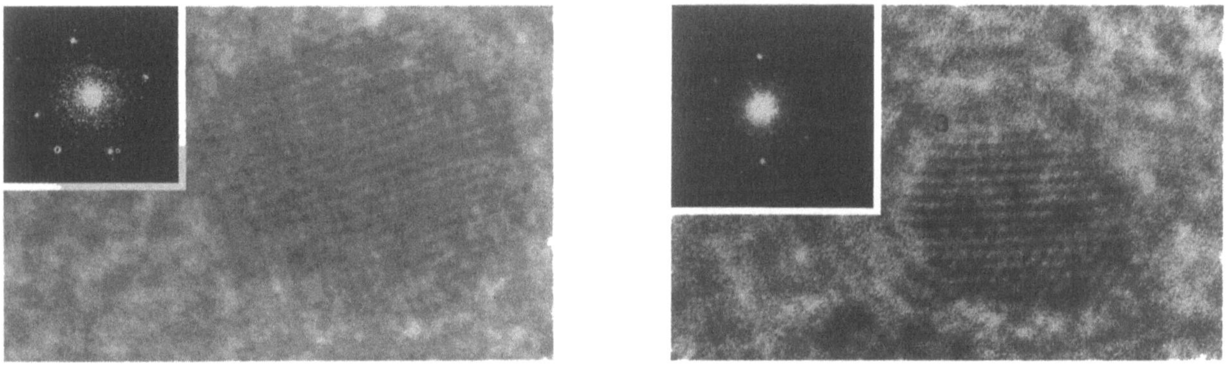

Fig. 3. HRTEM images of Pd particles (100) (on the left) and (110) oriented (on the right)

by several authors using various techniques [7–9]. The opposite trend is observed for the other metals.

The diffraction pattern for the particles supported on the carbon layer indicates that some particles are (110) oriented (see Fig. 2). Dark field observations confirm that only a small proportion of particles have this orientation. The epitaxy is better for particles deposited at 400°C than at 300°C and in the former case the proportion of (110) oriented particles is low (see Fig. 2).

These 2 orientations are confirmed by direct imaging of the atomic planes. (100) oriented particles show two perpendicular sets of (200) fringes and (110) particles of two sets of (111) fringes oriented at 70° each other as confirmed by optical diffractograms (see Fig. 3). Lattice planes fringes are clearly visible only from a diameter of 1.8 nm and sometimes for the smallest particles only one set of planes is visible. Two deposits prepared at 300°C and 400°C have been extensively examined in HRTEM. Figure 4 shows the size histograms of the two collections of particles. Particles deposited at higher temperature are larger and fewer. Some of these particles are presented on Fig. 5. Table 1 resumes the HRTEM observations for these two collections of Pd particles. It is noteworthy that for the 300°C deposit among the identified particles (except for 1% of simple twins) only perfect fcc particles are observed. The table seems to show that the number of (100) particles is only slightly larger than that of (110) particles, contrary to diffraction pattern and dark field observations. This is explained by the fact that (100) orientation is more difficult to image because the (200) lattice fringes are narrower than (111) one's . The higher proportion of unidentified particles for the 300°C deposit is due to the smaller size of the particles.

The lattice parameter cannot be measured with accuracy from the HRTEM images but it is possible to see distortions in the lattice. As an example Fig. 6 shows a (100) oriented particle of about 3 nm in diameter having a set of (200) planes expanded relatively to the other. Furthermore this expansion increases towards the edge of the particle. Such an anisotropy has been observed by HRTEM on Pd/TiO_2 [9] and by Moiré fringes on Al/MgO [10].

4. Discussion

The most important result of these observations is that on the 300°C deposit practically all particles have a perfect fcc structure while for the 400°C deposit 38% of the observed particles (or 55% of the identified ones)

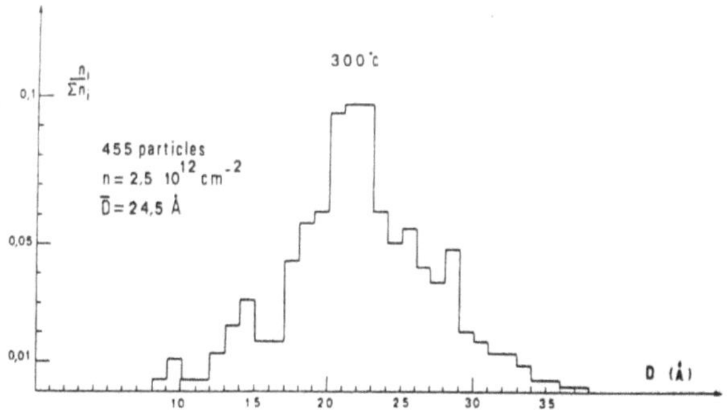

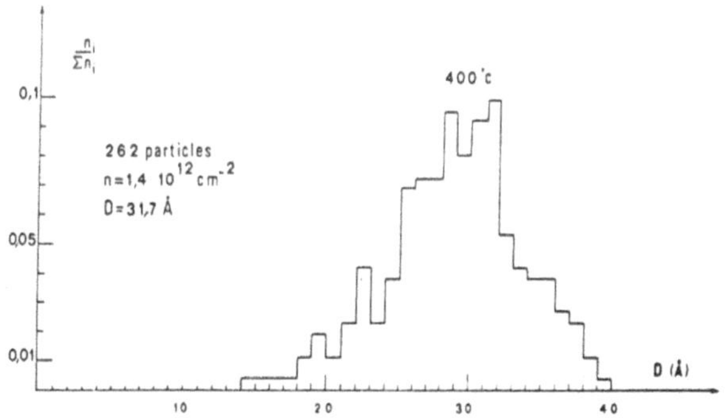

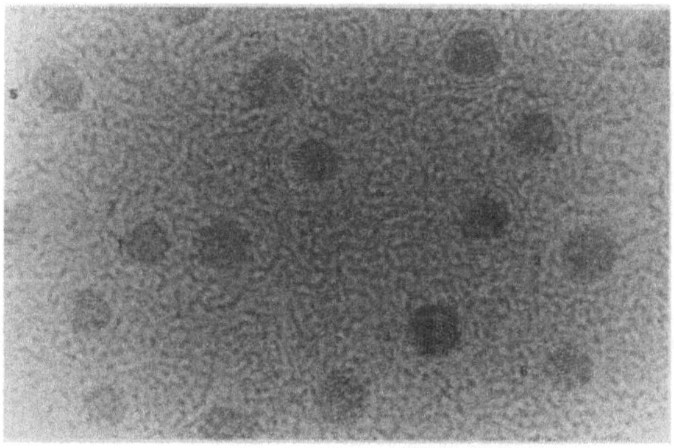

Fig. 5. HRTEM image of Pd particles from the 400°C deposit. (1) (100) particle, (2) (110) particle, (3) (110) twinned particle, (4–5) MTP (decahedra), (6) MTP

Table 1. Percentage of different orientations for the 2 deposits

Deposition Temperature	(100)	(110)	Twins	Unknown
300°C	9	8	1	82
400 °C	16	15	38	31

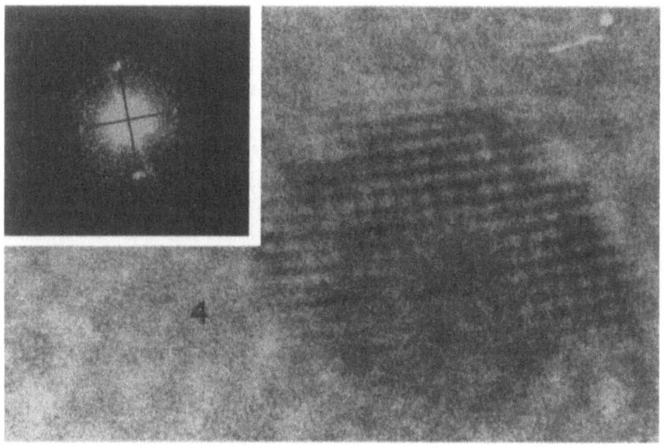

Fig. 6. (100) oriented Pd particle with a distorded lattice

are twinned. It can be argued that for the 300°C deposit a large number of unidentified particles are twinned. This is unlikely because most twinned particles on the 400°C deposit are in the larger size range of the distribution while unidentified particles are always in the smallest sizes. Furthermore the few defective particles observed on the 300°C deposit are simple twins and on the 400°C deposit roughly 70% of the defective particles are MTP with fivefold symmetry. However as the two size-histograms (see Fig. 4) overlap we cannot ruled out that MTP are favoured at higher temperature.

Taking into account the size limit (in our observations) to image atomic lattice planes we conclude that in the range 1.8 nm–2.5 nm small Pd particles have a perfect fcc structure and that MTP appear at large sizes. A same conclusion has been drawn by Monosmith and Cowley [11] from electron microdiffraction of isolated small gold particles. More recently Jacobs and Schryvers [9] observed by HRTEM that Pd particles on TiO_2, in the size range 1.5–4.0 nm are fcc single crystals. Finally Balerna et al [12] have deduced from EXAFS measurements that gold particles containing down to 50 atoms have an fcc structure.

All these findings ruled out the hypothesis of the formation of MTP from a 13-atomic icosahedron [4]. Thus MTP must have a postnucleatory origin. We suggest that the MTP result from coalescence events of small fcc clusters migrating on the substrate. In fact it is established that gold clusters smaller than 1.5 nm are mobile on NaCl and coalesce at room temperature [13]. This assumption is supported by observation of gold clusters prepared by gas evaporation technique and deposited on carbon film. Small clusters are perfectly fcc and large coalesced particles are MTP [14]. Higher the temperature, faster is the migration-coalescence mechanism [15].

Thus the Pd clusters should nucleate as perfect fcc crystal and could migrate on a given distance depending on the substrate temperature, when the density of clusters is sufficiently large the coalescence will occur resulting as MTP. Meanwhile another limiting factor of the dynamical coalescence is the epitaxy. If the cluster is in perfect registry with the substrate it becomes immobile [16].

As we have shown the smaller clusters are not twinned but they reveal distortions from the bulk fcc lattice: an expansion increasing as the cluster size drops. The opposite trends is observed for other metals and explained by an effect of the surface stress [10]. In the case of Pd clusters the same assumption holds if the surface stress has the opposite sign. However in the present case the oxidation (or bulk diffusion of oxygen) could explain that expansion as suggested recently [9].

5. Conclusion

The structure of small Pd clusters deposited on (100) MgO is obtained by lattice planes imaging. Particles with a diameter in the range 1.5–2.5 nm are perfectly fcc but the lattice is expanded relatively to the bulk. Larger clusters are often twinned and a large proportion of these defective particles have a five-fold symmetry. These results support a post-nucleatory origin of the MTP.

References

1. Bond, G.C.: Surf. Sci. **156**, 966 (1985)
2. Anderson, J.R.: Sci. Prog. **69**, 461 (1985)
3. Farges, J., de Feraudy, M.F., Raoult, B., Torchet, G.: J. Chem. Phys. **78**, 5067 (1983)
4. Gillet, M.: Surf. Sci. **67**, 139 (1977)
5. Burton, J.J.: Catal. Rev. Sci. Eng. **9**, 209 (1974)
6. Reichelt, K.: J. Cryst. Growth **19**, 258 (1973)
7. Heinemann, K., Poppa, H.: Surf. Sci. **156**, 265 (1985)
8. De Crescenzi, M., Diociaiuti, M., Lozzi, L., Picozzi, P., Santucci, S.: Phys. Rev. B **35**, 5997 (1987)
9. Jacobs, J.W., Schryvers, D.: J. Catal. **103**, 436 (1987)
10. Woltersdorf, J., Nepijko, A.S., Pippel, E.: Surf. Sci. **106**, 64 (1981)
11. Monosmith, W.B., Cowley, J.M.: Ultramicroscopy **12**, 177 (1984)
12. Balerna, A., Bernieri, E., Picozzi, P., Reale, A., Santucci, S., Burattini, E., Mobilio, S.: Phys. Rev. B **31**, 5058 (1985)
13. Henry, C.R., Chapon, C., Mutaftschiev, B.: Thin Solid Films **46**, 157 (1977)
14. Gao, P., Gleiter, H.: Acta Metall. **35**, 1571 (1987)
15. Chapon, C., Henry, C.R.: Surf. Sci. **106**, 152 (1981)
16. Métois, J.J., Zanghi, J.C., Erre, R., Kern, R.: Thin Solid Films **22**, 331 (1974)

Z. Phys. D – Atoms, Molecules and Clusters 12, 149–152 (1989)

Atoms, Molecules
Zeitschrift für Physik D and Clusters
© Springer-Verlag 1989

Size distribution of Bi clusters deposits on amorphous carbon substrates

F. Santos Aires, M. Treilleux, G. Fuchs, A. Hoareau, P. Melinon, and B. Cabaud

Départment de Physique des Matériaux. Associé au C.N.R.S. (U.A. 172) 43 Bd du 11 Novembre 1918, F-69622 Villeurbanne Cedex, France

Received 5 July 1988

Bismuth clusters are produced by the inert gas condensation technique. The cluster beam is analysed by time of flight mass spectrometry and is simultaneously deposited on an amorphous carbon film. We compare two kinds of deposition: molecular beam and free clusters deposition (18 Å mean size). Cluster deposition exhibits a larger mean particle size on the support. The two kinds of deposition may be correlated with the Schmeisser's law.

PACS: 81.15.G

1. Introduction

During the last decade, numerous investigations dealing with the growth of small metallic clusters on various substrates have been published in literature. The particles are generally formed on the substrate by vapodeposition [1, 2, 3, 4, 5]. A second approach is first to form clusters in the gas phase and then deposit the clusters themselves on the substrate [6, 7, 8]. In this case, the problem is to know if the mean size of clusters after deposition is different from the mean size of clusters in the beam.

The aim of this paper is to make a comparison between sizes distributions of bismuth clusters obtained by the two methods which are based on different physical effects and which are totally independent of each other.

2. Experimental

The bismuth clusters are generated by the gas aggregation technique in a source similar to that developed by Sattler et al. [9]. Details of source performances have been previously reported [10]. Briefly, the metal (99.995% bismuth grade) is vaporized in a helium atmosphere (99.99% purity) where clusters growth occurs under hight supersaturation conditions. Finally, bismuth clusters are carried by the helium jet.

The clusters beam is ionized by electron impact and then analysed by a time-of-flight mass spectrometer. Previous studies have shown that mass discrimination effects are negligible up to 20,000 a.m.u. In addition a translation sample holder can be inserted in the beam before the ionization chamber. This device is equipped with electron microscopy copper grids and with a quartz oscillator to monitor the intensity of the cluster beam. All the depositions are performed at room temperature in a background pressure better than 10^{-6} torr. A calibration of the quartz thickness monitor has been performed by comparison with Rutherford backscattering measurements (RBS) leading to relative precision of the flux determination better than 20%. The electron microscopy copper grids (Balzers 1000 mesh grids) are covered through electric arc technique with amorphous carbon films of typical thickness 40 Å. The cluster deposits are investigated using a conventional electron microscope (TEM Philips EM 300) at a typical magnification of 67,000. Particles with a size smaller than 12 Å can not be resolved by the TEM.

3. Results and discussion

Two kinds of samples were prepared and observed by TEM. Firstly the helium pressure in the cluster source was put to zero; in this case the source may be compared to a conventional effusive cell and the molecular beam consists predominantly of Bi_2 molecules. The mass spectrum corresponding to this mode is shown in Fig. 1a. Secondly the helium pressure was adjusted to produce bismuth clusters with a mean size of about 18 Å (Fig. 1b) The proportion of dimers and monomers is estimated to be lower than 5% of the total cluster species in this case. On the other hand the contribution of no detectable large clusters in our spectrometer is sufficiently low to be negligible. The distribution of the ionized species can be considered as a good representation of the neutral clusters in the beam.

3.1. Molecular beam deposition

The nucleation of atoms on support depends on the preparative methods conditions (impinging atoms flux, substrate temperature, position of the substrate to the vapour source, etc.). For all this set of experiments, the impinging atoms flux was fixed at 10^{13} atoms $cm^{-2} s^{-1}$. Micrographs of bismuth depositions are shown in Fig. 2 and the deduced size histograms are given in Fig. 3. The size histograms are obtained by measuring the diameter of 400 to 700 particles for each micrograph. From the size histograms, we deduce the particles number density n and the mean particle

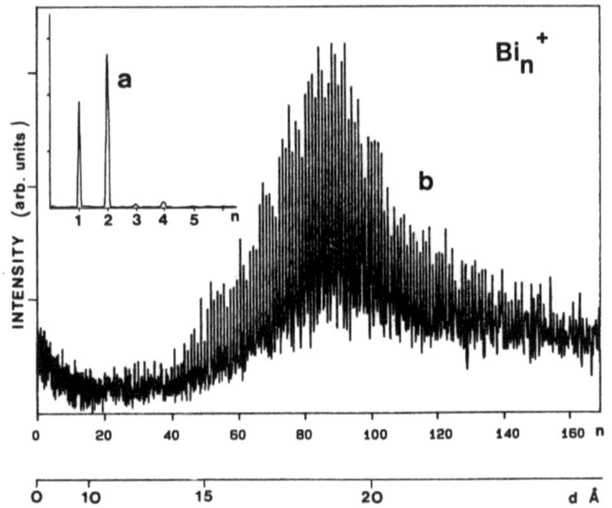

Fig. 1. a Mass distribution in molecular beam at 15 eV electron energy. The observed peak of monomer is essentially issued of fragmention of the dimer. **b** Mass distribution in the cluster beam at 24 eV electron energy and 1 torr of helium pressure

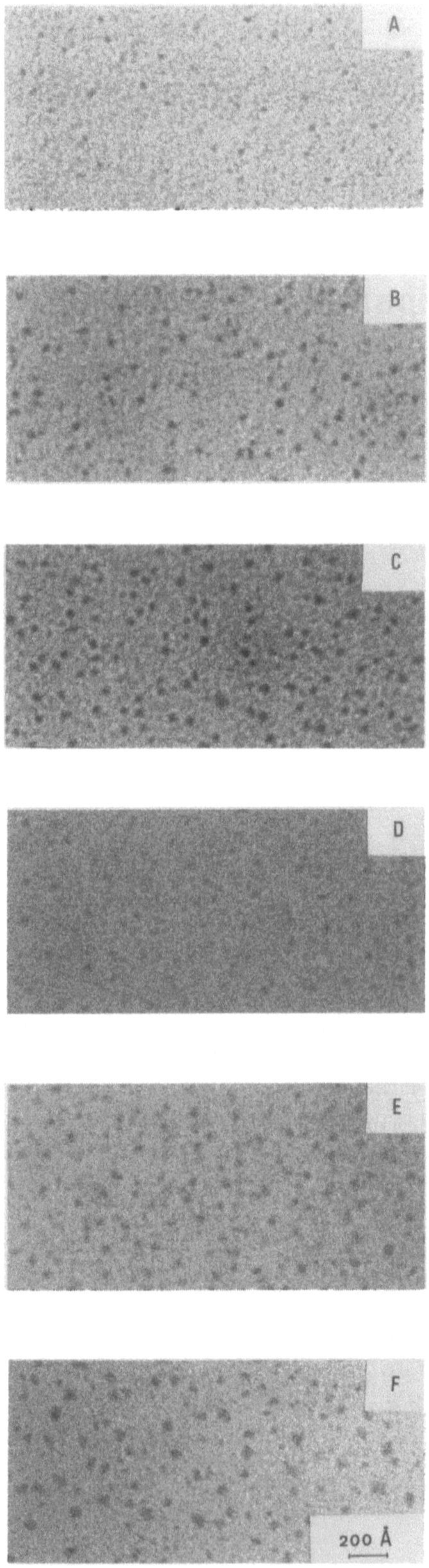

Fig. 2. A, B, C Micrographs of bismuth molecular beam depositions. The Coverage is respectively 1.3×10^{14} atoms cm^{-2}, 2.7×10^{14} atoms cm^{-2}, 5.1×10^{14} atoms cm^{-2}. **D, E, F** Micrographs of free clusters depositions. The Coverage is respectively 1.5×10^{14} atoms cm^{-2}, 2.9×10^{14} atoms cm^{-2}, 5.8×10^{14} atoms cm^{-2}

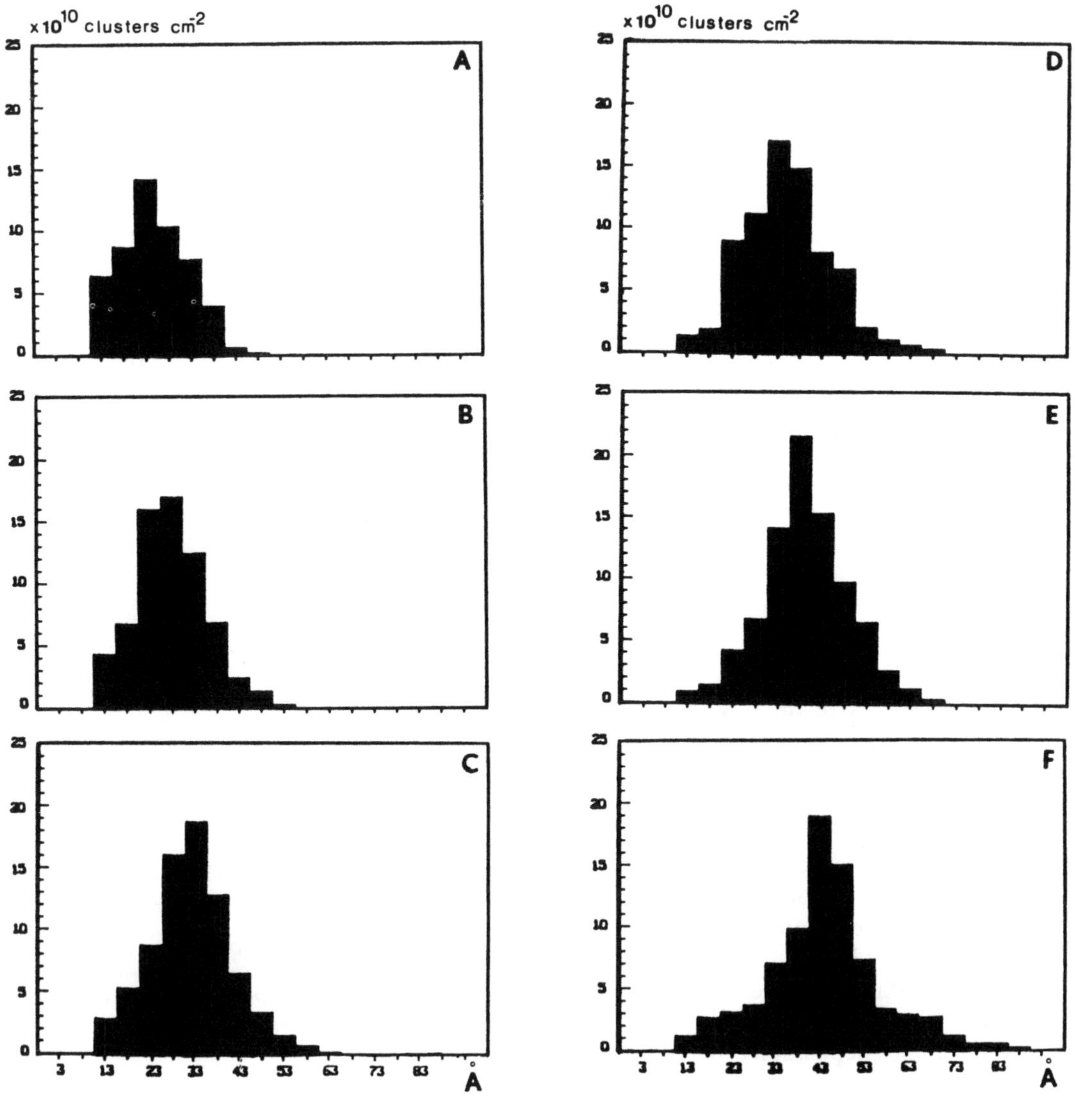

Fig. 3A–F Histograms corresponding to micrographs of Fig. 2 (respectively **A, B, C, D, E, F**)

diameter $\bar{d}$, using a normal distribution. Simulation with a Log-normal distribution give only a small shift in the results. Coalescence phenomenon is probably low in our experimental conditions [11].

3.2. Clusters deposition

For these experiments the equivalent impinging atoms flux is 2×10^{13} atoms $cm^{-2} s^{-1}$. The micrographs of free clusters depositions corresponding to the mass distribution in Fig. 1b are shown in Fig. 2. Size histograms deduced from these observations are given on Fig. 3. Coalescence phenomenon seems quite low [11].

3.3. Discussion

According to the work of Schmeisser [12] the mean particle diameter of clusters growing on a substrate by vapodeposition of atoms is proportional to the square root of the coverage θ (atoms cm^{-2}). Figure 4 shows the variations of the mean particle diameter versus the square root of the coverage for the two kinds of deposits. The first point to be noticed is that the linear dependence versus $\theta^{1/2}$ seems also valid for clusters depositions. This could be a proof of a similar growth process in the two cases. For a same coverage the cluster deposition exhibits a larger mean particle size than the molecular one. For the cluster deposition the extrapolated value of the mean diameter at null coverage is 18 Å in fairly good agreement with the mass

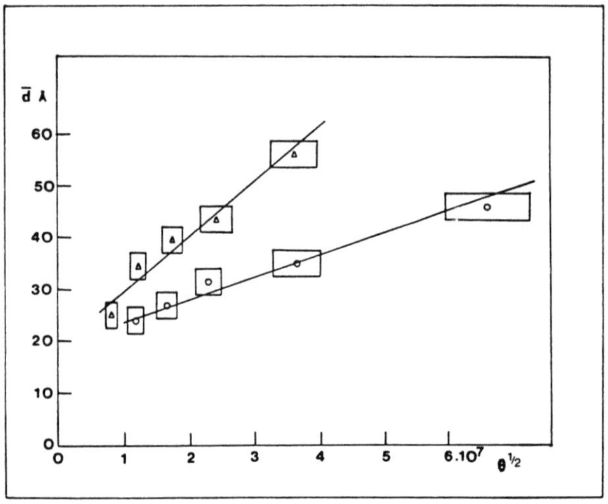

Fig. 4. Evolution of the mean diameter $\bar{d}$ of the observed particles as a function of the square root of the coverage θ
Δ: free clusters deposition
○: molecular beam deposition

spectra size distribution. However for the molecular beam the extrapolated value is far above the dimer diameter. In our experiments, particles of diameter less than 12 Å in the micrographs remain undetected which would result in an overrating of the larger clusters.

The number of nuclei at saturation n is of the same magnitude order for the two kinds of deposition (5×10^{11} up to 8×10^{11} clusters cm^{-2} in our experimental conditions). These numbers are two orders of magnitude greater than those given by Poppa [13] for depositions of bismuth molecular beam obtained with other experimental conditions. As a matter of fact the substrate temperature in the Poppa's depositions was more elevated (590 K).

4. Conclusion

Small bismuth clusters were grown on an amorphous carbon substrate using a molecular beam (Bi_2) or a cluster beam ($\bar{d} = 18$ Å). The main feature of this work was to elucidate the manner in which atoms or clusters impinging on the substrate can agglomerate to form larger particles.

It can be concluded that in the case where the impinging flux of clusters is sufficiently low (10^{13}atoms $cm^2 s^{-1}$) the observed mean size of particles on the substrate is only two or three times wider than the initial cluster size distribution in the beam. Therefore it is likely that diffusion of the incoming particles on the substrate is quite low in this case, and that the growth could be done around preferential sites of the surface. This conclusion is reinforced by the fact that the density number of particles observed on the substrate is the same for the molecular beam or the cluster beam. However for high impinging flux of clusters (3×10^{14} atoms $cm^{-2} s^{-1}$) and a mean size of about 10 Å, Treilleux et al. [14] have observed larger particles (~ 70 Å).

References

1 Hamilton, J.F., Logel, P.C.: Thin Solid Films, **23**, 89 (1974)
2 Lewis, B.: Surf. Sci. **21**, 289 (1970)
3. Chapon, C., Henry, C.R., Chemam, A.: Surf. Sci. **162**, 747 (1985)
4 Poppa, H., Moorhead, D., Heinemann, K.: Thin Solid Films **128**, 251 (1985)
5. Masson, A., Bellamy, B., Hadj Romdhane, Y., Che, M., Roulet, H., Dufour, G: Surf. Sci. **173**, 479 (1986)
6. Urban, J., Tesche, B.: Thin Solid Films **137**, 325 (1986)
7. Bowles, R.S., Park, S.B., Otsuka, N., Andres, R.P.: Mol. Catal. **20**, 279 (1983)
8. Sattler, K.: Physics and chemistry of small clusters. Jena, P Rao, B K., Khanna, S.N. (eds.) NATO ASI Series B: Physics Vol. 158, pp. 713–726. New York: Plenum Press 1987 (in cooperation with NATO Scientific Affairs Division)
9. Sattler, K., Muhlbach, J., Recknagel, E.: Phys. Rev. Lett. **45**, 821 (1980)
10. Broyer, M., Cabaud, B., Hoareau, A., Melinon, P., Rayane, D., Tribollet, D: Mol. Phys. **62** (n°3), 559 (1987)
11. Comsa, G.H.: Proceedings of the first ISSPIC Lyon 1976
12. Schmeisser, H.: Thin Solid Films **22**, 83 (1974)
13. Poppa, H.: J. Appl. Phys. **38** (No. 10), 3883 (1967)
14. Treilleux, M., Fuchs, G., Santos Aires, F., Mélinon, P., Hoareau, A., Cabaud, B.· 2. Phys. D–Atoms, Molecules and Clusters **12** (1989) (this issue)

Z. Phys. D – Atoms, Molecules and Clusters 12, 153–156 (1989)

Atoms, Molecules and Clusters
Zeitschrift für Physik D
© Springer-Verlag 1989

Stresses induced in the deposit substrate interface during particles growth

C. Chefi[1], **F. Hila**[2], **and M. Gillet**[1]

[1] Laboratoire de Microscopie et Diffractions Electroniques, Faculté des Sciences et Techniques de Saint-Jérôme UA 797, F-13397 Marseille Cedex 13, France

[2] Laboratoire de Microscopie Electronique, Département de Physique, Faculté des Sciences de Tunis, Tunis, Tunisie

Received 5 July 1988; final version 30 September 1988

We study stresses induced in the deposit substrate interface during a metal deposit growth. The stresses cause strains in the substrate and the substrate deformation is measured by an optical method. This technique enables the observation of strain variations in situ during the deposit growth. We have investigated the stresses in the case of Au and Cu vapor, deposited under vacuum on Al_2O_3, and the dependence of these stresses on the deposit mean thickness. The stresses are compressive and they increase during the first stages of growth. They decrease and become tensile when coalescence occurs.

PACS: 81.15

Introduction

Many properties of thin deposits, particularly thin protective layers, are influenced by the stresses which develop during the growth of the deposit on its support. These stresses depend on the difference in thermal dilatation between the support and the layer as well as the mode of growth [1–3], structure and deposit/support parameter differences [4], support and deposit mechanical properties [5, 6], impurities and imperfections at the interface [7].

Our aim is to study the stresses developing on an alumina support during the growth of a metallic deposit of Au and Cu. In this paper, we study simultaneously support strains and deposit morphology during the deposit growth. On the alumina support, the metallic deposit grows in tri-dimensional particles (Volmer Weber growth type). Particle density and shape are determined by Electron Microscopy and strains in the deposit-support system are related to the deposit morphology.

Experimental

Alumina supports are prepared from a sheet of aluminium 20 μm in thickness. An alumina layer of about 200 Å is obtained on the aluminium support by oxidation in air at 550°C for 24 h.

For electron microscope observation, the alumina layer is removed from its support by evaporation of aluminium at 700°C. Electron microscope observations show that the alumina layer is a continuous polycrystalline layer with grains 0.1 to 0.6 μ in diameter. Electron diffraction reveals the existence of 2 alumina phases γ and δ – with a strongly prevailing γ phase.

Deposits are obtained by evaporation in vacuum (10^{-6} torr) with a Knudsen cell. Evaporation rates are approximately 6 Å/mn for Au and 15 Å/mn for Cu.

Before each evaporation, alumina is degassed by heating at 300°C for 1 h. Auger spectroscopy analysis shows that the only surface impurity is a small quantity of carbon. Au and Cu deposits are made at different temperatures ($T = 20°C$, $T = 100°C$ and $T = 300°C$).

During metal deposition the stresses cause strains in the substrate. For strains measurements, the substrate is a rectangular strip (length $L = 43$ mm, width $l = 5$ mm, thickness $e = 20\,\mu$). One of its ends is fixed. The displacement of the free end is measured by an optical method. The stresses S occurring in the support at the deposit/support interface are either tensile or compressive depending on the sign of the displacement δ. Conventionally, $\delta < 0$ if the displacement occurs in the direction of the evaporation flux. S is compressive if $\delta > 0$, tensile if $\delta < 0$.

Results

Figure 1a, b show the dependence of the displacement δ on the mean deposit thickness h for various temperatures of the alumina substrate ($T = 20°C$, 100°C and 300°C). Figure 1a corresponds to the gold deposit and Fig. 1b to the copper deposit. We can in both cases distinguish two domains of thickness concerning the displacement δ – one domain where δ increases and is positive and another domain where δ decreases to become negative (except for Au at $T = 300°C$).

Electron microscopy observations show that in all cases the deposit grows as three-dimensional particles. The particles morphology and density are largely affected by the coalescence. Generally, the lower the deposit temperature, the faster the coalescence, and this coalescence is faster on copper than on gold. As an illustration, we present micrographs of Au deposits (Fig. 2a–d) and Cu deposits (Fig. 3a–d) for two deposit thickness and temperatures.

Figure 4a and b give the ratio of the support surface are covered with the deposit (deposit coverage θ) as a function of the mean deposit thickness for gold and copper respectively, in both cases for the different substrate temperatures. These curves illustrate the degree of coalescence as a function of the mean thickness for gold and copper deposit at different substrate temperatures.

The comparison of the results obtained in Fig. 1a and b concerning the displacements δ with those of Fig. 4a and b concerning the deposit coverage θ shows that the nature and importance of interfacial strains are related to the deposit coverage θ. In each experiment for gold and copper deposits, the maximum of δ corresponds to the same coverage θ. For gold the maximum of strain occurs for $\theta = 0.75$ and for copper the maximum of strain occurs for $\theta = 0.95$. We can also remark that the strains correspond to compressive stresses in substrate in the fast stage of growth, when

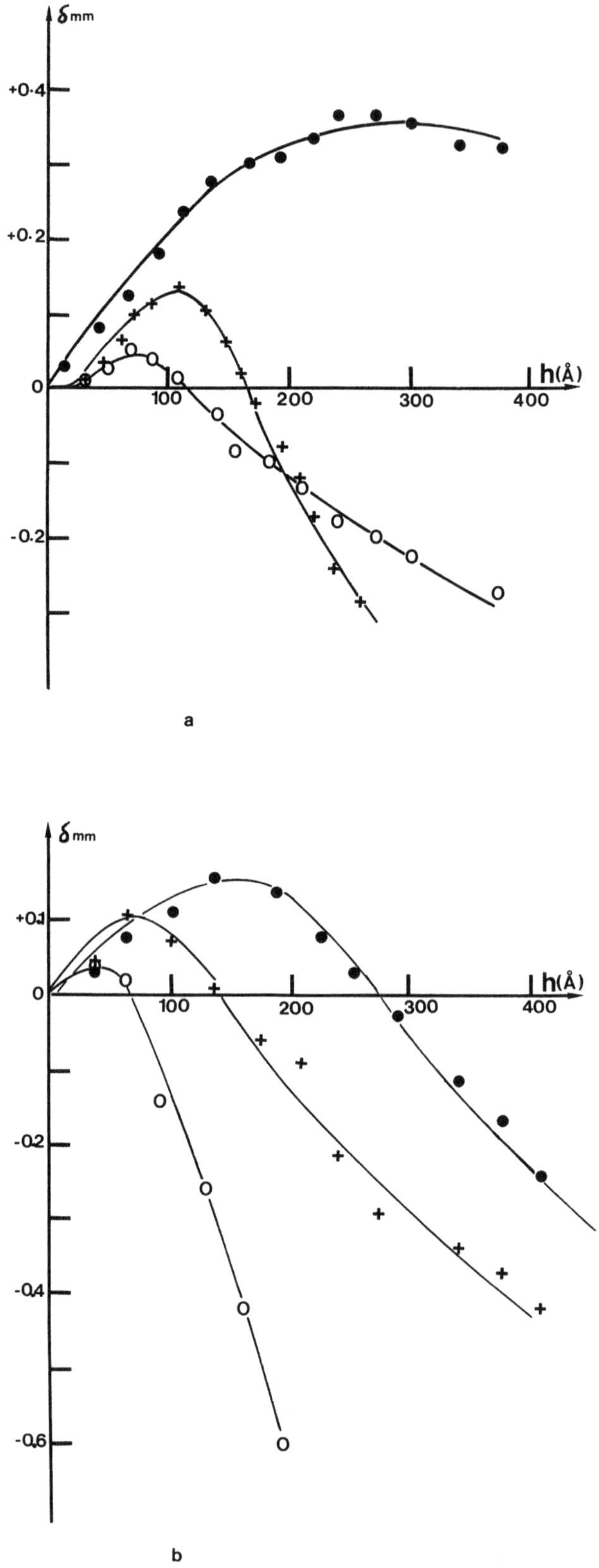

Fig. 1. **a** Dependence of the displacement δ on deposit thickness h for Au/alumina: ○ $T = 20°C$, + $T = 100°C$, ● $T = 300°C$. **b** Dependence of the displacement δ on deposit thickness h for Cu/alumina: ○ $T = 20°C$, + $T = 100°C$, ● $T = 300°C$

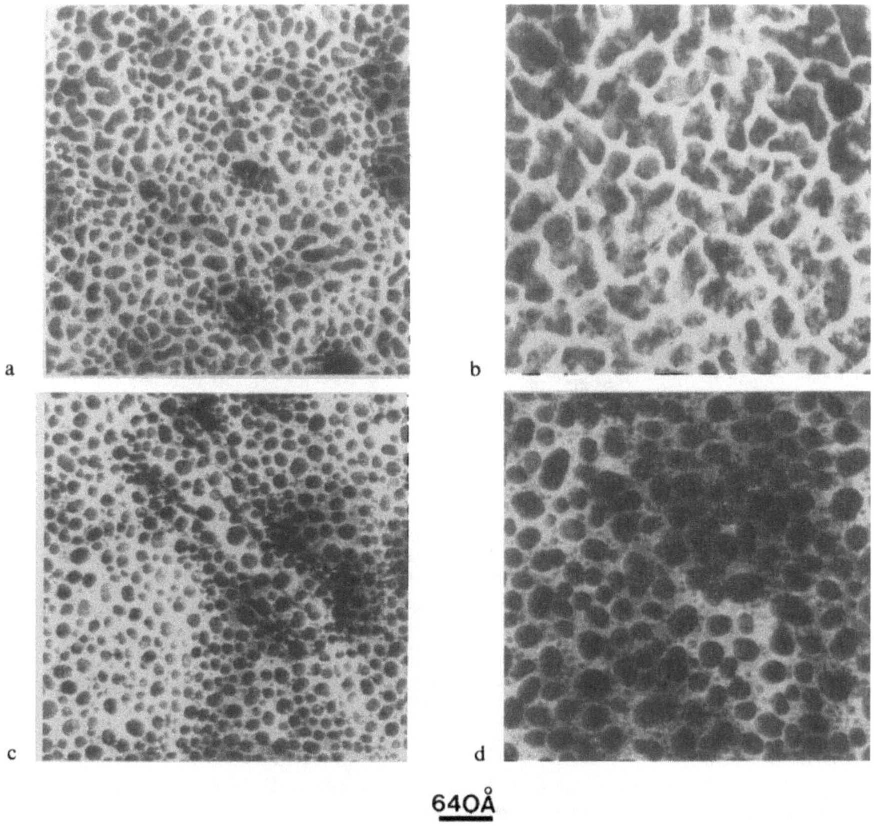

Fig. 2a–d. Transmission Electron micrographs of Au/alumina. **a** $T=20°C$, $h=20$ Å, **b** $T=20°C$, $h=50$ Å; **c** $T=300°C$, $h=20$ Å; **d** $T=300°C$, $h=45$ Å

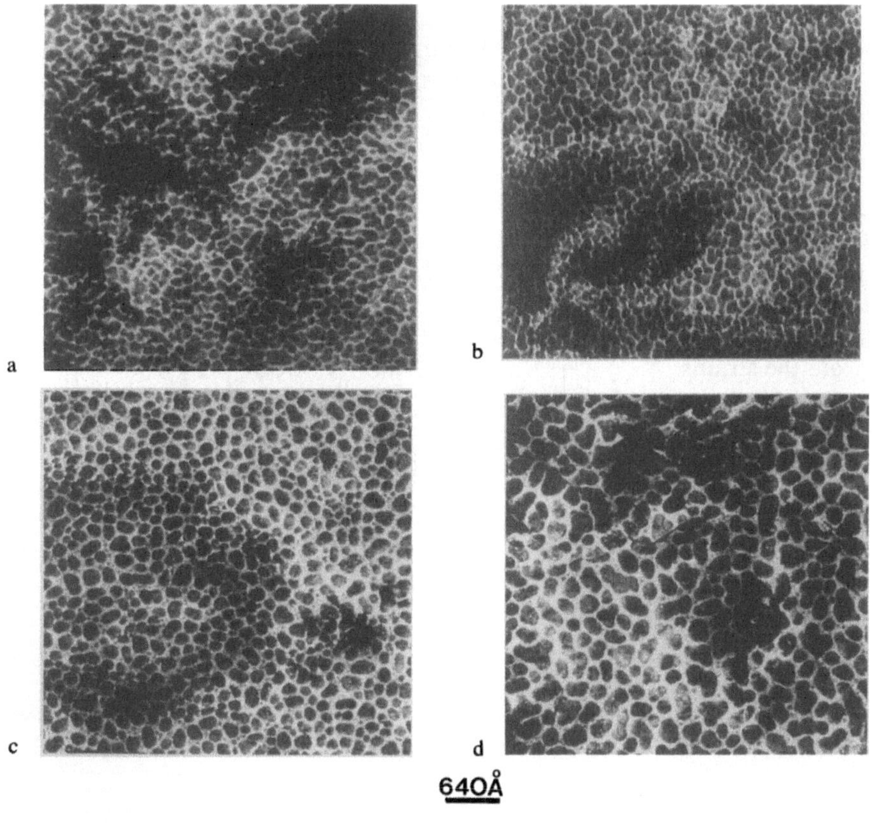

Fig. 3a–d. Transmission Electron micrographs of Cu/alumina. **a** $T=20°C$, $h=40$ Å; **b** $T=20°C$, $h=50$ Å; **c** $T=300°C$, $h=40$ Å, **d** $T=300°C$, $h=50$ Å

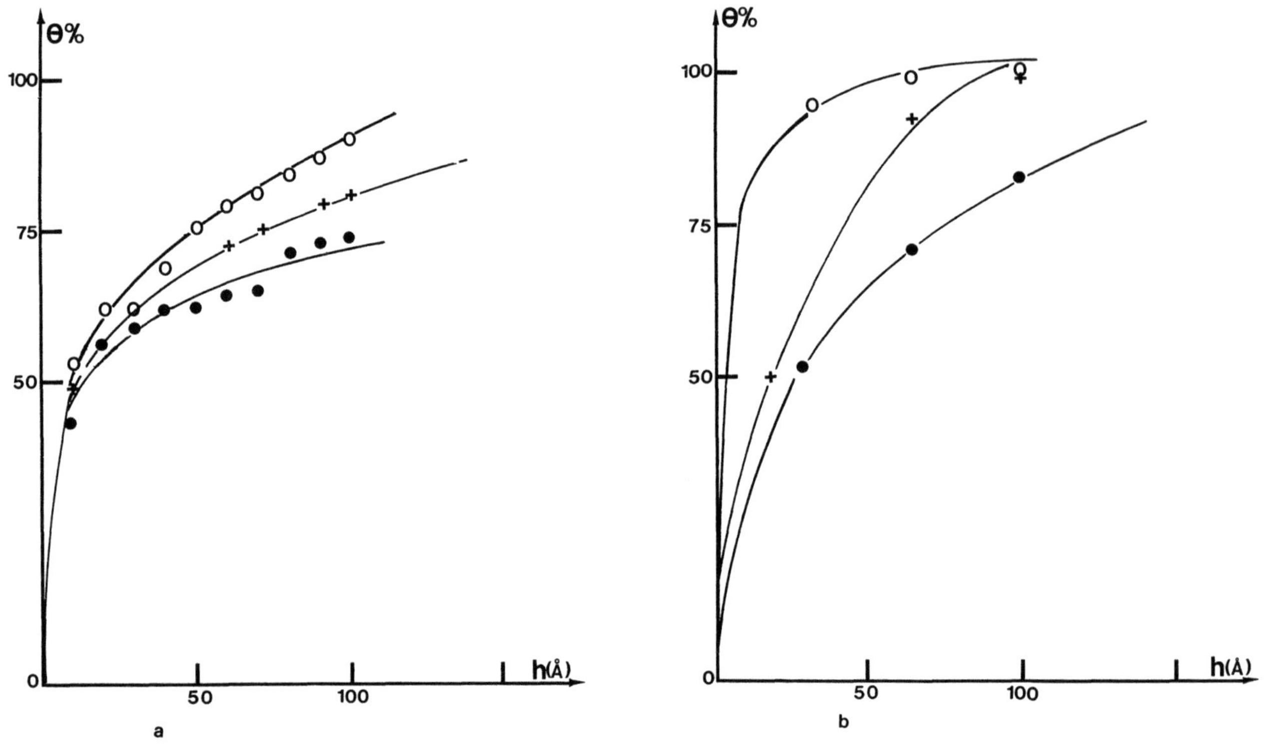

Fig. 4. **a** Dependence of deposit coverage θ on deposit thickness h for Au/alumina: ○ $T=20°C$, + $T=100°C$, ● $T=300°C$ **b** Dependence of deposit coverage θ on deposit thickness h for Cu/alumina: ○ $T=20°C$, + $T=100°C$, ● $T=300°C$

the deposit is discontinuous $\theta < 1$. The stresses become tensile for $\theta \simeq 1$ for a continuous deposit film.

Conclusion

The strain developed in a substrate (Al_2O_3) during a deposit growth (Au and Cu) depend on the deposit morphology. In the considered cases, we have a three-dimensional growth mode. The electron microscopy studies of the particle shape and density for various substrate temperature allowed us to relate the strains occurring in the substrate to the morphology of the deposit. These strains are related to the deposit coverage θ which depends on the growth conditions. They correspond to compressive stresses and increase in the first stage of growth. They decrease when the coalescence is important and become tensile for a continuous film.

References

1. Maissel, L., Glang, R.: Handbook of thin film technology, pp. 3–1 to 3–8 and 12–30 to 12–50. New York: Macgraw Hill 1970
2. Hass, G.: Physics of thin films, pp. 61–130 and 211–274. New York: Academic Press 1966
3. Blanckburn, H., Campbell, D.S.: Philos. Mag. **8**, 823 (1963)
4. Jesser, W.A., Matthews, J.W.: Philos. Mag. **17**, 461 (1968)
5. Johnson, R.M., Block, R.L.: Acta Metall. **16**, 831 (1968)
6. Koelher, J.S.: Phys. Rev. B **2**, 547 (1970)
7. Livesay, R.B., Starke E.A.: Acta Metall. **21**, 247 (1973)

Z. Phys. D – Atoms, Molecules and Clusters 12, 157–160 (1989)

Atoms, Molecules
Zeitschrift für Physik D and Clusters
© Springer-Verlag 1989

Nm-sized crystallites embedded in single crystalline films of magnesium oxide

K. Mihama and N. Tanaka
Department of Applied Physics, Faculty of Engineering, Nagoya University, Nagoya 464-01, Japan

Received 5 July 1988; final version 30 September 1988

Composite films of magnesium oxide embedding metals such as gold, iron and chromium are prepared by the simultaneous deposition technique onto cleavage surfaces of sodium chloride and characterized by high resolution electron microscopy. These metals grow epitaxially and uniformly inside magnesium oxide films as crystallites of a few nm in size. The thickness of the crystallites is estimated to be a few atomic layers. The observation of isolated single atoms or atomic clusters is attempted under an off-Bragg condition. For composite films of iron and magnesium oxide, a heat treatment at 500 °C for two hours brings about the growth of γ-iron crystallites. A strain field at the periphery of the γ-iron crystallites is analysed. For composite films of chromium and magnesium oxide, chromium crystallites of around 1 nm in size grow epitaxially inside magnesium oxide films with two kinds of orientation.

PACS: 61.14; 81.15.G; 81.40

1. Introduction

Composite films of metal and ceramics are of interest recently as functional materials. Since these films have been prepared by a sputtering technique in most cases, the structure is to be polycrystalline or amorphous [1]. To clarify the structure in connection with the physical properties, nm-sized isolated metal crystallites embedded inside single crystalline magnesium oxide films are prepared and characterized by high resolution electron microscopy [2, 3, 4].

In the present paper, some more details on structures for the metal crystallites in the composite films are described for gold, iron and chromium.

2. Experimental

Composite films of metal crystallites embedded in magnesium oxide were prepared by a simultaneous deposition technique. Details of the procedure were described in the previous papers [2, 3, 4]. Air-cleaved sodium chloride surfaces were used as growing substrates. Heat treatment was carried out for as-grown composite films at 500°C to 1000°C in vacuum of about 10^{-5} Pa.

The composite films thus prepared were observed by a high voltage electron microscope (H-1250ST) operated at 1000 kV and an electron microscope (JEM-2000FX) operated at 200 kV.

3. Results and discussion

For composite films of gold and magnesium oxide, gold grows epitaxially inside single crystalline magnesium oxide films with a few nm in size.

Figure 1 shows an electron micrograph of a gold-magnesium oxide film. The fringe-spacing observed at gold crystallites is measured to be 0.20 nm and 0.21 nm, depending on the amount of defocus. The result has been interpreted in terms of a selective imaging, the thickness of gold crystallites being estimated to be a few atomic layers [2].

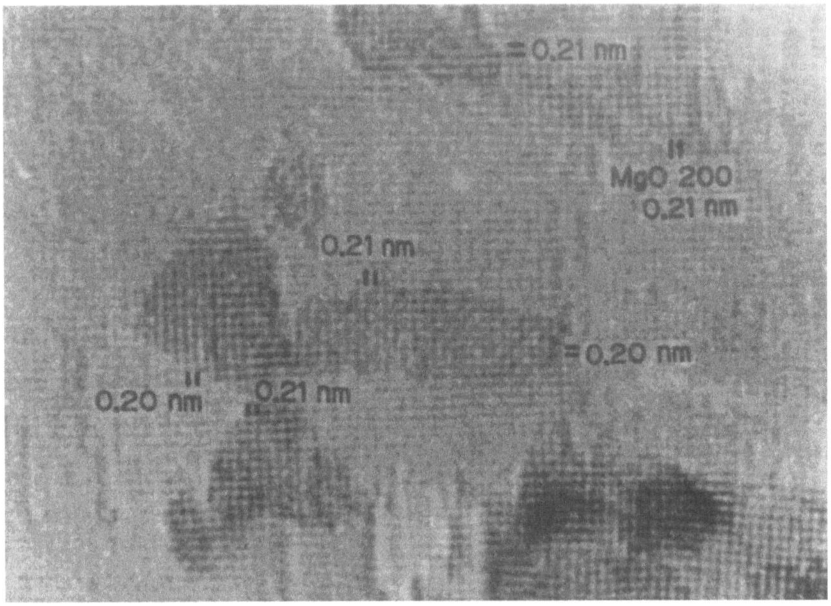

Fig. 1. Lattice image of gold crystallites embedded in a single crystalline magnesium oxide film

The existence of atomic clusters or isolated single atoms in the composite film may be expected with the present deposition technique. Figure 2 shows an electron micrographs and the corresponding diffraction pattern, the micrograph being taken with [$\bar{1}$0827] incidence, nearly [$\bar{1}$13] incidence for eliminating lattice fringes of magnesium oxide. In the micrograph, the lattice fringes due to magnesium oxide are almost disappeared. In the region of gold crystallites indicated by an arrow A, lattice fringes with a direction of [110] are observed, but in the region indicated by two arrows A and B, lattice fringes with two directions, [110] and [111] are observed. The difference of the appearance of lattice fringes is due to the difference of the thickness. Figure 3 shows atomic projections along the [$\bar{1}$13] direction for one to four atomic layers. In Fig. 3(a), atomic rows to the directions of [110] and [111] are observed for the monoatomic layer, but atomic rows with (111) spacing is accentuated with increasing the thickness. Since the electron micrograph reflects a projected potential of the specimen when the thickness is extremely thin, the micrograph of Fig. 2 may be compared with the projections shown in Fig. 3. Accordingly, the regions indicated by A and B with two directional lattice fringes correspond to the regions of monoatomic layer. Isolated black dots with similar contrast in some region as indicated by a white arrow. The dots can be detected in a narrow range of defocus around the Scherzer defocus. From the contrast of around 15% as well as the contrast behavior, they may be assigned as isolated gold atoms in magnesium oxide film [5, 6].

For composite films of iron and magnesium oxide, α-iron crystallites of around 1 nm in size grow epitaxially inside the magnesium oxide film with two kinds of orientation, (01$\bar{1}$) $[100]_{Fe}$//(001) $[100]_{MgO}$ and (001) $[1\bar{1}0]_{Fe}$//(001) $[100]_{MgO}$ [3]. The heat treatment for the as-grown films at 500°C for 1 h brings about a grain growth of α-iron up to about 4 nm and a growth of γ-iron for 2 h, the orientation relation between γ-iron and magnesium oxide being (001) $[100]_{\gamma-Fe}$// (001) $[100]_{MgO}$. The transformation to γ-iron at 500 °C may be due to the smallness of γ-iron as well as embedding in the magnesium oxide matrix. The lattice relation between α-iron and γ-iron is ($\bar{1}$10) $[001]_{\alpha}$//(010)$[001]_{\gamma}$, which is different from the Kurjumov-Sachs's [7]; in the present transformation, the lattice displacement is performed so as to have the epitaxial relation with magnesium oxide.

For γ-iron crystallites, the moiré fringes of 1.36 nm spacing formed by the (200) reflections of magnesium oxide and γ-iron are observed. The moiré fringes expand outward at the periphery of the crystallite with (010) systematic reflection, for example, and, then the moiré fringes of the same crystallite shrink with (100) systematic reflection. The result suggests that γ-iron crystallites have an asymmetric strain field. A characteristic two-dimensional streaks found in the diffraction patterns can be confirmed by a nm diffraction technique as well as the dark field technique with

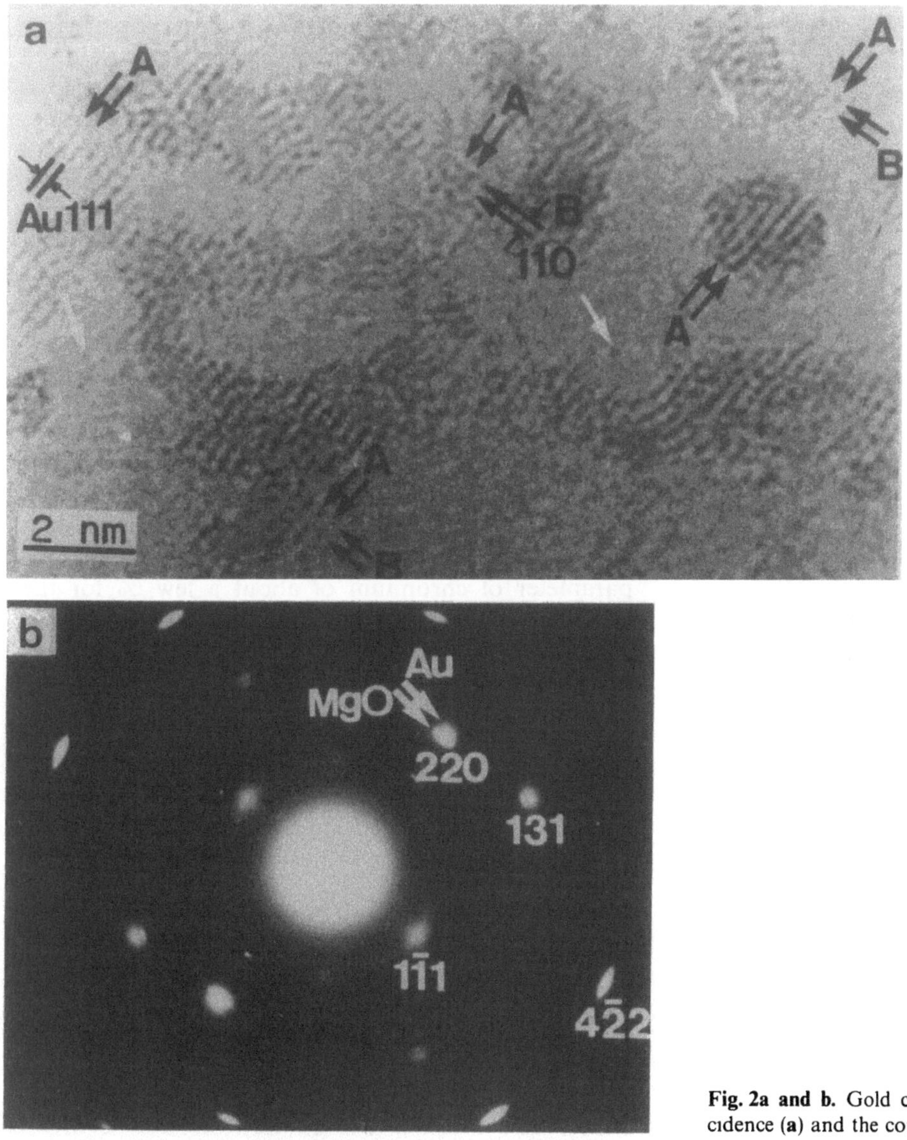

Fig. 2a and b. Gold crystallites taken with [$\bar{1}$13] electron incidence (a) and the corresponding diffraction pattern (b)

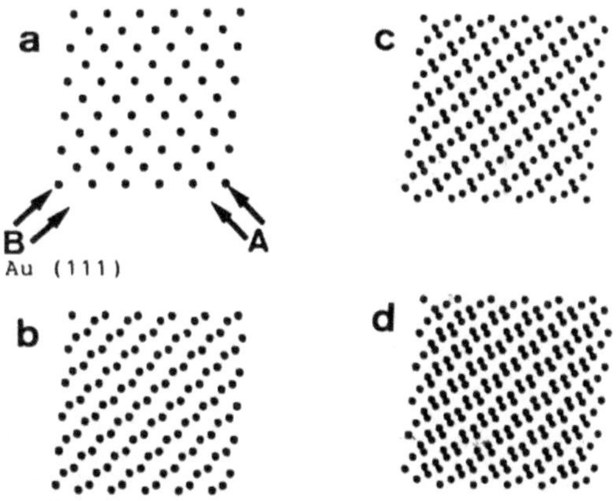

Fig. 3a-d. Atomic projections of gold atoms in a f.c.c. lattice along [$\bar{1}$13] direction for one (a), two (b), three (c) and four atomic layers

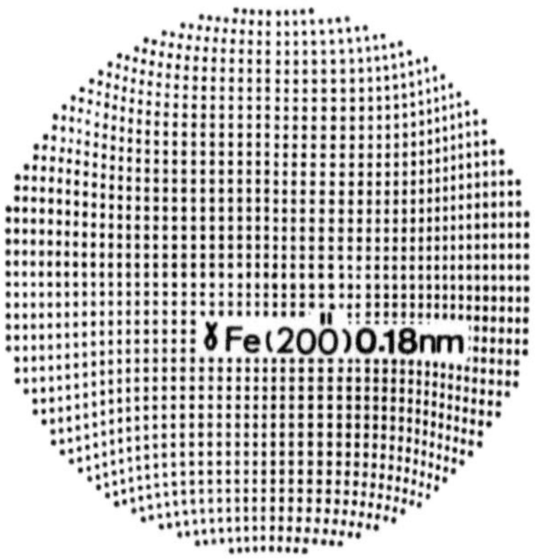

Fig. 4. Atomic model of a γ-iron crystallite with two-dimensional strain at the periphery

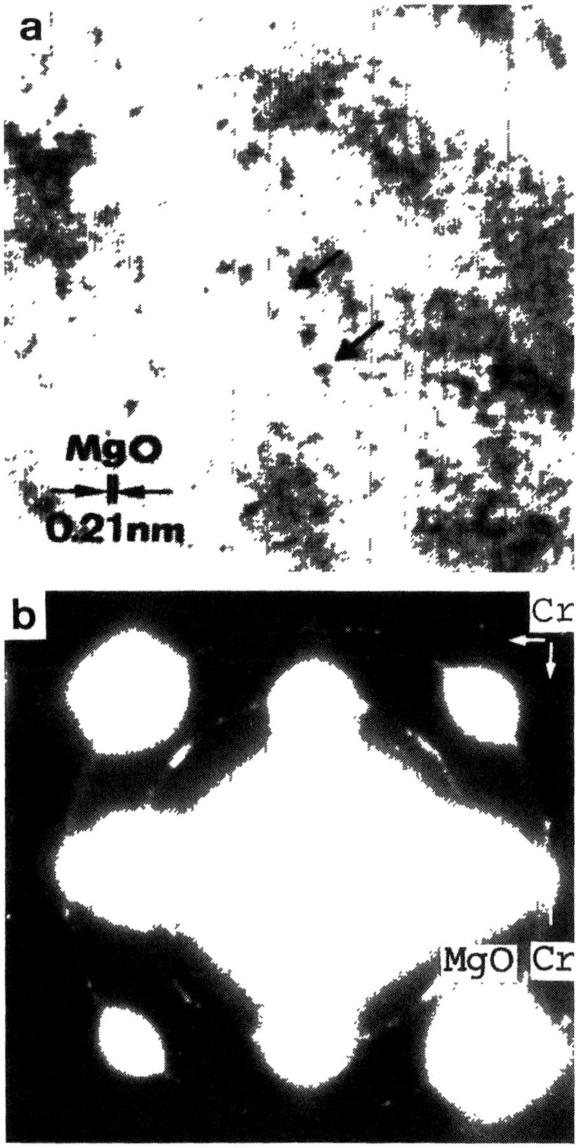

Fig. 5a and b. Composite film of chromium and magnesium oxide (a) and the diffraction pattern (b)

which the streaks arise from the region at the periphery of the crystallite [8]. Summarizing the above results, a model of strain field of γ-iron crystallites is proposed as shown in Fig. 4, in which the (020) lattice planes expand at the periphery so as to accommodate with those of magnesium oxide and the (200) lattice planes shrink so as to keep the volume of unit cell constant. In the model, the expansion of γ-iron crystallite at the periphery amounts to about 15%. The as-grown composite films of iron and magnesium oxide show a spin-glass character at low temperature. The present result may arise from the embedding effect of small size of iron crystallites inside the magnesium oxide film.

For composite films of chromium and magnesium oxide, chromium grows as b.c.c.-crystallites of around 1 nm in size with the two kinds of epitaxial orientation, the same as for composite films of iron and magnesium oxide, as shown in Fig. 5. An expansion of lattice parameter of chromium of about a few % for the accommodation of magnesium oxide lattice is detected. The heat treatment at 1000 °C for a few minutes brings about a formation of the spinel structure, Mg_2CrO_4.

References

1. Fan, J.C.C., Henrich, V.E.: J. Appl. Phys. **45**, 3742 (1974)
2. Nagao, M., Tanaka, N., Mihama, K.: Jpn. J. Appl. Phys. **25**, L215 (1986)
3. Nagao, M., Tanaka, N., Mihama, K.: Jpn. J. Appl. Phys. **25**, L614 (1986)
4. Nagao, M., Tanaka, N., Mihama, K.: Jpn. J. Appl. Phys. **26**, L215 (1987)
5. Mihama, K, Tanaka, N.: J. Electron Microsc. **25**, 65 (1976)
6. Nakano, K., Tanaka. N., Mihama, K.: Jpn. J. Appl. Phys. **19**, L445 (1980)
7. Kurdjumov, G., Sachs, G.: Z. Phys. **64**, 325 (1930)
8. Tanaka, N., Nagao, M., Yoshizaki, F., Mihama, K.: J E M Tech. (1989) (in print)

Z. Phys. D – Atoms, Molecules and Clusters 12, 161–165 (1989)

Atoms, Molecules
Zeitschrift für Physik D and Clusters
© Springer-Verlag 1989

Small clusters: between dynamics and thermodynamics

R.S. Berry
Department of Chemistry and The James Franck Institute, The University of Chicago, Chicago, IL 60637, USA

Received 6 July 1988

The relation between equilibrium properties and dynamical properties, and between the two kinds of descriptions, is explored by examining the dynamics of isomerization of argon clusters. The same general subject, from the viewpoint of ergodicity and chaos is examined through the fractal dimension of the trajectory in phase space and the Kolmogorov entropy.

PACS: 36.40+d

1. Introduction

Small clusters offer a special kind of insight into the structure of matter and into the nature of the theoretical descriptions we impose. On the first aspect, we can use the powerful techniques that have been developed for accurate treatment of atoms and small molecules, extending them to systems much richer in structure and dynamics than the pristine species for which they were first intended. We can address issues long though to be relevant solely in the province where N, the number of particles in a system, goes to infinity, yet maintain the control characteristic of the methods for small systems. This is leading us to new insights into how properties of bulk matter evolve from molecular properties, for example how structures of solids evolve from the structures of clusters, and how the electrical conductivity we associate with metals.

The second aspect of what clusters offer, insights into the conceptual foundations of our descriptions, is the main concern of this discussion. In particular, clusters offer such wide ranges of time scales for various kinds of phenomena that we can use them to clarify our notions of equilibrium and dynamical processes, and the relationships between the measurement process and how we think of equilibrium and dynamics. While this may seem more philosophical than scientific when stated in such general terms, the real importance of such considerations is in their application to specific scientific problems, particularly regarding what we consider significant information about a system and how we interpret it. Here we shall examine two aspects of small clusters relevant to the issue, both dealing with how a cluster explores the space available to it. The first is the question of how the shape of a potential surface governs the way the cluster's structure evolves in time, and how we can infer the most important information about the surface; we shall use Ar_7 as a principal illustration for this. The second is the question of how the levels of ergodicity and chaos evolve as the energy of a cluster increases. To illustrate this topic, we shall have to restrict ourselves to Ar_3 at present.

2. The dynamics of a cluster

Small clusters can exhibit well-distinguished solid-like and liquid-like forms, although not all clusters need show such sharp distinctions [1]. When the two forms may coexist, they display a dynamic equilibrium in which an ensemble of clusters contains some liquid-like and some solid-like clusters, in a ratio governed by a chemical equilibrium constant, the exponential of the difference between the (temperature dependent) free energies of the two forms. Any individual cluster spends an interval of time in one form, then switches to the other for some time, thus passing back and forth between the two. If the system is ergodic, as we normally expect, then the ratio of the amounts of the two forms at any instant is, within fluctuations, the same as the ratio of times any one cluster spends in the two forms.

One further phenomenon that small clusters may show is a finite band of temperature and energy within which the solid-like and liquid-like forms coexist [2]. We expect such "finite band coexistence" to be accompanied by sharp but unequal freezing and melting temperatures, T_f and T_m. The former is the lower bound for thermodynamic stability of the liquid-like form; the latter is the upper bound for the thermodynamic stability of the solid-like form. Evidence from simulations shows that argon clusters of some but not all sizes can be expected to show such sharp bands of coexistence [2]. Of the Ar_N clusters that have been studied by simulation, those that exhibit phase coexistence have $N=6, 7, 9, 11, 13, 15$ and 19. Those that do not have $N=8$, 14, 17, 20, 22, 26 and 33. Recent experimental data support the concept, both regarding a sort of surface melting of argon atoms clustered on a single molecule of carbazole [3] and especially of conventional bulk melting of argon custers containing a single benzene molecule [4]. The latter experiments indicate that $Ar_NC_6H_6$ clusters exhibit coexistence for all N from 17 through 24 under the specific conditions used by Hahn and Whetten.

The first question to address here is why some Ar_N clusters do not exhibit a finite band of phase coexistence while others do. Coexistence can be detected in several ways from simulations; the most useful and unambiguous is the appearance, in the history of a molecular dynamics simulation, of a bimodal distribution, either of (short-range) mean temperatures in an isoergic simulation or of (short-range) mean potential energies in an isothermal simulation. So long as the short-range averages are, at least in molecular dynamics (MD), over intervals long enough to include a few vibrations, bimodal distributions indicate two coexisting forms and average properties can be computed by examining separately the contributions to each of the peaks in the distribution. Under such circumstances, we can compute equilibrium properties for each form of the cluster and predict what each form should show in an experiment – for example, what line shapes each might show.

But not all clusters exhibit bimodal distributions; why? One reason that bimodal distributions may not appear is the first that makes us confront what equilibrium means. It may be that the clusters of a particular size pass between the solid and liquid forms faster than the intervals over which we take our short-time averages. In such a case, we would find unimodal distributions of mean temperatures in our simulations and no evidence for anything but gradual softening with energy or temperature. This can be achieved for any simulation by extending the interval for "short-time" averages until it is significantly longer than the mean time spent in either phase. Likewise, one can test for this phenomenon by reducing the interval for short-time averaging down to as brief as a single vibrational period. In our own simulations, we have most frequently used 500 steps of 10^{-14} s each as our averaging interval, about 3 breathing vibrations of the Ar clusters. Reducing this to only 150 steps did not reveal any bimodal behavior for any of the clusters, e.g. $N=8$, which show only unimodal behavior with 500-step averaging [5]. Hence we can rule out this explanation for unimodality of the specific argon clusters mentioned two paragraphs previously.

A second explanation of absence of a sharply bounded band of coexistence is the shape of the potential surface itself. It may be that the potential minima and saddles simply permit a gradually increasing exploration of configuration space as the energy or temperature of the cluster increases. This, at present, seems to be the best explanation of the behavior of Ar_8, Ar_{14}, and most of the others in the unimodal club [5].

A third explanation, which is very possibly appropriate for Ar_{33}, is that it contains so many atoms that the freezing and melting temperatures are closer than the resolution of any calculations thus far, so that the bimodal region simply has not yet been observed. We guess that as N gets large, T_f and T_m must converge to the same temperature T_e, the temperature at which the chemical potential is the same in both forms and the equilibrium constant is unity. We know relatively little about how these converge with N, only that the coexistence range for Ar_{19} is about 12°K and that no bimodality has been seen for Ar_{33}. This remains an important aspect of the problem still to be studied.

Two general inferences we can make, particularly from the insight given to us by simulations, are:

a) that a cluster – or a molecule – is *nonrigid* if it passes among potential wells on a time scale short compared with the time scale of our observations, and

b) that a cluster is *liquid* if it passes among many potential wells on a time scale roughly comparable with its vibrational frequency within one well.

It will probably be useful to distinguish these two. Of course we create a potential apparent paradox of liquids that are rigid, if we make fast observations, as with electron diffraction, but that is no real paradox because we need only specify the time scale with respect to which the system is rigid. The latter, the characterization of a liquid is extremely important because of what we intuitively want the concept of liquid to be. A liquid is a form of matter characterized by its compliance; one can put one's finger on ice, but only into water. It is not possible to put one's finger on water because even with the weakest force, the liquid complies – unless, of course, the force is applied much faster than the natural compliance time of the liquid, as every beginning diver learns.

Now we turn to the manner in which a cluster passes from one configuration to another. A solid cluster is confined to a single potential well, where it executes nearly harmonic vibrations. A liquid cluster can pass among a wide variety of potential wells which give rise, in general, to permutations of the atoms among the sites within the cluster, at least insofar as that concept has meaning for equilibrium geometries. The liquid cluster may or may not pass through the same potential wells as those that correspond to the solid forms of the cluster. If the cluster has a fixed energy enough above that of the saddles connecting solid-like forms with the liquid-like regions, then only the liquid-like form of the cluster is thermodynamically stable and the cluster may pass through solid-like regions with ease. In the coexistence range, the cluster spends much longer intervals in its solid-like wells than in any of the wells in its liquid-like regions.

This behavior is illustrated by the study of Ar_7 by Amar and Berry [6], using the quenching method of Stillinger and Weber [7]. The Ar_7 cluster has four kinds of potential minima, corresponding to geometries of 1) a pentagonal bipyramid, 2), and octahedron with one surface decorated, 3) a tetrahedron with three surfaces decorated, and 4) a "skew" structure which can be considered a tetrahedron with two faces and one edge decorated. The energies of the minima, the number of permutational isomers of each and the barriers [8] between each kind of minimum are given in Table 1.

The total isomerization rate for Ar_7 increases from 0.2 ns^{-1} at an energy of −14.4494, to 1.00 at −14.2898, to 2.01 at −14.1739, to 4.48 at −14.0202, to 8.36 at −13.8203, to 26.56 at −13.4352, to 63.7 at −13.0635. The isomerization rates give us considerable insight into how long one must look, either with molecular dynamics simulations or in following experimental heating curves, to see equilibrium frequencies of appearance of different regions of configuration space. The time scale required to see equilibrium distributions of time intervals in each region may be considerably longer, but need not be.

Table 1. Energies of potential minima, permutational degeneracies and barriers between minima for the cluster Ar_7. Energies are in units of the well depth for Lennard-Jones "argon", $1.67\text{–}10^{-21}$J. No additional barriers have been found

Structure	Energy	Permutational degeneracy	Barrier energy [8]
1.	−16.505	504	1-2: −15.44 1-3: −15.03 1-4: −15.02
2.	−15.935	1080	2-2. −15.10 2-3: −15.32 2-4: −15.28
3	−15.593	1080	
4.	−15.533	5040	

The distributions of residence times are especially revealing regarding solid-liquid coexistence. For energies up to at least −13.8203, all the durations in structures above the lowest are well below 200 ps, while the distribution of times in the pentagonal bipyramid extends into the thousands of ps except for the highest energy, where it extends only a little above 1000 ps in the observed runs. This range includes all the energies studied in the coexistence range. At −13.0635, in the liquid range, the times in the wells around the higher structures extended to about 40 ps and for the lowest-energy well, to just over 200 ps. In other words, the time scale separability for solid-like and liquid-like regions breaks down when the solid disappears, but it holds when the two forms coexist.

The Ar_7 cluster was studied with clusters of other sizes by Beck and Berry [5], also using quenching methods [7]. This study concentrated more than that of Ref. 6 on the mechanisms of passage from one well to another and how the geometries affected whether liquid-like regions of configuration space exist at all for particular clusters. For example Ar_{14} has no such well-defined liquid-like region. Rather, there are many low-lying minima in a sort of succession of energies like a staircase, so that as the energy increases, the cluster finds a smoothly growing volume of its configuration space to explore.

3. Ergodicity and chaos and how they grow

The other path we investigate to see how a cluster explores the space available to it is to quantify two characteristics that indicate the extent of this exploration [9, 10]. The first indicator is the degree of ergodicity, i.e. the degree to which the trajectory of a phase point fills the space energetically available to it. The second is the degree of chaos, the rate at which trajectories initially very close depart from each other. It is important to recognize the difference between these two. For Hamiltonian systems, we know of no connection between them. For dissipative systems, there is only a conjecture that the two are related.

To quantify ergodicity, we use the fractal dimension, precisely the "correlation dimension", of the trajectory of the phase point describing the cluster's evolution in phase space. This may be calculated by at least two algorithms; we used that of Grassberger and

Procaccia[11]. To quantify the degree of chaos we examined the full spectrum of Liapunov exponents as well as the sum of the positive exponents, which has been proven to be equal to the K-entropy [12].

The expectations for these quantities are easy to present, at least qualitatively. The fractal dimension for a perfect set of s uncoupled harmonic oscillators should be exactly s, corresponding to the one-dimensional periodic motion of each in its phase space. A cluster of N particles described in its own internal coordinate system either without rotation or in a rotating system has $3N$-6 internal vibrational modes and hence that many dimensions, at minimum, for its trajectory in phase space. The largest dimension we might expect is the total dimension of the phase space, $6N$, less the number of constraints. There are ten of the latter, three for the position of the center of mass, three for the conserved linear momentum, three for the conserved angular momentum and one for the conservation of energy, ten in all. Hence we expect a maximum dimensionality of $6N$-10 for the most "ergodic" trajectories. The largest cluster for which we have yet carried out such analyses is Ar_3. We therefore expect that the fractal dimension of the trajectory will be near 3 at low energies where the system is much like three harmonic oscillators, and perhaps as high as 8 at high energies.

The Kolmogorov entropy and Liapunov exponents are much more difficult to estimate from qualitative considerations, but we would naively expect them to increase with the energy. In particular, we might expect large increases to accompany melting.

One point must be made here regarding ergodicity and fractal dimension. If a system is ergodic at all, it must, at sufficiently long times, have an integral dimension. However for finite times corresponding to particular separable scales, nonintegral dimensions can be found and have been conjectured to be useful for 2-dimensional systems to characterize the degree of ergodicity attained by the system on that particular time scale. We have found [9, 10] that such separability does occur for Ar_3 and that the fractal dimension of the trajectory can be computed by the Grassberger-Procaccia algorithm for a wide range of energies. On this basis, we suggest that in general it may be valuable to consider fractal dimensions as characteristics of trajectories which can be associated with particular time scales, and therefore to compute such fractal dimensions over various time scales and corresponding length scales when there are several such scales involved in a problem.

A parallel point regarding the K-entropy and Liapunov exponents is that they can be considered characteristic of particular regions, even to the extent of being local properties of a phase space. This becomes important as we examine the energy dependence of the K-entropy of the Ar_3 cluster.

Now we turn briefly to the properties of Ar_3 itself. While this cluster does not have enough degrees of freedom or potential minima to exhibit a solid-liquid equilibrium like those of Ar_7 and Ar_{13} it does have one distantly analogous property helpful to us here. The equilibrium geometry of Ar_3 is an equilateral triangle, so there is only one potential minimum. The linear geometry of this species is a saddle on the energy surface, so there are three equivalent saddles whose energies lie between -0.70 and -0.56. Below the saddle energy, an Ar_3 cluster must of course reside in its well, undergoing relatively small-amplitude motions. Above the saddle energy, the cluster may pass through the saddle and return to the well by a large-amplitude motion. In fact, at energies below the saddle, the power spectrum of Ar_3 has almost no intensity at zero frequency, corresponding to no diffusion; above, the power spectrum has a nonvanishing contribution at zero frequency and diffusion can occur. This change is a sort of phase transition, which it is useful to think of as somewhat solid-to-liquid like.

Table 2 shows the fractal dimensions and K-entropies for Ar_3 at various energies and equivalent temperatures. The important point to note is that both these characteristic quantities increase monotonically up to the saddle energy and then decrease! The interpretation of this behavior is now clear; it remains to be determined whether Ar_3 is in any way a model for other, larger clusters. The larger clusters may well be able to pass from one region to another, across saddles, while continuing to increase the dimensionality and K-entropy associated with all their other degrees of freedom. For Ar_3, this is not possible, just because of the limited number of degrees of freedom.

The explanation, then, for Ar_3, is this. Below the saddle energy, the motion of the cluster within its well becomes more and more ergodic and chaotic as the energy increases. At the saddle energy, the cluster is able to penetrate all the way into the saddle and

Table 2. Dimensions (D_2) and K-entropies (K_2) for Ar_3 as functions of energy and the corresponding effective temperature. Where two values are given, they correspond to two different scales of correlation. (Taken from Ref. 9)

Energy	Temperature	D_2	K_2
-0.98	2.12	3.1	—
-0.97	4.15	3.3	1.8
-0.91	10.31	3.5	—
-0.83	18.19	5.9	5.6
-0.70	28.44	2.5, 7.6	4.4, 10.2
-0.56	30.54	2.6, 5.3	4.6, 7.3
-0.47	36.51	6.2	9.0

through, but at the cost of almost all its kinetic energy. This kinetic energy is what enables the trajectories to be ergodic and chaotic in the deep well. In the saddle region, at energies not very far above the saddle energy itself, all the trajectories correspond to slow motions along flow lines across the saddle. Hence the trajectories remain nearly parallel there, and hardly diverge at all. Moreover because the speeds are low in the saddle region, the cluster's phase point spends a long time there. The algorithms used to generate Table 2 produce the time averages of the fractal dimension and K-entropy. As a result the values reflect the rather large contribution to the average of the saddle region.

By separating the trajectories into portions associated with motion in the well from portions associated with the saddle, it has been possible to demonstrate that the reductions in the two indices are indeed due to the two kinds of contributions, from the highly disordered well and the much more ordered saddle.

This research was supported by a Grant from the National Science Foundation.

References

1. This transition is surveyed in many aspects in the recent volume Large Finite Systems, The Jerusalem Symposia on Quantum Chemistry and Biochemistry, Vol. 20, Jortner, J., Pullman, A., Pullman, B. (eds.) Dordrecht: Riedel 1987
2. Berry, R.S., Beck, T.L., Davis, H.L., Jellinek, J.: In: Evolution of Size effects in Chemical Dynamics, Part 2, Advances in Chemical Physics, Vol. LXX 1988, p. 75. New York: Wiley.
3. Bösiger, J., Leutwyler, S.: Phys. Rev. Lett. **59**, 1895 (1987)
4. Hahn M.Y., Whetten, R.L.: Phys. Rev. Lett. (1988)
5. Beck, T.L., Berry, R.S.: J. Chem. Phys. **88**, 3910 (1988)
6. Amar, F., Berry, R.S.: J. Chem. Phys. **85**, 5943 (1986)
7. Stillinger, F.H., Weber, T.A.: Phys. Rev. A **25**, 978 (1982); see also Weber, T.A., Stillinger, F.H.: J. Chem. Phys. **87**, 2833 (1983); Stillinger, F.H., Weber, T.A., Phys. Rev. A **28**, 2408 (1983)
8. Davis, H.L., Beck, T.L., Stephen Berry, R.: Chem. Phys. Lett. **147**, 13 (1988)
9. Beck, T.L., Leitner, D.M., Berry, R.S.: J. Chem. Phys. (1988)
10. Leitner, D.M., Beck, T.L., Berry, R.S.: J. Chem. Phys. (submitted, 1989)
11. Grassberger, P., Procaccia, I.: Phys. Rev. Lett. **50**, 346 (1983); ibid., Physica D **9**, 189 (1983); ibid., Phys. Rev. A **28**, 2591 (1983)
12. Pesin, Y.B.: Math. Surv. **32** (4), 55, (1977)
13. Bennetin, G., Casati, D., Galgani, L., Giorgilli, A., Sironi, L.: Phys. Lett. A **118**, 325 (1986)

Z. Phys. D – Atoms, Molecules and Clusters 12, 167–171 (1989)

Atoms, Molecules
Zeitschrift für Physik D and Clusters
© Springer-Verlag 1989

Dynamic and quantum size effects in molecular clusters

J. Jortner[1], U. Even[1], N. Ben-Horin[1], D. Scharf[1], R.N. Barnett[2], and U. Landman[2]

[1] School of Chemistry, Raymond and Beverly Sackler Faculty of Exact Sciences, Tel Aviv University, 69978 Tel-Aviv, Israel

[2] School of Physics, Georgia Institute of Technology, Atlanta, GA 30332, USA

Received 15 August 1988

The evolution of thermodynamic, dynamic and quantum size effects in clusters is expected to contribute towards the merging between microscopic and macroscopic points of view in molecular, surface and bulk phenomena.

PACS: 36.40; 64.70

I. Introduction

Ten years ago when cluster physics and chemistry were at their embryonic stages, Friedel stated in his concluding remarks at the first IMSPIC: "The study of clean isolated clusters can be viewed as a theoretician's dream" [1]. This dream has come true! The outstanding progress achieved in this field stems from two directions. First, the advent of supersonic and cluster beams had a remarkable impact on experimental progress in this field. Second, computer classical and quantum mechanical simulations are providing a conceptual framework for the understanding of the evolution of quantum, thermodynamic, dynamic and chemical size effects in clusters [2, 3]. We shall be concerned with the elucidation of some of these size effects in clean, isolated molecular clusters, contributing to the unified description of the interrelationship between molecular, surface and bulk phenomena in large finite systems.

II. Isomerization and "melting" of clusters

The nature of phase transformations in finite systems [4–7] is a subject of considerable conceptual interest in the context of the interrelationship between dynamic and thermodynamic size effects. Extensive computer simulation studies were performed for rare-gas clusters [4, 6, 7]. For these clusters the interaction potentials are "soft" resulting in a large number of structural isomers. For example, the Ar_{13} cluster is characterized by 988 isomeric structures, the lowest energy one corresponding to the icosahedron [8]. In alkali-halide clusters the strong interaction potentials result in a more distinct character and a smaller number of isomers. We have undertaken molecular dynamics (MD) simulations to study the energetics, structure and dynamics of alkali-halide clusters of variable sizes $(NaCl)_n$, for $n = 4$, 16 and 108, over a wide temperature range [5]. At the lower end of the size spectrum, the phase space of the system is characterized by a small number of stable configurations (solid isomers) between which the system transforms in a diffusionless manner, with temperature dependent rates and branching rations. This is evident from the caloric curve (total energy per particle vs kinetic energy per particle) for the $n = 4$ cluster (Fig. 1), which corresponds to molecular isomerization between the cube and ring conformations. This "small" system does not show soft modes and mass diffusion (Fig. 1) and does not exhibit melting. As the system size is increased, the number of accessible conformers increases leading to a hierarchial kinetics of isomerization events, which exhibits itself as a broadening of the transition region. Hierarchial isomerization is exhibited for $n = 16$ (Fig. 2). The lower temperature break corresponds to diffusionless isomerization between rigid isomers, while higher temperature breaks are accompanied by the gradual appearance of soft modes and diffusion

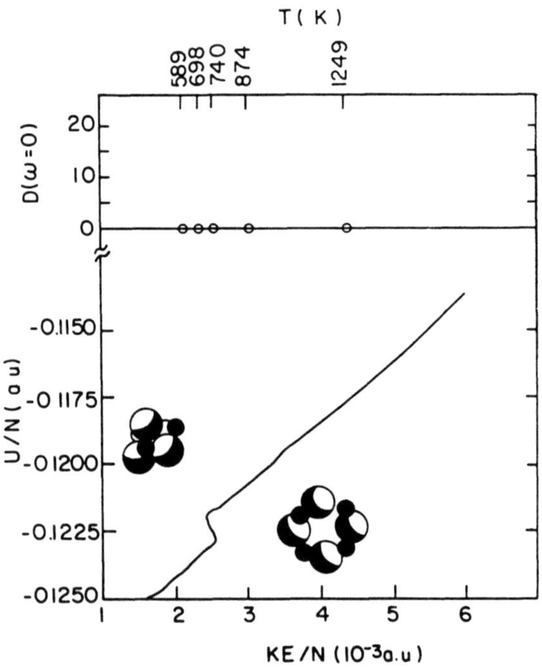

Fig. 1. Caloric curve (total energy per particle vs kinetic energy per particle) for Na_4Cl_4. The upper (O) data mark the density of states $D(\omega)$ at $\omega = 0$ (in units of $6.85 \times 10^{14}\ s^{-1}$), which show $D(\omega = 0) = 0$ for all temperatures. Note sharp molecular isomerization transition

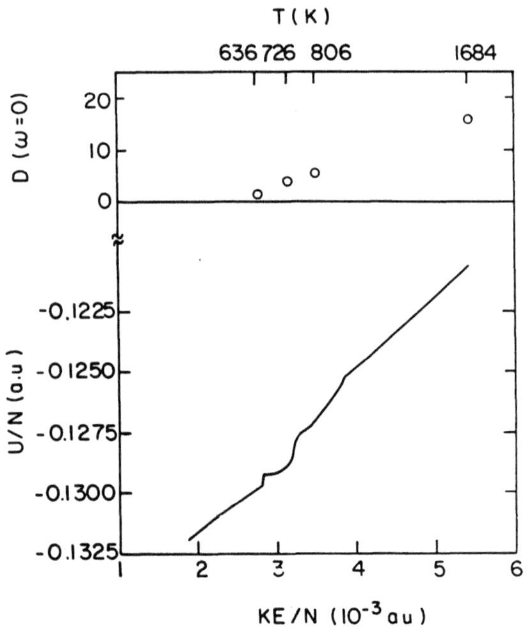

Fig. 2. Caloric curve and zero-frequency density of states for $Na_{16}Cl_{16}$. Note hierarchial isomerization, accompanied by the development of diffusion at higher temperatures. Notations same as in Fig. 1

characteristics (Fig. 2). For these small clusters coexistence is between solid isomers rather than interphase (solid-liquid) coexistence. The latter develops for clusters of sufficient size, characterized by a dense spectrum of accessible states, separated by thermally surmountable barriers [4–7]. Under these circumstances conventional melting is observed as a sharp transition, the separation of time-scales for inter-well and intra-well dynamics ceases and true solid-liquid coexistence is found. These studies, in conjunction with other computer simulations [9–11], reveal that the "melting" of clusters may be synonymous with isomerization between a multitude of isomers. No universal features of the interaction strength and of cluster size on multistate isomerization were yet elucidated.

III. Experimental interrogation of cluster isomerization

Experimental probing of cluster isomerization can rely on structural determinations [12], infrared and Raman spectroscopy [5b], and the measurements of ionization potentials [5b]. Most interesting is the interrogation of vibrational [10, 13] or electronic [5b, 14, 15] excitations of a foreign molecule in a heterocluster, which acts as a probe for the cluster nuclear dynamics. Interrogation of cluster isomerization can rely on vibrational or electronic spectroscopy of a guest molecule in a heterocluster, which acts as a probe for the nuclear dynamics of the cluster. Early spectroscopic studies of $SF_6 \cdot Ar_N$ [13] and tetracene $\cdot Ar_N$ [14] clusters were not size-specific, being hampered by "trivial" inhomogeneous broadening originating from the cumulative absorption of clusters of different size. Size-selective cluster spectroscopy is essential to explore the "intrinsic" inhomogeneous broadening, which originates from different isomers of the same size. Leutwyler and Bosinger [11] have applied the resonant two-photon one-color ionization (R2P1CI) method to study carbazole $\cdot Ar_N$ clusters, while we [15] have utilized the resonant two-photon two-color ionization (R2P2CI) technique for the study of $M \cdot R_N$ clusters (M ≡ tetracene, perylene, anthracene and dicloroanthracene, R = Ar, Kr). Extensive data were obtained for mass-selected clusters of dicloroanthracene (DCA) with Ar and Kr clusters ionized at the low ($\leq 30\ cm^{-1}$) excess vibrational energy of the parent ion, thus precluding any fragmentation and reflecting the dynamics of these clusters at approximately constant temperature (vibrational temperature 20–30 K).

The $S_0 \to S_1$ transition of mass selected $DCA \cdot Kr_N$ heteroclusters (Fig. 3) exhibits diverse spectroscopic phenomena with increasing N. Small clusters (N = 1–4) show a sharp electronic origin together with the intermolecular perpendicular vibration of the R(Ar, Kr) atom with respect to the microsurface ($\omega = 43\ cm^{-1}$ for DCA·Ar and $\omega = 40\ cm^{-1}$ for

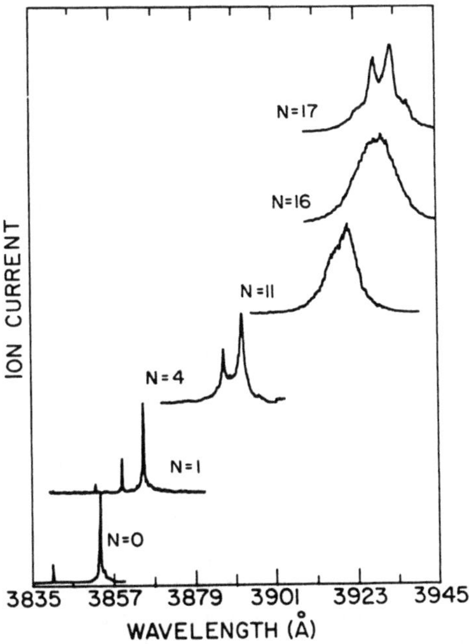

Fig. 3. Optical absorption spectra of mass-selected DCA · Kr_N clusters in pulsed supersonic jets of He (30%) and Ne (70%) obtained by resonant two-photon two-color ionization in conjuction with time of flight mass spectrometry

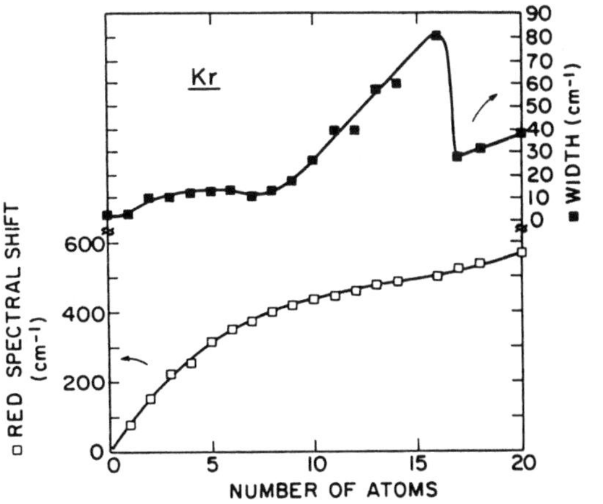

Fig. 4. Size dependence of the red spectral shift and the width (FWHM) of the prominent lowest energy spectral feature of the $S_0 \rightarrow S_1$ transition of DCA · Kr_N clusters

DCA · Kr). Each of the clusters exhibits a single prominent structure. Medium sized DCA · Kr_N clusters ($N = 5$–10) exhibit two qualitative changes: (A) $N = 5$. The appearance of a complex, but well resolved, structure, indicating coexistence of several isomers. These distinct isomers do not communicate on the time scale of the experiment (B) $N = 10$. The disappearance of the spectral structure and the appearance of "slushlike" broadening. This is assigned to a wetting-nonwetting transition, due to a transformation from a 2D to 3D structure [9, 11]. Transitions (A) and (B) were previously considered by Leutwyler and Bösinger [11].

A novel phenomenon on the onset of line narrowing with increasing N was observed for DCA · Kr_{17} clusters (Figs 3 and 4). In the range $N = 10$–16 the spectra of DCA · Kr_N are structureless and broad (Figs. 3 and 4). At $N = 17$ a discrete well-resolved spectral feature abruptly appears, being accompanied by pronounced line narrowing (Figs. 3 and 4), while in this range the red spectral shift exhibits a monotonous slow increase with increasing N. A qualitative similar effect is exhibited for DCA · Ar_N at $N = 20$. The onset of line narrowing is attributed to the nonrigid (disordered) to the rigid multistate isomerization in the 3D heterocluster, which is achieved by increasing N at (approximately) a constant temperature. The onset temperature for the rigid-nonrigid transition in a weakly interacting cluster is expected to increase with increasing N. Accordingly, the increase of N at constant temperature will result in the nonrigid-rigid "collective" isomerization. This transition is expected to be universal for molecular clusters and constitutes the microscopic analog of the melting transition in macroscopic systems.

We shall now proceed from classical dynamic and thermodynamic size effects to quantum size effects.

IV. Excess electrons in polar molecular clusters

Studies of electron localization in finite clusters of polar molecules, e.g., $(H_2O)_n$ and $(NH_3)_n$, open new avenues for probing the size effects of energetics and mechanisms of solvation, pertaining to the questions of the modes of electron localization and the minimal cluster sizes which sustain bound states of an excess electron. While single water and ammonia molecules do not attach an excess electron, weakly bound $(H_2O)_n^-$ ($n = 2, 5, 6$) clusters were detected and strongly bound, negatively charged free water [16–18] $(H_2O)_n^-$, and ammonia [16c], $(NH_3)_m^-$ clusters have recently been observed for $n \geq 11$ and $m \geq 35$, respectively, with accurate values of the vertical binding energies for the $(H_2O)_n^-$ cluster for $n = 2$, 5, 6, and $11 \leq n \leq 40$ being obtained via photoelectron spectroscopy [18]. These systems provide a theoretical challenge for the elucidation of distinct surface and interior localization modes in finite systems and the delicate dependence of their stability on the nature of the electron-molecule and interatomic interactions.

Employing the recently developed quantum path-integral molecular dynamics method (QUPID), which is based on the path-integral formulation of a quantum statistical mechanics [19], and an isomorphism between the quantum problem and a classical one in conjunction with a newly developed electron molecule pseudopotential [20], we have studied [20] the microscopic energetics, structure and stability of electron attachment to clusters of $(H_2O)_n$ and $(NH_3)_n$ for a wide range of cluster sizes and temperatures. In these studies the molecular degrees of freedom are treated classically (no significant isotopic dependence was observed) employing known intra- and intermolecular interactions for water [21] and ammonia [22], and the electron molecule interaction was described via a pseudopotential, constructed [20] in the spirit of the density functional method, consisting of Coulomb, polarization, exclusion and exchange contributions. The energetics of the systems can be expressed in terms of the electron vertical and adiabatic binding energies, EVBE and EABE, respectively, where EVBE is the electron ionization energy of the negatively charged cluster with no cluster reorganization and EABE = EVBE + E_c, where E_c is the cluster reorganization energy, i.e., the difference between the equilibrium intra- and intermolecular energies of the negatively charged cluster and the corresponding neutral. Energetic stability is inferred from the magnitude and sign of EABE, which at the limit of $n \to \infty$ is the heat of electron solvation in the bulk, and with EABE < 0 corresponding to an energetically stable bound state.

The adiabatic binding energies in $(H_2O)_n^-$ (Fig. 5), reveal that for small to intermediate size $(H_2O)_n$ ($8 < n < 64$) clusters stable well-bound electron localization occurs via surface states (SS), which do not constitute the precursors of the celebrated solvated electron. We have established the ubiquity of surface excess electron states on clusters which are exhibited in $(He)_n^-$ [5b], some alkali-halide clusters, i.e., $[Na_{13}Cl_{14}]^+ + e$ [23] and $(H_2O)_n^-$ ($n < 64$). For large $H_2O)_n^-$ clusters, localization in interior states (IS) requires cluster sizes $n \geq 64$. The electron charge distributions of these bound states, which are characterized by the radius of gyration R_g, varies from $R_g \sim 6a_0$ for the SS of the $n = 12$ cluster to $R_g \sim 4.0\,a_0$ of the IS for $n = 128$. Furthermore, the EVBE's of the SS states for the $8 \leq n \leq 32$ water clusters are in quantitative agreement with the measured values [18] (Fig. 6). The measurement of the transition from surface to internal localization upon increasing cluster size, signified by a marked increase in EVBE, (Fig. 6), is left as an experimental challenge.

In contrast to the case of the $(H_2O)_n^-$ cluster, we observed that the onset of stable electron attachment to ammonia clusters occurs via internal localization, requiring $n \gtrsim 32$ molecules, and is not preceded by well-bound surface localized states for smaller clusters. In addition, attempts to bind an electron via surface states to $n > 32$ ammonia clusters resulted in a transition to the internally localized states. The interplay between the competing effects of the interaction of the electron with the molecular constituents, the kinetic energy of localization and the intermolecular interactions (underlying cluster reorganization) precludes the formation of stable well-bounded surface states in

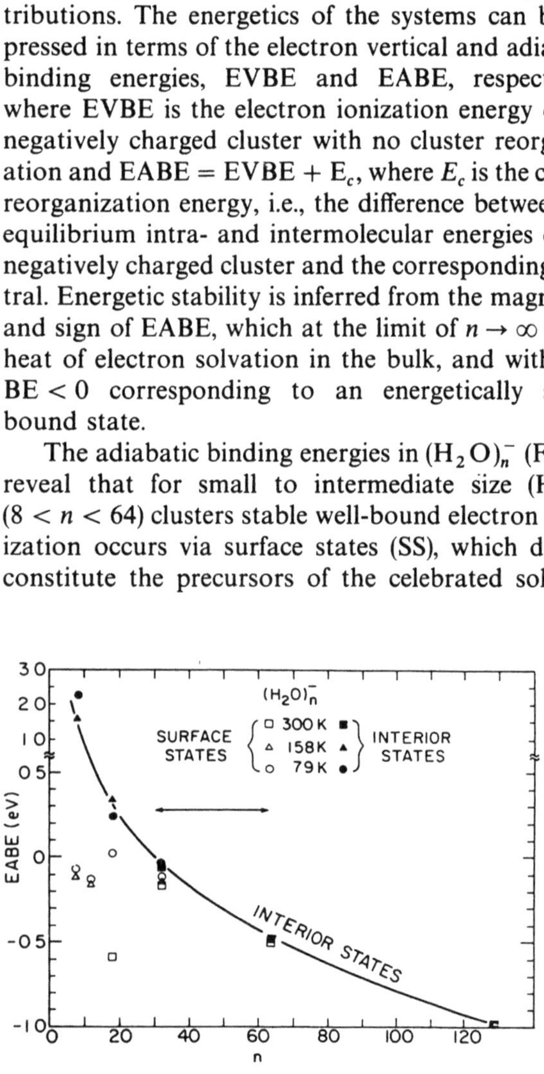

Fig. 5. The size dependence of the EABE in $(H_2O)_n^-$ ($n = 8-128$) clusters. Open symbols refer to surface states while black symbols refer to interior states. (The results for the interior state of the $(H_2O)_8^-$ were obtained for a static configuration of the molecules). Energetically stable surface states are exhibited for medium sized clusters ($8 \leq n \leq 30$), and for $n = 32$ and 64 the energies of surface and interior states are about equal, while for large ($n = 128$) clusters the interior state is energetically stable. The region of crossover from surface to interior states is marked by a horizontal double arrow

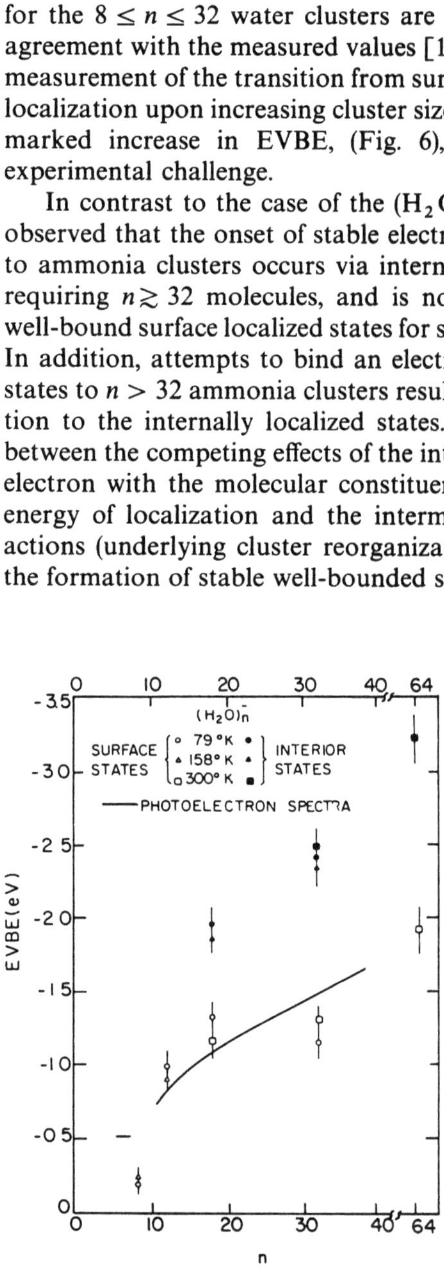

Fig. 6. The size dependence of EVBE of $(H_2O)_n^-$ ($n = 8-64$) clusters. Open symbols refer to surface states while black symbols refer to interior states. The experimental photoelectron detachment data of K. Bowen (reference 18) for $n = 11-40$ are in good agreement with the theoretical data for surface states

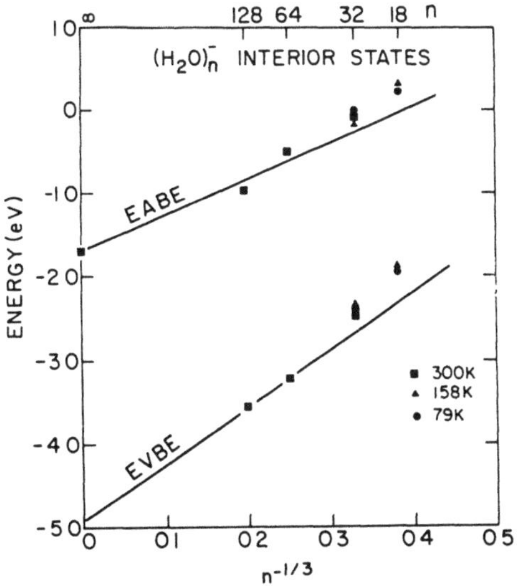

Fig. 7. The dependence of the vertical and adiabatic binding energies of the interior excess electron state in $(H_2O)_n^-$ clusters on $n^{-1/3}$. Upper and lower sets of points represent EABE and EVBE data, respectively at different temperatures as indicated on the figure. The black square for EABE at $n = \infty$ corresponds to the experimental heat of solution of an electron in water. The upper straight line corresponds to $(\text{EABE/eV}) = -1.70 + 4.48\, n^{-1/3}$ while the lower straight line corresponds to $(\text{EVBE/eV}) = -4.90 + 6.80\, n^{-1/3}$. The slopes of these lines are in perfect agreement with the predictions of the dielectric model

ammonia clusters, resulting in the onset of excess electron binding via internal localization for $n \gtrsim 32$, which is in agreement with the facts of life [16c].

The energies of the internal states of both materials exhibit linear dependence versus $n^{-1/3}$ (Fig. 7). An adaptation [20c, e] of a dielectric continuum model [24], where the molecular cluster is described by a dielectric sphere of radius $\bar{R} = r_s n^{-1/3}$ and where r_s is the (mean) radius of the molecular constituent, yields EABE $(\bar{R})$ = EVBE $(\infty) + An^{-1/3}$ and EVBE $(\bar{R})$ = EABE $(\infty) + Bn^{-1/3}$, where $A = e^2/2r_s\,(1 - D_s^{-1})$ and $B = (e^2/2r_s)\,(1 + D_{op}^{-1} - 2D_s^{-1})$ with D_s and D_{op} being the static and optical dielectric constants of the material, respectively. The values of the predicted slopes A and B from the dielectric model agree with those obtained from the simulations results (Fig. 7). Furthermore, an extrapolation of the EABE to $n \to \infty$ yields values in agreement with current experimental estimates (-1.7 eV and -1.1 eV for water and ammonia, respectively) for the bulk heats of solution in these materials. The quantitative analysis of the size dependence of EABE and AVBE for internal states establishes a continuous transition between microscopic solvation effects in finite systems and in the macroscopic polar fluids.

This research was supported by the U.S. DOE under Grant No. FG05-86ER45234 (to U.L.), the U.S.-Israel Binational Science Foundation Grant No 85-00361 (to J.J) and the Commission for Basic Research of the Israel Academy of Sciences (to U.E.).

References

1. Friedel, J.: J. Phys. (Paris) C2, **38**, 1 (1977)
2. Elemental and molecular clusters. Benedek, G., Martin, T.P., Pacchioni, G. (eds.). Berlin, Heidelberg, New York: Springer 1988
3. Small finite systems. Jortner, J., Pullman, A., Pullman, B. (eds.), Dordrecht: B. Reidel 1987
4. (a) Berry, R.S., Jellinek, J., Natason, G.: Phys. Rev. A **30**, 919 (1984); (b) Jellinek, J., Beck, T.L., Berry, R.S.: J. Chem. Phys **84**, 2783 (1986); (c) Beck, T.L., Jellinek, J., Berry, R.S.: J. Chem Phys. **87**, 545 (1987); (d) Beck, T.L., Berry, R S.: J. Chem Phys **88**, 3910 (1988)
5. (a) Luo, J. Landman, U., Jortner, J.: In: Physics and chemistry of small clusters. Jena, P., Rao, B.K., Khanna, S.N. (eds.), NATO Series B: **158**, New York: Plenum, 1987; (b) Jortner, J., Scharf, D., Landman, U.: Elemental and molecular clusters. Beneder, G. Martin, T.P., Pacchioni, G. (eds.) p. 148. Berlin, Heidelberg, New York: Springer 1988
6. Honeycutt, J.D., Andersen, H.C.: J. Phys. Chem. **91**, 4950 (1987)
7. (a) Blaisten-Barojas, E., Andersen, H.C.: Surf. Sci. **156**, 548 (1985); (b) Blaisten-Barojas, E., Garzon, I.L., Avalos-Borja, M.. Phys. Rev. **36**, 8447 (1987)
8. Hoare, M.R., McInnes, J.A.: Adv. Phys. **32**, 791 (1983)
9. Blaisten-Barojas, E.: Elemental and molecular clusters. Beneder, G. Martin, T.P., Pacchioni, G. (eds.), p. 102. Berlin, Heidelberg, New York: Springer 1988
10. (a) Eichenauer, D., Le Roy, R.J.: Phys. Rev. Lett. **57**, 2920 (1986); (b) Eichenauer, D., Le Roy, R.J.: Small finite systems. Jortner, J., Pullman, A., Pullman, B. (eds.) p. 153. Dordrecht: B. Reidel 1987; (c) Eichenauer, D., Le Roy, R.J.: J. Chem. Phys. **88**, 2898 (1988)
11. (a) Bösinger, J., Leutwyler, S.: Phys. Rev. Lett. **59**, 1895 (1987); (b) Leutwyler, S., Bösinger, J.: Small Finite Systems, Jortner, J., Pullman, A., Pullman, B. (eds.) p. 165. Dordrecht: B. Reidel 1987; (c) Bösinger, J., Leutwyler, S.: J. Chem. Phys. (in press); (d) Leutwyler, S., Bösinger, J.: J. Chem. Phys. (in press)
12. Farges, J., de Feraudy, M.F., Raoult, B., Torchet, J.: J. Chem. Phys. **78**, 5067 (1983)
13. Gough, T.E., Knight, D.G., Scoles, G.: Chem. Phys. Lett. **97**, 155 (1983)
14. Amirav, A., Even, U., Jortner, J.: J. Phys. Chem. **86**, 3345 (1982)
15. Even, U., Ben-Horin, N., Jortner, J.: Phys. Ref. Lett. (in press)
16. (a) Haberland, H. Langosch, H., Schindler, H.G., Worsnop, D.R.: Surf. Sci. **156**, 157 (1985); (b) Haberland, H., Schindler, H.G., Worsnop, D.R.: J. Chem. Phys. **81**, 3742 (1984); (c) Haberland, H., Schindler, H.G., Worsnop D.R.: Ber. Bunsenges. Phys. Chem. **88**, 270 (1984)
17. (a) Knapp, M., Echt, O., Kreisle, D., Recknagel, E.: J. Chem. Phys. **85**, 636 (1986); (b) Knapp, M., Echt, O., Kreisle, D., Recknagel, E.: J. Phys. Chem. **91**, 2601 (1987)
18. Bowen, K.H.: Z. Phys. D – Atoms, Molecules and Clusters **12** (1989) (this issue)
19. For a recent review see: Berne, B.J., Thirumalai, D.: Annu. Rev. Phys. Chem. **37**, 401 (1986)
20. (a) Landman, U., Barnett, R.N., Cleveland, C.L., Scharf, D., Jortner, J.: J. Phys. Chem. **91**, 4890 (1987); (b) Barnett, R.N., Landman, U., Cleveland, C.L., Jortner, J.: Phys. Rev. Lett. **59**, 811 (1987); (c) Barnett, R.N., Landman, U., Cleveland, C.L., Jortner, J.: J. Chem. Phys. **88**, 4421, 4429 (1988); (d) Barnett, R.N., Landman, U., Jortner, J.: Chem. Phys. Lett. **145**, 382 (1988); (e) Barnett, R.N., Landman, U., Kestner, N.R., Jortner, J.: J. Chem. Phys. **88**, 6670 (1988)
21. (a) Reimers, J.R., Watts, R.O., Klein, M.L.: Chem. Phys. **64**, 95 (1982); (b) Reimers, J.R., Watts, R.O.: ibid. **85**, 83 (1984)
22. Hinchliffe, A., Bounds, D.G., Klein, M.L., McDonald, I.R., Righini, R.: J. Chem. Phys. **74**, 1211 (1981)
23. Landman, U., Scharf, D., Jortner, J.: Phys. Rev. Lett. **54**, 1860 (1985)
24. Jortner, J.: J. Chem. Phys. **30**, 839 (1959)

Z. Phys. D – Atoms, Molecules and Clusters 12, 173–175 (1989)

Atoms, Molecules and Clusters
Zeitschrift für Physik D
© Springer-Verlag 1989

Computer simulations of photoionization

B. Wassermann and T.P. Martin
Max-Planck-Institut fur Festkörperforschung Heisenbergstrasse 1, D-7000 Stuttgart 80, Federal Republic of Germany

Received 5 July 1988

The Energy Distribution of Ionizing Transitions (EDIT) for $(Cs_2O)_n$ clusters has been calculated using the technique of molecular dynamics. Isomerization, thermal atomic motion and the electronic density of states all contribute to the shape of these curves.

PACS: 33.80.E

I. Introduction

Two closely related techniques are now being used to provide information concerning the electronic properties of mass selected clusters. One technique measures the wavelength dependence of the photoionization cross-section (photoionization spectroscopy). The other technique provides data on the energy distribution of the emitted electrons after photoionization at a fixed wavelength (photoelectron spectroscopy). The interpretation of these experiments is made complicated by the thermal motion of the atoms, isomerization and the multiplicity of ionizing transitions. It is, however, possible to sort out this complicated behaviour in some cases with the help of computer simulations. In this paper we will generate photoionization spectra for $(Cs_2O)_n$ clusters using the technique of molecular dynamics.

The calculational procedure is depicted schematically in Fig. 1. A pair potential is used to define a total potential energy surface in configuration space. Every conceivable arrangement of atoms in the cluster is represented by a point on this surface. Of course, this cannot be a complished in the two-dimensional space shown here. Configuration space is multidimensional, containing as many dimensions as the cluster has vibrational degrees of freedom. The molecular dynamics method allows us to follow the motion of the atoms with time. This motion can be described by a path on the total energy surface. Associated with each point on the total energy surface, (that is with each atomic arrangement) is a corresponding ionization energy. The ionization energy can also be represented by a surface in configuration space. Actually, there will be several surfaces corresponding to ionization to the ground state and to excited states of the ion. The path on the total energy surface maps into a path on the ionization energy surface which in turn defines an energy distribution of ionizing transitions (EDIT). It is this distribution which is measured in an experiment. Peaks in the distribution may occur in regions of high symmetry and high degeneracy, or in flat regions of the surface where the ionization energy is insensitive to atomic arrangement.

2. Interaction between ions

Ideally, the structure of all clusters should be determined using quantum mechanical methods [1–6]. We believe that for the Cs-O system the removal of an electron from a neutral stoichiometric cluster can be meaningfully modeled without the use of quantum mechanics. Because of the spherical, closed shell character of alkali metal and oxygen ions, the interactions appear to be well-described by a classical, two-body potential [7, 8]. Since charge is localized around the atoms, it is meaningful to describe the ionization process as a sudden neutralization of one of the anions.

The simplest two-body interaction capable of describing ionic bonding contains only two terms, a

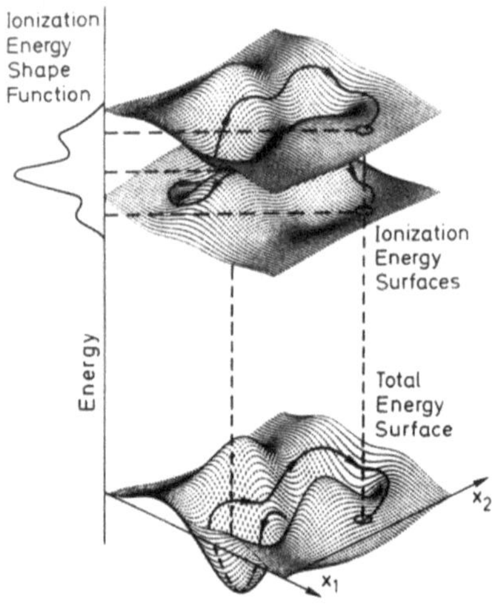

Fig. 1. The motion of the atoms in a cluster is represented by a path on the total energy surface. This path, when projected onto the ionization energy surface, defines an experimentally accessible EDIT (Energy Distribution of Ionizing Transitions) curve

point ion Coulomb and a Born-Mayer repulsive contribution:

$$V_{ij} = Z_i Z_j / r_{ij} + A \exp(-r_{ij}\beta)$$

The charge on an oxygen ion is not uniquely defined. It depends on the oxygen ion's environment. It is thought to be about -1.8 in the compound $Cs_{11}O_3$ [8]. However, this compound provides the oxygen atoms with a cesium-rich environment. The oxygen charge in our clusters is chosen to be -1.6 and the charge on cesium to be $+0.8$. The remaining parameters A and β have been fitted to the vibrational frequency of the antisymmetric mode of matrix isolated Cs_2O molecules and to the interatomic distance in Cs_2O crystals. The values used are $A = 6.2 \times 10^{-9}$ ergs and $\beta = 3.22\ \text{A}^{-1}$.

The energy required to ionize a cluster is the sum of the energy necessary to remove an electron from an isolated negative ion, i.e. the electron affinity, plus the interaction of the electron with all other ions in the cluster. In our simple model this last contribution is just the interaction between point charges,

$$IE_i = EA_i + \sum_j Z_i Z_j / r_{ij}.$$

Neglected here is the short range repulsive interaction between the electron and the occupied orbitals of neighbouring ions. This interaction would tend to lower the ionization energy. We have also neglected the dipolar interaction between polarizable ions and the electron removed.

3. Results

Consider the specific cluster $Cs_{10}O_5$. Ideally, the ensemble averages should be independent of starting configuration. Therefore, at $t = 0$ we assume the cluster is in the shallow local minimum corresponding to a chain configuration, i.e. oxygen ions are bridged with two Cs ions. Velocities corresponding to a temperature of 350 K are assigned to the ions during a brief heating period.

According to our model this cluster will have five distinct ionization energies determined by the electric field at each of the oxygen ions. The ionization energies are continually shifting as the atoms move with time. Figure 2 shows two sets of ionization energies, the instantaneous values for the cluster shown and also the values averaged starting from the first time step.

Due to its high symmetry, the chain at zero temperature has only three distinct instantaneous ionization energies. Even at higher temperatures the five non-degenerate energies are closely grouped together. Although the chain configuration is not particularly stable, it has a large catchment area. More than 20,000 time steps are required before a new local minimum is reached, B. Here a knot forms on one end of the chain. This knot seems to act as a seed for further isomerization, because a short time later the cluster condenses into a much more stable "planar" configuration, C. Since this condensation is isoenergetic, the

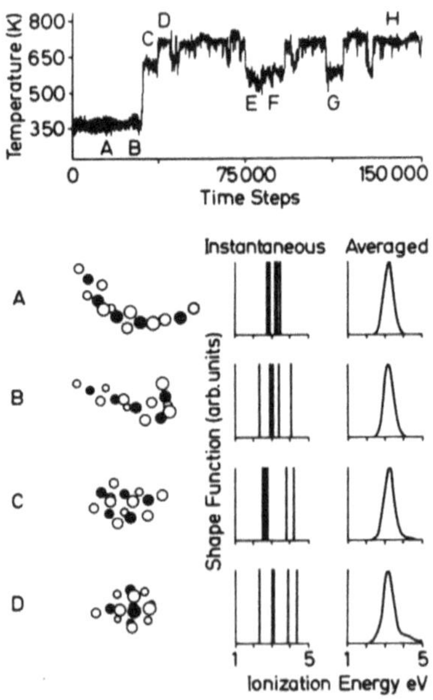

Fig. 2. Results of a molecular dynamics run for a $Cs_{10}O_5$ cluster Shown are cluster shape, temperature, and ionization energies at various instants in time. The time step is 2.4×10^{-15} s. Total time elapsed is only 0.4 nanosec

cluster heats up to about 600 K. That is, as the cluster reaches a new deeper local minimum on the potential energy surface, its kinetic energy increases, thus keeping the total energy constant. Shortly thereafter the cluster reaches an even deeper minimum *D* with a corresponding increase in temperature. Local minimum *D* corresponds to the first three dimensional configuration. Actually, there exists a set of three dimensional configurations having nearly the same binding energy. However, the cluster has too much kinetic energy to remain long in this set of deep minima. At E, for example, it returns to a "planar" form.

References

1. Swepston, P.N., Sellers, H.L., Schäfer, L.: J. Chem. Phys. **74**, 2372 (1981)
2. Landman, U., Scharf, D., Jortner, J.: Phys. Rev. Lett. **54**, 1860 (1985)
3. Galli, G., Andreoni, W., Tosi, M.P.: Phys. Rev. A **34**, 3580 (1986)
4. Dunlap, B.I.: J. Chem. Phys. **84**, 5611 (1986); J. Phys. Chem. **90**, 5524 (1986)
5. Scharf, D., Landman, U., Jortner, J.: J. Chem. Phys. **87**, 2716 (1987)
6. Sonil, K.K., Jordan, K.D.: J. Phys. Chem. **91**, 1710 (1987)
7. Martin, T.P.: Phys. Rep. **95**, 167 (1983)
8. Martin, T.P., Stolz, H.-J., Ebbinghaus, JG., Simon, A.: J. Chem. Phys. **70**, 1096 (1979)

Z. Phys. D – Atoms, Molecules and Clusters 12, 177–180 (1989)

Atoms, Molecules
Zeitschrift für Physik D and Clusters
© Springer-Verlag 1989

Rotating clusters: centrifugal distortion, isomerization, fragmentation

D.H. Li and J. Jellinek
Chemistry Division, Argonne National Laboratory, Argonne, IL 60439, USA

Received 7 July 1988

The effect of the overall rotation, as characterized by the total angular momentum, on the structure of an Ar_{13} cluster is studied by using a new technique for separation of the rotational motion from vibrations. No a priori limitation on the degree of nonrigidity of the cluster is imposed. The different structures for different total angular momenta of the cluster are arrived at by a thermal quenching procedure which preserves the overall rotation. It is shown that the centrifugal distortion is anisotropic and it may lead to isomerization transitions and even fragmentation of the cluster.

PACS: 82.30.Qt; 36.40.+d; 03.20.+i

I. Introduction

The different aspects of isomerization and phase change transitions in small clusters have been given considerable attention in the past few years; for recent reviews see [1]. As a result, a new picture of these phenomena is emerging which greatly enhances our understanding of both the structural and the dynamical properties of cluster systems. A serious shortcoming of the existing dynamical studies, however, has to be pointed out: virtually all of them assume that the total angular momentum of the cluster is zero, i.e., that the cluster is nonrotating. The reason for imposing this restriction lies primarily in the lack of a general approach which would be capable of treating the overall rotation in an arbitrary nonrigid system. It turns out, however, that the rotational motion has a pronounced and quite intricate effect [2, 3] and by no means is it allowed to be neglected a priori.

Recently we have suggested [4] a completely general procedure for separation of the overall rotation from the vibrational ("internal," thermal) motion in any N-body system, however floppy. In this communication we report results of application of this procedure to an Ar_{13} cluster. The specific phenomenon studied is that of centrifugal distortion of the rotating cluster as a function of its total angular momentum. This distortion, as will be shown, is anisotropic and it may lead to structural changes and even fragmentation of the cluster. In section II we recap the theoretical basis of our separation scheme. In section III the computational procedure, the results, and their discussion are presented. A brief summary is given in conclusion.

II. Separation of the overall rotation

Consider a classical N-body system evolving in accordance with Hamilton equations of motion. If its coordinates and momenta in the center of mass laboratory-oriented frame are $\{\vec{r}_i\}$ and $\{\vec{p}_i\}$, $i=1, \ldots, N$, respectively, then at any instant t one can define the tensor of inertia $\vec{I}(t)$ as a 3×3 matrix with matrix elements

$$I_{\alpha\alpha}(t)=\sum_i m_i(r_i^2(t)-\alpha_i^2(t)),$$

$$I_{\alpha\beta}(t)=-\sum_i m_i\alpha_i(t)\beta_i(t);\ \alpha, \beta(\neq\alpha)=x, y, z, \qquad (1)$$

where m_i is the mass of the i-th particle and x, y, z are the components of the vector $\vec{r}$. Knowing the total angular momentum $\vec{L}$ of the cluster,

$$\vec{L} = \sum_i \vec{r}_i(t) \times \vec{p}_i(t), \qquad (2)$$

which is a constant of motion, we can define the instantaneous angular velocity of the overall rotation of the body as a vector $\vec{\omega}(t)$ satisfying the equation

$$\vec{L} = \vec{I}(t) \cdot \vec{\omega}(t). \qquad (3)$$

The instantaneous energy of the overall rotation can be defined as

$$E_{\vec{L}}^{Rot}(t) = \tfrac{1}{2}\vec{\omega}(t) \cdot \vec{L}, \qquad (4)$$

and the total energy E^{Tot} of the body can be partitioned

$$E^{Tot} = E^{Th} + E_{\vec{L}}^{Rot} + V. \qquad (5)$$

In (5) $V = V\{\vec{r}_i\}$ is the potential energy and E^{Th} is the thermal ("internal," vibrational) energy defined as

$$E^{Th} = \sum_i (\vec{p}_i - m_i \vec{\omega} \times \vec{r}_i)^2 / 2m_i. \qquad (6)$$

That the partitioning (5), which completely separates the energy of rotation from that of vibrations, can always be performed is proved in [4]. Taking into account that $\vec{I}$, $\vec{\omega}$ and thus $E_{\vec{L}}^{Rot}$ can be considered as functions of $\{\vec{r}_i\}$ [cf. (1), (3), and (4)] we can rewrite (5) as

$$E^{Tot} = E^{Th} + V_{\vec{L}}^{Ad}, \qquad (7)$$

where

$$V_{\vec{L}}^{Ad}(\{\vec{r}_i\}) = E_{\vec{L}}^{Rot}(\{\vec{r}_i\}) + V(\{\vec{r}_i\}) \qquad (8)$$

is the $\vec{L}$-dependent "rotationally adiabatic potential surface" [4]. By letting $E^{Th} \rightarrow 0$ one "freezes" the system into an essentially rigid (nonvibrating) body. Note, however, that this thermal quenching procedure is distinctly different from the traditional one ($\vec{p}_i \rightarrow 0$) in that it preserves the overall rotation of the system ($\vec{L}$ = const.) leading to equilibrium structures of rotating bodies.

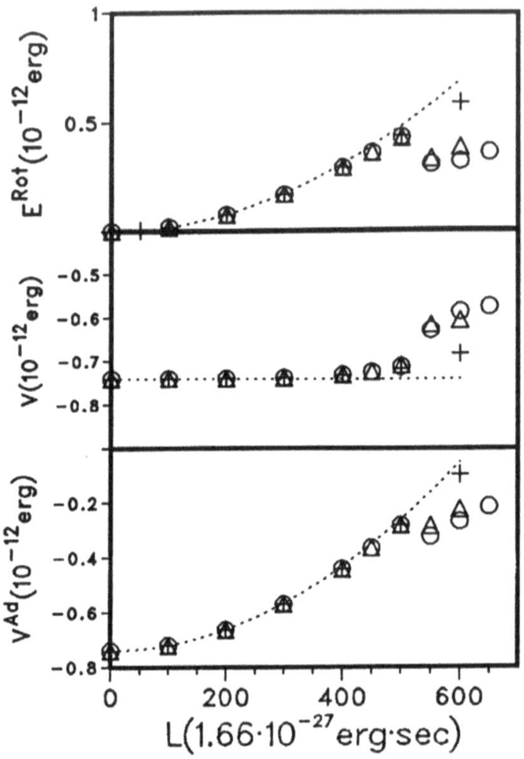

Fig. 1. E^{Rot}, V, and V^{Ad} (see text) of the thermally quenched initially ($\vec{L}=0$) spherical top isomer as a function of L. The dotted line corresponds to the rigid icosahedron. Symbols for the Lennard-Jones cluster rotating around: C_5-axis (+); C_3-axis (○); C_2-axis (Δ)

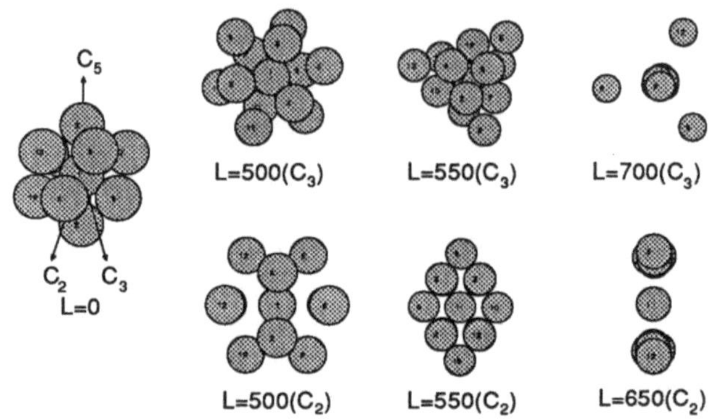

Fig. 2. Different structures of the initially spherical top isomer as functions of L. For $\vec{L} \neq 0$ the rotation axes are perpendicular to the plane of the pictures and cross the center of mass Their symmetry is indicated in parenthesis

III. The computational procedure, results, and discussion

We restrict our consideration here to cases where $\vec{\omega}$ is directed along one of the principal axes of the system since in these cases, as will be discussed in detail elsewhere [3], the thermal quenching can be performed most efficiently. Two different isomers of the initially nonrotating Lennard-Jones Ar_{13} cluster (see Figs. 2 and 4) were used in simulations. For a chosen nonzero $\vec{L} = \vec{L}_0$, directed along a selected principal axis, the same angular velocity $\vec{\omega}_0$ is assigned to all atoms of the cluster; $\vec{\omega}_0$ is calculated from (3) using $\vec{L} = \vec{L}_0$ and $\vec{I}(t=0)$ corresponding to the equilibrium geometry of a nonrotating isomer. This places the cluster onto the $V_{\vec{L}_0}^{Ad}$ surface. "Kicked" by the surface

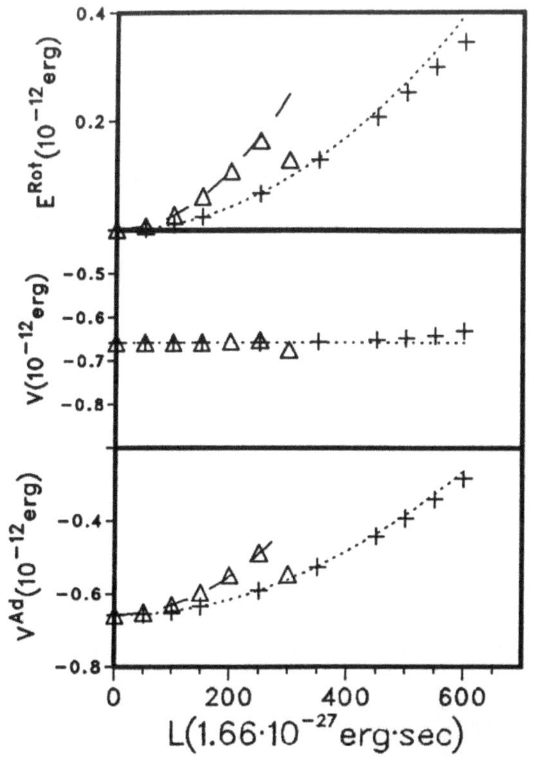

Fig. 3. The same as Fig. 1 but for initially ($\vec{L}=0$) asymmetric top isomer The dotted and dashed lines correspond to the rigid cluster rotating around the C_2- and C_1-axis, respectively. Symbols for the Lennard-Jones cluster rotating around: C_2-axis (+); C_1-axis (Δ)

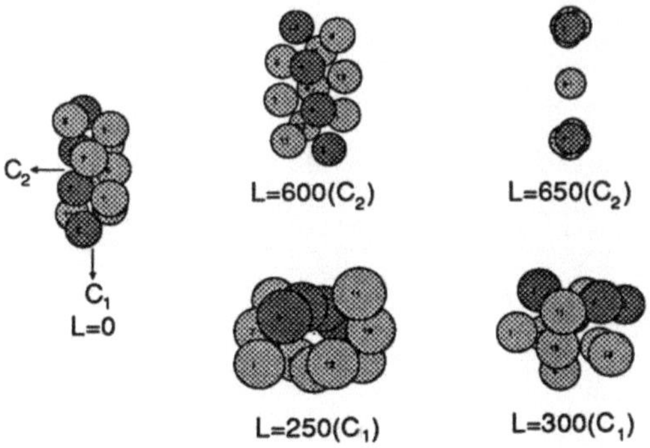

Fig. 4. The same as Fig. 2 but for initially asymmetric top isomer

the cluster is allowed to follow its Hamiltonian evolution for 2.5 psec during which part of the rotational energy transforms into thermal energy. After 2.5 psec the thermal energy is extracted which places the cluster into a lower lying point of the $V^{Ad}_{\vec{L}_0}$ surface. The 2.5 psec time evolution and subsequent thermal quenching steps are repeated until the difference between two consecutive points on the $V^{Ad}_{\vec{L}_0}$ surface visited by the cluster is less than $10^{-5}\%$. When this is achieved the cluster is deemed thermally quenched (i.e., it found a local minimum of $V^{Ad}_{\vec{L}_0}$) and its $E^{Rot}_{\vec{L}_0}$, $V(\vec{L}_0)$ and $V^{Ad}_{\vec{L}_0}$ are recorded. Then a new $\vec{L}=\vec{L}_1>\vec{L}_0$ directed along the same principal axis is considered and a new angular velocity $\vec{\omega}_1$ is assigned to atoms; $\vec{\omega}_1$ is calculated from (3) using $\vec{L}=\vec{L}_1$ and $\vec{I}(t)$ corresponding to the cluster in the minimum of $V^{Ad}_{\vec{L}_0}$. The cluster is placed now onto the $V^{Ad}_{\vec{L}_1}$ surface. The 2.5 psec time-evolutions followed by thermal quenches are repeated until the cluster finds a local minimum of the $V^{Ad}_{\vec{L}_1}$ surface. Then $E^{Rot}_{\vec{L}_1}$, $V(\vec{L}_1)$, and $V^{Ad}_{\vec{L}_1}$ corresponding to this minimum are recorded and so on. In this fashion the dependence of E^{Rot}, V, and V^{Ad} of the thermally quenched clusters on $\vec{L}$ directed along a selected principal axis is obtained. This dependence indicates centrifugal distortion, isomerization transition and even fragmentation of the cluster as a result of its rotation (see Figs. 1–4). The reason for the frequent quenching (every 2.5 psec) is to allow for study of the effect of rotation on a selected geometry (isomer) of the cluster before it has time to wander away from this geometry (at higher energies even nonrotating clusters can undergo spontaneous isomerization transitions [5]).

In Fig. 1 the results for the highly symmetrical icosahedral (at $\vec{L}=0$) isomer are shown. Initially a spherical top, it has three different axes of rotational symmetry: a C_5-axis through a surface atom and the center atom, a C_3-axis through the center of a face and the center atom, and a C_2-axis through the center of an edge and the center atom. The dotted line is obtained assuming that the icosahedron is a true rigid body. The deviations from the dotted line are the centrifugal distortions of the Lennard-Jones cluster characteristic for rotations around each of the three different symmetry axes. It is seen that the rotation around the C_5-axis is considerably more resistant to centrifugal distortions than rotations around the C_3- and C_2-axes, although below $L=500$ (in units of $1.66 \cdot 10^{-27}$ erg · sec) the distortions remain moderate and almost the same for all three axes. The quenches at $L=550$ produce new isomers for the C_3- and C_2-axes (cf. the abrupt changes in the graphs) which eventually fragment at $L=650$ for the C_2-axis and at $L=700$ for the C_3-axis. The pictures of the thermally quenched clusters for the characteristic cases are shown in Fig. 2.

In Fig. 3 we display the results for rotations around two principal axes of a less symmetric (asymmetric top at $\vec{L}=0$) isomer of Ar_{13}; one of the axes corresponds to the largest moment of inertia and has C_2 symmetry while the other is that with the smallest moment of inertia and has C_1 symmetry. Again, only moderate centrifugal distortions from the true rigid bodies (dotted and dashed lines) corresponding to the two different axes are observed before the cluster undergoes structural changes. These changes are detected as a result of thermal quenching at $L=300$ for the C_1-axis and at $L=650$ for the C_2-axis. In the new structure at $L=300$ the $\vec{\omega}$ vector turned out to be substantially out of the C_1 principal axis and only a partial thermal quenching could be achieved in a reasonable time. Interestingly, rotation around the C_2-axis resulted in fragmentation of the cluster at $L=650$ without a preceding isomerization transition. The pictures of characteristic cases for the asymmetric top isomer are shown in Fig. 4. Both Figs. 1 and 3 indicate that the lower the degree of symmetry of the cluster with respect to the rotation axis the more readily it undergoes isomerization transition(s) and/or fragmentation. The changes in the structure of the cluster induced by its rotation obey a "conservation of symmetry" law as is attested by Figs. 2 and 4. Rotation around a C_n-axis may create only isomers or a system of fragments of a cluster which have the same n-fold rotational symmetry. A note of caution should be made here: round-off errors can easily cause violation of this law in actual computer simulations unless these are performed with due care.

IV. Summary

Using a new scheme for separation of the overall rotation and a thermal quenching technique preserving the total angular momentum of the system we have studied the effect of rotation on equilibrium structures of a Lennard-Jones Ar_{13} cluster. The effect is that of anisotropic centrifugal distortion, isomerization and fragmentation of the cluster. More symmetric structures exhibit higher stability vis a vis rotation. If a change in the structure occurs it obeys the "symmetry conservation" law.

We thank Dr. W.F. Hoffman for using his program to produce the cluster pictures. This work was supported by the U.S. Department of Energy, Office of Basic Energy Sciences, Division of Chemical Sciences, under Contract W-31-109-Eng-38.

References

1. Advances in Chemical Physics, Vol. 70, Pts. 1 and 2. Prigogine, I., Rice, S.A. (eds.). New York: Wiley 1988
2. Jellinek, J., Jasien, P.G.: In: Structure of small molecules and atoms. Naaman, R., Vager, Z. (eds.), p. 39. New York: Plenum (1988)
3. Li, D.H., Jellinek, J.: (to be published)
4. Jellinek, J., Li, D.H.: Phys. Rev. Lett. **62**, 241 (1989)
5. Jellinek, J., Beck, T.L., Berry, R.S.: J. Chem. Phys. **84**, 2783 (1986)

Z. Phys. D – Atoms, Molecules and Clusters 12, 181–183 (1989)

Atoms, Molecules
Zeitschrift für Physik D and Clusters
© Springer-Verlag 1989

More on the melting of Lennard-Jones clusters

I.L. Garzón[1], M. Avalos-Borja[1], and E. Blaisten-Barojas[2,]*

[1] Instituto de Física, Universidad Nacional Autónoma de México, Apartado Postal 2681, 22800 Ensenada, Baja California, México

[2] Department of Physics, Virginia Commonwealth University, 1020 West Main Street, Richmond, VA 23284-2000, USA

Received 5 July 1988, final version 30 September 1988

The melting of 13-atom clusters interacting via Lennard-Jones potentials has been revisited using molecular dynamics coupled to steepest descent quenches. A procedure was devised to account for the fraction of times the global and local minima of the potential energy surface are accessed during a long trajectory. This quantity presents a sigmoid shape. A phenomenological model of melting is given in terms of a correlated walk that maps the short time excursions among the global and local minima in configuration space. Comparison between the simulation results and the theoretical model shows that the melting transition is well described in terms of the temperature changes of the fraction of high energy minima accessed during the cluster trajectory. Cooperativity is clear from the S shape of this quantity, i.e., the access to a local minimum favours the access to other local minima.

PACS: 36.40.+d; 64.70.Dv

Although the melting transition in Lennard-Jones clusters was studied several years ago [1], many features are not yet understood. Recent molecular dynamics simulations [2, 3, 4] have shown that very long time averages are needed in order to compensate for the large thermodynamic fluctuations characteristic of systems with a few atoms. Melting in finite size systems is commonly described in terms of a property that changes abruptly with temperature between two extreme values [5]. The Ising and related models have been applied with success to a number of finite systems such as polymers or biological molecules [5, 6]. We show here that this model can also be applied to clusters. A procedure is devised to measure the fraction of times the global or the local minima of the potential energy are accessed during a long trajectory. This quantity changes abruptly with temperature – at low temperatures the system is always found to access the global minimum, whereas at high temperatures the system accesses the local minima. At the transition temperature the system is found to access half of the times the global minimum and half of the times the local minima. Extensive molecular dynamics simulations were performed on the 13-atom cluster to measure this quantity. Furthermore, we give a phenomenological description of the temperature behavior of this quantity based on a correlated walk model [7].

The system is a 13-atom cluster where the atoms interact via Lennard-Jones potentials with constants ε, σ. The configuration space of this cluster is well known [8]. Namely, the global minimum of the potential energy is the icosahedron and there are at least 987 local minima corresponding to stable isomers. The binding energy of these conformations can be cast into a distribution where the icosahedron stands well detached by a gap of 2.85ε from the isomers with lower binding energies. We will refer to this gap as ΔV.

Constant energy molecular dynamics was used by solving the Newton equations of motion with a time step of 0.01τ, $\tau=\sqrt{mr_0^2/\varepsilon}$. The adopted units of energy

* On Sabbatical leave from Instituto de Física, Universidad Nacional Autónoma de México, A. Postal 20-364, México 01000, D.F

and distance were ε and $r_0 = 2^{1/6}\sigma$. Temperature was defined as $kT = 2\langle E_{kin}\rangle/(3N-6)$, with $\langle E_{kin}\rangle$ being the average kinetic energy without the center of mass and angular momentum contributions. It will be given in units of ε/k (k = Boltzmann's constant).

The computer experiment was started from a frozen icosahedral cluster. Next, the cluster was heated in a steplike manner [4]. The resulting total energy per particle as a function of temperature is shown in Fig. 1. In reaching the loop-shaped region of the curve, the cluster is hot enough to rearrange its bonds. In the course of a highly excited asymmetric vibration, the bond between one of the surface atoms and the central atom breaks. This process gives rise to the first isomer above ΔV, i.e., a structure where the atom that pops out decorates the face opposite to the hole left behind [8]. Now it is easier for the cluster to rearrange its atoms in the many other mechanically stable isomers with binding energies above the gap ΔV. Therefore in the loop shaped region of Fig. 1 the cluster is accessing other minima besides the icosahedron.

Our overall approach is to devise a measure of the way the cluster accesses all these isomers in configuration space and how this process changes with temperature. To fulfil this goal, each of the states labeled *a* to *f* in Fig. 1 was propagated for 300τ more in the following way. Every 5τ the molecular dynamics run was stopped and a quench following steepest descent paths drove the system to one of the minima of the potential energy V. A particular minima is visited every 5τ on the course of the 300τ trajectory. These visits were collected in histograms as shown in Fig. 2 for the temperatures labeled *a–f* in Fig. 1. At low temperatures the icosahedron (peak of low potential energy) is visited more frequently than the isomers above ΔV. As temperature increases, vibrations are

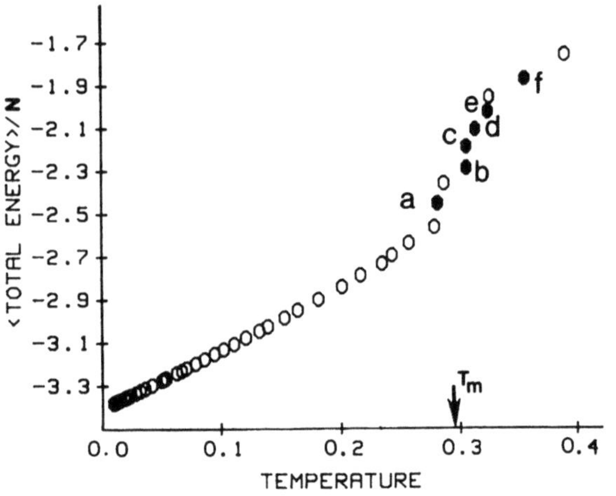

Fig. 1. Total energy as a function of temperature for the 13-atom cluster

excited allowing the cluster to restructure more easily from one isomeric form into another. Therefore, it is clear that temperature favors the access to minima above the gap. The fraction $\langle f\rangle$ of minima above the gap visited out of 60 visits changes abruptly from zero at low temperatures to one at high temperatures.

Let us consider a model cluster and speak of an intact icosahedron or broken icosahedron during a length of observation segmented in n pieces. An intact icosahedron is realized when the global minimum is visited, whereas a broken icosahedron occurs when any local minima above the gap ΔV is visited. The state of the system in the jth segment of time will be represented by μ_j ($\mu_j = 1$ for broken icosahedron and $\mu_j = -1$ for intact icosahedron). Thus, the number of broken icosahedra is $N_B\{\mu\} = \frac{1}{2}\Sigma_{j=1}^{n}(1+\mu_j)$ in a given configuration $\{\mu\} = \{\mu_1 \ldots \mu_n\}$. An ordered configuration is obtained when all μ's are -1. On the contrary, disorder increases the more 1's appear in a given configuration $\{\mu\}$. Further, let us propose a model energy in terms of the melting temperature T_m, the gap ΔV and δ the mean nearest-neighbor energy spacing between the potential energy of those local minima above ΔV:

$$E(T) = -B(T)\sum_{i=1}^{n}\mu_i - K\sum_{i=1}^{n}\mu_i\mu_{i+1} \qquad (1)$$

where $B(T) = (T-T_m)\Delta V/T$ and $K = \sigma\Delta V/\delta$. Here σ is a parameter. The partition function is easily solved in terms of the eigenvalues of the transfer matrix [6] such that the fraction of broken icosahedra is

$$\langle f\rangle = \frac{N_B}{n} = \frac{1}{2}\left(1 + \frac{\sinh(B/kT)}{\sqrt{e^{-4K/kT} + \sinh^2(B/kT)}}\right) \qquad (2)$$

$B(T)$ has the meaning of binding in the system and K stands for the correlation between neighboring segments. When $K\to 0$, either because $\sigma\to 0$ or $\Delta V\to 0$, all isomeric forms are uncorrelated in time. We recover the random model in which $\langle f\rangle$ changes smoothly with temperature. When K is large the quantity $\langle f\rangle$ becomes a step function, i.e., $\langle f\rangle = 1$ when $B>0$ and $\langle f\rangle = 0$ when $B<0$. The bulk limit is obtained when $\delta\to 0$, thus $K\to\infty$.

Now we refer to the computer experiment and assign values to the parameters as supported by the simulation. Thus, $\Delta V = 0.220$, $T = 0.3$, $\delta = 0.011$ and $\sigma = 0.022$. Substituting these values in Eqs. 1, 2, the fraction of broken icosahedra can be calculated analytically. The resulting curve is shown in Fig. 3 (continuous line) and compared to both the experimental evidence coming from the simulation (dots) and to the random model (dashed line). The agreement between theory and experiment is striking considering the simplicity of the model.

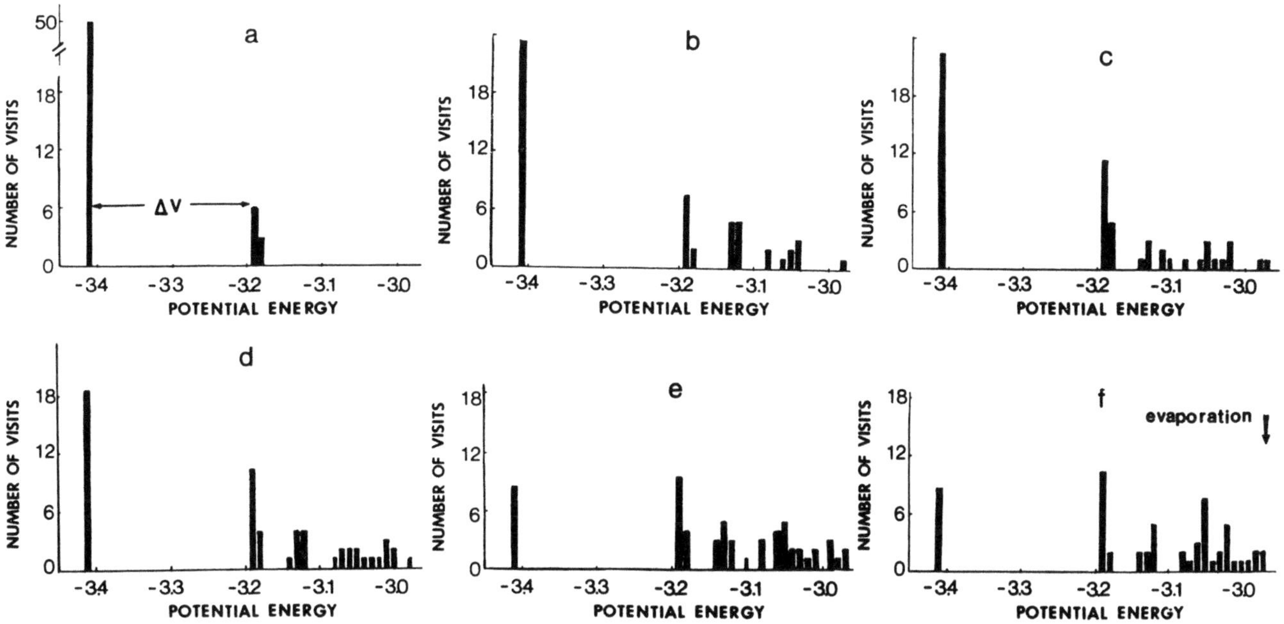

Fig. 2. Number of visits to the potential energy minima for the states labeled $a-f$ in Fig. 1

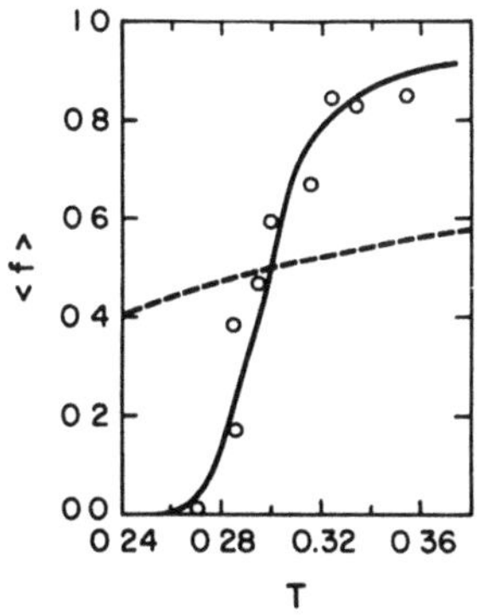

Fig. 3. Fraction of visited minima as a function of temperature

This interpretation of melting in clusters is different to what has been suggested up to now [1, 2, 3]. The approach is phenomenological but instructive. It equates the structural information obtained in n snapshots along a trajectory in phase space with a polymer containing N_B broken icosahedra and $n-N_B$ intact icosahedra. In this model clusters can be classified into three classes. The first class refers to cluster where there is an energy gap ΔV between the binding energy of the most stable isomer and that of the many other isomers. Also, the distribution of binding energies of these isomers is such that the mean nearest-neighbor energy spacing δ is very small compared to ΔV. Melting is observed in this class of clusters. The second class refers to clusters where there is no energy gap and therefore $\Delta V \gg \delta$. In this case we could speak of an amorphous cluster and melting is not observed. Clusters in the third class have only few, well known isomeric forms. This case can be understood on the basis of isomerization.

This work was supported by CONACyT grant PCEXCNA-050838. EBB acknowledges partial support by ARO grant DAAG 29-85-K-0244.

References

1. McGinty, D.J.: J. Chem. Phys. **58**, 4733 (1973); Kristensen, W.D., Jensen, E.J., Cotterill, R.M.J.: J. Chem. Phys. **60**, 4161 (1974); Briant, C.L., Burton, J.J.: J. Chem. Phys. **63**, 2045 (1975); Etters, R.D., Kaelberg, J.B.: Phys. Rev. A **11**, 1068 (1975); Kaelberer, J.B., Etters, R.D.: J. Chem. Phys. **66**, 3233 (1977); Nauchitel, V.V., Perstin, A.J.: Mol. Phys. **40**, 1341 (1980); Quirke N., Sheng P.: Chem. Phys. Lett. **110**, 63 (1984); Berry, R S., Jellinek, J., Natason, G.: Phys. Rev. A **30**, 919 (1984)
2. Jellinek, J., Beck, T.L., Berry, R.S.: J. Chem. Phys. **84**, 2783 (1986); Beck, T.L., Berry, R.S.: J. Chem. Phys. **88**, 3910 (1988)
3. Honeycutt, J.D., Anderson, H.C.: J. Phys. Chem. **91**, 4950 (1987)
4. Blaisten-Barojas, E., Garzon, I.L., Avalos-Borja, M.: Phys. Rev. B **36**, 8447 (1987); Blaisten-Barojas, E.: In: Elemental and Molecular Clusters. Benedek, G., Martin, T.P., Pacchioni, G. (eds.). Springer Series in Materials Science. Vol. 6, pp. 106–124. Berlin-Heidelberg-New York: Springer 1988
5. Hill, T.: Thermodynamics of Small Systems. Vol. 1, Ch. 2. New York: Benjamin 1970
6. Thompson, C.J.: Mathematical Statistical Mechanics, Ch. 7. Princeton, N.J.: Princeton University Press 1972
7. Fujita, S., Blaisten-Barojas, E., Torres, M., Godoy, S.. J. Chem Phys. **75**, 3097 (1981)
8. Hoare, M.R.: Adv. Chem. Phys. **40**, 49 (1979); Oksuz, I.: Surf. Sci. **122**, L585 (1982); Hoare, M.R., McInnes, J.A.: Adv. Phys. **32**, 791 (1983)

Z. Phys. D – Atoms, Molecules and Clusters 12, 185–187 (1989)

Atoms, Molecules
Zeitschrift für Physik D and Clusters
© Springer-Verlag 1989

Thermal decay of Lennard-Jones clusters

I. L. Garzón and M. Avalos-Borja

Instituto de Física, Universidad Nacional Autónoma de México, Apartado Postal 2681, 22800 Ensenada, Baja California, México

The decay mechanisms of argon clusters have been studied using Molecular Dynamics simulations and Lennard-Jones potentials. Heating up processes were applied to Ar_{13} up to temperatures in the melting region. In this range of temperatures large fluctuations in the mean kinetic energy of the system are present and a sequential evaporation is observed. The thermal decay of these aggregates occurs in a time scale of nanoseconds.

PACS: 36.40.+d.

The study of decay mechanisms and lifetime of charged clusters is an interesting field of research, in part by its importance in the interpretation of the mass spectrum obtained during a nozzle beam expansion experiment. Theoretical [1] and experimental [2, 3] approaches have been applied to understand these aspects of cluster physics. A model for the processes occurring in rare-gas clusters after ionization has been proposed by Haberland [4]. In this model the positive charge, initially delocalized in the whole cluster, localizes producing a dimer ion formation. This process produces a hot cluster that cools by evaporation in a 10^{-10}–10^{-8} seconds time scale. Experimental evidence supporting this model for small clusters ($n<10$) is given by the experiments of Buck and Meyer [5].

Recently the metastable decay of single and multiple charged cluster have been observed in argon cluster ions [2, 6]. A long time sequential decay or evaporation loss of only one monomer in the microsecond time regime has been recently reported for charged argon clusters [6]. These observations contrast with the Molecular Dynamics (MD) calculations [7] that predict multiple decay reactions in charged Van der Waals clusters in a picosecond time regime.

Other MD simulations of isolated [8] and adsorbed [9] neutral Lennard- Jones (LJ) clusters show that a "melting" transition occurs in these small systems before they evaporate. Infrared photodepletion spectroscopy of SF_6-Ar_n clusters [10] and resonant two-photon ionization spectroscopy of carbazole-Ar_n clusters [11] have been used to detect the phase transition in these finite systems.

In this communication we suggest that the long time sequential decay of charged clusters observed by Scheier and Märk [6] is another effect of the melting transition predicted by the MD simulations [8, 9] and it is not only an effect related with the electric charge present in the cluster. In the melting region, solid and liquid-like structures coexist in a range of temperatures and large thermal fluctuations produce the cluster decay in a time scale much longer than the picosecond multiple decay predicted for charged clusters [7].

Constant energy MD simulations of the Ar_{13} neutral cluster have been done to estimate its lifetime as a function of the internal temperature. The calculations show that when the cluster is heated up to the melting region, it decays in a time scale of nanoseconds. In this range of temperatures the cluster has enough energy to visit many of the available local minima of the surface potential thus suffering structural changes. At these temperatures the fluctuations in kinetic energy increase and eventually one or more atoms acquire enough energy to escape from the cluster. This atomic evaporation corresponds to the sequential decay observed by Scheier and Märk [6]. However, this is not a privilege of the charged clusters, but a phenomenon associated with the high value of the thermal fluctuations when the cluster is at the melting region.

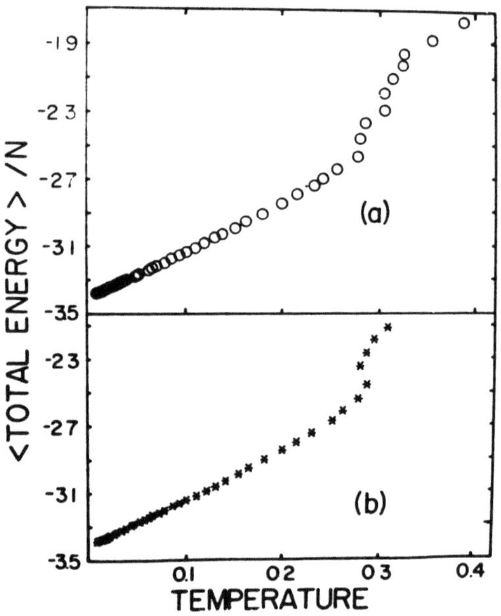

Fig. 1a and b. Caloric curve of Ar_{13} cluster. A LJ potential and the Verlet algorithm with a time step of .01 in reduced units were used. **a)** short time average. **b)** long time average.

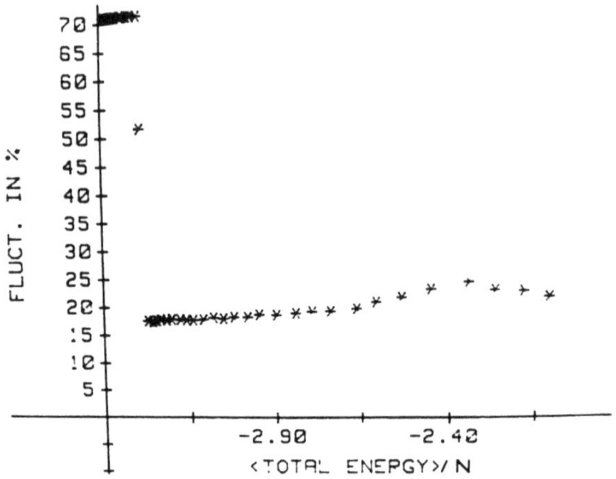

Fig. 2. Fluctuations in the kinetic energy as function of the total energy

Figure 1 shows the average total energy per atom as a function of the internal temperature [9] as obtained when two different heating up processes were done to the Ar_{13} cluster. In Fig. 1. a each circle was obtained by averaging over 3000 time steps while each point in Fig. 1b is an average over 10^5 time steps. For argon clusters the time step is 2.14×10^{-14} seconds. The simulations were stopped when one or more atoms escaped from the cluster. At low temperatures both curves coincide thus indicating that the solid-like phase is stable, at least in a time scale of nanoseconds. However, the long time MD simulations (fig. 1.b) show that the Ar_{13} cluster is "stable" only up to temperatures lower than 0.31 in reduced units. At higher temperatures it will decay emiting one or more atoms. The curve 1. a shows that the cluster is stable, in liquid-like phase, up to temperatures of 0.39 but in a much shorter time scale.

Figure 2 shows the fluctuations in kinetic energy as function of the mean total energy. The fluctuations are defined in the usual way as the square root of the variance divided by the average. For total energies per atom lower than -3.28 the fluctuations in kinetic energy are about 70% corresponding to the fluctuations of a set of harmonic oscillators vibrating in the breathing mode. This mode is favoured because the central atom is initially at rest. At higher energies the fluctuations remain constant at 17% corresponding to the thermodynamic limit value of $f^{-1/2}$ where f is the number of degrees of freedom. At total energies per atom equal to -2.4 there is an increase in the thermal fluctuations which produce the melting-like transition and the thermal decay.

The present work shows that a neutral Ar_{13} cluster in the coexistence region decays in a time scale of nanoseconds. This thermal decay is consequence of the large fluctuations in the kinetic energy of the system at temperatures in the melting region. This thermal evaporation differs from the much faster evaporation produced by the relaxation of a charged cluster when a dimer-ion formation [7] occurs after the ionization of the neutral cluster. We suggest that in the Scheier and Märk [6] experiments both effects are present, however the long time sequential decay could be associated mainly to the thermal decay. This interpretation indicates that the "melting" and the thermal fluctuations in finite systems can be observable effects in the mass spectrometry experiments. However, this thermal decay is only the last step of a long time history of processes which the cluster undergoes since its production in the nozzle chamber to its detection in the mass spectrometer.

We thank Prof. T.D. Märk for useful discussions during the realization of this work.

References

1. Klots, C.E.: J. Chem. Phys. **83**, 5854 (1985); Z. Phys. D – Atoms, Molecules and Clusters **5**, 83 (1987)
2. Märk, T.D.: Int. J. Mass Spectrom. Ion Phys. **79**, 1 (1987)
3. Campana, J.E.: Mass Spectrom. Rev. **6**, 395 (1987)

4. Haberland, H.: Surf. Sci. **156**, 305 (1985)
5. Buck, U., Meyer, H.: Phys. Rev. Lett. **52**, 109 (1984); Ber. Bunsenges. Phys. Chem. **88**, 254 (1984); Surf. Sci. **156**, 275 (1985)
6. Scheier, P., Märk, T.D.: Phys. Rev. Lett. **59**, 1813 (1987)
7. Soler, J.M., Saenz, J.J., Garcia, N., Echt, O.: Chem. Phys. Lett. **109**, 71 (1984); Saenz, J.J., Soler, J.M., Garcia, N.: Chem. Phys. Lett. **114**, 15 (1985)
8. Jellinek, J., Beck, T.L., Berry, R.S.: J. Chem. Phys. **84**, 2783 (1986); Berry, R.S., Beck, T.L., Davis, H.L.: Adv Chem. Phys. **70**, 75 (1988)
9. Garzón, I.L., Avalos, M., Blaisten-Barojas, E.: Proceedings of the International Symposium on the Physics and Chemistry of Small Clusters, NATO ASI Series, Jena, P., Rao, B.K., Khanna, S.N. (eds.) pp. 193–197. New York: Plenum Press 1987; Blaisten-Barojas, E., Garzón, I.L., Avalos-Borja, M.: Phys. Rev. B **36**, 8447 (1987)
10. Gough, T.E., Knight, D.G., Scoles, G.: Chem. Phys. Lett. **97**, 155 (1983); Eichenauer, D., Le Roy, R.J.: Phys. Rev. Lett. **57**, 2920 (1986)
11. Bösiger, J., Leutwyler, S.: Phys. Rev. Lett. **59**, 1895 (1987)

Z. Phys. D – Atoms, Molecules and Clusters 12, 189–191 (1989)

Atoms, Molecules and Clusters
Zeitschrift für Physik D
© Springer-Verlag 1989

Dynamics of transition-metal clusters

S. Sawada[1] **and S. Sugano**[2]

[1] Fundamental Research Laboratories, NEC Corporation, 4-11 Miyazaki, Miyamae-ku, Kawasaki 213, Japan
[2] The Institute for Solid state Physics, The University of Tokyo, 1-1-7 Roppongi, Minato-ku, Tokyo 106, Japan

Received 5 July 1988

The atomic structure and thermodynamic properties of transition-metal 6- and 7-atom clusters are investigated using the molecular dynamics method, where Gupta's potential taking into account many-body interaction is employed. The caloric and the structural fluctuations are studied. The "fluctuating state" is found for $N=6$ in the region of the temperature near and below the melting point, where clusters undergo structural transition from one isomer to others without making any topological change. The fluctuating state differs from the "coexistence state" found in Ar clusters [1] i.e. the former involves no liquid state. In the liquid state the motion of atom-permutation occurs besides the breathing motion. On the other hand, the fluctuating state is not found for $N=7$ but only the motion of atom-permutation in the liquid state. The coexistence state is found in both cases of 6- and 7-atom clusters. We also discuss a possibility of larger clusters displaying the fluctuating state.

PACS: 36.40; 64.60

1. Molecular dynamics of 6-atom cluster

To simulate motion of transiton-metal clusters, we use Gupta's model potential taking into account many-body interaction [2],

$$V=(1/2)\cdot U\sum_j \{A\sum_i \exp(-p(r_{ij}-r_0)) - [\sum_i \exp(-2q(r_{ij}-r_0))]^{1/2}\}, \quad (1.1)$$

where r_{ij} is the distance between the i-th and the j-th atoms, r_0 the lattice constant of the bulk crystal (fcc), $A=0.101035$, $p=9\cdot r_0^{-1}$ and $q=3\cdot r_0^{-1}$. Bulk cohesive energy is given by $E_{bulk}=1.7674\cdot U$. We use r_0, E_{bulk} and $r_0\cdot(m/E_{bulk})^{-1/2}$ (the order of the period of the atomic vibration) as the units of distance, energy and time, respectively, where m is the atomic mass.

The search for minima of the potential energy surface of the 6-atom cluster clarifies the existence of two kinds of isomers, octahedron (OCT) and tripyramid (TP) corresponding to the global minimum and the local minimum, respectively. These are illustrated in Fig. 1.

We have calculated the caloric curve and found that the transition from the solid state to the liquid state is not sharp and the intermediate fluctuating state appears in the temperature near and below the melting temperature [3]. The fluctuating state looks like the coexistence state of Jellinek, Beck and Berry [1], in which the distributions of the short-time averaged kinetic energy shows a bimodal form, but differs from it in several points. We have found that in the present case the cluster displays a bimodal distribution of the short-time averaged kinetic energy both in the fluctuating state and at a low energies in the liquid state. The meaning of the fluctuating state will be clarified in what follows.

To get information about the temporal structure of the cluster, we introduce nearest neighbor distance r_n ($=1.2$) and calculate an adjacency matrix A, whose elements A_{ij} is 1 for $|r_i-r_j|<r_n$ and 0 otherwise. We define distance index d in the matrix space by

$$d=\sum_i\sum_j \{A_{ij}-(A_{oct})_{ij}\}^2/2, \quad (1.2)$$

where A_{oct} is A for OCT. It should be noted that there are OCT isomers obtained by atom permutation

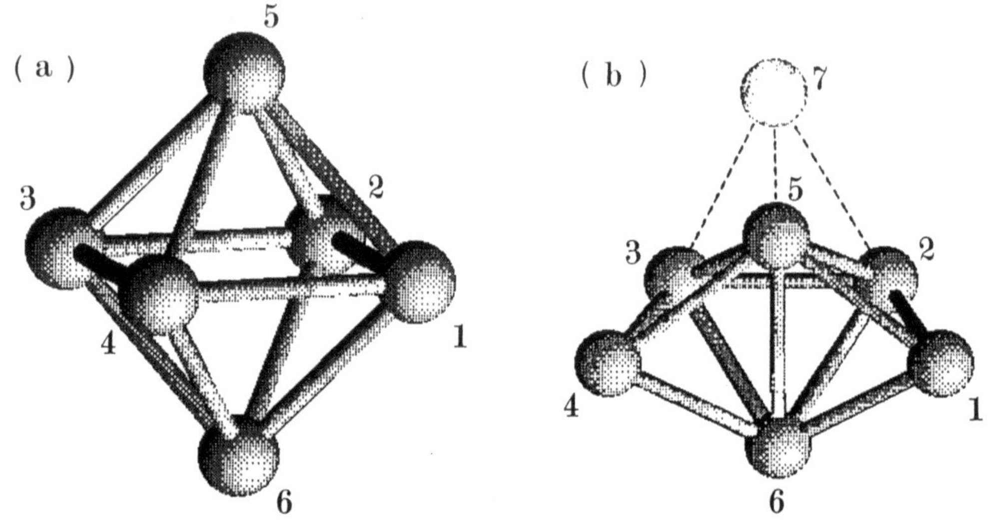

Fig. 1a and b. 6-atom clusters: (**a**) Octahedron (OCT), (**b**) tripyramid (TP) with dummy atom 7

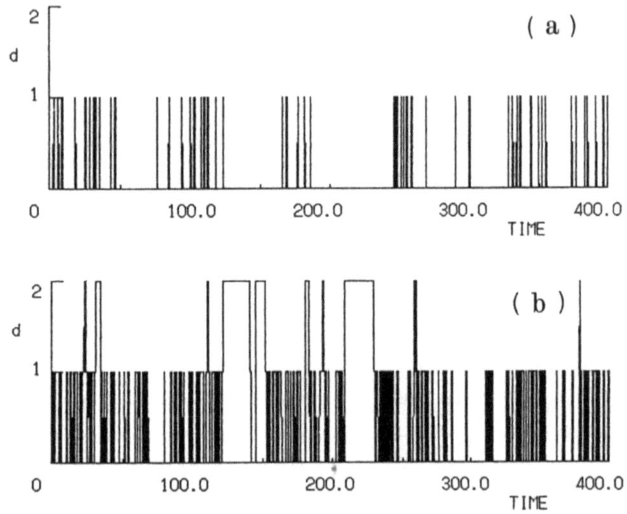

Fig. 2a and b. The time evolution of d in (**a**) the solid state (total energy E = −0.725) and (**b**) the fluctuating state (E = −0.710)

giving different A_{oct}'s. At first, (1.2) is calculated using A_{oct} for the initial OCT.

The time evolution of d is shown in Fig. 2a and b for the solid and fluctuating states, respectively. Values $d=0$ and 2 correspond to OCT and TP, respectively, and $d=1$ to intermediate structures between OCT and TP. It is seen that in the solid state the motion is limited to vibration around OCT. On the other hand, in the fluctuating state, the transition between OCT and TP takes place continually and the system vibrates in some period around OCT and TP. It was found by tracing the time evolution of A that OCT transforms to TP by the following motion of atoms: atom 1 and 4 (see Fig. 1a) are separated from each other and atom 5 and 6 approach each other. TP is transformed to OCT by just the reversed motion. This may be called "breathing" motion, where no exchange of the relative positions of the atoms occurs.

The time evolution of d in the liquid state is shown in Fig. 3a. It is seen that d takes values greater than 2. It is found that around $t=25$ the transition from one TP isomer to another (atom-permuted) TP isomer with $d=4$ is made by the cooperative motion of two atoms: atom 2 moves into the position of dummy atom 7 and atom 1 into that of atom 2 (see Fig. 1b). The succeeding breathing motion makes the transition from TP with $d=4$ to OCT with $d=6$, which is an atom-permuted OCT isomer different from the initial one. Substituting A_{oct} of the new atom-permuted OCT for A_{oct} in (1.2), time evolution of index d is re-calculated. The result is shown in Fig. 3b. It is seen that the system exhibits the breathing motion around the new OCT for a while (up to $t=230$) and then suddenly begins the motion of atom-permutation again. It is found that, as the total energy increases, the motion of atom-permutation takes place more frequently.

We have calculated the rms fluctuation in the bond length and found that it jumps up abruptly at the boundary between the fluctuating and the liquid states but not at the boundary between the solid and the fluctuating states. We have also calculated the diffusion coefficient and found that it is nearly equal to zero in both the solid and the fluctuating state and increases gradually in the liquid state as the total energy in-

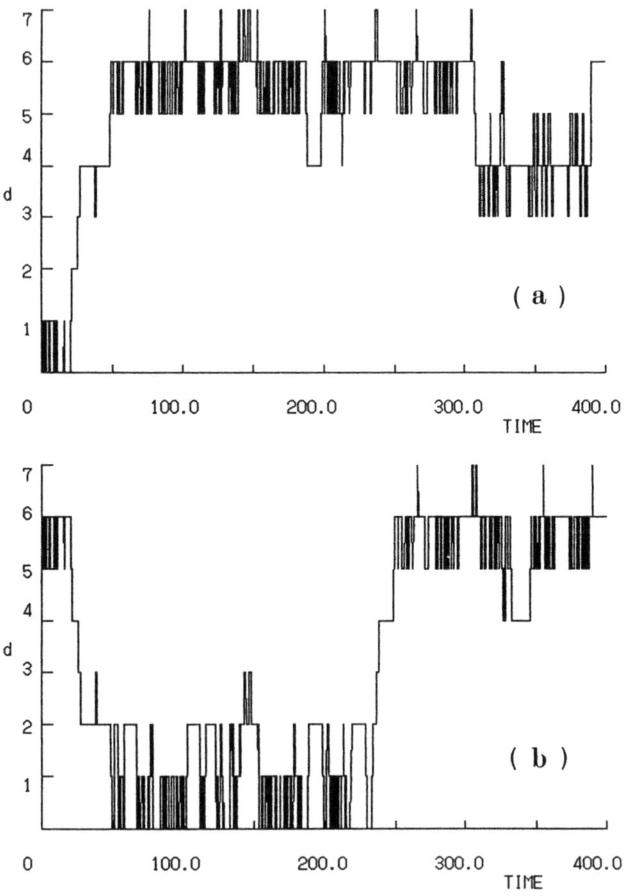

Fig. 3a and b. The time evolution of d in the liquid state ($E = -0.700$) calculated by using (1.2) with A_{oct} for (**a**) the initial OCT isomer and (**b**) the one obtained after the atom permutation

creases. Consequently, there is no soft-mode instability in the fluctuating state in contrast with the liquid state characterized by the existence of soft modes.

3. Molecular dynamics 7-atom cluster

The search for minima of the potential energy surface for the 7-atom cluster gives four kinds of isomers; pentagonal bipyramid (PBP), octahedron plus one (OCT+1), skewed arrangement (Skew) and incomplete stellated tetrahedron (IST) in the increasing order of potential energy.

In this case, no fluctuating states have been found. In the low energy region the motion is limited to the vibration around PBP. When the total energy exceeds a critical value, the transition from PBP to OCT+1 occurs. At the same time the fluctuation of the bond length abruptly increases. We consider this state to be the liquid state.

The reason why no fluctuating states appear is that the atomic motions inducing the transition between PBP and OCT+1 change the relative positions of atoms, i.e. the cluster motion going back and forth between PBP and OCT+1 is not a breathing motion.

We have also found that a transition-metal 7-atom cluster displays a bimodal distribution of the short-time averaged kinetic energy at low energies in the liquid state similarly to the case of a 7-atom Ar cluster [1].

4. Discussion

Existence of the fluctuating state similar to that of a 6-atom cluster can be expected for larger clusters. Suppose that there is the steepest descent path, which corresponds to breathing motion of the cluster, between two neighboring minima on the potential energy surface. In fact such a path can be found between icosahedron and cubooctahedron for $N = 13$, 55, 137, 309 and so on [3]. This motion is, so to speak, the collective mode of the cluster motion. There are other modes corresponding to individual atomic vibration. Only when the collective mode gains energy larger than the activation energy from the other modes, does the transition from one minimum to another take place. As N increases, the probability of this transition occuring decreases, so that the solid-liquid transition gets sharper and the region of the fluctuating state narrower. The molecular dynamics for larger clusters and the analysis of its results according to the above idea will lead to a good understanding of the structural fluctuation of metal clusters observed with the electron microscopic [4] and give a microscopic theory in contrast with the one using the continuum model [5]. Such a simulation is now going on.

A part of this work was presented by one of the authors in the NEC Symposium on Microclusters [3]. The authors acknowledge Dr. S. Saito, Dr. S. Ohnishi and Professor R.S. Berry for their valuable discussions and comments.

References

1. Jellinek, J., Beck, T.L., Berry, R.S.: J. Chem. Phys. **84**, 2783 (1986); Amar, F.G., Berry, R.S.: J. Chem. Phys. **85**, 5943 (1986); Beck, T., Jellinek, J., Berry, R.S.: J. Chem. Phys. **87**, 545 (1987)
2. Gupta, R.P.: Phys. Rev. B **23**, 6265 (1981); Gordon, M.B., Cyrot-Lackmann, F., Desjonqueres, M.C.: Surf. Sci. **80**, 159 (1979); Tomanek, D., Mukherjee, S., Bennemann, K.H.: Phys. Rev. B **28**, 665 (1983)
3. Sawada, S.: In: Microclusters. Sugano, S., Nishina, Y., Ohnishi, S. (eds.), p. 211. Berlin, Heidelberg, New York: Springer 1987
4. Iijima, S., Ichihashi, T.: Phys. Rev. Lett. **56**, 616 (1986)
5. Ajaya, P.M., Marks, L.D.: Phys. Rev. Lett. **60**, 585 (1988)

Z. Phys. D – Atoms, Molecules and Clusters 12, 193–197 (1989)

Atoms, Molecules
Zeitschrift für Physik D and Clusters
© Springer-Verlag 1989

The thermal motion of gold atoms in metal cluster compounds as determined by Mössbauer and specific heat

H.H.A. Smit, R.C. Thiel, and L.J. de Jongh

Kamerlingh Onnes Laboratorium der Rijksuniversiteit te Leiden, Postbus 9506 NL-2300 RA Leiden, The Netherlands

Received 6 June 1988

^{197}Au Mössbauer effect spectroscopy and specific heat measurements have been performed as a function of temperature on three gold polynuclear cluster compounds, $[Au_9(PPh_3)_8](NO_3)_3$, $Au_{11}(PPh_3)_7(SCN)_3$, and $Au_{55}(PPh_3)_{12}Cl_6$. The Mössbauer data yield information on the vibrational motions of the various distinguishable Au sites, as well as on the motion of the clusters as a whole. The Mössbauer and the specific heat data are successfully described by a superposition of inter- and intra-cluster vibrations. The latter are determined by calculating numerically the normal modes of vibration of the metal cores.

PACS: 33.40; 63.20; 65.40

1. Introduction

During the last decade chemists have made considerable progress in synthesizing polynuclear metal cluster compounds [1, 2]. Ru, Rh, Pt, and Au metal cluster compounds with a nuclearity of 55 have been reported [3, 4]. Metal cluster compounds are chemically synthesized materials, composed of molecules, which are made up of a metal core that is surrounded by a shell of ligands. The ligand shell stabilizes the metal core and prevents the metal cores from coalescing. A big advantage in comparison to e.g. colloidal metal particles is that a cluster compound is formed out of completely identical clusters. Moreover, in contrast to mass spectroscopic experiments on clusters formed in beams, large amounts of material can be supplied, often in the order of grams. Finally, one is able to study their properties as a function of cluster size. Fundamental questions arise when the particle size becomes so small that the bulk properties are lost (quantum size effects). Moreover, the fraction of surface atoms is very large. Consequently, they are model systems for surface effects and may be of interest for the understanding of the properties of small metal particle catalysts.

The Mössbauer Effect Spectroscopy (M.E.S.) provides an ideal microscopic tool to probe the chemical environment of the metal atoms in cluster compounds. The metal core is composed of Au atoms occupying inequivalent sites, which can be distinguished on the basis of their isomer shift (I.S.) and quadrupole splitting (Q.S.). In addition, by studying the temperature dependence and the absorption intensities of the M.E.S. spectra, information on the local Debye-Waller factors (*f*-factors, Mössbauer fractions, or recoilfree fractions) can be obtained. These possibilities appear to be of importance in connection with the problem of the size-dependent phonon spectra of very small metal particles.

Moreover the specific heat of the metal cluster compounds is measured and provides us independent information on the vibrational density of states of the metal cluster compounds. A mathematic model is presented that takes into account both intra- and inter-cluster vibrations. From this model both the Mössbauer and the specific heat results are fitted using only two adjustable parameters.

2. Metal cluster compounds

Three metal cluster compounds have been investigated, $[Au_9(PPh_3)_8](NO_3)_3$, $Au_{11}(PPh_3)_7(SCN)_3$, and $Au_{55}(PPh_3)_{12}Cl_6$. They are all chemically stable at room temperature and are not air sensitive. They

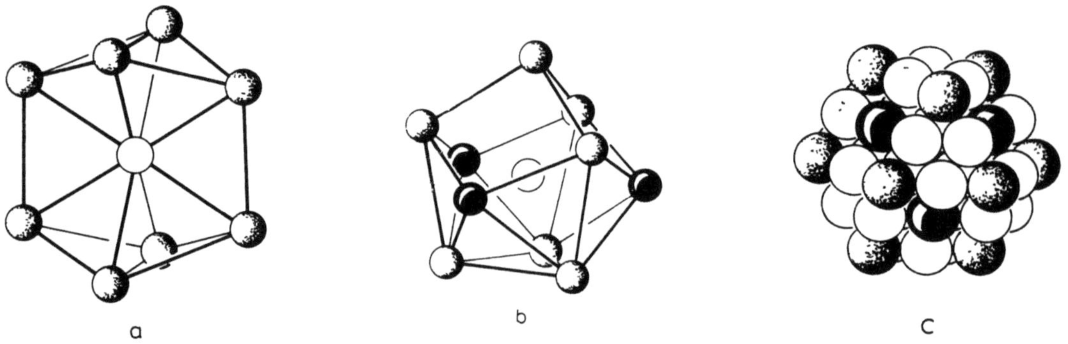

Fig. 1a–c. A schematic representation of the metal cores of the Au_9 (**a**), Au_{11} (**b**) and the Au_{55} (**c**) cluster. The P bonded gold surface atoms are the gray spheres, while the Cl or SCN bonded gold atoms are the black spheres. The unbonded gold surface atoms are slightly shaded and the inner-core atoms are blank

will be denoted by Au_9, Au_{11}, and Au_{55} in what follows.

A sketch of the metal core of $[Au_9(PPh_3)_8](NO_3)_3$ can be found in Fig. 1a. The central gold atom is bonded to eight peripheral gold atoms. Three mutually perpendicular two-fold symmetry axes, as found by Bellon et al. [5], leave only two inequivalent surface Au sites that are both bonded to a PPh_3 ligand.

A sketch of the metal core of $Au_{11}(PPh_3)_7(SCN)_3$ is shown in Fig. 1b. The three-fold symmetry axis running through the top and center Au atom (see Fig. 1b) leaves five different gold sites [6]. One site is the center gold atom, another the SCN bonded surface site, and further, three crystallographically different sites are bonded to a PPh_3 ligand.

The Au metal cluster compound with the largest nuclearity number known until now, is $Au_{55}(PPh_3)_{12}Cl_6$. The cuboctahedral structure proposed [3] for its metal core (see Fig. 1c) has only 5 distinct gold sites, because of its high symmetry. The central gold atom and the 12 gold atoms in the first shell, although strictly speaking occupying different sites, have the same nearest neighbour coordination. These 13 atoms will be referred to as inner-core atoms. Six surface sites are bonded to a chlorine ion, 12 surface sites are bonded to the phosphorus ion of a PPh_3 ligand, and 24 gold surface atoms are not bonded to a ligand and will be referred to as unbonded surface atoms in what follows. Based on the atomic radius of gold, the "diameter" of the metal core is roughly 15 Å. In contrast to the two crystalline metal cluster compounds described above, Au_{55} forms an amorphous solid.

3. Mössbauer and specific heat results

As an illustration we show the spectra of $Au_{55}(PPh_3)_{12}Cl_6$, obtained in the temperature range of 1.25 to 30 K, in Fig. 2. The fit procedure was as follows. First the spectra taken at 4.2 K, which have a very good signal to noise ratio, were analysed with the smallest number of subspectra that still produced an

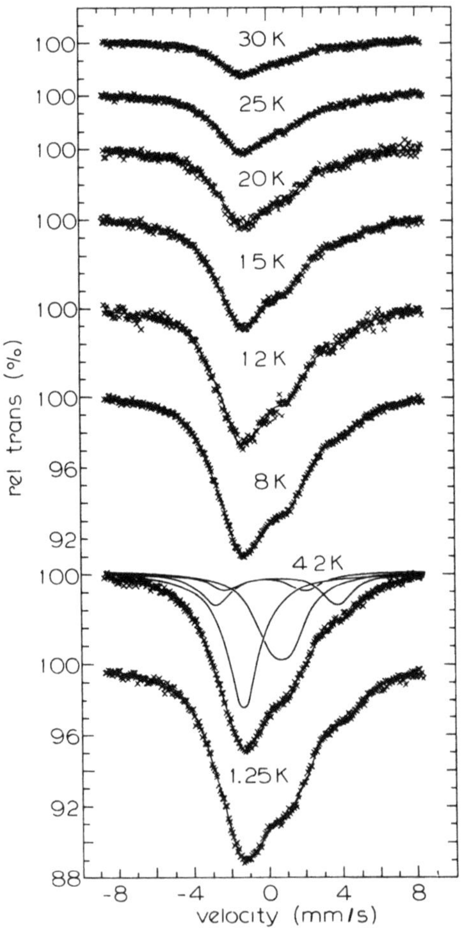

Fig. 2. Spectra taken at various temperatures for the Au_{55}, fitted with the hyperfine parameters as given in Table 1

acceptable fit, using a transmission integral. Subsequently the spectra measured at higher temperatures were analysed. In these refinements, the peak positions as found in the 4.2 K spectra were fixed and only the intensity ratios of the subspectra were adjusted. The values obtained for the I.S. and the Q.S. of the distinguishable subspectra can be found in Table 1. The assignments as indicated in Table 1 are based on a vast body of previous Au Mössbauer work on gold cluster compounds [7, 8, 9].

It should be noted that in the case of Au_9 and Au_{11} two different subspectra were found having Mössbauer parameters indicative for a PPh_3 bonded Au site. Since we were unable to assign these two subspectra to the inequivalent Au sites as found by X-ray measurements we have added their intensity for the discussion of the Debye-Waller factors. In Au_9 the contribution of the central Au atom apparently has Mössbauer parameters comparable to those of the peripheral Au atoms, since it is not observed separately.

The specific heat of the $Au_{11}(PPh_3)_7(SCN)_3$ has been measured by the square wave excitation (SWEX) method [10] between 0.5 and 20 K. The main reason for choosing this method was that only a small amount of material was available. The experimental specific heat of the Au_{11} is shown in Fig. 3a. The specific heat

Table 1. The I.S. (relative to Au in Pt at 4.2 K) and the Q.S. in mm/s for the different Au-sites

	Au_9 I.S.	Au_9 Q.S.	Au_{11} I.S.	Au_{11} Q.S.	Au_{55} I.S.	Au_{55} Q.S.
Au core			2.5	0.0*	−1.4	0.0*
Au unb. surf.					0.3	1.4
Au-SCN			0.8	4.6		
Au-Cl					0.1	4.4
$Au\text{-}PPh_3$ a	2.1	7.0	1.3	7.6	0.6	7.1
b	2.2	5.7	1.3	6.2		

The values with an asterisk were constrained during the refinement

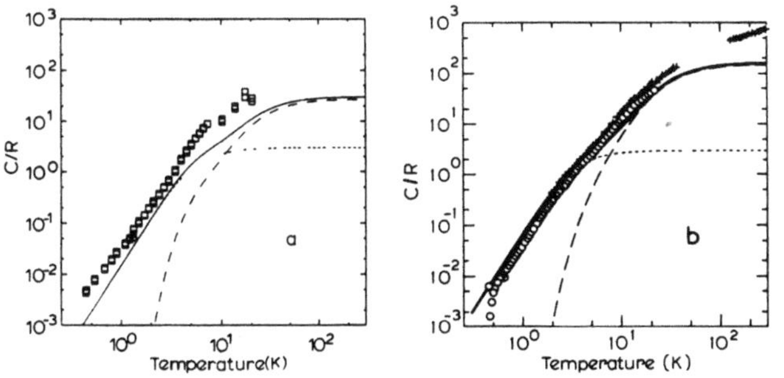

Fig. 3a and b. Specific heat of the Au_{11} (**a**) and the Au_{55} (**b**) cluster compound. The fit (*solid line*) is the addition of the inter- (*dotted line*) and intra-cluster (*broken line*) vibrational contributions

of $Au_{55}(PPh_3)_{12}Cl_6$ has been measured in our laboratory by means of an adiabatic heat pulse technique between 0.5 and 20 K. Additional data between the 2 and 35 K and between 120 and 300 K had been measured by Gmelin at the Max Planck Institute in Stuttgart, who kindly provided these data. These various data are combined and plotted in Fig. 3b.

4. Model for the vibrational density of states

From Fig. 2, which shows the spectra of the Au_{55} cluster compound as a function of temperature, the following three features can be noticed. The first is the drastic decrease in the absorption intensities as the temperature is increased. The second is that there is no visual change in the shape of the spectra as a function of temperature, and finally the intensity ratios found do not correspond to the relative occupancy ratios, giving different absolute f-factors for the different sites. The results for the absolute Debye-Waller factors for each gold site are plotted in Fig. 4 versus temperature. The difference in f-factors of the Au sites must be related to intra-cluster vibrations and results then from a difference in bonding stiffness for each Au site. However, the shape of the spectra in Fig. 2 is identical over the whole temperature range, indicating that these intra-cluster vibrations do not change the relative intensity ratios of the different Au sites over the measured temperature range. By contrast, the total intensity of the spectrum decreases drastically. This must therefore result from the centre of mass vibrations of the clusters, referred to as inter-cluster vibrations.

Assuming that the inter- and intra-cluster vibrations are decoupled we can write

$$\langle x^2 \rangle_{tot} = \langle x^2 \rangle_{inter} + \langle x^2 \rangle_{intra}, \tag{1}$$

where $\langle x^2 \rangle$ is the mean squared vibrational amplitude. This means that we can simply multiply the f-factors for the inter- and intra-cluster vibrations $f(T) = f_{inter}(T) \times f^{i}_{intra}(T)$. In order to calculate the intra-cluster vibrations we have used a simple brute force method [11] to calculate numerically all the normal modes of the metal core. We introduce identical massless springs between nearest neighbour atoms. The influence of the ligands are neglected, since the gold atoms in the core are at least a factor five heavier than the atoms in the ligands. Moreover, bondstretching modes of vibrations in the ligands and between the gold atoms and the ligands are usually found at several hundreds of cm^{-1} [12] and thus have no influence at the experimental temperatures. The eigenfunctions and energies of the normal modes can be found by solving $3N$ coupled linear differential equations. The

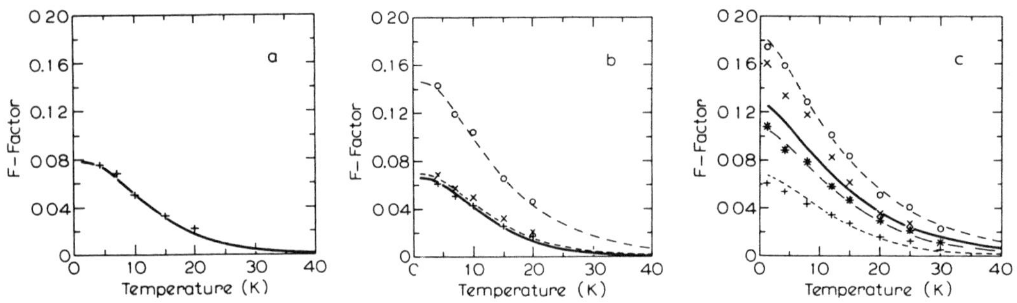

Fig. 4a–c. The temperature dependence of the Mössbauer fraction of the distinguishable gold sites of Au_9 (**a**), Au_{11} (**b**), and Au_{55} (**c**); (o) core, (x) Cl bonded, (★) unbonded surface, and (+) PPh_3 bonded Au atoms

only adjustable parameter for the intra-cluster vibrations is the force constant, F_{ij}, between nearest neighbours. The Mössbauer fraction $f_i(T)$ at the different gold sites i can be calculated using a modified Einstein function [13]

$$f_i(T) = \exp\left\{\frac{-E_\gamma^2}{6m_i c^2} \sum_{k=1}^{3N-6} \sum_{j=3i+1}^{3i+3} \frac{A_{jk}^2}{\hbar\Omega_k} \times\left(1+\frac{2}{\exp(\hbar\Omega_k/k_B T)-1}\right)\right\}, \quad (2)$$

where E_γ is the energy of the γ-ray, m is the mass of an Au atom, and N is the number of atoms per cluster. A_{jk} is the amplitude of the displacement of an atom at site i from equilibrium due to normal mode k with energy Ω_k and is obtained from the corresponding eigenfunction. The standard Einstein approximation can be used to calculate the contributions of the normal modes to the specific heat [14]. For more details of the calculation of the intra-cluster vibrations see [15, 16].

The inter-cluster vibrations will be calculated using the basic Debye scheme [13], where the phonon density of states is assumed to be proportional to the energy squared up to a cutoff energy, Ω_D. For the mass one should substitute the mass of the entire cluster. Clearly, the free parameter for the inter-cluster vibrations is the Debye temperature, and this was constrained to be equal for all the different gold sites of one particular cluster compound. For the contribution of the inter-cluster vibrations to the specific heat the standard Debye function has been used [14].

First we have fitted these two parameters to the Mössbauer fractions as shown by the lines in Fig. 4. The results for these parameters are given in Table 2. The curves fit the measurements quite accurately. Especially for the Au_{11} and the Au_{55} it is striking how well the model predicts the experimental ratio of Debye-Waller factors. Now we will use this model and the obtained parameters of Table 2 to calculate the specific heat contribution of the metal cores.

Table 2. The Debye temperature used to fit the inter-cluster vibrations and the force constant F_{ij} at 2.88 Å with which the intra-cluster vibrations are calculated

	Au_9	Au_{11}	Au_{55}
F_{ij}(N/m)	16 (2)	21 (2)	30 (2)
θ_D(K)	25 (3)	25 (3)	15 (1)

The calculated specific heats resulting from the intra-cluster vibrations using the force constants given in Table 2 are represented by the broken lines in Fig. 3. They fall off exponentially if the temperature becomes smaller than the energy of the lowest normal mode (i.e. 27 K and 23 K for Au_{11} and Au_{55} respectively). In the high temperature limit the intra-cluster specific heat C/R is equal to $3N-6$. The inter-cluster specific heat is calculated using the Debye approximation and is shown by the dotted lines in Fig. 3. At the highest temperatures, in the Dulong and Petit limit, the inter-cluster specific heat, C/R, amounts to 3, which corresponds to the three translational degrees of freedom. At the lowest temperatures the total specific heat is completely dominated by this contribution. Here the specific heat follows the commonly found third power of the temperature, characteristic for phonons in a three-dimensional continuous medium. This effective three-dimensional medium approximation is valid as long as the wave lengths of the phonons are larger than a characteristic length [17, 18], determined in our materials by the cluster size. The full curves in Fig. 3, which represent the sum of inter- and intra-cluster vibrational specific heats, follow the data very well indeed. The fact that these curves are not fitted to the data, but calculated using the parameters resulting from the M.E.S. measurements, gives credence to the proposed model.

It may be noticed that the specific heat contribution due to the electrons can be neglected in metal cluster compounds, since the energy splittings at the highest occupied molecular orbital are large compared to the temperature [19].

The experimental inter-cluster Debye temperatures θ_D^{inter}, as given in Table 2, can be related to earlier work on colloidal gold particles embedded in a gelatine matrix [20]. In that study, temperature dependent Mössbauer spectra were taken on three samples with average particle diameters of 42, 62, and 176 Å. For the inter-cluster vibrations Debye temperatures of 5.4, 3.3, and 1.1 K have been found respectively. In the Debye scheme the cutoff energy, Ω_D, varies with $(F/M)^{\frac{1}{2}}$, where F is the force constant and M the mass of the particle. An increase in mass indeed gives a decrease in Debye temperature for the inter-particle or -cluster vibrations (see above and Table 2).

In Table 2 the force constants are given with which the intra-cluster vibrations are calculated and fitted to the M.E.S. data. One can compare these to the bulk value. The compressibility of bulk gold yields a force constant of 34 N/m, if one assumes massless springs between nearest neighbours. This force constant has the same order of magnitude as those obtained experimentally. We think these force constants must be seen as empirical parameters and we do not want to draw further quantitative information from them because of the crudity of the model used.

5. Conclusion

In conclusion, we stipulate that the intra-cluster vibrations of metal cluster compounds up to at least 55 atoms cannot be described by models starting from a macroscopic view point. On the other hand, the microscopic approach as introduced in this paper is not only capable of explaining the intra-cluster vibrations as observed by a microscopic technique (M.E.S.), but also predicts the results obtained by a macroscopic technique (specific heat) very well indeed up to relatively high temperatures.

We are indebted to Prof. Dr. G. Schmid and Dr. N. Klein of the Universität Essen and Dr. J.J. Bour and R. Kanter of the Katholieke Universiteit Nijmegen for preparing the gold cluster compounds. The present work is part of the research program of the "Stichting voor Fundamenteel Onderzoek der Materie" (Foundation for Fundamental research on Matter) and was made possible by financial support from the Nederlandse Organisatie voor Zuiver-Wetenschappelijk Onderzoek (Netherlands Organization for the Advancement of Pure Research). The investigations are sponsored by the Leiden Materials Science Centre (Werkgroep Fundamenteel Materialen Onderzoek). The research program is also sponsored by the Commission of the European Communities.

References

1. Chini, P.: Gazz. Chim. Ital. **109**, 225 (1979)
2. Steggerda, J.J., Bour, J.J., van der Velden, J.W.A.: Rec. J. Royal. Neth. Chem. Soc. **101**, 164 (1982)
3. Schmid, G., Pfeil, R., Boese, R., Bandermann, F., Meyer, S., Calis, G.H.M., van der Velden, J.W.A.: Chem. Ber. **114**, 3634 (1981)
4. Schmid, G., Huster, W.: Z. Naturforsch. **41b**, 1028 (1986)
5. Bellon, P.L., Cariati, F., Manassero, M., Naldini, L., Sansoni, M.: J.C.S., Chem. Commun. 1423 (1971)
6. Bellon, P.L., Manassero, M., Sansoni, M.: J.C.S., Dalton Trans. 1481 (1972)
7. Vollenbroek, F.A., van der Velden, J.W.A., Bour, J.J., Trooster, J.M.: Recl. J. Royal Neth. Chem. Soc. **100**, 375 (1981)
8. Bartunik, H.D., Potzel, W., Mössbauer, R.L., Kaindl, G.: Z. Phys. **240**, 1 (1970)
9. Bos, W., Kanters, R.P.F., van Halen, C.J., Bosman, W.P., Behm, H., Smits, J.M.M., Beurskens, P.T., Bour, J.J.: J. Organometallic Chem. **307**, 385 (1986)
10. Regelsberger, M., Wernhardt, R., Rosenberg, M.: J. Phys. E. **19**, 525 (1986)
11. Wilson, E.B., Decius, J.C., Cross, P.C.: Molecular vibrations. New York: McGraw-Hill 1955
12. Greenwood, N.N., Ross, E.J.F.: Index of vibrational spectra of organic and inorganic metallic compounds. London: Butterworth 1975
13. Wegener, H.: Der Mössbauer Effect und seine Anwendung in Physik und Chemie. Mannheim: Bibliographisches Institut 1966
14. Kittel, C.: Introduction to Solid state physics, 5th Edn. New York: J. Wiley and Sons 1976
15. Smit, H.H.A., Nugteren, P.R., Thiel, R.C., de Jongh, L.J.: Physica B **153**, 33–52 (1988)
16. Smit, H.H.A.: Ph.D. Thesis, Rijksuniversiteit Leiden, 1988
17. Derrida, B., Orbach, R., Yu, Kin-Wah.: Phys. Rev. B **29**, 6645 (1984)
18. Rosenberg, H.M.: Phys. Rev. Lett. **54**, 704 (1985)
19. Mingos, D.M.P.: J.C.S., Dalton Trans. 1163 (1976)
20. Viegers, M.P.A., Trooster, J.M.: Phys. Rev. B **15**, 72 (1977)

Z. Phys. D – Atoms, Molecules and Clusters 12, 199–203 (1989)

Atoms, Molecules
Zeitschrift für Physik D and Clusters
© Springer-Verlag 1989

Stability of alkali-atom clusters

C. Bréchignac, Ph. Cahuzac, J. Leygnier, R. Pflaum*, J. Ph. Roux, and J. Weiner**
Laboratoire Aimé Cotton, CNRS II, Bâtiment 505, Campus d'Orsay, 91405 Orsay Cedex, France

Received 4 July 1988

A detailed study of the stability of alkali-atom clusters is presented and discussed. Free clusters are produced in an unseeded adiabatic expansion and ionized by a pulsed laser beam. A tandem time-of-flight system provides a size selection of ionized species and mass spectrometry of fragments. We have investigated for sodium and potassium clusters containing up to fifty atoms: the metastable decay following ionization, the photo-induced dissociation, the collisional induced dissociation and the collisional charge exchange. Experimental evidence is shown for dissociative process involving evaporation of monomers and dimers. This is explained by the statistical unimolecular theory and by the energetics of the systems.

PACS: 36.40

I. Introduction

The stability of clusters is present everywhere in cluster studies. From a fundamental point of view, the stability is related to important questions. How does a cluster hold itself together? How is dissipated an excess of energy deposited in that species? To which extent the knowledge of the dissociation processes reveals their geometrical or electronic structure? How is related the reactivity to the stability? etc . . .

To answer these questions, a numerous of experiments have been devoted to the study of fragmentation processes following photonic [1–12] or collisional [13–16] interactions. They are mostly performed on size-selected ionized species. However they concern various kind of clusters and various experimental techniques, and the comparison between results is difficult. We present here extensive results obtained under similar and well controlled experimental conditions on the dissociation processes of alkali-atom clusters. For size-selected species, in the size-range $n=3$–50, we have investigated the metastable decay following ionization, the photo-induced dissociation, the collisional induced dissociation and the collisional charge-exchange. Evidence is shown that evaporation of monomers (or dimers) governs the fragmentation of these species. This is well explained by considering both the energetics and the dynamics of the processes.

II. Experiment

The experimental apparatus is shown in Fig. 1. Neutral alkali-atom clusters are produced by an unseeded adiabatic expansion. They are ionized by a pulsed laser beam focused within the two first plates of an ion focusing and acceleration system. Accelerated ions, at typically 3 keV, are spatially resolved in a Time-Of-Flight Mass Spectrometer (TOFMS). Ion packets are selected due to two deflecting plates. They are then propagating through an interaction region, followed by a second TOFMS for the mass spectrometry of fragments.

A wide mass distribution of ionized clusters is prepared as follows. The adiabatic expansion through

* Permanent address: Universität Konstanz, Fakultät für Physik, Bücklestrasse 13, Postfach 5560, D-7750, Konstanz-1, Federal Republic of Germany
** Permanent address: Department of Chemistry, University of Maryland, College Park, MD 20742, USA

a conical nozzle provides a pseudo-gaussian distribution of neutral aggregates [17] centered on $n \sim 60$. They are ionized with a photon energy above their ionization limit and below the dissociation energy. At high laser fluence, excitation of the ions produces a fast dissociation during the laser pulse. Subsequent re-excitation of fragment ions shifts the ion distribution towards the smaller masses, whose relative intensities reflect the relative ion stabilities [17].

III. Evaporation following a photo-excitation: metastable decay and photo-induced dissociation

1. Metastable decay

The procedure to form small ionized masses gives rise to hot species which evaporate partly within the accelerating plates, partly in the TOF1, depending on their internal energy. Those which decompose in the field-free region give fragments which propagate in the same center-of-mass. Residual parent ions and fragments are mass discriminated in the second TOF by using a retarding potential (Fig. 1). Examples of such a metastable decay for Na_n^+ clusters are shown on Fig. 2. The even-numbered atom species evaporate always a monomer whereas the odd-numbered ones evaporate either a monomer or a dimer or both. This odd-even alternation vanishes at large masses.

From calculations of *atomization energies* at 0 K, now available for some small alkali-atom clusters [18–21], we compare the energy change for the evaporation of a monomer or of a dimer from the ionized parent X_n^+ (X stands for Na or K in the following). One finds that the observed channel always proceeds according to the minimum of energy consumption. The

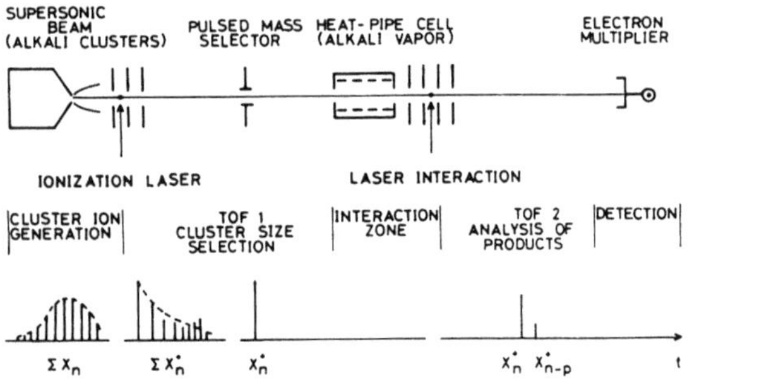

Fig. 1. Experimental set-up. For experiments which deal with photon interactions, the heat-pipe cell is cooled so that the alkali-vapor density is negligible. The multigrids system of the interaction zone may be used either as a simple retarding potential or as decelerating-accelerating system (see the text). At the bottom are schematized the successive cluster distributions from the neutral species as formed in the adiabatic expansion, ΣX_n, up to an example of dissociation

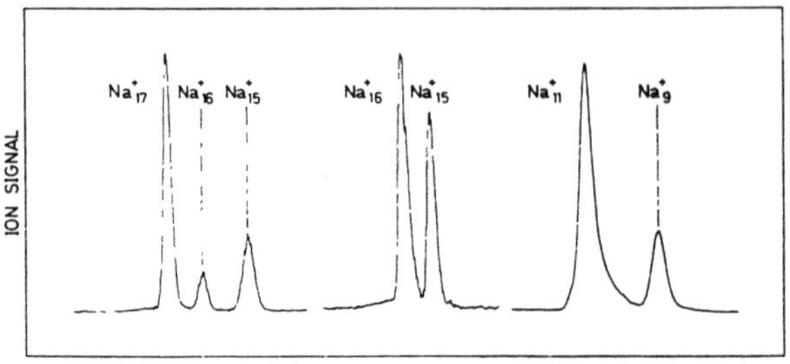

Fig. 2. Some examples of metastable decays corresponding to the three possible situations: evaporation of a monomer and of a dimer, evaporation of a monomer, evaporation of a dimer

odd-even alternation is then related to the relative stabilities [10, 11].

Our experimental procedure prepares ionized clusters X_n^+ which have an internal energy range determined by the time-window of the ionizing region. The TOF1 allows us to detect the metastable decay of those clusters having a decay constant θ_n of the order of Δt_n, the time window due to the TOF1 (10–60 μs). Following a model satisfying detailed balance [22], we learn from the *statistical unimolecular theory* that:

$$\theta_n^{-1} = v^3(s-1)\,8\pi mgS\,\frac{(E^* - D_n^+)^{s-2}}{E^{*s-1}} \tag{1}$$

where v is a characteristic vibrational frequency ($10^{12} - 10^{13}\,s^{-1}$), $s = 3n-6$ is the number of vibrational degrees of freedom, m is the reduced mass, g is the evaporative channel degeneracy and S is the geometrical cross section. E^* is the total internal energy of species n having D_n^+ as a binding energy. These X_n^+ clusters are generated by precursors X_{n+1}^+ (or X_{n+2}^+) having an internal energy $E^* + D_{n+1}^+$ and a decay constant given by (1). Assuming a slow evolution of D_n^+ with n, (1) implies that θ_{n+1} is several orders of magnitude smaller than θ_n. Therefore the precursor $n+1$ decays in a very short time within the two first plates of the accelerating system. Under these conditions the energy distribution of the X_n^+ clusters when entering the TOF1 is well approximated by:

$$P_n(E^*) = \left(1 - \exp\frac{\delta t_{n+1}}{\theta_{n+1}(E^* + D_{n+1}^+)}\right)\cdot \exp - \frac{\delta t_n}{\theta_n(E^*)} \tag{2}$$

where δt_n is the time-of-flight of the n-mer cluster within the accelerating plates (around 1 μs). Numerical calculations show that $P_n(E^*)$ is roughly rectangular with a width of the order of D_n^+. The survival fraction of these species entering the decelerating potential is:

$$Q_n(E^*) = P_n(E^*) \times \exp - \frac{\Delta t_n}{\theta_n(E^*)} \tag{3}$$

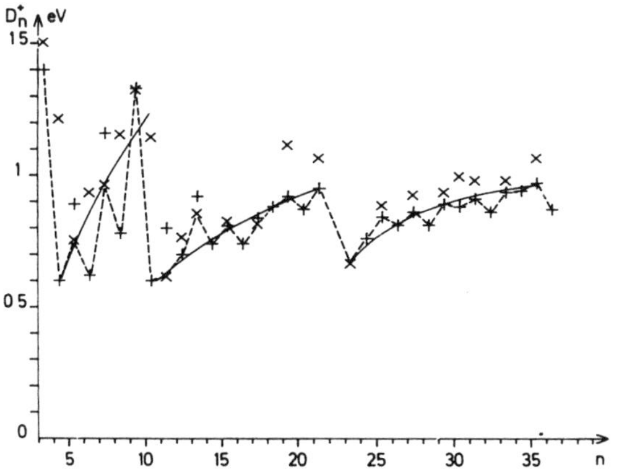

Fig. 3. Binding energies of ions deduced from the observed dissociation rates: × for dimer evaporation; + for monomer evaporation. The dashed line connects the predominant channels. The continuous line schematizes an electronic-shell structure behavior

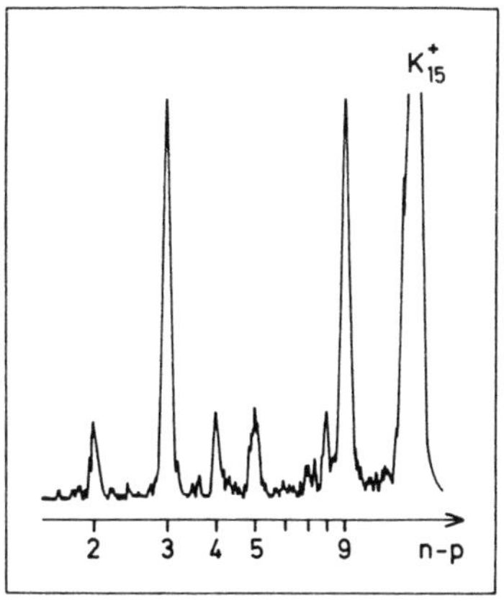

Fig. 4. Photodissociation of K_{15}^+ with a photon energy $h\nu = 3.67$ eV, at relatively high laser fluence. Two successive evaporative sequences are visible:

$K_{15}^+ \xrightarrow{h\nu} K_9^+ + 6K \xrightarrow{h\nu} K_3^+ + 6K$. The time duration of each sequence is much shorter than the residence time between the electrostatic plates and shorter than the laser pulse duration (6 ns)

and then the dissociation rate that we observe is:

$$F_n = \int_0^\infty [P_n(E^*) - Q_n(E^*)] dE^* / \int_0^\infty P_n(E^*) dE^* \qquad (4)$$

Taking D_n^+ as parameters we have deduced a set of D_n^+ values (Fig. 3) leading to the best fit between the calculated and observed F_n values.

The remarkable feature is the saw-toothed behavior of the D_n^+ values. This is a new confirmation for the validity of the jellium model applied to alkali-atom clusters [23, 24]. Moreover the very large sensitivity of θ_n on the ratio E^*/D_n^+ explains the observation of only one channel of evaporation, the simultaneous observation of a monomer and of a dimer corresponding to quasi-degenerate channels. In particular, no fission is expected for these species following the idea that a statistical behavior strongly favors the lowest dissociative channel.

2. Photo-induced evaporation

An illustration of the dissociation dynamics may be found in the photodissociation study of the size-selected ion clusters X_n^+. For this purpose the clusters are decelerated around 2 keV (Fig. 1) and then irradiated by a second pulsed laser at right angles to the cluster ion beam. Some fractions of the cluster ions dissociates. The residual parent ions and charged fragments are re-accelerated and mass dispersed in the TOF2. At low laser fluence one observes essentially only one charged fragment X_{n-p}^+. The integer p is roughly independent on n, but is proportional to the photon energy [12] and satisfies the relation $p \approx h\nu/\overline{D_n^+}$ where $\overline{D_n^+}$ is the average binding energy in the range n, $n-p$. This indicates that the photodissociation is an evaporative process in agreement with the satistical model described in paragraph 1. Indeed, the relatively hot and metastable ions X_n^+ are heated again, suddenly, by the pulses of the second laser. Their internal energy $E^* + h\nu$ leads to an extremely short decay time according to (1) for successive evaporations, at least up to the total burning of the photon energy (Fig. 4). At high laser fluence several sequences of evaporation can be observed due either to the absorption of another photon by the X_{n-p}^+, X_{n-2p}^+ ... fragments or to a multiphoton absorption process in the parent itself [12].

IV. Collisional interaction: evaporative dissociation and charge exchange

Collisional energy transfer is another way to put energy on the cluster. We have studied the response of the cluster ions X_n^+ to the collisional interaction with an alkali-vapor for a fixed high laboratory energy (3 keV). The heat-pipe kind absorbing cell (Fig. 1) is 20 cm long and works at low density so that clusters undergo only one collision during their crossing with the vapor.

1. Ionized potassium clusters and sodium vapor collisions: energy transfer

Our experimental set-up used with the retarding potential configuration (Fig. 1) gives essentially the residual parent ions and the mass spectrum of charged

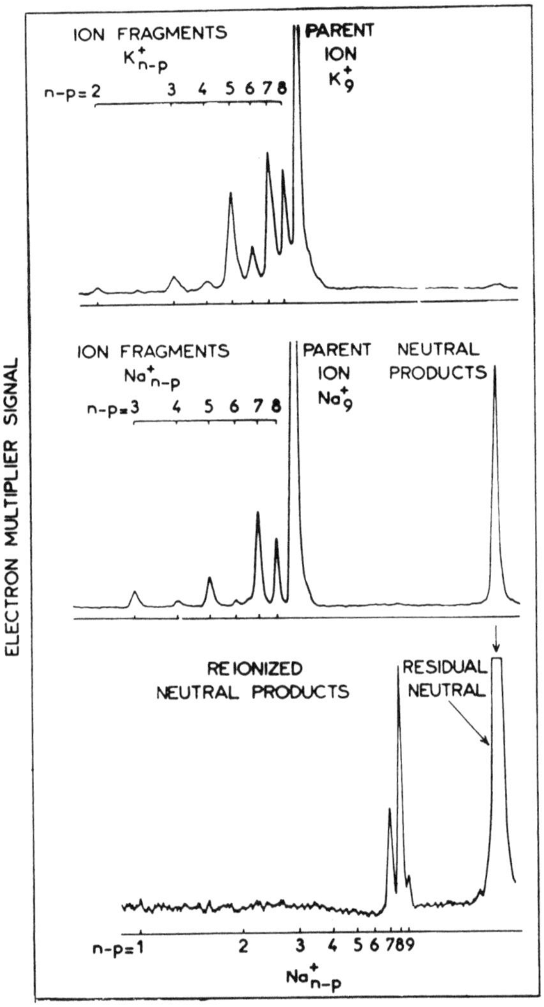

Fig. 5. Comparison of fragments observed for the $K_9^+ + Na$ interaction (*upper trace*) and those observed for the $Na_9^+ + Cs$ interaction (*medium trace*). The vanishingly small amount of neutrals visible in the $K_9^+ + Na$ correspond to the neutral monomer and dimer evaporated during the dissociation. These light particles are mostly ejected off the beam axis away from the detector acceptance angle. The lower trace shows the reionization spectrum of the neutralized K_9 parent when all the incident ions are rejected

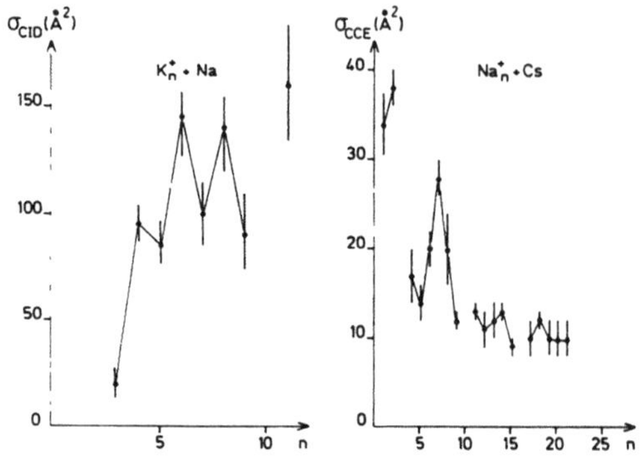

Fig. 6. Magnitude and size evolution of collisional induced cross-section σ_{CID} and collisional charge exchange one σ_{CCE}

fragments for the process $K_n^+ + Na \rightarrow K_{n-p}^+ + pK + Na$ (Fig. 5). The collisional-induced dissociation cross-section σ_{CID} can be deduced from the relative intensities of the residual parent and the fragments vs. the sodium density (Fig. 6). The increase of σ_{CID} when n increases is in agreement with the picture of a collective excitation of valence electrons [25, 26]. The odd-even alternation reflects the larger polarizability of species having an unpaired electron.

Looking at the dynamics of the interaction, a fraction ΔE of the relative kinetic energy (in the center of mass) is transferred to K_n^+ and is added to the initial internal energy E*. A subsequent dissociation takes place, according to equation (1). In contrast with the photo-induced dissociation, for which the added energy is well defined, ΔE varies with the impact parameter. The sequence of charged fragments (Fig. 5) results from a sum of contributions due to all the possible values of the impact parameter. The smallest observed fragment X_{n-p}^+ represents the last step of the dissociation allowed during the transient time between the cell and the retarding field plates (3–30 µs). The corresponding energy burning is found around 8 eV.

2. Ionized sodium clusters and cesium vapor interaction: charge exchange

For the collisional partners chosen here, the ionization potentials are relatively close [27] and an efficient neutralization by charge transfer is expected [28, 29]. This is indicated in the mass-spectrum of fragments as shown Fig. 5 for the $Na_9^+ + Cs$ collision. The remarkable change against the $K_9^+ + Na$ case, is the occurrence of a large signal corresponding to a neutral packet. Such a signal, vanishingly small with K_n^+ and Na partners increases when the cesium density increases. It must be attributed to large neutral species resulting from charge-exchange. The mass composition of this neutral packet is revealed as follows: using the deceleration-acceleration configuration we first sweep away the incident ions and then re-ionize the neutrals by the second pulsed laser beam. At low laser fluence and for a near-threshold ionization, the mass spectra due to this post-collisional ionization, present a narrow distribution of at most three masses centered on Na_n or Na_{n-1} (Fig. 5). This shows that a small dissociation is associated to the charge exchange, explained by a small increase in the internal energy of the neutralized parent. From the relative intensities of incident ions and neutral packets vs. the cesium density one deduces a collisional charge exchange cross-section σ_{CCE} (Fig. 6). The rather slow decline of σ_{CCE} as a function of n and its magnitude differs markely from the cross-section associated to the collisional dissociation. In the conventional picture of a charge transfer occurring during the crossing of two diabatic electronic states, the Landau–Zener–Stückelberg formulation gives the cross-section for the process $Na_n^+ + Cs \rightarrow Na_n + Cs^+$. The magnitude and cluster size behavior are consistent with our observations. This model implies a relatively long range interaction process (10 to 15 Å).

V. Conclusion

We have showed that following the electronic excitation of alkali-atom clusters a stabilization of these species occurs via evaporative cooling corresponding to (successive) ejections of monomers or dimers. Despite the fact that the observed dynamics is explained by a simple RRK statistical model other experiments are needed to refine the analysis such as the determination of the kinetic energy released by the fragments. The efficient charge-exchange provides a source of size-selected neutral aggregates suggesting further experiments. Among them, the photoionization spectroscopy of these neutrals should give new insight on the internal energy of these species by comparing the stability of neutrals to cluster ions. These works are in progress in our laboratory.

References

1. Echt, O., Kreisle, D., Knapp, M., Recknagel, E.: Chem. Phys. Lett. **108**, 401 (1984)
2. Bloomfield, L.A., Freeman, R.R., Brown, W.L.: Phys. Rev. Lett. **54**, 2246 (1985)
3. Alexander, M.L., Johnson, M.A., Levinger, N.E., Lineberger, W.C.: Phys. Rev. Lett. **57**, 976 (1986)
4. Stace, A.J.: J. Chem. Phys. **85**, 5774 (1986)
5. Begemann, W., Meiwes-Broer, K.H., Lutz, H.O.: Phys. Rev. Lett. **56**, 2248 (1986)
6. Brucat, P.J., Zheng, L.S., Pettiette, C.L., Yang, S., Smalley, R.E.: J. Chem. Phys. **84**, 3078 (1986)
7. Kamke, W., Kamke, B., Kiefl, H.U., Hertel, I.V.: J. Chem. Phys. **84**, 1325 (1986)
8. Scheier, P., Märk, T.D.: Phys. Rev. Lett. **59**, 1813 (1987)
9. Geusic, M.E., Jarrold, M.F., McIlrath, T.J., Freeman, R.R., Brown W.R.: J. Chem. Phys. **86**, 3862 (1987)
10. Bréchignac, C., Cahuzac, Ph., Roux, J. Ph., Pavolini, D., Spiegelmann, F.: J. Chem. Phys. **87**, 5694 (1987)
11. Bréchignac, C., Cahuzac, Ph., Pflaum, R., Roux, J. Ph.: J. Chem. Phys. **88**, 3732 (1988)
12. Bréchignac, C., Cahuzac, Ph., Roux, J. Ph.: J. Chem. Phys. **88**, 3022 (1988)
13. Kirchner, N.J., Bowers, M.T.: J. Chem. Phys. **86**, 1301 (1987)
14. Freas, R.B., Dunlap, B.I., Waite, B.A., Capana, J.E.: J. Chem. Phys. **86**, 1276 (1987)
15. Jarrold, M.F., Bower, J.E., Kraus, J.S.: J. Chem. Phys. **86**, 3876 (1987)
16. Hanley, L., Ruatta, S.A., Anderson, S.L.: J. Chem. Phys. **87**, 260 (1987)
17. Bréchignac, C., Cahuzac, Ph., Roux, J. Ph.: J. Chem. Phys. **87**, 229 (1987)
18. Boustani, I., Pewestorf, W., Fantucci, P., Bonacic-Koutecky, V., Koutecky, J.: Phys. Rev. B **35**, 9437 (1987)
19. Wang, Y., George, T.F., Lindsay, D.M., Beri, A.C.: J. Chem. Phys. **86**, 3493 (1987); Lindsay, D.M., Wang, Y., George, T.F.: J. Chem. Phys. **86**, 3500 (1987)
20. Spiegelmann, F., Pavolini, D.: J. Chem. Phys. **89**, 4954 (1988)
21. Bonacic-Koutecky, V., Boustani, I., Guest, M., Koutecky, J.: J. Chem. Phys. **89**, 4861 (1988)
22. Engelking, P.C.: J. Chem. Phys. **87**, 936 (1987)
23. Knight, W.D. Clemenger, K., De Heer, W.A., Saunders, W.A., Chou, M.Y., Cohen, M.L.: Phys. Rev. Lett. **52**, 2141 (1984)
24. Eckardt, W.: Phys. Rev. B **29**, 1558 (1984)
25. De Heer, W.A., Selby, K., Kresin, V., Masuri, J., Vollmer, M., Chatelain, A., Knight, W.D.: Phys. Rev. Lett. **59**, 1805 (1987)
26. Eckardt, W.: Phys. Rev. B **36**, 4483 (1987)
27. Kappes, M.M., Schär, M., Röthlisberger, U., Yeretzian, C., Schumacher, E.: Chem. Phys. Lett. **143**, 251 (1988)
28. Abshagen, M., Kowalski, J., Meyber, M., Zu Putlitz, G., Träger, F., Well, J.: Europhys. Lett. **5**, 13 (1988)
29. Begemann, W., Dreihöfer, S., Ganteför, G., Siekmann, H.R., Meiwes-Broer, K.H., Lutz, H.O.: Elemental and molecular clusters, pp. 230–253. Berlin, Heidelberg, New York: Springer 1988

Z. Phys. D – Atoms, Molecules and Clusters 12, 205–208 (1989)

Atoms, Molecules and Clusters
Zeitschrift für Physik D
© Springer-Verlag 1989

Jellium-model calculation for monomer and dimer decays of some potassium clusters

Susumu Saito* **and Marvin L. Cohen**

Department of Physics, University of California, and Materials and Chemical Sciences Division, Lawrence Berkeley Laboratory, Berkeley, CA 94720, USA

Received 5 July 1988; final version 30 September 1988

We have studied several decay processes of potassium clusters and found that a dimer-decay mechanism can explain the observed lowest abundance of K_{10} in the K_n mass spectra. Total-energy curves for decay processes are calculated using a jellium-background model for positive-ion cores and the local-spin-density-functional approximation for valence electrons. The energy-barrier height for a dimer decay of K_{10} from the energy-minimum point is found to be 0.18 eV, which is a reasonable magnitude for the decay to take place thermally in the experiment. The monomer decay of K_9 and the dimer decay of K_{11}, which are expected to be the most favorable decays of K_9 and K_{11}, are found to have high barriers. Monomer and dimer decays of K_8 are also studied and the monomer decay is found to be more favorable, in accord with the high-nozzle-temperature mass spectrum.

PACS: 36.40; 31.20

I. Introduction

Shell structure of valence electrons in alkali-metal-atom clusters has been studied extensively [1] since Knight et al. [2] reported the electronic shell structure of neutral sodium clusters produced by a supersonic nozzle source. Experimental mass spectra for Na_n [1–3] show peaks at $n=8$, 20, 40, 58, etc. ("magic numbers"), which are considered the relatively stable clusters and are found to correspond to the closed-shell configurations of valence electrons in a spherical potential well. A spherical jellium-background model, where positive-ion cores are replaced by the positively charged sphere with constant density, has been used for sodium clusters successfully to explain their electronic shell structure [1, 2, 4–7]. The electronic-energy-level sequence in that model is

$$1s, 1p, 1d, 2s, 1f, 2p, 1g, \ldots \tag{1}$$

Experimental magic-number clusters are found to correspond to closed-shell electronic structures such as $1s^2 1p^6$, $1s^2 1p^6 1d^{10} 2s^2$, etc.

In the case of neutral potassium clusters, supersonic-nozzle-source mass spectra [1, 3] show peaks at similar n values. They also show, however, some unique and interesting features, that is, clear dips at 10, 22, 42, and 60 (Fig. 1). These "anti-magic-numbers" cannot be explained by the static theory. Neither the total energies per atom nor the second differences of the energy with respect to the cluster size in the spherical jellium-background model (both of which clearly show the high stabilities of magic-number clusters [1, 6, 8]) indicate the poor stabilities of the anti-magic-number clusters. The anti-magic-numbers have, however, a simple relationship to magic numbers. They are displaced from magic numbers by two. This relationship suggests that anti-magic-number clusters tend to decay into magic-number clusters and dimers:

$$K_{n+2} \rightarrow K_n + K_2 \tag{2}$$

* Permanent address: Fundamental Research Laboratories, NEC Corporation, Miyazaki, Miyamae-ku, Kawasaki 213, Japan

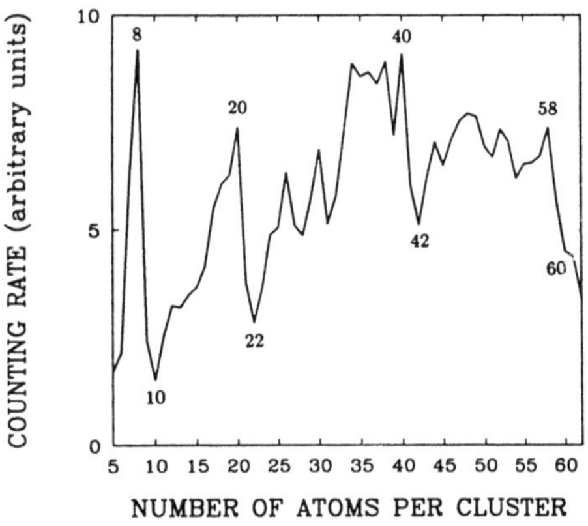

Fig. 1. Peak heights for K_n in a supersonic-nozzle-source mass spectrum (data from de Heer *et al.* [1])

where n is a magic number. Such decays may take place thermally in the experiment and make the abundance of K_{n+2} poor. Actually, when the temperature of the nozzle is higher, the experimental mass spectrum shows peaks at 7 and 18 instead of 8 and 20 [1], indicating the presence of thermal decays of potassium clusters.

In our previous paper [8], we studied potassium clusters using the jellium-sphere background model. In that work, we calculated the electronic structures of K_n using the local-spin-density-functional (LSDF) method. Reaction-energy calculations for fission processes of K_n reveal that there are several possible exoergic fissions. We have pointed out that K clusters which have exoergic fission channels are unabundant in the experimental mass spectra. This correlation indicates the important role of thermal fissions in K_n mass spectra. We have also found that most of the exoergic fissions produce dimers, K_2, as one of the daughter clusters. Such processes are referred to as dimer decays since some similarities to the α decay of the nucleus have been pointed out. Total-energy curves for several decay processes have been reported and the dimer decay has been proposed for the origin of anti-magic-number, 10.

In this paper, we will investigate several decay processes of potassium clusters in terms of the total-energy curves for the processes and examine the origin of the anti-magic-number 10. The decay processes studied are:

$$K_8 \to K_7 + K_1 \quad (3)$$

$$K_8 \to K_6 + K_2 \quad (4)$$

$$K_9 \to K_8 + K_1 \quad (5)$$

$$K_{10} \to K_8 + K_2 \quad (6)$$

$$K_{11} \to K_9 + K_2 \quad (7)$$

In these processes, we will consider minimum-spin-polarization K_n for daughter clusters, that is, K_n with $M=0$ and 1 for even and odd n, respectively ($M=N_\uparrow - N_\downarrow$, where N_σ is a number of σ-spin electrons), since the even-odd alternation observed in the ionization-potential experiment for K_n [1] indicates that they have minimum-spin configurations [8]. Monomer and dimer decays of K_8 ((3) and (4)) are studied to examine whether the monomer decay is the favorable decay channel of K_8, as expected from the high-nozzle-temperature experiment. A monomer decay of K_9 (process (5)) is the most favorable decay channel of K_9 from the viewpoint of the reaction energy calculated by the jellium-sphere-background model with the LSDF approximation for valence electrons [8]. Similarly, dimer decays of K_{10} and K_{11} ((6) and (7)) are also the most favorable decay channels of K_{10} and K_{11} and both decays are exoergic processes [8].

In the present work, two overlapping jellium spheres [8, 9] are used for the positive background (Fig. 2). In each process, the total energy of the system is calculated for several d values (defined in Fig. 2), from $R_n - R_m$ to larger than $R_{n-m} + R_m$ (where $R_l = l^{1/3} r_s$ is the radius of the jellium sphere for K_l), and then interpolated. The electronic structure of the system is calculated by the LSDF method. A more detailed description of the calculation is given in reference 8.

II. Results

In Fig. 3, the calculated total-energy curves for monomer and dimer decays of K_8 are shown. In the monomer-decay process, two different spin-polarization cases ($M=0$ and 2) are studied. In the case of no spin polarization ($M=0$), the total energy increases monotonically in the decay process. On the other hand, the

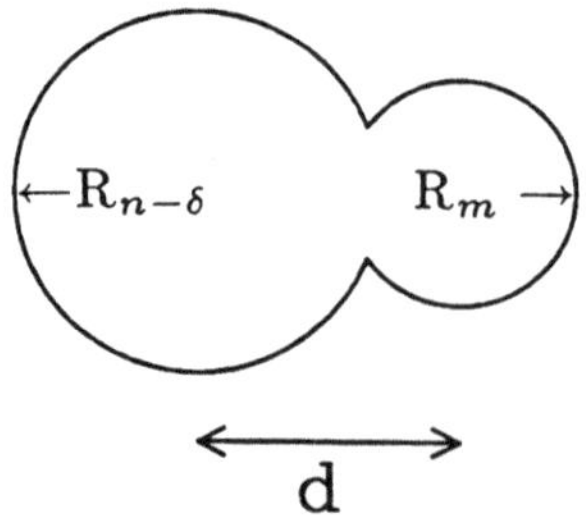

Fig. 2. Jellium background for m-mer decay process of K_n ($K_n \to K_{n-m} + K_m$). The density is uniform ($[4\pi r_s^3/3]^{-1}$, $r_s = 4.86$ a.u.) including the overlapping-sphere region. The daughter-sphere radius R_m is always kept $m^{1/3} r_s$. The mother-sphere radius $R_{n-\delta}$ ($0 \le \delta \le m$) depends on d, the distance between two sphere centers, and is so determined that the total positive charge equals n

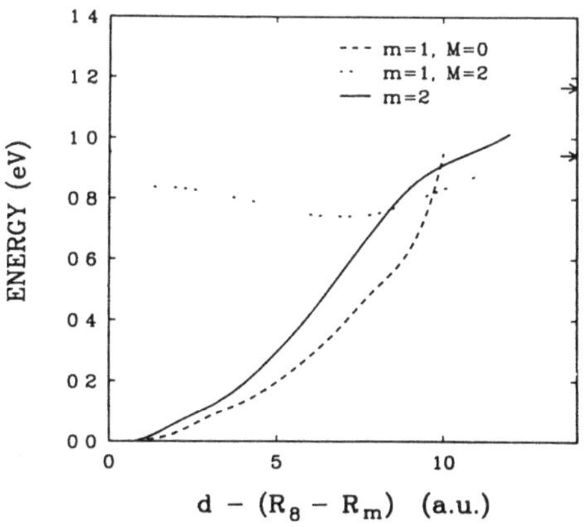

Fig. 3. Total-energy curves for monomer ($m=1$) and dimer ($m=2$) decays of K_8. Energies are measured from the ground-state energy of K_8. The expected ground-state energy of the monomer-decay system at $d=\infty$ (the sum of the total energies of K_7 ($N_\uparrow=4, N_\downarrow=3$) and K_1 ($N_\uparrow=1, N_\downarrow=0$)) is shown by the lower arrow. The sum of the energies of K_6 ($N_\uparrow=N_\downarrow=3$) and K_2 ($N_\uparrow=N_\downarrow=1$) is indicated by the upper arrow. $d-(R_8-R_m)=(R_{8-m}+R_m)-(R_8-R_m)$ is the value where the jellium background for the m-mer-decay system becomes two complete spheres. These are 9.30 and 11.36 a.u. for $m=1$ and 2, respectively

monomer-decay total-energy curve with a spin polarization ($M=2$) decreases initially and increases after a minimum point. At $d \leq R_8-R_1$, the system equals K_8 and the ground state is a closed-shell ($M=0$) K_8, and the spin-polarized state ($M=2$) corresponds to an excited state. The energy difference between the two states decreases in the decay process and two total-energy curves finally cross each other with the energy lower than the expected ground-state value of $d=\infty$. Hence, the energy-barrier height, which will determine the stability of the cluster against the decay, for the monomer decay of K_8 is considered the absolute value of the reaction energy of the process (3), that is, the difference between the energy of K_8 and the sum of the ground-state energies of K_7 and K_1, which is 0.94 eV.

In the dimer-decay process of K_8, we study only the $M=0$ case since the spin-polarized ($M=2$) system will lead to $K_6(N_\uparrow=4,\ N_\downarrow=2)+K_2(N_\uparrow=N_\downarrow=1)$, which includes the high-spin-state ($M=2$) K_6. The total-energy curve for the dimer decay of K_8 shows a monotonically increasing behaviour and approaches the $d=\infty$ value. The energy-barrier height for a dimer decay is considered to be the difference between the initial (K_8) and the final (K_6+K_2) energies, which is 1.17 eV and is higher than the monomer-decay barrier. Therefore, the monomer decay of K_8 is more favorable than the dimer decay. This result is consistent with the high-nozzle-temperature experiment mentioned above (150 K higher than the usual experiment) [1], where the mass spectrum for K_n shows a peak at $n=7$, indicating that the monomer decay is a favorable decay channel of K_8. Although the barrier height for the monomer decay is much higher than the nozzle temperature (about 1000 K) in the experiment, many K_8 clusters will possess enough kinetic energy for the decay to occur within the cluster flight time in the high-nozzle-temperature experiment. (Rough estimates for the thermal-evaporation rates for Na_n have been given and the flight time is shown to be enough long for the evaporation, or the monomer decay, to take place for Na_n at a temperature around 500 K [10]).

Total-energy curves for the decay processes (5)–(7) are shown in Fig. 4. Spin polarizations considered are $M=1$, 0, and 1, for the monomer decay of K_9, the dimer decay of K_{10}, and the dimer decay of K_{11}, respectively. These three curves have some common features. The energy of each system decreases initially in the decay process and then increases before the jellium background becomes two separated spheres. The energy-barrier height for each process can be defined as the difference between the minimum energy of the system and that at $d=\infty$. The results are 0.25, 0.18, and 0.25 eV for the monomer decay of K_9, the dimer decay of K_{10}, and the dimer decay of K_{11}, respectively. Therefore, the dimer decay of K_{10} is the most favorable decay process among the three. The temperature of the nozzle in the experiment is about the half of the barrier height for the dimer decay of K_{10}. Hence, most K_{10} clusters will possess enough kinetic energy for the dimer decay to take place within the flight time because the energy-barrier height for this case is much lower than the monomer decay of K_8. Since the three decay processes are the most favorable decay processes of K_9, K_{10}, and K_{11}, the present result shows that K_{10} is the most unstable cluster among the three.

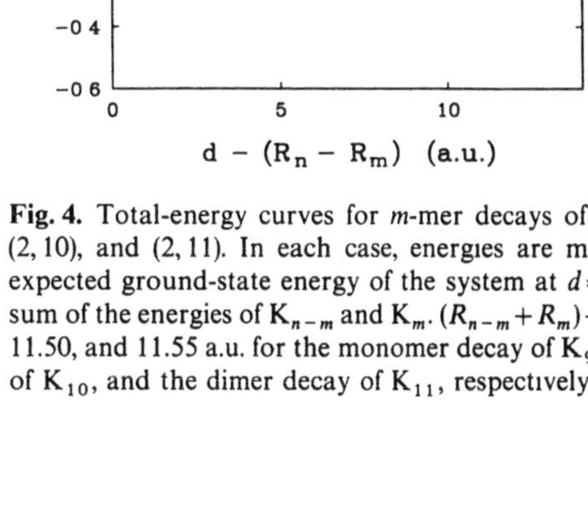

Fig. 4. Total-energy curves for m-mer decays of K_n: $(m,n)=(1,9)$, (2, 10), and (2, 11). In each case, energies are measured from the expected ground-state energy of the system at $d=\infty$, which is the sum of the energies of K_{n-m} and K_m. $(R_{n-m}+R_m)-(R_n-R_m)=9.33$, 11.50, and 11.55 a.u. for the monomer decay of K_9, the dimer decay of K_{10}, and the dimer decay of K_{11}, respectively

III. Discussion

The present work has shown that the dimer decay of K_{10} has the lowest energy barrier among the most favorable decay processes of K_9, K_{10}, and K_{11}, verifying that the dimer decay is a major cause of the anti-magic-number, 10. Since the abundances in K_n mass spectrum (Fig. 1) around the anti-magic-numbers (10, 22, 42, and 60) have very similar behavior, we expect that the dimer decays are also the origin of other anti-magic-numbers. The present work has also shown that the monomer decay is a favorable decay channel of K_8, in accord with the experimental high-nozzle-temperature mass spectrum.

It can be seen from Figs. 3 and 4 that the total-energy curves for monomer and dimer decays are classified into two groups: one with a monotonically increasing behavior (monomer decay of K_8 with $M=0$ and the dimer decay of K_8) and the other with an initial decrease and a final increase (monomer decay of K_8 with $M=2$, monomer decay of K_9, dimer decay of K_{10}, and dimer decay of K_{11}). In the former decays, the initial system is K_8 with $M=0$, which has a closed-shell electronic configuration. On the other hand, in the latter decays, all the initial electronic configurations are open. Therefore, the energy gain at the initial stage of the decay in the latter group can be considered to be an effect similar to the Jahn-Teller effect. Here the decay process lifts the degeneracy of the highest-occupied level which is partially occupied in the initial system. Hence, the electronic shell structure of the mother cluster plays a decisive role in the decay process, especially at the initial stage of the decay.

In the former group, the energy-barrier height for the decay is given by the absolute value of the reaction energy. On the other hand, in the latter group, the energy-barrier height does not have a simple relation to the reaction energy. Actually the monomer decay of K_9 and the dimer decay of K_{11} have almost the same barrier height even though they are endoergic and exoergic processes, respectively. In the latter group, however, we can give a qualitative estimation for the barrier height if we consider the energy barrier for a m-mer decay of K_n as the energy gain of the fusion process

$$K_{n-m}+K_m \rightarrow K_n \qquad (8)$$

A jellium-dimer-background model calculation for sodium clusters has revealed that the binding energy of $(Na_4)_2$, which is the initial stage of the fusion of two open-shell clusters, is considerably larger than the closed-shell case, $(Na_8)_2$ [11]. Therefore, a fusion of closed-shell clusters is less favorable and the energy gain will be smaller than other types of fusions. The total-energy curve for the fusion process of K_8 and K_2, that is, the curve for the dimer decay of K_{10}, also has a shallow minimum compared to others. Hence, the electronic shell structures of daughter clusters in the decay process play an important role at the final stage of the process for decays of open-shell clusters.

In summary, the present work is based on a model of overlapping jellium spheres for the positive-ion cores of K_n clusters. This model has enabled us to calculate the total-energy curves for decay processes of these clusters. Although the model is idealized and its applicability is limited to simple-metal clusters, it is a useful model since it is much more difficult to calculate these decay processes by treating the constituent atoms of a cluster separately.

We are grateful to Kathy Selby, Michael Vollmer, and Jun Masui for their useful discussions. The work of one of us (S.S.) is supported by NEC Human Resources International (Tokyo, Japan). This work was supported by National Science Foundation Grant No. DMR8319024 and by the Director, Office of Energy Research, Office of Basic Energy Sciences, Materials Sciences Division of the U.S. Department of Energy under Contract No. DE-AC03-76SF00098. CRAY computer time was provided by the U.S. Office of Energy Research of the Department of Energy at the National Magnetic Fusion Energy Computer Center.

References

1. de Heer, W.A., Knight, W.D., Chou, M.Y., Cohen, M.L.: Solid State Phys. Ehrenreich, H., Seitz, F., Turnbull, D. (eds.), vol. 40, pp. 93–181. New York: Academic Press 1987
2. Knight, W.D., Clemenger, K., de Heer, W.A., Saunders, W.A., Chou, M.Y., Cohen, M.L.: Phys. Rev. Lett. **52**, 2141 (1984)
3. Knight, W.D., de Heer, W.A., Clemenger, K., Saunders, W.A.: Solid State Commun. **53**, 445 (1985)
4. Martins, J.L., Car, R., Buttet, J.: Surf. Sci. **106**, 265 (1981)
5. Ekardt, W.: Phys. Rev. B **29**, 1558 (1984)
6. Chou, M.Y., Cleland, A., Cohen, M.L.: Solid State Commun. **52**, 645 (1984)
7. Ishii, Y., Ohnishi, S., Sugano, S.: Phys. Rev. B **33**, 5271 (1986)
8. Saito, S., Cohen, M.L.: Phys. Rev. B **38**, 1123 (1988)
9. Saito, S., Ohnishi, S.: Phys. Rev. Lett. **59**, 190 (1987)
10. de Heer, W.A., Selby, K., Kresin, V., Masui, J., Vollmer, M., Châtelain, A., Knight, W.D.: Phys. Rev. Lett. **59**, 1805 (1987)
11. Ishii, Y., Saito, S., Ohnishi, S.: Z. Phys. D – Atoms, Molecules, and Clusters **7**, 289 (1987)

Z. Phys. D – Atoms, Molecules and Clusters 12, 209–211 (1989)

Atoms, Molecules
Zeitschrift für Physik D and Clusters
© Springer-Verlag 1989

Theoretical study of the stability of X_N^n ($n=-1, 0, +1, +2$; X = Ag, Cu; $N \leq 25$) clusters as a function of size using a non-local density functional formalism

A. Rubio, L.C. Balbás, and A. Vega

Departamento de Fisica Teórica, Universidad de Valladolid, Valladolid, Spain

Received 7 July 1988; final version 30 September 1988

The spherical jellium model and self-consistent Weighted Density Approximation (WDA) to density functional theory have been used to study the stability of X_N^n ($n=-1, 0, +1, +2$; X = Ag, Cu; $N \leq 25$) clusters. The calculated magic numbers coincide with the observed ones. The first (IP1) and second (IP2) ionization potentials of Ag_N and Cu_N as a function of size show the typical oscillations induced by the electronic shell-filling effect. IP1 of Cu_N is about 0.5 eV higher than IP1 of Ag_N in the range studied ($N \leq 25$). For both Cu_N and Ag_N, IP1 appears to converge well towards the respective experimental values of the work function. The use of WDA allows us to obtain bound negative clusters of small size or with a nearly empty external shell, which is not possible using the Local Density Approximation (LDA) [1, 2]. However the electron affinity of X_N clusters obtained as the difference of energies of the neutral and the negatively charged clusters, becomes negative for $N=2$, 3 and 8 (very close to zero for $N=8$), revealing that WDA needs further refinements.

PACS: 36.40; 31.20

Introduction

In a recent paper [3] we have used the jellium model and self-consistent density functional theory to study the stability of Ag_N^{+2}, Ag_N^+, Ag_N, Ag_N^- and Na_N^- clusters. Calculations using the Local-Density-Approximation (LDA) for exchange and correlation effects have shown that:

i) The magic numbers of X_N^+ clusters (X = noble metal atom) produced as secondary ions in the experiments of Katakuse et al. [4] can be explained by a simple model in which the atomic 4*s*, 5*s* or 6*s* (for Cu, Ag and Au respectively) electrons move in a spherically symmetric effective potential in the cluster, giving rise to cluster electronic shells. When the number of atoms in the cluster is such that these shells become filled then the cluster is particularly stable with respect to clusters with adjacent sizes. The picture is, in fact, similar to the standard picture used to explain the magic number of alkaline clusters [1, 2].

ii) The same model explains the magic numbers [5] of Ag_N^- clusters. However, the LDA is unable to give solutions for the negative clusters of small size or with a nearly empty external electronic shell. This is due to the incorrect asymptotic behaviour of the exchange and correlation potential V_{xc}, provided by the LDA (in this theory $V_{xc}^{\mathrm{LDA}}(\bar{r})$ decreases exponentially at large r instead of the correct r^{-1} behaviour).

In [1] we also performed exploratory calculations for small Na_N^- clusters using a non-local Weighted-Density-Approximation (WDA) which improves the large distance behaviour of V_{xc}. The WDA solved the stability problem of the LDA giving rise to bound negative clusters even for nearly empty external shells. On the other hand the trends shown by the LDA calculations seemed to be preserved by the

WDA. In this paper we look further into this question by performing calculations, using the WDA, to obtain the stability properties of X_N^n ($n=-1, 0, +1, +2$; $N \leq 25$; X=Ag, Cu) clusters.

Method

The clusters are considered as a uniform distribution of positive charge (representing the ionic background) with spherical shape ($n_+(r)=n_0$ for $r<R$ and $n_+(r)=0$ for $r>R$) and a distribution of (valence) electronic charge, $n(\bar{r})$, calculated self-consistently by solving the one-electron equations of the density functional formalism

$$[-\tfrac{1}{2}\nabla^2 + V_H(\bar{r}) + V_{xc}^{\mathrm{WDA}}(\bar{r})]\,\Phi_i(\bar{r}) = \varepsilon_i \Phi_i(\bar{r}) \tag{1}$$

$$n(\bar{r}) = \sum_1 |\Phi_i(\bar{r})|^2 \tag{2}$$

(in atomic units). $V_H(\bar{r})$ is the electrostatic Hartree potential due to the positive background and to the electronic cloud, and V_{xc}^{WDA} is the exchange-correlation potential. The radius R of the cluster is linked to the number of atoms by the relation $4\pi R^3/3 = N\Omega$, where the volume per atom, Ω, is assumed to be the same as in the bulk metal. V_{xc}^{WDA} is obtained from

$$V_{xc}^{\mathrm{WDA}}(\bar{r}) = \frac{\delta E_{xc}^{\mathrm{WDA}}[n]}{\delta n(\bar{r})} \tag{3}$$

where

$$E_{xc}^{\mathrm{WDA}}[n] = \frac{1}{2}\int \frac{n(\bar{r})n(\bar{r}')}{|\bar{r}-\bar{r}'|} G^{\mathrm{WDA}}(\bar{r},\bar{r}';\, n^{\mathrm{WDA}}(\bar{r}))\,d\bar{r}\,d\bar{r}' \tag{4}$$

is the exchange-correlation energy, and the weighted density parameter n^{WDA} appearing in G^{WDA} is fixed for each $\bar{r}$ through the sum rule

$$\int n(\bar{r}')G^{\mathrm{WDA}}(\bar{r},\bar{r}';\, n^{\mathrm{WDA}}(\bar{r}))\,d\bar{r}' = -1 \tag{5}$$

The function $G^{\mathrm{WDA}}(\bar{r},\bar{r}')$ used in this work is taken from the work of Chacón and Tarazona [6], and it provides self-consistently the fundamental properties of the homogeneous electron gas. This function has been tested in several atoms [7] giving similar or better results than those obtained using the Langreth-Mehl [8] approximation to V_{xc}.

Equations (1)–(4) are solved self-consistently for X_N^n ($n=-1, 0, +1, +2$; $N \leq 25$; X=Ag, Cu) clusters, giving the density $n(\bar{r})$ and from this $n(\bar{r})$ we obtain the total energy of the clusters. This is not possible using the V_{xc}^{LDA} potential instead of V_{xc}^{WDA} in (1) for X_N^- clusters with $N \leq 5$ and $N=8, 9, 10, 20, 21, 22$, X=Ag, Cu.

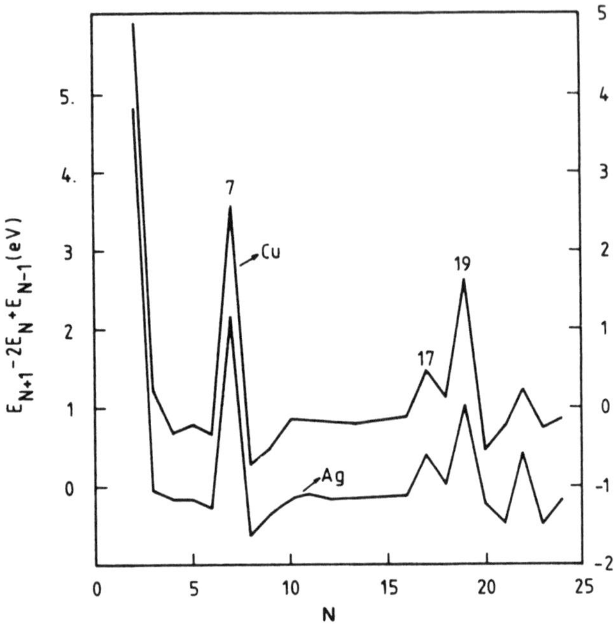

Fig. 1. Second derivative of the total energy, $S(N) = E_{N+1} - 2E_N + E_{N-1}$ as a function of size N, for Ag_N^- and Cu_N^- clusters. The scale for $S(N)$ to the left corresponds to Ag_N^- (*lower curve*) and the scale to the right, corresponds to $S(N)$ for Cu_N^- (*upper curve*)

Results

The quantity $S(N) = E_{N+1} - 2E_N + E_{N-1}$ as a function of N, where E_N, E_{N+1}, E_{N-1} are the total energies of clusters with N, $N+1$ and $N-1$ atoms respectively, must have a peak if a cluster with N atoms is more stable relative to those with one atom more and one atom less. The stability peaks for all the systems studied in this work correspond to clusters with filled electronic shells, the ordering of electronic levels being 1*s*, 1*p*, 1*d*, 2*s*, 1*f*... This behaviour is analogous to that previously reported [3] for Ag_N, Ag_N^+ and Ag_N^{+2} using the LDA. Now we confirm the same behaviour also for Ag_N^- clusters in the range $N \leq 25$ using the WDA. The same trend appears for Cu_N^n ($n=-1, 0, +1, +2$) clusters.

In Fig. 1 we show the functions $S(N)$ for Ag_N^- and Cu_N^-. The $S(N)$ function peaks at a number of atoms $N = 7$, 17 and 19 corresponding to 8, 18 and 20 electrons, which complete the electronic shells 1*p*, 1*d* and 2*s* respectively. We note a reduced shell-closing effect for X_{19}^+ and X_{17}^- (clusters with 18 electrons) with respect of X_{21}^+ and X_{19}^- (clusters with 20 electrons) respectively, in agreement with the observed trend [3]. This fact is a consequence of the competition in the occupation of the orbitals 1*d* and 2*s*, which is more remarkable in the WDA than in the LDA. With respect to the first and the second ionization potentials of Ag_N and Cu_N we find the typical oscillations induced by the electronic

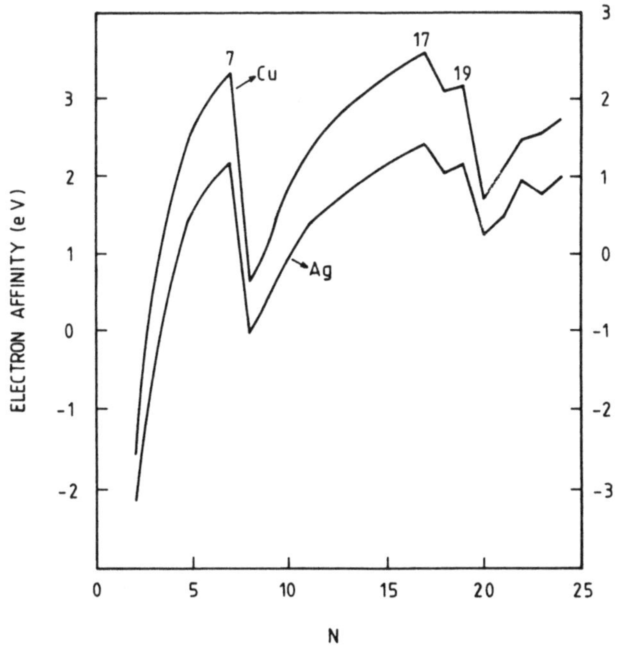

Fig. 2. Electron affinity of Ag_N (*lower curve*) and Cu_N (*upper curve*) clusters as a function of size. The scale to the left corresponds to Ag_N clusters and the scale to the right to Cu_N clusters

shell-filling effect. The absolute values of IP1 and IP2 are now a little bit smaller than those previously reported for Ag_N using the LDA. IP1 of Cu_N is about 0.5 eV higher than IP1 of Ag_N in the range studied ($N \leq 25$). For both Cu_N and Ag_N, IP1 appears to converge well towards the respective experimental values of the work function (4.26 eV for Ag and 4.65 for Cu).

For a cluster with N atoms the electron affinity $EA(X_N)$, is defined as $EA(X_N) = E(X_N) - E(X_N^-)$, where $E(X_N)$ and $E(X_N^-)$ are the total energies of clusters with N electrons and $N+1$ electrons, respectively. EA as a function of N is given in Fig. 2 for Ag_N and Cu_N. We note in both cases that for $N=2$, 3 and 8 the clusters have negative EA (very close to zero for $N=8$); that means that they are unstable with respect to dissociation into the neutral cluster and one free electron. This fact points to remaining small defficiencies in the WDA with respect to the long-range behaviour of the exchange-correlation potential (V_{xc}^{WDA} goes to zero as $(2r)^{-1}$ instead of the correct r^{-1} asymptotic behaviour). We remark that the failure to obtain stable negative ions of small clusters is not due to the spherical jellium model because the same problem appears for the atomic negative ions, and the long range behaviour of the external potential provided by the jellium model is coulombic as in the atomic case. The same argument is valid for an elliptic jellium model.

The study of the fragmentation of X_N^{2+} and X_N^+ (X = Ag, Cu) clusters reveals the same deficiencies as in the LDA calculations of [1], arising as a consequence of the jellium model.

Work supported by DGICYT (Grant Pr-86-0654-c02). SPAIN. We are grateful for very useful discussions with Professors J. A. Alonso and G. Borstel.

References

1. Knight, K.D., Clemenger, K., de Heer, A.W., Saunders, A.W., Chou, M.Y., Cohen, M.L.: Phys. Rev. Lett. **52**, 2141 (1985)
2. Eckardt, W.: Phys. Rev. B **29**, 1558 (1984); Surf. Sci. **152**, 180 (1985)
3. Balbás, L.C., Rubio, A., Alonso, J.A., Borstel, G.: Chem. Phys. **120**, 239 (1988)
4. Katakuse, I., Ichihara, T., Fujita, Y., Matsuo, T., Sakurai, T., Matsuda, H.: Int. J. Mass. Spectrom. Ion Proc. **67**, 229 (1985)
5. Katakuse, I., Ichihara, T., Fujita, Y., Matsuo, T., Sakurai, T., Matsuda, H.: Int. J. Mass. Spectrom. Ion Proc. **69**, 109 (1986)
6. Chacón, E., Tarazona, P.: Phys. Rev. B **37**, 4013 (1988); B **37**, 4020 (1988)
7. Balbás, L.C., Rubio, A., Alonso, J.A., Borstel, G.: J. Chem. Phys. (in print)
8. Langreth, D.C., Mehl, M.J.: Phys. Rev. B **28**, 1809 (1983)

Z. Phys. D – Atoms, Molecules and Clusters 12, 213–215 (1989)

Atoms, Molecules and Clusters
Zeitschrift für Physik D
© Springer-Verlag 1989

Theory of fragmentation of metal clusters

S. Sugano[1,2], A. Tamura[2,3], and Y. Ishii[2]

[1] Institute for Theoretical Physics, University of California, Santa Barbara, CA 93106, USA
[2] Institute for Solid State Physics, University of Tokyo, Roppongi, Minato-ku, Tokyo 106, Japan
[3] The Saitama Institute of Technology, Okabemachi, Ohsato-Gun, Saitama 369-02, Japan

Received 8 July 1988

The theory of shell correction developed for discussing fissing of heavy nuclei is applied to symmetric fragmentation of charged metal clusters. Assuming the effective potential of an anisotropic harmonic oscillator, we calculate the shell correction. We also calculate the energy of deformed charged droplets in the two-dimensional parameter space describing the deformation of droplets such as elongation and neck formation. The symmetric fragmentation of microclusters of alkali and noble metals is discussed.

PACS: 36.40.+d

1. Introduction

In the present paper, we deal with metal clusters containing a few tens through a few hundreds atoms, e.g. microclusters. Since both microclusters and heavy nuclei are interacting systems consisting of the finite number of particles, we may expect some similarities [1] in their physical properties in spite of the great difference in the interactions among the particles: *(i)* In heavy nuclei, the particle density is independent of the system size and the total energy is given as the sum of the volume and surface energies (saturation property). This property is naturally realized in clusters where the radii of core ions determine the maximum particle density. *(ii)* Shell structure arises from quantized single-particle states in an effective potential, which is usually assumed to be spherical for nuclei. Recent observations [2, 3] suggest that valence electrons of alkali and noble metal clusters are regarded as moving in a spherical potential. *(iii)* Theory of surface shape oscillation of heavy nuclei has been applied to analysis of vibrational spectrum of small particles [4].

One of the authors has proposed [1] a possible application of the theory of shell correction, which has been developed for fission of heavy nuclei by Strutinsky [5], to fragmentation of metal clusters. This is a subject of this paper.

2. Shell correction

The theory of shell correction [5] takes into account the shell structure and the large deformation. The theory is based on the fact that the asymptotic part of the total energy, E_{tot} is proportional to the number of particles, N, as given by the liquid-drop model and the energy of the shell oscillation is a small correction of the order of $N^{1/3}$.

According to the theory of shell correction, E_{tot} is given as

$$E_{tot} = E_{LD} + \Delta E_{shell}. \tag{1}$$

E_{LD} is the energy of a charged liquid droplet given as

$$E_{LD} = -\varepsilon_v N + \varepsilon_s N^{2/3} + E_{Coul}, \tag{2}$$

where the first, second and third terms are the volume, surface and Coulomb energies, respectively. The shell correction ΔE_{shell} is calculated as

$$\Delta E_{shell} = E_{shell} - \tilde{E}_{shell}, \tag{3}$$

where E_{shell} is the sum of single-particle orbital energies and $\tilde{E}_{shell}$ the averaged contribution from the single particle motion.

If the effective potential for single-particle motion is assumed to be of an anisotropic harmonic oscillator as

$$V(\mathbf{r})=\tfrac{1}{2}m\omega_0^2\,[e^{\gamma}(x^2+y^2)+e^{-2\gamma}z^2], \quad (4)$$

an equal-energy-surface of the effective potential, which is given by $V(\mathbf{r})=E$, is a spheroid where the ratio of the length of the rotational axis to that of the axis perpendicular to the rotational one is given as exp $(3\gamma/2)$. The volume of the spheroid, $4\pi R_0^3/3$, is invariant under the deformation if E is fixed. The shell correction for the effective potential in (4) may be calculated analytically [6]. Then $\tilde{E}_{shell}$ is expressed in a power series of $N^{-2/3}$ as

$$\tilde{E}_{shell}=\hbar\omega_0\,[\tfrac{1}{4}(3N)^{4/3}+\tfrac{1}{24}(2e^{\gamma}+e^{-2\gamma})(3N)^{2/3}+\ldots], \quad (5)$$

$$\hbar\omega_0=\frac{2\cdot 3^{2/3}}{r_s^2}[3N)^{-1/3}+\tfrac{1}{12}(2e^{\gamma}+e^{-2\gamma})(3N)^{-1} + \ldots], \quad \text{(a.u.)} \quad (6)$$

where $R_0=r_sN^{1/3}$ with the Wigner-Seitz radius r_s.

Figure 1 shows the calculated shell correction by using (3)–(6) for 46 valence electrons with $r_s=3.93$ a.u. corresponding to Na_{50}^{4+} cluster. The horizontal axis is given by $C=\exp(\gamma)$.

3. Symmetric fission of charged droplet

Let us consider a liquid droplet with charge Ze and surface tension σ. As we are interested in the stability of metal clusters, the droplet is assumed to be a perfect conductor where all the charges are at the surface. It should be stressed here that we do not necessarily assume liquid-like structures for spatial configuration of metal ions in a cluster [7]. In what follows, we deal with only the symmetric fission of a metal cluster, which is favored when Z is large. The asymmetric fission may be dealt with, for example, following the prescription in reference [8]. The asymmetry arises from both E_{LD} and ΔE_{shell}.

The symmetric fission reduces the total energy of such a droplet if the fissility parameter, $f=E_{Coul}^{(0)}/2E_{surf}^{(0)}$ where $E_{Coul}^{(0)}=(Ze)^2/2R_0$ and $E_{surf}^{(0)}=4\pi R_0^2\sigma$, satisfies the inequality

$$f>2^{-1/3}(2^{1/3}+1)^{-1}\sim 0.351. \quad (7)$$

We examine the stability of a liquid droplet in the two dimensional parameter space spanned by C and B, which are defined so that the shape of the droplet is given by the following equation [8]

$$\frac{z^2}{R^2}+\frac{x^2+y^2}{AR^2+Bz^2}=1, \quad (8)$$

where $R=CR_0$ and $A=C^{-3}-B/5$. If $B=0$, the shape is a prolate spheroid for $0<A<1$, and an oblate for $A>1$. Non-vanishing B gives the shapes illustrated in Fig. 2. The deformation energy is given by the sum of the surface and Coulomb energies. It can be shown that the spherical shape is stable, or at least metastable, as log as the fissility parameter f is less than unity. This and (7) suggest that a saddle point for symmetric fission may exist if

$$0.137<\frac{Z^2}{N}<0.392. \quad (9)$$

for parameters appropriate to sodium.

The calculation of the Coulomb energy needs an elaborate computer work. Figure 3 shows the contour map of the deformation energy of a droplet corre-

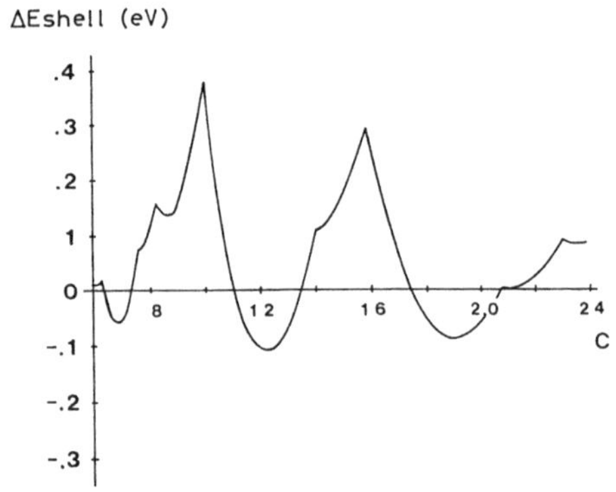

Fig. 1. Shell correction ΔE_{shell} versus deformation parameter C for the metal cluster with $N=50$ and $r_s=3.93$ a.u. corresponding to Na_{50}^{4+}

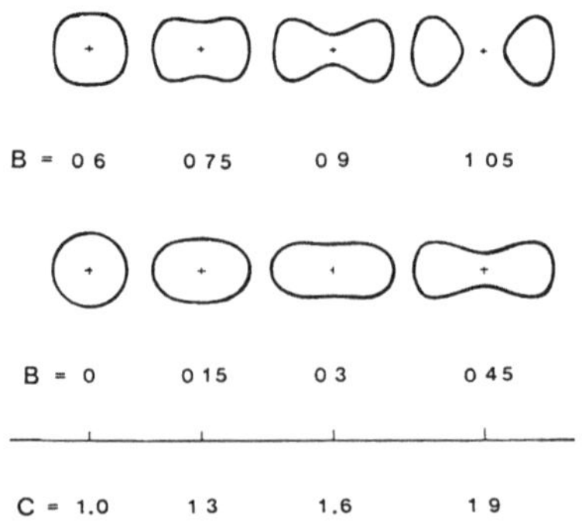

Fig. 2. Cluster shapes and deformation parameters

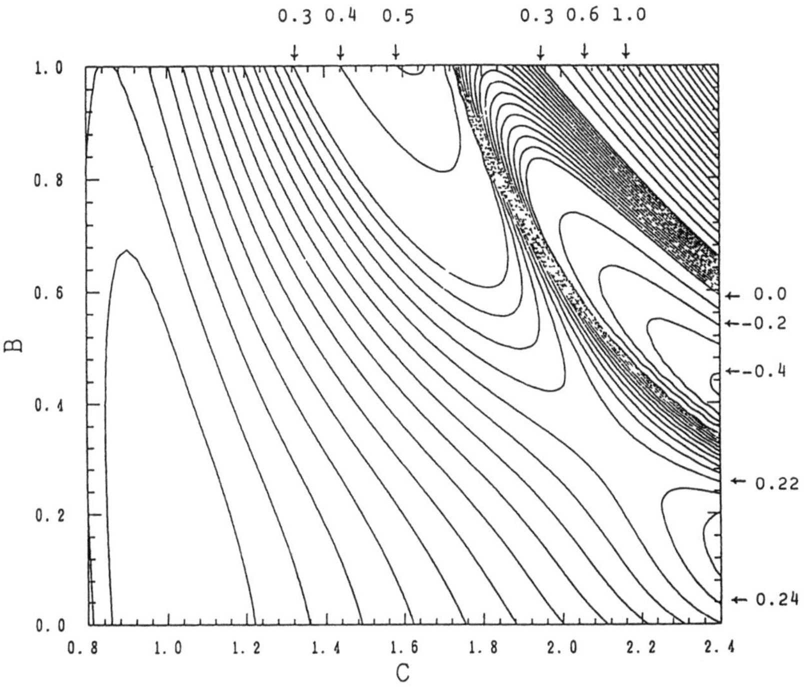

Fig. 3. Contour map of deformation energy of a charged metal droplet corresponding to Na_{50}^{4+}. ($\sigma = 200$ dyn/cm, $r_s = 3.93$ a.u.). The energy is given in eV

sponding to Na_{50}^{4+}. A saddle point is seen near the point, $C \sim 2.1$ and $B \sim 0.3$.

4. Discussion

In order to discuss the process of symmetric fission of Na_{50}^{4+}, we have to calculate the shell correction in the two dimensional space of deformation parameters, C and B, and add it to the contour map of Fig. 3. This calculation is in order and will be published elsewhere. From Figs. 1 and 3, we see that the magnitude of the shell correction is comparable to the height of the hills and the saddle point of the deformation energy. Thus, the importance of the shell effect in the fission of metal clusters is foreseen, although the use of the harmonic oscillator potential would probably overestimate the shell correction.

One of the authors (S.S.) thanks W. Kohn and J. Friedel for their stimulating discussions at ITP.

References

1. Sugano, S.: Microclusters. Sugano, S., Nishina, Y., Ohnishi, S. (eds.), p. 226. Berlin, Heidelberg, New York: Springer 1987
2. Knight, W.D., Clemenger, K., de Heer, W.A., Saunders, W.A., Chou, G.Y., Cohen, M.L.: Phys. Rev. Lett. **52**, 2141 (1984)
3. Katakuse, I., Ichihara, T., Fujita, Y., Matsuo, T., Sakurai, T., Matsuda, H.: Int. J. Mass Spectrosco. Ion Phys. **67**, 229 (1985)
4. Tamura, A., Ichinokawa, T.: Surf. Sci. **136**, 437 (1984)
5. Strutinsky, V.M.: Nucl. Phys. A **95**, 420 (1967); ibid A **122**, 1 (1968)
6. Bhaduri, R.K., Ross, C.K.: Phys. Rev. Lett. **27**, 606 (1971)
7. Ishii, Y.: Solid State Commun. **61**, 227 (1987)
8. Brack, M., Damgaard, J., Jensen, A.S., Pauli, H.C., Strutinsky, V.M., Wong, C.Y.: Rev. Mod. Phys. **44**, 320 (1972)

Z. Phys. D – Atoms, Molecules and Clusters 12, 217–219 (1989)

Atoms, Molecules and Clusters
Zeitschrift für Physik D
© Springer-Verlag 1989

Evaporation of tetramers in Sb_{4n} clusters and conditions for the formation of Sb_{2n+1} clusters

D. Rayane[1], P. Melinon[2], B. Cabaud[2], A. Hoareau[2], B. Tribollet[1], and M. Broyer[1]

[1] Laboratoire de Spectrométrie Ionique et Moléculaire (associé au C.N.R.S.), bât. 205
[2] Département de Physique des Matériaux (associé au C.N.R.S.), bât. 203, Université Lyon 1,
43 Bd du 11 Novembre 1918, F-69622 Villeurbanne Cedex, France

Received 6 July 1988

Antimony clusters are produced by the inert gas condensation technique. They are found to be built from Sb_4 units. The fragmentation by evaporation of Sb_4 units is studied as a function of the excess energy in the cluster. By this way the binding energy of the Sb_4 units in the cluster is found to be about 1.5 eV, well below the binding energy of a Sb atom in the bulk and in Sb_4 ($\simeq 3$ eV). The evolution of ionization potentials of Sb_{4n} clusters confirms that their structure is probably non metallic. Finally the possible metastable character of this Sb_{4n} structure is discussed.

PACS: 35.20.Vf; 36.40.+d

Introduction

With recent developments in clusters physics general trends begin to arise in electronic properties of alkaline, divalent and trivalent clusters. They are dominated respectively by the shell model [1], the van der Waals metal transition [2], and the *p* bonding [3]. For group IV and V metals the situation is presently less clear. Group V metals are particularly interesting because they are semi metals in the bulk, and moreover some of them (Sb and As) have in the bulk allotropic structures which are not metallic. Therefore the evolution of the properties of such clusters as a function of the size should be very interesting. We present in this paper recent results obtained on Antimony clusters. As already demonstrated by Sattler et al. [4], Sb_n clusters are built from Sb_4 units. By studying the evaporation phenomena, we are able to measure the binding energy of Sb_4 units (1.5 eV). The ionization potential measurements demonstrate that the structure of Sb_{4n} clusters is not the metallic one. Finally when the inert gas temperature is higher, Sb_{2n+1} clusters are more intense than Sb_{4n} clusters. This probably indicates that Sb_{4n} clusters are metastable.

Experimental set up

The Antimony clusters are generated by the inert gas condensation technique. The Antimony metal heated at about 550°C is condensed in helium vapour (few Torr) cooled to Liquid Nitrogen temperature. In some experiments, the Liquid Nitrogen circulation is switched off and the Helium temperature is controlled by the Helium pressure ($pV=nRT$), the flow rate remaining constant. After a 10^{-4} Torr differentially pumped chamber, the cluster beam passes into the high vacuum chamber (10^{-7} Torr) where the clusters are ionized by electron impact. The electron currents are typically 20–40 μA. The energy calibration is provided by the He beam and by the Sb_4 molecule in a free expansion. The mass analysis is performed by time of flight (T.O.F.) mass spectrometry.

Results and discussion

Figure 1 shows typical mass spectra observed on Antimony clusters. As already noted, they are built from Sb_4 units, at least when the He condensation gas

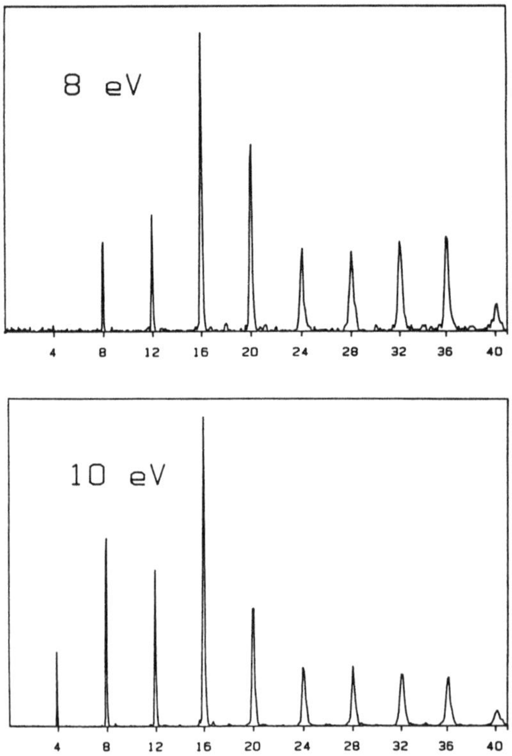

Fig. 1. Mass spectra of Antimony clusters for two different ionization energies. **a** E = 8 eV. **b** E = 10 eV. On both spectra only Sb_{4n} clusters appear. Sb_4 appears only for E = 10 eV (see text)

is cooled at low temperature. By varying the nucleation conditions, namely the Helium pressure, it is possible to shift the mass spectra from low masses to larger ones. By this method, we can select the size of the smallest clusters which exist in the molecular beam. In the experiment of Fig. 1, the smallest cluster is Sb_8 (Fig. 1a). At higher electron energy, Sb_4 appears in the spectrum (Fig. 1b). The most probable process is the evaporation of a Sb_4 unit when excess energy is released in the cluster by the ionization. To confirm this process, appearance potentials of Sb_{4n} clusters are systematically measured in different nucleation conditions. Figure 2 illustrates the obtained results: five series of measurements have been performed where the smallest clusters are respectively Sb_{20}, Sb_{16}, Sb_{12}, Sb_8 and Sb_4. For example, when Sb_{20} is the smallest, an additional energy is necessary to obtain Sb_{16} and so on. The appearance potentials of Fig. 2 support a successive evaporation mechanism for the fragmentation:

$$Sb_{4n} + e \rightarrow (Sb_{4n}^{+})^{*} + 2e$$

$$\hookrightarrow (Sb_{4n-4}^{+})^{*} + Sb_4$$

$$\hookrightarrow Sb_{4n-8}^{+} + Sb_4 \ldots.$$

From Fig. 2, the binding energy of Sb_4 units in the Sb_{4n} clusters can be measured to be about 1.5 eV. This binding energy between Sb_4 units is well below the binding energy of an atom in Sb_4 ($\simeq 3$ eV) or of an atom in the bulk ($\simeq 2.7$ eV). This means that the structure and electronic properties of small Antimony clusters produced by the present method are very

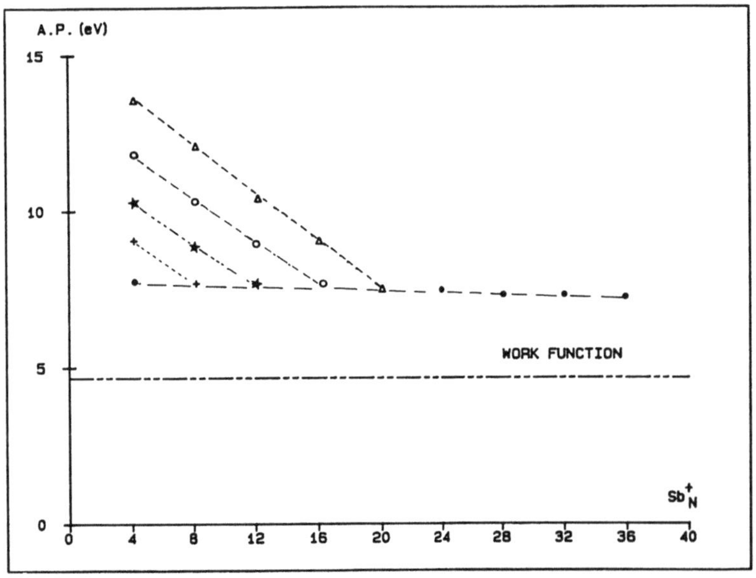

Fig. 2. Appearance potentials of Sb_N^+ ($N = 4n$) clusters in different nucleation conditions:

Series 1:	△	$p_{He} = 0.5$ T	Sb_{20} is the smallest cluster in the beam
Series 2:	○	$p_{He} = 1$ T	Sb_{16} is the smallest
Series 3:	★	$p_{He} = 1.5$ T	Sb_{12} is the smallest
Series 4:	+	$p_{He} = 2$ T	Sb_8 is the smallest
Series 5:	●	$p_{He} = 3$ T	Sb_4 is the smallest

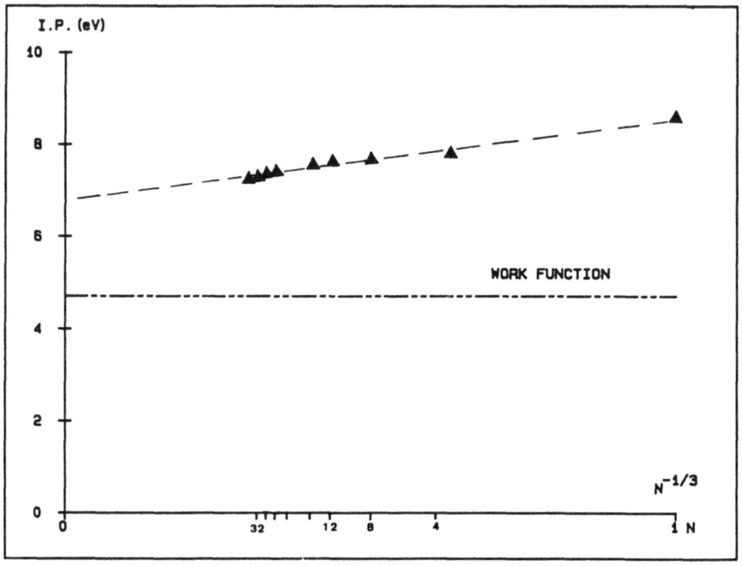

Fig. 3. Ionization potential (I.P.) of Sb_N clusters as a function of $N^{-1/3}$. The measured I.P. correspond to $N = 1, 4, 8, 12, 16, 20, 24, 28, 32, 36, 40$. The extrapolated I.P. value does not correspond to the bulk value

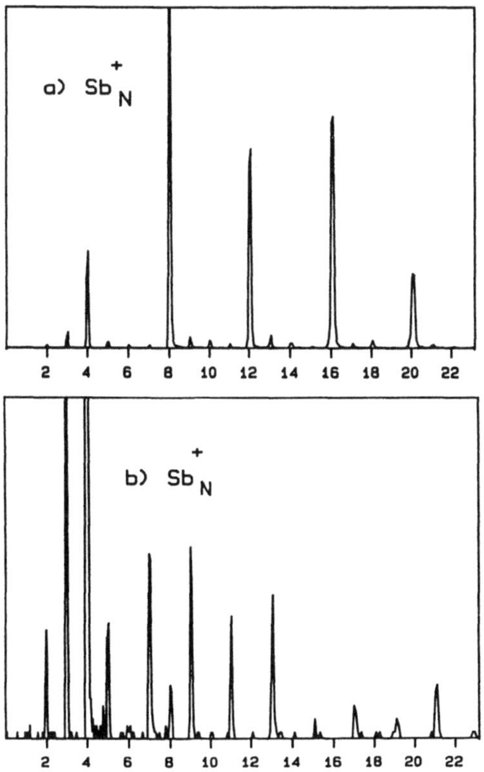

Fig. 4. Comparison between mass spectra of Sb_N clusters obtained with two different temperature of the Helium gas condensation **a** $T = 100$–150 K. **b** $T = 250$ K

different from the bulk structure. This is confirmed by the ionization potential measurements of Sb_{4n} clusters performed when these clusters are present in the beam before the ionization. Figure 3 shows the extrapolation of these ionization potentials as a function of $n^{-1/3}$. The extrapolated values (6.8 eV) is very different from the work function in the bulk which is 4.7 eV.

The question arises if this Sb_4 units structure is the most stable and if it is only obtained because in our experiment the Antimony vapour (at 550°C) contains almost only the Sb_4 molecule. An indication for this is given by experiments performed with a laser vaporization source [5] where the nucleation conditions are completely different and where the mass spectra do not exhibit a Sb_{4n} structure. In fact, in our experiment we have changed the Helium gas temperature and at higher condensation temperature (250 K) the mass spectra are completely different and only Sb_{2n+1} clusters appear at low electron energy (Fig. 4). As nucleation is a competition between evaporation and aggregation, our result indicates that the evaporation in Sb_{2n+1} clusters is less efficient than in Sb_{4n} clusters. Experiments are in progress in order best characterize Sb_{2n+1} clusters. However the present results show that the Sb_4 units structure is non metallic and could be metastable. Comparisons with theoretical calculations would be very interesting.

References

1. Knight, W.D., Klemenger, K., de Heer, W.A., Saunders, W.A., Chou, M.Y., Cohen, M.L.: Phys. Rev. Lett. **52**, 2141 (1984)
2. Bréchignac, C., Broyer, M., Cahuzac, Ph., Delacrétaz, G., Labastie, P., Wolf, J.P., Wöste, L.: Phys. Rev. Lett. **60**, 275 (1988)
3. Upton, T.H.: J. Chem. Phys. **86**, 7054 (1987); Rayane, D., Melinon, P., Cabaud, B., Hoareau, A., Khardi, S., Tribollet, B., Broyer, M.: J. Chem. Phys. **90**, 3295 (1989)
4. Sattler, K., Mühlbach, J., Pfau, P., Recknagel, E.: Phys. Lett. A **87**, 418 (1982)
5. Geusic, M.E., Freeman, R.R., Duncan, M.A.: J. Chem. Phys. **88**, 163 (1988)

Z. Phys. D – Atoms, Molecules and Clusters 12, 221–224 (1989)

Atoms, Molecules
Zeitschrift für Physik D and Clusters
© Springer-Verlag 1989

Role of fragmentation processes in the liquid metal ion source production of aggregates

J. Van de Walle and P. Joyes

Laboratoire de Physique des Solides, Bât. 510, Université Paris-Sud, F-91405 Orsay, France

Received 5 July 1988; final version 30 September 1988

We analyse the role of the fragmentation processes in the liquid metal ion source technique from a theoretical and an experimental point of view. We show that the "double peak" structure of the energy distribution of Sn_7^+ ions is due to the fact that there are two kinds of ions. On the one hand, ions which are directly emitted from the tip; on the other hand, ions produced by fragmentation at some distance (100–1000 Å) from the tip. We show that the existence of a peak for the second kind of ions strongly depends on the electric field mapping. We also deduce from the energy separation of the two peaks a life-time of the mother particles. The value obtained is compared to an evaluation made with the RRKM theory. The charge and mass of the mother particles are also discussed.

PACS: 36.40; 79.00

I. Introduction

The liquid metal ion source (LMIS) seems to be a very promising technique for producing intense currents of metallic aggregates with relatively large sizes. In this paper we intend to report on some features of the LMIS emission of tin and cerium aggregates.

The main question which will be examined is the role of fragmentation processes in the LMIS beam of charged particles. This subject has already been analysed for other techniques of aggregate production as, for example, the secondary ion mass spectrometry (SIMS) [1].

Our plan is the following. Firstly, we describe the various kinds of reaction which may take place in the LMIS beam and change its composition. Secondly, we give experimental results about the energy distributions of ions which tend to confirm the previous description.

II. The various steps in the LMIS beam evolution

Let us first recall the main lines of the experimental conditions. A tungsten tip is wetted by a molten metal X and a very strong electric field is applied at the tip summit. Then an intense emission of charged particles takes place. The sign of the charges is positive if the electric field is directed outward from the tip.

We give now what we expect to be the general features of the LMIS emission. As we will see some of these characteristics, but not all, are directly confirmed by experiment.

Immediately after the emission the beam is made of monoatomic, X^+, or polyatomic, X_n^+, singly charged ions and of droplets X_N^{p+}.

1) The first reaction which occurs at about one angström from the tip is a post-ionization reaction in which a part of the X^+ and X_n^+ particles become doubly or triply charged. This process has been studied by Kingham [2] for X^+ but the study remains to be done for X_n^+.

2) A fraction of the X_N^{p+} droplets is overcharged (we will write them $(X_N^{p+})^*$ and explodes at some distance from the tip. The life-time is relatively short $\tau \sim 5\,10^{-11}$ s and the range of distances from the tip where these reactions occur will be discussed later. During these reactions X_n^+ daughter particles are emitted which can be observed separately from the X_n^+

ions emitted directly by the tip. This is the central subject of the paper.

3) Other fragmentation reactions occur before the arrival of the particles at the collector ($t \sim 10^{-4}$ s). These reactions are not of the same nature as those of the preceding chapter. They can be written:

$$(X_n^{p+})^* \rightarrow X_{n-v}^{p+} + X_v$$

where one may note that now the emitted daughter particle X_v is neutral. For such reactions the needed excitation energy is smaller than in reactions of process 2.

Let us mention that reactions with emission of a neutral fragment have already been studied in SIMS experiments [1]. For example, the intensities $I(Cu_n^+)$ emitted from a copper target depend on the collecting time after emission. A strong decomposition of Cu_n^+ secondary ions is observed between times 10^{-8} s and 10^{-6} s for $n \geqslant 4$.

III. Observation of the two LMIS X_n species

In Figs 1–3 we give the energy distributions of Sn_7^+, Sn_2^+ and Ce_2^+ ions. These curves are indexed by the difference of potentials, V, between the tip and the cathode and by the total current, I, emitted from the tip. At constant V, the current I can be increased by decreasing the D tip-cathode distance. We will see that the important physical parameter is the electric field $\mathbf{E} = E(x)\mathbf{u}_x$ (x is the distance from the tip on the tip axis, $0 \leqslant x \leqslant D$). We will call E_0 the electric field at the tip summit, $E_0 = E(x=0)$.

On Figs. 2 and 3 we observe only one peak whereas two peaks appear on Fig. 1. This last structure can be explained by the reactions described in chapter II.

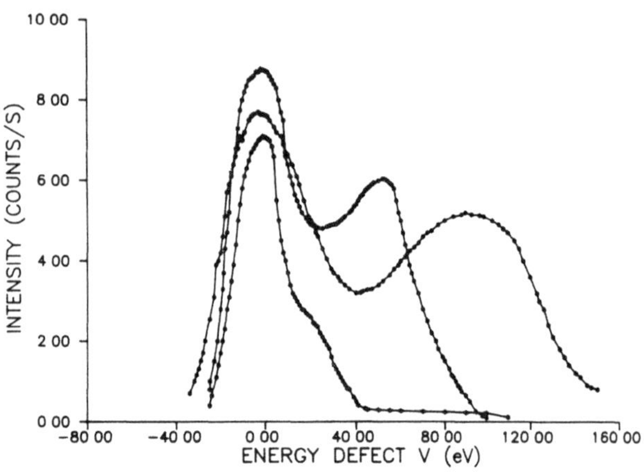

Fig. 1. Energy distribution of Sn_7^+ aggregates for a tip-cathode voltage of 5 kV and three I total currents: $I = 5$ μA is the curve with one peak: $I = 10$ μA (30 μA) the curve with a second peak at about 50 eV (100 eV)

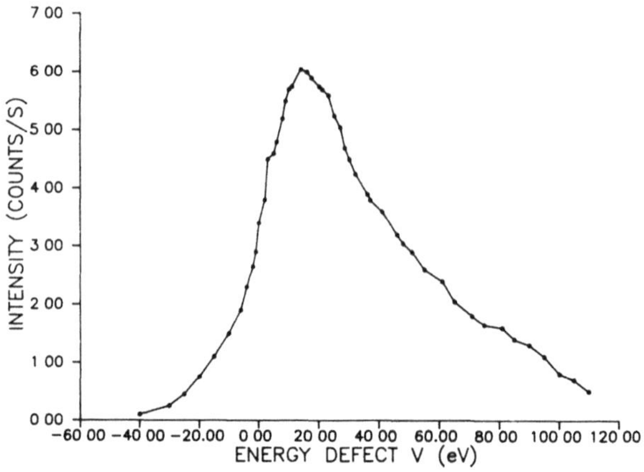

Fig. 2. Energy distribution of Sn_2^+ aggregates with a tip-cathode voltage of 5 kV and a total current $I = 30$ μA

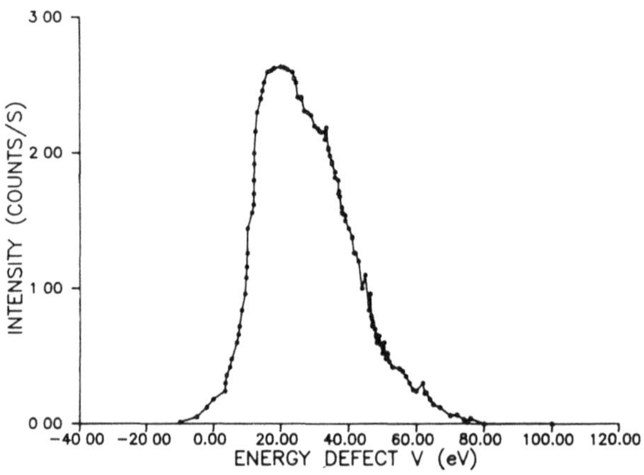

Fig. 3. Energy distribution of Ce_2^+ aggregates with a tip-cathode voltage of 5 kV and a total current $I = 10$ μA

Indeed there are two kinds of X_n^+ ions. On the one hand, ions, that we will call $(X_n^+)_a$, which are directly emitted from the tip and which have escaped from the post-ionization reaction. These ions are accelerated over the whole tip-cathode distance and have the maximum kinetic energy: $|e|$ V. On the other hand ions, $(X_n^+)_b$, produced by fragmentation and which are accelerated only over a part of the tip-cathode distance. Clearly these ions will have an energy defect, it is not obvious however that they can give rise to a resolved peak in the energy distribution and this point must be examined carefully.

We will first assume that the lifetime of the mother droplets is $\tau \sim 5\,10^{-11}$ s (a value which will be discussed later), by using an exponential rate of reaction:

$$1 - \exp(-t/\tau)$$

it is easy to see that almost all the reactions ($\sim 70\%$) occur between $t_1 = 10^{-11}$ and $t_2 = 10^{-10}$ s. We will then assume that the specific charge of the mother particle is about $Q/M \sim 4\,10^4$ MKSA (an order of magnitude which will also be reexamined later).

The next factor to be fixed is the electric field profile $E(x)$ along the tip-cathode axis. To our knowledge this quantity is not given with precision in the literature. In [3] it is shown that for a Ga source it varies relatively slowly over a distance $x \leqslant 50$ Å around an average value close to E_0. It is also noteworthy that when the total current I increases the average value of $E(x)$, for $x \leqslant 50$ Å, increases.

After the preceding region close to the tip, the field must decrease rapidly because its average value $\bar{E}$ for the whole tip-cathode distance, $x \leqslant D$, is much smaller ($\bar{E} \sim 10^6$–10^7 V/m) than $E_0 \sim 10^{10}$ V/m.

Now, if we suppose that the $E(x)$ profile can be approached by a two-step function we will write for the case $V = 5$ kV, $I = 10\,\mu$A (which corresponds to $D = 0.8$ mm [4]): $E(x) = E_0 = 10^{10}$ V/m for $x < 50$ Å and $E(x) \sim \bar{E} \sim 6\,10^6$ V/m for $50\,Å < x < 0.8$ mm. By using these values and the value of Q/M given before it is easy to show that the largest part of the mother particles explosions occurs between distances $d(t_1) = d_1 = 150$ Å and $d(t_2) = d_2 = 2050$ Å. The corresponding energy deficits are $V(0) - V(d_1) = 50.06\,V$, $V(0) - V(d_2) = 51.2$ V. It appears therefore that the larger part of the fragmentation events occurs in a narrow band of potential values. As a consequence, if we neglect the kinetic energy taken by the daughter particle during fragmentation, a relatively narrow peak (~ 1.14 eV) corresponding to $(X_n^+)_b$ particles should appear in the energy distribution at an energy defect of 50.63 eV.

Similarly, in the case $V = 5$ kV, $I = 30\,\mu$A (which corresponds to $D = 0.4$ mm [4]) we will take $E(x) = E_0 \sim 2\,10^{10}$ V/m for $x < 50$ Å and $E(x) \sim \bar{E} \sim 12\,10^6$ V/m for $50\,Å < x < 0.4$ mm. The increase of E in the $x \leqslant 50$ Å region is in agreement with the results of reference [3]. By using these values we obtain $d_1 = 230$ Å, $d_2 = 2850$ Å; $V(0) - V(d_1) = 100.2$ V, $V(0) - V(d_2) = 103.4$ V and hence a peak at 101.8 eV with a width of ~ 3.2 eV. If we compare to the first situation we see that the two experimental trends: shift of the energy defect toward large values and width increase when I increases are well reproduced. The fact that the observed widths are larger than the calculated widths is due to our approximations on the simplified field configuration and to our neglect of the kinetic energy communicated to $(X_n^+)_b$ during explosion.

The energy defects given by the previous estimations have to be multiplied by a factor $(1 - 7.p/N) \sim 0.6$ which takes into account the kinetic energy of the mother particle during explosion. The estimated energy defects become ~ 30.5 eV($I = 10\,\mu$A) and ~ 61 eV ($I = 30\,\mu$A). These values are smaller than the observed values but the order of magnitude is reproduced.

Before discussing the other experimental results let us emphasize that in our model the appearance of the second peak is strongly dependent on the steepness of the $E(x)$ decrease in the $x \sim 50$ Å region. Qualitatively, one may say that a slower decrease would give a broader peak and at the limit of a very slow decrease the peak may disappear. The calculation should be made with a more precise $E(x)$ function which is not available yet. Such a calculation may also lead to a change of the τ and Q/M values.

The other interesting characteristic of Figs. 2 and 3 is the "one peak" structure of Sn_2^+ and Ce_2^+. As it seems to be the case for tin, the examination of the experimental results for other elements (Ge [4], Au [5], Bi [6]) shows that the second peak only appears for n values larger than ~ 5. Though we have not yet obtain all the Ce_n^+ energy distributions one may think that this element follows the same behaviour.

Let us mention that the slow (linear) decrease at high energies might suggest that some Sn_2^+ dimers are formed by fragmentation but, anyway, with a probability much smaller than Sn_7^+.

The general reason why charged daughter particles are mainly produced when $n \gtrsim 5$ can be found in energetic arguments. When n decreases though the number of bonds which must be broken for the X_n^+ emission decreases (and thus the energy to be furnished decreases too) the electrostatic energy $e^2/2R(n)$ (where R is the average radius of the aggregate) increases and this last term is certainly preponderant. As a consequence no daughter aggregates with $n \lesssim 5$ are formed. Let us mention that the study of $(CO_2)_N^{3+}$ fragmentation [7] leads to a similar conclusion since the smallest daughter ion is $(CO_2)_5^+$.

IV. Discussion of the droplet parameters

Let us first discuss our choice of the specific charge. We have taken the value $Q/M \sim 4\,10^4$ MKSA from a study of the specific charge distribution in the LMIS emission from a Pb-Bi alloy [8]. This curve exhibits a relatively narrow minimum for this Q/M value which is interpreted [9] as due to the explosion, and therefore the disappearance, of the corresponding charged droplets.

If now we consider that the charge density on the tip $\sigma = \varepsilon_0 E_0$ is conserved on the droplets their charge is $Q = 4\pi\varepsilon_0 R^2 E_0$ and we obtain an other value of $Q/M = 3\varepsilon_0 E_0/\rho R$ where ρ is the mass per unit volume. By taking $E_0 = 10^{10}$ V/m and $\rho = 5.75\,10^3$ MKSA we see that a value of $R \sim 10$ Å leads us to a Q/M value close to the specific charge used (exactly $Q/M \sim 4.6\,10^4$

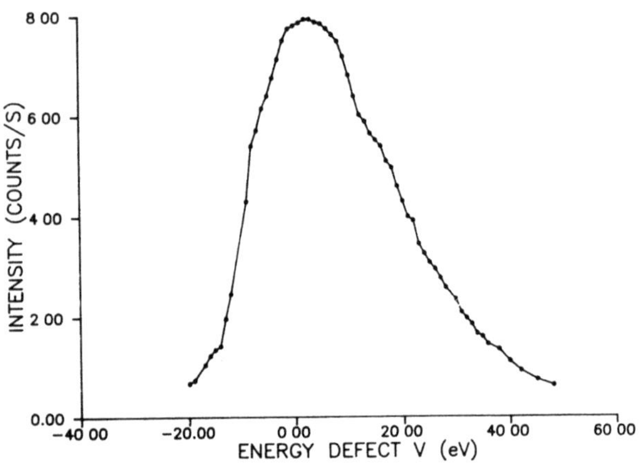

Fig. 4. Energy distribution of Sn_{25}^{3+} aggregates with a tip-cathode voltage of 5 kV and a total current $I = 30\,\mu A$

MKSA). To this R value corresponds $N \sim 122$ atoms and $p \sim 7$ elementary charges.

It is possible to discuss the lifetime value. The excitation energy E* of the droplets is made of an electrostatic part $Q^2/8\pi\varepsilon_0 R \sim 35.4$ eV and a thermal part $NkT \sim 10$ eV (for $T \sim 10^3$ °K), this gives E* ∼ 45 eV. When we use a RRKM theory [10] with an activation energy of 2.75 eV we obtain $\tau \sim 4.9\,10^{-11}$ s a value very close to the life-time used before.

V. Multiply charged ions

It remains to examine the energy distribution of multiply charged ions, an example is given Fig. 4. Two conclusions may be drawn from this curve which is characteristic of multiply charged species (Ge [4], Au [5], Bi [6]).

(1) There is only one peak with no energy defect which shows that the aggregates are formed in the immediate vicinity (∼ 1 Å) or at the summit of the tip. Our experimental results do not allow us to choose between these two possibilities however we think that, by analogy with monoatomic species, the first solution (immediate vicinity) is the most likely.

(2) No multiply charged ion is formed by the droplet fragmentation. As for the absence of production of X_n^+ with $n \lesssim 5$ this is certainly forbidden by the electrostatic energy required in the formation of a multiply charged aggregate. The conclusions of a recent theoretical study are in agreement with this idea since in the deexcitation paths of Li_n^{3+} (with $n \sim 7$) the lower activation energies are obtained when the total charge is shared on three particles [11].

VI. Conclusion

This paper shows that the formation of the LMIS aggregates is strongly influenced by the fragmentation processes. As we have seen the experimental discussion depends on the electric field mapping and the main effort in the future will be to get this quantity with the best possible precision. It is noteworthy that the various parameters fixed in this paper and particularly the relatively short lifetime of the metastable droplets which allow to interpret the experimental results are coherent with independent theoretical considerations.

References

1. Begeman, W., Meives-Broer, K.H., Lutz, H.O.: Phys. Rev. Lett. **56**, 2248 (1986)
2. Kingham, D.R.: Surf. Sci. **116**, 273 (1982)
3. Swanson, L.W., Kingham, D.R.: Appl. Phys. A **41**, 223 (1986)
4. Joyes, P., Van de Walle, J.: J. Phys. Paris **47**, 821 (1986)
5. Sudraud, P., Colliex, C., Van de Walle, J.: J. Phys. Lett. (Paris) **40**, L207 (1979)
6. Van de Walle, J., Joyes, P.: Phys. Rev. B **35**, 5509 (1987)
7. Kreisle, D., Echt, O., Knapp, M., Recknagel, E., Leiter, K., Märk, T.D., Saenz, J.J., Soler, J.M.: Phys. Rev. Lett. **56**, 1551 (1986)
8. Krohn, V.E.: J. Appl. Phys. **45**, 1144 (1980)
9. Thompson, S.P.: Vacuum **34**, 223 (1984)
10. Chait, B.T.: Int. J. Mass Spectrom. Ion Phys. **53**, 227 (1983)
11. Rao, B.K., Jena, P., Manninen, M., Nieminen, R.M.: Phys. Rev. Lett. **58**, 1188 (1987)

Z. Phys. D – Atoms, Molecules and Clusters 12, 225–227 (1989)

Atoms, Molecules and Clusters
Zeitschrift für Physik D
© Springer-Verlag 1989

Cluster cations ejected from liquid metal ion source: alkali metals (Li, Na) and group IV elements (Si, Ge, Sn, Pb)

Y. Saito[1] **and T. Noda**[2]

[1] Department of Applied Physics, Nagoya University, Nagoya 464-01, Japan
[2] Department of Electrical and Electronic Engineering, Toyohashi University of Technology, Toyohashi 440, Japan

Received 5 July 1988; final version 30 September 1988

Cluster abundance of Li_n^+ ($n \leq 19$), Na_n^+ ($n \leq 25$), Si_n^{z+} ($n \leq 8$ for $z=1$, $3 \leq n \leq 7$ for $z=2$), Ge_n^{z+} ($n \leq 11$ for $z=1$, $3 \leq n \leq 9$ for $z=2$, $n=4$ for $z=3$), Sn_n^{z+} ($n \leq 7$ for $z=1$, $3 \leq n \leq 9$ for $z=2$, $n=4$ for $z=3$) and Pb_n^{z+} ($n \leq 6$ for $z=1$, $5 \leq n \leq 7$ for $z=2$) ejected from a liquid metal ion source has been investigated by mass spectrometry. The abundance spectra of alkali metal clusters showed distinct maxima and steps at $n=3, 7, 9, 13$ and 19 for Li, and at $n=3, 5, 11, 13$ and 19 for Na. Mass spectra of Si, Ge and Sn clusters were very similar each other, showing intensity drops after $n=4$ and 6 (and also $n=10$ for Ge) for singly charged clusters. The magic numbers observed are discussed in terms of stability of charged clusters.

PACS: 36.40; 79.70; 35.20.X

1. Introduction

Mass spectra often display ion abundance anomalies where the intensity of a cluster with specific size (magic number) is more abundant than its neighbors [1, 2]. Since a secondary ion source or liquid metal ion source generates "born ionized" clusters, it is usually accepted that mass spectra reflect directly relative abundance and stability of cluster ions that were formed in the source and survived during flight time [3, 4]. The purpose of the present study is to produce "born ionized" clusters of alkali metals and Group IV elements by a liquid metal ion source (LMIS), and to investigate the origin of the magic numbers.

2. Experimental

The LMIS used in the present experiment is equipped with a closed-type reservoir for source material and with an emitter tip, which are heated by electron bombardment. The detailed structure has already been described elsewhere [5]. Pure metals were used as the source materials. The mass distributions of ions from the LMIS were measured by a single focusing magnetic analyzer except for Si in which a Wien filter was used.

3. Results and discussion

3.1. Alkali metals (Li, Na)

Figure 1a and b show relative ion intensity of Li_n^+ and Na_n^+ ejected from the LMIS, respectively. For Li, singly charged clusters containing up to 19 atoms were distinctly observed, but multiply charged ones were not detected. Ion intensity as a function of cluster size showed peaks or steps at $n=3, 7, 9, 13$ and 19. For Na, on the other hand, singly charged clusters containing up to 25 atoms were observed, and distinct steps and local maxima were found at $n=3, 5, 11, 13$ and 19. Intensity of $n=21$ was slightly high compared to its neighbors, but the intensity variation around it is not so distinct as that around the other magic numbers.

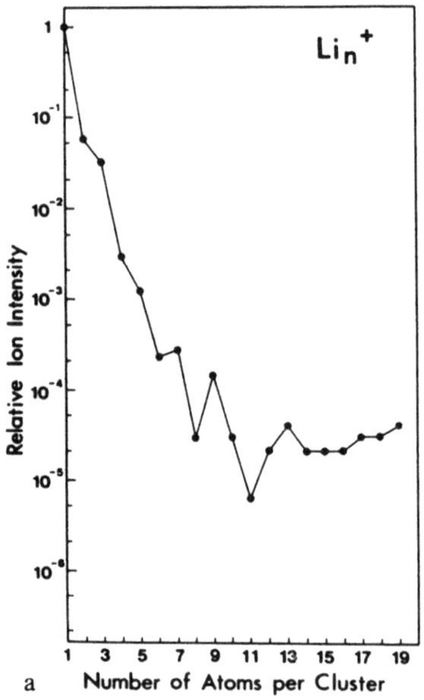

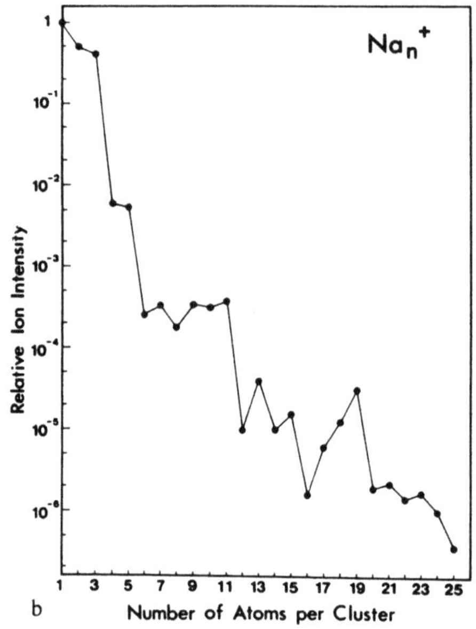

Fig. 1. **a** Relative ion intensity of Li_n^+ ejected from LMIS. The acceleration voltage V_{ac} is 7 kV, and the total emission current I_t, 50 μA. **b** Na_n^+. $V_{ac} = 4.0$ kV, and $I_t = 40$ μA

Concerning multicharged clusters, only Na_7^{2+} is identified. The odd-even alternation in the intensity variation is prominently found for Na in a region from $n=5$ up to 19, clusters with an odd number of atoms showing higher intensity. This alternation shows a significant contribution of the spin-pairing energy to the total energy of these cluster cations.

Prominent features in the ion abundance found in the present experiment, the peaks or steps and the odd-even alternation, correlate well with the variation in bond dissociation energies calculated for cationic Li [6] and Na [7] clusters. This coincidence indicates that the anomalies in abundance variation are attributed to cluster stability against fragmentation.

According to the electronic shell model [2], cationic alkali clusters of $n=3$, 9, 19, 21, 35, . . . have increased stability. The model can only partially explain the observed magic numbers, indicating the importance of geometric structure in predicting stability and other physical properties of metallic clusters.

3.2. *Group IV elements (Si, Ge, Sn, Pb)*

Figure 2 shows a mass spectrum of Ge_n^{z+}. Not only singly charged clusters but also multicharged ones are observed. The size of singly charged Ge_n^+ clusters ranges up to $n=11$. Abundance curve of the singly charged clusters showed discontinuous variations (steps) at $n=4$, 6, and 10. There has so far been a report by Van de Walle and Joyes [8] on Ge clusters from

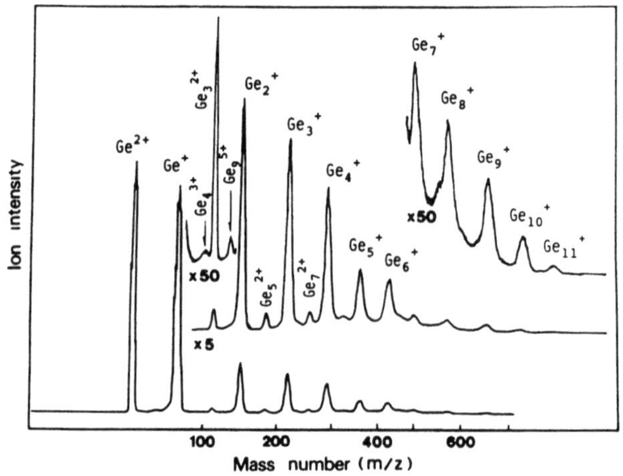

Fig. 2. Mass spectrum of Ge_n^{z+}. $V_{ac} = 3.8$ kV, and $I_t = 20$ μA

LMIS. They were concerned mainly with multiply-charged big clusters Ge_n^{z+} ($z=3$, $n \leq 88$), but did not reveal the anomalous intensity variations of singly-charged clusters. In the present study, we focused our attention on magic numbers of singly-charged clusters. Since the experimental apparatus and procedure are not described at all in their report, the reason why the edges of intensity drop are not observed in their mass spectra is not clear.

For Si and Sn, clusters only up to $n=8$ and $n=7$, respectively, were observed [9]. Mass spectra of Si and

Sn clusters showed features very similar to those of Ge: Intensity variations of singly charged Si and Sn clusters also showed steps at $n=4$ and 6. These magic clusters coincide with those revealed by photofragmentation studies for Si and Ge clusters [10]. This supports the hypothesis that the magic numbers in mass spectra from LMIS are attributed to the size dependent stability of the clusters. It should be noted that several studies on Sn LMIS [11, 12] have already been reported, and that the trend of cluster abundance spectra shown in the literature agree with the present results. On the other hand, emission of Si ions from pure Si LMIS was reported for the first time by our group [9, 13].

The resemblance of Sn to Si and Ge in cluster abundance spectra suggests that covalent nature of bonding is dominant in Sn clusters as well. On the other hand, Pb_n^+ did not show any discontinuous variation in the intensity curve within an observed size range ($n \leq 6$). This trend is also found in mass spectra from Pb LMIS reported by Komuro [12]. The difference in cluster abundance spectra between Pb and the other elements in Group IV reflects types of bonding in cluster; metallic for Pb [14], and covalent for Si, Ge and Sn.

Finally, the critical sizes of detectable multicharged clusters M_n^{z+} ($z \geq 2$) were as follows; Si_3^{2+}, Ge_3^{2+}, Sn_3^{2+} and Pb_5^{2+} for $z=2$, Ge_4^{3+} and Sn_4^{3+} for $z=3$. For $z=2$, the critical sizes here agree with those reported so far [15, 8, 11], except for Pb. Although Joyes and Van de Walle [16] have cited Pb_3^{2+} as a critical size, Pb_3^{2+} was not detected in our experiment. For $z=3$, the critical size of Ge agrees with that already reported [8], but that of Sn found here is much smaller than that ($Sn_{19(\text{or } 13)}^{3+}$) reported by Joyes and Van de Walle [16]. Studies on stability of multicharged clusters would provide much knowledge on cluster geometry and binding between atoms. Discussion on these critical sizes will appear elsewhere.

References

1. Mühlbach, J., Pfau, P., Sattler, K., Recknagel, E.: Z. Phys. B-Condensed Matter **47**, 233 (1982)
2. Knight. W.D., Clemenger, K., de Heer, W.A., Saunders, W.A., Chou, M.Y., Cohen, M.L.: Phys. Rev. Lett. **52**, 2141 (1984)
3. Katakuse, I., Ichihara, T., Fujita, Y., Matsuo, T., Sakurai, T., Matsuda, H.: Int. J. Mass Spectrom. Ion Proc. **67**, 229 (1985)
4. Joyes, P., Sudraud, P.: Surf. Sci. **156**, 451 (1985)
5. Noda, T., Okutani, T., Yagi, K., Tamura, H., Okano, H., Watanabe, H.: Int. J. Mass Spectrom. Ion Phys. **46**, 15 (1983)
6. Boustani, I., Pewestorf, W., Fantucci, P., Bonačić-Koutecký, V., Koutecký, J.: Phys. Rev. B **35**, 9437 (1987)
7. Rao, B.K., Khanna, S.N., Jena, P.: Phys. Rev. B **36**, 953 (1987)
8. Van de Walle, J., Joyes, P.: Phys. Rev. B **32**, 8381 (1985)
9. Minami, K., Saito, Y., Noda, T.: Rapid Commun. Mass Spectrom. **2**, 115 (1988)
10. Brown, W.L., Freeman, R.R., Raghavachari, K., Schluter, M.: Science **235**, 860 (1987)
11. Dixon, A., Colliex, C., Ohana, R., Sudraud, P., Van de Walle, J.: Phys. Rev. Lett. **46**, 865 (1981)
12. Komuro, M.: In: Proc. Inter. Ion Engineering Congress, -IS-IAT'83-, Kyoto (1983), Takagi, T. (ed.), pp. 337–348
13. Noda, T., Okutani, T., Tamura, H.: Oyo Butsuri **54**, 935 (1985) [in Japanese]
14. Saito, Y., Yamauchi, K., Mihama, K., Noda, T.: Jpn. J. Appl. Phys. **21**, L396 (1982)
15. Tsong, T.T.: Appl. Phys. Lett. **45**, 1149 (1984)
16. Joyes, P., Van de Walle, J.: J. Phys. B **18**, 3805 (1985)

Z. Phys. D – Atoms, Molecules and Clusters 12, 229–233 (1989)

Atoms, Molecules and Clusters
Zeitschrift für Physik D
© Springer-Verlag 1989

Sputtered metal and silicon cluster ions: collision-induced fragmentation and neutralization

W. Begemann, R. Hector, Y.Y. Liu, J. Tiggesbäumker, K.H. Meiwes-Broer, and H.O. Lutz
Fakultät für Physik, Universität Bielefeld, D-4800 Bielefeld 1, Federal Republic of Germany

Received 7 July 1988; final version 30 September 1988

Mass separated metal and silicon cluster ion beams $M_n^{+,-}$ are produced by sputtering and undergo fragmenting and/or neutralizing collisions at different kinetic energies (100–1800 eV) in Ar and SF_6. Fragment patterns induced by rare gas collisions open a way to determine ionization potentials and electron affinities of clusters. These values are compared to known experimental and theoretical data. For negatively charged clusters the absorption in gas targets is mainly due to neutralization, the cross sections varying with cluster material, number of atoms and collision partner from 10 Å^2 to about 50 Å^2.

PACS: 36.40.+d; 79.20.Nc; 35.20.Vf; 34.70.+e

Introduction

The description of cluster properties as a function of their size is an important experimental and theoretical challenge. The resulting need for intense, mass-selected cluster beams is hampered by the fact that cluster sources create broad mass distributions; therefore, mass selection becomes necessary, usually performed by ion-optical methods. As a result, most experiments with reliable mass identification have been performed with charged clusters. However, experiments are underway to generate also size selected neutral clusters or at least neutral beams with a narrow mass distribution [1–4].

An interesting experimental method to investigate the composition and structures of unknown compounds is the study of collision-induced fragmentation (CIF). Since more than 20 years CIF is used to analyze the structures of organic ions. First results on the fragmentation behaviour of cluster ions (H_n^+) are known since that time as well [5]. More recently, various research groups have applied this technique to mass-selected metal clusters within a large energy scale [6–11]. In the work to be reported here, we have studied fragmentation and neutralization in collisions between ionic metal and silicon mass-selected clusters and gas targets. The combination of high neutralization and low fragmentation cross sections (as found for Al_n^- and Si_n^-) indicate a method to produce neutral mass-selected cluster beams.

Experimental

Figure 1 shows the five-stage differentially pumped UHV-machine used for collision experiments in the kinetic energy range of 10–4000 eV [3, 6]. Cluster ions of both polarities are sputtered by 10–25 keV Xe^+ bombardment of different bulk materials; the ions are accelerated and focussed through a Wien filter to produce a mass selected continuous cluster beam. Collision experiments have been performed either without deceleration ($E_{kin} = 1.8$ keV) in a collision chamber of length 1.35 m, or inside a gas cell (3 cm long) at energies 100 eV $< E_{kin} <$ 1.8 keV after the cluster ions have been decelerated by a lens system. Product ions are separated in forward direction by a retarding field energy analyzer in front of detector 1, or by TOF-mass spectrometry in a reflectron (detector 2).

Results and discussion

Fragmentation

In a first experiment we have investigated the interaction of accelerated sputtered cluster ions with rare

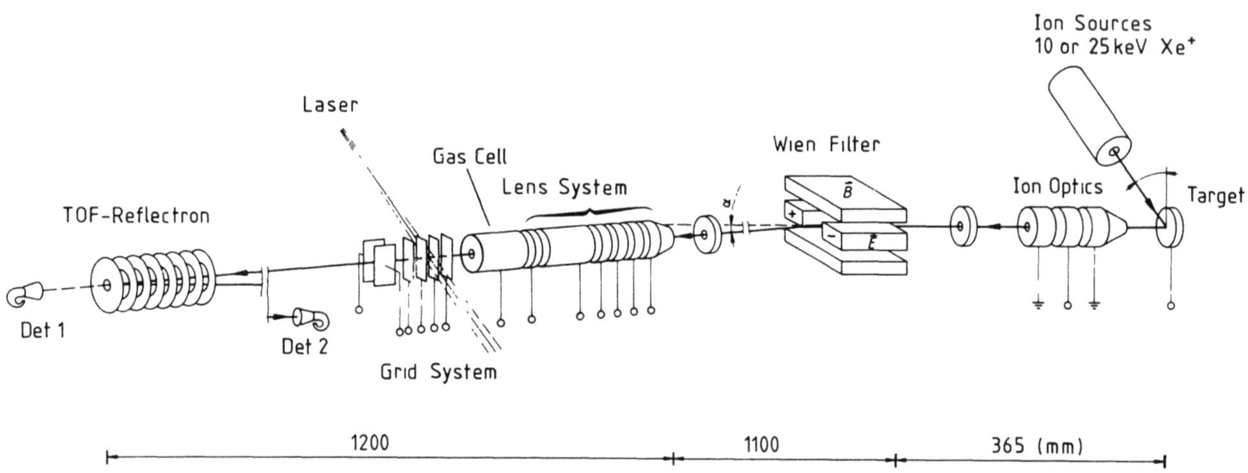

Fig. 1. Experimental arrangement for collision-induced fragmentation and neutralization of mass selected sputtered cluster ions

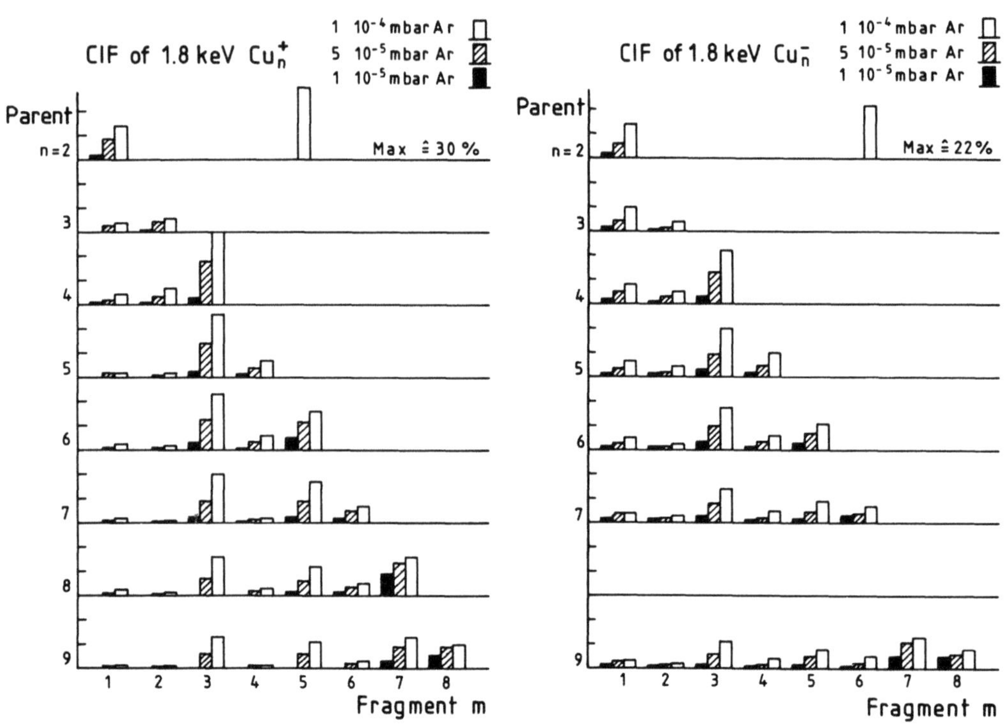

Fig. 2. Fragment ion intensities I_m/I_0 of 1.8 keV $Cu_n^{+,-}$ in Ar

gases using a collision length of 1.35 m. As examples, Fig. 2 shows the fragment ion intensities I_m of 1.8 keV $Cu_n^{+,-}$ in an argon target at different gas pressures. The signal is normalized to the sum I_0 of transmitted stable parent ions I_p and all fragment ions ΣI_m detected under small angles in forward direction. With this procedure, the influence of cluster beam absorption due to scattering and neutralization is minimized.

Even-odd alternations in the fragment ions abundances $Cu_m^{+,-}$ match the n-dependence in the mass spectra of sputtered $Cu_n^{+,-}$ and their in metastable decay behaviour [3, 12]. Cu cluster ions with an odd number of atoms and hence paired electrons show higher stabilities compared to neighbouring cluster masses.

When alternative fragment channels are open, those with the lowest fragmentation energies $D_n^{+,-}$ will be dominant. For decay processes $M_n^{+,-} \rightarrow M_m^{+,-} + M_{n-m}$ the required energies $D_n^{+,-}(M_m^{+,-}, M_{n-m})$ are roughly

$$D_n^+(M_m^+, M_{n-m}) = D_n(M_m, M_{n-m}) + IP(M_m) - IP(M_n) \quad (1)$$

$$D_n^-(M_m^-, M_{n-m}) = D_n(M_m, M_{n-m}) - EA(M_m) + EA(M_n) \quad (2)$$

Table 1. Aluminium and silicon cluster ionization potentials and electron affinities, obtained by the evaluation of fragmentation patterns. For comparison the data of other authors are given

		Al_2	Al_3	Al_4	Al_5	Al_6
IP (Al_n)	this work	5.5±0.1[a]	6.1±0.1	6.5±0.1	6.4±0.1	6.5±0.1
	(9) exp.	5.5	6.3	6.5	6.4	6.5
	(17) exp.	6.0–6.42	6.42–6.5	6.5–7.87	6.42–6.5	6.0–6.42
	(18) theor.	6.02	6.31	6.46	5.57	6.55
EA (Al_n)	this work	1.1±0.2	1.4±0.2	1.0±0.2	1.3±0.2	
	(19) exp.	1.0±0.2	1.4±0.2	2.0±0.2	2.1±0.2	
	(18) theor.	0.09	1.43	1.5	1.91	

	Si_2	Si_3	Si_4	Si_5	Si_6	Si_7
IP (Si_n)						
this work	8.0±0.3	7.7±0.3	7.5±0.3	7.8±0.2	7.7±0.2	7.5±0.2
(20) exp.	>7.87	>7.87	>7.87	>7.87	>7.87	>7.87
(21) theor.	7.5	7.9	7.6	7.8	7.5	
EA (Si_n)						
this work	1.9±0.3	2.0±0.3	1.7±0.3	1.6±0.2		
(22) exp.		2.0	1.8	2.5		

[a] The errors represent the experimental noise. Uncertainties which may originate from the approximation of ⟨E⟩ (6) are not included

with fragmentation energy D_i, ionization potential $IP(M_i)$ and electron affinity $EA(M_i)$ of the neutral cluster M_i. The correlation leads to preferred ionic products with low ionization potential IP or high electron affinity EA, respectively, as confirmed for Cu_n (cf. Fig. 2), Al_n and Si_n [3].

From the relative fragment ion intensities more quantitative information about IP- and EA-values can be extracted, as has been shown by Jarrold et al. [9] for the IPs of Al_n. If the fragment channels

$$\text{(a)}\quad M_n^+ \xrightarrow{D_{n,a}^+} M_{n-m}^+ + M_m$$

$$\text{(b)}\quad M_n^+ \xrightarrow{D_{n,b}^+} M_m^+ + M_{n-m}$$

are observed simultaneously under single collision conditions, the difference of the corresponding fragmentation energies is calculated from (1) as:

$$D_{n,b}^+ - D_{n,a}^+ = D_n^+(M_m^+, M_{n-m}) - D_n^+(M_{n-m}^+, M_m) \tag{3a}$$

$$= IP(M_m) - IP(M_{n-m}) \tag{3b}$$

Under the assumption that fragmentation occurs statistically and that the transition states for the $M_{n-m}^+ + M_m$ and $M_m^+ + M_{n-m}$ channels are similar, the following relation between fragmentation energies and fragment ion intensities can be found [9].

$$\ln\left(\frac{I(M_{n-m}^+)}{I(M_m^+)}\right) \approx \frac{D_{n,b}^+ - D_{n,a}^+}{k \cdot T} \tag{4a}$$

$$= \frac{IP(M_m) - IP(M_{n-m})}{k \cdot T} \tag{4b}$$

with the Boltzmann-factor k and the cluster's internal temperature T after the collision. In the same manner

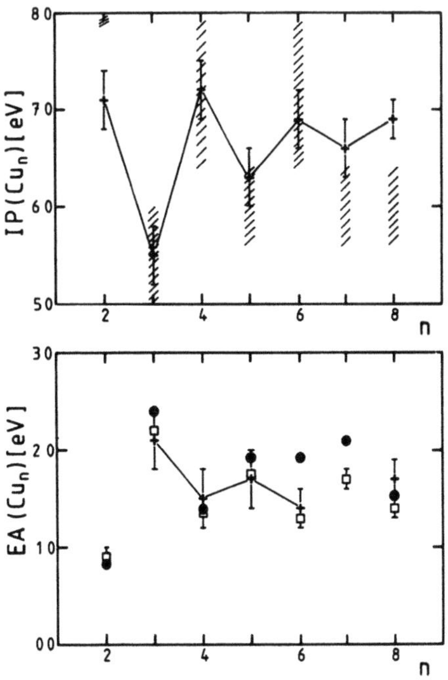

Fig. 3. Ionization potentials and electron affinities (+) for copper clusters, extracted from the fragmentation patterns of Fig. 2. For comparison, values of other authors are given as well:
(≡): photoionization bracketing [13, 14],
(□): photoelectron thresholds [15], and
(•) : other photoelectron data [16]

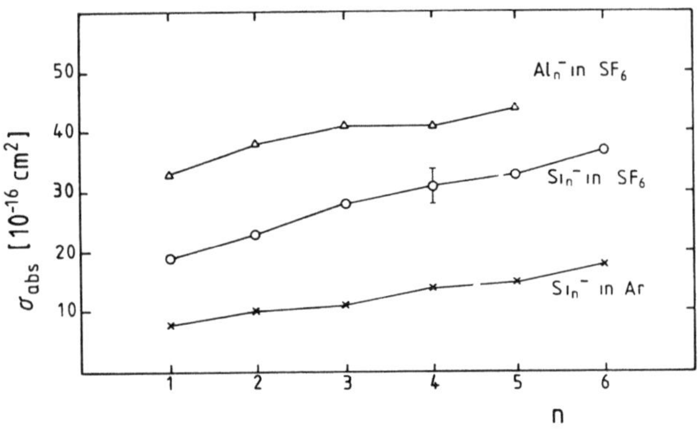

Fig. 4. Absorption cross sections σ_{abs} of 1.8 keV Si_n^- (in Ar and SF_6) and Al_n^- (in SF_6). The main contributions to σ_{abs} are supposed to originate from neutralization of the anions

we write

$$\ln\left(\frac{I(M_{n-m}^-)}{I(M_m^-)}\right) \approx \frac{EA(M_{n-m}) - EA(M_m)}{k \cdot T} \tag{5}$$

for the decay of negatively charged clusters M_n^-. The average internal energy $\langle E \rangle$ of an excited cluster distributed over $n-3$ internal modes can be estimated from

$$T \approx \frac{\langle E \rangle}{(3n-3)k} \tag{6}$$

Approximate $\langle E \rangle$ values for trimers can be calculated from (4b) or (5) with experimental fragment ion intensities and known IP- and EA-values of monomers and dimers; for the larger clusters the trimer internal energy is taken. With this estimate our IP- and EA-data in Fig. 3 and Table 1 were extracted from fragment patterns of 1.8 keV $Al_n^{+,-}$, $Si_n^{+,-}$ and $Cu_n^{+,-}$ in argon and compared to published experimental and theoretical values. Although this method is based on rough assumptions, it leads to fairly reasonable results, as shown, e.g., by the appearance of even-odd alternations in IP(Cu_n) and EA(Cu_n).

Neutralization

Besides fragmentation, competing collision-induced processes are conceivable, especially for negatively charged clusters. In earlier scattering experiments in argon targets we found low fragmentation cross sections for Si_n^- (<5 Å^2, $n=2$–6) if compared to Si_n^+. Cluster absorption (by neutralization or elastic scattering), on the other hand, turned out to proceed with cross sections σ_{abs} which are for negatively charged silicon clusters twice as large as for positive ones. This behaviour suggested effective neutralization of Si_n^- in Ar. More recent measurements of Al_n^- and Si_n^- cluster ions in Ar and SF_6 target gases corroborated these earlier results. Figure 4 shows absolute cross sections σ_{abs} for Si_n^- and Al_n^-. The values increase with n, an SF_6 target resulting in a higher cross section than Ar.

In addition, the neutral particle production was studied as function of target gas pressure. Since the detection efficiencies for neutral clusters are not known, only relative neutralization cross sections σ_{neu} have been determined. σ_{neu} and σ_{abs} show the same n-dependence, indicating that cluster anion absorption is mainly due to neutralization. This is further evidenced by the detection of negative reaction products; as an example TOF-spectra of negatively charged products created in low energy collisions of Si_4^- in SF_6 are given in Fig. 5. The right part compiles reaction channels observed for Si_n^- collisions. Only negatively charged fragments of SF_6^-, and no Si_m^- ($m<n$) have been detected; since the degree of fragmentation for neutralized Si_n is not known yet, neutral clusters possibly including fragments are indicated by "Si_n".

A large charge transfer rate has also been found in Si_n^--reactions with WF_6 at thermal collision energies [23], there the dominating channel yielded WF_6^- for all clusters with n = 1–6.

Although the properties of the neutralized cluster-species have not been studied yet, we suppose that neutralization of negatively charged clusters with such high cross sections is a "soft" method for producing mass-identified neutral cluster beams. Apparently, in contrast to charge exchange of positively charged clusters, electronic excess energy in cluster anion neutralization is mainly transferred to the (dissociating) SF_6-molecules.

Conclusion

CIF-experiments with cluster ions allow the evaluation of ionization potentials and electron affinities. By comparison of fragment ion intensities $I(M_1^{+,-})$ and $I(M_{n-1}^{+,-})$ produced from the same parent ion species $M_n^{+,-}$, differences between known atomic IP, EA and that of M_{n-1} clusters can be calculated. Resulting values of small Al_n, Si_n and Cu_n have been compared to other experimental and theoretical data. In addition, neutralization and absorption of Al_n^- and Si_n^- has been investigated. Charge transfer cross sections are found to lie in the region of up to 50 Å^2; such effective neutralization processes may open a way to create mass-selected neutral cluster beams.

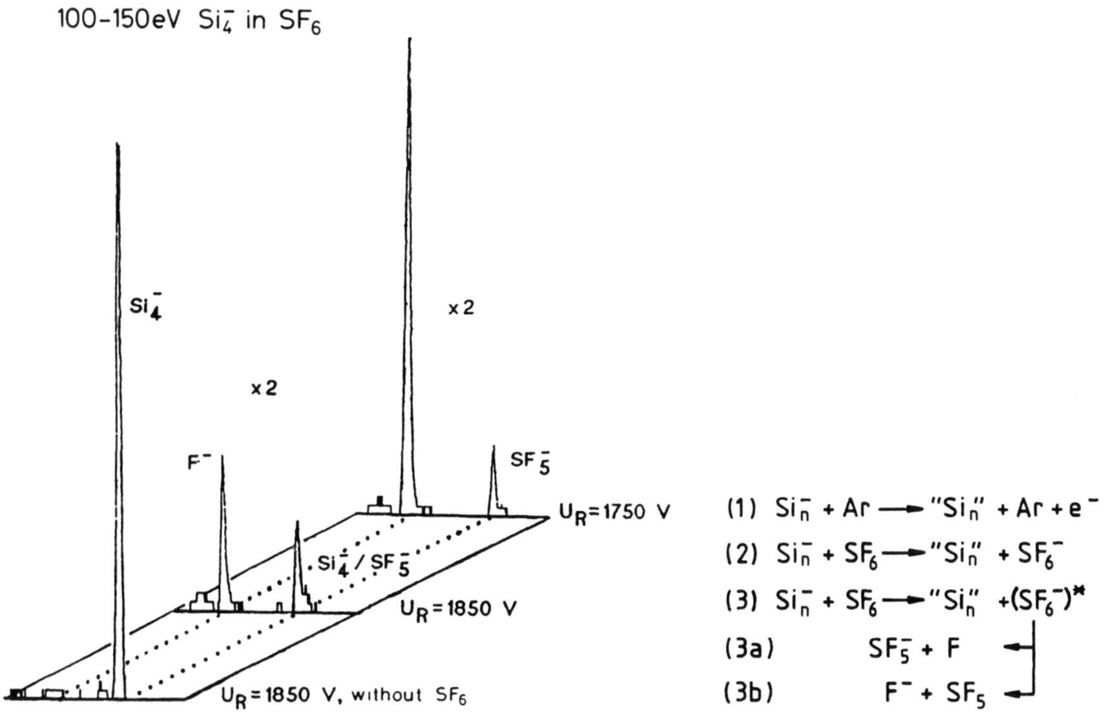

Fig. 5. *First trace:* Reflectron-TOF mass spectrum of 100–150 eV Si_4^- after deceleration from 1.8 keV. *Second trace:* Analysis of all negatively charged masses with $\sim 10^{-4}$ mbar SF_6 in the gas cell. *Third trace:* After lowering of the reflection potential U_R the parent mass (Si_4^-) vanishes; the spectrum contains only reaction products (F^- and SF_5^-). The right part displays all reaction channels which have been observed in Si_n^-–collisions. "Si_n" represents neutralized clusters and possibly includes their (neutral) fragments

References

1. Buck, U., Meyer, H.: Phys. Rev. Lett. **52**, 109 (1984)
2. Abshagen, M., Kowalski, J., Meyberg, M., zu Putlitz, G., Träger, F., Well, J.: Europhys. Lett. **5**, 13 (1988)
3. Begemann, W., Dreihöfer, S., Ganteför, G., Siekmann, H.R., Meiwes-Broer, K.H., Lutz, H.O.: In: Elemental and molecular clusters. Martin, T.P. (ed.) Springer Series in Materials Science, Vol. 6, p. 230. Berlin, Heidelberg, New York: Springer 1988
4. Bréchignac, C., Cahuzac, Ph., Leygmier, J., Pflaum, R., Weiner, J: Phys. Rev. Lett. **61**, 314 (1988)
5. van Lumig, A., Reuss, J.: Int. J. Mass Spectron. Ion Phys. **25**, 137 (1977); **27**, 197 (1978)
6. Begemann, W., Dreihöfer, S., Meiwes-Broer, K.H., Lutz, H.O.: Z. Phys. D–Atoms, Molecules and Clusters **3**, 183 (1986)
7. Ervin, E., Loh, S.K., Aristov, N., Armentrout, P.B.: J. Phys. Chem. **87**, 3593 (1983); Loh, S.K., Li Lian, Hales, D.A., Armentrout, P.B.: J. Chem. Phys. **89**, 610 (1988)
8. Ruatta, S.A., Hanley, L., Anderson, S.L.: Chem. Phys. Lett. **137**, 5 (1987)
9. Jarrold, M.F., Bower, J.E., Kraus, J.S.: J. Chem. Phys. **86**, 3876 (1987)
10. Jarrold, M.F., Bower, J.E.: J. Chem. Phys. **87**, 1610 (1987)
11. Begemann, W., Dreihöfer, S., Meiwes-Broer, K.H., Lutz, H.O.: In: Physics and chemistry of small clusters, NATO ASI Series B, Physics Vol. 158, p. 269. New York: Plenum Press 1987
12. Begemann, W., Meiwes-Broer, K.H., Lutz, H.O.: Phys. Rev. Lett. **56**, 2248 (1986)
13. Powers, D.E., Hansen, S.G., Geusic, M.E., Michalopoulos, D.L., Smalley, R.E.: J. Chem. Phys. **78**, 2866 (1983)
14. Morse, M.D., Hopkins, J.B., Langridge-Smith, P.R.R., Smalley, R.E.: J. Chem. Phys. **79**, 5316 (1983)
15. Ganteför, G.: PhD thesis, Bielefeld 1989
16. Leopold, D.G., Ho. J., Lineberger, W.C.: J. Chem. Phys. **86**, 1715 (1987)
17. Upton, T.H., Cox, D.M., Kaldor, A.: In: Physics and chemistry of small clusters, NATO ASI Series B, Physics, Vol. 158, p. 755. New York: Plenum Press 1987
18. Upton, T.H.: J. Chem. Phys. **86**, 7054 (1987)
19. Ganteför, G., Meiwes-Broer, K.H., Lutz, H.O.: Phys. Rev. A **37**, 2716 (1988)
20. Trevor, D.J., Cox, D.M., Reichmann, K.C., Brickmann, R.O., Kaldor, A.: J. Phys. Chem. **91**, 2598 (1987)
21. Raghavachari, K., Logovinsky, V.: Phys. Rev. Lett. **55**, 2855 (1985)
22. Cheshnovsky, O., Yang, S.H., Pettiette, C.L., Craycraft, M.J., Liu, Y., Smalley, R.E.: Chem. Phys. Lett. **138**, 119 (1987)
23. Reents Jr., W.D., Mandich, M.L., Bondybey, V.E.: Chem. Phys. Lett. **131**, 1 (1986)

Z. Phys. D – Atoms, Molecules and Clusters 12, 235–236 (1989)

Atoms, Molecules and Clusters
Zeitschrift für Physik D
© Springer-Verlag 1989

Enrichment of Zintl cluster ions

D. Schild, R. Pflaum, G. Riefer, and E. Recknagel
Fakultät für Physik, Universität Konstanz, D-7750 Konstanz, Federal Republic of Germany

Received 5 July 1988

Compound cluster ions of heavy post-transition atoms (In, Pb, As, Sb, Bi) are enriched by electron induced dissociation if they are analogous to the Zintl polyanions Pb_5^{2-} and Pb_9^{4-} with respect to the atom and valence electron number.

PACS: 36.40; 34.80.G

Binary clusters containing heavy post-transition elements are generated in a double oven source by inert gas condensation and probed by electron impact ionization. Mass analysis by time of flight spectrometry shows high abundances of special cluster configurations indicating their enhanced stability. High electron energy ionization populates stable cluster ions by fragmentation of larger clusters.

Cluster ions containing group IVa/Va atoms show high abundances for combinations of atoms which are analogous with the Zintl polyanions Pb_5^{2-} and Pb_9^{4-} [1] with respect to the total number of atoms and valence electrons in the clusters: $(Pb_2As_3)^+$ [2], $(Pb_2Sb_3)^+$ [3, 4, 5], $(Pb_4As_5)^+$ [2], $(Pb_4Sb_5)^+$ [3, 4, 5] (Fig. 1). To test if the stability of the Zintl cluster analogues is general to the clusters containing atoms of heavy post-transition elements, the Bi-In system was chosen since both Zintl cluster ions can be formed ($(Bi_4In)^+$[2], $(Bi_7In_2)^+$ [2]). The experimental result shows that equal numbers of atoms and valence electrons perform enhanced stability of the cluster analogues (Fig. 1).

The ion abundance of the Zintl cluster analogues reaches up to 10% of the total ion signal depending on the cluster distribution and electron energy. The electron energy giving maximum yield depends on the compound cluster system. Bi-In cluster analogues are populated more effectively at lower electron energies than Pb-Sb cluster analogues.

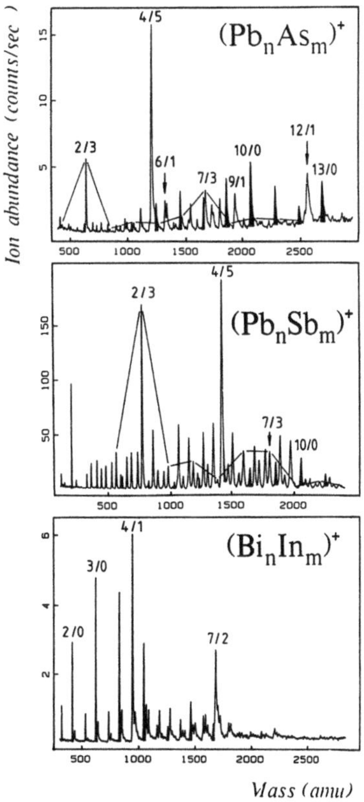

Fig. 1. Mass spectra of Pb-As, Pb-Sb and Bi-In clusters at 80 eV. Filled peaks correspond to pure lead or bismuth clusters, respectively. Fragmentation enriches Zintl cluster ions. They are observed with the highest abundance of the compound clusters in each case

The fact that Zintl clusters are observed in various compound cluster systems shows the intrinsic stability of certain configurations both in complex chemistry and cluster physics. In contrast to alkali clusters, the number of atoms and valence electrons is decisive for the stability of Zintl clusters.

The Deutsche Forschungsgemeinschaft is acknowledged for financial support.

References

1. Corbett, J.D.: Chem. Rev. **85**, 383 (1985)
2. Schild, D., Pflaum, R., Riefer, G., Recknagel, E.: Z. Phys. D – Atoms, Molecules and Clusters **10**, 329–335 (1988)
3. Schild, D., Pflaum, R., Sattler, K., Recknagel, E.: J. Phys. Chem. **91**, 2649 (1987)
4. Wheeler, R.G., LaiHing, K., Wilson, W.L., Allen, J.D., King, R.B., Duncan, M.A.: J. Am. Chem. Soc. **108**, 8101 (1986)
5. Wheeler, R.G., LaiHing, K., Wilson, W.L., Duncan, M.A.: J. Chem. Phys. **88**, 2831 (1988)

Z. Phys. D – Atoms, Molecules and Clusters 12, 237–239 (1989)

Atoms, Molecules
Zeitschrift für Physik D and Clusters
© Springer-Verlag 1989

Enrichment and segregation in alkali heteroclusters

M.J. López, A. Mañanes*, J.A. Alonso, and M.P. Iñiguez
Departamento de Física Teórica, Universidad de Valladolid, Valladolid, Spain

Received 5 July 1988; final version 30 September 1988

A spherical average pseudopotential method (SAPS) is used to investigate some properties of compound alkali clusters. The effect that the substitution of a Sodium atom by a Lithium atom in a Na_n cluster has on the stabilities and geometries is studied for $n \leqslant 21$. We have found that substitution is always energetically possible. On the other hand equiatomic $Na_n Cs_n$ clusters are considered in the size range $n \leqslant 16$. We find a strong segregation effect of the Cesium atoms towards the cluster surface. This agrees with what happens in liquid $Na_x Cs_{1-x}$ alloys.

PACS: 31.20; 36.40

Introduction

The study of alkali heteroclusters opens new ways to get insight into the correlation between features in the mass spectra obtained in supersonic expansions and the relative stabilities of clusters of different size [1, 2, 3]. Some ab initio results exist for small size clusters [4]. On the other hand, the scope of the applicability of the shell model [5] has also been discussed [1, 6, 7].

A spherically averaged pseudopotential method has been used by us [8] to study stabilities of pure clusters in a size range (more than about 10 atoms) that offers computational difficulties for ab initio calculations. In this work we use the same method to study the total energies and geometries of $LiNa_{n-1}$ ($n \leqslant 21$) and Na_nCs_n ($n \leqslant 16$) heteroclusters.

Method

For a starting geometry, characterized by the set of ionic positions $\{\vec{R}_i\}$, the electron density is obtained by integrating the Kohn-Sham equations [9] for the valence electrons in the external potential formed by spherically averaging the total ionic potential around the center of ionic charge of the cluster. An Ashcroft's pseudopotential is used to describe the individual ions, the empty core radii being 1.61 a.u., 1.74 a.u. and 2.74 a.u. for Li, Na and Cs respectively. Keeping now frozen the electron density obtained in the above calculation, we minimize the cluster energy by relaxing the ionic positions $\{\vec{R}_i\}$ in the steepest descent direction. For the new geometry we recalculate the spherical average of the potential and the new ground state electron density and the process is iterated until convergence is achieved and a local energy minimum reached. The process is repeated a number of times starting with different random configurations to be sure that the absolute minimum is reached.

Li Na_{n-1} clusters

We first report on the geometries found for $n \leqslant 21$. In the range $n = 2$–6 all atoms are located on a spherical layer with no atom at its centre. Then the central site is occupied by a Li atom for $n = 7$–10 and by a Na atom for $n = 11$–19. Li Na_{19} and Li Na_{20} have two atomic shells with two and three Na atoms respectively in the inner one and the rest in the outer shell. The inner and outer shell radii are 2.7 a.u. (3.3) and 8.0 a.u. (8.2) for $n = 20$ (21). We now compare with the geometries obtained for pure Na_n clusters by the same method [8].

* On leave from Universidad de Cantabria, Santander, Spain

In Na_n there is no central atom for $n \leqslant 8$. Then a central atom plus an external shell is the geometry for $n = 9$–19. Na_{20} and Na_{21} have an inner shell with two and four atoms respectively and the rest in an outer shell. Fig. 1 shows the heat effect ΔH for the substitution of a Na atom by a Li atom in a pure Na_n cluster

$$Na_n + Li + \Delta H \rightarrow Li\,Na_{n-1} + Na$$

For comparison we have included in Fig. 1 the results of a jellium calculation (cf. [7]) in which the Li impurity is located at the cluster center and modelled by an ionic density higher than that corresponding to the Na host. The size evolution of the geometry and electronic structure accounts for the size oscillations in ΔH. In the jellium model, the pronounced increase of ΔH between $n = 8$ and $n = 18$ is associated to the filling of an electronic shell with high angular momentum (1*d*). On the other hand, the drastic drop between $n = 18$ and $n = 20$ occurs for the filling of an electronic shell with low angular momentum (2*s*). This is understandable due to the central location of the more attractive Li atom. In the SAPS calculation there are geometrical effects superimposed on the electronic oscillations. A minimum at $n = 7$ occurs because Li Na_6 is the first cluster with a central Li atom and, as we mentioned above, the central site in Na_7 is still empty. The minimum at Li Na_8 is due to the 2*s* orbital (its electronic configuration is $1s^2\,1p^6\,2s$). There is a lack of oscillations for $11 \leqslant n \leqslant 20$ because in both pure Na_n and Li Na_{n-1} clusters the innermost atoms are Na in this size range. The calculated ΔH in the SAPS method is negative, which leads to the conclusion that the Li/Na substitution reaction is always energetically favourable. This fact might be the way to explain the Li enrichment in $Li_x\,Na_{n-x}$ clusters observed when Li and Na vapours are coexpanded in a supersonic beam [2].

$Na_n\,Cs_n$ clusters

We have analysed the total energy and the geometry of $Na_n\,Cs_n$ for $n \leqslant 16$. The general trend of the total energy per atom reproduces the electronic shell structure features, that is, clusters with filled electronic shells become specially stable. This occurs for $n = 4$ and $n = 10$ which correspond to clusters with a total number of electrons $N = 2n$ closing their 1*p* and 1*d* shells respectively. The eigenvalues for the 2*s* and 1*d* orbitals follow the geometry in the following way. In the range $6 \leqslant n \leqslant 11$ ($12 \leqslant N \leqslant 22$) the central atom is Sodium and the cluster 2*s* level is below the 1*d*. Instead, for $12 \leqslant n \leqslant 16$ Cesium is the central atom, and the 2*s* level is above the 1*d*. The reversal between the 2*s* and 1*d* levels has permitted the magic numbers of other alkali heteroclusters to be understood [1, 6, 7]. The main interest in our study of $Na_n\,Cs_n$ clusters is concerned with segregation effects. These effects arise in our calculation solely from the different empty-core radius of the Na and Cs pseudopotentials. Na is more attractive for the valence electrons. Na atoms will be located at regions of higher electron density (internal regions) while Cs segregates to the lower electron density surface region. MonteCarlo simulations of Na–Cs liquid–vapour interfaces show that the surface is almost pure Cs and the next inner layer is dominated by Na [10]. In Fig. 2 we show the atomic radial distributions for $Na_4\,Cs_4$ and $Na_{10}\,Cs_{10}$ which are representative of the overall trend in the whole size range studied. There are two homoatomic layers, the inner one of Na atoms and the outer one of Cs. All of the Na layer-Cs layer radial distances lie within 4.4 ± 0.6 a.u. The width of the layers, measured as the standard deviation of the atomic radial distributions, σ, increases with *n* and is always smaller for the Cs layer. 2σ ranges from 0.15 a.u. (0.0 a.u.) to 1.6 a.u.

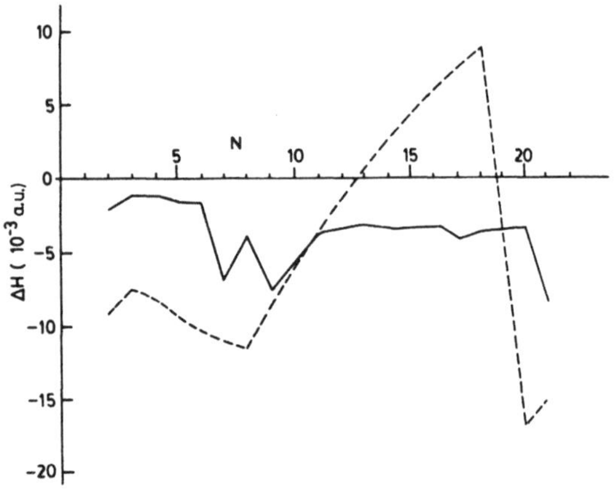

Fig. 1. Heat of substitution of a Na atom by a Li atom in Na_n clusters versus the number of atoms *n*. ——— SAPS and --- jellium

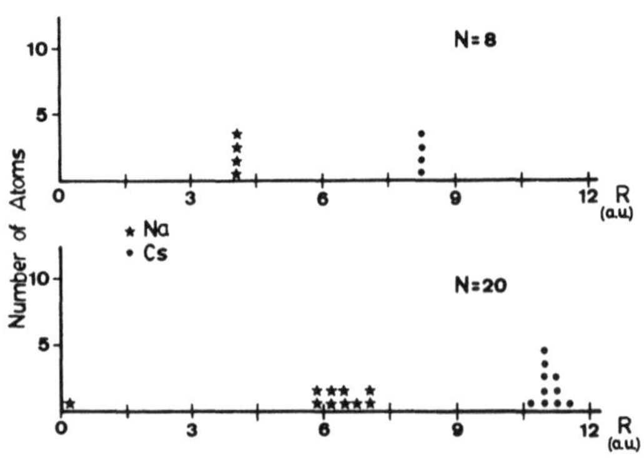

Fig. 2. Radial distribution of atoms, showing segregation effects in $Na_{n/2}Cs_{n/2}$. The symbols ★ and ∗ denote Na and Cs atoms respectively

(1 a.u.) for the Na (Cs) layers from $n = 2$ to $n = 16$. By considering that the nearest Na–Cs pair is at 7 a.u. (being the most probable pair distance of 7.7 a.u.) we induce that a Na-Cs pair, each atom in its layer, is never in the same radial direction.

Work supported by DGICYT (Grant Pr-86-0654-C02)

References

1. Kappes, M.M., Radi, P., Schär, M., Schumacher, E.: Chem. Phys. Lett. **119**, 11 (1985)
2. Kappes, M.M., Schär, M., Schumacher, E.: J. Phys. Chem. **91**, 658 (1987)
3. Brechignac, C., Cahuzac, P., Roux, J.P.: In: Elemental and Molecular Clusters, Springer Series in Materials Science, Vol. 6, p. 254. Berlin, Heidelberg, New York: Springer 1988
4. Rao, B.K., Khanna, S.N., Jena, P.: Z. Phys. D – Atoms, Molecules and Clusters **3**, 219 (1986)
5. de Heer, W.A., Knight, W.D., Chou, M.Y., Cohen, M.L.: Solid St. Phys. **40**, 93 (1987)
6. Zhang, S.B., Cohen, M.L., Chou, M.Y.: Phys. Rev. B **36**, 3455 (1987)
7. Baladrón, C., Alonso, J.A., Iñiguez, M.P.: J. Phys. F **17**, L197 (1987)
8. Iñiguez, M.P., Lopez, M.J., Alonso, J.A., Soler, J.M.: Z. Phys. D – Atoms, Molecules and Clusters **11**, 163 (1989)
9. Kohn, W., Sham, L.J.: Phys. Rev. **140**, A 1133 (1965)
10. Rice, S.A.: Z. Phys. Chem. **156**, 53 (1988)

Z. Phys. D – Atoms, Molecules and Clusters 12, 241–244 (1989)

Atoms, Molecules and Clusters
Zeitschrift für Physik D
© Springer-Verlag 1989

Metal-insulator transitions in microclusters? Mixed sodium/sodium chloride aggregates

S. Pollack, C.R.C. Wang, and M.M. Kappes
Department of Chemistry, Northwestern University, Evanston, IL 60204, USA

Received 5 July 1988

Photoionization measurements on cluster beams obtained by coexpanding sodium and sodium chloride vapors allow inferences regarding the thermodynamic stability of a number of mixed metal/salt aggregates. We discuss the possibility of observing composition dependent metal/insulator transitions among clusters with a fixed number of monomer units.

PACS: 36.40; 33.80.E; 34.80.G

I. Introduction

The large differences in per atom binding energy between small neutral clusters of s^2 metals and bulk have been explained in terms of a metal/non-metal transition which is thought to occur at intermediate particle size [1–4]. According to this argument, weakly van der Waals binding microclusters become "metallic" as the number of atoms increases and the initially empty *p*-electron derived "band" overlaps the filled *s*-"band". There have been several experimental probes of the concept for mercury clusters: ionization potential determinations [1], spectroscopic characterization of autoionization line shapes via synchrotron irradiation [2] and determinations of lower bounds on particle size required to stabilize two positive charges within the same cluster against Coulomb explosion (related to polarisability and shielding) [3]. These measurements imply that the transition occurs at 40 ± 20 atoms and has a gradual rather than dramatic particle size dependence [1–3].

Do metal/non-metal transitions in individual particles occur for all elements which are metallic in bulk? If the criterion is valence electron localization, simple metals do not qualify: recent measurements are consistent with "metallic" bonding and full valence electron delocalization *down to the dimer level* [4]. Clusters of *p*-, *d*- and *f*-electron metals represent intermediate cases, which have yet to be fully understood [5].

There is another class of clusters which might manifest "metal insulator transitions": Two-component aggregates consisting of monomer units which are respectively metallic and insulating in bulk. Now, composition in addition to particle size will determine the nature of bonding. In this study we report some preliminary measurements obtained for sodium/sodium halide mixed aggregates.

II. Experimental

Experiments were performed in a newly constructed molecular beam apparatus which will be described in detail elsewhere. Clusters were generated in continuous supersonic beams by coexpansion of sodium and sodium halide vapors from a high temperature oven (30 degree conical nozzle, 0.6 mm throat diameter). Neutral species were characterized by mass analysis of ions formed after electron impact or UV photoionization using an Extrel *C*-50 quadrupole mass filter mounted perpendicular to the molecular beam. Ions were detected with a DeTech 305A continuous dynode electron multiplier.

For electron impact studies, electron kinetic energies were kept below 7.5 eV, to reduce ion fragmentation. While average electron energies were calibrated against the ionization potentials of Na_2 and Na_3 [6], energy spreads could not be determined, but are assumed to be less than an electron volt based on broadening of appearance thresholds.

In photoionization, we made use of focused radiation provided by a 1 kW Xe arc lamp (PTI) mounted perpendicular to both cluster beam and mass spectrometer.

III. Results and discussion

Figures 1–4 contain, respectively, typical data and summaries of the relative intensity distributions for major ions observed in electron impact (EI) and photoionization (PI) probes of sodium/sodium chloride beams. The differences observed between EI and PI measurements are due to: (i) the inherently lower sensitivity and larger fragmentation cross sections associated with 7.5 eV electron impact [7] versus photoionization and (ii) the presence of some neutrals (Na_3Cl_2 and Na_5Cl_2) with ionization potentials beyond the arc lamp range. Analogous mass spectra have been obtained in recent Na/NaBr studies.

Probes of sodium beams under similar conditions yield mass spectra which reflect *neutral* cluster stabilities in the case of near threshold photoionization (large signal decreases after Na_8^+ and Na_{20}^+) and *ion* stabilities for electron impact (drop after Na_9^+) [8]. Can we draw any inferences regarding the relative stabilities of halogenated species in Na/NaCl beams? Ionization

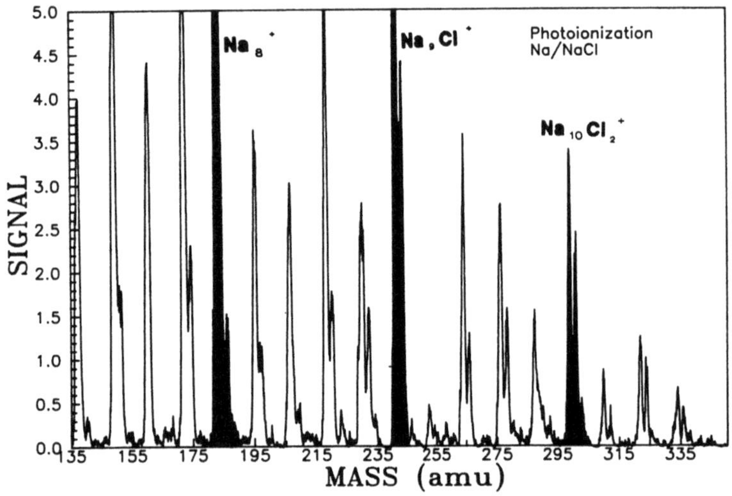

Fig. 1. Photoionization mass spectrum obtained by Xe-arc lamp irradiation of a Na/NaCl cluster beam (T_{oven} = 800°C)

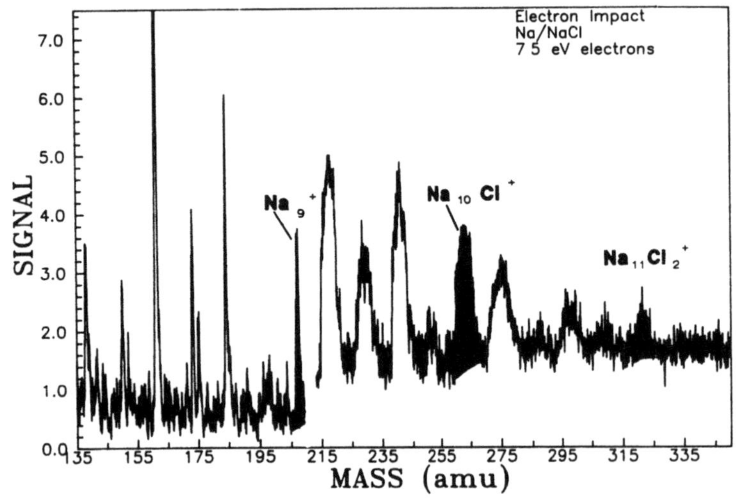

Fig. 2. Composite of partial mass scans obtained upon electron impact ionization (7.5 eV) of neutrals generated by Na/NaCl coexpansion (T_{oven} = 800°C)

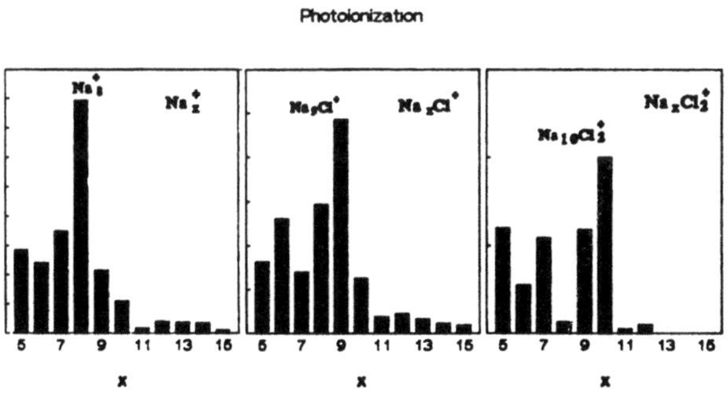

Fig. 3. Typical intensity distributions for Na_x^+, Na_xCl^+ and $Na_xCl_2^+$ ions observed upon photoionization of Na/NaCl cluster beams. From these data we infer relative stability islands at Na_8, Na_9Cl and $Na_{10}Cl_2$

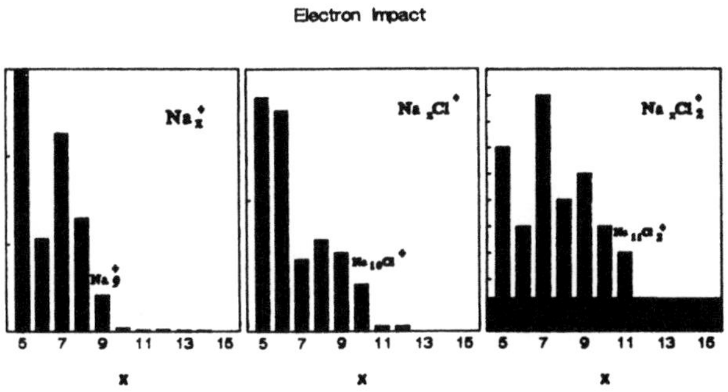

Fig. 4. Typical intensity distributions for Na_x^+, Na_xCl^+ and $Na_xCl_2^+$ ions observed upon electron impact ionization of Na/NaCl cluster beams. These data are a measure of ion stabilities due to extensive fragmentation associated with 7.5 eV electron kinetic energies. We infer relative stability maxima at Na_9^+, $Na_{10}Cl^+$ and $Na_{11}Cl_2^+$.

and fragmentation cross sections are unknown. However, if treated like sodium mass spectra, the data imply Na_xCl/Na_xCl_2 stability islands at $Na_9Cl/Na_{10}Cl_2$ and $Na_{10}Cl^+/Na_{11}Cl_2^+$ for neutral and charged clusters, respectively. Considering the low sodium chloride to sodium metal vapor pressure ratios present in the oven source, the strong signals observed for mono- and bichlorinated clusters suggest that sodium chloride monomers may be bound more strongly to these clusters than sodium atoms, throughout the composition and size range studied.

Bulk alkali/alkali halide solutions are generally characterized by miscibility gaps with high consolute temperatures being necessary to achieve continuous solubility [9]. Metal/insulator transitions corresponding to large, composition dependent changes in those observables related to electron localization, occur in such high temperature melts, as the concentration of the alkali halide is increased. The details are dependent on the observable. For electrical conductivity the transition is continuous, with the critical composition (typically halide rich) depending on the definition of what constitutes "metallic" conductivity [10]. In contrast, neutron diffraction [11] and optical reflectivity [12] studies locate a metal/insulator transition in the metal rich regime.

Experiments on alkali halide ($(MX)_x$) and alkali (M_x) clusters have established that respectively ionic or metallic bonding occurs down to the microcluster regime [8, 15]. What happens at intermediate compositions? For low internal temperatures and sufficiently large clusters, the molecular equivalent to bulk segregation may occur. Analogous segregation may in fact have been observed in a recent beam study of the $Cs_x(Cs_2O)_y$ system [13]. For $Na_x(NaCl)_y$ microclusters ($x+y<13$), "phase separation" is not possible and a stepwise transition from bonding dominated by valence electron delocalization to bonding dominated by electrostatics may occur. Clearly for the smallest species, we are pushing the bulk/cluster analogy too far. However it is interesting to consider this idea in an intermediate size range. There are several experimentally accessible observables which can be used as rough diagnostics for valence electron delocalization: *(a)* ionization potentials conforming to the predictions of a classical electrostatic droplet model; *(b)* stability maxima at characteristic valence electron counts and *(c)* optical spectra (*F* centers versus collective dipole oscillations?) [14].

The preliminary data presented above suggests that $Na_8(NaCl)$, $Na_8(NaCl)_2$, $Na_9(NaCl)^+$ and $Na_9(NaCl)_2^+$, still manifest extensive valence electron delocalization in that stability maxima are observed among Na_xCl/Na_xCl_2 and $Na_xCl^+/Na_xCl_2^+$ clusters at electron counts ($Na \equiv +1$; $Cl \equiv -1$) consistent with observations on pure alkali aggregates [16]. We are presently extending these experiments in two directions: (i) generation of mixed species with higher halide contents through reactive coexpansion of alkalis with halogen molecules and (ii) optical spectroscopy via size selective ion depletion measurements.

IV. References

1. Rademann, K., Kaiser, B., Even, U., Hensel, F.: Phys. Rev. Lett. **59**, 2319 (1987)
2. Brechignac, C., Broyer, M., Cahuzac, P., Delacretaz, G., Labastie, P., Wolf, J., Wöste, L.: Phys. Rev. Lett. **60**, 275 (1988)
3. Brechignac, C., Broyer, M., Cahuzac, P., Delacretaz, G., Labastie, P., Wöste, L.: Chem. Phys. Lett. **133**, 45 (1987)
4. Kappes, M., Schär, M., Radi, P., Schumacher, E.: J. Chem. Phys. **84**, 1863 (1986)
5. See for example: Niemann, W., Malzfeldt, W., Rabe, P., Haensel, R., Lubcke, M.: Phys. Rev. B **35**, 1099 (1987)
6. Kappes, M. Schär, M., Röthlisberger, U., Yeretzian, C., Schumacher, E.: Chem. Phys. Lett. **143**, 251 (1988)

7. See for example: Hermann, A.: Ph.D. Thesis, University of Bern, 1978
8. Kappes, M.: Chem. Rev. **88**, 369 (1988)
9. Bredig, M.: In: Molten salt chemistry. Blander, M. (ed.). New York: Interscience 1964
10. Warren, W.: In: The metallic and nonmetallic states of matter. Edwards, P., Rao, C. (eds.). London: Taylor and Francis 1985
11. Jal, J., Mathieu, C., DuPuy, J., Chieux, P., Sadoc, A.: Z. Phys. Chem. N.F. **156**, 189 (1988)
12. Freyland, W., Gasbade, K., Heyer, H., Pfeiffer, E.: J. Phys. Chem. **88**, 3745 (1984)
13. Bergmann, T., Limberger, H., Martin, T.: Phys. Rev. Lett. **60**, 1767 (1988)
14. de Heer, W., Selby, K., Kresin, V., Masui, J., Vollmer, M., Chatelain A., Knight W.: Phys. Rev. Lett. **59**, 1805 (1987)
15. Pflaum, R., Pfau, P., Sattler, K., Recknagel, E.: Surf. Sci. **156**, 165 (1985)
16. Knight, W., Clemenger, K., de Heer, W., Saunders, W., Chou, M., Cohen, M.: Phys. Rev. Lett. **52**, 2141 (1984)

Z. Phys. D – Atoms, Molecules and Clusters 12, 245–247 (1989)

Atoms, Molecules
Zeitschrift für Physik D and Clusters
© Springer-Verlag 1989

Dissociation of small metal particles induced by surface plasmon excitation with laser light

W. Hoheisel, U. Schulte, M. Vollmer, R. Weidenauer, and F. Träger

Physikalisches Institut der Universität Heidelberg, Philosophenweg 12, D-6900 Heidelberg 1, Federal Republic of Germany

Received 6 July 1988

This paper reports the dissociation of supported metal clusters after illumination with low intensity visible laser light. The process is non-thermal and exhibits a resonant dependence on the laser frequency. This indicates that a surface plasmon excitation preceeds the dissociation.

PACS: 36.40.+d; 33.80.Gj

Photodissociation of clusters has recently found a continuously growing interest [e.g. 1–14]. On the one hand there is a fundamental interest in the basic processes that promote dissociation/fragmentation of free clusters after interaction with electromagnetic radiation. On the other hand, research in this field is stimulated by surface processes, especially photodesorption [15, 16] involving supported clusters. Furthermore there are interesting applications of cluster photodissociation such as measurements of the velocity distributions of cluster beams [17]. One of the basic questions of photofragmentation studies concerns the dynamics of the process, i.e. whether dissociation results from fission or is due to evaporation of e.g. neutral monomers. Knowledge of the primary step of excitation is of significant importance to understand the underlying mechanism. Recently it was found that the excitation of surface plasma resonances in sodium clusters [7, 14] is the primary step for the subsequent dissociation. In the present paper new results concerning desorption of metal atoms from supported sodium clusters [14, 18] are reported and the relevance of the effect for the production of monodisperse clusters on surfaces is discussed. Emphasis will be on the nonthermal ejection mechanism after the collective electronic excitation and a comparison to thermal dissociation of clusters in beams [7, 19, 20].

The experimental set up (see Fig. 1) has been described in detail elsewhere [14, 21, 22]. Briefly, sodium clusters are formed on a LiF single crystal in UHV via surface diffusion and nucleation at defects. The average cluster size and density are determined by scattering experiments [21]. A cw argon or krypton ion laser beam can be directed onto the sample and passes the transparent crystal. Part of the incident light, however, is absorbed and causes the desorption of atoms, which are detected with a quadrupole mass spectrometer. The counting rate as a function of time is stored in a microcomputer. For time-of-flight measurements of the desorbing atoms, the cw lasers are either chopped to produce pulses of 5 μs duration or a pulsed Nd:YAG or an excimer laser pumped dye laser with ns pulses is used.

Desorption of atoms can be easily detected even for laser intensities as low as 40 mW/cm^2. No threshold is observed. The desorption rate depends linearly on the light intensity, but decreases with illumination time. A determination of the integral desorption rate at a given light intensity and a measurement of the intensity of the transmitted light gives a total quantum efficiency of about 10^{-5}. Since the measured extinction is not only due to absorption, but also to scattering, this number constitutes a lower limit. The time-of-flight measurements reveal energies of the desorbing atoms of more than 1 eV. The signal strongly depends on the laser frequency for a given mean cluster radius and shows a resonant behavior. Similarly, the desorption rate as a function of the particle size at fixed wavelength shows a resonance (see Fig. 2). Desorption of ions is not observed.

As can be seen from Fig. 2, illumination with blue light of $\lambda = 458$ nm results in a resonance of the desorption rate which peaks around 30 nm cluster size whereas green light with $\lambda = 514$ nm gives a resonance for clusters with sizes of about 45 to 50 nm. Red light shifts the resonance to even larger clusters. A comparison of our results (Fig. 2) with the optical absorption spectra of sodium clusters in beams [7, 19, 20, 23, 24], and matrices [25, 26] shows that the wavelength dependence of the process results from surface plasmon excitation. The linear dependence of the desorption rate on the laser power suggests that dissociation occurs as a result of a single absorption/deexcitation cycle and therefore is a true quantum process.

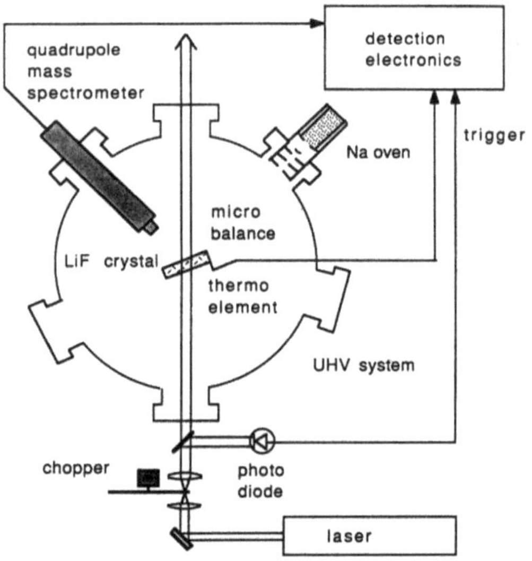

Fig. 1. Experimental arrangement for the study of laser induced desorption processes from supported sodium clusters

The underlying mechanism is non-thermal. This is supported by the high kinetic energy of the ejected atoms, by the independence of the velocity distribution from the light intensity, and by the fact that no threshold for the desorption can be detected. In addition, the energy deposited in the transparent substrate or in a particle cannot produce a temperature rise sufficiently high for thermal evaporation [22]. Consequently, direct heating of the substrate or coupling of the electronic excitation to internal vibrational modes cannot explain the experimental data. It seems that the dissociation occurs after decay of the collective electronic excitation into a localized, repulsive single electron state. The only other experiment where dissociation of clusters after excitation of surface plasmons was reported is for small sodium clusters in beams [7, 19, 20]. It was argued that a simple thermal evaporation model is sufficient to explain the depletion of the illuminated clusters in the beam. This does not contradict the nonthermal dissociation process described here particularly since the experimental conditions are appreciably different. In the beam experiment, very small clusters ($N < 40$ atoms/cluster) were investigated, whereas clusters with around 10^6 atoms are studied in the present work. Due to this size difference, which has

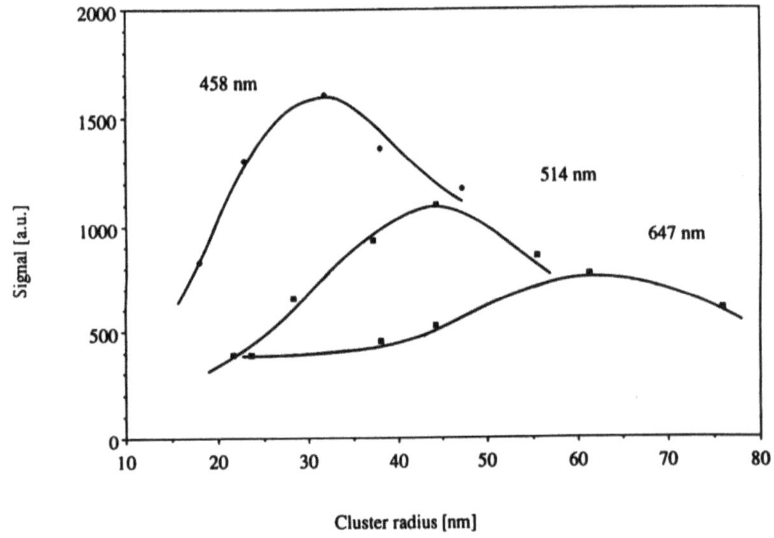

Fig. 2. Dependence of the laser induced desorption signal on the mean cluster radius for fixed laser wavelengths of 458 nm, 514 nm and 647 nm

a pronounced influence on the heat capacity, it seems quite reasonable, that the dissociation of small clusters can be governed by thermal evaporation. On the contrary supported clusters lose heat to the substrate, which constitutes an additional cooling mechanism. Also, the quantum efficiency of 10^{-5} indicates that the observed nonthermal dissociation has to compete with other relaxation channels, especially the above mentioned heat transfer to the substrate.

The shift of the resonance as a function of cluster size (see Fig. 2) can be used to change the cluster size distribution on the surface. The clusters have a size distribution the width of which is typically one half of the value of the mean particle diameter [22]. However, for many experiments and applications it would be convenient to have a very narrow distribution on a surface. The reported size dependence of the laser induced dissociation, i.e. desorption of single atoms from the surface of a cluster could help to narrow the size distributions. Illumination with red light for example should preferably dissociate the larger clusters, thus leading to a narrowing of the size distribution.

Besides this interest, a deeper understanding of laser-induced desorption is of considerable importance for other applied areas like material ablation or laser induced etching (e.g. [27–29]). From the scientific point of view, the reported desorption mechanism also sheds light on how the energy is released by a small metal particle after a collective electron excitation, a phenomenon that parallels the emission of neutrons from atomic nuclei if a giant dipole oscillation is excited [30]. Very little is known about the energy dissipation after electronic or vibrational excitation in small particles and the coupling of the different degrees of freedom. This, however, is of major importance for a detailed and fundamental understanding of such aggregates intermediate between atoms and molecules on the one side and the condensed state on the other [31–33].

We wish to thank Klaus Jungmann for his work on the time of flight measurements and on valuable discussions.

References

1. Märk, T.D., Castleman, Jr., A.W.: Adv. Atomic Molec. Phys. **20**, 65 (1985)
2. Bloomfield, L.A., Freeman, R.R., Brown, W.L.: Phys. Rev. Lett. **54**, 2246 (1985)
3. Brucat, P.J., Zheng, L.-S., Petiette, C.L., Yang, S., Smalley, R.E.: J. Chem. Phys. **84**, 3078 (1986)
4. Alexander, M.L., Johnson, M.A., Levinger, N.E., Lineberger, W.C.: Phys. Rev. Lett. **57**, 976 (1986)
5. Broyer, M., Delacrétaz, G., Labastie, P., Wolf, J.P., Wöste, L.: Phys. Rev. Lett. **57**, 1851 (1986)
6. Geusic, M.E., Jarrold, M.F., McIlrath, T.J., Bloomfield, L.A., Freeman, R.R., Brown, W.L.: Z. Phys. D – Atoms, Molecules and Clusters **3**, 309 (1986)
7. de Heer, W.A., Selby, K., Kresin, V., Masui, J., Vollmer, M., Châtelain, A., Knight, W.D.: Phys. Rev. Lett. **59**, 1805 (1987)
8. Buck, U., Huisken, F., Lauenstein, Ch., Meyer, H., Sroka, R.: J. Chem. Phys. **87**, 6276 (1987)
9. Pflaum, R., Sattler, K., Recknagel, E.: Chem. Phys. Lett. **138**, 6 (1987)
10. O'Brien, S.C., Heath, J.R., Curl, R.F., Smalley, R.E.: J. Chem. Phys. **88**, 220 (1988)
11. Geusic, M.E., Freeman, R.R., Duncan, M.A.: J. Chem. Phys. **88**, 163 (1988)
12. Zhang, Q.-L., Liu, Y., Curl, R.F., Tittel, F.K., Smalley, R.E.: J. Chem. Phys. **88**, 1670 (1988)
13. Bréchignac, C., Cahuzac, Ph., Roux, J.-Ph.: J. Chem. Phys. **88**, 3022 (1988)
14. Hoheisel, W., Jungmann, K., Vollmer, M., Weidenauer, R., Träger, F.: Phys. Rev. Lett. **60**, 1649 (1988)
15. Chuang, T.J.: Surf. Sci. Rep. **3**, 1 (1983)
16. Desorption induced by electronic transitions, DIET II. Brenig, W., Menzel, D. (eds.). Berlin, Heidelberg, New York: Springer 1985
17. Vollmer, M., Selby, K., Kresin, V., Masui, J., Kruger, M., Knight, W.D.: **59**, 1965 Rev. Sci. Instr. (1988)
18. Hoheisel, W., Schulte, U., Vollmer, M., Weidenauer, R., Träger, F.: **36**, 664 Appl. Surf. Sci. (1989)
19. Selby, K., Vollmer, M., Masui, J., Kresin, V., de Heer, W.A., Knight, W.D.: Z. Phys. D–Atoms, Molecules and Clusters **12** (1989) (this issue)
20. Selby, K., Vollmer, M., Masui, J., Kresin, V., de Heer, W.A., Knight, W.D.: Phys. Rev. B (submitted)
21. Vollmer, M., Träger, F.: Z. Phys. D–Atoms, Molecules and Clusters **3**, 291 (1986)
22. Vollmer, M., Träger, F., Surf. Sci. **187**, 445 (1987)
23. Mann, D.M., Broida, H.P.: J. Appl. Phys. **44**, 4950 (1973)
24. Hecht, J.: J. Appl. Phys. **50**, 7186 (1979)
25. Smithard, M.A., Tran, M.Q.: Helv. Phys. Acta **46**, 869 (1974)
26. Abe, H., Charlé, K.-P., Tesche, B., Schulze, W.: Chem. Phys. **68**, 137 (1982)
27. Kelly, R., Cuomo, J.J., Leary, P.A., Rothenberg, J.E., Braren, B.E., Aliotta, C.F.: Nucl. Instrum. Meth. B **9**, 329 (1985)
28. Srinivasan, R.: Science **234**, 559 (1986)
29. Bäuerle, D.: Chemical processing with lasers. Springer Series in Materials Sciences. Berlin, Heidelberg, New York: Springer 1986
30. Bertsch, G.F., Bortignon, P.F., Broglia, R.A.: Rev. Mod. Phys. **55**, 287 (1983)
31. Metal clusters. Träger, F., zu Putlitz, G. (eds.) Berlin, Heidelberg, New York: Springer 1986
32. Physics and chemistry of small clusters. Jena, P., Rao, B.K., Khanna, S.N. (eds.). New York: Plenum 1987
33. Elemental and molecular clusters. Benedek, G., Martin, T.P., Pacchioni, G. (eds.) Springer Series in Materials Sciences, Vol. 6. Berlin, Heidelberg, New York: Springer 1988

Z. Phys. D – Atoms, Molecules and Clusters 12, 249–251 (1989)

Atoms, Molecules
Zeitschrift für Physik D and Clusters
© Springer-Verlag 1989

Sodium halide clusters: evidence for direct electronic fragmentation

R. Pflaum and E. Recknagel
Universität Konstanz, D-7750 Konstanz, Federal Republic of Germany

Received 5 July 1988

Sodium halide clusters are generated by inert gas condensation and probed by electron and photon induced dissociation. The dependence of fragmentation on electron energy and initial cluster temperature indicates a direct electronic fragmentation process without an intermediate state of vibrationally hot clusters.

PACS: 36.40; 34.80; 33.80

Introduction

Cluster fragmentation links the observable ion abundance to cluster stability which is determined by both, geometric and electronic structure of the clusters. For alkali halide clusters, bonding is determined by the Coulomb force between alkali cations and halogen anions [1]. Since the overlap between adjacent inert gas like electron shells is small [2], the cluster potential is not fundamentally changed by addition or desorption of single molecules and, cluster stability is mainly determined by the cluster's geometrical structure. For comparison, stability of alkali metal clusters, is mainly determined by the electronic structure, since each atom contributes its valence electron to the cluster's molecular orbitals. Therefore, the most stable clusters are the ones with closed electron shells [3]. Results of photofragmentation are consistent with a close coupling of electronic and vibrational energy within these clusters: for each photon of energy E rapid desorption of E/E_B atoms is observed, where E_B denotes the evaporation energy of one alkali metal atom from the cluster [4].

The first informations about the structure of alkali halide clusters came from the dissociation of cluster ions. An enhanced stability of clusters which can be represented by cuboids, i.e. small fractions of the rock salt lattice, is found whether the energy for dissociation remains from cluster generation or is transferred during ionization [5, 6, 7] and whether the ions are positively or negatively charged [5, 6]. Dissociation of hot cluster ions produced by ion sputtering [5] or laser vaporization [6] as well as electron induced dissociation (E.I.D.) of cold clusters produced by inert gas condensation [7] show enhanced stability for cluster *ions* composed of $3 \times 3 \times 3$, $3 \times 3 \times 5$, $3 \times 5 \times 5$, ... single ions (cf. Fig. 1a for E.I.D.).

The structure of neutral sodium halide clusters is accessible by multiphoton induced dissociation (M.I.D.) which occurs during the rise time of the laser pulse when clusters are ionized by multiphoton absorption, simply because excitation of admolecules requests less energy than does ionization. In this case, the enhanced cluster sizes are the ones that can be represented by *neutral* cuboids possessing equal numbers of anions and cations [8].

An explanation for the similarity of these results might be that electronic energy is converted into vibrational energy by radiationless transitions – there never is a lack of collision partners for an atom within a cluster – and that from then on dissociation proceeds identically. However, there are observations which do not fit into this simple picture. It will be shown that there is evidence for decoupling of electronic excitation and vibrational heating of sodium halide clusters.

Experiment

Neutral sodium halide clusters are produced by inert gas condensation of sodium halide vapor (partial pressure about 0.1 Torr) in cold helium gas (1 Torr, 70

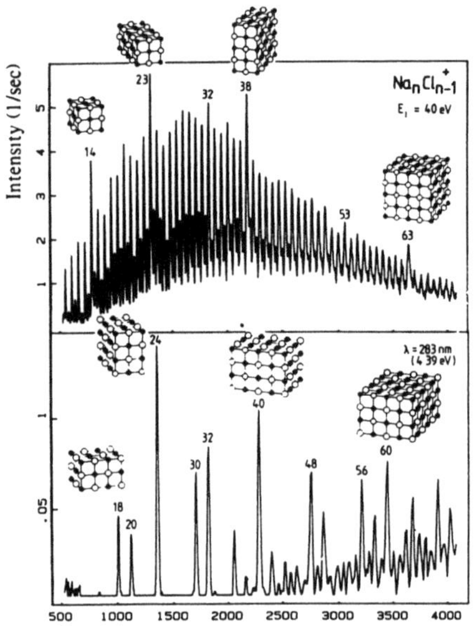

Fig. 1. Mass spectra of sodium chloride clusters condensed in helium at 300 K and ionized by electron impact (above) or multiphoton absorption (below)

– 300 K). The relation of partial pressures applied provides an efficient cooling since each cluster undergoes more than 100 collisions with helium atoms between two monomer additions. After collimation into high vacuum the clusters are ionized by electron impact or multiphoton absorption and detected by a time of flight mass spectrometer. The time scale crucial for fragmentation of ions before mass analysis, depends on the cluster mass and lies between 1 and 3 microseconds for clusters reported here. Dissociation before ionization, induced by (multi-) photon absorption, is restricted to the rise time of the laser pulse, given by about 5 ns.

Results and discussion

The first indication for direct electronic fragmentation comes from multiphoton ionization of sodium chloride clusters, which was reported in detail in [8]. Mass spectra taken from warm clusters (condensed in helium at room temperature) show intensity maxima for cluster sizes which can be represented by cuboids *before* ionization, whereas M.I.D. depopulates other clusters by more than 90% (Fig. 1), independent of the laser wavelength applied [8]. However, the resulting ions cannot have been heated considerably: no population of the ion cuboids at $n = 14$ and 23 is observed although the time scale for observable dissociation of ions is more than a hundred times longer than the one for dissociation before ionization (cf. above) during which the observed intensity anomalies are produced.

Since the momentum transfer by electron impact is negligible with respect to the transfer of electronic excitation energy, the mechanism of E.I.D. might be similar to the one of M.I.D. Electron ionization experiments were performed with a focus on vibrational heating which, for alkali halide clusters, can be detected by halogen desorption. We remember that ionization of an alkali halide cluster preferentially works via neutralization of one of the cluster's halogen anions [3, 7]. The resulting halogen atom is loosely bound by polarization and desorbs at much lower vibrational temperature than, for instance, admolecules. The vanishing of stoichiometric $(NaCl)_n^+$ lines therefore detects an increased vibrational temperature, while the intensity enhancement of cuboid cluster ions (Fig. 1) marks the onset of E.I.D.

The second indication for direct fragmentation, this time during E.I.D., comes from the onset of fragmentation monitored in Fig. 2 by the relative abundance of cuboid cluster ions. The threshold for E.I.D. is not shifted to higher energies for clusters condensed at liquid nitrogen temperature, although the temperature difference, added up over the $6n$-9 vibrational degrees of freedom of, for instance, $(Na_{38}Cl_{37})^+$, amounts to more than 4 eV. If thermalization of electronic excess energy were the main reason

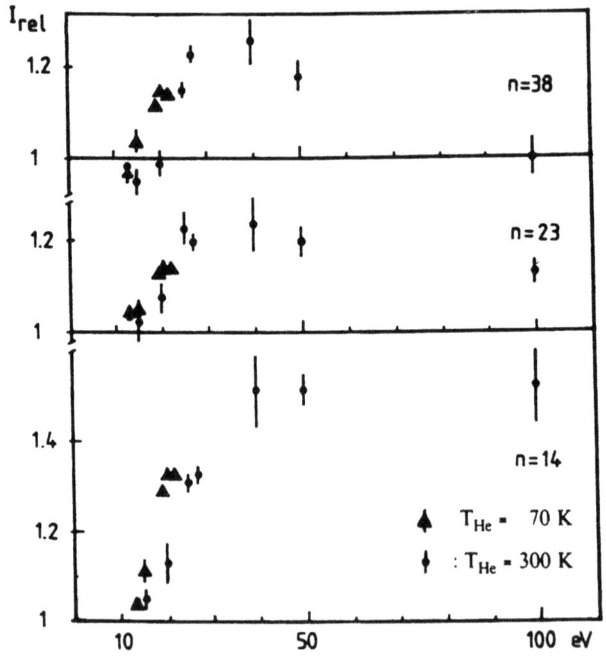

Fig. 2. Relative line intensity of the cluster ions $(Na_nCl_{n-1})^+$ ($n = 14, 23$ and 38) with respect to the line intensity of $n-2, \ldots n+2$, versus the electron energy

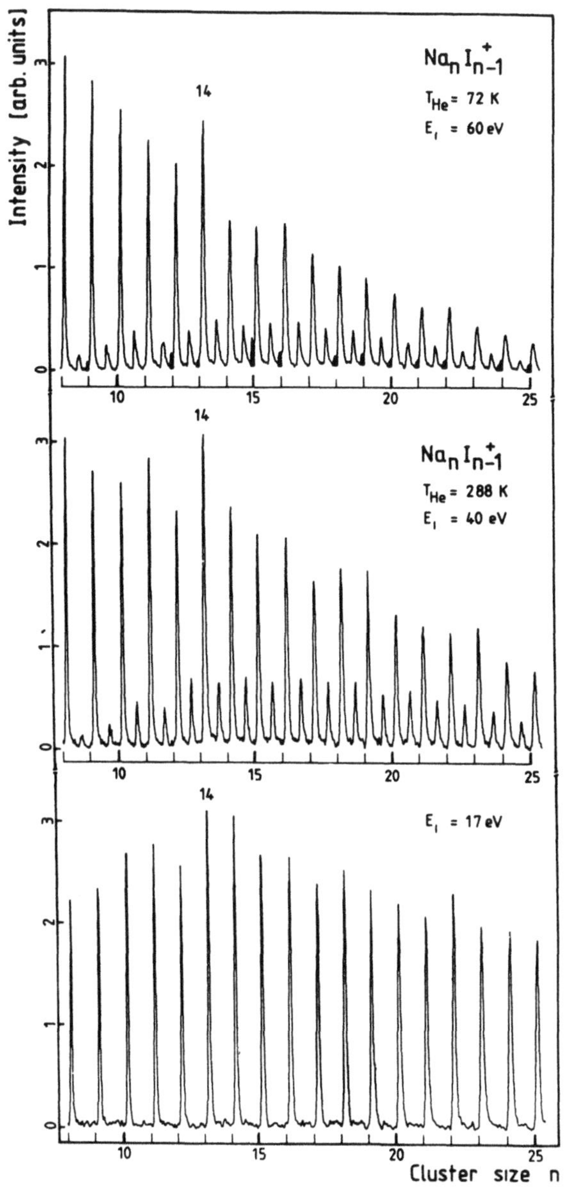

Fig. 3a–c. Mass spectra of NaI clusters condensed in helium at 70 K **a** or 300 K **b, c** and ionized by electron impact (electron energy E_I). Lines of stoichiometric ions (**a**) are filled black

for fragmentation, the corresponding shift would be observed since the energy width of the electron gun applied is less than 1 eV.

Another indication for E.I.D. without an intermediate state of a vibrationally hot cluster ion is displayed in Fig. 3. Even after ionization by 60 eV electrons, which is high above the thresholds for E.I.D. *and* double ionization (⇒ intermediate lines in Figs. 3a, b), we find lines of stoichiometric $(NaI)_n^+$ cluster ions. Warm clusters, on the contrary, strictly lose one halogen, whether low (*c*) or high (*b*) energies are applied.

The mechanism which explains both, direct E.I.D. and M.I.D., is a straightforward consequence of a molecular property of alkali halides: upon electronic excitation, the molecule changes into the dissociation continuum of a van-der-Waals state [9], represented by two neutral atoms. For a molecule on a cluster surface, the electronic field of the cluster provides a weak bonding of neutral atoms, by polarization. Therefore, neutralized atoms resulting from an electronic excitation easily desorb from the cluster without vibrational heating of the residual cluster fragment.

The authors wish to acknowledge financial support by the Deutsche Forschungsgemeinschaft.

References

1. Martin, T.P.: Phys. Rep. **95**, 167 (1983)
2. Jennison, D.R., Kunz, A.B.: Phys. Rev. B **13**, 5597 (1976)
3. Knight, W.D., Clemenger, K., Heer, W.A. de, Saunders, W.A., Chou, M.Y., Cohen, M.L.: Phys. Rev. Lett. **52**, 2141 (1984)
4. Brechignac, C., Cahuzac, Ph., Roux, J.-Ph.: J. Phys. Chem. **88**, 3022 (1988)
5. Ens, W., Beavis, R., Standing, K.G.: Phys. Rev. Lett. **50**, 27 (1983), and references therein
6. Conover, C.W.S., Yang, Y.A., Bloomfield, L.A.: Proceedings of the Symposium on Atomic and Surface Physics, La Plagne, France, p. 288 (1988)
7. Pflaum, R., Pfau, P., Sattler, K., Recknagel, E.: Surf. Sci. **156**, 165 (1985)
8. Pflaum, R., Sattler, K., Recknagel, E.: Chem. Phys. Lett. **138**, 8 (1987)
9. Levi, H., Dissertation, Frankfurt/Main (1934) Swaminathan, P.K., Laaksonen, A., Corongiu, G., Clementi, E.: J. Chem. Phys. **84**, 867 (1986)

Z. Phys. D – Atoms, Molecules and Clusters 12, 253–261 (1989)

Atoms, Molecules and Clusters
Zeitschrift für Physik D
© Springer-Verlag 1989

Ionization dynamics of van-der-Waals clusters

A. Ding
Bereich Strahlenchemie, Hahn-Meitner-Institut, Berlin, Germany

Received 9 July 1988; final version 30 September 1988

The ionization process of homogeneous and heterogeneous van-der-Waals clusters has been investigated using various ionization methods (electron bombardment, charge exchange, photoionization methods), and different analyzing techniques. Direct and indirect ionization processes can be distinguished in the experiments from the shape of the ionization curve which depends on the type of cluster. These features appear differently in homogeneous and heterogeneous systems: Homogeneous systems exhibit characteristic ionization efficiency curves where the direct ionization path appears as a sudden increase in the ionization efficiency while the indirect transition gives rise to a long drawn out tail extending to the true ionization threshold. In heterogeneous clusters the indirect ionization path proceeds via excited states of the component with the larger ionization potential and subsequent energy transfer to the other component. These transitions are shifted and broadened depending on the type of internal interaction. Conclusions are drawn concerning the geometry and the interaction potential inside the cluster. The resolution of the TEPICO (Threshold Electron Photo Ion Coincidence) experiments makes it possible to determine the kinetic energy release of the fragments. It is shown that the results are related to the stabilities of the cluster ions involved in the fragmentation chain. Results are presented for pure rare gas clusters (Ar_n, Kr_n, Xe_n) and for mixed systems (Ar_nO_{2m}, Ar_nXe, Kr_nXe, $(CH_4)_nNe$).

PACS: 33.65, 33.80E, 36.40

Introduction

Ionization of clusters plays a significant role in cluster research because it is the only general technique which allows size dependent analysis of free cluster beams [1]. Other techniques are severely limited to special cases or only applicable to very small clusters [2]. Ionization induced fragmentation goes along with most ionization processes of clusters [3, 4]. This effect is particularly pronounced in weakly bound van-der-Waals systems but occurs frequently also in systems with stronger bonds like metal clusters [5].

The measured mass distribution of a free cluster beam is the convolution of the original neutral parent distribution with the ionization and fragmentation probabilities of each cluster which strongly depends on the mass and the internal excitation of the particular cluster. As a general rule strongly bound clusters will decay less easily than clusters with weak bonds, the bonding being a non monotonic function of the cluster size. This is the reason why stable cluster ions appear more pronounced in cluster mass spectra (magic numbers) [6].

The charge distribution in a cluster changes upon ionization inducing a change in geometry. This leads to the release of internal energy which normally causes fragmentation of the system[1].

[1] A similar behaviour is found when complex (organic) molecules are ionized; mass spectra taken of such molecules show a large degree of fragmentation which may be reduced by applying the right ionization method

Ionization proceeds via a direct or an indirect transition to the ionic state, the former resembling a direct transition from the state of the neutral cluster (not necessarily in its vibrational ground state) into one of the ionic states of the system (paths 1 and 2 in Fig. 1) [7]. It is governed by the Franck-Condon principle; i.e. the geometry change during the ionization process will be small leaving the ion in a vibrationally excited state, if the equilibrium geometry of the ion is different from that of its neutral precursor as is usually the case. The photoionization efficiency increases strongly above the energy which corresponds to the direct transition into the continuum state [8]. Because of the large number of internal degrees of freedom this is a rather smooth function of photon energy (as compared to the step function which is found in the case of photoionization of atoms). The amount of internal excitation depends on the magnitude of the ionization induced geometry changes and is much larger for homogeneous than for heterogeneous clusters. This is caused by the formation of strongly bound ionic substructures (dimer-, trimer-, and tetramer-ions) within the homogeneous cluster ion [3, 9], which introduce large changes in binding energy and geometry[2]. If the binding energy in the cluster ion is small – for example in the case of van-der-Waals clusters – this will lead to evaporation of cluster components or to fragmentation.

Indirect ionization proceeds via at least one intermediate excited state of the neutral cluster which lies above its lowest ionization potential. It is followed by an autoionization process, whereby the internal electronic energy is used to remove one electron from the system. The excitation spectrum is therefore dominated by the excited neutral state and may exhibit narrow resonances (path 3a in Fig. 1). Except for very small clusters the ionization probability of the indirect process is usually lower than that of the direct one because of the larger number of steps involved and additional competing processes which could quench the excited neutral state without forming an ion (path 3b in Fig. 1).

These effects appear very pronounced in binary mixed cluster systems consisting of atoms or molecules with large differences in ionization potential. Because of the correspondence of states [10], the component with the lowest ionization potential correlates with the lowest ionic state of the cluster. It is lower than that of the separated components, an effect caused by the binding in the cluster. Above this state lie neutral excited states correlating with the Rydberg states of the other component (i.e. the component with the higher ionization potential).

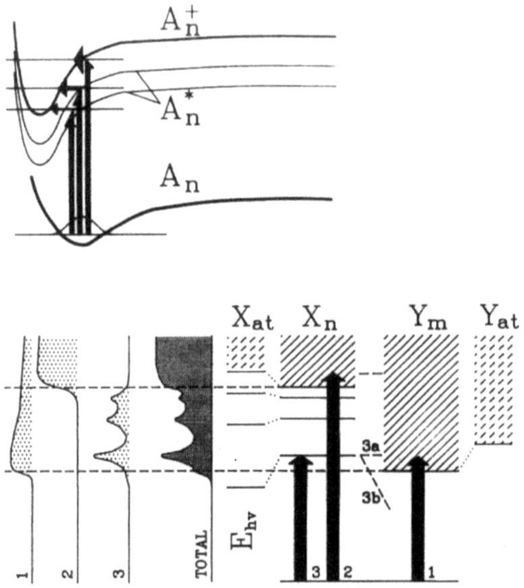

Fig. 1. *Bottom*: Schematic diagram of the energy levels of a mixed cluster X_nY_m (*right*) and its ionization efficiency curve (*left*). The energy levels (the ionization continuum is hatched) of the cluster components (X_n,Y_m) are shifted in respect to the atomic ones (X_{at}, Y_{at}). Only such excitations which lie above the lowest ionic level (– –) will contribute to ionization. The hole will be finally trapped in the lowest energy level possible, i.e. at the ionization threshold of Y_m. *Top*: Schematic diagram of the energy levels of a homogeneous cluster. A transition into the ionic state is possible through a Rydberg state with a very small possibility, depending on the geometric differences between the neutral and the ionic potential surfaces

Such processes occur also in homogeneous cluster systems exhibiting several sequences of energy levels converging to different ionic states: Rydberg states above the lowest ionization potential are excited, and in turn decay into ionic states thus ejecting an electron. The ionization efficiency curves should display a rich fine structure as a function of photon energy. Because of insufficient wavelength resolution they have so far only been observed in small systems.

The effect of the ionization process will be discussed with respect to the primary excitation and ionization process, the subsequent fragmentation of the cluster, and the resulting cluster ion mass distribution featuring particular abundant cluster ions.

Experimental

Experiments have been performed in which different ionization procedures have been applied to homogeneous and heterogeneous van-der-Waals clusters and their mass spectra measured.

[2] This is similar to the behaviour of diatomic van der Waals ions: Homonuclear dimer ions (like He_2^+, Ar_2^+, $(O_2)_2^+$, ...) have a much greater dissociation energy than the heteronuclear compounds (like $HeAr^+$, $ArXe^+$, ArO_2^+)

Table 1. Ionization procedures applied in cluster research (e_p: primary, e_s: ejected electron; C^+: cluster ion)

	Ionizing particle	Energy partitioned between	Primary energy
Electron bombardment	e^-	$e_p + e_s + C^+$	variable
Charge exchange	I^+	$e_s + C^+$	fixed
Standard PI	$h\nu$	$e_s + C^+$	variable
TEPICO	$h\nu$	C^+	variable

Van-der-Waals clusters have been generated by expanding a high pressure gas or a gas mixture from a small orifice into the vacuum. The ionization methods used are listed in Table 1. A more detailed description of the experiment is given [11–14].

Results and discussion

A. The argon-oxygen system

Figures 2–4 show ionization efficiency curves for mixed argon-oxygen clusters $(O_2)_n^+$, $Ar(O_2)_n^+$, and $Ar_2(O_2)_n^+$ produced by photoionization of a cluster beam which has been generated by expanding an Ar-O_2 mixture. All spectra except that of ArO_2 exhibit sharp resonances at 12.05 eV and 13.6 eV and a sharp increase in intensity between 15 eV and 16 eV. Moreover, spectra of oxygen clusters formed by expanding pure oxygen did not show any resonance structure nor the sharp increase around 15.5 eV [15]. Therefore it has to be assumed that the detected structures and resonances are caused by the admixture of Ar.

Comparison with absorption spectra of solid Ar (top left in Figs. 3b, 4) show a striking similarity. The structures in the solid Ar spectra are interpreted as excitons, i.e. collective excitations of states which correlate with the first excited ^1P-state of Ar. We conclude, therefore, that the resonance structure observed is caused by the excitation of the Ar component in the cluster, its state correlating to the ^1P state of the Ar atom. The blue shift observed is similar to that found for excitons in solid Ar [16].

These structures are not only visible in the spectra of mixed $Ar_n(O_2)_m^+$ cluster ions, but also in those of pure $(O_2)_m^+$. This implies that the pure $(O_2)_m^+$ ions are fragments from a neutral precursor which must have contained Ar atoms:

$$Ar_n(O_2)_m + h\nu \rightarrow Ar_n(O_2)_m^{+*} \rightarrow (O_2)_k^+ + (m-k) \times O_2 + n \cdot Ar \quad (1)$$

As has been pointed out the charge will travel to the particle with the lowest ionization potential, i.e. finally the charge will not be located on one of the Ar atoms. It has to be assumed that the Ar atom is more loosely bound in such a cluster than the O_2-molecules and is, therefore, the first component evaporated, hence increasing the oxygen content of the cluster.

Some of the curves deviate from the described picture. Smaller cluster ions (O_4^+, O_6^+, $Ar \cdot O_2^+$) show additional or shifted resonances, particularly at low expansion pressures, i.e. for cluster distributions containing only small clusters; ArO_2 does not show the resonance at 12.05 eV at all (c.f. Fig. 3a). This can be related to the small binding energy of that molecule. The resonance energy is not sufficient for ionization and subsequent fragmentation of larger clusters nor for the direct ionization of ArO_2 because of unfavourable Franck-Condon factors. Measurements varying the expansion conditions show an increase of the resonance structure in respect to the direct ionization, when reducing the stagnation pressure, i.e. when producing distributions of smaller clusters. We explain this by quenching processes of the neutral excited cluster without the formation of an ion (path 3*b* in Fig. 1).

B. The system Ar-Xe

Similar resonances are observed for mixed Ar-Xe clusters [7]. Figure 6 shows the cluster ion intensity as a function of wavelength for several selected cluster ions. Sharp resonances occur at a photon energy of $E_{h\nu} = 11.95$ eV which is attributed to the excitation of the Ar component in the cluster. This resonance, whose nature is similar to the ones observed for the argon-oxygen system, lies closer to the atomic value than that of the $Ar_n(O_2)_m^+$-system implying a weaker interaction within the cluster.

The form of the resonance depends slightly on the cluster ion under investigation: clusters of $Ar_nXe^+ (n < 10)$ show a single resonance peak (c.f. Ar_7Xe^+ in Fig. 6) while the resonance line of larger cluster ions exhibits a shoulder extending to smaller photon energies. This is interpreted as being due to different positions of the excited Ar-atom in respect to the Xe-ion: the main resonance is caused by Ar-atoms in the first shell of the cluster. They are the nearest neighbours to the central Xe-ion and show, therefore, a stronger interaction. The shoulder is caused by the excitation of Ar atoms in the second shell, which are more loosely bound, exhibiting a smaller energy deviation compared to the atomic value. This effect is additional evidence for ionization induced fragmentation. From this it is concluded that the precursors of the Ar_nXe^+ $(n > 10)$ are to be fragmentation products of larger neutral clusters consisting of at least two shells

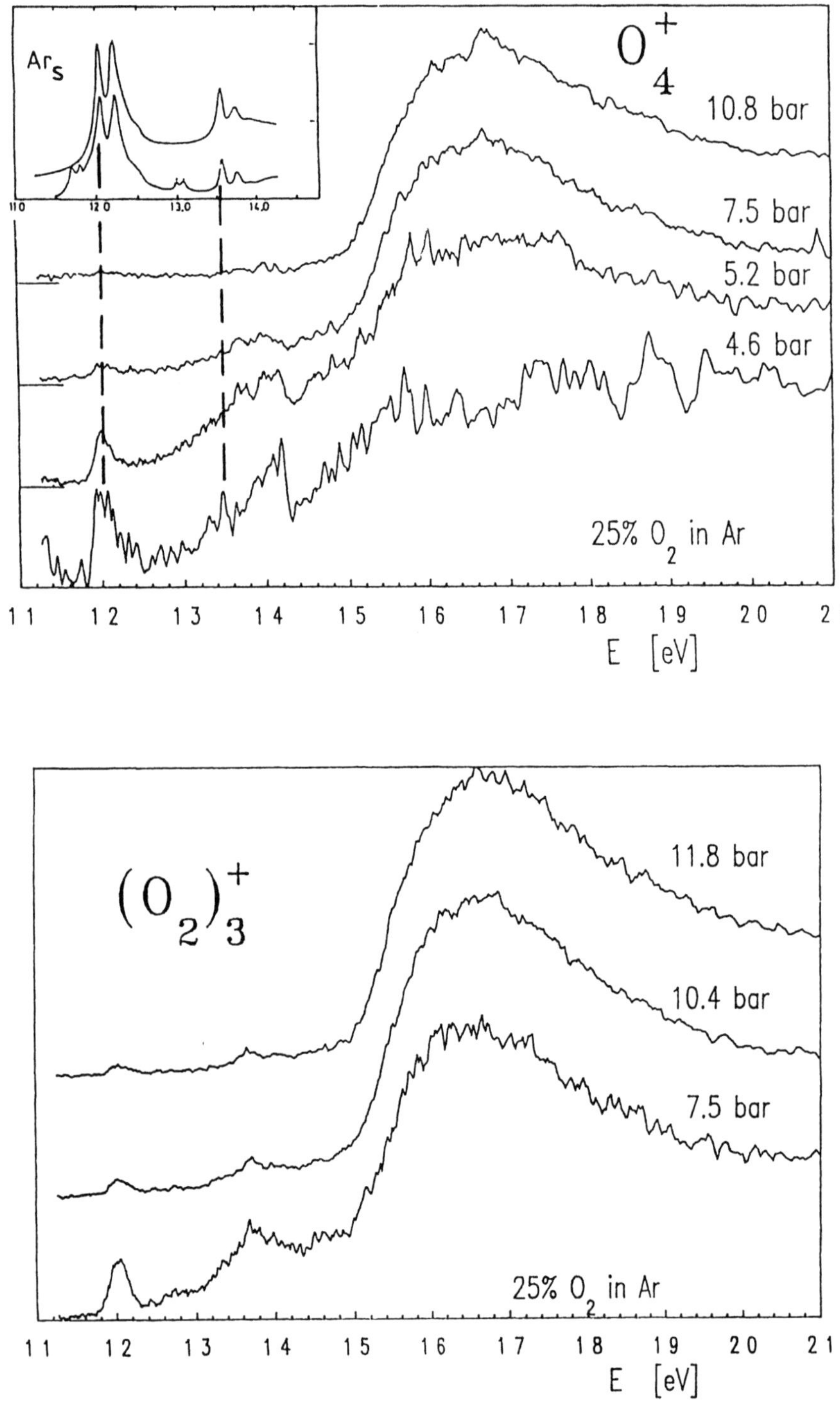

Fig. 2. Ionization efficiency curves for pure oxygen clusters (O_4^+, O_6^+) as a function of photon energy E and stagnation pressure. The clusters have been produced by expansion of an argon-oxygen mixture through a 30 μm nozzle. The larger clusters show resonances at 12.05 eV and 13.6 eV, independent of the composition of the cluster ions, the resonance is shifted for the small clusters. Absorption spectra of solid argon (Ar_s) displaying exciton resonances are shown in the top left corner for comparison [16]. The upper one of the two curves is taken with solid Ar covered by a thin layer of Kr and displays only bulk excitons, the lower curve is obtained from pure solid Ar and displays both bulk and surface resonances

($Ar_nXe(n > 12)$), while the main precursor for the Ar_7Xe^+ ion is the neutral Ar_nXe with only one Ar-shell filled ($8 \leqslant n \leqslant 12$). It is probable that $n = 12$, because neutral clusters with one shell completely filled are also extremely stable and should be particularly abundant in the neutral beam.

C. Other systems

Similar resonances have been observed in other systems. The system Ne-CH_4 exhibits resonances caused by the excitation of the Ne component into its first excited state (^1P at 16.85 eV) [17]. The corresponding

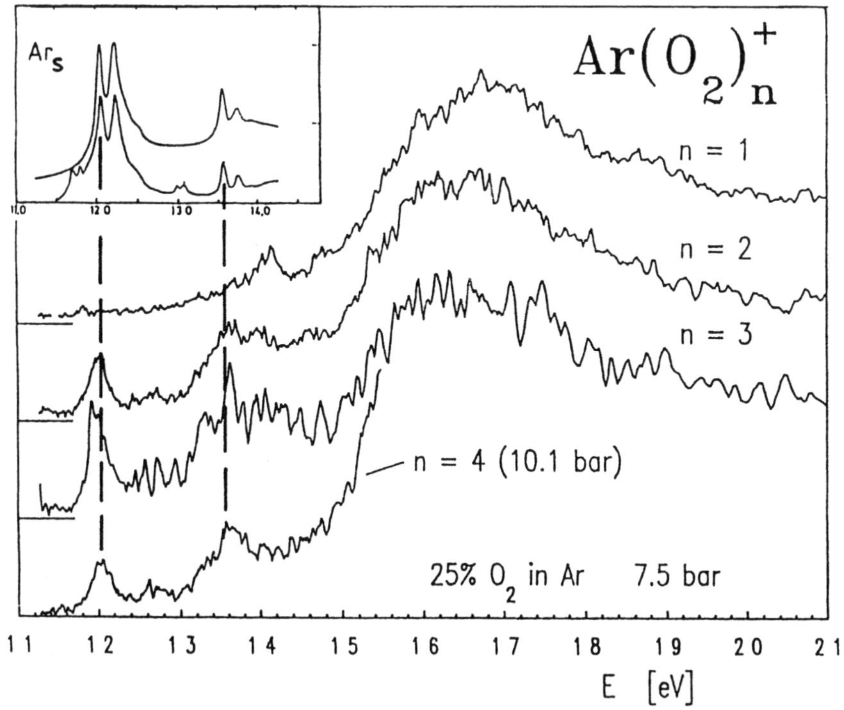

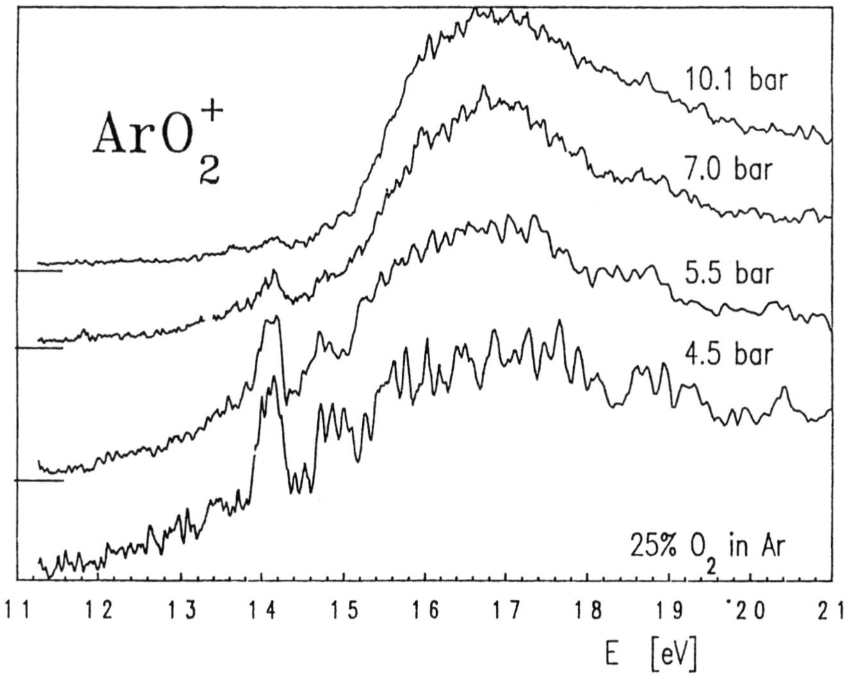

Fig. 3. Ionization efficiency curves for pure mixed oxygen-argon clusters containing one Ar-atom ($Ar \cdot O_{2m}$) as a function of photon energy and stagnation pressure. The clusters have been produced by expansion of an argon-oxygen mixture through a 30 μm nozzle. Exciton resonances are displayed in the top left corner for comparison (a detailed description is given in the caption of Fig. 2)

resonance for the Ar-CH_4 system cannot be observed because of the high ionization potential of the CH_4. Both show however the direct ionization path.

Resonances have also been observed in systems consisting of rare gases mixed with organic molecules. This has been called Penning-ionization [18], implying the transfer of one electron to the excited atom and the subsequent removal of its highest excited electron according to the scheme

$$A \cdot B + h\nu \rightarrow A^*B \rightarrow A^-B^+ \rightarrow A \cdot B^+ + e^- \quad (2)$$

The ejected electron has a relatively high kinetic energy. The TEPICO-method excludes such processes as it is only sensitive to low energy electrons. Systems

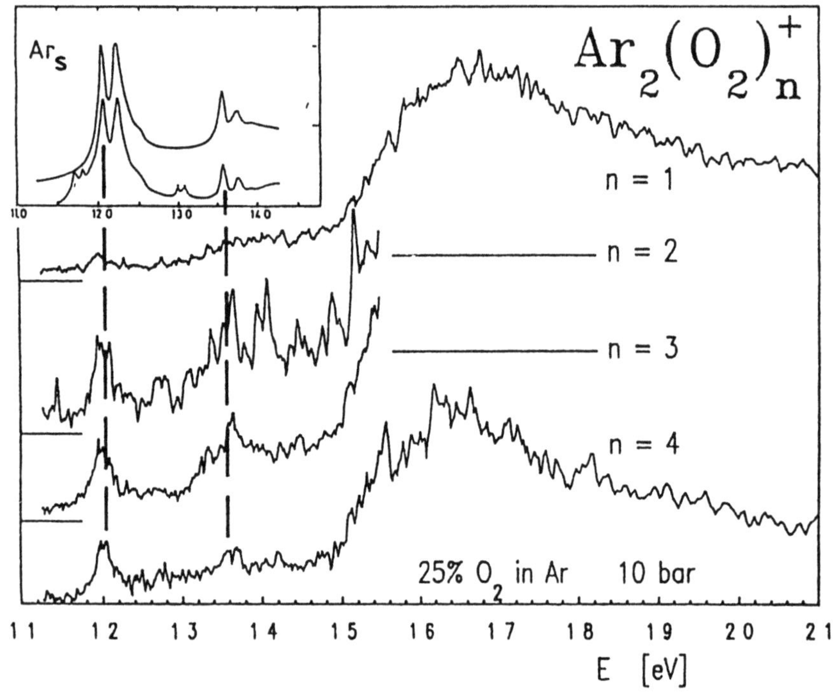

Fig. 4. Ionization efficiency curves for pure mixed oxygen-argon clusters containing two Ar-atoms (Ar_2O_{2m}) as a function of photon energy and stagnation pressure. The clusters have been produced by expansion of an argon-oxygen mixture through a 30 μm nozzle. Exciton resonances are displayed in the top left corner for comparison (a detailed description is given in the caption of Fig. 2)

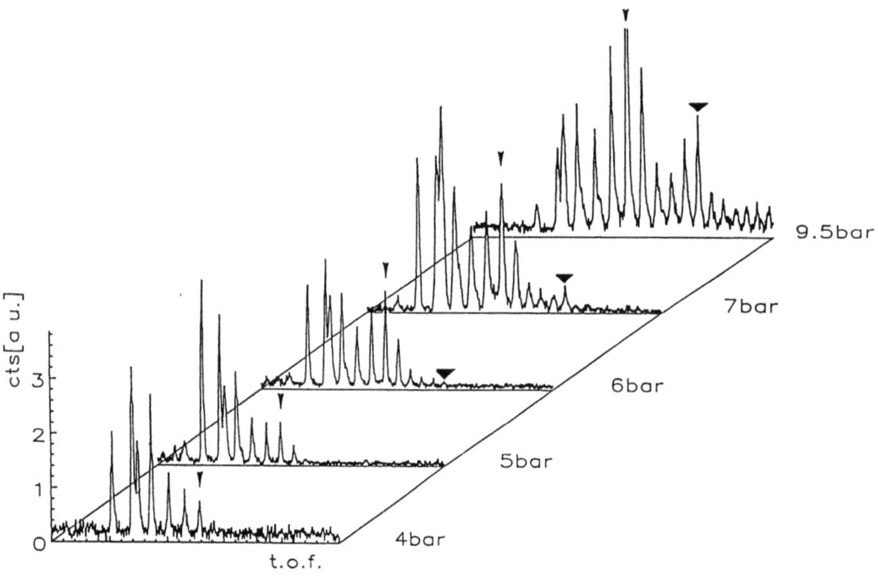

Fig. 5. Time-of-flight spectra of Ar_nXe^+ cluster ions as a function of stagnation pressure generated by an expansion of a gas mixture of 0.4% Xe in Ar through a 30 μm nozzle. The shape of the $Ar_{12}Xe^+$ peak consisting of a narrow and a broad distribution should be noted

like Ar_nXe which have been measured by this method exhibit therefore a different energy transfer mechanism.

Fragmentation spectroscopy of rare gas clusters

Time-of-flight mass spectrometry can be employed not only to measure the final cluster ion distribution but also the initial kinetic energy of the mass resolved ions. In conjunction with the TEPICO method this is an excellent tool to determine the distribution of energy among the different components of the cluster.

For these experiments, both a two field and a three field time-of-flight arrangement have been used. The flight time of an ion in the present set-up is given by the following equations (v_0: initial velocity, m: ion mass, e: elementary charge, d: length of field free region)

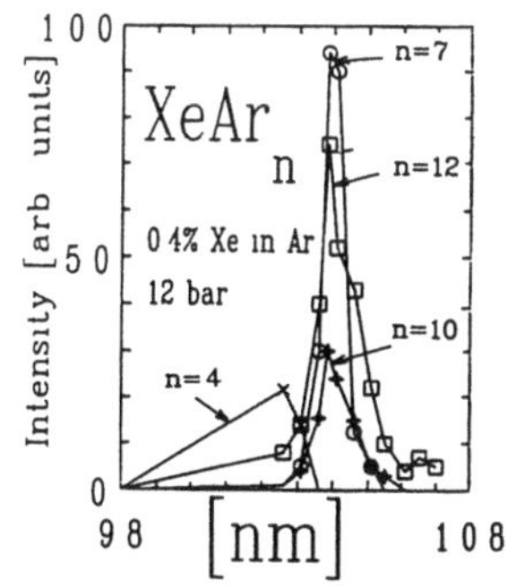

Fig. 6. Relative intensities for some selected Ar_nXe^+-cluster-ions as a function of photon energy

2-field arrangement: $$t_I = \frac{m}{eE}(v_t - v_0) + \frac{d}{v_t} \quad (3)$$

E: field strength and

d: length of the acceleration field $$v_t = \frac{2b}{m} + v_0^2$$

3-field arrangement:

$$t_{II} = \frac{m}{eE_1}(v_{tt} - v_t) + \frac{m}{eE_2}(v_{tt} - v_t) + \frac{d}{v_{tt}} \quad (4)$$

E_1, E_2: field strengths and

d_1, d_2: $$v_t = \frac{2b_1}{m} + v_0^2$$

length of the first and

second acceleration field $$v_{tt} = \frac{2b_2}{m} + v_t^2$$

and depends on the ion mass, the acceleration fields, and also on the initial velocity of the particle shortly after ionization. Rapid fragmentation ($t < 50\ ns$) will therefore lead to a symmetric broadening of the lines (c.f. Fig. 7). The peak shape and width may be used to determine the amount of fragment energy released. If the life time of the internally excited cluster ion is in the order of the acceleration time, the arrival time of the fragment ion will lie between that of the parent ion and that of an ion which has immediately fragmented after ionization. Because of the statistical distributions of lifetimes this will lead to an unsymmetric broadening of the ion peak with an exponential shape. It is similar to the appearance of metastable peaks in a normal mass spectrum [19]. It has to be pointed out that this classification in metastable and non-metastable compounds is to a certain degree arbitrary and depends on the experimental conditions. Unbroadened ion peaks can be seen only in exceptional cases which are attributed to long lived (or even stable) cluster ions not decaying during the acceleration time.

Examples of such fragmentation processes are presented for mixed and homogeneous cluster systems:

The TEPICO-spectra of Ar_nXe^+ cluster ions show a line broadening which depends on the photon energy [7]. This is particularly pronounced for the $Ar_{12}Xe^+$-peak. As shown in Fig. 7 this peak consists of a narrow and a broad line. This is interpreted as the occurrence of an unfragmented metastable cluster ion and one which represents the end of a fragmentation chain and displays therefore kinetic energy release:

$$Ar_{12}Xe + h\nu \rightarrow Ar_{12}Xe^+_{metastable} \quad (5a)$$

$$Ar_nXe\,(n > 12) + h\nu \rightarrow Ar_{12}Xe^+ + (n-12)Ar + E_{kin} \quad (5b)$$

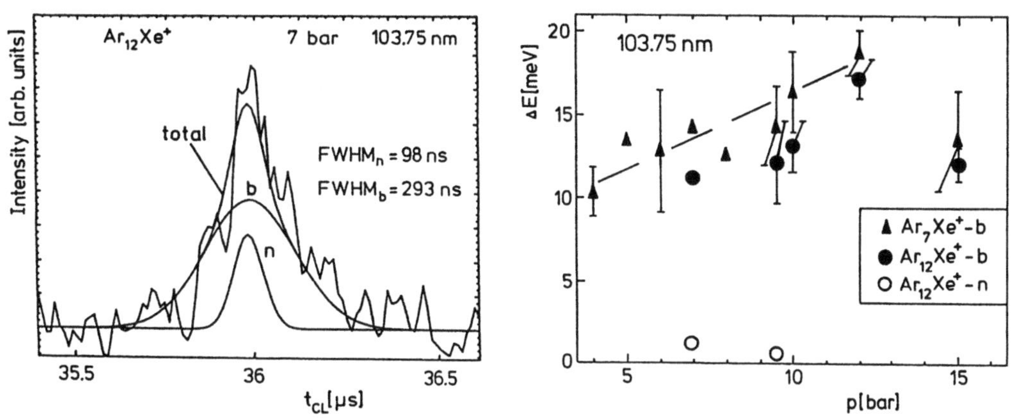

Fig. 7. *Left*: Line analysis of the $Ar_{12}Xe^+$ cluster. *Right*: Fragmentation energy release ΔE as a function of pressure for Ar_7Xe^+ and $Ar_{12}Xe^+$. Symbols *n* and *b* denote the narrow and the broad components of the cluster peaks, respectively. The *broken lines* indicate the average release energy for the Ar_7Xe^+ clusters. At low pressures ($p = 6$ bar) $Ar_{12}Xe$ is no longer produced. At high pressure ($p = 10$ bar) the narrow component cannot be seen because of too large fragment ion intensity

The contribution displaying the narrow line width is only seen under certain experimental conditions, i.e. if the distribution of the neutral clusters does not contain too many large particles (small expansion pressure) *and* if the photon energy is low. Other clusters like Ar_7Xe^+ show either large line broadening or do not appear at all, which classifies them as fragmentation products stemming from larger clusters. This is strong evidence that ionization is usually accompanied by fragmentation.

A very detailed study has been performed to investigate the fragmentation of homogeneous rare gas clusters [20] as a function of photon energy and expansion conditions (stagnation pressure and temperature). Fragmentation is a very sensitive function of photon energy and causes significant kinetic energy release. At energies just below the ionization potential of the Ar-atom (15.76 eV; 78.8 nm) three particular cluster ions (Ar_2^+, Ar_{19}^+, and Ar_{24}^+) carry a substantial kinetic energy release of appr. 17 meV (see Fig. 8). These ions are interpreted as representing the end of a fragmentation chain, because their relative stability is large compared to the other members of the chain. The abundance of these particular cluster ions has also been observed with conventional mass spectrometry. The experiments show that – under the given experimental conditions – fragmentation of small Ar_n^+ ($n < 10$) clusters is very extensive having the dimer ion as the ultimate fragmentation product. This picture changes significantly at lower photon energies: at photon energies of 14.94 eV, maxima of the kinetic energy release appear for Ar_3^+, Ar_{15}^+, Ar_{18}^+, and Ar_{21}^+ implying that fragmentation has become less violent. Fragmentation of small Ar_n-clusters ($n < 10$) now terminates at Ar_3^+, which is the second most stable cluster ion in this series. Fragmentation is still strong, the lifetimes of the respective precursors are, however, much longer. The kinetic energy is not released instantaneously and does not cause symmetric line broadening; instead, the precursors appear as metastable ions.

Lowering the photon energy further to 14.76 eV, the kinetic energy release decreases further and is nearly independent of cluster size. Instead, a large number of metastables with rather long lifetimes appear.

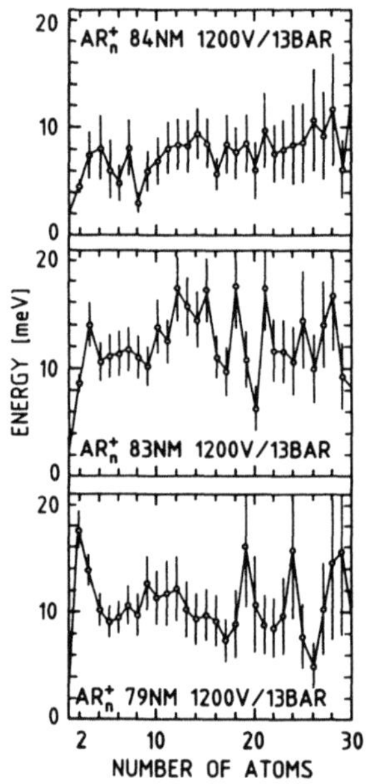

Fig. 8. Kinetic energy release from the fragmentation of Ar-clusters as a function of cluster size and photon wavelength

Mass spectra of van-der-Waals clusters

Mass spectra of clusters are determined by the neutral cluster distribution as well as the excitation and fragmentation probabilities for a specific cluster. If the internal excitation is sufficiently strong, the mass spectrum will be dominated by the fragmentation process. The most stable cluster ions will then be the most abundant ones in the mass spectrum. It is, therefore, not surprising that the resulting mass distributions change only little when using different ionization techniques.

In the case of Ar-clusters all three previously mentioned techniques – ionization by photons, electron bombardment or charge transfer – have been applied and produce nearly identical results. It is noteworthy to mention in this context that electron bombardment ionization is equivalent to photoionization with nearly white light, as the interacting electrons generate an instantaneous change of the electric field at the location of the particle. This corresponds to a spectrum with a large frequency interval.

Charge transfer ionization also behaves similarly to photoionization [12]: the energy dissipated is the difference of the ionization energy of the incoming particle and that of the cluster under attack. Large differences lead to profound internal excitation and therefore to heavy fragmentation (c.f. Fig. 9).

Photoionizing the cluster using photon energies near the ionization threshold proceeds via an indirect ionization path, as the direct transition is energetically no longer accessible. This expresses itself in a very low ionization efficiency. Indirect transitions proceed via high lying Rydberg states, the transition probability governed by the Franck-Condon factors between this and the ionic state.

As the binding energy increases with cluster size, larger clusters are formed by direct ionization while

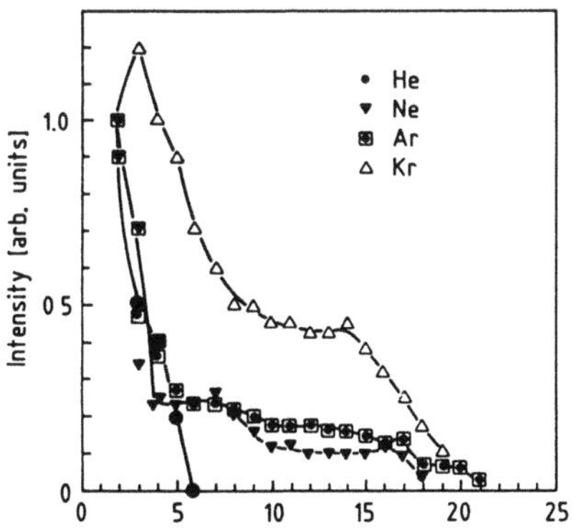

Fig. 9. Ionization of Ar-clusters by charge transfer from low energy (50 eV) rare gas ions. The different spectra are normalized to the Ar ion

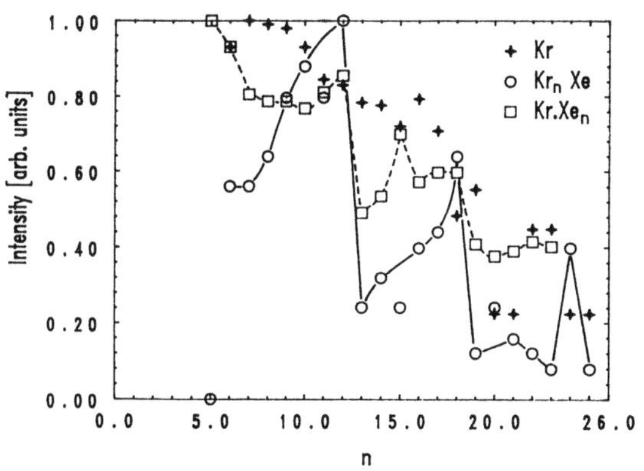

Fig. 10. Ionization of Kr$_n$Xe and Xe$_n$Kr by electron bombardment (E_{el} = 25 eV). The spectra show large intensity drop above $n = 12$ and $n = 18$, which is explained by a geometry where the ion is surrounded with particles of the second species. This is found for a great number of heterogeneous systems

the smaller ones are generated with nearly equal intensity through the indirect channel. Going to smaller ionization energies these probabilities decrease exponentially depending on the overlap of the excited Rydberg states with the corresponding ones of the cluster ion.

If geometry differences between neutral and ionic clusters are not too large as in the case of heterogeneous clusters, the picture is even simpler: when direct ionization prevails fragmentation will dominate the measured ion mass distribution. Cluster ions of the type AB_n^+ $(IP(A) < IP(B))$ favour stable ions of the form AB_{12}^+, AB_{18}^+. These masses are found abundantly in the mass spectrum independent of the ionization technique (see Fig. 10). In the cases where exothermic ion-molecule reactions are possible within the cluster, the large amount of internal energy released will smooth out the ionic distributions.

The assistance with the experiments of Dr. R.A. Cassidy, Dr. J. Valldorf, Dr. E. Holub-Krappe, Mr. G. Ganteför and Mr. J. Lehmann is greatfully acknowledged. The author thanks the staff of the Berlin Synchrotron Facility BESSY for the technical support of the photoionization experiments.

References

1. Märk, T.D., Castleman, Jr., A.W.: Adv. Atom. Molec. Phys. **20**, 65 (1985); In: Gaseous ion chemistry and mass spectrometry. Futrell, J.H. (ed.) New York: Wiley 1986
2. Buck, U., Meyer, H.: J. Chem. Phys. **84**, 4854 (1986); Abshagen, M., Kowalski, J., Meyberg, M., zu Putlitz, G., Träger, F., Well, J.: Europhys. Lett. **5**, 13 (1988)
3. Haberland, H.: Surf. Sci. **156**, 305 (1985)
4. Ding, A., Futrell, J.H., Cassidy, R.A., Cordis, L., Heßlich, J.: Surf. Sci. **156**, 282 (1985)
5. Begemann, W., Dreihöfer, S., Meiwes-Broer, H.W., Lutz, H.O.: Z. Phys. D-Atoms, Molecules and Clusters **3**, 183 (1986)
6. Echt, O., Sattler, K., Recknagel, E.: Phys. Rev. Lett. **47**, 1121 (1981)
7. Holub-Krappe, E., Ganteför, G., Bröker, G., Ding, A.: Z. Phys. D-Atoms, Molecules and Clusters **10**, 319 (1988)
8. Berkowitz, J.: Photoabsorption, photoionisation and photoelectron spectroscopy. New York: Academic Press 1979
9. Heßlich, J., Kuntz, P.J.: Z. Phys. D - Atoms, Molecules and Clusters **2**, 251 (1986); Kuntz, P.J., Valldorf, J.: Z. Phys. D - Atoms, Molecules and Clusters **8**, 195 (1988); Amarouche, M., Duvand, G., Malrieu, J.P.: J. Chem. Phys. **88**, 1010 (1988)
10. Kebarle, P., Davidson, W.R., French, M., Cumming, J.B., McMahon, T.B.: Faraday Discuss. Chem. Soc. **64**, 220 (1977)
11. Ding, A., Hesslich, J.: Chem. Phys. Lett. **94**, 54 (1983) Cassidy, R.A., Hesslich, J., Miller, R.E., Kuen, I., Ding, A.: Proc. of the SASP'84 **3**, 222 (1984)
12. Ding, A., Futrell, J.H., Cassidy, R.A.: Proc. SASP'84 **3**, 237 (1984)
13. Ding, A., Cassidy, R.A., Cordis, L., Lampe, F.W.: J. Chem. Phys. **83**, 3426 (1985)
14. Cordis, L., Ganteför, G., Heßlich, J., Ding, A.: Z. Phys. D - Atoms, Molecules and Clusters **3**, 323 (1986)
15. Linn, S.H., Ono, Y., Ng, C.Y.: J. Chem. Phys. **74**, 3348 (1981)
16. Schwentner, N., Koch, E.-E., Jortner, J.: Electronic excitations in condensed rare gases, p. 33. Berlin, Heidelberg, New York: Springer 1985
17. Ding, A., Cassidy, R.A., Futrell, J.H., Cordis, L.: J. Phys. Chem. **91**, 2562 (1987)
18. Kamke, W., Kamke, B., Kiefl, H.U., Hertel, I.V.: Chem. Phys. Lett. **122**, 356 (1985)
19. Beynon, J.H.: Mass spectrometry and its applications to organic chemistry. Amsterdam: Elsevier 1960
20. Ganteför, G., Holub-Krappe, E., Bröker, G., Ding, A.: (to be published)

Z. Phys. D – Atoms, Molecules and Clusters 12, 263–267 (1989)

Atoms, Molecules and Clusters
Zeitschrift für Physik D
© Springer-Verlag 1989

Metastable decay of rare gas cluster ions: mechanism and kinetics

T.D. Märk
Institut für Ionenphysik, Leopold Franzens Universität, Technikerstrasse 25, A-6020 Innsbruck, Austria

Received 7 July 1988; final version 30 September 1988

Metastable decay of cluster ions has been discovered only recently. It was noted that one has to take this metastable decay into account when using mass spectrometry to probe neutral clusters, because ion abundance anomalies in mass spectra of rare gas and molecular clusters are caused by delayed metastable evaporation of monomers following ion production. Moreover, it was found that *(i)* the individual metastable reaction rates k depend strongly on cluster size and cluster ion production pathways and that *(ii)* there exists experimental evidence ($k = k(t)$) and a theoretical prediction that a given mass selected cluster ion generated by electron impact ionization of a nozzle expansion beam will comprise a range of metastable decay rates. In addition, it was discovered that metastable Ar cluster ions which lose two monomers in the μs time regime decay via sequential decay series $Ar_n^{+*} \rightarrow Ar_{n-1}^{+*} \rightarrow Ar_{n-2}^{+*}$ with cluster sizes $7 \leq n \leq 10$ and $n = 3$ (similar results were obtained recently in case of N_2 cluster ions). Conversely, the dominant metastable decay channel of Ar_4^{+*} into Ar_2^+ was found to proceed predominantly via a single step fissioning process.

PACS: 36.40; 34.80.G

1. Introduction

Metastable decay of various cluster ions (e.g. rare gas, molecular, ionic, metal) has been discovered only recently [1–4]. It was noted quite early [3] that one has to take this metastable decay into account when using mass spectrometry to probe neutral cluster beams, because ion abundance anomalies in mass spectra of rare gas [5–7], molecular [8], ionic [9] and metallic [10] clusters are at least in part caused by delayed metastable evaporation of monomers following ion production. Properties of the metastable decay of cluster ions have been recently the subject of several experimental and theoretical studies, including the metastable decay mechanism, the release of kinetic energy, and the metastable reaction rates [11].

In particular, it was found that (i) the individual metastable reaction rates depend strongly on cluster size [2–10, 13] and cluster ion production pathways [3, 6, 14] and that (ii) there exists experimental evidence [15] and a theoretical prediction [16] that a given mass selected cluster ion generated by electron impact ionization of a nozzle expansion beam will comprise a range of metastable decay rates. Moreover, we discovered [17, 18] recently the occurrence of sequential metastable decay series $Ar_n^{+*} \rightarrow Ar_{n-1}^{+*} \rightarrow Ar_{n-2}^{+*}$, with $n = 3$ and $7 \leqslant n \leqslant 10$. Conversely, the dominant metastable decay channel of Ar_4^{+*} into Ar_2^+ was found to proceed predominantly via a single step fissioning process [17].

The present progress report summarizes recent results on this subject obtained in our laboratory for rare gas cluster ions including also related studies with molecular cluster ions [19].

2. Experimental

The supersonic beam/electron impact ionization/mass spectrometer system used here has been described in detail previously [6]. A schematic view of the apparatus (including the addition of the laser facility) is provided by Fig. 1 in [18]. Neutral rare gas clusters are produced by expanding the gas under study at a constant temperature (typically between 0 and $-180°C$) and constant pressure (typically between 1 and 10 bar) through a 20 μm nozzle. The ensuing supersonic beam passes a skimmer (differential pumping stage) and is crossed 10 cm downstream at right angles by an electron beam of variable energy. Cluster ions thereby produced are *(i)* extracted at right angles from the ionization region, *(ii)* accelerated by an acceleration voltage (typically between 0.5 and 3 kV), *(iii)* analyzed in a 90° magnetic and 90° electric sector field, and *(iv)* detected with an electron multiplier. Mass spectra obtained with different expansion conditions strongly depend on the stagnation temperature and stagnation pressure, i.e. lower temperature and higher pressure yield distributions with an increased abundance of larger clusters.

The neutral cluster distribution (as diagnosed indirectly [6] via the above mentioned cluster ion mass spectra) can be changed dramatically with help of an Ar ion laser (Spectra Physics 164, 5 Watt) being fired in the direction of the neutral cluster beam. In this irradiation mode there exists a collinear interaction of the (mechanically chopped) laser beam with the cluster beam leading to extensive photofragmentation of the neutral clusters, i.e. up to 80% depletion for large cluster sizes [20].

Essential to the present unimolecular dissociation measurements (which go beyond the simple metastable ion detection) is the existence *(i)* of a first field-free region between the end of the acceleration region and the beginning of the magnetic sector field (length: 41 cm) and *(ii)* of a second field-free region between the end of the magnetic sector field and the beginning of the electric sector field (length: 25 cm). A possible metastable decay of an ion m_1^+ (produced in the ion source) into m_2^+ can be detected either *(i)* in the first field free-region by tuning the magnetic sector field to a nominal $m^* = m_2^2/m_1$ and the electric sector field to an electric field $E^* = m_2 E/m_1$, or *(ii)* in the second field-free region by tuning the magnetic sector field to m_1^+ and the electric sector field to $E^* = m_2 E/m_1$ (with E being the correct electric field to detect m_1^+). In the present study we employed an additional experimental procedure in order to discover and study the occurrence of successive metastable decay series [17–19]. In this alternative mode of operation we detect a possible metastable decay of m_1^+ into m_2^+ in the first field-free region followed by a metastable decay of this m_2^+ into m_3^+ in the second field-free region. This is accomplished by tuning the magnetic sector field to a nominal mass $m^* = m_2^2/m_1$ and the electric sector field to an electric field $E^* = m_3 E/m_1$. In this case only those m_3^+ ions will arrive at the detector of our mass spectrometer which are produced by two successive decays in the first and second field-free regions, respectively.

Some of the metastable reactions occurring in the first field-free region may be contaminated by collision induced dissociations due to the presence of background gas [21]. However, by studying the unimolecular dissociations as a function of pressure in the first field free region (see for instance examples given in [6, 17]), it is possible to distinguish between truly metastable dissociation (whose occurrence is independent from the background pressure) and collision induced dissociations. Careful analysis of this pressure dependence yields the metastable abundance fraction present in the first field free region (for details of this analysis see [6, 17]). Moreover, using the appropriate kinetic rate equations it is possible to derive the corresponding metastable dissociation rates k(I) in the first field free region and k(II) in the second field free region. In order to study the temporal evolution of this rate in greater detail (in particular to obtain a continuous set of k values connecting the two time windows available) we have varied the time t_1 at which the ions enter the field free region by varying the ion source extraction voltage and acceleration voltage (between 3 kV and 0.5 kV) [22].

3. Results

3.1. Metastable dissociation reactions

Using the technique discussed above we were able to detect metastable loss of *one* monomer in the first or in the second field free region and metastable loss of *two* monomers in the first or in the second field free region for Ne_n^+, Ar_n^+ and $(N_2)_n^+$ (with n up to ~ 80). In the latter case it is not possible to distinguish in the experiment between a simultaneous evaporative decay or a sequential decay. Of particular interest, however, is the definitive observation of a sequential reaction $X_n^{+*} \xrightarrow{I} X_{n-1}^{+*} \xrightarrow{II} X_{n-2}^+$, with $X_n = Ar$ and $n = 3$, 7 to 10 [17, 18]. This constitutes the first experimental evidence for the occurrence of sequential decay of metastable atomic cluster ions. Similar decay sequences were also discovered very recently to be present for nitrogen cluster ions [19], i.e. $(N_2)_n^{+*} \to (N_2)_{n-1}^{+*} \to \ldots \to N_2^+$, with $n = 20, 6-3$.

Conversely, Ar_4^+, which shows a particularly strong tendency to lose two monomers in the metastable time regime, was shown to decay exclusively via a single step fissioning process $Ar_4^+ \rightarrow Ar_2^+ + 2Ar$ [17].

3.2. Metastable reaction rates for the various dissociation reactions

Table 1 gives some of the derived metastable dissociation rates for Ar_n^{+*} cluster ions decaying in the first and second field free region. Several conclusions can be drawn from these data:

(i) Reaction rates $k_{n,n-1}$ (I) deduced for reaction (1) is smaller than $k_{n,n-1}$ (I) deduced for reaction (3). This difference is caused by assuming in the analysis of reaction (1) that the Ar_{n-1}^+ ions produced in the first field free region via reaction (1) are stable for the rest of the flight time. Therefore, previous analysis of such data in the literature, which have neglected the additional decay (as evidenced by reaction (3)) give too low values of the decay rates.
(ii) The metastable decay rates obtained in the sequential decay of reaction (3) are very different, e.g., $k_{3,2}$ (I) $= 45\ s^{-1} \ll k_{2,1}$ (II) $= 680\ s^{-1}$. This result is in contrast to observation for Ar_7^+ through Ar_{10}^+. It indicates that for Ar_3^+ different metastable decay mechanism may be operative (e.g. electronic predissociation, barrier tunneling) than in the case of larger cluster ions (vibrational predissociation). In case of larger clusters each successive decomposition is accompanied by additional cooling of the remaining complex (thus leading to a lowering of the rate for further evaporation). The situation in case of $(N_2)_n^{+*}$ is even more complicated, see [19, 23].

Moreover, Märk et al. [6, 7, 25] have recently measured reaction rates $k_{n,n-1}$ (I) for reaction (1) (see Table 1) as a function of cluster size n in Ne and Ar up to $n = 80$. Besides a general increase of the reaction rate with cluster size (which has been nicely accounted for by Klots using the concept of an evaporative ensemble [16, 24]), we have observed anomalous reaction rates corresponding to abundance anomalies (magic numbers) in the observed mass spectra (see similar results [5] in Xe). It was possible to conclude from these results that the occurrence of magic numbers in rare gas cluster spectra is at least in part caused by the intrinsic instability of the cluster ions.

3.3. Apparent time dependence of metastable reaction rates

Klots [16] noted recently that a given mass-selected cluster ion generated in a nozzle expansion source will comprise a broad range of internal energies. The range of evaporative metastable fragmentation rates, k, thus subtended should be large leading to an apparent time dependence of the rates. In order to confirm this prediction, we varied recently the time, t_1, at which

Table 1. Reaction rates k(s^{-1}) for metastable dissociation in the first and second field free region. t_1: time at which the ions enter the first field free region; t_2: total flight time from ionization to detection. Acceleration voltage 3 kV. Electron energy: 180 eV. Stagnation pressure: 2.5 bar Ar. Stagnation temperature: −130°C

	Times		Reaction (1)*	Reaction (2)*	Reaction (3)*	
Ion	t_1(μs)	t_2(μs)	$k_{n,n-1}$(I)	$k_{n,n-2}$(I)	$k_{n,n-1}$(I)	$k_{n-1,n-2}$(II)
Ar_3^+	3.2	22	35	20	45	680
Ar_4^+	3.6	26	100	1800	not observed	
Ar_7^+	4.8	34	490	15	890	150
Ar_8^+	5.2	36.5	650	10	940	110
Ar_9^+	5.5	39	700	15	1250	180
Ar_{10}^+	5.8	41	3900	25	4030	65

* Reaction (1): $Ar_n^{+*} \xrightarrow{I} Ar_{n-1}^+ + Ar$ (first field free region)

* Reaction (2): $Ar_n^{+*} \xrightarrow{I} Ar_{n-2}^+ + 2Ar$ (first field free region)

* Reaction (3): $Ar_n^{+*} \xrightarrow{k_{n,n-1}(I)} Ar_{n-1}^{+*} + Ar$ (first field free region)

$\xrightarrow{k_{n-1}, k_{n-2}\ (II)} Ar_{n-2}^+ + Ar$ (second field free region)

Ne_n^+ cluster ions ($10 \leqslant n \leqslant 60$) enter the first field free region and found that the reaction rates derived change by more than a factor of 10 if the acceleration voltage is changed from 3 to 1 kV [15] (see also Refs. [10, 26]). In contrast, N_2O trimer ions produced by photoionization were found by Hertel and coworkers [27] to decay with a single metastable rate, presumably because these ions were produced with a fairly narrow range of internal energies. In order to follow up these observations we studied recently [18, 25] in detail the time dependence of the metastable reaction rate of small Ar_n cluster ions (with $n = 3, 4, 10$) produced by electron impact ionization, i.e. $k_{n,n-1}$ in both field free regions for different flight times (acceleration voltages). It was found that there is a strong dependence of the apparent measured reaction rate on the time elapsed since production of the ion [28]. This is the first experimental proof that such small metastable cluster ions cannot be assigned a unique lifetime.

3.4. *Dependence of metastable reaction rates on the ion production conditions: Ar_3^{+}* and Ar_4^{+}**

In agreement with a previous observation in our laboratory [3, 6] metastable reaction rates of Ar_3^{+*} and Ar_4^{+*} increase with increasing stagnation gas temperature above a specific temperature. Moreover, it was found that the reaction rates show a minimum and that there is an increase at very low temperatures. In addition, the reaction rates also increase with increasing pressure (at least at higher temperatures). For details see [8, 25].

In order to interpret this dependence on the expansion conditions, it is (according to Ref. [6]) necessary to correlate used expansion conditions with the corresponding neutral cluster properties and size distributions. At low temperatures and/or high pressures we find Ar_n cluster distributions up to at least $n = 300$ and more [29], whereas at higher temperatures and/or lower pressures the abundance of larger clusters decreases drastically. It can therefore be concluded that, at low temperature (and/or high pressure), most of the Ar_3^+ (or Ar_4^+) observed are produced via dissociative ionization of Ar_n clusters with $n \gg 3$ (or 4), whereas at higher temperatures (and/or lower pressures) most of the Ar_3^+ ion observed are produced via direct ionization of neutral Ar_3 with only small contributions from dissociative ionization of larger Ar_n. In the latter case, vertical Franck Condon transitions from the neutral Ar_3 (in triangular configuration) to the linear Ar_3^+ are likely to result in highly excited ion states leading to much stronger predissociation (via electronic predissociation and/or tunneling through rotational barriers) than in the former case where Ar_3^+ ions are the result of a dissociative ionization process. Clearly, a possible influence of the internal energy (vibrational and rotational temperature) of the neutral precursors cannot be excluded in the list of possible effects to be responsible for the observed temperature dependence of k.

Moreover, in order to shed new light on this dependence we have used a recent method to increase drastically the vibrational and rotational temperature of neutral van der Waals clusters [20], i.e. interaction of visible laser light with neutral clusters. According to [20] Raman scattering on the Strokes side is tentatively suggested to occur in this case and to be responsible for the strong photodissociation observed. Here, we have measured the metastable reaction rates under the same expansion conditions as above, however, using visible laser photodissociation to change the neutral cluster beam properties (temperature, distribution). It was found [18] that in this case a dramatic increase of the metastable reaction rates can be observed (up to one order of magnitude) yielding values of up to 100 000 s^{-1} in the μs time regime. This is clear proof that the neutral precursor distribution and temperature plays a significant role in determining the final ion state distribution which in turn is responsible for the differing metastable reaction rates.

Work partially supported by the Österreichischer Fonds zur Förderung der Wissenschaftlichen Forschung. It is a pleasure to acknowledge the important contributions to the presented results by several coworkers as mentioned in the corresponding references given throughout the paper.

References

1. Stace, A.J., Shukla, A.K.: Int. J. Mass Spectrom. Ion Phys. **36**, 119 (1980)
2. Sunner, J., Kebarle, P.: J. Phys. Chem. **85**, 327 (1981)
3. Stephan, K., Märk, T.D.: Chem. Phys. Lett. **90**, 51 (1982)
4. Katakuse, I., Ichihara, T., Fujita, Y., Matsuo, T., Sakuai, T., Matsuda, H.: Int. J. Mass Spectrom. Ion Proc. **67**, 229 (1985)
5. Kreisle, D., Echt, O., Knapp, M., Recknagel, E.: Phys. Rev. A **33**, 768 (1986)
6. Märk, T.D., Scheier, P., Leiter, K., Ritter, W., Stephan, K., Stamatovic, A.: Int. J. Mass Spectrom. Ion Proc. **74**, 281 (1986)
7. Märk, T.D., Scheier, P.: Chem. Phys. Lett. **137**, 245 (1987)
8. Echt, O., Kreisle, D., Knapp, M., Recknagel, E.: Chem. Phys. Lett. **108**, 401 (1984)
9. Ens, W., Beavis, R., Standing, K.G.: Phys. Rev. Lett. **50**, 27 (1983)
10. Begemann, W., Meiwes-Broer, K.H., Lutz, H.O.: Phys. Rev. Lett. **56**, 2248 (1986)
11. A summary on this subject has been recently given in Ref. 12, p. 30 to 42
12. Märk, T.D.: Int. J. Mass Spectrom. Ion Proc. **79**, 1 (1987)
13. Stace, A.J., Moore, C.: Chem. Phys. Lett. **96**, 80 (1983)
14. Stephan, K., Märk, T.D.: Phys. Rev. A **32**, 1447 (1985)
15. Märk, T.D., Scheier, P.: J. Chem. Phys. **87**, 1456 (1987)
16. Klots, C.E.: J. Chem. Phys. **83**, 5854 (1985)
17. Scheier, P., Märk, T.D.: Phys. Rev. Lett. **59**, 1813 (1987)

18. Scheier, P., Stamatovic, A., Märk, T.D.: J. Chem. Phys. **89**, 295 (1988)
19. Scheier, P., Märk, T.D.: Chem. Phys. Lett. **148**, 393 (1988)
20. Stamatovic, A., Howorka, F., Scheier, P., Märk, T.D.: Chem. Phys. Lett. **145**, 95, 581 (1988); Stamatovic, A., Scheier, P., Howorka, F., Märk, T.D.: Int. J. Mass Spectrom. Ion Proc. **83**, R1 (1988); Märk, T.D., Stamatovic, A., Howorka, F., Scheier, P.: Chem. Phys. Lett. (1988) (in press); Acta Phys. Slov. (1988) (in press); Märk, T.D.: Nuovo Cimento (1988) (in press); Stamatovic, A., Howorka, F., Märk, T.D.: Chem. Phys. Lett. (1988) (in press); Stamatovic, A., Howorka, F., Scheier, P., Märk, T.D.: J. Chem. Phys. (1988) (in press)
21. The background gas pressure in the second field free region is too low to induce appreciable amounts of collision induced dissociations
22. This also involves inevitably a change of the sampling time window, a fact which has to be taken into account in the derivation of the respective reaction rate *k*
23. Magnera, T.F., David, D.E., Michl, J.: Chem. Phys. Lett. **123**, 327 (1986); Leisner, T., Echt, O., Kandler, O., Miehle, W., Xue-Jian, Y., Recknagel, E.: Chem. Phys. Lett. **148**, 386 (1988)
24. Klots, C.E.: Private communication, 1988
25. Märk, T.D., Scheier, P.: (unpublished)
26. Echt, O., Cook, M.C., Castleman, Jr. A.W.: Chem. Phys. Lett. **135**, 229 (1987)
27. Kamke, W., Kamke, B., Kiefl, H.U., Hertel, I.V.: J. Chem. Phys. **84**, 1325 (1986)
28. It is especially gratifying to observe that there is a good agreement between the *k* values determined in the first field free region (where collision induced dissociation has to be subtracted from the metastable signal) and in the second field free region. This agreement excludes also artefacts from different extraction and transmission efficiencies
29. Scheier, P., Märk, T.D.: Int. J. Mass Spectrom. Ion. Proc. **76**, R11 (1987)

Z. Phys. D – Atoms, Molecules and Clusters 12, 269–271 (1989)

Atoms, Molecules and Clusters
Zeitschrift für Physik D
© Springer-Verlag 1989

Intra-cluster ion-molecule reactions and photodissociation of molecular clusters

N. Nishi and H. Shinohara
Institute for Molecular Science, Myodaiji, Okazaki 444, Japan

Received 7 July 1988

Intra-cluster ion-molecule reactions in benzene clusters and benzene-water binary clusters were observed in Resonance-Enhanced Multi-Photon Ionization (REMPI) Time-of-Flight (TOF) mass spectra. Condensation reactions of the excited dimer unit or higher cluster unit are found to occur within clusters.

PACS: 33.80.E; 36.40

Introduction

Photoionization or electron impact ionization of molecular clusters often induces the reaction of the primary excited ion with an adjacent molecule in the clusters. Proton transfer reaction in pure water and ammonia clusters is a simple and typical example [1–3]. This process is assisted by the excess energy deposited in clusters on ionization. Recently, Schriver et al. reported a new example of intracluster ion-molecule reaction in benzene clusters: an initially formed $C_6H_6^+$ attacks a neighboring benzene of cluster according to the reaction [4].

$$C_6H_6^+ + C_6H_6 \rightarrow C_7H_7^+ + C_5H_5, \qquad \Delta H = 0.63 \text{ eV}. \quad (1)$$

They used the resonant two-photon ionization technique to selectively excite benzene clusters via the $^1B_{1u}$ second excited state (6.42 eV; ArF laser) of benzene. Here we report that the benzene clusters and benzene-H_2O and benzene-CCl_4 binary clusters showed drastic intracluster ion-molecule reactions.

Experimental

Benzene and its mixed clusters are formed by bubbling argon (or helium) through room temperature liquid benzene or solution mixtures, respectively, at nozzle stagnation pressures ranging from 500 to 1500 Torr. Essential feature of the apparatus has been given elsewhere [5]. The mass spectrometer is a Mamyrin-Lubman type linear reflectron TOF mass spectrometer (R.M. Jordan Co. Ltd.) [6]. With a field free drift length of 53.6 cm, this spectrometer guarantees unit mass resolution up to m/e = 250.

Results and discussion

a) Resonance enhanced 2-photon ionization time-of-flight (RE2PI) mass spectra of benzene clusters via S_1 and S_2 states

TOF mass spectra of benzene cluster beams showed a change of the pattern depending on the resonant electronic states. Figure 1 exhibits a comparison of the RE2PI TOF mass spectra with the 243.1 nm ($S_1(^1B_{2u})$ or monomer) and 211.0 nm ($S_2(^1B_{1u})$ of monomer) excitation under the same laser power condition (63 μJ/pulse). Upon excitation to the S_1 state, the spectrum showed a famous fragmentation patterns of monomer ion [7, 8]. Excitation at the onset of the second absorption, however, provided a spectrum with an entirely different feature (Fig. 1B). The intensity of fragment ions was greatly diminished to such an extent that except the $C_4H_4^+$ ion the fragment ions observed

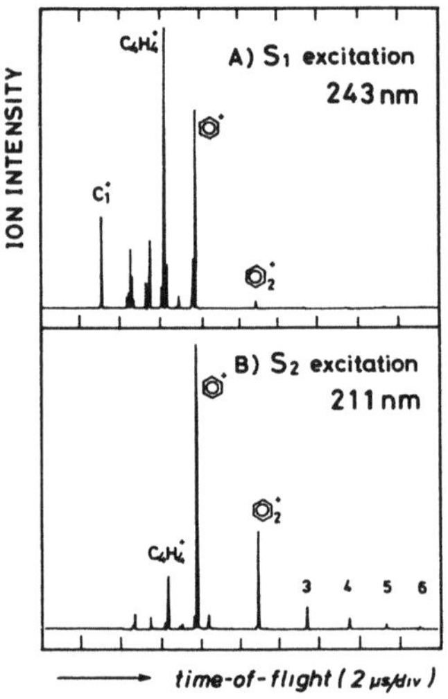

Fig. 1A and B. Comparison of RE2PI-TOF mass spectra of benzene clusters via the $S_1(^1B_{2u})$ excitation at 243 nm (A) with the $S_2(^1B_{1u})$ state at 211 nm (B). The conditions: room temperature benzene seeded in Ar (900 Torr), 40°C nozzle temperature and 63 μJ/pulse laser intensity

in Fig. 1A are almost missing. On the other hand, the cluster ions, $(C_6H_6)_n^+$, up to the heptamer were very much enhanced compared with the upper spectrum. As discussed frequently [9–12], the excited benzene dimer is expected to form a sandwich excimer. According to the theoretical calculation by Vala et al. [13], $^1B_{1u}$ and $^1E_{1u}$ states of benzene are very much stabilized due to Coulombic interaction between positive and negative molecular ion states (charge-resonance states). The $^1B_{1u}$ (S_2) state of Benzene is the so-called V state [14], which is described only by ionic structures in the simple VB theory. In contrast, the $^1B_{2u}(S_1)$ state does not show such charge polarization but in the higher $^1E_{1u}$ state the contribution of the charge polarization becomes much salient. The theoretical calculation suggests that the observed excited state of the dimer is an excimer state stabilized by the configuration mixing of charge-resonance and neutral-excitation states. We observed that the $^1B_{1u}$ stabilization energy is larger than 6 kcal/mol. This is attributed to the ionic nature of the excited state. This nature of the "S_2" excited state makes the resonant excitation via the "S_2" state very efficient because of large Franck-Condon factors between the "S_2" and the ion states.

b) Extensive fragmentation of benzene cluster ions due to additional multiphoton excitation

Resonant two-photon ionization of a benzene dimer unit appeared to be very efficient because of the ionic nature of the excimer state and the consequent advantage for the ionization from the excited state. The transparency of the monomer absorption around the onset of the "S_2"←S_0 cluster absorption is quite suited to extract the nature of the dimer state. We also found that the two-color ionization through the "S_2" state of the clusters shows similar effectiveness of the ionization even for the delay time of 3 ns between the two pulses. This clearly demonstrates that the relaxation to the lowest excited singlet state of the excimer was achieved before the second photon ionization. This stability of the excited clusters and the cluster ions enabled us to observe the extensive photofragmentation within the clusters. Figure 2 exhibits the various photofragment ions larger than parent benzene. The applied laser power at 213 nm was 400 μJ/pulse. Formation of $C_7H_7^+$ was already reported by Schriver et al. [5] who demonstrated that the two-photon ionization via the $^1B_{1u}$ state is capable of inducing the intracluster reaction (1). The $C_7H_7^+$ ion is also seen in the spectrum taken with a lower laser energy of 120 μJ/pulse (Fig. 1B). The newly appeared large product ions showed a common feature on the number of hydrogen (y) relative to the number of carbon (x). Namely, $y=x\pm1$ or $y=x-2$ with some exceptions. For C_{10} and C_{12} compounds, one can find dehydrogenated species with $y=x-i$ $(i=1\sim4)$. This kind of dehydrogenated ions was also found by Squire and Bernstein for toluene clusters [15]. The atomic composition of the product ions suggests the following condensation reactions in the clusters.

$$(C_6H_6)_n \xrightarrow{mh\nu} [(C_6H_6)_n^+]^* + C_xH_y^+ + C_{6(n-k)-x}H_{6(n-k)-y} + kC_6H_6, \qquad (2)$$

where k is the number of evaporated benzene

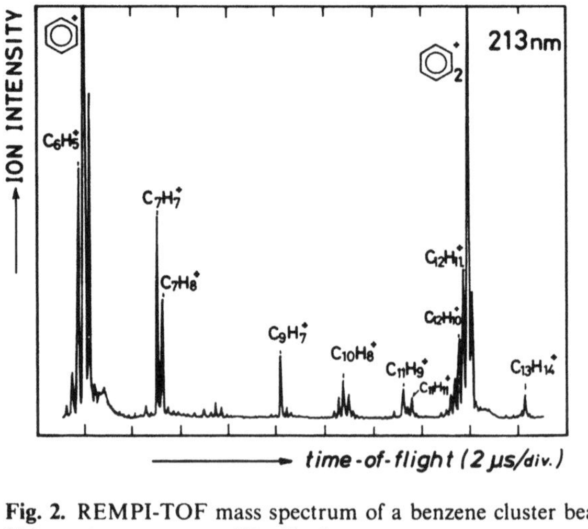

Fig. 2. REMPI-TOF mass spectrum of a benzene cluster beam at 213 nm. Laser power = 400 μJ/pulse

molecule(s) (most probably k = 1). Reaction (2) must be only possible for the ions $C_xH_y^+$ which are energetically much stable compared with the parent cluster ions. A typical example is $C_7H_7^+$. It is probable that the solvated tropylium ions, $C_7H_7^+(C_6H_6)_n$, are the parents of the newly observed ions.

c) Intra-cluster ion-molecule reactions in benzene-water mixed clusters

Benzene-water mixed clusters showed ion-molecule reactions inside of the hydrate clusters. Figure 3 shows the time-of-flight mass spectrum of the mixed cluster beams excited at 261.3 nm. At least two series of water progressions are seen in the spectrum, which originate from mass 91 ($C_7H_7^+$; tropylium ion) and mass 95 ($C_6H_5OH_2^+$). One more interesting feature is the appearance of protonated benzene dimer and trimer ions, although these ions are not accompanied with the water sequence. The reaction schemes will be expressed as follows.

$$C_6H_6\cdot(H_2O)_n \xrightarrow{m\cdot h\nu} [C_6H_6^+(H_2O)_n]^* \longrightarrow C_6H_5OH_2^+(H_2O)_{n-s} + H + s\cdot H_2O, \quad (3)$$

$$(C_6H_6)_2(H_2O)_n \xrightarrow{m\; h\nu} [(C_6H_6)_2^+(H_2O)_n]^* \longrightarrow C_7H_7^+(H_2O)_{n-t} + C_5H_5 + t\cdot H_2O, \quad (4)$$

$$(C_6H_6)_p(H_2O)_n \xrightarrow{m\cdot h\nu} [(C_6H_6)_p^+(H_2O)_n]^* \longrightarrow (C_6H_6)_{p-q}H^+ + OH + q\cdot C_6H_6 + (n-1)H_2O. \quad (5)$$

Benzene-(carbon tetrachloride) mixed cluster beams also produced large ions such as $C_6H_6Cl^+$, $C_7H_6Cl^+$, $C_6H_6Cl_2^+$, and $C_7H_6Cl_2^+$ through REMPI at 209 nm. The ionization of the mixed clusters fragmented a carbon tetrachloride molecule selectively inside of the clusters.

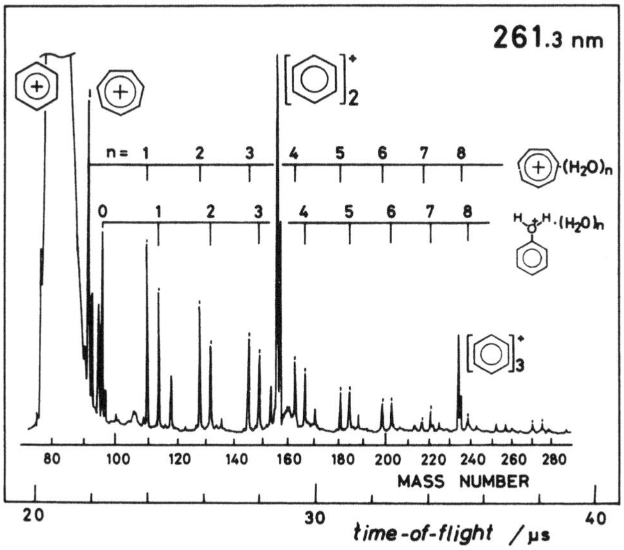

Fig. 3. REMPI time-of-flight mass spectrum of a benzene-water mixed cluster beam at 261.3 nm. The cluster beam was produced by bubbling Ar gas flow (350 Torr) through a benzene-water (4:1) mixture. Laser power ≈ 1 mJ/pulse

References

1. Shinohara, H., Nishi, N.: Chem. Phys. Lett. **87**, 561 (1982)
2. Choo, K.Y., Shinohara, H., Nishi, N.: Chem. Phys. Lett. **95**, 102 (1983)
3. Shinohara, H., Nishi, N.: J. Chem. Phys. **83**, 1939 (1985)
4. Schriver, K.E., Camarena, A.M., Hahn, M.Y., Paguia, A.J., Whetten, R.L.: J. Phys. Chem. **91**, 1786 (1987)
5. Shinohara, H., Nishi, N.: Chem. Phys. Lett. **141**, 292 (1987)
6. Lubman, D.M., Bell, W.E., Kronick, M.N.: Anal. Chem. **55**, 1437 (1983)
7. Zandee, L., Bernstein, R.B.: J. Chem. Phys. **71**, 1359 (1979)
8. Boesl, U., Neusser, H.J., Schlag, E.W.: J. Chem. Phys. **72**, 4327 (1980)
9. Hopkins, J.B., Powers, D.E., Smalley, R.E.: J. Phys. Chem. **85**, 3739 (1981)
10. Langridge-Smith, P.R.R., Brumbaugh, D.V., Haynam, C.A., Levy, D.H.: J. Phys. Chem. **85**, 3742 (1981)
11. Law, K.S., Schaver, M., Bernstein, E.R.: J. Chem. Phys. **81**, 4871 (1984)
12. Bärnsen, K.O., Selzle, H.L., Schlag, E.W.: J. Chem. Phys. **85**, 1726 (1986)
13. Vala, M.T., Hillier, I.H., Rice, S.A., Jortner, J.J.: Chem. Phys. **44**, 23 (1966)
14. Mulliken, R.S.: Chem. Phys. Lett. **25**, 305 (1974)
15. Squire, D.W., Bernstein, R.B.: J. Phys. Chem. **88**, 4944 (1984)

Z. Phys. D – Atoms, Molecules and Clusters 12, 273–274 (1989)

Atoms, Molecules
Zeitschrift für Physik D and Clusters
© Springer-Verlag 1989

Magic numbers of large rare gas clusters

W. Miehle, O. Echt*, O. Kandler, T. Leisner, and E. Recknagel
Physics Department, Universität Konstanz, Postfach 5560, D-7750 Konstanz, Federal Republic of Germany

Received 5 July 1988

The superior stability of closed-shell icosahedral structures is evident from size distributions of argon, krypton and xenon cluster ions in the size range $100 \lesssim n \lesssim 1000$.

PACS: 36.40.+d; 35.20.Wg; 61.25.Bi

1. Introduction

Magic numbers are frequently observed in mass spectra of atomic clusters, if a beam of neutral clusters is ionized by electron impact or photoionization. In case of xenon clusters, most of these numbers may be related to particularly stable icosahedral structures with complete coordination layers or sublayers [1, 2]. However, the agreement with anomalies being observed for clusters of He [3], Ne [4], Ar [5, 6], and Kr [6] is rather poor in the size range $n \lesssim 70$. In contrast, distributions of argon clusters, grown around ions, show intensity anomalies which mostly agree with those in xenon spectra [7]. Moreover, observed magic numbers at $n = 13$, 55, 147, ~ 309, and ~ 561 agree with numbers expected for complete-shell icosahedra. Very recently, it was noted [8] that mass spectra of argon clusters, born neutral, also show many features in common with the above-mentioned Xe and Ar spectra, if the size n becomes larger than 71 [5].

In this contribution we show that large clusters of Ar, Kr and Xe, born neutral and ionized by electron impact, do indeed tend to exhibit identical magic numbers which are related to icosahedral structures. The agreement is particularly striking for $n \gtrsim 230$. A full account of these results will be published elsewhere.

2. Experiment

Neutral clusters are grown in a supersonic, non-seeded jet; they are subsequently ionized by a pulsed electron beam. Generation of multiply charged clusters is avoided by keeping the electron energy slightly below the corresponding threshold. The size distribution of the cluster ions is analyzed in a time-of-flight mass spectrometer equipped with a reflectron. More details may be found elsewhere [9, 10].

3. Results and discussion

We have obtained mass-resolved spectra up to $n \sim 565$ for Ar, up to $n \sim 440$ for Kr, and up to $n \sim 240$ for Xe. The non-resolved part of the spectra extends beyond $n \sim 1000$ in all cases. The apparent resolution of Kr and Xe clusters suffers from the occurrence of several natural isotopes. Figure 1 displays a representative spectrum of Kr_n^+, $45 \leq n \leq 325$. Pronounced intensity anomalies are observed, most notably at $n = 147$ and 309, where closure of the 3rd and 4th icosahedral shell is expected to occur. Many additional anomalies are observed at, or close to, n-values which can be related to icosahedral packings with some adjacent faces being completely unfilled [7, 11].

The anomalies in argon and xenon spectra are strikingly similar to the krypton spectra, and to the distribution of argon clusters born ionized [7], for $n \gtrsim 130$. Concerning argon clusters, born neutral, intensity anomalies are often displaced by $\Delta n = +1$ in comparison to the other spectra, but they all agree with each other for $n \gtrsim 230$. In particular, closure of the 5th icosahedral shell is clearly indicated by an intensity drop at $n = 561$ ($n \sim 561$ for Kr and Xe). Another drop at $n \sim 923$ agrees with the expected stability drop upon closure of the 6th icosahedral shell.

Thus charged rare gas clusters appear to require at least ~ 1000 atoms before transition to the fcc

* Present address: University of Copenhagen, Niels Bohr Institute, Tandem Accelerator Laboratory, DK-4000 Roskilde, Denmark

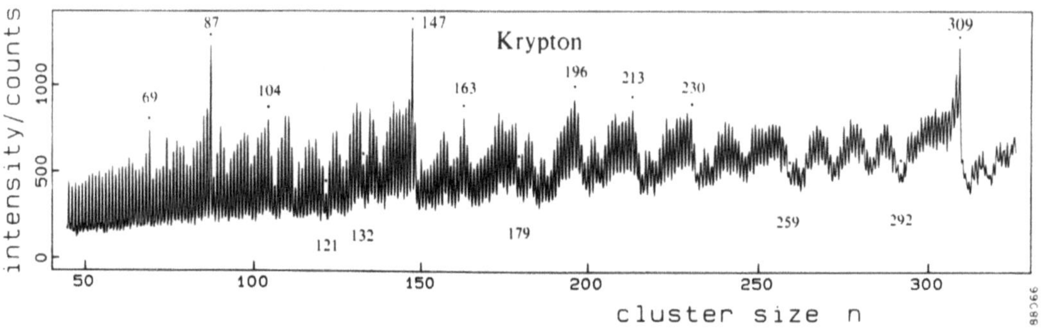

Fig. 1. Time-of-flight mass spectrum of krypton clusters

structure of bulk rare gases occurs. In comparison, Honeycutt and Andersen calculate that neutral Lennard-Jones clusters require at least 5000 atoms for the transition to occur [12]. Based on electron diffraction data on neutral Ar clusters, Lee and Stein place the location of the transition at $1500 \leq n \leq 3500$ [13], while Farges and colleagues, based on similar measurements, report a crossover from icosahedral to fcc structure at a size of $n \sim 750$ [14].

In conclusion, our results confirm recent calculations which indicate that the geometric structure of sufficiently large rare gas clusters will hardly be affected by the charge state or by the detailed structure of the cluster core [15]. Furthermore, it is interesting to note that molecular van der Waals clusters are also expected to arrange in icosahedral structures [16], and that size distributions of $(CO)_n^+$, recently obtained in our laboratory, do indeed exhibit intensity anomalies related to complete-shell icosahedra [10].

One of us (O.E.) gratefully acknowledges the stimulating atmosphere at the Niels Bohr Institute, where this manuscript was written. This work was financially supported by the Deutsche Forschungsgemeinschaft

References

1. Echt, O., Sattler, K., Recknagel, E.: Phys. Rev. Lett. **47**, 1121 (1981); Echt, O., Reyes Flotte, A., Knapp, M., Sattler, K., Recknagel, E.: Ber. Bunsenges. Phys. Chem. **86**, 860 (1982)
2. Kreisle, D., Echt, O., Knapp, M., Recknagel, E.: Phys. Rev. A **33**, 768 (1986)
3. Stephens, P.W., King, J.G.: Phys. Rev. Lett. **51**, 1538 (1983)
4. Märk, T.D., Scheier, P.: Chem. Phys. Lett. **137**, 245 (1987)
5. Scheier, P., Märk, T.D.: Int. J. Mass Spectrom. Ion Proc. **76**, R 11 (1987)
6. Ding, A., Hesslich, J.: Chem. Phys. Lett. **94**, 54 (1983)
7. Harris, I.A., Kidwell, R.S., Northby, J.A.: Phys. Rev. Lett. **53**, 2390 (1984); Harris, I.A., Norman, K.A., Mullkern, R.V., Northby, J.A: Chem. Phys. Lett. **130**, 316 (1986)
8. Märk, T.D.: Int. J. Mass Spectrom. Ion Proc. **79**, 1 (1987)
9. Knapp, M., Echt, O., Kreisle, D., Recknagel, E.: J. Phys. Chem. **91**, 2601 (1987); Leisner, T., Echt, O., Kandler, O., Yan, Xue-Jian, Recknagel, E.: Chem. Phys. Lett. **148**, 386 (1988)
10. Kandler, O., Leisner, T., Echt, O., Recknagel, E.: (Otto Stern Centennial Issue) Z. Phys. D – Atoms, Molecules and Clusters **D10**, 291 (1989)
11. Northby, J.A.: J. Chem. Phys. **87**, 6166 (1987)
12. Honeycutt, J.D., Andersen, H.C.: J. Phys. Chem. **91**, 4950 (1987)
13. Lee, J.W., Stein, G.D.: J. Phys. Chem. **91**, 2450 (1987)
14. Farges, J., de Feraudy, M.F., Raoult, B., Torchet, G.: J. Chem. Phys. **84**, 3491 (1986); Farges, J., de Feraudy, M.F., Raoult, B., Torchet, G.: In: Large finite systems. Jortner, J., Pullman, A., Pullman, B. (eds.), p. 113. Dordrecht: Reidel, D. 1987
15. Böhmer, H.-U., Peyerimhoff, S.D.: Z. Phys. D – Atoms, Molecules and Clusters **8**, 91 (1988); ibid, **11**, 239 (1989)
16. Van de Waal, B.W.: J. Chem. Phys. **86**, 5660 (1987)

Z. Phys. D – Atoms, Molecules and Clusters 12, 275–277 (1989)

Atoms, Molecules
Zeitschrift für Physik D and Clusters
© Springer-Verlag 1989

Fragmentation of neutral van der Waals clusters with visible laser light: a new variant of the Raman effect?

A. Stamatovic[2], F. Howorka[1], P. Scheier[1], and T.D. Märk[1]

[1] Institut für Ionenphysik, Leopold Franzens Universität, Technikerstrasse 25, A-6020 Innsbruck, Austria
[2] Department of Physics and Meteorology, PMF Beograd, Yugoslavia

Received 8 July 1988; final version 30 September 1988

We have observed strong photodissociation (using visible laser light) of neutral van der Waals clusters (Ar, N_2, O_2, CO_2, SO_2, NH_3) produced by supersonic expansion and detected by electron ionization/mass spectrometer. Several tests were performed, all of them supporting this surprising discovery. We suggest that Raman induced photodissociation (RIP) is responsible for this phenomenon. This first observation of Raman induced photodissociation provides a new technique for the study of neutral van der Waals clusters.

PACS: 33.80.G

1. Introduction

The infrared photodissociation of *neutral* molecular clusters has been studied for a large variety of species, including van der Waals (vdW) and hydrogen bonded systems [1]. In these experiments, a single infrared photon is absorbed in some well-defined (monomer) vibrational mode within a small cluster. If this photon energy is larger than the enthalpy of evaporation, the resulting energy transfer process leads ultimately to dissociation of the cluster. The photodissociation of cluster *ions* with infrared and/or visible light is currently an area of intense research [2]. In these studies, a single visible photon is thought to be absorbed in a well defined ion state above the dissociation limit leading (via a repulsive curve) to dissociation of the cluster ions. Conversely, to the best of our knowledge, no experimental information on the photodissociation of neutral van der Waals clusters with visible photons has been reported in the literature [3]. This might be due to the fact that absorbing electronic states lie energetically beyond the visible range and that therefore no effect has to be expected.

Here we report the first such study, i.e. the interaction of neutral Ar, N_2, O_2, CO_2, SO_2 and NH_3 [6–9] clusters with visible laser light at around 500 nm. The observed effects are very strong, leading in some cases to the depletion of more than 80% of the target clusters. This finding at first appears curious, because these clusters do not possess any optically allowed transitions in this wavelength region. We tentatively interpret this surprising effect by Raman scattering: Stokes transitions induced by Raman scattering lead to the excitation of low-lying vibrational states and subsequently to unimolecular dissociation.

2. Experimental

The supersonic beam/electron impact ionization source/mass spectrometer system used has been described previously [10]; the only difference is the addition of a Spectra Physics 164 argon ion laser with a nominal power of 5 Watt (all lines). Neutral clusters are formed by expanding several bar of the gas under study at various stagnation gas temperatures through a 20 μm nozzle. The cluster beam proceeds through a skimmer into a differentially pumped inner chamber. 10 cm downstream from the nozzle, the cluster beam is crossed perpendicular by an electron beam. Cluster ions thereby produced are extracted at right angles from the ionization region and analyzed in a 90° magnetic sector field followed by a 90° electric sector field. The cluster beam was irradiated co-axially over the entire flight path by a counterpropagating *cw* visible laser beam. In this arrangement, the van der Waals complexes were bathed in the radiation field for a period of appr. 200 μs, corresponding to the flight time from the nozzle to the ionization regime. The laser light was mechanically chopped at 13 Hz; thus photodissociation of the neutral clusters could be detected as modulation of cluster ion abundances in the mass spectrum.

3. Results and discussion

Photofragmentation spectra are recorded with the use of a lock-in amplifier synchronized to the chopper. The spectrum thus obtained gives directly the difference in cluster ion abundance with and without laser on (due to the photodissociation of the respective neutral precursors). It was found that for larger clusters depletion of the original cluster ion signal prevails (negative signal in the difference spectrum), whereas for smaller clusters growth due to fragments from larger clusters outweighs loss by fragmentation (positive signal in the difference spectrum). Moreover, the turning point between net loss or net gain is strongly depending on the stagnation conditions (that is on the neutral cluster distribution [10]), e.g., in case of Ar_n clusters at a stagnation temperature of $T = -110°C$ the turning point occurs at $n \sim 7$, at $T = -120°C$ at $n \sim 12$, and $T = -165°C$ at $n \sim 20$ [7]. It is interesting to note, that we have observed such net loss peaks in Ar up to cluster size $n = 150$ [7]. Furthermore, photofragmentation of the neutrals shows up not only as modulation of the singly charged cluster ion mass spectrum, but also as loss peaks in the difference spectrum of doubly charged cluster ions (e.g. as observed in Ar and NH_3).

A series of measurements was performed to corroborate the existence [11] of this photofragmentation process with visible laser light and to gain information about the nature of the dissociation reaction. First, we studied the photodissociation yield as a function of laser focus position. The yield reaches a minimum if the laser focus lies close to the ionization region. This is clear experimental evidence in favor of photoionization of the neutral clusters, excluding photoionization of the ions.

We also observed that the intensity of the neutral cluster beam (as measured by the total mass flux determined with an ionization gauge) decreases by more than 20% if there is good overlap between the chopped laser beam and the neutral cluster beam. This indicates that neutral dissociation products fly out of the cluster beam and are pumped away in the differentially pumped inner chamber.

As a further test, we examined different irradiation geometries. For instance, if the ionized clusters are crossed at right angles with the laser beam in the ion source extraction region, the obtained spectra look very different (i.e. all ions exhibit depletion), and the depletion is a factor of 300 smaller than in the other case. In this irradiation mode only ions are dissociated, because neutral clusters are not present in appreciable quantities at this position of the ion source. This indicates that ion dissociation is negligible in the other irradiation mode. Moreover, further tests also exclude any other sources (e.g. photoelectrons at the surfaces) for the observed phenomenon.

There remains the question of the mechanism of this visible photodissociation. As there are no allowed optical transitions the only conceivable inelastic interaction process is Raman scattering. Raman scattering on the Stokes side of the scattered line leads to vibrational and rotational excitations of the electronic ground state of the target systems. Recently, Godfried and Silvera [2] observed for the first time rotational and vibrational Raman transitions (spectra) for the lowest two vibrational states of the argon dimer. It was suggested [12], that the $v = 2$ and 3 states of Ar_2 are metastable with respect to tunnelling through a rotational barrier; the latter state is even strongly predissociating (leading to linewidth broadening in the Raman spectrum). Excitation of this kind to strongly predissociated states by Raman scattering is tentatively suggested as a possible mechanism of the presently observed photodissociation of van der Waals clusters (this is also in accordance with the linear dependence of the photodissociation yield on laser power observed here in all cases). Raman induced photodissociation (RIP), observed here for the first time, seems to be possible for van der Waals clusters, because of the extremely weak binding energies (for instance 12 meV between two Ar atoms and 80 meV for the binding energy per atom in an Ar crystal).

Moreover, we have just changed the experimental set up, such that the laser beam no longer has to pass through the ion source and no longer can interact with the skimmer and nozzle surface in order to interact with the beam of neutral clusters. In this new set up the neutral clusters are crossed at right angles with the laser. First results with Ar clusters confirm the present observations [13].

In conclusion, we have observed strong depletion of neutral van der Waals clusters induced by visible laser light. Several tests were performed to establish this discovery, all of them supporting the existence of this remarkable effect. Raman induced photodissociation is suggested to be responsible for this observation.

Work partially supported by the Österr. Fonds zur Förderung der Wissenschaftlichen Forschung.

References

1. Gentry, W.R.: Am. Chem. Soc. Symp. Ser. **263**, 289 (1984); Janda, K.C.: Adv. Chem. Phys. **60**, 201 (1985); Miller, R.E.: J. Phys. Chem. **90**, 3301 (1986)
2. Johnson, M.A., Lineberger, W.C.: Pulsed Methods for Cluster Ion Spectroscopy. In: Techniques for the study of gas ion molecule reactions. Saunders, W.H., Farrar, J.M. (eds.). New York: Wiley (in press)
3. Except for these cases where electronic states of the neutral can be reached by visible photons, e.g. van der Waals complexes of iodine [4] or sodium [5]
4. Levy, D.H.: Adv Chem. Phys. **47**, 323 (1981)
5. Broyer, M., Delacretaz, G., Labastie, P., Wolf, J.P., Wöste, L.: Phys. Rev. Lett. **57**, 1851 (1986)
6. Stamatovic, A., Howorka, F., Scheier, P., Märk, T.D.: Chem. Phys. Lett. **145**, 95, 581 (1988)
7. Stamatovic, A., Scheier, P., Howorka, F., Märk, T.D.: Int. J. Mass Spectrom. Ion Proc. Rapid Commun 83, R1 (1988)
8. Märk, T.D., Stamatovic, A., Howorka, F., Scheier, P.: Chem. Phys. Lett. **149**, 230 (1988), Acta Phys. Solv. (1989) (in press)
9. Märk, T.D., Stamatovic, A., Howorka, F., Scheier, P., Walder, G.: Nuovo Cimento, (1989) (in press)
10. Märk, T.D., Scheier, P., Ritter, W., Stephan, K., Stamatovic, A.: Int. J. Mass Spectrom. Ion Proc. **74**, 281 (1986)
11. In particular, it is necessary to demonstrate that the present observations are due to photodissociation of the neutral clusters and not due to the photodissociation of ionized clusters [7, 8]
12. Godfried, H.P., Silvera, I.F.: Phys. Rev. Lett. **48**, 1332 (1982); Phys. Rev. A **27**, 3008, 3019 (1983)
13. Walder, G., Howorka, F., Stamatovic, A., Rozsa, K., Höpfel, R.A., Märk, T.D.: Proc. 12th Intern. Symp. Molecular Beams, Perugia, 1989.

Z. Phys. D – Atoms, Molecules and Clusters 12, 279–281 (1989)

Atoms, Molecules
Zeitschrift für Physik D and Clusters
© Springer-Verlag 1989

Multiply charged cluster ions of Ar, Kr, Xe, N_2, O_2, CO_2, SO_2 and NH_3: Production mechanism, appearance size and appearance energy

T.D. Märk[1], P. Scheier[1], M. Lezius[1], G. Walder[1], and A. Stamatovic[2]

[1] Institut f. Ionenphysik, Leopold Franzens Universität, Technikerstrasse 25, A-6020 Innsbruck, Austria
[2] Department of Physics/Meteorology, PMF Beograd, Yugoslavia

Received 8 July 1988; final version 30 September 1988

Clusters of Ar, Kr, Xe, N_2, O_2, CO_2, SO_2 and NH_3 formed by supersonic nozzle expansion have been studied by electron impact ionization mass spectrometry (up to 15000 amu). Besides mass spectra of singly charged ions showing the characteristic anomalous distributions, we have in particular investigated the properties of multiply charged cluster ions. Critical appearance sizes of doubly and triply charged cluster ions, n_2 and n_3 respectively, found in the present study confirm recent theoretical predictions about n_3/n_2 and their dependence on the properties of the cluster constituents. The appearance energies of multiply charged cluster ions determined are shifted way below the appearance energies of the respective monomer ions. These huge red shifts together with the observed linear threshold laws and large maximum ionization efficiencies indicate that multiply charged cluster ions are produced by sequential single ionization events of one incoming electron at different cluster sites. Furthermore, we have also obtained for the first time clear evidence that (for electron energies above the appearance energy of doubly charged ions) an appreciable amount of singly charged (also fragment) ions is produced via Coulomb explosion of unstable doubly charged ions in the ion source.

PACS: 36.40; 07.75

1. Introduction

A particularly interesting subject in the growing field of cluster science is the investigation of the production mechanism and of the stability of multiply charged cluster cations. Such multiply charged cluster ions X_n^{z+} are usually detected in a mass spectrum only if their size is larger than a characteristic critical appearance size, n_z. In case of van der Waals clusters there exists a large body of reliable determinations of the critical appearance size n_2 for doubly charged ions. Some systematic trends for the dependence of n_2 on the properties of the gases under study have emerged (see the comprehensive reviews given in [1, 2]).

Conversely, for triply (and higher) charged cluster ions there exist less data. Moreover, these data are less accurate due to the fact that critical appearance sizes have been deduced with insufficient mass resolution from apparent steps in the unresolved mass spectra. Recently, we have been able to improve the mass range and resolution of a double focussing sector field mass spectrometer [3] allowing the unambiguous determination of the appearance size of triply charged cluster ions. Moreover, due to the high sensitivity of this instrument it became possible to determine reliable appearance energies and threshold laws of multiply charged ions of van der Waals clusters.

In the present investigation we have determined appearance sizes and appearance energies of doubly and triply charged ions of Ar, Kr, Xe, N_2, O_2, CO_2, SO_2 and NH_3 clusters (see also [4–7]). These values and the shape and magnitude of the ionization cross section curves give valuable insight into the electron impact ionization mechanism of van der Waals clusters. Moreover, a systematic and thorough study of the electron impact ionization cross section curves of singly charged cluster ions (below the appearance size of doubly charged cluster ions) gives for the first time clear evidence that for electron energies above the appearance energy of doubly charged ions an appreciable fraction of singly charged cluster ions is produced via Coulomb explosion of metastable doubly charged ions in the ion source (i.e. at times of $<10^{-6}$ s since production) [6, 7].

2. Experimental

The supersonic beam – electron impact ion source – mass spectrometer system used has been described in detail previously [8]. In short, it consists of a modified three electrode open type ion source and high resolution double focussing sector mass spectrometer (reversed geometry) with increased mass range (up to 15000 amu [3]). A neutral cluster beam is produced by expanding the gas under study at high pressures and low temperatures through a 20 μm nozzle. The ensuing supersonic beam passes a skimmer and is crossed at right angles with an electron beam of variable electron energy (up to 180 eV; 0.2 to 0.5 eV FWHM) and variable electron current (up to 0.5 mA). Cluster ions produced are *(i)* extracted at right angles from the ionization region, *(ii)* accelerated (up to 3 kV), *(iii)* analyzed in a 90° magnetic sector field followed by a 90° electric sector field and *(iv)* postaccelerated and detected with a CuBe electron multiplier.

3. Results and discussion

Mass spectra of Ar, Kr, Xe, N_2, O_2, SO_2 and NH_3 clusters show above certain size limits additional peaks between singly charged parent and fragment ions. Due to the mass-to-charge ratios and the appearance energies deduced (see below) these ion series can be attributed to multiply charged cluster ions. The critical appearance sizes n_2 and n_3 deduced are shown in Fig. 1 together with values for other van der Waals clusters studied previously [10–13].

Several theoretical approaches have been used recently in order to elucidate and predict the occurrence of critical appearance sizes. All of them are based on some kind of liquid drop approximation [10, 11, 14]. A very simple and useful relation for n_2 has been recently derived by Echt [15] using a two term (electric and surface energy) liquid drop model. Following this model, we have plotted here in Fig. 1 the appearance sizes n_2 and n_3 versus $10000/T_c \cdot \sqrt[3]{v}$

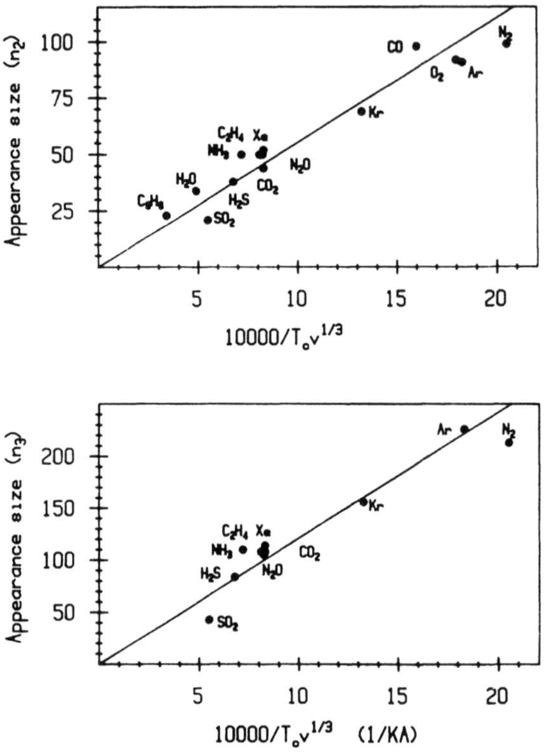

Fig. 1. Appearance size n_2 and n_3 versus one over the critical temperature T_c times the cubic root of the molecular volume v for doubly and triply charged ions of Ar, Kr, Xe, N_2, O_2, CO_2, SO_2 and NH_3 van der Waals clusters, respectively. Values for v and T_c have been taken from [9] and previously measured n_2 and n_3 values for CO_2 also from [10], for Xe, N_2O and H_2S from the measurements of Echt et al. [11], for CO from [12] and for C_6H_6 from [13]. There exist additional data on n_2 not shown here on various hydrocarbons (see [11])

(T_c, critical temperature; v, molecular volume). The available data appear to lie in these plots on a line through the origin. Moreover, Bennemann and coworkers [16] using a simple electronic theory predicted a ratio for the critical appearance size for van der Waals clusters on $n_3/n_2 = 2.3$ (see also a value of 2.25 predicted in [14]). The averaged ratio for all experimental determinations (Fig. 1) with $n_3/n_2 \cong 2.1$ agrees rather well with these predictions.

In order to aid the identification of multiply charged cluster ions in the mass spectra and in order to shed light on the ionization mechanism it is interesting to study the energy dependence of the various cluster ion abundances close to threshold. An overriding conclusion from these studies (for more detail see [4–7]) is that the onsets of doubly and triply charged cluster ions are shifted way below the onsets of the respective monomer ions. These large red shifts cannot be attributed to solvation effects. We therefore interpret the present results, i.e. that the appearance energy of doubly charged van der Waals clusters is about two times the appearance energy of singly charged clusters and the appearance energy of triply charged clusters is about three times the appearance energy of singly charged clusters as evidence for the occurrence of multiple sequential electron collision processes in the ionization event. For instance, doubly charged clusters are produced by two sequential single ionization events of one incoming electron at two different monomers within the cluster. Such a multiple step sequential ionization mechanism also explains the large abundance of the doubly and triply charged cluster ions observed. These large abundances cannot be explained by multiple ionization in a single collision event at one monomer, because cross sections for doubly charged monomer ions are about a factor of 10^2 to 10^3 smaller than those for the respective singly charged monomer ions [17]. Moreover, another argument in favor of this multiple step sequential ionization mechanism is the linear shape of the ionization cross section function (see discussion in [7]).

Work partially supported by the Österreichischer Fonds zur Förderung der Wissenschaftlichen Forschung.

References

1. Echt, O.: In: Electronic and atomic collisions. Gilbody, H.B., Newell, W.R., Read, F.H., Smith, A.C.H. (eds.) Amsterdam: North Holland 1988; Märk, T.D.: In: Book of Invited Papers, 6th Symp Elem. Reak. Nizk. Plazma, Lukac, P. (ed.), p. 142. Jelsava-Hradok CSSR 1986
2. Märk, T.D.: Int. J. Mass Spectrom. Ion Proc. **79**, 1 (1987)
3. Scheier, P., Märk, T.D.: Int. J. Mass Spectrom. Ion Proc. **76**, R11 (1987)
4. Scheier, P., Märk, T.D.: Chem Phys. Lett. **136**, 423 (1987)
5. Scheier, P., Stamatovic, A., Märk, T.D.: J. Chem. Phys. **88**, 4289 (1988)
6. Scheier, P., Stamatovic, A., Märk, T.D.: Chem Phys. Lett. **144**, 119 (1988)
7. Scheier, P., Walder, G., Stamatovic, A., Märk, T.D.: Chem. Phys Lett. **150**, 222 (1989); J. Chem. Phys. **90**, 1288 (1989)
8. Märk, T.D., Scheier, P., Leiter, K., Ritter, W., Stephan, K., Stamatovic, A.: Int. J. Mass Spectrom. Ion Proc. **74**, 281 (1986)
9. Weast, R.C.: Handbook of Chemistry and Physics. Cleveland, OH: Chemical Rubber Company 1974
10. Kreisle, D., Echt, O., Knapp, M., Recknagel, E., Leiter, K., Märk, T.D., Saenz, J.J., Soler, J.M.: Phys. Rev. Lett. **56**, 1551 (1986)
11. Echt, O., Kreisle, D., Recknagel, E., Saenz, J.J., Casero, R., Soler, J.M.: Phys. Rev. A **38**, 3236 (1988)
12. Kandler, O., Leisner, T., Echt, O., Recknagel, E.: Z. Phys. D – Atoms, Molecules and Clusters **10**, 295 (1988)
13. Stace, A.J., Bernhard, D.M., Crooks, J.J., Reid, K.L.: Mol. Phys. **60**, 671 (1987)
14. Casero, R., Saenz, J.J., Soler, J.M.: Phys. Rev. A **37**, 1401 (1988) and references therein
15. Echt, O.: In: Contr. Symp. Atomic Surface Physics, Howorka, F., Lindinger, W., Märk, T.D. (eds.), p. 272. Obertraun, Austria 1986
16. Tomanek, D., Mukherjee, S., Bennemann, K.H.: Phys. Rev. B **28**, 665 (1983)
17. Märk, T.D., Dunn, G.H.: Electron impact ionization. New York Wien: Springer 1985

Z. Phys. D – Atoms, Molecules and Clusters 12, 283–284 (1989)

Atoms, Molecules and Clusters
Zeitschrift für Physik D
© Springer-Verlag 1989

Quantized losses from charged nitrogen clusters

T. Leisner, O. Echt*, O. Kandler, Yan Xue-Jian, and E. Recknagel
Fakultät für Physik Universität Konstanz, D-7750 Konstanz, Federal Republic of Germany

Received 5 July 1988

Metastable decay of $(N_2)_n^+$, formed in a supersonic jet and ionized by electron impact, has been analyzed for $n \leq 50$. The probability for decay of $(N_2)_n^+$ into $(N_2)_{n-x}^+$, plotted versus x, exhibits pronounced oscillations. The "period" of these oscillations increases with increasing precursor size n, but converges to an average value of approximately 4.7 beyond $n=25$.

PACS: 36.40.td; 34.80.G; 35.20.Wg

1. Introduction

It is well-known that van der Waals clusters, ionized by either electron impact or by multi-photon absorption or by electron attachment, are prone to prompt and delayed decomposition [1]. The excess energy which drives these reactions arises from the large difference between vertical and adiabatic ionization potential or, in other words, from substantial structural reorganization in these systems. Energy equilibration, however, proceeds rapidly; most of the excitation energy will be converted into ejection of monomers within less than $\simeq 1\,\mu s$, before analysis of unimolecular reactions can be accomplished. Hence, the probability for metastable loss of more than one monomer, measured in the time regime $t \geq 1\,\mu s$, is, in general, very low.

In this contribution we present measurements on nitrogen clusters of size $n \leq 50$, which are found to decay in a very peculiar manner [2]. We have employed a novel technique in order to determine the probability of all competing reaction channels. Our findings, together with related results obtained by Märk and coworkers [3], strongly support a model previously proposed by Michl and coworkers [4]. Those authors invoked slow, radiationless decay of intramolecular vibrational excitation in order to explain the appearance of so-called "magic losses" from sputtered nitrogen cluster ions.

* Present address: University of Copenhagen, Niels Bohr Institute, Tandem Accelerator Laboratory, DK-4000 Roskilde, Denmark

2. Experiment

Clusters of nitrogen are generated in a supersonic, non-seeded jet. The collimated cluster beam is crossed by a pulsed electron beam; the size-to-charge ratio of the cluster ions is analyzed in a time-of-flight mass spectrometer. Details have been given elsewhere [5].

Analysis of all possible decay channels is accomplished by using two electrostatic mirrors: The first one, kept at a potential U_1, reflects all ions of energy $E_{kin} < e_0 U_1$ towards the ion detector (the drift tube is at ground potential, the ions start at a potential $U_0 \simeq 2\,kV$, and all cluster ions are singly charged [6]). A second mirror, at potential $U_2 (U_2 < U_1)$, is located just in front of this detector. Hence, we combine time-of-flight mass spectrometry with differential energy analysis. In order to be detected, the ions have to fulfill the inequality

$$U_2/U_0 < n_f/n_i < U_1/U_0 \qquad (1)$$

(n_i and n_f are the initial and final cluster size, respectively). In other words, the decay channels of all precursor ions may be analyzed if one records a set of, say, 40 mass spectra with $U_1 - U_2 = U_0/40$, and varying U_2 from 0 to U_0 in increments of $U_0/40$.

3. Results and discussion

The probability for the net reaction[1]

$$(N_2)_n^+ \rightarrow (N_2)_{n-x}^+ + xN_2 \qquad (2)$$

measured on the time scale $1.6\sqrt{n}\ \mu s \leq t \leq 18\sqrt{n}\ \mu s$ with respect to ionization, has been plotted in Fig. 1. Each desorption trace exhibits pronounced oscillations. The "period" ranges from $\Delta x = 2$ to 5, it increases with increasing precursor size. A more refined analysis reveals that Δx changes rapidly from 3.0 to 4.7 over a narrow range of "momentary"[2] cluster size $m = 20$ to 25, but stays constant for larger m [2]. These observations indicate that the excess energy associated with the loss of x monomers is released into the cluster in multiples of a well-defined quantum, and that the binding energy of nitrogen cluster ions has converged rather abruptly to a constant value (presumably the bulk value of condensed nitrogen) at cluster size 25. A quantitative determination of the binding energy would require knowledge not only of the energy quantum (cf. below), but also of the effective cluster temperature [7].

The following model, originally proposed by Michl and coworkers [4], offers a consistent explanation for our findings: If intramolecular vibrations in one or more neutral N_2 molecules within the cluster are excited as a direct or indirect consequence of the ionization event, one will have a source of large ($E_0 = 0.29$ eV [8]), long-lived energy quanta. Once such a localized excitation is transferred into the cluster, a bunch of N_2 molecules will be ejected rapidly; the release of several quanta within the relevant experimental time window causes the oscillatory decay pattern.

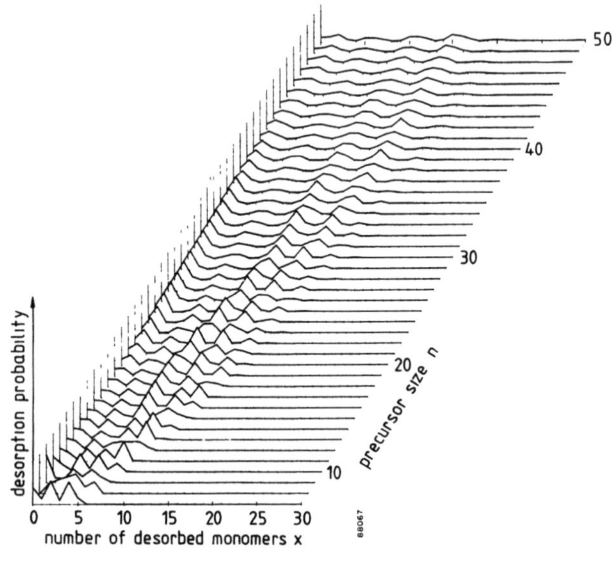

Fig. 1. Probability for metastable decay of $(N_2)_n^+$ into $(N_2)_{n-x}^+$ for precursor size $7 \leq n \leq 50$. The ordinate is marked in increments of 10%

The lifetime of vibrational quanta in isolated N_2 is exceedingly large. Even in a clean matrix, vibrational relaxation proceeds on the time scale of seconds to minutes [9–11]. Explanation of our observations does, indeed, require a strongly enhanced coupling of the intramolecular vibration to the cluster lattice, because our experimental time scale is 10 µs to 100 µs with respect to ionization. It is likely that the molecular ion in the cluster, presumably N_4^+, acts as an efficient mediator between the intra- and intermolecular vibrations of N_2.

We have searched for similar phenomena in other van der Waals clusters. CO appeared to be a promising candidate, because it also features vibrational quanta being considerably larger than the binding energy of the bulk, and a long vibrational lifetime [9, 11, 12]. The decay channels of $(CO)_n^+$, however, did not indicate delayed relaxation of vibrational energy quanta. Other observations pertaining to CO clusters have been published elsewhere [13].

We appreciate stimulating discussions with Profs. S. Bjørnholm, E.E. Ferguson, and T.D. Märk. This work was financially supported by the Deutsche Forschungsgemeinschaft.

References

1. Echt, O.: In: Elemental and molecular clusters. Benedek, G., Martin, T.P., Pacchioni, G., (eds.). p. 263. Springer Series in Materials Science, Vol. 6. Berlin, Heidelberg, New York: Springer 1988
2. Leisner, T., Echt, O., Kandler, O., Xue-Jian, Yan, Recknagel, E.: Chem. Phys. Lett. **148**, 386 (1988)
3. Scheier, P., Märk, T.D.: Chem. Phys. Lett. **148**, 393 (1988); and contribution to this conference
4. Magnera, T.C., David, D.E., Michl, J.: Chem. Phys. Lett. **123**, 327 (1986)
5. Knapp, M., Echt, O., Kreisle, D., Recknagel, E.: J. Phys. Chem. **91**, 2601 (1987)
6. Echt, O., Kreisle, D., Recknagel, E., Saenz, J.J., Casero, R., Soler, J.M.: Phys. Rev. A (1988); Scheier, P., Stamatovic, A., Märk, T.D.: J. Chem. Phys. (1987) (in press)
7. Engelking, P.C.: Chem. Phys. Lett. **139**, 6 (1987)
8. Radzig, A.A., Smirnov, B.M.: Reference data on atoms, molecules and ions. Springer Series in Chemical Physics, Vol. 31. Berlin, Heidelberg, New York: Springer 1985
9. Ewing, G.E.: J. Phys. Chem. **91**, 4662 (1987)
10. Chandler, D.W., Ewing, G.E.: J. Chem. Phys. **73**, 4904 (1980)
11. Legay, F.: Vibrational Relaxation in Matrices, In: Chemical and biochemical applications of lasers. Moore, C.B. (ed.), Vol. 2, p. 43. New York: Academic Press 1977
12. Bahrdt, J., Schwentner, N.: J. Chem. Phys. **88**, 2869 (1988)
13. Kandler, O., Leisner, T., Echt, O., Recknagel, E.: Z. Phys. D Atoms, Molecules and Clusters **10**, 295 (1989)

[1] The identity of the neutral products cannot be determined with our technique; but Märk and coworkers have shown that $(N_2)_n^+$ decays by sequential loss of monomers [3]

[2] See [2] for a definition

Z. Phys. D – Atoms, Molecules and Clusters 12, 285–287 (1989)

Atoms, Molecules and Clusters
Zeitschrift für Physik D
© Springer-Verlag 1989

Observation and investigation of metastable decay series of $(N_2)_n^+$ cluster ions

T.D. Märk and P. Scheier

Institut für Ionenphysik, Leopold Franzens Universität, Technikerstrasse 25, A-6020 Innsbruck, Austria

Received 8 July 1988

N_2 cluster ions are produced by electron impact ionization of a supersonic N_2 cluster beam and analyzed with a double focussing sector field mass spectrometer. It is found that metastable N_2 cluster ions lose more than one N_2 molecule in the µs time regime and decay predominantly via sequential series $(N_2)_n^{+*} \to (N_2)_{n-1}^{+*} \to \cdots \to N_2^+$, evaporating a single monomer in each of these successive decay steps. The metastable decay rates determined in detail for cluster sizes $2 \le n \le 6$ and $n = 20$ lie between 1 and $10^6\,s^{-1}$. These rates *(i)* depend strongly on the time elapsed after ion formation and on the respective parent cluster ion size, and *(ii)* exhibit a quasiperiodic pattern in magnitude.

PACS: 36.40; 34.80.G

1. Introduction

The electron impact ionization of van der Waals clusters has been the subject of considerable recent interest [1, 2]. In particular, the production and stability of rare gas cluster ions has been studied extensively [1]. In contrast, other van der Waals cluster ions have not been studied in great detail and the present study therefore is directed to the stability of N_2 cluster ions. Previous results on the electron impact ionization and mass spectrometry of N_2 clusters are reviewed in [3].

Stephan et al. [4] were the first to report the occurrence of metastable decay reactions involving N_2 clusters, e.g. that of $(N_2)_2^{+*}$ into $N_2^+ + N_2$. In a recent experimental study on the stability of $(N_2)_n^+$ by Michl and coworkers [5] cluster ions were formed by Ar^+ sputtering of frozen N_2. Three types of metastable decay were observed, i.e. *(i)* loss of a single N_2 molecule (attributed to statistical predissociation), *(ii)* nearly total disintegration (attributed to radiationless decay of an electronically excited N_2 state), and *(iii)* losses of very specific (magic) numbers (more than one) of N_2 molecules (attributed to N_2 ($v=1$) vibrational relaxation; see also [6]). Moreover, Echt and coworkers [7] studied the metastable decay of $(N_2)_n^+$ produced by electron impact ionization of a supersonic cluster beam using a TOF reflectron apparatus. They found in accordance with the previous results that $(N_2)_n^+$ ions decay via loss of one and/or more N_2 molecules with certain 'magic' numbers of N_2 being preferentially lost.

The present investigation is a natural extension of these previous studies [4–7] in that we determine the dynamics and kinetics of metastable $(N_2)_n^{+*}$ ion dissociations. We present the first experimental evidence of sequential decay series in metastable molecular cluster ions in the µs time regime [8]. N_2 cluster ions are produced by electron impact ionization of a supersonic N_2 cluster beam and analyzed by the innovative use of a double focussing sector field mass spectrometer, reversed geometry (see similar studies on Ar_n^+ [9]). It is found that metastable N_2 cluster ions lose more than one N_2 molecules in the µs time region and decay predominantly via sequential series $(N_2)_n^{+*} \to (N_2)_{n-1}^{+*} \to \cdots \to N_2^+$, evaporating a single monomer in each of these successive steps with widely differing rates. For instance, $(N_2)_4^{+*}$ decays sequently via $(N_2)_3^+$ and $(N_2)_2^+$ to N_2^+ with the following decay rates in the first field free region $\sim 10^4$, $\sim 10^6$ and $\sim 1560\,s^{-1}$, respectively. These rates depend strongly on the original parent ion size and on the time elapsed since formation of these cluster ions. The latter is in accordance with previous experimental and theoretical results obtained for rare gas cluster ions [1, 10, 11].

2. Experimental technique

The supersonic beam/electron impact ionization/mass spectrometer system used has been described in detail previously [12]. Neutral nitrogen clusters are produced by expanding pure N_2 gas at a constant temperature (typically at appr. $-170°C$) and a constant pressure (typically between 1 and 2 bar) through a 20 μm nozzle. The ensuing supersonic beam passes a skimmer and is crossed 10 cm downstream at right angles by an electron beam. Cluster ions produced are *(i)* extracted at right angles from the ionization region, *(ii)* accelerated by an acceleration voltage, and *(iii)* analyzed in a 90° magnetic and 90° electric sector field.

Essential to the present measurements is the existence *(i)* of a field-free region between the end of the acceleration region and the beginning of the magnetic sector and *(ii)* of a second field-free region between the end of the magnetic sector field and the beginning of the electric sector field. A possible metastable decay of an ion m_1^+ into m_2^+ can be detected either *(i)* in the first field free region by tuning the magnetic sector field to a nominal $m^* = m_2^2/m_1$ and electric sector field to $E^* = m_2 E/m_1$, or *(ii)* in the second field-free region by tuning the magnetic sector field to m_1^+ and the electric sector field to $E^* = m_2 E/m_1$ (with E being the correct sector field to detect m_1^+). This corresponds to the usual operating mode for the detection of metastable decay of polyatomic ions. Here we employed an additional experimental procedure [9]. In this alternative mode of operation we detect a possible metastable decay of m_1^+ into m_2^+ in the first field-free region followed by a metastable decay of this m_2^+ into m_3^+ in the second field-free region. This is accomplished by tuning the magnetic sector field to a nominal mass $m^* = m_2^2/m_1$ and the electric sector field to $E^* = m_3 E/m_1$. In this case only those m_3^+ ions will be detected which are produced by successive decays in the first and second field-free regions, respectively.

Some of these metastable reactions occurring in the first field free region may be contaminated by collision induced dissociations due to the presence of background gas. However, by studying the unimolecular dissociations as a function of pressure in the first field free region, it is possible to distinguish between truly metastable dissociations (whose occurrence is independent from the background pressure) and collision induced dissociations. It was found, however, that the fragment ion fractions measured here are within the experimental error independent of this pressure. This is due to the fact that in presence of a strong metastable decay possible collision induced dissociations have only a negligible influence on the measured fragment ion fractions in this pressure range. Moreover, it is interesting to point out that metastable fractions measured here are in excellent agreement with results obtained by Echt et al. [7] using a different experimental setup.

3. Results and discussion

We have studied metastable decay series originating from nitrogen cluster ions $(N_2)_n^{+*}$ with $n = 3, 4, 5, 6$ and 20 (for $n = 2$ see [4]). Except for $(N_2)_{20}^+$, we have measured the metastable ion fractions of the decay series at two different ion acceleration voltages in the ion source (3 and 1.5 kV, respectively). This allows the determination of the respective reaction rates k (details of this analysis are described in [8] and [9]) at two different times (since ion production) in each field free region. Table 1 gives as an example the obtained results for $(N_2)_3^{+*}$. We have also in turn calculated with help of the derived reaction rates k the corresponding metastable ion fractions. These calculated fractions are in excellent agreement with the experimental ion fractions. In case of the trimer the agreement is especially good. The situation gets worse with the number of decay channels to be resolved.

An overriding conclusion of the present investigation is that all meastable decay reactions (e.g. 3 to 1, 4 to 1, 5 to 1, 6 to 2 and 20 to 5) observed here proceed predominantly via sequential single monomer evaporations in the μs time regime. This constitutes the first experimental evidence for the occurrence of sequential decay series of molecular cluster ions. Similar decay series – involving only two evaporative steps – were

Table 1. Metastable reaction rates k (in s^{-1}) for the sequential decay series $(N_2)_3^+ \xrightarrow{k_{3,2}} (N_2)_2^+ \xrightarrow{k_{2,1}} N_2^+$ as determined in the first field free region, k(I), and second field free region, k(II) for two different ion acceleration voltages U (3 and 1.5 kV), respectively. The corresponding flight times from ionization to the middle of the first field free region and of the second field free region are 5.1 and 13.1 μs in case of $U = 3$ kV and 7.2 and 18.6 μs in case of $U = 1.5$ kV, respectively

Reaction rate	$U = 3$ kV	$U = 1.5$ kV
$k_{3,2}$ (I)	10 800	9 900
$k_{2,1}$ (I)	2 000	1 850
$k_{3,2}$ (II)	3 200	2 400
$k_{2,1}$ (II)	620	590

reported so far only for Ar_3^+ and Ar_7^+ through Ar_{10}^+ [9, 13].

Moreover, the metastable rates obtained here decrease as a function of time elapsed since the formation of the cluster ions. It has to be concluded in accordance with recent conjectures by Klots [10] that $(N_2)_n^{+*}$ parent ions are produced with a broad range of internal energies leading to a broad range of metastable fragmentation rates. This in turn yields the observed time dependence of the measured (apparent) reaction rates. Moreover, not only parent ions are present in the analyzed ion beam with a broad range of internal energies, but this broad energy distribution is passed on to the ensuing daughter ions within a metastable decay series.

Furthermore, individual reaction rates within a sequential decay series vary widely, i.e. small rates are followed by large rates and viceversa. This is in contrast to observations [9] for $Ar_{n=7\text{ to }10}^+$ cluster ions where metastable reaction rates were observed to decrease in a sequential series as a result of the cooling which accompanies each decomposition. This indicates that for the N_2 cluster ions a different metastable decay mechanism may be operative (i.e. similar to the case of Ar_3^{+*} where $k_{3,2} \ll k_{2,1}$ [13]). As a matter of fact it can be seen that the magnitude of an individual reaction rate within a decay series is not depending on the specific cluster ion size of the daughter ion but rather on the size of the parent ion. For instance, if we consider the decay of $(N_2)_4^+$ into $(N_2)_3^+$ and subsequently into $(N_2)_2^+$, it was found that in case of $(N_2)_6^{+*}$ as parent ion $k_{4,3} \ll k_{3,2}$, whereas in case of $(N_2)_5^{+*}$, $k_{4,3} \gg k_{3,2}$ and in case of $(N_2)_4^+$, $k_{4,3} \ll k_{3,2}$. This indicates that the observed metastable rates within a metastable ion series of N_2 cluster ions do not mirror the intrinsic stability of the corresponding (daughter) cluster ions but that the relative magnitude of these rates is depending on the parent ion and a decay mechanism initiated in this ion.

Work partially supported by the Österreichischer Fonds zur Förderung der Wissenschaftlichen Forschung.

References

1. Märk, T.D.: Int. J. Mass Spectrom. Ion Proc. **79**, 1–59 (1987)
2. Sattler, K.: Curr. Top. Mater. Sci. **12**, 1 (1985); Stace, A.J.: Mass Spectrom. **9**, 96 (1987)
3. Scheier, P., Stamatovic, A., Märk, T.D.: J. Chem. Phys. **88**, 4289 (1988)
4. Stephan, K., Märk, T.D., Castleman, Jr., A.W.: J. Chem. Phys. **78**, 2953 (1983)
5. Magnera, T.F., David, D.E., Tian, R., Stulik, D., Michl, J.: J. Am. Chem. Soc. **106**, 5040 (1984); Magnera, T.F., David, D.E., Michl, J.: Chem. Phys. Lett. **123**, 327 (1986)
6. Engelking, P.C.: Chem. Phys. Lett. **139**, 6 (1987)
7. Leisner, T., Echt, O., Kandler, O., Kreisle, D., Miehle, M.: Jahresbericht 1987; Nukleare Festkörperphysik. Konstanz, p. 107; Leisner, T., Echt, O., Kandler, O., Miehle, W., Xue-Jian, Y., Recknagel, E.: Chem. Phys. Lett. **148**, 386 (1988)
8. Scheier, P., Märk, T.D.: Chem. Phys. Lett. **148**, 393 (1988)
9. Scheier, P., Märk, T.D.: Phys. Rev. Lett. **59**, 1813 (1987)
10. Klots, C.E.: J. Chem. Phys. **83**, 5854 (1985)
11. Märk, T.D., Scheier, P.: J. Chem. Phys. **87**, 1456 (1987); Chem. Phys. Lett. **137**, 245 (1987)
12. Märk, T.D., Scheier, P., Leiter, K., Ritter, W., Stephan, K., Stamatovic, A.: Int. J. Mass Spectrom. Ion Proc. **74**, 281 (1986)
13. Scheier, P., Stamatovic, A., Märk, T.D.: J. Chem. Phys. **89**, 295 (1988)

Z. Phys. D – Atoms, Molecules and Clusters 12, 289–290 (1989)

Atoms, Molecules
Zeitschrift für Physik D and Clusters
© Springer-Verlag 1989

Electron attachment to clusters composed of closed shell, hydrogen containing molecules

H. Haberland, C. Ludewigt, and T. Richter
Fakultät für Physik der Universität Freiburg, Hermann-Herder-Strasse 3, D-7800 Freiburg i.Br., Federal Republic of Germany

Received 6 July 1988

Electron attachment to clusters of HCl, H_2O, D_2O, $C_2H_4(OH)_2$, NH_3, and ND_3 is compared. The clusters are produced in a seeded supersonic beam traversing a sustained gas discharge. For the first four molecules $N_{min} = 2$ is the lowest number of molecules necessary to support a bound negative ion state. A much larger N_{min} and an isotope effect is observed for ammonia, $N_{min}(NH_3) = 35$, and $N_{min}(ND_3) = 41$.

PACS: 36.40.+d; 82.20.TR; 87.55.−d

Introduction

Closed shell atoms and molecules have no bound negative ion states in general, whereas the solid or liquid of the same species do often have a positive electron affinity. A transition must therefore occur as a function of cluster size from an unbound to a bound level for the extra electron. A series of experiments intended to elucidate this transition has been performed in this laboratory [1–5] and elsewhere [6, 7]. This paper gives a short summary of the results obtained and reports new data on HCl and ND_3.

The interaction of a slow electron with closed shell atoms or molecules often leads to resonances at or near zero energy. This resonance must lower its energy as a function of cluster size, and at some N_{min} it becomes a bound negative ion state. Earlier measurements have shown, that $N_{min} = 2$ for H_2O, D_2O, $C_2H_4(OH)_2$. For ammonia an unexpected large N_{min} of 35 was observed [2–5]. This large difference in the minimal number of molecules necessary to bind an extra electron, is not really understood today, although some progress has been made [8–10]. In performing experiments with other hydrogen containing, closed shell clusters, it was hoped that light could be shed on this problem.

Experiment

The experimental setup is similar to the one used earlier to generate negatively charged clusters [2–5]. A mixture of 1 to 5% of the molecule in question diluted in helium or argon is expanded from a sonic nozzle into vacuum, as shown in Fig. 1. Inlet pressures of 15 to 40 bars, temperatures of 250 to 400 K, and nozzle diameters of 50 to 80 μm are typical. Between nozzle and skimmer a heated filament is used to generate electrons in order to sustain an electric glow discharge in front of the nozzle. Typical discharge parameters are: $U = 30$ V, $I = 30$ mA. Due to the electron emission from the heated filament the electron temperature and the applied voltages are much lower than for an unsustained discharge. The low electron temperature enhances negative cluster ion formation, and the low voltage reduces sputter damage on nozzle and skimmer. The electromagnet seen in Fig. 1 collimates the slow electrons, but more important, prevents their escape through the skimmer, thus reducing the background. The skimmer is graphite coated, and normally kept within one Volt of the grounded nozzle. A simple ion optics is used to form a beam, which is mass analysed using a quadrupole mass filter. Individual ions are counted.

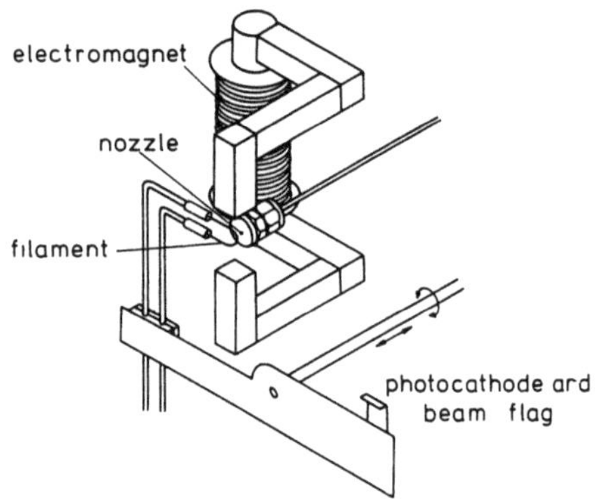

Fig. 1. Source for negatively charged clusters. A mixture of He or Ar with a small amount of the molecule to be clustered goes through a 3 mm stainless steel pipe, closed at the end by a pierced (50 to 80 µm) "Swagelock" end cap. The electron emission from the heated filament sustains a glow gas discharge in front of the nozzle. An electromagnet enhances the electron density, and more important, impedes electrons from going through the skimmer into the detection chamber, thus reducing the background significantly

Table 1. Minimal number of molecules in some hydrogen bonded clusters, necessary to support an extra bound electronic state. See text for discussion

Molecule	N_{min}
HCl	2
H_2O	2
D_2O	2
$C_2H_4(OH)_2$	2
NH_3	35
ND_3	41

Discussion

We envision the negative cluster ion formation to proceed as follows. The slow electrons from the gas discharge attach to the growing clusters, which are "hot" due to the heat of condensation liberated in the clustering process. The cluster ions cool by evaporation of monomers and rare gas collisions.

As will be discussed elsewhere in more detail, N_{min} (HCl) = 2 and $N_{min}(ND_3) = 41$ is observed on the time scale of this experiment (10^{-4} s). This places HCl in the same category as H_2O and D_2O as can be seen from Table 1. No published calculations exist for $(HCl)_2^-$. For the kindred case of $(H_2O)_2^-$, calculations [9, 10] are in reasonable agreement with field [3] and photodetachment [7] experiments. The existence of the $(HCl)_2^-$ has been inferred earlier from reactions in γ-irradiated high pressure HCl [11]. The large value of N_{min} and its isotope effect for NH_3 and ND_3 is not understood in detail. An attempt to apply the theory of Stampfli and Bennemann [8], which correctly predicts N_{min} for NH_3, to H_2O and HCl gave N_{min} values above 20, indicating that this theory is not applicable in these cases [12].

The temperature of the clusters is difficult to assess. If the structure and binding of the observed $Ar_n(HCl)_2^-$, $n = 1$ or 2, was known, an estimate of the upper limit of the temperature would be possible. No $He(HCl)_2^-$ was ever observed. Whether this is due to a temperature effect or, as postulated for $(H_2O)_2^-$, to the repulsive interaction with the delocalised extra electron [3], remains to be determined.

An experimental reason for the isotope effect in ammonia might be a different temperature of NH_3 and ND_3 clusters. In spite of heavy seeding no argon attached to these large cluster ions was ever observed. A temperature effect on N_{min} for ammonia can therefore not be excluded, although it is considered to be unlikely. The only alternative seems to be breakdown of the Born-Oppenheimer approximation, which is equally difficult to accept. More experimental and theoretical work is needed to clarify this situation. A more detailed account of this work is given in [13].

This work was supported by the Deutsche Forschungsgemeinshaft.

References

1. Armbruster, M., Haberland, H., Schindler, H.G.: Phys. Rev. Lett. **47**, 323 (1981)
2. Haberland, H., Ludewigt, C., Schindler, H.G., Worsnop, D.R.: Surf. Sci. **156**, 157 (1985)
3. Haberland, H., Ludewigt, C., Schindler, H.G., Worsnop, D.R.: Phys. Rev. **A36**, 967 (1987)
4. Haberland, H.: In: Fundamental processes of atomic dynamics, Nato ASI Series, Briggs, J.S., Kleinpoppen, H., Lutz, H.O. (eds.). New York: Plenum 1988
5. Haberland, H.: In: Large finite systems. Jortner, J. Pullman, A., Pullman, B. (eds.) Dordrecht: D. Reidel 1987
6. Kondow, T.: J. Phys. Chem. **91**, 1307 (1987)
7. Lee, G.H., Eaton, J., Ludewigt, C., Haberland, H., Bowen, K.H.: J. Chem. Phys. (to be published)
8. Stampfli, P., Bennemann, K.H.: Phys. Rev. Lett. **58**, 2635 (1987)
9. Barnett, R.N., Landmann, U., Cleveland, C.L., Jortner, J.: Phys. Rev. Lett. **56**, 2635 (1987)
10. Wallqvist, A., Thirumalai, D., Berne, B.J.: J. Chem. Phys. **85**, 15883 (1986)
11. Nagra, S.S., Armstrong, D.A.: Can. J. Chem. **54**, 3580 (1976)
12. Richter, Th.: Diplomarbeit Freiburg 1987 (unpublished)
13. Haberland, H.: In: 107. Enrico Fermi School, *The Chemical Physics of Atomic and Molecular Clusters*, Varenna 1988 (accepted for publication)

Z. Phys. D – Atoms, Molecules and Clusters 12, 291–292 (1989)

Atoms, Molecules
Zeitschrift für Physik D and Clusters
© Springer-Verlag 1989

A mechanism of electron attachment to small clusters

T. Kondow, T. Nagata, and K. Kuchitsu*
Department of Chemistry, Faculty of Science, The University of Tokyo, Bunkyo-ku, Tokyo 113, Japan

Received 5 July 1988; final version 30 September 1988

Van der Waals clusters of various molecules were collisionally ionized by high-Rydberg rare gas atoms and slow electrons. Negative cluster ions thus produced were detected by mass spectroscopy. The ionization mechanism were investigated by measurements of the size- and the energy-dependences of the electron attachment cross sections.

PACS: 36.40; 34.50.H; 34.80.D

1. Introduction

When a high-Rydberg rare gas atom, Rg**, or a very slow electron is admitted to a cluster having a positive vertical electron affinity, negative cluster ions are produced as a result of electron transfer from Rg** or attachment of the slow electron. The formation of the negative cluster ions can be explained in general by the following electron attachment and relaxation processes [1]:

$$(M)_m \xrightarrow{Rg^{**}/e} (M)_m^{-*}, \quad (1)$$

$$(M)_m^{-*} \longrightarrow (M)_n^- + (m-n)M, \quad (2)$$

$$\longrightarrow \text{intracluster chemical reactions}, \quad (3)$$

where $(M)_m$ represents a cluster of molecule M with its size, m. The experimental features of the mass spectra evidently depend on the properties of the component molecule, such as the electron affinity, the dipole moment, and the polarizability [1]. This fact indicates that the affinity states and the electron-vibration coupling in the cluster play an important role in the electron attachment process (process (1)). In the present study, the size- and the energy-dependences of the electron attachment cross sections for several van der Waals clusters were obtained from the analysis of the mass spectra of the negative cluster ions produced. Vibrational modes coupled with the incoming electron can be identified with the aid of a theory given by Tsukada et al. [2].

2. Experimental

The experimental details are almost identical with those reported previously [3], except for the location of a triple-grid ionization source where the collision occurs. The source was placed just after the collimation aperture of the beam, and cluster ions formed in the source were focused and steered by a lens system onto the entrance aperture of the quadrupole mass spectrometer. The signal-to-noise ratio was improved by about one order of magnitude.

In addition to the ion source of a gas-cell type, a crossed-beam negative ion source was designed and constructed. This ion source facilitates the use of a state-selected Rg** beam. The design of the ion source is shown in Fig. 1. A beam of Rg** was produced by electron bombardment of 50 eV on a pulsed beam of Rg atoms with a 500 μs duration at a stagnation pressure of 1–2 atm. A series of grids were mounted in front of the electron gun in order to eliminate charged particles, especially stray electrons. The beam of Rg** was allowed to collide orthogonally with a neutral cluster beam at the center of the acceleration grids of a time-of-flight mass spectrometer and negative ions

* Present address: The Technological University of Nagaoka, Nagaoka, Niigata 940–21, Japan

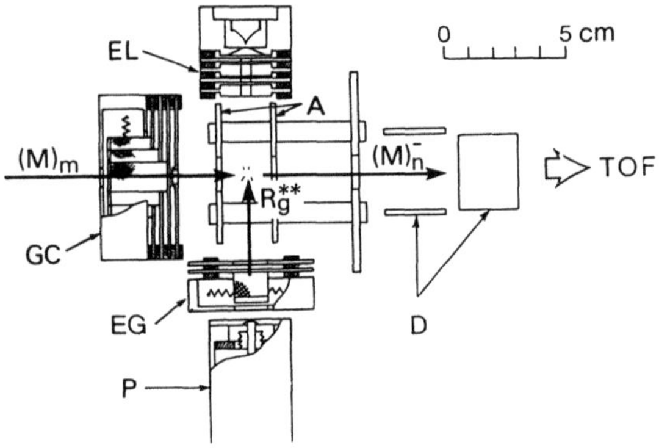

Fig. 1. A crossed-beam negative ion source. A: acceleration grids; D: deflectors; EG: concentric electron gun; EL: electron gun for production of positive ions; GC: triple-grid ion source of gas-cell type; P: pulsed nozzle

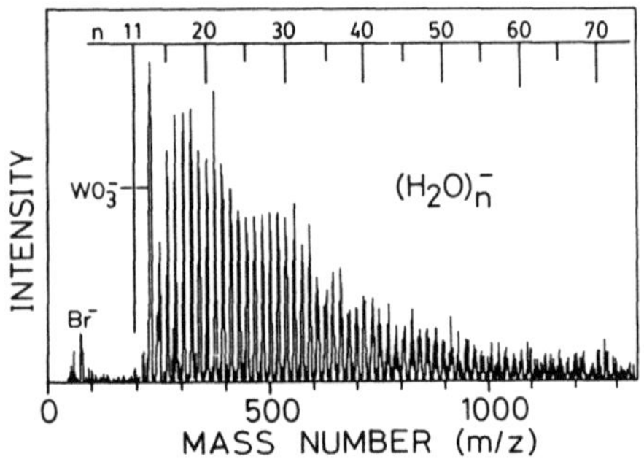

Fig. 2. Mass spectrum of $(H_2O)_n^-$ produced from $(H_2O)_m$ in collision with Kr**. The threshold size is 11

were produced. The ions were accelerated up to 1.75 keV in the direction parallel to the incident neutral beam with a delay of several microseconds after the formation of the ions. After traveling a flight path of ≈ 175 cm, the ions were post-accelerated up to 20 keV and hit on a conversion dynode. Secondary positive ions from the dynode were detected by an electron multiplier. A finite spatial distribution of the ions at the collision region limited a mass-resolution less than 100.

3. Results and discussion

In the formation of negative cluster ions, $(NH_3)_m$, $(H_2O)_m$, $(CH_3CN)_m$ etc., no extensive evaporation appears to occur as estimated from the energetics of the ionization processes, if the excess energy is distributed statistically among the internal degrees of freedom of the product ions; the excess energy is generated by the stabilization of the cluster systems after the electron attachment.

The cross section for the formation of $(NH_3)_n^-$ is much larger than those given by electron impact (2.5 and 5 eV). Since the cluster ion, $(NH_3)_n^-$, starts at $n = 37$, the vertical electron affinity of $(NH_3)_m$ turns to be positive at this threshold size. On the other hand, the vertical electron affinity of $(ND_3)_m$ is positive for m larger than 45 because of the onset of $(ND_3)_n^-$ at $n = 43$. Similar results were obtained for the other clusters. The threshold size and the shape of the size-distribution for $(NH_3)_n^-$ differ greatly from those for $(ND_3)_n^-$. These differences originate from a marked isotope effect in the cross section for the electron attachment to the cluster systems. An analysis of the isotope effect based on the theory of Tsukada et al. [2] reveals that the incoming electron is coupled with the motion of the hydrogen atom associated with hydrogen bonds and is captured in the clusters. A similar isotope effect has been reported by Haberland et al. [4], where their threshold sizes are different from the present results.

Figure 2 shows a typical mass spectrum of $(H_2O)_n^-$ produced from $(H_2O)_m$ in collision with Kr**. No ion other than $(H_2O)_n^-$ is observed. As shown in Fig. 2, the threshold size for H_2O is 11; in the case of D_2O, the threshold size is 12. It was found that the features of the spectrum of H_2O were different from those for D_2O in a higher mass range. This result indicates that there exists an isotope effect in the electron attachment cross sections. Furthermore, several intensity anomalies were observed in the mass spectra, especially prominent in the spectrum for D_2O; the intensities at $n = 28$, 52, 54, 56 and 58 were weak while those at $n = 51$, 53, 55, 57 and 59 were strong. These intensity anomalies, i.e. magic numbers, can be related to the stability of $(H_2O)_m$. The same threshold sizes and intensity anomalies are observed in the mass spectra obtained by electron impact on water clusters [5].

References

1. Kondow, T.: J. Phys. Chem. **91**, 1307 (1986)
2. Tsukada, M., Shima, N., Tsuneyuki, S., Kageshima, H., Kondow, T.: J. Chem. Phys. **87**, 3927 (1987)
3. Kondow, T., Kuchitsu, K.: Mass Spectrosc. **34**, 339 (1986)
4. Haberland, H., Ludewigt, C., Richter, T.: Z. Phys. D–Atoms, Molecules and Clusters **12** (1989) (this issue)
5. Knapp, M., Echt, O., Kreisle, D., Recknagel, E.: J. Phys. Chem. **91**, 2601 (1987)

Z. Phys. D – Atoms, Molecules and Clusters 12, 293–295 (1989)

Atoms, Molecules
Zeitschrift für Physik D and Clusters
© Springer-Verlag 1989

Electron bombardment fragmentation of size selected molecular clusters

U. Buck, J. Kesper, Ch. Lauenstein, M. Tolle, and M. Winter
Max-Planck-Institut für Strömungsforschung, D-3400 Göttingen, Federal Republic of Germany

Received 5 July 1988

$(CO_2)_n$, $(NO)_n$ and $(NH_3)_n$ clusters are generated in a supersonic molecular beam and size selected by scattering from an He beam. By measurements of angular dependent mass spectra, TOF distributions and the angular dependence of the scattered signal quantitative information on the fragmentation probability by electron impact is derived. The van der Waals systems $(CO_2)_n$ and $(NO)_n$ appear only at masses which are simply multiples of the monomer mass. The preferred cluster ion is the monomer ion for all investigated cluster sizes with $n = 2$ to 4. The fragment pattern for the quasi-hydrogen bonded $(NH_3)_n$-cluster shows, beside a large number of fragment masses, a preference for protonated ions. The results are explained in terms of simple models based on the structural change from the neutral to the ionized configuration and the fragmentation pattern of the monomer followed by ion-molecule reactions.

PACS: 36.40; 34.40; 34.80.G

1. Introduction

During the last years great interest has been devoted to the study of weakly bound neutral complexes. Unfortunately, the severe fragmentation following the most ionisation processes can influence the correlation between neutral and ionic cluster intensities dramatically, resulting in large uncertainties of the derived data for neutral species. In some experiments it was tried to derive information about the fragmentation pattern from studies of the time evolution of the cluster ions after the ionization process [1–5]. Theoretical studies [6, 7], however, show that the evaporation process happens in the nano-second scale. One possible method is to utilize the different kinematical behavior of clusters of different size in a scattering process [8, 9]. Using this method in the present contribution, fragmentation analysis of small van der Waals bond clusters $(NO)_n$, $(CO_2)_n$, and of the quasi-hydrogen bonded $(NH_3)_n$-cluster is carried out.

2. Experimental

As described in earlier papers [8, 9] the scattering analysis of cluster beams enables us to find the correlation between detected cluster ions and their neutral parents independent of the cluster size distribution and the fragmentation in the ion source. The method is based on the kinematically different behaviour of different cluster sizes in the scattering process with light He-atoms. The angular area in which different clusters are scattered decreases with increasing cluster size.

The two used crossed beam machines have been described elsewhere [10, 11, 12]. Both machines work on the same principle. Briefly, in the scattering chamber the two beams, which are generated as nozzle beams in two differentially pumped chambers, intersect each other at an angle of 90°. To allow angular resolved measurements, the detection system consisting of electron bombardment ionizer, quadrupole, and particle multiplier rotates with respect to the beams around the scattering center. The clusters are generated by using the seeded beam technique, using He or Ne as the carrier gas. The target beam was in all cases a pure He beam. The energy of the ionizing electrons is 70 eV and the transmission of the quadrupole is carefully adjusted to the known fragmentation of CF_2ClBr.

3. Results and discussion

A detailed derivation of the formula to calculate the fragmentation from the measured TOF-distributions and angular dependence is given in previous papers [13]. The fragmentation of one cluster size n detected at the ion mass k is given by

$$f_{nk} = X_{nk} S_k / \sum_k X_{nk} S_k$$

where the X_{nk} is the relative contribution of the cluster size n in a TOF-spectrum at mass k and S_k is the total scattered intensity at one mass k. The results for the CO_2 and NO clusters are shown in Table 1. Both systems show for all examined cluster sizes a strong preference of the monomer ion as the product of the ionisation. Only small contributions are measured at the mass of the dimer. Only for CO_2-clusters small probabilities appear for detection of the cluster at the parent mass. The NO-Clusters with $n \leq 4$ appear only on the monomer and dimer mass. The explanation which stands behind these results is the change in geometry and thus energetics [14] when going from the neutral cluster in a Franck-Condon like transition to the ionized cluster. After vertical ionisation, the charge preferably localizes in the dimer and subsequent energy transfer from the highly vibrationally excited dimer to the rest of the cluster leads to fragmentation. In the cases CO_2 and NO, additional fragmentation by evaporation of another monomer occurs. For the NH_3-clusters, the results are shown in Fig. 1. The dimer ion at mass $m = 18$ amu (NH_4^+) is the dominating fragment. But there are some more differences compared to CO_2 and NO. First the fragment

Table 1. Fragmentation probabilities of electron impact ionisation f_{nk} for $(CO_2)_n$ and $(NO)_n$-clusters

	CO_2			NO	
n	f_{n1}	f_{n2}	f_{n3}	f_{n1}	f_{n2}
2	0.91	0.09	—	0.92	0.08
3	0.86	0.13	0.01	0.94	0.06
4	0.85	0.14	0.01[a]	0.97	0.03

[a] all $f_{4,k}$ are more qualitative, $f_{4,4} < 0.01$

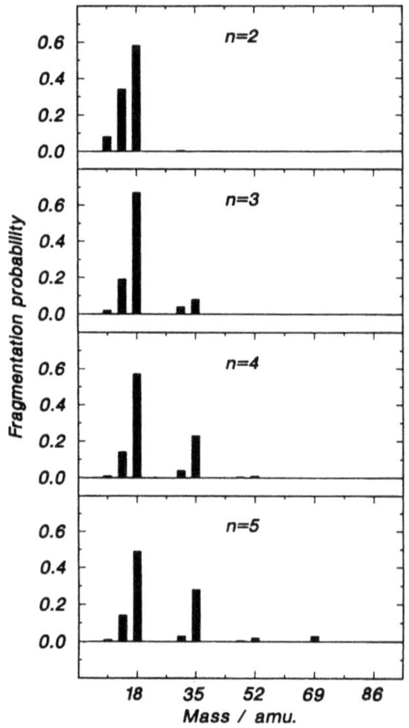

Fig. 1. Fragmentation probabilities for $(NH_3)_n$-clusters up to $n = 5$ versus detection mass for an electron energy of 70 eV

masses are not only simply multiples of the monomer mass. The favored channels are the protonated ions which can be described by the formula $(NH_3)_{n-x} H^+$ where x is between 1 and $n-2$. Additional probabilities exist for the appearance of $(NH_3)_{n-x}(NH_2)^+$. The preference of the protonated channel is in good agreement with recent ab initio calculations [15], which favor the NH_4^+ channel for high excess energies. The appearance of ions containing NH_2^+ and NH_3^+ is well explainable with fast exothermic ion molecule reactions following the initial ionisation and fragmentation of the monomer. Various reported data for the monomer fragmentation [16–18] give a strong fragmentation to NH_2^+ and weak to NH^+ for the monomer. From the ratio of the monomer fragmentation we would expect a ratio for NH_4^+/NH_3^+ of 1.25 and NH_4^+/NH_2^+ of about 10 for the dimer, which is in agreement with the experimental results 1.7 and 7.3, respectively. For increasing cluster size these channels seem to become less important, the probability for the appearance of NH_2^+ as a fragment from a cluster with $n = 5$ is only $f_{5,16} = 0.01$. Very recent work of Garvey and Bernstein [19] on ammonia clusters claimed the presence of ions containing NH_5^+. Carefully measured mass spectra which contain clusters up to $n = 10$, greater clusters being excluded by the scattering process, had also contributions of complexes which seem to contain NH_5^+. Tests with ND_3 showed, however, no complexes containing ND_5^+. So we consider that water impurities, which are probably contained in the steel bottle, form these peaks.

References

1. Futrell, J.H., Stephan, K., Märk, T.P.: J. Chem. Phys. **76**, 5893 (1982)
2. Stephan, K., Märk, T.P.: Int. J. Mass. Spectrom. Ion Phys. **47**, 195 (1983)
3. Echt, O., Kreisle, D., Knapp, M., Recknagel, E.: Chem. Phys. Lett. **108**, 401 (1984)
4. Echt, O., Dao, P.D., Morgan, S., Castleman, A.W.: J. Chem. Phys. **82**, 4076 (1985)
5. Kamke, W., Kamke, B., Kiefl, H.U., Hertel, I.V.: J. Chem. Phys. **84**, 1325 (1986)
6. Saenz, J.J., Soler, J.M., Garcia, N.: Surf. Sci. **156**, 121 (1985)
7. Polymeropoulos, E.E., Brickmann, J.: Surf. Sci. **156**, 563 (1985)
8. Buck, U., Meyer, H.: Phys. Rev. Lett. **86**, 439 (1984)
9. Buck, U., Meyer, H.: J. Chem. Phys. **84**, 4854 (1986)
10. Buck, U., Huisken, F., Schleusener, J., Schaefer, J.: J. Chem. Phys. **72**, 1512 (1980)
11. Andres, J., Buck, U., Huisken, F., Schleusener, J., Torello, F.: J. Chem. Phys. **73**, 5620 (1980)
12. Buck, U., Hoffmann, G., Kesper, J., Otten, D., Winter, M.: Chem. Phys. **126**, 159 (1988)
13. Buck, U.: J. Phys. Chem. **92**, 1023 (1988)
14. Haberland, H.: Surf. Sci. **156**, 305 (1985); In: Physics and chemistry of small clusters. Jena, P., Rao, B.K., Khanna, S.N. (eds.) p. 667. New York: Plenum 1987
15. Tomoda, S., Kimura, K.: Chem. Phys. Lett. **121**, 159 (1985)
16. Crowe, A., McConbey, J.W.: Int. J. Mass Spectrom. Ion Phys. **24**, 181 (1977)
17. Märk, T.D., Egger, F., Cheret, M.: J. Chem. Phys. **67**, 3795 (1977)
18. Bederski, K., Wojcik, L., Adamczyk, B.: Int. J. Mass Spectrom. Ion Phys. **35**, 171 (1980).
19. Garvey, J.F., Bernstein, R.B.: Chem. Phys. Lett. **143**, 13 (1988)

Z. Phys. D – Atoms, Molecules and Clusters 12, 297–301 (1989)

Atoms, Molecules
Zeitschrift für Physik D and Clusters
© Springer-Verlag 1989

Kinetic energy release of clusters after multiphoton ionisation with time of flight and electrostatic analyser techniques

J.D. Coffman, M. Gallmann, and I.V. Hertel
Fakultät für Physik der Albert Ludwigs Universität Freiburg, Hermann-Herder-Strasse 3, D-7800 Freiburg, Federal Republic of Germany

Received 5 July 1988

A new technique is presented which allows direct observation of initial kinetic energies in multiphoton ionisation–fragmentation processes of molecules and clusters and provides an unambiguous determination of metastable decay channels. Results are presented for the unimolecular loss of a monomer from clusters $(C_6H_6)_8^+$ to $(C_6H_6)_{12}^+$ and for the reaction $C_6H_6^+ \rightarrow C_4H_4^+ + C_2H_2$. We also observe a significant amount of probably collision induced fragmentation processes $(C_6H_6)_n^+ \rightarrow (C_6H_6)_{n-x}^+ + (C_6H_6)_x$ with x much larger than 1.

PACS: 36.40; 32.80.K

1. Introduction

Resonant enhanced multiphoton ionisation (REMPI) provides a high degree of state selectivity in molecular and cluster mass spectrometry. Even so the mass spectra obtained via multiphoton ionisation, usually in conjunction with time of flight (TOF) detection, are often very confused by the large amount of fragmentation. The new technique which we introduce here uses kinetic energy analysis of the TOF mass spectra and thus allows one to elucidate fragmentation pathways (at least for ions with fragmentation lifetimes in the μs to 50 μs time domain) by determination of the energy loss due to fragmentation in the field free drift region. The second and main advantage of this technique is the ability to measure energy broadening in metastable mass peaks due to decomposition in flight [1]. The fragmentation energy is amplified as a consequence of a transformation from a center of mass to a lab system and allows thus an easy and direct observation of initial kinetic energies while the REMPI ion source can produce ions with very well defined excess energies and lifetimes [2]. In this paper we demonstrate the complete isolation of prompt and metastable benzene clusters by an energy window and present preliminary measurements of the average kinetic energy release for the unimolecular decomposition of the benzene monomer and benzene clusters.

2. Experimental

The traditional approach to metastable mass spectroscopy for many years has been to use an electron impact ionisation source followed by a field free drift region, a magnetic sector mass analyser, and finally an electrostatic energy analyser [3]–[5]. Our approach (Fig. 1) employs a REMPI ion source with a Wiley, McLaren [6] extraction field, a field free drift zone and a double hemispherical electrostatic analyser. We have the option of measuring a normal TOF through an alignment hole or an energy resolved TOF through the hemispheres. Stable ions pass through the condensers with the transmission energy U_t, typically — 100 eV in our set up, giving an energy resolution of ~ 1.2 eV. Ions with excess internal energies above various dissociation thresholds will undergo unimolecular decay and thus lose mass according to $m_0^+ \rightarrow m^+ + \delta m$. If a parent ion of mass m_0^+ decays rapidly in the neighborhood of the laser focus its' daughter ion of mass m^+ will be detected at the same energy as the stable, unfragmented ions (prompt TOF). If on the other hand

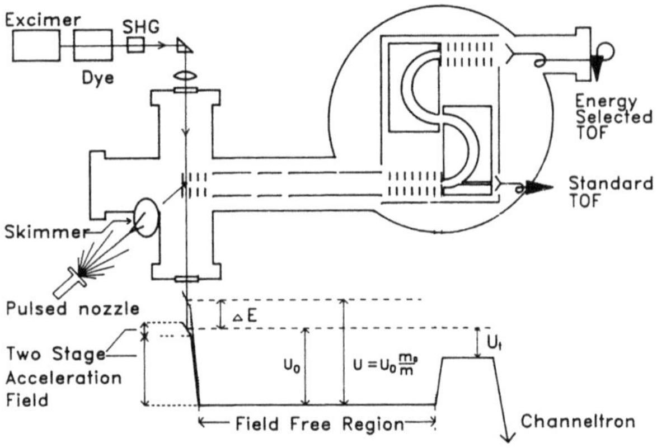

Fig. 1. Experimental setup for kinetic energy analysis of TOF spectra. A pulsed supersonic beam passes through a 2 mm skimmer and intersects a focussed ultraviolet laser beam at right angles. Stable ions produced via 1C-REMPI are accelerated to the energy U_0 of the field free region and pass through the energy analyser which is set to a fixed transmission energy. Ions decaying in the field free region lose mass according to $m_0^+ \rightarrow m^+ + \delta m$ and suffer an energy loss equal to $(\delta m/m_0)\, U_0$. These metastable ions are detected by raising the potential at the laser focus to $(m_0/m)\, U_0$

the parent ion decays in the field free region the daughter ion produced loses kinetic energy equivalent to $U_{loss} = U_0 \cdot \delta m/m_0$ where U_0 is the accelerating potential between the laser focus and the field free zone. To detect these daughter ions U_0 is raised to $U^\dagger = U_0 \cdot m_0/m$. By increasing the extraction and acceleration fields proportionally the Wiley–McLaren spatial focussing condition and the flight times of the ions are held constant, i.e. the flight times for a specific ion mass always the same regardless of being promptly or metastably formed and independent of the mass loss encountered. Ion kinetic energy scans (IKES) give us two valuable pieces of information. (I) The energy loss between the prompt and metastable peak gives us the mass loss in the drift region and (II) the broadened metastable peak allows us to determine the average amount of kinetic energy ε_f released [1]:

$$\varepsilon_f = \frac{(\delta E)^2 m_0^2}{16\, m\, \delta m (U^\dagger)^2} \tag{1}$$

where δE is the effective broadening of the metastable peak in the energy loss scan

$$(\delta E)^2 = (\delta U^\dagger)^2 - (m_0/m)^2 (\delta U_0)^2 \tag{2}$$

Equation (2) accounts for instrumental broadening of energy peaks, δU_0 describing the width of the unfragmented prompt peak mainly due to the chosen energy resolution setting.

The benzene clusters are produced by a pulsed General Valve nozzle with a diameter of 0.8 mm. Argon gas at pressures ranging from 2 to 8 bar has been employed as carrier gas. The cooled gas beam passes through a skimmer of diameter 2 mm and intersects a focussed UV pulsed laser beam. The present experiment has been performed with a single laser frequency (1C-REMPI) set to the much studied vibronic transition $^1B_{2u}$, 6_0^1 at $\lambda = 259$ nm. At this wavelength we find strong metastable ion peaks in the benzene fragmentation mass spectra in good agreement with Boesl et al. [2, 7]. For metastable cluster ion production the laser is tuned off resonance to $\lambda = 259.2$ nm which reduces the monomer ionization rate and thus helps to eliminate unwanted space charge effects within the laser focus. Typical intensities are 200 μJ/pulse focused onto 100 to 500 μm. The laser is a Lambda Physics Excimer pumped dye laser with either a KDP or a BBO doubling crystal.

3. Results

Figure 2 shows the measured TOF spectra as a function of the acceleration energy close to U_0, i.e. for small energy losses. We see immediately the large difference between the REMPI energy loss spectra of benzene molecules and clusters. The benzene dimer exhibits no sign of single H-loss whereas of the monomer signal roughly 50% are found in the $C_6H_6^+$ prompt- and 50% in the $C_6H_5^+$ metastable-channel. The metastable $C_6H_5^+$ peak is slightly broader than the parent monomer or dimer peaks. This is to be expected if one studies (1). It represents a kinetic energy amplification which is mass dependent. A fifty percent mass loss brings the maximum amplification. Electron impact ionisation of this process points to very small kinetic energy releases for H-losses in general. The mechanism here being that of simple bond breaking with little or no reverse activation energy involved.

The high selectivity of our experiment for discriminating different fragmentation channels is demonstrated in Fig. 3. The TOF spectrum (a) shows the promptly fragmented and stable ions with all metastables filtered out. The most reproducible of the small nonintegral cluster peaks beginning above $n = 3$ may be assigned to $Ar\text{–}(C_6H_6)_n^+$ mixed clusters. Metastable loss of a benzene (Bz) monomer from these $Ar\text{–}Bz_n^+$ mixed clusters has also been observed (not shown in Fig. 3). The other small non integral cluster ion peaks could be due to internal fragmentation of one of the molecules within the cluster. However, mixed clusters due to trace impurities could also be responsible for these peaks. Figure 3(b) shows a completely isolated metastable transition from $n = 8$ to $n = 7$. By comparison with Fig. 3(a) we can see that the fraction of metastable to prompt $n = 7$ is roughly 40%. The high production rate of metastable clusters between $n = 7$

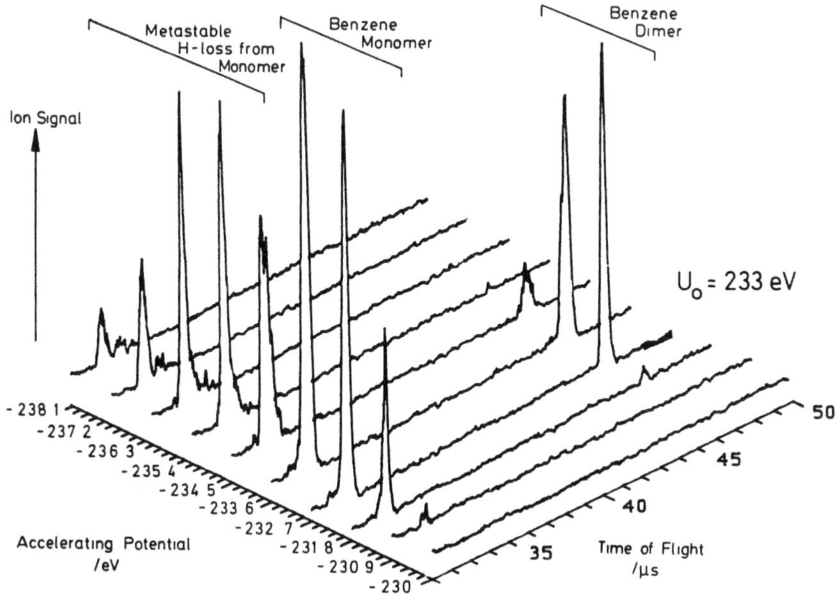

Fig. 2. Energy scan of the benzene monomer and dimer. The monomer exhibits a second slightly broadened satellite peak due to the metastable loss of an H atom. The dimer ion exhibits no signs of fragmentation within one of it monomer subunits

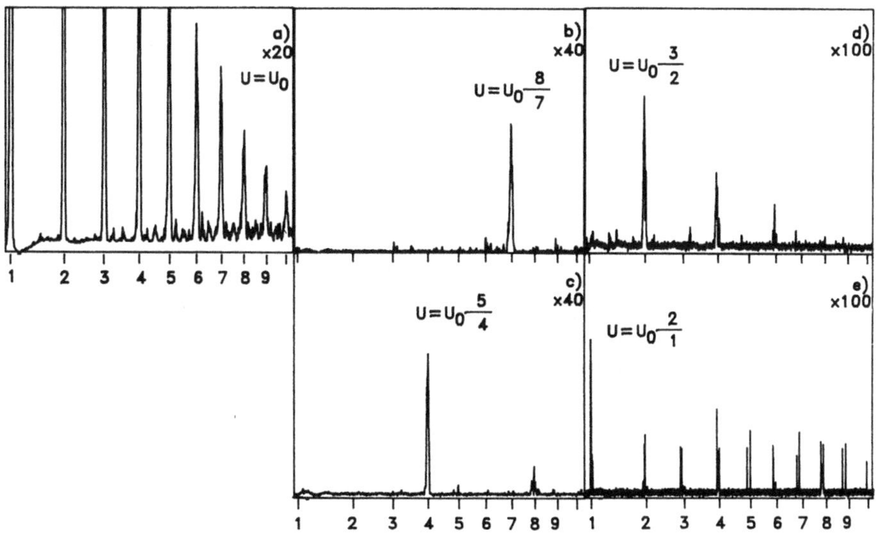

Fig. 3. Metastable channels of fragmenting clusters are fully isolated by the transmission energy window of the mass spectrometer. Each TOF spectrum represents a specific fractional mass loss as indicated by the numbers $U = U_0 \cdot (m_0/m)$ given in the figures. It should also be noted that the metastable mass peaks have the same flight times as their corresponding prompts

to 13 makes them good candidates for energy broadening studies. Figure 3(c) isolates the reactions $n = 5 \rightarrow 4$ and $n = 10 \rightarrow 8$. Figure 3(d) shows reactions $3 \rightarrow 2$, $6 \rightarrow 4$, and $9 \rightarrow 6$ while Fig. 3(e) shows reactions $2 \rightarrow 1$, $4 \rightarrow 2$, $6 \rightarrow 3$, $\ldots 20 \rightarrow 10$. Metastable mass losses of $\delta m/m_0 = 0.5$ have been observed in this way for parent cluster ion sizes larger than $n = 34$. Kiermeier et al. [8] have seen metastable peaks corresponding to the loss of up to 4 monomer subunits only for parent cluster ions up to $n = 6$ in a reflectron mass spectrometer. They suggest that the loss of larger neutral fragments is less probable for larger ionized clusters due to higher threshold energies. If this description is correct then the large mass losses seen here must be attributed to collisional induced decompositions or else three photon processes. An alternative description would be that these very large mass losses did not occur by sequential monomer loss, rather by the splitting off of large neutral bound clusters, processes which should have much lower threshold energies. Background pressure tests on these large mass loss reactions are inconclusive due to the low signal. Tests on the single molecule evaporations such as $n = 8 \rightarrow 7$ with a second pulsed nozzle being used to better mimic background conditions showed no significant increase in daughter ion signals due to collisions.

Measurements of energy broadened metastable peaks for benzene molecules and clusters have also been carried out. In Fig. 4(a) the $3h\nu$ ionisation fragmentation of C_6H_6 to $C_4H_4^+$ shows large broadening in the IKES corresponding to a dissociation energy release of 80 meV. The energy width amplification from prompt to metastable is favourable in this reaction and therefore the choice of the prompt peak to represent instrumental broadening is not critical. Figure 4(b) displays the observed kinetic energy broadening in the case of a typical metastable cluster. Here the broadening is not nearly as pronounced as in Fig. 4(a) due to the unfavourable mass loss ratio in (1). Therefore much care must be taken in the selection of a prompt peak which properly represents instrumental effects. The prompt and metastable peaks of $(C_6H_6)_{10}^+$ are compared first. Owing to the high degree of metastability we believe that the prompt peak is asymmetric due to decay processes in the 10^{-7} s regime. This asymmetry is directly related to the ratio of prompt to metastable and disappears in the case of smaller clusters where the prompt mass peak is dominant (see insert). If the metastable cluster peak shown here were also asymmetric then we would be led to the conclusion that sequential monomer loss was present and the prompt $(C_6H_6)_{10}^+$ or the parent $(C_6H_6)_{11}^+$ would be the appropriate choice to represent instrumental broadening. Since no asymmetry was observed in the metastable peak and since the loss of two benzene monomers has a much lower probability we

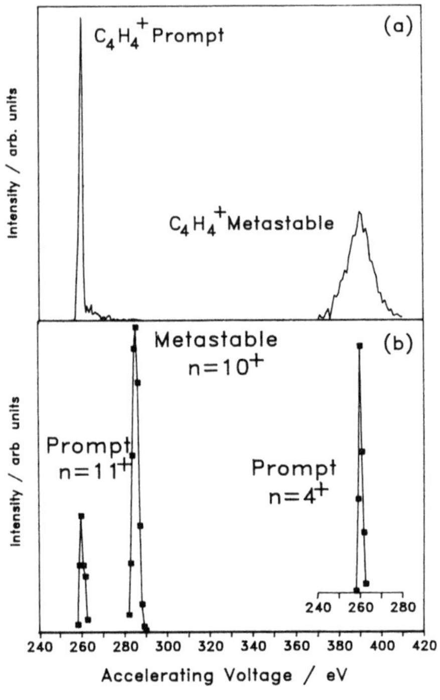

Fig. 4a and b. Measurement of energy broadening in metastable peaks. The increase in the widths of the metastables in comparison to the prompt peaks is due solely to initial kinetic energy release from fragmentations in flight. **a** the unimolecular reaction $C_6H_6^+ \rightarrow C_4H_4^+ + C_2H_2$ exhibits strong broadening. **b** Broadened metastable peak due to loss of a monomer in a benzene cluster

Table 1. Average kinetic energy release for the monomer loss of benzene clusters 7 to 11 $(C_6H_6)_n^+ \rightarrow (C_6H_6)_{n-1}^+ + C_6H_6$ and for the fragmentation $C_6H_6^+ \rightarrow C_4H_4^+ + C_2H_2$

$n \rightarrow n-1$	ε_f/meV	E_{excess}/eV
$8 \rightarrow 7$	13 ± 5	< 0.6
$9 \rightarrow 8$	16 ± 4	< 0.6
$10 \rightarrow 9$	21 ± 5	< 0.6
$11 \rightarrow 10$	25 ± 5	< 0.6
$12 \rightarrow 11$	26 ± 6	< 0.6
for comparison:		
$C_6H_6^+ \rightarrow C_4H_4^+ + C_2H_2$		
	80 ± 10	> 0.2 < 0.5

feel justified in using an unbroadened prompt cluster peak such as the $(C_6H_4)^+$ as a representative of instrumental broadening. Table 1 gives the average kinetic energy release of metastable clusters as evaluated from our data by eqs. (1) and (2) using this procedure. Also given are the estimated excess energies above the dissociation thresholds. The $2h\nu$ energy at $\lambda = 259$ nm corresponds to an energy of 9.57 eV. The ionisation potentials of the clusters listed are believed to be slightly lower than those of the dimer (8.86 eV) and trimer (8.85 eV) [8]. Calculations of neutral binding energies [9] suggest that removal of a monomer in this cluster size range requires 0.2 to 0.3 eV. The energy removed from the complex by the photoelectron is at present unknown. Therefore an upper limit of the excess energy is taken to be 0.6 eV. For the molecular fragmentation reaction assuming ladder switching to low lying vibrational states of the benzene ion an upper and lower bound to the excess energy content of 0.5 and 0.2 eV respectively may be assigned to the present experiment.

4. Conclusion

Direct ion kinetic energy measurements were carried out for metastable decay processes of molecules and molecular clusters following multiphoton ionisation. 1C-REMPI allows selection of small cluster ions via vibrationless intermediate states [8, 10]. A 2C-REMPI ion source in conjunction with our new time of flight and electrostatic analyser techniques will make it possible for the first time to measure the kinetic energy release of clusters with well defined variable excess energies in combination with intermediate state selectivity. Such experiments will provide additional information on the detailed mechanisms (i.e. statistical partitioning of excess energy [4, 5], tunneling) which are intrinsically present in the mass spectroscopy of cluster distributions.

References

1. Ottinger, Ch.: Z. Naturforsch. **20a**, 1229 (1965)
2. Kühlewind, H., Neusser, H.J., Schlag, E.W.: J. Chem. Phys. **88**, 6105 (1984)
3. Cooks, R.G., Beynon, J.H., Caprioli, R.M., Lester, G.R.: Metastable ions. Amsterdam: Elsevier 1973
4. Stace, A.J., Shukla, A.K., Chem. Phys. Lett. **85**, 157 (1981)
5. Illies, A.J., Jarrold, M.F., Bowers, M.T.: Int. J Mass Spectrom. Ion Phys. **47**, 93 (1983)
6. Wiley, W.C., McLaren, I.H.: Rev. Sci. Instrum. **26**, 1150 (1955)
7. Boesl, U., Neusser, H.J., Weinkauf, R., Schlag, E.W.: J. Chem. Phys. **86**, 4857 (1982)
8. Kiermeier, A., Ernstberger, B., Neusser, H.J., Schlag, E.W.: Z. Phys. D–Atoms, Molecules and Clusters **10**, 311 (1988)
9. Van De Waal, B.W.: Chem. Phys. Lett. **123**, 69 (1986)
10. Hopkins, J.B., Powers, D.E., Smalley, R.E.: J. Chem. Phys. **85**, 3739 (1981)

Z. Phys. D – Atoms, Molecules and Clusters 12, 303–305 (1989)

Atoms, Molecules and Clusters
Zeitschrift für Physik D
© Springer-Verlag 1989

Interactions of electrons with SF_6: ionization and attachment

T.D. Märk,[1] A. Stamatovic[2], and P. Scheier[1]

[1] Institut für Ionenphysik, Technikerstrasse 25, A-6020 Innsbruck, Austria
[2] Dept. Physics/Meteorology, PMF Beograd, Yugoslavia

Received 8 July 1988; final version 30 September 1988

Electron attachment and electron impact ionization of SF_6 clusters have been investigated quantitatively in a molecular beam/electron ion source/mass spectrometer system as a function of electron energy E ($0 \leqslant E \leqslant 180$ eV) and as a function of cluster size.

PACS: 36.40; 34.80.G

1. Introduction

Much attention has been recently directed towards the interaction of electrons with van der Waals clusters [1]. The present investigation is devoted to the production and properties of positive and negative cluster ions produced by the interaction of electrons with a cluster beam consisting of $(SF_6)_n$.

Mitsuke et al. [2] and Kondow [3] have recently reported that formation of negative SF_6 cluster ions via collisions with Rydberg atoms is more effective than free electron attachment. Mitsuke et al. [2] used however in their experiments free electrons with an average energy of 2.5 eV. It is well known that, with the exception of N_2O [4], other van der Waals clusters exhibit (irrespective of the monomer behavior) huge resonances close to zero electron energy leading to highly efficient and nondestructive attachment [4–8]. In the case of SF_6 already the monomer is produced only by a sharp resonance near zero electron energy [9]. Hence we suspected that Misuke et al. [2] were operating their electron source off resonance and therefore were obtaining a much weaker signal in case of free electron attachment than in Rydberg atom collisions. The present study thus extends the electron energy range used down to appr. zero electron energy [10]. A strong zero energy resonance in the attachment cross section leading to the formation of $(SF_6)_n^-$ was discovered [11].

Another question concerns the absence of stoichiometric $(SF_6)_n^+$ cluster ions in positive ion spectra as observed by Echt et al. [12]. The present study confirms this observation in case of SF_6 clusters, i.e. positive $(SF_6)_n$ cluster spectra contain only $(SF_6)_n \cdot SF_x^+$ fragment ion series with $0 \leqslant x \leqslant 5$ and in addition $(SF_6)_n\ S_2F_5^+$ ions formed by internal ion molecule reactions. According to [11] positive ion mass spectra of mixed SF_6/Ar clusters include ion series of non-stoichiometric fragment ions $(Ar_n \cdot SF_5)^+$, $(Ar_n \cdot SF_6 \cdot SF_5)^+$, $(Ar_n (SF_6)_2\ SF_5)^+$ and, in addition, stoichiometric ions comprised of $(SF_6\ Ar_n)^+$ and $((SF_6)_2\ Ar_n)^+$. These stoichiometric ions are only present with $n \geqslant 3$.

2. Experimental

The crossed molecular beam/electron ionization/mass spectrometer system has been described in detail previously [5, 6, 13]. Neutral SF_6 clusters are formed by expanding typically 10 to 350 Torr SF_6 through a 20 μm nozzle into vacuum. The nozzle and gas temperature in this study is typically kept constant at around −55°C. The supersonic gas beam passes a skimmer and is crossed in the ion source at right angles by an electron beam of variable energy E ($0 \leqslant E \leqslant 180$ eV). After electron attachment or electron ionization of the neutral clusters the resulting anions or cations are extracted at right angles with respect to the neutral cluster beam and the electron beam by a weak electric extraction field. After extraction the ions are accelerated, pass a field free region and are then analyzed in a double focussing sector field mass spectrometer (reversed geometry). The electron energy scale was found to be linear (under identical experimental conditions in the ion source) and was calibrated previously with fragment anions in O_2, CO_2 and SO_2 [5, 6, 8]. In the present study SF_6^- and SF_5^- produced by electron attachment to SF_6 were used as reference standards.

3. Results and discussion

3.1. Positive SF_6 cluster ions

Positive ion mass spectra of neat SF_6 clusters have been studied in detail by Echt et al. [12] up to appr. $n=40$. They reported the occurrence of non-stoichiometric $(SF_6)_n \cdot SF_x^+$ ions, with x from 5 to 1 with decreasing intensities in that order (similar to the situation in the monomer). The present study of neat SF_6 clusters shows in accordance with this earlier study only non-stoichiometric ions. Besides the ions observed by Echt et al. [12] we find also an ion series with $x=0$ corresponding to $(SF_6)_nS^+$. Moreover, at a mass of 159 amu we are able to observe ions consisting of two sulphur and five fluorine atoms. Such an ion cannot be produced by a simple dissociative electron impact ionization process of one cluster moiety. It either has to be produced by an internal ion molecule reaction after the ionization event or by the occurrence of two sequential dissociative ionization events of the incoming electron within the cluster. There is evidence in the spectra for the existence of larger homologs of this ion.

3.2. Negative SF_6 cluster ions

Foremost among polyatomic anions which have been studied extensively in recent years is SF_6 [9]. Electron attachment to SF_6 with thermal electrons yields mainly SF_6^{-*} with $\tau > 10^{-5}$ s. The cross section for dissociative attachment yielding SF_5^- is much smaller and peaks at slightly larger electron energies (appr. 0.35 eV). Electron attachment to neat SF_6 clusters with appr. thermal electrons gives rise only to stoichiometric $(SF_6)_n^-$ ions. This is in accordance with similar studies in Konstanz [4, 10]. Attachment to mixed Ar/SF_6 clusters with appr. thermal electrons yields also only stoichiometric ion series consisting of $Ar_n \cdot SF_6^-$, $Ar_n(SF_6)_2^-$ and $Ar_n(SF_6)_3^-$ ions (see [11]). These results indicate that the dissociative channel into SF_5^- present in the monomer is efficiently quenched in case of the cluster target. A similar result has been recently reported for SO_2 clusters [8, 14]. Moreover, in contrast to the cluster cations the intensity of the mass peaks of the cluster anions depends strongly on the electron energy, i.e. decreasing sharply above thermal energy.

The dramatic change of the negative mass spectrum as a function of electron energy is due to the resonance like structure of the attachment cross section functions involved. It was found that for instance $(SF_6)_2^-$ and $(SF_6)_3^-$ cluster ions have approximately the same energy dependence as SF_6^- from the cold beam, i.e. showing a resonance like structure close to zero electron energy though all of them slightly blue shifted with respect to the SF_6^- resonance produced from thermal SF_6 monomers. It is interesting to note that the observed blue shift and broadening of the cross section peaks is in accordance with similar studies in Konstanz [4, 10]. The apparent blue shift and the absence of any SF_5^- cluster homologs may be indicative that part of the $(SF_6)_n^-$ cluster ions are produced via the energetically higher lying SF_6^{-*} channel [11].

Finally, the observed near zero energy resonance like structure of the cross section of $(SF_6)_n^-$ cluster ions explains a recent observation of Mitsuke et al. [2, 3]. They reported that formation of SF_6 cluster anions by collisional electron transfer proceeds with higher probability than via free electron attachment. They used, however, for comparison in their experiment electrons of an average energy of 2.5 eV. Thus the cluster ions produced in their experiment via free electron attachment were produced off resonance. Judging from the presently observed shape of the attachment cross section, negative cluster ion production via free electron attachment is as efficient (if electron energies are used close to 0 eV) as bound electron transfer. This statement holds for all of the systems (e.g., O_2, CO_2, H_2O and SO_2 clusters [5–8]) exhibiting the recently discovered "zero energy resonance" for electron attachment. Moreover, attachment of thermal free electrons as well as electron transfer from high Rydberg atoms essentially release the same excitational energy (i.e. the adiabatic electron affinity), therefore possible fragmentation in the attachment process will be similar [4].

Work partially supported by the Österreichischer Fonds zur Förderung der Wissenschaftlichen Forschung.

References

1. Märk, T.D.: Int. J. Mass Spectrom. Ion Proc. **79**, 1 (1987)
2. Mitsuke, K., Kondow, T., Kuchitsu, K.: J. Phys. Chem. **90**, 1522 (1986)
3. Kondow, T.: J. Phys. Chem. **91**, 1307 (1987)
4. Knapp, M., Echt, O., Kreisle, D., Märk, T.D., Recknagel, E.: In:

Physics and chemistry of small clusters. Jena, P., Rao, B.K., Khanna, S.N. (eds.), pp. 693–698. New York: Plenum 1987
5. Märk, T.D., Leiter, K., Ritter, W., Stamatovic, A.: Phys. Rev. Lett. **55**, 2559 (1985); Int. J. Mass Spectrom. Ion Proc **74**, 265 (1986)
6. Stamatovic, A., Leiter, K., Ritter, W., Stephan, K., Märk, T.D.: J. Chem. Phys. **83**, 2942 (1985)
7. Knapp, M., Echt, O., Kreisle, D., Recknagel, E.: J Chem. Phys. **85**, 636 (1986)
8. Stamatovic, A., Scheier, P., Märk, T.D.: Z. Physik D – Atoms Molecules and Clusters **6**, 351 (1987)
9. Christophorou, L.G., Mc Conkle, D.L., Christodoulides, A.A.: In: Electron molecule interactions and their application. Christophorou, L.G. (ed.), pp. 477–617. Orlando: Academic Press 1984
10. A similar study on electron attachment to SF_6 clusters has been made simultaneously to our study by the Konstanz group [4]. Their results confirm the present data
11. Stamatovic, A., Scheier, P., Märk, T.D.: J. Chem. Phys. **88**, 6884 (1988)
12. Echt, O., Reyes Flotte, A., Knapp, M., Sattler, K., Recknagel, E.: Ber. Bunsenges. Phys. Chem. **86**, 860 (1982)
13. Märk, T.D., Scheier, P., Ritter, W., Leiter, K., Stephan, K., Stamatovic, A.: Int. J. Mass Spectrom. Ion Proc. **74**, 281 (1986)
14. Märk, T.D., Scheier, P., Stamatovic, A.: Chem. Phys. Lett. **136**, 177 (1987)

Z. Phys. D - Atoms, Molecules and Clusters 12, 307–314 (1989)

Atoms, Molecules
Zeitschrift für Physik D and Clusters
© Springer-Verlag 1989

General properties of the electronic structure of alkali metal clusters and Ia-IIa mixed clusters*

P. Fantucci[1], V. Bonačić-Koutecký[2], and J. Koutecký[2]

[1] Dipartimento di Chimica Inorganica e Metallorganica, Centro CNR, Universita di Milano, Via Venezian, 21, I-20133 Milano, Italy

[2] Freie Universität Berlin, Institut für Physikalische und Theoretische Chemie, Takustrasse 3, D-1000 Berlin 33, Germany

Received 9 July 1988

The results of the systematic ab-initio CI investigation of neutral and charged Li_n, Na_n, $BeLi_k$ and $MgNa_k$ clusters are summarized and analyzed. The general characteristic features of the electronic structure are pointed out: *a)* The participation of the atomic orbitals, which are empty in Ia and IIa metal atoms, allows for a higher valency of these atoms in clusters. *b)* Jahn-Teller and pseudo-Jahn-Teller effects strongly influence the electronic and geometric structure of clusters. *c)* Deformations of cluster geometry can lead to biradicaloid structures with higher spin multiplicity in their ground states. *d)* The peculiarities of the electronic structures of clusters can be deduced from the presence of many "surface" atoms. The theoretical results agree with experimental data presently available and they are useful for interpretation of the experimental findings.

PACS: 31.20.D; 31.30; 36.40

Introduction

The investigation of alkali metal clusters with quantum chemical methods can substantially contribute to the general understanding of the electronic structure of small metal clusters. Since Ia metal atom has only one valence electron, the rules governing the electron interaction in clusters are relatively simple. The application of sophisticated theoretical methods makes it possible to discover specific cluster properties which are absent in molecules or solids. Due to the simplicity of the electronic structure of alkali metal clusters the theory yields predictions of several observables which are experimentally accessible.

Therefore, the results [1–9] obtained from the ab initio all-electron CI investigation of small Li_n, Li_n^+, Li_n^-, Na_n, Na_n^+ clusters will be discussed in this contribution from an unified point of view. Furthermore, general concepts gained from the study of alkali metal clusters will be used and adequately extended for the understanding of the properties of mixed $BeLi_k$, $BeLi_k^+$ [10], $MgNa_k$ and $MgNa_k^+$ [11] clusters.

Comments on methods

The geometries of neutral and charged alkali metal clusters and mixed Ia-IIa clusters have been optimized using an analytical gradient method [12] in the framework of all electron Hartree-Fock procedure. Relatively small AO basis sets can be used for a rough estimate of cluster geometry provided that they include *p* functions. Therefore, a minimal basis set augmented by one *p*-function has been used for Li_n since it was shown that extension of this basis set does not substantially change the calculated properties of clusters [1]. In the case of Na_n (9*s*, 4*p*) AO basis set

* Dedicated to Prof. Rudolf Zahradník on the occasion of his 60th birthday

contracted to (3s, 1p) and augmented by one diffuse p function with exponent $\alpha_p = 0.065$ has been used [4, 5]. Careful investigations of some cluster properties with larger AO basis sets: (10s, 4p/3s, 2p) and (14s, 8p, 1d/6s, 5p, 1d) for Li and Na, respectively, has confirmed again the reliability of the relatively small basis. Note, that the minimal basis set for Li augmented by the p function (6s, 1p/2s, 1p) does not suffer from large basis set superposition error (BSSE) (e.g. BSSE(Li_4) = 0.37 kcal/mol).

Standard split valence basis sets 3-21G, 6-31G, 6-31G* of increasing flexibility have been used for neutral and cationic $BeLi_k$ clusters. The search of optimal geometries has been carried out for $BeLi_k$ ($k<4$) employing all three basis sets, but for larger $BeLi_k$ (>3) as well as for $BeLi_k^+$ clusters only the smallest 3-21G basis has been used, since the results obtained with different basis sets do not differ significantly.

For neutral and cationic $MgNa_k$ clusters the relatively large AO basis set (9s, 5p, 1d/4s, 3p, 1d) has been used for $k = 1$–6. For $k \geq 6$ the d functions on Na atoms have been omitted.

The multireference double-excitation configuration interaction (MRD-CI) [13a] energies or the multireference single and double direct CI (MR-SD) [13b] energies have been calculated for the optimized Hartree-Fock geometries. Only valence electrons have been correlated. In the case of MRD-CI calculations the Davidson estimate of the full CI energy has been computed.

Electronic and geometric structure of alkali metal clusters

The topologies of Hartree-Fock optimized geometries of lithium and sodium clusters are shown in Fig. 1 next to the binding energies per atom obtained from the MRD-CI procedure. The atomization energies per atoms (E_b/n) as functions of the number of atoms in Li_n, Li_n^+ and Li_n^- clusters (Fig. 1a) and in Na_n and Na_n^+ clusters (Fig. 1b) exhibit a generally increasing tendency with the increasing nuclearity n for all five investigated classes of alkali metal clusters. The E_b/n functions exhibit oscillations for even-odd number of valence electrons n_e. These oscillations which favour energetically the clusters with an even number of electrons n_e can be clearly depicted from Fig. 2 where the second differences of the cluster energies defined as,

$$\Delta^2 E_n = E_{n+1} + E_{n-1} - 2E_n$$

are plotted as functions of n. Well developed maxima for Li_2, Na_2, Li_3^+ and Na_3^+ ($n_e = 2$), for Li_8, Na_8, Li_9^+ and Na_9^+ ($n_e = 8$), for Li_4 and Na_4 ($n_e = 4$), as well as for Li_7^+ and Na_7^+ ($n_e = 6$) are noticeable. The quantity $\Delta^2 E_n$ can be interpreted as the difference between the energies of the fragmentation $M_n \rightarrow M_{n-1} + M_1$ for the Na_n or Li_n cluster and the fragmentation $M_{n+1} \rightarrow M_n + M_1$ for the Na_{n+1} or Li_{n+1} cluster.

The very parallel behaviour of the E_b/n functions for Na_n and Li_n as well as for Na_n^+ and Li_n^+ and the similarities in cluster geometries (c.f. topologies shown in Fig. 1) demonstrates that the alkali metal cluster properties depend dominantly upon the number of the valence electrons. The absolute values of the calculated binding energy per atom are substantially smaller for Na_n than for Li_n clusters. For comparison, E_b/n functions for sodium clusters obtained with the local spin density method (LSD) by Martins et al. [14–16] are also drawn in Fig. 3. The larger values of the atomization energies per atom calculated with the LSD procedure confirm that the electron density methods using local potentials strongly overestimate the strength of the interatomic interactions.

The evolution from planar towards three-dimensional cluster shapes with increasing nuclearity n is striking for neutral and cationic alkali metal clusters. The planar forms of Li_n and Na_n clusters ($n<7$) are slightly deformed sections of (111) plane of the *fcc* (or hcp) lattice while the three-dimensional forms for $n = 7$, 8, 9 and 10 can be approximately considered as steps in tetrahedral cluster growth. The interatomic distances in small alkali metal clusters do not differ appreciably from each other and the Mulliken atomic charges are relatively small. For several cases the geometries of the cationic and mainly the anionic Li_n^- clusters are quite different from the shapes of neutral clusters. These differences in cluster shapes are due not only to the different number of electrons but also to the preference of such cluster geometries for which the polarization of the electron cloud is able to partly compensate the unfavourable influence of the Coulomb repulsion.

The vertical and adiabatic ionization potentials (IP) of Li_n and Na_n clusters as functions of cluster size are reported in Fig. 4, where the vertical IP's are compared with the experimental appearance potential of Na_n following the suggestion of Kappes et al. [17]. Three features are characteristic for the general form of this dependence: *i)* the overall decreasing trend which is in agreement with the classical spherical droplet model, *ii)* the even-odd oscillations showing higher stability of clusters with even number of electrons, *iii)* an especially large decrease of the ionization potential values for nuclearities 3 and 9 with respect to those for 2 and 8, respectively. The theory predicts very similar values for ionization potentials of Na_4, Na_6 and Na_8. The parallel behaviour of characteristic features of the calculated ionization potentials and experimentally estimated appearance potentials is evident. The values of the theoretical ionization potentials for Na_n clusters

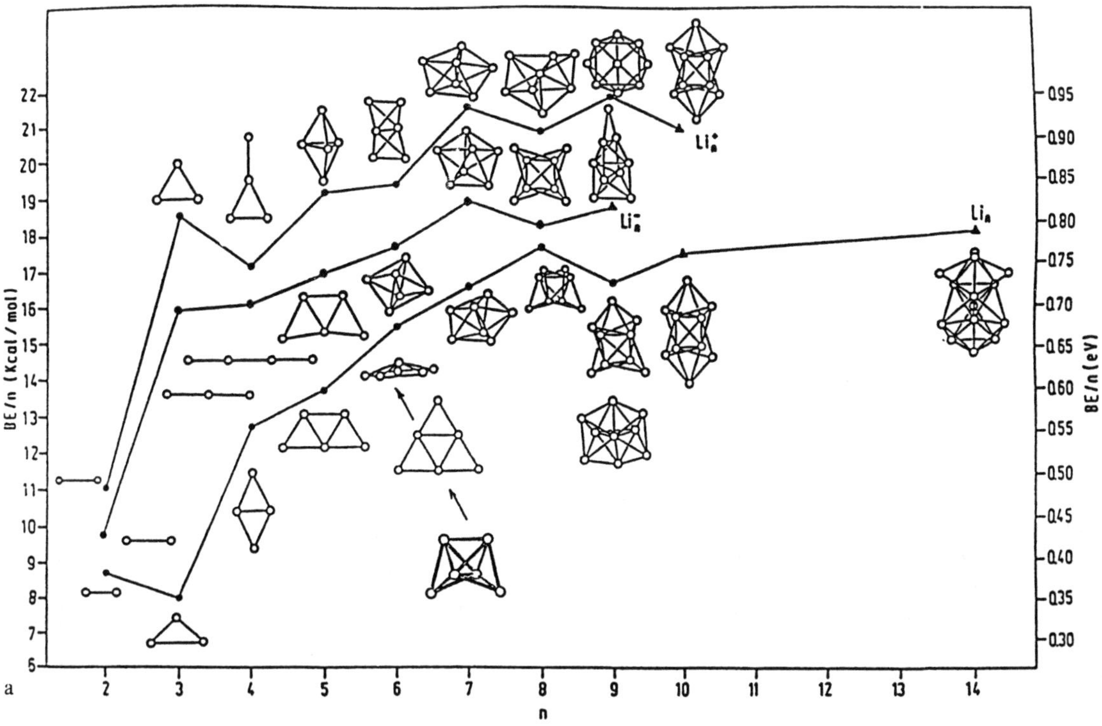

a

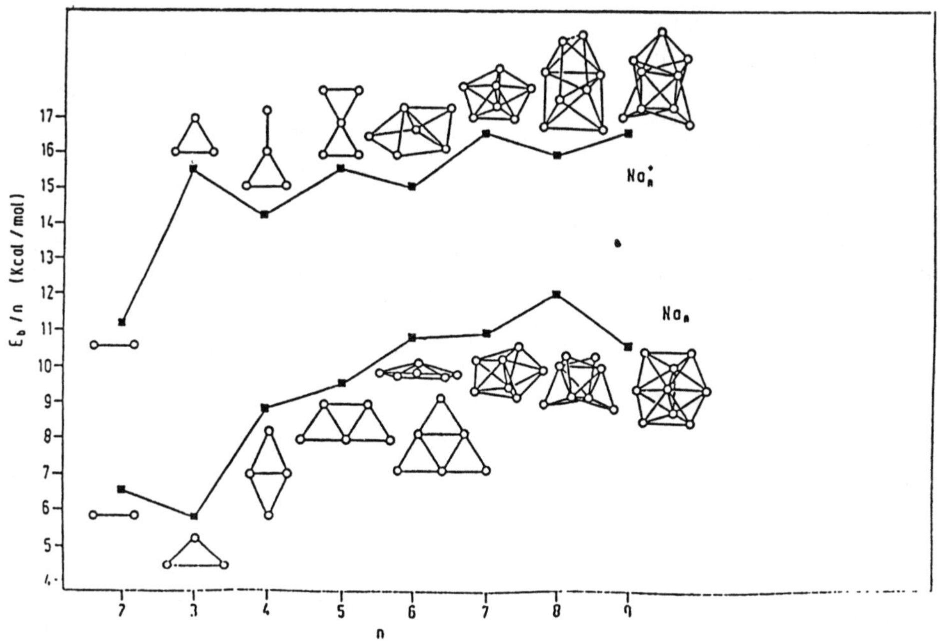

b

Fig. 1a and b. The binding energies per atom E_b/n of **a** neutral cationic and anionic lithium clusters n and **b** of neutral and cationic sodium clusters as functions of nuclearity n. The cluster energies are calculated with the MRD-CI procedure and with the Davidson corrections. The topologies of the SCF optimized clusters are schematically drawn

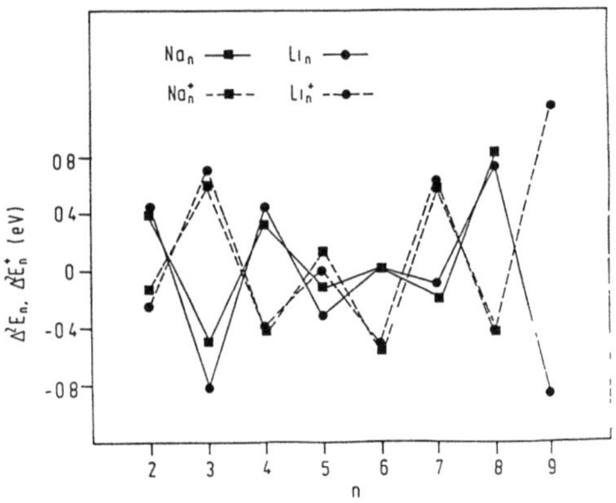

Fig. 2. The second differences Δ^2E_n and $\Delta^2E_n^+$ for neutral (——) and cationic (---) clusters as functions of the number of atoms n (●) lithium and (■) sodium

are, slightly lower in comparison with the experimental values. This is due to the fact that the correlation effects between the core and valence shells of Na atoms have not been appropriately considered. According to our preliminary results, if the core-polarization effects are taken into account, a quantitative agreement with the experiment can be reached.

The dependence of electron affinity on the nuclearity n for Li_n clusters determined with ab initio methods exhibit remarkable similarity with the experimentally obtained electron affinity for Cu_n clusters [18] (c.f. Fig. 5). The calculated absolute values must be taken with caution due to methodological difficulties.

The fragmentation energies for the two channels: $Na_n^+ \rightarrow Na_{n-1}^+ + Na_1$ ($\delta_1E_n^+$) and $Na_n^+ \rightarrow Na_{n-2}^+ + Na_2$ ($\delta_2E_n^+$) are drawn in Fig. 6. For Na_5^+, Na_7^+ and Na_9^+ the fragmentation through the second channel is energetically favourable. This prediction is in agreement with the experimental findings of Bréchignac et al. [19].

Mixed neutral and cationic $BeLi_k$ and $MgNa_k$ clusters

The optimal geometries of the mixed clusters have been determined with the analytical gradient method in the framework of the Hartree-Fock SCF procedure [12]. The energies of clusters for these geometries have been calculated employing the MRD-CI [13a] procedure for $BeLi_k$ and the multireference direct CI (MR-SD) [13b] method for $MgNa_k$.

The topologies of the optimal geometries for $BeLi_k$ and $BeLi_k^+$ clusters are schematically shown in Fig. 7. The Be atom is in the center of the cluster surrounded by Li atoms in all cases except for $BeLi_3$. The atomization energies per atom for neutral and cationic

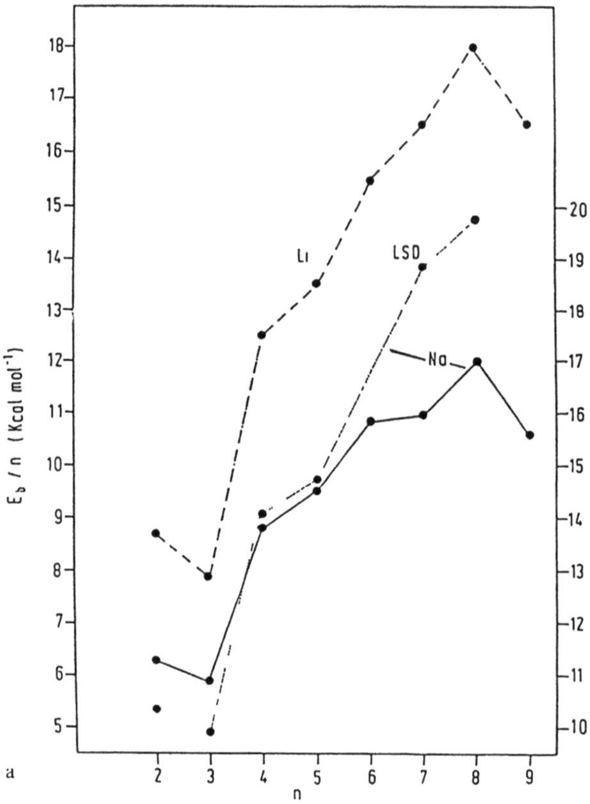

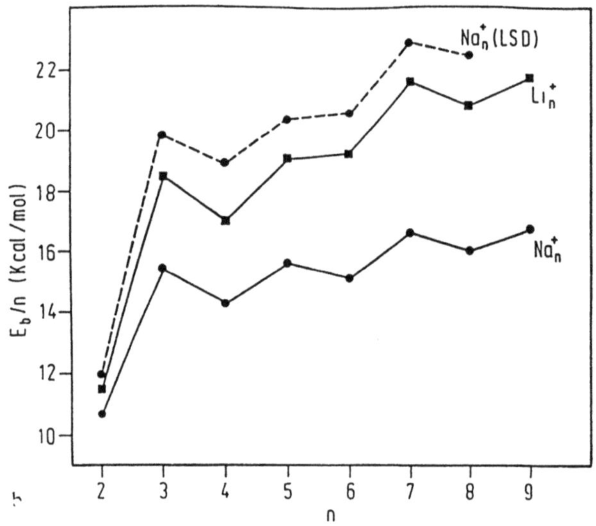

Fig. 3. a Comparison of the binding energies per atom E_b/n as functions of n for neutral Li and Na clusters and for cationic clusters. The broken (-.-.-) and solid (——) line labels all-electron MRD-CI results for Li and Na clusters, respectively. The energies are determined for the corresponding HF optimal geometries. The dotted curve labels LSD results for Na clusters for which the scale on the right-hand side of Fig. 2 is valid. **b** Comparison of the binding energies per atom E_b/n as function of n for the cationic Na_n^+ and Li_n^+ clusters. The full lines (——) label the all-electron MRD-CI results and the dashed line (---) the LSD results

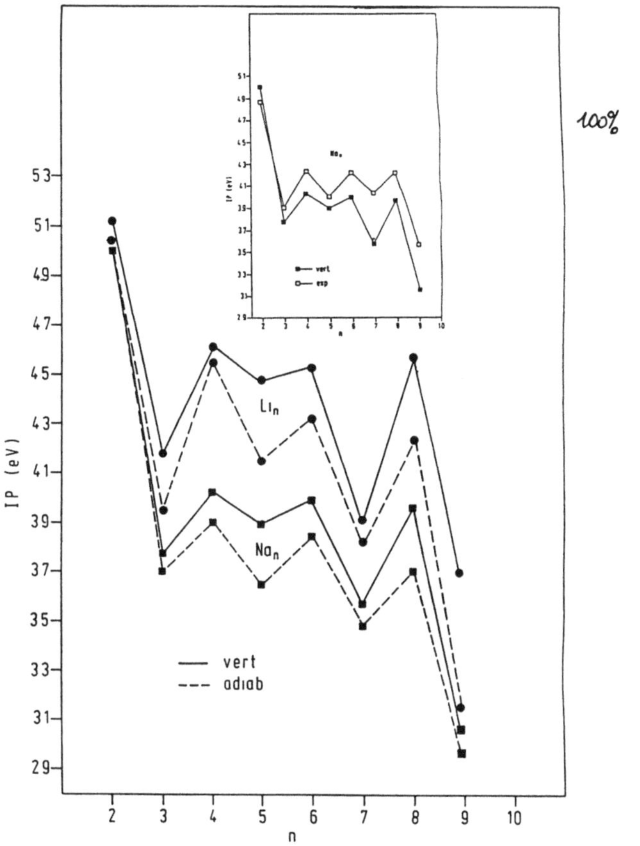

Fig. 4. Vertical (——) and adiabatic (---) ionization potentials (in eV) of Li (●) and sodium (■) clusters. The comparison of calculated vertical IP's for Na_n with experimental neutral data from Ref. 17 are given in the box

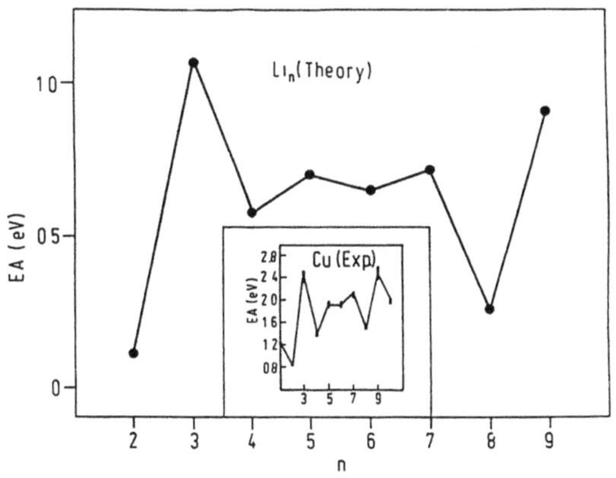

Fig. 5. The adiabatic electron affinities of Li_n clusters as function of the nuclearity n. The experimental data for Cu_n from Ref. 19 are given in the box

$BeLi_k$ clusters drawn in Fig. 7 increase with the cluster size until the number of valence electrons is 8. E_b/n remains nearly constant for $BeLi_6$, $BeLi_7$, $BeLi_8$, $BeLi_9$. In the case of cationic $BeLi_k^+$ clusters there is a decrease in E_b^+/n from $n_e=8$ to $n_e=9$.

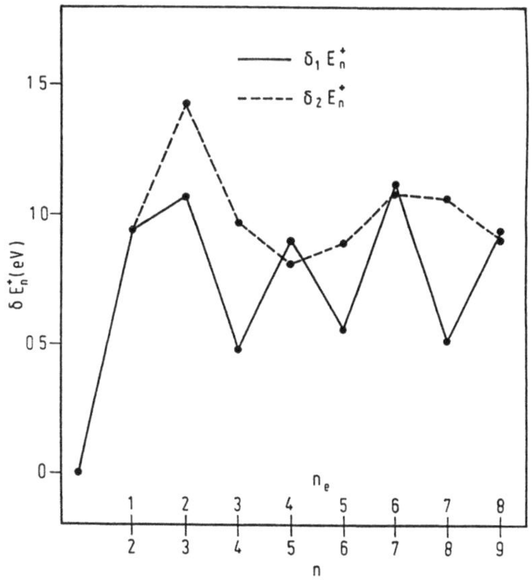

Fig. 6. Dissociation energies $\delta_1 E_n^+$ for the channel $Na_n^+ \rightarrow Na_{n-1}^+ + Na_1$ and $\delta_2 E_n^+$ for the channel $Na_n^+ \rightarrow Na_{n-2}^+ + Na_2$

The neutral and cationic $MgNa_k$ clusters exhibit striking similarities to $BeLi_k$ clusters concerning geometries and other properties. There are some minor differences for small clusters. For comparison the E_b/n are shown in Fig. 8 for neutral and cationic $MgNa_k$ and $BeLi_k$ clusters.

The second differences in energies exhibit very pronounced maxima for $n_e=8$ as shown in Fig. 9.

The vertical and adiabatic ionization potentials of $BeLi_k$ clusters (Fig. 10) are slowly decreasing functions of k for $k<9$ with usual even odd oscillations except of vertical IP for $BeLi_3$. This exception can be understood as due to the anomalous shape of $BeLi_3$. Both vertical and adiabatic IP's decrease strongly for $n_e=9$ with respect to $n_e=8$.

Discussion and conclusion

Let us summarize the results of the investigation of alkali metal and Ia-IIa mixed clusters.

The theoretically predicted geometries of alkali metal clusters exhibit some common characteristics.

1) The calculated interatomic nearest neighbour distances in Li_n clusters are fairly close to the experimental value found in the fcc Li crystal (3.1 Å) and do not exhibit large deviations from each other. The calculated interatomic distances in Na_n clusters do not show large fluctuations and those computed for Na_8 (T_d) cluster are quite close to the value observed in Na crystal (~3.7 Å).

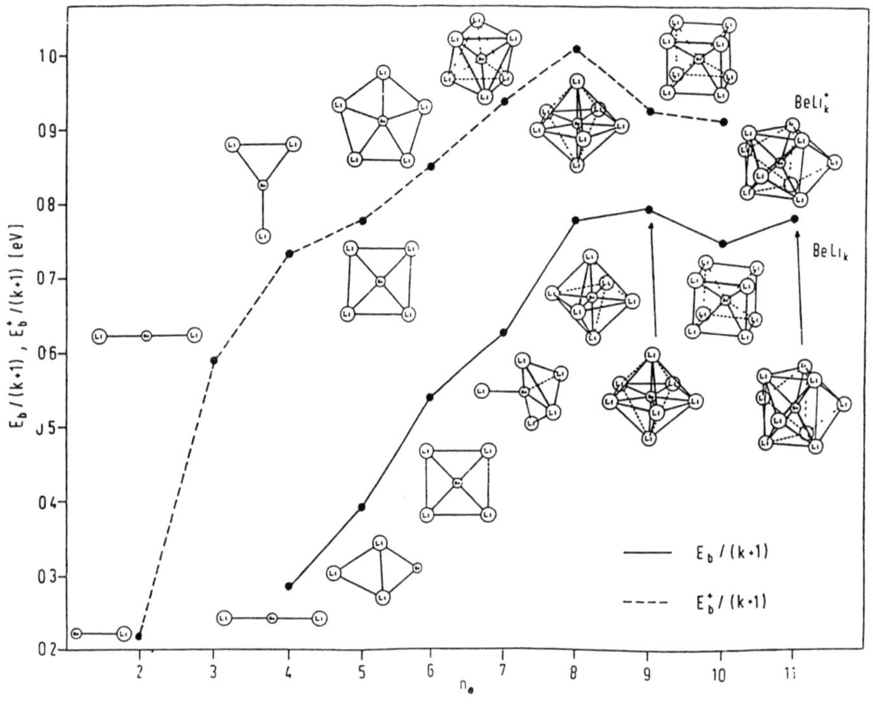

Fig. 7. The binding energies $E_b/(k+1)$ of neutral $BeLi_k$ and $E_b^+/(k+1)$ of cationic $BeLi_n^+$ clusters as functions of number of valence electrons n_e. The 3-21G AO basis set and MRD-CI procedure have been employed. The geometries of the SCF-optimized clusters are shown

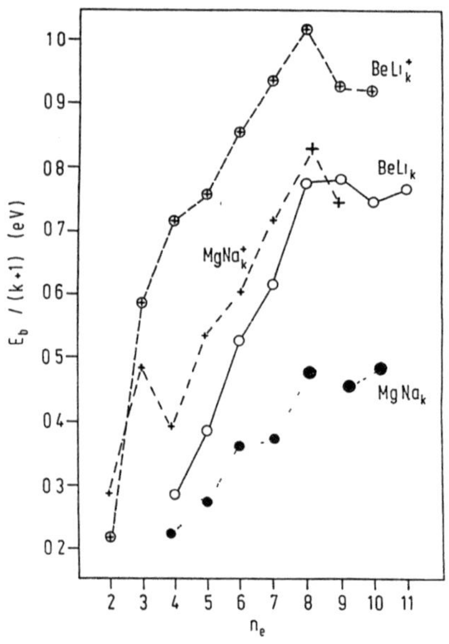

Fig. 8. The comparison of the atomization energies per atom $E_b/(k+1)$ of neutral $BeLi_k$ and cationic $BeLi_k^+$ clusters (cf. Fig. 7) with the $E_b/(k+1)$ of neutral $MgNa_k$ and cationic clusters $MgNa_k^+$. For $MgNa_k$ with $k=1-6$ the $4s3p1d$ basis set was used for Mg and Na and for $k \geq 6$ the d function on Na atoms has been omitted. The CI energies for $k=7, 8$ are shifted for the energy difference obtained employing smaller and larger basis set for $MgNa_6$. The multi-reference single and double direct CI procedure has been employed for investigation of $MgNa_k$ clusters

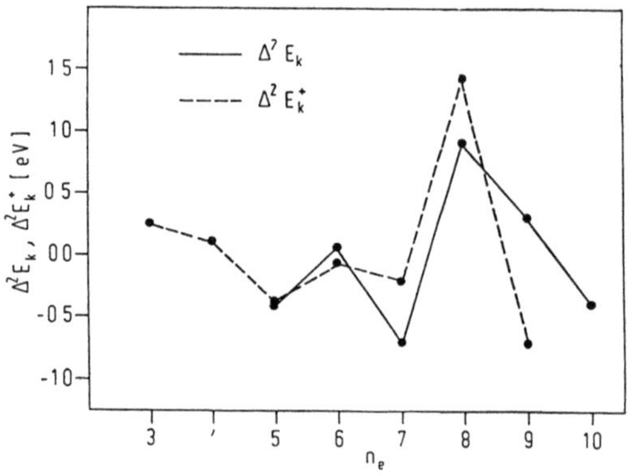

Fig. 9. Second differences $\Delta^2 E_k$ for neutral $BeLi_k$ and $\Delta^2 E_n^+$ for cationic $BeLi_k^+$ clusters as functions of number of valence electrons n_e

2) The reliability of the absolute values of our calculated equilibrium distances should not be overestimated since they can vary with the AO basis set as well as with the applied treatment of electron correlation. However, the small fluctuations of the interatomic distances and their rough agreement with the experimental data of bulk metals show that in this respect the character of bonding in clusters and in crystals is similar.

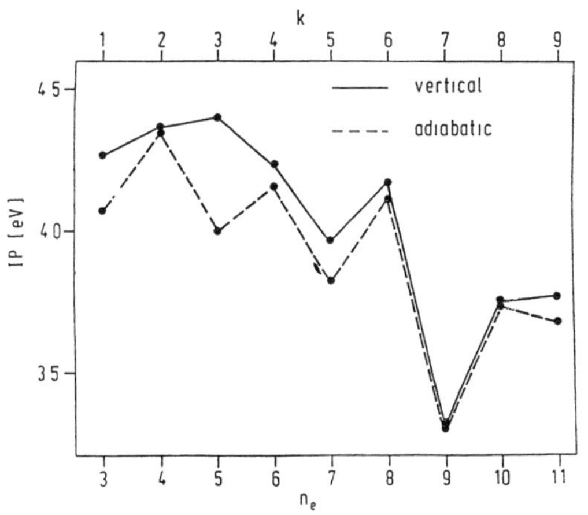

Fig. 10. Adiabatic and vertical ionization potentials of mixed $BeLi_k$ clusters calculated with the MRD-CI procedure and the 3-21G basis set

3) The geometries of alkali metal clusters exhibit some features which are very different from the crystal properties. The tendency towards compactness which maximizes the number of interatomic interactions leads to structures with tetrahedral building elements for clusters with the nuclearity $n \geq 6$.

4) For very small clusters $n \leq 6$ the pseudo-Jahn-Teller effect can lead to planar structures with highest possible compactness i.e. deformed segments of (111) plane in fcc (or hcp) crystal lattice.

5) The geometries predicted for "medium" size clusters $(6 < n < 14)$ which are built from tetrahedral subunits must undergo a change in their character for further increasing *n*, either through introduction of larger irregularities in shape or by incorporation of some octahedra. Only in this way, a transition to any crystal lattice with translational symmetry can be achieved.

6) The geometries of small neutral and charged clusters can differ appreciably.

7) The mixed Ia-IIa clusters have the tendency to place the IIa atom in the cluster center.

8) In summary, cluster geometries are evidently due to a compromise between the interatomic interactions similar to those in crystals and the influence of large number of "surface" atoms (dangling bonds).

The electronic structure of the alkali metal clusters and Ia-IIa mixed clusters is influenced by geometrical structures and vice versa. Several general rules can be formulated:

1) Only the inclusion of *p*-type AO's in the AO basis set gives rise to the stability of clusters. These functions which are of no importance in the ground state of the alkali metal atoms allow then to reach a higher valency and consequently more "chemical" bonds with neighbouring atoms in cluster. The connection with the overlap of *s*-band and *p*-band in *sp* metals is evident. For example, the very stable Li_4 and Li_8 clusters loose completely their stability if the *p*-AO basis functions are not used.

2) The Mulliken net atomic charges in clusters are quite small if the electron correlation is properly considered. The deviation from charge neutrality in Li_n clusters is usually smaller than 0.1 electron. The atomic centers with smaller number of nearest neighbours are negatively charged. The Mulliken population of *p*-orbitals are larger for atoms with larger number of nearest neighbours.

3) The mentioned pseudo-Jahn-Teller effects which can cause the deformations of highly symmetrical cluster forms arise if the degeneracy of one-electron functions required by the dimension of the irreducible representations of the symmetry group can not be achieved, because of the unsufficient number of electrons. For example, a deviation from the T_d symmetry occurs for Li_4 and the most stable tetramer has only the D_{2h} symmetry.

4) The described deformations do not occur if the number of electrons is sufficient to fill the degenerate (or nearly degenerate) one-electron functions and such a cluster can be especially symmetrical and stable.

5) The lower stability of Na_n clusters with respect to Li_n clusters can be at least partly understood as due to the weaker hybridization of *s* and *p* AO's in the former case.

6) The success of simplified methods [20, 21] in describing some cluster properties is mainly due to the very general character of the rules governing the electronic structure of clusters. The more sophisticated quantum chemical approach shows the limits of the applicability of these approximate methods and yields reliable predictions of many cluster properties.

The theoretical statements concerning the cluster properties can be summarized as follows:

1) The abundances observed in the experimental recording devices must be interpreted with caution because both the stability of clusters and the fragmentation processes should be taken into account. The theory shows that the large atomization energy (E_b/n) is not necessarily accompanied with the large differences of the fragmentation energies of cluster with *n* atoms and $(n+1)$ atoms $(\Delta^2 E_n)$.

Experimental investigation of mass spectrum of Li_n^+ clusters emitted from liquid metal ion source [23] exhibit large drops in the intensity of peaks after Li_7^+, Li_9^+, Li_{13}^+ and Li_{19}^+. First two "magic" numbers ($n_e=6$, 8) are in agreement with quantum chemical predictions (Δ^2E_7, Δ^2E_9 large). The further two ($n=13$, 19) could be interpreted [22] as due to large stability of compact icosahedric structures.

2) The theoretical predictions of the quantity Δ^2E_n and $\Delta^2E_n^+$ for neutral and cationic alkali metal clusters support the interpretation that the large abundances with even or odd nuclearities depend on the energies of the photons or electrons used to ionize originally neutral clusters.

3) The calculated ionization potentials of alkali metal clusters generally decrease with cluster size but the stabilizing effects of closing electron shells as well as the even-odd oscillations are very important. The theory predicts, very characteristic dependence of the ionization potentials upon the size of alkali metal clusters which seems to agree in details with the observed appearance potentials. Especially, the fact that the ionization potentials of Na_6, Na_7 and Na_8 have very similar values can be well understood.

4) The peculiarities of the cluster fragmentation are due to the even-odd oscillations of the cluster stability with the nuclearity: Some cationic clusters with the odd nuclearity should fragment yielding a Ia dimer.

5) The observed properties of the mixed Ia-IIa clusters can be explained as due to the above mentioned rules obtained for homogeneous Ia clusters. In particular, the abundances of Ia-IIa clusters with 8 and 10 valence electrons agree with those criteria of cluster stability.

According to our opinion one important aspect of the bonding in Ia and IIa metal clusters is the "activation" of the atomic orbitals which are without importance for the electronic structure of atoms. The participation of ionic structure in the VB sense is not the most important factor of the interatomic interactions in cluster. The large number of "surface" atoms present in clusters necessarily modify drastically the character of binding in cluster with respect to solid.

This work has been partly supported by the Sfb 6 "Structure and Dynamics of Interfaces", Sfb 337 "Energy- and Charge-Transfer in Molecular Aggregates", and by the CNR "Italian National Research Council".

References

1. Boustani, I., Pewestorf, W., Fantucci, P., Bonačić-Koutecký, V., Koutecký, J.: Phys. Rev. B **35**, 9437 (1987)
2. Koutecký, J., Fantucci, P.: Z. Phys. D – Atoms, Molecules and Clusters **3**, 147 (1986)
3. Koutecký, J., Fantucci, P.: Chem. Rev. **86**, 538 (1986)
4. Bonačic-Koutecký, V., Fantucci, P., Koutecký, J.: Phys. Rev. B **37**, 4369 (1988)
5. Bonačić-Koutecký, V., Boustani, I., Guest, M.F., Koutecký, J.: J. Chem. Phys. **89**, 4861 (1988)
6. Koutecký, J., Plavšić, D., Döhnert, D.: Croat. Chim. Acta **56**, 451 (1983)
7. Fantucci, P., Balzarini, P.: J. Mol. Catal. **4**, 337 (1978)
8. Boustani, I., Koutecký, J.: J. Chem. Phys. **88**, 5657 (1988)
9. Beckmann, H.-O., Koutecký, J., Botschwina, P., Meyer, W.: Chem. Phys. Lett. **67**, 119 (1979); Beckmann, H.-O., Koutecký, J., Bonačić-Koutecký, V.: J. Chem. Phys. **73**, 5182 (1980); Plavšić, D., Koutecký, J., Pacchioni, G., Bonačić-Koutecký, V.: J. Phys. Chem. **87**, 1096 (1983); Fantucci, P., Koutecký, J., Pacchioni, G.: J. Chem. Phys. **80**, 325 (1984); Koutecký, J., Pacchioni, G.: Ber. Bunsenges. Phys. Chem. **88**, 323 (1984); Koutecký, J., Pacchioni, G., Jeung, G.-H., Hass, E.C.: Surf. Sci. **156**, 650 (1985)
10. Pewestorf, W., Bonačić-Koutecký, V., Koutecký, J.: J. Chem. Phys. **89**, 5794 (1988)
11. Fantucci, P., Bonačić-Koutecký, V., Koutecký, J.: J. Chem. Phys (in press)
12. Guest, M.F., Kendrick, J.: CCP1/86/1; Computational Science Group, Daresbury Laboratory
13. a) Buenker, R.J.: In: Studies in physical and theoretical chemistry. Carbo, R.T. (ed.), Vol. 21, pp. 17–34. Current aspects of quantum chemistry, Amsterdam: Elsevier 1982; Buenker, R.J., Phillips, R.A.: J. Mol. Struct. Theochem. **123**, 291 (1985); b) Saunders, V.R., van Leuthe, J.H.: Mol. Phys. **48**, 923 (1983)
14. Martins, J.L., Buttet, J., Car, R.: Phys. Rev. B **31**, 1804 (1985)
15. Martins, J.L., Car, R., Buttet, J.. J. Chem. Phys. **78**, 5646 (1983)
16. Martins, J.L., Buttet, J., Car, R.: Ber. Bunsenges. Phys. Chem. **88**, 239 (1984)
17. Kappes, M.M., Schär, M., Röthlisberger, U., Yeretzlav, Ch., Schumacher, E.: Chem. Phys. Lett. **143**, 251 (1988)
18. Leopold, D.G., Ho, J., Lineberger, W.C.: J. Chem. Phys. **86**, 1715 (1987)
19. Bréchignac, C., Cahuzac, Ph.: Z. Phys. D – Atoms. Molecules and Clusters **3**, 121 (1986); Bréchignac, C., Cahuzac, Ph., Pavolini, D., Spiegelman, F.: J. Chem. Phys. **87**, 5694 (1987)
20. Wang, Y., George, T.F., Lindsay, D.H., Beri, A.C.: J. Chem. Phys. **86**, 3493 (1987); Lindsay, D.M., Wang, Y., George, T.F.: J. Chem. Phys. **86**, 3500 (1987)
21. Knight, W.D., Clemenger, K., de Heer, W.A., Saunders, W.A., Chou, M.Y., Cohen, M.L.: Phys. Rev. Lett. **55**, 5391 (1986)
22. Saito, Y., Watanabe, M., Hagiwara, T., Nishigaki, S., Noda, T.: J. Appl. Phys. (Jpn) **27**, 424 (1988)

Z. Phys. D – Atoms, Molecules and Clusters 12, 315–318 (1989)

Atoms, Molecules
Zeitschrift für Physik D and Clusters
© Springer-Verlag 1989

Metallic free clusters

W.D. Knight
Department of Physics, University of California, Berkeley, CA 94720, USA

Received 23 July 1988

Simple models and corresponding experimental results are presented concerning single particle and collective motion in metal clusters. Relative abundances, photo-electron spectra, ionization potentials, polarizabilities, and plasma resonances are discussed in relation to the shell model, and suggestions are made for future experiments.

PACS: 36.40.+d; 31.20.–d; 33.20.Kf

I. Introduction

The first significant study of free metal clusters [1] was made about 20 years ago, when ionization potentials were determined for several alkali metal clusters containing up to eight atoms. The significant advantages of examining metal clusters in the molecular beam free of any significant interactions with other clusters or with the background gas were recognized then, and a considerable body of literature has accumulated in the past two decades. The reader may consult [2–6] for surveys of the pertinent literature.

At the present time much information is available concerning the production and properties of free metal clusters of a number of metals containing up to a few thousand atoms. The general aim has been to learn how the properties of clusters change as they develop toward the characteristic bulk structures, and to learn whether a unique "metal cluster structure" exists. Over the years many experiments were performed on embedded clusters, however the observed properties are not characteristic of the pure free clusters. Clusters deposited on minimally interacting substrates such as carbon are not free. In fact, one of the promising developments to be expected in surface physics is comparison of the properties of size selected free metal clusters with identical supported clusters.

This paper will briefly survey current knowledge concerning the properties of simple metal clusters in relation to the electronic shell model [2], and will emphasize points where further experimental work is needed to improve our understanding of these systems. In general we will consider a system to be metallic if the valence electrons are delocalized with respect to the ion cores, and can participate in charge transport processes, such as conduction, screening external fields, or oscillating collectively as in plasmon excitation. Electron-phonon interactions may result in conversion of electronic energy to heat.

II. Abundances of metal clusters in beams

The simplest shell model [2] for metal clusters assumes that the delocalized valence electrons all move in the same potential (mean field approximation). The eigenstates for such a system may be found in the first approximation by assuming a finite ellipsoidal potential well of the Woods-Saxon type. The eigenvalues may be found and the total electron energy minimized with respect to the shapes. The most stable shape is spherical for clusters containing $N = 8$, 20, 40, 58 and 92 atoms of an alkali metal. Because of the high degeneracy associated with spherical symmetry, these clusters are characterized by a relatively large energy gap between the highest filled state and the next unoccupied state, and they appear to be especially stable relative to neighboring clusters. Other clusters are found to be ellipsoidal with characteristic degeneracies and related shell closings predicted to occur at for example $N = 26$, 30, 34, 38. Energy levels may be plotted for spheroidal clusters as a function of deformation parameter on the Clemenger-Nilsson diagram [7], which indicates highest occupied levels and deformations which are zero for spherical clusters.

The degeneracies are associated not only with shell closings in the heirarchy of eigenvalues, but also with features in various properties plotted as a function of N for a sequence of clusters. In this model the electronic structure is directly related to the geometrical shape of the delocalized electron distribution, and is independent of the interatomic geometry usually depicted in chemical bonding. The energy gap between the closed shell and the next unoccupied state is a measure of the stability of a cluster relative to the next cluster in the sequence. This is reflected in the mass abundance spectra of a sequence of clusters.

In mass spectra [2] for Na, K, all of the above spherical and ellipsoidal shell closings are marked by relatively high abundances. Also, in mixed Na($N-1$)K [2] clusters all of the closings through $N=40$ are observed. This is an indication of the dominance of the electronic structure of the system and the basic geometry of the delocalized electron distribution. It is predicted that clusters $N=9$, 10 are prolate spheroids; 11, 12, and 13 are triaxial ellipsoids, 14 is an oblate ellipsoid; 15, 16, and 17 are triaxial; 18 is oblate, etc. In these predictions there is good agreement between simple potential well theory and a fully self-consistent treatment [8]. Experimental evidence is consistent with the predictions for shapes.

Neutral sodium and potassium mass spectra have been observed up to $N \sim 100$. It is not difficult to extend this range, and further shell closings will be observed. When the energy gap in the level system becomes comparable with kT, it is expected [9] that thermal fluctuations may moderate the abrupt changes associated with shell closings. Observation would be facilitated by cooling the observed clusters. Further experiments are needed.

The patterns observed in the neutral alkali clusters are also seen in positive cluster ions of the noble metals [10, 11], where the respective shell closings are observed for the corresponding electron numbers (20 electrons for the +21 ion, etc.). The ellipsoidal shell closings have not been reported for the noble metal cluster ions. The odd-even patterns [10, 11] in the noble metal ion sequences are unusually strong, and appear to persist to cluster sizes approaching $N=100$. Some features in the noble metal mass spectra suggest that some clusters may show octupole distortions [12]. Experiments are needed for a larger variety of elements, and for a range of temperatures to distinguish behavior of neutral and ionized clusters.

Some features consistent with the shell model appear in the divalent cluster ions of Zn and Cd [13], but a number of features remain to be explained. The absence of odd-electron systems for these clusters is significant. Among the trivalent elements Al, Ga, and In have been investigated. Of these, Al [14] shows definite shell effects, while for Ga and In [15] it is questionable. Further investigation of these materials should give some clues as to when and how the valence electron energies cease to be dominant. The mass spectrum of Pb [16] fails to show features which are consistent with the simple shell model described above. Similarly, although definite patterns appear in the mass spectra of transition metal clusters, they seem not to be simply related to the shell structure which is derived principally from delocalized electrons. For these metals, it is probably necessary to include the effects of d-electrons. More experiments are needed.

The relative mass abundance spectra for clusters appear to give a good account of the relative stabilities, and the more direct measurement of the binding energies is in good agreement with abundance spectra and the shellmodel [17]. Further studies of fragmentation [18, 19, 20] add valuable information on cluster stabilities and the role of temperature in the processes.

The interesting question of metal-insulator transitions is raised for example in the behavior of Hg clusters, where evidence indicates metallic behavior above size range N ~ 20–70 [21, 22]. The evidence comes from two quite different experiments involving ionization potential and inner core ionizations respectively. Further experiments along these lines will be valuable in filling out the picture for the not-so-simple clusters. Relevant experiments involving mass abundance spectra, polarizability, and plasma resonance should help in clarifying the processes which take place.

III. Some properties of metal clusters

While the abundance spectra and stabilities reflect the underlying structure of metal clusters, there is further interest in the behavior as shown in other properties. In this section we will discuss some known cluster properties, and relate them to the shell structure. Thus far we have discussed indirect evidence for discrete levels in the electronic structure.

Photoelectron spectroscopy

Direct evidence for the existence of levels is found in several experiments, in particular the measurement of photoelectron spectra. From these experiments [23–26] one can determine the location of highest occupied states as well as more strongly bound states, and higher excited states. These appear as distinct peaks in the observed kinetic energy spectra of electrons emitted following photon absorption. In an advanced version [27] of these experiments coincidence techniques identify the emitted electrons with the respective parent clusters. Electron affinities and energy gap values have been obtained by these methods for the alkali and noble metal clusters. The results appear to be consistent with the shell model. It is expected that further experiments will extend the results to a wider range of elements and clusters and give information on fine structure and temperature effects. In particular, the splitting of levels [25] could be investigated and related to the residual crystal fields and the order of level filling as clusters grow in size.

Ionization potentials

The electron affinities and ionization potentials can be determined in photoelectron spectroscopy experiments. It is important to compare these experiments in detail with directly obtained ionization potentials. Information is available on IPs for a number of metal clusters, but few measurements extend over a wide range, say for $N \sim 100$. Since shell effects are clearly observed for potassium [28] over this range, it is important to obtain accurate IP values for other metals over the same range for comparison. Detailed correlation of IPs and stability is an important feature in relating properties to shell structure. Some developments are needed in experimental techniques and data analysis to refine values for IP, and to understand the physics of threshold and post-threshold behavior of the photoionization efficiency curves. Recent experiments show the effects of temperature and fragmentation during ionization [18, 29]. From these it appears that except for small effects the IP is not directly dependent on temperature. Further investigation of these effects will be important.

The problem of the overall progression of the IP, in a long sequence of clusters as bulk behavior is approached, also appears to require further study. A general trend in which IP varies inversely as the cluster radius is usually observed. However, the coefficient of the $1/R$ term depends on the metal, and also on the range of cluster size. Thus, while the IP is one of the important properties which characterizes a cluster species, there remain a large number of questions, not the least of which is a significant discrepancy between experimental and theoretical values [3].

Static polarizability

The polarizability measures the capacity of a system to screen out external fields. Results on Na and K [31] indicate that metallic screening sets in at least as early as $N=3$. Measurements of the polarizability combined with ellipsoidal depolarizing factors facilitates determination of the probable shapes of metal clusters. Spherical clusters $N=8$ and 20 show lower polarizabilities than neighboring open shell systems. Different experimental values were obtained for cluster $N=20$ [31] under conditions which suggest that the polarizability is dependent on temperature. This important question deserves further experimental investigation. Since the experimental deflection measurement is sensitive to cluster beam velocity, the measurement techniques should be improved to permit direct measurement of polarizability on individual clusters which are both mass and velocity selected. The measurements should include a number of elements and a wide range of cluster sizes.

Since metallic behavior is associated with screening, polarizability measurements, e.g. on Hg, should give some valuable information regarding the metal-insulator transition. It is also worthwhile to point out that the polarization experiments involve a deflection sensitivity which on the average is nearly independent of mass. The measurement may therefore be applied to clusters of any size, and therefore is particularly suitable for investigating the development of properties as cluster size increases toward the bulk limit. In general the polarizability experiments should be valuable in answering questions concerning metallic behavior. As yet, no experimental evidence has been found for the existence of permanent dipole moments in metal clusters. However, the question will not be settled until a series of careful measurements has been made.

IV. Collective electron motion in metal clusters

A system which screens should also exhibit collective oscillations. In larger metal clusters, such oscillations have been called Mie resonances [32]. These effects have recently been observed [33] in Na clusters for $N=3$–70, and the important fact that the phenomenon can occur even for very small systems, i.e. $N=3$–8, is established. Present indications [34] are that the "spherical" cluster $N=8$ shows a single resonance peak, and that "prolate" clusters $N=9$ and 10 show resonance structures consistent with the expected double peaks. However, there are experimental gaps in the measured spectra. Until the experiments investigate a wider frequency range continuously, questions will remain. For example: are the detailed predictions of shapes fulfilled for a wide range of clusters; is the oscillator strength correct; can single particle peaks [35] be observed; what are the functional shapes of the resonance curves; what are the damping mechanisms; what is the temperature dependence of the damping?

The experiments should be extended to include a wide range of conducting, semi-conducting, and non-conducting neutral and ionized clusters. The accuracy of the assumption that static and dynamic polarizabilities are approximately equal [35] should be tested. Since the dipole resonance depends mainly on the density of electrons, the size dependence should not be strong, and the specific size dependence of the resonance frequency should be tested directly. Preliminary indications are that the size dependence is not strong for small clusters. Some size dependence of resonance frequency has been reported for larger particles [36]. Experiments should be undertaken to bridge the gap between the size ranges which have been studied, remembering that the larger clusters have mainly [37] been studied in matrix.

There are reports of definite size dependence [36] of plasma resonance damping for larger clusters, and extension of the cluster beam experiments to large clusters should be undertaken in order to sort out temperature, surface collision, and resistive loss effects as cluster size increases toward the bulk. Further areas for extended investigations are in the multipole resonances, for example the breathing and quadrupole modes.

The study of plasmons in clusters appears to be one of the fruitful areas for cluster study in the near future. It provides direct means for investigating metallic behavior, the interaction between photons and cluster electrons, and interaction between the cluster electrons and vibrational modes. More generally the study of cluster plasma resonances permits some evaluation of interactions between single particle and collective modes in the Fermi liquid. Finally, as we mentioned in the introduction, supported clusters are not free. The nature of cluster-matrix surface interactions may be studied directly by a series of experiments which compare plasmon behavior in mass selected free clusters with the same clusters in selected supports.

This work has been partly supported by the U.S. National Science Foundation under Grant #DMR 86-15246. I am grateful to Dr. Sven Bjornholm and the Neils Bohr Institute, Copenhagen, Denmark, for their hospitality during the spring of 1988 while this manuscript was in preparation.

References

1. Robbins, E.J., Leckenby, R.E., Willis, P.: Adv. Phys. **16**, 739 (1967)
2. DeHeer, Walt A., Knight, W.D., Chou, M.Y., Cohen, M.L.: Solid State Phys. **40**, 93 (1987)
3. Cohen, M.L., Chou, M.Y., Knight, W.D., deHeer, W.A.: J. Phys. Chem. **91**, 3141 (1987)
4. Koutecky, J., Fantucci, P.: Chem. Rev. **86**, 539 (1986)
5. Benedek, G., Martin, T.P.: Elemental and Molecular Clusters, Proceedings of the Ettore Majorana Summer School, Erice 1987. Berlin Heidelberg New York: Springer 1988
6. Sugano, S., Nishina, Y., Onishi, S., (eds.) Microclusters. Berlin Heidelberg New York: Springer 1987
7. Clemenger, K.: Phys. Rev. B 32, 1359 (1985)
8. Ekardt, W., Penzar, Z.: Phys. Rev. B **38**, 4273 (1988)
9. Clemenger, K.: PhD Thesis, University of California, Berkeley, 1986
10. Katakuse, I., Ichihara, T., Fujita, Y., Matsuo, T., Sakurai, T., Matsuda, H.: Int. J. Mass Spectrom. Ion Proc. **67**, 229 (1985)
11. Joyes, P.: Surf. Sci. **106**, 272 (1981)
12. Nishioka, H.: private communication
13. Katakuse, I., Ichihara, T., Fujita, Y., Matsuo, T, Matsuda, T.: Int. J. Mass Spectrom. Ion Proc. **69**, 109 (1986)
14. Devienne, F.M., Roustan, J.-C.: Org. Mass Spectrom. **17**, 173 (1982); Jarrold, M.F., Bower, J.E., Kraus, J.S., J. Chem. Phys. **86**, 3876 (1987)
15. Martin, T.P.: in Ref. 5
16. Mühlbach, J., Sattler, K., Pfau, P., Recknagel, E.: Phys. Lett. A 87A, 415 (1982); for a discussion of transition metal clusters, see Whetten, R.L., Cox, D.M., Trevor, D.J., Kaldor, A.: Surf. Sci. **156**, 8 (1985)
17. Bréchignac, C., Cahuzac, Ph., Leyginier, J., Weiner, J.: J. Chem. Phys. **90**, 1492 (1989)
18. Bréchignac, C., Cahuzac, Ph., Roux, J.-Ph: J. Chem. Phys. **88**, 3022 (1988)
19. Kappes, M.M., Schär, Schumacher, E., Vayloyan, A.: Z. Phys. D-Atoms, Molecules and Clusters **5**, 359 (1987)
20. Begemann, W., Dreihöfer, S., Ganteför, G., Siekmann, H.R., Meiwes-Broer, K.H., Lutz, H.O.: in Ref. 5
21. Rademann, K. Kaiser, B., Even, U., Hensel, F.: Phys. Rev. Lett. **59**, 2319 (1987)
22. Bréchignac, C., Cahuzac, Ph., Labastic, P., Wolf, J.P., Wöste, L.: Phys. Rev. Lett. **60**, 275 (1988)
23. Kasdan, A., Lineberger, W.C.: Phys. Rev. A 10, 1658 (1974)
24. Cheshnovksy, I.O., Yang, S.H., Pettiette, C.L., Craycraft, M.J., Liu, Y., Smalley, R.E.: Chem. Phys. Lett. **138**, 119 (1987); see also Cheshnovsky, I.O., Smalley, R.E: Z. Phys. D–Atoms Molecules and Clusters **12** (1989) (this issue)
25. Bowen, K.: Z. Phys. D–Atoms, Molecules and Clusters **12** (1989) (this issue)
26. Meiwes-Broer, K.H.: Z. Phys. D–Atoms, Molecules and Clusters **12** (1989) (this issue)
27. Rademann, K., Kaiser, B., Hensel, F.: Z. Phys. D–Atoms, Molecules and Clusters **12** (1989) (this issue)
28. Saunders, Winston A., Clemenger, Keith, deHeer, Walt A., Knight, W.D.: Phys. Rev. **B 32**, 1366 (1985)
29. Cahuzac, Ph.: Z. Phys. D–Atoms, Molecules and Clusters **12** (1989) (this issue)
30. Chou, M.Y., Cleland, A., Cohen, M.L.: Phys. Lett. **A 113**, 420 (1986)
31. Knight, W.D., Clemenger, K., deHeer, W.A., Saunders, W.A.: Phys. Rev. **B 31**, 2539 (1985)
32. Mie, G.: Ann. Physik **25**, 377 (1908)
33. deHeer, Walt A., Selby, K., Kresin, V. Masui, J., Vollmer, M., Châtelain, A., Knight, W.D.: Phys Rev. Lett. **59**, 1805 (1987)
34. Selby, K., Vollmer, M., Masui, J., Kresin, V., Kruger, M., DeHeer, W., Knight, W.D.: Z. Phys. D–Atoms, Molecules and Clusters **12** (1989) (this issue)
35. Ekardt, W.: Phys. Rev. B **31**, 6360 (1985)
36. Genzel, L., Martin, T.P., Kreibig, U.: Z. Physik B–Condensed Matter **21**, 339 (1975)
37. Mann, D.M., Broida, H.P.: J. Appl. Phys. **44**, 4950 (1973)

Z. Phys. D – Atoms, Molecules and Clusters 12, 319–326 (1989)

Atoms, Molecules
Zeitschrift für Physik D and Clusters
© Springer-Verlag 1989

Electronic structure of metal clusters

G.K. Wertheim
AT&T Bell Laboratories, Murray Hill, NJ 07974, USA

Received 5 July 1988; final version 30 September 1988

Photoemission spectra of valence electrons in metal clusters, together with threshold ionization potential measurements, provide a coherent picture of the development of the electronic structure from the isolated atom to the large metallic cluster. An insulator-metal transition occurs at an intermediate cluster size, which serves to define the boundary between small and large clusters. Although the outer electrons may be delocalized over the entire cluster, a small cluster remains insulating until the density of states near the Fermi level exceeds 1/kT. In large clusters, with increasing cluster size, the band structure approaches that of the bulk metal. However, the bands remain significantly narrowed even in a 1000-atom cluster, giving an indication of the importance of long-range order.

The core-electron binding-energy shifts of supported metal clusters depend on changes in the band structure in the initial state, as well as on various final-state effects, including changes in core hole screening and the coulomb energy of the final-state charge. For cluster supported on amorphous carbon, this macroscopic coulomb shift is often dominant, as evidenced by the parallel shifts of the core-electron binding energy and the Fermi edge. Auger data confirm that final-state effects dominate in cluster of Sn and some other metals. Surface atom core-level shifts provide a valuable guide to the contributions of initial-state changes in band structure to cluster core-electron binding energy shifts, especially for Au and Pt. The available data indicate that the shift observed in supported, metallic clusters arise largely from the charge left on the cluster by photoemission. As the metal-insulator transition is approached from above, metallic screening is suppressed and the shift is determined by the local environment.

PACS: 79.60; 36.40

Introduction

The fundamental goal of cluster studies is to elucidate the size-dependent atomic and electronic structure of clusters ranging in size from the isolated atom to the bulk metal. The structural and electronic properties are, of course, intimately coupled. The electronic structure depends on the spatial arrangement of the atoms, and the equilibrium spatial structure depends on the ability of the resulting electronic bands to accommodate the outer electrons. First principles calculations to determine the equilibrium structure of few-atom clusters have been carried out for small Na [1] and Si [2] clusters, among others.

Many of the interesting phenomena that occur in clusters arise from the transition from bonds to bands with increasing cluster size. For elements that are metals in the bulk there will be a transition from insulator to metal with increasing cluster size. Since this phenomenon occurs in all the systems discussed here, it is important to have a clear understanding of what constitutes metallic behavior. Phenomenologically a metal is defined by its ability to conduct an electric current and by the positive temperature coefficient of its resistivity. These properties depend on the presence of a conduction band intersected by the Fermi level, which make it possible to excite electron-hole pairs of all energies, no matter how small. This is the fundamental criterion which defines a metal. All the properties which are unique to metals depend on such a partially filled band. The presence of electronic states delocalized over all the atoms of a cluster is a necessary, but not a sufficient condition to make it a metal.

The transition from insulator to metal takes place when the density of electronic states in the vicinity of the Fermi level is high enough so that the levels overlap by virtue of the thermal width [3]. The critical cluster size for the transition to the metallic state, estimated on this basis, falls in the vicinity of 100 atoms for an alkali metal at room temperature. For compact Al clusters the development of the conduction band has been calculated for clusters of 13, 55, and 147 atoms [4], showing a rapid convergence toward a bulk-like conduction band.

Cluster growth

Most of the data to be discussed here were obtained with clusters grown on a substrate, typically sputter-cleaned amorphous carbon. The metal atoms are usually deposited from a Knudsen cell onto a room temperature substrate, and have sufficient mobility to nucleate $\sim 3 \times 10^{12}$ clusters/cm^2 during the early stages of the deposition [5, 6]. The number of atoms in the average cluster is obtained by dividing the coverage by the density of nucleation sites. The average volume and radius can then be readily obtained. Continued deposition results in growth of these clusters, leading ultimately to coalescence into a contiguous metallic layer, but one which typically has holes. Since the cohesion of the cluster atoms is much greater than their adhesion to the substrate, the clusters will be compact and will tend to have a faceted, spherical shape. The mass of each cluster is proportional to the amount of material deposited per unit area of the substrate, up to the point where coalescence becomes significant. The cluster diameter consequently varies as the cube root of the coverage, Θ. Small regions of this relationship have been approximated by a linear behavior, but the average cluster diameter is never proportional to the coverage. Confirmation of the growth and coalescence of clusters was recently obtained from the coverage-dependence of the core electron binding energy and signal intensity [7].

Electron spectroscopy of clusters

We begin by considering the contribution of electron spectroscopy to our understanding of the electronic structure of clusters.

1. Threshold ionization potential

The smallest photon energy capable of ejecting an electron from a cluster locates the highest occupied molecular orbital (HOMO) relative to the vacuum level. Measurements on small alkali metals clusters show a rapid initial drop of this threshold ionization potential (IP) from the single atom to the few-atom clusters, followed by a continued slow decrease in many-atom clusters [8–10]. These results are in good accord with a shell-model calculations in which the nuclear and inner electronic charge is treated as jellium [11, 12]. The outer electrons are delocalized over the entire cluster, but have discrete energy levels and exhibit shell-closure effects (magic numbers) analogous to those of the nuclear shell model. These are in good accord with experimental observations. The model also reproduces the even-odd fluctuations found in alkali atom clusters. For the range of cluster sizes typically studied in beam experiments, the clusters are not metallic by the criterion established above. The combination of measurement and theory for the few-atom cluster leads to a simple picture for the behavior of the outer electrons, which are destined to become the conduction electrons in the larger, metallic cluster.

In large metallic clusters the threshold ionization potential approaches the work function of the bulk metal from above with increasing cluster size. The difference between the threshold IP of a free cluster and the work function of the bulk metal is equal to the coulomb energy $e^2/2r$ of the charged cluster. (The cluster has unit positive charge in the final state due to the emission of the electron.) An image-charge model had earlier led to the conclusion that this term has the form $3e^2/8r$ [13, 14], but an analysis based on energies is more reliable. For small metallic clusters, with band structures significantly different from that of the bulk metal, there will also be changes in the chemical potential and in the surface dipole layer [15].

In general then, as we form clusters starting with the isolated atom, progressing through the molecular and metallic cluster to the bulk metal, the threshold ionization potential decreases from the first ionization potential of the free atom of the work function of the bulk metal. The states of the outer electrons are delocalized but remain discrete until the density of states becomes so high that they overlap by virtue of their thermal width, forming a band. If this band is not fully occupied metallic behavior obtains. The threshold IP follows a $1/r$ law in the metallic region. The size below which this law breaks down may be identified with the metal-insulator transition. These simple measurements combined with suitable theory provide a great deal of insight into the behavior of the outermost electrons of clusters of metal atoms. For more detailed information about the electronic structure we must turn to a more sophisticated spectroscopy.

2. Photoemission spectroscopy of valence bands

In order to observe the electronic structure in greater detail one determines the kinetic energy distribution of the electrons excited by photons with energy well above the IP threshold. In spite of the multitude of phenomena that modify the details of this spectrum, it corresponds in first order to the distribution of occupied electronic states, both core and valence.

The most immediate application is the measurement of valence band density of states. Such data have consistently shown a narrowing of the bulk metal bonds, when the metal is in the form of small clusters [16–21], Fig. 1. The measured band widths are typically monotonically related to the cluster size, and to parameters like the average coordination number, which provide a quantitative measure of the difference between a cluster and the bulk metal. In model calculations for simple metals the bandwidth is proportional to the square root of the coordination number [22]. The band narrowing at the surface of metals is in rough agreement with this prediction, so that one might expect it to be applicable to large metallic clusters, which one can visualize as a core of bulk-like metal surrounded by a surface-like layer. However, one should not extrapolate this "law" from the large metallic cluster all the way to the molecular clusters with small average coordination numbers.

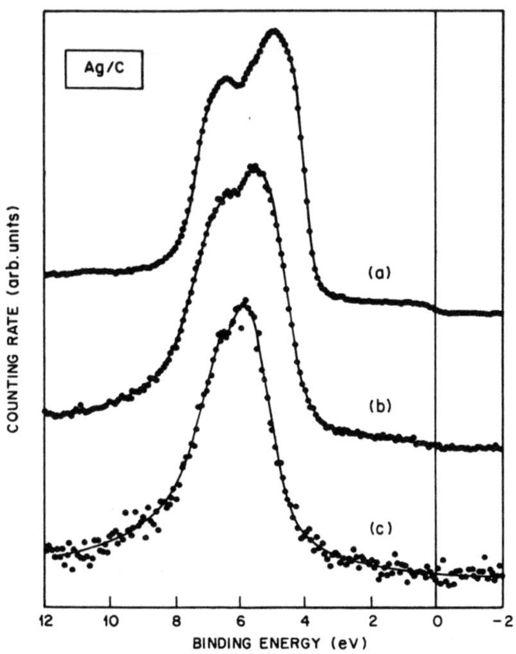

Fig. 1. X-ray photoelectron spectra of the valence bands of bulk silver, and Ag clusters at coverages of 1.0×10^{15} and 2.5×10^{14} atoms/sm^2 on amorphous carbon. The substrate signal has been subtracted. (from Ref. 27)

X-ray photoelectron spectroscopy of the valence bands of random-sized supported clusters shows not only an over-all narrowing, but also a loss of the detailed features characteristic of the bulk solid. Since this effect is seen in quite large clusters, it is not readily explained by the presence of different cluster sizes, but must be due to the loss of long-range periodicity. Recent measurements with relatively small, size-selected supported Au clusters [23] show the *5d* band approaching a broadened spin-orbit doublet. It is not yet clear whether this broadening should be attributed to *5d-5d* overlap or to the inequivalent atomic sites in the cluster. Significant progress in valence band spectroscopy has also been made in pulsed laser-excited photoemission from small, free Cu clusters, in which individual 4*s* electron states were resolved [24].

Under favorable conditions, photoemission can provide more detailed information about the structure of the valence bands. For example, for a planar polycrystalline surface, the band structure of the surface atoms can be separated from that of the bulk simply by changing the take-off angle [25]. It is unfortunately impossible to apply this techniques to supported clusters. All we can obtain is a total density of valence band states for all the atoms in the cluster. Nevertheless, the information obtained from surface layers on bulk crystals is relevant to the interpretation of cluster properties. For a 100 atom cluster, with a diameter typically near 10Å, 80 percent of the atoms are part of the surface. An appreciable fraction of these are edge atoms, which should have properties even further removed from those of the bulk than the surface atoms. To a low order of approximation one might then expect to be able to predict the changes observed in the valence band of large clusters in terms of those found in the bulk and at the surface of macroscopic solids, using the fraction of bulk and surface atoms in the cluster as a guide.

3. Photoemission from core levels

Core-electron spectra are more easily obtained by XPS than valence band spectra, because the lines are narrower and the core electrons have larger cross sections. A great deal of work focusing on the core-electron binding-energy shift has been reported for clusters of simple, noble, transition, and rare earth metals. The remarkable fact is that in every case the core-electron binding energy is larger in the cluster than in the metal,

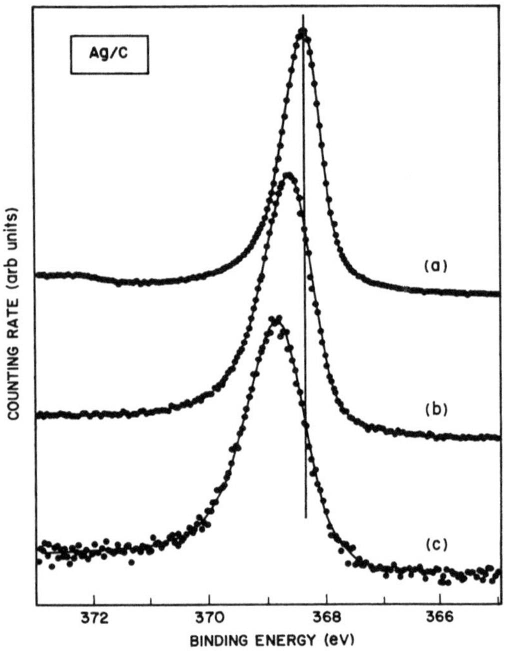

Fig. 2. The Ag $3d_{5/2}$ core electron spectra of samples shown in Fig. 1. (from Ref. 27)

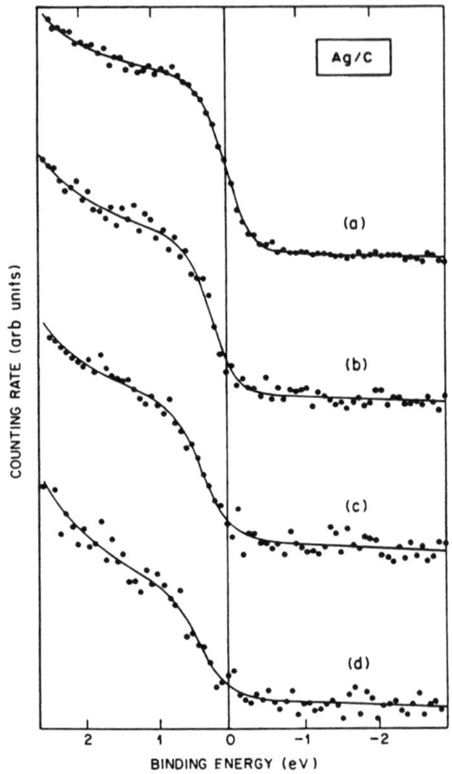

Fig. 3. Shift of the Fermi edge of small supported Ag clusters relative to that of the amorphous carbon substrate. (from Ref. 27)

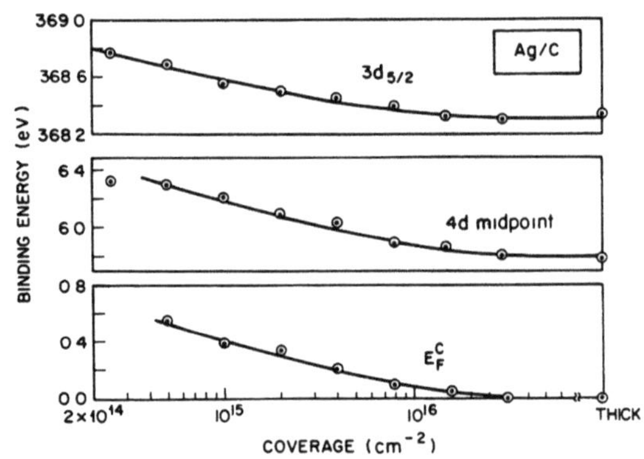

Fig. 4. Comparison of Fermi edge, valence band, and core level shifts of supported Ag clusters. (from Ref. 27)

increasing monotonically with decreasing cluster size, as illustrated in Fig. 2. This is traceable simply to the fact that photoemission may leave even a supported cluster with unit positive charge in the final state [26]. However, depending on cluster size, other mechanisms will also contribute. Included here are (1) a reduction of the screening in the final state, and (2) changes in band structure resulting in a redistribution of charge between orbitally-distinct bands in the initial state. For supported clusters the substrate interaction will also have to be considered.

For large, *isolated* metal clusters the analysis is simple. The core hole is screened by the conduction electrons much as it is in the bulk metal. The resulting charge-deficit in the conduction band appears at the surface of the cluster, making the final result quite similar to that encountered in the discussion of the threshold IP. The energy of the final state is increased by $e^2/2r$ relative to that of the infinite metal, correspondingly reducing the kinetic energy of the photoelectron, and increasing the core-electron binding energy. Confirmation that this mechanism is dominantly responsible for the core-level shift comes from the comparison of the shifts for the core electron and the cluster Fermi level [26, 27], see Figs. 3 and 4. It should be noted that for the smaller clusters the shift of the 3*d* electrons is smaller than the Fermi edge shift. This is due to changes in band structure that add a small negative components, i.e., a shift with the same sign as the surface atom core-level shift of silver. This is in accord with the increase of the surface to volume ratio with decreasing cluster size. The band structure contribution to the shift should in general have the same sign as the surface-atom core-level shift, e.g. positive for transition metals with less than half-filled *d*-shell, and negative for those with more than half-filled shell.

For *supported* metal clusters the energy of the final state of the system is reduced to about 1/2 by the formation of an image of the charged cluster in the substrate. For grounded clusters, e.g. clusters supported on a metallic substrate, this shift will vanish [26, 28].

In small, metallic clusters conduction electron screening will be suppressed if the coulomb energy of the charged cluster ($e^2/2r$) exceeds the energy reduction afforded by the core-hole screening process [28–30]. Consequently, the macroscopic coulomb shift will never exceed the relaxation energy in the bulk metal. Metals with a small density of states at E_F, and hence small small screening energies, will have small cluster core-electron binding-energy shifts.

In smaller, non-metallic clusters conduction electron screening is replaced by the polarization of the neighboring atoms or even by the transfer of a valence electron to the atom with the core hole. The final state is more localized, and no longer depends strongly on cluster size. In these clusters the screening will be a function of the neighbor environment (coordination number), so that one may expect inhomogeneous broadening of the core-electron photoemission line. In this regime the core-electron binding-energy shift is not equal to the change in threshold IP.

Theoretical calculations of core-electron binding energy shifts have so far been reported only for small, nonmetallic cubical Li clusters [31]. They show the expected decrease in binding energy with increasing cluster size, and demonstrate that atoms with different local environment have distinct core-electron binding energies. The calculated range of energies is, however, much larger than one would find in real clusters, because the model includes corner atoms with very low coordination number [32]. These shifts are described as having an initial state origin, while those measured in our experiments are clearly due to final state effects. The authors of [31] see this as a fundamental contradiction. However, experiment and theory are concerned with different regimes of cluster size, metallic vs. molecular, so that they are not directly comparable.

4. Auger electron spectroscopy

The kinetic energy of Auger transitions between core levels depends on the initial state band structure and the final state relaxation. The same contributions appear in the core-electron binding energy obtained by photoemission. In the Auger process the relaxation term has a coefficient of 3, because the initial state has one core hole and is singly charged, while the final state has two and is doubly charged. This makes it possible to combine core-electron binding energy shifts with Auger kinetic energy shifts to separate the initial and final state contributions [33]. Core-valence-valence Auger transitions cannot be used in this analysis, because the final state screening of a valence hole is inherently different from that of a core hole. In this analysis, the effect of charge left on a cluster by photoemission is not distinguishable from that of final state relaxation [34]. Limited attempts to use this formalism to identify the initial and final state contributions to core-electron binding energy shift indicate that final state effects dominate [20, 34].

Case studies

We will now examine some of the available data for a number of different metals to illustrate and expand on the points raised above. We begin with a particularly simple case.

1. Tin

In Sn the conduction band consist of 5*s* and 5*p* states, with limited opportunity for charge flow between bands with greatly different radial distribution. The magnitude of the surface atom core level shift is much less that 0.1 eV. The 3*d* core electron binding energy Sn clusters supported on amorphous carbon follows the 1/r behavior expected on the basis of the final state charge [7, 29, 35], see Fig. 5. No information about the electronic structure is consequently obtained from

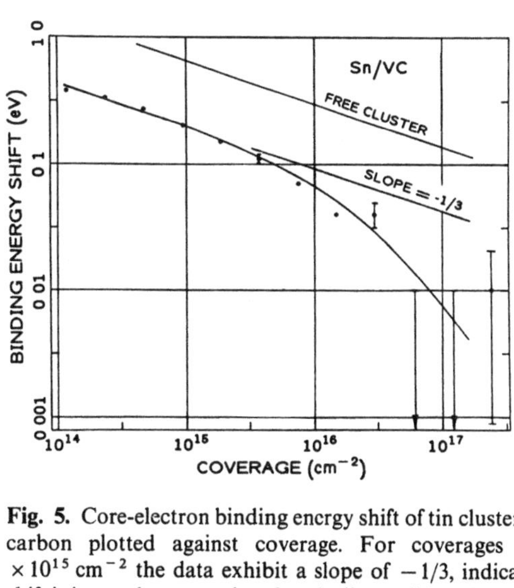

Fig. 5. Core-electron binding energy shift of tin clusters on vitreous carbon plotted against coverage. For coverages less than 3 $\times 10^{15}$ cm^{-2} the data exhibit a slope of $-1/3$, indicating that the shift is inversely proportional to the linear dimension of the clusters. At higher coverage the shift drops rapidly below this line indicating that the clusters have begun to coalesce. The coulomb shift for free clusters of the same diameter is also indicated. (from Ref. 7)

core level studies, but the results confirm that the final state charge is the source of the observed core level shifts. Auger data support the final-state nature of the shift [34].

2. Silver

In Ag the 4*d* band lies in the interval from 4 to 7.5 eV below the Fermi level. The density of states at E_F is much smaller than in Sn. The surface atom core level shift is small, i.e., −0.08 eV [25]. As we have already seen in Fig. 1, the Ag 4*d* valence band narrows significantly and remains well below E_F as the cluster size is decreased. The 4*d* band, consequently, remains filled, so that there is no flow of charge between the 4*d* and 5*s* bands. The largest observed 3*d* cluster core-level shift is smaller than that in Sn, reflecting the smaller density of states. The observations are fully accounted for by the cluster charge [7, 27].

3. Gold

In Au the 5*d* band lies in the interval between 2 and 7.5 eV below E_F. The surface-atom core-level shift is −0.4 eV, i.e. toward smaller binding energy [25]. The 5*d* band is narrowed at the surface, and exhibits a centroid shift comparable to the core-level shift. In the clusters the *d*-band is also narrowed, but the shifts are toward larger binding energy with decreasing cluster size. The data do not conform to the behavior of Sn or Ag when plotted as in Figs. 4. A reasonable interpretation is that initial-state effects produces a shift toward smaller binding energy, i.e., with the same sign as the surface atom shift, which competes with the effect of the cluster charge, resulting in a more complicated size-dependent shift. No attempt has so far been made to separate these two contributions.

Recent experiments with size-selected Au clusters [23] have given band widths comparable to those of random-sized clusters with the same average diameter, indicating that the width is an inherent property of the individual cluster, and not due to the size distribution. This is reasonable if the width is due to residual *d*-*d* interactions. The structure of the valence band spectra is no more detailed than that of random-sized clusters. Core electron spectra, Fig. 6, also exhibit shifts and broadenings comparable to those of random-sized clusters. The broadening of the core levels is, however, much smaller than that of the valence band, and probably reflects the inequivalence of the atoms within each cluster. The interaction with the substrate undoubtedly makes an important contribution to this inequivalence of the various atoms in the cluster.

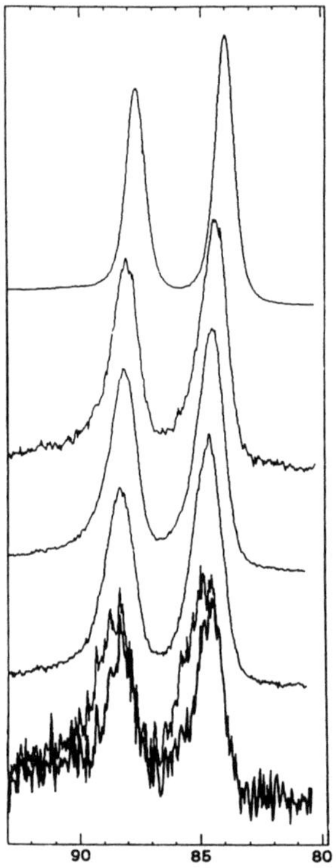

Fig. 6. XPS spectra of the Au 4f electrons of size-selected, supported Au clusters. The upper curve is for bulk gold, those below for clusters of 33, 27, 7, and "5" atoms. (from Ref. 23)

4. Palladium

In Pd the 4*d* band is intersected by the Fermi level. There would consequently appear to be ample opportunity for charge redistribution between it and the 5*s* conduction band. However, charge flow between the bands can only come about by the motion of the two band relative to each other [27]. Narrowing of the almost filled 4*d* band will move the top edge away from the vacuum level. Narrowing of the almost empty 5*s* band will move its charge toward the vacuum level. The net result of the narrowing of both bands will be a transfer of charge into the 4*d* band moving the system toward the $4d^{10}$ configuration of the free Pd atom. The position of the Fermi level after band-narrowing will be largely determined by the motion of the top of the high-density-of-states 4*d* band. On the basis of these considerations one would expect a negative surface-atom core-level shift, but this has not been experimentally established. Our own attempts to measure the shift were unsuccessful, and indicated only that it is much smaller than that of Au.

The narrowing of the 4*d* valence band in Pd clusters has been known for many years [16, 17]. The valence-band data also show that the *d* band moves

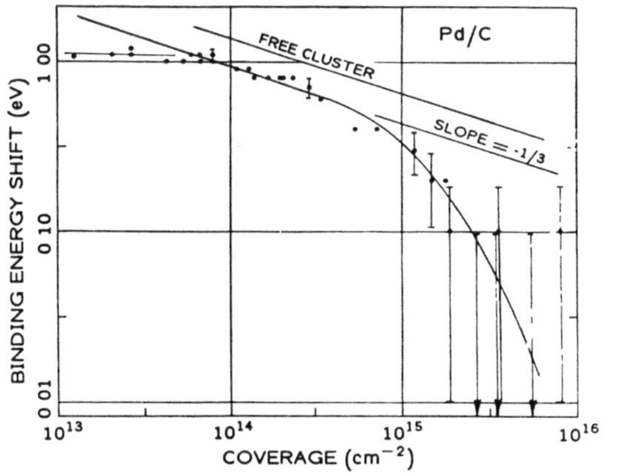

Fig. 7. Core-electron binding energy shift for Pd clusters, from Ref. 37, replotted as in Fig. 3. Note the saturation of the shift below a coverage of 10^{14} cm^{-2}, where the average clusters contains fewer than 30 atoms. (from Ref.7)

away from the Fermi level of the substrate with decreasing cluster size. If it were not for the effect of the final state charge, the separation of the *d*-band from E_F would be a clear indication of a transition to an insulating state. (A filled 4*d* band implies an empty 5*s* band.) The core-level shift toward larger binding energy with decreasing cluster size is also well established [20, 27, 36, 37]. The sign of the shift is opposite to that expected from the increased *d*-band occupancy, and is again best explained in terms of the final-state charge. The 3*d* binding energy shift, Fig. 7, conforms quite well to the predictions of the cluster charge model, lending considerable weight to that interpretation, and confirming that the band structure contribution to the shift is small. The data are unique in clearly showing a leveling off in the limit of small clusters, presumably corresponding to the transition to the non-metallic state in which the 5*s* band is empty.

Auger data have been interpreted as supporting an initial state origin for the Pd cluster shift [37]. However, since this work is based on a final state with two valence band holes, the conclusion is invalid [34].

5. *Platinum*

Platinum, the 5*d* analog of Pd, exhibits similar behavior. XPS data show that the valence band narrows and loses some of its barely resolved structure in small clusters. The loss of the sharp peak at E_F has been interpreted as indicating a loss of occupied 5*d* density [20], but is more likely due simply to the change in the band structure of a finite cluster. The size dependence of the core-electron binding energy shift conforms quite well to the predictions of the final-state cluster charge model [7], although a significant contribution from changes in band structure should be present.

Conclusions

Threshold ionization potentials provide only limited information about the development of the electronic structure of clusters. For small, insulating clusters the threshold IP can be interpreted in terms of the shell-model. For large, metallic clusters it reflects the coulomb energy of the charged final state. Valence band photoemission spectra have clearly demonstrated the narrowing of the bands in metallic clusters. More detailed information about valence bands is only now beginning to emerge from the study of size-selected clusters, both free and supported. Core-level binding energy shifts are dominated by the effects of the charged final state, and shed little light on the electronic structure. Currently major interest is focused on the metal-insulator transition, which is expected for clusters in the 7 to 10Å range, and appears to have been observed for supported Pd clusters.

I am indebted to S.B. DiCenzo for permission to use the unpublished results in Fig. 6, as well as for a critical reading of this manuscript.

References

1. Martins, J.L., Buttet J., Car, R.: Phys. Rev. Lett. **53**, 655 (1984)
2. Raghavachari, K., Logovinsky, V.: Phys. Rev. Lett. **55**, 2853 (1985); see also Blaisten-Barojas, E., Levesque, D.: Phys. Rev. B. **34**, 3910 (1986)
3. Kubo, R., Kawabata, A., Kobayashi, S.: Ann. Rev. Mater. Sci. **14**, 49 (1984)
4. Casula, F., Andreoni, W., Maschke, K.: J. Phys. C. **19**, 5155 (1986)
5. Hamilton, J.F., Logel, P.C.: Thin Solid Films **16**, 49 (1973); **23**, 89 (1974)
6. DiCenzo, S.B., Wertheim, G.K.: Comments Solid State Phys. **11**, 203 (1985)
7. Wertheim, G.K., DiCenzo, S.B.: Phys. Rev. B **37**, 844 (1988)
8. Rohlfing, E.A., Cox, D.M., Kaldor, A.: Chem. Phys. Lett. **99**, 161 (1983)
9. Powers, D., Hansen, S.G., Geusic, M.E., Michalopolous, D.L., Smalley, R.E.: J. Chem. Phys. **78**, 2866 (1983)
10. Knight, W.D., Clemenger, K., deHeer, W.A., Saunders, W.A., Chou, M.Y., Cohen, M.L.: Phys. Rev. Lett. **52**, 2141 (1984)
11. Ekardt, W.: Phys. Rev. B. **29**, 1558 (1984)
12. Beck, D.E.: Solid State Commun. **49**, 381 (1984)
13. Smith, J.M.: AIAA J. **3**, 648 (1965)
14. Wood, D.M.: Phys. Rev. Lett. **46**, 749 (1981)
15. Plieth, W.J.: Surf. Sci. **156**, 530 (1985)
16. Unwin, R., Bradshaw, A.M: Chem. Phys. Lett. **58**, 58 (1978); Takasu, Y., Unwin, R., Tesche, B., Bradshaw, A.M., Grunze, A.M.: Surf. Sci. **77**, 219 (1978)
17. Egelhoff, W.F., Tibbetts, G.G.: Solid State Commun. **29**, 53 (1979); Phys. Rev. B **19**, 5028 (1979)
18. Oberli, L., Monot, R., Mathieu, H.J., Landolt, D., Buttet, J.: Surf. Sci. **106**, 301 (1981)

19. Lee, S.T., Apai, G., Mason, M.G., Benbow, R., Hurych, Z.: Phys. Rev. B **23**, 505 (1981)
20. Mason, M.G.: Phys. Rev. B **27**, 748 (1983), and references therein
21. Colbert, J., Zangwill, A., Strongin, M., Krummacher, S.: Phys. Rev. B **27**, 1378 (1983)
22. Cyrot-Lackmann, F.: Adv. Phys. **16**, 393 (1967); J. Phys. Chem. Solids **29**, 235 (1968)
23. DiCenzo, S.B., Berry, S., Hartford, E.: Phys. Rev. B **38**, 8465 (1988)
24. Smalley, R.E.: Bull. Am. Phys. Soc. **33**, 642 (1988)
25. Citrin, P.H., Wertheim, G.K., Baer, Y.: Phys. Rev. Lett. **41**, 1425 (1978); Phys. Rev. B **27**, 3160 (1983)
26. Wertheim, G.K., DiCenzo, S.B., Youngquist, S.E.: Phys. Rev. Lett. **51**, 3210 (1983)
27. Wertheim, G.K., DiCenzo, S.B., Buchanan, D.N.E.: Phys. Rev. B **33**, 5384 (1986)
28. Wertheim, G.K., DiCenzo, S.B., Buchanan, D.N.E., Bennett, P.A.: Solid State Commun. **53**, 377 (1985)
29. Cini, M.: Surf. Sci. **62**, 148 (1977)
30. Castellani, N.J., Leroy, D.B., Lambrecht, W.: Chem. Phys. **95**, 459 (1985)
31. Parmigiani, F., Kay, E., Bagus. P.S., Nelin, C.J.: J. Electron Spectrosc. Relat. Phenom. **36**, 257 (1985)
32. DiCenzo, S.B., Wertheim, G.K.: J. Electron Spectrosc. Relat. Phenom. **43**, c7 (1987)
33. Wagner, C.D.: Faraday Discuss. Chem. Soc. **60**, 291 (1975)
34. Wertheim, G.K.: Phys. Rev. B **36**, 9559 (1987)
35. Cheung, T.T.P.: Chem. Phys. Lett. **110**, 219 (1984)
36. Cheung, T.T.P.: Surf. Sci. **127**, L129 (1983); ibid. **140**, 151 (1984)
37. Kohiki, S., Ikeda, S.: Phys. Rev. B. **34**. 3786 (1986)

Z. Phys. D – Atoms, Molecules and Clusters 12, 327–332 (1989)

Atoms, Molecules
Zeitschrift für Physik D and Clusters
© Springer-Verlag 1989

Electronic properties of small platinum particles as seen by NMR

J.J. van der Klink
Institut de Physique Expérimentale, Ecole Polytechnique Fédérale de Lausanne, PHB-Ecublens, CH-1015 Lausanne, Switzerland

Received 9 July 1988; final version 30 September 1988

No doubt partially because of their catalytic activity, small platinum particles on mineral supports have been extensively studied by NMR during recent years. Here we will discuss these data in relation to the expected electronic properties of small metal particles. Experimentally, no quantum size effects are found, but the local densities of state at the Fermi energy show considerable site-dependence when going from the surface to the interior of the particle. Such effects have been theoretically predicted for an infinite film.

PACS: 76.60.Cg; 36.40.+d

This paper focuses on magnetism-related NMR effects that arise through the hyperfine coupling between the electronic and nuclear spins: in bulk metals, the static part of the hyperfine coupling results in the Knight shift K and the dynamic part in the Korringa relation $T_1^{-1} \propto T$ (the spin-lattice relaxation rate is proportional to the temperature). We will restrict the discussion mainly to the case of ^{195}Pt NMR for which detailed data have become available since the first studies by C.P. Slichter's group, reported at the Lausanne edition of this conference [1].

In theoretical descriptions of the magnetic quantum size effect (QSE) one imagines a collection of particles all containing the same number of atoms; the size-parameter is expressed as

$$\delta = 4\,E_F/3N \tag{1}$$

where N is the number of "conduction" electrons in the particle, and E_F the Fermi energy of the bulk, measured from the bottom of the conduction band. Whereas all particles are assumed to have the same N, the set of one-electron energy values $\{E_i\}$ is thought to vary from particle to particle. For magnetic (and most other) problems, only levels within a few kT from E_F are important. Give the level index 0 to the Fermi level (or, more precisely, in the canonical ensemble to the highest occupied level at T=0): $E_0 \equiv E_F$. Then the distribution of the $\{E_i\}$ around E_0 is characterized by a number of probability densities: $P_n(x)$, where

$$P_n\left(\frac{\Delta}{\delta}\right) \mathrm{d}\left(\frac{\Delta}{\delta}\right) \tag{2}$$

is the fraction of particles that has the n-th level (counted from E_0 either to higher or to lower energies) such that

$$\frac{\Delta}{\delta} < \frac{|E_n - E_0|}{\delta} < \frac{\Delta + \mathrm{d}\Delta}{\delta} \tag{3}$$

By definition, for any probability density:

$$\int_0^\infty P_n\left(\frac{\Delta}{\delta}\right) \mathrm{d}\left(\frac{\Delta}{\delta}\right) = 1 \tag{4}$$

The fraction of particles with *any* level in that interval is given by:

$$R\left(\frac{\Delta}{\delta}\right) \mathrm{d}\left(\frac{\Delta}{\delta}\right) = \sum_{n=1}^{\infty} P_n\left(\frac{\Delta}{\delta}\right) \mathrm{d}\left(\frac{\Delta}{\delta}\right) \tag{5}$$

(called the two-level correlation function) with

$$\lim_{x \to \infty} R(x) = 1 \tag{6}$$

Let us first look at the susceptibility at very low temperatures, such that

$$(E_0 - E_{-1})_j \gg \mathrm{kT} \quad \text{and} \quad (E_1 - E_0)_j \gg \mathrm{kT} \tag{7}$$

Define

$$(E_1 - E_{-1})_j = 2\Delta_j \tag{8}$$

where the index j refers to the j-th particle in our collection. Its susceptibility at these low temperatures is:

$$\chi_j/\chi_{\text{Pauli}} = \delta/2\,\text{kT} \tag{9}$$

for an "odd" particle in both the canonical and grand canonical ensembles; and

$$\chi_j/\chi_{\text{Pauli}} = (4\,\delta/\text{kT})\exp(-\Delta_j/\text{kT}) \tag{10}$$

for an "even" particle in the canonical ensemble; due to a difference in definition of the Fermi level the grand canonical result is obtained by replacing (kT) with (2 kT).

If we neglect "local" effects within the particle we would say that the Knight shift of the NMR signal in this particle is:

$$\text{K}_j/\text{K}_\infty = \chi_j/\chi_{\text{Pauli}} \tag{11}$$

(where K_∞ is the bulk Knight shift).

The total NMR signal from the complete collection of particles is a superposition of signals with different Δ_j. At these low temperatures, the "odd" particle signal will remain sharp, since it does not depend on Δ_j; but $\text{P}_1(\Delta/\delta)$ will broaden the line from the "even" particles, so that in theory, $\text{P}_1(\Delta/\delta)$ can be determined from the low-temperature even-particle NMR signal. In practice, size distributions (that affect δ) will mask this effect, and also broaden the "odd" particle signal.

It is clear from the above that the collection of $P_n(x)$ (with $n = 1, 2, \ldots$) describes the distribution of the energy levels of the zero-field Hamiltonian, and that this distribution is hardly more "fundamental" than a size distribution. Any given particle j is thought to have a well-defined, time-independent energy level structure $\{E_i\}_j$. We apply the reasoning of statistical mechanics to an ensemble of particles all strictly identical to the one under consideration. The results depend on $\{E_i\}_j$. If experimentally we measure a collection of non-identical particles (having the same N or not), we obtain a superposition of results pertaining to each of the $\{E_i\}_j$ present in the sample. The nature of the superposition depends on the type of measurement: with a Faraday balance we simply obtain the sum of the forces exerted by the magnetic field gradient on all of the particles. In CESR, different $\{E_i\}_j$ will in principle lead to a spectrum of signals with different g, one signal for each particle. In NMR, we obtain a spectrum consisting of one signal from each distinct site in each particle (these "local" effects are not included in the simple reasoning leading to (11)).

As an illustration of such local effects, Fig. 1 shows a ^{195}Pt NMR spectrum (absorption derivative) in a very slightly alloyed (7×10^{-4} atom fraction Mo) bulk platinum sample, obtained by Weisman and Knight in 1968 [2]. The overwhelming majority of platinum nuclei resonates at around 1.14 G/kHz (off-scale in the figure) as in a pure sample, but those platinums that happen to be nearest-neighbours to a Mo impurity resonate at around 1.127 G/kHz; and second- and third-neighbours are visible in between. Apparently, the conduction electron spin density (or equivalently the local density of states (LDOS) at the Fermi level) on neighbouring sites to an impurity is different from that in "bulk" sites.

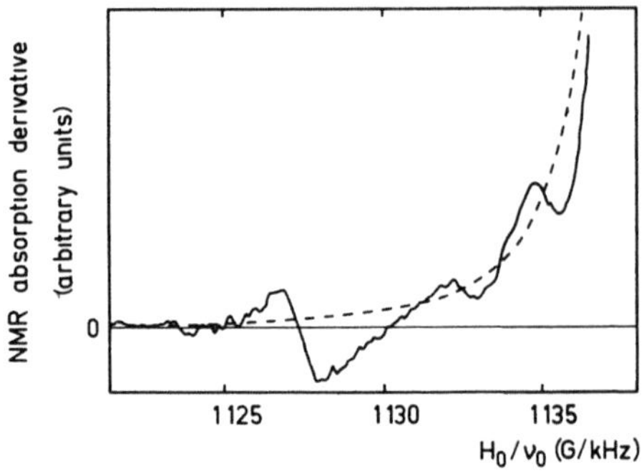

Fig. 1. Absorption derivative NMR spectrum of ^{195}Pt in a very slightly alloyed bulk Pt sample. Most of the ^{195}Pt nuclei resonate at about 1.14 G/kHz, off-scale to the right. Those ^{195}Pt that are first neighbours to the alloying impurity resonate around 1.127 G/kHz, and second- and third-neighbors give separate resonances as well, showing variations in the local densities of states at the Fermi level with distance from the impurity. Data from Ref. 2

As measured by the Faraday method, χ_0 of such alloys is lower than that of the bulk [3]: the NMR shows that this change can be considered as "localised" around the impurity (a diminished d-like LDOS). An immediate question is now: if the small-particle susceptibility is different from that of the bulk, will the difference also be "localised"?

A preliminary answer to the question is found in the NMR spectra shown in Fig. 2. Figure 2a represents data from the early work by Slichter and coworkers in platinum catalysts prepared by impregnation of η-alumina [4]. Figure 2b is a spectrum obtained in our laboratory for a catalyst prepared by ion exchange (a method related, but not identical, to impregnation) on titania [5]. As inferred from size distributions measured on electron micrographs, both these samples have a dispersion (fraction of atoms on the surface) around 60%; and in both cases the peak at the low-field end in Fig. 2 has a relative area of about 60%. This correlation has also been found for other values of dispersion: so we may assume that indeed there is "local" information in these spectra.

We find no detectable variation of the gross features of these spectra with temperature in the range

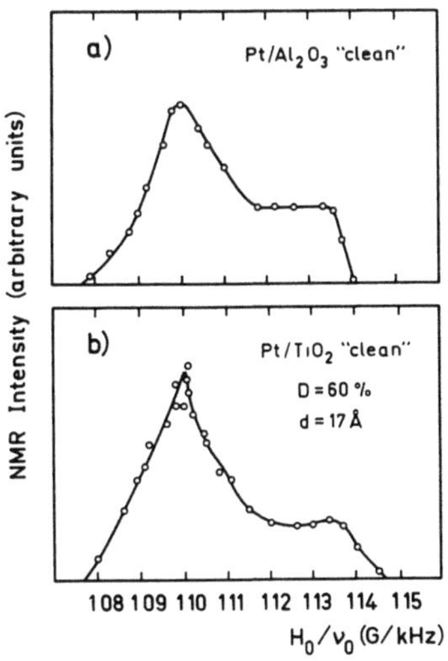

Fig. 2.a. One of the first ^{195}Pt NMR spectra of a platinum catalyst, obtained by C.P. Slichter's group. These absorption spectra are extremely broad, and are obtained from point-by-point spin-echo measurements. They show that the ^{195}Pt nuclei are in a wide diversity of electronic environments, resonating in general at a lower value of field than do bulk platinum samples (cf. Fig. 1). Data from Ref. 4. **b.** Similar spectra are obtained for Pt catalysts prepared in a different way on a different substrate: the particle size distribution here is comparable to that in **a**, and so is the spectrum. Data from Ref. 5

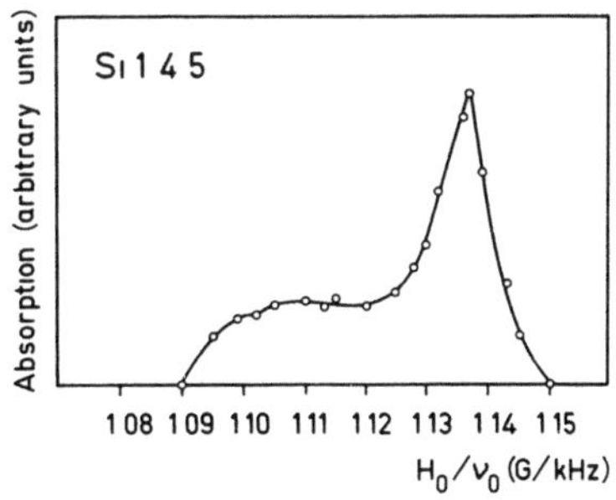

Fig. 3. ^{195}Pt NMR spectrum of particles prepared in the same way as those in Ref. 6a, but that have been deposited on silica powder instead of embedded in gelatin. Mean particle diameter 48 Å: note that more nuclei are in a "bulk-like" environment, compared to the samples in Fig. 2. Data from Ref. 7

studied (from 80 K down to 10 K), so the susceptibilities that determine them apparently behave in a Pauli-like way. In the case of alloys mentioned earlier, the appearance of NMR signals in the low field region is accompanied by a reduction in the overall susceptibility χ_0: in small Pt particles prepared by a third method (citrate-reduction) well-known in catalyst preparation, a nearly temperature-independent susceptibility was initially found at temperatures between 50 and 200 K, with a value $\chi/\chi_{\mathrm{bulk}}=0.25$ [6a]. It is now believed however that improved sample preparations lead to $\chi/\chi_{\mathrm{bulk}}\approx 1$ in this temperature region [6b] and that the initial results are atypical for this kind of preparation. Combined NMR and susceptibility experiments on the same sample would help in clarifying this situation. For the moment, we show in Fig. 3 the NMR spectrum of a silica-supported catalyst, prepared by the citrate reduction method [7]. The difference in the high field/low field intensity ratio as compared to Fig. 2 can be explained completely by a difference in particle size: more atoms are in a bulk-like environment. There are no indications that the method of preparation itself affects the NMR spectrum.

Marzke et al. [6] also reported an attempt to observe ^{195}Pt NMR in their sample; but they were unsuccessful. This has been considered as an indication of inhibition of the Korringa nuclear spin-lattice relaxation mechanism by the quantum size effect (although in hindsight it seems more likely that they were experimentally not prepared to observe extremely broad resonances as in Fig. 2a, that were discovered only several years later [1]). The Korringa mechanism is due to the non-diagonal $(I_+S_- + I_-S_+)$ part of the hyperfine interaction $\mathbf{I}\cdot\mathbf{S}$ between electron spin **S** and nuclear spin **I**. Therefore the relaxation mechanism involves a simultaneous flip-flop of a nuclear and an electron spin, changing their combined Zeeman energy by a quantity $|\hbar\omega_{\mathrm{electron}} - \hbar\omega_{\mathrm{nucleus}}|$. In magnetic fields used for NMR this energy corresponds to between 1 and 10 K on a temperature scale. Energy conservation then requires that the electron spin flip is accompanied by a scattering into an orbital wavefunction whose energy in zero field is different from the energy of the initial orbital wavefunction. In order that this scattering be possible we need first of all that an energy level at the required energy-separation does indeed exist, and next that the corresponding spin-orbital has a thermal probability of being empty. The latter condition restricts our discussion to states around the Fermi level, as usual in magnetic phenomena.

In a bulk sample, the spacing δ between the one-electron energy levels is small compared to their width w caused by lifetime broadening (scattering by electron-phonon interactions), resulting in a continuous density of states, that is (more or less) constant around E_F. The existence of the energy levels required for energy conservation in the nuclear spin-lattice relaxation is therefore assured.

In a small particle $w \ll \delta$, and in the general case the required energy levels are not available. In a sample characterized by a two-level correlation function $R(x)$ (cf. (5)) there is a fraction of the particles

$$R\left(\frac{\hbar\omega}{\delta}\right)\frac{w}{\delta}$$

that has a suitable level centered between $(\hbar\omega)$ and $(\hbar\omega$

$+w$), assumed to be broadened so as to have a non-zero intensity at exactly $h\omega$. (The actual value of the relaxation time will of course vary according to whether at distance $h\omega$ we are in the "maximum" or in the "wings" of the broadened level). Because $w \ll \delta$, and R is of the order of 1 at most, the Korringa relaxation mechanism is completely inhibited for a fraction $(\delta - w)/\delta \approx 1$ of the nuclei in the total small-particle sample; and this independent of the temperature of the experiment, as long as $\delta \ll w$ stays valid.

Therefore, to show the existence of a magnetic QSE by NMR, two results are necessary, both at *high* temperatures ($kT > \delta$):

i) the signal is centered at the bulk Knight shift ($kT > \delta$)

ii) the Korringa relation $T_1^{-1} \propto T$ is violated at any temperature ($kT > \delta$; although it then will remain true also at $kT < \delta$) and the values of T_1^{-1} are considerably less than in the bulk at the same temperature.

Note that these experiments completely avoid arguments concerning the "kind of ensemble", or "effect of spin-orbit coupling" that are current in the literature for low-T results. It is astonishing that they have never been performed with a positive (for demonstration of QSE, that is) result.

A suppression of the Korringa relation and a decrease in spin-lattice relaxation rates by four orders of magnitude at 1 K have been observed by Brom and coworkers [8] in platinum carbonyl cluster compounds of nuclearity 26 and 38 (Fig. 4). Their linepositions (Fig. 4a) are very roughly at zero Knight shift (as appropriate in most QSE theories for a metal with an even number of electrons per atom at very low temperature), independent of temperature between 1 and 4.2 K. The QSE effect would require the lines to move to the full Knight shift at "sufficiently" high temperatures, but the equipment used in this work could not be operated above 4.2 K.

In the case of platinum catalysts, it has been found that the spin-lattice relaxation rate T_1^{-1} varies across the spectrum. Its value at the position of "bulk" platinum in the "high-field" peak equals that in the metal and follows Korringa's relation. At the low-field peak (the "surface peak") its value depends on the surface treatment. For "clean", hydrogen-covered or CO-covered surfaces Korringa relations are found (Fig. 5a from our laboratory [7] and Fig. 5b from Jean-Philippe Ansermet's thesis [9]), and there are indications that the same is true for a surface exposed to ambient air [10]. These observations show that the level spacing around the Fermi energy in these catalyst particles is quasi-continuous ($w/\delta \geq 1$). In bulk platinum $D(E_F) = 24.5$ states per Rydberg and per atom; estimating $\delta^{-1} = N\,D(E_F)$ for a particle of N atoms, one finds $\delta = 6$ K for $N = 1000$ or a diameter $d = 30$ Å. The usual estimate for w is much less than 1 K below liquid nitrogen temperatures. Clearly either (or both) estimates are wrong; most likely that for δ, since

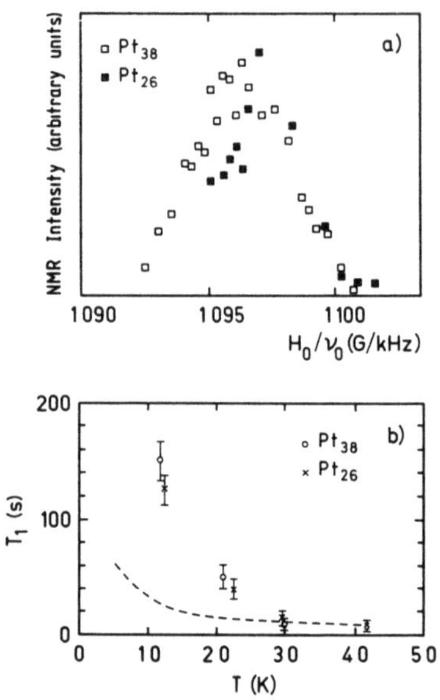

Fig. 4a and b. ^{195}Pt NMR lineshapes **a** and spin-lattice relaxation time as a function of temperature **b** for two platinum carbonyl cluster compounds of nuclearity 26 and 38. The lineposition is in the zero Knight shift region (although not centered at $K=0$), and the spin-lattice relaxation is not due to the Korringa mechanism. Data from Ref. 8.

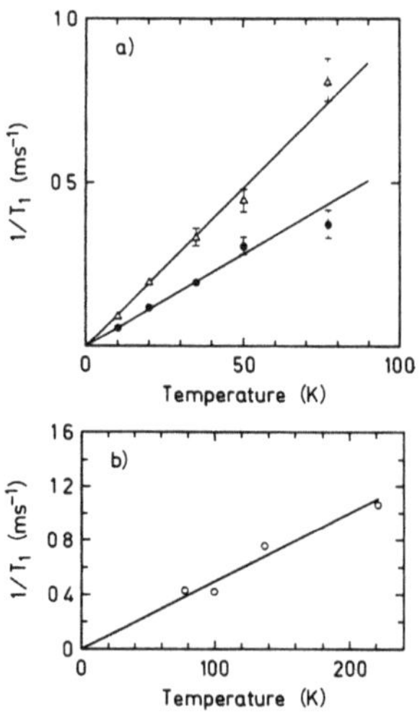

Fig. 5a and b. Korringa relationship for the temperature variation of the spin-lattice relaxation rate of the "surface peak" (region of 1.10 G/kHz in Fig. 2) for a clean (*open triangles*) and hydrogen-covered (*closed circles*) platinum on titania catalyst **a** and for a CO-covered platinum on alumina catalyst **b**. Data in **a** from Ref. 7, in **b** from Ref. 9 (with permission of the Department of Physics, University of Illinois at Urbana-Champaign)

one finds an unmodified spin-lattice relaxation rate with respect to the bulk metal at the "bulk" position [5, 11].

As explained in more detail in a paper at this conference [5], it appears that the NMR quantities K and T_1^{-1} in small platinum particles are mainly governed by surface/volume ratio and by surface treatment. Recently, we have proposed [7] to analyse χ, K and T_1^{-1} with equations and parameters that work well for bulk platinum, but allowing local values for the *s*-like and *d*-like densities of state $D_s(E_F)$ and $D_d(E_F)$. The equations incorporate separate Stoner enhancements for *s*-like and *d*-like electrons, but neglect interactions between bands. Without going here into the details of such calculations, we show in Fig. 6 the resulting contours of constant susceptibility, constant Knight shift and constant relaxation rate, with $D_s(E_F)$ and $D_d(E_F)$ as parameters.

As one would expect, the middle panel of Fig. 6 shows that a given position in the spectrum (a given value of K) can be due to atoms in many different environments (many different combinations of $D_s(E_F)$ and $D_d(E_F)$); and it is likely that all of these combinations will indeed occur. According to the lower panel however, each environment will be characterized by a different relaxation rate. As a result, the experimental relaxation curves are expected to be non-exponential. To actually observe such non-exponentiality however requires signal-to-noise ratios that may be difficult to achieve on this kind of samples. In the following we will reason as if each measured relaxation time resulted from a truly single-exponential decay curve (for a somewhat different analysis, see the paper presented by Jean-Pierre Bucher at this conference [5]). The combined information (signal intensity as function of position) $+(T_1^{-1}$ as a function of position) then results in a histogram of local susceptibility as in Fig. 7. The labelling of the horizontal axis includes an assumed core diamagnetic contribution of -30×10^{-6} emu/mole; therefore the paramagnetic susceptibility varies over a factor of five from (surface) site to (bulk) site. In an infinite five-layer slab of Pt(111) the paramagnetic

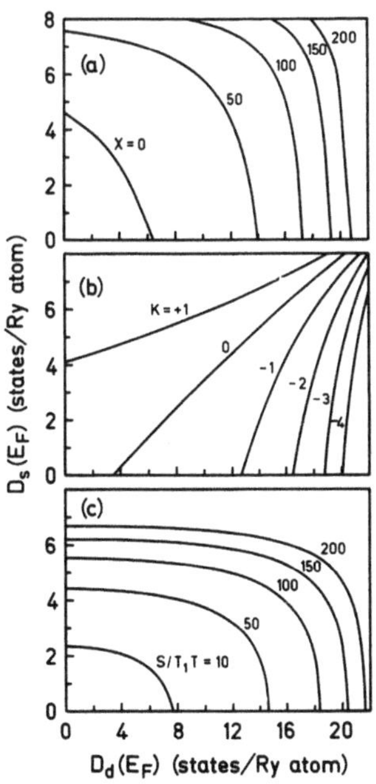

Fig. 6a–c. Contours of constant susceptibility (in units of 10^{-6} emu/mole) **a** Knight shift (in percent) **b** and spin-lattice relaxation rate (units 10^{-6}) **c** as function of the *s*-like and *d*-like total densities of states for platinum. For bulk platinum we assume $D_s(E_F) = 4.08\ Ry^{-1}\ atom^{-1}$ and $D_d(E_F) = 20.4\ Ry^{-1}\ atom^{-1}$, leading to $\chi = 205\times10^{-6}$ emu mole^{-1}, $K = -3.44\%$ and $S/T_1T = 1.87\times10^{-4}$ (where $S = 5.59\times10^{-6}$ sK). Such contours allow to determine a "local susceptibility" from panel **a** if the relaxation time (panel **c**) and Knight shift (panel **b**) are known. From Ref. 7

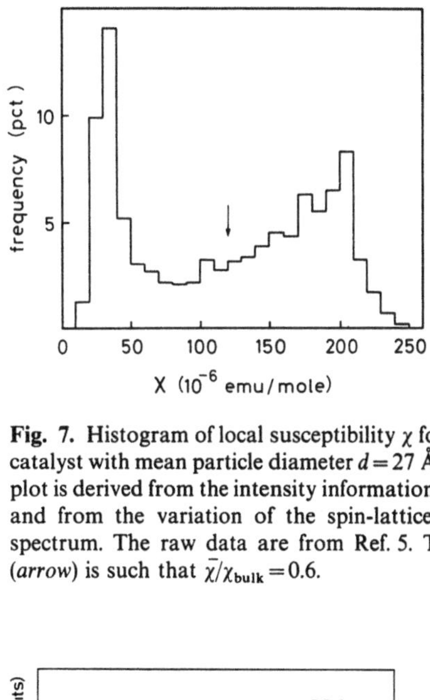

Fig. 7. Histogram of local susceptibility χ for a platinum on titania catalyst with mean particle diameter $d = 27$ Å (dispersion 36%). This plot is derived from the intensity information in the NMR spectrum, and from the variation of the spin-lattice relaxation across the spectrum. The raw data are from Ref. 5. The average value of χ (*arrow*) is such that $\bar{\chi}/\chi_{bulk} = 0.6$.

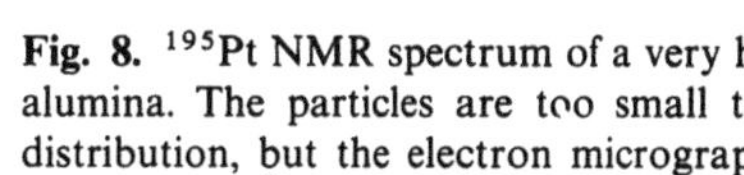

Fig. 8. ^{195}Pt NMR spectrum of a very high dispersion catalyst on alumina. The particles are too small to measure a reliable size distribution, but the electron micrographs indicate a typical diameter $d = 10$ Å, with essentially no particles having $d > 20$ Å. We would expect a sample with such an NMR spectrum to have a temperature-independent susceptibility χ, with χ considerably below the bulk value. Sample courtesy of R. Prins, then at Technical University Eindhoven. Spectrum from Ref. 13

susceptibility has been calculated as 285, 190 and 142×10^{-6} emu/mole for the central, subsurface and surface layers, showing the same trend towards a diminished susceptibility at the surface [12].

We expect magnetometer experiments to find the center of gravity of this distribution, at $\chi_0/\chi_{bulk} = 0.6$. To obtain still lower values for the average susceptibility, one would need particles sufficiently small that the NMR spectrum peaks strongly at the low-field side, for example such as shown in Fig. 8 [13].

In conclusion, NMR data indicate that the susceptibility of small (10–30 Å) platinum particles is less than that of the bulk, but remains constant at practical temperatures. The occurrence of a Korringa-like nuclear spin-lattice relaxation shows that the level spacing near E_F is small, so no QSE will occur for this size of particles.

It is a pleasure to acknowledge the cooperation of all the authors whose work has been shown in the figures, and among them my former student Jean-Pierre Bucher. Part of his work was supported by the Swiss "Commission d'Encouragement de la Recherche Scientifique".

References

1. Slichter, C.P.: Surf. Sci. **106**, 382 (1981)
2. Weisman I.D., Knight, W.D.: Phys. Rev. **169**, 373 (1968)
3. Inoue, N., Sugawara, T.: J. Phys. Soc. Jpn **45**, 450 (1978)
4. Rhodes, H.E., Wang, P.-K., Stokes, H.T., Slichter, C.P., Sinfelt, J.H.: Phys. Rev. B. **26**, 3559 (1982)
5. Bucher, J.-P., van der Klink, J.J., Buttet, J., Graetzel, M.: Surf. Sci. (in press)

6a. Marzke, R.F., Glausinger, W.S., Bayard, M.: Solid State Commun. **18**, 1025 (1976)

6b. Marzke, R.F.: Private communication; Van Rheenen, P.: Ph.D. Thesis, Arizona State University (1981)

7. Bucher, J.-P.: Ph.D. Thesis, Ecole Polytechnique Fédérale de Lausanne (1988)
8. Pronk, B.J.; Brom, H.B., Ceriotti, A., Longoni, G.: Solid State Commun. **64**, 7 (1987)
9. Ansermet, J.-Ph.: Ph.D. Thesis, University of Illinois, Urbana Champaign, 1985 (unpublished)
10. Rhodes, H.E., Wang, P.-K., Makowka, C.D., Rudaz, S.L., Stokes, H.T., Slichter, C.P., Sinfelt, J.H.: Phys. Rev. B 26, 3569 (1982)
11. Makowka, C.D., Slichter, C.P., Sinfelt, J.H.: Phys. Rev. B 31, 5663 (1985)
12. Weinert, M., Freeman, A.J.: Phys. Rev. B 28, 6262 (1983)
13. Bucher, J.-P., Buttet, J., van der Klink, J.J., Graetzel, M. Newson, E., Truong, T.B.: J. Mol. Catal. **43**, 213 (1987)

Z. Phys. D – Atoms, Molecules and Clusters 12, 333–339 (1989)

Atoms, Molecules and Clusters
Zeitschrift für Physik D
© Springer-Verlag 1989

Investigation of local chemical and electronic properties of small particles with EELS point analysis and image energy filtering in a STEM

D. Ugarte and C. Colliex

Laboratoire de Physique des Solides associé au CNRS, Bâtiment 510, Université Paris-Sud, F-91405 Orsay, France

Received 5 July 1988

The finely focused electron beam of a STEM microscope provides a very powerful probe for the investigation of the local chemical and electronic properties of small particles. After a general description of different techniques which have been developed to take benefit of the Electron Energy Loss Signal (EELS) in this configuration, we describe two experiments under progress. The first one concerns clusters of iron oxide particles in a magnetic superlattice, the second one deals with silicon spheres. These examples involve quite different excitations which are respectively more characteristic of the properties of the bulk (core loss edges) or of the surface (plasmon modes).

PACS: 61.16.Di; 71.45.Gm; 79.20.H

I. Introduction

As a consequence of its general design, the STEM (Scanning Transmission Electron Microscope) is well adapted to the study of very local specimen properties [1, 2, 3]. First, its field emission gun and 3-lens focusing optics delivers on the sample entrance surface a subnanometer probe of high energy (100 keV) incident electrons which can be fixed or rastered over a specimen feature [4]. The next important aspect of STEM is its unique and efficient strategy for detecting simultaneously many different signals. On the transmitted beam it can be either the microdiffraction pattern on a 2D detector, or the beam elastically scattered to large angles within an annular detector and the electron energy loss spectrum (EELS) through a magnetic spectrometer. A Si-Li detector introduced within the objective pole pieces close to the specimen collects and analyzes the secondary X-ray emission. All these signals can be used to visualize, identify and study the structural, chemical and electronic properties of small particles which are supported on thin layers or embedded within a thin matrix. A final major aspect of STEM technique lies in its easy and natural coupling to digital systems. Computer driven units control data acquisition, store and process a posteriori the enormous amount of results in order to improve the extraction of useful information [5].

The purpose of the present paper is to describe present studies which illustrate the great EELS capabilities of the Orsay Vacuum Generators HB501 dedicated STEM for tackling problems involving small particles.

II. EELS point analysis and image energy filtering in a STEM

Two methods have generally been used to acquire and display EELS data in an electron microscope environment [6]. In the point analysis mode, one selects a specimen feature, i.e. by focusing the incident beam on a given particle or at a well defined impact parameter from its surface, and records the corresponding electron energy loss spectrum, that is the distribution of the measured intensity along the energy loss scale. This approach is very useful to measure the energy of an excitation process as a function of the particle size [7, 8, 9] or of the beam position with respect to the particle [10, 11, 12]. The second technique is to select a given energy loss window and record the specific image which is made with only the electrons having suffered this energy loss; these are 2D energy filtered images. Acquisition and processing techniques have been developed to extract from sequences of energy filtered images recorded at different energy losses, the characteristic contribution of a plasmon line or of a core loss edge superposed over a non characteristic background [13, 14]. With this method, one can study the variation of excitation probabilities or of cross sections as a function of particle size and beam position and examples can also be found in [10, 11, 12, 15].

In the EELS spectrum, the useful information is contained in different domains. For small particles, most of the work to now has been devoted to the study of plasmon excitations; these modes imply collective oscillations of the quasi-free electron gas, for instance in Al, Ga and Sn in which materials bulk and surface

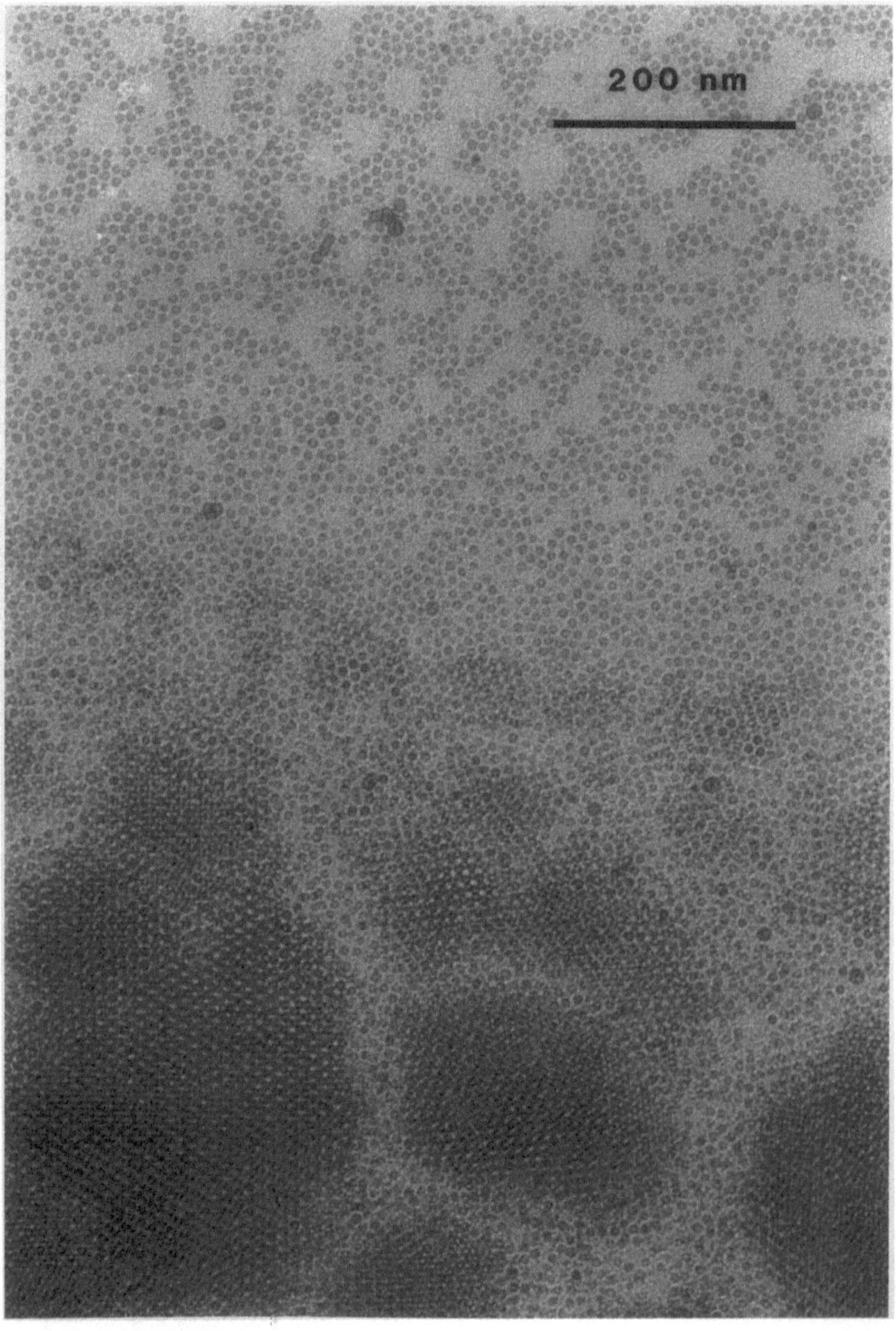

Fig. 1. Conventional electron microscopy images of the magnetic superlattice of iron oxide particles. There exists a juxtaposition of 3D crystal type arrangements and of 2D random distributions. The contrast is mostly phase contrast which is very dependent on focusing conditions

modes have been identified. A typical energy loss lies then between 5 to 25 eV, depending on the electron density. The second category of useful information is in the core edges which relate unambiguously the signal to the presence of a given species: this method has therefore been advertized for its microanalytical capabilities. Very recently, its very high sensitivity has been demonstrated on subnanometer clusters--discrimination between two elements and identification at the single atom level [16, 17]. Moreover the detailed profile of the spectrum on the core threshold and above it, is very sensitive to the local density of vacant electron states with given quantum symmetry and to the local crystal symmetry, similarly to XANES and EXAFS structures on X-ray absorption curves. The great advantage of EELS techniques is that this type of analysis can be made on an object down to the nanometer size in best cases. It is therefore possible to anticipate its application on a single cluster: for instance, we have already shown the change of the Pd M_{45} edge for particles of decreasing size, from 5 to 2 nm, corresponding to an increased oxidation level [18]. Further examples of work in progress are shown in the next two sections.

III. A magnetic super lattice made of small iron oxide particles (work in collaboration with M. Bentzon, Technical University of Denmark, Lyngby)

Assemblies of small iron oxide particles are obtained from the oxidation of a magnetic fluid, produced by thermal decomposition of $Fe(CO)_5$ in Declin-carrier liquid – in the presence of oleic acid – surfactant – [19]. The individual particles have an average diameter of 6.9 nm with a very narrow standard deviation ($\sigma = 0.4$ nm). As shown in Fig. 1, they aggregate to form large (0.1 – 1 μm) three dimensional superlattice crystals, together with single layers of randomly distributed particles. In any case the distance between two adjacent spheres is $\simeq 9$ nm and it was never possible to observe two particles in contact. It suggests that they are covered with a 1–2 nm thick layer of surfactant molecules.

EELS spectroscopy has been performed over large areas of supercrystals (200×200 nm^2) to have a good signal to noise ratio on the O-K edge at 532 eV and the Fe-L_{23} edge at 705 eV (see Fig. 2). The oxygen/iron atomic ratio is determined from the ratio of the integrated signals after background subtraction, taking into account reliable calculated values for the cross sections [20]. We obtain O/Fe $\simeq 1.7 \pm 0.1$, which agrees satisfactory with the trivalent iron oxide ratio (1.5), the more as one can expect an excess of oxygen

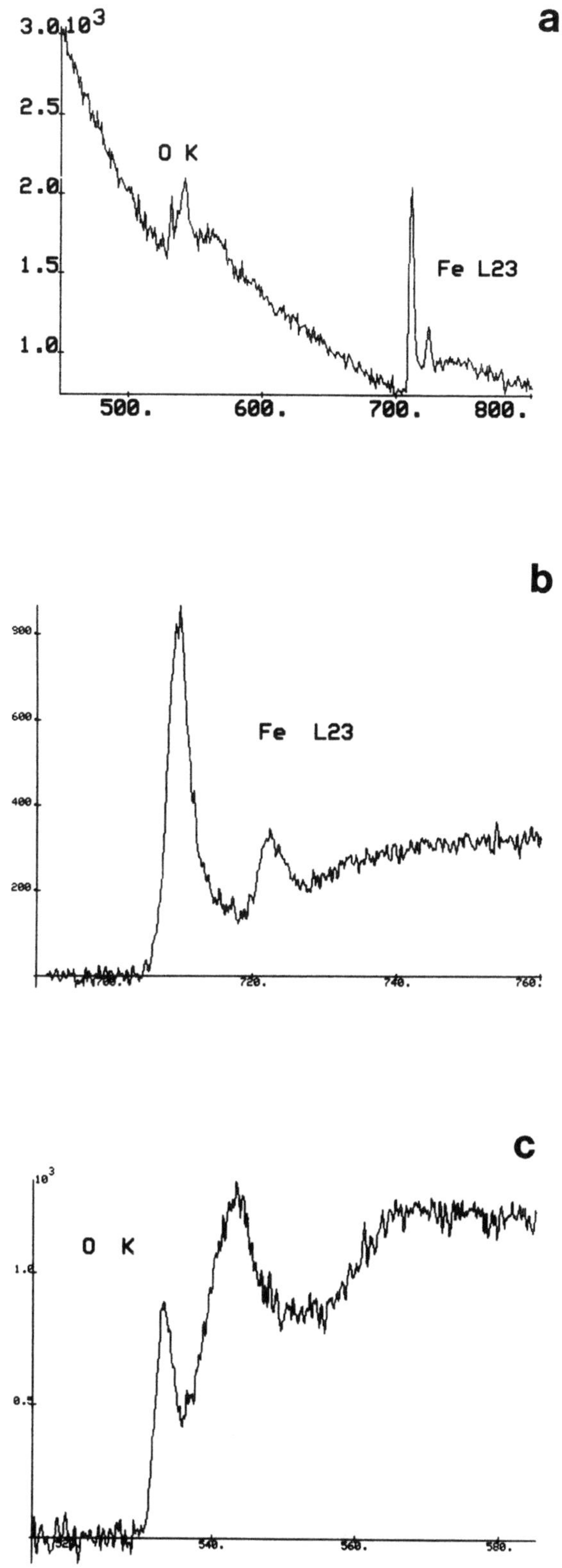

Fig. 2a–c. Electron energy loss spectra recorded over a collection of iron oxide particles: **a** O-K and Fe-L_{23} edges used for quantitative microanalysis; **b** detail of Fe-L_{23} edge after background subtraction displaying the white lines (dose $\simeq 4 \times 10^7$ e$^-$/nm^2); **c** detail of O-K edge after background subtraction (dose $\simeq 4 \times 10^7$ e$^-$/nm^2)

due to the unavoidable presence of oxygen-rich surfactant molecules. The fine structure on the edges are then compared with previously recorded standard profiles on FeO, α-Fe_2O_3, γ-Fe_2O_3, Fe_3O_4 [21] and confirm the above determination, as a consequence of several similarities. The white-line ratio, which corresponds to the intensity ratio of the two prominant resonance lines L_3 and L_2 on the iron edge, is estimated of the order 6.3, after a search of the best decomposition into its different components – see [22]. This has to be compared with the previously measured 6.5 and 6.8 values, respectively for the alpha and gamma phases. For the oxygen K edge, the prepeak at 535 eV is also an evidence of the occurrence of the trivalent phase, but the XANES structure at $\simeq 550$ eV is not clear enough for an unambiguous discrimination between the two structures. Nevertheless, it can be assumed that the hematite α-Fe_2O_3 phase is dominant.

However, for high primary doses, one notices an evolution in the shape of the O-K edge, with a change towards the FeO fine structure. This is due to electron-beam induced reduction of iron oxide, and this effect must be taken into account when improved spatial resolution characterization is searched. It is the case in the series of energy filtered images recorded below and above the iron M_{23} edge at 55 eV, in order to evaluate the homogeneity of individual particles. A few examples are shown in Fig. 3 which exhibits a pair of annular dark field images and Fe maps with a 10 min. irradiation period between them. The contrast in the annular dark field micrograph is proportional to the mass thickness for such thin objects and the intensity in the Fe map varies linearly with the iron absolute number of atoms. This behavior is an evidence of an inhomogeneous radial distribution within single particles and may be due to a less dense (maybe carbonaceous) core in the centre. It is also enhanced by radiation damage, with hints of gradual oxide reduction on the outer surface. Similarly to other situations involving oxidized particles of catalytic interest, many beam-induced effects prevent from a clear identification of the components when ultra high spatial resolution is required [23].

IV. Clusters of silicon spheres

Up to now, mostly metals with low melting point such as Al, Ga, Sn have been crystallized with spherical shapes. They constitute very nice objects for the comparison of the experimental results with theoretical models in the spherical geometry [24–29]. Iijima has reported the growth of fine particles of silicon prepared by gas evaporation [30]: they are nearly spherical, their sizes are in the range 20–200 nm in diameter and they have a tendency to coagulate when collected for

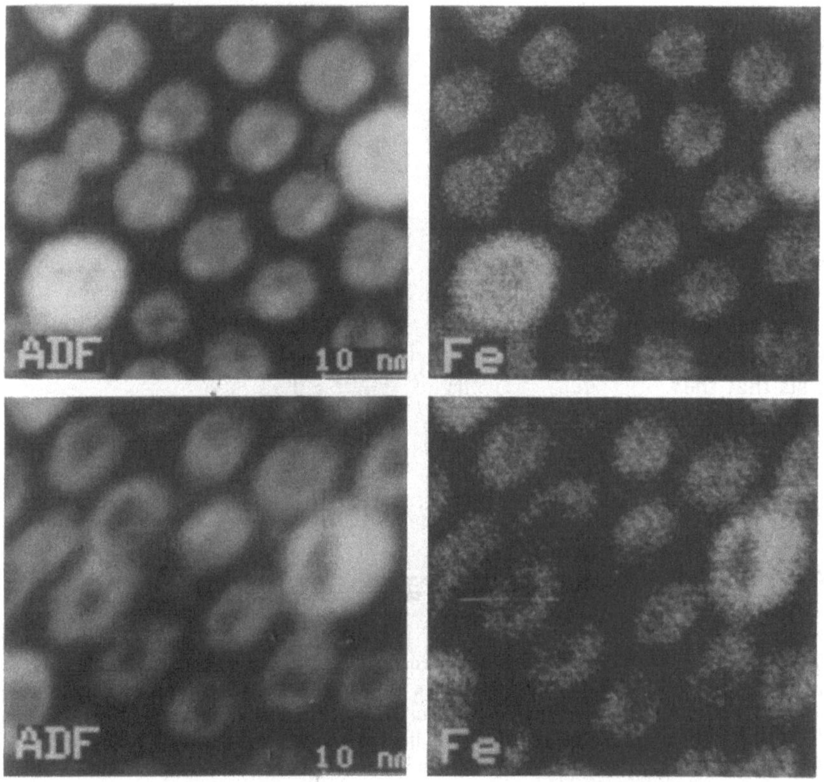

Fig. 3. Annular dark field and characteristic iron micrographs recorded after different irradiation doses; the first one is $\simeq 6 \times 10^7$ e^-/nm^2 and the second one is about 5 times higher than the first one Notice the increased contrast in the second set of images and the elongated shapes which are due to a slight specimen drift For the Fe map, one has recorded three energy filtered images at 43, 52 eV before the edge M_{23} for estimating the background contribution in each pixel, the last one after the edge at 57 eV contains the characteristic signal superposed on top of the background

electron microscopy observation, as shown in Fig. 4. In many cases, one can find situations with a smaller sphere on the surface of a bigger one. A few of them display dark contrast as lines or bands running across them close to the center: they are due to planar growth faults. The particles are deposited on holey carbon films and the EELS studies are carried out on particles hanging over the holes in order to eliminate the carbon background. They are all covered with a 2 to 6 nm thick oxide layer, during the atmospheric exposition while preparing the sample for the electron microscope work. Moreover a very thin contamination layer about 1 to 2 nm thin covers the most external surface.

A model object therefore consists of a Si core, with two outside shells, an intermediate one of SiO_x and the outer most one of C. It has not been considered either by Batson [31] or by Wang and Cowley [9] who deal respectively with bispherical situations or with embedded geometries. A few experimental results are gathered in Fig. 5 which contains a spectrum recorded over a cluster of spheres and a series of energy filtered images at all noticeable features seen in the low loss range. From our previous knowledge for such cases, we can make the following comments:

- the 18 eV is the bulk plasmon for Si;
- the 9 eV loss is the surface mode on the Si-SiO_x boundary and is very similar to the interface loss at 8 eV reported by Howie and Milne [32] on silicon-silica interfaces seen parallel to the beam. The fact that it also exhibits a weak contrast on the outside surface of the Si oxide has been interpreted by Batson [10] in terms of coupling between the major metal/oxide interface oscillation and the oxide surface one;
- the 4 eV loss is also a very localized excitation because it induces a sharp contrast in the image. This second excitation at an energy loss about one half of the main surface mode was interpreted by Batson [37] as a bispherical mode. Our experimental observation does not confirm his own ones, because our intensity distribution is azimuthally very regular and

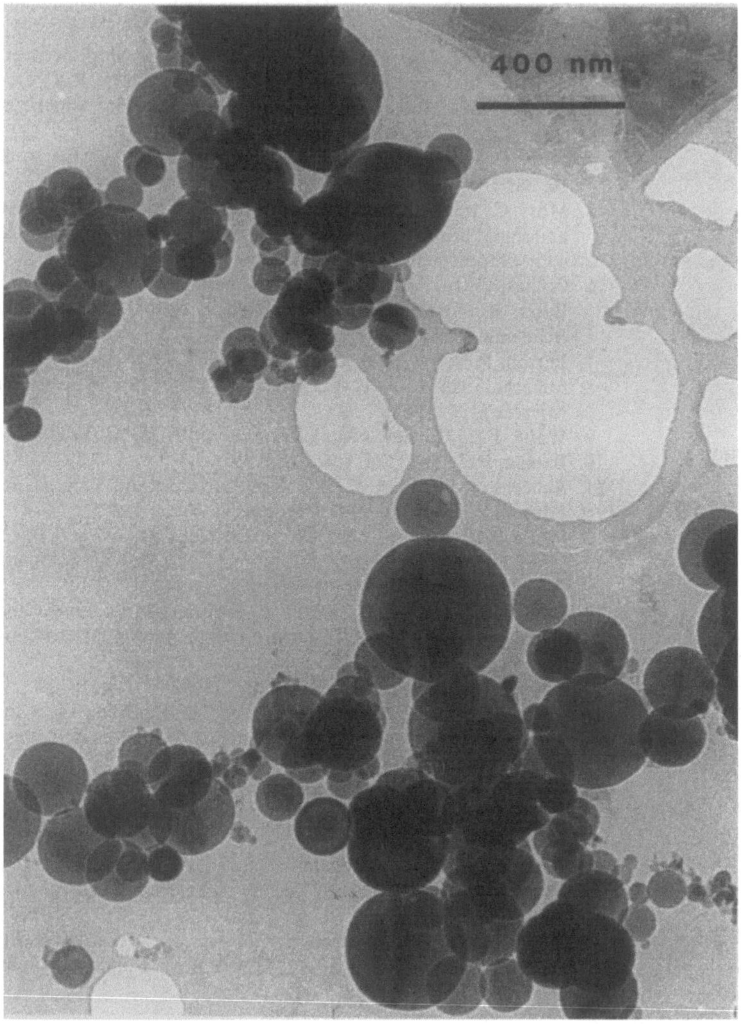

Fig. 4. Conventional electron microscopy image of a cluster of silicon spheres

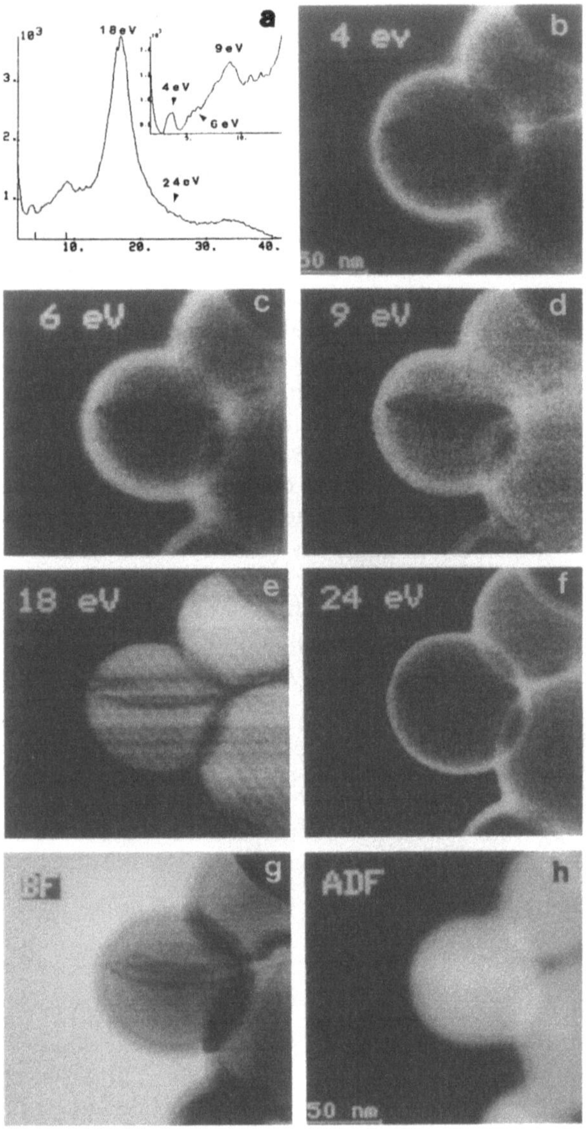

Fig. 5. a Low energy loss spectrum recorded on a collection of spheres with the major features of interest in the 4-25 eV range. **b** to **h** family of images on a sphere attached to other spheres–see text for discussion.

does not display a clear dipolar behavior. Consequently, our suggestion is to connect it to the secondary boundary between Si oxide and amorphous carbon;

- the last 6 eV loss is not as well localized because this image is more diffuse between the two previous ones. We can rule out the occurrence of the higher ℓ-modes for the carbon-SiO_x interface because they would be even more localized than the dipole one [33]. It may be due to the superposition of different more delocalized contributions, among which bulk interband edges in the SiO_x layer and carbon plasmon loss.

As a conclusion, these semiconductor spheres will provide a stringent test for classical electrodynamical descriptions once the satisfactory model has been worked out. More extensive experimental studies and calculations are in progress. Generally it may be emphasized that surface plasmon modes produce a high spatial resolution technique for the visualization and study of the electronic properties of surface and interfaces on small particles.

Thanks are due to M. Bentzon and S. Iijima who provided the specimens for this work. The help of M. Tencé has been very efficient for the use of the digital control of the STEM. Private communications with T. Manoubi, who has made his spectra available to us before publication, is also acknowledged.

References

1. Crewe, A.V.: In: Electron microscopy in material science. Valdre, U., Zichichi, A. (eds.), p. 162.New York: Academic Press 1971
2. Colliex, C., Mory, C.: In: Quantitative electron microscopy, Chapman, J.N., Craven, A.J.(eds.), pp. 149–216.Scottish Universities Summer School in Physics 1984
3. Cowley, J.M.: In: Catalytic materials: relationship between structure and reactivity. Whyte, T.E., Dalla Betta, R.A., Derouane, E.G., Baker, R.T.K. (eds.), pp. 353–366. American Chemical Society Symposium Series n° 248 1984
4. Mory, C., Tencé, M., Colliex, C.: J. Microsc. Spectrosc. Electron. **10**, 381 (1985)
5. Mory, C., Bonnet, N., Colliex, C., Kohl, H., Tencé, M.: In: Image and signal processing in electron microscopy. Scanning Microscopy (Supplement) **2**, 329 (1988)
6. Colliex, C.: In: Advances in Optical and Electron Microscopy, Barer, R., Cosslett, V.E. (eds.), Vol. 9, pp. 65–177. London: Academic Press 1984
7. Batson, P.E.: Sol. State Commun. **34**, 477 (1980)
8. Acheche, M., Colliex, C., Trebbia, P.: Scanning Electron Microsc.I, 25 (1986)
9. Wang, Z.L., Cowley, J.M.: Ultramicroscopy **23**, 97 (1987)
10. Batson, P.E.: Surf. Sci. **156**, 720 (1985)
11. Achèche, M., Colliex, C., Kohl, H., Nourtier, A., Trebbia, P.: Ultramicroscopy **20**, 99 (1986)
12. Wang, Z.L., Cowley, J.M.: Ultramicroscopy **21**, 347 (1987)
13. Jeanguillaume, C., Tencé, M., Trebbia, P., Colliex. C.: Scanning Electron Microsc. **II**, 745 (1983)
14. Bonnet, N., Colliex, C., Mory, C., Tencé, M.: In: Image and signal processing in electron microscopy scanning. Microscopy (Supplement) **2**, 351 (1988)
15. Batson, P.E.: Ultramicroscopy **9**, 277 (1982)
16. Colliex, C., Mory, C.: Proceedings of the 46th Meeting of the Electron Microscopy Society of America, San Francisco Press, p. 522, 1988
17. Mory, C., Colliex, C.: Ultramicroscopy **28**, 334 (1989)
18. Colliex, C.: In: Proceedings of 4th Conf.on Solid State Analysis, Karl Marx Stadt, RDA, 1987
19. Bentzon, M.D., Wonterghem, J., Thölen, A.: Proc.45th Meeting of the Electron Microscopy Society of America, San Francisco Press, 1987
20. Egerton, R.F.: In: Electron energy loss spectroscopy in the electron microscope, pp. 262–277. New York, London: Plenum Press 1986

21. Ortiz, C., Manoubi, T., Colliex, C.: In: Proceedings of the International Conference on Magnetism. J. Phys. (Paris) (1989) (in press); Manoubi, T.: Private communication 1988
22. Manoubi, T., Tencé, M. and Colliex, C.: Ultramicroscopy **28**, 49 (1989)
23. Wang, Z.L., Colliex, C., Paul-Boncour, V., Percheron-Guegan, A., Achard, J.C., Barrault, J.: J. Catal. **105**, 120 (1987)
24. Fujimoto, F., Komaki, K.: J.Phys.Soc.Japan **25**, 1679 (1968)
25. Schmeits, M.: J. Phys. C **14**, 1203 (1981)
26. Ruppin, R.: J. Phys. Chem. Sol. **39**, 233 (1978)
27. Kohl, H.: Ultramicroscopy, **11**, 53 (1983)
28. Barberan, N., Bausells, J.: Phys. Rev. B **31**, 6354 (1985)
29. Ferrell, T.L., Echenique, P.M.: Phys. Rev. Letters **55**, 1526 (1985)
30. Iijima, S.: Jn. J. Appl. Phys. **26**, 357 (1987)
31. Batson, P.E.: Phys. Rev. Letters **49**, 936 (1982)
32. Howie, A., Milne, R.H.: Ultramicroscopy **18**, 427 (1985)
33. Colliex, C.: Ultramicroscopy **18**, 131 (1985)

Z. Phys. D – Atoms, Molecules and Clusters 12, 341–346 (1989)

Atoms, Molecules and Clusters
Zeitschrift für Physik D
© Springer-Verlag 1989

Electronic correlation and effective interactions in small alkali clusters

F. Spiegelmann, P. Blaise, J.P. Malrieu, and D. Maynau

Laboratoire de Physique Quantique, (UA 505 CNRS), Université Paul Sabatier, 118 Route de Narbonne, F-31062 Toulouse, France

Received 5 July 1988; final version 30 September 1988

The crucial importance of correlation effects versus delocalization, and their nature in small Alkali clusters is analysed from an ab-initio point of view through a detailed investigation of the Li_2 dimer. The role of the external correlation (provided by extended basis sets and large Configuration Interaction calculations) is shown to lower the energy of ionic configurations and to increase their weight in the electronic wavefunction, increasing simultaneously the importance of delocalization versus internal correlation within the s-band. Effective interactions are determined from accurate diabatic calculations on dimers and transfered to clusters via an effective hamiltonian spanned by s orthogonal orbitals. Although not including explicitely the p-band, this model provides results in good agreement with ab-initio calculations on Lithium clusters.

PACS: 31.20.D; 36.40

I. Introduction: strong or weak correlation, s band or extended basis sets?

There has been a number of systematic theoretical studies concerning both neutral and charged alkali clusters [1–18]. Within the Born-Oppenheimer approximation and for clusters involving a number of atoms $n<10$, ab-initio CI calculations [7–11] can be considered as providing reliable results (geometries, stabilities and ionization potentials). Indeed they make no a-priori hypothesis on the nature of the electronic wavefunction, the hamiltonian used is exact, and although they might underestimate the binding energies, because of the limitation of the basis set, their results are in reasonable agreement with experimental data, essentially available on sodium and potassium clusters. The first realistic calculations on Alkali clusters were done with the local Spin Density (LSD) method and representing correlation effects through a monoelectronic functional [5, 6], and they proved to be in good agreement with the posterior ab-initio results.

As concerns approximate methods, one should mention the failure of the Diatomics in Molecules (DIM) attempts to treat Alkali clusters beyond trimers [12–16]. Alternatively the simplest conceivable model, namely the empirical Hückel topological hamiltonian [17] happens to predict correctly the ground state topologies as well as the binding energies using as parameter the binding energy of the dimer.

While the optimal structures and stabilities are thus now fairly well characterized from a theoretical point of view, some fundamental questions concerning the nature of the electronic wavefunction in small Alkali clusters have not yet been answered or clearly formulated. Indeed the simultaneous success of the most simple method (Hückel) and the most sophisticated treatment (ab-initio CI calculations) raises a contradiction since:

(i) Ab-initio calculations show that extended basis sets are needed (at least $s+p$ basis sets), that the monoelectronic SCF level is not sufficient to account for the structure of alkali clusters and that large CI

calculations are necessary, thus suggesting that correlation is important.

(ii) The Hückel model is a simple monoelectronic MO description, completely neglecting the electronic correlation. Moreover it is formulated in the pure basis of s atomic orbitals.

In order to solve this dilemna, one should be able to answer the following questions which all concern the fundamental nature of the electronic wavefunction in Alkali clusters:

(i) Can Alkali clusters be represented as s-electron clusters and to which extent is this representation valid?

(ii) What is the respective importance of correlation versus delocalisation in these clusters? In a Valence-Bond description, this question is often formulated using the Hubbard hamiltonian [19, 20]:

$$H_H = \sum_{i \neq j} t_{ij} a_i^+ a_j + \sum_i U_i a_i^+ a_i^+ a_i a_i$$

where t_{ij} is the hopping integral allowing for delocalisation and U_i is the effective repulsion of two electrons with opposite spin when they occur to be on the same atom i, yielding correlation effects. Materials are generally classified according to their delocalisation versus correlation ratio measured by $|t|/U$ and it is interesting to check from an ab-initio point of view whether the hypothesis of weak correlation ($U \simeq 0$) usually assumed in such metals and in the Hückel description is grounded or not.

(iii) Besides these correlation effects within the s-band (hereafter called the internal correlation), one should consider the so-called external correlation [21] provided by excitations outside the valence shell. What is the relative importance of those two aspects of correlation and what is their interplay in the structure of small Alkali clusters?

In this paper, we shall support the following theses:

(i) the role of external correlation is prominent and the implication of p orbitals (and in general of extended orbital basis sets) consists in providing angular (and radial) correlation of the ionic configurations in the VB meaning. This role is more crucial than hybridization and mixing between s and p bands, at least for small clusters.

(ii) this effect can be incorporated in an effective way through a dressing of the interactions in the s-band. This means that the problem may be reduced (projected) into a model space within the s-band using effective integrals. If this is done properly, the essential effect of external correlation is to lower the energy of the ionic forms and thus to decrease the value of the U parameters in the Hubbard model.

II. Physics of the dimer and effective interactions

We will first consider the neutral dimer in the minimal VB description of the s-band (i.e. using the exact Li 2s orbitals eventually orthogonalized). There are two neutral configurations $s_a\bar{s}_b$ and $\bar{s}_a s_b$, and two ionic configurations $s_a\bar{s}_a$ and $s_b\bar{s}_b$. The CI matrix is as follows:

$$\begin{array}{c} \\ s_a\bar{s}_b \\ s_b\bar{s}_a \\ s_a\bar{s}_a \\ s_a\bar{s}_b \end{array}
\begin{array}{c} \begin{array}{cccc} s_a\bar{s}_b & s_b\bar{s}_a & s_a\bar{s}_a & s_b\bar{s}_b \end{array} \\ \begin{pmatrix} h_n & K_n & t & t \\ K_n & h_n & t & t \\ t & t & h_i & K_i \\ t & t & K_i & h_i \end{pmatrix} \end{array}$$

h_n and h_i may be considered as representing the diabatic potentials correlated to the neutral dissociation Li + Li or to dissociation into the ion pair $Li^+ + Li^-$. The Hubbard model for a dimer consists in neglecting the exchange integrals. $t(R)$ then corresponds to the transfer integral while $U(R)$ is the difference between the neutral and ionic potentials:

$$U(R) = h_i(R) - h_n(R)$$

The limit $U(R) = 0$ in which neutral and ionic configurations are degenerate corresponds to the Hückel model and the whole physics is then controlled by the transfer integral and yields an adiabatic ground state at $2t$. Alternatively, in the limit $U \gg t$, correlation is strong, the ionic configurations are very high and may be neglected or treated pertubatively, which is the hypothesis of magnetic models or DIM models in which effective exchange integrals between neutral configurations regulate the spin order. The frontier between those two worlds is given by the critical value $|t|/U = 0.25$ which is easily obtained considering the convergence radius of a perturbation series expressed in either scheme.

Figures 1–3 show the results of an ab-initio valence calculation of those integrals (dotted lines) using core effective potentials. The neutral potential h_n is repulsive while the ionic potential h_i is attractive due to the $1/R$ coulombic attraction. The diagonalisation of the above matrix yields four adiabatic states, two singlets ${}^1\Sigma_g^+$ which exhibit neutral/ionic mixing, one triplet ${}^3\Sigma_\mu^+$ which is purely neutral and a singlet ${}^1\Sigma_\mu^+$ which is purely ionic. The bound character of the ${}^1\Sigma_g^+$ ground state of neutral Alkali dimers may thus be considered as resulting from the interaction of a repulsive neutral state strongly coupled with an ionic state. However the quantitative description of the Li_2 dimer obtained in this minimal CI representation is rather poor ($R_e \simeq 6.0$ bohrs and $D_e \simeq .2$ eV while the experimental values are $R_e = 5.05$ bohrs and $D_e = 1.05$ eV). One may also notice that while the asymptotic limit Li + Li is obviously well accounted for, the dissociation of the ion

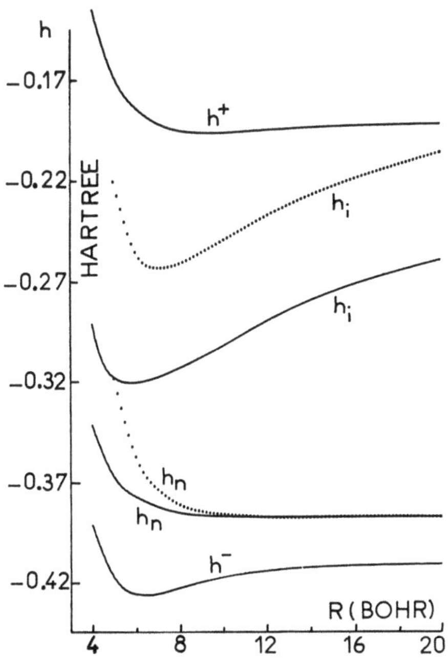

Fig. 1. Interaction potentials for Li_2 (h_n and h_i), Li_2^+ (h^+) and Li_2^- (h^-)
.... minimal representation
———— effective representation

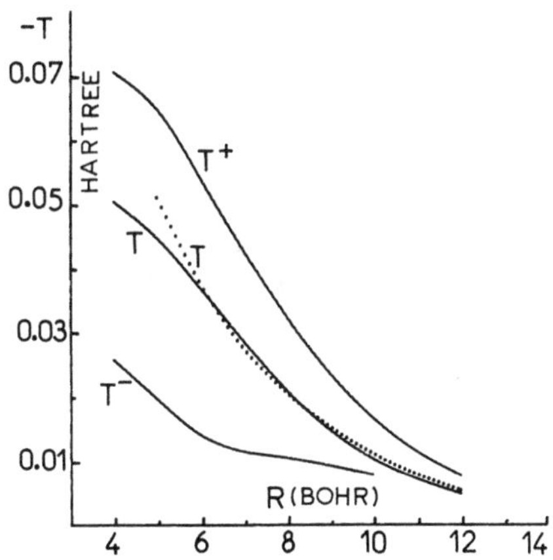

Fig. 2. Transfer integrals for the Alkali dimers Li_2 (t), Li_2^+ (t^+) and Li_2^- (t^-)
.... minimal representation
———— effective representation

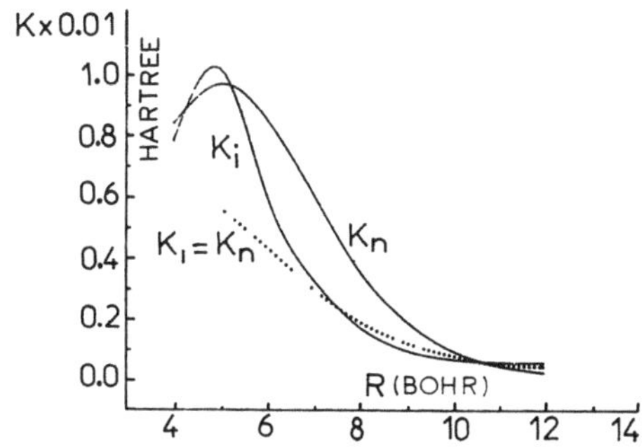

Fig. 3. Exchange integrals for the Alkali dimer Li_2 (K_n and K_i)
.... minimal representation
———— effective representation

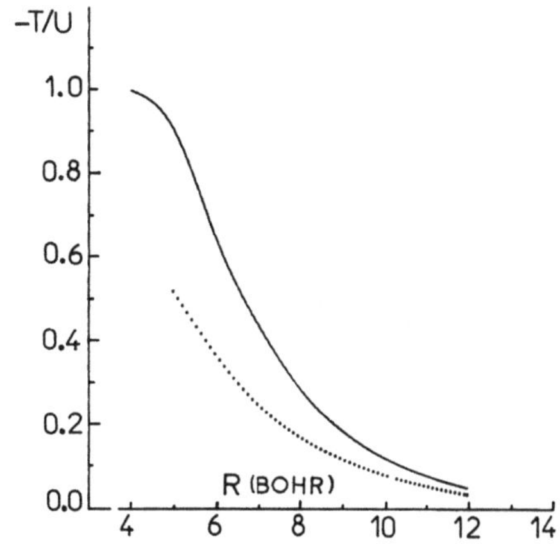

Fig. 4. $|t|/U$ ratio in Li_2
.... minimal representation
———— effective representation

pair $Li^+ + Li^-$ lies much too high due to the bad representation of Li^- with a single determinant $s_a\bar{s}_a$ and the s orbitals of the Li atom (the error is 0.04788 Hartree on the electron affinity). The $|t|/U$ ratio at this level of accuracy is plotted in Fig. 4 (dotted line).

In order to check the role played by extended basis sets the same analysis has been achieved using a valence $7s/4p/1d$ gaussian orbital basis set and performing full CI (exact result with core pseudopotentials in the above basis). In those new conditions, the Li_2 equilibrium distance becomes $R_e = 5.07$ bohrs, the dissociation energy is 1.03 eV and the electron affinity of the Lithium atom comes to 0.61 eV instead of 0.62 experimentally. A diabatisation process was used in order to provide an effective hamiltonian which has the same basic meaning as the matrix of the minimal representation but which includes exactly the correlation effects in the extended basis set. This diabatization scheme will be explained in details elsewhere [22] and essentially consists of a rotation of the lowest adiabatic vectors issued from the CI, maximizing their overlap with the reference configurations of the minimal representation [23]. The use of the Theory of Intermediate Effective Hamiltonians [27] then allows to determine a 4×4 representation, the diagonalisation of which provides exact eigenvalues for the two lowest states, ${}^1\Sigma_g^+$ and ${}^3\Sigma_u^+$.

Those effective hamiltonian H^{eff} matrix elements (full lines in Figs. 1 to 3) differ substantially from the minimal representation. The position of the diabatic neutral states h_n is lowered due to:

(i) some weak s–p hybridization effects resulting from single excitations in Valence Bond description

and which would equivalently appear in a Hartree-Fock calculation

(ii) An improved representation of the repulsion because of the better flexibility of the basis set.

(iii) The inclusion of dispersion forces

However, the dominant feature is the very important lowering of the ionic states (h_i). The new curve is also displaced to somewhat shorter distances and remain roughly parallel to its representation in the minimal basis set. The vertical shift lies in the range 0.04–0.06 Hartree eV depending on the internuclear distance and this shift corresponds to the error which is made on the electron affinity of Li in the minimal representation. The t value is slightly reduced at small distances (Fig. 2) while K_n and K_i are now slightly different and increased (Fig. 3).

The most significant result is that the $|t|/U$ ratio obtained in this accurate description is doubled with respect to its minimal description value. Thus the effect of dressing the hamiltonian in the s-band is essentially to lower the ionic forms and diminish the U term in the Hubbard hamitonian and this is the action of external correlation. In a paradoxical way, the external correlation reinforces the importance of delocalization versus internal correlation. This solves the apparent contradiction mentioned in the introduction and show that:

(i) Hartree-Fock representation is a MO delocalized picture but includes no correlation at all and thus fails

(ii) The essential role of large basis sets in CI calculations is to account for the external correlation, thus enhancing the delocalized character of the electronic wavefunction

(iii) The Hückel model is a MO delocalized model. It is *a*-priori non-correlated but since the t parameter is taken empirically, it contains implicitly the external correlation effects which lack in Hartree-Fock description.

Effective integrals spanned in a s-representation can also be extracted from accurate calculations on Li_2^+ (h^+ and t^+) or Li_2^- (h^- and t^-). All of them are plotted in Figs. 1 and 2 (full lines).

III. Relevance of *s*-effective hamiltonians for clusters

Are the previous conclusions still valid in the case of clusters? The abundance of ionic character in electronic wavefunctions of small clusters was first mentionned in order to interpret Electron-Spin Resonance (ESR) results [28]. It is conforted by electronic density maps or Mulliken population analysis following ab-initio calculations which exhibit significant charge transfers. This aspect has also been illustrated by Maynau and Malrieu [18] who projected CI wavefunctions on non-orthogonal VB wavefunctions and found large overlaps with the ionic configurations. This explains the failure of correlated models such as DIM and magnetic models which were grounded on a representation in the subspace of neutral forms.

This section introduces an effective hamiltonian for Alkali clusters. It is spanned by the space of VB Configurations $|K>$ in the s-band built using a set of orthogonalized AO's that do not need to be defined explicitly. This space involves neutral configurations, singly ionic configurations and multi-ionic configurations. The energy of any of these configurations is defined by a sum of effective two-body potentials running on atom pairs (i, j):

$$H_{KK}^{\text{eff}} = \sum_{i \neq j} \mu_{ij} V(R_{ij})$$

where $V(R_{ij})$ is taken equal to one of the effective diabatic potentials $h_n(R_{ij})$, $h_i(R_{ij})$, $h^+(R_{ij})$, $h^-(R_{ij})$, $h^{++}(R_{ij})$ or $h^{--}(R_{ij})$ according to the electron count and distribution $n_{ij} = n_i + n_j$ on the (i, j) pair. $h^{++}(R_{ij})$ and $h^{--}(R_{ij})$ are respectively the $Li^+ + Li^+$ and the $Li^- + Li^-$ interactions which are taken as $1/R_{ij}$. μ_{ij} is 1 if $n_i \neq n_j$ and 1/2 in the case $n_i = n_j$. The off diagonal-elements H_{KI}^{eff} are non-zero only if:

(i) configurations K and L differ by one electron transfer on a pair (i, j). H_{KL}^{eff} is then equal to a transfer integral $t(R_{ij})$, $t^+(R_{ij})$ or $t^-(R_{ij})$ according to electron counts n_i and n_j

(ii) configurations K and L differ by two differences on a pair (i, j). H_{KL}^{eff} is then equal to $K_n(R_{ij})$ if the difference implies a spin alternation or $K_i(R_{ij})$ if the difference results from a double charge transfer within the pair.

This model is a generalisation of a distance-dependant Hubbard hamiltonian adding exchange integrals and extracted from accurate ab-initio calculations. The diagonalisation of this hamiltonian defines the energies of the states of the clusters and optimisation procedures can be used to search the energy minimum of the ground state surface for n-atom cluster labelled hereafter $E(n)$.

The equilibrium geometries thus obtained are compared in Fig. 5 with the ab-initio results of Boustani et al [8] for Li_n clusters until the heptamer. All the shapes are identical and in general, close quantitative agreement is observed as concerns the internuclear distances. The binding energies per atom

$$e(n) = -(E(n) - nE(1))/n$$

are given in Fig. 6. While showing the same trend, they are systematically larger than those of the ab-initio calculations which were performed in a $2s/1p$ basis set

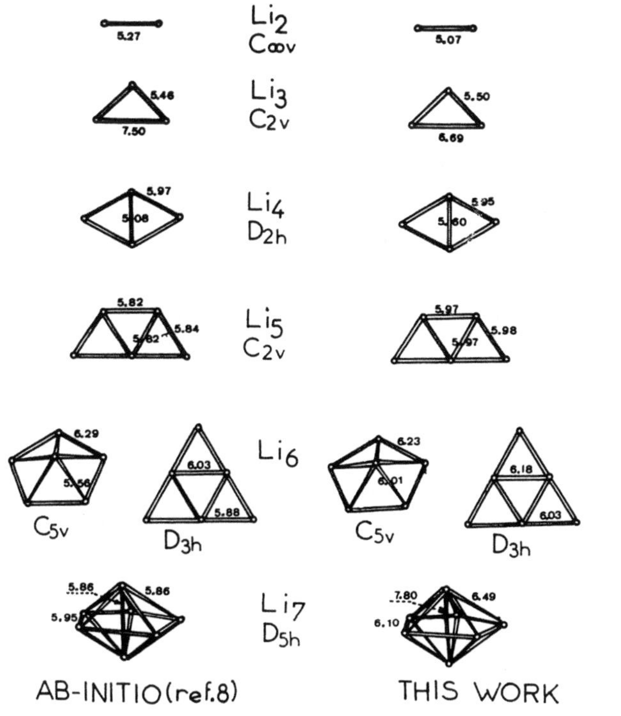

Fig. 5. Equilibrium geometries of small Lithium clusters compared with the ab-initio results of Boustani et al [8]

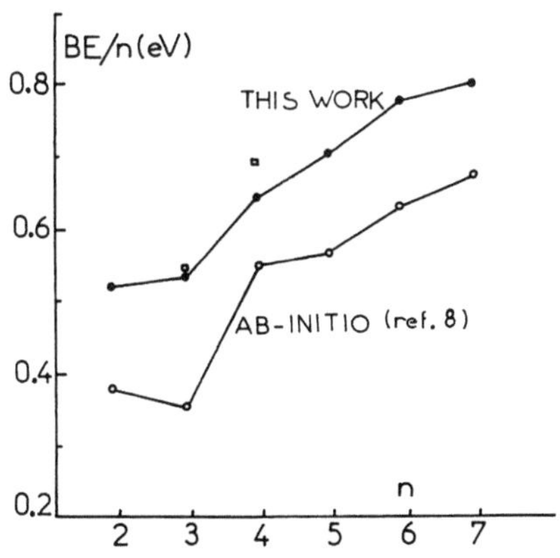

Fig. 6. Binding energies per atom of small lithium clusters compared with the ab-initio results of Boustani et al [8]. The squares are accurate ab-initio results for the trimer [25] and for the tetramer [26]

and thus probably underestimate the (external) correlation energy.

Some apparently minor but still significant differences with ab-initio results must however be quoted. The obtuse angle of the tetramer rhombus is 123.5° instead of 130° [26], which corresponds to a longer small diagonal. In the same way the distance between the two extreme atoms of the heptamer pentagonal bipyramid is too long in our model and the two atoms are not bound. Those differences are indications of the limits of the present model, since the p orbitals are not meant to restrict their action in two-body interactions but may also be implied in characteristic many-body interactions which happen to be significant in specific geometrical situations. For instance the D_{2h} tetramer MO description is $1a_1^2 1b_2^2$, the $1a_1$ orbital is totally bonding, while the $1b_2$ is anti-bonding on the long diagonal. Both orbitals belong essentially to the s-band, however $1b_2$ shows significant mixing with the $2b_2$ MO which is spanned by p orbitals directed along the same axis and centered on the atoms of the short diagonal; this mixing diminishes the antibonding character and stabilizes the $1b_2$ orbital. The stabilization is obviously related to the initial energy difference between the two mixing orbitals. The energy of p-band MO's can be lowered in some conditions by collective interactions; it is the case on the D_{5h} heptamer where there is again a significant mixing [8] between the HOMO which is antibonding between the two summits of the pyramid and a π orbital involving the p-orbitals of the pentagonal basis. The π MO energy is lowered due to cyclic conjugation effects and this enhances its mixing with the s-band anti-bonding MO which yields a shortening of the central axis. This importance of p mixing on Li_7 is also supported by ESR experiments [28].

In conclusion, the most important part of the effects due to p orbitals (weak axial hybridization, dispersion energy, repulsion and essentially angular correlation) may be incorporated in a s-model through two-body effective interactions and couplings. The p collective effects are lacking but the magnitude of their contribution for Li_4 can be estimated to 0.25 eV out of a total binding energy of 2.80 eV [26]. The general concordance of the present work with ab-initio results shows that our model includes the essential of the physics governing the structure of small Alkali clusters and proves that delocalization within the s-band plays the dominant role.

References

1. Knight, W.D., Clemenger, K., De Heer, W.A., Saunders, W.A., Chou, M., Cohen, M.L.: Phys. Rev. Lett. **52**, 2141 (1984)
2. Eckardt, M.: Phys. Rev. Lett. **52**, 1925 (1984)
3. Ishü, Y., Ohnishi, S., Sugano, S.: Phys. Rev. B **31**, 1804 (1985)
4. For a review, see Koutecky, J., Fantucci, P.: Chem. Rev. **86**, 539 (1986); Elemental and molecular clusters. Springer Series in Material Sciences, Vol. 6. Benedek, G., Martin, T.P., Pacchioni, G., (eds.). Berlin, Heidelberg, New York: Springer 1988; Metal Bonding and interactions in high temperature systems with emphasis on alkali metals. Gole, J.L., Stwalley, W.C., (eds.). Washington D.C.: A.C.S. 1982
5. Flad, J., Stoll, H., Preuss, H.: J. Chem. Phys. **71**, 3042 (1979); Flad, J., Igel, G., Dolg, M., Stoll, H., Preuss, H.: Chem. Phys. **75**, 331 (1983)

6. Martins, J.L., Car, R., Buttet, J.: J. Chem. Phys. **78**, 5646 (1983); Martins, J.L., Buttet, J., Car, R.: Phys. Rev. B **31**, 1804 (1985)
7. Rao, B.K., Jena, P.: Phys. Rev. B **31**, 1804 (1985)
8. Boustani, I., Pewestorf, W., Fantucci, P., Bonacic-Koutecky, V., Koutecky, J.: Phys. Rev. B **35**, 9437 (1987)
9. Bonacic-Koutecky, V., Fantucci, P., Koutecky, J.: Phys. Rev. B **37**, 4369 (1988)
10. Brechignac, C., Cahuzac, Ph., Roux, J.Ph., Pavolini, D., Spiegelmann, F.: Chem. Phys. **87**, 5694 (1987)
11. Pavolini, D., Spiegelmann, F.: J. Chem. Phys (in press)
12. Companion, A.L., Stieble, D.J., Starshak, A.J.: J. Chem. Phys. **49**, 3637 (1968)
13. Companion, A.L.: Chem. Phys. Lett. **56**, 500 (1978)
14. Pickup, B.T., Byer, W., Brown, S.: Mol. Phys. **23**, 1189 (1972)
15. Richstmeier, S.C., Dixon, D.A., Gole, J.L.: J. Phys. Chem. **86**, 3942 (1982)
16. Varandas, J.C., Morais, V.M.F.: Mol. Phys. **47**, 1241 (1982); Varandas, J.C., Morais, V.M.F., Pais, A.A.C.C.: Mol. Phys. **58**, 285 (1985)
17. Wang, Y., Georges, T.F., Lindsay, D.M., Beri, A.C.: J. Chem. Phys. **86**, 3493 (1987); Lindsay, D.M., Wang, Y., Georges, T.F.: **86**, 3500 (1987)
18. Maynau, D., Malrieu, J.P.: J. Chem. Phys. **88**, 3163 (1988)
19. Hubbard, J.: Proc. Soc. London A **276**, 238 (1963)
20. For an analysis in small clusters, see: Friedel, J.: In: Elemental and molecular clusters. Springer Series in Material Sciences, Vol. 6, p. 12. Berlin, Heidelberg, New York: Springer 1988
21. Spiegelmann, F., Malrieu, J.P., Zurru, J.P., Maynau, D.: J. Chim. Phys. **83**, 69 (1986) and references therein
22. Blaise, P., Spiegelmann, F., Malrieu, J.P., Maynau, D.: (to be published)
23. Cimiraglia, R., Malrieu, J.P., Persico, M., Spiegelmann, F.: **18**, 3073 (1985)
24. Komiha, N., Daudey, J.P., Malrieu, J.P.: J. Phys. B **20**, 4375 (1987); Rajzman, M., Spiegelman, F., Malrieu, J.P.: J. Chem. Phys. **89**, 433 (1988)
25. Pavolini, D., Spiegelmann, F.: J. Chem. Phys. **87**, 2854 (1987)
26. Bonacic-Koutecky, V., Fantucci, P., Koutecky, J.: Chem. Phys. Lett. **146**, 518 (1988)
27. Evangelisti, S., Daudey, J.P., Malrieu, J.P.: Phys. Rev. A **35**, 4930 (1987)
28. Lindsay, D.M., Herschbach, D.R., Kurian, A.L.: Mol. Phys. **32**, 1199 (1976); Garland, D.A., Lindsay, M.: J. Chem. Phys. **80**, 10 (1984)

Z. Phys. D – Atoms, Molecules and Clusters 12, 347–351 (1989)

Atoms, Molecules
Zeitschrift für Physik D and Clusters
© Springer-Verlag 1989

New methods for the calculation of the electronic structure of clusters: application for Na_{13} to Na_{147}

J.L. Martins

Department of Chemical Engineering and Materials Science, University of Minnesota, Minneapolis, MN 55455, USA

Received 5 July 1988

We present a review of recent developments in the field of electronic structure calculations and of their application to the ab-initio quantum mechanical studies of the properties of clusters. The possibilities of the new computational methods are illustrated by the calculation of the binding energy and level structure of sodium clusters in the size range of Na_{13} and Na_{147}.

PACS: 31.20.Ef; 36.40.+d

Introduction

The ab-initio quantum mechanical calculations of the electronic structure of clusters face two main problems: (i) We are usually interested in the evolution of cluster properties with size, but the computational difficulty of a problem increases rapidly with the size of the system, and will exceed any finite computer resources. (ii) Contrary to most studies of molecules and solids, the geometry of clusters are usually unknown, and should be, in principle, predicted by minimizing the calculated total energies with respect to cluster geometry. Realistic calculations of the electronic structure of clusters have been limited to very small sizes, three to ten atoms tipically, and only in very few cases was the optimization of the cluster geometry attempted. For clusters larger than a dozen atoms, theoretical calculations of the electronic structure had to make some drastic simplifying assumption about the shape of the ionic potential, the shape of the wavefunctions, or the values of the hamiltonian matrix elements, an undesirable feature if we want a realistic description of the cluster properties.

In a discussion of the efficiency of different computational schemes for the study of clusters it is crucial to consider the asymptotic dependence of the computing time τ on the number of atoms n present in the cluster, this dependence can usually be expressed for a reasonable range of cluster size in the form,

$$\tau \simeq A n^{\gamma},$$

where the prefactor A will depend on the intrinsic computer spead and details of the algorithm, but the more important exponent γ will be a characteristic of the computational scheme. For the traditional computational schemes, we have that $\gamma=3$ for the local density calculations, $\gamma=4$ for the Hartree-Fock method, and $\gamma=5$ for configuration interaction calculations. This dependence of the computing time, τ, on large powers of the number of atoms explains the problems found by the cluster theorists in extending their calculations to larger cluster sizes.

Recent progresses in electronic structure calculation methods [1, 2] are particularly interesting for cluster applications, because they have smaller power dependencies of the computing time on the number atoms as compared to the traditional methods. The molecular dynamics method of Car and Parrinello [1] for local density calculations, and related methods [3], have the computing time dependence, $\tau \propto n^2 \log n$, and the pseudospectral formulation of the Hartree-Fock method[2] has the computing time dependence $\tau \propto n^3$.

The Car and Parrinello method is specially suited for cluster calculations, because it allows the simultaneous calculation of the electronic structure, and the search for the equilibrium structure of the cluster, improving on both the botlenecks of the electronic structure calculations.

Calculations for Si and Se clusters [4, 5] using the molecular dynamics method have been recently published, showing how powerful the method is for cluster simulations, in particular for finding the equilibrium geometries. We will present here some new results for medium sized sodium clusters, up to Na_{147}, using a similar computational method[3] that should be more efficient for fixed geometry calculations, but is less efficient if the equilibrium geometry of the cluster has to be searched.

In the following sections, we will review briefly the possibilities of developing algorithms with small values of the exponent γ, and stress the similarities between the molecular dynamics method in local density theory and the pseudospectral Hartree-Fock method: both represent the wavefunctions in a finite basis set – plane waves for Car and Parrinello, gaussians orbitals for the pseudospectral method – and in a grid in real space. This dual representation of the wavefunction is used extensively in our recently developed dual space pseudopotential formalism [3]. Applications to Na_n clusters will illustrate this last method.

New computational methods

Let us consider first the origin of the cubic dependence of the computing time on the number of atoms for the traditional local density calculations. In the local density method we must solve a one electron Schrödinger equation in an effective potential V_{eff} that contains (within the local density approximation) all the relevant many-body interactions,

$$H = \tfrac{1}{2}\nabla^2 + V_{\text{eff}}(r), \tag{1a}$$

$$H\psi(r) = \varepsilon\psi(r). \tag{1b}$$

If we expand the wavefunction in a basis set $\{\phi_i(r)\}$ of functions,

$$\psi(r) = \sum_{i=1}^{N} a_i \phi_i(r),$$

the differential Schrödinger equation is transformed into a matrix eigenvalue equation

$$\sum_{j=1}^{N} H_{ij} a_j = \varepsilon a_i, \tag{2}$$

where,

$$H_{ij} = \int \phi_i^*(r) H \phi_j(r)\, \mathrm{d}^3 r. \tag{3}$$

For a constant numerical accuracy, the number N of functions in the basis set is proportional to the number of atoms in the cluster, $N \propto n$, and the computing time required to diagonalize a matrix with a Givens-Householder type of algorithm is proportional to the cube of the matrix size, $\tau \propto N^3$, whence the cubic dependence of computing time on the cube of the number of atoms, $\tau \propto n^3$. Using iterative diagonalization schemes, one can improve the prefactor A of the required computing time if only a few eigensolutions of (2) are required, but the dependence on the number of atoms will still be cubic. The basic operation of an iterative scheme is the calculation of the product of the Hamiltonian by a trial wavefunction, $\sum_{j=1}^{N} H_{ij} a_j$, requiring N^2 multiplications, and it is possible to calculate a single eigenvector within a computing time $\tau \propto n^2$. Since the number of eigenvectors that must be calculated is equal to the number of occupied states and therefore proportional to the number of atoms, the iterative procedure also has $\gamma = 3$. The matrix H_{ij} contains N^2 elements, and the number of operations needed to calculate each of them is, at least, proportional to the number of atoms, as can be seen by considering the spatial extent of the integral defining the matrix H_{ij} (3). The calculation of the Hamiltonian matrix will thus require a computing time $\tau \propto n^3$.

From the above discussion it would seem that the exponent $\gamma = 3$ is the best we can hope for in the local density scheme, so we will give two counter-examples with $\gamma = 2$. The first example is obtained by considering a basis set, $\{\phi_i(r)\}$, of atomic wavefunctions with a finite range, i.e. functions that are identically equal to zero outside a sphere centered on a atom. For large clusters the number of non-zero elements of H_{ij} will be proportional to the number of atoms, n, and with the techniques for sparse matrices it is possible to develop a electronic structure calculation scheme with $\tau \simeq An^2$. The second counter-example is obtained if we solve the Schrödinger equation (1) by a finite elements method. The wavefunctions would be represented in a real space mesh of points, $\{r_j\}$, where the number of points is proportional to the number of atoms. The computing time for finding one eigensolution of the finite differences, using, for example, a relaxation method, is proportional to the number of mesh points, and therefore the computing time for calculating the wavefunctions for all the occupied states will be proportional to the square of the number of atoms. Neither of these two methods has been popular because the prefactor A in the approximate expression for the computing time can be very large. In the case of the finite range basis set the problem is that the calculation of many center molecular integrals over arbitrarily shaped functions is extremely difficult; in the case of the finite differences equation, one problem is the Coulomb singularity of the potential at the nucleus, and the other problem is the absence of the useful error cancellations that occur in basis set expansion methods and other schemes based on variational principles [2].

The two methods discussed above have $\gamma=2$ because they recognize the local character of the differential operators in the Schrödinger equation (1). This local character can be used advantageously by representing the wavefunction simultaneously in a basis set and in a mesh of real space points. The one particle effective Hamiltonian, H is the sum of a kinetic energy operator, diagonal in momentum (reciprocal) space, and a potential energy operator, diagonal in real (coordinate or direct) space. The work required to calculate the action of each of the operators on a wavefunction scales with the number of atoms in a cluster, if the representation where the operator is diagonal is used, and it scales with the square of the number of atoms otherwise. If we could find an efficient and accurate transformation from one representation to the other, we would be able to use each operator in its diagonal representation.

In a plane wave basis set the fast Fourier transform (FFT) provides an efficient method for calculating the value of a function at N points from the N coefficients of its plane wave expansion, and vice-versa [1, 3]. To calculate the action of the effective Hamiltonian H on a vector ψ, represented, for example, in momentum space by N components $\psi(G)$, we obtain easily the result of the action of the kinetic energy operator, $\frac{1}{2}G^2\psi(G)$; the action of the potential energy operator can be obtained by calculating $\psi(r_j)$ in a regularly spaced grid in real space using the FFT, evaluating the potential on the same grid, $V_{\text{eff}}(r_j)$, calculating the product, $(V_{\text{eff}}\psi)(r_j)=V_{\text{eff}}(r_j)\psi(r_j)$, and using the FFT to transform $V_{\text{eff}}\psi$ back to momentum space obtaining $(V_{\text{eff}}\psi)(G)$. The FFT algorithm requires only $O(N\log N)$ floating point operations, and therefore the calculation of the product $H\psi$ scales with $n\log n$ and not with n^2. A fundamental property of the FFT procedure is that no accuracy is lost by using the FFT scheme with respect to the straightforward matrix multiplication.

The FFT based matrix multiplication can be used with different iterative schemes for solving (2). In the original implementation [1], a molecular dynamics algorithm was used, which allowed to solve simultaneously the electronic structure problem, and to simulate the dynamics of the ions. Recently we integrated [3] the FFT matrix multiplication scheme with a direct inversion of the iterative subspace scheme, (the method used for the calculations presented in the next section) and a conjugated gradient scheme was proposed [6] by Štich et al.

The methods just described for solving the Schrödinger equation (1) belong to the class of spectral methods. A pseudospectral method has been proposed recently [2] for solving the Hartree-Fock equations following similar ideas. In the Roothaan Hartree-Fock scheme the wavefunctions are expanded in a basis set, $\{\phi_i\}$, and the Fock matrix is represented on that basis, requiring the computation of the complicated four center integrals,

$$\langle ij|kl\rangle=\int\int\frac{\phi_i^*(r)\phi_j^*(r')\phi_k(r)\phi_l(r')}{|r-r'|}\,d^3r\,d^3r'.$$

The total number of these electrostatic integrals is proportional to n^4 and therefore the computing time dependence on the number of atoms is $\tau\propto n^4$. The pseudospectral method starts by choosing a set of N points, $\{r_j\}$, such that the matrix,

$$R_{ij}=\phi_i(r_j),$$

is not singular. An wave function is then represented either by its basis set expansion a_i, where $\psi(r)=\Sigma_{i=1}^N a_i\phi_i(r)$, or by its value in the real space grid, $b_j=\psi(r_j)$. The change of representation is then described by the matrix R_{ij},

$$b_j=\sum_{i=1}^{N} a_i R_{ij},$$

and we have the choice of calculating any operator in the most practical representation. For example the kinetic energy operator can be calculated in the basis set representation, while the Hartree operator can be calculated in the grid point representation. The cumbersome four center integrals are not needed in the grid point representation, for example the Hartree potential at the grid point r_j requires the integral,

$$\int\frac{\phi_i^*(r)\phi_i(r)}{|r-r_j|}\,d^3r.$$

The computing time of the pseudospectral Hartree-Fock method increases only with the cube of the number of atoms, and is numerically accurate and stable as demonstrated by the first applications [2].

Application of Na_n clusters

Although our computational method was developed for electronic structure calcultions of periodic solids, it can be applied to clusters by using a supercell method, i.e. by using a periodic lattice of clusters, with enough empty space between clusters. Using supercell techniques we are able to study the isolated atom, very small clusters, medium sized clusters, the surface and the bulk solid, within the same physical and numerical approximation. It is therefore an appropriate method to study the transition from atoms to clusters to solids.

We studied a series of sodium clusters with a fixed geometry, corresponding to fragments of an fcc lattice of increasing size with a cubo-octahedral geometry and containing 13, 55, and 147 atoms; we also calculated the electronic structure for the isolated atom and fcc bulk sodium. Since we were interested in general cluster properties and not on the properties of a given cluster, we did not optimize the geometries, in particular, the interatomic distance were kept constant, and the distance between clusters was only two lattice constants, sufficiently large for the interaction between neighboring clusters to be small although not negligeable. By the same reason that we are interested in trends, we should not worry that bulk Na is not fcc (though the energy of the fcc structure is practically identical to the equilibrium bcc structure), or that medium sized clusters should not be expected to look like fcc crystal fragments. The calculations were done using a first-principles local pseudopotential, and a kinetic energy cutoff for the plane wave expansion of 5 Rydberg, corresponding to more than 14000 basis functions in our calculation of the properties of Na_{147}.

The binding energies of the fcc clusters Na, Na_{13}, Na_{55}, Na_{147}, and Na_{∞}, are plotted in Fig. 1 as a function of the $-1/3$ power of the number of atoms, i.e. versus the inverse of their "radius". The almost linear behaviour of the binding energy with the inverse radius displayed by the data is to be expected if the main contribution to the lowering of the binding energy per atom is a surface effect. The data suggests that surface energy based arguments (drop models) could be valid even for very small sizes.

There is a strong experimental and theoretical evidence for an underlying electronic spherical shell structure in sodium clusters [7, 8]. The shell structure is very pronounced in the abundance mass spectrum, but is much weaker in the measured ionization potentials. Real clusters are not spherical, and an important question is how the shell structure is modified when the spherical degeneracies are lifted by the lowering of the symmetry. An example of the effects of lowering the symmetry is shown in Fig. 2, where the level structure of the 55 atom cubo-octahedron sodium cluster is compared with the level structure of the 53 atom cluster obtained by removing two of the corner atoms. The cubo-octahedron cluster has a cubic symmetry, and has several double and triple degenerate levels. Some of the larger gaps between the levels can be associated with the gaps of a spherical cluster, showing that the reduction of the symmetry from spherical to cubic does not destroy entirely the underlying spherical structure. The 53 atom cluster has a low symmetry and none of its electronic levels is degenerate. The gaps between levels are smaller but we can still identify some of the larger gaps with the corresponding spherical gaps. This illustrates how some of the effects of the

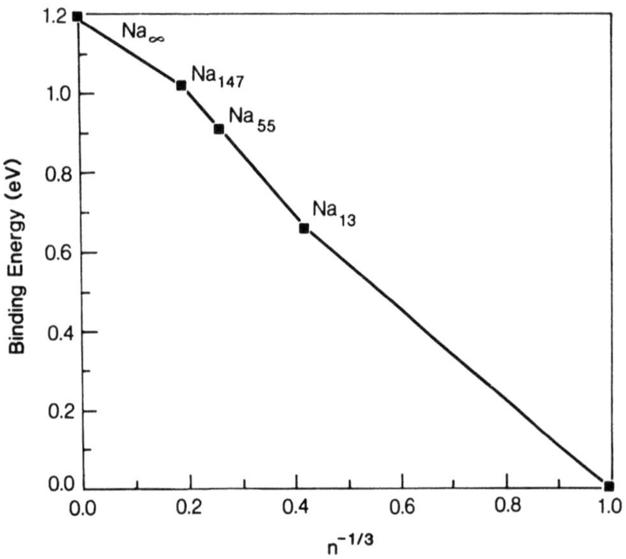

Fig. 1. Calculated binding energy of sodium clusters as a function of the number of atoms in the cluster. The squares correspond to the bulk, cubo-octahedral clusters with 147, 55 and 13 atoms, and the isolated atom. The line was drawn as guide to the eye

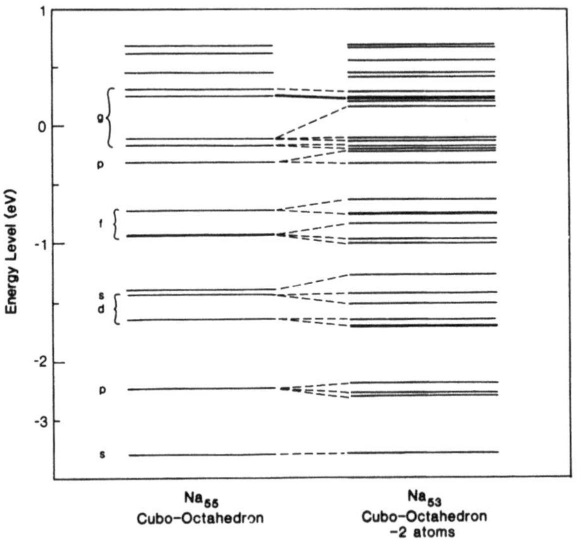

Fig. 2. Level structure for a highly symmetric 55 atom cluster compared with a low symmetry 53 atom cluster. See text for the discussion

spherical shell structure are preserved by non-spherical clusters.

Conclusions

We have reviewed recent progresses [1, 2] in the calculation of the electronic property of clusters using the local density approximation or the Hartree-Fock method. The use of a dual representation of the wavefunctions in a basis set expansion, and in a grid in real space, is an important element in the development of new computational methods that are more efficient than the traditional basis set expansions.

Using a recently developed ab-initio pseudopotential local-density scheme [3], we performed realistic calculations for medium sized sodium clusters, up to 147 atoms, an order of magnitude larger than those studied previously with traditional ab-initio computational schemes. When the optimization of the geometry of the clusters is important – and in clusters this is the rule – the recently developed molecular dynamics method of Car and Parrinello should be extremely useful as demonstrated by the recent calculations [4, 5] for Si and Se clusters.

The calculations for sodium clusters were supported by a computer time grant from the Minnesota Supercomputer Institute

References

1. Car, R., Parrinello, M.: Phys. Rev. Lett. **55**, 2471 (1985)
2. Friesner, R.A.: Chem. Phys. Lett. **116**, 39 (1985); Friesner, R.A.: J. Chem. Phys. **86**, 3522 (1987)
3. Martins, J.L., Cohen, M.L.: Phys. Rev. B **37**, 6134 (1988)
4. Ballone, P., Andreoni, W., Car, R., Parrinello, M.: Phys. Rev. Lett. **60**, 271 (1988)
5. Hohl, D., Jones, R.O., Car, R., Parrinello, M.: Chem. Phys. Lett **139**, 540 (1987)
6. Štich, I., Car, R., Parrinello, M.: Phys. Rev. B **39**, 4997 (1989)
7. Knight, W.D., Clemenger, K., De Heer, W.A., Saunders, W., Chou, M.Y., Cohen, M.L.: Phys. Rev. Lett. **31**, 1804 (1985)
8. Martins, J.L., Buttet, J., Car, R.: Phys. Rev. B **31**, 1804 (1985)

Z. Phys. D – Atoms, Molecules and Clusters 12, 353–356 (1989)

Atoms, Molecules
Zeitschrift für Physik D and Clusters
© Springer-Verlag 1989

Static electric dipole polarizability of small sodium aggregates

I. Moullet[1], José Luís Martins[2], F. Reuse,[1] and J. Buttet[1]

[1] Institut de Physique Expérimentale, Ecole Polytechnique Fédérale de Lausanne, CH-1015 Lausanne, Switzerland

[2] Department of Chemical Engineering and Materials Science, University of Minnesota, Minneapolis, Minnesota 55455, USA

Received 5 July 1988; final version 30 September 1988

We present pseudopotential local-spin-density calculations of the static electric polarizability of sodium dimers and trimers and their respective cations. The electronic polarizabilities are obtained from self-consistent calculations in the presence of an external electric field, which is kept sufficiently small to avoid non-linear effects. The calculated polarizability tensor has a strong anisotropy directly related to the geometric and electronic structures of the molecules, the anisotropy being larger for the neutral clusters. The polarizabilities are averaged over the vibrational motion and rotations of the aggregates in order to be compared with the experimental measurements. The obtained values show an improvement in the agreement with experiment with respect to the values calculated in the spherical approximation.

PACS: 31.20.Sy; 31.90.+s

Introduction

The static electric polarizabilities of sodium clusters with sizes up to 40 atoms were recently obtained by Knight et al. [1] by measuring the deflection of a molecular beam in the presence of a static electric field. Their results show that the polarizability follows the classical scaling with the cluster volume. The general trend of the polarizabilities per atom normalized to the atomic polarizability is a slow decrease towards the bulk value of 0.4 with a superimposed fine structure that should be explained by the cluster's electronic and geometric structure. Smaller relative values of the normalized polarizability are found for the magic number clusters which have a larger abundance in the observed mass spectrum [2] and which correspond to the closed shells of the spherical models of simple metal clusters [2, 3]. The fine structure observed between two shell closings cannot be explained within the spherical models and reflects the deviations from "sphericity" of the true cluster geometry.

The response of an alkali cluster to electric fields have been the subject of several recent calculations, most of them using the spherical jellium approximation and the local density formalism [4, 5, 6, 7, 8]. These studies show that the polarizability of a jellium sphere is enhanced with respect to the classical value obtained using the bulk polarizability. This behaviour is attributed to an extension of the spill out of the electronic charge density across the classical spherical surface. The spherical jellium calculations consistently underestimate the experimental polarizability by about 20% and are unable to reproduce the fine structure of the cluster polarizability, since they do not take into account the actual geometry of the cluster.

We have performed pseudopotential local-spin-density (LSD) calculations of the static electric properties of Na_2 and Na_3 molecules. The effects of the underlying geometric structure is taken into account without any shape approximation and we include corrections due to the vibrational motion of the ions. Our calculated values underestimate the experimental values by 9 to 18%, the order of magnitude of the error attributed to the self interaction correction of the LSD is 10% [8].

Theoretical and computational method

The computational method has been previously described in detail [9, 10, 11]. The exchange and correlation effects are treated in the local-spin-density approximation of the density functional formalism [12] using the Ceperley-Alder correlation [13], the effect of the core on the valence electrons is described by a first principles pseudopotential [14].

The interaction with a uniform static electric field **E** is obtained by adding a term of the form **E.r** to the self-consistent Kohn-Sham equations[1]. The geometrical structure of the perturbed clusters are found by minimizing the total energy, including the dipolar term, and the polarizability tensor is derived from the linear relation of the induced dipole moment to the electric field for small value of **E**,

$$\mathbf{p}_{\text{ind}} = \mathbf{p}_{\text{tot}} - \boldsymbol{\mu} = -\int(\rho(\mathbf{r}) - \rho^0(\mathbf{r}))\,d\mathbf{r} + \sum_j Z_j(\mathbf{R}_j - \mathbf{R}_j^0)$$
$$= \alpha\mathbf{E} + O(\mathbf{E}^2) \tag{1}$$

where $\rho(\mathbf{r})$ is the valence electron density, Z_j and $\mathbf{R}_j$ are the charge and position of the jth core, and α the polarizability tensor. The superscript 0 indicates the values in absence of perturbation. Unless specifically indicated, atomic units [a.u.] are used in this work.

The electric field values we use are in the range of $5\ 10^6\ \text{Vm}^{-1}$ up to $5\ 10^8\ \text{Vm}^{-1}$ (10^{-5}–10^{-3} a.u.). They are sufficiently large to avoid numerical errors but sufficiently small that all effects are linear.

In order to include the effects of molecular vibrations on the polarizabilities we average the electronic properties over the molecular vibrational ground state. The equation for the ionic vibrations in the adiabatic approximation is

$$H_N = [T_N + U_e(Q, \mathbf{E})]\chi(Q) = E\chi(Q), \tag{2}$$

where Q are the normal coordinates and the electric field dependent adiabatic potential is given by

$$U_e(Q, \mathbf{E}) = U_e^0(Q) - \boldsymbol{\mu}(Q)\mathbf{E} - \tfrac{1}{2}\alpha(Q)\mathbf{E}^2 + O(\mathbf{E}^3). \tag{3}$$

Expanding both $\boldsymbol{\mu}$ and α in terms of the normal coordinates, the field dependent nuclear hamiltonian is

$$H_N = H_N^0 + A(\mathbf{E}) + \sum_k B_k(\mathbf{E})Q_k + \frac{1}{2}\sum_{k,l} C_{kl}(\mathbf{E})Q_kQ_l$$
$$+ O(Q^3), \tag{4}$$

where,

$$A(\mathbf{E}) = -\boldsymbol{\mu}\mathbf{E} - \tfrac{1}{2}\alpha\mathbf{E}^2 \tag{5.a}$$

$$B_k(\mathbf{E}) = -\left(\frac{\partial\boldsymbol{\mu}}{\partial Q_k}\right)\mathbf{E} - \frac{1}{2}\left(\frac{\partial\alpha}{\partial Q_k}\right)_0\mathbf{E}^2 \tag{5.b}$$

$$C_{kl}(\mathbf{E}) = -\left(\frac{\partial^2\boldsymbol{\mu}}{\partial Q_k\partial Q_l}\right)_0\mathbf{E} - \frac{1}{2}\left(\frac{\partial^2\alpha}{\partial Q_k\partial Q_l}\right)_0\mathbf{E}^2. \tag{5.c}$$

Treating the A, B and C terms in first order perturbation theory and using the harmonic approximation for the vibrational wavefunction gives the vibrationally averaged polarizability,

$$\langle\alpha\rangle = \alpha + \frac{1}{4}\sum_k \frac{1}{m_k\omega_k}\left(\frac{\partial^2\alpha}{\partial Q_k^2}\right)_0, \tag{6}$$

where m_k and ω_k are, respectively, the mass and the frequency of the kth normal mode.

Because of high experimental temperatures, the electric field is too low to align the molecule along one preferred direction and the measured values are the polarizabilities averaged over the orientations of the clusters. Thus, we defined the mean averaged polarizability,

$$\langle\bar{\alpha}\rangle = \tfrac{1}{3}(\langle\alpha_{xx}\rangle + \langle\alpha_{yy}\rangle + \langle\alpha_{zz}\rangle). \tag{7}$$

Results and discussion

To test the frozen core approximation, we have compared the polarizability of a sodium atom in both all-electron and pseudopotential calculations (Table 1), and verified that the polarization of the ionic core can be neglected. We also reported in Table 1 the result obtained in the spherical jellium model [15] which underestimates the atomic polarizability of 18%, versus our underestimate of 11%.

In order to calculate the induced dipole moment of a molecule that has not an inversion symmetry, we must determine its permanent dipole moment. All realistic calculations of the equilibrium structure of Na_3 predict that it is an "obtuse" isosceles triangle and that the Born Oppenheimer (BO) surface is strongly anharmonic and has saddle points corresponding to an "acute" isosceles geometry [11][2]. These two ex-

Table 1. Atomic polarizability in Å^3

α_{ae}	α_{ps}	$\alpha^a_{\text{jellium}}$	α^b_{exp}
21.2	21.0	19.4	23.6

[a] From [15]
[b] From [1].

[1] Since we use a localized basis set the asymptotic behaviour of the potential is irrelevant for our calculations

[2] The "obtuse" geometry corresponds to an isosceles triangle with an apex angle larger than 60°

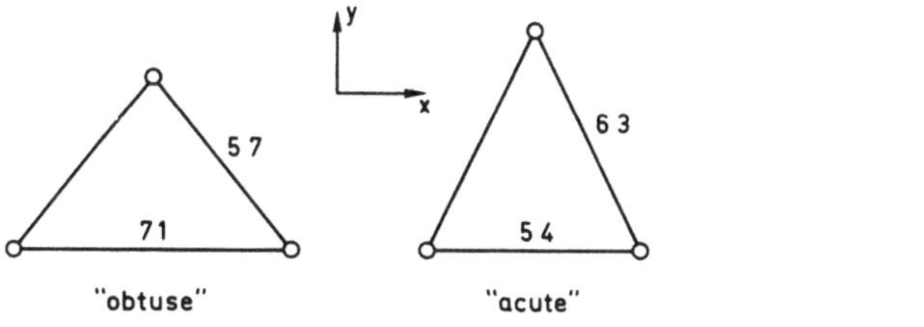

Fig. 1. Calculated equilibrium and saddle point geometries for Na_3

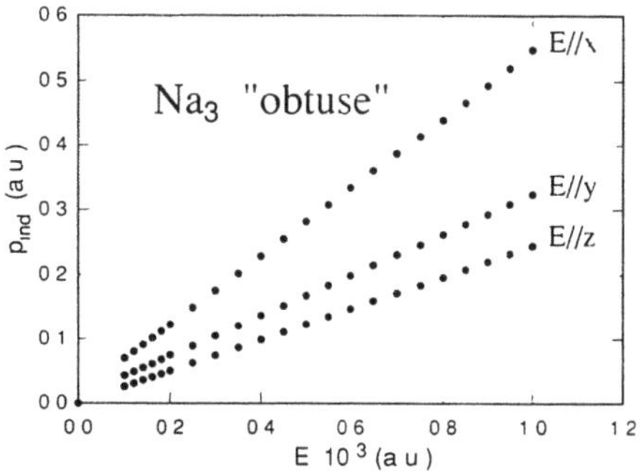

Fig. 2. Induced dipole moment versus the electric field for the equilibrium geometry of Na_3. The electric field is oriented along the three principal axis of the molecule

tremal configurations are shown in Fig. 1. By symmetry the permanent moment will be oriented along the C_2 axis of the molecule (the y axis in the Fig. 1). In both cases, the major contribution to the permanent dipole moment $\boldsymbol{\mu}$ is found to come from the non-bonding state while the electronic charge of the bonding state seems to be equally distributed among the three atoms. For the equilibrium structure the permanent dipole moment value is two orders of magnitude smaller than the value obtained for the saddle point, which is -0.26 a.u..

As previously mentioned (1) the polarizability is evaluated from the dependence of the induced dipole moment on the electric field. The induced polarization is calculated for several values of **E** oriented along each of the principal axis of the molecule. Our numerical results are reported in Fig. 2 for the equilibrium geometry of Na_3 showing a linear behaviour of the induced charge density on the electric field. The slopes of the lines give directly the diagonal elements of the polarizability tensor, these calculated values are reported in Table 2 for the neutral and ionized dimers and trimers. For our choice of coordinate system (see Fig. 1), the non diagonal elements of the polarizability tensor are zero. The calculated values of the electric polarizability show a strong anisotropy for the neutral clusters (see also Fig. 2). Qualitatively this difference can be explained if we interpret the polarizability as a measure of the charge extension in each direction. A simple perturbation scheme based on a LCAO analysis of the molecular orbitals shows that for Na_2 the perpendicular component results from the coupling of the highest occupied state (HOMO) with some π-type molecular orbital higher in energy while the parallel polarizability is due to the interaction with the LUMO (lowest unoccupied molecular orbital), the σ anti-bonding orbital.

A similar analysis for Na_3 explains the difference between the components of the polarizability parallel (α_{xx}, α_{yy}), and perpendicular (α_{zz}) to the molecular plane. Examining the contribution to the polarizability of each occupied state, we have seen that the non-bonding state is the easiest to polarize. Compared with the results for the neutral clusters, the ionized clusters are less anisotropic, a trend found previously for the lithium dimer [16]. This is due to the reduction of the total valence charge (we remove one electron) and to the fact that the remaining electrons are more tightly bounded.

We calculated the diagonal tensor elements of α in several points of the BO surface near the equilibrium for the neutral clusters and calculated the averaged polarizabilities using (6). The results for the averaged quantities are reported in Table 3 for both the dimer and trimer. For the perpendicular component of the dimer polarizability the average over the ground state vibrations results in an increase of about 7%, while the parallel component, which is proportional to the interatomic distance, is not affected by the vibrations. In the case of Na_3 we see an increase of 7% and 17% for the polarizability in the molecular plane while the perpendicular component decreases by 20%. The atomic vibrations tend to reinforce the anisotropy and their contribution is not negligeable.

Table 2. Electronic molecular polarizabilities in Å^3

			Na_3		
	Na_2	Na_2^+	"obtuse"	"acute"	Na_3^+
α_{xx}	26.1	10.8	78.8	50.6	25.0
α_{yy}	26.1	10.8	46.6	74.3	23.9
α_{zz}	47.2	17.9	36.0	34.3	17.0
$\bar{\alpha}$	33.1	13.2	53.8	53.1	22.0

Table 3. Averaged molecular polarizabilities in Å^3

	Na_2	Na_3
$\langle\alpha_{xx}\rangle$	27.9	88.1
$\langle\alpha_{yy}\rangle$	27.9	54.5
$\langle\alpha_{zz}\rangle$	47.2	28.9
$\langle\bar{\alpha}\rangle$	34.4	57.2
$\alpha^a_{jellium}$	30.3	53.6
α^b_{exp}	37.9	69.8

[a] From data of [7]
[b] From [1]

We report in the fourth row of Table 3 the rotational averaged polarizabilities of the dimer and trimer as defined by (7). The corresponding pure electronic quantities are in the last row of Table 2.

When the values of $\langle\bar{\alpha}\rangle$ are compared with the experimental results, we find that our calculations underestimate the polarizabilities by about 9% for Na_2 and 18% for Na_3. The LSD approximation is expected to underestimate the true polarizabilities by about 10% according to the calculations of Stampfli and Bennemann [8] who introduced self-interaction correction in their spherical jellium calculations. The discrepancy between our calculated values of the polarizability for Na and Na_2 and the experimental results can be attributed to the deficiencies of the LSD approximation, the discrepancy for the trimer is however too large. We suspect that contributions of the molecular vibrations to the static polarizability that were neglected in the approximations leading to eq. 6 (e.g. the harmonic approximation for the vibrations is very doubtful in the case of the trimer) are responsible for the remaining discrepancy between theory and experiment.

The work reported in this paper was partially supported by the Swiss National Science Foundation through grant # 2.454-0.87.

References

1. Knight, W.D., Clemenger, K., de Heer, W.A., Saunders, W.: Phys. Rev. B **31**, 2539 (1985)
2. Knight, W.D., Clemenger, K., de Heer, W.A., Saunders, W., Chou, M.Y., Cohen, M.L.: Phys. Rev. Lett. **52**, 2141 (1984)
3. de Heer, W.A., Knight, W.D., Chou, M.Y., Cohen, M.L.: Solid State Phys. **40**, 93 (1987)
4. Snider, D.R., Sorbello, R.S.: Phys. Rev. B **28**, 5702 (1983)
5. Beck, D.E.: Phys. Rev. B **30**, 6935 (1984)
6. Ekardt, W.: Phys. Rev. Lett. **52**, 1925 (1984)
7. Manninen, M., Nieminen, R.M., Puska, M.J.: Phys. Rev. B **33**, 4289 (1986)
8. Stampfli, P., Bennemann, K.H.: In: Physics and chemistry of small clusters. Jena, P., Rao, B.K., Khanna, S.N. (eds.), p. 473. New York: Plenum Press, 1987
9. Martins, J.L.: Calcul des propriétés structurales et électroniques d'agrégats de metaux alcalins. Ph.D. Thesis, Ecole Polytechnique fédérale de Lausanne (1984)
10. Martins, J.L., Buttet, J., Car, R.: Phys. Rev. B **31**, 1804 (1984)
11. Martins, J.L., Car, R., Buttet, J.: J. Chem. Phys. **78**, 5646 (1983)
12. Kohn, W., Vashishta, P.: In: Theory of the inhomogeneous electron gas. Lundqvist, S., March, N.H. (eds.), p. 79. New York: Plenum Press, 1983
13. Ceperley, D.M., Alder, B.J.: Phys. Rev. Lett. **45**, 566 (1980)
14. Bachelet, G.B., Hamann, D.R., Schlüter, M.: Phys. Rev. B **26**, 4199 (1982)
15. Puska, M.J., Nieminen, R.M., Manninen, M.: Phys. Rev. B **31**, 3486 (1985)
16. Bishop, D.M., Pouchan, C.: J. Chem. Phys. **80**, 789 (1984)
17. Bishop, D.M., Lam, B., Epstein, S.T.: J. Chem. Phys. **88**, 337 (1988)

Z. Phys. D – Atoms, Molecules and Clusters 12, 357–359 (1989)

Atoms, Molecules
Zeitschrift für Physik D and Clusters
© Springer-Verlag 1989

Electronic structure and stability of charged beryllium clusters

S.N. Khanna[1], **F. Reuse**,[2] **and J. Buttet**[2]

[1] Physics Department, Virginia Commonwealth University, Richmond, VA 23284-2000, USA
[2] Institut de Physique Expérimentale, Ecole Polytechnique Fédérale de Lausanne, CH-1015 Lausanne, Switzerland

Received 6 July 1988; final version 30 September 1988

Electronic structure studies on neutral, singly and doubly ionized Be_n clusters ($n \leq 5$) have been carried out in order to investigate the stability and observability of charged clusters. Our studies employ wave function expansion in terms of gaussian type orbitals and have been carried out within local spin density formalism. It is shown that although small doubly ionized clusters are unstable, they are protected from fragmentation by energy barriers. We illustrate this explicitly for trimers by presenting a Born-Oppenheimer surface of Be_3, Be_3^+ and Be_3^{++}. It is argued that depending on their geometries, the observable doubly charged clusters can be generated through a one or two photons ionization. We also present results on the distribution of "hole charge" in doubly ionized clusters and show that a small cluster exhibits metallic like behaviour in regard to distribution of missing electronic charge.

PACS. 31.20.Di; 36.40.+d.

Recently there has been a lot of interest in the studies of charged clusters. This interest is sparked both by the fact that the cluster detection often proceeds via ionization and detection of charged species and by the underlying fundamental physics. Singly ionized clusters are mostly stable. However, instability sets in when a cluster is multiply ionized as the Coulomb repulsion between charges acts against the cohesion and the cluster can fragment. This phenomenon called the "Coulomb Explosion" has been of considerable experimental and theoretical investigation. One of the central issues has been the minimum or the critical cluster size, n_c, for which a multiply ionized cluster can be formed.

The purpose of this paper is to focus on the problem of doubly ionized metallic clusters. Both the theoretical and the experimental situation in this area are far from clear. Early experiments [1] had reported a critical size $n_c = 31$ for doubly charged lead clusters while in recent experiments [2] doubly charged lead trimers have been observed. Same situation holds for silver clusters [3]. Several other doubly charged dimers and trimers have also been observed. Recently Hg_5^{++} clusters ionized both by electron impact and single photon ionization have been observed [4].

Basically the problem of doubly charged clusters can be described by considering the evolution of interaction as the two singly ionized fragments are brought together from large separation. At distances large compared to the cluster sizes, the dominant force is the coulomb repulsion. However, as the singly charged fragments come closer, the electronic forces due to overlap and polarization of electronic charge densities start to become important. Situations can then arise when the electronic forces dominate over the simple Coulomb repulsion and lead to attractive interaction. Under these circumstances the doubly charged cluster can exhibit a deep metastable minima protected by potential barriers in its configuration space [5]. If appropriate experimental situations could thus be created so as to generate the charged cluster in this minimum, it would have long lifetime and could thus be observable. In this paper, we demonstrate the existence of such minima and show how the observability of doubly charged clusters in certain experiments only (e.g. case of Pb_3^{++}), are linked to the location of these minima. We also consider the charge localization in doubly charged clusters and show that even a small cluster exhibits metallic like tendencies in regard to distribution of "hole charge". We have carried out ab-initio self-consistent local spin density (LSD) calculations on neutral, singly and doubly charged Beryllium clusters having up to 5 atoms. The choice of Beryllium is motivated by the low number of electrons and by the interest in Group II elements. As is well known, these elements have weakly bound molecules while there is metallic bonding in the bulk phase. The theoretical studies were done by representing molecular orbitals as sum of gaussian type orbitals (9*s*, 3*p*) centered on atomic sites [6]. The exchange correlation contributions were treated within a LSD scheme and the calculations were carried out at the all electron level. The basis sets employed in our studies reproduced the atomic ionization potentials and excitation energies to within 3% of experimental values.

Our calculations predict the ground state geometries of neutral trimer, tetramer and pentamer to be equilateral triangle, tetrahedron, and distorted triangular bypyramid respectively. These are in agreement with previous C.I. calculations [7]. The corresponding geometries for singly ionized clusters were found to be linear for trimer, distorted tetrahedron for tetramer and a distorted triangular bipyramid for pentamer. The doubly ionized clusters (for $n \leq 5$) all had linear chains as minimum energy configuration. However, we found a centered triangular geometry for Be_4^{++} and distorted triangular bypyramid geometry for Be_5^{++} to be respectively 0.5 eV and 0.6 eV higher than the corresponding linear structures.

In Fig. 1 we show the total energy of Be_3^{++} linear structure as a function of separation between Be_2^+ and Be^+ fragments. It is seen that at large distances, the interaction is coulombian. However as the fragments come together, the overlap of electronic charges, leads to the development of cohesion or attractive forces. We notice that although Be_3^{++} is energetically unstable, it is protected by a potential barrier of a height of about 1.4 eV. A simple WKB approximation shows that the life-time at low temperatures is much longer than the observation time. In principal Be_3^{++} is thus observable provided it could be created in the potential well. To examine such a possibility we show in Fig. 2 a Born-Oppenheimer (BO) surfaces of Be_3, Be_3^{++} and Be_3^{++} for isoceles triangular configurations. Starting from a Be_3 in its ground state, it is unlikely that a observable Be_3^{++} can be formed in a single ionization process since such a ionized cluster will reach an

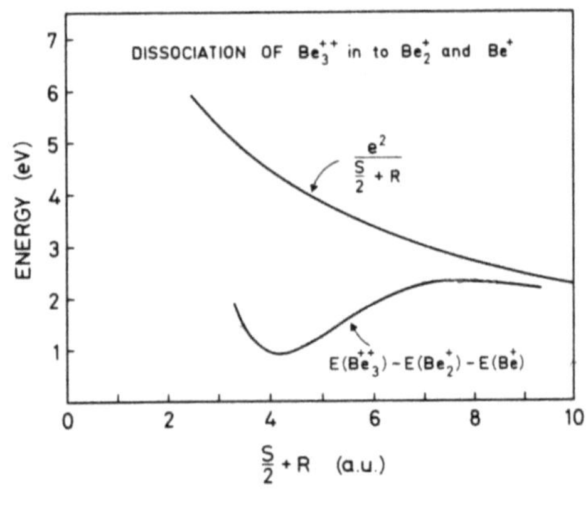

Fig. 1. Binding energy of a linear Be_3^{++} as a function of distance between the center of Be_2^+ (size *S*) and Be^+ (distant *R* from dimer end)

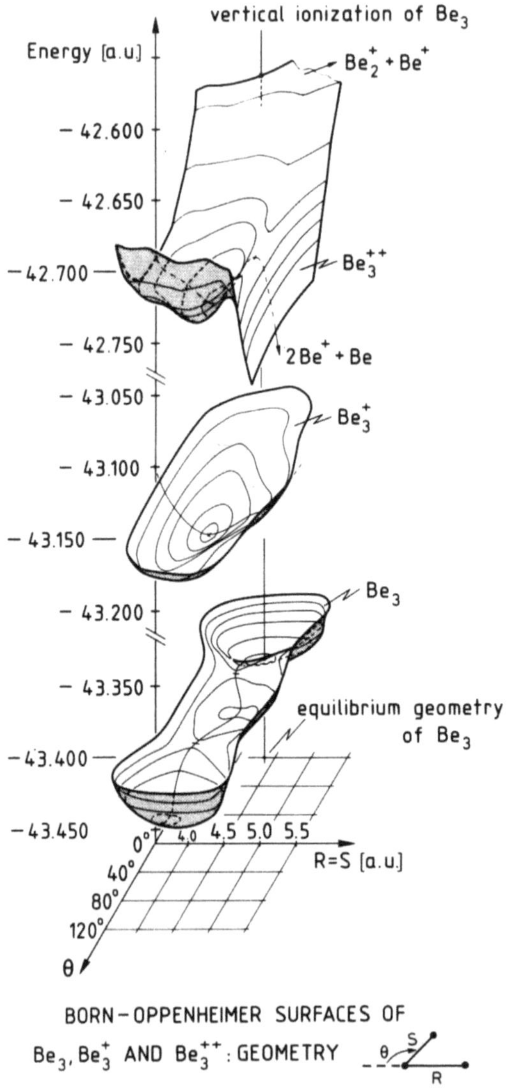

Fig. 2. Born-Oppenheimer surfaces of singly and doubly ionized and neutral Be_3 corresponding to various isoceles triangular configurations

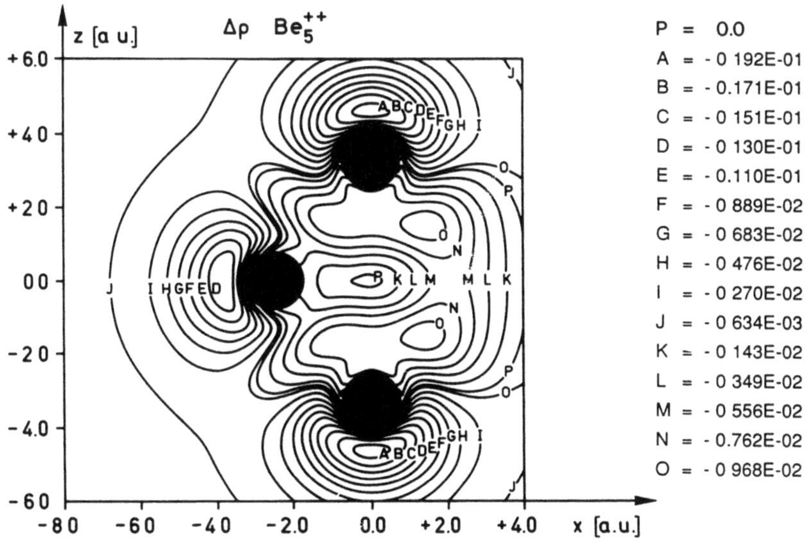

Fig. 3. Difference charge density representing the difference between electronic charge density in Be_5^{++} minus the charge density obtained by superposing neutral Be atoms. The figure corresponds to the vertical symmetric plane of the distorted triangle-bipyramid configuration

energy higher than its dissociation threshold. It is however possible to form Be_3^{++} in the metastable state in a two-step ionization process if enough time is given to singly ionized cluster to reach its equilibrium geometry since both singly and doubly ionized clusters have similar configuration.

Although it is difficult to make such a detailed analysis for Be_4^{++} and Be_5^{++}, it should be possible to observe them if they could be prepared in low energy configurations because of the existence of potential barriers. For Be_5^{++}, the existence of a triangular bypyramid structure suggests that it should be possible to generate Be_5^{++} in a single photon experiment from Be_5 because of their similar geometries. These considerations could account for the observation of Hg_5^{++} clusters.

Finally we illustrate the behavior of small clusters in regard to distribution of hole charges in particular, we show that the simple picture where one supposes hole charges in doubly ionized clusters to be localized at opposite ends is too simple [8]. In Fig. 3 we have shown the difference charge density in the vertical plane of the distorted triangular-bypyramid. The difference charge density corresponds to electronic charge density of Be_5^{++} from which we have subtracted the charge density obtained by superposing neutral atoms. It is seen that part of the holes are localized at the outermost atoms of the bipyramid, but there are also sizable holes localized outside the other three atoms. Same behaviour is observed for linear structure where the holes are localized at the end atoms as well as in torus shaped charge depletions around inner atoms. It is as if the cluster was behaving like a metal.

To sum up, the present work shows that the potential barriers play a crucial role in the observability of small charged clusters. This is in contrasts to works where one discusses clusters stability only in terms of initial and final state energies [9]. An analysis of results on singly and doubly ionized geometries suggests that although the minimum energy configurations of doubly ionized clusters are linear up to $n \leq 5$, three dimensional structures should stabilise at higher sizes.

This work was supported by the Department of Energy (DE-FGO5-87ER-45316)

References

1. Sattler, K., Muhlback, J., Echt, O., Pfau, P., Recknagel, E.: Phys. Rev. Lett. **47**, 160 (1981)
2. Hoareau, A., Melinon, P., Cabaud, B., Rayne, D., Tribollet, B., Broyer, M.: Chem. Phys. Lett **143**, 602 (1988)
3. Schulze, W., Winter, B., Urban, J.: Z. Phys. D – Atoms, Molecules and Clusters **4**, 379 (1987)
4. Brechignac, C., Broyer, M., Cahuzac, Ph., Delacretaz, G., Labastie, P., Woste, L.: Chem. Phys. Lett. **133**, 45 (1987)
5. Durant, G., Daudey, J.P., Malrieu, J.P.: J. Phys. **47**, 1335 (1986); Liu, F., Press, M.R., Khanna, S.N., Jena, P.: Phys. Rev. Lett. **59**, 2562 (1987)
6. Martins, J.L., Buttet, J., Car, R.: Phys. Rev. B **31**, 1804 (1985)
7. Marino, M.M., Ermler, W.C.: J. Chem. Phys. **86**, 6283 (1987)
8. Tomanek, D., Mukherjee, S., Bennemann, K.H.: Phys. Rev. B. **28**, 665 (1983)
9. Rao, B.K., Jena, P., Manninen, M., Nieminen, R.M.: Phys. Rev. Lett. **58**, 1188 (1987)

Z. Phys. D – Atoms, Molecules and Clusters 12, 361–363 (1989)

Atoms, Molecules and Clusters
Zeitschrift für Physik D
© Springer-Verlag 1989

Magnetism and local order

Feng Liu, M.R. Press, S.N. Khanna, and P. Jena
Physics Department, Virginia Commonwealth University, Richmond, VA 23284-2000, USA

Received 6 July 1988; final version 30 September 1988

A simple electronic structure scheme based upon a combination of the molecular clusters and tight binding formulation is proposed that is not only quantitative but also enables electronic structure studies in a computationally efficient manner. As a first application, we use the new scheme to study the magnetic moment of Fe and Ni in a wide range of environments where the local coordination of the magnetic atom changes from two to as many as twelve. This is achieved by considering configurations such as linear chain, crystallographic surfaces of various orientations, thin films and bulk. The magnetic moment is shown to increase with decreasing coordination due to narrowing of electron bands. Its implications for the magnetic moments of small Fe and Ni clusters are discussed.

PACS: 71.20; 36.20.K; 31.20; 73.20; 75.20.H

Metallic clusters are now known to exhibit electronic and magnetic behavior different from their crystalline counterparts and evolving with cluster size [1]. Due to advances in experimental techniques, it is possible to perform experiments on size selected clusters and measure quantities such as core-shifts, ionization potentials, reactivity and magnetic moment of size selected clusters. A theoretical understanding of these experiments clearly requires electronic structure studies on clusters and during the past few years, a lot of theoretical effort has been devoted to these studies. The available electronic structure techniques for clusters can be broadly grouped into two categories. On the one hand, the development of ab initio self-consistent techniques like configuration-interaction [2] or the local density functional methods with Gaussian [3] or numerical basis such as Discrete Variational Method (DVM) [4] permit an ab initio study of cluster electronic structure. However, these schemes are computationally heavy and even with extensive computational resources one is limited to clusters having few dozen atoms. On the other hand and to circumvent these limitations, semi-empirical and approximate methods such as tight-binding [5] or effective medium theories [6] have been developed. These methods by their very nature are not quantitative and can only yield qualitative trends. For example, in the tight-binding model, the matrix elements of the tight-binding hamiltonian are *fitted* to some known Band-Structure data. The transferability of these parameters to reduced dimensions such as clusters and their dependence on the band structure chosen remain a nagging problem.

The purpose of this paper is to propose a new electronic structure scheme based upon a combination of the conventional tight-binding approach [5] and the ab initio self-consistent linear combination of atomic orbitals, molecular orbital technique (SCF-LCAO-MO) [4]. We *calculate* the various overlap matrix elements appearing in the tight-binding method from SCF-LCAO-MO studies of the diatomic molecules. The new scheme contrasts the conventional tight-binding approach in that there are no adjustable parameters and thus has an ab initio character. As a first application, we use the new scheme to address the problem of variation of magnetic moments of Fe and Ni atoms in a variety of local environments. The basic motivation of these studies is firstly to compare the results of the new scheme with the available experimental and other first-principles studies and secondly to show that the local magnetic moments of the atoms are primarily related to their local surroundings and establish this relationship. Our studies constitute a first step towards an understanding of the recent experiments [7] on the magnetic moment of small clusters which indicate an enhancement of magnetic moment of Fe_n ($n \leq 25$) clusters with respect to their bulk value.

Consider a system described by a one electron hamiltonian, H, in atomic units,

$$H = -\frac{1}{2}\nabla^2 + \int \frac{\rho(r')\,d^3r'}{|r-r'|} + \sum_j \frac{Z_j}{|r-R_j|} + V^{\sigma}_{xc}(r) \quad (1)$$

Here, j is the site index. The second and third term represent, respectively, the electronic and the nuclear contributions to the energy. The last term is the exchange-correlation potential for the spin σ approximated by the Von-Barth-Hedin approximation to local spin density functional. The wave function ψ for the system is expressed in terms of linear combination of atomic orbitals (LCAO) $|j, \alpha\rangle$ localized at site j. α represents the spin-orbital index.

$$\psi = \sum_{j\alpha} C_{j\alpha} |j, \alpha\rangle \quad (2)$$

$C_{j\alpha}$ are variational parameters to be determined by a solution of Rayleigh-Ritz equation,

$$(\underline{H} - E\,\underline{S})\underline{C} = 0 \quad (3)$$

Here $\underline{H}$ and $\underline{S}$ are hamiltonian and overlap matrices evaluated numerically as weighted sums over a set of points $\mathbf{r}_k$ [4]. The matrix equations are solved self consistently to determine $C_{j\alpha}$.

The electronic structure of the system is described in terms of local density of states $n_i(E)$ with i as site index. The moment and the continued fraction technique relate $n_i(E)$ to its moments μ^i_n which are given by [5]

$$\mu^i_n = \frac{1}{N} \sum_{\substack{i;j;k\ldots \\ \lambda;\mu;\nu\ldots}} \langle i, \lambda|H|j, \mu\rangle \langle j, \mu|H \ldots H|\, i, \lambda\rangle \quad (4)$$

where N is the number of atoms.

Given a set of orbitals $|i, \lambda\rangle$, a calculation of μ_n requires a knowledge of $\langle i, \lambda|H|j, \mu\rangle$. In this work, we *calculate* these elements explicitly from the SCF-LCAO-MO method described in (1)–(3). We have used the DVM to calculate these elements from a dimer calculation with bond length same as for the system under study. The choice of DVM is motivated by the fact that the basis functions $|i, \lambda\rangle$ are nearly orthonormal, a feature which simplifies the moment approach. Our studies demonstrate that the spin-polarized hamiltonian matrix elements obtained via a dimer calculation are transferable to a wide range of systems with differing environments. Electronic structure studies were carried out for Ni and Fe in a number of configurations: linear chain, ⟨100⟩ and ⟨110⟩ surfaces, slabs containing 5, 7 and 9 layers of atoms, and bulk. The choice of these wide range of systems was motivated by the availability of accurate FLAPW [7] results as well as a test for the applicability of the scheme to diverse environments. For the moment calculation, 4*s* and 3*d* orbitals were used and the continued fraction was constructed from twenty-two moments.

Our calculations yield bulk moments of Ni and Fe to be 0.6 μ_B and 2.5 μ_B respectivley in good agreement with the corresponding experimental values of 0.6 μ_B and 2.2 μ_B. For linear chain, our calculated value of 1.0 μ_B for Ni and 3.2 μ_B for Fe are in good agreement with the corresponding FLAPW values [8] of 1.1 μ_B and 3.3 μ_B. For a 5-layer Ni(100) slab, our studies predict a moment enhanced by 4.3% over central layer again in close agreement of the FLAPW results of 5.2% of Jepsen et al. [9] and 5.8% of Krakauer et al. [10].

Small clusters contain atoms with varying local surroundings and it is of interest to consider variation in moments with local environment. To illustrate such a relation, we show in Fig. 1, the deviation, $\Delta\mu = \mu - \mu_B$, in the moment μ from its Bulk value of μ_B as a function of cordination number in various configurations considered in this work. One notices that a decreasing coordination enhances local moment. It is related to the fact that a reduction in coordination narrows the band width which reduces overlap between majority and minority bands and thus leads to promotion of minority spins to majority bands. This is

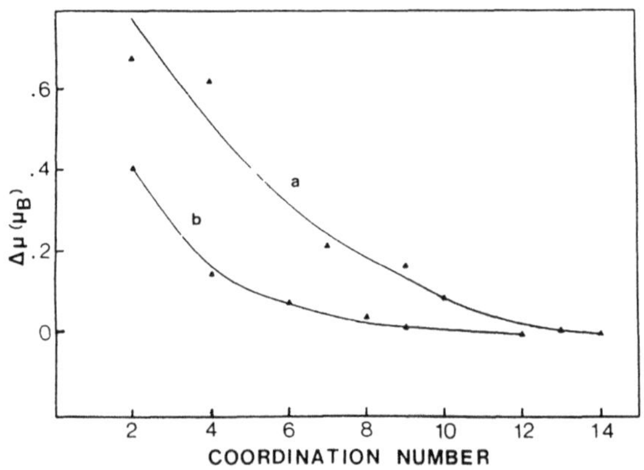

Fig. 1. Deviation from the bulk magnetic moment, $\Delta\mu$, in Fe and Ni as a function of nearest coordination number (in various structures). For Fe which has BCC structure, the coordination number in bulk includes nearest and next nearest atoms. A coordination number of 13 for Fe was obtained by creating a monovacancy and looking at atoms surrounding the vacancy. *(a)* and *(b)* correspond to Fe and Ni respectively. The triangles are the actual calculated points. Smooth lines are drawn simply to guide the eye

in agreement with other electronic structure studies on small Ni and Fe clusters (having 13 and 19 atoms) which indicate an enhancement of moment in these clusters [11]. This is also in agreement with experiments on Fe clusters [7] which indicate an enhancement of moment in small Fe clusters. We would like to emphasize that in reality, the change in magnetic moment with coordination represents only one part of the change in cluster moments with size [12]. It is now known that interparticle distances in small metallic clusters are few percent smaller than in bulk. A contraction in interparticle spacing results in an increase in band-widths which can produce moment changes in opposite direction to the effect produced by reduced coordination. An application of the scheme developed here to calculation of actual cluster moments requires total energy calculations to determine cluster geometry. We are now in the process of developing it and will be presented in later publications.

This work was supported by the Department of Energy (DE-FG05-87ER45316).

References

1. Jena, P., Rao, B.K., Khanna, S.N.: Physics and chemistry of small clusters. New York: Plenum 1987
2. Hehre, W.J., Radom, L., Schleyer, P.v.R., Pople, J.A.: Ab initio molecular orbital theory. New York: Wiley 1986
3. Martins, J.L., Buttet, J., Car, R.: Phys. Rev. **B31**, 1804 (1985)
4. Ellis, D.E.: In handbook on the physics and chemistry of the actinides. Freeman, A.J., Lander, G.H. (eds.), p. 1 Amsterdam, New York: Elsevier 1985
5. Cyrot-Lackmann, F., Khanna, S.N.: In: Excitations in disordered systems. Thorp, M.F. (ed.), p. 59. New York: Plenum 1982
6. Norskov, J.K., Lang, N.D.: Phys. Rev. B **21**, 2136 (1980)
7. Cox, D.M., Trevor, D.J., Whetten, R.L., Rohlfing, E.A., Kaldor, A.: Phys. Rev. B **32**, 7290 (1985)
8. Weinert, M., Freeman, A.J.: J. Magn. Magn. Mater. Dig. **38**, 23 (1983)
9. Jepsen, O., Madson, J., Anderson, O.K.: Phys. Rev. B **26**, 2790 (1982)
10. Krakauer, H., Freeman, A.J., Wimmer, E.: Phys. Rev. B **28**, 610 (1983)
11. Lee, K., Callaway, J., Kwong, K., Tang, R., Ziegler, A.: Phys. Rev. B **31**, 1796 (1985)
12. Pastor, G.M., Dorantes-Davila, J., Bennemann, K.H.: Physica B **149**, 22 (1988)

Z. Phys. D – Atoms, Molecules and Clusters 12, 365–367 (1989)

Atoms, Molecules and Clusters
Zeitschrift für Physik D
© Springer-Verlag 1989

Calculation of the size dependence of the ionization energy and autoionization energy of Hg-clusters

G.M. Pastor, P. Stampfli, and K.H. Bennemann
Institut für Theoretische Physik der Freien Universität Berlin, Arnimallee 14, D-1000 Berlin 33, Germany

Received 5 July 1988; final version 30 September 1988

To study the transition from van der Waals to metallic bonding we calculate the size dependence of the ionization energy and $5d \to 6p$ autoionization energy of Hg_n-clusters using a parametrized LCAO model. Our results are in good qualitative agreement with experiment. Comparison with experimental results suggests that electron correlations play an important role for the transition from localized (van der Waals-like) to delocalized (covalent or metallic) electronic states occuring in Hg_n at $n \simeq 13$–19.

PACS: 36.40.+d

The size dependence of the properties of clusters, resulting from the evolution of the electronic structure as we go from the atom to the bulk, is one of the central problems in cluster research. Furthermore, the progress in this field will also improve our understanding of solids, surfaces and interfaces. Interesting size dependent properties are expected for Hg-clusters since the Hg-atom has a closed shell electronic configuration in contrast to metallic bulk-Hg. Indeed, Tomanek et al. [1] pointed out several years ago that an insulator to metal transition should take place in Hg_n for $20 \lesssim n \lesssim 50$. This physical picture seems to be confirmed by recent experimental results [2, 3]. In previous papers [4] we showed that electron delocalization should occur in Hg_n-clusters with n larger than the critical size $n_{cr} \simeq 13$–19. However, our calculations [4] yield for $n \simeq n_{cr}$ a considerable gap Δ_n between occupied and unoccupied states ($\Delta_{19} \simeq 1.8$ eV). Thus, no metallic but rather covalent bonding is expected for this cluster size range. Metallic like behaviour occurs for $n \gtrsim 80$, when the s- and p-band overlap [4]. Here we discuss results for the size dependence of the ionization potential and $5d \to 6p$ autoionization energy of Hg-clusters obtained by using a simple tight-binding electronic theory [4]. Comparison with experiment [2, 3] suggests that electron correlations play an important role for determining the size dependent transition from localized to delocalized electronic states.

We consider the tight-binding Hamiltonian for s- and p-electrons:

$$H = \sum_{i,a} \varepsilon_a c_{i\alpha}^+ c_{i\alpha} + \sum_{\substack{i \neq j \\ \alpha,\beta}} t_{ij}^{\alpha\beta} c_{i\alpha}^+ c_{j\beta} + \sum_{\substack{i,j \\ \alpha,\beta}} U_{ij}^{\alpha\beta} n_{i\alpha} n_{j\beta}, \quad (1)$$

where $c_{i\alpha}^+$, $c_{i\alpha}$ refer to the usual creation and annihilation operator of an electron on site i and orbital $\alpha(\alpha = s, \ p_x, p_y, p_z)$ and $n_{i\alpha} = c_{i\alpha}^+ c_{i\alpha}$ is the number operator.

The ionization potential I_n is calculated from:

$$I_n = -\varepsilon_F(n) + \frac{e^2}{2R}, \quad (2)$$

where $\varepsilon_F(n)$ refers to the energy of the highest occupied state in Hg_n and $R = (3v/4\pi)^{1/3} n^{1/3}$ is the cluster radius (v refers to the volume per atom). Equation (2) can be derived by assuming that the shape of the total density of states (DOS) of Hg_n remains unchanged upon ionization. This is a reasonable approximation at least for large clusters for which the changes in the electronic density ρ, and thus the changes in the potential, caused by ionization are small and localized at the cluster surface ($\Delta\rho \propto R^{-2}$). Furthermore, we performed self-consistent calculations for both Hg_n and

Hg_n^+ (e.g. for $n=4$, 13) and obtained results similar to those derived from (2). Physically, the first term in (2) takes into account the changes in the electronic structure with increasing cluster size due to *sp*-hybridization and band formation. The second term accounts for the shift of the electronic levels of Hg_n^+ with respect to those of neutral Hg_n occurring upon electron emission.

The $5d \rightarrow 6p$ autoionization energy is given by

$$\delta_n = \varepsilon_{6p} - \varepsilon_{5d} - \varepsilon_{pd}. \qquad (3)$$

Here, ε_{6p} refers to the energy of the excited 6*p*-electron, initially in the 5*d*-state, and ε_{pd} to its interaction energy with the remaining 5*d*-hole. For localized electrons ($n < n_{cr}$) the excitation is like a Frenkel exciton and thus we use the ansatz $\varepsilon_{6p} = \varepsilon_{6p}(\text{atom}) - \alpha z_n$. The shift αz_n results from the polarization of the atoms surrounding the one in which absorption occurs. z_n refers to the average coordination number. The electron-hole interaction energy is approximately independent of cluster size ($n < n_{cr}$). Therefore for $n < n_{cr}$, $\delta_n \simeq \delta_1 - \alpha z_n$. The constant α is fitted to the observed [2] red shift in Hg_2. For delocalized electrons ($n > n_{cr}$) the excitation is like a Mott–Wannier exciton. Then, ε_{6p} is given by the energy of the lowest unoccupied *p*-like state ($\varepsilon_{6p} \equiv \varepsilon_{\text{LUMO}}(n)$), and $\varepsilon_{pd} \simeq -e^2/\kappa R$. Here $\kappa = 1 + 4\pi\chi$ stands for the size dependent dielectric constant. We approximate the susceptibility χ by $\chi = (z_n/z_b)\,(\gamma/v)\,(\varepsilon_p - \varepsilon_s)/\Delta_n$ where γ refers to the atomic polarizability of Hg, Δ_n to the gap between occupied and unoccupied states, and z_b to the bulk coordination number [5].

The parameters used for the calculations were determined as follows. The *sp*-promotion energy $\varepsilon_p - \varepsilon_s = 5.8$ eV is obtained from the atomic $6s^2 \rightarrow 6s6p$ transition averaging between the ^3P and ^1P transitions [6]. For the hopping integrals we use the relation [7]: $t_{ss\sigma} = -1.32t$, $t_{sp\sigma} = 1.42t$, $t_{pp\sigma} = 2.22t$, $t_{pp\pi} = -0.63t$, and set $t = 0.44$ eV such that the number of *p*-electrons per atom $n_p = \int_{-\infty}^{\varepsilon_F} d\varepsilon N_p(\varepsilon)$ obtained in APW calculations for bulk fcc-Hg [8] is reproduced. In this way the bulk-DOS calculated within tight-binding approximation [4] is in good agreement with the results shown in Ref. [8]. The position of the 6s-level (ε_s) and the 5*d*-level (ε_{5d}) with respect to vacuum are fitted to the bulk work function $\phi = 4.5$ eV and bulk $5d \rightarrow 6p$ transition energy [9], respectively.

In Fig. 1 results are given for the size dependence of the ionization energy I_n of Hg_n. For bulk-like clusters ($n \gtrsim 80$–135) the position of the Fermi energy $\varepsilon_F(n)$ with respect to the *s*-level ε_s is approximately the same as in the bulk ($\varepsilon_F(n) \simeq \varepsilon_F(\text{bulk})$) and thus $I_n \simeq \phi + e^2/2R$. For $n < 80$ a gap Δ_n opens between occupied *s*-like and unoccupied *p*-like states. We have $\varepsilon_F(n) < \varepsilon_F(\text{bulk})$, where now $\varepsilon_F(n)$ refers to the energy of the highest occupied state. Therefore, for $n < 80$, $I_n > \phi + e^2/2R$ in

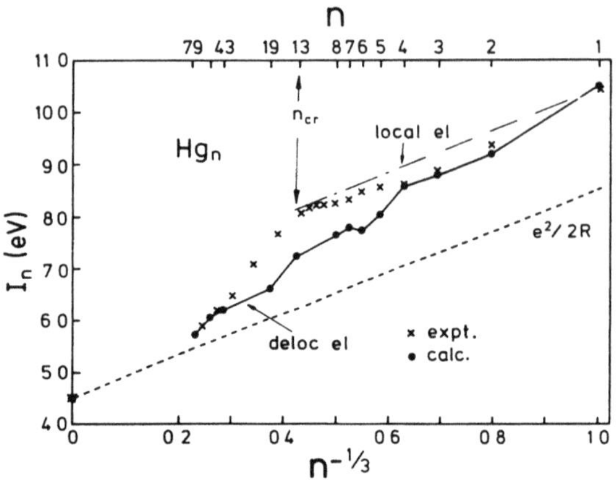

Fig. 1. The size dependence of the ionization energy I_n of Hg_n-clusters. The calculated results (*full line*) were obtained using $I_n = -\varepsilon_F(n) + e^2/2R$, where $\varepsilon_F(n)$ refers to the energy of the highest occupied state calculated from (1) in the Hartree-Fock approximation, and R to the cluster radius. The effects of strong correlations, leading to electron localization in Hg_n with $n < n_{cr}$, are simulated by 'blocking' the hopping integrals ($t_{ij} = 0$)

contrast to alkali-metal clusters which follow, approximately, the R^{-1}-law to much smaller cluster sizes. This can be understood recalling that alkali-metal clusters with one valence electron per atom have a half-filled band and thus $\varepsilon_F(n) \simeq \varepsilon_s$ even though the band width W_s reduces with decreasing cluster size ($W_s \propto \sqrt{z_n}$, z_n = effective coordination number). For Hg_n with $n < 80$ we have a full, mainly *s*-like, band and thus the reduction of W_s causes I_n to increase beyond the electrostatic contribution $e^2/2R$ (see Fig. 1). However note, we underestimate the reduction of W_s in small Hg_n clusters with $n \lesssim 20$, since we neglected electron correlations (beyond Hartree-Fock approximation), which tend to suppress electron hopping in small Hg-clusters. In Hg_n with $n < n_{cr}$ the Coulomb interactions U_{ij} amongst the electrons favor the highly correlated van der Waals state where electrons are localized instead of the covalent bonded state where charge fluctuations are necessarily present [4]. This situation can be simulated by setting $t_{ij} = 0$ for Hg_n with $n < n_{cr}$. Clearly, one obtains as expected $\varepsilon_F(n) \simeq \varepsilon_s(\text{atom})$ and thus $I_n = -\varepsilon_s + e^2/2R$, in good agreement with the experimental results for $n \leq 13$. Comparison with experiment suggests therefore that electron localization occurs in Hg_n with $n < n_{cr} \simeq 13$.

In Fig. 2 results are shown for the size dependence of the $5d \rightarrow 6p$ autoionization resonance of Hg_n-clusters. For small clusters $n \lesssim 19$ good agreement with experiment is obtained with our ansatz $\delta_n \simeq \delta_1 - \alpha z_n$, indicating that the excited 6*p*-electron occupies an atomic-like state and that electrons are localized (Frenkel exciton). For larger clusters the experimental observations are close to the results obtained with our

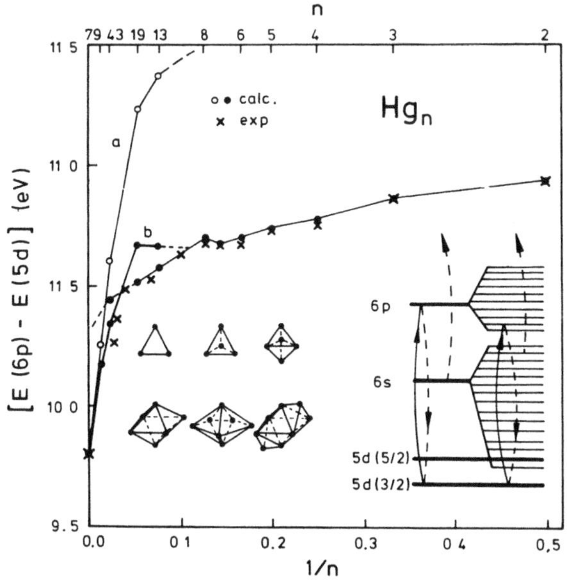

Fig. 2. The size dependence of the $(5d^{10}6s^2(^1S_0) \rightarrow 5d^9 6s^2(D_{3/2})6p)$ autoionization resonance calculated from $\delta_n = \varepsilon_{6p} - \varepsilon_{5d} + \varepsilon_{pd}$. For $n \lesssim 19$ the transition is atomic-like (Frenkel exciton). Thus, we use $\delta_n = \delta_1 - \alpha z_n$. For $n \gtrsim 19$ the excited 6*p*-electron occupies a delocalized state (Mott-Wannier exciton). Thus, $\varepsilon_{6p} = \varepsilon_{\mathrm{LUMO}}$ and $\varepsilon_{pd} \simeq e^2/\kappa R$. The electron-hole interaction ε_{pd} corrects curve a to b. The inset figures show the most stable structures for $n \leq 8$

tight-binding calculations. Thus, the transition is like a Mott-Wannier exciton with the excited 6*p*-electron occupying one of the low lying delocalized states above ε_F. These two types of transitions are illustrated in the inset of Fig. 2. From these results we conclude that electron delocalization and band formation occurs in Hg_n-clusters consisting of approximately 19 atoms, in good agreement with our results for I_n.

In summary, by comparing our results for I_n and δ_n with experiment we conclude that electron localization due to strong correlations occurs in Hg_n-clusters with $n < n_{cr} \simeq 13$–19 atoms. Clearly, for $n < n_{cr}$ van der Waals like bonding should dominate, as has been already shown by comparing the cohesive energy resulting from pairwise van der Waals-like bonding with that due to covalent bonding [4]. Furthermore, electron delocalization is expected to occur for $\tau \lesssim \Delta t$, where $\tau \simeq \hbar / W_s$ is the life time of an *s*-electron hole at an atom and $\Delta t \simeq \hbar / \Delta \mathrm{E}$ is the time of an electron charge fluctuation involving the energy $\Delta \mathrm{E} \simeq \mathrm{C} - e^2/d$. C is the Coulomb energy of an *s*-electron at a Hg-atom ($\mathrm{C} \simeq 7.0$ eV, $e^2/d \simeq 4.5$ eV). Thus, one may estimate n_{cr} from $W_s(n_{\mathrm{cr}}) \simeq \mathrm{C} - e^2/d$. In this way we estimate $n_{cr} \simeq 13$.

A more detailed discussion of the size dependence of I_n, δ_n and also the cohesive energy $E_{\mathrm{coh}}(n)$ and the gap Δ_n between occupied and unoccupied states can be found in [4]. Presently, we attempt to include the important correlations in our calculations. This should properly describe the blocking of the hoppings t_{ij}, and thus lead to a sharper transition from van der Waals to covalent to metallic Hg-clusters.

References

1. Tománek, D., Mukherjee, S., Bennemann, K.H.: Phys. Rev. B **28**, 665 (1983)
2. Bréchignac, C., Broyer, M., Cahuzac, Ph., Delacretaz, G., Labastie, P., Wöste, L.: Chem. Phys. Lett. **120**, 559 (1985); Phys. Rev. Lett. **60**, 275 (1988)
3. Rademann, K., Kaiser, B., Even, U., Hensel, F.: Phys. Rev. Lett. **59**, 2319 (1987)
4. Pastor, G.M., Stampfli, P., Bennemann, K.H.: Phys. Scr. **38**, 623 (1988); Europhys. Lett. **7**, 419 (1988)
5. The expression for χ yields proper limiting values: no screening ($\kappa = 1$) for an isolated bond and perfect screening ($\kappa = \infty$) for a metal. Furthermore, χ gives for bulk semiconductors results in reasonable agreement with experiment.
6. Moore, C.E.: Atomic energy levels, Nat. Bur. Stand. Circ. No. 467 (1958)
7. Harrison, W.A.: Phys. Rev. B **27**, 3592 (1983)
8. Mattheiss, L.F., Warren, W.W.: Phys. Rev. B **16**, 624 (1977)
9. Svensson, S., Martensson, N., Basilier, E., Malmqvist, P.A., Gelius, U., Siegbahn, K.: J. Electron Spectry. **9**, 51 (1976)

Z. Phys. D – Atoms, Molecules and Clusters 12, 369–371 (1989)

Atoms, Molecules
Zeitschrift für Physik D and Clusters
© Springer-Verlag 1989

Shell structure of electronic states of Cu clusters

T. Yamaguchi and N. Fujima
Faculty of Engineering, Shizuoka University, Hamamatsu 432, Japan

Received 5 July 1988; final version 30 September 1988

Electronic states of icosahedral Cu_{13} and Cu_{12} clusters were calculated by the DV-Xα method which took into account of many-electron effects. The valence 3$\underline{d}$ orbitals are localized within atom and isolated from the valence 4$\underline{s}$ orbitals extended over the whole cluster. The characteristic feature of the states of $\underline{s}$ valence electrons of Cu_{13} cluster well corresponds to that of one-electron picture i.e. the shell model of the 3-dimensional isotropic harmonic oscillator potential. Electronic states of valence levels of Cu_{12} cluster which have no central atom fairly well correspond to those of a combined potential of the harmonic oscillator and a 3-dimensional Gaussian potential barrier at the center of the cluster.

PACS: 71.45.Nt; 61.50.Em; 21.60.Cs

I. Introduction

Mass spectroscopy of ion cluster beam of simple metals such as Na shows the existence of magic numbers, 8, 18, 20, 34, 40, 58, . . . [1]. That is, clusters composed of atoms of magic numbers are relatively more stable than those of other numbers. The experiment of ion cluster beam of noble metals such as Cu shows the existence of similar magic numbers [2]. This suggests that the shell model in the nuclear physics holds good in noble metal clusters. In the conventional solid state physics of noble metals [3], the valence band has the localized $\underline{d}$ electrons as well as the extended $\underline{s}$ electrons and the $\underline{s}$–$\underline{d}$ interaction is not negligible. The picture of valence electrons is far from that of the free electron. Then, it is marvellous that the shell model holds in noble metal clusters. Note that the number of valence electrons giving the magic number is not eleven but one per atom.

Based on the jellium model, electronic states of simple metal clusters are theoretically obtained, for example, for the monovalent Na, the divalent Mg and the trivalent Al [4]. For Na clusters, the number of atoms $\underline{N}$ is equal to the total number of valence electrons $\underline{Z}$. Peaks of the second difference of the total energy are found at shell closing numbers. For Mg clusters, N is equal to $Z/2$, and positions of the peaks are shifted to one half of their previous $\underline{N}$ numbers. For Al clusters, N is equal to $Z/3$, and peaks occur at clusters with numbers of total valence electrons close to the shell closing numbers. For instance, the peaks at $N = 13$ ($Z = 39$) and $N = 19$ ($Z = 57$) are related to the shell closing numbers 40 and 58, respectively.

In this paper, using Slater's discrete-variational exchange-alpha (DV-Xα) method [5] where many electron effect is taken into account, we calculate the electronic states of icosahedral (Ih) Cu_{13} and Cu_{12} clusters. We discuss the calculated results in terms of the shell model based on the one electron picture.

II. Electronic states

Figure 1(b) and (e) show the energy levels of valence electrons of the Ih-Cu_{13} and Cu_{12} clusters calculated by the DV-Xα method, taking into account of the 4p excited state as well as the valence 3$\underline{d}$ and 4$\underline{s}$ orbitals of atoms. Here, the distance between the central atom and a surrounding atom of the Ih-Cu_{13} cluster is the bond length in crystal $\underline{b} = 4.83$ a.u. Then, the distance between neighboring atoms in the Ih-Cu_{12} cluster is longer by 5% than $\underline{b}$.

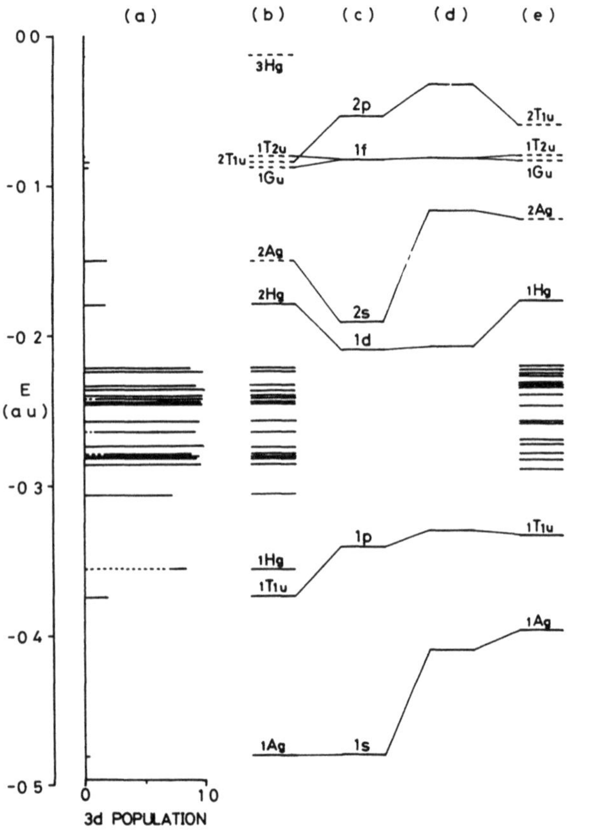

Fig. 1. The *d*-character of valence levels of Ih-Cu_{13} cluster and energy levels of Ih-Cu_{13} and Cu_{12} clusters.
a. Mulliken charge analysis of the *d*-character of valence levels of Ih-Cu_{13} cluster calculated by the DV-Xα method. *Dotted and full* lines represent the *d*-character of the central and the surrounding atoms, respectively.
b. Energy levels of Ih-Cu_{13} cluster by the DV-Xα method.
c. Energy levels of Ih-Cu_{13} cluster by the shell model.
d. Energy levels of Ih-Cu_{12} cluster by the shell model to which the 3-dimensional Gaussian potential barrier is added in the central region of the cluster.
e. Energy levels of Ih-Cu_{12} cluster by the DV-Xα method

Many levels are nearly degenerate in the region around -0.3 a.u. There are 17 and 16 levels for the Ih-Cu_{13} and Cu_{12} clusters, respectively. Mulliken charge analysis shows that most of them are composed of more than 90% 3d orbitals, as shown in Fig. 1(a). Among the 17 levels of Cu_{13} cluster, the lowest 1H_g level is isolated from other levels. It has 72% 3d character of the central atom and it bonds weakly with the 3d levels of surrounding 12 atoms. The remaining 16 levels correspond to 16 levels of the Ih-Cu_{12} cluster. These facts are consistent with the group theory which says that the d orbital of the central atom has an H_g representation and that the linear combinations of 12d orbitals of surrounding atoms have an A_g, a T_{1g}, a T_{2g}, two G_g, three H_g, two T_{1u}, two T_{2u}, two G_u and two H_u representations. The energy width of 16 levels is about 0.07 a.u. which is comparable to the d band width of Cu metal bulk. On the contrary, Mulliken charge analysis also clarifies that other levels are composed of mainly the 4s and 4p orbitals.

Assuming that the valence 4s electrons of our Cu_{13} cluster feel little the core potential of atoms, we compare the above results with those calculated by the shell model. The energy level of the shell model is given by

$$E_{nl} = (n + 3/2)\omega - U_0 l(l + 1) + V_0, \tag{1}$$

where n and l denote the major-shell and subshell, respectively, and ω, U_0 and V_0 are fitting parameters. The least-square-fit to the result calculated by the DV-Xα method gives the energy levels shown in Fig. 1(c). We can see that the shell model reproduces fairly well the energy levels by the DV-Xα method except the nearly degenerate levels.

The A_g levels of the Ih-Cu_{12} clusters in Fig. 1(e) have drastically higher energies than those of the Ih-Cu_{13} clusters in Fig. 1(b), and the degenerate levels have nearly the same energies. The A_g level in the Ih symmetry is the s level in the shell model. Note that the s level has density distribution in the region near the origin of the coordinate. Physically, we assume a 3-dimensional Gaussian potential-barrier for the Cu_{12} cluster in addition to the harmonic oscillator potential for the Cu_{13} cluster. We obtain energy levels of the perturbation of this potential barrier shown in Fig. 1(d). We can see that our model reproduces fairly well the energy levels by the DV-Xα method except the nearly degenerate levels.

One may think that the existence of magic numbers cannot be inferred from an analysis of highly symmetric 12- and 13-atoms clusters, because the cluster stability requires a calculation of total energies for various cluster geometries. Furthermore, the optimal cluster geometry may differ for different numbers of atoms [6]. The calculation shows that the four- and five-fold degenerate levels of the icosahedral cluster in Fig. 1 are split into two in the octahedral symmetry. But, the energy splittings are small compared with those in Fig. 1. The gross feature of energy levels is similar to that in Fig. 1. Thus, the one-to-one correspondence between energy levels by the DV-Xα method and the shell model gives the conclusion that the shell model is applicable to Cu clusters, since the s valence electrons are distributed nearly spherically and the localized d electrons of Cu clusters do not mix with the extended s electrons. Our calculation gives a theoretical foundation of magic number in noble metal clusters.

References

1. Knight, W.D., Clemenger, K., de Heer, W.A., Saunders, W.A., Chou, M.Y., Cohen, M.L.: Phys. Rev. Lett. **52**, 2141 (1984)
2. Katakuse, I., Ichihara, T., Fujita, Y., Matsuo, T., Sakurai, T., Matsuda, H.: Int. J. Mass Spectrom. Ion Phys. **67**, 229 (1985)
3. Kittel, C.: Introduction to solid state physics, 6th Edn. New York: John Wiley & Sons, 1986
4. Chou, M.Y., Cohen, M.L.: Phys. Lett. A **113**, 420 (1986)
5. Slater, J.C.: The calculation of molecular orbitals. New York: John Wiley & Sons, 1979
6. Devienne, F.M., Roustan, J.C.: Org. Mass Spectrom. **17**, 173 (1982)

Z. Phys. D – Atoms, Molecules and Clusters 12, 373–376 (1989)

Atoms, Molecules and Clusters
Zeitschrift für Physik D
© Springer-Verlag 1989

The nickel-group IV molecules NiC, NiSi, and NiGe

I. Shim[1] and K.A. Gingerich[2]

[1] Chemical Physics, Chemistry Department B, The Technical University of Denmark, DTH 301, DK-2800 Lyngby, Denmark
[2] Department of Chemistry, Texas A&M University, College Station, TX 77843, USA

Received 5 July 1988; final version 30 September 1988

All electron ab initio calculations have been applied to elucidate the electronic states and the nature of the chemical bonds in the molecules NiC, NiSi, and NiGe. The calculations have revealed that the ground states of all three molecules are $^1\Sigma^+$, but due to the open $3d$ shell of the Ni atom the molecules have many low-lying electronic states. The NiC molecule is strongly polar, and the low-lying electronic states have been identified as those arising when the angular momenta of the ^3F_g Ni^+ ion are coupled to the angular momenta of the $^4S_uC^-$ anion. The chemical bond in the NiC molecule has triple bond character due to the valence bond couplings between the Ni $4s$ and $3d\pi$ electrons and the C $2p$ electrons. The chemical bonds in the molecules NiSi and NiGe are very much alike; they are double bonds composed of one σ and one π bond. The σ bond is due to the doubly occupied delocalized molecular orbital composed of the Ni $4s$ orbital and the Si $3p\sigma$ or the Ge $4p\sigma$ orbital. The π bond originates from the valence bond coupling between the localized hole in the Ni $3d\pi$ orbital and the valence $p\pi$ electron of Si or Ge.

PACS: 31.20.D; 31.30

I. Introduction

Investigations of the physical and chemical properties of transition metal containing molecules are of considerable current interest [1–4]. In continuation of our theoretical and experimental investigations of the Pd group IV molecules, PdC [5, 6], PdSi [7], and PdGe [8] we have also undertaken such investigations for the analogous Ni group IV molecules, NiC [9], NiSi and NiGe [10]. It is expected that the basic knowledge of the chemical bonding and the electronic properties in small molecular units containing Pd and Ni will aid the understanding of catalytic processes.

In the present work we compare the chemical bonds as well as the low-lying electronic states of the Ni group IV molecules. By comparing also with the Ni containing transition metal dimers, Ni_2 [11, 4], NiCu [12], and NiFe [13], we extract certain characteristic features of the Ni atom as it participates in the formation of the chemical bonds.

The electronic ground states and the low-lying excited states of the Ni group IV molecules have been investigated by performing all electron ab initio Hartree-Fock (HF) and configuration interaction (CI) calculations. The HF calculations have been performed in the Hartree-Fock-Roothaan formalism [14]. The integrals have been calculated using the program MOLECULE [15]. For the HF calculations we have utilized the ALCHEMY program system [16], and the CI calculations have been performed using ALCHEMY in conjunction with the program ENERGY [17] for generating the symbolic energy expressions.

II. Calculational details

The basis sets consisted of Gaussian type functions, and for all the atoms, Ni, C, Si, and Ge, they are of better than double zeta quality. For Ni and Ge we have added a diffuse d function to the modified basis set of Wachter's [18] and Huzinaga's [19], respectively. The primitive basis sets for Ni and Ge (14s, 11p, 6d) have been contracted to (8s, 6p, 3d). The basis set for the Si atom is essentially Veillard's (12s, 9p) basis [20], but augmented with a d polarization function. Likewise, for the C atom, we have utilized Huzinaga's (10s, 6p) basis [21] augmented with a d polarization function. The primitive basis sets have been contracted to (6s, 4p, 1d) for Si and to (4s, 3p, 1d) for C. The details of the basis sets as well as results of atomic HF calculations have been described elsewhere [9].

In order to determine the electronic ground states of the Ni-group IV molecules, NiC, NiSi, and NiGe, we have performed HF calculations of 13 different electronic states for each of the molecules. However, it turned out that it was not possible to predict the electronic ground states of the molecules on basis of the HF calculations, and therefore we have also performed CI calculations. The CI calculations allowed full reorganization within the valence shells of the molecules, i.e. the 3d and 4s orbitals of Ni and the valence s and p orbitals of the group IV atoms. These calculations allow the molecules to dissociate correctly into ground term atoms. The calculations have been carried out in the subgroup C_{2v}, of the full symmetry group $C_{\infty v}$, and the number of configurations included to describe the electronic ground states of the molecules amount to 1308. The final wave functions of the molecules have been analysed by performing Mulliken population analyses.

III. The Ni atom in chemical bonding

According to Moore's tables [22] the lowest lying terms of the Ni atom, i.e. ^{3}F$(3d)^8$ $(4s)^2$ and ^{3}D$(3d)^9$ $(4s)^1$, are practically degenerate. Although the Ni atom in either of these two low-lying configurations is capable of forming chemical bonds, it appears that the Ni atom preferably binds in the $(3d)^9$ $(4s)^1$ configuration, because the open 3d shell in the $(3d)^8$ $(4s)^2$ configuration is screened by the doubly occupied 4s orbital. This is the case for the metallic as well as the intermetallic molecules Ni_2 [4, 11], NiCu [12], and NiFe [13].

In molecules where the chemical bond is formed between a Ni atom and a more electronegative atom, the bonding can originate from the $(3d)^8$ $(4s)^2$ configuration of Ni. In this case the electronegative atom gives rise to a charge transfer, thereby enabling the bonding interaction with the 4s orbital or the 4s–4p hybrid orbital of Ni. In the molecule NiH, the configurations $(3d)^8$ $(4s)^2$ and $(3d)^9$ $(4s)^1$ mix considerably resulting in a population of 8.66 electrons in the 3d shell of Ni [23]. As described in the present work the more electronegative C atom causes a considerable charge transfer in the NiC molecule resulting in a gross atomic charge of +0.40e on Ni, and the population of the 3d orbital of Ni has been determined as 8.13e. The molecules NiSi and NiGe are isovalent with the NiC molecule, but for both the NiSi and the NiGe molecule the charge transfer is small, if not negligible, and the chemical bond in each molecule originates from the $(3d)^9$ $(4s)^1$ configuration of Ni.

IV. The NiC molecule

The electronic ground state of the NiC molecule is predicted as $^1\Sigma^+$, originating mainly from the $(3d)^8$ configuration of Ni. Due to the electronegative C atom the molecule becomes appreciably polar, and at the internuclear distance 3.60 a.u., which is close to the equilibrium distance 3.57 a.u. of the molecule, the gross atomic charge on Ni is +0.40e. This gives rise to a strong electrostatic interaction between the atomic fragments. In addition, the electrons in the Ni 4s, 4p, and 3$d\pi$ orbitals valence bond couple to the C 2p electrons resulting in a chemical bond in the NiC molecule that has triple bond character.

From Table 1 it is noted that the molecule has 12 electronic states with T_e values in the energy range 0.00–0.61 eV, i.e. the singlet, triplet and quintet states of the symmetries Σ^+, Π, Δ, and Φ. In all these states the C atom has two electrons with parallel coupled spins in the 2$p\pi$ orbital. In the Σ^+ states the Ni atom has two electrons in the 3$d\pi$ orbitals. In the Π states, the Ni atom is associated with a (3$d\pi$, 3$d\sigma$) hole-pair, in the Δ states with a (3$d\delta$, 3$d\sigma$) hole-pair, and finally in the Φ states with a (3$d\delta$, 3$d\pi$) hole-pair. Within the Ni atom, the spins of the partly filled orbitals are high-spin coupled. The molecular states arise by coupling of the triplets of each atomic fragment to the singlets, triplets and quintets of the NiC molecule. Thus, all the low-lying electronic states are dominated by the high-spin coupling within each atomic fragment, and the molecular states can be regarded as arising from antiferromagnetic couplings of the high-spin coupled fragments.

The Ni 4s and 4p orbitals from a hybrid that tends to reduce the charge in the internuclear region. In all the above-mentioned states, the hybrid orbital contributes to a σ bond through valence bond coupling to the C 2$p\sigma$ orbital. The decoupling of the hybrid orbital and the C 2$p\sigma$ orbital gives rise to the states $^7\Sigma^+$, $^7\Pi$, $^7\Delta$, and $^7\Phi$ with T_e values of 0.90, 0.84, 0.79, and 1.03 eV, respectively. This indicates that the formation of the bond between the Ni 4σ, 4p hybrid and the C 2$p\sigma$

Table 1. Low-lying electronic states of the Ni-group IV molecules NiC, NiSi, and NiGe. Together with each grouping of molecular states the Table shows the approximate occupations associated with the partly occupied *d* subshell of Ni and with the partly occupied $p\pi$ valence orbitals of the group IV atoms

Molecule	Electronic states	Transition energy (eV) T_e	Open shell occupation Ni	Open shell occupation Group IV atom
NiC	$^1\Sigma^+$, $^3\Sigma^+$, $^5\Sigma^+$	0.00, 0.24, 0.52	$(3d\pi)^2$	$(2p\pi)^2$
	$^1\Pi$, $^3\Pi$, $^5\Pi$	0.19, 0.25, 0.46	$(3d\pi)^1$ $(3d\sigma)^1$	$(2p\pi)^2$
	$^1\Phi$, $^3\Phi$, $^5\Phi$	0.36, 0.43, 0.61	$(3d\delta)^1)$ $(3d\pi)^1$	$(2p\pi)^2$
	$^3\Delta$, $^1\Delta$, $^5\Delta$	0.36, 0.37, 0.39	$(3d\delta)^1$ $(3d\sigma)^1$	$(2p\pi)^2$
NiSi	$^1\Sigma^+$, $^3\Sigma^+$, $^3\Delta$, $^1\Delta$, $^1\Sigma^-$, $^3\Sigma^-$	0.00, 0.54, 0.61, 0.65, 0.67, 0.68	$(3d\pi)^3$	$(3p\pi)^1$
	$^1\Phi$, $^3\Pi$, $^3\Phi$, $^1\Pi$	0.35, 0.35, 0.36, 0.38	$(3d\delta)^3$	$(3p\pi)^1$
	$^5\Delta$, $^3\Delta$	0.48, 0.72	$(3d\delta)^3$ $(4s\sigma)^1$	$(3p\pi)^2$
	$^3\Pi$, $^1\Pi$	0.55, 0.59	$(3d\sigma)^1$	$(3p\pi)^1$
NiGe	$^1\Sigma^+$, $^3\Sigma^+$, $^3\Delta$, $^1\Delta$, $^1\Sigma^-$, $^3\Sigma^-$	0.00, 0.52, 0.62, 0.63, 0.65, 0.66	$(3d\pi)^3$	$(4p\pi)^1$
	$^1\Phi$, $^3\Pi$, $^3\Phi$, $^1\Pi$	0.34, 0.35, 0.36, 0.37	$(3d\delta)^3$	$(4p\pi)^1$
	$^1\Pi$, $^3\Pi$	0.57, 0.60	$(3d\sigma)^1$	$(4pn)^1$
	$^5\Delta$, $^3\Delta$	0.69, 0.86	$(3d\delta)^3$ $(4s\sigma)^1$	$4p\pi)^2$

orbital results in an energy gain of approximately 0.4 eV.

As a whole the low-lying electronic states of the NiC molecule as to the numbers and the symmetries are those derived by coupling of the angular momenta of the monopositive ^3F_g Ni^+ ion with those of the $^4S_uC^-$ anion.

V. The molecules NiSi and NiGe

As part of the molecules NiSi and NiGe, the Ni atom essentially retains the $(3d)^9$ $(4s)^1$ configuration. The electronic ground states of both molecules are predicted as $^1\Sigma^+$. The chemical bonds in the NiSi and NiGe molecules are of strikingly similar natures. Thus, the chemical bond in each molecule is a double bond composed of one σ and one π bond. The σ bond is a conventional chemical bond due to the doubly occupied delocalized molecular orbital composed of the Ni 4*s* orbital and the Si $2p\sigma$ or the Ge $3p\sigma$ orbital. From Table 1 it is noted that the Ni atom has three electrons in the $3d\pi$ subshell in the $^1\Sigma^+$ ground state as well as in the low-lying states $^3\Sigma^+$, $^3\Delta$, $^1\Delta$, $^1\Sigma^-$, and $^3\Sigma^-$. In the $^1\Sigma^+$ ground state this gives rise to a $3d\pi$ lone pair on Ni as well as to the π bond, since the third $3d\pi$ electron on Ni is valence bond coupled to the Si $3p\pi$ or to the Ge $4p\pi$ electron.

In addition, Table 1 reveals that the low-lying electronic states in the molecules NiSi and NiGe are of identical symmetries, and that they are all associated with definite symmetries of the hole in the 3*d* shell of the Ni atom. In all the low-lying electronic states of NiSi and NiGe the Ni 4*s* orbital and the Si $3p\sigma$ or the Ge $4p\sigma$ orbital combine to form a delocalized molecular orbital. This orbital is doubly occupied at the equilibrium internuclear distances of the molecules in all the low-lying electronic states except the states $^5\Delta$ and $2^3\Delta$. The many low-lying electronic states in the molecules NiSi and NiGe arise from the valence bond coupling of the Si $3p\pi$ or the Ge $4p\pi$ electron with the hole in the 3*d* shell of the Ni atom.

VI. Conclusions

In all the Ni containing molecules investigated except NiC, the Ni 4*s* orbital forms part of a delocalized molecular orbital which is doubly occupied in the electronic ground states. In the metal molecules, Ni_2 NiCu, and NiFe, the 3*d* electrons on the different nuclei interact through a Heisenberg exchange coupling that results in very small energy splittings between the potential energy curves. In the Ni-group IV molecules the Ni 3*d* electrons contribute significantly to the formation of the chemical bonds through the valence bond coupling with the valence *p* electrons of the group IV atoms.

The authors appreciate the support by a NATO Grant No. RG. 85/0448 for international collaboration in research and by the Center of Energy and Mineral Resources of the Texas A&M University.

References

1. Faraday Symp. R Soc. Chem. No. 14, Diatomic Metals and Metallic Clusters (1980)

2. Shim, I.: K. Dan. Vidensk. Selsk. Mat.-Fys. Medd. **41**, 147 (1985)
3. Morse, M.D.: Chem. Rev. **86**, 1049 (1986)
4. Jena, P., Rao, B.K., Khanna, S.N. (eds.): Physics and Chemistry of Small Clusters. NATO ASI Series. Series B· Physics 158. New York: Plenum Press 1987
5. Shim, I., Gingerich, K.A.: J. Chem. Phys. **76**, 3833 (1982)
6. Shim, I., Gingerich, K.A.: Surf. Sci. **156**, 623 (1985)
7. Shim. I., Kingcade, J.E., Jr., Gingerich, K.A.: Z. Phys. D – Atoms, Molecules and Clusters **7**, 261 (1987)
8. Shim, I., Kingcade, J.E., Jr., Gingerich, K A.: J. Chem. Phys. **85**, 6629 (1986)
9. Shim, I.: In: Understanding molecular properties. Avery, J., Dahl, J.P., Hansen, A.E. (eds.), pp 555–583. Holland: Reidel 1987
10. Shim,I., Kingcade, J.E., Jr., Gingerich. K.A.: J.Chem. Phys. (in press)
11. Shim, I., Johansen, H., Dahl, J.P.: Int. J. Quantum Chem. **15**, 311 (1979)
12. Shim, I.: Theor. Chim. Acta (Berlin) **54**, 113 (1980)
13. Shim, I.: Theor. Chim. Acta (Berlin) **59**, 413 (1981)
14. Roothaan, C.C.J.: Rev. Mod. Phys. **32**, 179 (1960)
15. Almolöf, J.: In: Proceedings of the Second Seminar on Computational Problems in Quantum Chemistry. München: Max Planck Institut 1973
16. The ALCHEMY program system has been written at IBM Research Laboratory in San José, CA Bagus, P.S., Liu, B., Yoshimine, M., McLean, A.D.
17. Sarma, C.R., Rettrup, S.: Theor. Chim. Acta (Berlin) **46**, 63 (1977); Rettrup, S., Sarma, C.R.: Theor. Chim. Acta **46**, 73 (1977)
18. Wachters, A.J.H.: J. Chem. Phys. **52**, 1033 (1970)
19. Huzinaga, S.: J. Chem. Phys. **66**, 4245 (1977)
20. Veillard, A.: Theor. Chim. Acta (Berlin) **12**, 405 (1968)
21. Huzinaga, S.: J. Chem. Phys. **42**, 1293 (1965)
22. Moore, C.E.: Nat. Bur. Stand. Circ. No 467. Washington, D.C.: U.S. GPO 1952
23. Walch, S.P., Bauschlicher, J.C.W., Langhoff, S.R.: J. Chem. Phys. **83**, 5351 (1985)

Z. Phys. D – Atoms, Molecules and Clusters 12, 377–380 (1989)

Atoms, Molecules and Clusters
Zeitschrift für Physik D
© Springer-Verlag 1989

First-principles calculations of K-edge X-ray absorption near edge spectra of some transition metal dimers

B. Wästberg[1], A. Rosén[1], and D.E. Ellis[2]

[1] Department of Physics, Chalmers University of Technology and, University of Göteborg, S-412 96 Göteborg, Sweden

[2] Department of Physics and Astronomy and Materials Research Center, Northwestern University, Evanston, IL 60201, USA

Received 5 July 1988; final version 30 September 1988

Theoretical X-ray absorption near-edge spectra have been evaluated for different bond lengths of Mn_2, Co_2 and Ni_2 using a modified Xα scattered-wave procedure.

PACS: 31.20.E; 78.70

I. Introduction

The physics and chemistry of small transition metal clusters are today a research area of increasing importance for understanding heterogeneous catalysis and other surface processes [1–3]. Studies of the reactivity of for example clusters of cobalt, nickel and niobium and other transition metals with H_2, N_2 and CO, have shown very pronounced variations in the reaction rate with the cluster size [4] as achieved with the new laser vaporization technique [5, 6]. This behaviour indicates that the size, the geometrical shape of the clusters and other properties depending on the size and the geometry, as the electronic structure, are important for the reactivity.

Research in this field means by necessity determination of the geometrical shape of the clusters by employing techniques sensitive for changes in geometry and bond length. Although optical spectroscopy [7] has been used on clusters in gas phase and matrix isolated clusters, knowledge about structure is at present very limited. In spite of intense use of optical spectroscopy in gas phase, spectra of clusters larger than the dimers have to our knowledge only been reported for a few system as Na_3 and Cu_3 [8, 9]. In parallel with achievements in the experimental field, theoretical methods of different sophistications have been applied to calculate electronic properties and predict stable geometries by total energy minimization [10, 11].

Other methods of interest are X-ray spectroscopy using the extended X-ray absorption fine structure, EXAFS and X-ray absorption near edge structure XANES [12–15]. In a related work [16] we examined how theoretical near threshold XANES-spectra of Ni_2 and Ni_3 depends on changes in bond length and geometrical arrangement as a first step to investigate the applicability of XANES for larger clusters. In this work we present a comparison between dimers of Mn, Co and Ni calculated for some different bond lengths.

II. Method

The XANES region ranges from about 10 eV below up to 50–60 eV above the threshold. The K-edge XANES spectra is dominated by two processes, *a)* transition from an initial core state to an eigenstate of the molecular hamiltonian. *b)* transition from an initial core state to a continum state which is perturbed from the backscattering field of the neighbouring atoms. For description of the near threshold structure the multiple scattering theory is usually needed.

The molecular SCF potential was calculated using local spin density (LSD) methods [18, 19] with the exchange correlation potential by von Barth and

Hedin [20]. The molecular wavefunctions were constructed of atomic numerical basis functions using the LCAO method. The cross-sections were calculated with a modified multiple scattering code [21] employing the muffin tin potential calculated from the SCF molecular potential by angular averaging. The final state continuum wave functions were normalized by requiring that the wave function asymptotically approach a Coulomb wave plus incoming Coulomb spherical waves at large distances.

III. Results and discussion

Generally the electronic structure of transition metal dimers and larger molecules are very complicated with a number of states in a narrow energy range. The calculated valence energy levels for majority (up) and minority (down) spin of Mn_2, Co_2 and Ni_2 are shown in Fig 1. The internuclear distances used for Mn_2 and Co_2 in Fig. 1 are taken to be close to a theoretical estimate of the ground state bond length [22] since no reliable experimental value exists [23]. Three different internuclear distances are presented for Ni_2 with the experimental bond length at 4.2 au [24]. The symmetry used in the calculations was $C_{\infty v}$. The spin difference vary as 5.0 for Mn_2, 2.3 for Co_2 to 1.0 for Ni_2 ($r = 4.2$ au). The energy of the constant intermediate muffin-tin potential used in the cross-section calculation is also indicated in the figure. The constant muffin-tin energy vary strongly from a value for a weak interaction potential as for Mn_2, $r = 4.72$ au, of $V_{con} = -3.3/-2.5$ eV to a value for Ni_2, $r = 4.2$ au, of $V_{con} = -11.8/-11.6$ eV, where the two values represent the spin up and spin down potential, respectively.

In our previous paper [16] we presented theoretical XANES spectra for Ni_2 and Ni_3. These spectra are characterized by the "shape resonances" [25] occurring 0–15 eV above the threshold and their dependence on geometrical parameters. Figure 2 shows the calculated continuum part of the XANES cross-section of Ni_2 for three different bond lengths and their partial cross-sections of σ and π symmetry. The variations in the shape and position of these resonances for Mn_2 and Co_2 are compared with Ni_2 as shown in Fig. 3. The continuum cross sections of the dimers are obviously sensitive to changes in the bond length as they induce variations in the molecular potential. Our

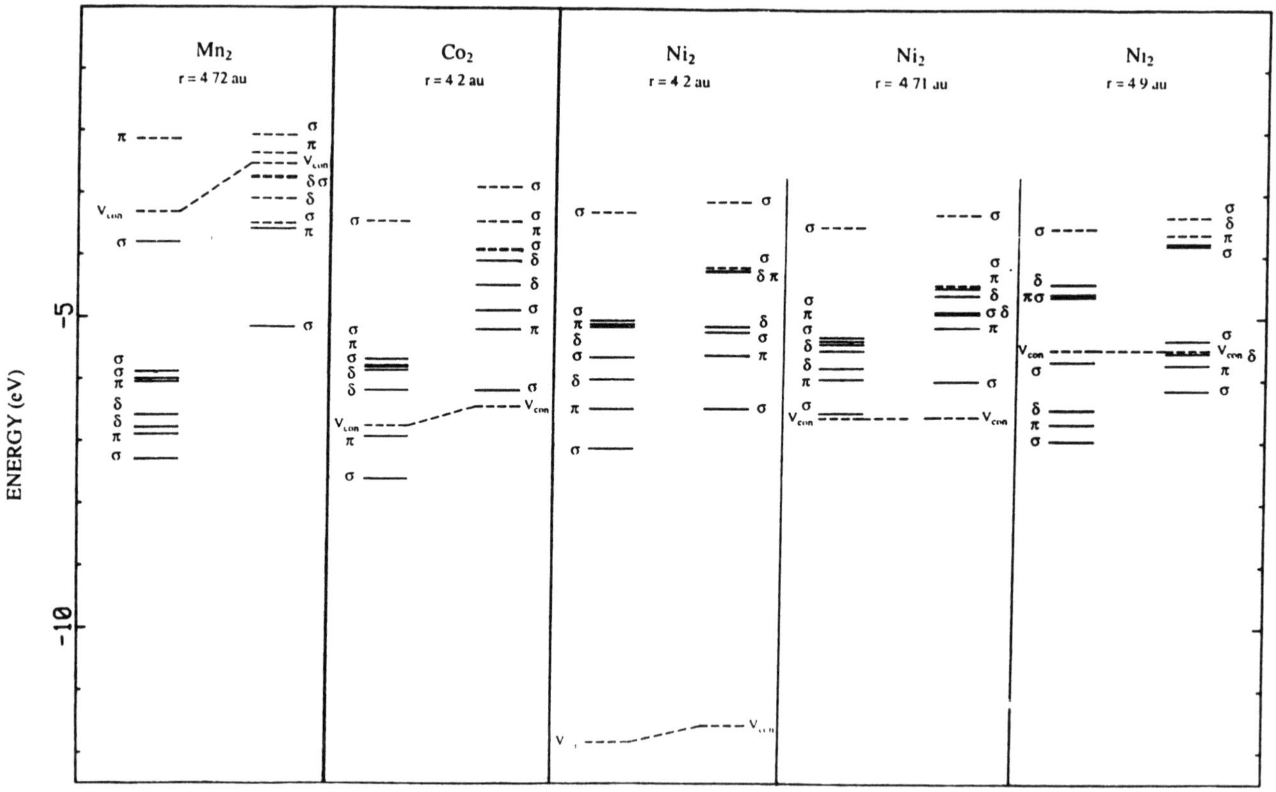

Fig. 1. Energy level diagram of Mn_2 with the bond length $r = 4.72$ au, Co_2 with $r = 4.2$ au and Ni_2 with $r = 4.2$ au, $r = 4.71$ au and $r = 4.9$ au. The levels to left are those with majority (*up*) spin and the minority (*down*) spin to right. Occupied levels are marked with solid lines and unoccupied with dashed lines. V_{con} is the energy of the intermediate muffin-tin potential used in the cross-section calculations

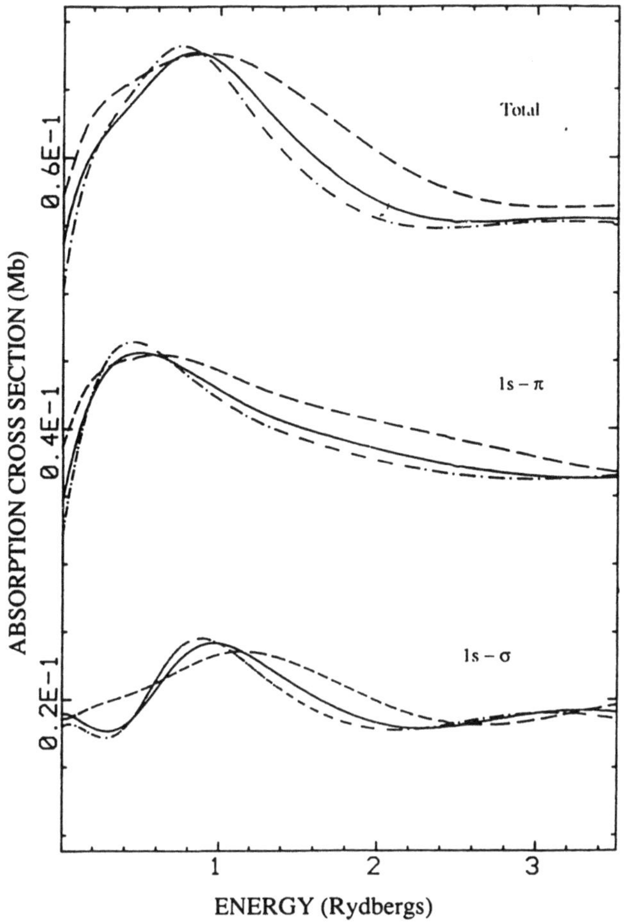

Fig. 2. The continum part of the theoretical Ni_2 *K*-edge XANES spectra calculated for the following lengths; **a** the experimentally determined bond length for the dimer, 4.2 au (*dashed line*)
b nearest neighbour bond length in the solid 4.71 au (*solid line*).
c Chosen bond length of 4.9 au. (*point dashed line*). The figure shows the total cross section and partial cross sections for the two dipole allowed channels of π and σ symmetry

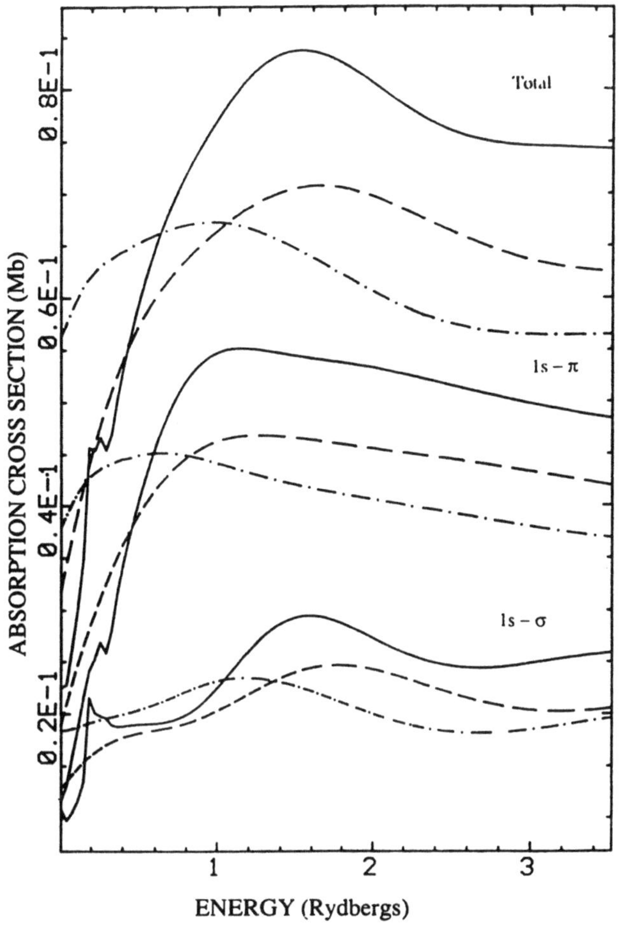

Fig. 3. Comparison between the continuum part of theoretical K-edge XANES spectra calculated for Mn_2 with $r = 4.72$ au (*solid line*), Co_2 with $r = 4.2$ au (*dashed line*) and Ni_2 with $r = 4.2$ au (*point-dashed line*). The cross-sections are also shown for the two dipole allowed channels of π and σ symmetry

earlier calculations show that symmetry resolved cross sections contain more pronounced variations as a function of bond length which may indicate that an angular resolved experiment, if it is possible to perform, could give more information.

In summary we believe that XANES could be a method for analyzing the structure of metal clusters. The sensitivity of the molecular potential as exemplified in the dependence of variations in the interatomic distance of the dimers as well as in the apex angle of a trimer as Ni_3 [16] show the applicability of XANES for information about the geometrical shape of larger clusters.

We are grateful for financial support from STU, The National Swedish Board for Technical Development through the research program in MICRONICS (contract 87-03403P) and STUF, Engineering Research Council of STU (contract 86-03067P) and from NFR, the Swedish Natural Science Research Council (contract F-FU 2560-128).

References

1. Metal Clusters, Proceedings of an International Symposium, Heidelberg, West Germany 1986. (eds.) Trager, F., zu Putlitz, G.: Z. Phys. D–Atoms, Molecules and Clusters **3**, 99 (1986)
2. International Symposium on the Physics and Chemistry of Small Clusters. Richmond Virginia, Oct 28–Nov 1, 1986. Jena, P., Rao, B.K., Khanna, S.N (eds.) New York: Plenum 1987
3. Metal Clusters, Moskowitz, M. (ed.). New York: Wiley 1986
4. Cox, D.M., Zakin, M.R., Kaldor, A.: In· Elemental and molecular clusters. Benedek, G, Martin, T.P., Pacchioni, G. (eds.) Berlin, Heidelberg, New York· Springer 1988

5. Hopkins, J.B., Langridge-Smith, P.R.R., Morse, M.D., Smalley, R.E.: J. Chem. Phys. **78**, 1627 (1983)
6. Kaldor, A., Cox, D.M., Trevor, D.J., Zakim, M.R.: In Ref. 1, p. 195
7. Moskowitz, J.M.: In Ref. 3, p. 131
8. Herrmann, A., Hoffmann, M., Leutwyler, S., Schumacher, E., Wöste, L.: Chem. Phys. Lett. **62**, 216 (1979)
9. Morse, M.D., Hopkins, J.B., Langridge-Smith. P.R.R., Smalley, R.E.: J. Chem. Phys. **79**, 5316 (1983)
10. Walch, S.P., Bauschlicher, C.W.: In: Comparison of ab initio quantum chemistry with experiments for small molecules. the State of the art. Bartlett, J. (ed.). Dordrecht. Holland: D. Reidel 1985
11. Salahub, D.R., Roos, B., Jones, B.: In: Ab initio methods in quantum chemistry. Advances in chemical physics, Vol LXVII and LXIX. New York: Wiley 1987
12. Bart, J.C.J.: Adv. Catal. **34**, 203 (1986)
13. Montano, P.A., Shenoy, G.K.: Solid State Commun. **35**, 53 (1980)
14. Purdum, H., Montano, P.A., Shenoy, G.K., Morrison, T.: Phys. Rev. B **25**, 4412 (1982)
15. Montano, P.A., Schulze, W., Tesche, B., Shenoy, G.K., Morrison T.I.: Phys. Rev. B **30**, 672 (1984)
16. Wästberg, B., Rosén, A., Ellis, D.E.: Z. Phys. D–Atoms, Molecules and Clusters 12 (1989) (this issue)
17. Kohn, W., Sham, L.J.: Phys. Rev. **140**, A1133 (1965)
18. Rosén, A., Ellis, D.E., Adachi, H., Averill, F.W.: J. Chem. Phys. **65**, 3629 (1973)
19. Delley, B., Ellis, D.E.: J. Chem. Phys. **76**, 1949 (1982)
20. von Barth, U., Hedin, L.: J. Phys. C **5**, 1629 (1972)
21. Chou, S.-H., Kutzler, F.W., Ellis, D.E. Shenoy, G.K., Morrison, T.I., Montano, P.A.: Phys. Rev. B **31**, 1069 (1985)
22. D.R. Salahub in Ref. 11.
23. Huber, K.P., Herzberg, G.: Constants of diatomic molecules. New York: Van Nostrand Reinhold 1979
24. Morse, M.D., Hansen, G.P., Langridge-Smith, P.R.R., Zheng, L.-S., Geusic, M.E., Michalopoulos, D.L., Smalley, R.E.: J. Chem. Phys. **80**, 5400 (1984)
25. Dehmer, J.L., Dill, D.: Phys. Rev. Lett. **35**, 213 (1975); J. Chem. Phys. **65**, 5327 (1976)

Z. Phys. D – Atoms, Molecules and Clusters 12, 381–386 (1989)

Atoms, Molecules and Clusters
Zeitschrift für Physik D
© Springer-Verlag 1989

Compared electronic structure of negative ions $M_pC_n^-$: I. Normal elements in Hückel theory

M. Leleyter*

Groupe de Physique Théorique, Faculté des Sciences, 33 rue Saint-Leu, F-80039 Amiens Cedex, France

Received 5 July 1988

Negative cluster ions $M_pC_n^-$ (M normal element, $n < 10$, $p = 1$–4) produced by various experimental techniques from carbides show in their emission intensities a very strong even-odd effect according to the parity of the carbon atom number n. This is in particular the case when M = N, F, Cl ($p = 1$), M = H, Al, Si, S ($p = 1$, 2) or M = B ($p = 1$–4). The largest intensities of $M_pC_n^-$ ions always take place for even n except in the cases of NC_n^-, $B_2C_n^-$ and $Al_2C_n^-$, for which the maxima of emission occur for odd n. This oscillating behaviour corresponds to alternations in the stability of the clusters which are mainly due to the fact that, in Pitzer and Clementi model (linear chains in the sp hybridization within the framework of Hückel theory), the HOMO (highest occupied molecular orbital) of the clusters lies in a double degenerate π level band: a cluster with a complete HOMO is always more stable than a cluster with a nearly empty HOMO. This result involves that the total number of π electrons is the main factor governing the parity of the stability alternations. Accordingly, since the knowledge of the π electron number requires the determination of the σ electron number too, these alternations enable us to infer a very likely electronic structure of the ions.

PACS: 31.20.N; 36.40; 79.20.N.

Negative cluster ions $M_pC_n^-$ (M normal element, $n < 10$, $p = 1$–4) produced by secondary ionic emission (SIMS) or laser ionization mass spectrometry (LIMS) from various carbides [1–4] or by surface ionization of organic compounds [5] generally show in their emission intensities a very pronounced even-odd effect as a function of the carbon atom number n. This is in particular the case when M = N, F, Cl ($p = 1$), M = H, Al, Si, S ($p = 1, 2$) and when M = B ($p = 1$–4). The largest intensities of $M_pC_n^-$ ions always take place for even n except in the cases of NC_n^-, $B_2C_n^-$ and $Al_2C_n^-$, for which the maxima of emission occur for odd n (Figs. 1, 2).

As the intensity alternations are observed with various experimental techniques, it means that the parity effects are due to the clusters themselves and not to the emission mechanisms. It is well-known that such an alternating behaviour can be interpreted from our "correspondence rule" [1, 6] which reads that the ions which give the most intense emission currents are just the most stable ones, so that the alternations in the emission intensities according to the parity of n correspond to alternations in the stabilities of the ions and are therefore strongly related to their electronic structures.

We interpret the intensity alternations from the Pitzer and Clementi model (P.C.) within the framework of Hückel theory: $M_pC_n^-$ are assumed to be linear (case of sp hybridization) with the M atoms attached to one or the 2 ends of C_n chains. The assumption of a linear C_n chain is pertinent since recent electron spin resonance experiments on C_6 molecules indicate "cumulene type" bonding as ... $=C=C=C=$ ... [7] with unpaired spins in $p\pi$ orbitals. This result fairly

* Also Laboratoire de Physique des Solides, associé au CNRS (LA 2), Batiment 510, Université Paris-Sud, F-91405 Orsay Cedex, France

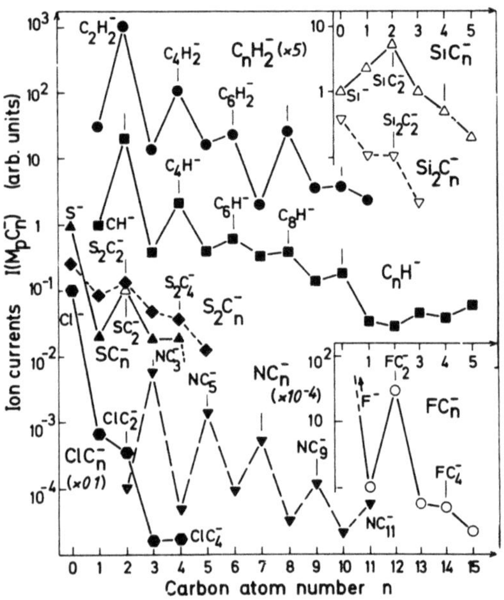

Fig. 1. Relative intensities of $M_pC_n^-$ ions versus n, the carbon atom number. FC_n^-, ClC_n^- (Ref. 1), $H_pC_n^-$ (Ref. 6) (primary ions 6.5 keV Ar^+), $Si_pC_n^-$ (Ref. 2) and $S_pC_n^-$ ions (Ref. 3) are observed by SIMS. NC_n^- (Ref. 5) are produced by surface ionization of organic compounds. All the maxima occur for even n (local maxima for ClC_n^- and $Si_pC_n^-$) except for NC_n^- ions (and for HC_n^- with $n > 11$), the maxima of which take place for odd n

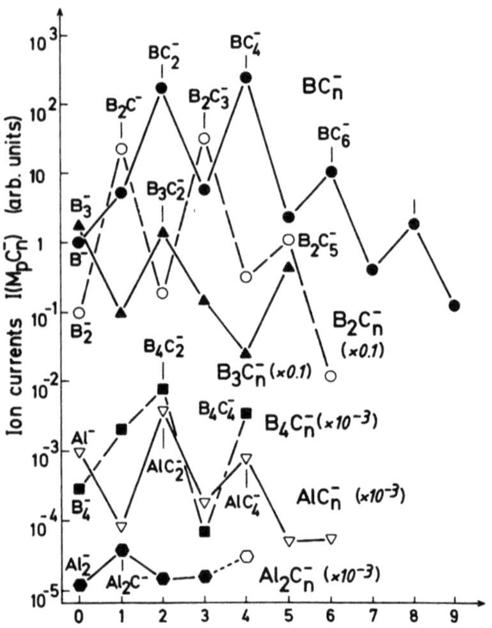

Fig. 2. Relative intensities versus n, the carbon atom number, of $M_pC_n^-$ ions obtained from carbides of III-A elements: $Al_pC_n^-$ ions formed in SIMS experiments (Ref. 1, primary beam 6.5 keV Ar^+ ions), $B_pC_n^-$ ions in laser mass spectra (Ref. 4). Maxima of AlC_n^-, $B_pC_n^-$ ($p = 1$, 4 and 3 with $n < 4$) occur for even n, whereas those of $Al_2C_n^-$ (for $n < 4$, because the intensity of $Al_2C_4^-$ ions may probably be perturbed by foreign ions) and $B_pC_n^-$ ions ($p = 2$, $p = 3$ with $n \geq 3$) occur for odd n

well agrees with our model in which the alternations arise from the fact that the highest occupied molecular orbital (HOMO or also "Fermi level") of the linear chain falls in a double degenerate π molecular orbital band. This fact has already been pointed out for C_n with various theoretical methods (ab initio calculations, semi-empirical CNDO approximations, etc . . .).

Now, HOMO may be either half-filled or complete with 4 electrons and a cluster with a complete (or nearly complete) HOMO is always more stable than a cluster with half-filled HOMO. Therefore the stabilities of the ions depend markedly on whether HOMO is full or nearly empty. The consequence is that the total number of π electrons is the main factor governing the parity of the alternations.

Case of MC_n^- ions

According to the above model, NC_n^- ions must have a filled HOMO if n is odd whereas the other MC_n^- ions (M = H, B, F, Al, Cl) must have a filled HOMO for even n. However the determination of the π electron number involves the knowledge of the bonding σ level band of the cluster.

The structure of MC_n^- ions is similar enough to that of C_n^- chain recalled in Table 1. For C_n^- chain, there are in particular 2 "surface levels" (noted σ_t) due to the dangling bonds on the 2 terminal C atoms. When the extra M atom is added, because of the M–C bonding, one of the σ_t levels of C_n^- chain combines with the $p\sigma_M$ atomic orbital (A.O.) of the M atom (except for H atom in the case of which it is obviously the s AO which combines with the σ_t level) to give hybrid orbitals: one σ_B(M–C) bonding orbital (filled) and one σ_{AB}(M–C) antibonding one (empty) (Table 1). The $s\sigma$ AO of M atom remains almost only localized on M (except for H atom). It is obvious that, in order that the complete HOMO corresponds to an even value of n, there must be $2n$ or $2n+4$ π electrons whereas NC_n^- ions will have complete HOMO for odd n with $2n+2$ π electrons. One can see that it is actually the case if the $s\sigma_M$ level is filled.

The electronic structures of MC_n^- ions are given in Tables 1 and 2 and the structures of π bands are sketched in Figs. 3 and 4 for BC_n^- and for NC_n^- (or HC_{n+1}^- or ClC_{n-1}^-) with filled or half-filled HOMO according to the parity of n (and the number of valence electrons); we take the usual notations of Hückel theory and the further assumption that the hopping integral β_π has the same value for C–C and for M–C bonds. In Figs. 3, 4, the dotted lines and empty circles exhibit the positions of the π electron barycentre; the lower the barycentre is in the energy scale, the

Table 1. Electronic structures of the bonding σ and π MO (molecular orbital) bands of MC_n^- ions when M is a 1st or 2nd row element (H, B, N, F)

1st and 2nd row elements

MC_n^- ions	C_n^-	HC_n^-	BC_n^-	NC_n^-	FC_n^-
valence elec.	4n+1	4n+2	4n+4	4n+6	4n+8
occupied σ MO's	← n − 1 (C_n) →				
			← + sσ_M →		
		← + 1 σ_B(M−C) →			
	+ 2σ_t	← + 1 σ_t →			
total σ occ.	n+1	n+1	n+2		
π MO's	← n (C_n) →				
			← + pπ_M →		
total	n		n+1		
number of π electrons	2n−1	2n	2n	2n+2	2n+4
Max. of stab for n	even			odd	even

The structure of C_n^- is recalled for comparison with its $n-1$ bonding σ MO's and the $2\sigma_t$ (see the text). σ_B(M–C) is the bonding σ MO which arises from the combination of one terminal MO, σ_t, of C_n chain, with the $p\sigma$ AO (atomic orbital) of M atom (it exists the corresponding antibonding MO, σ_{AB}(M–C), which is not given here since it is empty and does not play any role in the filling of HOMO). $s\sigma_M$ holds for the σ MO which is almost localized on M atom (it does not appear for H atom since it has combined with one of the σ_t MO's to give σ_B(H–C)). $p\pi_M$ holds for a π MO especially localized on M atom and arising from a p AO of M atom. In the bottom of the table are recapitulated the number of π electrons, deduced from the structure given in the table, and the parity of n which gives the maxima of relative stabilities Δ_n^-

Table 2. Electronic structures of the bonding σ and π MO bands of MC_n^- ions when M is a 3rd row element (Al, Si, S, Cl)

3rd row elements

MC_n^- ions	AlC_n^-	SiC_n^-	SC_n^-	ClC_n^-
valence elec.	4n+4	4n+5	4n+7	4n+8
occupied σ MO's	← n − 1 (C_n) →			
	+ sσ_{Al}	-	+ sσ_S	+ sσ_{Cl}
	← + 1 σ_B(M−C) →			
	← + 1σ_t →			
total σ occ.	n+2	n+1	n+2	
π MO's	← n (C_n) →			
	← + pπ_M + dπ_M →			
total	n+2			
π electrons	2n	2n+3		2n+4
Max. of stab. for n	even			

The notations are the same ones as in Table 1. $d\pi_M$ holds for a π MO especially localized on M and arising from a d AO of M atom

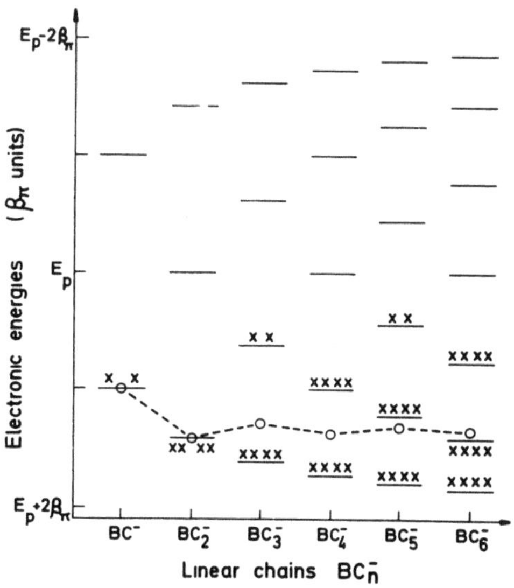

Fig. 3. Electronic structure of the π level band of the linear chains BC_n^- versus the carbon atom number n. The electronic energies are given in β_π units where β_π is the usual hopping integral of Hückel theory; its value is assumed to be the same one for C–C and M–C bonds The electronic population of the levels is represented by crosses. The *empty circles* and *dotted lines* give the positions of the π electron barycentre against n. HOMO is complete for even n, which corresponds to the lower positions of the barycentre in the energy scale and to more stable aggregates. E_p is the atomic energy of the p level of the carbon atom and lies in the middle of the π band

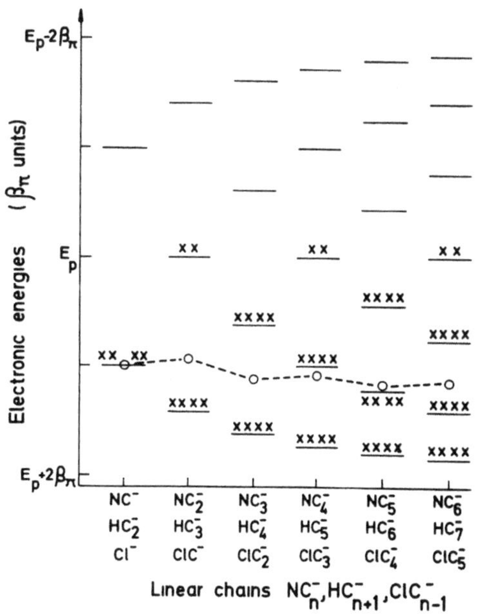

Fig. 4. Electronic structure of the π level band of the linear chains NC_n^- versus the carbon atom number n. This sketch is also valid for HC_{n+1}^- and ClC_{n-1}^- chains HOMO is then complete for odd n with corresponding lower positions of the barycentre

more stable the BC_n^- (or NC_n^-) ions are, which is the case for even n in Fig. 3 (or odd n in Fig. 4).

In Table 3, we give the quantities $\Delta_n^- = |E_\pi^-(n) - E_\pi^-(n-1)|$ obtained from the corresponding values

Table 3. Relative stabilities $\Delta_n^- = |E_\pi^-(n) - E_\pi^-(n-1)|$ in β_π units of MC_n^- ions, that is the increases in Hückel energies of the π electrons when one goes from MC_{n-1}^- ion to MC_n^- ion for M = B, N, F, Al, Si, S.

	Δ_n^-			
n	BC_n^- FC_n^-	NC_n^- ($= HC_{n+1}^-$, ClC_{n-1}^-)	AlC_n^-	SiC_n^- SC_n^-
1	2.0	4.0	2.828	2.657
2	3.657	1.657	3.644	2.669
3	2.051	3.287	2.456	2.602
4	3.220	1.984	3.268	2.603
5	2.158	3.048	2.384	2.579
6	3.024	2.135	3.066	2.577
7	2.230	2.924		
8	2.914	2.219		
9	2.283	2.852		
10	2.846	2.227		
11	2.320	2.802		

The maxima are underlined: they take place when n is odd for NC_n^- ions and when n is even in the other cases

of the energies $|E_\pi^-|$ of the π electrons (in β_π units) for the various MC_n^- ions. These quantities (with underlined maxima) measure the increases of the π electron energies of the ions when one extra carbon atom and one extra C–C bond are added to the MC_{n-1}^- cluster, and have the interest of making more evident that the energy gain is effectively larger when one goes from BC_{2k-1}^- to BC_{2k}^-, that is $\Delta_{2k}^- > \Delta_{2k\pm1}^-$ (or from NC_{2k}^- to NC_{2k+1}^- with $\Delta_{2k}^- < \Delta_{2k+1}^-$) than in the opposite case, which fairly well agrees with the data of Figs. 1 and 2.

Case of $M_2C_n^-$ ions

As Fig. 2 shows, the experimental curves for $B_2C_n^-$ and $Al_2C_n^-$ exhibit an alternation inversion with respect to the BC_n^- and AlC_n^- ions, with intensity maxima for odd n, contrary to the other $M_2C_n^-$ ions which have their emission maxima for even n as the corresponding MC_n^- ions.

The electronic structures of $M_2C_n^-$ ions are summed up in Table 4. There are 2 possible isomers, $M_2C_n^-$ and MC_nM^- ions. In the symmetrical isomers, the second σ_t level disappears too to give rise to a second σ_B(M–C) MO (molecular orbital) whereas in the disymmetrical cases, it appears a new bonding level σ_B(M–M) between the M atoms built from $s\sigma_M$ AO's. In the 2 cases, it remains one $s\sigma_M$ MO almost only localized on terminal M atoms. The structure of the π band of $B_2C_n^-$ ions is sketched in Fig. 5 and we have still shown the positions of the π electron barycentre which make clear that the "odd" ions are more stable than the "even" ones. HOMO is no more a "closed shell" since the total number of valence electrons is odd. However, it can be seen from Fig. 5 that even when HOMO is only nearly filled (3 electrons), the corresponding cluster is nevertheless more stable than when HOMO only accommodates one π electron.

Let us notice that for $H_2C_n^-$, the symmetrical isomers would lead to $2n+1$ π electrons with maximum stabilities for odd n (Table 4), contrary to the experimental data. Only the disymmetrical isomers agree with the curve in Fig. 1.

In the other $M_2C_n^-$ cases, the π number which is deduced from Table 4 gives a result in good agreement

Table 4. Electronic structures of the bonding σ and π MO bands of $M_2C_n^-$ ions for the 2 kinds of isomers.

$M_2C_n^-$	$H_2{>}C_n^-$	HC_nH^-	$B_2C_n^-$	BC_nB^-	$Al_2C_n^-$	AlC_nAl^-	$Si_2C_n^-$	SiC_nSi^-	$S_2C_n^-$	SC_nS^-
valence el.	4n+3		4n+7		4n+7		4n+9		4n+13	
occ. σ MO's	← n-1 (C_n) →									
$s\sigma_M$			1	2	1	2	1	2	1	2
σ_B(M-C)	2		1	2	1	2	1	2	1	2
σ_B(M-M)	-		1	-	1	-	1	-	1	-
σ_t	1	0	1	0	1	0	1	0	1	0
total σ occ.	n+2	n+1	n+3							
π MO's	n		n+2		n+4					
π electrons	2n-1	2n+1	2n+1				2n+3		2n+7	
Max. stab. for n	even	~~odd~~	odd				even			

The notations are those of Tables 1 and 2. σ_B(M – M) is the bonding σ MO which appears in the disymmetrical isomers M – M – C_n^- and which is mainly localized on the 2 M atoms.
In this case, it still remains one σ_t MO on the free end of C_n chain. The number of π MO's is only $n+2$ for $B_2C_n^-$ since there are only the 2 $p\pi_B$ in addition to the n π MO's of C_n, whereas it is $n+4$ for the 3rd row elements because of the 2 $d\pi_M$ MO's of M atom. Let us remark that HC_nH^- ions give the wrong parity of the stability alternations, contrary to the disymmetrical isomers

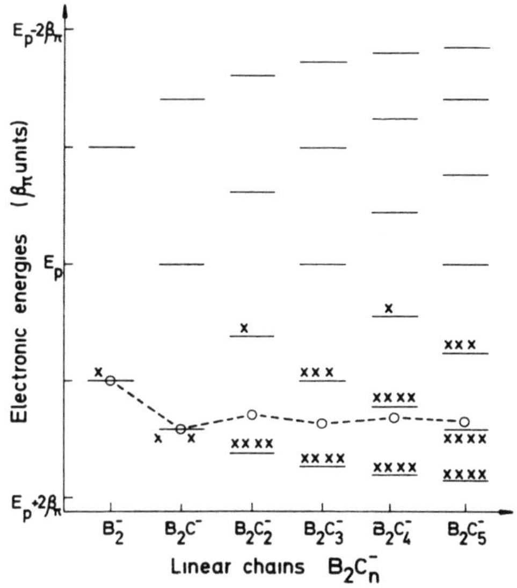

Fig. 5. Electronic structure of the π level band of the linear chains $B_2C_n^-$ or BC_nB^- versus the carbon atom number n Now HOMO is no more complete. However, when HOMO is almost complete (3 electrons if n is odd), the barycentre is lower in the energy scale (whence the cluster is more stable) than when HOMO is nearly empty (1 electron if n is even)

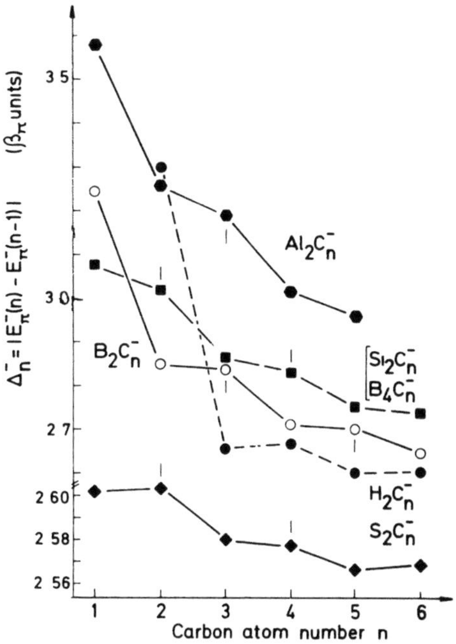

Fig. 6. Relative stabilities, versus the carbon atom number n, $\Delta_n^- = |E_\pi^-(n) - E_\pi^-(n-1)|$ (in β_π units), that is differences between the π electronic energies $E_\pi^-(n)$ and $E_\pi^-(n-1)$ of 2 consecutive linear chains $M_pC_n^-$ for M = H, B, Al, Si, S ($p = 2$) and for B ($p = 4$). The local maxima of the curves show the maxima of the relative stabilities of the various clusters. $B_2C_n^-$ and $Al_2C_n^-$ are then more stable for odd n than for even n, contrary to the other ions

with the curves of Figs. 1 and 2 as it can be seen from Fig. 6, in which we have drawn the values as a function of n of Δ_n^- (in β_π units) calculated according to the structures given in Table 4 (and with the further assumption that β_π keeps the same value for M–M bonds as for M–C or C–C bonds). One can observe local maxima of Δ_n^- for odd n in the cases of $B_2C_n^-$ and $Al_2C_n^-$ ions and for even n in the other cases.

Our model can be easily extended to $B_3C_n^-$ and $B_4C_n^-$ ions by assuming that the clusters are still linear chains.

$B_3C_n^-$ ions have $4n+10$ valence electrons and 2 possible isomers $B_3C_n^-$ and $B_2C_nB^-$. According to the above discussion, the occupied σ MO's of $B_3C_n^-$ are, besides the $n-1$ σ MO's of C_n, 1 σ_B(B–C) + 2 σ_B(B–B) + 1 remaining σ_t, that is $n+3$ σ MO's in all, and consequently it remains $2n+4$ π electrons to be shared in $n+3$ π MO's, which leads to a complete HOMO for even n, in agreement with the experimental results for $n < 4$. In the same way, for $B_2C_nB^-$ ions, there are only $n+2$ σ MO's ($n-1+2$ σ_B(B–C) + 1 σ_B(B–B)) whence $2n+6$ π electrons (complete HOMO for odd n). If the $s\sigma_B$ MO mainly localized on the terminal B atoms (1 for $B_3C_n^-$ and 2 for $B_2C_nB^-$) are assumed to be filled, one is lead to $n+4$ σ MO's for the 2 isomers whence $2n+2$ π electrons (complete HOMO for odd n). So, the only case where HOMO is completely filled for even n is this of $B_3C_n^-$ ions with an empty $s\sigma_B$ MO. The 3 other cases are in agreement with the second part of the experimental curve, when $n > 3$.

$B_4C_n^-$ ions ($4n+13$ electrons) can have 3 linear isomers: $B_4C_n^-$, $B_3C_nB^-$ and the symmetrical $B_2C_nB_2^-$ ions. The 4 boron atoms give rise, in the same way as above, to σ_B(B–C) and σ_B(B–B) MO's (and to the corresponding antibonding ones too, which lie above the π band). We have thus for $B_4C_n^-$ ions, $n+5$ bonding σ MO's ($n-1+1$ σ_B(B–C) + 3 σ_B(B–B) + 1 σ_t + 1 $s\sigma_B$) whence $2n+3$ π electrons in $n+4$ π MO's (the same situation as for $Si_2C_n^-$) with nearly complete HOMO for even n, and, according to Fig. 6, maximum stabilities for even n as in the experiment. For the 2 other isomers, there are $n+5$ σ MO's too ($n-1$ + 2 σ_B(B–C) + 2 σ_B(B–B) + 2 $s\sigma_B$), thus the same result as for $B_4C_n^-$ ions. If $s\sigma_B$ MO's are assumed to be empty, they only have $n+3$ σ MO's, whence $2n+7$ π electrons as in the case of $S_2C_n^-$, that is still a result consistent with the experimental data. On the contrary, $B_4C_n^-$ ions have this time $n+4$ σ bonding MO's and $2n+5$ π electrons with nearly complete HOMO for odd n, which does not agree with experiment.

So we have seen that, whenever experimental results on $M_pC_n^-$ ions, where M is a normal element belonging to almost all the columns (I and III to VII) of the periodic table, present very strong parity effects, it is possible, within the very simple model of the linear chain (sp hybridization) in Hückel approximation, to deduce a very likely electronic structure of the corresponding clusters and to analyse the roles played by the various orbitals of the extra M atoms, especially the role of the $s\sigma_M$ AO's; it appears indeed, according to the above discussions, that their filling has an important influence on the HOMO filling as a function of the n parity. To sum up, we can conclude that the alternations in the emission of $M_pC_n^-$ ions are a good reflection of their electronic structures and a means of "measuring" their relative stabilities. Finally, our model enables us to get a better understanding of the M–C bonding between carbon and a normal element atom.

References

1. Leleyter, M., Joyes, P.: ISSPIC 3 (Berlin 1984). Surf. Sci. **156**, 800 (1985)
2. Honig, R.E.: Adv. Mass Spectrom. **2**, 25 (1962)
3. Orth, R.G., Jonkmann, H.T., Michl, J.: Int. J. Mass Spectr. Ion Phys. **43**, 41 (1982)
4. Becker, S., Dietze, H.J.: Int. J. Mass. Spectr. Ion Proc.: **82**, 287 (1988)
5. Zandberg, E.Y., Paleev, V.I.: Sov. Phys. Tech. Phys. **17**, 671 (1972)
6. Leleyter, M.: Thèse de Doctorat d'Etat (1975) Orsay (France)
7. Van Zee, R.J., Ferrante, R.F., Zeringue, K.J., Weltner, W. Jr., J. Chem. Phys. **88**, 3465 (1988)

Z. Phys. D – Atoms, Molecules and Clusters 12, 387–390 (1989)

Atoms, Molecules
Zeitschrift für Physik D and Clusters
© Springer-Verlag 1989

Electronic structure of spheroidal metal containing carbon shells: study of the LaC_{60} and C_{60} clusters and their ions within the local density approximation

A. Rosén and B. Wästberg

Department of Physics, Chalmers University of Technology, and University of Göteborg, S-412 96 Göteborg, Sweden

Received 5 July 1988; final version 30 September 1988

Ionization thresholds and electron affinities have been calculated within the Local Density Approximation (LDA) for the neutral, positively and negatively charged clusters of LaC_{60} and C_{60}. The evaluated energies are found to be in good agreement with available experimental data. More accurate measurements are however necessary to verify the suggested spheroidal cage structure for these molecules.

PACS: 31.20.N; 36.40

I. Introduction

Since the fascinating discovery of the remarkable stable molecule C_{60} [1] a number of experimental and theoretical papers have appeared [2–4]. The suggested icosahedral cage structure has caused a widespread debate about the structure of C_{60} but provides a simple explanation of the remarkable high stability and chemical inertness of C_{60}. Shortly after the first observation of C_{60}, experimental evidence was presented [5] for a new class of organo – metallic clusters consisting of a metal atom bound to a C_{60} cluster, i.e. an MC_{60} molecule, where M was the La atom in the first experiment and K or Cs in more recent works [6]. The new molecule was produced by laser vaporization of a $LaCl_3$ – impregnated graphite disc using the conventional laser vaporization technique [7]. It was suggested that the metal atom was located within the spheroidal carbon cage of the C_{60} molecule although other suggestions have appeared [8]. This organometallic molecule was found to have a similar stability as observed for the C_{60} molecule. A metal atom chemically bound in such a carbon cage would consist of a positively or negatively charged atom with surrounding carbon atoms of opposite charge depending on the charge transfer. Such a molecule has great similarities to the "super-atom" concept proposed in connection with the production of semi-conductor hetero-structures [9–11]. This type of quazi-atomic system could be realized as a spherical core doped with donors and a surrounding impurity-free matrix of greater electron affinity. The donors are ionized under certain conditions, resulting in the release of electrons to form a shell of negative charge outside the core. The electrons in this almost spherical Coulomb potential will be quantized with respect to the l quantum number but occupy electronic states with different ordering and magnitude of the eigenvalues compared with an atom. Therefore LaC_{60} could be considered as a "super-atom" or "super-antiatom" depending on the sign and magnitude of the charge transfer, with an electronic structure similar to the C_{60} molecule.

Some time ago we applied the CNDO/S method [12] to calculate the ionization energy, electron affinity and energies for optical transitions of C_{60} [13]. This calculation has recently been extended to evaluation of the ionization thresholds and electron affinities for C_{60}^-, C_{60} and C_{60}^+ [14] and LaC_{60} [15] within the local density approximation (LDA) [16]. Generally a very good agreement has been obtained between available experimental values of the ionization thresholds, electron affinities and optical transitions for these systems [17–19]. In this work we will summarise the previous work with inclusion of some new results.

II. Theory

The one-electron equations appearing in the LDA scheme were solved for the exchange correlation potential of von Barth and Hedin [20] by using the Discrete Variational Method (DVM) [21] and the Self Consistent Charge (SCC) approximation [22]. The basis functions were generated with an atomic LDA code for the configurations of C . . . $2s^2 2p^2$ and La . . . $5s^2 5p^6 4f^0 5d^1 6s^2 6p^0$. For increasing the flexibility of the basis, additional basis functions were added from the C^{2+} ion, 2*s* and 2*p*, and from the La^{1+} ion, 6*s* and 6*p*. The occupation of the carbon 1*s* core and the core of La atom was kept constant within the "frozen core" approximation while other basis functions were treated self-consistently. The calculations were performed for the alternating bond lengths [23] and in C_s symmetry, giving groups of levels of gerade and ungerade symmetry corresponding to the groups of levels in the I_h symmetry.

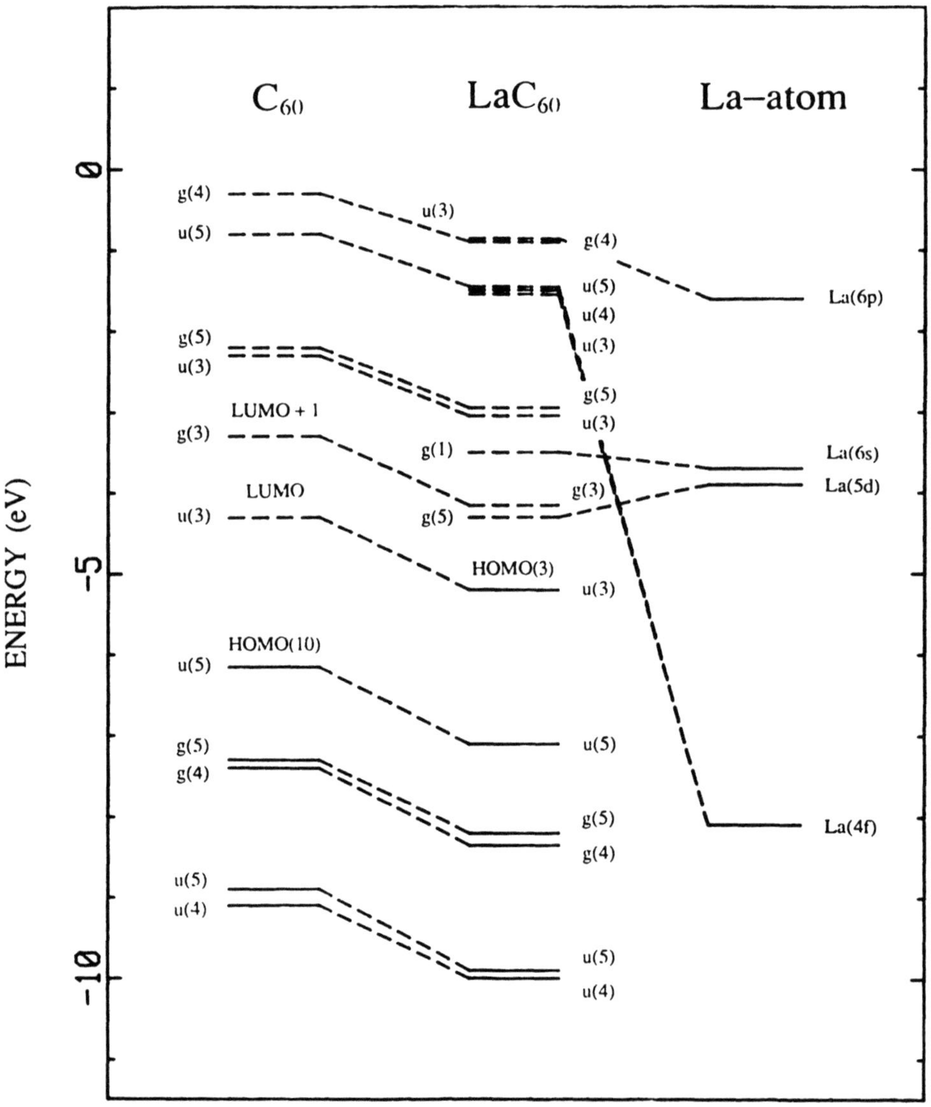

Fig. 1. Comparison of the one-electron molecular eigenvalues for C_{60}, LaC_{60} and the La atom. The symmetry of the levels and degeneracy divided by two are given as g(*n*) and u(*n*). The occupied levels are denoted by *solid lines* while *dashed lines* denote the unoccupied ones. Occupation numbers of the HOMO levels are given in parenthesis

Table 1a. Comparison of experimental and theoretical values of ionization thresholds and electron affinities in eV for C_{60}^+, C_{60} and C_{60}^-. The energies are calculated using the von Barth & Hedin scheme

	C_{60}^+		C_{60}		C_{60}^-	
	Theor	Exp	Theor	Exp	Theor	Exp
ε_i (LUMO+1)	−6.7		−3.3			
ε_i (LUMO)	−7.6		−4.3		−0.1	
ε_i (HOMO)	−9.4		−6.1		−1.1	
Ionization threshold	10.8		7.8	6.4–7.9	2.7	2.6–2.8
Electron affinity	7.8	6.4–7.9	2.7	2.6–2.8	−0.4	

Table 1b. LaC_{60}^+, LaC_{60} and LaC_{60}^-. The energies are calculated using the von Barth & Hedin scheme

	LaC_{60}^+		LaC_{60}		LaC_{60}^-	
	Theor	Exp	Theor	Exp	Theor	Exp
ε_i (LUMO+1)	−7.8		−4.2		−1.9	
ε_i (LUMO)	−7.8		−4.3		−1.9	
ε_i (HOMO)	−8.3		−5.2		−2.4	
Ionization threshold	10.0		6.9	<6.4	3.8	
Electron affinity	6.9	<6.4	3.8		0.8	

III. Results

Results of the ground-state calculations of C_{60} and LaC_{60} are presented in Fig. 1 and compared with the atomic eigenvalues of La. The levels are assigned as $g(n)$ or $u(n)$ corresponding to gerade or ungerade symmetry. n denotes the degeneracy of the particular level divided by two. The molecular and atomic origin of the levels in LaC_{60} has been identified through a density of state analysis and denoted by dashed lines to the levels in C_{60} and the La atom, respectively. Occupation numbers are given for the HOMO levels in parenthesis. The HOMO levels in LaC_{60} are of carbon character which may explain the similarities in reactivity for these molecules. The LUMO level is of gerade symmetry and originates from the La 5*d* state. The molecular levels have also been examined through a Mulliken analysis which gave a gross positive charge of 2.85 for the La atom and a total negative charge of 2.85 for the surrounding carbon atoms. The molecule LaC_{60} can therefore be considered as a "super-atom".

Eigenvalues for the HOMO, LUMO and LUMO +1 levels from the ground-state calculations are presented in Table 1. For comparison with available experimental data, the ionization threshold and electron affinity were calculated using the transition state procedure [24, 25]. Our value for the ionization threshold for C_{60} is in good agreement with the experimental values and our earlier theoretical values [13] as well as with a value of 7.92 eV obtained in a large-scale Hartree-Fock calculation performed by Lüthi and Almlöf [26]. Their calculation of the electron affinity of 0.80 eV is, however, quite different from our value of 2.7 eV which is in good agreement with the experimental value of 2.6-2.8 eV [18]. Our prediction of an ionization threshold of 6.9 eV for LaC_{60} is somewhat too high compared with the approximate experimental value of less than 6.4 eV [19]. The trend of the change in the ionization threshold between C_{60} and LaC_{60} is however correct. The geometry of LaC_{60}, in the form of a "super atom" with the La atom in the centre of the spheroidal C_{60} cage, is therefore quite plausible. Further measurements of a more precise value of the ionization threshold and the electron affinity for LaC_{60} should, however, be of great interest as a check on these calculations, as well as to obtain further data to resolve the question concerning the structure of LaC_{60}.

We are grateful for financial support from STU, The National Swedish Board for Technical Development through the research program in MICRONICS (contract 87-03403P) and STUF, Engineering Research Council of STU (contract 86-03067P) and from NFR, The Swedish Natural Science Research Council (contract F-FU 2560-128).

References

1. Kroto, H.W., Heath, J.R., O'Brien, S.C.O., Curl, R.F., Smalley, R.E.: Nature **318**, 162 (1985)
2. Smalley, R.E.: NASA Conference on Carbon in the Galaxy, November 1987, NASA Conference Report. This paper gives an up to date overview of the field
3. Klein, D.J., Seitz, W.A., Schmalz, T.G.: Nature **323**, 703 (1986)
4. Kroto, H.W.: Nature, **329**, 529 (1987) This paper gives an overview of the discussion of the stability of carbon clusters

5. Heath, J.R., Brien, S.C.O., Zhang, Q., Liu, Y., Curl, R.F., Kroto, H.W., Tittel, F.K., Smalley, R.E.: J. Am. Chem. Soc. **107**, 7779 (1980)
6. Weiss, F.D., Elkind, J.L., O'Brien, S.C., Curl, R.F., Smalley, R.E.: J. Am. Chem. Soc. **110**, 4464 (1988)
7. Hopkins, J.B., Langridge-Smith, P.R.R., Morse, M.D., Smalley, R.E.: J. Chem. Phys. **78**, 1627 (1983)
8. Cox, D.M., Trevor, D.J., Reichmann, K.C., Kaldor, A.: J. Am. Chem. Soc. **108**, 2457 (1986)
9. Watanabe, H., Ionshita, T.: Optoelectronics Dev. **1**, 33 (1986)
10. Ionshita, T., Ohnishi, S., Oshiyama, A.: Phys. Rev. Lett. **57**, 2560 (1986)
11. Watanabe, H.: The Physics and Fabrication of Microstructures and Microdevices, Kelly, M.J., Weisbuch, C. (eds.). Berlin, Heidelberg, New York: Springer 1986
12. Del Bene, J., Jaffé, H.H.: Chem. Phys. **48**, 1807, 4050 (1968)
13. Larsson, S., Volosov, A., Rosén, A.: Chem. Phys. Lett. **137**, 501 (1987)
14. Rosén, A., Wästberg, B.: J. Chem. Phys. **90**, 2525 (1989)
15. Rosén, A., Wästberg, B.: J. Am. Chem. Soc. **110**, 8701 (1988)
16. Kohn, W., Sham, L.J.: Phys. Rev. **A140**, 1133 (1965)
17. Cox, D.M., Trevor, D.J., Reichmann, K.C., Kaldor, A.: J. Am. Chem. Soc. **108**, 2457 (1986)
18. Yang, S.H., Pettitte, C.L., Conceicao, J., Cheshnovsky, O., Smalley, R.E.: Chem. Phys. Lett **139**, 233 (1987)
19. Heath, J.R., Curl, R.F., Smalley, R.E.: J. Chem. Phys. **87**, 4236 (1987)
20. von Barth, U., Hedin, L.: J. Phys. C **5**, 1629 (1972)
21. Ellis, D.E., Painter, G.S.: Phys. Rev. B **2**, 2887 (1970)
22. Rosén, A., Ellis, D.E.: Adachi, H., Averill, F.: J. Chem. Phys. **65**, 3629 (1976)
23. Newton, M.D., Stanton, R.E.. J. Am. Chem. Soc. **108**, 2469–2470 (1986)
24. Slater, J.C.: The self-consistent field for molecules and solids. New York: McGraw-Hill 1974
25. Slater, J.C.: Adv. Quantum Chem. **6**, 1 (1972)
26. Lüthi, H.P., Almlöf, J.: Chem. Phys. Lett. **135**, 357 (1987)

Z Phys. D – Atoms, Molecules and Clusters 12, 391–394 (1989)

Atoms, Molecules
Zeitschrift für Physik D and Clusters
© Springer-Verlag 1989

Molecular orbital calculations on transition metal clusters

M.C. dos Santos[1,*] and J. Albino O. de Aguiar[2,]**

[1] Facultés Universitaires Notre-Dame de la Paix, Laboratoire de Chimie Théorique Appliquée, B-5000 Namur, Belgium
[2] Kamerlingh Onnes Laboratorium der Rijksuniversiteit Leiden Postbus 9506, NL-2300 RA Leiden, The Netherlands

Received 5 July 1988

Transition metal clusters have been subject of experimental and theoretical interest due to their catalytic activity, as well as their unusual physical properties. Semi-empirical extended Hückel molecular orbital calculations are presented for a series of small metal clusters with nuclearity ranging from 3 to 6. Naked and carbonylated clusters of Fe, Ru, and Os are studied. The charge transfer between ligands shell and metal core is found to be a function of nuclearity, CO coordination and the chemical species forming the bare cluster. The observed magnetic properties of these systems are discussed in terms of their electronic structure and CO-metal charge transfer.

PACS: 31.20Pv; 36.40+d

Polynuclear transition metal cluster compounds have received considerable attention not only due to their catalytic activity but also because of their unusual chemical and physical properties [1, 2, 3, 4]. These macromolecules, formed by a core with n metal atoms surrounded by a ligand shell, provide also a good model of small metallic crystallites covered with suitable molecules, e.g., carbon monoxide and phosphines. Thus by studying the ligand-metal core interaction in such class of compounds one may be able to draw comparisons between the physics of these systems and that of chemisorbed molecules on metal surfaces.

A remarkable aspect of metal cluster compounds concerns the change in their magnetic properties as a function of cluster size and ligand coordination. For example, it was experimentally shown that the magnetic susceptibilities of diamagnetic Os [5] and Pt [6] carbonyl clusters increase with cluster nuclearity, and that high nuclearity Pt and Ni clusters are paramagnetic [6, 7]. Recently one of the authors has shown that also low nuclearity Ru clusters [8] are paramagnetic. This is in marked contrast with the behavior of Os clusters where paramagnetism appears only in a ten metal atoms particle [9]. The role played by the ligands and cluster nuclearity in determining the magnetic properties of these systems has been theoretically investigated. Recent molecular orbital calculations on Ni clusters show that their magnetic moment increases with nuclearity [10, 11], and that carbonylation [10, 11, 12] or hydrogenation [13] of a naked paramagnetic cluster can reduce or even suppress its paramagnetism.

In this work, we present semi-empirical molecular orbital calculations based on the extended Hückel theory (EHT) [14] for a series of small naked and carbonylated metal clusters of Fe, Ru and Os containing 3 to 6 metal atoms. Within this very qualitative approach, properties like magnetic moments cannot be accessed. Instead, we analyze charge modifications on the metallic core due to carbonylation of the naked cluster, and search for trends that could allow us to

* Permanent address Departamento de Quimica Fundamental, Universidade Federal de Pernambuco, 50739 Recife, PE, Brasil
** Permanent address: Departamento de Fisica, Universidade Federal de Pernambuco, 50739 Recife, PE, Brasil

relate features of the chemical bond with the appearance of paramagnetism. This procedure is similar to that adopted in an EHT study of CO adsorption on metal surfaces [15], where the correct picture of the change from molecular to dissociative chemisorption on the first row transition metals is reproduced. It should also be stressed that the numerical problem of dealing with macromolecules is far from trivial. For instance, the smallest carbonyl cluster treated here has 144 electrons. In this context, the EHT approach represents a reasonable compromise between computational effort and quality of the results.

We have used the standard EHT technique to calculate the electronic structures of naked and carbonylated clusters. The appropriate EHT parameters for molecular orbital calculations were taken from the literature [16], and are listed in Table 1. Bond lengths (also presented in Table 1) and bond angles adopted here are averages of the available experimental data on carbonyl clusters [2, 17]. Calculations on both the naked and carbonylated clusters have been performed for the structures shown in Fig. 1. To facilitate comparisons, we have chosen the same geometries for the three metals, even though some of them may not correspond to the most stable structure for a specific metal. We believe this is not going to affect significantly the results at the EHT level. The clusters with six metal atoms were calculated as doubly negative ions. Their real structures do actually correspond to that shown in Fig. 1 [2]. In order to better compare the spectra of the different metal clusters, we have calculated the electronic structure of the ligand shell, in

Table 1. EHT parameters. Energies are given in eV. ζ_1 and ζ_2 are atomic orbital exponents and C_1 and C_2 are coefficients of the double-zeta expansion of *d* orbitals

Atom	orbital	H_{ii}	ζ_1	ζ_2	C_1	C_2
C	2*s*	−21.4	1.55			
	2*p*	−11.4	1.325			
O	2*s*	−32.3	2.20			
	2*p*	−14.8	1.975			
	4*s*	−9.17	1.90			
Fe	4*p*	−5.37	1.90			
	3*d*	−12.70	5.35	2.0	0.5505	0.6260
	5*s*	−8.60	2.078			
Ru	5*p*	−4.30	2.043			
	4*d*	−12.20	5.378	2.303	0.5340	0.6365
	6*s*	−8.17	2.452			
Os	6*p*	−4.81	2.429			
	5*d*	−11.84	5.571	2.416	0.6372	0.5598

Bond lengths: Fe-Fe 2.56 Å, Ru-Ru 2.85 Å, Os-Os 2.88 Å, Fe-C 1.82 Å, Ru-C 1.92 Å, Os-C 1.95 Å, C-O 1.15 Å

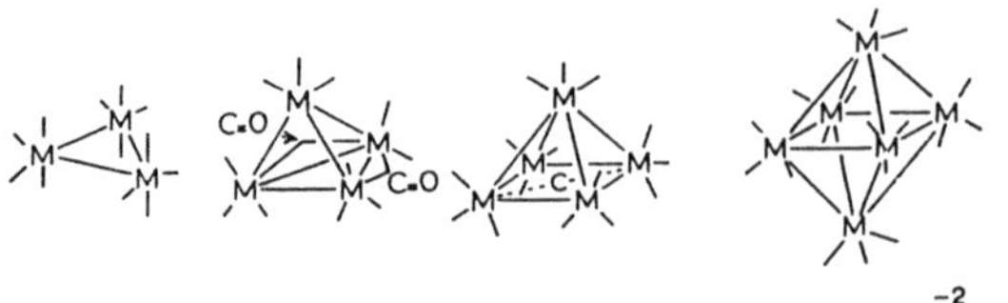

Fig. 1. Structures of carbonyl clusters (M = Fe, Ru, and Os). In $M_3(CO)_{12}$ two bridging carbonyls were adopted for M = Fe. Note two bridging CO for the $M_4(CO)_{13}$ cluster

the same geometry adopted for Ru carbonyl clusters, but without the metal atoms.

We present in Fig. 2 (a) to (d) the one-electron levels for the ligands, naked, and carbonylated clusters with nuclearity 3 to 6, respectively. Since the larger clusters loose their high symmetry when the ligands are attached, we have preferred to relate groups of levels to the original CO molecular orbitals or to the metal atomic orbitals instead of their usual group classification. From the figures, the following features are observed: first, as one coordinates more and more ligands, the 1π, 5σ and 2π bands broaden and start to slightly mix for $(CO)_{15}$. The energy gap between the 5σ and 2π bands decreases in the series. The same occurs to the *d* metal bands, which are wider for larger naked clusters, but in a less extent when compared to the ligands bands. We also note an increase in the density of states on the top of the *d* bands with increasing nuclearity. Contrary to the tendency observed for the bulk metals, we obtain a shift of the Fermi level to higher energies when moving from Fe to Os. This is expected since we have adopted EHT parameters suitable for cluster calculations and also the bulk crystalline structures are more compact than those of the clusters.

While the naked structures M_4 and M_6^{-2} were found to be open shell systems for the three metals, a closed shell configuration was found for all carbonylated clusters. In these last systems one observes a tendency of the HUMO-LUMO gap to decrease with increasing nuclearity, even though some oscillations of this property appears to be present. Carbonylation leads to the stabilization of the first excited states, which approach more and more the LUMO level when the nuclearity is increased. The well known fluxionality [2] of both the metal core and the ligands might very well lead to an open shell configuration if, e.g., the ligands reach a position that stabilizes the low lying excited states. The charge modifications on the metal as one considers the interaction of a bare cluster with the ligands can also give indications about the expected magnetic behavior. In Table 2 we list the average charge variation in the orbital population on the metal atoms of the carbonyl clusters with respect to

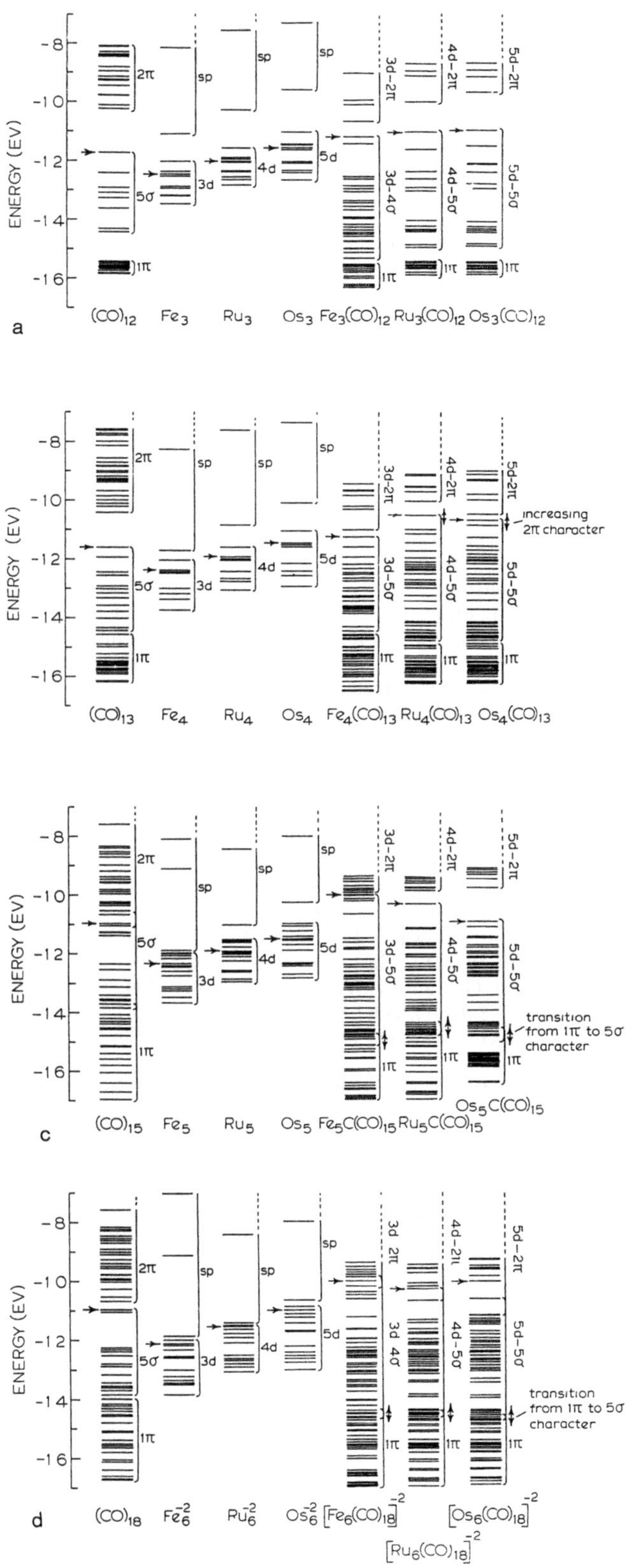

Fig. 2. One-electron energy levels for the ligands shell, naked and carbonylated clusters of Fe, Ru and Os. The HOMO levels are indicated by the horizontal arrows. **a** $(CO)_{12}$, M_3 and $M_3(CO)_{12}$; **b** $(CO)_{13}$, M_4 and $M_4(CO)_{13}$; **c** $(CO)_{15}$, M_5 and $M_5C(CO)_{15}$; **d** $(CO)_{18}$, M_6^{-2} and $M_6(CO)_{18}^{-2}$

Table 2. Average charge variations in orbital populations on the metal atom due to carbonylation of a naked metal cluster. $\Delta\varepsilon$ is the HOMO-LUMO energy gap in the carbonyl cluster, given in electron-Volts

Atom	orbital	nuclearity 3	4	5	6
	Δs	0.28	0.25	0.21	−0.06
Fe	Δp	0.64	0.67	0.82	0.76
	Δd	−0.46	−0.38	−0.02	0.19
	total	0.46	0.54	1.01	0.89
	$\Delta\varepsilon$	0.53	0.25	0.08	0.14
	Δs	0.33	0.28	0.28	0.28
Ru	Δp	0.87	0.60	0.70	0.66
	Δd	−1.04	−0.80	−0.95	−1.02
	total	0.16	0.08	0.03	−0.08
	$\Delta\varepsilon$	1.03	0.49	0.46	0.10
	Δs	0.32	0.25	0.25	0.26
Os	Δp	0.96	0.71	1.08	0.76
	Δd	−1.46	−1.28	−1.57	−1.45
	total	−0.18	−0.32	−0.25	−0.43
	$\Delta\varepsilon$	1.30	0 20	1.16	0.18

the bare cluster, as well as the evolution of the HOMO-LUMO gap, as a function of nuclearity. The conclusions to be drawn from this table are similar to those found in the chemisorption problem [15]. The strong variation on the *s* and *p* populations are an indication of hybridization of the *d* bands since in the bare clusters these orbitals contribute with only a few percent to the charge density. Iron clusters interact with CO molecules mainly by a $3d$-5σ bond, where the ligands transfer charge from the 5σ states to the low lying $3d$ levels. In truth, the last occupied levels in the carbonylated Fe clusters present some 2π component, but it is clear that the 5σ charge donation from the ligands overcompensates the metal back donation to 2π levels. For Ru clusters the 5σ transfer is nearly equivalent to the 2π back donation, thus the metal atoms remain approximately neutral. In Os clusters we find the opposite situation as that found for Fe ones. Charge transfer from the metal to the ligands is more important resulting in positively charged Os atoms. This is the natural conclusion when we compare the relative positions of the *d* levels for the naked clusters versus the ligands 5σ and 2π bands. As the HOMO level in the naked clusters increases in energy from Fe to Os, a stronger interaction with the ligands 2π levels is expected in the series. Magnetic properties are concerned with charge transfer mechanism, as one sees from the results obtained for Ni clusters [10]. It was shown that the magnetic moment of a Ni cluster decreases when the number of coordinated CO is increased, and at the same time the charge transfer from the metal to the ligand is increased. Therefore, we would expect Fe clusters to possess distinct magnetic behavior as compared to Os carbonyls.

However, charge transfer does not seem to be the only point to influence the magnetic properties. The possibility of finding open shell structures is also determined by the energy separation between the ground state (in our calculation it is always diamagnetic) and the low lying excited paramagnetic states as well as the mechanisms, e.g., fluxionality, that could promote the molecule to such open shell structures. This analysis is out of the scope of our qualitative study. Nevertheless, we would predict an increase in the probability of reaching a high spin state due to the reduction of the HOMO-LUMO gap and the accumulation of excited states near the gap for larger clusters.

We have presented EHT calculations on the electronic structure of naked and carbonylated clusters of Fe, Ru and Os with metal nuclearity ranging from 3 to 6. Our results indicate that the ligand-metal chemical bond involves a 5σ charge transfer from CO to the metal and a back donation from the d metal band to the 2π ligands band. In Fe clusters the former process dominates (the metal core has negative net charge) while the opposite situation occurs for Os clusters (the metal core is left positively charged by the interaction with CO shell). Ru clusters appear as the intermediate case, where charge transfer from ligands to metal core nearly compensates back donation. For the three metals it is observed a reduction of the HOMO-LUMO gap with increasing nuclearity, Fe clusters having smaller gaps than the corresponding Os systems.

It is still to be investigated how these trends in charge transfer mechanism couple with the magnetic properties, where an accurate treatment of the electron-electron interaction is expected to play an important role. Nevertheless, the observed trend might indicate diverse magnetic properties for Fe, Ru, and Os carbonyl clusters of same nuclearity. In this respect it will be particularly interesting to follow the evolution of the charge transfer for high nuclearity Os clusters, e.g., decanuclear Os carbonyl which is known to present paramagnetism [9].

Financial support of the Brazilian Agencies CAPES and CNPq are gratefully acknowledged. One of us (M.C.S.) thanks the Laboratoire de Chimie Théorique Appliquée for hospitality, the National Fund for Scientific Research (Belgium), IBM Belgium and the Facultés Universitaires Notre-Dame de la Paix for the use of the Namur Scientific Computing Facility.

References

1. Chini, P.: Gazz. Chim. Ital. **109**, 225 (1979)
2. Johnson, B.F.G. (ed.): Transition Metal Clusters. New York: Wiley-Interscience 1980
3. Schmid, G.: Struct. Bonding Berlin **62**, 51–85 (1985)
4. de Jongh, L.J.: In: Organic and Inorganic Low-Dimensional Crystalline Materials, Delhaes, P., Drillon, M. (eds.) pp. 231–241. New York: Plenum 1987
5. Johnson, D.C., Benfield, R.E., Edwards, P.P., Nelson, W.J.H., Vargas, M.D.: Nature **314**, 231 (1985)
6. Teo, B.K., DiSalvo, F.J., Waszczak, J.V., Longoni, G., Cerriotti, A.: Inorg. Chem. **25**, 2262 (1986)
7. Pronk, B.J., Brom, H.B., de Jongh, L.J., Longoni, G., Cerriotti, A.: Solid State Commun. **59**, 349 (1986)
8. de Aguiar, J. Albino O., Mess, A., Darriet, J., de Jongh, L.J., Drake, S.R., Edwards, P.P., Johnson, B.F.G., Lewis, J.: Solid State Commun. **66**, 913 (1988)
9. Benfield, R.E., Edwards, P.P., Stacy, A.M.: J. Chem. Soc. Chem. Commun. **10**, 525 (1982)
10. Pacchioni, G., Fantucci, P.: Chem. Phys. Lett. **134**, 407 (1987)
11. Raatz, F., Salahub, D.R.: Surf. Sci. **176**, 219 (1986)
12. Holland, G.F., Ellis, D., Trogler, W.C.: J. Chem. Phys. **83**, 3507 (1985)
13. Fournier, R., Salahub, D.R.: Int. J. Quantum Chem. **29**, 1077 (1986)
14. Hoffmann, R.: J. Chem. Phys. **39**, 1397 (1963)
15. Sung, S-S., Hoffmann, R.: J. Am. Chem. Soc. **107**, 578 (1985)
16. Evans, D.G.: Inorg. Chem. **25**, 4602 (1986); Jorgensen, K.A., Hoffmann, R.: J. Am. Chem. Soc. **108**, 1867 (1986); Summerville, R.H., Hoffmann, R.: J. Am. Chem. Soc. **98**, 7240 (1976)
17. Plummer, E.W., Salaneck, W.R., Muller, J.S.: Phys. Rev. B **18**, 1673 (1978)

Z Phys. D – Atoms, Molecules and Clusters 12, 395–397 (1989)

Atoms, Molecules
Zeitschrift für Physik D and Clusters
© Springer-Verlag 1989

Bonding in nickel cluster carbonyls

G. Pacchioni and P. Fantucci

Dipartimento di Chimica Inorganica e Metallorganica, Centro CNR, Universita' di Milano, via Venezian 21 I-20133 Milano, Italy

Received 4 July 1988

Semiempirical (INDO) and ab initio valence only calculations of the electronic structure of Ni_n and $Ni_n\ (CO)_m$ clusters support the idea that the bonding with the CO ligands destabilizes the 4*s*-like MOs of the Ni_n fragment thus eliminating their contribution to the metal-metal bonding in the carbonylated form. This process is accompanied by a complete quenching of the magnetic moment of the bare cluster.

PACS: 36.40; 31.20.R

Introduction

Inorganic clusters with ligands have been proposed as a bridge between organometallic and surface chemistry [1]. This idea is based on the hypothesis that similar electronic mechanisms govern the bonding of small molecules on surfaces and on clusters. However, the nature of the metal-metal bonding and in particular the role of metal *d* electrons in transition-metal organometallic clusters is far from being completely understood and the assumed similarity with the bonding in bulk metals has still to be proved. Qualitative arguments [2] support the view of a metal-metal bond arising entirely from interactions involving the metal *d* orbitals in bare and in carbonylated metal clusters. It has been suggested [2a] that the role of the ligands is basically that of inducing a considerable *d-p* hybridization, thus maximizing metal-metal and metal-ligand overlaps.

Only recently it became possible to perform more rigorous quantum-mechanical studies [3] on organometallic clusters because of the difficulties connected with the treatment of transition metal atoms. These studies [3] provide first evidences of the very different metal-metal interaction in clusters with ligands (in particular CO) with respect to the bare counterparts. It is now clear that in small naked clusters of atoms of the first transition series (e.g. Fe, Co, Ni) the bonding originates mainly from 4*s*-4*s* interactions [4–6]. The 3*d* electrons belong to orbitals strongly contracted around the metal centers and characterized by small interatomic 3*d*-3*d* overlap. The appearance of typical ferromagnetic spin ordering [4–6] is the consequence of the weak coupling between unpaired electrons on different centers.

In a preliminary investigation performed at semiempirical level of theory [7] we have shown that the effect of adding CO ligands to a bare paramagnetic Ni_n cluster is that of reducing the number of unpaired electrons per atom leading to low-spin (for coordinatively unsaturated) or diamagnetic (for coordinatively saturated) $XNi_n\ (CO)_m$ clusters.

Here we present an extension of our previous INDO investigation focused on the metal-metal bond changes caused by the carbonylation. In addition, we further support our conclusions reporting the results of preliminary non empirical Effective Core Potential (ECP) calculations on Ni_3 and $Ni_3(CO)_6$ systems.

Results of semiempirical INDO calculations

Previous INDO investigations on Ni_3, Ni_5, $Ni_3(CO)_x$ and $Ni_5(CO)_y$ systems [7] have been extended to Ni_6 and $Ni_6(CO)_{12}$ clusters (Fig. 1). The details of the computational procedure are given elsewhere [7]. The

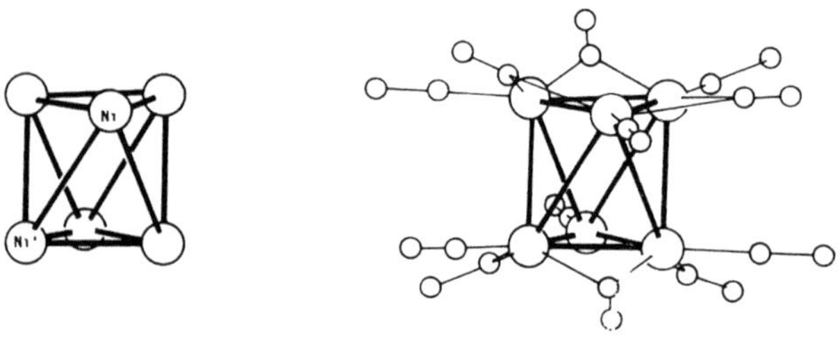

Fig. 1. Geometry of Ni_6 and $Ni_6(CO)_{12}$ clusters

geometry of the carbonylated system has been derived from the experimental X-ray structure of $[Ni_6(CO)_{12}]^{2-}$ [8]. For the bare Ni_6 cluster the same geometry of the metal core of $Ni_6(CO)_{12}$ has been assumed. We found for Ni_6 at least six states in an energy range of less than 2 eV; all these states are paramagnetic containing two to six unpaired electrons. An attempt of precisely defining the ground state of Ni_6 on the basis of a semiempirical method can be questionable. Nevertheless, it is significative that the present method gives the lowest closed shell singlet state lying at about 3.3 eV above the lowest paramagnetic state ($^3A_{2u}$). The singlet state corresponding to three $3d^{10}$ Ni atoms lies 8 eV above the ground state, i.e. much higher than the dissociation limit.

The existence of several low-lying states, a typical feature of clusters of metal atoms [9], originates from $3d \rightarrow 3d$ and $3d \rightarrow 4s$ excitations. Accurate ab initio CI calculations on Ni_3 [5] have shown the existence of four almost degenerate states which differ only for the position of the $3d$-holes, i.e. $3d\delta$ or $3d\pi$. When we consider the $Ni_6(CO)_{12}$ cluster, we found that the ground state is a $^1A_{1g}$ well separated from the lowest paramagnetic configuration ($^3A_{2u}$ ($a_{2u} \rightarrow a_{1g}$ LUMO), $\Delta E = 2.2$ eV). The a_{1g} LUMO in $Ni_6(CO)_{12}$ is the bonding combination of the 4s Ni orbitals and the $^3A_{2u}$ state arises indeed from a $3d \rightarrow 4s$ excitation. The change in the magnetic moment due to the bonding with the CO ligands reflects the electronic rearrangements occurred in the metal fragment. To follow these modifications we adopted three criteria (Table 1): *a)* the average atomic Ni population, *b)* the values of the Wiberg indices [10] (the sum of the squares of the elements in the interatomic blocks of the one-particle density matrix, roughly proportional to the bond order) and *c)* the Ni-Ni two-center energy contribution obtained by partitioning the total energy of the system as sum of one- and two-center terms: $E_{tot} = \Sigma_a E_a + \Sigma'_{ab} E_{ab}$.

The effect of the carbonylation of the Ni_6 cluster is that to change the atomic Ni configuration from $4s^{0.3}$ $3d^{9.7}$ (in the triplet state) to $4s^{0.0}$ $3d^{9.4}$. The reduced 4s population is due to the destabilization of the 4s-derived MOs caused by repulsion of the 4s and CO 5σ charge distributions (Pauli repulsion) [11], with the consequent promotion of the Ni electrons from 4s to 3d. The 4s-4s contribution to the metal-metal bonding is therefore completely removed in the carbonylated cluster. All the fifteen 3d-like MOs are doubly occupied in the $Ni_6(CO)_{12}$ complex, but the Ni 3d population is only 9.4 because of the metal-CO back-donation mechanism. The electronic redistribution following the interaction with the CO ligands has a direct effect on the strength of the metal-metal bond. This is shown by the values of the two-center contributions to the total energy, E_{Ni-Ni} and by the Wiberg indices: both these quantities decrease significantly on going from the naked to the carbonylated form (Table 1).

Table 1. Analysis of the strength of the metal-metal bond in Ni_6 and $Ni_6(CO)_{12}$ clusters (INDO results)

	$Ni_6(D_{3h})$	$Ni_6(CO)_{12}(D_{3h})$
Ground state	$^3A_{2u}$	$^1A_{1g}$
Total energy (au)	−233.5032	−492.6723
De/n (eV)	1.15[a]	1.43[b]
average Ni configuration	$4s^{0.33}\,3d^{9.66}$	$4s^{0.03}\,3d^{9.38}$
E_{Ni-Ni} (au)	−0.241	−0.133
$E_{Ni-Ni'}$ (au)	−0.093	−0.028
Wiberg index Ni–Ni	0.229	0.026
Wiberg index Ni–Ni′	0.114	0.005

[a] Defined as $[6E(Ni)-E(Ni_6)]/6$
[b] Defined as $[(12E(CO)+E(Ni_6))-E(Ni_6(CO)_{12})]/12$

Results of ECP calculations

These conclusions, deduced from semiempirical calculations, need to be substantiated by more rigorous quantum-mechanical approaches. For this reason we performed ab initio ECP Hartree-Fock calculations on Ni_3 and $Ni_3(CO)_6$ clusters. The calculations use the non-relativistic ECP developed by Hay and Wadt [12] which replaces the Ne core but leaves the 3s, 3p, 3d, 4s and 4p as valence orbitals. The Ni basis set is the same as reported by Hay and Wadt [12], except that the most diffuse p function has been eliminated because it originates nearly-linear dependency problems when combined with the analogous functions on different Ni atoms. The basis for Ni was contracted to [3s2p2d]. C and O atoms where treated at all electron level adopting a minimal basis set of Gaussian type orbitals [7s3p/2s1p] [13].

The Ni_3 was considered in a fixed equilateral triangular geometry with r(Ni-Ni) = 2.38 Å, the same distance observed in the $Ni_3(CO)_6$ subunit of the $[Ni_6(CO)_{12}]^{2-}$ cluster [8]. As discussed by Basch et al. [4] and more recently by Walch [5], the dominant configurations for this geometry arise from three Ni atoms in $3d^9\ 4s^1$ configuration. The 4s-electrons distribution is $4s(a_1')^2\ 4s(e')^1$ and the lowest configuration is a quintet with three unpaired electrons localized in the 3d-shell. We also find that a Ni $3d^9\ 4s^1$ quintet with localized 3d-holes is favoured, but we encounter convergency problems when we try to calculate the $4s(a_1')^2$ $4s(e')^1$ state and we only succeed to calculate the energy of the $4s(a_1')^2\ 4p(a_2'')^1$ configuration, where the out-of-plane combination of Ni $4p_z$ orbitals is occupied instead of the $4s(e')$ MO. The total energy of this configuration is higher than the sum of the energies of three separated Ni atoms (− 504.54276 a.u.) even when the Ni-Ni distance is optimized ($E(Ni_3) = -504.53163$ a.u. for $r_e = 2.83$ Å); the average Ni atomic configuration is $(3s, 4s)^{2.50}\ (3p, 4p)^{6.50}\ (3d)^{9.0}$.

When we compute the $Ni_3(CO)_6$ cluster we find that the lowest state is a singlet bound by 24.2 kcal/mol with respect to 3 Ni atoms and 6 CO molecules. The average Ni atomic population in this case is $(3s, 4s)^{1.96}\ (3p, 4p)^{6.63}\ (3d)^{9.06}$, with a net charge + 0.35 on each Ni atom. With respect to the analogous INDO calculation on $Ni_3(CO)_6$ [7] we observe the same tendency toward a destabilization of the 4s-like levels (which are empty in the carbonylated form), the presence of a set of completely filled 3d-like MOs, a large donation of charge from the metal to the CO ligands (the INDO net charge on Ni is + 0.38), but a different population of the 4p Ni orbitals. The ECP calculation suggests that the 4s-4p hybridization becomes important upon formation of the carbonylated complex, a feature not present in INDO calculations because an artificial destabilization of the 4p atomic orbitals was introduced in the parametrization. However, it is apparent that the 4p orbital contribution to the M-CO bonding is not small and a new INDO parametrization accounting for this behaviour is planned. The involvement of the 4p orbital is of great importance when the formation of the dianions is considered because of the competition between $4s(a_1')$ and $4p_z(a_2'')$ MOs as orbital acceptors on the neutral fragments [14].

Conclusions

ECP calculations on $Ni_3(CO)_6$ have confirmed the general tendency deduced from qualitative semiempirical investigations, i.e., the bonding with the CO ligands destabilizes the 4s-derived MOs of the Ni_n fragment thus eliminating their bonding contribution to the metal-metal interaction. The reduced strength of the metal-metal bond is shown by the decreased bond order (as measured by the Wiberg indices) and Ni–Ni two-center energy contributions to the total energy of the systems. A third, more direct evidence of a weak metal-metal bonding should come from a comparison of the total electron densities in Ni_3 and $Ni_3(CO)_6$ obtained from ECP calculations; further work is planned in this direction. The existence of regions of very low electron density in correspondence of the metal-metal bond axis has been already suggested for the $Fe_2(CO)_9$ dinuclear complex [15].

References

1. Muetterties, E.L., Rhodin, T.N., Band, E., Bruker, C.F., Pretzer, W.: Chem. Rev. **79**, 91 (1979)
2. a) Evans, D.G.: Inorg. Chem. **25**, 4602 (1986)
   b) Woolley, G. Inorg. Chem. **27**, 430 (1988)
3. Holland, G.F., Ellis, D.E., Trogler, W.C.: J. Chem. Phys. **83**, 3507 (1985)
4. Basch, H., Newton, M.D., Moskowitz, J.W.: J. Chem. Phys. **73**, 4492 (1980)
5. Walch, S.P.: J. Chem. Phys. **86**, 5082 (1987)
6. Lee, K., Callaway, J., Dahr, S.: Phys. Rev. B **30**, 1724 (1984)
7. Pacchioni, G., Fantucci, P.: Chem. Phys. Lett. **134**, 407 (1987)
8. Calabrese, J.L., Dahl, L.F., Cavalieri, A., Chini, P., Longoni, G., Martinengo, S.: J. Am. Chem. Soc. **96**, 2616 (1974)
9. Koutecky, J., Fantucci, P.: Chem. Rev. **86**, 539 (1986)
10. Wiberg, K.W.: Tetrahedron **24**, 1083 (1968)
11. Bagus, P.S., Nelin, C.J., Bauschlicher, C.: Phys. Rev. B **28**, 5423 (1983)
12. Hay, P.J., Wadt, W.R.: J. Chem. Phys. **82**, 299 (1985)
13. Van Duijneveldt, F.: IBM Tech. Report RJ 945, 1971
14. Underwood, D.J., Hoffmann, R., Tatsumi, K., Nakamura, A., Yanamoto, Y.: J. Am. Chem. Soc. **107**, 5968 (1985)
15. MacDougall, P.J.: Ph.D. Thesis, McMaster University, Hamilton (Canada), 1989.

Z. Phys. D – Atoms, Molecules and Clusters 12, 399–400 (1989)

Atoms, Molecules
Zeitschrift für Physik D and Clusters
© Springer-Verlag 1989

NMR in aluminum small particles in high magnetic field

S. Kobayashi, K. Wago, and F. Komori
Department of Physics, Faculty of Science, University of Tokyo, 7-3-1 Hongo, Bunkyo-ku, Tokyo, Japan 113

Received 5 July 1988; final version 30 September 1988

The Knight shift of ^{27}Al in small Al particles was measured in wide ranges of temperature and magnetic field, in order to study the magnetic field dependence of the universarity class of random hamiltonian governing the conduction electrons. The samples with diameters 80 A and 27 A were investigated. We observed the change in the symmetry of the hamiltonian in high magnetic field as the recovery of the spin paramagnetic susceptibility in 80 A particles.

PACS: 75.20

The energy levels of conduction electrons in small particles are correlated according to the universality classes of the hamiltonian. For light metals as Al, the universality class is orthogonal and the distribution of levels in represented by well known Wigner's distribution function which has a peak around the mean value of the level spacing. Because of this peak, thermodynamic properties behave as activation type, and consequently the spin paramagnetic susceptibility in the particles with even electrons die out at temperatures lower than the mean level spacing [1].

However, the level correlation works within a single spin band. If the up and down spin bands are shifted oppositely by magnetic field so much that the Zeeman energy exceeds the mean level spacing, the correlation between the levels of opposite spins near the Fermi level will be lost, and the level distribution becomes Poisson type, although the correlation within the spin subbands is retained.

We previously reported the NMR measurement of the spin susceptibility in small particles of Cu in high magnetic fields. We observed a recovery of the Knight shift with the increase of the magnetic field. But in case of copper particles, it was rather hard to observe the clear recovery because the spin-orbit interaction is so large that the residual shift is large even in low magnetic field [2]. At the same time in Cu, the spin states are no more the good eigenstates, and the levels are correlated even in high fields.

In the present work, we used the aluminum small particles, in which the spin-orbit interaction is negligibly small. The only complication is that the aluminum metal exhibits the superconductivity. However, the range of magnetic field interesting for the present study is high enough to suppress superconductivity completely. Thus we can discuss the level correlation effect being independent of the superconductivity.

We measured the NMR Knight shift of Al particles in the vacuum-evaporated samples. This method had not been adopted so far for aluminum because it forms continuous film even with average thickness as small as 10 A. We overcame this difficulty by increasing the temperature of a substrate up to 200 C. The samples were formed on SiO film supported on Capton film in the vacuum of about 1×10^{-6} Torr. We repeated the evaporation of SiO and aluminum many times to obtain sufficient amount of aluminum particles for the NMR measurements. Two samples were prepared. The distribution of the particle size was measured on electron micrograph of a specimen which was prepared on thin carbon film with the same condition for the NMR sample. The measured diameter and the average thickness of the aluminum film suggested that the particles are in oblate shape. We prepared two samples with mean diameter of 27 A (Sample A) and 80 A (Sample B) in which the mean level spacings are estimated to be 65 K and 2.5 K respectively. The range of temperature and magnetic field were 0.5–1.5 K and

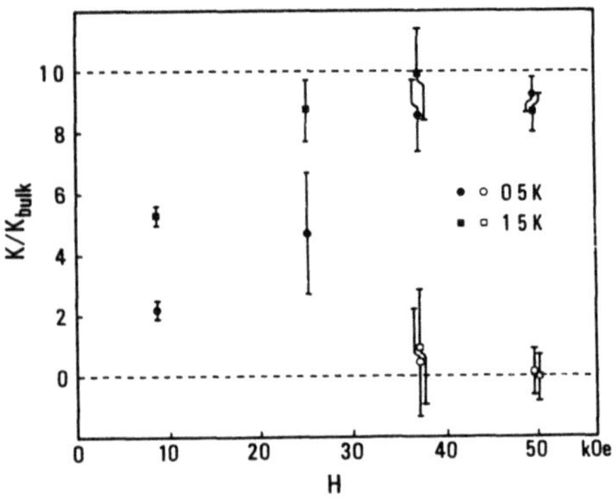

Fig. 1. Magnetic field dependence of the Knight shift of Al Sample B (d = 80A) normalized to the bulk value in normal state. Signals shown with open symbols are assigned to those from aluminum oxide. In lower fields, metallic and oxide signals are not resolved

8.7–49 kOe. The critical magnetic field of superconductivity due to the paramagnetic pair breaking mechanism in these particles is expected to be about 22 kOe [3].

In Sample A, the Knight shift is almost zero and independent of the temperature and the field. We attribute the absence of the shift in the field higher than 22 kOe to the level discreteness because the Zeeman energy in the highest field is still too small to destroy the level correlation. In the field lower than 22 kOe, both the superconductivity and the level discreteness kill the shift. The signal from oxide, if exists, overlaps at the same zero shift position. The signal from particles with odd electrons is supposed to spread widely with large shifts.

On the other hand, two NMR lines were observed in Sample B. One line has zero Knight shift in all measured fields, and the other has the field dependent Knight shift. The field dependence of the Knight shift of two lines at 0.5 K and 1.5 K are shown in Fig. 1. We assign the line which depends on the field to the NMR signal from the metallic part of the particles (metallic line), and the one which is independent of the field to the one from the oxide on the surface of the particles. The reasons for this assignment are as follows. There is no choice for the field dependent one to be assigned to the metallic signal. The field independent one has two possibilities. One is that from superconducting particles and the other is that from insulating compound. However, the highest field is sufficiently higher than the superconducting critical field which was experimentally confirmed [3], and therefore no superconducting aluminum can survive. Moreover, the line width is considerably narrower for the field independent line, indicating the absence of the second order quadrupole broadening arising from the Friedel oscillation of conduction electrons.

At the lowest temperature the Knight shift of the metallic line is almost zero in 8.7 kOe and very close to the bulk value in 49 kOe, in accordance with the above expectation, that the level correlation will be lost in high magnetic fields. The Zeeman energy in 49 kOe is about 6.4 K while the mean level spacing in 80 A Al is about 2.5 K.

The signal from particles with odd electrons cannot be detected in low fields because it is expected to be very broad due to its Curie like nature of the spin susceptibility. But in high fields, it surely merges into the observed metallic line because the distinction between odd and even loses its sense.

At 1.5 K, the shift of the metallic line is larger than those at 0.5 K. This increase of the shift can be explained as the combined effects of level discreteness with the mean level spacing of 2.5 K, and the superconductivity with the transition temperature of 1.2 K, both of which are moderated as the temperature is raised to 1.5 K, although the superconductivity is less effective in high fields.

We did not expected the oxides in our samples because they were prepared in vacuum. The intensity ratio of metallic line to the insulating one in Sample B is roughly 3:1. An only possible explanation is that our vacuum is not high enough to prevent the surface oxidation. This should be checked by improving the vacuum system.

References

1. Halperin, W.P.: Rev. Mod. Phys. **58**, 533 (1986)
2. Kobayashi, S., Katsumoto, K.: J. Phys. Soc. Jpn. **56**, 2256 (1987)
3. Nomura, K., Kobayashi, S., Sasaki, W.: J. Phys. Soc. Jpn. **48**, 37 (1980); The data of magnetic field dependence of the Knight shift given in this reference (Fig. 1) are very much different to the present ones. The difference arises from the difference in the NMR methods. The present one is a steady state method, while in [3], they used a pulse method. By the latter, a metallic line and a line of oxide cannot be resolved, giving an incorrect value of the Knight shift. In addition, the samples in ref. [3] are prepared by vacuum evaporation method and inevitably contain more oxide than the present samples. The detailed discussion will be given elsewhere (to be submitted to J. Phys. Soc. Jpn.).

Z. Phys. D – Atoms, Molecules and Clusters 12, 401–403 (1989)

Atoms, Molecules
Zeitschrift für Physik D and Clusters
© Springer-Verlag 1989

Size dependence of the paramagnetic susceptibility of Mg and Zn clusters

S. Pasche and J.-P. Borel
Institut de Physique Expérimentale, Ecole Polytechnique Fédérale de Lausanne, CH-1015 Lausanne, Switzerland

Received 5 July 1988; final version 30 September 1988

The paramagnetic susceptibility of small divalent clusters (Mg and Zn) trapped in Xe and Adamantane matrices has been studied with EPR spectroscopy. This study represents an observation of the transition between the atomic behaviour and the extended solid. The signal of clusters >15 Å as a function of temperatures shows that their susceptibility has a Curie-low behaviour. These results are contrary to the quantum size effects models of Kubo [1] and Mühlschlegel [2].

PACS: 36.40; 33.35.Ex

Introduction

The spin susceptibility of small metallic particles is strongly dependent on the distribution of the conduction electron energy levels. Several models have been reported in order to calculate the partition function from which the spin susceptibility is deduced. First the quantum size effects model predicted by Kubo [1] and Mühlschlegel [2] suggest an equal energy level spacing. The average level spacing near Fermi surface is given by $\delta = 1/\mathscr{D}(\varepsilon_F)$ where $\mathscr{D}(\varepsilon_F)$ is the electronic density of states. This model according to Denton et al. [2] predicts for $kT < \delta$ a vanishing susceptibility when $T \to 0$ for even particles, and a Curie-law behaviour for odd ones.

Contrary to this model, more recent calculations proposed by Bucher and van der Klink [3] using a spherical jellium model show that χ varies as $1/T$, for odd and even particles with an unfilled shell.

Other studies of the polarizability on sodium reported by Knight, de Heer and Saunders [4] are in agreement with the ellipsoïdal jellium model elaborated by Clemenger [5].

Having so different models, this work on the paramagnetic susceptibility, can be considered as an important element, to confirm or invalidate one or another of these theories.

Experiments

In order to measure the susceptibility of even particles, we have chosen two pure divalent metals: Mg and Zn (60 *N*).

The particles are evaporated at 450°C and 550°C in a good conventional vacuum (10^{-7} Torr). The nucleation of clusters is produced in a flow of Xe gas at room temperature. The clusters are respectively isolated in a Xe and Adamantane ($C_{10}H_{16}$) matrix at 60 K. The sample is then collected in a quartz tube, in the same apparatus.

The quartz tubes are supplied with a special quartz-Pyrex-Pyrex-Copper connection. This allows to crimp the tubes instead of welding them. Such a technique avoids the usual temperature elevation of the samples.

The paramagnetic susceptibility is measured in a Varian spectrometer E102 in a continuous helium flow cryostat. The size distribution of the particles is measured by electron-microscopy using Philips EM 300 and EM 430 microscopes.

Results

The ESR measurements of the Zn clusters susceptibility show that χ is strongly dependent on size. In Fig. 1

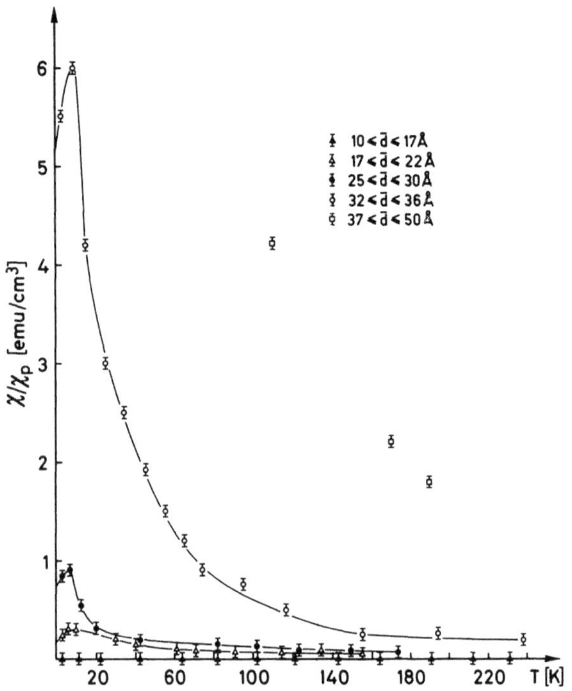

Fig. 1. Normalized susceptibility of small Zn clusters as a function of temperature

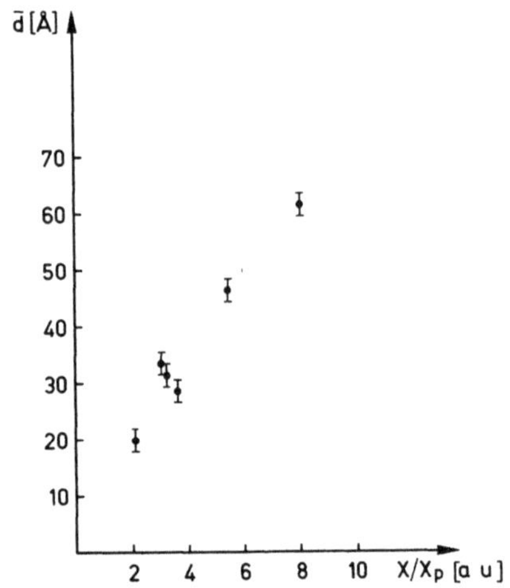

Fig. 2. Magnetic susceptibility as a function of average size diameter

the susceptibility has been normalized to χ_{Pauli} as a function of T. We find here for small particles, no ESR signal is observed up to room temperature.

When *d* (*d* is the average size diameter) increases, the ESR signal increases χ varies as 1/T. Furthermore, at high temperatures, χ remains less than χ_{Pauli}, and $\chi/\chi_{\mathrm{Pauli}}$ tends to 1 with increasing size particles.

To put this relationship into evidence we have ploted in Fig. 2 χ/χ_{P} as a function of the average diameter. Although our results are approximate (χ_{p} is different from the absolute χ_{Pauli}. χ_{p} represents the constant paramagnetic susceptibility observed at high temperature); it is clear that χ is proportional to particles size.

This behaviour is not expected from Kubo like models; however it can be explained by an ellipsoïdal shell model.

(1) For small particles, the degeneracy is lifted by the geometrical distortion. The energy level spacing is much larger than kT, resulting in the fact that no susceptibility is observed in the region of 0 to 200 K. For very small particles, one expects temperatures at least of order 400 K, before any susceptibility could be observed.

(2) For the larger particles, the shape is approximately spherical, and the energy levels are degenerate at Fermi-level. This results a Curie-law behaviour of χ. However small irregularities of the surface, slightly lift the degeneracy implying a decreasing χ at low temperatures.

Similar results have been observed on Mg clusters trapped in a Xe matrix (Fig. 3). In a previous paper [7] we have suggested that Curie-law behaviour of the susceptibility could be due to electrons transitions in the matrix. Taking into account the differences of the Pauling electronegativity (1.6 for Zn, and 1.2 for Mg) and the different matrices employed; Fig. 3 shows that the χ behaviour is independent of the choice of metal and matrix.

Oxidation effects

In order to get sure that our results are not due to any contamination effects, we have deliberately introduced

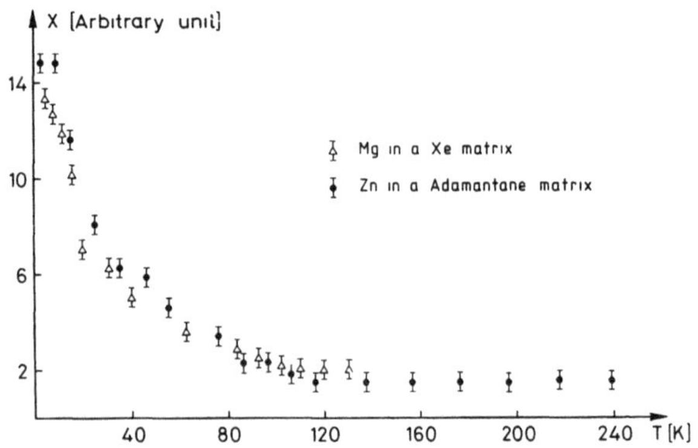

Fig. 3. Susceptibility of Mg and Zn clusters, trapped resp. in a Xe and $C_{10}H_{16}$ matrix. The average size diameter is 32 Å for Zn, and 35 Å for Mg

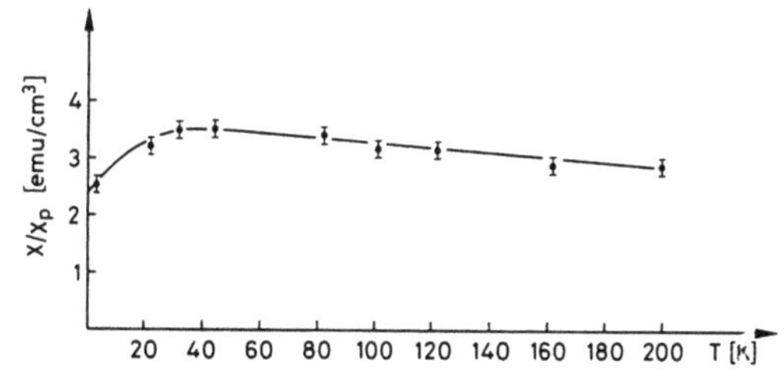

Fig. 4. Normalized paramagnetic susceptibility of Zn clusters, contaminated with 20% of oxygen. The average size diameter $d = 27$ Å

20% of pure oxygen, in the Xe flow necessary for the nucleation. The ESR measurements (Fig. 4) are quite different of the results obtained in (Fig. 1). The susceptibility increases up to 30 K and then becomes nearly constant for higher temperatures. In previous experiments on Mg clusters [6], such contamination effects may have lead to a misinterpretation of the results. The χ behaviour under oxidation conditions looks similar to the one predicted by the quantum size effect. In our experiments oxidation effects are eliminated by evaporating, collecting, and sealing the samples within the same UHV (10^{-7} torr) chamber.

Conclusion

Our study on Mg and Zn clusters, show that the paramagnetic susceptibility depend strongly on size and shape of the particles. It varies like a Curie-law for particles > 17 Å, while for very small ones no ESR signal is observed. These results can be qualitatively explained by an ellipsoïdal jellium model, where the energy levels are degenerate at Fermi-level in larger particles, but with decreasing cluster size, ellipsoïdal distorsion lift the degenerecy.

We would like to thank Dr. de Heer for his invaluable discussions, and Dr. J.J. van der Klink for his suggestions. This work is partially supported by the Swiss National Found.

References

1. Kubo, R.: J. Phys. Soc. Jpn **17**, 973 (1962)
2. Denton, R., Mühlschlegel, B., Scalapino, D.J.: Phys. Rev. B **7**, 3589 (1973)
3. Bucher, J.P., van der Klink, J.J.: Solid State Comm. **65**, 27 (1988)
4. Knight, W.D., Clemenger, K., de Heer, W.A., Saunders, W.A.: Phys. Rev. B **31**, 2539 (1985)
5. Clemenger, K.: Phys. Rev. B **32**, 1359 (1985)
6. Kimura, K., Bandow, S.: Phys. Rev. B **37**, 4473 (1988)
7. Pasche, S., Borel, J.P.: Solid State Commun. **58**, 865 (1986)

Z. Phys. D – Atoms, Molecules and Clusters 12, 405–409 (1989)

Atoms, Molecules
Zeitschrift für Physik D and Clusters
© Springer-Verlag 1989

Photoemission from tin and lead cluster anions

G. Ganteför, M. Gausa, K.H. Meiwes-Broer, and H.O. Lutz
Fakultät für Physik, Universität Bielefeld, D-4800 Bielefeld 1, Federal Republic of Germany

Received 18 August 1988

Photoelectrons from mass-identified jet-cooled tin and lead cluster anions (Sn_n^-, Pb_n^-) are detached by ultraviolet laser light ($h\nu = 3.68$ eV). The photoelectron energy spectra give the detachment energies of ground state cluster anions (electron affinities) as well as excitation energies of neutral clusters in the geometry of the anions. The energy spectra for Sn_n^- are dominated by flat thresholds with an n-dependence similar to that of other group IV clusters. In contrast, for Pb_n^- we find pronounced narrow lines close to threshold, generally followed by a 0.3–1.4 eV gap which indicates closed-shell behaviour of Pb_n^- for nearly all n.

PACS: 36.40.+d; 33.60.Cv; 79.60.−i

I. Introduction

The physics of unsupported clusters has overcome the phase where mainly mass spectra and their – often speculative – interpretations were of main interest. With the development of intense ion and anion sources it is now possible to perform chemical or spectroscopic investigations on mass selected isolated clusters in a beam [1, 2]. When dealing with metal clusters classical spectroscopic methods like flame or absorption spectroscopy, laser-induced fluorescence or resonant two-photon ionization are mostly sentenced to fail since in general electronic excitation energies are redistributed into vibrational degrees of freedom before light emission would occur or before a second photon can ionize the cluster. This rapid energy redistribution, as well as the low densities of free mass-selected cluster beams, have reduced much of the general instrumentarium of atomic, molecular and surface physics in its significance for cluster research. One technique which has recently been shown to remain a powerful tool is photoelectron spectroscopy (PES). Within a short time a large amount of photoelectron data of mainly metal or group IV clusters have become available (see, e.g., [3–6]). Although theory is lacking behind the experimental progress a number of physical properties can already be extracted from such PES data. In this paper PE-spectra of tin and lead clusters will be presented; although the elements Sn and Pb are chemically similar, PES reveals the different nature of these group IV clusters.

II. Experimental

The experimental set-up has been discussed elsewhere [5, 6]. In short, metal cluster anions are produced in a laser vaporization source without an additional ionization agent. After cooling in a (pulsed) supersonic jet the anions are mass-separated in a time-of-flight (TOF) mass spectrometer collinearly to the source direction. Under ultrahigh vacuum conditions fixed-frequency light of an ultraviolet laser detaches electrons from mass-identified anion bunches. The electron energies are analyzed by recording their flight times in a magnetic-bottle TOF electron spectrometer [7] perpendicular to the cluster beam direction. The energy resolution $\Delta E/E$ depends on a variety of experimental parameters, in particular the magnetic field configuration, the ionization volume and the anion's velocity (see [5]). In the spectra presented here $\frac{\Delta E}{E} \approx 0.05$ was achieved. Monomer anion photoelectron spectra serve to calibrate the electron spectrometer.

III. Results and discussion

The cluster ion beam composition is studied by TOF mass analysis yielding broad, uniform intensity distributions for (positively and negatively) charged metal clusters [5, 6, 8]. Figure 1a gives as example the mass spectrum of Sn_n^- where no intensity anomalies appear. Such structureless spectra can be produced for all metals we studied so far; we conclude that the source operates at aggregation conditions where the cluster ion growth is not determined by differences in cluster binding energies. A change of critical source parameters (e.g., detuning of the laser and source pulse overlap) can lead to more structured spectra. Figure 1b gives as example the mass spectrum of Pb_n^- where Pb_7^- and Pb_{10}^- appear with enhanced intensity. In addition, an intensity step is observed at Pb_{15}^-. Further detuning will lead to an enhancement of these features and to an overall decrease of cluster ion intensity. The appearance of irregularities might originate from cluster growth where the partial pressure of the He seeding gas during aggregation is too low for an effective transport of the heat of condensation. In this case

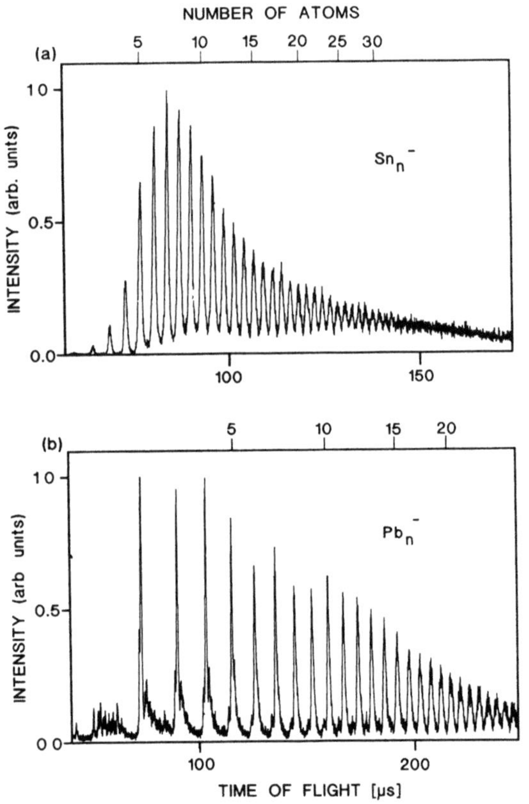

Fig. 1a and b. Time-of-flight mass spectra of jet-cooled Sn_n^- (*upper trace*) and Pb_n^- (*lower trace*). The anions are produced in a laser vaporization source (without additional ionization) and analyzed collinearly to the source direction. For Pb_n^- the source parameters are slightly detuned in order to produce "magic numbers"

condensation heats the clusters and can lead to a decomposition before sufficient stabilization in collisions with He would occur. "Magic numbers" result, reflecting differences in cluster binding energies. It is interesting to note that the pronounced lines in Pb_n^- spectrum at $n=7$, 10, 15 also appear at the same positions in Pb_n^+ spectra after electron impact ionization [9] or photoionization [10]. Therefore, simple electron counting rules or the jellium model cannot explain these numbers. Other source operation conditions may produce other pronounced lines, e.g., Pb_{12}^-. Negatively charged 13-atom clusters, in contrast to positive ones, do not appear with enhanced intensity at all. This finding sheds some doubt on the interpretation based on positive cluster abundances according to which Pb_{13} is built in an icosahedral structure. Instead, the close-packing model should be extended by detailed quantum-chemical considerations in which the influence of electronic symmetries is included.

To investigate the electronic structure of the cluster anions, photoelectrons have been detached by pulsed laser light ($h\nu = 3.68$ eV). Resulting electron flight times are converted numerically into electron binding energies. Figure 2a, b gives such energy distribution curves (EDC) of photoemitted electrons for tin and lead cluster anions. All cluster anions detach at photon energies of <3 eV. For each cluster a characteristic EDC is obtained, starting with a more or less pronounced threshold. Such spectra reflect the distribution of transitions from the anionic ground states into the ground states of the neutralized clusters (in the geometry of the anions) or into their excited states: Distinct narrow lines near threshold (see Table 1) are interpreted as vertical transitions into the respective neutral electronic ground state (electron affinities); peaks at the high energy sides of the first lines belong to excitations into higher electronic neutral states, i.e. the energetic positions of the electronic states are revealed directly by photoelectron spectroscopy. In contrast to optical spectroscopy, however, a larger amount of states is accessible since the only important "selection rule" is the change of the spin multiplicity by one.

Most of the **tin** cluster EDCs do not show narrow peaks near threshold. Therefore, PE-thresholds (as marked by bars in Fig. 2, see also Table 1) are rough estimates of tin electron affinities. A similar threshold behaviour is observed also for a number of other metal clusters [5, 6]. One reason might be a strong change in geometry upon ionization resulting in a poor overlap between low vibrational states. This would lead to a heating of the neutralized cluster and perhaps to its dissociation. Although anion photofragmentation might be possible in some cases [11] we do not expect such a strong change in geometry upon neutralization

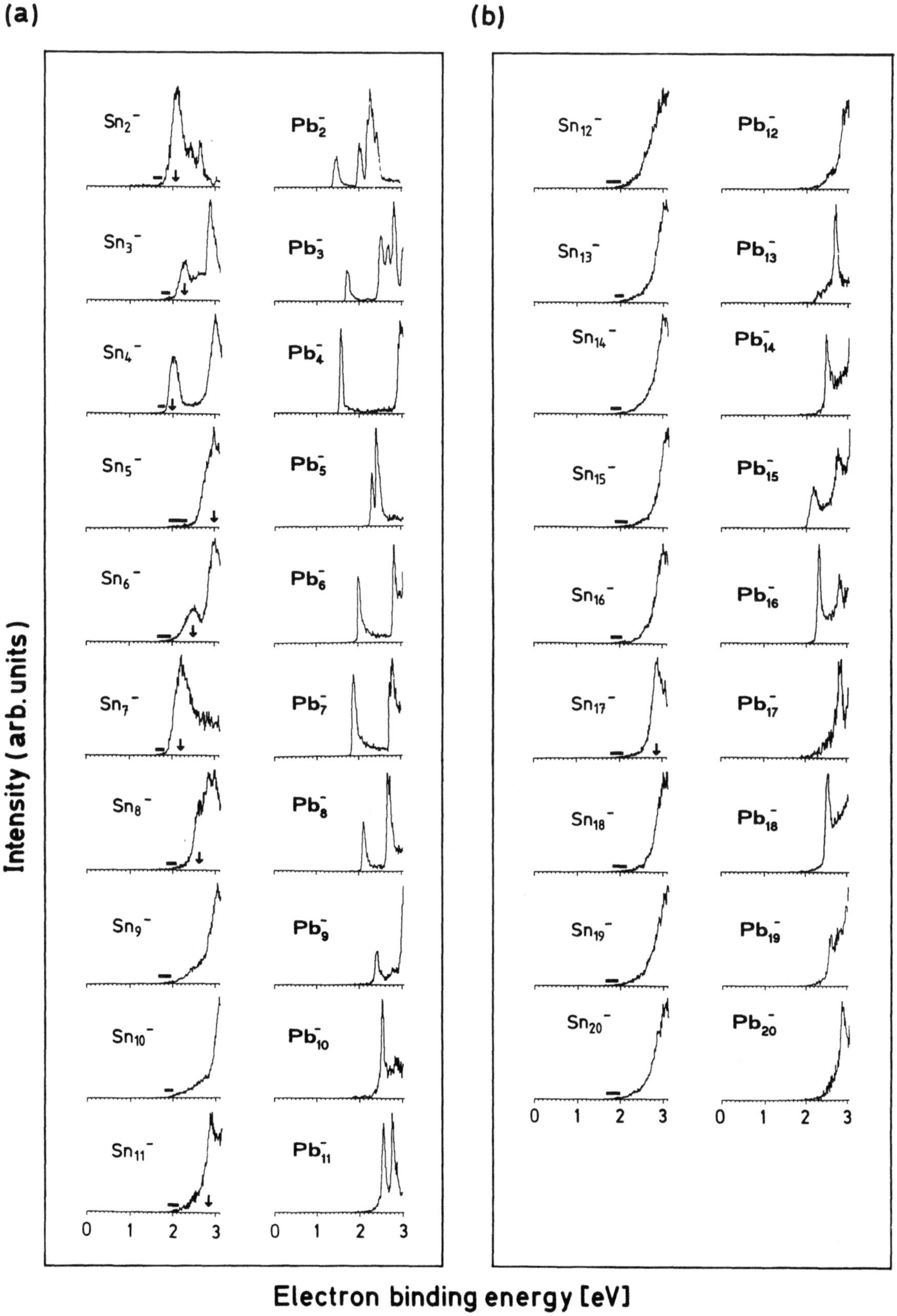

Fig. 2a and b. Photoelectron energy spectra of tin and lead cluster anions at a detachment energy of $h\nu = 3.68$ eV. At this photon energy no background correction is necessary. For Sn_n^- the thresholds are marked by horizontal bars. Pb_n^- show pronounced peaks near threshold

Table 1. Photoelectron threshold energies (for Sn_n^-) and first maximum positions (for Pb_n^-) as estimates of cluster electron affinities (in eV)

n	2	3	4	5	6	7	8	9	10	11
Sn_n^- threshold	1.7	1.8	1.7	2.2	1.8	1.8	2.0	1.8	1.9	2.0
Pb_n^- 1st maximum	1.45	1.70	1.55	2.35	2.05	1.95	2.15	2.45	2.55	2.55
n	12	13	14	15	16	17	18	19	20	
Sn_n^- threshold	1.9	2.0	1.9	2.1	2.0	2.0	2.1	1.9	1.9	
Pb_n^- 1st maximum	(2.6)	2.3	2.50	2.2	2.30	(2.85)	(2.55)	2.6	(2.6)	

Table 2. Excitation energies E of neutral lead cluster in the geometry of the anions, relative to the neutral ground state

n	2	3	4	5	6		
E (eV)	0.55, 0.80, > 1.5	0.80, 0.95, 1.10, ≥ 1.30	1.40	0.10, > 0.70	0.80, > 1.0		
n	7	8	9	10	11	12	13
E (eV)	0.90, > 1.05	0.60, > 0.80	≥ 0.60	> 0.50	0.20, > 0.45	(0.3)	0.4, > 0.80
n	14	15	16	17	18	19	20
E (eV)	≥ 0.55	0.60, ≥ 0.90	0.50, > 0.70	(> 0.20)	(> 0.45)	(≥ 0.40)	(> 0.15)

for most of the clusters. Strong isomerization is another possible origin for broad lines and flat thresholds as it had been demonstrated recently with alkali-oxide clusters [12]. Different source conditions should alter the degree of isomerization. Up to now, however, we found no significant change of the PES upon source parameter adjustments.

Comparing the Sn_n data with EDCs of silicon and germanium, $n \leq 12$ [13], we find similar spectra within the range of overlap. The observed photodetachment thresholds of Si_n^- and Ge_n^- are essentially the same, whereas the values of Sn_n^- are shifted to lower binding energies by about 0.5 eV (see Table 1). Only for $n = 10$ we find a strong qualitative difference: The Si_{10}^- spectrum has a low-energy hump which – according to the jellium model [14–16] – shows that neutral Si_{10} is a closed shell cluster; in Ge_{10}^- such a feature is missing. Also Sn_{10}^- (see Fig. 2) has no hump indicating that neutral Sn_{10} is an open shell cluster, in contrast to Si_{10}.

Photoelectron spectra from **lead** cluster anions, on the other hand, are different to those of Si_n^-, Ge_n^-, and Sn_n^-. For most n we find a defined threshold with one or two narrow lines followed by a distinct gap. The energetic positions of neutral excited states (see Table 2) vary irregularly with n. The existence of a gap indicates that the corresponding neutral cluster is a closed shell species with well separated electronically excited states; in view of this, Pb_n are closed shell clusters for nearly all n observed so far. The narrow lines clearly indicate that isomerization in lead clusters is weak or absent. The gap widths are large and reach 1.4 eV in the case of Pb_4^-. This is the largest gap we observed so far; it is even larger than that of Ag_8^- [6] which – as an alkali-like cluster with nine valence electrons – has one excess electron after a shell closing according to the jellium model. The huge gap for lead at $n = 4$ and the low PE-threshold could be interpreted as an $1p$ shell closing. In this case each s^2p^2-atom gives two electrons to the valence band. A similar behaviour would then also be expected for $n = 10$ ($2s$-closing). The width of the Pb_{10}^- band gap cannot be determined as the employed photon energy is too low. The electron affinity, on the other hand, is relatively high, in contrast to the expectation of the jellium model. Instead, Pb_{15}^- has a low electron affinity with a pronounced gap indicating that this cluster appears in a special closed-shell configuration with high symmetry. As noted earlier, in the mass spectra of Pb_n^- the line at $n = 15$ can be produced with enhanced intensity. We conclude that the simple jellium picture, which has been modelled for monovalent metal clusters, does not work reliably for lead clusters. Instead, it appears to be necessary to consider also the influence of the cluster's geometry.

A complete discussion of the Sn_n and Pb_n photoelectron spectra goes beyond the scope of this communication. In conclusion, we find a strong similarity between semiconductor and tin clusters which may be a hint that Sn_n are not free-electron-like species. Bulk electronic bands of group IV diamond structure species (i.e., Si, Ge, Sn) are nearly identical with a decreasing band gap width going from Si to Ge and Sn. Bulk lead, on the other hand, is a free-electron-like metal; its photoelectron spectrum reveals two *p*-bands near the Fermi level and an *s*-band at higher binding energies [17]. The cluster photoelectron spectra have to evolve to these bands.

References

1. Cf. Physics and chemistry of small clusters. NATO ASI Ser. B **158**, 755 (1987)
2. Cf. Elemental and molecular clusters. Springer Series in Materials Science, Vol. 6. Berlin, Heidelberg, New York: Springer 1988
3. Leopold, D.G., Ho, J.H., Lineberger, W.C.: J. Chem. Phys. **86**, 1715 (1987)
4. Pettiette, C.L., Yang, S.H., Craycraft, M.J., Conceicao, J., Laaksonen, R.T., Cheshnovsky, O., Smalley, R.E.: J. Chem. Phys. **88**, 5377 (1988)
5. Ganteför, G., Gausa, M., Meiwes-Broer, K.H., Lutz, H.O.: Z. Phys. D–Atoms, Molecules and Clusters **9**, 253 (1988)
6. Ganteför, G., Gausa, M., Meiwes-Broer, K.H., Lutz, H.O.: Faraday Discuss. Chem. Soc. **86**, (1989)
7. Kruit, P., Read, F.H.: J. Phys. E **16**, 313 (1983)
8. Begemann, W., Dreihöfer, S., Ganteför, G., Siekmann, H.R., Meiwes-Broer, K.H., Lutz, H.O.: Elemental and molecular clusters. Springer Series in Materials Science, Vol. 6, p. 230. Berlin, Heidelberg, New York: Springer 1988
9. Sattler, K., Mühlbach, J., Echt, O., Pfau, P., Recknagel, E Phys. Rev. Lett. **47**, 160 (1981)
10. LaiHing, K., Wheeler, R.G., Wilson, W.L., Duncan, M.A.: J. Chem. Phys. **87**, 3401 (1987)
11. Saunders, W.A., Fayet, P., Wöste, L.: Phys. Rev A (in press) (1989)
12. Wassermann, B., Martin, T.P.: J. Chem. Phys. (in press)
13. Cheshnovsky, O., Yang, S.H., Pettiette, C.L., Craycraft, M.J., Liu, Y., Smalley, R.E.: Chem. Phys. Lett. **138**, 119 (1987)
14. Martins, J.L., Car, R., Buttet, J.: Surf. Sci. **106**, 265 (1981)
15. Ekardt, W.: Ber. Bunsenges. Phys. Chem. **88**, 289 (1984)
16. Knight, W.D., Clemenger, K., de Heer, W.A., Saunders, W.A., Chou, M.Y., Cohen, M.L.: Phys. Rev. Lett. **52**, 2141 (1984)
17. Jezequel, G., Barski, A., Steiner, P., Solal, F., Roubin, P., Pinchaux, R., Petroff, Y.: Phys. Rev. B **30**, 4833 (1984)

Z. Phys. D – Atoms, Molecules and Clusters 12, 411–413 (1989)

Atoms, Molecules
Zeitschrift für Physik D and Clusters
© Springer-Verlag 1989

Electronic and chemical properties of indium clusters

D. Rayane[1], P. Melinon[2], B. Cabaud[2], A. Hoareau[2], S. Khardi[1], B.Tribollet[1], and M.Broyer[1]
[1] Laboratoire de Spectrométrie Ionique et Moléculaire (associé au C.N.R.S.), Bât 205
[2] Départment de Physique des Matériaux (associé au C.N.R.S.), Bât. 203
Université Lyon I, 43 Bd. du 11 Novembre 1918, F-69622 Villeurbanne Cedex, France

Received 7 July 1988

Indium clusters are produced by the inert gas condensation technique. The ionization potentials are found higher for small clusters than for the Indium atom. This is explained by the p character of the bonding as in Aluminium. Doubly charge clusters are also observed and fragmentation processes discussed. Finally small Indium clusters $3 < n < 9$ are found very reactive with hydrocarbon.

PACS: 35.20.Vf; 36.40.+d

Introduction

Recent experimental [1, 2, 3] and theoretical studies of Aluminium clusters have shown that clusters of trivalent metals have particular properties. As discussed by Upton [1] in a theoretical paper on Aluminium, the bonding in such clusters is dominated by the p orbitals. One of the most important consequence is the evolution of the Ionization Potentials as a function of the size. They do not follow the $n^{-1/3}$universal law predicted by the classical model for the work function of a conducting sphere. On the contrary the calculated ionization potentials for Al_n clusters ($2 \leqslant n \leqslant 6$) are higher than for the Al atom. These results are confirmed by recent experimental measurements [2, 3]. Concerning the chemical properties, the chemisorption of Hydrogen on Aluminium clusters exhibits a remarkable sensitivity to the cluster size. Al_6 is the smallest cluster for which chemisorption of H_2 is observed and for $n > 6$, the reactivity decreases with increasing clustter size. In the present paper, we report experimental results obtained on Indium clusters which have a similar atomic structure. Ionization potentials are measured and fragmentation processes are discussed. The ionization thresholds are also observed for doubly charged clusters. Moreover Indium clusters are found to be very reactive with hydrocarbon and this reactivity is strongly size dependent.

Experimental set up

The Indium clusters are generated by the gas aggregation technique [4]. The Indium metal heated at about 1000°C is condensed in helium vapour (few Torr) cooled to Liquid Nitrogen temperature. After a 10^{-4} Torr differentially pumped chamber, the cluster beam passes into the high vacuum chamber of the mass spectrometer (10^{-7} Torr) where they are ionized by electron impact. The electron currents are typically 20–40 μA. The energy calibration is provided by an Ar or He beam, or by the In_2 molecule. The mass analysis is performed by time of flight (T.O.F.) mass spectrometry.

Results and discussion

Figure 1 shows typical mass spectra of Indium clusters recorded at low and high electron energy. Doubly charged clusters appear on the spectrum for $n \geqslant 15$. In order to measure the ionization potentials of Indium clusters, the electron energy is varied in steps of 0.25 eV from 5 to 12 eV and in steps of 1 eV from 12 to 30 eV. At each electron energy value, the whole mass spectrum is recorded. Figure 2 shows the evolution of the ionization potentials as a function of the cluster size. Despite the lack of precision of the measurements

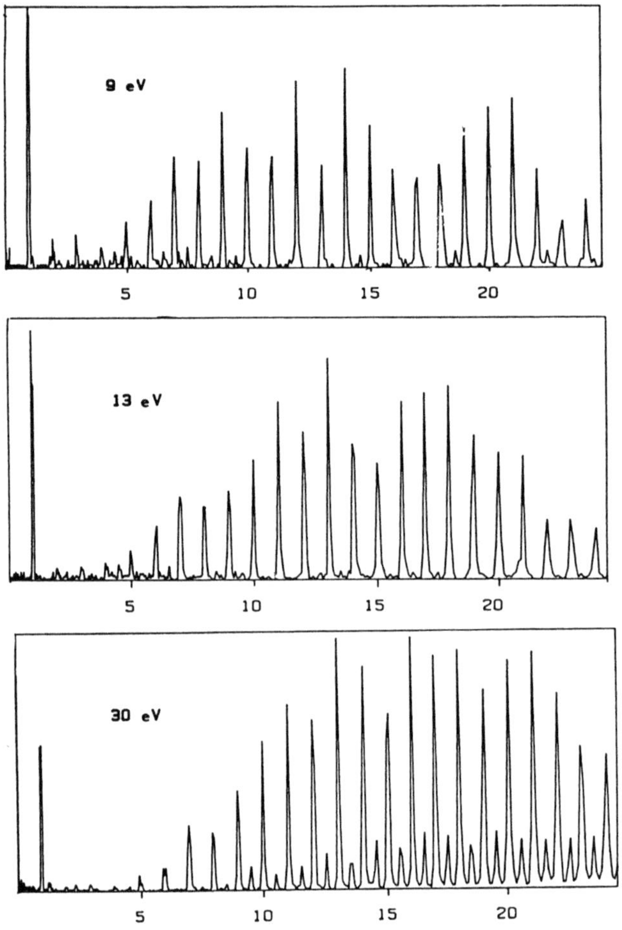

Fig. 1. Mass spectra of Indium clusters as a function of the electron energy

($\pm$0.25 eV), the general trends are quite clear: the ionization potentials do not decrease as a function of the clusters size and they do not follow the prediction of the classical model of the work function of a conducting sphere ($W_R = W_0 + (3e^2/8R)$). The curve corresponding to this classical model is also drawn in Fig. 2. Since the Indium atom has an atomic structure ($5s^2 5p$) similar to that of the Aluminium, the evolution of the ionization potentials can be interpreted by similar arguments [1]: the p band is only 1/6 filled and the cluster Fermi level is deeper relative to the atom. At large cluster size this effect decreases because s and p states become mixed. In Indium, complications arise from the spin-orbit interaction.

As illustrated by Fig. 1, the mass distribution seems essentially unaffected by the ionizing electron energy. This behavior is very different to that observed on Lead [5] and Antimony [6] clusters where the evaporation phenomena strongly shift the mass distribution to low masses at high electron energy. This probably means that the fragmentation of small ionized Indium clusters occurs mainly by evaporation of the ionized atom. This fragmentation pattern has been also observed in collision induced dissociation of Aluminium ion clusters [3] and may be explained by the fact that the ionization potential is lower for the atom than for the small clusters.

Doubly charged clusters are observed for $n \gtrsim 17$ (Fig. 1) and the corresponding ionization potential is measured to be 15 ± 2 eV. The shape of the ionization threshold is not linear and this makes the extrapolation very difficult. The intensity of these doubly

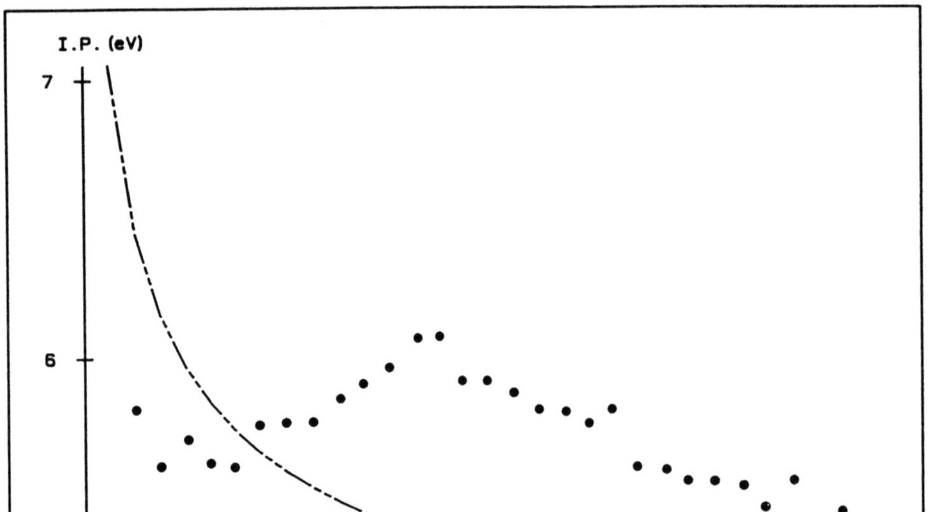

Fig. 2. Ionization potentials of Indium clusters as a function of the size. The dashed line corresponds to the curve $W = W_0 + (3e^2/8R)$.

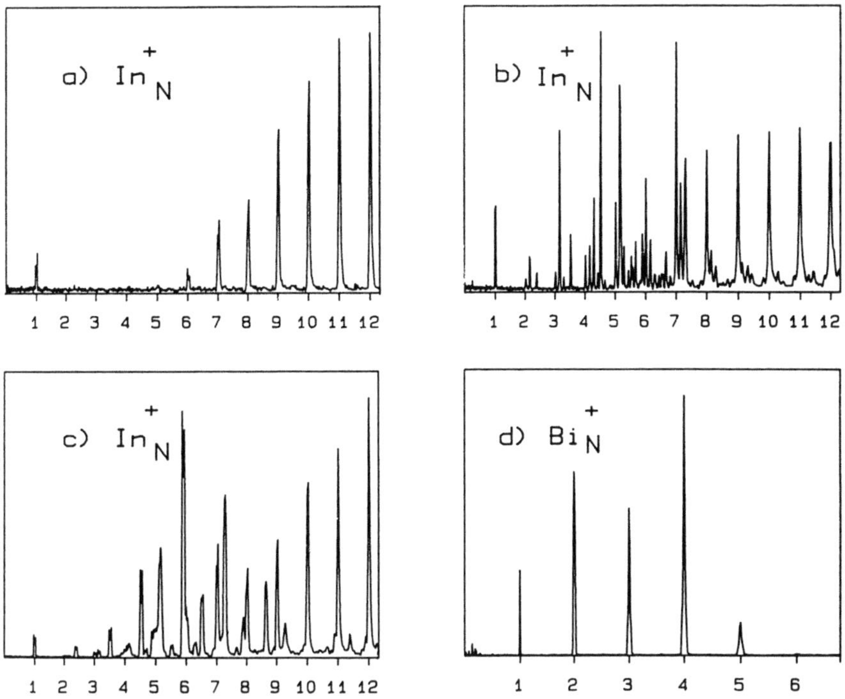

Fig. 3a–d. Reactivity of Indium clusters with hydrocarbon. **a)** Indium clusters with cold Helium (100–150 K). **b)** Indium clusters with Helium at 200 K and with Butane. **c)** Indium clusters with Helium at 250 K. **d)** Bismuth clusters with Helium at 200 K and with Butane. Despite different intensities of the various peaks, **b** and **c** are similar. **d** shows the absence of reactivity of Bismuth clusters with hydrocarbons

charged remains always small, even at an electron energy larger than 30 eV. This behavior is very different of that of Mercury clusters [7] or large Lead clusters ($n > 30$) [5] and rather similar to the behavior [5] of small Lead clusters ($7 < n < 27$). This is probably due to the importance of Coulomb explosion in the dissociation phenomena.

Finally, we have found that Indium clusters are very reactive with hydrocarbon. In current mass spectra of Indium clusters, impurities are frequently observed in the mass region $4 < n < 10$ (Fig. 1). The most intense complex $(In)_n$ X corresponds to X = 56–58. It suggests that X may be hydrocarbons C_4H_8 or C_4H_{10}. The intensity of these peaks is strongly enhanced, when the temperature of Helium inert gas is increased (Fig. 3), because the hydrocarbon impurities are no more condensed on the walls of the condensation chamber. This effect and these kind of peaks have been never observed in other clusters (Bi, Sb, Pb, Ag, Ba) produced by the same method. In order to assign the hydrocarbon responsible of these phenomena, Butane (C_4H_{10}) has been injected in the condensation chamber. The results are shown on Fig. 3 and clearly demonstrate the intense reactivity of hydrocarbons with Indium clusters. This reactivity is strongly size specific and is intense for $3 < n < 9$. The most intense peaks are $(In)_nX$ with $X = CH_4$ and C_4H_{10}. As a comparison, the same experiment, namely the injection of butane, has been performed on Bismuth clusters and no peaks of Bi_nX complex are observed. New experiments are under investigation in our laboratory to best characterize this chemical reactivity.

References

1. Upton, T.H.:. J. Chem. Phys. **86**, 7054 (1987)
2. Upton, T.H., Cox, D.M., Kaldor, A.: Physics and chemistry of small clusters. Jena, P., Rao, B.K., Khanna, S.N.: (eds), pp. 755–768. NATO ASI Series 1987
3. Jarrold, M.F., Bower, J.E., Kraus, J.S.: J. Chem. Phys. **86**, 3876 (1987); Hanley, L., Ruatta, S.A., Anderson, S.L.: J. Chem. Phys. **87**, 260 (1987)
4. Sattler, K., Mühlbach, J., Recknagel, E.: Phys. Rev. Lett. **45**, 821 (1980)
5. Hoareau, A., Melinon, P., Cabaud, B., Rayane, D., Tribollet, B., Broyer, M.: Chem. Phys. Lett. **143**, 602 (1988)
6. Rayane, D., Melinon, P., Cabaud, B., Hoareau, A., Tribollet, B., Broyer, M.: Z. Phys. D – Atoms, Molecules and Clusters (this issue) (1989)
7. Bréchignac, C., Broyer, M., Cahuzac, Ph., Delacrétaz, G., Labastie, P., Wolf, J.P., Wöste, L.: Chem. Phys. Lett. **118**, 174 (1985)

Z. Phys. D – Atoms, Molecules and Clusters 12, 415–416 (1989)

Atoms, Molecules
Zeitschrift für Physik D and Clusters
© Springer-Verlag 1989

Anomalous nuclear spin-lattice relaxation in copper small particles

T. Goto, F. Komori, and S. Kobayashi
Department of Physics, Faculty of Science, University of Tokyo 7-3-1, Hongo, Bunkyo-Ku, Tokyo 113, Japan

Received 5 July 1988; final version 30 September 1988

Anomalously long nuclear spin-lattice relaxation time is expected in small metallic particles when the life time width of the conduction electron energy levels is smaller than their mean spacing. A new method for setting nuclear magnetization to zero is developed in order to measure the relaxation time in Cu small particles. The anomaly is observed in the relaxation of ^{63}Cu in the particles with the diameters smaller than 100 A at temperatures below 0.5 K.

PACS: 76.60.-k

1. Introduction

In small metallic particles, the nuclear spin-lattice relaxation time is expected to be very long when the life time width of the conduction electron energy level is smaller than the mean spacing δ; it becomes difficult to satisfy the condition of the energy conservation in the process of the Korringa type nuclear spin relaxation.

In this system, there are large quadrupole fields arising from the Friedel oscillation for nuclear species with $I > 1/2$. This causes large NMR width, and consequently, it is hard to obtain uniform increase of spin temperature for relaxation measurement by rf excitation. Actually, the reported recovery of NMR signal in Cu small particles [1] was strongly non-exponential, although an anomalously long relaxation tail was observed.

To avoid this difficulty, we have developed a new method to set the nuclear magnetization to zero. We have studied T_1 of ^{63}Cu in copper small particles with diameter of few tens A at temperatures between 30 and 500 mK. We choose copper, because its NMR properties have been extensively studied [1, 2] and because it is free from superconductivity which also strongly modifies T_1 and conceals the effect.

2. Experimental

In preparing small particles, we used vaccum evaporation method. This is based on the fact that the early stage of evaporation produces small island-shaped particles well dispersed on the evaporation substrate. The source copper metal with five nines purity was evaporated from electrically-heated crucible. In order to obtain an amount of particles sufficient to observe NMR signals, we repeated the successive evaporation of SiO and the metal on Mylar substrate. These SiO layers prevent copper particles from oxidization.

We prepared three samples with different mean diameters by controlling the amount of copper for each evaporation. The size distribution was determined by electron micrographs. The parameters of samples are listed in Table 1.

The field dependence of the Knight shift of Sample 1 and 2 at low temperature was reported [2], and the results agreed with the expectation based on the effects of discrete levels with spin-orbit interaction.

In the experiments below 250 mK, the copper particles were put in the mixing chamber of a dilution refrigerator, which provided temperatures down to 30 mK. For the measurements at 0.5 K, we used the simple ^{3}He cryostat, where the particles were immersed in pure ^{3}He.

The nuclear spin-lattice relaxation time of ^{63}Cu in copper small particles was measured by monitoring the recovery of the NMR signal after depolarizing the spins. The signal was detected by Q-meter method with field sweep below 250 mK. The amplitude of rf field was kept very small to prevent the saturation of resonance.

Table 1. List of samples

No.	1	2	3
$d(A)$	100	60	30
$\bar{\delta}(K)$	2.4	11.3	90.5

d: average diameter of particles.
$\bar{\delta}$: mean spacing of kinetic energy levels of conduction electrons for the particle with the average size d: explicitly defined as $3/4\pi D(E_E)d^3$, where $D(E_F)$ is the density of the states per spin at the Fermi level, and is calculated from the specific heat of electrons

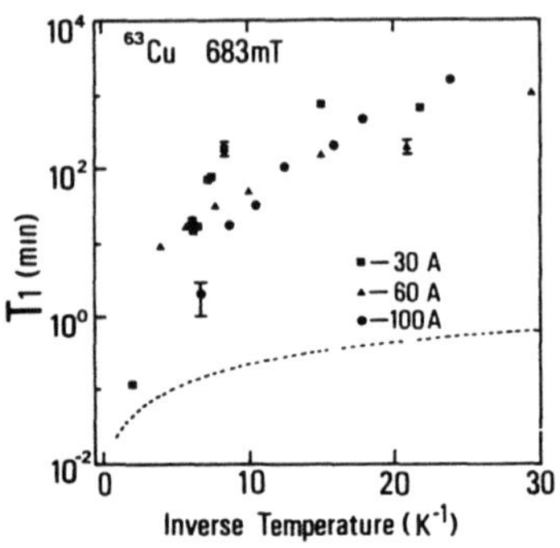

Fig. 1. The nuclear spin-lattice relaxation time T_1 of ^{63}Cu in small particles as a function of temperature. The *dotted line* indicates the Korringa's relation, $T_1 T = 1.27$ K.s

In a previous paper [3], we have shown that the nuclear polarization, which is built up in high magnetic field, can be easily lost when the field is once lowered so much that the quadrupole broadening exceeds the Zeeman energy. This depolarization takes place in the time much shorter than T_1 in a fixed high field.

We made use of this phenomenon to set the initial condition for present measurement of T_1, that is, we kept the field at zero for several minutes, and suddenly increased it to 683 mT.

The transient time for switching the field was 1 min, which was slow enough to prevent the eddy current heating. The resonance intensity was detected as a function of time, and fitted to a single exponential to obtain T_1. For the measurement at 0.5 K, where high field T_1 is shorter than a minute, we employed pulsed NMR. Echo height was measured as a function of the time after the field switching to the resonance field from zero in 10 s.

3. Results and discussion

The relaxation was found to be single exponential within experimental accuracy being in clear contrast with that observed in ref. [1]. This assures that the spectral diffusion accompanied with incomplete saturation does not take place.

As shown in Fig. 1, T_1 increases rapidly when the temperature goes down, and levels off at around 100 mK. Both the magnitude and the temperature dependence of T_1 deviate far from Korringa's relation. Below 100 mK, T_1 is larger than the Korringa's value by three or four orders of magnitude. The relaxation time also depends on the mean particle size, and is longer for the smaller particles.

In order to confirm that the observed relaxations are intrinsic to copper nuclei in the particles, we measured the T_1 of ^{63}Cu in multilayers of continuous film prepared by the same method as that for the particles. The thickness of each film is 100 A. At temperatures down to 50 mK, no deviation from the Korringa value was observed.

In the measured temperature range, $k_B T$ is much smaller than δ. The life time width of the levels which arises from inelastic processes such as electron-phonon scattering is also expected to be smaller than δ [4]. Therefore, the long T_1 can be ascribed to the discreteness of electron levels.

The leveling off of T_1 below 100 mK implies that there is additional relaxation path such as magnetic centers, whose effect is almost independent of temperature and of particle size.

4. Conclusion

By using a new method for T_1 measurement in inhomogeneously wide resonance, we observed the anomalously long T_1 in small particles at very low temperature. The origin of the anomaly is attributed to the discreteness of the electron levels.

We thank Dr S. Katsumoto who provided us with a part of samples.

References

1 Kobayashi, S.: J. Phys. Paris Colloq. **38**, C2-121 (1977)
2. Kobayashi, S., Katsumoto, S.: J. Phys. Soc. Jpn. **56**, 2256 (1987)
3. Komori, F., Goto, T., Kobayashi, S.: J. Phys. Soc. Jpn. **55**, 2133 (1986)
4. For example: Komori, F., Kobayashi, S., Prog. Theor. Phys. Suppl. **84**, 224, (1985)

Z. Phys. D – Atoms, Molecules and Clusters 12, 417–420 (1989)

Atoms, Molecules and Clusters
Zeitschrift für Physik D
© Springer-Verlag 1989

Core edge energy loss studies of Pd clusters on graphite

M.De Crescenzi*, L. Lozzi, P. Picozzi, and S. Santucci
Dipartimento di Fisica Università dell' Aquila, I-67100 L'Aquila, Italy

Received 6 July 1988; final version 30 September 1988

In order to verify that the d-band of Pd clusters moves towards higher binding energies when the cluster size decreases, we have carried out energy loss measurements of the $M_{4,5}$ core edge of Pd clusters with different diameters evaporated on polycristalline graphite. Generally the shifts have been attributed in the past both to initial and or final state effects. In the last case one could not observe variations with our technique which probes the partial unfilled density of states. Our experimental results show an enhancement of the absorption and a narrowing of the lineshape with increasing cluster size. We attribute the observed effects to a genuine change in the electronic structure which occurs in the initial state of the cluster. Comparative studies of this effect have been performed following the evolution of the $M_{4,5}$ VV Auger levels and the collective excitations in the range 0–30 eV far from the elastic peak.

PACS: 79.20; 73.20

Introduction

The electronic and structural properties of small metal clusters deposited on inert substrates have prompted great interest in the last few years for their technological implications (heterogeneous catalysis [1], optical devices [2] etc) and for fundamental aspects [3, 4].

A number of spectroscopic techniques like Auger [5, 6], XPS [7, 8] and XANES [9, 10], have been applied to understand the evolution of the empty and full states above and below the Fermi level respectively. Many authors [11–14] have reported a sizeable shift towards higher binding energies of the core levels when the cluster size decreases.

In the past these shifts have been attributed to initial-state effects [11, 14] (variation of the electronic structure as a function of the size of the metal cluster which approaches the atom like behaviour) and to final state effects [8, 13] (electrostatic charging of the cluster which is not neutralized by the free electrons of the substrate). In particular this last effect should shift all energies by $e^2/2r$, where r is the cluster radius, resulting in an apparent shift of the whole electronic structure of the cluster (valence band and core level) with respect to that of the bulk.

In addition discontinuous metal thin films deposited on metallic substrates like noble metals on Al (111) [15], Pd on Aluminium or Silver [16] and Cu on Zinc [17] show an important shift of the valence band towards higher binding energies and a detectable narrowing of the lineshape. The two effects are of the same order of magnitude.

This suggests that both initial and final state effects should be present when the metallic clusters are deposited on the insulating substrates. In order to discriminate one effect from the other we have evaporated Pd clusters on clean polycristalline graphite. We have chosen graphite because it ensures, due to its low reactivity with metals, an island formation. Furthermore as no core level shifts have been observed for different metal clusters [12, 14] so no sizeable charging effect should be invoked.

* Permanent address: Dipartimento di Fisica, Università di Roma II, "Tor Vergata", I-00 173 Roma, Italy

The purpose of this work is to assess the real amount of change of the electronic valence band of Pd clusters decreasing film thickness. The result obtained in the $M_{4,5}VV$ Auger transition clearly shows that also for Pd deposited on graphite there is a shift of about 1 eV of the centroid of the valence band towards higher energies. This effect is confirmed by the concomitant energy loss measurements of the $M_{4,5}$ core level which probe mainly the empty density of states (DOS) above E_F. As a consequence, a depression of the *d* valence band with respect to E_F, as reported by photoemission experiments, should produce detectable variations on the onset and lineshape of the absorption edge.

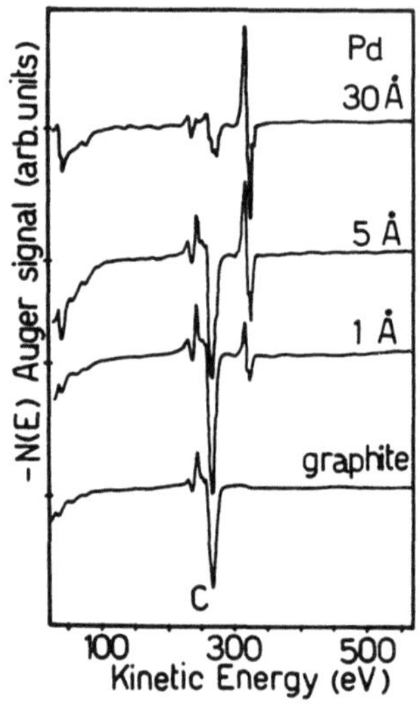

Fig. 1. Auger spectra of different Pd thicknesses (1, 5 and 30 Å) on clean graphite

Experiment

Clusters were prepared by evaporating palladium (99.9% pure) from a tantalum boat onto a clean polycristalline graphite substrate at room temperature in a UHV chamber with a base pressure in the 10^{-10}-torr range. The cleanliness of graphite has been checked by Auger measurements showing, within the sensitivity of this technique ($\lesssim 1\%$), the absence of any contaminants. The nominal thicknesses were monitored by an Inficon quartz microbalance.

The vacuum chamber was equipped with a Riber single pass cylindrical mirror analyzer (CMA) for electron spectroscopy measurements. The energy resolution $\Delta E/E$ was about 0.3%. A primary-electron energy of about 2000 eV was used with a beam current of 1 μA. Signals were detected with a lock-in amplifier recording the first derivative of the electron yield distribution [dN(E)/dE].

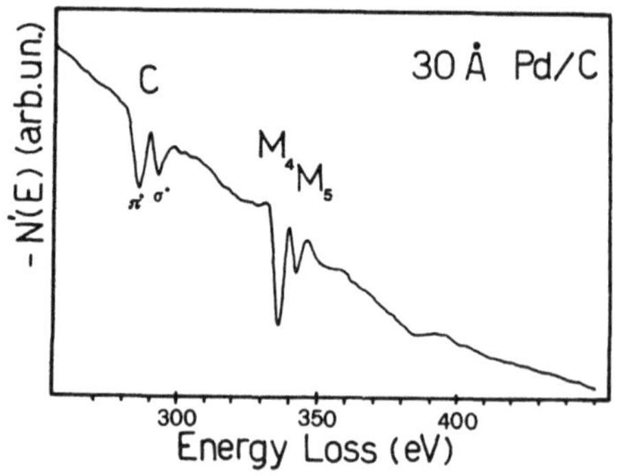

Fig. 2. First derivative of the electron yield in a tipical ELS spectra in the region between 250 and 450 eV showing the structures related both to the K-edge of graphite substrate and the $M_{4,5}$ core edge of Pd.

Results and discussion

Figure 1 shows the Auger spectra of Pd nominal thicknesses on graphite. We always note the presence of both features of graphite and palladium ensuring the discontinuity of the films except for the thicker films (> 20 Å), because of the overlapping of C and Pd Auger features.

Figure 2 shows a typical ELS spectra in a region between 250 and 450 eV energy loss. We note the structures related both to the graphite substrate at 284 and 290 eV (π^* and σ^* transitions) and the $M_{4,5}$ core edge at 335 and 340 eV respectively. As the spectrum has been obtained using a 30 Å film thickness, the ELS technique, in this case, shows an unambiguously discontinuity of the film better than the Auger spectroscopy. Figure 3a displays a magnification of the Pd $M_{4,5}$ VV Auger lines for the various nominal thicknesses. We note a large shift towards lower kinetic energies of both M_4VV and M_5VV structures going from 30 Å to 1 Å evaporation. The observed shift amounts to ~1 eV for the 1 Å nominal thickness deposit. No shift has been observed for depositions with nominal thickness greater than 30 Å. These results are in good agreement with previous investigation of Pd on amorphous carbon as reported by Kohiki and Ikeda [14].

The Fig. 3b displays the numerical integration of the Auger lineshipe of Fig. 3a. We do not note a detectable broadening as observed for other metallic clusters like copper deposited on graphite [6].

Recent XPS preliminary results of Pd on polycristalline graphite do not show appreciable 3*d* ($M_{4,5}$) core level shift and an important shift of the *d*-valence band centroid in good agreement with Egheloff et al

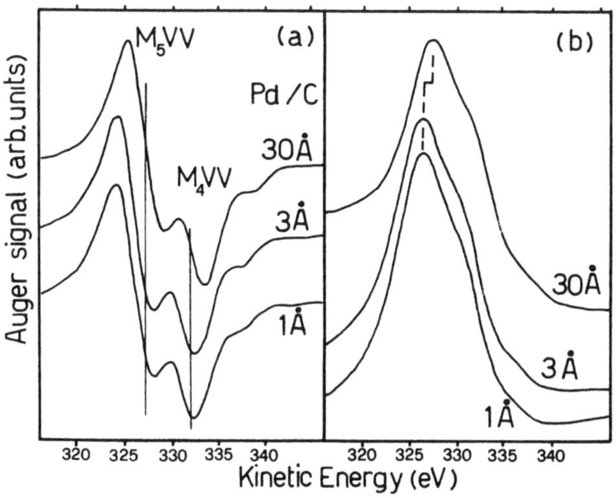

Fig. 3. a Magnification of the Pd $M_{4,5}$ VV Auger line of different Pd thicknesses.
b Numerical integration of the Auger lineshapes of **a**

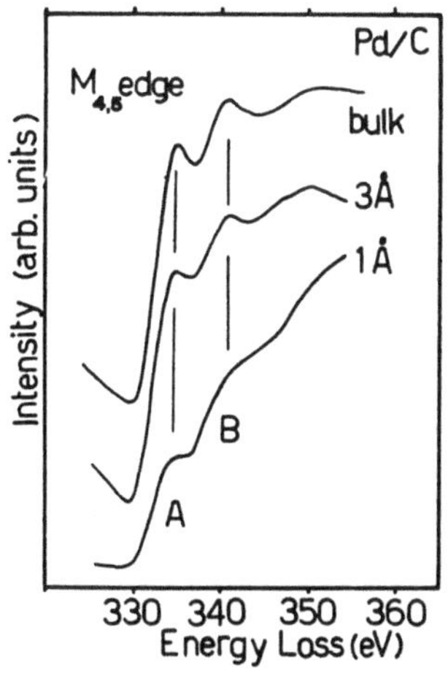

Fig. 4. Numerical integration of EELS spectra around $M_{4,5}$ edge for different thicknesses of Pd graphite

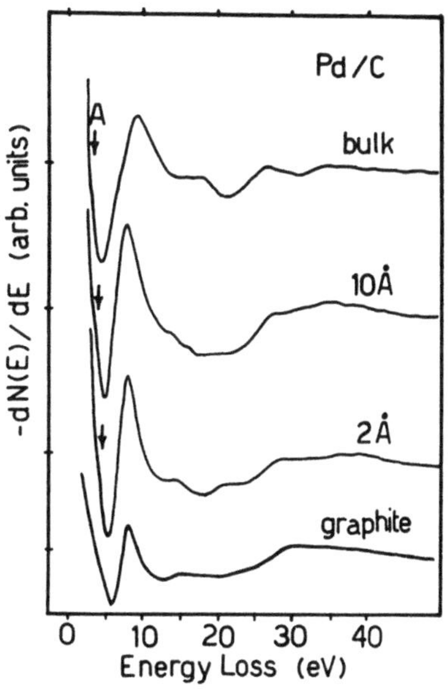

Fig. 5. Electron energy loss spectra of different Pd coverages on clean graphite. The primary beam energy was 400 eV

[12]. On the basis of the comparative analysis of XPS and Auger spectra we can rule out final state effects caused by the photoionization of the clusters. In our case, as the polycristalline graphite behaves like a metallic substrate the hole created by the photoemission process is completely screened by the flowing charge supplied by the substrate. We consider therefore the observed shift in the Auger Pd spectra as a genuine effect caused mainly by the change of the valence band of the Pd clusters.

Furthermore, in order to probe the change of the electronic structure above E_F, we have used the energy loss measurements above the $M_{4,5}$ ionization level. The ELS core edge energy is that required to excite an electron from a core level to the first unoccupied orbital just above E_F. This state corresponds to the screened or relaxed energy, requiring no further screening charge from the support so that extratomic relaxation should be relatively unimportant.

Figure 4 displays the numerical integration of the ELS spectra for different nominal thicknesses. We note that for the bulk the features A and B related to the *d*-density of states, are sharpened while the same features evolve towards a smooth lineshape decreasing the Pd thickness. This is a clear fingerprint of the lowering of the *d* density of states with respect of the E_F which leaves less empty states available to the excited core electron.

The change of the *d* band has also been studied by energy loss measurements involving valence and conduction bands (Fig. 5). The main features of Pd bulk have been carefully interpreted in terms of interband transitions [18, 19]. In particular the features A at about 4 eV should involve transitions from the full *d* states to the first empty states above E_F. Following the electron band structure calculated by Lasser and Smith [19], this transitions involve initial states with band indices from 1 to 4 towards final band with index 5. The A structure shifts by about 1 eV on going from the bulk to the smallest cluster. As a result an increased energy loss of this transition indicates a lowering of the *d* band respect to E_F. This shift is of the same order of magnitude of that reported in Fig. 2 as obtained from Auger data.

Other transitions located at higher energy loss follow the same behaviour. Unfortunately the intense transition located around 7 eV in the bulk is masked by the carbon π^* transition of the substrate. However Henry and Poppa [20] performing ELS measurements of Pd on MgO substrate, where the Pd 7 eV structure does not overlap with the substrate features, have observed a sizeable shift towards higher energy loss decreasing the cluster size.

In conclusion we can confirm that in Pd discontinuous films the changes in the valence band structure (narrowing and lowering of the *d* band with respect to E_F) are sizeable and evolve continuously towards higher binding energies for decreasing cluster size. These effects are not masked by charging (final state effects) because graphite sufficiently screens the core level involved in the Auger process. Moreover our conclusions are strongly supported by ELS experiment for which electrostatic charging effects cannot be invoked.

The authors are grateful to O. Consorte for his technical assistance.

References

1. Sinfelt, J.: Rev. Mod. Phys. **51**, 569 (1979)
2. Granqvist, C.G.: J. Phys. C **41**, 247 (1981)
3. Delley, B., Ellis, E., Freeman, A.J., Baerends, E.J., Post, D.: Phys. Rev. B.**27**, 2132 (1983)
4. Baetzold, R.C.: Surf. Sci. **106**, 243 (1981)
5. Oberli, L., Monot, R., Mathieu, H.J., Landolt, D., Buttet, J.. Surf. Sci. **106**, 301 (1981)
6. De Crescenzi, M., Diociaiuti, M., Lozzi, L., Picozzi, P., Santucci, S., Battistoni, C., Mattogno, G.: Surf. Sci. **178**, 282 (1986)
7. Mason, M.G., Gerenser, L.J., Lee, S.T.: Phys. Rev. Lett. **39**, 288 (1986)
8. Wertheim, G.K., Di Censo, S.B., Buchanan, D.N.E.: Phys. Rev. B. **33**, 5384 (1986)
9. De Crescenzi, M., Diociaiuti, M., Picozzi, P., Santucci, S.: Phys. Rev. B. **34**, 4344 (1986)
10. Balerna, A., Bernieri, E., Picozzi, P., Reale, A., Santucci, S., Burattini, E., Mobilio, S.: Phys. Rev. B. **31**, 5085 (1985)
11. Mason, M.G.: Phys. Rev. B. **27**, 748 (1983)
12. Egheloff, Jr, W.F., Tibbets, G.G.: Solid State Commun. **29**, 53 (1979)
13. Wertheim, G.K., Di Cenzo, S.B., Buchanan, D.N.E., Bennett, P.A.: Solid State Commun. **53**, 377 (1985)
14. Kohiki, S., Ikeda, S.: Phys. Rev. B. **34**, 3786 (1986)
15. Egheloff, Jr, W.F.: J. Vac. Sci. Technol. **20**, 668 (1982)
16. Frick, B., Jacobi, K.: Surf. Sci. **178**, 907 (1986)
17. Abbati, I., Braicovich, L., Bertoni, C.M., Calandra, C., Manghi, F.: Phys. Rev. Lett. **40**, 469 (1978)
18. Chiarello, G., Colavita, E., De Crescenzi, M., Nannarone, S.: Phys. Rev. B. **29**, 4878 (1984)
19. Lasser, R., Smith, N.V.: Phys. Rev. B. **25**, 806 (1982)
20. Henry, C.R. Poppa, H.: Z. Phys. D – Atoms, Molecules and Clusters **12** (1989) (this issue)

Z. Phys. D – Atoms, Molecules and Clusters 12, 421–424 (1989)

Atoms, Molecules and Clusters
Zeitschrift für Physik D
© Springer-Verlag 1989

Auger and electron energy loss spectroscopies of small palladium particles supported on (111) oriented MgO films

C.R. Henry* and H. Poppa

IBM Almaden Research Center, 650 Harry Road, San-José, CA 95120 USA and Department of Chemical Engineering, Stanford University, CA 94305 USA

Received 5 July 1988; final version 30 September 1988

The most common tool used to characterize supported metal clusters is the transmission electron microscope. The main advantage of TEM is the combination of high (lateral) resolution imaging with electron diffraction. However the TEM observations are usually made ex-situ i.e. UHV deposited clusters have to be exposed to the atmosphere during transfer to the TEM. This could be a severe limitation for very small reactive clusters. This paper demonstrates that electron spectroscopies can provide in-situ information on the cluster growth (AES), on the electronic structure (ELS) and on the local atomic order (SEELFS) of the clusters. These techniques were applied to Pd clusters of varying size ($\sim$10–200 Å) vapor-deposited on thin (111) MgO support films under UHV conditions. An expansion of the lattice and a shift of the loss peaks towards higher energies are observed with decreasing particle size.

PACS: 79.20; 73.20; 61.14

1. Introduction

There is continued interest in systematic studies of UHV grown supported metal clusters and small particles. One important goal of this research is progress in the understanding of the origin of the catalytic properties of small transition metal particles. For this it is necessary to understand the relationship between the atomic structure, the electronic structure and the reactivity of the particles [1] and a Pd/MgO model catalyst system has been studied with a combination of several surface and structure sensitive techniques. The atomic structure of the particles is studied ex-situ by Transmission Electron Microscopy (TEM) and in situ by Surface Electron Energy Loss Fine Structure (SEELFS), the electronic structure by Energy Loss Spectroscopy (ELS) and the reactivity by Temperature Programmed Desorption (TPD) [2]. This paper focuses on some of the more recent electron spectroscopy measurements and should be read in conjunction with [2].

2. Experimental

The specimens are prepared and characterized in a multi-chamber UHV system. Thin films of MgO are deposited from an electron-gun evaporator on a UHV cleaved and dehydroxylated mica substrate [3]. Deposition at 400°C at a rate of 8–13 $\times 10^{13}$ cm^{-2} s^{-1} resulted in (111) epitaxial MgO films [2]. The palladium particles are prepared by subsequent deposition at 325°C substrate temperature at a rate of 3–5 10^{13} cm^{-2} s^{-1}. The equivalent thickness of the deposits is monitored by an oscillating quartz crystal. A single pass Varian CMA was used for the electron spectroscopy. For AES the primary electron energy was 1500 eV with a beam current of 2 μA. For ELS the beam energy was in the range 100–500 eV with a beam

* Permanent address: CRMC2–CNRS, Campus de Luminy, case 913, F-13288 Marseille Cedex 09, France

current of 1 μA. During the SEELFS measurement a primary energy of 2000 eV, a beam current of 3 μA, a modulation of 10 V and a time constant for the lock-in amplifier of 10 s (for a corresponding sweep time of 15 min) were utilized. Data acquisition on 2048 data points per spectrum was performed with the help of the IBM PC interfaced to the lock-in amplifier. At the end of the experiment the mica samples were thinned by microcleavage for TEM observation.

3. Results

3.1. AES

The growth of the Pd particles was followed by monitoring the palladium and oxygen Auger peaks. At low coverages the intensity of the Pd peak increased linearly with deposition time indicating a quasi two-dimensional growth, later the growth became more 3-dimensional.

3.2 TEM

Electron diffraction indicated that the Pd particles are (111) epitaxially oriented on the MgO. In bright field the particles have an hexagonal outline [2]. The mean diameter of the particles and their number density are given in Table 1 as a function of the amount of deposited palladium. For the 5 Hz deposit very small cluster were present but inhomogeneities of the deposit indicated that some particles were lost and other have coalesced during the preparation of the specimen for the TEM; for this deposit the mean size has therefore been estimated. For the 20 Hz Pd deposit, no TEM observations are available and, therefore, the mean diameter is estimated from a $t^{0.5}$ time growth law as observed on Pd/MgO(100) [4]. For the 200 Hz deposit the diameter is larger than given by the $t^{0.5}$ law because of coalescence (indicated by the decrease of the number density of clusters).

3.3 ELS

Figure 1 shows a set of ELS spectra obtained for different Pd thicknesses at a primary energy of E_p = 500 eV. For the continuous Pd film, loss peaks (obtained from the $d^2\ N(E)/dE^2$ curves) were obvious at 7.4, 16.1, 23.6, 32.1 and 43.6 eV. Shifts towards higher energies were observed as the size of the particle decreases. In the case of the 7.4 eV loss, the shift (up to $\Delta E \simeq 1.8$ eV) is clearly visible in the figure. When the primary energy (E_p) decreased the loss energies did not change but their intensities droped. For $E_p < 300$ eV, only the 7.4 eV peak remains visible indicating the surface character of this loss [5]. The height of the loss peaks increased (at a higher rate for the 7.4 eV peak) with the amount of deposited Pd. For the thinnest Pd deposits ($\Delta f \leqslant 50$ Hz) MgO losses appeared clearly.

Table 1. Mean size D and number density n of particles deduced from TEM as a function of the deposited Pd amount

Δf (Hz)	5	10	20	100	200	400
D (Å)	$\leqslant 15$	20	30	70	340	∞
n (cm^{-2})	—	3.1×10^{12}	—	1.1×10^{12}	$4\ 10^{10}$	—

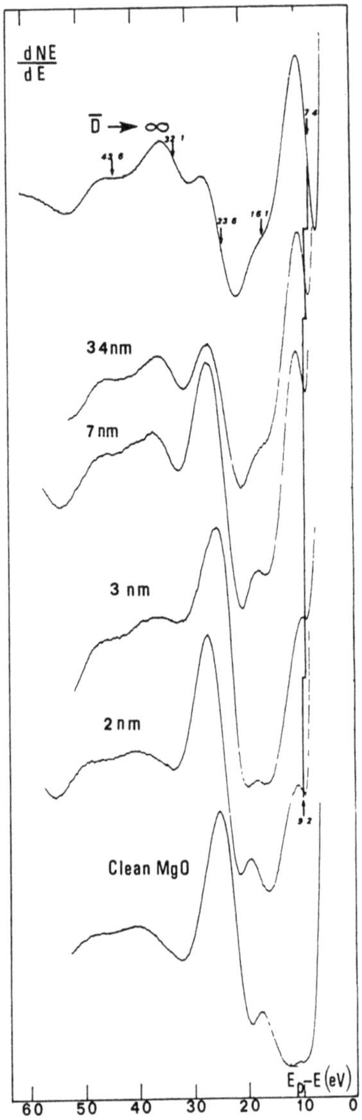

Fig. 1. ELS spectra from a continuous Pd film, from small Pd particles with different mean diameters and from a clean MgO substrate, taken at a primary energy of 500 eV. The shift of the plasmon loss peak is indicated

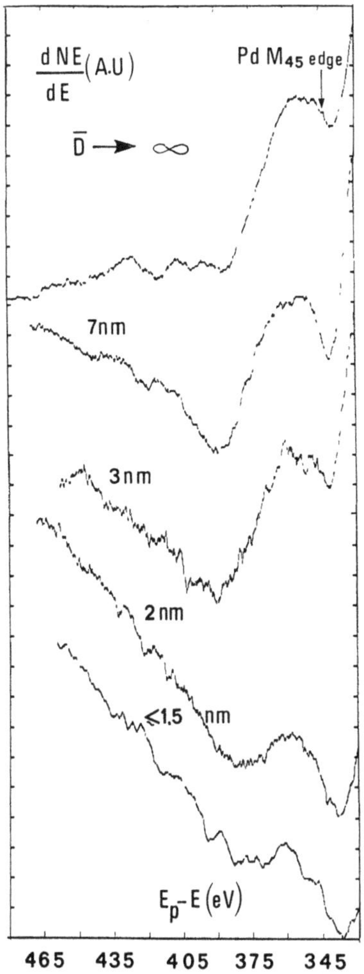

Fig. 2. ELS spectra above the Pd M45 edge for a continuous (111) Pd film and for small particles of different sizes. The actual noise is displayed on the spectra except for the 2 smallest sizes for which the spectra have been numerically smoothed

3.4 SEELFS

Figure 2 shows a set of energy loss spectra above the Pd M45 edge taken for different Pd particles mean sizes. The position of the edge varied between 340 and 350 eV. After the edge, intensity oscillations were clearly visible for the continuous Pd film (400 Hz). The spectra were noisier for the thinner deposits and the oscillations were more difficult to see. It is noteworthy that the shape of the background changed appreciably with the size of the Pd particles. De Crescenzi [6] has shown that local order information can be obtained from the oscillations above such an edge by using an EXAFS-like formalism. The radial distribution of our Pd clusters was calculated in this way and Fig. 3 displays distributions obtained for a continuous film and for a deposit with 30 Å clusters. Without a phase shift correction, the nearest neighbors distance for the continuous film appeared at 2.58 Å vs. 2.76 for the bulk; others peaks can be seen at 3.77, 4.65 and 5.97 Å. The first peak at 0.82 Å is due to the Ramsawer-Townsend modulation of the backscattering amplitude [7]. The peak positions found here are close to ones obtained in EXAFS [8], however in our case the peak near 4 Å is more developed.

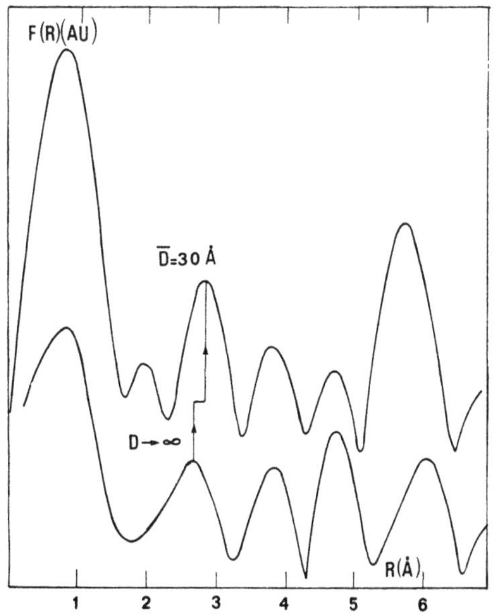

Fig. 3. Fourier transform of the SEELFS signals for a continuous Pd film and for 3 nm particles

For the small particles the first neighbour distance is located at 2.82 Å indicating a definite expansion of 9%. For even smaller particles the radial distributions tended to show larger expansions but the present noise level of the spectra is too high to permit reliable estimates.

4. Discussion

The particles growth can be monitored in situ by AES. With this technique one can deduce the growth mode (*2D*/*3D*) and determine an integral value related to the amount of metal deposited. The particle size distribution can only be determined unambiguously by TEM. Thus the sample had to be air exposed and thinned in this case and this operation can influence the particle size distribution (particularly for the smallest sizes $D < 15$ Å). To overcome this problem it is necessary to determine through systematic measurement the nucleation and growth law for the system studied [4].

The atomistic structure of the particles is conveniently investigated by HRTEM [9] but for the same reasons as above it may be necessary to obtain additional informations by in-situ measurements. SEELFS is an in-situ method and give the local atomic order like EXAFS. Present measurements show an expansion of approximately 9% for 3 nm Pd particles. By EXAFS, a smaller expansion of 3% has recently been obtained [8] for 1.4 nm Pd clusters on carbon. For 2.5–3 nm Pd particles on MgO [9–11] and on mica [12–13], electron diffraction indicates an expansion of up to 4%. HRTEM observations for Pd particles on TiO_2 in the size range 2-5 nm showed an expansion of up to 9%, which increased to 15% after heating in air [14]. This result for Pd particles is contrary to the results for other metals that were found contracted [6, 13], which was interpreted as due to surface stress [6]. For Pd a very large expansion is always found which is not clearly correlated with size. Taking into account the effect of heating in oxygen Jacobs and Schryvers [14] suggest that oxygen might diffuse into the Pd lattice (already at room temperature) and cause the expansion. For the present in situ experiments, the effect of oxygen diffusion can also not be eliminated because it has been demonstrated that oxygen from the MgO film can diffuse towards the support surface [2]. Clearly further experiments with the same technique and different substrates are needed.

ELS measurements indicate loss energies close to Pd single crystals [15] and films [5, 16]. From optical constants measurements, surface and volume plasmons occur at 7.3 and 7.6 eV respectively [17]. These two features are not separated in our case and may correspond to the 7.4 loss. Other losses have been interpreted as interband transitions [5, 17]. It is tempting to explain the shift of the plasmon peak with particle size by a size effect as calculated by Wang and Cowley for Al/Al_2O_3 particles smaller than 4 nm [18]. However shifts, towards higher energy are also observed for the other loss peaks. Alternatively these shifts might be explained by a continuous evolution of the *d*-band as suggested by De Crescenzi et al. [19]. Indeed, photoemission results for Pd clusters on carbon substrate indicate a positive shift of the electron binding energy, as the cluster size decreases, of about 1.3 eV [20] which is the same order of magnitude as the shifts observed presently in the loss features.

5. Conclusion

By a combination of TEM, AES, ELS and SEELFS techniques the evolution of the atomistic and electronic structure of Pd particles supported on (111) MgO epitaxial films has been studied as a function of the particle size. Two important results are found with decreasing partilce size: 1/ an increase of the lattice parameter and 2/ a shift of the energy loss features (mainly the plasmon peak).

This work was performed at Stanford University and was supported by the IBM Research Division/Almaden Research Centre and by CNRS (France).

References

1. Poppa, H.: Vacuum **34**, 1081 (1984)
2. Henry, C.R., Poppa, H.: J. Vac. Sci. Technol. A **6**, 1113 (1988)
3. Koch, R., Poppa, H.: J. Vac. Sci. Technol. A **5**, 1845 (1985)
4. Chapon, C., Henry, C.R., Chemam, A.: Surf. Sci. **162**, 747 (1985)
5. Vankar, V.D., Vook, R.W.: Surf. Sci. **131**, 468 (1983)
6. De Crescenzi, M., Diociaiuti, M., Lozzi, L., Picozzi, P., Santucci, S.: Phys. Rev. B **35**, 5997 (1987)
7. Teo, B.K., Lee, P.A.: J. Am. Chem. Soc. **101**, 2815 (1979)
8. De Crescenzi, M., Diociaiuti, M., Picozzi, P., Santucci, S.: Phys. Rev. B **34**, 4334 (1986)
9. Henry, C.R., Chapon, C., Penisson, J.M., Nihoul, G.: Z. Phys. D – Atoms, Molecules and Clusters **12** (1989) (this issue)
10. Heinemann, K., Osaka, T., Poppa, H., Avalos-Borja, J: J. Catal. **83**, 61 (1983)
11. Renou, A., Rudra, A.: Surf. Sci. **156**, 69 (1985)
12. Anton, R., Poppa, H.: Proc. Xth Int. Congr. Electron Microscopy Hamburg 1982, p. 509.
13. Heinemann, K., Poppa, H.: Surf. Sci. **156**, 265 (1985)
14. Jacobs, J.W., Schryvers, D.: J. Catal. **103**, 436 (1987)
15. Legaré, P., Holl, Y., Maire, G.. Solid State Commun. **31**, 307 (1979)
16. Liu, D.R.: Philos. Mag. B **57**, 619 (1988)
17. Vehse, R.C., Arakawa, E.T.: Phys. Rev. B **1**, 517 (1970)
18. Wang, Z.L., Cowley, J.M.: Ultramicroscopy **23**, 97 (1987)
19. De Crescenzi, M., Lozzi, L., Picozzi, P., Santucci, S.: Z. Phys. D – Atoms, Molecules and Clusters **12** (1989) (this issue)
20. Kohiki, S., Ikeda, S.: Phys. Rev. B **34**, 3786 (1986)

Z. Phys. D – Atoms, Molecules and Clusters 12, 425–429 (1989)

Atoms, Molecules
Zeitschrift für Physik D and Clusters
© Springer-Verlag 1989

Electron distribution of small Ni particles and Ce ions in X and Y zeolites related to their catalytic properties

F. Vergand[1], H. Roulet[1], G. Dufour[1], C. Bonnelle[1], E. Ramaroson[2], J.F. Tempere[2], M.F. Guilleux[2], and D. Delafosse[2]

[1] Laboratoire de Chimie Physique, Université Pierre et Marie Curie, UA 176 CNRS, 11, rue Pierre et Marie Curie, F-75231 Paris Cedex 05, France

[2] Laboratoire de Réactivité de Surface et Structure, UA 1106 CNRS, 4, Place Jussieu, F-75232 Paris Cedex 05, France

Received 5 July 1988; final version 30 September 1988

Catalytic behaviour of nickel particles smaller than 1.3 nm, encaged in X and Y zeolites, in the presence of Ce^{3+}, is related to their electronic properties. In general the rates of reaction are lower than those of a standard catalyst made of 2.5 nm nickel particles on SiO_2; this is attributed to the strong localization of the Ni $3d$ states due to the size effect. However a high rate, more than twice the standard, is found for Ni in X zeolite, in relation with a Ni $3d^{10}$ configuration. On the other hand the fragmentation of butane is deeper with Ni in Y support than in X support and this corresponds to a high density of Ni d states at the Fermi level.

PACS: 79.60; 36.40

Zeolite containing small metal particles are well known industrial catalysts. Their properties depend on the zeolite acidity; addition of rare earth ions also influences their reactivity particularly concerning hydrocarbon chemistry. In this work we show that catalytic behaviour of nickel particles ($\lesssim 100$ atoms) in X and Y zeolitic support containing cerium ions can be correlated with their electronic structure.

Experiment

Nickel-cerium zeolites Y and X were prepared by conventional ion exchange techniques, contacting dilute solutions of Ni^{2+} and Ce^{3+} nitrates with Linde NaY and NaX zeolites. These precursor samples were reduced by a flow (5 $dm^3 h^{-1}$) of molecular hydrogen at 623 K. The extent of reduction and the size of nickel particles were determined by magnetic measurements [1, 2]. Hydrogen chemisorption experiments were performed at room temperature and 100 torr gas pressure. Thermodesorption measurements showed that hydrogen remains adsorbed on the particles after reduction. In order to demonstrate the role of cerium, one sample with Gd instead of Ce was also studied. Sample data are reported in Table 1. Data for nickel particles on SiO_2 [3] are recalled as a reference.

The catalytic activities and selectivities in the $CO + H_2$ reaction were measured in a flow system as reported in [4]; they are given for the temperature of 573 K. Kinetic experiments on the butane hydrogenolysis were carried out at 523 K on a dynamical differential microreactor described in [5].

Ce $3d$ and Ni $2p$ core levels were studied by X-ray photoelectron spectroscopy (XPS) whereas Ni $3d$ valence distributions were analysed by X-ray emission spectroscopy induced by electrons (EXES).

For the XPS measurements, a standard X-ray source ($\mathrm{Al}\,K\alpha_{1,2}, \hbar\omega = 1486.6$ eV) was used. The samples being electrical insulators we got rid of any charge effect by referencing all binding energies to the C_{1s} at 285 eV. The nickel-cerium zeolite powders were transported from the reactor under vacuum into tubes and introduced into the spectrometer under argon atmosphere using a glove box.

For the EXES experiments we used a bent crystal spectrometer which has a resolution around 10^{-4} in the considered range. We analysed the Ni Lα emission which corresponds to $3d \rightarrow 2p_{3/2}$ transition. X-ray emission was induced by electrons whose incident energy, 1300 eV, was chosen so as to probe a thickness of about 350 Å in the zeolitic support. The spectral 3*d* densities of states were obtained for nickel particles by subtracting, from the overall spectrum of each sample, its own Ni^{2+} spectrum obtained from precursor stage [6].

Results and discussion

The catalytic activity of the samples regarding $CO + H_2$ reaction and butane hydrogenolysis are listed in Table 2.

In the $CO + H_2$ reaction the catalytic selectivity is measured by the CH_4 percentage in products of reaction (C_1) and by the ratio of the number of C_2H_6 molecules on the total number of CH_4 and C_2H_6 molecules $C_2/(C_1 + C_2)$. In butane hydrogenolysis the catalytic selectivity is defined by the ratio r_{C1}/r_t, where r_{C1} is the rate of production of methane due to the complete cracking of the butane molecule ($C_4H_{10} + 3H_2 \rightarrow 4CH_4$), and r_t is the total butane hydrogenolysis rate.

According to Table 2: *i* – the rates measured with the nickel particles in zeolites are smaller than the ones of nickel on SiO_2, except for NiCeX in CO hydrogenation: therefore nickel particles in NiCeX favours the CO dissociation; *ii* – the selectivity in the CH_4 production is greater with NiCeY than with NiCeX: for nickel particles in NiCeY, interaction with adsorbed butane leads to an enhanced degree of fragmentation of the molecule.

Figure 1 shows the XPS spectra of Ce 3*d* Ni 2*p* for NiCeY sample in the precursor stage (Fig. 1a), in the reduced stage (Fig. 1b), and after $CO + H_2$ reaction (Fig. 1c).

Table 1. Sample data

Sample	Composition	wt %	α	D(nm)	N	N_S/N	H/Ni^0
NiCeY	$Ni_5Ce_6Na_{28}$ $(AlO_2)_{56}(SiO_2)_{136}$ $250H_2O$	Ni^{2+} 1.87 Ce^{3+} 5.84 Na^+ 3.88	0.80	1.3	100	0.76	0.05
NiCeX	$Ni_{11}Ce_{17}Na_9H_4$ $(AlO_2)_{86}(SiO_2)_{106}$ $264H_2O$	Ni^{2+} 3.52 Ce^{3+} 13.29 Na^+ 1.21	0.54	1.2	80	0.78	0.15
NiGdX	$Ni_{13}Gd_{17}Na_4H_5$ $(AlO_2)_{86}(SiO_2)_{106}$ $264H_2O$	Ni^{2+} 4.52 Gd^{3+} 15.36 Na^+ 0.64	0.42	1.0	50	0.85	0.00
$NiSiO_2$		Ni^{2+} 4.55	1	2.5		0.49	0.40

α: degree of reduction
D: mean diameter of nickel particles
N: total number of Ni atoms per particle assuming a spherical shape and the specific gravity of bulk Ni
N_S: number of surface atoms per particle
H/Ni^0: ratio of the number of adsorbed hydrogen atoms on the total number of nickel atoms

Table 2. Catalytic activity

	$CO + H_2$ Reaction			Butane hydrogenolysis	
		Selectivity			
Sample	Rate × 10^9 mol $m^{-2}s^{-1}$	C_1 %	$C_2/(C_1+C_2)$ %	Rate × 10^9 mol $m^{-2}s^{-1}$	Select. r_{C1}/r_t
NiCeX	360	96	4	110	0.09
NiCeY	(253)	(92.1)	(0.06)	126	0.20
$NiSiO_2$	170	99	1	380	0.47
	(327)	(95.4)	(0.03)		

$CO + H_2$ reaction: total pressure = 1 atm (values in brackets refer to 10 bars); $H_2/CO = 3$; T = 573 K; total flow = 3.6 $dm^3 h^{-1}$
Butane hydrogenolysis: total pressure = 1 atm; T = 523 K

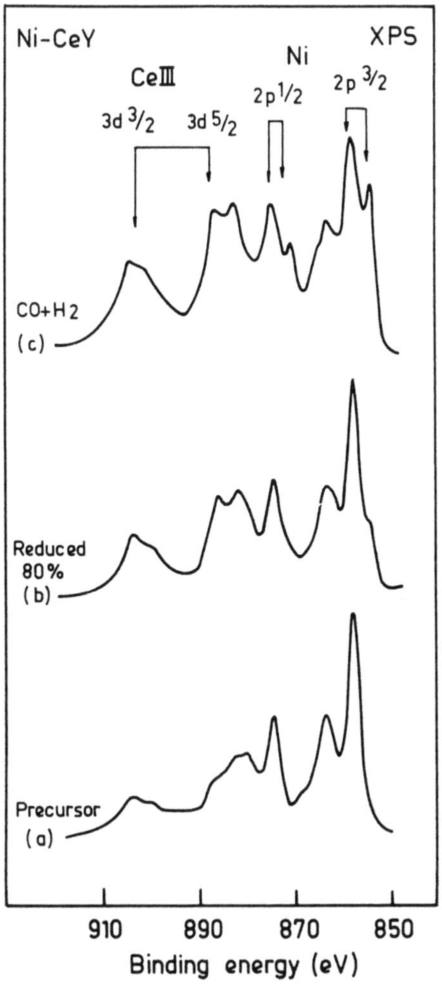

Fig. 1. XPS spectra of NiCeY

In the precursor spectrum, a structure is seen at 905.4 eV. Its position corresponds to Ce $3d_{3/2}$ of Ce III species obtained by Praline et al. [7]. The spin orbit splitting of Ce $3d$ is 18.5 eV and Ce $3d_{5/2}$ peak is present at 886.9 eV mixed with Ni peaks. After reduction, and after $CO+H_2$ reaction the Ce III peak intensity increases but no contribution of Ce IV species is seen. The Ni $2p$ peaks are observed with a spin orbit splitting of 17.0 eV. In the precursor stage the Ni $2p_{3/2}$ line appears at 857.2 eV, with its satellite shifted by +6 eV; this Ni $2p_{3/2}$ binding energy is almost identical to the value 857.4 eV obtained for NiF_2 [8], revealing the presence of Ni^{2+} states in this sample.

After reduction a shoulder becomes visible on the low binding energy side of Ni $2p_{3/2}$ peak, and this new structure is enhanced largely after $CO+H_2$ reaction. This new peak has a binding energy of 853.4 eV i.e. 0.8 eV higher than that of bulk metal (852.6 eV). By referring to Citrin and Wertheim review on core level shifts in metallic clusters [9] this peak can be attributed to the small Ni particles. Such a shift could also arise from NiO but as this new structure increases when reduction progresses it cannot be due to an oxidation effect.

Figure 2 shows the XPS spectra for the Ce $3d$ Ni $2p$ region in the precursor (Fig. 2a) and the reduced (Fig. 2b) NiCeX samples. These spectra are rather similar to those obtained with NiCeY. As in NiCeY no contribution of CeIV is visible.

Changes in XPS intensities can be due to a more or less important amount of each species at the surface of the zeolite grains. Indeed the various treatments that the samples suffer, exchange, reduction, methanation, cause the ion migration. But the total EXES spectra show the contributions of Ni^{2+} and of Ni particles in the correct proportion relative to the degree of reduction.

Ni $3d$ spectral densities of states in particles (solid line) and in bulk metal (dotted line) are reported in Figs. 3 and 4 respectively for NiCeY and NiCeX. Normalized intensities are plotted versus binding energy relative to the Fermi level. They are very different. NiCeY shows a maximum in coincidence with that of bulk metal and a strong resonant emission [10] as

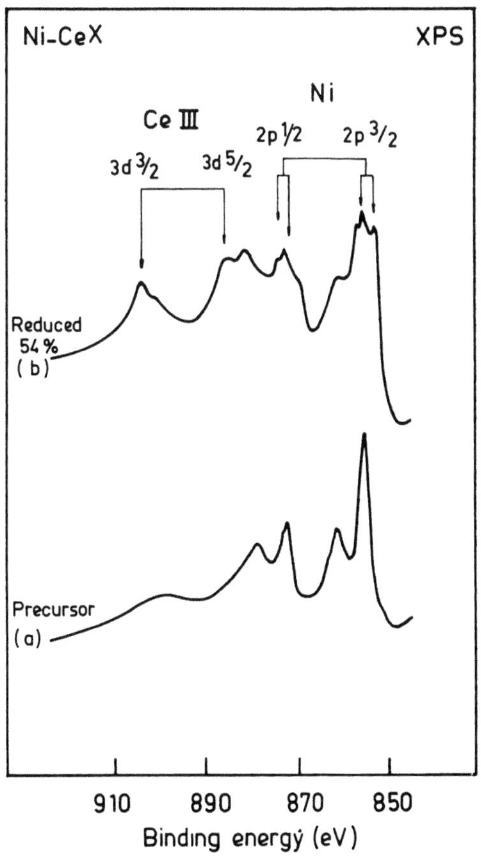

Fig. 2. XPS spectra of NiCeX

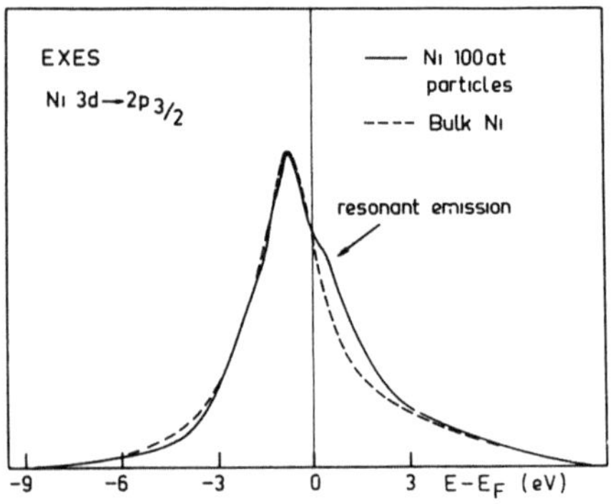

Fig. 3. EXES spectrum of NiCeY compared to the bulk

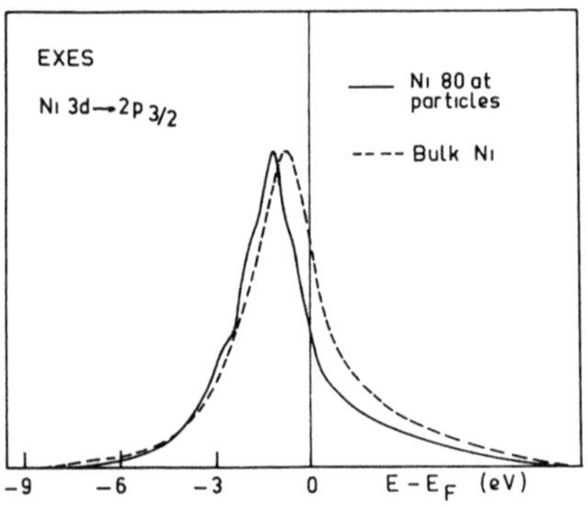

Fig. 4. EXES spectrum of NiCeX compared to the bulk

proved by the enhanced intensity in the range 0 to 2 eV above the Fermi level. On the contrary NiCeX has its maximum shifted towards higher binding energy compared to bulk metal and does not show any resonant emission. The density of states at the Fermi level is much larger for NiCeY than for NiCeX.

A common feature of the particle spectra is the presence of structures which do not exist in bulk metal. As already reported [11–13] these structures are observed when the number of atoms is of the order or less than one hundred per particle. They are due to a size effect. The decrease of the mean coordination number leads to a decrease in the overlaping between the *d* orbitals of neighbouring nickel atoms and then to a strong localization of the *d* states. This leads to a structured valence band. If empty *d* orbitals are present in the ground state of the particle, a $2p^5 3d^{10}$ excited state can be created; its radiative decay gives rise to a resonant emission.

In NiCeX structures are also present showing that the states are strongly localized, but the shift of the maximum towards higher binding energies and the absence of resonance indicate that empty *d* states are completely missing. This confirms our previous assumption of a nearly $3d^{10}$ configuration in NiCeX [14]. Such a configuration could be stabilized by a small amount of negative charge at the surface of particles. We had suggested that this could be due to Ce^{3+} electrodonation towards Ni with Ce^{4+} formation. Such a species cannot be observed by XPS in the studied samples because of the small number of cerium ions that are concerned. Indeed, substituting Gd^{3+} to Ce^{3+} in X zeolite makes less contracted the Ni3*d* spectral distribution and lower than in NiCeX the density of states at the Fermi level. These results prove the existence of the Ni–Ce^{3+} interaction.

Conclusion

Spectral density of states like catalytic activity are very sensitive to the zeolitic support acidity and to the presence of Ce^{3+} ions. Both properties can be closely correlated.

The decrease of catalytic activity, as observed between particles in zeolite and standard $NiSiO_2$, is attributed to a size effect. Indeed, for 2.5 nm particles such as $NiSiO_2$ the spectral 3*d* density of states is almost similar to that of bulk metal [15, 16]. Smaller particles have more localized 3*d* states and, in the absence of particular process like Ce^{3+}–Ni interaction in NiCeX, they lead to low rates of reaction.

In the case of NiCeX, the rate of the $CO+H_2$ reaction is more than twice that of the standard $NiSiO_2$. Because X zeolite is less electroacceptor than Y zeolite, Ni–Ce interaction can take place; it leads to an electron enrichment of the nickel particles that agrees with the nearly $3d^{10}$ configuration observed. The excess of charge explains that CO dissociation is favoured. For NiCeY we observed a higher density of states at the Fermi level than for NiCeX; we propose that it can explain the better adsorption and higher degree of fragmentation of butane.

References

1. Weiss, P., Forrer, R.: Ann. Phys. (Paris) **5**, 153 (1926)
2. Selwood, P.W.: Chemisorption and magnetization. New York: Academic Press 1975
3. Dalmon, J.A., Martin, G.A.: Proceedings of the 7th International Congress on Catalysis, Tokyo (1980), Part A, p. 402. Amsterdam: Elsevier 1980
4. Sauvion, G.N., Tempère, J.F., Guilleux, M.F., Delafosse, D.: J. Chim. Phys. **81**, 133 (1984)
5. Sauvion, G.N., Guilleux, M.F., Tempère, J.F., Delafosse, D.: Stud. Surf. Sci. Catal. **12**, 229 (1982)
6. Iraqi, B.: Thesis Paris (1987)
7. Praline, G., Roel, B.E., Hance, R.L., Lee, H.I., White, J.M.: J. Electron Spectrosc. Related Phenom. **21**, 17 (1980)

8. Wagner, C.D., Riggs, W.M., Davis, L.E., Moulder, J.F., Muilenberg, G.E.: Handbook of X-ray photoelectron spectroscopy, physical electronics division. Eden Prairie, MN: Perkin-Elmer Corp. 1979
9. Citrin, P.H., Wertheim, G.K.: Phys. Rev. B **27**, 3176 (1983)
10. Bonnelle, C.: Resonant X-ray emission spectroscopy in solids. Bonnelle, C., Mande, C. (eds.), p. 104. Oxford, New York: Pergamon Press 1982
11. Fargues, D., Vergand, F., Belin, E., Bonnelle, C., Olivier, D., Bonneviot, L., Che, M.: Surf. Sci. **106**, 239 (1981)
12. Fargues, D., Vergand, F., Bonnelle, C.: Surf. Sci. **156**, 590 (1985)
13. Vergand, F., Fargues, D., Bonnelle, C.: J. Phys. **48**, C9- 923 (1987)
14. Vergand, F., Iraqi, B., Bonnelle, C., Ramaroson, E., Guilleux, M.F., Tempère, J.F., Delafosse, D.: J. Chim. Phys. **83**, 619 (1986)
15. Vergand, F., Fargues, D., Belin, E., Bonnelle, C.: J. Phys. F. **11**, 1887 (1981)
16. Fargues, D., Vergand, F., Bonnelle, C.: J. Phys. F **14**, 1897 (1984)

Z. Phys. D – Atoms, Molecules and Clusters 12, 431–434 (1989)

Atoms, Molecules
Zeitschrift für Physik D and Clusters
© Springer-Verlag 1989

Vacuum-UV photoemission-photoion coincidence spectroscopy of neutral metal atom clusters

K. Rademann, B. Kaiser, T. Rech, and F. Hensel

Fach Physikalische Chemie, Philipps-Universität Marburg, Hans-Meerwein-Strasse, D-3550 Marburg, Federal Republic of Germany

Received 5 July 1988; final version 30 September 1988

A newly developed photoion-photoelectron Vacuum-UV coincidence spectrometer has been coupled to a supersonic metal atom cluster beam source and has been used to investigate the electronic structure of isolated mercury clusters in the size range from 1 to 110 at several selected discrete excitation energies between 11.3 and 7.1 eV. Excitation of the van der Waals cluster Hg_{10} at the center of the strong $D_{3/2}$-autoionization line at 10.7 eV yields a photoelectron kinetic energy distribution between 0 and 2.5 eV indicating the population of Hg_{10} ionic states, which are also accessible by threshold ionization.

PACS: 35.20.V; 36.20; 36.40

Introduction

The electronic structure of molecular beam isolated metal clusters is for several compelling reasons of great current interest [1–4]. One particularly strong driving force in studying electronic properties of metal atom clusters has been the fact, that there must be a size dependent transition from nonmetallic behaviour in small aggregates to metallic cohesion in larger clusters. As for mercury clusters, such a transition was predicted theoretically [5, 6] to occur at small sizes of about 30 atoms and was observed experimentally by energy resolved mass spectrometry [7] as well as by autoionization mass spectroscopy of the innershell excitation of the 5*d* electrons [8]. This transition is intimately related to the size dependent hybridization [9] of the fully occupied 6*s* and the empty 6*p* levels of the mercury atom. To study the evolution of the *sp*-valence band structure in greater detail, photoelectron spectroscopy would be a most suitable technique to employ. However, as clusters in general appear as a broad distribution in a molecular beam, the size of the cluster has to be determined experimentally, for instance by scattering processes in a crossed molecular beam [10] or by a photoion-photoelectron coincidence detection scheme, which was first applied by Poliakoff et al. [11] several years ago to explore rare gas dimers and trimers in a free jet. It is the main purpose of this paper to show the applicability of the coincidence technique to neutral mercury clusters in a molecular beam.

Experimental

The photoelectron photoion cluster beam spectrometer will be described in greater detail elsewhere [12]. Briefly, the experimental set-up consists of a conventional three chamber molecular beam apparatus used to produce an intense, seeded or non-seeded metal atom cluster beam from a resistively heated stainless steel oven equipped with mechanically drilled nozzles of different size and geometry. The doubly skimmed molecular beam is irradiated 130 mm downstream the nozzle by the pulsed and monochromatized output of an intense vacuum-UV ns flashlamp operated with nitrogen as a free running relaxation oscillator at high repetition rates between 10 and 100 kHz. These light pulses allow for the detection of photoelectrons and photoions by high transmission time-of-flight spectrometers. The mass analyser is a conventional Wiley-McLaren device [13] combined with a modified Daly [14] detector. Electric acceleration fields are applied to the mass spectrometer immediately after the detection of a single photoelectron. The photoelectron spectrometer is an advanced magnetic type 4π-acceptance time-of-flight spectrometer based on the ideas of Kruit and Read [15] and Cheshnovski et al. [16]. Electrons with kinetic energies between 0.1 and several eV are generated in an electrically field free ionization volume of about 5 mm^3. Their trajectories are parallelized by an inhomogenous static magnetic field of 0.5 kG and guided by a homogenous magnetic field of 10 G through a 0.67 m drift tube towards a scintillation detector. The energy resolution of the *pe*-spectrometer depends on the ratio of the two magnetic fields, the length of the drift tube, the ionization volume, the time duration of the ionizing light pulses and the energy resolution of the monochromator, which was set to be 0.1 eV at 10 eV photon energy. An overall energy resolution of $dE/E = 0.1$ at 1 eV kinetic energy was achieved in the present set-up. The data acquisition electronics and coincidence timing was described briefly [17].

Results and discussion

In Figs. 1 and 2 photoelectron-photoion coincidence spectra are displayed for different mercury clusters (or clusters grouped together in a predetermined time window of the mass spectrometer). Three different coincidence spectra excited at a photon energy of 11.1 eV are shown in Fig. 1. In Fig. 2 the excitation energy has been changed slightly to 10.7 eV. The photoelectrons are recorded in a time window of 4000 ns in coincidence with the corresponding cations as indicated. The cluster beam was prepared by coexpanding a mercury-argon mixture through a 450 μm diameter conically shaped 60° nozzle at a temperature of 300°C and a total stagnation pressure of 700 mbar. Under these expansion conditions mercury clusters up to size number 35 were present in the molecular beam. At a flashlamp repetition rate of about 15 kHz only 500 ion-electron pairs per sec were generated and a coincidence counting rate between 1 and 100 s^{-1}, depending on the width of the mass gate, was observed. The *pe*-coincidence spectrum for the mercury atom is exhibited at a kinetic energy of 0.6 eV. This single peak corresponds to the removal of a 6s valence electron and is also observed in the conventional He-I gas phase photoelectron spectrum. The coincidence spectrum for the cluster group 2–25 exhibits photoelectrons with a kinetic energy distribution between 0 and 4.1 eV. This result is consistent with an appearance potential of Hg_{25} of 7 eV. At the excitation energy employed in this experiment most of the mercury ions

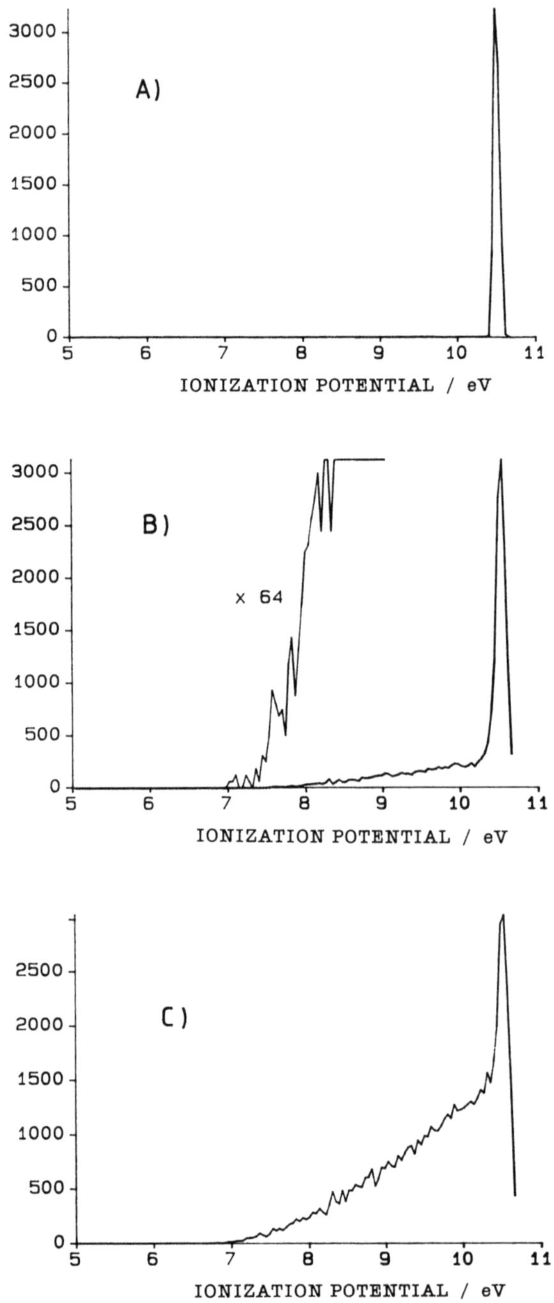

Fig. 1A–C. Photoelectron-Photoion Coincidence spectra recorded consecutively on a cluster beam containing mercury aggregates up to Hg_{35}. Excitation energy $E = 11.1$ eV. **A** Coincidence spectrum for atomic Hg; **B** Coincidence spectrum for mercury particles Hg_1 to Hg_{25}. **C** Coincidence spectrum for mercury clusters Hg_2 to Hg_{25}

are generated predominantly by autoionization of the $D_{3/2}$ line rather than by direct ionization. This becomes more obvious in the case of the photoelectron-photoion coincidence spectrum for a single cluster mass. In Fig. 2 a coincidence spectrum for Hg_{10} is shown. At the given photon energy of 10.7 eV Hg_{10} is excited in the center of the $D_{3/2}$ line [8]. This neutral

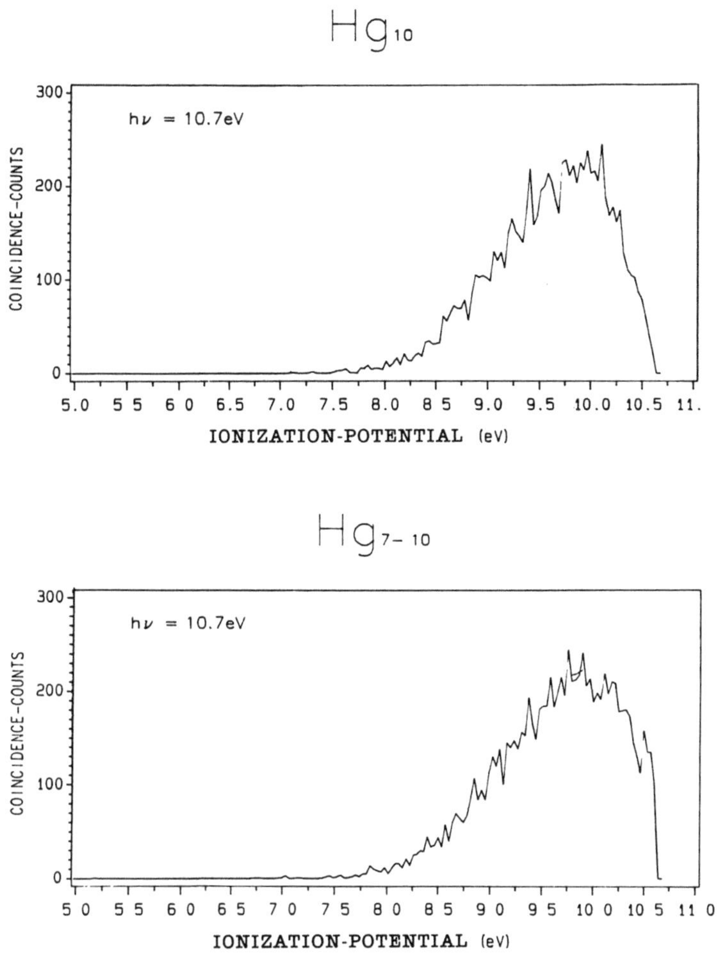

Fig. 2. Photoelectron-Photoion coincidence spectrum for Hg_{10} and Hg_{7-10}. The excitation energy 10.7 eV is predominantly absorbed by the $D_{3/2}$ autoionization line

superexcited state undergoes a rapid autoionization process because of a strong coupling to the ionization continuum. From the onset of the photoemission curve a maximum kinetic energy of 2.5 eV can be extracted which gives an appearance energy for the neutral Hg_{10} of 8.2 eV. This value is in good agreement with the vertical ionization potential reported by Cabaud et al. [18]. Similar observation were made for larger clusters up to size number 23.

Under different expansion conditions at 330°C larger mercury clusters up to size number 110 were present in the molecular and observed in the mass spectrum. Photoemission spectra have been recorded for this fixed mass distribution at three different photon energies 7.1, 8.3 and 11.3 eV. The maximum kinetic energy observed in the photoelectron time-of-flight spectra is consecutively shifted to larger values of 1.7, 2.9 and 5.9 eV respectively. These results were used to assign a vertical ionization potential of 5.4 eV to the largest mercury cluster Hg_{110} observed in the mass spectrum. Further coincidence measurements on selected clusters in the size range between 30 and 100 atoms indicated a size dependent change in the overall shape of the photoemission spectra which is attributed to an increase of the density of states at the ionization threshold. An increase of the total density of states is expected theoretically [9] and should occur because of the overlap of fully occupied 6*s* and empty 6*p* states in this size range.

In conclusion, we have shown for the first time that the photoelectron-photoion coincidence technique can be applied to neutral metal atom clusters prepared in a supersonic molecular beam. The coincidence spectrometer was used to study the photoemission of mercury clusters of different sizes and at various discrete excitation energies between 11.3 and 7.1 eV. Vertical ionization potentials, which were extracted from the photoemission data, are in perfect agreement with the results obtained from energy resolved mass spectrometry [7].

Financial assistance from the Deutsche Forschungsgemeinschaft is gratefully acknowledged. We thank Professor U. Even for valuable help and discussions. One of us (B.K.) gratefully acknowledges a grant from the Land Hessen.

References

1. Proceedings of the International Symposium on the Physics and Chemistry of Small Clusters, Jena, P., Rao, B.K., Khanna, S.N. (eds.). NATO ASI Series B: 158. New York, London: Plenum Press 1987
2. Metal Clusters, Träger, F., zu Putlitz, G (eds.) Berlin-Heidelberg-New York: Springer 1986
3. Elemental and molecular clusters Benedek, G., Martin, T.P., Pacchioni, G. (eds.). Berlin, Heidelberg, New York: Springer 1988
4. Microclusters, Sugano, S., Nishina, Y., Ohnishi, S. (eds.). Berlin, Heidelberg, New York: Springer 1987
5. Tomanek, D., Mukherjee, S., Bennemann, K.H.: Phys. Rev. B **28**, 665 (1983)
6. Miedema, A.R., Dorleijn, J.W.F.: Philos. Mag. B **43**, 251 (1981)
7. Rademann, K., Kaiser, B., Even, U., Hensel, F.: Phys. Rev. Lett. **59**, 2319 (1987)
8. Brechignac, C., Broyer, M., Cahuzac, Ph., Delacretaz, G., Labastie, P., Wolf, J.P., Wöste, L.: Phys. Rev. Lett. **60**, 275 (1988)
9. Pastor, G.M., Stampfli, P., Bennemann, K.H.: Preprint
10. Buck, U.: J. Phys. Chem. **92**, 1023 (1988)
11. Poliakoff, E.D., Dehmer, P.M., Dehmer, J.L., Stockbauer, R.: J. Chem. Phys. **76**, 5214 (1882)
12. Rademann, K.: Ber. Bunsenges. Phys. Chem. (to be published)
13. Wiley, W.C., McLaren, I.H.: Rev. Sci. Instrum. **26**, 1150 (1955)
14. Daly, N.R.: Rev. Sci. Instrum. **31**, 264 (1960)
15. Kruit, P., Read, F.H.: J. Phys. E **16**, 313 (1983)
16. Cheshnovsky, O., Yang, S.H., Pettiette, C.L., Craycraft, M.J., Smalley, R.E.: Rev. Sci. Instrum. **58**, 2131 (1987)
17. Rademann, K., Kaiser, B., Rech, T., Hensel, F.: Z. Phys. Chem. N.F. (submitted)
18. Cabaud, B., Hoareau, A., Melinon, P.: J. Phys. D **13**, 1831 (1981)

Z. Phys. D – Atoms, Molecules and Clusters 12, 435–437 (1989)

Atoms, Molecules
Zeitschrift für Physik D and Clusters
© Springer-Verlag 1989

Fluorescence of rare gas clusters excited with synchrotron radiation

J. Stapelfeldt, J. Wörmer, G. Zimmerer, and T. Möller
II. Institut für Exp. Physik, Universität Hamburg, D-4000 Hamburg, Federal Republic of Germany

Received 5 July 1988; final version 30 September 1988

A cluster beam experiment for fluorescence measurements on rare gas clusters has been built up. First results on the evolution of energy levels of neutral krypton clusters with a cluster size between 200–8500 atoms/cluster are reported. The well known bulk excitons of solid krypton do not merge into the first atomic lines and appear only in clusters larger than 300 atoms/cluster. Smaller clusters absorb only at energies which fit well with surface excitons of the solid.

PACS: 36.40; 33.20; 78.40

1. Introduction

In the past a large number of studies concerning the geometrical structure [1] of rare gas clusters, their stability [2] and ionization potential [3] were performed leading to a detailed knowledge of these properties. On the contrary, information on electronic properties of *neutral* rare gas clusters is nearly missing because of various experimental difficulties. In the past, only one group performed fluorescence experiments on rare gas cluster beams with electron excitation [4]. Among others, the authors observed phase transitions as a function of cluster size.

The aim of our study is twofold, namely:

(a) determination of energy levels of unrelaxed clusters,

(b) investigation of the relaxation processes in electronically excited clusters.

For this purpose, a supersonic beam experiment has been installed behind the high flux beamline Superlumi [5] at HASYLAB (DESY, Hamburg). At present, fluorescence of clusters is excited selectively with spectrally selected synchrotron radiation and monitored without further spectral analysis as a function of excitation wavelength. The cluster beam experiment is a quite natural extension of former investigation of rare gas molecules and solids [6].

In this contribution we present first results concerning electronic excitation of krypton clusters. It is tried to correlate the cluster excitations with the well-known excitations of krypton atoms, molecules, and of solid krypton. It should be pointed out that our data remarkably differ from reflection measurements on dense, gaseous krypton [7]. This indicates that the variation of cluster size and variation of gas density are quite different approaches to bridge the gap between atoms and solids.

2. Experiment

The measurements were carried out at the experimental station Superlumi at HASYLAB, Hamburg [8]. Spectrally selected synchrotron radiation with a band pass of 0.05 . . . 0.25 nm was used to excite the cluster beam in the wavelength range 110 nm to 150 nm.

The cluster beam set-up consists of two vacuum chambers. The first one contains an interchangeable temperature controlled nozzle. It is pumped down to $10^{-2} \ldots 10^{-3}$ mbar with a booster pump. The cluster beam enters the main chamber through an interchangeable skimmer and then crosses the focussed photon beam. At a band pass of 0.25 nm, the photon flux in the intersection region is about $10^{11} \ldots 10^{12}$ photons per second. The main chamber is pumped down to a pressure between 4×10^{-4} and 1×10^{-8} mbar with a cryopump.

Most of the measurements were performed with a conical nozzle (inner diameter 80 μm, opening cone $2\theta = 25°$, length 15 mm) which offers much higher cluster densities than a simple orifice. An optimal overlap with the optical beam was achieved with a relatively large skimmer (diameter 2.5 mm). The stagnation pressure ranges between 0.3 and 5 bar at room temperature. 99.99% Kr and Xe gas was used. The mean cluster size was determined using the concept of corresponding beams.

The fluorescence light was collected perpendicular to both beams with a large spherical mirror and focussed on the CsTe cathode of a channel-plate detector with MgF_2 windows. The signal was recorded with the single-photon counting method. In addition, first preliminary time resolved measurements were performed taking advantage of the time structure of synchrotron radiation. The counting rates ranged between a few hundred and $\sim 10^5$ cps.

From the stagnation pressures given in Fig. 2, the mean cluster sizes in the beam were estimated using the concept of corresponding beams [13]. Hagena introduced a dimensionless parameter, Γ^*, which is independent from the kind of rare gas atoms in a cluster beam, and which describes the degree of condensation. For the calculation of Γ^* we took an equivalent nozzle diameter of 670 μm obtained from a comparison with a simple orifice of 80 μm. The results of electron diffraction of Farges et al [14] on Ar clusters were used for calibration. Then the mean cluster size $\bar{N}$ (in atoms per cluster) is related to Γ^* in the following way,

$$N = 1.6 \cdot 10^{-4} \cdot \Gamma^{*1.95} \tag{1}$$

This formula describes the results of Farges et al. [14] with a deviation of less than 50% in the range $20 \leq \bar{N} \leq 5000$. With this calibration we get the values of $\bar{N}$ given in Fig. 2. The diameters of the clusters range between 18 Å and 70 Å.

3. Results

Figure 1 shows a typical fluorescence excitation spectrum of a Kr cluster beam (stagnation pressure 3.24 bar). It consists of the atomic resonance lines ($5s\,(3/2)_1$, $5s'\,(1/2)_1$) and features which are characteristic for clusters. The width of the atomic lines is mainly caused by the band pass of excitation (0.05 nm). All the other features are much broader.

Except for the resonance lines, the cluster beam is optically thin. From solid state [9] and from gas cell experiments [10] it is known that the quantum efficiency is approximately constant in the range of excitation of Fig. 1. Extrapolating this result to clusters, the fluorescence excitation spectra then display the absorption coefficient in relative units.

The rather broad maxima in the vicinity of the resonance lines are similar to excitonic excitations of solid Kr. This can be seen from Fig. 2 (upper part)

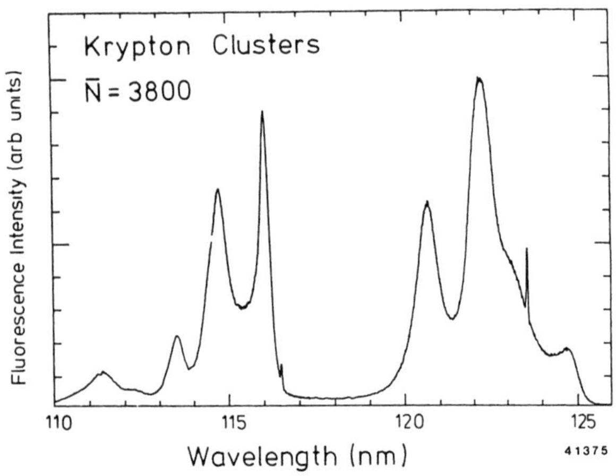

Fig. 1. Fluorescence excitation spectrum ($\lambda_{fl} = 110$ nm-250 nm) of krypton clusters recorded with a band pass of 0.05 nm. (Nozzle parameters: conical nozzle $d = 80$ μm, $2\theta = 25°$, $P_0 = 3.24$ bar $T_0 = 300$ K) The estimated cluster size is 3800 Atoms/cluster. The sharp lines at 123.6 nm and 116.5 nm correspond to the lowest atomic transitions

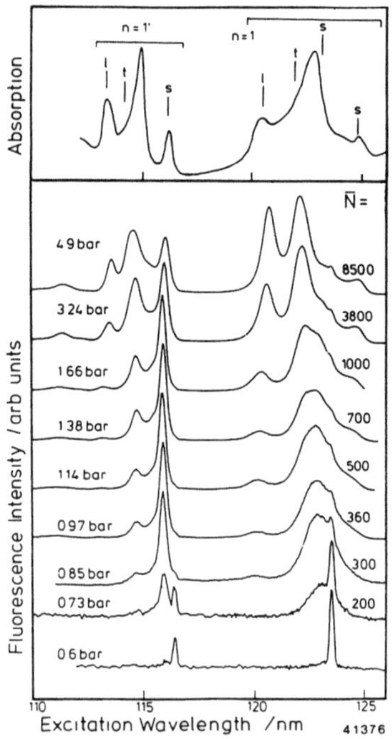

Fig. 2. Fluorescence excitation spectra ($\lambda_{fl} = 110$ nm–250 nm) of krypton clusters as a function of stagnation pressure (cluster size $\bar{N}$ in atoms/cluster). For comparison an absorption spectrum of a thin film (30 Å) of solid krypton taken under grazing incedence ($\alpha = 73°$) is included in the upper part. Additionally energy positions of bulk and surface excitations of the solid [16] are marked

which shows an absorption spectrum of solid Kr [11]. The evolution of these features with stagnation pressure is shown in Fig. 2 (lower part). At a pressure of 0.6 bar, only the resonance lines are visible. With increasing stagnation pressure, two broad, blue shifted bands appear. Molecular, red shifted bands [12] at ~ 125 nm $(X^1\Sigma_g^+ \rightarrow A^3\Sigma_u^+)$, ~ 124 nm $(X^1\Sigma_g^+ \rightarrow B^1\Sigma_u^+)$, and at ~ 117 nm $(X^1\Sigma_g^+ \rightarrow C^1\Sigma_u^+)$ are not observed in the range of pressure investigated. With further increasing stagnation pressure, two additional bands on the high energy side of the resonance lines appear and finally a fourth band at the low energy side of the $5s$ $(3/2)_1$ resonance line. At the highest pressures, in addition, some weak structures around 111 nm are established. The energetic position of some of the features are unaffected, others shift as a function of stagnation pressure.

4. Discussion

As the results have been obtained very recently, only some preliminary remarks and aspects of interpretation can be given here. For the largest clusters observed ($\bar{N} \approx 8500$) nearly a 1:1 correspondence to the absorption of a thin solid Kr film (Fig. 2, upper part) is found (neglecting small deviations of the energetic position). The features at, e.g., 124.7 nm, 123.1 nm and 115.8 nm correspond to surface excitons [15], whereas the pronounced maxima at 122.1 nm and 120.6 nm are tentatively ascribed to the transverse and the longitudinal $n=1$ exciton, the maxima at 114.6 nm and 113.5 nm to the transverse and longitudinal $n=1'$ exciton. It is evident, that the bulk $n=1$ and $n=1'$ excitons do not merge into the atomic lines with decreasing cluster size but are absent in small clusters and show up from a certain cluster size already near to their energetic positions in very large clusters. As an estimate of the threshold we tentatively give $\bar{N} \approx 300$ atoms/cluster.

The smallest clusters yield only broad structures which energetically overlap with the surface excitons of the solid phase. However, at a cluster size of $\bar{N} = 100$–200 atoms/cluster, a discrimination between surface and bulk excitations seems to be inappropriate. At least the features at 123.1 nm and 115.8 nm which show up in large clusters and which correspond to surface excitons of solid Kr, are correlated with the excitation of the smallest clusters. Concerning the long-wavelength peak at 124.7 nm ($\bar{N} = 8500$) it is not clear whether it merges into the tail of the broad 123.1 nm peak for $\bar{N} \approx 200$ clusters or whether it needs a minimum cluster size from which on it shows up.

The faint maxima at 112.3 nm seem to be correlated with $n=2$ surface excitations of solid Kr. Only the largest clusters yield some evidence of $n=2$ bulk excitons which is entirely reasonable in view of the large Wannier radius (~ 10 Å) of $n=2$ excitons in the solid and the respective cluster size. Figure 2 shows, that the peak position of some of the maxima slightly depends on the cluster size. Both, blue- and red shifts are observed with increasing $\bar{N}$.

In the near future, the experiments will be extended to larger and smaller clusters as well as to higher excitation energies. In view of the large counting rates, spectral analysis of cluster fluorescence seems an achievable goal. Then the dynamics of clusters will be investigated in detail. A TOF spectrometer for mass analysis will be attached to the experimental set-up in the near future. For a better understanding of the various excitations observed, the electronic structure of rare gas clusters should be calculated by theorists.

Note added in proof

The average cluster size has been reduced to the dimer. (Stapelfeldt, J., Wörmer, J., Möller, T.: Phys. Rev. Lett. **62**, 98 (1989)

The work was supported by the Bundesminister für Forschung und Technologie of the Fed. Rep. Germany (contract 05 305 AX BO TP7).

References

1. Farges, J., de Feraudy, M.F., Raoult, B., Torchet, G.: J. Chem. Phys. **78**, 5067 (1983)
2. Echt, O., Sattler, K., Recknagel, E.: Phys. Rev. Lett. **47**, 1121 (1981); Ding, A., Hesslich, J., Chem. Phys. Lett. **94**, 54 (1983)
3. Poliakoff, E.D., Dehmer, P.M., Dehmer, J.L., Stockbauer, R.: J. Chem. Phys. **75**, 1568 (1981); Cordis, L., Ganteför, G., Hesslich, J., Ding, A.: Z. Phys. D – Atoms, Molecules and Clusters **3**, 323 (1987)
4. Verkhovtseva, E.T., Bondarenko, E.A., Doronin, Yu.S: Chem. Phys. Lett. **140**, 181 (1987); Verkhovtseva, E.T., Bondarenko, E.A., Yaramenko, V.I., Doronin, Yu.S.: Chem. Phys. Lett. **97**, 483 (1983)
5. Wilcke, H., Böhmer, W., Haensel, R., Schwendtner, N.: Nucl. Instrum. Methods **208**, 835 (1983)
6. Roick, E., Gaethke, R., Gürtler, P., Woodruff, T.O., Zimmerer, G.: J. Phys. C **17**, 945 (1984); Möller, T., Stapelfeldt, J., Behland, M., Zimmerer, G.: Chem. Phys. Lett. **117**, 301 (1985)
7. Laporte, P., Subtil, J.L., Reininger, R., Saile, V., Bernsdorff, S., Steinberger, I.T.: Phys. Rev. B **35**, 6270 (1987)
8. Möller, T., Gürtler, P., Roick, E., Zimmerer, G.: NIMPR A **246**, 461 (1986)
9. Roick, E.: Ph.D Thesis, DESY F41 internal report 84-07, Hamburg 1984
10. Brodmann, R., Zimmerer, G., Hahn, U.: Chem. Phys. Lett. **41**, 164 (1976)
11. Saile, V.: Appl. Opt. **19**, 4115 (1980)
12. La Rocque, P.E., Lipson, R.H., Hermann, P.R., Stoicheff, B.P.: J. Chem. Phys. **84**, 6627 (1986)
13. Hagena, O.F.: Z. Phys. D – Atoms, Molecules and Clusters **4**, 291 (1987)
14. Farges, J., Feraudy, M.F., Raoult, B., Torchet, G.: J. Chem. Phys. **84**, 3491 (1986)
15. Saile, V., Skibowski, M., Gürtler, P., Koch, E.E., Konzevnikov, A.: Phys. Rev. Lett. **37**, 305 (1976)
16. Saile, V.: Ph.D. Thesis, München (1976)

Z. Phys. D – Atoms, Molecules and Clusters 12, 439–442 (1989)

Atoms, Molecules
Zeitschrift für Physik D and Clusters
© Springer-Verlag 1989

Ionization energies of cesium and cesium oxide clusters

H.G. Limberger and T.P. Martin

Max-Planck-Institut für Festkörperforschung, D-7000 Stuttgart 80, Federal Republic of Germany

Received 5 July 1988; final version 30 September 1988

The vertical ionization potentials of 7 cesium and 86 oxidized cesium clusters were determined using the technique of photoionization mass spectrometry. The spectra were obtained using a tunable *cw* dye laser for clusters in a mass range 1 to 2024 amu. The vertical ionization potentials (IP) are presented as a function of size and composition. The ionization energies of cesium clusters, Cs_n, decrease with cluster size. Unusually low IP were observed for the enneamer, Cs_9, and for the cesium monoxide $Cs_{11}O$. With increasing oxidation of the cesium metal clusters the IP decreases (suboxides) reaches a minimum at $Cs(Cs_2O)_n$ and then increases (superoxides).

PACS: 3380E

Introduction

Photoionization efficiency spectra (PIE) and one-photon ionization potentials (IP) have been reported for transition metal clusters composed of iron [1], vanadium, and niobium [2] as well as for many homonuclear and heteronuclear alkali clusters [3–7] and some small alkali monoxide clusters [8]. We have measured PI spectra of oxidized cesium clusters from Cs_n up to Cs_nO_m with $m \leq 2n-1$ [9–11] using a *cw* Ar^+-laser pumped tunable dye laser. Ionization with long wavelength continuous dye laser radiation is possible, because cesium atoms have the lowest ionization energy among the alkali atoms and oxidized cesium clusters are, astonishingly, even more easily ionized. Compared to a pulsed dye laser system this continuous system has the advantage of minimizing peak power and therefore fragmentation due to multiphoton ionizations.

Experiment

The cesium oxide clusters were produced by inert gas quenching [9, 10]. By allowing the elements cesium and oxygen to react in the vapour phase, it is possible to vary the composition of Cs-O clusters. Using a mass spectrometer this composition can be determined. The apparatus is described in more detail in previous publications [9–11]. Briefly the cluster beam apparatus consists of three parts, the cluster condensation cell, the differentially pumped region, and the high vacuum chamber. Within the cell, cesium was evaporated from an oven into cold He-gas containing the oxygen. The cesium oxide clusters condensed out of the supersaturated vapour and passed through a 2.3 mm diameter hole. After passing through a skimmer the clusters entered a high vacuum chamber containing the quadrupole mass spectrometer. There they were ionized in a *one-photon process* by laser photons and mass analyzed.

Results and discussion

PI spectra

Tuning the dye laser wavelength we recorded the frequency dependence of the ion count rate for 86 Cs_nO_m and 7 Cs_n clusters. Ion counting experiments have a large dynamical range, usable signals vary from 1 to 10^5 cps. Therefore, data have to be presented on a logarithmic scale if information is not to be lost, even though such a presentation tends to suppress detailed but reproducible structure. Figure 1 shows a few examples of PI spectra for Cs_n and Cs_nO_m clusters as

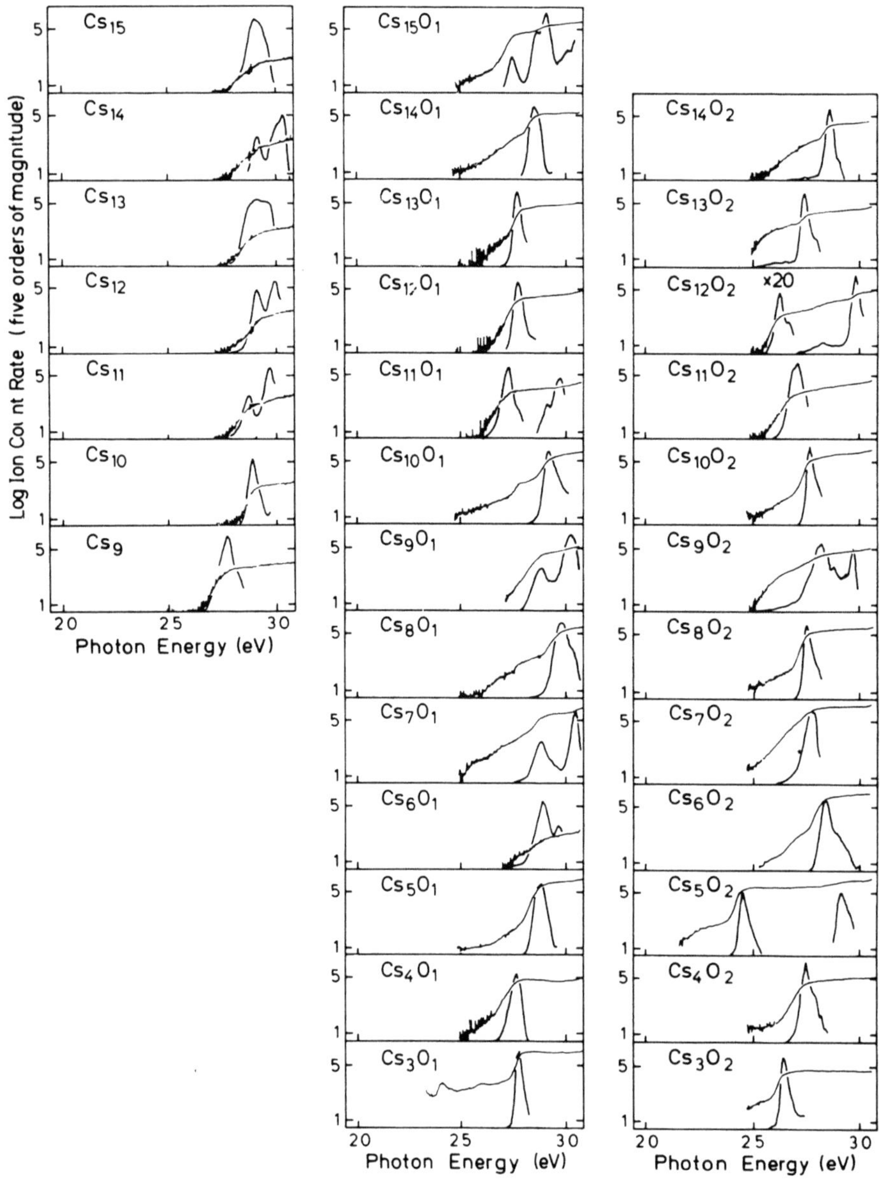

Fig. 1. A few examples of PI spectra (logarithmic) for Cs and Cs-O clusters The derivative of the spectra are linearly drawn

well as the numerically calculated derivatives of these spectra (PID). The IP of the metal clusters with $n \leq 8$ should lie above the maximum dye laser energy (3.1 eV).

How can the ionization potential be extracted from such spectra? The interpretation of PI spectra is made difficult by their richness in spectral features. It is not exceptional to find that PI spectra have more than one well-defined ionization energy. Possible explanations are: *1)* Different final ionic states require different energies for ionization. *2)* The cluster has several isomeric forms, each with it's own threshold. *3)* The cluster ion observed can also be produced as a fragment resulting from ionization of a large cluster.

Assuming that different electronic states appear as distinct peaks in the PID spectrum, we applied the model of displaced harmonic oscillators to interpret each observed peak as arising from one electronic transition from the neutral to the ion. In this case the maximum in the photoionization derivative spectrum (PID) corresponds to the vertical ionization potential and the width of the peak includes the effect of geometry changing, temperature, and vibrational frequency [12].

Ionization potentials of Cs_nO_m clusters

From the measurements of PI spectra in the threshold region, first ionization potentials were obtained showing interesting features not only as a function of size but also the influence of the composition.

The cesium metal clusters Cs_n exhibit a clear drop in the IP as a function of size from $n=8$ to $n=9$. The IP's of clusters with $n \leq 8$ lie above 3.1 eV, whereas for $9 \leq n \leq 15$ the ionization energies can be rather precisely determined from the obtained PI spectra and range from 2.78 eV to 2.93 eV. Therefore, the gap has at least a magnitude of about 0.31 eV. The relatively high IP of Cs_8 indicates a closed shell structure for the octamer which is also observed for other alkali clusters as sodium and potassium [5, 6, 13–15].

The oxidized cesium clusters, Cs_nO_m, show two behaviors: 1) Strong even-odd alternation for clusters containing more oxygen atoms than the pure oxide cluster $(Cs_2O)_m$, i.e. $\frac{1}{2} < \frac{m}{n} < 1$, as a function of cesium atoms per cluster. Cs_nO_m clusters with n odd contain one additional cesium atom, i.e. an unpaired electron, whereas in clusters with n even all electrons are paired, explaining the higher stability and IP. 2) The IP of a 9 electron cluster, $Cs_{11}O$, is lower than that of an 8 electron cluster, $Cs_{10}O$. The observed relatively low IP for $Cs_{11}O$ is somewhat surprising. In suboxide clusters only the oxide anion, O^{2-}, and an excess of Cs atoms are present. Because of the relatively high ionicity, the oxygen anion binds two cesium atoms. Thus 9 free electrons remain from the sea of 11 delocalized valence electrons. Recent measurements with the same cluster source, but a TOF mass spectrometer, showed that $Cs_{2n+z}O_n$ clusters with $z=8$, 18, 34, 58 and 92 have unusually high ionization energies. The anomalies are well-defined for clusters containing one to seven oxygen atoms. The indicated values of z corresponds to the number of free electrons in closed shells of angular momentum. Thus in $Cs_{11}O$ the $1d$ shell will be opened [16].

Will the ionization energy of a Cs cluster increase or decrease when one or more oxygen atoms are added? Figure 2 clearly shows that most oxidized cesium clusters have lower ionization energies than the bare metal clusters. In fact, oxidized cesium has perhaps the lowest work function of any known material, a property utilized for decades in the construction of photocathodes sensitive in the infrared. With increasing oxidation of the cesium clusters, Cs_n, the IP decreases (suboxides), reaches the minimum at $Cs(Cs_2O)_n$ and increases for superoxides. The decrease of ionization energy in the case of suboxide clusters could be explained with a somewhat modified shell model. In a cesium suboxide cluster the oxygen atom could be described as a barrier within the potential well. Therefore, the free electrons are constrained to move in a smaller volume as compared to the bare metal cluster. As a consequence of the Heisenberg uncertainty principle, confinement of the free valence electrons in a smaller space will be associated with an increase of their momentum, and hence, of their energy

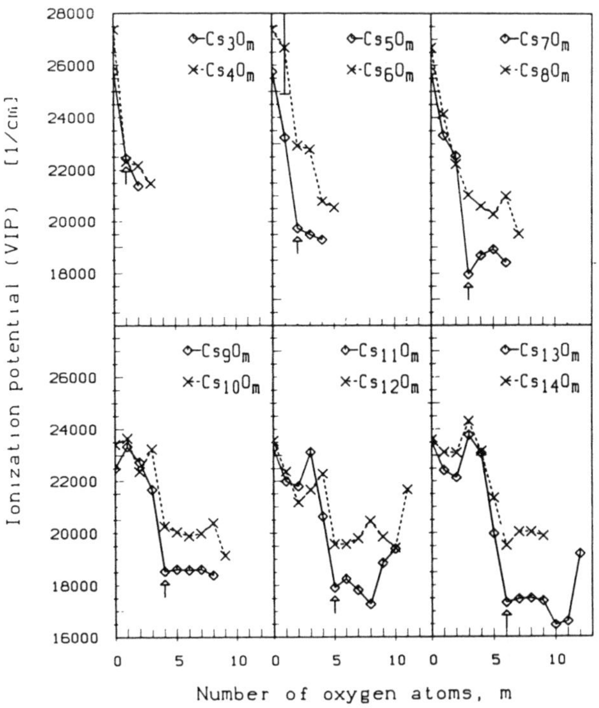

Fig. 2. Ionization potentials of Cs_nO_m clusters as a function of m

[17]. Burt and Heine [18] explained the low work function of bulk cesium suboxides [19] in just this way.

The oxygen dependence shows that the minimum in the ionization energy is reached at $Cs_2(Cs_2O)_m$ for clusters with an even number of cesium atoms per cluster or for an odd number at the 'doped' oxide $Cs(Cs_2O)_m$. 'Doped' oxide clusters containing one excess cesium atom show the lowest IP. The IP difference between 'doped' and stoichiometric cesium oxide clusters is approximately 220 cm^{-1}, possibly increasing with cluster size. Optical transmission measurements on thin films of Cs_2O indicate that the normal oxide is a semiconductor with a band gap of above 2.0 eV. If the material is prepared under conditions which result in an excess of cesium, measurements of the activation energy of electrical conductivity indicate the presence of donor states 0.25 eV below the conduction band [20]. In analogy with the bulk properties, the excess cesium in an oxide cluster may act as a donor state well above the highest occupied orbital (top of the valence band) of the normal oxide, resulting in an appreciable lowering of the ionization energy (work function).

Cesium oxide layers adsorbed on various substrates reduce the work function of the material. The discussion about the cesium oxygen concentration in the range of the minimal work function still continues: Is it cesium suboxide as a layer [21, 22], a doped cesium oxide, $Cs(Cs_2O)$ [23, 24], or, an oxygen-rich cesium, Cs_2O_2? Under the assumption that cluster properties transform with increasing cluster size into the bulk properties, it seems to be clear that doped cesium oxide should have the lowest work function.

References

1. Rohlfing, E.A., Cox, D.M., Kaldor, A.: J. Chem. Phys. **81**, 3846 (1984)
2. Cox, D.M., Zakin, M.R., Kaldor, A.: In. Elemental and molecular clusters. Benedek, G., Martin, T.P., Pacchioni, G. (eds.) Ser. Mat. Science, Vol. 6, p. 329; Berlin, Heidelberg, New York: Springer 1988
3. Kappes, M.M., Schär, M., Röthlisberger, U., Yeretzian, C., Schumacher, E.: Chem. Phys. Lett. **143**, 251 (1988)
4. Herrmann, A., Schumacher, E., Woste, L.: J. Chem. Phys. **68**, 2327 (1978)
5. Bréchignac, C., Cahuzac, Ph.: Z. Phys. D. – Atoms Molecules, and Clusters **3**, 121 (1986)
6. Saunders, W.A., Clemenger, K., de Heer, W.A., Knight, W.D.: Phys. Rev. B **32**, 1366 (1985)
7. Peterson, K.I., Dao, P.D., Farley, R.W., Castleman, A.W., Jr.: J. Chem. Phys. **80**, 1780 (1984)
8. Dao, P.D., Peterson, K.I., Castleman, A.W., Jr.: J. Chem. Phys. **80**, 563 (1984)
9. Martin, T.P.. Angew. Chem., Int. Ed. Engl. **25**, 197 (1986)
10. Martin, T.P.: J. Chem. Phys., **81**, 4426 (1984)
11. Martin, T.P., Limberger, H.G.: In: Elemental and molecular clusters, Ser. Mat. Science. Benedek, G., Martin, T.P., Pacchioni, G. (eds.), Vol. 6, p. 350. Berlin, Heidelberg, New York: Springer 1988
12. Limberger, H.G., Martin, T.P.: J. Chem. Phys. **90**, 2979 (1989)
13. Cohen, M.L., Chou, M.Y., Knight, W.D., Saunders, W.A.: J. Phys. Chem. **91**, 3141 (1987)
14. Martins, J.L., Car, R., Buttet, J.: Surf. Sci. **106**, 265 (1981)
15. Ekardt, W.: Phys. Rev. B **29**, 1558 (1984)
16. Bergmann, T., Martin, T.P., Limberger, H.G.: Phys. Rev. Lett. **60**, 1767 (1988)
17. Woratschek, B., Ertl, G., Küppers, J., Sesselmann, W. Haberland, H.: Phys. Rev. Lett. **57**, 1484 (1986)
18. Burt, M.G., Heine, V.: J. Phys. C **11**, 961 (1978)
19. Woratschek, B., Sesselmann, W., Küppers, J., Ertl, G., Haberland, H.: J. Chem. Phys. **86**, 2411 (1987)
20. Davey, J.E.: J. Appl. Phys. **28**, 1031 (1957)
21. Ebbinghaus, G., Braun, W., Simon, A., Berresheim, K.: Phys. Rev. Lett. **37**, 1770 (1976)
22. Bates, C.W., Jr.: Phys. Rev. Lett. **47**, 204 (1981)
23. Uebbing, J.J., James, L.W.: J. Appl. Phys. **41**, 4505 (1970)
24. Phillips, C.C., Hughes, A.E., Sibbett, W.: J. Phys. D. **17**, 611 (1984)

Z. Phys. D – Atoms, Molecules and Clusters 12, 443–444 (1989)

Atoms, Molecules and Clusters
Zeitschrift für Physik D
© Springer-Verlag 1989

Electronic shell structure in Cs-O clusters

T. Bergmann and T.P. Martin
Max-Planck-Institut für Festkörperforschung, Heisenbergstrasse 1, D-7000 Stuttgart 80, Federal Republic of Germany

Received 5 July 1988

Unusually high ionization energies have been observed for Cs-O clusters having certain sizes and composition, namely for $Cs_{2n+z}O_n$ with $z = 8, 18, 34, 58$ and 92. The anomalies are well-defined for clusters containing from 1 to 7 oxygen atoms. The indicated values of z are identical to the number of electrons in closed shells of angular momentum.

PACS: 07.75; 36.40

1. Introduction

Recently, it has been shown that the electronic properties of alkali metal clusters can be described by using a spherical potential not merely for individual atoms, but for the cluster as a whole [1–10]. In this paper we will present evidence that such a model may also be applicable to compound clusters composed of cesium and oxygen atoms.

2. Experiment

The experiments have been carried out using a pulsed dye laser and a time-of-flight mass spectrometer. This system has a mass range of over 20,000 amu with a mass resolution of 30,000. However, due to the high peak power levels, the laser must be strongly attenuated to suppress multiphoton processes. The laser power was, of course, reduced to levels for which the ion signals were linearly proportional to photon flux. A time-of-flight mass spectrum of oxygen-deficient clusters from 0–15000 amu contains more than one thousand mass peaks. Although each of these mass peaks is well resolved in the experiments, the limited resolution of a printed page prevents us from presenting complete mass spectra. A partial spectrum, from 4000–6000, is shown in Fig. 1. The lines fall in bunches. Each bunch contains a fixed number of Cs atoms with varying numbers of oxygen atoms. We have connected with a solid line those peaks corresponding to clusters containing a fixed number (two) of oxygen atoms. Notice the steep increase in intensity between $Cs_{38}O_2$ and $Cs_{39}O_2$.

Mass spectra, however, yield only very indirect information about the stability and electronic properties of clusters. Therefore, we have studied how the intensity of each mass peak varies with laser wavelength. At the end of an experiment 1000 photoionization spectra are stored in the computer, one for each cluster observed. It is hardly possible to analyze all these spectra individually. Therefore, it was necessary to program the computer to further distill the large amount of data. This was done in the following way: The data points comprising each photoionization spectrum were fitted with a small number of error functions. The first significant peak in the derivative of the fitting function was assumed to define the vertical ionization energy. The ionization energies for a small set of clusters is shown in Fig. 2. Notice that clusters with composition $Cs_{2n+59}O_n$ can be ionized with photons of lower energy than $Cs_{2n+58}O_n$. Such a step was also observed for clusters with composition $Cs_{2n+z}O_n$ for z: $8 \rightarrow 9$, $18 \rightarrow 19$, $20 \rightarrow 21$, $34 \rightarrow 35$, $58 \rightarrow 59$, $92 \rightarrow 93$.

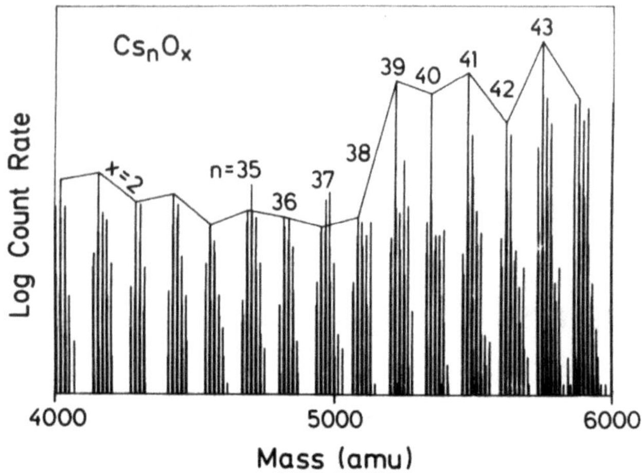

Fig. 1. Portion of a high resolution time-of-flight mass spectrum of Cs-O clusters photoionized with 480 nm laser light. The mass peaks of clusters containing two oxygen atoms have been connected

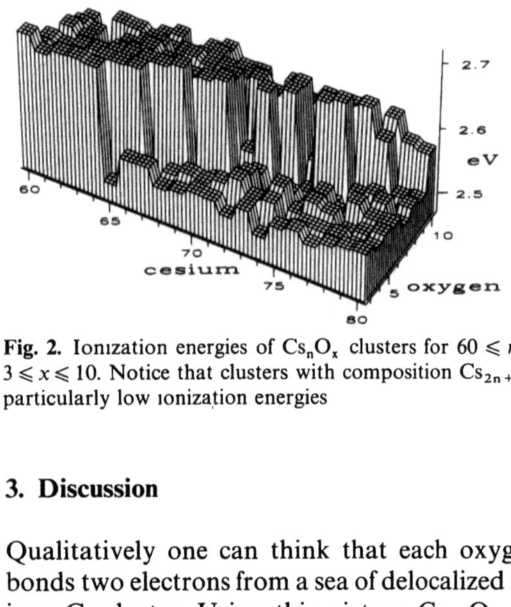

Fig. 2. Ionization energies of Cs_nO_x clusters for $60 \leqslant n \leqslant 80$ and $3 \leqslant x \leqslant 10$. Notice that clusters with composition $Cs_{2n+59}O_n$ have particularly low ionization energies

3. Discussion

Qualitatively one can think that each oxygen atom bonds two electrons from a sea of delocalized electrons in a Cs-cluster. Using this picture $Cs_{38}O_2$, $Cs_{40}O_3$, $Cs_{42}O_4$, etc. would be said to contain 34 delocalized electrons. Our photoionization spectra show that such clusters have high ionization energies. The clusters $Cs_{39}O_2$, $Cs_{41}O_3$, $Cs_{43}O_4$, etc. contain 35 electrons and are observed to have relatively low ionization energies. By choosing the laser photon energy to lie between the sets of high and low ionization energies, we have also been able to demonstrate [11], by means of mass spectra, that clusters containing 8, 18, 34, 58 and 92 delocalized electrons are much more difficult to ionize than the corresponding clusters containing 1 more electron.

These results are consistent with the following simple interpretation. Clusters with composition $Cs_{2n+z}O_n$ have z delocalized electrons which can apparently be described as moving in a spherical potential, just as in pure alkali metal clusters. The orbitals are defined by angular momentum *l* and the observed anomalies in the mass spectra correspond to electron configurations: (1*s*, 1*p*); (1*s*, 1*p*, 1*d*); (1*s*, 1*p*, 1*d*, 2*s*, 1*f*); (1*s*, 1*p*, 1*d*, 2*s*, 1*f*, 2*p*, 1*g*); (1*s*, 1*p*, 1*d*, 2*s*, 1*f*, 2*p*, 1*g*, 2*d*, 1*h*, 3*s*).

References

1. Martins, J.L., Car, R., Buttet, J.: Surf. Sci. **106**, 265 (1981)
2. Ekardt, W.: Ber. Bunsenges. Phys. Chem. **88**, 289 (1984)
3. Knight, W.D., Clemenger, K., Heer, W.A., Saunders, W.A., Chou, M.Y., Cohen, M.L.: Phys. Rev. Lett. **52**, 2141 (1984)
4. Begemann, W., Dreihofer, S., Meiwes-Broer, K.H., Lutz, H.O.: Z. Phys. D. – Atoms, Molecules and Cluster **3**, 183 (1986)
5. Katakuse, I., Ichihara, I., Fujita, Y., Matsuo, T., Sakurai, T., Matsuda, T.: Int. J. Mass Spectrom. Ion Proc. **67**, 229 (1985)
6. Herrmann, A., Schumacher, E., Wöste, L.: J. Chem. Phys. **68**, 2327 (1978).
7. Kappes, M.M., Kunz, R.W., Schumacher, E.: Chem. Phys. Lett. **91**, 413 (1982)
8. Peterson, K.I., Dao, P.D., Farley, R.W., Castleman, A.W., Jr.: J. Chem. Phys. **80**, 1780 (1984)
9. Brèchignac, C., Cahuzac, Ph., Roux, J.-Ph.: Chem. Phys. Lett. **127**, 445 (1986)
10. Saunders, W.A., Clemenger, K., de Heer, W.A., Knight, W.D.: Phys. Rev. B **32**, 1366 (1985)
11. Bergmann, T., Limberger, H., Martin, T.P.: Phys. Rev. Lett **60**, 1767 (1988)

Z. Phys. D – Atoms, Molecules and Clusters 12, 445–450 (1989)

Atoms, Molecules
Zeitschrift für Physik D and Clusters
© Springer-Verlag 1989

Physical properties of high-nuclearity metal cluster compounds

L.J. de Jongh[1], J. Albino O. de Aguiar[1,*], H.B. Brom[1], G. Longoni[2], J.M. van Ruitenbeek[1], G. Schmid[3], H.H.A. Smit[1], M.P.J. van Staveren[1], and R.C. Thiel[1]

[1] Kamerlingh Onnes Laboratorium, Rijksuniversiteit Leiden, Postbus 9506, NL-2300 RA Leiden, The Netherlands
[2] Dipartimento di Chimica Fisica ed Inorganica, Viale del Risorgimento, 4 40136 Bologna, Italia
[3] Institut für Anorganische Chemie, Universität Essen, Universitätsstrasse, S-7, D-4300 Essen, Federal Republic of Germany

Received 21 July 1988; final version 30 September 1988

Metal cluster compounds are composed of large macromolecules, which consist of a metal core (cluster) containing a certain number ($n=6$–560) of metal atoms, to which core a shell of ligands is coordinated. They provide excellent model systems for an assembly of identical metal clusters, embedded in a dielectric matrix. We discuss a number of physical properties of these materials.

PACS: 36.40; 75.20

1. Introduction

Polynuclear metal cluster compounds provide an interesting class of model systems for small metal particles with chemisorbed molecules [1–4]. They are composed of large macromolecules consisting of a metal core of a certain number (n) of metal atoms, to which core a "shell" of ligands is coordinated. Some representative examples are shown in Fig. 1. In going from compound to compound the type of metal atom can be varied, as well as the number n of atoms in the metal core. For Au, Pt, Ru and Rh, the maximum value of n reached at this moment is already $n=55$, namely in $Au_{55}(PPh_3)_{12}Cl_6$ and related Pt, Ru and Rh cluster compounds. For Pd a metal core as large as $n=560$ has recently been reported [5, 6]. The advantages of these materials for fundamental research in small metal particle physics are clearly: *(i)* complete homogeneity of the metal cluster size in each sample; *(ii)* possibility to study the properties as a function of cluster size by comparing compounds with different n; *(iii)* availability of macroscopically large (0.01–10 g) samples for various physical experiments.

Problems of fundamental interest in the physics of small metal particles are: the transition from nonmetallic to metallic behavior as the cluster size is increased, the relation with surface physics and chemistry, the magnetic and the (super)conducting properties, the phonon spectra etc. In our group at Leiden we have recently started a systematic study of the physical properties of these polynuclear metal cluster compounds [7–12]. We combine the measurements of macroscopic thermodynamic variables like the magnetization, susceptibility, specific heat and electrical conductivity, with microscopic probes like neutron scattering and ESR, NMR, and Mössbauer Effect studies on the electronic and nuclear spins of the metal atoms. For instance some of the materials show interesting low-gap semiconducting behaviour [8], with strong non-Ohmic effects in the conductivity at low temperatures. With NMR it is possible to probe the metal nuclei for possible signs of metallic behaviour in e.g. the nuclear spin lattice relaxation, or in the position of the resonance frequencies (Knight-shift) [9]. Mössbauer effect represents also a microscopic probe by which the chemical and physical properties of the metal-core atoms can be studied [10]. Since most of these results are available in the literature and/or reported in the other papers of our group at this conference, we shall concentrate here on the magnetic

* Permanent address: Department de Fisica, Universidade Federal de Pernambuco, 50739 Recife, Brasil

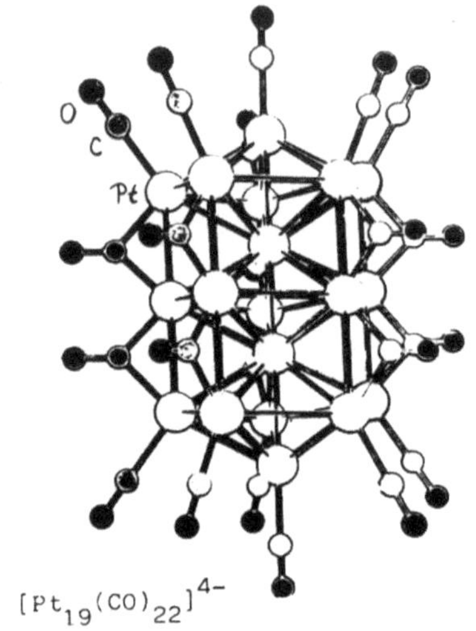

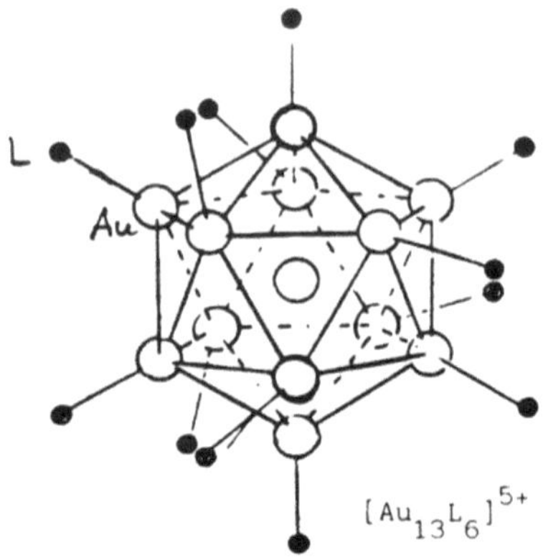

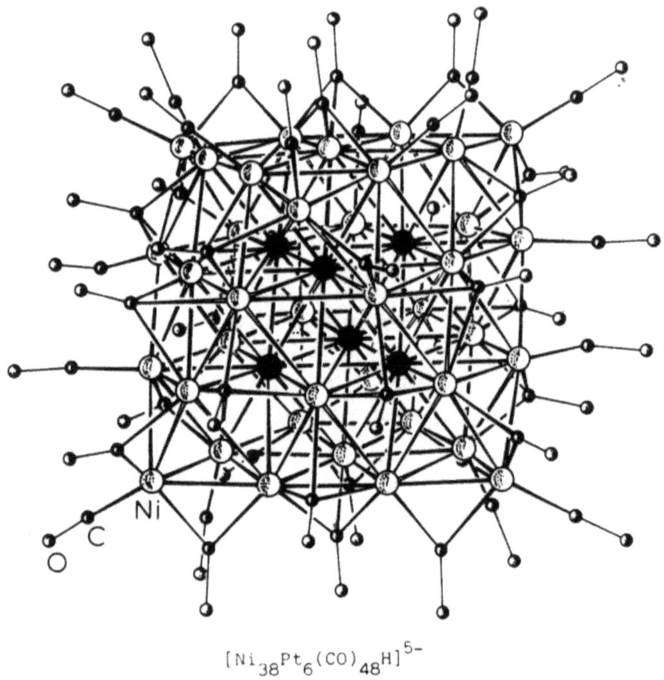

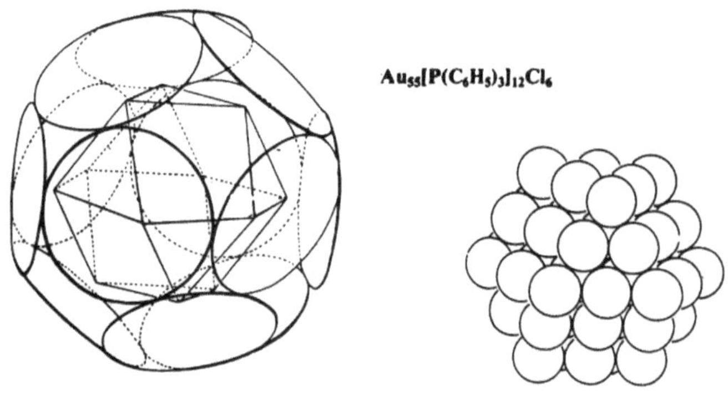

Fig. 1. Some examples of metal cluster compounds containing 19 Pt atoms, 13 Au atoms, 38 Ni and 6 Pt atoms, and 55 metal atoms where the metal can be Au, Pt, Ru, Rh or Co

measurements. The importance of these is that the electronic properties of the cluster molecules should be directly reflected in the behaviour of the susceptibility and the high-field magnetization.

2. Magnetic experiments

Starting with the low-field magnetic susceptibility, we recall that previously [7] we have reported the temperature dependence of the magnetic susceptibility of polynuclear carbonyl clusters, namely $[Pt_{38}(CO)_{44}H_2](PPN)_2$ and a number of Ni clusters, both large (Ni_{34}, Ni_{38}) and small (Ni_9, Ni_{12}). Independently, Teo et al. [13] have published susceptibility data on a large series of Pt carbonyl clusters, ranging from Pt_6 up to Pt_{38}. In all cases the temperature dependence of the susceptibility can be described by the formula

$$\chi(T) = \chi_{dia} + \chi_0 + C_0/(T - \theta) \quad (1)$$

Here χ_{dia} is the negative diamagnetic susceptibility, which is temperature independent, χ_0 is a positive temperature independent contribution and the last term is a contribution in the form of a Curie–Weiss law that dominates the behaviour at low temperatures ($\lesssim 10$ K). The effective moments deduced from the experimental Curie constants C_0 (assuming a spin value $S = \frac{1}{2}$) are quite small for the Pt clusters (0.1–0.5 μ_B per cluster), are only slightly larger for the small Ni clusters ($\simeq 1$ μ_B/cluster), and increase to 4–9 μ_B/cluster for the large Ni clusters. For comparison, we recall that in bulk Ni metal the magnetic moment is about 0.6 μ_B/atom, which would correspond to 20–23 μ_B/cluster for Ni_{34} and Ni_{38}. This evidences a strong reduction of the magnetic moment with respect to the bulk.

Similar results have been obtained for the "Schmid-clusters" [14]. For the clusters of the nonmagnetic metals $Au_{55}[P(C_6H_5)_3]_{12}Cl_6$, $Rh_{55}[P(C_6H_5)_3]_{12}Cl_6$, $Ru_{55}[P(t\text{-}Bu)_3]_{12}Cl_{20}$ and $Pt_{55}[As(tert\text{-}C_4H_9)_3]_{12}Cl_{20}$ the susceptibility is low-valued, although in all cases a weak contribution of the Curie–Weiss form can be clearly distinguished at low temperatures. As an example Fig. 2 shows the results for the Ru_{55} cluster. For the Co_{55} cluster in $Co_{55}(PMe_3)_{12}Cl_{20}$ the susceptibility is considerably higher, although the value found for the magnetic moment of about 5 μ_B/cluster is still very small as compared to the bulk Co metal (in Co metal the moment is 1.7 μ_B/atom, which would give 95 μ_B/cluster for the Co_{55} cluster).

It thus appears that in all these metal cluster compounds there is a certain fraction of unpaired spin density residing on the cluster, the relative amount of which depends on whether the metal atom would be magnetic or nonmagnetic in the bulk, on the cluster size, etc. Also the value of temperature-independent term χ_0 depends on the type of metal atom. We remind the reader of earlier studies on Osmium cluster compounds [15], which likewise show a Curie-term in the susceptibility for cluster with a nuclearity of 10 Os atoms. For a series of small ruthenium carbonyl clusters (Ru_3, Ru_6, . . .) that we are currently investigating [11] in collaboration with B.F.G. Johnson and P.P. Edwards (Univ. of Cambridge), our susceptibility data show again the typical diverging contribution at low temperatures in almost all cases. Some of these data are shown in Fig. 3. We note in particular the deviating behaviour the $Ru_3(CO)_{12}$ cluster, which shows no unpaired spin density (no Curie tail). In fact, whereas all the "magnetic" Ru clusters appear in the form of black, powdered samples, the $Ru_3(CO)_{12}$ cluster forms the exception in that it is a lightly coloured compound. This appears to be a strong indication that the observed moments are indeed intrinsic, i.e. they are not due to magnetic impurities.

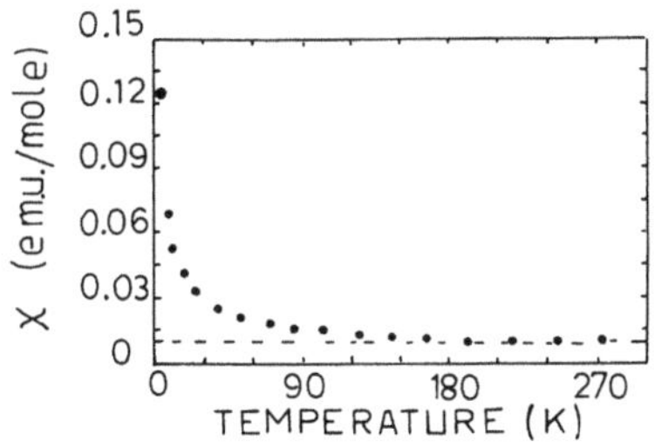

Fig. 2. Magnetic susceptibility versus temperature for the Ru_{55} Schmid-cluster. The data have been corrected for an estimated contribution of $\chi_{dia} = -0.0025$ emu/mole for the diamagnetism of the ligand molecules

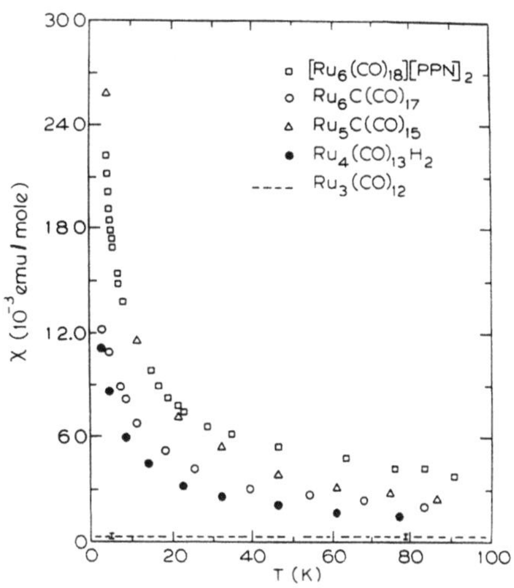

Fig. 3. Susceptibility curves for small Ru-clusters; the data have been corrected for the estimated diamagnetism of the ligands (CO and PPN)

The fact that even such small clusters as Ru_4 and Ru_5 carry a certain amount of unpaired spin density (about 1 μ_B per cluster, similar as for the Ru_{55} "Schmid-cluster"), seems to be important from the point of view of theories for chemical bonding, in which it is generally assumed that the simple electron counting models would suffice for such small clusters.

As already hinted at in the above, we have obviously been worried about the possible presence of magnetic impurities in the samples. Repeating the measurements on different samples of the same material, however, yielded essentially the same results. Only in some cases where the compounds were air-sensitive differences have been observed that could be brought down to partial decomposition of the material, with the associated formation of ultra small free metal particles. These effects have been avoided by keeping the materials and sample holders constantly in nitrogen atmosphere.

Strong evidence that the magnetic behaviour cannot be due to paramagnetic impurities of the usual kind is obtained by measurement of the magnetization in very high magnetic fields. In Fig. 4 the results for the Co_{55} "Schmid-cluster" are shown up to 30 tesla at $T = 4.2$ K [14]. At this low temperature a paramagnetic impurity magnetization should saturate in fields of 2–4 tesla already. By contrast after an initial more rapid increase the magnetization of Co_{55} keeps increasing up to the highest fields in a gradual, almost linear fashion. Previously we have reported similar behaviour for the large Ni clusters [7]. These experiments were performed in the high-field pulse magnet at the Kamerlingh Onnes Laboratory [16]. Finally, Fig. 5a shows that also in the small Ru clusters the magnetization shows no sign of saturation in comparatively high fields of 5 T. When compared to the results shown for one of the large Ni clusters of Fig. 5b, it appears that the behaviour is quite similar except that the magnetization per cluster is a factor of 25 larger in the Ni compound. Note also that the temperature dependent behaviour found indicates that we

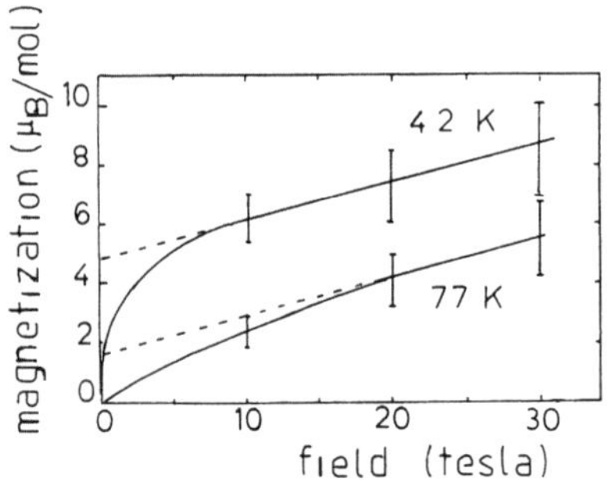

Fig. 4. High-field magnetization curves at $T = 4.2$ K and $T = 77$ K for the Co_{55} Schmid-cluster. (pulsed-field data)

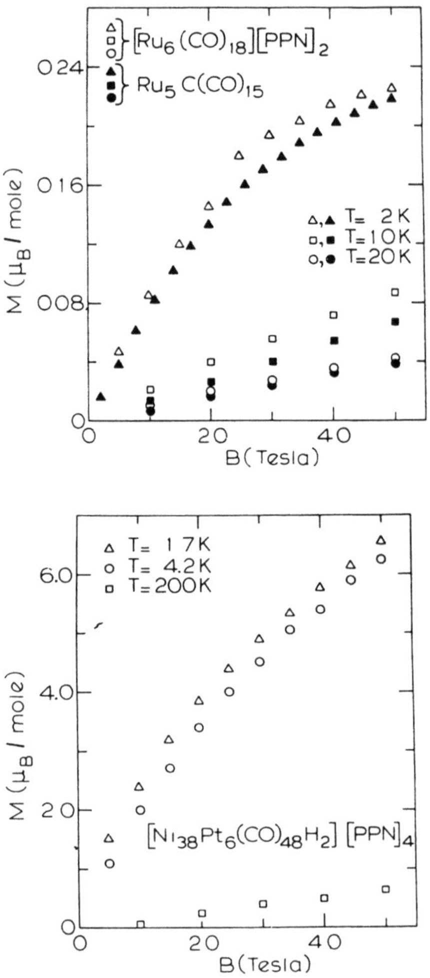

Fig. 5. Magnetometer data for two small Ru-clusters and for a large Ni-cluster

are dealing here with thermal (de)population of levels as far as the low-field part of the magnetization is concerned. From Fig. 4 on the other hand, one would conclude that the high-field contribution may be only weakly dependent on temperature.

Detailed theoretical calculations for the observed magnetic behaviour are still lacking at present. Qualitative explanations run along the following lines. For bare (naked) small metal particles experimental evidence as well as Molecular Orbital calculations [17–19] indicate that the magnetic moment per cluster should roughly correspond to what is estimated on basis of the per atom value for the bulk metal. The moments appear to be due to unpaired spins of electrons in a set of nearly degenerate energy levels close to the Fermi (HOMO-)level. At the same time, experimental results [20] for the effects on the magnetic moment of metal surface atoms due to adsorption of molecules (CO, H_2) as well as M.O. calculations [21–23] for such effects on magnetic clusters, clearly indicate a severe reduction of the magnetic moment. The extremely strong reductions of the magnetic moments of the large Ni and Co clusters with respect to the bulk values may thus be understood from the fact that the majority of the metal atoms are at the surface, where they are ligated to the ligand molecules. According to the M.O. calculations the effect of the ligands is to destabilize one or more levels that are occupied for the bare cluster. The electrons of the destabilized orbital(s) pair-off the originally unpaired spins near the Fermi-level, thereby reducing the total magnetic moment.

It thus appears that the effect of ligand coordination to the metal core has strong analogies with the crystal-field effects on single metal atoms due to their surrounding ligands. These effects form an extensively studied part of the theory of transition metal compounds. Depending on the symmetry and strength of the ligand field the single magnetic metal atom can be in a high spin or in a low-spin state. We are planning to extend these theories to clusters of more than one metal atom.

The appearance of unpaired spin density on the clusters of the nonmagnetic metals may be similarly understood. By the effect of the ligand field the energy level structure will profoundly be changed. The ensuing redistribution of the available electrons may then very well result in some small amount of unpaired spin density. The noninteger values found for the magnetic moments per cluster may be explained in terms of *s-p* hybridization, similar as in the band picture of magnetic metals. The behaviour of the high-field magnetization has strong analogy with the Stoner model for a magnetic metal, where the sub-bands of spin-up and spin-down levels are shifted with respect to one another, and full saturation is thus only reached for a field value equal to the bandwidth. In the metal cluster the energy level structure is discrete due to the finite size, but one can nevertheless distinguish "sub-bands" of spin-up and spin-down levels, as e.g. in the M.O.-calculations. We would therefore attribute the high-field linear part of the magnetization curves to a combination of the transfer from spin-down to spin-up levels by the field, plus the orbital (Van Vleck) paramagnetic contribution arising from field-induced admixtures of the excited levels into the ground state. Both contributions should be independent of temperature in the experimental range. The initial more rapid increase of the magnetization can be ascribed to saturation of the same unpaired spins that are responsible for the Curie–Weiss terms in the magnetic susceptibility. The unpaired spin density can also be seen in ESR experiments on e.g. the Pt carbonyl clusters. The ESR signal has a *g*-value slightly larger than the free-electron one and a temperature dependence that mirrors that of the magnetic susceptibility [9]. Lastly, we remark that similar strong reductions of the moment of a magnetic metal atom due to finite size effects appear to have been observed by Sinfelt in his magnetization studies of Ni catalysts [24].

The work in Leiden is part of the research program of the "Stichting voor Fundamenteel Onderzoek der Materie" (Foundation for Fundamental Research on Matter) and was made possible by financial support form the "Nederlands Organisatie voor Zuiver-Wetenschappelijk Onderzoek" (Netherlands Organization for the Advancement of Pure Research). The investigations are sponsored by the Leiden Materials Science Group ("Werkgroep Fundamental Materialen Onderzoek"). The research program is also sponsored by the Commission of the European Communities.

References

1. Chini, P.: Gaz. Chim. Ital. **109**, 225–240 (1979)
2. Transition metal clusters. Johnson, B.F.G. (ed.). New York: Wiley 1980
3. Schmid, G.: Struct. Bonding Berlin **62**, 51–85 (1985)
4. de Jongh, L.J., Brom, H.B., Longoni, G., Nugteren, P.R., Pronk, B.J., Schmid, G., Smit, H.H.A., van Staveren, M.P.J., Thiel, R.C.: In: Physics and chemistry of small clusters. Jena, P., Rao, B.K., Khanna, S.N. (eds.) p. 807. New York: Plenum (1986); de Jongh, L.J.: In: Organic and inorganic low-dimensional crystalline materials. Delhaes, P., Drillon, M. (eds.), 231, New York: Plenum 1987
5. Vargaftik, M.N., Zagorodnikov, V.P., Stolyarov, I.P., Moiseev, I.I., Likholobov, V.A., Kochubey, D.I., Chuvilin, A.L., Zaikovsky, V.I., Zamaraev, K.I., Timofeeva, G.I.: J. Chem. Soc. Chem. Commun. 937 (1985)
6. Schmid, G.: Nachr Chem. Tech. Lab. **34**, 249 (1987)
7. Pronk, B.J., Brom, H.B., de Jongh, L.J., Longoni, G., Ceriotti, A.: Solid State Commun. **59**, 349 (1986)
8. van Staveren, M.P.J., Brom, H.B., de Jongh, L.J., Schmid, G.: Solid State Commun. **60**, 319 (1986); Proceedings of the 4th International Symposium on Small Particles and Inorganic Clusters, Aix-en-Provence, July 1988
9. Pronk, B.J., Brom, H.B., Longoni, G., Ceriotti, A.: Solid State Commun. **64**, 7 (1987); van Staveren, M.P.J., Brom, H.B., de Jongh, L.J., Schmid, G.: Proceedings of the 4th International Symposium on Small Particles and Inorganic Clusters, Aix-en-Provence, July 1988
10. Smit, H.H.A., Thiel, R.C., de Jongh, L.J., Schmid, G., Klein, N.: Solid State Commun. **65**, 915 (1988); Physica B (to be published) Proceedings of the 4th International Symposium on Small Particles and Inorganic Clusters, Aix-en-Provence, July 1988
11 Albino O. de Aguiar, J., Mees, A., Darriet, J., de Jongh, L.J., Drake, S., Edwards, P.P., Johnson, B.F.G., Lewis, J.: Solid State Commun. **66**, 913 (1988)
12. Albino O. de Aguiar, J., Brom, H.B., de Jongh, L.J., Schmid, G.: Proceedings of the 4th International Symposium on Small Particles and Inorganic Clusters, Aix-en-Provence, July 1988
13. Teo, B.K., Disalvo, F.J., Waszczak, J.W., Longoni, G., Ceriotti, A.. Inorg. Chem. 25:2262 (1986)
14. Mees, A., Darriet, J., Albino O. de Aguiar, J., Jurgens, M.J.G.M.: unpublished York
15. Benfield, R.E., Edwards, P.P., Stacey, A.M.: J. Chem. Soc. Chem. Commun. 525 (1982)
16. Jordaan, H.A., Wolf, R., de Klerk, D.: Physica **69**, 129 (1973)

17. Messmer, R.P., Knudson, S.K., Johnson, K.H., Diamon, J.B., Yang, C.Y.: Phys. Rev. B **13**, 1396 (1976)
18. Yang, C.Y., Johnson, K.H., Salahub, D.R., Kaspar, J., Messmer, R.P.: Phys. Rev. B **27**, 5673 (1981)
19. Lee, K., Callaway, J., Dhar, S.: Phys. Rev. B **30**, 11724 (1984); Lee, K., Callaway, J., Kwong, K., Tang, R., Ziegler, A.: Phys. Rev. B **31**, 1796 (1985)
20. See e.g. Feigerle, C.S., Seiler, A., Peña, J.L., Celotta, R.J., Pierce, D.T.: Phys. Rev. Lett. **56**, 2207 (1986)
21. Salahub, D.R., Raatz, F.: Int. J. Quantum Chem., Symp. **18**, 173 (1984); Surf. Sci. **146**, L609 (1984)
22. Holland, G.F., Ellis, D.E., Trogler, W.C.: J. Chem. Phys. **83**, 3507 (1985)
23. Pacchioni, G., Fantucci, P.: Chem. Phys. Lett. **134**, 497 (1987); Proceedings of the 4th International Symposium on Small Particles and Inorganic Clusters, Aix-en-Provence, July 1988
24. Sinfelt, J.H.: Catal. Rev. **3**, 175 (1969); CRC Crit. Rev. Sol. State Sci. 311 (1974); Rev. Mod. Phys. **51**, 3 (1979)

Z. Phys. D – Atoms, Molecules and Clusters 12, 451–452 (1989)

Atoms, Molecules
Zeitschrift für Physik D and Clusters
© Springer-Verlag 1989

NMR of metal-core bonded ^{31}P in polynuclear metal cluster compounds

M.P.J. van Staveren[1], H.B. Brom[1], L.J. de Jongh[1], and G. Schmid[2]

[1] Kamerlingh Onnes Laboratorium, Rijksuniversiteit Leiden Postbus 9506, NL-2300 RA Leiden, The Netherlands

[2] Institut fur Anorganische Chemie, Universität Essen Universitätsstrasse 5–7, D-4300 Essen 1, Federal Republic of Germany

Received 5 July 1988

In this paper we present measurements of nuclear relaxation times of ^{31}P in the polynuclear cluster compounds $Au_{55}(PPh_3)_{12}Cl_6$ and $Ru_{55}(P(t\text{-}Bu)_3)_{12}Cl_{20}$. Above 15 K the data can be described as a Korringa process, while below 15 K the relaxation time appears to be thermally activated.

PACS: 76.60; 36.40

NMR is a valuable tool to study the properties of small (metal) particles. By examining lineshapes, chemical/Knight shifts and relaxation times one can get information about the environment of a nucleus inside the particle or on its surface. In general, one expects that the continuous energy bands of a "bulk" metal particle will split up into a discrete distribution of levels upon decreasing the size of the particle. This so-called Quantum Size Effect will be reflected in the quantities determined by NMR [1]. In particular, the magnetic relaxation time will increase with respect to the bulk due to the decrease of the number of electronic states at the fermilevel that couple to the nuclei by contact or dipolar interactions. If the average level distance is denoted by d then one may expect an activated temperature dependence of the relaxation time T_1 as soon as $k_B T < d$.

In this paper we present preliminary results for the longitudinal relaxation times as a function of temperature of the ^{31}P nuclei in $Au_{55}(PPh_3)_{12}Cl_6$ and $Ru_{55}(P(t\text{-}Bu)_3)_{12}Cl_{20}$. The molecules of these compounds consist of a metal atom core surrounded by a shell of ligands [2]. The metal atom cores have a diameter of approximately 13 Å, making them comparable in size to the small metal particles that can be obtained by sputtering, deposition and condensation techniques [1]. In contrast to the particles produced by these techniques, however, the cores of the cluster molecules are strictly uniform throughout a sample. As part of a general research program on polynuclear cluster compounds in our group [3, 4, 5] we have used NMR to study the microscopic properties of these compounds.

The P atoms of the Au_{55} and Ru_{55} compounds are bonded directly to the metal atom cores, the Ph and t-Bu groups being bonded to the P atoms. This suggests that studying ^{31}P-NMR in cluster molecules may be analogous to studying for example [6] the ^{13}C-NMR of CO molecules chemisorbed on the small Pt particles of a Pt catalyst. In that case it has been found that the delocalized electron states of the metal particles mix with the discrete states of the adsorbate, giving rise to a T_1 that is inversely proportional to the temperature for large enough particle size. Such a temperature dependence is characteristic for relaxation by contact with conduction electrons [7].

The P relaxation times were determined by measuring the magnetization recovery as a function of time after saturation. The T_1's were found to increase with decreasing temperature. In Fig. 1 we have plotted $T_1 T$ for the Ru_{55} compound as a function of temperature. It is seen that above about 15 K the product $T_1 T$ is more or less constant at about 450 sK. This value is rather large (Ansermet et al. [6] found 45 sK for ^{13}C in CO on Pt, while bulk metals have Korringa constants in the range 1–15 sK). If the data above 15 K are

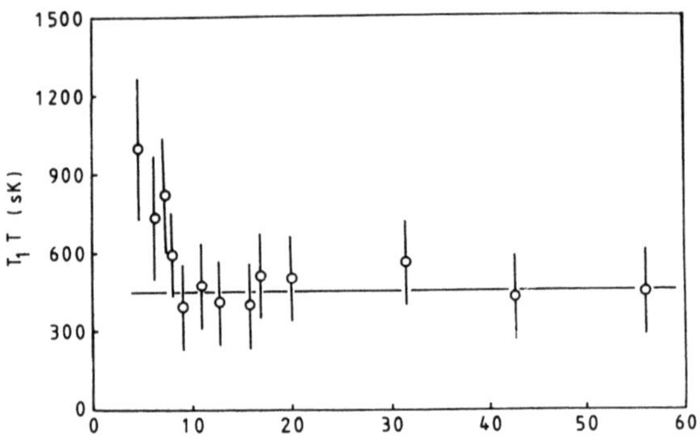

Fig. 1. T_1T as a function of temperature for $Ru_{55}(P(t\text{-}Bu)_3)_{12}Cl_{20}$

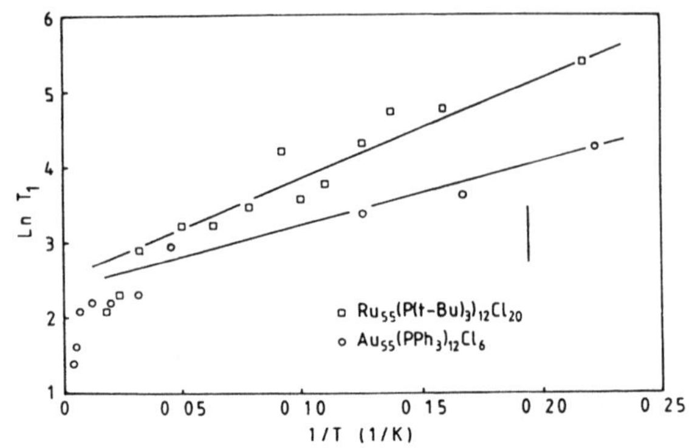

Fig. 2. Logarithm of T_1 as a function of $1/T$

interpreted in terms of a Korringa process, this would indicate a very low density of electrons at the ^{31}P site, since in this case Korringa's constant is inversely proportional to the square of the density of states that are coupled to the ^{31}P nuclei [7]. However, at lower temperatures an upturn to higher values is observed, indicating a T_1 that rises faster than $1/T$. In Fig. 2 we show that the T_1's below 15 K of both the Ru_{55} and the Au_{55} compounds may be described as $T_1 \propto \exp(T_0/T)$ with $T_0 \approx 15$ K for Ru_{55} and 9 K for Au_{55}. The apparent transition from a Korringa-like to an activated behaviour might be interpreted as a sign of a Quantum Size Effect, but at this point it can not be ruled out that the ^{31}P relaxation is due to e.g. paramagnetic fluctuations or unpaired electron spin density on the cluster. Work is in progress to see if there is any correlation between the magnetic susceptibilities of the compounds and the observed relaxation times.

This work is part of the research program of the Stichting voor Fundamenteel Onderzoek der Materie (FOM) and was made possible by financial support from the Nederlandse Organisatie voor Wetenschappelijk Onderzoek (NWO). This work is also part of the research program of the Leiden Materials Science Centre (WFMO) and is supported by the Commission for the European Communities.

References

1. Halperin W.P.: Rev. Mod. Phys. **58**, 533 (1986)
2. Schmid G.: Struct. Bonding Berlin **62**, 51–85 (1985)
3. de Jongh L.J.: In: Organic and inorganic low-dimensional crystalline materials. Delhaas, P., Drillon, M. (eds.), p. 231. New York: Plenum 1987 and references cited therein
4. Pronk B.J., Brom, H.B., Ceriotti, A., Longoni, G.: Solid State Commun. **64**, 7 (1987)
5. Smit H.H.A., Thiel, R.C., de Jongh, L.J., Schmid, G., Klein, N.: Solid State Com. **65**, 915 (1988)
6. Ansermet J.P., Po-Kang Wang, Schlichter, C.P., Sinfelt, J.H.: Phys. Rev. B **37**, 1417 (1988)
7. Schlichter, C.P.: Principles of magnetic resonance, 2nd Edn. Berlin, Heidelberg, New York: Springer 1980

Z. Phys. D – Atoms, Molecules and Clusters 12, 453–455 (1989)

Atoms, Molecules
Zeitschrift für Physik D and Clusters
© Springer-Verlag 1989

Magnetic properties and EPR spectroscopy of molecular metal clusters

R.E. Benfield
Chemical Laboratory, University of Kent, Canterbury CT2 7NH, UK

Received 5 July 1988; final version 30 September 1988

The magnetic properties of molecular metal cluster compounds resemble those of small metal particles in the metametallic size regime. Even-electron metal carbonyl clusters with 10 or more metal atoms are paramagnetic, because their frontier orbital separations of less than 1 eV lead to high-spin electronic configurations. The rhodium cluster $[Rh_{17}S_2(CO)_{32}]^{3-}$ gives EPR below 200 K with $g=2.04$, the first example of this type of paramagnetism in an even-electron carbonyl cluster of this 4*d* metal. Its spectral parameters are compared with those of osmium carbonyl clusters and some significant differences highlighted. Attempts have also been made to generate radical cations from lower-nuclearity, diamagnetic molecular clusters such as $Rh_6(CO)_{16}$ by chemical oxidation in sulphuric acid. An EPR active species ($g=2.09$) believed to be $[Rh_6(CO)_{16}]^+$ has been obtained.

PACS: 33.30; 36.40

1. Introduction

There is increasing interest in the physical properties of transition metal carbonyl cluster molecules, which are good structural models for small metal particles bearing adsorbed carbon monoxide ligands [1]. Their structures have been successfully rationalised by theories based on skeletal electron counting [2–4], which assign the electrons in pairs to closed shells of molecular orbitals. Nonetheless, even-electron carbonyl clusters containing 10 or more atoms of osmium [5], 9 of nickel [6] or 6 of platinum [6, 7] have been found to be in fact weakly Curie paramagnetic. This "anomalous" paramagnetism has been extensively studied in osmium clusters by EPR spectroscopy [8–12].

We have now extended these EPR studies to rhodium carbonyl clusters. $[Rh_{17}S_2(CO)_{32}]^{3-}$ is the first example of an even-electron carbonyl cluster of this 4*d* metal to show this type of paramagnetism. The generation of a cluster radical from the diamagnetic molecule $Rh_6(CO)_{16}$ is also described.

2. EPR of $[Rh_{17}S_2(CO)_{32}]^{3-}$

$[Rh_{17}S_2(CO)_{32}]^{3-}$ has an axially symmetric structure with four stacked squares of rhodium atoms enclosing an interstitial S-Rh-S alignment [13]. The EPR spectrum of its solid $[N(PPh_3)_2]^+$ salt is shown in Fig. 1. It is instructive to compare the form and temperature-dependence of this spectrum with that of $H_2Os_{10}C(CO)_{24}$ [8, 9].

The g-value of $[Rh_{17}S_2(CO)_{32}]^{3-}$ was 2.040 ± 0.002. The smaller g-shift from free spin than in osmium clusters ($g = 2.280$ in $H_2Os_{10}C(CO)_{24}$) arises from the weaker spin-orbit coupling in rhodium. The lineshape of the spectrum reflected the axial symmetry of the $[Rh_{17}S_2(CO)_{32}]^{3-}$ cluster. Additional features arising from hyperfine coupling to ^{103}Rh nuclei were not significantly better resolved by the use of lower modulation or second derivative spectra. The derivative peak-to-peak linewidth ΔHpp of the main resonance was 83 ± 2 Gauss, implying an electronic spin-spin relaxation time T_2 of the order of 10^{-9} s, similar to that found in osmium carbonyl clusters [8, 9, 11].

The g-value, lineshape and resolution of the EPR of $[Rh_{17}S_2(CO)_{32}]^{3-}$ were essentially unchanged between 4 and 150 K. ΔHpp was constant between 4 and 75 K, but increased slightly at higher temperatures. This contrasts with the EPR linewidth of $H_2Os_{10}C(CO)_{24}$, which increases linearly with temperature between 20 K and 120 K [8, 9], and confirms that the spectrum of $[Rh_{17}S_2(CO)_{32}]^{3-}$ is inhomogeneously broadened by unresolved hyperfine coupling to

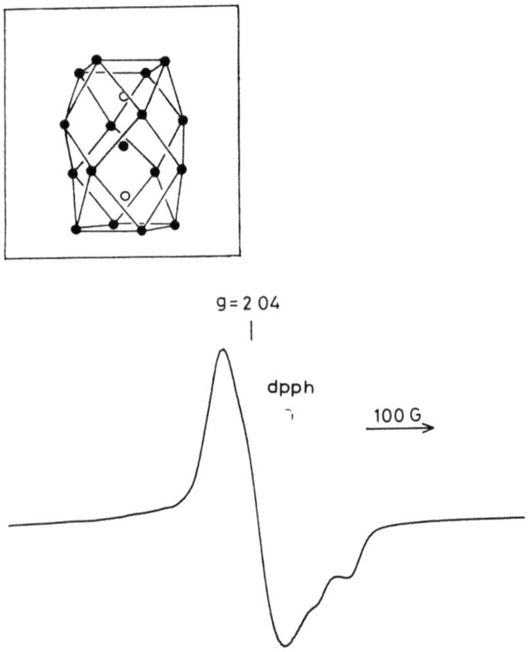

Fig. 1. EPR spectrum (JEOL PE-1-X, X-band, 9.23 GHz) of polycrystalline $[Rh_{17}S_2(CO)_{32}]^{3-}[N(PPh_3)_2]_3^+$ at 10 K. Microwave power 10 mW, modulation 20 G; $g=2.04$, $\Delta H_{pp}=83$ G. Inset· cluster structure of $[Rh_{17}S_2(CO)_{32}]^{3-}$. ● rhodium; ○ sulphur

^{103}Rh nuclei. There are important implications for the degree of delocalisation and *s*-orbital hybridisation component of the spin density in the Rh_{17} cluster.

The EPR intensity of $[Rh_{17}S_2(CO)_{32}]^{3-}$ obeyed the Curie law up to 75 K. At higher temperatures, the intensity reduced more than given by a simple reciprocal temperature-dependence, following instead the Curie-Weiss law with a Weiss temperature of about 80 K. This intensity variation is almost identical to that of $H_2Os_{10}C(CO)_{24}$, despite the completely different temperature-dependence of the spectral linewidths of the two clusters.

3. EPR of the radical cation $[Rh_6(CO)_{16}]^+$

Low-nuclearity metal carbonyl clusters are diamagnetic [14]. The molecule $Rh_6(CO)_{16}$ falls into this class. To probe the nature of the highest occupied molecular orbital of this octahedral metal cluster by EPR spectroscopy, we have attempted to generate the molecular cation radical by chemical oxidation in concentrated sulphuric acid. This classical method for the study of bonding in polyaromatic hydrocarbons [15] has been applied to $Fe(CO)_5$ [16], but not to metal carbonyl clusters, which may broadly be viewed as three-dimensional aromatic systems.

$Rh_6(CO)_{16}$ dissolved slowly in 98% sulphuric acid to give a brown solution. Although this was EPR silent

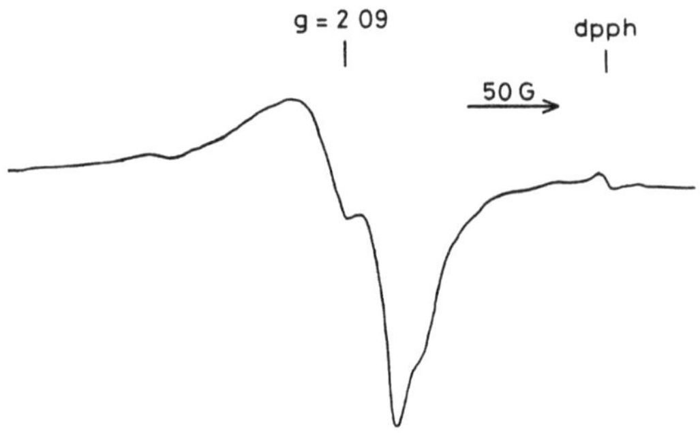

Fig. 2. EPR spectrum (9.19 GHz) of frozen $Rh_6(CO)_{16}/H_2SO_4$ solution at 50 K. Microwave power 8 mW, modulation 8 G; $g=2.09$, $\Delta H_{pp}=63$ G

at room temperature in a quartz capillary, freezing the solution led to observation of the spectrum in Fig. 2. The main resonance at $g = 2.09$ remained unsaturated at temperatures down to 20 K, behaviour typical of the EPR of metal clusters [8, 11], and is assigned to the cation radical $[Rh_6(CO)_{16}]^+$. Two less prominent resonances at $g = 2.00$ and $g = 1.94$ saturated at higher temperatures indicating much longer electronic relaxation times T_1, and are assigned to sulphur-based radicals or mononuclear Rh decomposition products.

Other spectroscopic methods also suggest that all the $Rh_6(CO)_{16}$ in the sulphuric acid solution was oxidised into radicals. No ^{13}C NMR signals could be detected, whereas carbonyl ligand resonances were observable from an extremely dilute solution of $Rh_6(CO)_{16}$ in $CDCl_3$. The UV-visible spectrum also differed completely from that of $Rh_6(CO)_{16}$ in $CHCl_3$.

The *g*-value of the $[Rh_6(CO)_{16}]^+$ resonance was 2.09 ± 0.01, intermediate between the 2.04 of $[Rh_{17}S_2(CO)_{32}]^{3-}$ and the 2.155 of the odd-electron cluster radical $[Rh_{12}C_2(CO)_{23}]^{3-}$ [17]. The spectrum showed partly resolved hyperfine coupling to ^{103}Rh nuclei, and the peak-to-peak derivative linewidth of 63 $\pm$ 2 Gauss was virtually temperature-independent. As in the case of $[Rh_{17}S_2(CO)_{32}]^{3-}$ this suggests inhomogeneous broadening, implying substantial delocalisation of the unpaired electron over the Rh_6 cluster. $[Rh_{12}C_2(CO)_{23}]^{3-}$ behaves differently, showing at 4.2 K a well-resolved *g*-tensor structure and no ^{103}Rh hyperfine splitting; this has been interpreted in terms of the unpaired electron being localised in *d* atomic orbitals [17].

4. Magnetism in molecular clusters and small metal particles

Even-electron osmium carbonyl clusters with fewer than 10 metal atoms are diamagnetic, with a temperature-independent paramagnetic (TIP) susceptibility component which increases with cluster nuclearity as the frontier orbital separation reduces [14]. For $Os_3(CO)_{12}$, the frontier orbital structure has been well-characterised by theoretical and experimental methods. The energy gap between occupied and unoccupied orbitals is 3 eV, comparable with that in octahedral low-spin complexes of the Co^{3+} ion, which are also diamagnetic with a TIP component [18].

Larger even-electron molecular clusters such as $H_2Os_{10}C(CO)_{24}$ and $[Rh_{17}S_2(CO)_{32}]^{3-}$ behave differently. Their weak Curie paramagnetism is quite different from both the diamagnetism with TIP of the lower-nuclearity clusters and the Pauli paramagnetism (also ideally temperature-independent) of bulk metals. This type of paramagnetism has now been observed in clusters of 3*d*, 4*d* and 5*d* metals.

Theoretical calculations predict that the large paramagnetic moments of bare metal particles are quenched by the coordination of ligands [19, 20]. Experimentally though, this effect is not complete in high-nuclearity carbonyl clusters, and a residual paramagnetic moment remains. It is thought that a high-spin electronic configuration is adopted in these clusters when the frontier orbital separation is small enough [8, 9, 10, 21], exactly as in the case of octahedral metal ion complexes ML_6 [22]. The computed energy gaps in decanuclear osmium carbonyl clusters are well under 1 eV [23]. The resulting magnetic properties are well understood in ML_6 complexes, where the unpaired electrons are essentially localised in *d* atomic orbitals on the central metal atom. The picture in molecular clusters is much more complex, because the hybridisation of the metal atomic orbitals and the degree of electron delocalisation are not known. EPR spectroscopy offers the key to this problem by helping to characterise the molecular orbitals containing the spin density.

There are strong resemblances between the magnetic properties of large cluster molecules and those of small metal particles in the size range where quantum-size effects are important. The term "metametallic" has been introduced to describe the distinctive behaviour of these systems [8, 9, 21]. For example, a well-established quantum size effect is the inhibition of electron spin relaxation so that EPR spectra of metal particles are narrower than the conduction electron spin resonances of bulk metals [24]. Exactly the same effect is observed in molecular metal clusters, where the electronic spin-spin relaxation times T_2 are several orders of magnitude longer than in bulk metals, and the spin-lattice relaxation times T_1 are substantially longer still [8, 11].

In studies of this type of effect, molecular metal clusters have many advantages over other metal particle systems. Experimental samples are completely homogeneous in cluster size and shape, with geometries and inter-atomic distances accurately known from X-ray crystallography. Additionally, the problem of charge neutrality [25, 26], which obscures the study of magnetic effects in metal particles, is overcome, since in molecular compounds we can be certain of the electron count in each cluster.

The award of a 1983 University Research Fellowship by the Royal Society is gratefully acknowledged. SERC and the Commission of the European Communities are also thanked for supporting the research programme.

References

1. Chini, P.: J. Organomet. Chem. **200**, 37 (1980)
2. Wade, K.: In: Transition metal clusters. Johnson, B.F.G. (ed.), p. 193. Chichester: Wiley 1980
3. Mingos, D.M.P.: Accounts Chem. Res. **17**, 311 (1984)
4. Teo, B.K., Longoni, G., Chung, F.R.K.: Inorg. Chem. **23**, 1257 (1984)
5. Benfield, R.E., Edwards, P.P., Stacy, A.M.: J. Chem. Soc. Chem. Commun 525 (1982)
6. Pronk, B.J., Brom, H.B., de Jongh, L.J., Longoni, G., Ceriotti, A.: Solid State Commun. **59**, 349 (1986)
7. Teo, B.K., DiSalvo, F.J., Waszczak, J.V., Longoni, G., Ceriotti, A.: Inorg. Chem. **25**, 2262 (1986)
8. Benfield, R.E.: J. Phys. Chem. **91**, 2712 (1987)
9. Benfield, R.E.: In: Physics and chemistry of small clusters. Jena, P., Rao, B.K., Khanna, S.N. (eds.), p. 401. New York: Plenum 1987
10. Drake, S.R., Edwards, P.P., Johnson, B.F.G., Lewis, J., Obertelli, S.D., Pyper, N.C.: Chem. Phys. Lett. **139**, 336 (1987)
11. Drake, S.R., Edwards, P.P., Johnson, B.F.G., Lewis, J., Obertelli, S.D., Pyper, N.C.: J. Chem. Soc. Chem. Commun., 1190 (1987)
12. Benfield, R.E.: J. Organomet. Chem., (in press, 1989)
13. Vidal, J.L., Fiato, R.A., Cosby, L.A., Pruett, R.L.: Inorg. Chem. **17**, 2574 (1978)
14. Johnson, D.C., Benfield, R.E., Edwards, P.P., Nelson, W.J.H., Vargas, M.D.: Nature **314**, 231 (1985)
15. Carrington, A., Dravnieks, F., Symons, M.C.R.: J. Chem. Soc. 947 (1959)
16. Lionel, T., Morton, J.R., Preston, K.F.: J Chem. Phys. **76**, 432 (1982)
17. Beringhelli, T., Morazzoni, F., Strumolo, D.: J. Organomet Chem. **236**, 109 (1982)
18. Ballhausen, C.J.: Introduction to ligand field theory. p. 147. New York: McGraw Hill 1962
19 Holland, G F, Ellis, D.E, Trogler, W C. J. Chem. Phys. **83**, 3507 (1985)
20 Pacchioni, G, Fantucci, P.. Chem. Phys. Lett. **134**, 407 (1987)
21 Jongh, L.J., de Brom, H.B., Longoni, G., Pronk, B.J., Schmid, G, Staveren, M.P.J. van: J. Chem. Res. 150 (1987)
22. Orgel, L.E.: Introduction to transition metal chemistry. 2nd Edn, Chap 3 London. Methuen 1966
23 Bullett, D.W.: Chem. Phys Lett. **135**, 373 (1987)
24 Ozin, G.A.: J. Am. Chem. Soc. **102**, 3301 (1980)
25. Henglein, A., Lilie, J.: J. Am. Chem. Soc. **103**, 1059 (1981)
26. Staveren, M.P.J. van, Brom, H.B., Jongh, L.J., de Ishii, Y: Phys Rev. B **35**, 7749 (1987)

Z. Phys. D – Atoms, Molecules and Clusters 12, 457–459 (1989)

Atoms, Molecules
Zeitschrift für Physik D and Clusters
© Springer-Verlag 1989

EPR on the high-nuclearity palladium cluster $Pd_{561}Phen_{36}O_{200}$

J. Albino O. de Aguiar,[1,*], H.B. Brom[1], L.J. de Jongh[1], and G. Schmid[2]

[1] Kamerlingh Onnes Laboratorium der Rijksuniversiteit Leiden, Postbus 9506, NL-2300 RA Leiden, Nederland

[2] Institut für Anorganische Chemie, Universität Essen, Universitätsstrasse 5-7, D-4300 Essen 1, Federal Republic of Germany

Received 5 July 1988; final version 30 September 1988

We present EPR measurements on the polynuclear metal-cluster compound $Pd_{561}Phen_{36}O_{200}$ for temperatures ranging from 300 K to 8 K. No temperature dependence was found in either the resonance field or the line-width. The signal intensity shows a Curie-like behavior. The effect of the ligand on this behavior is discussed. We also discuss the possibility that quantum size effects govern the observed temperature dependence of the line-width and resonance field.

PACS: 75.60.Jp; 36.40.+d; 76.30.-V

The study of small metal particles has received considerable attention in recent years not only due to their catalytic activity but also because of their unusual physical and chemical properties [1]. Nevertheless a precise characterization of these unusual properties has not yet been achieved due to the unavoidable spread of the particle diameter in the experimentally available samples. On the other hand polynuclear metal cluster compounds formed by a metal core surrounded by a shell of ligands [2] provide a model system of small, identical particles embedded in a dieletric solid [3]. Apart from the size uniformity of the metal core, the possibility of a systematic variation of this core size [2] gives the opportunity to study the physical and chemical properties of these molecular clusters as a function of cluster size. Furthermore one may be able to unravel their electronic structure, and also to draw comparisons with surface effects for bulk metals, as well as the problem of adsorption of molecules at metal surfaces.

As part of our systematic study [4–7] of the physical properties of metal cluster compounds, we present here preliminary results of Electron Paramagnetic Resonance (EPR) on the giant metal cluster $Pd_{561}Phen_{36}O_{200}$, hereafter denoted by Pd_{561}. The measurements were carried out with a Bruker ER 200 D spectrometer operating at 9.45 GHz and employing 100 kHz field modulation. The microwave power incident on the sample was kept constant at 20 mW. No saturation effect was observed on the line for microwave power up to 200 mW. The temperature was varied between 8 to 300 K with the aid of a flow cryostat, and was measured (within ± 1 K) using a chromel vs gold-0.07% iron thermocouple.

The Pd_{561} cluster is obtained by reduction of palladium acetate with H_2 in the presence of small amounts of Phen($C_{12}H_8N_2$) followed by oxidation in air [8]. From the half width of the X-ray lines one can conclude that the Pd atoms are arranged in a ccp structure having a diameter of 24.3 Å in the [111] direction. The 36 phen groups are accommodated on the surface of a cuboctahedron. The exact number of Pd is not precisely determined, but the X-ray experiments [8] suggest that it is very close to 561 corresponding to the idealized five-shell Pd cluster. Also the number of O as well as it localization is not precisely determined. In this work we consider the Pd_{561} cluster to have the formula $Pd_{561}Phen_{36}O_{200}$.

* Permanent address Departamento de Fisica, Universidade Federal de Pernambuco, 50739 Recife, Brasil

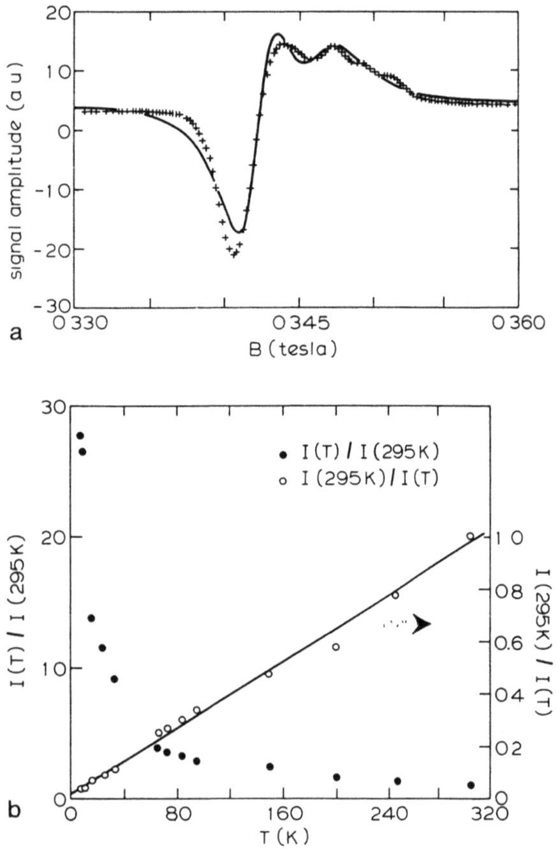

Fig. 1a. Typical EPR signal at 95 K (*crosses*), and fit (*continuous line*) as explained in the text. **b** Temperature dependence of the total intensity (●) of the fitted signal, and it inverse (○) normalized to the value at 295 K

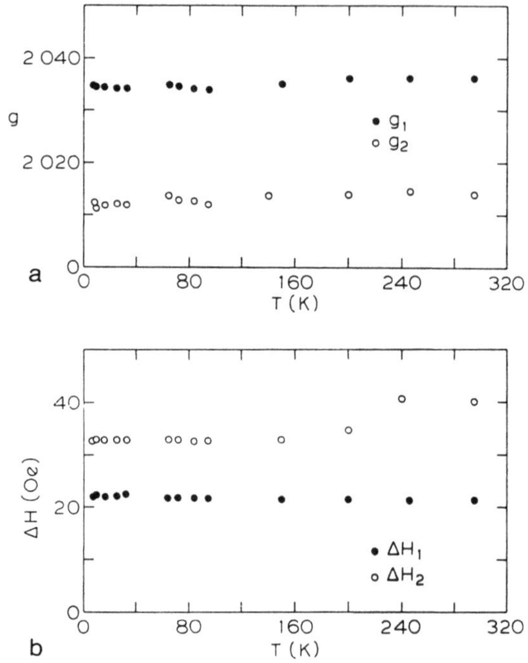

Fig. 2a and b. Temperature dependence of the *g*-values **a** and line-width **b** of the two components of the EPR signal (see text)

In Fig. 1a we show a typical EPR signal (crosses) of the Pd_{561} cluster at 95 K. Also shown is the simulated spectrum (continuous line) obtained from the superposition of two Lorentzian lines, which gives a reasonable fit to the experimental data. We will continue to present our data as being the result of two Lorentzian lines the origin of which will be discussed later. In Fig. 1b we present the temperature dependence of the total intensity, resulting from the sum of the two separate lines, and its inverse normalized to the value at 295 K. The intensity follows a Curie-Weiss law with a Curie-Weiss temperature $\Theta = -6$ K. Using a second sample, that gave essentially the same signal as that reported here (Fig. 1a), and DPPH as a standard a calibration of the signal intensity was made. We calculated that the EPR signal corresponds to 0.02 unpaired electrons per cluster, or one spin per 50 Pd_{561} clusters. Similar spin density was observed in small platinum microcrystals embedded in a gelatin matrix [9].

In Fig. 2 the temperature dependence of the *g*-values (Fig. 2a) is shown together with the line-width at half maximum (Fig. 2b) of the two lines used in the fit. We call the low- and high-field lines line 1 and 2 respectively. Apart from the high temperature ($T \geq 200$ K) data of the line-width of the high-field line, both the *g*-values and line-width of the two lines are temperature independent.

The paramagnetism observed on the Pd_{561} cluster seems to be a general feature of polynuclear metal cluster compounds. It was observed to occur both in high-nuclearity Ni and Pt clusters (metal core with 9 or more metals atoms) [4] and in low-nuclearity Ru clusters (metal core with 4 metal atoms or more) [6]. The origin of this paramagnetic behavior seems to come from the relative small gap between the frontier orbitals of the naked cluster, together with the charge transfer between the metal core and the ligands [4, 6]. The paramagnetic behavior of the Pd_{561} might have the same origin.

The theory of EPR absorption in metals was studied long ago by Dyson [10]. He has shown that for a metal particle with a diameter *d* less than the rf skin depth the line is Lorentzian with a temperature dependent line-width. The line-width is determined by the spin-orbit coupling as proposed by Elliot [11]. If one reduces further the particle diameter electron scattering by the particle surface starts to dominate the relaxation mechanism, and the theory predicts a line-width that diverges with decreasing particle diameter *d*. However, if the particle diameter is sufficiently decreased a situation is reached were the discreteness of the energy levels, the so-called quantum size effects, comes into play. Kawabata [12] studied in detail the influence of these effects on the EPR absorption. When the particles are sufficiently small, i.e. when they satisfy the Kawabata conditions, non-conservation of energy quenches the normal electron scattering process and one obtains a temperature independent Lorentzian EPR line-width. The same temperature behavior is obtained for the g-value.

We come now back to our fitting procedure. By using the resistivity value of bulk Pd we find that our sample diameter is much smaller than the rf skin depth ($\simeq \mu$m), therefore we feel justified to use Lorentzian lines to describe the EPR absorption curve. Furthermore, in attempting to describe the effects of the ligands on the EPR line shape we consider our cluster as being composed by two parts, namely, the inner part and the surface shell. The former is composed out of the first four Pd shells having a total of 309 metal atoms, the latter is constituted by the fifth surface metal shell with 252 Pd atoms, to which shell the ligands are attached. In this model the effect of the ligands is to create a site distinction among the Pd atoms and probably to give additional characteristics to the EPR line as for example the structure observed in the high field part of the spectrum. We attribute this structure to unresolved hyperfine coupling to the ^{14}N nuclei. As seen in Fig. 1a the fit obtained by using this shell model is quite reasonable. Another indication of the correctness of this fitting model is shown in Fig. 3, where we see a nice agreement for the temperature dependence of the intensity ratio of the two lines (black points) and the theoretical value for the ratio between the number of palladium atoms in the inner to that in the surface shell, v.i.z. 309/252 = 1.23 (dotted line). At the moment we do not have an explanation for the departure from the theoretical value, observed for $T \geq 200$ K. One possible explanation could be the extent of the electron delocalization on the cluster.

We now apply the above outlined EPR absorption theory to our case. By using the data published by Monod [13] for palladium we obtain that in order to satisfy the Kawabata conditions the mean level separation of a Pd cluster should be much larger than 4 mK. For our Pd_{561} cluster with a diameter of 24.3 Å we obtain, using the bulk Sommerfeld constant and molar volume, an average level spacing of 11.2 K. From the above considerations we conclude that the temperature independent behavior of the g-value and line-width (Figs. 2a and b) can be interpreted as a manifestation of quantum size effects.

One must be cautious in the attribution of an observation of quantum size effects [1]. Particulary in the present case one should look to other evidence

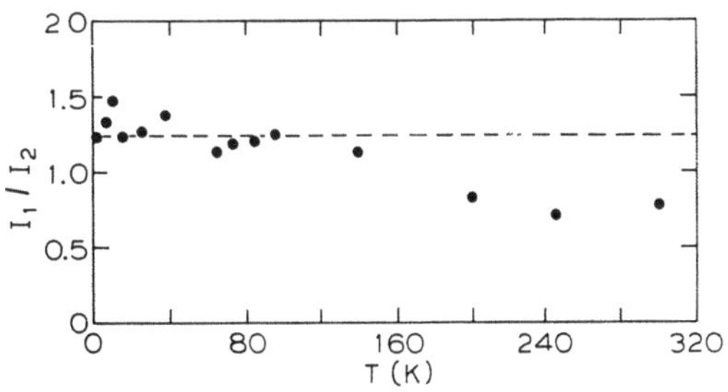

Fig. 3. Temperature dependence of the intensity ratio of the two components of the EPR signal (•), and ratio of the number of Pd atoms in the inner and outer shells (*dotted line*) of the metal core of the Pd_{561} cluster (see text)

predicted by the theory [12]. For example, according to Kawabata [12] the EPR line-width should vary linearly with the microwave frequency and with the square of the particle diameter. Another point to remark is the small number of spins per cluster (0.02). It can be seen as an "impurity" contribution arising for example from chemical doping or small crystallite formation. In our model we assume that no matter its origin, the spin still resides on a Pd cluster of the average size. Studies to clarify this point are under way and will be reported elsewhere.

This work is part of the research program of the Leiden Materials Science Center (WFMO), and is sponsored by the Commission of the European Communities. The authors are grateful to J. Felius for his assistance in the experiments. JAO de A thanks the Brazilian agency CNPq for financial support during his stay in Leiden, and the Kamerlingh Onnes Laboratorium for the hospitality offered to him.

References

1. Halperin, W.P.: Rev. Mod. Phys. **58**, 533 (1986)
2. Schmid, G.: Struct. Bonding Berlin **62**, 51–85 (1985)
3. de Jongh, L.J.: In: Organic and Inorganic Low-Dimensional Crystalline Materials, p. 231–241 Delhaes, P., Drillon, M., (eds) New York: Plenum 1987
4. Pronk, B.J., Brom, H.B., de Jongh, L.J., Longoni, G., Ceriotti, A.: Solid State Commun. **59**, 349 (1986)
5. Smit, H.H.A., Thiel, R.C., de Jongh, L.J., Schmid, G.: Solid State Commun. **65**, 915 (1988), and references therein.
6. de Aguiar, J., Albino, O., Mees, A., Darriet, J., de Jongh, L.J., Drake, S., Edwards, P.P, Johnson, B.F.G., Lewis, J.: Solid State Commun. **66**, 913 (1988)
7. de Aguiar, J., Albino, O., Jurgens, M.J.G.M., Schmid, G., de Jongh, L.J.: in J. Phys. (Paris) Proceedings of ICM/88 (to be published)
8. Schmid, G.: to be published.
9. Gordon, D.A., Marzke, R.F., Glausinger, W.S : J. Phys. C: **2**, 87 (1977)
10. Dyson, F.J.: Phys. Rev. **98**, 349 (1955)
11. Elliot, R.J.: Phys Rev. **96**, 266 (1954)
12. Kawabata, A.. J. Phys. Soc. Japan **29**, 902 (1970)
13. Monod, P.: J. Phys. C: **6**, 1473 (1978)

Z. Phys. D – Atoms, Molecules and Clusters 12, 461–464 (1989)

Atoms, Molecules
Zeitschrift für Physik D and Clusters
© Springer-Verlag 1989

Ac and dc electrical conductivity of polynuclear metal cluster compounds

M.P.J. van Staveren[1], **J.T. Moonen**[1], **H.B. Brom**[1], **L.J. de Jongh**[1], **and G. Schmid**[2]

[1] Kamerlingh Onnes Laboratorium der Rijksuniversiteit Leiden Postbus 9506, NL-2300 RA Leiden, The Netherlands

[2] Institut fur Anorganische Chemie, Universität Essen, Universitätsstrasse 5-7, D-4300 Essen 1, Federal Republic of Germany

Received 5 July 1988; final version 30 September 1988

Ac and dc electrical conductivity data are presented for two polynuclear metal cluster compounds $Au_{55}(PPh)_3)_{12}Cl_6$ and $Pd_{561}Phen_{36}O_{200}$. The temperature dependence of the low field dc conductivity increases with electrical field, the rate of increase becoming stronger with decreasing temperature. These results are compared with data for granular metal systems, and a close correspondence is observed. The conductivity is found to increase with frequency in analogy to what is observed in amorphous systems. A scaling law for low frequencies that has been recently proposed is shown to describe the data quite well.

PACS: 72.80

The study of small metal particles raises a lot of questions concerning the physics and chemistry of systems in the region between atom and bulk [1]. As the properties of systems in this transition region depend strongly on the size of the particles, it is important to reduce the spread in particle size as much as possible. Polynuclear metal clusters compounds provide ideal systems in this respect; the molecules consist of a well defined metal atom core surrounded by a ligand shell [2]. At present core sizes ranging up to 561 atoms for a giant Pd cluster compound have been realized [3].

As part of a general research program on the physical properties of these cluster compounds [4] we have investigated the electrical conductivity of several compounds as a function of temperature, frequency and electrical field strength.

In this paper we will present results for $Au_{55}(PPh_3)_{12}Cl_6$ and for $Pd_{561}Phen_{36}O_{200}$. The exact chemical formula for the Pd compound is not known as yet, but the combined results of x-ray experiments and chemical analysis strongly support the idea that the Pd core is in fact a five shell cluster containing (in the ideal case) 561 atoms.

The measurements were performed on pressed powder pellets with a thickness of about 0.3 mm. Thin gold wires attached to the pellets with Electrodag silverpaint served as contacts. The high impedances of the samples at lower temperatures made it necessary to employ a Keithley 617 electrometer in a two-point configuration for the dc-measurements. A four-point measurement on the Pd_{561} sample at a temperature high enough to allow it, gave results consistent with a two-point measurement, indicating that the influence of the contacts on the data is not important. For the ac-measurements we employed a Hewlett-Packard 4192A LF Impedance Analyzer capable of supplying measurement frequencies between 5 Hz and 13 MHz.

The low field dc conductivity σ_{dc} of the Au_{55} compound was found to be very well described by the relation [5]

$$\sigma_{dc}(T)=\sigma_0 \exp-(T_0/T)^{1/2} \tag{1}$$

with $\sigma_o=7.5$ (Ohm cm)$^{-1}$ and $T_0=57000$ K. This same temperature dependence of the low field σ_{dc} was observed for the Pd_{561} compound between 4 and 80 K, with $\sigma_0=49$ (Ohm cm)$^{-1}$ and $T_0=4500$ K. However, around 80 K a transition to a simple activated behaviour was seen with a gap of 0.02 eV. It is however possible to alternatively fit the data for both compounds over the entire temperature range with

$$\sigma_{dc}(T)=\tilde{\sigma}_0 T^{-1}\exp-(T_0/T)^{\frac{1}{2}} \tag{2}$$

as is shown in Fig. 1 for the Pd_{561} compound. The value of T_0 then increases to 5000 K for the Pd_{561} compound. The fact that a deviation from eq. 1 is only observed in the Pd_{561} compound may then be attributed to the low value of T_0 for this compound.

The deviation from the straight line at low temperatures can be described to non-Ohmic effects which (at fixed field) increase with decreasing temperature. This can be seen for the Pd_{561} compound in Fig. 2 where the field dependence of σ_{dc} is shown at several temperatures. Similar effects were observed in the Au_{55} sample [5].

The peculiar temperature dependence of σ_{dc} as described by (1) as well as the strong field dependence at lower temperatures has also been observed in granular metal systems [6] which are obtained by co-sputtering a metal and an insulator. At small metal volume concentrations these systems consist of small metal grains in a dielectric matrix. Electrical conduction in this regime is assumed to be due to the tunneling and/or hopping of charge carriers between the metal grains. It is generally assumed that the spread in particle sizes and shapes and in inter-particle distances will lead to a distribution of the energies of the charged grains and of the inter-particle hopping/tunneling probabilities.

It should be realized that the sizes of the metal cores in the Au_{55} (13 Å) and the Pd_{561} (24 Å) compounds are in the range of sizes observed in these granular systems. The similarity in microscopic physical appearance is thus obvious, with the important difference that the metal core size is absolutely uniform throughout the cluster samples.

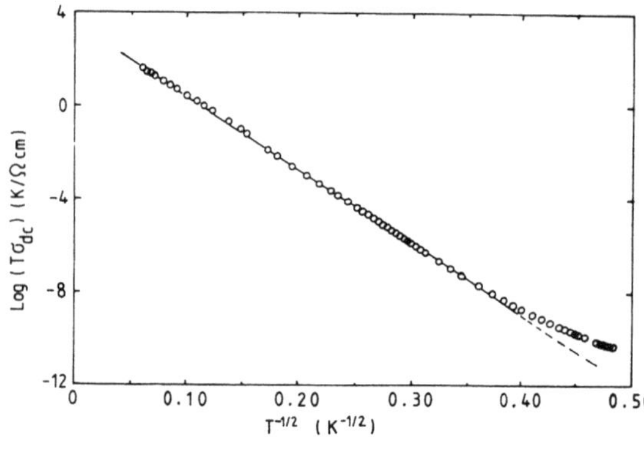

Fig. 1. The logarithm of $T\sigma_{dc}$ against $T^{-1/2}$ for $Pd_{561}Phen_{36}O_{200}$

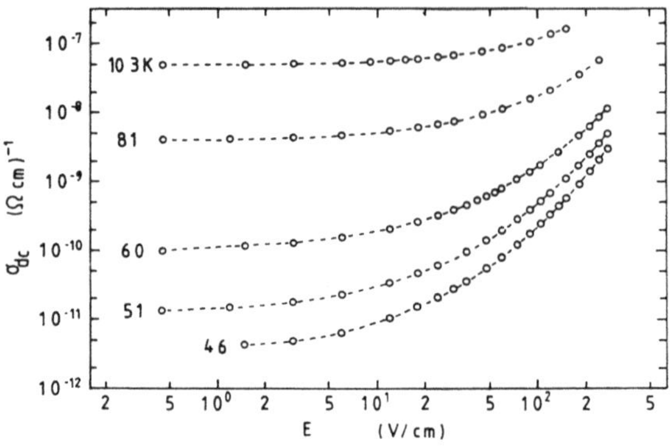

Fig. 2. The logarithm of σ_{dc} against electrical field strength E for $Pd_{561}Phen_{36}O_{200}$

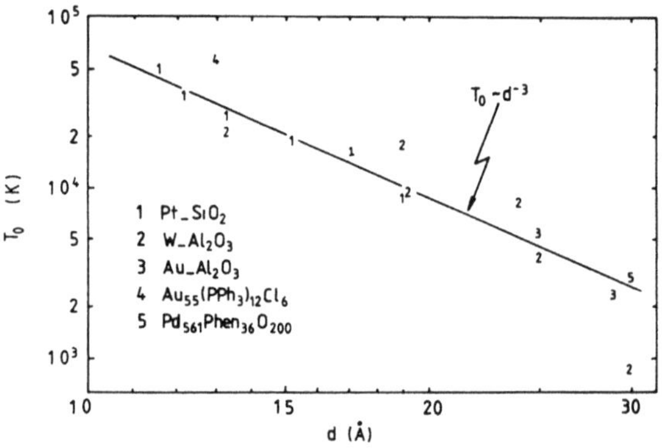

Fig. 3. Log-log plot of T_0 against particle size d

In Fig. 3 we have collected from the literature [6] the values of T_0 in (1) as a function of (mean) particle size for a number of granular metal systems, together with the values of T_0 for the cluster compounds. It is seen that there is a correlation between T_0 and particle size, and that this same correlation seems to be obeyed by the cluster compounds. For particles that are small enough to adopt more or less spherical shapes, the value of T_0 seems to be inversely proportional to the particle volume, and not to the particle radius:

$$T_0 \propto 1/d^3 \tag{3}$$

where d is the particle diameter. The similarities in composition and physical behaviour justify the conclusion that the cluster compounds can be viewed as the first model systems for granular metals without a distribution of particle sizes (although a distribution in hopping/tunneling probabilities between neighbouring macromolecules may still be present since the materials are not crystalline).

The existing models (see e.g. [7, 8]) that are invoked to explain (1) all rely basically on Mott's concept [9] of variable range hopping. Assuming a density of localized states $g(\varepsilon)$ near the Fermi level E_F given by

$$g(\varepsilon)=N_0\varepsilon^p \tag{4}$$

where ε is measured from E_F, and p is a non-negative numerical factor, one obtains [10]

$$\sigma_{dc}(T)=CT^{-2x}\exp-(T_0/T)^x \tag{5}$$

with C a constant and $x=(p+1)/(p+4)$. In these models T_0 scales with the localization length a as

$$T_0 \propto a^{-3/(p+1)} \tag{6}$$

Different assumptions about the physical mechanisms governing the conductivity lead to different values of p. It is seen that taking $p=2$ in (5) leads to (2). This has been interpreted [8] as evidence for the existence of coulomb correlations, which have been shown to lead to $p=2$ [11]. To get the dependence of T_0 on particle size as described by (3), however, one has to take $p=0$ if one assumes that the particle diameter may be equated with the localization length. It thus seems that, if (3) is correct, one cannot describe the dc conductivity data on the basis of the above mentioned assumptions.

It has been recently suggested [12] that the temperature dependence of in granular systems may be due to hopping between states that are "superlocalized" due to the fractal geometry of an incipient infinite percolation cluster. From a.o. these assumptions one gets (3) for T_0 and (1) with an exponent of 3/7 instead of 1/2. We do not know, as yet, whether this approach is correct.

The frequency dependence of the conductivity of the Pd_{561} compound is shown in Fig. 4. It is seen that one can approximately write for the total conductivity at angular frequency ω

$$\sigma(\omega)=\sigma(0)+\gamma\omega^s \tag{7}$$

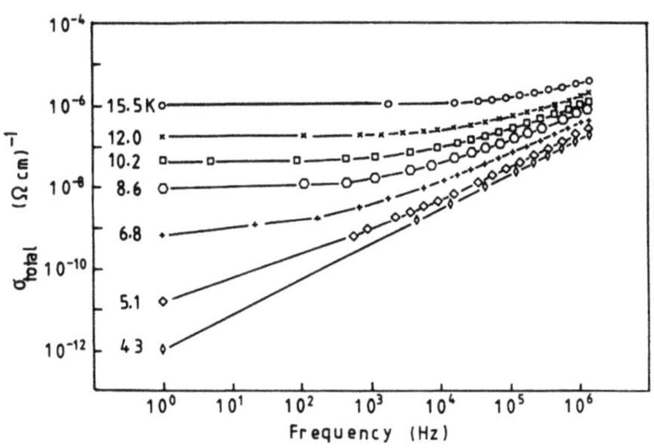

Fig. 4. Frequency dependence of the total conductivity of $Pd_{561}Phen_{36}O_{200}$ at several temperatures

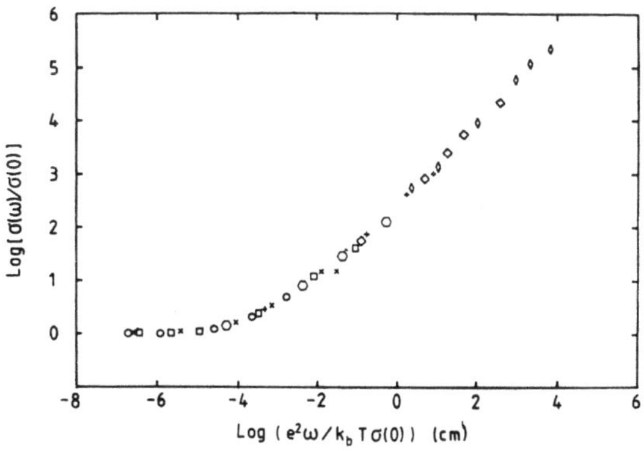

Fig. 5. Normalized total conductivity at angular frequency of $Pd_{561}Phen_{36}O_{200}$ as a function of the scaling variable $e^2\omega/k_BT\sigma(0)$

where γ is approximately constant. The exponent s varies somewhat with temperature and frequency. This same behaviour has been observed in a great number of amorphous semiconductors [13]. The increase of $\sigma(\omega)$ with frequency arises basically from the fact that the dc conductivity is solely due to current paths with an "infinite" extension, while at higher frequencies one also gets contribution from paths that are finite but longer than the distance travelled by a charge carrier during one half period of the alternating fields. In recent years there has emerged a unified theory for both the dc and ac conductivity of disordered hopping systems [14]. This theory is based on an Extended Pair Approximation (EPA) to Kirchhoff's equations for the equivalent resistor/capacitor network of Miller and Abrahams [15]. On the basis of this theory Summerfield [16] has proposed a low frequency scaling relationship for the ac-conductivity of amorphous hopping systems that can be described as

$$\sigma(\omega)/\sigma(0)=f(A\alpha e^2\omega/k_BT\sigma(0)) \tag{8}$$

where, for a particular system, A is a constant and α is the reciprocal localization length of the localized states between which the hopping occurs. It has been proposed that the quasi-universal function $f(x)$ might be written as

$$f(x)=1+x^s \tag{9}$$

with $s=0.725$ [16].

This scaling relation has been observed in a number of systems [15, 14]. In Fig. 5 we have plotted the ac data for the Pd_{561} compound in a way suggested by (8). It is seen that the scaling relationship is obeyed very well. It has been found that the Summerfield scaling also applies to the ac data of the Au_{55} compound. For Au_{55} we get $A\alpha=6.3\ 10^4\ \text{cm}^{-1}$ and $s=0.6$ while for Pd these values are $5\cdot10^3\ \text{cm}^{-1}$ and 0.7 resp. We would like to stress that, as far as we are aware, the present cluster compounds are the first systems to show both a

$T^{-\frac{1}{2}}$ behaviour for σ_{dc} (instead of T^{-1} or $T^{-\frac{1}{4}}$) and the Summerfield scaling law for σ_{ac}.

This work is part of the research program of the Stichting voor Fundamenteel Onderzoek der Materie (FOM) and was made possible by financial support from the Nederlandse Organisatie voor Wetenschappelijk Onderzoek (NWO). This work is also part of the research program of the Leiden Materials Science Center (WFMO) and is supported by the Commission for the European Communities.

References

1. Halperin, W.P.: Rev. Mod. Phys. **58**, 533 (1986)
2. Schmid, G.: Struct Bonding Berlin **62**, 51–85 (1985)
3. Schmid, G.: (to be published)
4. de Jongh L.J.: In: Organic and inorganic low-dimensional crystalline materials. Delhaas, P., Drillon, M. (eds.), p. 231. New York: Plenum 1987
5. van Staveren, M.P.J., Brom, H.B., de Jongh, L.J., Schmid, G.: Solid State Commun. **60**, 319 (1986)
6. Abeles, B., Ping Sheng, Coutts, M.D., Arie, Y.: Adv. Phys. **24**, 407 (1975)
7. Ping Sheng, Klafter, J.: Phys. Rev. B **27**, 2583 (1983)
8. Entin-Wohlmann O., Gefen, Y., Shapira, Y.: J. Phys. C: **16**, 1161 (1983)
9. Ambegaokar V., Halperin, B.I., Langer, J.S.: Phys. Rev. B: **4**, 2612 (1971)
10. Shklovskii, B.I., Efros, A.L.: In: Electronic properties of doped semiconductors. Cardona, M. (ed.). Berlin, Heidelberg, New York: Springer 1984
11. Efros, A.L., Shklovskii, B.I.: J. Phys. C: **8**, 149 (1975)
12. Deutscher G., Levy, Y., Souillard, B.: Europhys. Lett. **4**, 577 (1987)
13. Elliott, S.R.: Adv. Phys. **36**, 135 (1987)
14. Summerfield, S., Butcher, P.N.: J. Phys. C: **15**, 7003 (1982) J. Phys. C: **16**, 295 (1983)
15. Miller, A., Abrahams, E.: Phys. Rev. **120**, 745 (1960)
16. Summerfield, S.: Philos. Mag. **52**, 9 (1985)
17. Abboudy S., Fozooni, P., Mansfield, R., Lea, M.J.: Philos. Mag. Lett. **57**, 277 (1988)

Z. Phys. D – Atoms, Molecules and Clusters 12, 465–467 (1989)

Atoms, Molecules and Clusters
Zeitschrift für Physik D
© Springer-Verlag 1989

Electrical and magnetic properties of thin films containing metal clusters

C. Laurent* **and E. Kay**
IBM Almaden Research Center, 650, Harry Road, San Jose, CA 95120-6099, USA

Received 5 July 1988; final version 30 September 1988

A large variety of granular polymer thin films can be synthetized in a capacitively coupled coplanar diode radiofrequency (r.f.) discharge system in which an argon-monomer mixture is injected at low pressure (20 mTorr). This approach offers several advantages over the other techniques, e.g. easy control of the metal content in the film from a few % up to 100%. The d.c. electrical properties of gold containing plasma polymerized tetrafluoroethylene (PPTFE) and the magnetic behavior of cobalt containing plasma polymerized propane (PPP) are reported in this paper.

PACS: 73.90.+f; 75.70.−i; 77.50.+p.

Introduction

Dispersion of metallic particles in a polymeric matrix is interesting from several points of view. Basic studies such as the investigation of optical or electrical properties can be performed on these composite systems as a function of the metal dispersion. Applications of these materials in the field of optical filters, magnetic or optical recording, have been envisaged. Different techniques are available for the preparation of these films, e.g. plasma polymerization of organometallic molecules, metal evaporation and plasma polymerization, co-sputtering of metal and polymer, reactive etching and plasma polymerization in a capacitively coupled r.f. discharge. This latter technique, extensively explored by E. Kay et al. [1], relies on the concept of physical sputtering of a metal target by ion bombardment from a rare gas plasma. When sufficient halo or hydrocarbon monomer is injected into such a plasma, polymerization dominates at all the grounded surfaces whereas on the r.f. powered metal electrode physical sputtering can be made to dominate but only when this electrode is sufficiently negatively biased with respect to the plasma potential and the partial pressure of the organic monomer is sufficiently low.

If the discharge contains chemical etchant species as is the case, for example, in some fluorocarbon plasmas then ion bombardment enhanced "chemical etching" also contributes to the removal of material from the electrode surfaces. In the case of a chemically inert metal, e.g. gold, no chemical reactions are expected between the metal and the reactive species of the discharge. At metal volume fractions below that necessary for percolation gold is found in the polymeric matrix as almost spherical clusters which are isolated from one another throughout the matrix. The optical and electrical properties of such films have been extensively studied by Kay et al. [2, 3]. The behavior of gold containing PPTFE under high field is reported here. When the metallic target consists of a reactive metal, such as cobalt, then reactions in a fluorocarbon plasma and at the surface of the growing film between fluorine atoms and the metal lead to the formation of metal fluoride particles, CoF, CoF_2, within the organic matrix [4]. The electrical properties of such a composite are no longer controlled by simple percolation between metallic clusters in a uniform dielectric polymeric matrix. In order to avoid such complexity during plasma polymerization in the pre-

* Present address: LGET, Université Paul Sabatier, 118, Route de Narbonne, F-31062 Toulouse Cedex, France

sence of a reactive metal, a hydrocarbon instead of a fluorocarbon can be injected into the plasma, thereby getting rid of the very highly reactive fluorine atoms. As a result, spherical cobalt clusters are formed in the matrix. The magnetic properties of cobalt containing PPP are described below.

Experimental procedure

The films are prepared in a r.f. capacitively coupled diode system already described elsewhere [5]. The residual vacuum in the chamber is 10^{-6} Torr. A plasma is generated in a tetrafluoroethylene or propane monomer-argon mixture, at a total pressure of (20 mTorr). The typical gas flow is 20 $cm^3 min^{-1}$. The electrode to which the radio frequency (13.56 MHz) power is applied consists of a gold or a cobalt sputtering target whereas the grounded electrode holds the temperature controlled substrates (15°C).

The gases are characterized by means of a UTI type 100C quadrupole mass spectrometer. The rate of film deposition is measured with a frequency dependent quartz microbalance mounted near the substrate. Optical emission spectra of the plasma are recorded using an Optical Multichannel Analyzer (OMA II) system equipped with an intensified Si vidicon detector.

The crystallographic structure of the films is determined by X-Ray & Electron diffraction and the morphology by Electron Microscopy. Surface chemical analysis is determined by X-Ray photoelectron spectroscopy (XPS) using a Hewlett–Packard 5950B ESCA spectrometer.

Results

High field electrical properties of gold containing PPTFE

The films obtained showed a wide range of conductivities from insulating (low field resistivity $\rho > 10^{+16}$ Ω.cm) to metallic ($\rho \sim 3.10^{-6}$ Ω.cm) depending on the gas partial pressure ratio during synthesis (Fig. 1). The electrical percolation threshold lies at 38 ±5 vol.% Au. In the dielectric regime, composite film are essentially insulating, but the presence of metallic clusters introduces additional conduction mechanisms in comparison with metal-free polymers. As an example, the current-voltage characteristic (J-V) of a 28% gold containing sample is shown in Fig. 2. A reversible switching transition occurs for field threshold around

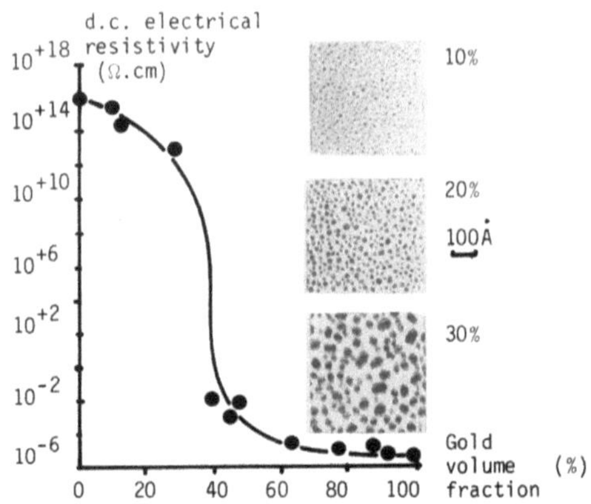

Fig. 1. Low field resistivity versus gold volume fraction, together with film microstructure in the dielectric regime

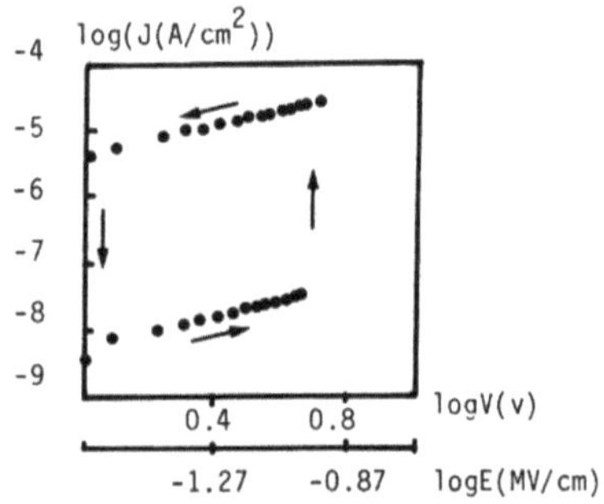

Fig. 2. Current voltage characteristic in a (log J, log V) plot for a 28 Vol. % Au. The switching transition takes place around 10^5 V/cm

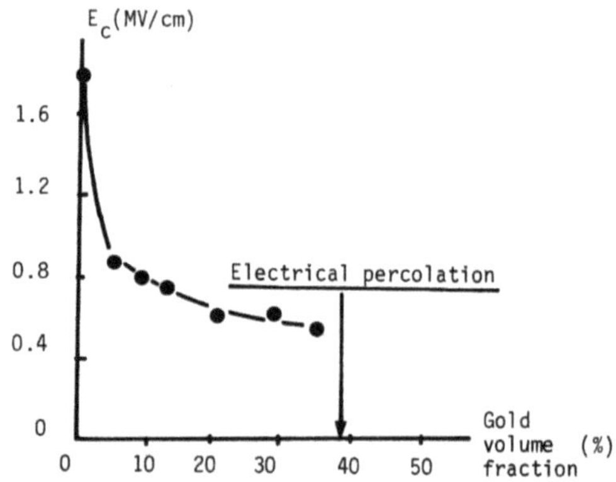

Fig. 3. Evolution of the specific breakdown field E_c versus gold volume fraction

10^5 V/cm. This phenomenon, only observed above ~10 Vol.% Au, is probably due to a transition from thermally induced tunneling type of conduction to a direct Fowler–Nordheim effect [6]. Under very high field, breakdown occurs in the films. Figure 3 shows the evolution of the specific breakdown field E_c calculated from statistical data of time to breakdown [7].

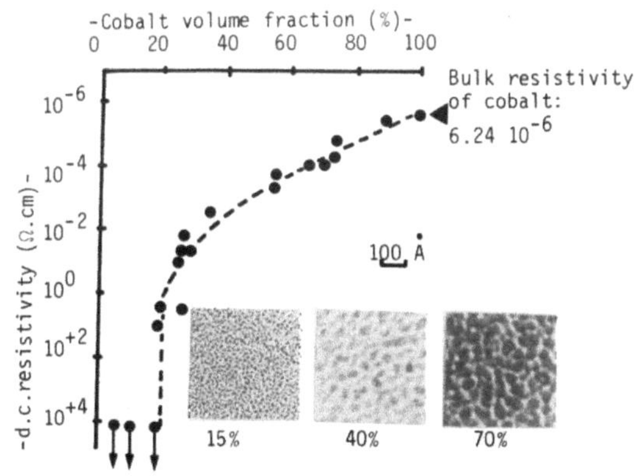

Fig. 4. d.c. resistivity versus cobalt volume fraction together with film microstructure

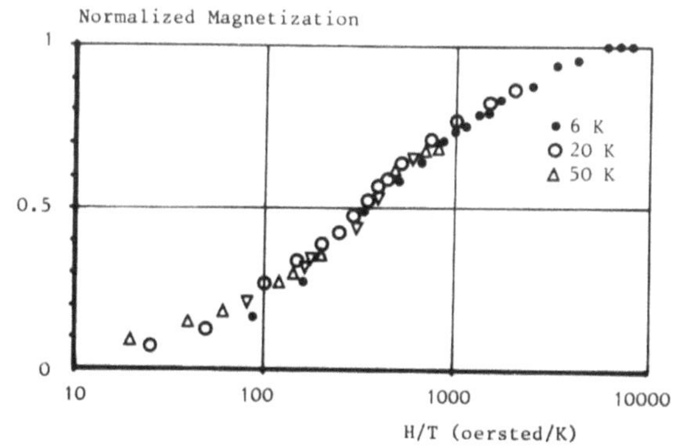

Fig. 5. Superparamagnetic behavior of Co clusters below electrical percolation (12 Vol. % Co sample)

After the initial decrease, the breakdown field changes slowly with an increasing concentration of gold particles. Surprisingly, no major changes take place until very near the electrical percolation.

Magnetic properties of cobalt containing PPP

The d.c. electrical conductivity of the system is controlled by a percolation threshold at ~22 Vol.% cobalt (Fig. 4). The onset of percolation is also reflected in the magnetic properties by the appearance of hysteresis in the magnetization curves and the onset of ferromagnetism.

Below percolation. The samples consist of isolated Co particles of a few angström radius as seen on Fig. 4. This is well below the estimated critical size (~150 Å radius) at which single magnetic domain state becomes energetically favored. The magnetic behavior of such a system depends on the relative magnitude of the thermal energy kT and the anisotropy barrier KV, where V is the volume of the cluster and K its magnetic anisotropy. When $kT \gg KV$, the system is superparamagnetic, i.e. the magnetic moments will relax in a very short time into a paramagnetic thermal equilibrium. Two salient features help identify the superparamagnetic thermal equilibrium. First the absence of magnetic hysteresis, second the dependence of the total magnetization of the system on $\mu H/kT$, where μ is the magnetic moment of a cluster, and H is the applied field. This is exactly what is observed experimentally as shown on Fig. 5. The system behaves superparamagnetically down to 6 K indicating a very small size of the clusters as revealed by electron microscopy.

Above percolation. A ferromagnetic ordering is observed through the appearance of hysteresis.

Conclusion

The ability to produce noble and transition metal clusters dispersed in a polymeric matrix by metal sputtering and plasma polymerization in a r.f. capacitively coupled diode reactor system has been demonstrated. The d.c. electrical properties and dielectric breakdown field values have been reported in the dielectric regime of gold containing plasma polymerized tetrafluoroethylene whereas magnetic behavior of cobalt containing plasma polymerized propane is discussed. The very unique feature of plasma polymerized matrix is its ability to soften or melt at reasonable annealing temperature. As a consequence, it is possible to change the microstructure of the metallic clusters by post deposition annealing.

One of us (C.L.) wants to express his gratitude to the IBM Almaden Research Center for granting a one year fellowship.

References

1. Kay, E., Dilks, A., Hetzler, U.: J. Macromol. Sci. Chem. A **12**, 1393 (1978)
2. Perrin, J., Despax, B., Hanchett, V., Kay, E.: J. Vac. Sci. Technol. A **4**, 46 (1986)
3. Perrin, J., Despax, B., Kay, E.: Phys. Rev. B, **32**, 719 (1985)
4. Kay, E., Hecq, M.: J. Vac. Sci. Technol. A **2**, 401 (1984)
5. Kay, E., Dilks, A.: J. Vac. Sci. Technol. **18**, 1 (1981)
6. Laurent, C., Kay, E., Souag, N.: J. Appl. Phys. **64**, 336 (1988)
7. Lacoste, R., Muhammad, A.: IEEE Trans. Elect. Insul., EI-**19**, 234 (1984)

Z. Phys. D – Atoms, Molecules and Clusters 12, 469–470 (1989)

Atoms, Molecules and Clusters
Zeitschrift für Physik D
© Springer-Verlag 1989

Optical absorption of small metallic particles

S. Tomiya and K. Kawamura

Department of Physics, Faculty of Sciences and Technology, Keio University, Hiyoshi 3-chome, Kohoku-ku, Yokohama, 223 Japan

Received 5 July 1988; final version 30 September 1988

In an extremely small metallic particle in which electron energies take discrete values, a light is absorbed by the surface plasmon which is not decayed by the surface scattering. The usually used relation between polarizability of a particle and electrical susceptibility does not hold.

PACS: 36.40; 78.00

1. Introduction

It is well-known [1] that a spherical small metallic particle exhibits an absorption band centered at the surface plasmon frequency ω_s with width of the order of v_F/D where D is diameter of the particle and v_F, the fermi velocity of the electron within it. Then, there arise the following questions about extremely small particles in which electronic energies take discrete values [2, 3]. *(i)* Does the absorption spectra still form a band whose width increases with decreasing diameter of particles? *(ii)* Do electrons move still collectively in such a small system or do they move individually?

2. Calculation

The absorption rate of an oscillating electric field is proportional to an imaginary part of electric polarizability which is expressed by the following Kubo formula [4];

$$\alpha(\omega) = -(i/\hbar)\int_0^\infty \langle[P_E(t), P_E]\rangle \exp(-i\omega t)\,dt \tag{1}$$

where P_E is a component of total dipole moment in the direction of an external electric field. The right-hand side of this expression is written by a linear combination of Green functions of particle-hole operators which satisfy the following equations in the random phase approximation;

$$\langle\langle a_m^+ a_n; P_E\rangle\rangle_\omega = G_{mn}^{(0)}(\omega)(P_E)_{nm} + \Sigma'_{rs} G_{mn}^{(0)}(\omega)\, V_{rnsm} \langle\langle a_r^+ a_s; P_E\rangle\rangle_\omega . \tag{2}$$

Here, $G_{mn}^{(0)}(\omega)$ are Green functions of independent electron-hole pairs, and V_{rnsm} are matrix elements of potential energy of Coulomb force between two electrons. The dash on the summation indicates that the terms with $r=m$ and/or $s=n$ should be omitted to exclude the self-interaction of each electron.

In some literatures, the polarizability of a particle is related to the electric susceptibility $\chi(\omega)$ of material of the particle. A similar expression could be obtained from (2), if (*i*) the self-interaction were included and (*ii*) the induced electric charge density were assumed to be localized exclusively on the surface. The resultant expression for the polarizability is

$$\alpha(\omega) = \chi(\omega)\,\Omega/(1+\chi(\omega)/3\varepsilon_0). \tag{3}$$

where Ω is volume of the particle, ε_0 is the dielectric constant of vacuum, and $\chi(\omega)$ is defined by

$$\chi(\omega) = \Sigma_{mn} |(P_E)_{mn}|^2\, G_{mn}^{(0)}(\omega)/\Omega. \tag{4}$$

Instead of employing this ordinary formalism, we have solved (2) numerically and the Green function on the right-hand side of (1) has been computed. As for the basis of the matrix elements, we have employed the wave functions for electrons trapped by an infinitely

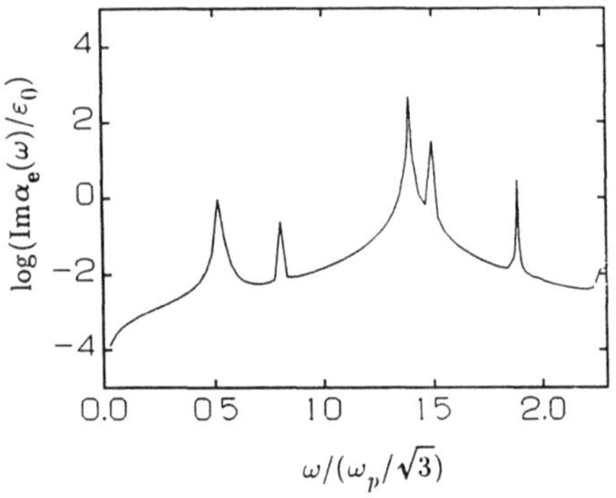

Fig. 1. The logarithmic plot of an imaginary part of electric polarizability against the frequency scaled by the surface plasmon frequency

deep potential well. We have used parameters for a silver particle with radius of 5 Å. The electron states whose energies are less than four times the fermi energy have been taken.

3. Electric polarizability

The imaginary part of the polarizability in a logarithmic scale is plotted in Fig. 1 against the frequency normalized to the surface plasmon frequency $\omega_s = \omega_p/\sqrt{3}$ where ω_p is the plasma frequeney. To display δ-function like spectra, a very small imaginary number is added to energies of a single electron which appear in the expression for $G^{(0)}_{mn}$.

There is essentially a single line which is accompanied by a very low satellite line. Although its position is shifted from ω_s toward the high frequency side, this plot implies that the oscillator strength is condensed on the *undamped* surface plasmon mode. Then, the answers to the questions at the beginning of the present paper is that the surface plasmon mode is well-defined in such an extremely small particle and the line broadening is not observed.

4. Discussion

Our highest peak exhibits the blue-shift from ω_s whereas recent numerical studies in the time dependent local density approximation predict the red-shift of the surface plasmon mode [5].

In contrast to larger particles which exhibit the absorption band [1] with width inversely proportional to the particle size, the absorption spectrum of an extremely small particle consists of a (few) discrete line(s). This suggests that it is wrong to argue that the surface scattering causes the broadening of absorption spectrum. As pointed out by Kawabata and Kubo [6], the line broadening is caused by transfer of energy from the collective mode to individual modes. Since energies of those individual modes are discrete in an extremely small particles, the absorption spectrum consists of discrete lines [7].

In our study, the polarizability has been computed also by using (3) which has been compared with that evaluated by the rigorous solution of (2). Both the functions exhibit much different features and the former is rather similar to $\chi(\omega)$ although the intensity of absorption changes so that the highest peak is the one which is the nearest to ω_s on the low frequency side. The usually employed expression (3) is wrong for small particles, since self-interaction is taken on deriving it. The influence of the self-interaction is not of quantitative nature such that oscillator strength is modified by an amount of $(1/N)$ (N; electron number), but we have confirmed by retaining the self-interaction term in (2) that it causes very small shift of the peak position. Further, the assumption that the induced charge density is localized on the surface is not sound; it oscillates in space with amplitude which decays gradually toward the center of a particle [7]. Details of such studies will be published elsewhere.

The present authors express their sincere thanks to Professor R. Kubo for stimulating discussion.

References

1. See for the review in this field Kreibig, U., Genzel, L.: Surf. Sci. **156**, 678 (1985)
2. Kubo, R.: J. Phys. Soc. Japan **17**, 975 (1962)
3. As for the quantum size effects in small particles see Halperin, W., P.: Rev. Mod. Phys. **58**, 533 (1986)
4. Kubo, R.: J. Phys. Soc. Japan **12**, 570 (1957)
5. Ekardt, W.: Phys. Rev. B **31**, 6360 (1985)
6. Kawabata, A., Kubo, R.: J. Phys. Soc. Japan **21**, 1767 (1966)
7. Penzar, Z., Šunjić, M.: Solid State Commun. **52**, 747 (1984)

Z Phys. D – Atoms, Molecules and Clusters 12, 471–475 (1989)

Atoms, Molecules and Clusters
Zeitschrift für Physik D
© Springer-Verlag 1989

The size dependent shift of the surface plasmon absorption band of small spherical metal particles

K.-P. Charlé, W. Schulze, and B. Winter*

Fritz-Haber-Institut der Max-Planck-Gesellschaft, Faradayweg 4-6, D-1000 Berlin 33, Germany

Received 9 July 1988

It has been previously established that the surface plasmon of small spherical silver particles, which are embedded in a noble gas matrix, shifts to higher energies (blue shift) as the mean diameter D of the particles decreases (100 Å > D > 20 Å). This blue shift has also been found for supported silver particles, and quite recently we observed it by elastic light scattering in the gas phase. This latter experiment proves unambiguously that the blue shift in small silver particles is not induced by interactions with the environment, the presence of which is clearly recognized in less inert matrices such as O_2 or CO. From self-consistent calculations of the surface response of planar jellium surfaces one would expect a red shift, which is also directly confirmed by a few calculations for selected jellium spheres. The contradiction between the observed blue shift for small particles and the predicted red shift for jellium spheres disappears, if one accounts for the *d*-electrons of silver in a very simple approximation.

PACS: 78.00; 42.20

Introduction

The size dependence of the surface plasmon of small metal particles is usually investigated by measuring the absorption of colloids of such particles. If the fillfactor of the colloid is sufficiently low (ratio of metal volume to colloid volume smaller than 10^{-3}) and if the particles are spherical and small enough ($R \leqq 100$Å), one observes a distinct absorption band due to surface plasmon or dipole resonance absorption. A calculation of the dipole resonance of metal particles within the framework of classical optics [1] shows that the position of the absorption band is determined by the classical resonance condition for spherical particles

$$\mathrm{Re}\,\varepsilon(\omega) + 2\varepsilon_{Ma} = 0 \tag{1}$$

where $\varepsilon(\omega)$ is the dielectric function of the bulk metal and ε_{Ma} is the dielectric constant of the (transparent) matrix, in which the particles are embedded. It should be noticed that the classical resonance condition (1) predicts a size independence of the surface plasmon for small particles. In order to prove, whether this prediction is correct, one has to prepare colloids of spherical metal particles with a well defined mean diameter D, i.e., the size distribution of the particles has to be sufficiently narrow. The gas aggregation technique (GAT) [2, 3] is ideally suited for this purpose, because it yields almost Gaussian size distributions with a dispersion σ/D of about 0.2 and mean diameter D below 100 Å. The colloids are then generated by condensing the metal particles grown in the gas phase together with the matrix gas on a cooled substrate (sapphire, 10 K). Figure 1 shows the absorption spectrum of Ag/Ar colloids with the mean diameter D of the Ag particles decreasing from bottom to top. In the upper two spectra the absorption by the Ag monomer dominates, and the particles are too small (D < 20 Å) for a size determination by electron microscopy [3]. The next to top spectrum contains already a slight indication of the surface plasmon absorption band,

* To whom all correspondence should be addressed

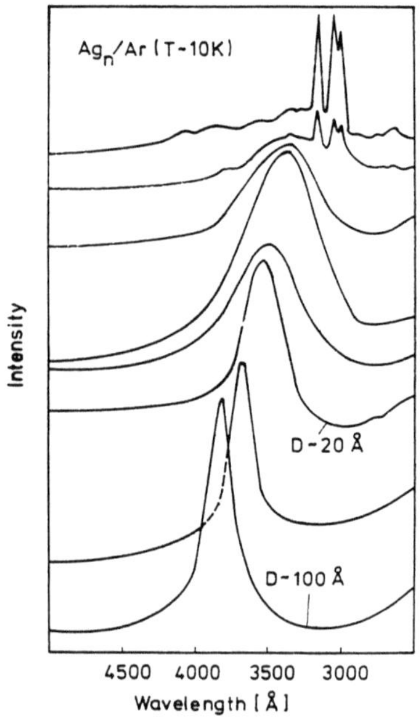

Fig. 1. Absorption spectra for Ag/Ar colloids. The mean diameter D of the spherical Ag particles decreases from bottom to top

which dominates the other spectra. For $D \triangleq 100$ Å (bottom) the position of the resonance agrees almost perfectly with the classical prediction (1). The same holds for the calculated full width at half maximum (FWHM) of the band [4]. However, as D gets smaller, the surface plasmon absorption band obviously shifts to the blue, and the FWHM increases. This size dependent shift and damping of the surface plasmon of small metal particles ($D < 100$ Å), which is not predicted by classical optics [1], is a general phenomenon commonly called 'the size effect'. The size effect as such has been observed in Ag colloids before [5–7]. The point of the results for Ag/Ar colloids shown in Fig. 1 is that they are radically different: First a blue shift was observed for the first time [2]. In [7] practically no shift was observed and in [5, 6] a red shift was found. Second, the FWHM increases significantly more slowly (about 1/3) with decreasing diameter than was observed in [5–7].

Experimental

Results

Since the theoretical predictions for the size effect were confusing as well, an attempt was made to clarify the situation experimentally. A natural starting point was the question, whether the observed differences are due to different matrices (glass [5, 7], silica gel [6]) and probably also due to different preparation techniques. Therefore the influence of the embedding matrix on the size effect was studied [8]. The preparation technique described above (GAT plus subsequent embedding) is perfectly suitable for such an investigation, because the Ag particles can be embedded in different matrices within the same experimental run. This way matrices can be directly compared with each other as shown in Fig. 2 (drawn out part of the curves), where CO, which is known to interact strongly with Ag, is compared with the inert Ar. In Ag/CO the blue shift disappears and the FWHM increases about 3 times faster than in Ag/Ar. Thus, the size effect in Ag/CO is very similar to that in Ag/glass [7]. The same kind of matrix influence has also been observed in Ag/NaCl and Ag/KCl (unpublished). On the other hand, in most matrices like the noble gases and N_2 the size effect is almost independent of the matrix. (In Ne there is a small quantitative change [8]). Even in the more reactive matrices O_2 and C_2H_4, where the expected matrix influence is revealed by a stronger increase of the FWHM, the peak shift is the same as in Ar. These observations demonstrate clearly that the blue shift is a property of the Ag particles, whereas a red shift or no shift and a stronger increase of the FWHM is a matrix effect. This conclusion is confirmed at least qualitatively by reflectance measurements for Ag particles on an alumina support [9] and by preliminary results for Rayleigh-Scattering in the gas phase, where any matrix or support is absent.

Another novel and remarkable result is the observation that the peak shift for Ag in inert and weakly interacting matrices (O_2, C_2H_4) depends linearly on the reciprocal mean diameter at least for $D \geqq 20$ Å, where the size distribution of the Ag particles could be

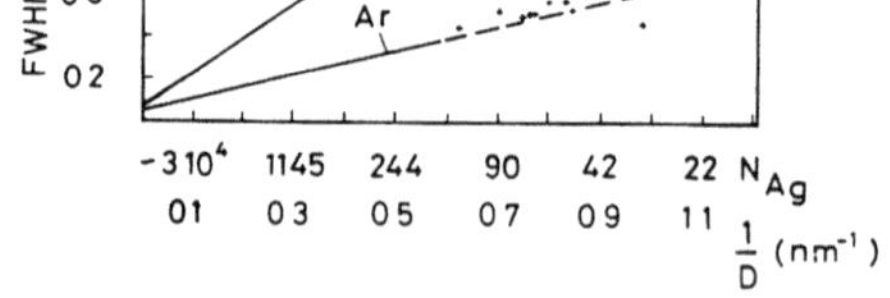

Fig. 2a and b. The position of the maximum (**a**) and the FWHM of the surface plasmon absorption band (**b**) in dependence of the reciprocal mean diameter of the particles for Ag/Ar and Ag/CO colloids. The number of atoms per particle is also indicated. For the dashed part of the curves and the crosses in **b** see text

determined in-situ with sufficient accuracy. As an example the linear shift is shown for Ag/Ar in Fig. 2a, drawn out part of the curve 'Ar'. The dashed part of this curve represents a conjecture for $D < 20$ Å, which was used the other way round to estimate the mean diameter from the peak position in cases such as represented by the upper spectra in Fig. 1. The crosses in Fig. 2b show that the measured FWHM coincide with the corresponding extrapolated FWHM curve. Admittedly, this proves only that the peak shift and the broadening of the absorption band both depend in the same way on the reciprocal mean diameter of the particles. Nevertheless, these results suggest strongly that clusters of a few ten Ag atoms still have a well defined surface plasmon. An investigation of this interesting question is underway.

The by now experimentally well established blue shift of the surface plasmon of Ag particles seems to be a serious problem for theory, which predicts a red shift for jellium (see below), the only tractable working model, which is free of questionable assumptions. In the discussion it will be shown that there is neither a qualitative contradiction nor a substantial quantitative disagreement between the measured shift of the surface plasmon frequency $\omega_{sp}(D)$ of spherical Ag particles and the jellium model, which is a good one for the conduction electrons in Ag. The theoretical background for the following treatment of the experimental data is the theory of the electromagnetic surface response of metals [10, 11] as developed by Feibelman for planar surfaces. For a spherical geometry this theory was subsequently formulated in [12, 13]. The result of interest in the present context is the new resonance condition

$$\mathrm{Re}[\varepsilon(\omega)] + 2\varepsilon_{Ma} + 2/R \times \mathrm{Re}[(\varepsilon(\omega) - \varepsilon_{Ma})\, d(\omega; 1/R)] = 0 \tag{2a}$$

for a sphere with radius R. The generally complex quantity d, which has the dimension of a length, is given by

$$d = d_{\perp} - d_{\parallel} \tag{3}$$

with $d_{\perp}$, $d_{\parallel}$ being the surface response functions for electric fields normal and parallel to the surface. These surface response functions account essentially for the non-locality of the electromagnetic response of metals, which cannot be neglected in strongly inhomogeneous regions such as interfaces, especially not for normal fields.

Since the $\mathrm{Im}[\varepsilon(\omega)]$ of Ag is very small compared to $|\mathrm{Re}[\varepsilon(\omega)]|$ and ε_{Ma} in the spectral region of interest (3.0 eV $\leqq \hbar\omega \leqq$ 3.7 eV) the somewhat simpler version

$$\varepsilon_1(\omega) + 2\varepsilon_{Ma} + 2/R[(\varepsilon_1(\omega) - \varepsilon_{Ma})\,\delta(\omega; 1/R)] = 0 \tag{2b}$$

with $\mathrm{Re}[\varepsilon(\omega)] = \varepsilon_1(\omega)$ and $\mathrm{Re}[d] = \delta$ is sufficiently accurate for an experimental determination of δ. It is emphasized that for this evaluation the actual $\varepsilon_1(\omega)$ for Ag [14] is used and not some free-electron surrogate, as is often done. In the results (symbols in Fig. 4 below) no statistically significant influence of the matrix (ε_{Ma}) and of the particle size (D) is found. Especially, there is no systematic trend (see below) for different ε_{Ma}. Therefore the experimental result in the frequency range 3.0 eV $\leqq \hbar\omega \leqq$ 3.6 eV can be represented with sufficient accuracy by

$$\delta(\omega) = 0.5\, r_s^{1/2}(1 - \omega^2/\omega_p^2)^{-1/2}\ \text{Å} \tag{4}$$

with $\hbar\omega_p$ = 3.78 eV being the experimental bulk plasmon energy of Ag [14] and $r_s = 3.04$ being the density parameter for the conduction electrons.

Discussion

Since the experimentally determined real part of the surface response of Ag particles turns out to be independent of the particle diameter within experimental error, it must be the same as for planar surfaces [13]. Indeed, the surface response functions $d_{\perp}$, $d_{\parallel}$ describe deviations from classical optics in a narrow range (a few Å) around the interface, where classically a jump of local dielectric properties is presumed. Therefore one has in the limit of large spheres or vanishing curvature $1/R$ of the surface:

$$\lim_{R\to\infty} d_{\text{sphere}} = d_{\text{plane}} \tag{5}$$

This relation can be directly verified (unpublished) at least within the framework of the semiclassical infinite barrier (SCIB) model and the hydrodynamic approximation (HA) for the non-local response of the conduction electrons as applied to spheres in [15]. For a detailed discussion of SCIB and HA and their application to a plane the reader is referred to [10, 11]. In Fig. 3 the experimental result (4) for Ag particles (dotted line) is compared with Re $[d_{\perp}(\omega/\omega_p) - d_{\parallel}(\omega/\omega_p)]$ as calculated for planar jellium surfaces. The dashed curve is the SCIB + HA result

$$\mathrm{Re}[d_{\perp} - d_{\parallel}] = 0.45\, r_s^{1/2}(1 - \omega^2/\omega_p^2)^{-1/2}\ \text{Å} \tag{6}$$

where damping in the HA has been neglected, and the full curve is the result of the self consistent calculation of Feibelman [10] for $r_s = 3$. Note that the bulk plasmon frequency

$$\omega_p^2 = 4\pi n e^2/m \tag{7}$$

($\hbar\omega_p$ = 9.0 eV for Ag) of jellium is used for these two curves, whereas for the experimental fit the actual plasmon frequency of Ag (see above) is used.

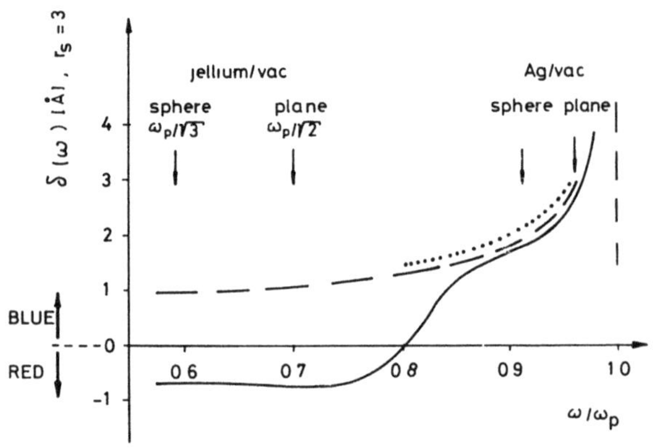

Fig. 3. Frequency dependence of the real part of the surface response, which is responsible for the shift of the surface plasmon frequency. The dotted curve is deduced from the experimental results for Ag particles in various matrices. Compare (4) in the text. *Dashed curve*: Theoretical result for a planar jellium/vacuum interface in the SCIB approximation plus HA (damping neglected). *Drawn out curve*: Result of a self consistent calculation for the same system redrawn from [10]

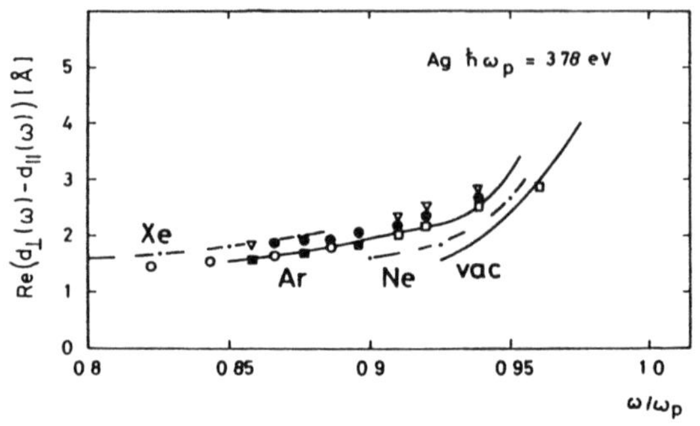

Fig. 4. Real part of the surface response of various Ag/dielectric interfaces. The symbols represent experimental results: *Open squares*: Ne, *filled circles*: Ar, *filled squares*: Kr, *open circles*: Xe, *open triangles*: O_2. The curves are calculated with the SCIB approximation plus HA (damping neglected)

As Fig. 3 shows, such a rescaled jellium approach, which has been suggested in [9, 13] for Ag, explains the observed blue shift very nicely. It reveals very clearly that the reason is the contribution $\varepsilon_0(\omega)$ of the *d* electrons to the real part of the dielectric function $\varepsilon(\omega)$, due to which the surface plasmon for both a sphere and a plane are very close to the bulk plasmon (arrows on the right). If the *d* electrons can be neglected like in the alkali metals, the surface plasmons are more separated from the bulk plasmon (arrows on the left), and one expects a red shift, which is predicted from the self consistent calculation [10]. This qualitative prediction is confirmed by corresponding calculations for jellium spheres ($r_s = 4$; [16]) and recent experiments on Na clusters [17] in the gas phase.

Although the scaled jellium approach [9, 13] gives a correct hint for the observed size effect in Ag particles, the good agreement between (4) and the rescaled formula (6) is fortuitous, because the latter is not correct. This can be shown by a SCIB calculation of $\mathrm{Re}(d_\perp - d_{||})$ for an Ag/dielectric interface, either along the lines described in [11, 10] for a plane or as described in [15] for a spherical interface. The Fourier-Transform of the non-local longitudinal bulk dielectric function of Ag is in the HA given by [11, 15]

$$\varepsilon_l(\omega, k) = \varepsilon_0(\omega) - (\omega_p^0)^2(\omega(\omega + i\gamma) - \beta^2 k^2)^{-1} \qquad (8)$$

with $\beta^2 = 0.6\, v_F^2$ (v_F being the Fermi velocity). The local contribution $\varepsilon_0(\omega)$ of the *d* electrons is determined from experimental data, which are described by the corresponding transversal function

$$\varepsilon_t(\omega) = \varepsilon_l(\omega, 0) \qquad (9)$$

using $\hbar\omega_p^0 = 9.2$ eV and damping is neglected ($\gamma = 0$) as before. The result depends now explicitly on $\varepsilon_0(\omega)$, which is frequency dependent, and on the dielectric constant of the matrix. Relevant examples for $\mathrm{Xe}(\varepsilon_{Ma} = 2.3)$, $\mathrm{Ar}(\varepsilon_{Ma} = 1.75)$, $\mathrm{Ne}(\varepsilon_{Ma} = 1.25)$ and vacuum are shown in Fig. 4 together with detailed experimental results for various matrices, which were summarized above in (4) or the dotted curve in Fig. 3. The rescaled jellium curve (dashed curve in Fig. 3) is very close to the curve for Ar in Fig. 4. Though the slight trend predicted by the correct calculations is not reproduced by the experimental results, the quantitative agreement is still satisfactory. In summary, it has been substantiated that the observed blue shift of the surface plasmon of spherical Ag particles ($D > 20$ Å) is essentially due to the peculiar bulk optical properties of silver, which are caused by its *d* electrons.

The continuing interest and support of Prof. H. Gerischer is gratefully acknowledged. The cluster project is supported by the Deutsche Fortschungsgemeinschaft in Sfb 337.

References

1. Mie, G.: Ann. Phys. **25**, 377 (1908)
2. Abe, H., Schulze, W., Tesche, B.: Chem. Phys. **47**, 95 (1980)
3. Frank, F., Schulze, W., Tesche, B., Urban, J., Winter, B.: Surf. Sci. **156**, 90 (1985)
4. Abe, H., Charlé, K.-P., Tesche, B., Schulze, W.: Chem. Phys. **68**, 137 (1982)
5. Ganière, J.D., Rechsteiner, R., Smithard, M.A.: Solid State Commun. **16**, 113 (1985)
6. Ascarelli, P., Cini, M.: Solid State Commun. **16**, 385 (1975)
7. Genzel, L., Martin, T.P., Kreibig, U.: Z. Phys. B – Condensed Matter **21**, 339 (1975)
8. Charlé, K.-P., Frank, F., Schulze, W.: Ber. Bunsenges. Phys. Chem. **88**, 350 (1984)

9. Borensztein, Y., De Andrès, P., Monreal, R., Lopez-Rios, T., Flores, F.: Phys. Rev. B **33**, 2828 (1986)
10. Feibelman, P.: Prog. Surf. Sci. **12**, 287 (1982)
11. Forstmann, F., Gerhardts, R.R.: Springer Tracts in Modern Physics, Vol. 109 (especially Chap. 5). Berlin, Heidelberg, New York: Springer 1986
12. Apell, P., Ljungbert, Å.: Phys. Scr. **26**, 113 (1982)
13. Apell, P., Ljungbert, Å.: Solid State Commun. **44**, 1367 (1982)
14. Irani, G.B., Huen, T., Wooten, F.: J. Opt. Soc. Am. **61**, 128 (1971)
15. Dasgupta, B.B., Fuchs, R.: Phys. Rev. B **24**, 554 (1981) (The version given in Sect. IV has been adopted, because it is consistent with SCIB for free electrons)
16. Ekardt, W.: Phys. Rev. B **31**, 6360 (1985)
17. de Heer, W.A., Selby, K., Kresin, V., Masui, J., Vollmer, M., Châtelain, A., Knight, W.D.: Phys. Rev. Lett. **59**, 1805 (1987)

Z. Phys. D – Atoms, Molecules and Clusters 12, 477–479 (1989)

Atoms, Molecules
Zeitschrift für Physik D and Clusters
© Springer-Verlag 1989

Investigation of surface plasma resonances in small sodium clusters

K. Selby, M. Vollmer*, J. Masui, V. Kresin, W.A. de Heer***, and W.D. Knight**

Department of Physics, University of California, Berkeley, CA 94720, USA

Received 5 July 1988

Photoabsorption cross sections of small neutral sodium clusters are measured at wavelengths from 452 to 604 nm. We find that surface plasma oscillations of the valence electrons dominate the photoabsorption spectra. The experimental results are consistent with predictions based on an extended Clemenger–Nilsson model, general sum rules, and the experimental static electric polarizabilities. For Na_8 a single broad resonance is observed, the peak wavelength is in excellent agreement with the model. The data for Na_9 and Na_{10} support the prediction of double resonances.

PACS: 36.40.+d, 33.80.Gj, 33.20.Kf

It is well known that metal clusters exhibit shell structure of single electron levels, with spherical shell closings occurring at 8, 20, 40, . . . valence electrons. Shell structure is observed in measurements of abundance spectra, ionization potentials, static electric polarizabilities [1], and photoelectron spectra [2]. Here we report recent progress in the investigation of photoabsorption spectra [3, 4]; an extension of cluster spectroscopy to collective oscillations of the valence electrons.

Measurement of cluster photoabsorption cross sections by longitudinal beam depletion spectroscopy has been described in detail elsewhere [3, 4]; a brief description follows. A supersonic beam of neutral sodium clusters is illuminated by a colinear and counterpropagating chopped laser beam. Photon absorption excites surface plasma oscillations, which have resonances at visible wavelengths. Excited clusters rapidly evaporate atoms or small aggregates. The subsequent recoil of the daughters causes their removal from the highly collimated beam, resulting in a reduced counting rate at the detector. The remaining clusters are photoionized, and then enter a quadrupole mass analyser (QMA). Ions of the chosen mass are steered into the detector.

The QMA is set to a specific mass, and the counting rate at the detector of clusters which were illuminated for the full duration of a laser pulse is determined. Between laser pulses, the counting rate of unilluminated clusters is also measured. The ratio of these two counting rates, r, and the measured photon flux determine the cluster photoabsorption cross section [3].

Cross sections have been measured for Na clusters with 2 to 21 atoms, at wavelengths from 452 to 515 nm, and from 584 to 604 nm. For a given wavelength and cluster size, $\ln r$ is found to be directly proportional to the laser energy per pulse, demonstrating that removal of a cluster from the beam is a single photon process. This relationship has been verified for clusters as large as 40 atoms [3, 4].

An extended Clemenger-Nilsson model [5] in conjunction with the measured static electric polarizabilities [1] has already been successful in describing the photoabsorption spectra of sodium clusters [3, 4]. We will briefly review the model below.

* Permanent address: Physikalisches Institut der Universität Heidelberg, D-6900 Heidelberg 1, Federal Republic of Germany
** Permanent address: Department of Physics, Cornell University, Ithaca, NY 14853, USA
*** Permanent address: Institut de Physique Expérimentale, Ecole Polytechnique Fédérale de Lausanne, PHB-Ecublens, CH-1015 Lausanne, Switzerland

Photoabsorption by small metal spheres was first investigated theoretically by Mie [6]. For a particle with a radius much smaller than the photon wavelength, the photoabsorption cross section, σ, is given by

$$\sigma = \frac{4\pi N e^2}{m_e c} \frac{\omega^2 \Gamma}{(\omega^2 - \omega_0^2)^2 + \omega^2 \Gamma^2}, \quad (1)$$

where N is the number of valence electrons, ω_0 is the resonance frequency, and Γ is the damping constant. Equation (1) assumes that all of the dipole oscillator strength is exhausted by the surface plasma resonance at ω_0, and the resonance frequency is given by (see, e.g., [7])

$$\omega_0^2 = \frac{N e^2}{m_e \alpha}, \quad (2)$$

where α is the static electric polarizability of the cluster. (Classically, α is the cube of the cluster radius. With a Wigner-Seitz density parameter $r_s = 4.0$ a.u., the classical resonance wavelength is 365 nm.)

The polarizability of a metal ellipsoid has a different value along each of the three axes; the photoabsorption spectrum is then the sum of three resonances of equal weight. The resonant frequencies are obtained by inserting each of the three polarizability components into (2), and (1) is easily generalized to give the photoabsorption cross section.

The extended Clemenger-Nilsson model is used to calculate the axial ratios of ellipsoidal metal clusters [3, 4]. The classical depolarization factors together with the experimentally determined average static electric polarizabilities [1] then give the three polarizability components. For closed shell clusters, the model predicts a spherical shape, and hence a single resonance peak.

The axial ratios of neutral sodium clusters have also been determined using a self-consistent spheroidal jellium model [8]. For those clusters which our ellipsoidal model finds to be spheroidal, we obtain axial ratios which are consistent with this calculation.

Figure 1 shows the measured photoabsorption cross section vs. wavelength for *(a)* Na_8, *(b)* Na_9, and *(c)* Na_{10}. In each figure, the results of our model are included, with $\Gamma = 0.15\ \omega_0$. For Na_8, the model predicts a single resonance at 478 ± 4 nm. Experimentally, a resonance is observed which is very close to the predicted position. For Na_9 and Na_{10}, the model gives double resonance peaks; the data support this prediction.

The spherical self-consistent jellium model has been used for theoretical investigation of surface plasma resonances in small metal clusters [7, 9]. The calculations give photoabsorption spectra consisting of several single-particle resonances and one surface plasma resonance. No predictions have been made of the widths. For Na_8, Beck [9] calculates that the surface plasma resonance is red shifted by 19% with respect to the classical value. Our data shows a red shift of about 25%, which is in agreement with (2), using the experimental static polarizability. We have not yet observed single-particle resonances.

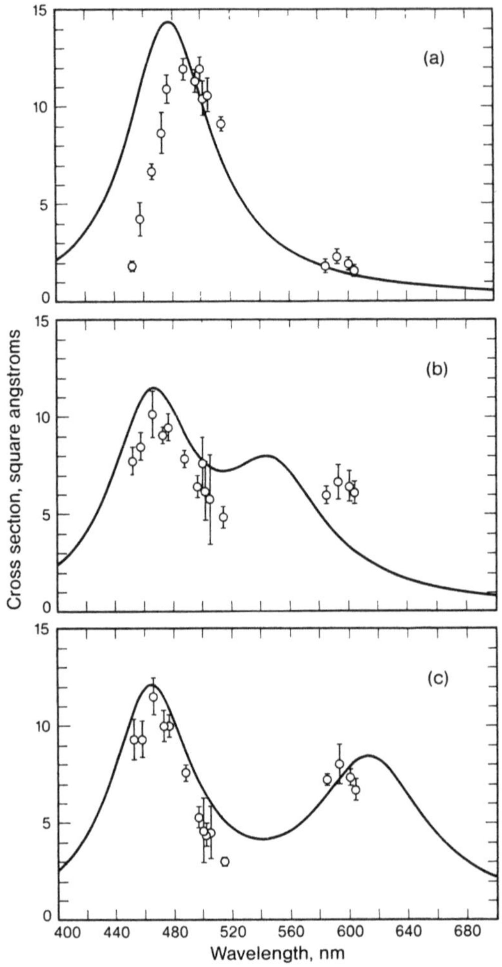

Fig. 1a–c. Experimental (circles) and theoretical (solid lines) photoabsorption cross sections for: **a** Na_8, **b** Na_9, and **c** Na_{10}. For the theoretical curves a damping factor of $\Gamma = 0.15\ \omega_0$ is used

There are striking similarities between our investigations of surface plasma resonances in clusters and the observed giant dipole resonances in nuclei, the collective motion of the protons with respect to the neutrons. Predictions of resonance frequencies and photoabsorption cross sections parallel our method (for a review, see [10]).

Recent experiments with neutral Na clusters on an LiF surface [11] show that resonant photodesorption of atoms occurs, which is attributed to single-photon excitation of surface plasma oscillations. For particles of 50 nm mean radius, the resonance is at 490 ± 5 nm with a width of 90 nm.

In summary, we report recent progress in the study of surface plasma resonances of mass selected neutral sodium clusters in a beam. The data are consistent with predictions based on the extended Clemenger-Nilsson model, general sum rules, and the measured static electric polarizabilities. Data for Na_8, Na_9, and Na_{10} support the prediction of a single resonance in the cross section for closed shell clusters and a double resonance for certain open shell clusters. For Na_8 the measured resonance wavelength is very close to the predicted value.

Future work will include measurement of the cross sections at other visible wavelengths. Resonance positions and widths will be determined, and the dependence of the damping on cluster size will be investigated.

This work was supported by the U.S. National Science Foundation under Grant No. DMR-86-15246. We thank Prof. André Châtelain for his participation in the experiment and for stimulating discussions. We are very grateful to Dr. Andrew Kung, Dr. Tetsuo Hadeishi, and Prof. Eugene Commins, for the loan of lasers, and thank Michael Kruger for assistance with the experiment. One of us (M.V.) gratefully acknowledges the support of the Deutsche Forschungsgemeinschaft.

References

1. de Heer, W.A., Knight, W.D., Chou, M.Y., Cohen, M.L.: In: Solid state physics. Ehrenreich, H., Turnbull, D. (eds.), Vol. 40, p. 39. New York: Academic Press 1987
2. Pettiette, C.L., Yang, S.H., Craycraft, M.J., Conceicao, J., Laaksonen, R.T., Cheshnovsky, O., Smalley, R.E.: J. Chem. Phys. **88**, 5377 (1988). Bowen, K.H.: Z. Phys. D - Atoms, Molecules and Clusters **12** (1989) (this issue); Rademann, K., Kaiser, B., Hensel, F.: Z. Phys. D - Atoms, Molecules and Clusters **12** (1989) (this issue)
3. de Heer, W.A., Selby, K., Kresin, V., Masui, J., Vollmer, M., Châtelain, A., Knight, W.D.: Phys. Rev. Lett. **59**, 1805 (1987)
4. Selby, K., Vollmer, M., Masui, J., Kresin, V., de Heer, W.A., Knight, W.D.: Phys. Rev. B (submitted)
5. Clemenger, K.: Phys. Rev. B **32**, 1359 (1985)
6. Mie, G.: Ann. Phys. **25**, 377 (1908)
7. Ekardt, W.: Phys. Rev. B **31**, 6360 (1985)
8. Ekardt, W., Penzar, Z.: Phys. Rev. B **38**, 4273 (1988)
9. Beck, D.E.: Phys. Rev. B **35**, 7325 (1987)
10. Berman, B.L., Fultz, S.C.: Rev. Mod. Phys. **47**, 713 (1975)
11. Hoheisel, W., Jungmann, K., Vollmer, M., Weidenauer, R., Träger, F.: Phys. Rev. Lett. **60**, 1649 (1988)

Z. Phys. D – Atoms, Molecules and Clusters 12, 481–484 (1989)

Atoms, Molecules
Zeitschrift für Physik D and Clusters
© Springer-Verlag 1989

Absorption spectroscopy of size-selected trimers and pentamers transition metal clusters isolated in rare gas solids: preliminary steps

J.C. Rivoal[1], C. Grisolia[1], J. Lignieres[1], D. Kreisle[2], P. Fayet[3,]*, and L. Wöste[3,]**

[1] Lab. Opt Phys., ER5-CNRS, 10 rue Vauquelin, F-75231 Paris Cedex 05, France
[2] Fakultät für Physik, Universität Konstanz, D-7750 Konstanz, Federal Republic of Germany
[3] Inst. Phys. Exp., EPFL, PHB-Ecublens, CH-1015 Lausanne, Switzerland

Received 8 July 1988; final version 30 September 1988

Size-selected positive ionic silver trimers and pentamers and nickel trimers have been co-deposited, at low kinetic energy, with rare gas (Kr) onto a cold transparent surface. They have been neutralized onto the surface by electrons of few eV, with an efficiency of 80 ± 20%. The optical absorption spectra of the resulting matrix-isolated neutral species have been recorded, with good sensitivity, in the UV-visible range. The evaporation of clusters, detected on the atom signal after deposition of Ni trimers of 20 eV, neutralized by electrons of 5 eV, is shown to be roughly 20%. This indicates that Ni_3 clusters are present in the matrix but their number is actually too low to be optically detected.

PACS: 36.40; 33.20

Introduction

The present interest in the thermodynamical, electronical or magnetical properties and the structure of small metallic aggregates, is due to their special reactivity and to the fact that they give rise to a family of materials having entirely novel properties.

In spite of the large amount of theoretical predictions regarding the structure, the stability, and also the ground state electronic properties of small neutral transition metal clusters, and in spite of a large number of experimental results, details of the electronic structure of these clusters are still lacking [1, 2]. In gas phase studies, the use of supersonic expansion decreases the rotational and vibrational energy of the particles but usually their temperature is not precisely known. When mass selection is effective, clusters are in well-characterized state but it is the properties of the ions which are under study. Matrix isolation technique allows the study of the electronic and vibrational properties of cold clusters [1, 2, 3] but the size of the observed species is not known with certainty [4, 5].

Since matrix isolation offers the possibility to accumulate substantial quantities of the desired species at a fixed low temperature and since a continuous high flux source of size-selected clusters was available [6], we decided to combine the advantages of the two techniques. Positive mass selected ionic clusters have been co-deposited with rare gas onto a cold transparent substrate and subsequently neutralized onto the surface with electrons. The optical properties of the trapped species have been investigated using UV-visible absorption.

Experimental

Our experimental arrangement, already described [6], is shown in Fig. 1. A cold reflex ion source (CORDIS) provides a primary Kr^+ ion beam of typically 20 KeV energy and 1 mA current. It is directed onto a water-cooled metal target. The emerging secondary ions are first focussed and then selected by mass in a triple quadrupole mass analyser (QMS). The total amount of selected ions is either measured using a Faraday plate

* Present address: Corporate Research EXXON Research and Engineering Co., Annandale, NJ 08801, USA
** Permanent address: C.A.L.-EPFL, PHB-Ecublens CH 1015 Lausanne, Switzerland

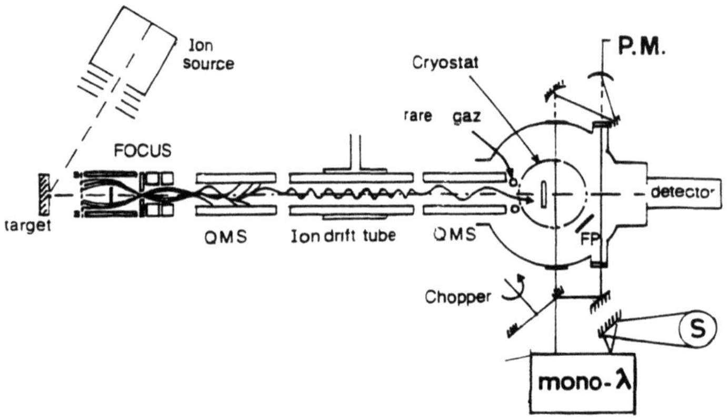

Fig. 1. Experimental set-up

(F.P.) attached to the rotating cryostat or enter the nitrogen guard of the cryostat. They are co-deposited onto an optical CaF_2 window, at low kinetic energy, with krypton gas, whose polarisation by the ions is believed to help the "soft-landing". Since the interaction cross-section of ions beam and electrons is very low in the gas phase we make the choice to try a neutralization onto the surface of the matrix. Such a process requires electrons with low kinetic energy to avoid either penetration in the volume of the matrix or the dissociation of the clusters during neutralization; e.g. Ni_2 is known to have a dissociation energy of 2.07 eV [7]. The co-deposition time was typically 10^3 s. The rare gas solid with its good thermal conductivity can possibly dissipate the excess energy of the neutral cluster created. Figure 2 shows the bottom part of the rotating cryostat. The tungsten filament creating the electrons is located inside the nitrogen guard, below the optical window. Since all the cryogenic parts are electrically grounded it is placed between two isolated tantalum plates. The electrons emerge from a slit (2×10 mm) in the top plate (grounded) and their kinetic energy can be varied by applying DC voltages to the filament and the bottom plate (repeller). The electron current can be measured by another plate located above the optical window.

The absorption spectrophotometer has been built in the Paris's laboratory. It is a standard double beam device (Fig. 1). The light created by a xenon arc source enters a grating home-made medium resolution monochromator. The optical output beam is chopped at 1.2 KHz and the two resultant beams (measure and reference) are collected after two reflections by a photomultiplier (P.M.). After electronic processing we record the log-ratio of the measure signal by the reference signal. The optical resolution can be varied between 1 and 12 A (6 A in our experiment) and the sensitivity is close to 5.10^{-4} in absorbance unit over the entire range (760–230 nm).

Amount of particles

An estimation of the superficial density of particles, needed to obtain the electronic absorption spectra of small clusters embedded in the rare gas matrix, is a prerequisite to our study. The oscillator strengths of the electric dipole allowed transitions are well known in the gas phase [8]. We can use them "to count" the centres as suggested by Dexter [9]. If we call N the total number of particles deposited on a surface S, formula 6-2 of [9] can be expressed by

$$N/S = 2.6 \cdot 10^{12}\ (n\sigma/\alpha^2 f)\langle A/\sigma\rangle_0 \tag{1}$$

– n is the index of refraction of the matrix, not precisely known; we will take $n = 1.4$.
– α is a coefficient relating the microscopic Lorentz local field to the macroscopic field $\alpha = (n^2 + 2)/3$, σ is the barycenter of the absorption band (in cm^{-1})
– $\langle A/\sigma\rangle_0$ is the zeroth order moment of the absorption band [10]
– f is the oscillator strength of the transition. The tables [8] give $gf = |g_1 f_{12}| = |g_2 f_{21}|$, g_1 et g_2 being respectively the statistical weights of the ground and excited states [11], then (1) can be rewritten

$$N/S = 2.09 \cdot 10^{12} (\sigma/gf) g_1 \langle A/\sigma\rangle_0 \tag{2}$$

The surface S need to be precised. If the area on which the particles are deposited is smaller than the surface of the optical probing beam we will consider the surface of the optical beam. If the particles are spreaded out on the optical window the surface will be the one of the window.

Results

The first step was to check the neutralization process and to make a calibration of the optical "counting" of the particles. We checked with the F.P located outside the nitrogen guard (Fig. 1) and with a F.P. placed in front of the window inside the nitrogen guard, that the number N_p of the particles detected was the same and that they were confined inside an area $S_p \leq 0.65\ cm^2$. Then S will be the surface of the optical beam, that is $S = 1.2 \pm 0.2\ cm^2$. A deposition of $N_p = 3.72 \cdot 10^{12}\ Ni^+$ ions neutralized with e^- onto a krypton matrix lead to a number of optically detected atoms of $N = 2.93 \cdot 10^{12}$, showing that the neutralization process is effective at $80\% \pm 20\%$. In a second step we have tried to make a "soft-landing" of the ions onto the matrix. This requires to decrease the kinetic energy of the ions, unfortunately at the expense of the amount of size-selected particles. The kinetic energy distribution has been measured before each run, with the F.P. by varying the potential of the last QMS. We define $E_c = E_{co} \mp \Delta E_c$, E_{co} being the mean kinetic energy and ΔE_c half the energy variation associated with a decrease from 90 to 10% of the maximum of ions collected. In all runs E_{co} was close to the potential energy difference between the target and the cryostat (grounded) and the maximum of ions was optimised by adjusting the focus lenses and the potentials of the three QMS. We were able to select all size of small nickel and silver positive clusters and we present in Table 1 some of our results on nickel trimer, silver trimer and silver pentamer. The maximum current of positive ions was close to 0.1 nA, in all runs, except in the second run on Ag_5^+ where it was only 0.06 nA. The voltage of the filament (Fig. 2) was kept at -8 V except in the third run on Ni_3^+ where it was lowered to -5 V. In all runs we recorded the absorption spectra over the entire UV-visible range. Fig. 3 and 4a) are parts of the spectra relevant for the determination of N in the present study. They cover the absorption ranges of the Ni and Ag atoms respectively isolated in krypton matrices, they are in good agreement with already published data [3, 4]. No significant absorption band can be detected outside these ranges. Since we were only able to detect atoms in the matrix, we define, in Table 1, the percentage of fragmentation (or evaporation) as the ratio of N, deduced from (2), by N_p times the size of the cluster. In Fig. 4b) and c) are also depicted the absorption and magnetic circular dichroism (MCD) spectra, obtained in Paris, after deposition of silver in krypton, the metal beam being generated with a Knudsen cell. The signal to noise ratio, in absorption (Fig. 4a and b) is the same but in MCD (Fig. 4c) it is improved by a factor roughly hundred [10]. We note that MCD, measured on the same sample, at the same time, with an applied field of only 1 Tesla, presents a sign characteristic of the nature of the excited state.

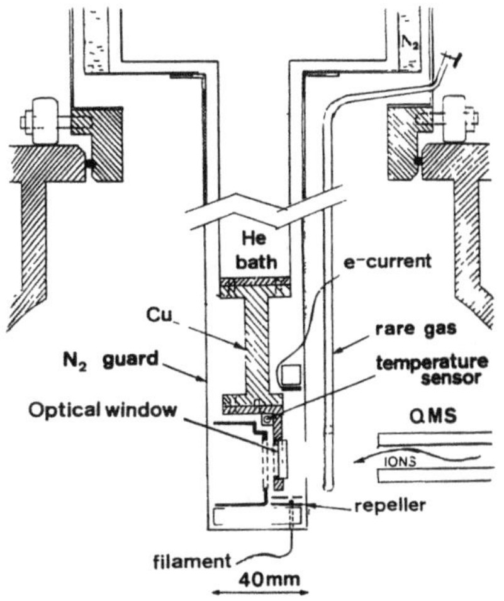

Fig. 2. Detail of the bottom part of the helium cryostat

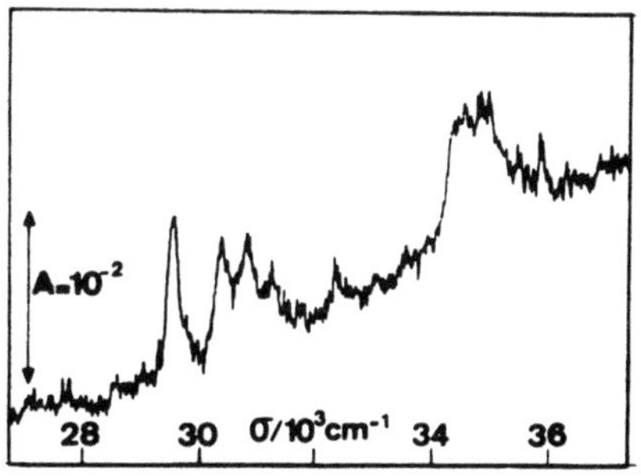

Fig. 3. Absorption spectrum of Ni atoms resulting from deposition of Ni trimers

Table 1

Selected cluster		Ag_3^+			Ag_5^+		Ni_3^+	
Kinetic energy of the ions Ec/eV		57 ± 14			57 ± 14		20 ± 10	
Time of the deposition (mn)		30			30		30	15
Maximum kinetic energy of the electrons (eV)		8			8		8	5
Number of depositions	1	2	3	1	2	1	2	3
Number of ions (N_p) detected by the F.P. ($/10^{12}$)	1.24	0.94	1.12	0.91	0.64	1.12	1.30	0.62
Number of atoms (N_0) optically detected ($/10^{12}$)	4.32	2.88	4.27	1.99	3.23	1.31	1.30	0.39
Percentage of fragmentation	121	106	133	53	101	39	34	21

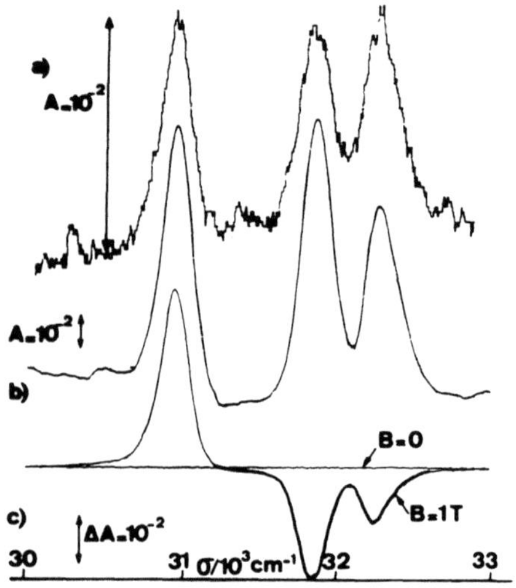

Fig. 4. Silver clusters

Discussion and conclusion

Table 1 shows that silver clusters evaporate when arriving onto the surface or during the neutralization process. The number of nickel atoms detected reduces when the mean kinetic energy was decreased to 20 eV. Moreover a further decrease in the kinetic energy of the electrons leads to only 21% of fragmentation. Assuming that the oscillator strengths of the electronic transitions of the remaining clusters are the same as the ones of the atoms, we think that the absorption signal due to the trimers is of the order of the experimental noise. One way to get the electronic spectra of the size-selected clusters is to increase the number of incident clusters; this is achieved now with the new experimental set-up installed by Lindsay et al. at the same institute [12]. As shown in Fig. 4 another way to increase the signal to noise ratio is to measure the MCD on the same sample. We have already choosen the first possibility and experiments are in progress using the new beam set-up and our neutralization and our optical system. This will insure a gain in sensitivity of at least ten. Further improvement is planed in the near future by measuring, at the same time, absorption and MCD spectra on the same sample.

One of us (J.C.R.) is indebted to J. Buttet and A. Chatelain who offer us the possibility to work at the Institut de Physique Expérimentale of the EPFL, and for their kind hospitality and financial support. We gratefully acknowledge the support of the Action incitative Europe of the CNRS (J.C.R.) and a grant from the Deutsche Forschungsgemeinschaft (D.K.). Technical assistance of R. Nahoum (Paris) and M. Dupuis (Lausanne) was kindly appreciated.

References

1. Proceedings of the 3rd International Meeting on Small Particles and Inorganic Clusters.: Surf. Sci. **156**, 1–1092 (1985)
2. Physics and chemistry of small clusters. NATO ASI Series B, Vol. 158, Jena, P., Rao, B.K., Khanna, S.N. (eds.). New York: Plenum Press 1986
3. Metal clusters. Moskovits, M. (ed.). New York: Wiley Interscience 1986

4a. Schultze, W., Becker, H.U., Abe, H.: Ber. Bunsenges. Phys Chem. **82**, 138 (1978) b. Schultze, W., Abe, H.: Diatomic metals and metallic clusters. Far. Symp. Roy. Soc. **14**, 87 (1980)

5a. Rivoal, J.C., ShaksEmampour, F., Zeringue, K.F., Vala, M.: Chem. Phys. Lett. **92**, 313 (1982)
b. Vala, M., Pyzalski, R., ShaksEmampour, F., Eyring, M., Pyka, F., Tipton, T., Rivoal, F.C.: J. Chem. Phys. **86**, 5951 (1987)

6. Fayet, P., Wöste, L.: Surf. Sci. **156**, 134 (1985)
7. Morse, M.D.: J. Chem. Phys. **80**, 5400 (1984)
8. Corliss, C.H., Bozman, W. R.: Nat. Bur. Stand. U.S. Monogr. **53**, 1 (1962)
9. Dexter, D.L.: In: Solid state physics. Seitz, F., Turnbull, D. (eds.), Vol. 6, p. 370. New York: Academic Press 1970
10. Rivoal, J.C., Briat, B.: Mol. Phys. **27**, 1081 (1974)
11. Di Bartolo, B. Optical Interactions in solids, Chap. 17. New York: John Wiley and Sons 1968
12. Lindsay, D.M., Meyer, F., Harbich, W.: Z. Phys. D – Atoms, Molecules and Clusters **12** (this issue) (1989)

Z. Phys. D – Atoms, Molecules and Clusters 12, 485–487 (1989)

Atoms, Molecules
Zeitschrift für Physik D and Clusters
© Springer-Verlag 1989

Fluorescence spectroscopy of laser vaporized species

F.W. Froben, J. Kolenda, and K. Möller
Institut fur Molekülphysik, Freie Universität Berlin, Arnimallee 14, D-1000 Berlin 33, Germany

Received 5 July 1988; final version 30 September 1988

The plasma produced by a Nd-Yag laser above a metal surface is measured by time resolved optical spectroscopy. The emission of atoms, ions and diatomics is observed for silver, copper and molybdenum. (Cu_2, $\omega e = 263.3\ \mathrm{cm}^{-1}$ X-state, $189.8\ \mathrm{cm}^{-1}$ A-state). From these small species clusters are formed with a size distribution between 10 and 80 Å diameter measured by electron microscopy.

PACS: 36.40.+D; 52.50.Jm; 32.20.Kf, Lg

1. Introduction

The method of laser vaporization has been used extensively since 1983 [1–3] for the production of clusters, and results have been obtained on size distribution, reactivity and laser induced fluorescence of the cold beam. But little is known on the hot plasma produced above the metal surface by the vaporizing laser pulse. Interest in this plasma is threefold.
1. The vapor above the solid contains atoms and ions, which can combine and form a cluster beam. The investigation of the plasma can contribute to the understanding of the cluster formation dynamics.
2. The plasma can be used as radiation source from the IR to the VUV and due to the time structure for time resolved spectroscopy [4–6].
3. The emission spectra of the plasma above the metal surface has been seen and analysed, but only atomic emission reported so far [7–8]. Here we present the first molecular spectra in this plasma [9]. This investigation may serve as an early view into the molecule and cluster formation process.

2. Experimental method

Contrary to the Smalley source where the combination of atoms occur in a channel and the cluster size is depending mainly on the gas pressure and the length and diameter of this channel, an open vaporization or a modified pick-up source was used [10]. The metal rod is mounted on a support inside a vacuum chamber at 10^{-6} Torr and can be moved freely up and down by a stepper motor and turned as in the Smalley source. The laser hits the rod somewhat of center so that the cosine distribution of the vaporizing atoms is about 30° to the beam direction. A pulse valve adjustable in distance and angle pushes the atoms and ions further into the forward direction and allows clustering by three-body collisions (Fig. 1). Ar, Kr or N_2 have been used as gases with a backing pressure of 8 atm. The pulse length can be varied between 100 μs and 1 ms and also the time interval between laser shot and gas pulse. The pressure in the vacuum chamber with the gas rises to 10^{-3} Torr. The gas changes the cluster size but reduces the emission intensity of the molecules, so the optical spectra were measured without adding gas. The cluster size distribution of this modified source is probably much wider than for the Smalley source, but there are some important advantages.
1. The gas volume around the metal rod does not change by the rod movement, resulting in less emission intensity fluctuations.
2. No problem with clogging of the channels, mainly the laser entrance channel.
3. No problem changing materials for vaporization. The emission of the plasma and of the beam is visible

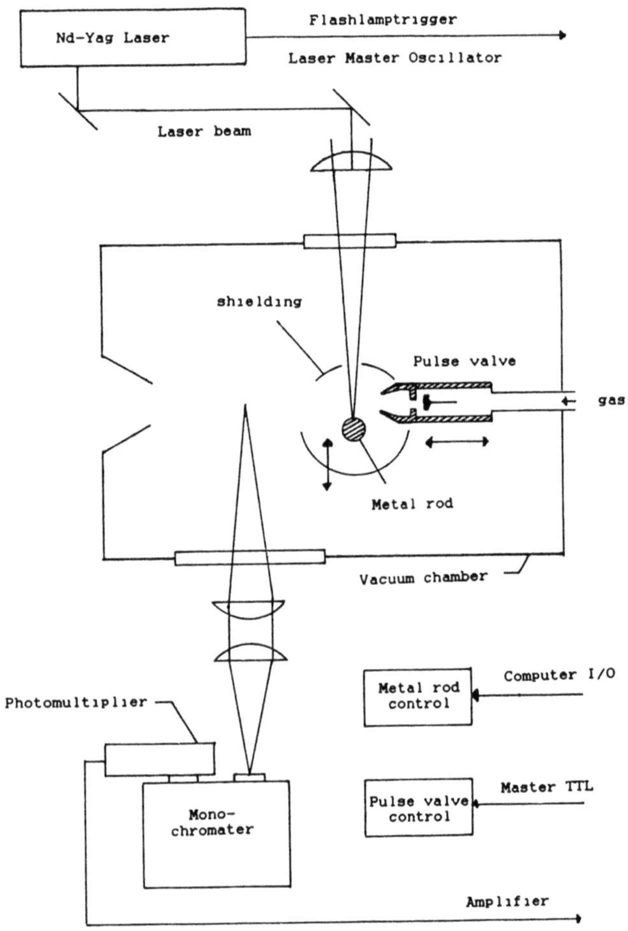

Fig. 1. Experimental set up

from top or side and an area of about 0.1 cm^3 is focussed on the entrance slit of a monochromator, amplified and recorded in a multi-channel analyser and stored with 1 μs resolution in a microcomputer. In principle it is simple to record a spectrum but because of the vast data it is a very time consuming process. Therefore it is difficult to recover the full time resolved spectrum out of the digital data.

3. Results and discussion

As an example of the results for pure metals, Fig. 2 shows the emission spectrum of the *A*-*X* transition of Cu_2. The position of the vibrational levels results in $\omega_e = 263.3$ for the *X*-state and 189.8 cm^{-1} for the *A*-state within experimental error of the published values [11]. The intensity of the vibrational bands change proportional to the laser power. Time resolved spectroscopy shows that the intensity distribution is highly nonthermal for the first 12 μs after the laser shot. The emission from the plasma and the beam above silver and molybdenum is measured also, showing atoms, ions and dimers. It is important to remark, that the intensity distribution, especially in the early stages of the emission, might be very different from the emission observed in an normal high temperature oven source. This is due to the fact, that the molecules are formed from highly excited atoms, and not from ground state atoms [9].

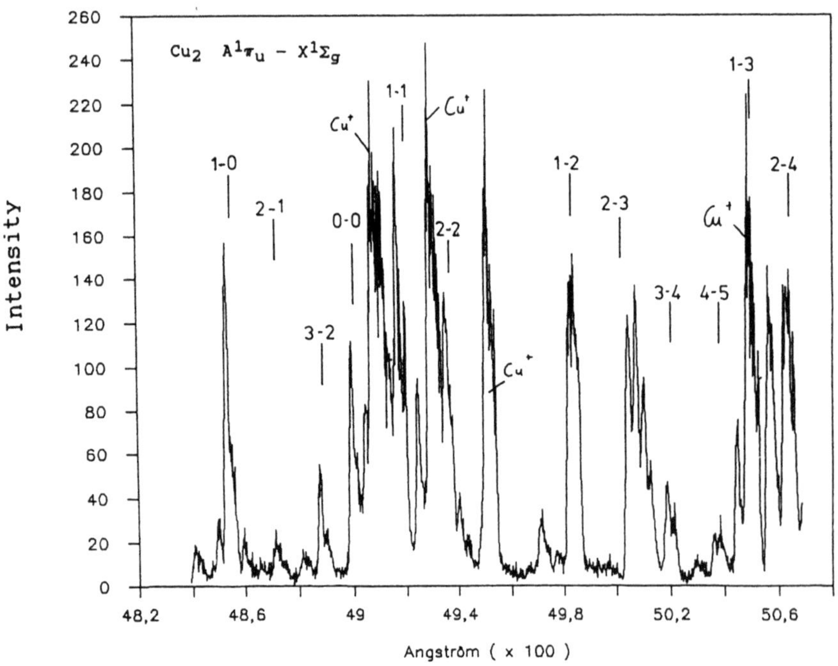

Fig. 2. Spectrum of *A*–*X* transition of Cu_2

Nevertheless, the emission spectrum is an important additional information on the early stages of molecule formation in the plasma and it may be possible to observe spectra of not yet known species in this non thermal source above cheramics or hetero nuclear materials. Further downstream clusters are collected on a carbon film supported by a copper net and measured by electron microscopy. The diameter is between 10 and 100 Å with a maximum of the size distribution at 40 Å. This is a much wider distribution than for the unmodified Smalley source, where under similar conditions the halfwidth of the distribution was between 10 and 20 Å. It is planned to complete this source with a TOF mass spectrometer in the near future.

This research was supported by the "Deutsche Forschungsgemeinschaft" Sfb 337 which is gratefully acknowledged. Thanks are due to Dr. B. Tesche form the Fritz Haber Institut of the Max-Planck-Society for the electron microscopy.

References

1. Hopkins, J.B., Langridge-Smith. P.R.R., Morse, M.D., Smalley, R.E.: J. Chem Phys. **78**, 1627 (1983); Dietz, T.G., Duncan, H.A., Liverman, H.G., Smalley, R.E.: J. Chem. Phys. **73**, 4816 (1980)
2. Bondybey, V.E., Schwartz, G.P., English, J.H.: J. Chem. Phys. **78**, 11 (1983)
3. Riley, S.J., Parks, E.K., Mao, C.R., Pobo, L.G , Wexler, S.: J. Phys. Chem. **86**, 3911 (1982)
4. Ueda, K., Sugawara, M., Shindoh, Y., Kaminishi, K., Namioka, T.: J. Spectrosc. Soc. Japan **34**, 366 (1985)
5. Caro, R.G., Wang, J.C., Young, J.F., Harris, S.E.: Optic News **11**, No. 39 (1985)
6. Adamson, A.W., Cimolino, M.C.: J. Phys. Chem. **88**, 488 (1984)
7. Kagawa, K., Yokoi, S., Nakajima, S.: Opt. Commun. **45**, 261 (1983)
8. Viswanathan, R., Hussla, I.: Springer Series in Chemical Physics Vol. 39, p. 148. Berlin, Heidelberg, New York: Springer 1984
9. Schulze, W., Winter, B., Möller, K., Froben, F.W.: High Temp. - High Pressure **20**, 331 (1988)
10. Schulz, C.P., Haugstätter, R., Tittes, H.U., Hertel, I.V.: Phys. Rev. Lett. **57**, 1703 (1986)
11. Kleman, B., Linkvist, S.: Ark. Fys. **8**, 333 (1954)

Z. Phys. D – Atoms, Molecules and Clusters 12, 489–492 (1989)

Atoms, Molecules
Zeitschrift für Physik D and Clusters
© Springer-Verlag 1989

UV-visible and Raman investigations of matrix isolated adducts between metal vapors of chromium and benzene: selective photo-irradiations effects

M. Dalibart and J. Derouault
L.S.M.C. (U.A. 124), Université de Bordeaux I F-33405 Talence-Cedex, France

Received 5 July 1988

The UV-visible absorption and the Raman spectra of chromium vapors trapped either in Ar-benzene mixtures or in neat benzene have been investigated. The data give evidence for the formation of the well-known dibenzene chromium sandwich complex but show also the formation of benzene complexes with the dimer Cr_2 and with the cluster Cr_3 photo-irradiations effects have been studied.

PACS: 78.00

Introduction

Our research group is engaged in the vibrational and UV-visible spectroscopic characterization of complexes between aromatic macrocyclic ligands such as cyclophans and dimers or very small aggregates of transition metals.

Some preliminary attempts have convinced us that the spectroscopic data will be consistently interpreted only on the basis of results related to smaller analogous compounds (complexes between benzene and chromium for example).

In a previous paper [1], we have demonstrated the existence of several complexes, and experimental conditions for spectroscopic (IR, Raman, and electronic) investigations and photo-irradiations studies have been defined and tested.

We present here a more detailed analysis of the Raman spectra and of the influence of photo-irradiation upon the formation of complexes chromium + benzene.

Experimental section

Reagents, apparatus for vaporization and spectrometers have been described elsewhere [1, 2].

Results and discussion

Infrared spectra

The infrared spectra of deposits Cr/Bz/Ar have been recorded with normal C_6H_6, with deuterated C_6D_6 and with a mixture $C_6H_6+C_6D_6$. The bands shifts and profiles clearly establish that the main species contains two benzene molecules [1].

This result confirms the previous assignment of $Cr(C_6H_6)_2$ but only very weak features may be assigned to other species than $Cr(C_6H_6)_2$ and numerous attempts to increase the proportion of these new species have failed. Furthermore, a deposit in neat Benzene does not yield any detectable spectrum at all.

UV-visible spectra

Our results (Fig. 1f) are in good agreement with those of previous papers [3] concerning the UV-visible spectra of Cr vapors co-condensed with Ar, except for the lack of the reported band at 480 nm. But, probably due to a lower isolation ratio, we observe a much higher relative concentration for the dimer Cr_2 at 462 nm.

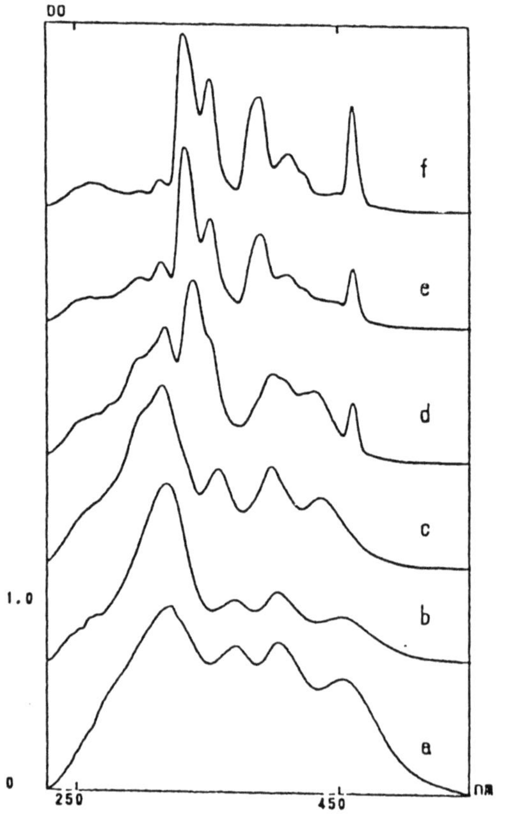

Fig. 1. UV-visible spectra of chromium vapors deposited at 12K (except for *b*: 20 K) *a*: Cr/5 Bz *b*: Cr/7 Bz *c*: Cr/.7 Bz/15 Ar *d*: Cr/1.4 Bz/30 Ar *e*: Cr/.7 Bz/40 Ar *f*: Cr/40 Ar

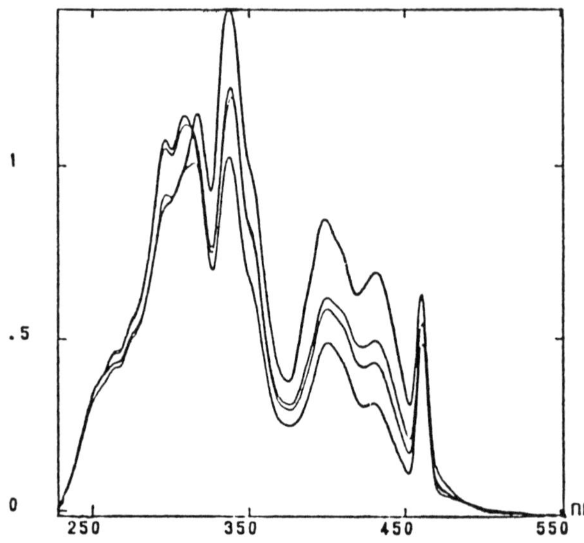

Fig. 2. UV-visible spectra of Cr in Bz+20 Ar after irradiations at 395 nm with times from 0 to 16 min.

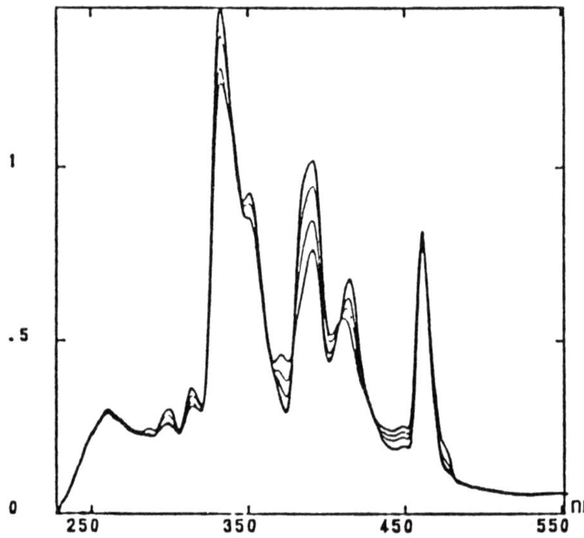

Fig. 3. UV-visible spectra of Cr in pure Ar after irradiations at 395 nm with times from 0 to 30 min.

Monomeric Cr atoms are responsible for the strong absorption at 330 nm. The band observed at 355 nm, which spontaneously decreases with time (even in darkness), is therefore due to an unstable species.

Introduction of benzene in the matrix (Fig. 1c–e) leads to the appearance of new bands in the 300 nm and 400 nm ranges: these bands are ascribed to the formation of complexes. When the amount of benzene increases, the bands due to Cr atoms (332 nm) and Cr_2 dimers (462 nm) tend to disappear. Co-deposition of Cr vapors and neat benzene (Fig. 1a, 1b) yields spectra with broad features and no free Cr atoms or dimers can be detected.

Each of the observed band for each deposit has been selectively irradiated for increasing durations. The spectra are deeply modified and their complete analysis, using factor analysis mathematical techniques [4] is still in progress. Figures 2 and 3 display the effect of selective irradiation at 395 nm with a 1 kW Xe lamp on two deposits:

—on a deposit of Cr in a pure Ar matrix, the irradiation induces the decrease of five bands: 313, 330, 350, 395 and 450 nm which are due to Cr atoms or Cr_2 dimer species. The counter-part is the increase of four others bands at 295, 370, 412 and 476 nm. The last one has been assigned by others authors either to a site effect for Cr_2 or to the trimer cluster. The three others ones are most probably due to site effects upon Cr and Cr_2 species.

—on a deposit of Cr in an Ar matrix with benzene, irradiation at the same wavelength induces a change in intensity of the band due to free Cr atoms, but the main effect is the decrease of all the complex bands in the 400–450 nm range and the increase of those in the

Table 1. Raman wavenumbers (cm^{-1}) of chromium vapors trapped in neat benzene obtained with different laser excitation lines

C_6H_6			C_6D_6			$C_6H_6+C_6D_6$			
676 nm	568 nm	488 nm	676 nm	568 nm	488 nm	676 nm	568 nm	488 nm	
					609 vS				A
608 S	607 S	606 S	607 w(R)	607 w(R)	—	607 S	607 S	608 S	C_6H_6
579 w(R)	582 w(R)	582 w(R)	579 S	578 S	580 S	580 S	579 S	579 S	C_6D_6
		571 m		546 vw	553 m	551 w			
510 vw	515 m	516 S		506 m	508 S			514 S	A
450 vw	456 vw	461 vw	455 vw	458 vw	458 m	465 vw		460 m	A
								430 m(T)	B
409 vw	410 w	409 S			403 S			405 S	A
	392 S(T)					392 m(T)			Cr_2
380 S		381 vw	380 vw(R)	379 vw(R)		380 m	382 w	386 m(T)	B_H
360 sh	356 w	356 w	357 w	356 w	360 w	355 w	355 w	365 w	$B_{HD}+C$
335 vw(R)	330 m		338 S	330 w		338 m	336 m	340 m(T)	B_D
		324 m			312 m		319 w	325 m(T)	A
312 w			312 w					315 m(T)	Cr_3
				306 vw		305 vw			
	275 m	273 sh		260 w	255 sh		267 m	275 w	$CrBz_2$
		256 S		246 vw	245 S			256 m 241 m	$A{:}Cr_xBz$
202 vw	202 vw	206 vS		199 vw	199 vS	204 vw		203 vS	A

S: Strong m: medium w: weak (R): Residue of isotopic partner coming from oil desorption from the debitmeter (T): line observed only in some cases according to deposition conditions
v: very sh: shoulder

300 nm area. The most probable interpretation is the dissociation of complexes of Cr_2 and the increase of complexes of Cr, especially $Cr(C_6H_6)_2$.

Raman spectra

The Raman spectra of deposits of chromium in neat benzene have been recorded with three different excitation laser lines: 676, 568 and 488 nm (Table 1). Some attempts have been made at 457 nm with the immediate destruction of the deposit as a consequence.

The spectra are very weak, due to the very low allowable excitation power (typically 15 mW at 488 nm), and numerous scans have to be averaged. Despite the poor signal to noise ratio, it is possible nevertheless to show very important properties:

* the spectra differ completely at 676 nm and at 488 nm, owing to different (pre-)resonance effects.

* lines coming from $Cr(C_6H_6)_2$ are weak on all recorded spectra, and the Cr_2 dimer may be detected in some cases, although the UV spectra had demonstrated its very low abundance.

* several different species can be characterized according to the variations upon wavelength excitation and upon deuteration:

A. Species of which the spectrum is excitated mainly by the 488 nm laser line. The isotopic shift of the line at 256 cm^{-1} suggests that it corresponds to a one benzene containing complex. Thus the *A* species may be described as $Cr_x(C_6H_6)$, and most probably $x=3$ since the 206 cm^{-1} line is very closed to a strong line of Cr_3 [5]. This assumption is supported also by the resonance Raman effect since the reported absorption band for Cr_3 is located at 477 nm [6].

B. Species mainly detected by excitation at 676 nm. The vicinity of the strongest line at 380 cm^{-1} with that of Cr_2 [5] suggests that *B* is a complex of the dimer, while the isotopic shifts establish that at least two benzene molecules are involved in such a complex — other species are detected in the 300 cm^{-1} range, but the available data do not allow a more accurate interpretation.

Conclusions

When trapped in benzene containing argon matrices or in neat benzene, the chromium atoms or the dimer molecules which exist in the metallic vapors may form several complexes with one or two ligands.

The infrared spectra confirm the main occurence of the well-known di-benzene chromium sandwich complex.

The UV-visible absorption spectra demonstrate the formation of several new complexes which are sensible to selective photo-irradiation.

The analysis or Raman spectra (including the influence of deuteration) and the comparison of the results with literature data concerning Cr in Ar matrices allows for the assignment of some lines to complexes of Cr_2 and Cr_3 clusters with one or two benzene molecules.

Future developments of this work will concern the ability to stabilize analogous complexes with some polyaromatic macrocycles like cyclophans.

References

1. Dalibart, M., Derouault, J., Labarrere, H., Tranquille, M.: J. Mol. Struct. **174**, 17 (1988)
2. Tranquille, M.: Imag. Chimie **36**, 18 (1980)
3. Pellin, M.J., Gruen, D.M.: J. Chem. Phys. **79**, 5887 (1983)
4. Dalibart, M., Spectra 2000. **116**, 69 (1986)
5. Dilella, D.P., Limm, W., Lipson, R.H., Moskovits, M., Taylor, K.V.: J. Chem. Phys. **77**, 5263 (1982)
6. Klotzbucher, W.E., Ozin, G.A.: J. Am. Chem. Soc. **100**, 2262 (1978)

Z. Phys. D – Atoms, Molecules and Clusters 12, 493–496 (1989)

Atoms, Molecules and Clusters
Zeitschrift für Physik D
© Springer-Verlag 1989

Nanosecond emission spectra in μm space of Si particles sputtered by laser irradiation

A. Kasuya and Y. Nishina
Institute for Materials Research, Tohoku University, Sendai, 980 Japan

Received 5 July 1988

Laser sputtering process from surfaces of Si and other Group IV elements are analyzed by space/time resolved spectroscopy. Our measurements demonstrate the capability of analyzing dynamical characteristics of cluster formation/fragmentation processes in μm and nanosecond resolution.

PACS: 33.22.-t; 52.50.Jm; 82.30.N

1. Introduction

Laser vaporization technique is used extensively in preparing atomic clusters of metals and semiconductors [1]. The process of cluster formation in this technique, involves desorption of surface atoms by laser sputtering followed by their clustering reactions through collision with inert gas. Hence, clusters are presumably formed during the collision process. There are, however, many experimental indications in which the desorbed particles are initially in the form of clusters prior to the collision with cooled gas. In the case of graphite, for example, C_n's with $n > 27$ have been observed by laser sputtering of graphite surface in vacuum [2]. Our previous time-of-flight measurements of emitted ions in Si also show the presence of Si clusters during the first 100 ns after laser excitation [3]. The present paper reports our analysis of laser desorption process immediately after laser excitation in terms of space and time-resolved spectra of light emitted from the sputtered particles from the target of silicon and other crystals of Group IV elements.

2. Experimental results

The KrF laser beam of its peak intensity up to 500 MW/cm^2 is incident at right angle to the target surface. The plume of emission is visible over the distance ranging from the irradiated spot at $z = 0$ on the target to about 1 mm off the surface. The emission over 5 mm length along the z direction normal to the surface is focused and projected along the entrance slit of a streak camera directly, or along that of a single monochromator of $f = 4$ connected to the streak camera. Figure 1a shows the streak image obtained by one laser shot on Si without the monochromator. The ordinate represents the time t measured from the onset of laser pulse which is indicated by the double-headed arrow. The intensity distribution in the z-t plane is represented by various dot patterns according to the intensity levels of emission as shown in the inset. The pattern shown in Fig. 1 indicates that intensity of emitted light has a maximum in the area a few tenths of a millimeter off the surface and at a time a few nanoseconds after the end of laser excitation. The region of the maximum intensity is narrow horizontally (in space) and elongated vertically (in time). Although the details of intensity distribution in the pattern fluctuate from pulse to pulse of measurements, yet the essential feature on the presence of a maximum intensity remains for each measurement. The result shows that the desorbed particles do not expand rapidly in vacuum right after being removed from the surface, but rather move together in one direction for the first few nanoseconds after their desorption. From

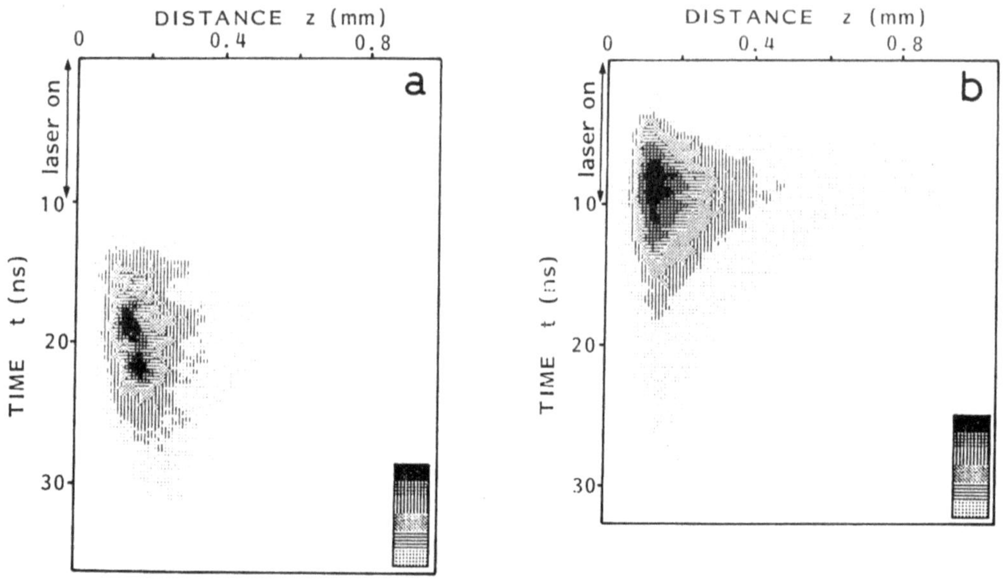

Fig. 1a and b. The time evolution of the intensity of the light emitted from the plume of laser desorbed particles toward the z direction normal to the surface of (**a**) Si and (**b**) graphite. The peak intensities of the pulsed KrF laser beam are 100 MW/cm^2 for Si and 30 MW/cm^2 for graphite, respectively

the trajectory of this high intensity region, the translational energy of desorbed particles is estimated to be in the order of 10 eV. Besides this trajectory, there is some area of intensity remaining in the region around 0.4 mm and 20 ns. Such an additional area indicates the presence of another component which moves faster (fast component) with respect to the main component mentioned above (slow component). The presence of the fast component is more evident in the case of Carbon atoms desorbed from graphite surface as shown in Fig. 1b. The patterns shown in Fig. 1 are the images of emission along the z-axis which pass through the center (maximum intensity) of the plume. The measurements along the off-axis of the plume show that the slow component remains in the pattern only within a few tenths of a millimeter below and above the z-axis. This measurement confirms that the slow component expands in a very narrow region in the z direction away from the surface, whereas the fast component expands more isotropically into vacuum. In order to find out whether the slow component is directed along the surface normal or along the direction of the incident laser beam, the direction of the plume is measured with the laser beam incident 40° tilted from the surface normal. The pattern of the plume is flatten in the z direction and elongated toward the laser beam in comparison with the case of the normal incidence. This pattern together with its streak image shows that the slow component points against the incident direction of the laser beam, whereas the fast component is spread in much wider directions with respect to the z-axis. The difference noted in the desorption characteristics between Si and C is that the fast component of C is more prominent, i.e., being more intense and distributed far in the z direction in comparison with the case of Si. On the contrary, the slow component is more intense and confined in a narrow region in Si than in C. Also, Fig. 1 indicates that the maximum in the light emission occurs much earlier in C than in Si.

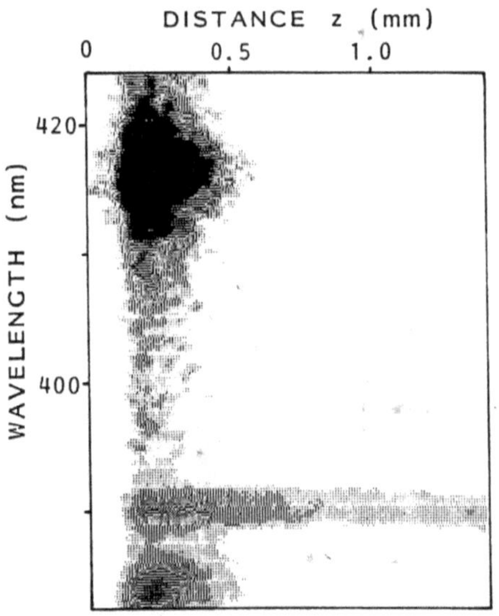

Fig. 2. The emission spectrum between 385 nm and 425 nm of the plume along the surface normal of Si

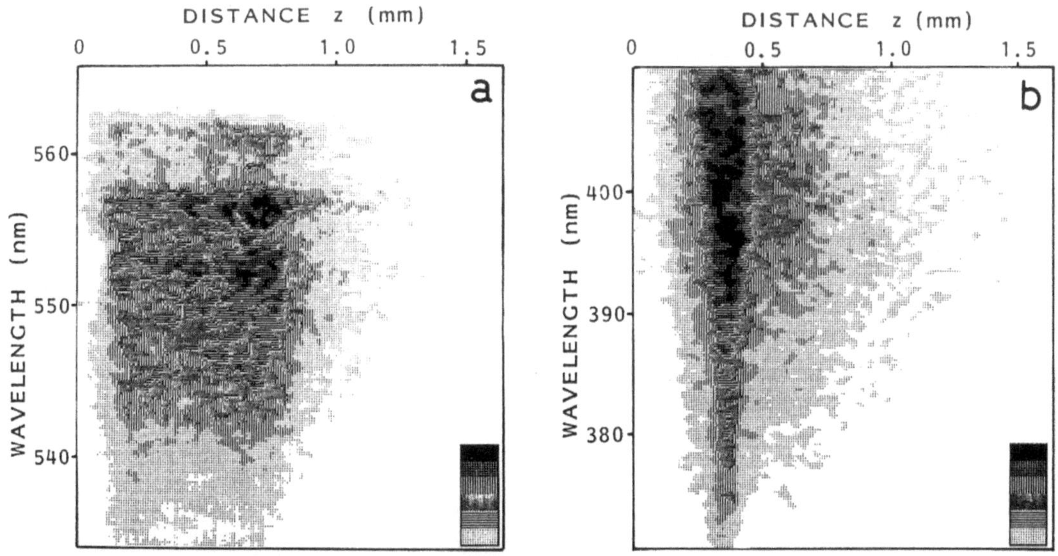

Fig. 3a and b. Emission spectra of the plume along the direction normal to the graphite surface, showing in (**a**) the 0-1 transition of the Swan band of C_2, and in (**b**) the 405 nm band of C_3

Similar measurements are also carried out for Ge and their results are essentially the same as for Si.

3. Discussion

Spectral characteristics of emitted light are analyzed in space and time for the slow component. The spectra in the range from 200 to 800 nm on Si consist of many lines as well as weak continua. The spectra from 385 nm to 425 nm is shown in Fig. 2. The line at 390 nm is identified as emission from mono-atomic neutrals (Si^0), and those at 386 nm and 416 nm are from their ions (Si^+). These emission lines are the transitions from the excited states at 5.1 eV, 10.1 eV and 9.8 eV, respectively. All neutral lines observed are relatively sharp and appear in space more than 3 mm off the surface. On the contrary, the iòn lines are broad near the surface and become narrow with increasing z, and disappear within 1 mm. As the laser intensity is increased, the spectrum extends far in the z direction and tends to be dominated by ionic lines. Band-like continua are dominantly distributed in the range from 300 to 600 nm. These bands exhibit fine structures composed of many sharp lines. The most dominant one around 410 nm coincides with the H-X transition reported on Si_2 [4]. Other bands may be interpreted as superposition of fine structures, indicating that they are due to Si_n with $n>2$. These bands are observed in position near the sample surface and disappear sharply beyond a few tenths of a millimeter along the z axis. In the case of graphite, spectral bands due to neutral molecules C_2 and C_3 are dominant as shown in Fig. 3. These bands are identified as Swan band of C_2 (the 0-1 pair of d-a transition) for Fig. 3a and "405 nm band" of C_3 for Fig. 3b [5]. The presence of the C_n cluster bands are consistent with the result of mass analysis on laser vaporization measurements where molecular species are dominant over atomic [2]. It is difficult to assign other continuous bands in C as well as in Si because of lack of both theoretical and experimental information on spectroscopic parameters at present for C_n with $n>3$, and Si_n with $n>2$.

It is clear from our measurements that particles are not desorbed from the surface by ordinary evaporation process. It is particularly evident for particles identified with the slow component in the pattern that they change their states continuously in tens of nanoseconds after laser excitation. For the first few tenths of a millimeter and a few nanoseconds, they move together in the direction toward the laser beam from the sample surface. Fine structures in the band emission from the desorbed particles in this region indicate that they are not in the form of individual atoms but are chemically bound, at least, partially in the form of clusters. In the case of Si, disappearence of quasi-continuous (molecular) spectra, in contrast with the persistent appearence of atomic lines beyond this region both in space and time indicate that the desorbed particles change their chemical states through the process $Si_n \rightarrow Si_{n-1} + Si^*$, where Si^* is the electronically excited Si. The observed emission of atomic lines comes from de-excitation of Si^* which is produced in the process of fragmentation in desorbed particles. This interpretation is consistent with our observation through light emission that molecules, ions and atoms expand in space and time, and follow similar and continuous trajectories in the z-t plane.

The authors would like to thank Professor C. Horie of Tohoku University and Professor K. Murakami of University of Tsukuba for valuable discussion. This work was supported in part by the scientific grant from the Mitsubishi Foundation.

References

1. See for example, Physics and chemistry of small particles. NATO ASI Series B158. Jena, P., Rao, B.K., Khanna, S.J.: (eds.) New York: Plenum 1987
2. Kasuya, A., Nishina, Y.: Phys. Rev. **B28**, 6571 (1983)
3. Kasuya, A., Nishina, Y.: Phys, Rev. Lett. **57**, 755 (1986)
4. Douglas, A.E.: Can J. Phys. **33**, 800 (1955)
5. Herzberg, G.: Spectra of diatomic molecules. New York: Van Nostrand Reinhold 1950

Z. Phys. D – Atoms, Molecules and Clusters 12, 497–500 (1989)

Atoms, Molecules
Zeitschrift für Physik D and Clusters
© Springer-Verlag 1989

Melting for lead particles of very small sizes; influence of surface phenomena

R. Garrigos, P. Cheyssac, and R. Kofman
Laboratoire de Physique de la Matière Condensée UA190 Parc Valrose, F-06034 Nice Cedex, France

Received 5 July 1988; final version 30 September 1988

Melting, solidification and undercooled state of lead particles of very small sizes (mean radius $\bar{r}$ lower than 100 Å) have been investigated through high sensitive optical measurements. A phenomenological interpretation of melting is proposed which is based both on thermodynamic size effect and surface phenomena.

PACS: 64.00; 78.00

Introduction

The molten state of matter has been extensively studied [1]. Despite numerous theoretical works [2, 3] or numerical simulations [4, 5], melting remains a complicated problem since various phenomena such as premelting effects are involved. Undercooling and solidification are also difficult to explain since the stability of the undercooled liquid and nucleation mechanisms are involved.

In the last few years new phase transition phenomena, induced by the presence of the surface, have been brought in evidence [6, 7]. The nature of this transition is not definitely established. Anyway, direct observations by HREM have shown the surface of a crystal to be an unstable region [8, 9] and surface loosening of small crystals has been observed at the oncoming of the melting region [10].

In this paper, we report about optical investigation of melting, undercooled state and solidification of lead particles.

In Sect. I, the principle of the experiment is reminded and experimental results are given.

In Sect. II, an interpretation of melting is proposed which takes account of thermodynamic size effect and surface phenomena.

I. Experiment: principle and results

Although the experimental procedure has thoroughly been given in a previous paper [11], we remind its main features here.

Lead particles of small diameter (100 to 500 Å) have been obtained by evaporation and condensation on alumina films in ultra-high vacuum and covered by another film of alumina. The size distribution of the particles, characterized by the mean radius $\bar{r}$, the standard deviation σ and the number N of particles per unit area, has been determined by analysis of electron microscopy pictures. A gaussian law is assumed for the radii of the particles in each distribution.

In the optical properties of a material, bulk properties and surface properties are involved. One way to obtain optical properties is to measure the reflectance of the sample. For a given material, the reflectance R of the solid phase differs from that of its liquid phase; then, by high sensitive reflectance measurements we are able to study the solid-liquid transition of the material. By using small particles embedded in a transparent medium, we insure a high degree of reproducibility along cycles of melting and solidification.

The reflectance R of the sample has been recorded as a function of temperature T for many cycles of

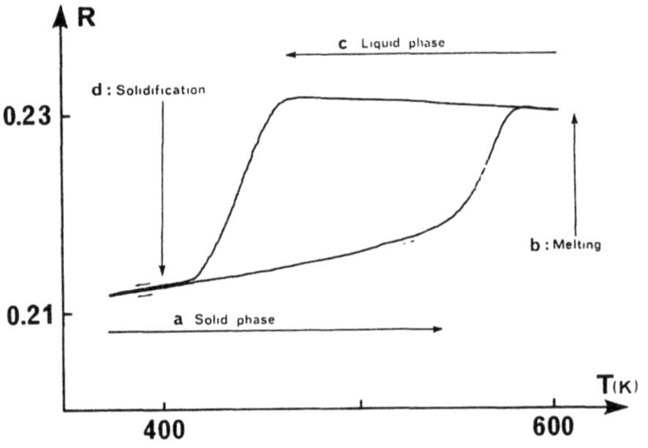

Fig. 1. Phenomenological description of the phase transition as revealed by reflectance measurements
- Region *a*, where *R* increases weakly as *T* increases: the particles are solid.
- Region *b*, where *R* increases strongly as *T* increases: the particles melt.
- Region *c*, where *R* varies very weakly as *T* decreases: the particles are liquid and supercooled for the major part of this region.
- Region *d*, where *R* is strongly decreasing as *T* decreases: the particles solidify

melting and solidification. A typical $R(T)$ plot is given in Fig. 1 in which the occuring phenomena can be distinguished.

In Fig. 2, plóts of $R(T)$ have been selected for five distributions of different mean radius. The striking feature is the progressive disappearance of hysteresis when the mean radius decreases.

II. Phenomenological interpretation of melting

II.1. Size effect

From classical thermodynamics, it is known that the melting temperature of a particle decreases with its size [12]. Such an effect can clearly be seen on Fig. 2. Furthermore, particles of various sizes are contained in each sample and this fact (polydispersion) makes the transition regions broader.

This thermodynamic size effect relates the melting temperature T_M of a particle to its size. This is expressed through macroscopic quantities: the radius r_s of the solid particle which is assumed to be spherical, its latent heat of fusion L, and the surface tensions γ (interfacial energies) between the solid or liquid particle and the surrounding medium. Several models have been proposed to describe this size effect $T_M(r_s)$. A review of these models is given in [13] and [14].

This experiment relates the optical properties of a sample and its solid-liquid transition. Then, one needs a description of the reflectance in terms of the optical characteristics of the substrate, the matrix and the lead particles.

Within the frame of an effective medium theory [15] the sample is equivalent to a e_0 thick layer of dielectric constant $\bar{\varepsilon}$ given by:

$$\bar{\varepsilon}=\varepsilon_m\frac{1-\Sigma_i(q_i-3\varepsilon_i A_i)}{1-\Sigma_i(q_i-3\varepsilon_m A_i)};\quad A_i=\frac{q_i}{\varepsilon_i+2\varepsilon_m} \tag{1}$$

This dielectric constant accounts for:

- the size distribution of the particles
- the total volume $q=q_l+q_s$ of particles per unit volume of the effective medium, q_s being the solid part and q_l the liquid part
- the dielectric constant ε_m of the matrix
- the dielectric constant ε_i ($i=s$ or l for solid or liquid phases) of lead particles with their temperature dependence

The reflectance R of the sample is given by Fresnel relations as a function of $\bar{\varepsilon}$ through ε_i, q_l and q_s [11].

Taking into account that, for a given distribution at T_M the particles of radii $r<r_s$ are liquid while those of radii $r>r_s$ are solid, we have calculated R as a function of T (Fig. 2). Numerical computation has been

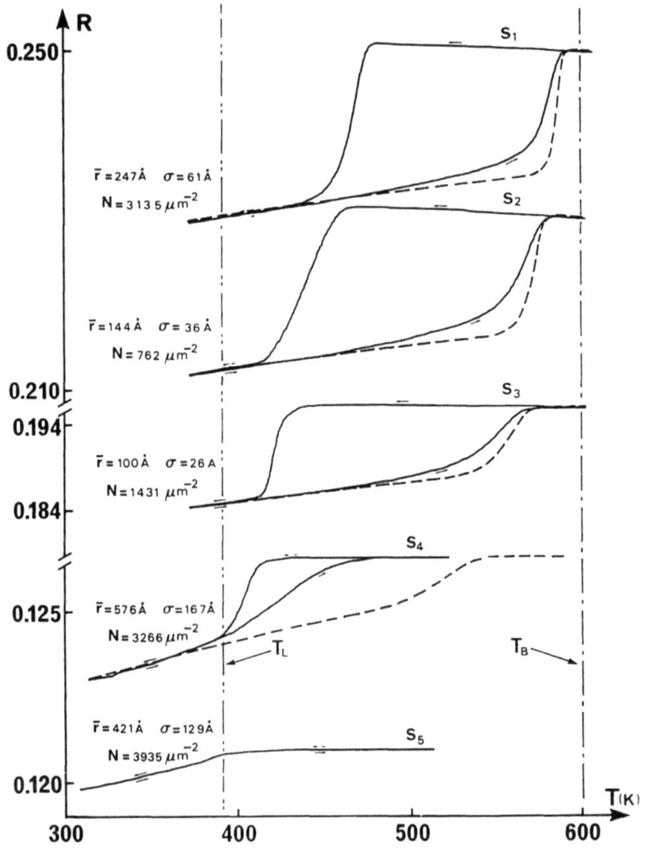

Fig. 2. Reflectance curves for five distributions of lead particles —— Experiment; $T_L \sim 392$ K is the maximum undercooling temperature of the liquid [13] ------- Theory (sample S_5 excepted)

performed in order to fit the values of the reflectance and its slopes both in the solid and liquid regions. Such a procedure allows determination of unknown parameters (such as the thickness e_0 of the effective medium) and adjust dubious values of other ones (such as the free electron plasma frequency ω_p for liquid particles). As expected, the value of e_0 varies in the same manner as the mean radius (from 434 Å for sample S_1 to 144 Å for sample S_4). The value of ω_p is 2.10^{16} rd.s^{-1} and is found to have a slight increase when the mean radius decreases (10% in the $\bar{r}$ explored range).

It can be seen that for distributions having the upper mean radii (samples S_1, S_2 and S_3) the model agrees with the experiment particularly at the temperature where the whole distribution is melted. Nevertheless, a discrepancy remains in the region of transition.

For distributions of lower mean radii (sample S_4 and "*a fortiori*" S_5) the thermodynamic model is no longer valid.

II.2. Surface phenomena

As reported in Fig. 2 the calculated transition width is not in agreement with the experimental one. In the upper model, polydispersion only is responsible of the width of the transition. It turns out that other contributions must be involved.

This model describes a first-order transition and is a macroscopic point of view which ignores premelting effects such as the occurrence of disorder at the surface of a particle. Such a crude model cannot reproduce the details of the whole results, particularly when the surface to volume ratio of the particle becomes appreciable (small particles).

In the present paper, we attempt to account for surface phenomena by assuming that, at the surface of each solid particle, there is a film of thickness δ in which disorder occurs (Fig. 3). Such a geometrical description has already been used by Wronski [16] for melting of small particles of tin. For reasons of simplicity, we will use the terminology of "surface melting" for effects arising at the surface of a material before it melts, although a real surface melting is not clearly established.

For calculations, we will consider that for a given distribution, δ depends on temperature and not on radius (δ is an average value of the thickness of the disordered layer, over the distribution). Furthermore, assuming this layer to have the same dielectric constant as the liquid, we can calculate the new expressions of q_s and q_l; i.e.:

$$q_s = \frac{4\pi N}{3e_0}\int_{r_s}^{\infty}(r-\delta)^3 p(r)\,\mathrm{d}r;\; p(r) = \frac{1}{\sigma\sqrt{2\pi}}\,\mathrm{e}^{-\left(\frac{r-\bar{r}}{\sigma\sqrt{2}}\right)^2} \qquad (2)$$

$$q_l = \frac{4\pi N}{3e_0}\int_0^{r_s} r^3 p(r)\,\mathrm{d}r + \frac{4\pi N}{3e_0}\int_{r_s}^{\infty}[r^3-(r-\delta)^3]p(r)\,\mathrm{d}r \qquad (3)$$

Then, every parameter being the same as used or determined in the previous section, we can obtain R and deduce δ in order to fit exactly theory on experiment. The results for samples S_1 to S_4 are given in figure 4 in which is also drawn the thickness of the liquid layer as calculated by Frenken et al. [6] for a lead single crystal.

It appears that δ increases with temperature, which is characteristic of a second-order transition. Furthermore, the lower $\bar{r}$, the higher δ. Then, we can reasonably consider melting as a series of two transitions when T increases:

– a second-order transition (order-disorder or surface melting) which starts at low temperature and predominates for the smaller particles.

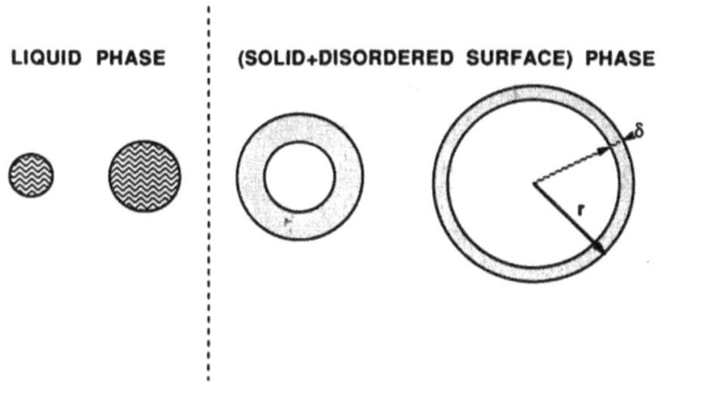

Fig. 3. Schematic representation of the particle distribution at a given temperature

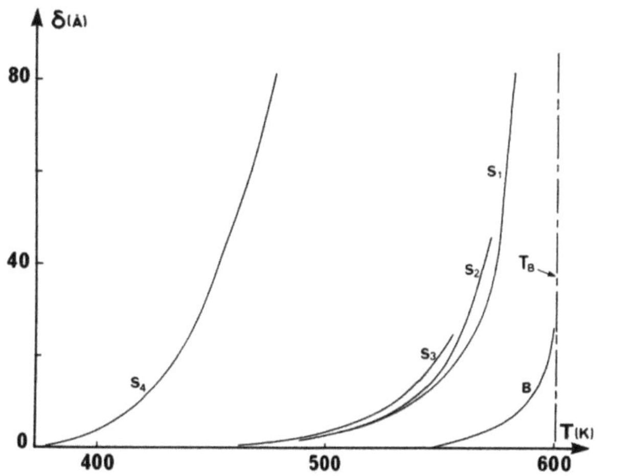

Fig. 4. Thickness of the disordered layer versus temperature for four distributions of lead particles (S_1, S_2, S_3, S_4) and for a bulk single crystal of lead (B)

– the usual solid-liquid first-order transition which takes place at T_M (T_B for the bulk) the contribution of which is weaker as the size of the particles decreases.

References

1. Ubbelohde A.R.: The molten state of matter. New York: J. Wiley 1978
2. Wood D.M., Ashcroft N.W.: Phys. Rev. B. **25**, 6255 (1982)
3. Haymet A.D.J., Oxtoby. D.W.: J. Chem. Phys **84**, 1769 (1986)
4. La Violette R.A., Stillinger F.H.: J. Chem. Phys **83**, 4079 (1985)
5. Stillinger F.H., Weber T.A.: Phys. Rev. A, **28**, 2408 (1983)
6. Frenken J.W.M., Maree P.M.J., van der Veen J.F.: Phys. Rev. B. **34**, 7506 (1986)
7. Prince K.C., Breuer U., Bonzel H.P.: Phys. Rev. Lett. **60**. 1146 (1988)
8. Iijima S., Ichihashi T.: Phys. Rev. Lett. **56**, 616 (1986)
9. Bovin J.O., Wallenberg R., Smith D.J.: Nature **317**, 47 (1985)
10. Lereah Y., Deutscher G., Kofman R.: Europhys. Lett. **8**, 53 (1989)
11. Garrigos R., Kofman R., Cheyssac P.. Europhys. Lett. **1**, 355 (1986)
12. Semenchenko V.K.: Surface phenomena in metals and alloys. New York: Pergamon Press 1961
13. Cheyssac P., Kofman R., Garrigos R.: Phys. Scr. **38**, 164 (1988)
14. Borel J.P.: Surf. Sci. **106**, 1 (1981)
15. Aspnes D.E.: Thin Solid Films **89**, 249 (1982)
16. Wronski C.R.M.: B. J. Appl. Phys. **18**, 1731 (1967)

Z. Phys. D – Atoms, Molecules and Clusters 12, 501–504 (1989)

Atoms, Molecules
Zeitschrift für Physik D and Clusters
© Springer-Verlag 1989

Nonlinear optics in quantum confined semiconductor particles

P. Horan and W. Blau

Department of Pure and Applied Physics, Trinity College, Dublin 2, Ireland

Received 5 July 1988; final version 30 September 1988

The field of nonlinear optics is dominated by the search for materials which show a large nonlinear optical response. Physical confinement of charge carriers, whether in 1, 2 or 3 dimensions, can greatly enhance the nonlinear optical behaviour. We have studied the nonlinear optical properties of small particles, in the size range 100-10Å which provide 3-D charge confinement. The larger particles are too big to show significant quantum confinement effects. Smaller particles show distinct confinement effects and a large optical nonlinearity.

PACS: 42.65

1. Introduction

The interaction of light with matter is usually characterized by several phenomena, such as light absorption, refraction, scattering and luminescence. These are regarded as properties of the material, independent of light intensity. However, for sufficiently large light intensities, such as are available with lasers, these optical characteristics become a function of intensity. This is the field of nonlinear optics [1].

The great activity in the area of nonlinear optics arises both from the interesting basic physics and the possible applications in the fields of optical computing and signal processing. Towards this end, materials with a large nonlinear response are very desirable i.e. materials which exhibit a large change in some characteristic optical property for a low light intensity. Most work has been done in semiconductors [2]. In the search for larger nonlinearities organic polymer materials show great promise [3]. Another interesting approach is the field of quantum confined semiconductor systems. When charge carriers are physically confined within a dimension comparable with the exciton radius r_{ex}, the energy structure is radically altered. This technique allows the engineering of novel optical materials.

Structures which produce confinement within one dimension i.e. where the electron or hole is free to move only within a plane but confined perpendicular to that plane, are known as quantum wells [4]. These have shown very interesting nonlinear properties. Confinement of the charge carriers in all three dimensions should show an enhanced nonlinear response [5]. Such structures are known as "quantum dots". The purpose of this paper is to investigate the nonlinear optical properties of such quantum dots. Towards this end we have studied several systems exhibiting different degrees of confinement. We observe an enhanced nonlinear response in the well confined samples.

2. The quantum dot system

By a quantum dot we refer only to particles retaining the bulk lattice structure. This imposes a lower size limit, since at some point rearrangement to a differentially packed "cluster" structure will occur so as to minimize the number of broken bonds. As an upper limit, particles must be small enough to provide charge confinement, typically smaller than the bulk exciton radius. This defines a range of approximately 20–200 Å.

The electronic behaviour can be simply modelled by the "particle-in-a box" problem. Consider an electron confined within a cubic volume of semiconductor

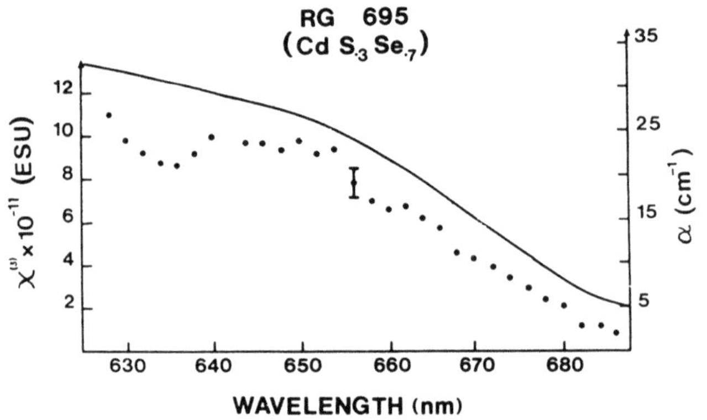

Fig. 1. Absorption coefficient ——— and measured third order susceptibility $\chi^{(3)}$ . . . . for Schott glass R G 695

of side length "a". The wave function must be zero outside and go to zero at the bounderies. The simplest solutions are sine' waves; i.e. only particular "standing waves" which satisfy the boundary conditions are selected from the conduction band continuum. The resulting energies are given by:

$$E_{nlm} = E_g + \frac{h^2 \pi^2}{2 m_e a^2}(n^2 + l^2 + m^2) \tag{1}$$

where E_g is the bulk semiconductor bandgap, m_e is the electron effective mass and $n, l, m = 1, 2, 3 \ldots$. These are the only allowed eigenstate energies, and the absorption spectrum should show a series of sharp absorption lines. It is because of this concentration of bulk oscillator strength into single spectral lines and small volumes that an enhanced nonlinear response is to be expected.

3. Experimental method

Linear optical spectra were taken using a Phillips SP8-200 UV/VIS double beam spectrophotometer. The nonlinear response was investigated by two methods. The variation of sample transmission with intensity was investigated using a nitrogen pumped dye laser system; PRA LN1000 high pressure nitrogen laser and PRA LN107 high resolution dye laser. This system gave 50–100 μJ pulses of ~500 ps duration, with a spectral bandwidth of ~5 Å. The same laser system was also used for Laser Induced Grating (L.I.G.) experiments [6]. In this technique two coherent beams are made to overlap at a small angle within the sample. The resulting interference pattern modulates the materials' absorption/refractive index. The pulses self-diffract from the created grating. The energy diffracted out depends on the ease with which the materials' properties may be altered. Measurement of the diffraction efficiency allows direct calculation of the effective third order susceptibility $\chi^{(3)}$, a figure of merit of the nonlinear response [7].

4. Cd S_x Se_{1-x} doped glass

Initial studies were carried out with semiconductor doped glasses. These are commercially available as

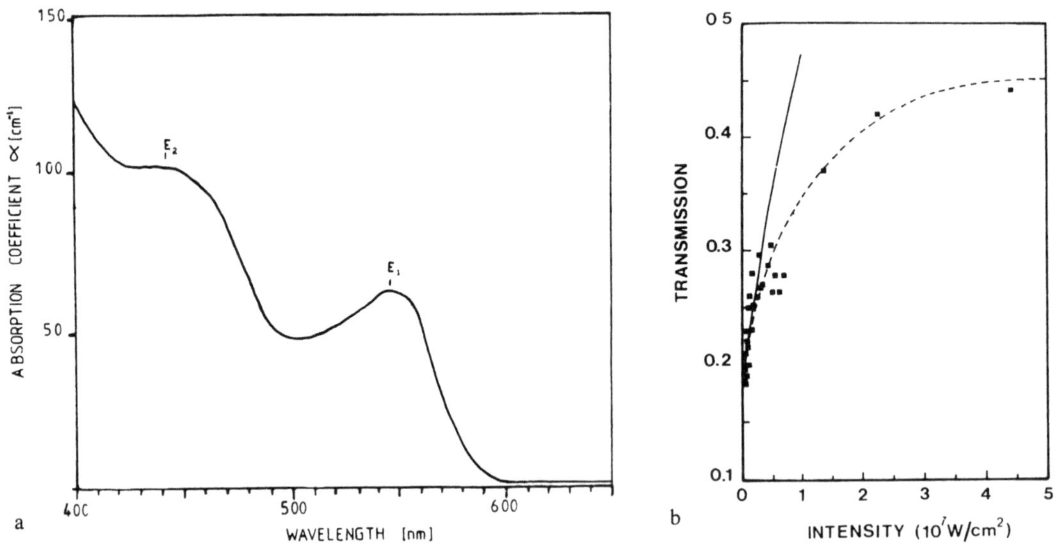

Fig. 2. **a** Absorption spectrum for the quantum confined glass (Corning 203 AME) showing the first and second transitions. **b** Measured transmission with intensity for the quantum confined glass 203 AME. The solid line is a fit to a simple two level model, the dashed line allows for an excited state absorption

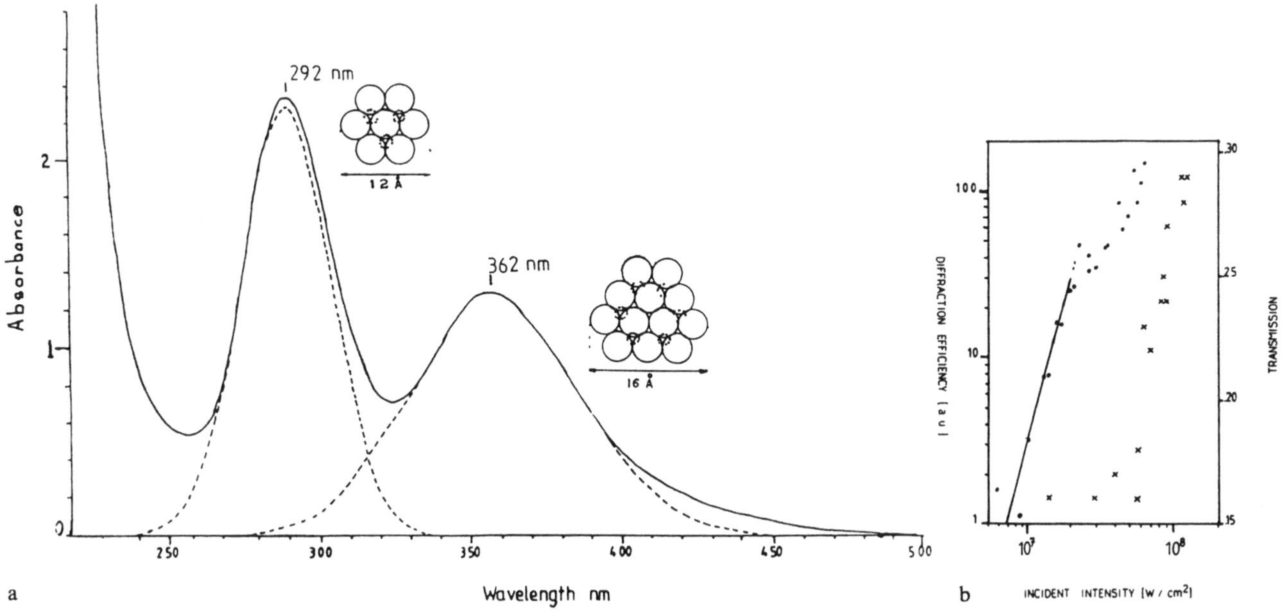

Fig. 3. a Absorption spectrum of colloidal BiI_3, showing the structure of the associated particles. **b** Intensity dependence of the transmission x(log-lin) and of the diffraction efficiency from a L.I.G. • (log-log) for BiI_3 colloid. The straight line fits the low intensity cubic behaviour of the diffracted signal

sharp cut-off high pass filters. They are formed by doping a borosilicate glass, which is transparent in the visible region, with Cd, S and Se and annealing at high temperatures ($\sim 700°C$). The microcrystallites grow by the process of Ostwald ripening to give particles of $\sim 100 \pm 30$ Å dimension with occasional bulk islands [8]. These are not small enough to provide strong confinement, the exciton radius in CdS is ~ 25 Å, and a sharp bulk-like band edge is observed in absorption. See Fig. 1.

We carried out wavelength resolved L.I.G. experiments on a series of glasses for a different ratio x [7]. Typical results are illustrated in Fig. 1, where we observed an effective $\chi^{(3)}$ proportional to absorption. A simple band-filling model has been proposed to account for this behaviour [9]. However, the effective $\chi^{(3)}$ is enhanced compared with the bulk due to surface effects and local field effects.

5. Quantum confined CdSe doped glass

For semiconductor doped glasses the particle size is determined by the annealing process. The commercial glasses used above were optimised to give a sharp absorption cut-off, rather than for particle size. By careful control of the annealing process a series of glasses were produced showing clear quantum confinement effects [10]. The absorption spectrum for one of these glasses (Corning 203 AME) is illustrated in Fig. 2a, the peak and shoulder corresponding to the first and second allowed eigenstates of the quantum confined system as given in (1). This spectrum corresponds to a particle size of 27 ± 3 Å.

We studied the intensity dependence of the transmission for this sample [11], the results are shown in Fig. 2(b). Fitting with a simple two level saturation model (solid curve) does not give a good fit. The change in transmission is not as large as expected, suggesting a second mechanism is at work. If we allow for an excited state absorption a much better fit to data is obtained (dashed curve). Higher levels of the quantum confinement series or conduction states in the glass could provide the necessary absorbing states. L.I.G. measurements give a calculated effective $\chi^{(3)}$ not greatly enhanced compared to the nonconfined samples of Sect. 4.

6. BiI_3 colloid

To further investigate well confined samples we prepared colloidal BiI_3 as described by Sandroff et al [12]; which gives a dilute colloidal suspension in acetonitrile, a highly nonpolar solvent which only absorbs light below 280 nm. This colloidal technique exploits the layered honeycomb structure of BiI_3. The colloid forms a single layer sandwich of bismuth between iodine and seems to be particularly stable for the first complete rings of 12 Å and 16 Å illustrated in Fig. 3a. The observed Gaussian absorption peaks are attributed to the first transition for these two particle sizes. The width of these peaks is probably due to a spread in particle sizes.

We carried out nonlinear transmission and L.I.G. measurements at the absorption maximum of 362 nm. The variation of transmission and diffraction efficiency with intensity are shown in Fig. 3b. We can see the onset of absorption saturation. At the same intensities the cubic dependence of diffraction efficiency is also seen to drop off, as expected for saturation behaviour. A large effective $\chi^{(3)}$ is also observed, of the same order as seen in the quantum confined glass.

7. Conclusions

Quantum confined systems show great promise for nonlinear optics. Further studies with more homogeneous samples will be necessary to gain a better understanding of these systems.

Note added in proof

Recent work by Wang and Herron (J. Phys. Chem. **91**, 5005 (1987)) has shown that the sharp features in the absorption spectrum of the BiI_3 colloid are due to molecular species I_3^-. In this light the optical nonlinearity has to be reinterpreted as being thermal in origin, perhaps augmented by photochemical changes.

We would like to thank D.W. Hall of Corning Glass Company who supplied the sample of quantum confined glass. Financial support for part of this work was provided by EOLAS, the Irish science and technology agency.

References

1. Shen, Y.R.: The principles of nonlinear optics. Berlin, Heidelberg, New York: Springer 1982
2. Horan, P., Blau, W.: Contemp. Phys. **28**, 59 (1987)
3. Zyss, J., Chemla, D.S.: Nonlinear optical properties of organic polymers, 2 Vols. New York: Academic Press 1987
4. Chemla, D.S., Miller, D.A.B.: J. Opt. Soc. Am. **B2**, 1155 (1985)
5. Schmitt-Rink, S., Miller, D.A.B., Chemla, D.S.: Phys. Rev. B. **35**, 813 (1987)
6. Eichler, H.J.: Laser induced dynamic gratings. Berlin, Heidelberg, New York: Springer 1986
7. Horan, P., Blau, W.B.: Semicond. Sci. Technol. **2**, 382 (1987)
8. Warnock, J., Awschalom D.D.: Phys. Rev. B. **32**, 5529 (1985)
9. Roussignol, P., Ricard, D., Lukasik, J., Flytzanis, C.: J. Opt. Soc. Am. **B4**, 5 (1987)
10. Hall, D.W., Borelli, N.F.: J Appl. Phys. (to be published)
11. Horan, P., Blau, W.: Phys. Rev. Lett. (submitted)
12. Sandroff, C.J., Wang, D.M.H.: J. Chem Phys. **85**, 5337 (1986)

Z. Phys. D – Atoms, Molecules and Clusters 12, 505–514 (1989)

Atoms, Molecules and Clusters
Zeitschrift für Physik D
© Springer-Verlag 1989

Many-cluster-systems: models of inhomogeneous matter

U. Kreibig, K. Fauth, M. Quinten, and D. Schönauer
FB 11-Physik – Universität des Saarlandes, D-6600 Saarbrücken, Federal Republic of Germany

Received 5 July 1988

First, some general remarks concerning macroscopic "cluster matter" are given. In the second part, three recent, mainly optically and electron-microscopically performed investigations are discussed which deal with special properties of noble metal cluster systems forming the building units of this kind of matter: *(1)* dressed Au-55 clusters, *(2)* electromagnetic coupling effects among coagulated clusters, *(3)* the transition towards compact inhomogeneous matter caused by coalescence of clusters.

PACS: 36.40; 42.20; 78.00; 81.20; 81.40; 82.70

I. Cluster matter

When inspecting the program of the present conference, one may notice two interesting aspects: first, the majority of the contributions is devoted to properties of the single cluster and, second, the aspect of technical applications of clusters is only little represented. Probably, both aspects are not independent: most technically and practically important cluster systems are forming macroscopic compact matter and consist of very large numbers of clusters which, usually are in interaction with each other. This special kind of material may be designated as "cluster-matter" or, less specifying, "inhomogeneous matter".

Macroscopic crystalline matter, "homogeneous matter", is usually built up by atomic/molecular building units which are ordered regularly with macroscopic correlation lengths of spatial order.

A vast number of compact materials – in special those of practical/technical importance – however consist of larger building units which may be identified as more or less well defined clusters, the sizes of which range from small molecular to large solid clusters, as crystalline particles or grains. Preferentially these building units are not in macroscopic regular order, instead there is statistical spatial and/or chemical disorder with limited topological correlation lengths.

Some arbitrarily chosen examples may illustrate the term "cluster-matter":

(1) Fine-grain-material (polycrystalline matter, small-domain-matter, nanocrystalline matter, ceramics, cluster-produced thin films, incommensurate phase materials etc.) Glasses (?); Quasicrystals (?)
(2) Cluster-matrix-systems (metal-nonmetal systems, granular superconductors including oxide interlayers, cluster tunnel junctions, cermets, photographic systems, dispersion colors etc.)
(3) Natural colloidal systems (interstellar dust, aerosols, hydrosols, clays etc.)
(4) Quantum-semiconductor devices ("quantum dots")

At a first glance the differences between these samples are more obvious than their common features. Yet, there are underlying systematics which are illustrated in Table 1.

The key effect of cluster matter is that the clusters do not behave like isolated single units but are essentially influenced by interaction effects both, among each other and with the embedding material (if existing). These interactions mainly determine the macroscopic physical properties of the material. They may be divided into interactions which are permanently present (v.d. Waals forces, elastic strain fields etc.) and those which are introduced only by the experiment (electric fields between neighboring particles when an electric voltage is applied; electromagnetic scattering fields during irradiation by light etc.)

Table 1. Classification of cluster matter

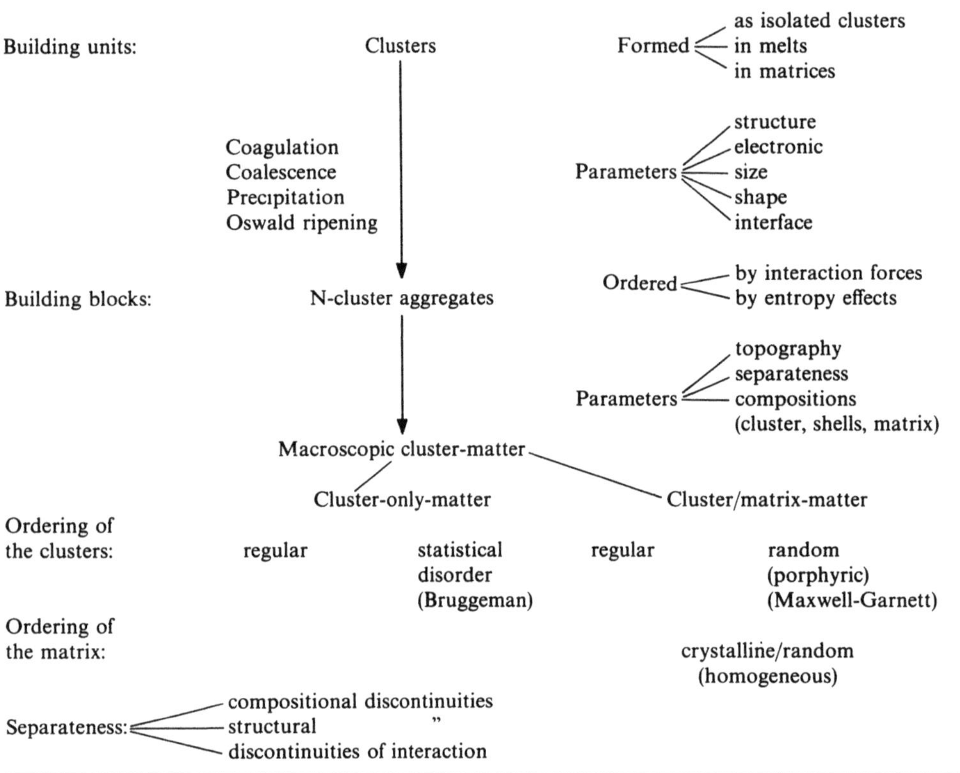

Macroscopic properties of cluster matter depend on parameters the number of which is larger than for homogeneous matter. They describe the single cluster, the near-order close to a given cluster and the far-order. They all have to be regarded in a detailed microscopic description, and several can be varied to tailor new materials with special, desired macroscopic properties.

A description may be based upon

(1) the building units (clusters, cluster shells, embedding media etc.), which are characterized by the material properties, the characteristic lengths (e.g. sizes), the geometrical shapes and the separateness (a measure of the formation of separating interfaces around each building unit), and
(2) the building blocks (clusters plus surface shells; fractal N-cluster aggregates in a matrix, etc.) being the areas in which clusters are arranged according to some given more or less strict order. They may be characterized by correlation lengths for spatial order, filling factors, the structure of geometrical ordering etc.

In the case of completely random systems, the characteristic length of *(1)* equals the correlation length of *(2)*. For ideal cluster crystals, on the other hand, the building block is the whole sample. In general, the characteristic lengths of the building units are larger and the correlation lengths of the building blocks are smaller than in homogeneous matter.

The property of separateness as introduced in Table 1 characterizes the differences between the cluster and its surrounding, e.g. due to differences of chemical composition and of interaction forces within the building units and with their surroundings. It is mainly these differences which lead to stable interfaces and discerns inhomogeneous cluster matter from homogeneous material. E.g. separateness is strong for metal clusters embedded in a rare gas matrix, but weak in a mineral glass with the building units of SiO_4-tetrahedra in the random network.

To conclude this rather general part: The most important applications of clusters are in macroscopic cluster matter. It is not sufficient, however, for understanding to refer to properties of single clusters only. Instead, the interactions in the many-cluster-systems and the cluster/matrix interfaces contribute essentially to macroscopic physical properties of bulk cluster material.

In the second part of this paper three special topics concerning cluster matter will be reported on. They are based on three papers of this volume [3, 7, 13]. First, recent results of a special and very interesting kind of building units are presented: the ligand-stabilized Au-55 clusters. Second, some implications are discussed of the first successful attempt to perform a quantitative spectral analysis of cluster matter, founded upon its microscopic details, in special the interaction effects. Third, the property of separateness, introduced above is treated as it is changed when isolated clusters successively form fine grain (or nanocrystalline) metals.

II. Building units of a cluster-matrix system: Au-55 clusters

55 is one of the structural magic numbers of densely packed spheres but not an electronic magic number as derived from the shell model [1]. Recently, Schmid [2] succeeded in producing Au-55 clusters by stabilizing them through chemical bonds in a metalorganic compound. The ligand layers surrounding the Au clusters prevent them from close approach, hence these clusters form the loosely packed cluster-matrix system with high separateness (Fig. 1).

Details of the clusters and their investigation are contained in the paper [3] of this volume; a general review of such chemically bound metal clusters is given by Mingos [4], also in this volume. The following will be devoted to the most interesting question: are these clusters still molecules or already solid particles?

The optical absorption spectroscopy of 6*s*-electron plasma excitations has proved to be an especially sensitive sensor for structural and electronic properties of Au clusters [5]. In solid state clusters the "conduction" electrons give rise to the very pronounced absorption band of particle plasmon polaritons which successively is smeared out and fades away when the cluster sizes are decreased, thus indicating quantum size effects [5]. These elementary or collective excitations are hybridized with electron-hole excitations which in the bulk gold form the 5*d*-6*s*, *p* interband transition edge, in the vicinity of the plasmon frequencies. These effects are shown in Fig. 2 which was obtained from earlier experiments on Au-clusters of 10^2 to 10^4 atoms which were embedded in a glass matrix. In these latter samples the particles were not chemically bound but only physisorbed to the glass [5].

For the Au-55 clusters one might expect, instead, a molecular many-line spectrum provided their properties are still molecule-like and the surrounding dielectric matter does not contribute to the line broadening too much.

The extended investigations of ligand stabilized metal clusters (Mingos et al; Ozin; etc. [6]) rather point to smeared out spectral structures in the case of Au-55. Therefore we measured the absorption spectrum cooling the samples down below 2 K with the aim of reducing broadening effects. Figure 3 shows one of the resulting spectra. Obviously there is, down to the experimental resolution of ± 0.5 nm, no hint of molecular line structures or shell model [1] structures in the investigated frequency range from 1.9 to 3.5 eV. On the other hand, the absorption clearly deviates from the spectra calculated with the Mie-Debye theory

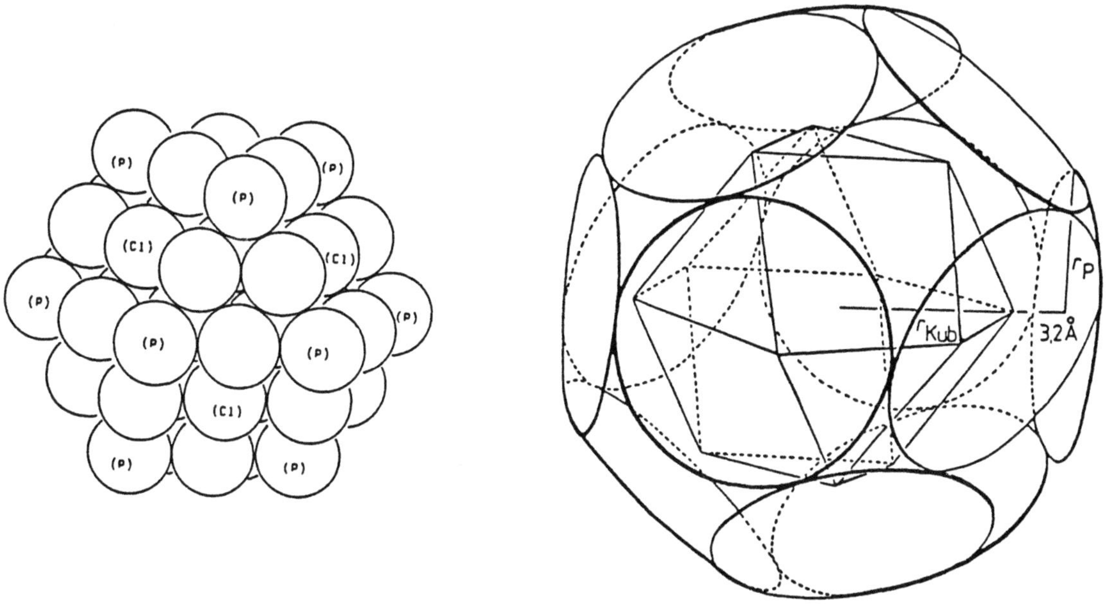

Fig. 1. Structure of the "naked" (*left*) and the "dressed" (*right*) Au_{55}-cluster (after [2])

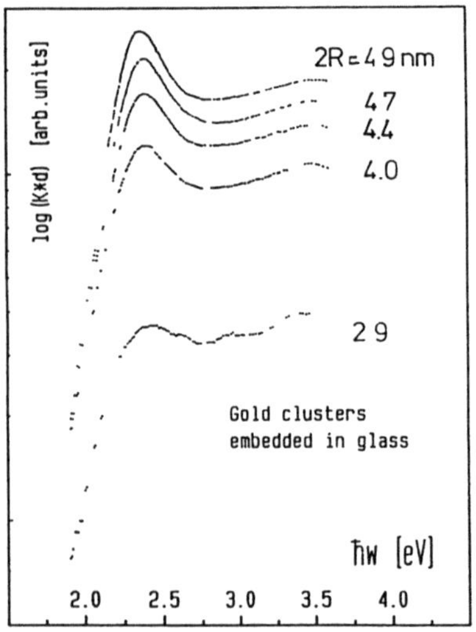

Fig. 2. Optical absorption of Au-clusters of various sizes embedded in a glass-matrix (after [5])

including quantum size effects, of the polarizability, which, as discussed in [3] of this volume, would have been expected if the clusters were solid state metallic.

In the investigated spectral region the absorption, instead, follows an exponential decrease without any hint of a plasma resonance. This is shown in Fig. 3. The wavelength dependence is given by $K \propto \exp(-3.44 \cdot \lambda[\mu m])$, both at room temperature and below, 2 K. Such dependencies were observed from "amorphous" materials. By slight heat-treating of the samples which probably leads to partial cluster growth (or, alternatively, to stronger cluster-cluster interactions), a weak 6*s*-electron plasmon structure could be produced, as described in [3], the spectra then resembling those of the previously investigated Au-cluster/glass system (Figs. 2, 3a).

This latter system has been analyzed in detail [5], including a Kramers Kronig analysis of the optical data of Fig. 2. It was proved that, beside the 6*s*-electron quantum size effects, it is, below, say 2500 atoms per cluster, the size dependence of the electron-hole excitations which contributes most to changes of the absorption spectrum. It was then shown (Fig. 4) that around a critical number of several hundred atoms per cluster the interband transition edge rather abruptly decreased its slope and was shifted towards lower frequencies. Analogous results were obtained by other authors for the interband edge of metal clusters. They also follow from UPS experiments the latter being interpreted as due to a decreasing spin-orbit splitting of the 5*d* electrons.

The size dependent 5*d*-6*s*,*p* electron-hole transition edge of Fig. 4 resembles the Au-55 spectra of Fig. 3, thus a tentative interpretation of the latter spectra can be given – though only very reservedly – as follows:

First, no marked collective 6*s*-electron excitations nor free electron shell model excitations [1] could be identified. It is open whether these electrons are more strongly localized close to their parent atoms or by the surrounding ligands. Second, the observed exponential decay of the absorption towards longer wavelengths is due to 5*d*-6*s*,*p* electron-hole transitions as a precursor of the interband transition edge of bulk gold. Fine structures were not resolved, possibly because of relaxation/lifetime effects and Stark effects caused by the surrounding ligand shells.

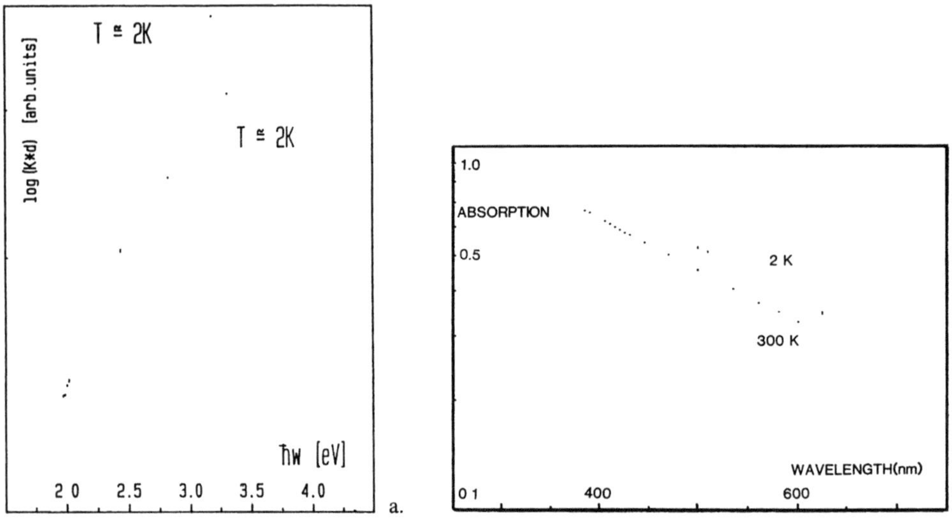

Fig. 3. a. Optical absorption of Au_{55}-clusters, dried after filtration and stabilization on a quartz substrate. (*lower curve*), and of 2 nm Au-clusters in a glass matrix (*upper curve*). Spectra taken at 2 K. **b.** Absorption of Au_{55}-clusters versus wavelength ($T = 300$ K; 2 K)

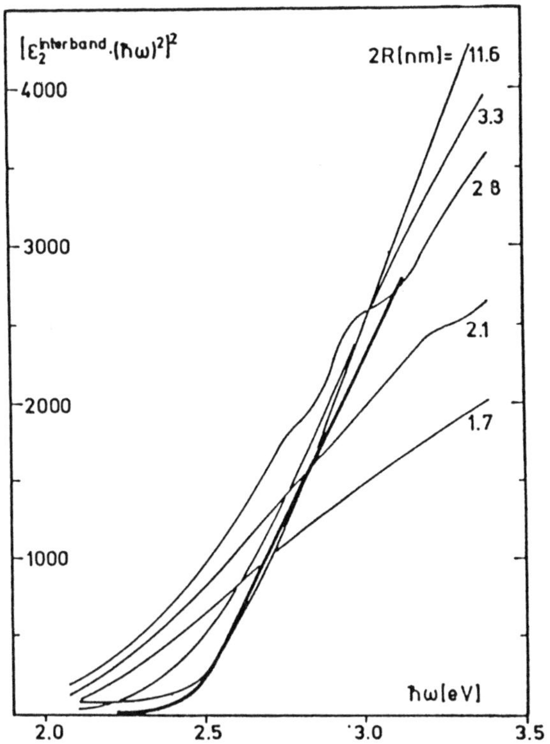

Fig. 4. Edge of the 5*d*-6 *s*, *p* interband transitions in bulk gold (*bold line*) and of Au clusters of various mean sizes 2R, embedded in glass. ε_2: imaginary part of the dielectric function. (after [5])

So, with all caution, it may be stated that these Au-55 clusters are below the molecule-solid state transition but that this transition does not show up in the optical spectra as a discrete "phase transition" but in rather continuous alterations. Perhaps, it does not make much sense to define generally a such transition without specifying the physical property which is expected to show it up. So, more precisely: the transition in the optical absorption spectrum is above the number of 55 atoms, provided, the ligand stabilized clusters behave similar to free clusters in view of this property.

III. Multipolar coupling in densely packed many-cluster-systems

This investigation, the details of which are presented in [7] of this volume, was performed on cluster matter composed of larger, solid state like gold clusters which could be systematically induced to form coagulation aggregates by allowing diffusion limited aggregation (DLA). Then, the separateness – as defined in sect. I – is reduced the particles being able to almost touch their neighbors. Yet, the particles are assumed to be electrically isolated by thin dielectric interlayers [12] in what we call coagulation aggregates. Hence, there is strict charge conservation for each cluster; tunnel processes are not regarded.

The aggregates were well separated in the samples, so the investigation is restricted to these building blocks of the macroscopic material and aggregate-aggregate interaction is excluded.

The optical excitation of the 6*s*-electron plasma is extremely sensitive to the topological structure details of such aggregates as has been demonstrated earlier [8]. Figure 5 shows for some arbitrarily chosen examples how, starting from the narrow absorption of the dipolar plasmon polariton of the single cluster, the absorption spectrum is markedly changed when, without changing the mean filling factor of the macroscopic sample, the particles are forced to concentrate in aggregates. Mainly two extinction bands appear in the spectra which are due to optically active plasmon coupling modes. (As will be shown in the following section, still broader spectra extending towards the IR and eventually resembling the spectra of continuous films are observed when coalescence occurs among the clusters.)

The changes of the extinction occurring in such former aggregates are due to electromagnetic interaction effects, i.e. to the superposition of near-field scattered and incident waves. In principle they are

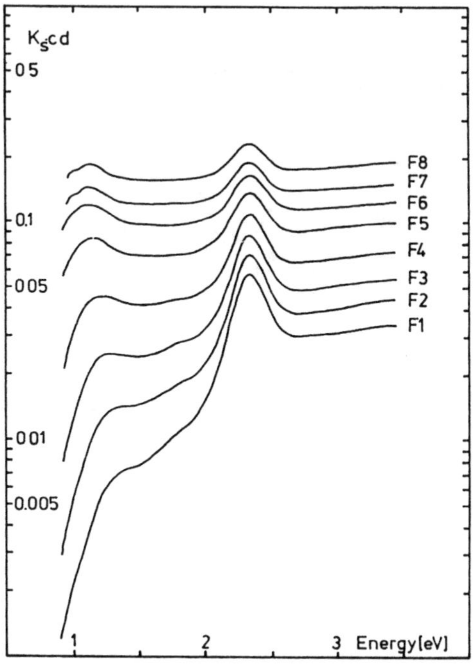

Fig. 5. Experimental spectra of optical extinction of a Au-cluster system (hydrosol) with various amounts of aggregation (increasing from F1 to F8). Mean particle size 2R = 30 nm

present in aggregates irrespective of the sizes of the clusters (e.g. also in aggregates of Au-55 clusters) since their strengths are mainly determined by the relative variable (cluster distance)/(cluster size). Yet, the optical effects are more pronounced due to the stronger single cluster plasmon band and spectral details can be better resolved in the case of larger particles. Then, the single particle band consists of, both, absorption and scattering loss contributions and of plasmon modes of different spherical multipole orders. The hierarchy of these excitations [9] is given in Fig. 6. Recently, contributions of the dipolar and the quadrupolar mode could be separated experimentally for the first time [9] by combining optical absorption and photothermal spectroscopy. Figure 7 gives an example for well separated silver particles of 19 and of 63 nm mean diameter.

The multipolar extinction spectra of the single particle are described by the electromagnetic theory of Mie, which predicts, for particles of less, say, 100 nm size that mainly the dipolar and the quadrupolar mode are excited. Recently, Rojas and Claro [10] published model calculations of the absorption of two coupled clusters approaching each other till to direct contact. They predict the markedly ehnanced excitation of

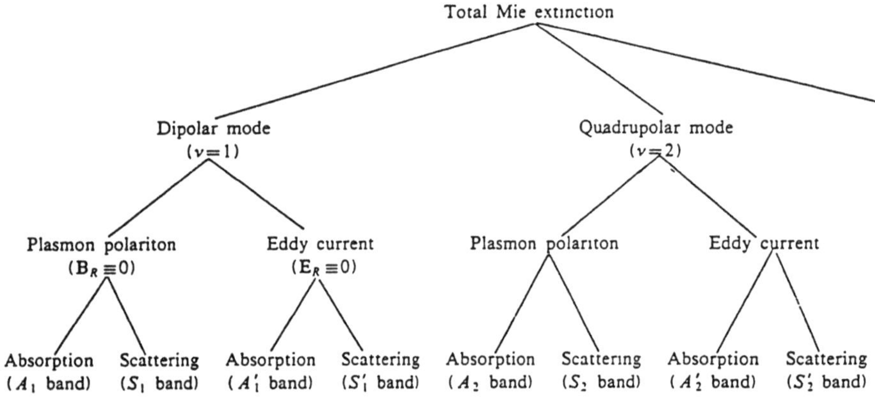

Fig. 6. Hierarchy of the plasma resonance extinction of small metal particles (after [9]). A_ν, S_ν: contributions of the ν-th multipole absorption and scattering. E_R, B_R: electric and magnetic field normal to the particle surface. Primes: magnetic contributions

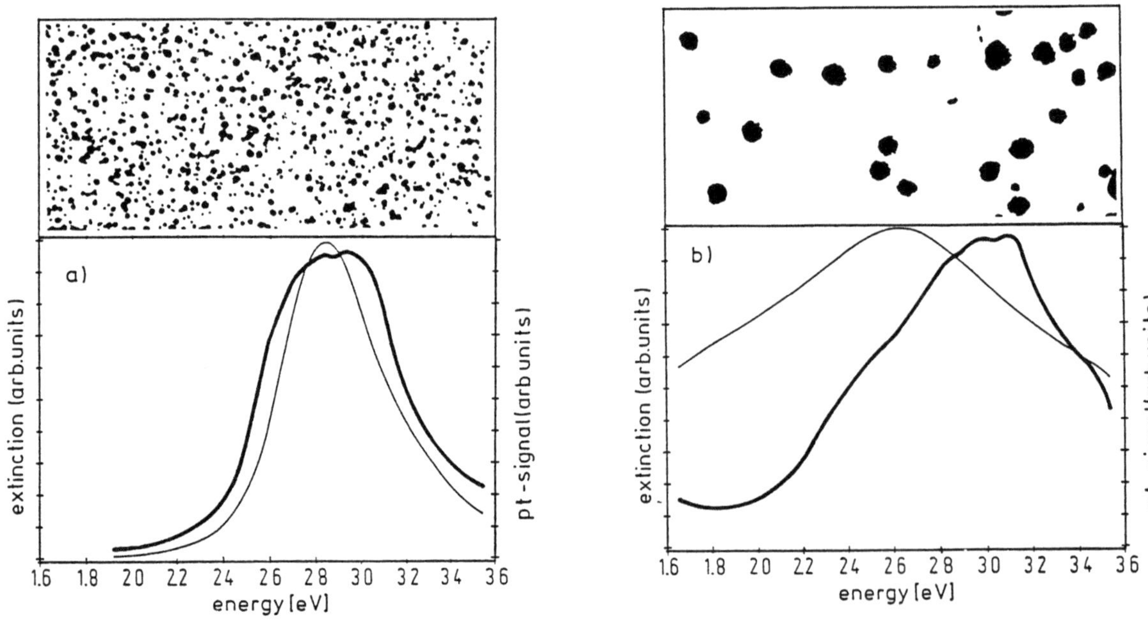

Fig. 7a and b. Optical extinction (*thin lines*) and photothermal signal (*bold lines*) of Ag-particles of 2R = 19 nm (*left*) and 2R = 63 nm (*right*). In the 63 nm particles the extinction is mainly due to the dipolar scattering, the photothermal signal mainly due to quadrupolar absorption (after [9]). The samples are shown at top

higher multipole modes due to the electromagnetic coupling fields between the two particles. In the limit of direct contact their number, i.e. the multipolar order, even diverges to infinity. At finite but small separations more than several ten modes should be important. As a consequence, large numbers of higher multipolar plasmon modes, shifting the overall absorption toward low frequencies, have to be taken into account to reproduce theoretically the experimental spectra.

This is in contradiction to our previous electrodynamic calculations [7] which yielded contributions to the extinction that are strongly decreasing with increasing polar order. This prevailed on us to perform a quantitatively comparing analysis of the measured extinction spectrum of a selected many-cluster sample by applying our earlier developed theoretical methods. Details of this analysis are given in [7]. Here only the implications for the theory of Claro will be presented.

The sample, used, consisted of small coagulation aggregates of 30 nm size Au clusters. The DLA-aggregates, composed of 2 to, at maximum, 20 individuals, were electron-microscopically analyzed and all observed kinds of aggregates were filed into the total of about 50 classes of similar but regular aggregates (Fig. 8).

The calculations were performed as described in [7], avoiding the approximations used commonly and by Claro as well. This means that retardation was included, both, for the single particle excitations and for the coupling fields. In a first step only dipolar excitations and dipolar coupling were treated and second also quadrupolar excitations and dipole-quadrupole and quadrupole-quadrupole interactions were taken into account. The extinction spectra were then computed for all of the aggregate classes of Fig. 8, including orientation averaging. Weighting these partial spectra by their relative numbers of observed aggregates and summing them all up, finally a calculated spectrum was obtained that could be quantitatively compared to the measured extinction. The results are shown in Fig. 9.

It is obvious that

- the dipolar contributions do not explain the experimental spectrum. Yet, almost quantitative correspondence is reached when quadrupolar excitations and interactions are fully considered.
- as a consequence the electromagnetic theory of direct multipolar interactions is quantitatively applicable. Essential contributions of higher multipole modes than the quadrupolar are not missing in the calculated spectrum and hence – vice versa – are not contained in the experimental spectrum.

These latter contributions appear to be overestimated in the Claro theory, probably owing to the quasistatic limit which was admitted. It might be mentioned that in view of the experimental details the limiting case of a vanishing particle-particle distance by keeping them individuals is unrealistic since in the absence of a separating interlayer the particles will spontaneously coalesce, forming new, larger building units as will be discussed in the following section.

IV. Changes of separateness: the formation of grain boundaries by coalescence

Though the separateness of the single clusters of the sample discussed in the preceding section was reduced during the aggregation process, since neighboring particles closely approach, the clusters remain still to

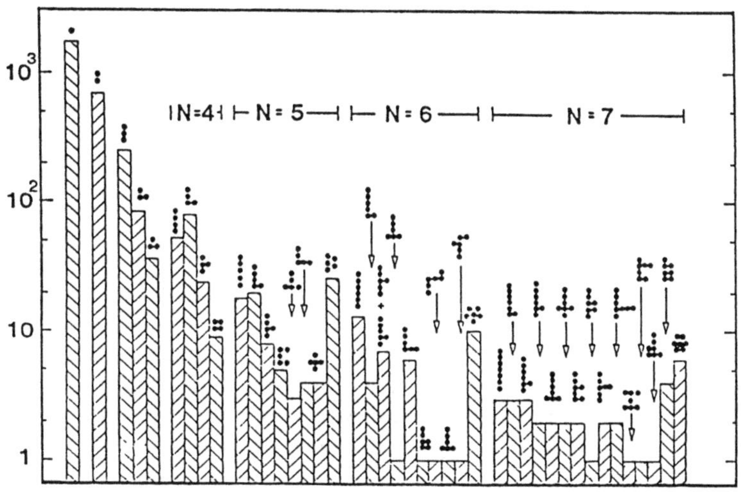

Fig. 8. Size and shape histogram of the examined sample: the cluster aggregates were allocated to classes of high symmetry aggregates drawn at top of each column. N: number of particles in the aggregate

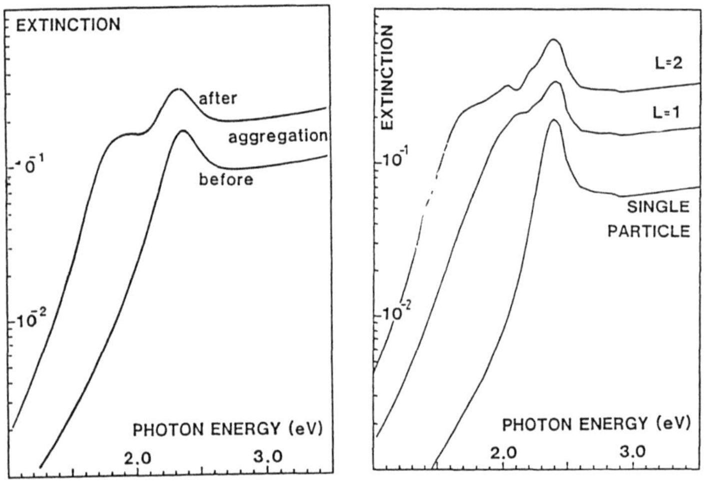

Fig. 9. Experimental (*left*) and computed (*right*) extinction spectra of the Au cluster sample before and after coagulation aggregation. $L=1$, 2: contributions of dipolar ($L=1$) and of dipolar and quadrupolar plasmons and interactions ($L=2$)

be individuals. If, however, coagulation turns to coalescence in the aggregate matter, grain boundaries are formed between these individuals at their connection points [11] and separateness drops markedly. Though electrical conductivity now extends over the whole aggregate, the aggregate does not act as a new, larger building unit, since the grain boundaries still contribute extra resistance due to the limited transmission probability for conduction electrons [12].

It is possible by chemical means or by heat treatment to introduce coalescence in samples of the kind discussed in the previous section. Details are given in the paper of Schönauer et al [13] of this volume. Samples of Au clusters in liquid and solid matrix were produced which formed coagulation chains of quasi-fractal structure. By treating them, coalescence occurred between part of the clusters, each newly formed grain boundary switching on electrical conductivity between the connected clusters. Thus, electrical percolation paths may be obtained with correlation lengths extending from two particle diameters up to the length of the macroscopic sample. Only in the latter case electrical percolation can be probed by applying an electric field. The precursor states of this (complete) percolation proved to be observable by optical spectroscopy. Figures 1 and 2 of paper [13] shows electron micrographs of the sample used. The coagulation and the coalescence links are clearly to be recognized.

The corresponding spectra of optical extinction in the spectral range of the plasma resonances were recorded, both, for the liquid and the solid samples before and after initiating coalescence (Fig. 4 of paper [13]). The strong changes due to the coalescence are obvious. It should be mentioned that at least in the case of solid embedding matrix the topological structure of the fractal aggregates remains completely unchanged during the coalescence treating. The coagulation aggregates show clearly the typical two-peak structure due to splitting of the plasmon peak by electromagnetic interaction [7, 13]. In the case of linear aggregate shapes they are ascribed to the longitudinal and transverse collective aggregate mode. After coalescence the low frequency, longitudinal, mode almost disappears and a new band develops about 0.5 eV lower. This latter is to be ascribed to the percolation paths as can be argued from the electromagnetic coupling theory described in the previous section.

For details may be referred to [13], again. In Fig. 5 of that paper it is demonstrated that for all shapes of coagulation aggregates of a given number of clusters it is the chain-like one that gives the largest peak splitting. The investigated sample only contained short chainlike strings as show the electron-micrographs, so the splitting of the linear chain gives an upper bound of splitting energies occurring in the sample. Figure 6 of paper [13] gives the dependence of the splitting on the length of the chain, calculated for various cluster diameters. It is essentially the longitudinal, low frequency mode that exhibits this effect. There is a cross-over from strong dependence of the longitudinal mode frequency on the length of the chain for short chains to a constant position of this mode, independent of chain length, for long chains, varying only with the size of the single clusters. Hence, there is a limiting low frequency $\hbar\omega_g$ for all possible coagulation aggregates of a given single cluster size. Its value is plotted in Fig. 10, it is almost reached by the used coagulation samples.

The low energy peaks of the coalescence samples clearly are beyond this limit, thus indicating that they are due to the coalescence structures. Vice versa the existence of strong absorption beyond $\hbar\omega_g$ gives evidence for coalescence in aggregated samples.

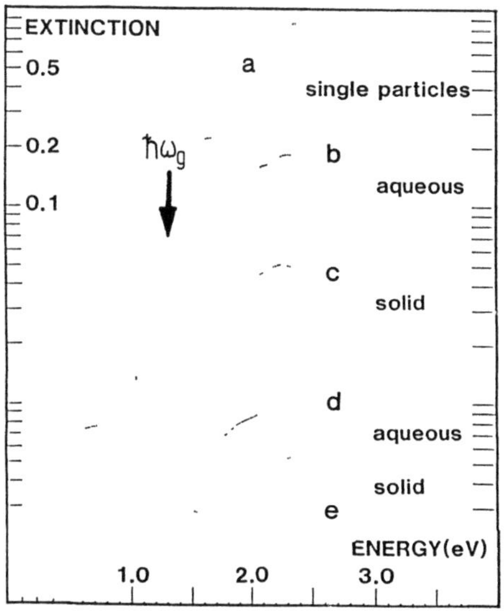

Fig. 10. Extinction spectra of Au clusters embedded in H_2O and in solid gelatin. *a* isolated Au clusters *b*, *c* after coagulation (fractal aggregates) *d*, *e* after coalescence. ω_g is the lower limit of the low frequency collective mode for coagulation aggregates of the given cluster size

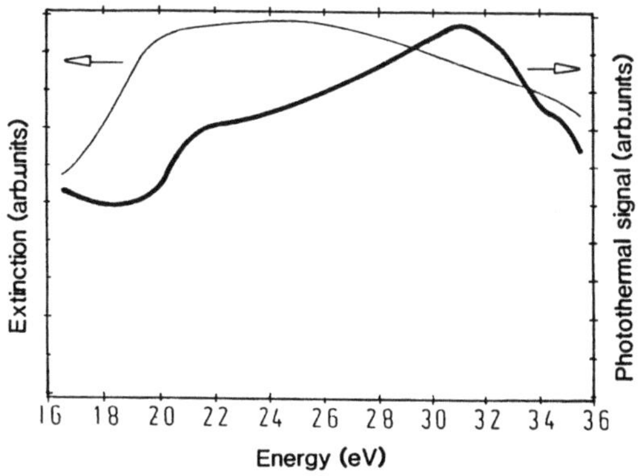

Fig. 11. Experimental extinction (=absorption+scattering) spectrum (thin line), and photothermal absorption spectrum (bold line) of densely packed Ag-particles. Particle diameter $\overline{2R}=10$ nm; filling factor ~ 0.4

No comparingly precise calculations are available for coalescence aggregates, mainly because of the complicated resulting structures of the building blocks. Estimations can be performed, however, replacing them by building units of regular shapes as cylinders or prolate ellipsoids. These particles exhibit two absorption peaks as can be shown by quasistatic calculations, the low frequency peak being shifted towards zero frequency for increasing eccentricity.

Comparing the peak positions of ellipsoidal units, as given in Fig. 7 of [13], with the measured coalescence peaks one finds a mean value for the eccentricity of ellipsoids substituting our percolation paths and what follows is a mean correlation length of about 10 single clusters describing the precursor states of percolation discussed above. This result roughly corresponds to the electron-microscopic analysis.

In conclusion, the detailed optical spectral analysis enables to distinguish between coagulation and coalescence structures in macroscopic samples of cluster matter of the investigated kind. Percolation structures can be identified of short correlation lengths which cannot be probed by electrical measurements.

The present results throw some light upon the strange extinction of densely packed cluster samples in the IR. However, detailed computations of the extinction of ellipsoids or cylinders including retardation are missing. They would be needed since recent investigations have shown that in such samples there is large amount of scattering beside the true absorption even if the included single clusters are too small to show scattering at all [9]. Scattering generally only occurs when retardation is regarded in the computations. This may be demonstrated by Fig. 11, where, both, optical absorption and photothermal spectra are shown for a densely packed cluster sample, the former being due to extinction while the latter only indicates true absorption contributions to the extinction.

Including the property of separateness, the model of cluster matter may be applied to the ample field of fine grain polycrystalline materials. E.g., nanocrystalline materials [14] appear as intermediate states of small clusters strongly linked by coalescence. Since particular emphasis is laid in this model on the structural unit of the cluster, investigations of the properties of the single free cluster are prerequisite. On the other hand, it is a promising field of cluster science to tailor directedly kinds of cluster matter with specific, desired macroscopic properties by varying the microscopic parameters.

References

1. Knight, W.D., Clemenger, K, de Heer, W.A., Saunders, W.A., Chou, M.Y., Cohen, M.L.: Phys. Rev. Lett. **52**, 2141 (1984) Ekardt, W.: Phys. Rev. B **29**, 1558 (1984)
2. Schmid, G., Pfeil, R., Boese, R., Bandermann, F., Meyer, S., Calis, G., v.d. Velden, J.: Chem. Ber. **114**, 3634 (1981) Schmid, G.: In: Structure and Bonding, Vol. 62. Berlin, Heidelberg, New York: Springer 1985
3. Fauth, K., Kreibig, U.: Z. Phys. D-Atoms Molecules and Clusters (this issue) (1989)
4. Mingos, D.M.P.: Z. Phys. D-Atoms, Molecules and Clusters (this issue) (1989)
5. Kreibig, U: J. Phys. **38**, 2–97 (1977); Solid State Commun. **28**, 767 (1978); In: Growth and props of metal clusters. Bourdon, J. (ed.), p. 371 Amsterdam: Elsevier 1980
6. Hall, K.P., Mingos, D.M.P.: Prog. Inorgan. Chem. **32**, 237

(1984); Ozin, G.A.: Faraday Symposia of the Chem. Society, Vol. 14, 7 (1980)
7. Quinten, M., Schönauer, D., Kreibig, U.: Z. Phys. D – Atoms, Molecules and Clusters (this issue) (1989)
8. Kreibig, U.: Z. Phys. D – Atoms, Molecules and Clusters **3**, 239 (1986)
9. Kreibig, U., Schmitz, B., Breuer, H.D.: Phys. Rev. B **36**, 5027 (1987)
10. Rojas, R., Claro, F.: Phys. Rev. B **34**, 3730 (1986)
11. Thölen, A.: Contributions of clusters physics to material science, Davenas, J., Rabette, P.M., (eds.), p. 601. Dordrecht: Nijhoff 1986
12. Kreibig, U.: Z. Phys. B – Condensed Matter **31**, 39 (1978)
13. Schönauer, D., Quinten, M., Kreibig, U.: Z. Phys. D – Atoms, Molecules and Clusters (this issue) (1989)
14. Birringer, R., Gleiter, H., Klein, H.P., Marquardt, P.: Phys. Lett **102A**, 365 (1984)

Z. Phys. D – Atoms, Molecules and Clusters 12, 515–520 (1989)

Atoms, Molecules
Zeitschrift für Physik D and Clusters
© Springer-Verlag 1989

Optical properties of systems containing Au_{55}-clusters

K. Fauth,[1] U. Kreibig,[1] and G. Schmid[2]

[1] FB 11-Physik-, Universität des Saarlandes, D-6600 Saarbrücken, Federal Republic of Germany
[2] Institut für Anorganische Chemie, Universität-GH-Essen, D-4300 Essen, Federal Republic of Germany

Received 5 June 1988; final version 30 September 1988

Gold cluster compounds were investigated which contain clusters of 55 gold atoms in each unit. The ligands stabilize these clusters and, hence, many-cluster systems can be prepared which mainly show the properties of the single dressed cluster. Experimental results of optical and electron-microscopical investigations are presented and shortly discussed in view of the question whether the clusters are molecular or solid-state like.

PACS: 33.20.K; 36.40; 42.20; 78.40; 82.70.

Introduction

Experiments on free clusters require macroscopic samples with large numbers of clusters. Up to now no method of providing such numbers of identical clusters has been found. Instead, clusters in macroscopic samples have distributed sizes and shapes. Clusters of uniform geometry can be obtained only if their number is small and/or the clusters are not free. The cluster number is reduced by cluster selection with TOF-mass-spectroscopy. The alternative way is to produce clusters which are bound in a regular molecular unit. This latter way has been very successfully investigated in chemistry, by building metal-organic complexes. A review is given by Mingos in this volume [1], previous reviews concerning noble metal clusters were presented by Hall and Mingos [2] and by Ozin [3].

Recently Schmid succeeded in preparing Au-55 clusters enveloped and stabilized by ligand shells [4]. The number 55 appears as a structural magic number when packings of spheres are considered. The closest electronic magic number following from the free electron shell model [5] is 58.

The obviously fundamental question for the electronic state of these clusters, -molecular or rather solid-state like-, appears to be still open, irrespective of a number of different experiments which were performed (Lit. given e.g. in [6]).

In the following, results are presented of an investigation combining electron-microscopic and optical experiments, the latter having been performed at 300 K and at low temperatures below 2 K.

Characterization of the single clusters

Schmid et al. developed two kinds of stabilized Au-55 clusters differing by their ligand shells. The according composition formulae are:

(1) $Au_{55}((C_6H_5)_3P)_{12}Cl_6$ and

(2) $Au_{55}((C_6H_5)_3PC_6H_4SO_3Na)_{12}Cl_6$.

The first can be dispersed in e.g. CH_2Cl_2, the latter in H_2O. Preparation and constitution of the metal-organic complexes are described in detail in a series of papers by Schmid and co-workers [6].

In the following the clusters dressed by their ligand shells will be regarded simply as a metal cluster embedded in an almost spherical dielectric shell (Fig. 1) with strong but otherwise unspecified cluster-matrix interaction.

The magic number of 55 appears both in icosahedral and cuboctahedral symmetry, which only slightly differ in their atomic positions. The latter of the structures is assumed to be preferred, as has been concluded from symmetry arguments for the ligands

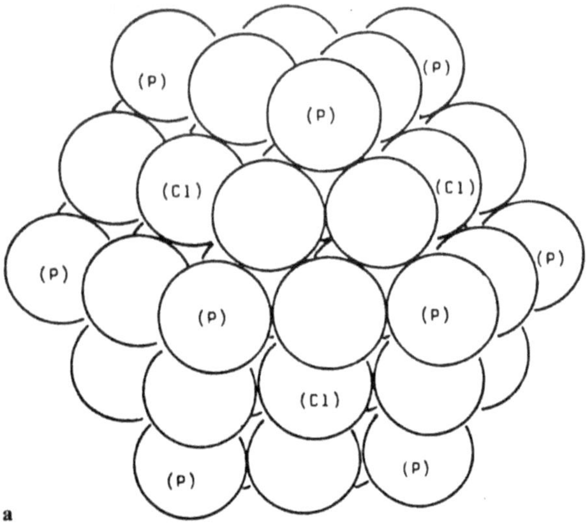

a

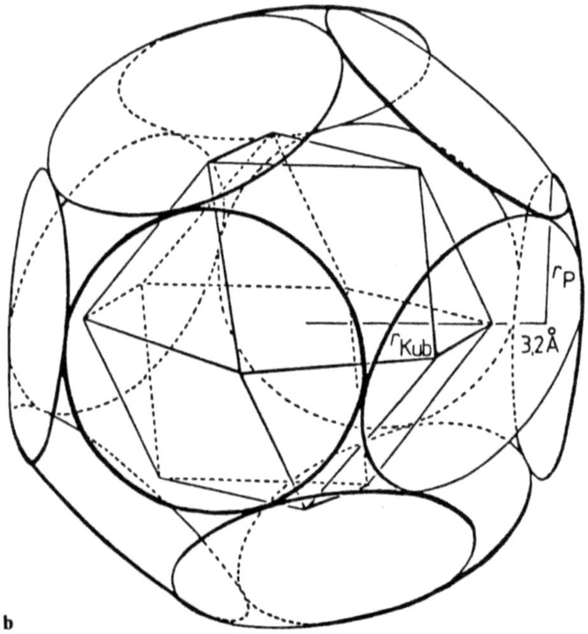

b

Fig. 1a and b. Au_{55}-clusters: **a**) metallic nucleus. **b**) dressed cluster with ligand structure

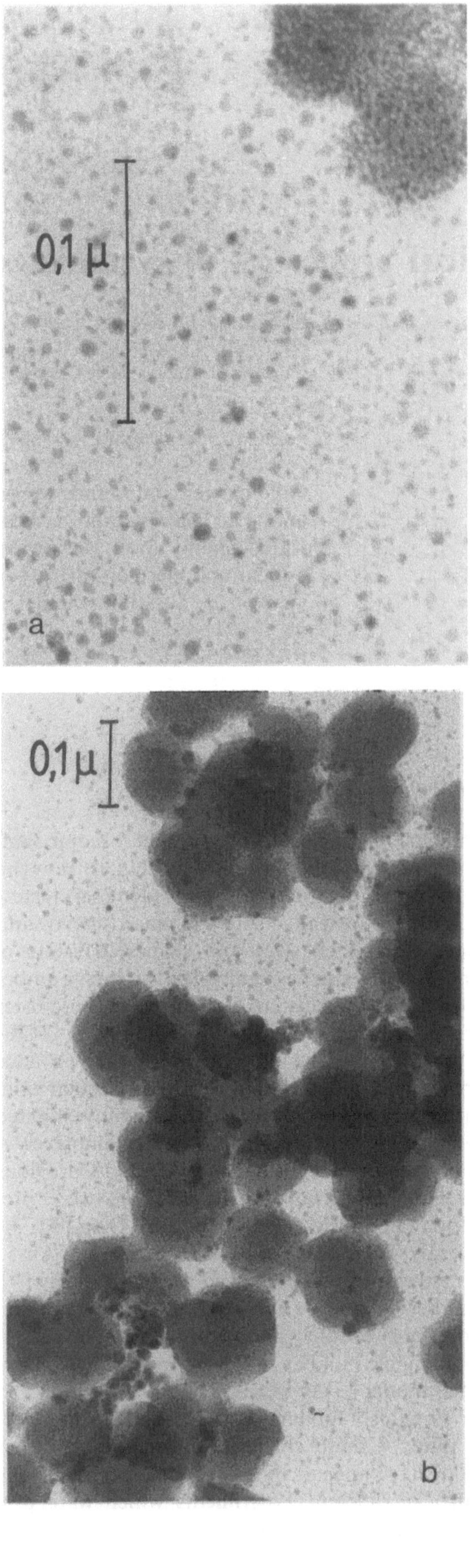

Fig. 2a and b. Electron micrographs of dried Au_{55}-cluster systems containing single systems, cluster aggregates and, in **b** also compact particles

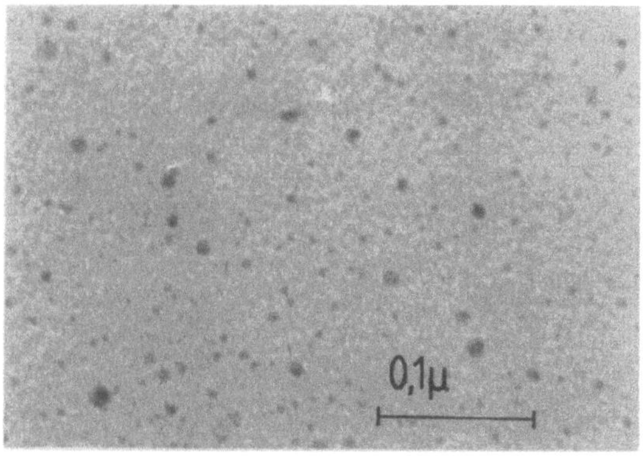

Fig. 3. Electron micrograph of a Au_{55}-cluster–system stabilized with gelatin before drying

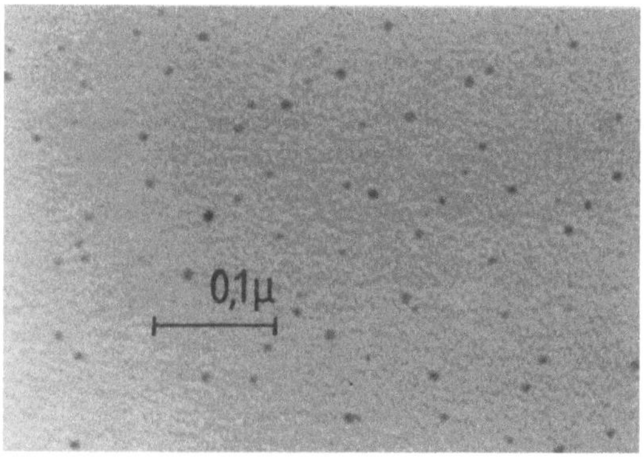

Fig. 4. Electron micrograph of a Au_{55}-cluster-system, filtered and stabilized before drying

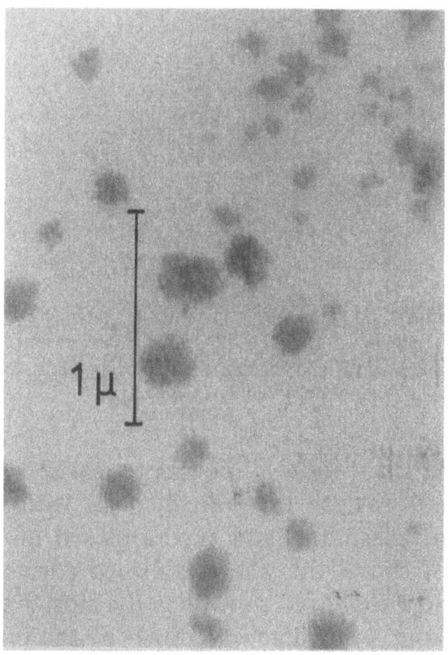

Fig. 5. Electron micrograph of a Au_{55}-cluster-system, heat-treated at 70°C for 30 min

[6]: there are 12 phosphine groups and 6 Cl-ions in the molecule and it is the cuboctahedron which has 12 corners and 6 centers of square planes at its surface. The diameter of the naked cluster is then $10 \times r_{Au} = 1.44$ nm, where r_{Au} is the radius of one Au unit. There is one center atom, surrounded by 12 next neighbors forming the first shell. Then follows the second shell of 42 atoms which builds the interface between the Au atoms and the ligands. Only the 13 inner atoms are surrounded completely by Au atoms. In the outer shell three different types of Au atoms are distinguished. 12 in the corners are bound chemically to the phosphine groups, 6 in the centers of the square planes are bound to Cl and the remaining 24 atoms are again in direct interaction only with other Au atoms. From the observed effect that the second shell can be stripped off leaving a naked Au_{13} cluster follows that the bonds to the ligands are strong.

Characterization of the cluster samples

Freshly produced substances, obtained as dry powders, were dispersed in their respective solvents and then dried on carbon foil covered copper grids to provide TEM investigations.

Solid samples for the optical experiments were produced by drying the liquids on quartz substrates. Figure 2 gives an example: the smallest visible point-structures are due to well separated dressed Au-55 clusters of about 2 nm diameter. Far more frequently, however, larger units are observed. Apparently, they accumulate in a rather narrow range of sizes between 4 and 8 nm. Moreover there are extended aggregates, one of them shown in Fig. 2a and several ones in Fig. 2b. It can be stated from the electron-microscopic contrast that these are thin plane layers. Careful inspection of their structure indicated a regular structure due to dense packing of the dressed clusters [7].

To exclude the formation of such large units we tried to stabilize the Au-55 in the liquid. This was done by gelatin for the aqueous system and by polystyrene for the CH_2Cl_2 system. The electron micrographs (Fig. 3) show that large aggregates now are missing, yet the smaller aggregates still are present with similar size distribution. We conclude that they were present already in the powder substances.

As a next step to select single clusters the dispersions were filtered through Celite and stabilized immediately afterwards. The micrograph of Fig. 4 confirms that larger aggregates were extracted, yet the smaller aggregates were still present. The smallest ones with about 4 nm diameter may be tetrahedra of 4 clusters, each. Similar structures may be guessed for the larger ones (55 in the 8 nm aggregates?)

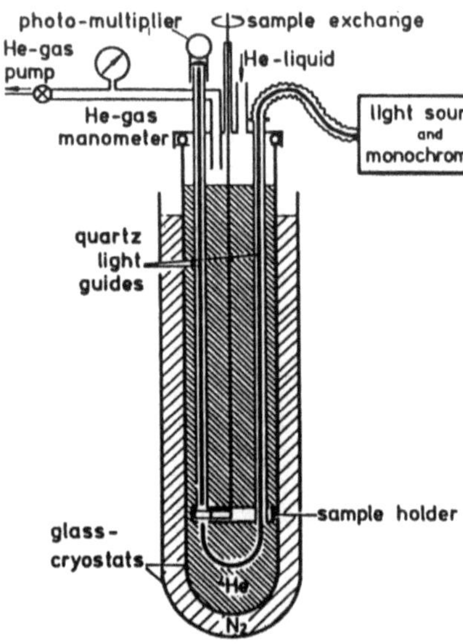

Fig. 6. Optical cryostat for $T \geq 1.5$ K

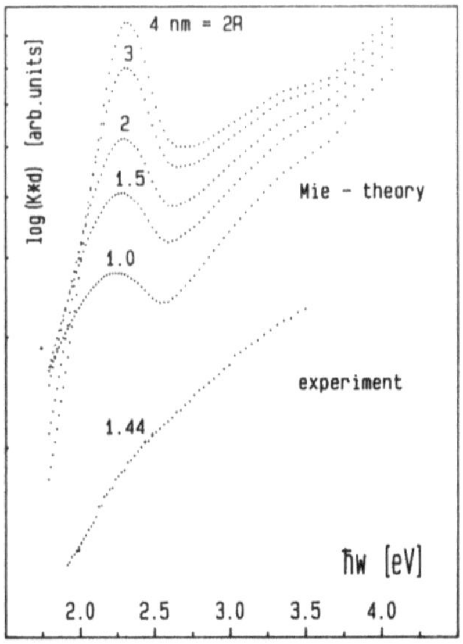

Fig. 7. Opt. absorption of solid state Au clusters of various sizes, calculated from Mie-theory (9), including quantum size effects

Further information about the tendency of the Au-55 to form larger aggregates were obtained by heat-treatment of the liquid samples (30 min, 70 °C). Figure 5 shows that very extended aggregates were formed exceeding 50 nm in their diameters.

Optical experiments and discussion

It is well known from larger Au-clusters [8] that the optical absorption around the dipolar plasmon excitation at about 2.5 eV is highly sensitive to structural and electronic properties.

Optical spectra of the samples described in the previous section were recorded in the spectral range from 1.4 to 4 eV, both, at 300 K and at 2 K. The type of low temperature equipment, used, was described earlier [8]. It is shown in Fig. 6.

In the following, arguments will be drawn from these spectra concerning the question for the electronic state of the Au-55.

At a first glance at the experimental spectra, Figs. 8, 9, 10, confirm the absence of any line structures as typical of simple molecules. This holds, to an experimental resolution of ± 0.5 nm, both for room temperature and for 2K. First, we therefore assume tentatively the clusters to be solid state like. Then, the spectra can be compared to theoretical spectra obtained from the electromagnetic theory of Mie and Debye [9]. In these calculations the dielectric function of bulk Au is incorporated which, to allow for the limited cluster sizes, was corrected due to quantum size effects [8]. The absorption of the theoretical spectra, Fig. 7, decreases toward lower frequencies with the narrow dipolar plasmon band superimposed at around 2.3 eV. It shows the well established effects of broadening and a slight red shift when the cluster size is decreased from 4 nm diameter down to the (only hypothetically chosen) size of 1 nm.

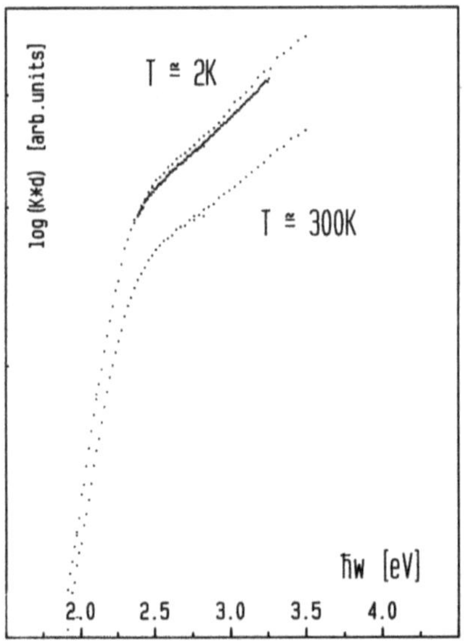

Fig. 8. Optical absorption spectra measured from stabilized Au_{55}-clusters. The two curves of the $T=2$ K-spectrum represent two subsequent measurements. The corresponding micrograph is shown in Fig. 3

None of these spectra compare well with the measured spectra of the sample stabilized in water,

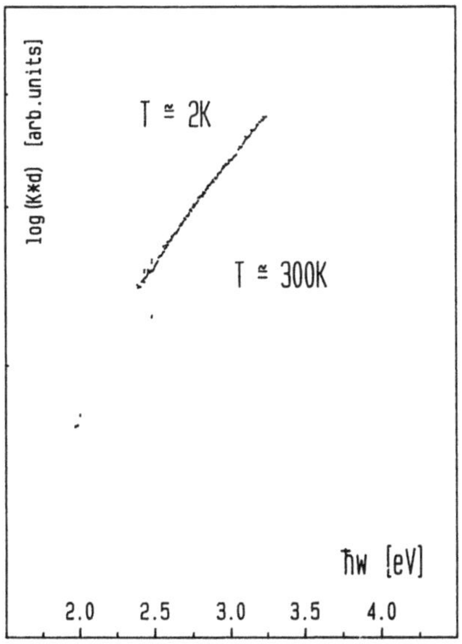

Fig. 9. Optical absorption spectra measured from filtered and stabilized Au_{55}-clusters. See Fig. 4. for micrograph

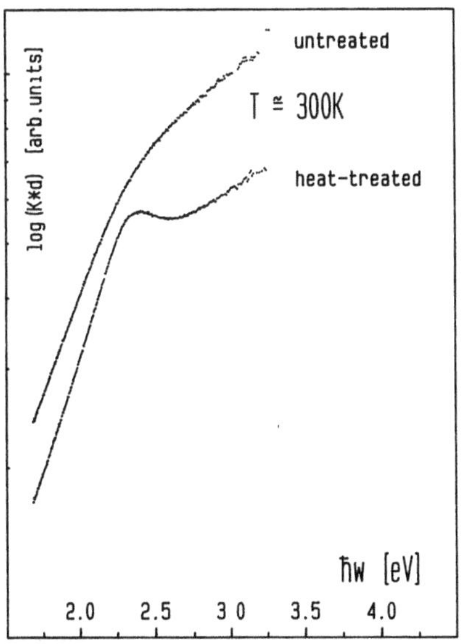

Fig. 10. Optical absorption spectra measured from untreated and heat-treated Au_{55}-cluster-systems. See Fig. 5. for micrograph

which is plotted in Fig. 8. There is an analogue decrease to low frequencies, yet, instead of the plasmon band there is only a bend around 2.5 eV. This structure is only slightly changed by cooling the samples below 2 K. Very similar spectra were obtained from CH_2Cl_2-samples and also from the solidified samples. After filtration, however, the recorded spectra (Fig. 9) deviate from the Mie spectra even more drastically[1]. Now a monotonous decrease of the absorption is left which, on the wavelength scale follows an exponential, $K \propto \exp(-3.44 \cdot \lambda\,(\mu m))$. This is true for both the 300K- and the 2K-spectrum, with the same value for the constant. It might be mentioned that exponential absorption edges are frequently observed for "amorphous" material. As a clear result we find that the Au-55 clusters do not behave as predicted for solid state particles by the extended Mie theory. It cannot yet be excluded that the strongly bound ligands cause additional damping of the plasmon making its detection difficult. Also the question remains for the implications of the observed tendency of the clusters to form larger, aggregated units.

First, if in these aggregates the clusters were compacted to solid gold by coalescence, we would expect pronounced plasmon peaks as previously observed for 4 to 8 nm particles [8]. There is no indication in the measured spectrum, hence, the aggregates consist of dressed Au-55 clusters, each cluster being protected from coalescence by its ligand shell. This is in agreement with the electron-microscopic findings.

Second, if the clusters were packed close together there should be direct electromagnetic coupling as described in [10] causing the broadening of spectral features. This effect depends on the relative parameter (center to center distance)/(cluster diameter) and is negligible for numerical values larger than, say, 1.2. In our samples, the 1.4 nm Au-55 clusters are separated by, at least twice the distance between Au and phosphine, that is 2×0.3 nm (the diameter of the dressed cluster is 2 nm). Hence, the above parameter amounts to $2/1.4 = 1.4$ and direct interactions can be excluded therefore. As a result, the Au-55 clusters act as well separated single units even if the dressed clusters are closely packed together. This holds for the filtrated samples but not as stringently for the untreated samples. As obvious from Fig. 2b there are large, obviously solid particles which were accumulated in the area of this micrograph, but very rare in the average. Only few of them would suffice to explain the bend that appears in the spectrum of Fig. 8 but not in Fig. 9 and which would then be due to slight contributions of Mie-like absorption to the otherwise exponential absorption of the majority of dressed Au-55 clusters. The formation of these solid particles is increased by heat-treating the samples: Fig. 10 shows a slight but well resolved plasmon peak. Compared to the computed spectra of Fig. 7 it is blue-shifted, pointing to small sizes.

However, at this point, the alternative explanation of slight changes of the majority of dressed Au-55 clusters cannot yet be excluded. Further investigations will concentrate on this problem. A tentative interpretation of the optical properties, supporting the molecular state of the Au-55 clusters, is given in [11] of this volume.

[1] Benfield et al. [12] do not observe the disappearance of the plasmon peak. This is probably due to a small amount of colloidal gold still present in their samples

References

1. Mingos, D.M.P.: Z. Phys. D – Atoms, Molecules and Clusters **12** (this issue) (1989)
2. Hall, K.P., Mingos, D.M.P.: Prog. Inorg. Chem. **32**, 237 (1984)
3. Ozin, G.A. Faraday Symp. R. Chem. Soc. **14**, 7 (1980)
4. Schmid, G., Pfeil, R., Boese, R., Bandermann, F., Meyer, S., Calis, G.H.M., van der Velden, J.W.A.: Chem. Ber. **114**, 3634 (1981)
5. Ekardt, W., Phys. Rev. B **29**, 1558 (1984); Knight, W.D., Clemenger, K., de Heer, W.A., Saunders, W.A., Chou, M.Y.: Phys. Rev. Lett. **52**, 2141 (1984)
6. Schmid, G.: In: Structure and bonding, Vol. 62. Berlin, Heidelberg, New York: Springer 1985
7. Schmid, G., Klein, N., Korste, L., Kreibig, U., Schönauer, D.: Polyhedron **7**, 605 (1988)
8. Kreibig, U.J. Phys. **38**, C2-97 (1977); Kreibig, U.: Growth and properties of metal clusters. Bourdon, J. (ed.), p. 371. Amsterdam: Elsevier 1980
9. Mie, G.: Ann. Phys. **25**, 377 (1908)
10. Quinten, M., Schönauer, D., Kreibig, U.: Z. Phys. D – Atoms, Molecules and Clusters **12** (this issue) (1989)
11. Kreibig, U., Fauth, K., Quinten, M., Schönauer, D.: Z. Phys. D – Atoms, Molecules and Clusters **12** (this issue) (1989)
12. Benfield, R.E., Creighton, J.A., Eadon, D.G.: Z. Phys. D – Atoms, Molecules and Clusters **12** (this issue) (1989)

Z Phys. D – Atoms, Molecules and Clusters 12, 521–525 (1989)

Atoms, Molecules
Zeitschrift für Physik D and Clusters
© Springer-Verlag 1989

Electronic excitations in many-particle systems: a quantitative analysis

M. Quinten, D. Schönauer, and U. Kreibig
Universität des Saarlandes, FB 11 Physik, D-6600 Saarbrücken, Federal Republic of Germany

Received 5 July 1988

Small particles, produced in usual experiments, commonly form many-particle systems. Interactions of various kinds among the particles influence the properties of such systems. We consider here the optical properties of noble metal particle systems with separated nearly spherical particles and with aggregates formed by induced coagulation of the single particles. In order to describe their optical extinction we apply electrodynamic interaction models on particles and aggregates. We perform a quantitative analysis of the extinction spectrum of one selected sample with the electrodynamic interaction model.

PACS: 41.70; 78.40; 82.70

I. Introduction

Plasmon polariton excitations in small metal particles are well described by Mie's scattering theory for electromagnetic waves [1]. Additional aspects arise when the particles form aggregates by lumping. Electromagnetic interactions among the particles of an aggregate result in new excitations. They can be described by the electromagnetic multipole scattering theory for the aggregate as a whole. We present here two contradicting versions of the same model.

We consider the optical properties of colloidal noble metal particle systems prepared with usual techniques. By inducing coagulation we achieve in systems where the well separated single particles of the colloids form aggregates of various shapes and sizes.

In this paper we present the optical properties of a selected sample of small Au-particles (mean diameter $2R = 38$ nm) before and after coagulation. Applying the electromagnetic theory on single particles and on aggregates we perform a quantitative comparison of experimental and theoretical results.

II. Experiment

The colloidal gold particles of our sample were prepared by chemical reduction applying Zsigmondy's preparation technique [5]. The resulting diluted colloidal system contained well separated, nearly spherical particles of mean diameter $2R = 38\ \text{nm} \pm 1.5\ \text{nm}$ (Fig. 1a and 1b). The measured optical extinction spectrum exhibited a plasmon polariton peak at 2.36 eV with a halfwidth of 0.16 eV and in addition extinction from the (5*d*-6*s*, *p*)-interband transitions (Fig. 1c).

Colloidal metal particles diluted in aqueous electrolytic solutions are known to be surrounded by clouds of ions forming the electric double layer [6]. This double layer causes a repulsive electrostatic potential V_R between particles, approximatively described by a screened Coulomb-potential [6]

$$V_R = V_0 \cdot \exp(-\kappa r)/r \qquad (1a)$$

κ is the Debye-Hückel parameter and depends on the ion concentrations n_i and valences z_i

$$\kappa = (2e_0^2/(\varepsilon\varepsilon_0 k_B T) \sum_i n_i z_i^2)^{1/2} \qquad (1b)$$

This repulsion counteracts the Van-der-Waals attraction between particles which can approximatively be described by [6]

$$V_A = A_0/r^2 \qquad (1c)$$

Particles will coagulate if the electrostatic repulsion becomes smaller than the Van-der-Waals attraction.

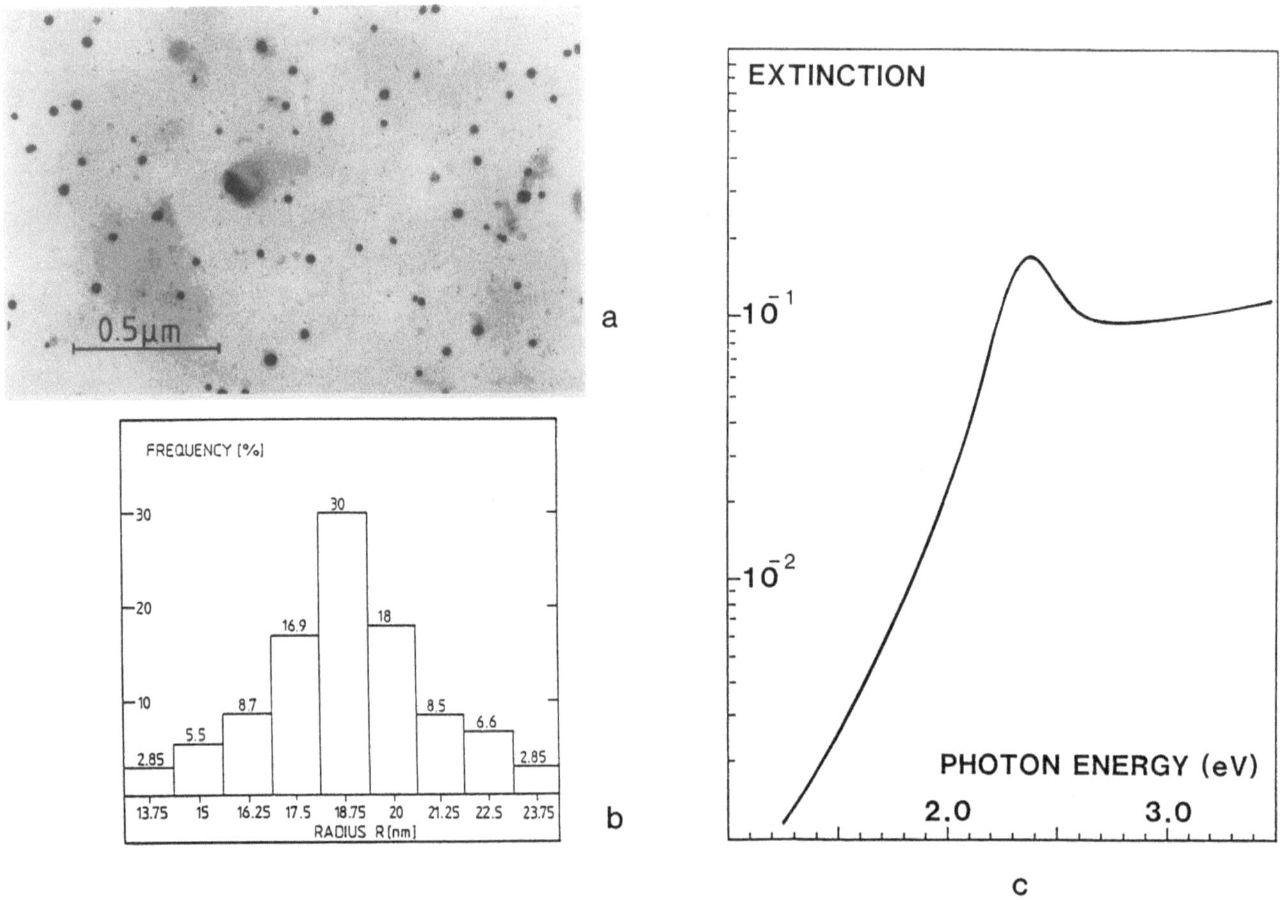

Fig. 1. a TEM-micrograph **b.** size histogram **c.** optical extinction spectrum of the single particles

This can experimentally be achieved by addition of a soluble salt to the colloid which causes κ to increase and V_R to decrease. Then the single particles tend to approach their neighbours during their diffusion and form aggregates of statistically varying shapes and sizes. Remaining ions, adsorbed at the particle surfaces, prevent direct metallic contact among the particles within an aggregate. The aggregation process can be stopped by addition of a stabilizing agent like gelatin which forms coverings around the particles.

In our sample aggregation was induced by addition of $CuSO_4$ up to a concentration of $1.5\ 10^{-4}$ –M. About 10 s after addition of $CuSO_4$ we stopped the aggregation process by addition of gelatin.

The optical extinction spectrum of the resulting sample is shown in Fig. 2a. Instead of the single particle Mie maximum there are two clearly separated maxima with an energy difference $\Delta E = 0.41$ eV. Their intensity ratio is about 2:1. For the TEM analysis of our sample (one selected micrograph is given in Fig. 2b) we counted the amount of aggregates with same given number of particles N. Introducing several kinds of highly symmetric aggregates for each N we classified the shapes of the recognized aggregates. The result for $N = 1$ to $N = 7$ particles per aggregate is given in Fig. 2c.

III. Theory of optical extinction

In this section we describe the calculations of extinction of electromagnetic waves by single spherical particles and by coagulation aggregates.

For single particles the optical extinction is given by Mie's theory [1]. Mie's theory bases on an multipolar expansion of the waves inside and outside

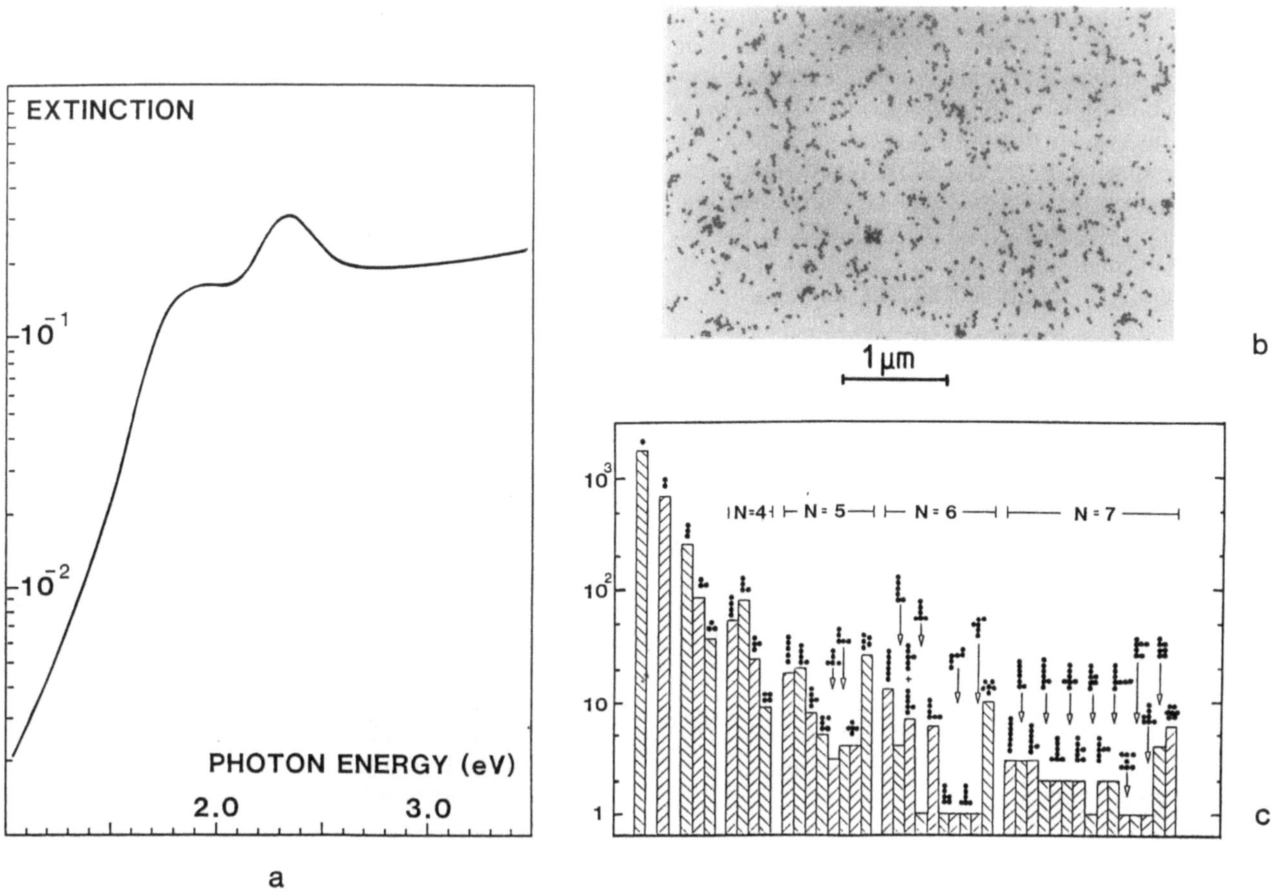

Fig. 2. a Optical extinction spectrum of the aggregated sample. **b** TEM-micrograph. **c** size and shape histogram

the particle into spherical harmonics $Y_{lm}(\theta, \phi)$ and Riccati-Bessel functions $\psi_1(kr)$ and $\zeta_1(kr)$. From Maxwell's boundary conditions at the surface of the sphere result two equations which combine the expansion coefficients c_{lm}, d_{lm} of the scattered wave and the coefficients a_{lm}, b_{lm} of the incident plane wave:

$$d_{lm}=\Delta_1 b_{lm}= -b_{lm}\cdot\frac{K\psi_l(KR)\psi_l'(kR)-k\psi_l'(KR)\psi_l(kR)}{K\psi_l(KR)\zeta_l'(kR)-k\psi_l'(KR)\zeta_l(kR)} \tag{2a}$$

$$c_{lm}=\Gamma_l a_{lm}= -a_{lm}\cdot\frac{k\psi_l(KR)\psi_l'(kR)-K\psi_l'(KR)\psi_l(kR)}{k\psi_l(KR)\zeta_l'(kR)-K\psi_l'(KR)\zeta_l(kR)} \tag{2b}$$

Δ_l and Γ_l are the electric and magnetic multipolar moment of a sphere. The prime denotes the first derivative of the functions. $k=\omega/c\cdot n_0$ is the wave-vector in the embedding medium and $K(\omega)=\omega/c\cdot N(\omega)$ is the complex wave-vector in the particle. Finally the extinction $E(\omega)$ of n spheres in a volume V is given by

$$E(\omega)=n/V2\pi/k^2\mathrm{Re}\sum_{l=1}^{\infty}\sum_{m=-1}^{l} a_{lm}^{*}c_{lm}+b_{lm}^{*}d_{lm}$$

$*$ = conjugated complex (3)

In order to extend Mie's theory on particle aggregates one has to take into account the scattered waves of all particles $j\neq i$, $j=1,\ldots,N$, of a N-particle aggregate as additional incident waves at all particles i, $i=1,\ldots,N$.

For this purpose the scattered waves of the particles j have to be transformed into incident waves at particle i. This transformation was given by Langbein [2] for two particles with a common z-axis. For arbitrary positions of the particles an additional transformation of the spherical harmonics under rotation is needed. This transformation was given by Jeffreys [3]. Gérardy and Ausloos [4] finally combined both transformations in order to generalize the transformation.

Applying Maxwell's boundary conditions on the system of incident and scattered waves one results in two linear sets of equations for each particle i:

$$d_{lm}(i) - \Delta_l(i) \cdot \sum_{j \neq i}^{N} \sum_{q=1}^{\infty} \sum_{p=-q}^{q} c_{qp}(j) B_{qplm} + d_{qp}(j) A_{qplm} = \Delta_l(i) b_{lm}(i) \tag{4a}$$

$$c_{lm}(i) - \Gamma_l(i) \cdot \sum_{j \neq i}^{N} \sum_{q=1}^{\infty} \sum_{p=-q}^{q} c_{pq}(j) A_{qplm} + d_{qp}(j) B_{qplr} = \Gamma_l(i) a_{lm}(i) \tag{4b}$$

The terms A_{qplm} and B_{qplm} describe the full transformation and are given as [4]

$$A_{qplm} = \exp(i(p-m)\phi_{ij}) \times \sum_{\lambda=-r}^{r} 0(l, m, \lambda, \theta_{ij})\, 0(q, p, \lambda, \theta_{ij})\, V_{ql}^{\lambda}(kR_{ij}) \tag{5a}$$

$$r = \min(q, l)$$

$$B_{qplm} = \exp(i(p-m)\phi_{ij}) \times \sum_{\lambda=-r}^{r} 0(l, m, \lambda, \theta_{ij})\, 0(q, p, \lambda, \theta_{ij})\, W_{ql}^{\lambda}(kR_{ij}) \tag{5b}$$

where R_{ij}, θ_{ij}, ϕ_{ij} are the relative coordinates of particle j respectively particle i. Finally the extinction of n N-particle aggregates in a volume V is given by

$$E(\omega) = n/V \cdot 2\pi/k^2 \operatorname{Re} \sum_{i=1}^{N} \sum_{l=1}^{\infty} \sum_{m=-l}^{l} a_{lm}^{*}(i) c_{lm}(i) + b_{lm}^{*}(i) d_{lm}(i) \tag{6}$$

Numerical computations of the extinction spectrum, however, are only possible when the infinite sum over q and l in (4) is truncated at a maximum multipole order $L = \max(q, l)$.

According to the signs of the terms V_{ql}^{λ} and W_{ql}^{λ} two contradicting versions of the presented model are in discussion. Gérardy and Ausloos argue that the terms V_{ql}^{λ}, W_{ql}^{λ} have the following "symmetry" (version 1):

$$V_{lq}^{\lambda} = (-1)^{q+l} V_{ql}^{\lambda}, \quad W_{lq}^{\lambda} = (-1)^{q+l} W_{ql}^{\lambda} \tag{7a}$$

which indicates a change in sign when $q+1$ is an odd number. Instead we suggest that these terms have always the *same* sign (version 2)

$$V_{lq}^{\lambda} = V_{ql}^{\lambda}, \quad W_{lq}^{\lambda} = W_{ql}^{\lambda} \tag{7b}$$

The different definitions of the symmetry of these terms strongly affect the computed extinction spectra. For both versions we computed the extinction of a linear chain with $N=3$ up to maximum multipole order $L=3$ (this upper limit was given by the accuracy of the numerical computations). The resulting peak energies are shown in Fig. 3. The peak positions for $L \to \infty$ were estimated by extrapolation. As clearly to be recognized from this figure the band splitting of version 1 converges alternating towards an energy difference ΔE which is lower for $L \to \infty$ than for $L=1$, in expression $\Delta E(L \to \infty) < \Delta E(L=1)$. In our version 2 ΔE instead converges monotonously towards a value $\Delta E(L \to \infty)$ which is higher than for $L=1$, $\Delta E(L \to \infty) > \Delta E(L=1)$.

IV. Comparison

With respect to the theoretical models for the extinction of single particles and of aggregates as described

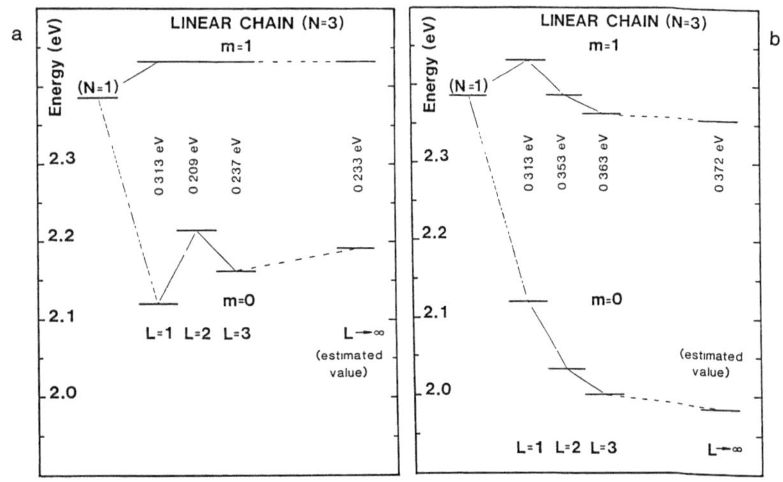

Fig. 3a and b. Band peak splitting for a linear chain with $N=3$ particles **a.** for computations respectively version 1 **b.** for computations respectively version 2

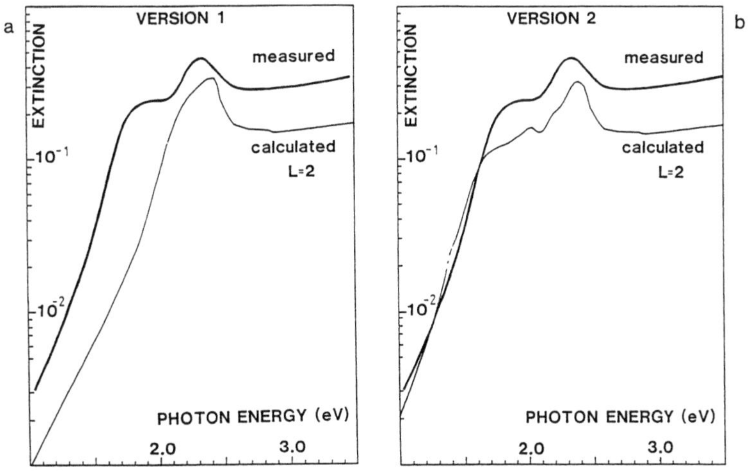

Fig. 4a and b. Comparison of computed and measured extinction spectrum of the selected sample. **a.** result for version 1. **b.** result for version 2. The computed spectra are shifted by a factor 0.5 for better separation

in Sect. III we performed a quantitative analysis of the measured extinction spectrum of our selected sample. For this purpose we made numerical computations on a PDP 11/44 computer for all high symmetric aggregate shapes shown in Fig. 2c and for both sign-versions with a maximum multipole order $L=2$. The resulting computed spectra of the different aggregates were weighted with their corresponding relative numbers in the sample and added up to the complete extinction spectrum. For both versions the computed spectrum is compared to the measured one as shown in Fig. 4.

Obviously the computations for version 1 do not fit the measured spectrum. However, the computations for version 2 fit the measured spectrum well. Remaining differences are believed to be mainly due to the choice of literature data for the dielectric function ε which do not fit quantitatively the particle material. For example, quantum size effects and grain boundary scattering in the particles modify ε. Therefore we plan an improvement by using data for ε derived by Kramers-Kronig analysis of the measured extinction of the single particles. Finally, we conclude that the model of *pure* electromagnetic interactions among the particles of an aggregate (what means that there is *not any charge transfer* due to metallic contacts or electron tunneling) yield a rather good quantitative agreement of the experiment with the theoretical description.

References

1. Mie, G.: Ann. Phys. **25**, 377 (1908)
2. Langbein, D.: Springer Tracts in Modern Physics, Vol. 72. Berlin, Heidelberg, New York: Springer 1974
3. Jeffreys, B.: Geophys. J. R. Soc. **10**, 141 (1965)
4. Gérardy, J.M., Ausloos, M.: Phys. Rev. B **25**, 4204 (1982)
5. Zsigmondy, R.: Das kolloide Gold. Leipzig: Deuticke 1925
6. Stauff, J.: Kolloidchemie. Berlin, Heidelberg, New York: Springer 1960

Z. Phys. D – Atoms, Molecules and Clusters 12, 527–532 (1989)

Atoms, Molecules
Zeitschrift für Physik D and Clusters
© Springer-Verlag 1989

Precursor-states of percolation in quasi-fractal many-particle-systems

D. Schönauer, M. Quinten, and U. Kreibig
FB 11-Physik, Universität des Saarlandes, D-6600 Saarbrücken, Federal Republic of Germany

Received 6 July 1988

Samples are described which consist of coagulated quasi-fractal structures of 20 nm gold-particles embedded in a nonconducting matrix. By varying coagulation into coalescence electrical percolation can be induced while all topological parameters are kept constant. The first steps of percolation were investigated from the optical spectra. From calculations a fundamental limit for the splitting of plasmon modes in coagulation aggregates of arbitrary size and shape was found. In coalescence aggregates this limit is markedly surpassed. The percolation correlation length can be estimated therefrom. In the present case it amounts to about 300 nm.

PACS: 78.00; 82.70

I. Introduction

A realization of percolation models are systems consisting on small metallic particles embedded in an insulating matrix. One example are the cermets (ceramics-metal-systems) of Au-particles produced in an Al_2O_3-matrix [1]. By increasing the mean volume fraction of the particles, the electrical *dc* conductivity of such samples can be varied from insulator behavior (10^{-6} $S \cdot m^{-1}$) up to metallic conductivity (10^4 $S \cdot m^{-1}$). Below the percolation threshold particle-particle tunneling assisted conductivity has been observed [2].

Probably the particles form statistical networks, sometimes assigned as fractal structures, and near the percolation threshold conducting paths are formed with metallic contacts between neighboring particles. The length of such paths, the correlation length, increases with increasing filling factor, and above the percolation threshold the lengths of conducting paths in the sample are macroscopic. Detailed investigations of such cermet systems are difficult since changing the filling factor means to influence at same time other structural parameters as particle sizes and topological distributions.

In the following a study is presented of a similar system, where all topological parameters are kept constant when the lengths of the conducting paths are changed.

Detailed information about this process can be obtained by optical means rather than by measuring *dc* conductivity since, then, the important scale for the length 1 of the conducting paths is given by the wavelength λ of the used light, instead of the macroscopic sample length, and, consequently the optical "percolation" occurring when $1 \simeq \lambda$ is a pre-state of the *dc*-percolation.

II. Sample preparation

(a) The many-particle-systems used for the present investigation are Au-Colloids which were produced as Zsigmondy hydro-sols [3] and subsequently caused to form coagulation aggregates by diffusion limited aggregation (DLA) in solution of proper chosen pH-value.

The topological arrangement of statistical network consisting of irregular chains of coagulated particles [4] is shown in Fig. 1. These structures were stabilized

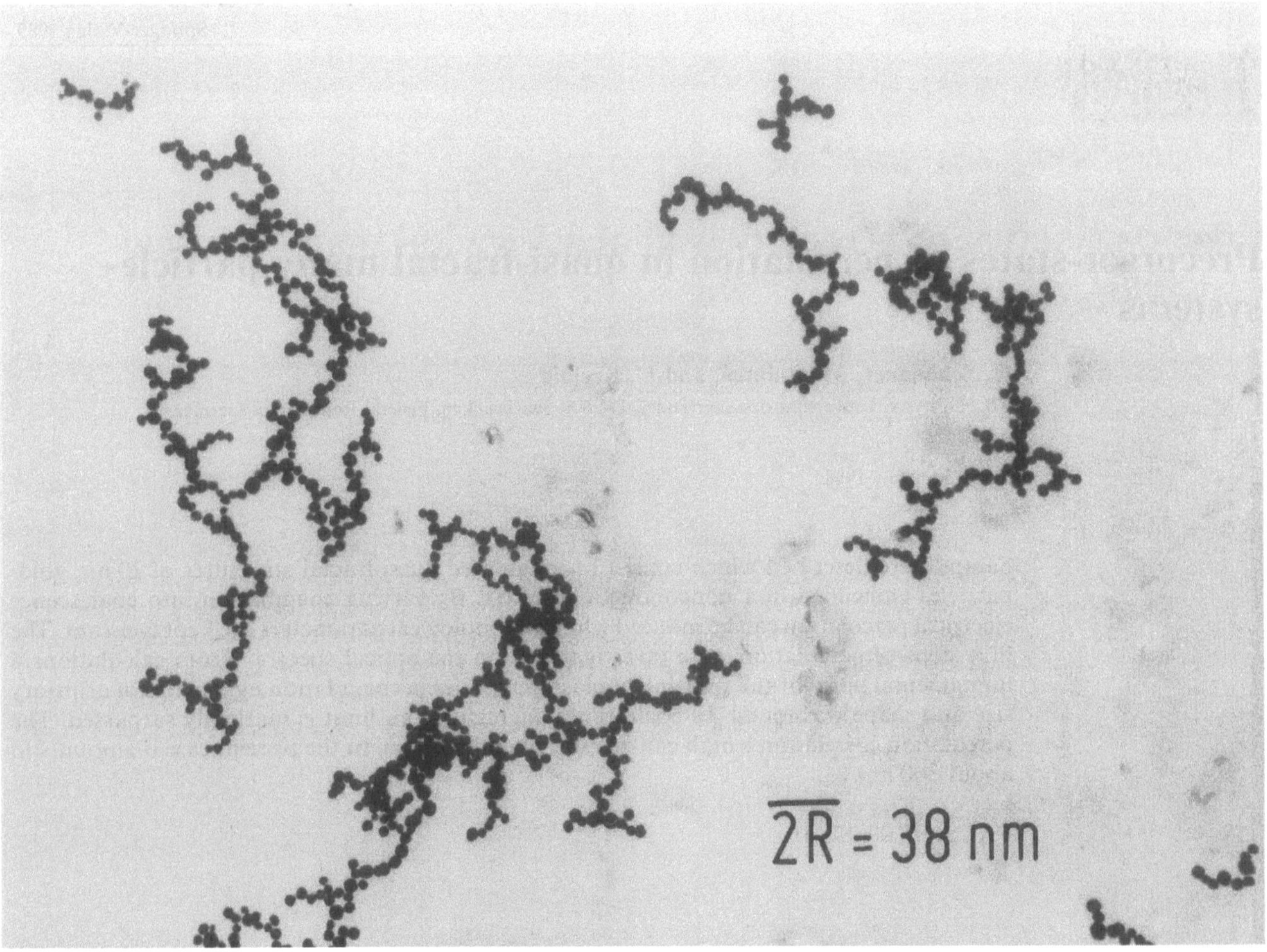

Fig. 1. Au-particle sample with quasi-fractal aggregates ($2R$ = 38 nm)

by adding gelatin and then investigated by measuring their optical extinction spectra. Part of these specimen were then dried into thin films. The 3-dimensional network existing in the aqueous solution were thus transformed into 2-dimensional ones, embedded in solid gelatin. Thin carbon foils were used as substrates and so, beside measuring the optical extinction, TEM micrographs could be performed of one and the same sample to analyse the topographic details.

The percolation aggregates, the compactness of which can vary between the dense 3-dimensional lump and, as the other extreme, the fractal network, are formed in two steps:

(b) Coagulation

Usually the particles in Au-Hydro-sols are prevented from aggregation by repulsive Debye-Hückel-forces coming from the ion double-layer at the particle surfaces and the adjacent ion cloud. These forces overcompensate the attractive van-der-Waals-forces. Addition of ions to the colloidal system influences this cloud and releases the aggregation.

This leads primarily to coagulation aggregates where neighboring particles are separated by thin ion shells at their surface and/or are stabilized in the secondary DLVO potential minimum [5]. Particles in coagulation aggregates are almost touching but are well separated individuals with their numbers of conducting electrons conserved. Electronic current along the aggregates is only possible via tunneling [2].

(c) Coalescence

By additional change of pH-value of the system and/or heat treatment neighboring particles grow together by building a common grain boundary, i.e. a neck between each other. This can take place, both, in the liquid and in the solid samples. For the latter it is ensured, that the formation of grain boundaries is the only effect and *no* changes of the sample topography occurs. The coalescence times vary between seconds and weeks.

The resulting grain-boundaries are clearly seen in the micrograph of Fig. 2. The large contact areas, the formation of which was described by Thölen [6], allow for direct electrical conductivity. In contrast to the coagulation aggregates the number of electrons in one particle is no longer conserved. As shows Fig. 2, there are percolation chains of roughly 5–10 particles which are separated by links of coagulation only. The length of these chains may be identified with the correlation length of the percolation path.

III. Optical analysis

(a) Coagulation aggregates

The optical extinction (i.e. absorption plus scattering losses) of such samples is determined, both, by properties of the individual structural units, which in coagulation areas are the particles and collective effects due to interaction among the particles densely packed in the aggregates. Figure 3 shows extinction spectra of a sample, where merely the topography of otherwise unchanged particles is varied. It is obvious that spectra are very sensitive to the details of the collective effects among the particles. They are at present subjects of intensive investigation [4, 7, 8, 9, 10].

The spectra are dominated by the plasmon-polariton-excitations which form one very intense band in the spectra of isolated Au-particles (Fig. 4, curve a), that is the aggregates split mainly into two maxima. This splitting is due to electromagnetical couplings via scattering near fields. For the case of linear chains they are attributed to excitations with the electric field parallel (low energy mode) and normal (high energy mode) to the chain axis. In all kinds of aggregates, it is in special the low frequency mode, that is extremely sensitive to changes of the topology. In Fig. 5 this is demonstrated with the calculated spectra of some examples all consisting of 5 particles. From this figure follows – and it is together a general result – that from all shapes of an N-particle-aggregate, it is the linear chain that results in the largest energetic difference $\Delta\hbar\omega$ between the two major modes. With increasing length of the N-particle-chain, there is a cross-over in the splitting energy from strong N-dependence to an asymptotic value $\Delta\hbar\omega_g$ which only depends on the radius of the single particle. Figure 6 shows this behavior, as calculated with the theory described by Quinten et al. in this volume. All retardation effects within the aggregates were taken into account and multipole effects were considered up to the quadrupole [4].

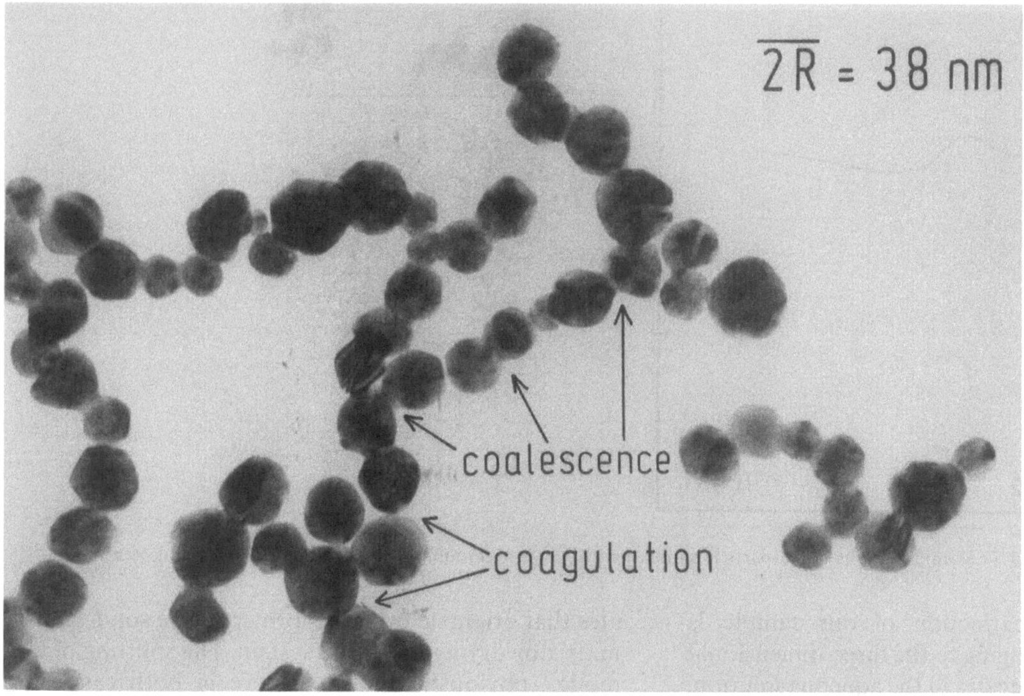

Fig. 2. Coagulation and Coalescence in a Au-particle-aggregate of the sample of Fig. 1

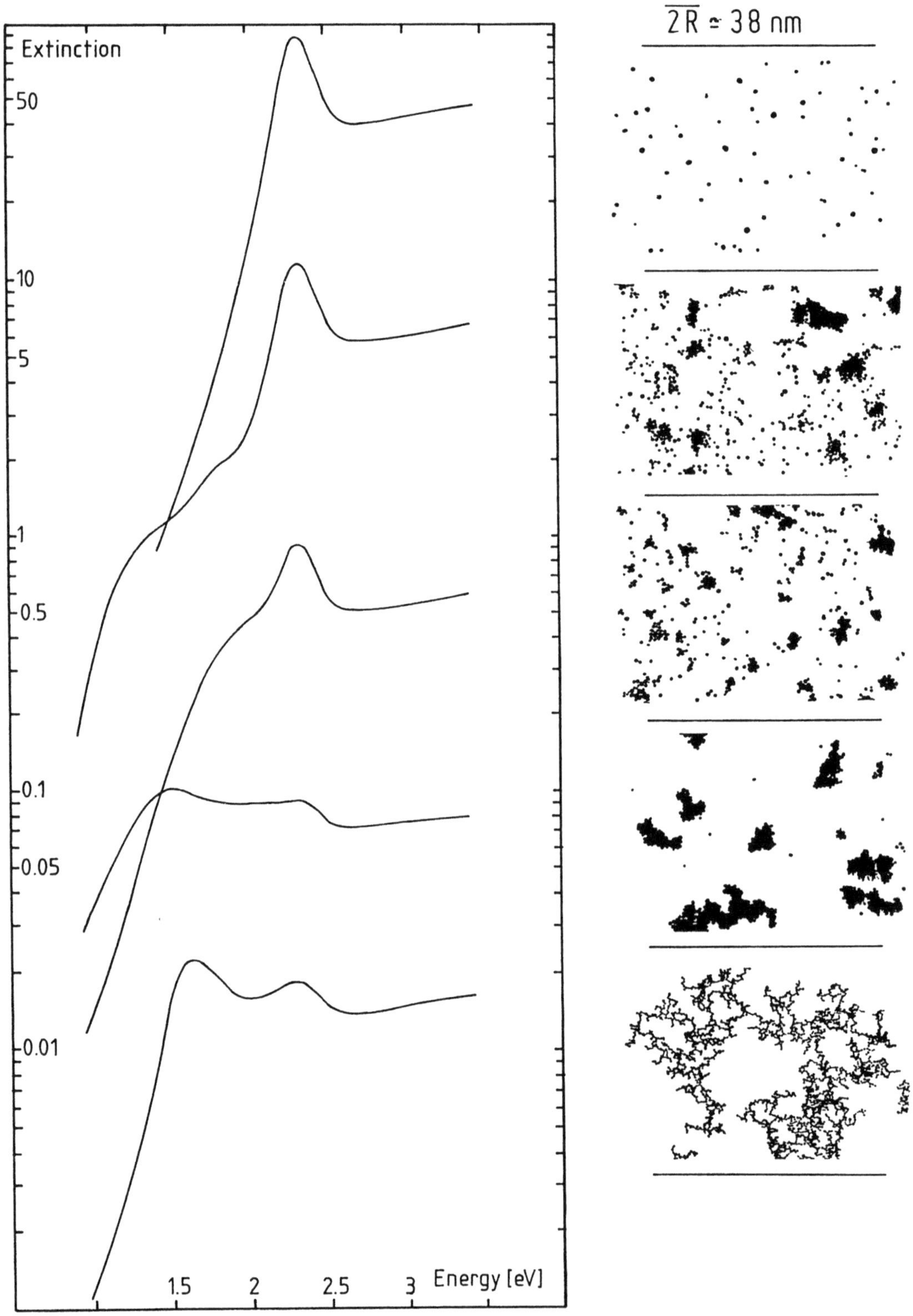

Fig. 3. Extinction spectra and TEM micrographs of Au-particle samples with varying amount of aggregation ($2R$ = 38 and 56 nm)

The experimental extinction of our sample is shown in Fig. 4, (b) belonging to the three-dimensional fractal coagulation aggregates in the aqueous solution, while (c) is appropriate to the two-dimensional aggregates that originate from the former with a solid gelatin matrix by drying the liquid system. The splitting of the modes, obviously, almost achieve in both cases the corresponding values of the limiting energy $\Delta\hbar\omega_g$ of

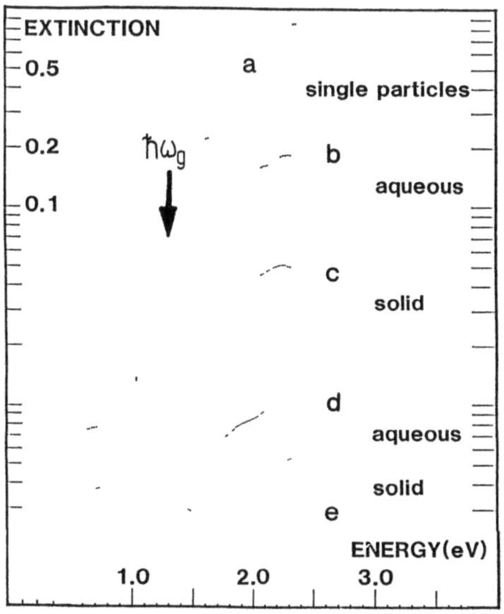

Fig. 4. Measured extinction spectra of the sample of Fig. 1 *a)* before aggregation, *b), d)* 3-dimensional aggregates, *c), e)* 2-dimensional aggregates

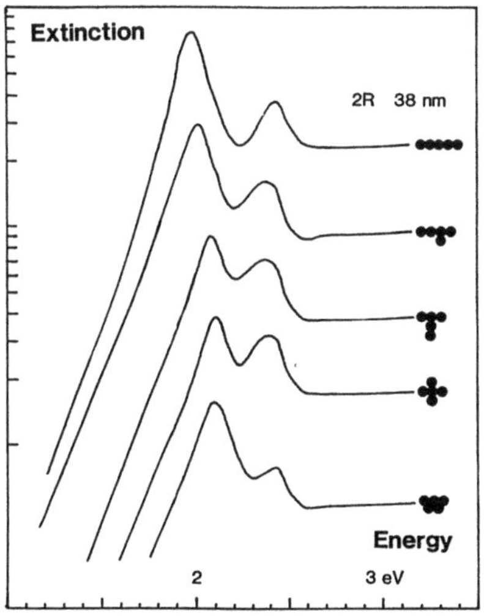

Fig. 5. Computed extinction spectra for 5 selected high-symmetric aggregates all containing 5 particles. The spectra are shifted vertically for arbitrary amounts

Fig. 6 which are marked in the spectra. This reflects the fact that the quasi-fractal aggregate structure essentially consists of chainlike segments.

(b) Coalescence aggregates

The transformation from coagulation to coalescence, which took place very slowly as was observed with the electronmicroscope, caused marked changes in the excitation spectra, Fig. 4: the spectra (d) and (e) were measured in the same way as (b) and (c) but two weeks later. They show an intense change: the low energy-peaks of spectra (b) and (c) have nearly disappeared and, instead, we find an extinction band which cannot be ascribed to a coagulation aggregate mode since it is clearly below $\hbar\omega_g$. The detailed micrograph, Fig. 2, shows that the coagulation transmuted in part into coalescence.

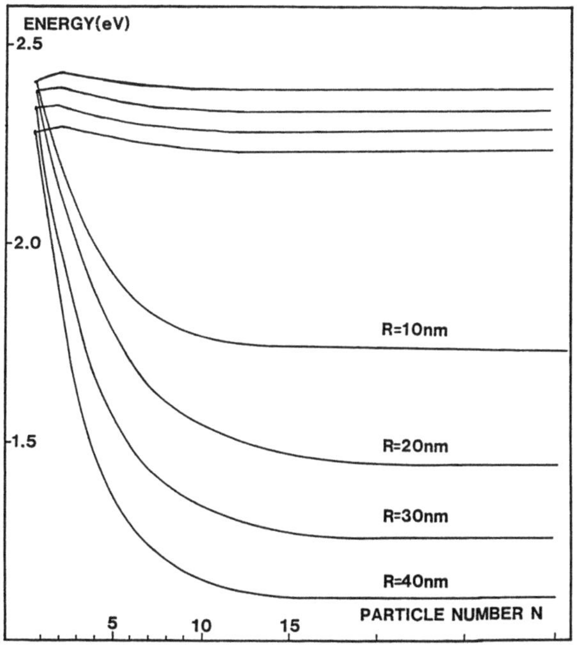

Fig. 6. Computed extinction peak positions of linear Au-particle-chains with varying number of particles N. R: particle radius. Calculations include retarded dipolar and quadrupolar excitations and coupling

The optical extinction of coalescence segments cannot be calculated numerically due to their irregular shape. However they may be replaced by simple models, as are the oblate ellipsoidal or the cylindric particles, the extinction of the latter can be calculated up to now only in the quasi-static case: in both models however, yield a two peak structure, the low-energy plasmon-polariton-mode shifting towards zero frequency with increasing excentricity of the ellipsoids or length of the cylinders (Fig. 7).

Comparing the position of the measured low-energetic peak in Fig. 4 with Fig. 7 it is possible, to estimate roughly the mean length of electrical conducting paths due to coalescence: for the spectra Fig. 4 (d) and (e) we determined approximately 9 particle-diameters. This is in good correspondence to the electron micrographic informations.

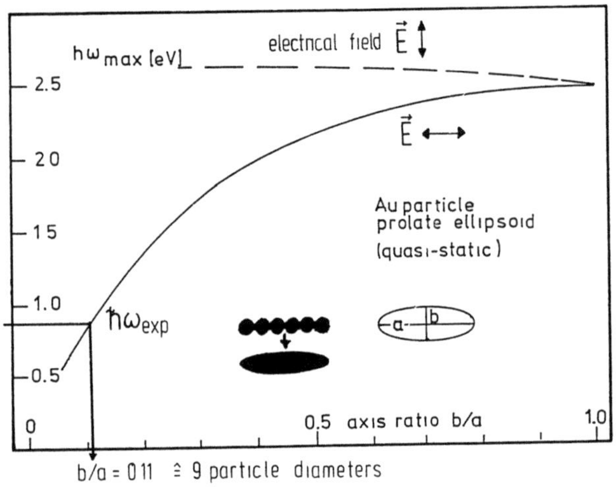

Fig. 7. Calculated extinction peak positions of ellipsoidal particles with varying excentricity. The low energetic excitation of Fig. 4 is drawn in and yield to a mean percolation length of the coalescent aggregates of about 9 particle diameters

IV. Results

1) Samples of small conducting particles embedded in an insulating matrix proved to be a proper model substance for experiments on percolation: By artificially changing coagulation into coalescence the lengths of conducting percolation paths could be varied, while all the other topological parameters were kept constant.

2) The optical plasma resonance extinction allowed to observe the formation of electrical percolation paths, yet not on the correlation length scale of the whole sample, as probed by static conductivity but on the smaller scale of the order of the wavelength of light. Such percolation segments may be interpreted as precursor-states of *dc*-electric percolation.

3) It should be possible analogously to observe the elementary process during the formation of nanocrystalline material [11], when starting with more compact coagulation aggregates.

References

1. Niklasson, G., Granqvist, C.G.: J. Appl. Phys. **55**, 3382 (1984)
2. Kreibig, U., Andersson, D., Niklasson, G., Granqvist, C.G.: Thin Solid Films **125**, 199 (1985)
3. Zsigmondy, R.: Das kolloide Gold. Leipzig: Akademische Verlagsgesellschaft 1925
4. Quinten, M., Kreibig, U.: Proceedings of the Symposium on Optical Particle Sizing: Theory & Practice. Rouen/France 1987
5. Stauff, J.: Kolloidchemie. Berlin-Heidelberg-New York: Springer 1960
6. Thölen, A.: In: Contributions of cluster physics to material science and technology. Davenas, J., Rabette, P., (eds.), p. 601. Dordrecht: Nijhoff, 1986
7. Kreibig, U., Quinten, M., Schönauer, D.: Phys. Scr. **34**, **T13**, 86 (1986)
8. Quinten, M., Kreibig, U.: Surf. Sci. **172**, 557 (1986)
9. Langbein, D.: Springer tracts in modern physics, Vol. 72. Berlin, Heidelberg, New York: Springer 1974
10. Gérardy, J., Ausloos, M.: Phys. Rev. B **27**, 6446 (1983)
11. Birringer, R., Gleiter, H., Klein, H.P., Marquardt, P.: Phys. Lett. **102A**, 365 (1984)

Z. Phys. D – Atoms, Molecules and Clusters 12, 533–536 (1989)

Atoms, Molecules and Clusters
Zeitschrift für Physik D
© Springer-Verlag 1989

The relationship of $Au_{55}(PPh_3)_{12}Cl_6$ to colloidal gold

R.E. Benfield[1], **J.A. Creighton**[1], **D.G. Eadon**[1], **and G. Schmid**[2]

[1] Chemical Laboratory, University of Kent, Canterbury CT2 7NH, U.K.

[2] Institut für Anorganische Chemie der Universität Essen, Universitätsstrasse 5–7, D-4300 Essen 1, Federal Republic of Germany

Received 5 July 1988; final version 30 September 1988

The electronic (UV-visible) spectrum of the molecular cluster $Au_{55}(PPh_3)_{12}Cl_6$ shows features corresponding to the 520 nm plasma resonance and the shorter-wavelength interband transition of colloidal gold. These absorptions differ qualitatively from the simpler one-electron transitions of lower-nuclearity cluster molecules. Differential scanning calorimetry has been used to measure the enthalpy of decomposition of $Au_{55}(PPh_3)_{12}Cl_6$. The Au-Au bonding appears to be substantially stronger than in bulk gold.

PACS: 33.20.K; 35.20.G; 36.40

1. Introduction

Molecular metal cluster compounds are good structural models for small metal particles bearing surface adsorbates. Evidence is emerging that they may also represent physical models [1].

The molecule $Au_{55}(PPh_3)_{12}Cl_6$ is believed to contain cuboctahedral clusters with three concentric close-packed shells of 1, 12 and 42 gold atoms [2, 3]. Its solid-state electrical conductivity resembles that of granular metals [4]. We have compared some of its spectroscopic and thermochemical properties with the well-characterised behaviour of colloidal gold.

2. Electronic spectrum of $Au_{55}(PPh_3)_{12}Cl_6$

The UV-visible spectrum of $Au_{55}(PPh_3)_{12}Cl_6$ in solution is shown in Fig. 1, with spectra of two lower-nuclearity clusters and of colloidal gold. It comprises three features: an intense peak at 229 nm, a broad band extending beyond 800 nm, and a weak shoulder at 510 nm.

The 229 nm peak is the $\pi \rightarrow \pi^*$ transition of the phenyl rings in the coordinated PPh_3 ligands [5], showing vibrational structure on the long wavelength side around 270 nm.

The broad absorption extending to long wavelengths (1.5–4.5 eV) differs from the discrete peaks in the spectra of the lower-nuclearity clusters. Instead, it resembles the absorption extending across the spectrum of colloidal gold. This arises from interband transitions, principally from the dense *5d* band into the *6s*, *6p* band just above the Fermi level [6]. In $Au_{55}(PPh_3)_{12}Cl_6$ the spectrum has developed significantly away from one-electron transitions towards the continuous absorption envelope of colloidal gold.

The 510 nm shoulder in the spectrum of $Au_{55}(PPh_3)_{12}Cl_6$ occurs at almost exactly the wavelength of the 520 nm plasma resonance of colloidal gold. This resonance is a collective oscillation of delocalised, freely mobile conduction electrons, induced by the radiation field at a frequency characteristic of the metal and the particle shape [7, 8]. We propose that the 510 nm feature represents a molecular plasma resonance in $Au_{55}(PPh_3)_{12}Cl_6$. Calculations of the absorption spectra of particles according to Mie theory [9] are valid down to particle radii of *ca.* 10 Å, roughly the size of the Au_{55} cluster, if corrections are applied for reduced electronic mean free path. The calculated and experimentally observed plasma resonance absorption of 10 Å gold particles is weaker than in larger colloids, but at the same frequency [9].

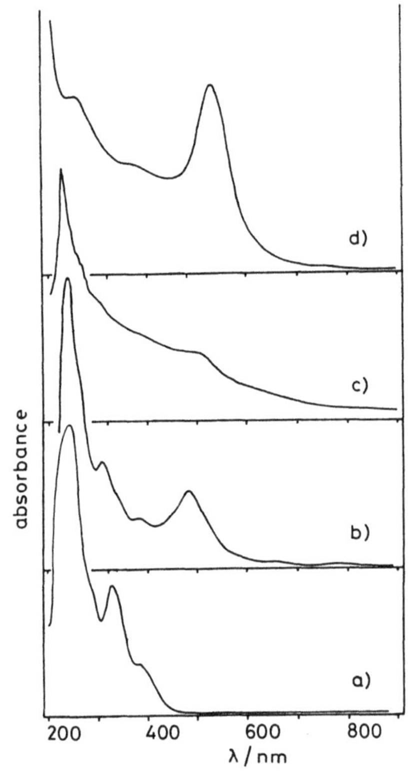

Fig. 1. UV-visible spectra (Shimadzu UV 240). *a)* $Os_3(CO)_{12}$ in CH_2Cl_2 solution *b)* $[Os_{10}C(CO)_{24}]^{2-}[N(PPh_3)_2]_2^+$ in CH_2Cl_2 solution *c)* $Au_{55}(PPh_3)_{12}Cl_6$ in CH_2Cl_2 solution *d)* Au colloid, 150–200 Å diameter, in aqueous solution

Thus the 510 nm feature has exactly the characteristics expected for the plasma resonance of a gold particle of 55 atoms.

A similar claim has been made [10] for a band at 495 nm which dominates the spectrum of the mixed-metal molecular cluster $Ag_{20}Au_{18}(PPh_3)_{12}Cl_{14}$. The much lower prominence of the band in this region of the $Au_{55}(PPh_3)_{12}Cl_6$ spectrum casts doubts on this assignment. The $Ag_{20}Au_{18}$ cluster contains fewer metal atoms, and is not a close-packed cluster but consists of three fused icosahedra [11]. It is unlikely that the plasma resonance absorption could dominate the spectrum of $Ag_{20}Au_{18}(PPh_3)_{12}Cl_{14}$ but be so weak in $Au_{55}(PPh_3)_{12}Cl_6$. Indeed, the overall appearance of the $Ag_{20}Au_{18}(PPh_3)_{12}Cl_{14}$ spectrum is similar to that of the lower-nuclearity cluster $[Os_{10}C(CO)_{24}]^{2-}$ (Fig. 1).

We are also critical of the use of Mie theory to calculate the plasma resonance absorption wavelength of the $Ag_{20}Au_{18}$ cluster as equal to that of a bimetallic silver-gold colloid of similar composition [10]. The plasma resonance frequency is sensitive to particle shape, reducing markedly with change from spherical to ellipsoidal shape by deformation or particle aggregation [8, 12, 13]. The $Ag_{20}Au_{18}$ cluster has the shape of three spheres aggregated into an oblate spheroid [11], and any plasma resonance absorption should occur at significantly longer wavelength than in a spherical alloy particle. The cuboctahedral shape of the Au_{55} cluster makes the calculation of its absorption wavelength more reliable.[1]

3. Synthetic relationships between gold colloids and molecular clusters

Molecular gold clusters are synthesised by reduction of $(PPh_3)AuCl$ in solution. Use of B_2H_6 as reducing agent complexes excess PPh_3 so that the stoichiometry $Au_{55}(PPh_3)_{12}Cl_6$ is attained [2, 3]. The reducing agent $NaBH_4$ yields instead smaller clusters with proportionately more phosphine ligands, such as $Au_{11}(PPh_3)_7Cl_3$ [16].

With no phosphines present, reduction of the gold salt $KAuCl_4$ in water or methanol with $NaBH_4$ at room temperature rapidly gives gold colloids of 100,000–250,000 atoms with a characteristic red colour (λ_{max}= 520 nm) [17]. If the milder reducing agent tri-sodium citrate is used, refluxing is necessary; in aqueous solution a shortlived intermediate blue colour is observed, and in methanol a more stable blue phase is obtained. We have demonstrated that these intermediate blue gold species are *not* analogous to $Au_{55}(PPh_3)_{12}Cl_6$. Their electronic spectra were quite different, showing strong absorption around 600 nm; and electron microscopy revealed large gold particles which were poorly shaped and not dense. The nature of the earliest stages of gold colloid nucleation remains unknown.

The synthesis of gold colloids in methanol was strongly influenced by addition of PPh_3 at different stages. Reduction of $KAuCl_4$ with $NaBH_4$ in the presence of PPh_3 gave a direct route to brown molecular $Au_{11}(PPh_3)_7Cl_3$. Addition of PPh_3 to the blue citrate-reduced colloid appeared to redissolve the colloid, decolourising the solution. However, nucleophiles such as PPh_3 or pyridine added to the stable red colloid caused aggregation of particles [13].

[1] Kreibig and co-workers believe that the 510 nm shoulder in the spectrum of $Au_{55}(PPh_3)_{12}Cl_6$ arises from cluster aggregates larger than 40 Å in diameter, and that the plasma resonance absorption in individual $Au_{55}(PPh_3)_{12}Cl_6$ molecules is too weak to be observed above the interband transition [14, 15]. If this is the case, our description of the 510 nm feature as a molecular plasma resonance is incorrect, but our doubts concerning the spectrum of $Ag_{20}Au_{18}(PPh_3)_{12}Cl_{14}$ are reinforced

It may prove possible to decompose $Au_{55}(PPh_3)_{12}Cl_6$ in solution in a controlled way into colloidal nuclei. Decomposition of $Au_{55}(PPh_3)_{12}Cl_6$ at an electrode gave a metallic phase composed of clusters of cuboctahedra [18]; this may be related to the non-dense intermediates formed during reductive colloid preparation.

4. Differential scanning calorimetry of $Au_{55}(PPh_3)_{12}Cl_6$

We have investigated the thermal decomposition of solid $Au_{55}(PPh_3)_{12}Cl_6$ (Fig. 2). The strong sharp exothermic decomposition at 156°C is important additional evidence for its identity as a stoichiometric molecular compound. The weaker endotherm at 201°C corresponds to the melting point of $(PPh_3)_2AuCl$, shown by NMR to be a decomposition product of $Au_{55}(PPh_3)_{12}Cl_6$ in solution [2, 3]. As no other thermal events were detected, decomposition in the solid state must follow the same route as in solution, to $(PPh_3)_2AuCl$ and metallic gold:

$$Au_{55}(PPh_3)_{12}Cl_6(s) \rightarrow 6(PPh_3)_2AuCl(s) + 49Au(s)$$

The decomposition enthalpy at 156 °C was 112 J/g, equal to 1590 kJ/mol of $Au_{55}(PPh_3)_{12}Cl_6$. We can estimate the relative bonding strengths of $Au_{55}(PPh_3)_{12}Cl_6$ and bulk metallic gold. In $Au_{55}(PPh_3)_{12}Cl_6$ there are 216 nearest-neighbour Au-Au interactions. After decomposition, there are 294; the number of other bonds stays the same, with 12 P-Au and 6 Cl-Au coordinate bonds on both sides of the equation. Assume first that these other bonds remain unchanged in strength. The decomposition exotherm represents the additional stability of 49 atoms of bulk gold compared to 55 gold atoms in the molecular cluster.

However, the picture changes if we calculate the binding energy per 2-centre Au-Au interaction. The atomisation energy of 49 moles of bulk gold is 18032 kJ; the total Au_{55} cluster binding energy is therefore 16442 kJ/mol of cluster. The resulting enthalpies per mole of nearest-neighbour Au-Au interactions are 61.3 kJ for bulk gold and 76.1 kJ for the Au_{55} cluster.

Although this calculation is clearly oversimplified, it indicates that the Au-Au interactions in $Au_{55}(PPh_3)_{12}Cl_6$ are substantially stronger than in bulk gold. The driving force for decomposition of the molecular cluster comes from the lower average coordination number of its metal atoms.

This conclusion is unaffected by possible changes in the other bond enthalpies on cluster decomposition. Bond strengths within the PPh_3 ligands are certain to be almost constant. With so few P-Au and Cl-Au bonds involved, even substantial changes in their enthalpies would only slightly alter the calculated Au-Au bonding strength in $Au_{55}(PPh_3)_{12}Cl_6$.

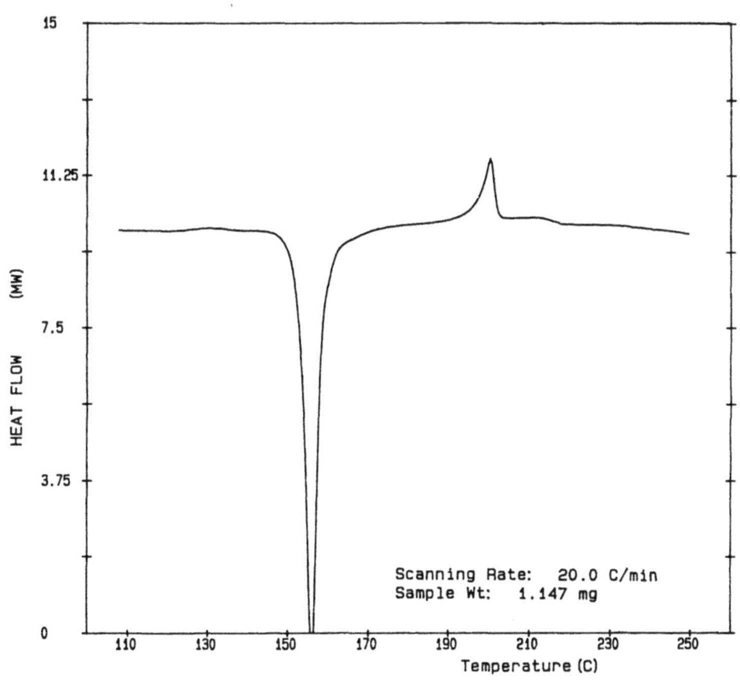

Fig. 2. Differential scanning calorimetry trace (Perkin Elmer DSC-7) of solid $Au_{55}(PPh_3)_{12}Cl_6$

Au-Au distances in the strongly bonded $Au_{55}(PPh_3)_{12}Cl_6$ have not been measured crystallographically, but seem certain to be on average shorter than the 2.884 Å of bulk gold. In lower-nuclearity gold clusters with two shells of metal atoms, Au-Au distances range between 2.6 Å and 3.2 Å; bonds on the cluster periphery are longer than those to interstitial atoms [19]. It remains to be seen whether $Au_{55}(PPh_3)_{12}Cl_6$, with three shells of gold atoms, follows a similar pattern.

5. Conclusion

The electronic spectrum of $Au_{55}(PPh_3)_{12}Cl_6$ differs from those of lower-nuclearity series of molecular clusters, more closely resembling that of colloidal gold. Its bonding appears stronger than in bulk gold. There are important implications for the electronic structure of the cluster. Evidence is mounting that $Au_{55}(PPh_3)_{12}Cl_6$ should not be regarded as merely a large cluster molecule, but more as a "molecular colloid" of gold.

This work was funded by the Royal Society, S.E.R.C. and Johnson Matthey. We also thank the Commission of the European Communities for an award under the "Stimulation Action-Laboratory Twinning" programme.

Note added in proof

We have now used EXAFS to measure the Au–Au distance in $Au_{55}(PPh_3)_{12}Cl_6$, it is 2.77 ± 0.02 Å

References

1. Benfield, R.E.: J. Phys. Chem. **91**, 2712 (1987)
2. Schmid, G., Pfeil, R., Boese, R., Bandermann, F., Meyer, S., Calis, G.H.M., van der Velden, J.W.A.: Chem. Ber. **114**, 3634 (1981)
3. Schmid, G.: Struct. Bonding (Berlin) **62**, 51 (1985)
4. van Staveren, M.P.J., Brom, H.B., de Jongh, L.J., Schmid, G.: Solid State Commun. **60**, 319 (1986)
5. Rao, C.N.R.: Ultra-violet and visible spectroscopy. 3rd Edn., Chap. 5. London: Butterworths 1975
6. Lässer, R., Smith, N.V., Benbow, R.L.: Phys. Rev. B **24**, 1895 (1981)
7. Kerker, M.: The scattering of light and other electromagnetic radiation. New York: Academic Press 1969
8. Eadon, D.G.: The visible absorption spectra of colloidal metal dispersions. Ph.D. thesis, University of Kent at Canterbury 1988
9. Kreibig, U.: J. Phys. (Paris) **38**, C2-97 (1977)
10. Teo, B.K., Keating, K., Kao, Y-H.: J. Am. Chem. Soc. **109**, 3494 (1987)
11. Teo, B.K., Hong, M., Zhang, H., Huang, D., Shi, X.: J. Chem. Soc. Chem. Commun. 204 (1988)
12. Papavassiliou, G.C.: Progr. Solid State Chem. **12**, 185 (1980)
13. Blatchford, C.G., Campbell, J.R., Creighton, J.A.: Surf. Sci. **120**, 435 (1982)
14. Fauth, K., Kreibig, U., Schmid, G.: Z. Phys. D. Atoms, Molecules and Clusters 12 (1989)
15. Kreibig, U., Genzel, L.: Surf. Sci. **156**, 678 (1985)
16. Cariati, F., Naldini, L.: Inorg. Chim. Acta **5**, 172 (1971)
17. Turkevich, J., Stevenson, P.C., Hillier, J.: Disc. Faraday Soc. **11**, 55 (1951)
18. Schmid, G., Klein, N.: Angew. Chem. **25**, 922 (1986)
19. Hall, K.P., Mingos, D.M.P.: Progr. Inorg. Chem. **32**, 237 (1984)

Z. Phys. D – Atoms, Molecules and Clusters 12, 537–541 (1989)

Atoms, Molecules
Zeitschrift für Physik D and Clusters
© Springer-Verlag 1989

Reactivity of free clusters*

S.J. Riley
Chemistry Division, Argonne National Laboratory, Argonne, IL 60439 USA

Received 6 July 1988

Recent studies of the reactions of free or isolated transition metal clusters with simple molecules will be reviewed. Cluster chemical reactions are carried out in a laser-vaporization cluster source coupled to a continuous-flow reactor. Reactions can be categorized as chemisorptive (or surface) reactions, and bulk reactions. Chemisorptive reactions can be characterized as *facile*, in which most cluster sizes of a given metal are equally reactive towards a particular reagent, or *activated*, in which strong dependence of reactivity on cluster size is found. Under the normal operating conditions of the cluster reactor, a reaction may be kinetically controlled or at equilibrium. Following chemisorption, adsorbate decomposition and product desorption may occur. Specific reactions to be discussed include the reactions of iron clusters with ammonia and with water and the reactions of nickel clusters with ammonia.

PACS: 82.30; 36.40

I. Introduction

Free or isolated clusters provide a fascinating arena for the study of fundamental chemical processes. In the case of metal clusters, numerous studies in recent years have shown that bare clusters often have unique chemical properties, and that these properties can change dramatically with cluster size [1]. The challenge of current and future studies of cluster chemistry is to understand these properties, to correlate them with electronic and geometrical structure, and to devise experimental methods to probe the sequence of steps that comprise the breaking and making of chemical bonds on cluster surfaces.

In a broad sense, the chemical reactions of isolated metal clusters can be categorized as either surface chemisorption reactions or bulk reactions. In the former, reactions involve the formation of surface-adsorbate bonds (either chemical or physical), and are generally characterized by curvature in a composition plot – the plot of the total number of adsorbed molecules at saturation versus cluster size [2]. From such data it is apparent that many clusters can be characterized as being "globular" in shape, that is, roughly spherical. Most of the detailed studies of the chemistry of isolated clusters have focused on reactions of the chemisorptive type. These studies include measurements of cluster reaction kinetics, thermodynamics and saturated product compositions.

Bulk reactions are characterized by a more extensive breakage of metal-metal bonds and the formation of a bulk compound having strong metal-nonmetal bonds. Early examples of this are the reactions of oxygen with clusters of such transition metals as iron, nickel, chromium and vanadium, reactions that lead essentially to metal oxide clusters [3]. In these reactions, a composition plot will approach linearity for larger clusters, with the slope giving the stoichiometry of the metal compound that is formed.

Chemisorption reactions of transition metal clusters are of particular interest, as they mimic and perhaps model the many studies of the chemistry of

* Work performed under the auspices of the Office of Basic Energy Sciences, Division of Chemical Sciences, U.S. Department of Energy, under Contract W-31-109-Eng-38

transition metal surfaces. Often the interaction between the adsorbate and the metal is fairly localized, so that only a few metal atoms in the vicinity of the adsorbate are involved in the interaction or are affected by it. Modeling surface or bulk metal properties using an ensemble of a few metal atoms is essentially modeling a cluster, so experimental data on cluster chemistry can have a direct input to theoretical descriptions of metal systems. Of course, clusters of transition metals are also obvious models for heterogeneous catalysts.

In the few years that the chemistry of isolated transition metal clusters has been studied, a few general trends have been noted. Chemisorption reactions generally fall into two categories: *activated* and *facile*. In the former, there is often a strong dependence of cluster reactivity on cluster size, at least for smaller clusters. There is usually some rate limiting step, often the breakage of a bond within the adsorbate molecule, that is very sensitive to cluster electronic properties and size. The classic example of such a reaction is that between iron clusters and hydrogen, in which orders of magnitude differences in rate constant are found over a small range in cluster size [4]. In facile reactions, all cluster sizes for a given metal-reactant system are generally very reactive. These reactions are often quite exothermic and/or represent cases where the adsorbate binds nondissociatively, so that adsorbate bond breakage is not required for reaction. The principal example of such reactions that will be discussed here is the reaction of iron clusters with ammonia.

In reactions in which adsorbate bonds are broken, there is the possibility of additional chemical steps that may lead to desorption of constituent species. The first reported cases of such reactions were the dehydrogenation of cyclohexane on platinum clusters [5] and, more recently, the dehydrogenation of benzene on niobium clusters [6]. We have seen similar processes in the reactions of water with iron clusters [7], and in the reactions of ethylene and ammonia with nickel and platinum clusters. In each case, loss of hydrogen from the clusters is seen. An understanding of the detailed sequence of reaction steps in such processes will help us understand the way metal surfaces can promote and affect chemical transformations.

II. Experimental

Numerous descriptions of the techniques for making and studying the chemical reactions of isolated transition metal clusters have been given, so a detailed description will not be given here. Briefly, in our experiments clusters are formed by pulsed laser vaporization of a metal target rod located in a continuous-flow reactor through which inert He carrier gas is flowed. If possible, isotopically enriched targets are used. Reagents enter the reactor at a point downstream of the target where cluster growth has terminated and the clusters have thermalized to the temperature of the carrier gas. Clusters and their reaction products expand out of a nozzle at the end of the flow-tube reactor (FTR) and are formed into a molecular beam. After passing through several stages of differential pumping, the clusters are detected in a laser-ionization time-of-flight (TOF) mass spectrometer. The general experimental procedure is to monitor changes in the intensities of peaks in the TOF spectrum as FTR conditions are changed. The most common condition to vary is the partial pressure of the reagent gas.

An important feature of our current FTR design is the ability to add reagent gas at different points along the flow tube. For the FTR arrangement used in the studies described here, there are two reagent addition ports that provide cluster – reagent interaction times of 1.1 ms and 0.23 ms for hydrogen addition. The two-port arrangement allows an independent variation of the interaction time and the reagent pressure. This is crucial for determining whether a given reaction is kinetically controlled or at equilibrium. If the distribution of products is independent of interaction time for a given reagent pressure, then the reactants are at equilibrium in the FTR. If, on the other hand, increasing the interaction time at constant pressure leads to products with more adsorbate molecules on them, then the reaction is kinetically controlled.

III. Examples of the chemical reactivity of free clusters

A. The reactions of iron clusters with ammonia

The iron cluster – ammonia reaction is fairly simple and straightforward – chemisorption is nondissociative – and the iron clusters' IPs decrease with increasing ammonia coverage. This assures us that we can photoionize all possible species produced in the reaction, so that we can follow the uptake of ammonia all the way from the first to the last chemisorption step. To quantify the ammonia uptake, we calculate the average number of adsorbates, $\bar{m}$, at a given ammonia partial pressure in the FTR [8]. A plot of $\bar{m}$, versus pressure for Fe_{55} is shown in Fig. 1. Note that the (logarithmic) ammonia pressure scale spans almost five orders of magnitude. At pressures below 3 μ, data is shown for different interaction times. That the data is distinctly different shows that the reaction is kinetically controlled for the first 12 or so ammonia molecules that bind to the cluster. Above 3 μ, the

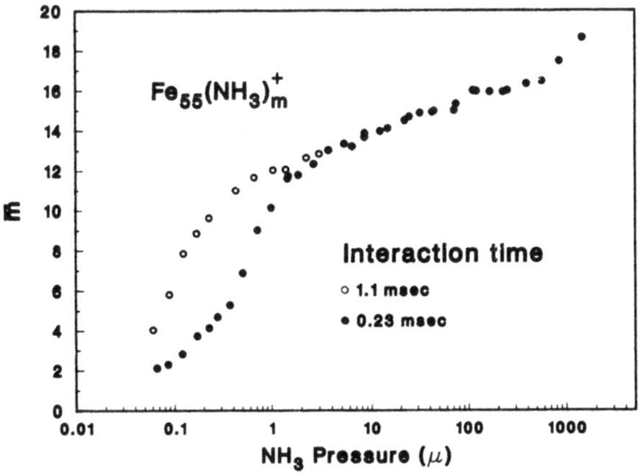

Fig. 1. Uptake plot for the reaction of Fe_{55} with ammonia. Above 3 μ pressure, only 0.23 msec data is shown

composition is independent of time, indicating equilibrium. In this region, there is a gradual, essentially linear, increase in coverage with log $P(NH_3)$ from 3 μ to a little over 100 μ. The data in this region can be analyzed to determine the thermodynamics of chemisorption, i.e., ammonia binding energies. At higher pressures, the uptake plot in Fig. 1 shows a short plateau followed by a further rise in $\bar{m}$. In general, a plateau in an $\bar{m}$ plot implies a large reduction in binding energy for the *next* adsorbate molecule to add to the cluster [8b]. In this case, we interpret the plateau as the filling of all of the *chemisorptive* sites on the Fe_{55} cluster. Thus, as can be seen in Fig. 1, Fe_{55} can accommodate 16 chemisorbed ammonias. The final rise at higher pressures we interpret as the onset of *physisorption.* At these very high ammonia partial pressures there is substantial cooling of the clusters in the expansion out of the nozzle. This, coupled with the increased cluster-ammonia collisions in the expansion, lead to the beginning of the formation of what is essentially a second, physisorbed layer.

Data similar to that in Fig. 1 has been collected for iron clusters from Fe_2 to Fe_{85}. In general, the results show similar features: a rapid rise in the kinetically controlled region, a more gradual increase in the equilibrium region, and a saturation plateau of varying extent. Additional plateaus at lower pressures are seen for a few clusters, and can be used to infer something about cluster structure and the nature of the adsorption sites. For example, for Fe_{13}, we see a dramatic plateau at $Fe_{13}(NH_3)_4$ that extends for an order of magnitude change in ammonia pressure [9]. At higher pressures, another plateau is seen at $Fe_{13}(NH_3)_6$. This suggests that Fe_{13} has four equivalent sites that strongly bind ammonia, and two more that are more weakly binding. If we adopt the usual idea that ammonia binds most strongly to atop sites [10], both the fcc and bcc structures become good candidates for the structure of Fe_{13}. In a few cases, such as Fe_{15}, the uptake curve is essentially featureless, with no plateaus or large changes in slope evident [8b]. This we interpret as an indication of a cluster having all chemisorption sites equivalent. For a 15-atom cluster this is nicely satisfied by bcc packing. Thus we see that purely chemical studies can tell us something about cluster structure.

B. The reactions of iron clusters with water

The chemisorption of water on iron clusters can lead to adsorbate decomposition and H_2 desorption, but the process shows a dramatic isotope effect. H_2O undergoes decomposition and loss of H_2 from the clusters, while no D_2 loss is seen for D_2O. We will consider the D_2O reaction in which there is no desorption, because it illustrates one important experimental complication in studies of this type, and what we can do to account for it.

The reaction between iron clusters and D_2O under the normal FTR conditions is an equilibrium one. This is shown by product spectra that are independent of reagent interaction time at constant reagent pressure. For a series of equilibria such as

$$Fe_n + D_2O \rightleftharpoons Fe_n(D_2O);$$
$$Fe_n(D_2O) + D_2O \rightleftharpoons Fe_n(D_2O)_2; \text{ etc.}$$

we expect there to be an equilibrium relationship between the densities of the species in the FTR. For example, the ratio $[Fe_n(D_2O)]/[Fe]$ should equal $K[D_2O]$, where K is the equilibrium constant for the first addition reaction. If we now assume that ion intensities I_m in the mass spectra are proportional to the neutral densities $[Fe_n(D_2O)_m]$ in the FTR, then we would expect a similar relationship between integrated ion signals and the D_2O pressure in the FTR. In fact, as is shown in Fig. 2 for Fe_{20}, a plot of I_1/I_0 vs D_2O pressure is not linear, but shows deviation at higher pressures. Interestingly, if the second ratio, I_2/I_1, is plotted vs D_2O pressure, the curve is linear.

The reason for this discrepancy is that some fraction of the clusters absorb a second photon in the photoionization step. The excess energy thus deposited in the clusters causes the desorption of adsorbates. For iron clusters that have just a few (perhaps no more than three) D_2O molecules on them, this energy will be sufficient to desorb all of the molecules, yielding bare cluster ion signal. As the D_2O pressure increases, relatively more ion intensity appears as bare cluster signal (I_0), and the I_1/I_0 ratio deviates below linearity. Since the absorption of the second photon causes the desorption of all D_2O molecules, the I_2/I_1 ratio will

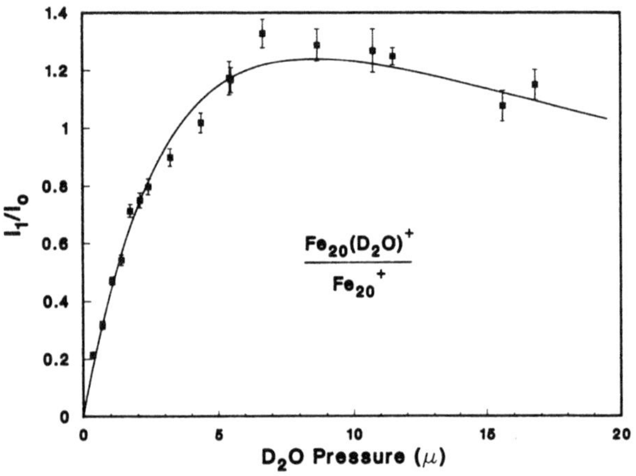

Fig. 2. Plot of the ratio of $Fe_{20}(D_2O)^+$ ion signal to $Fe_{20}{}^+$ ion signal as a function of D_2O pressure in the FTR. Error bars are determined from a statistical analysis of the data. The smooth curve is calculated by modeling the two-photon absorption process that leads to the curvature of the data

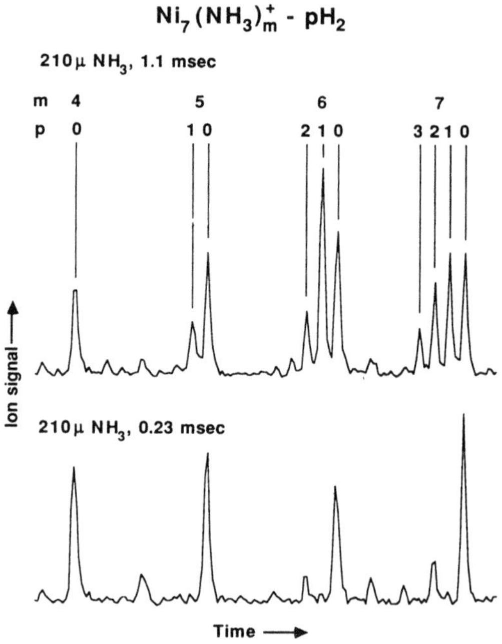

Fig. 3. The Ni_7 mass spectra recorded for reaction with ammonia at the partial pressures indicated and for the reaction times shown. In the upper spectrum the number of H_2 molecules desorbed from each product is indicated

remain linear because both the second and first products are fractionally depleted to roughly the same extent.

A coupled kinetic model has been developed to put this picture on a more quantitative basis. The details and results will appear in a future publication [7]. The prediction of the modeling procedure for the $Fe_{20}(D_2O)/Fe_{20}$ case is shown by the smooth curve in Fig. 2. One result of this procedure is that an estimation of the probability for two-photon absorption can be made, a quantity not always easy to measure. We find that at moderate ionizing laser fluences (~ 1.5 mJ/cm^2) about 20–25% of iron clusters absorb a second photon. The assumptions that have to be made to relate observed ion intensities to actual neutral cluster distributions when multiphoton processes are involved have been considered [11].

C. *The reactions of nickel clusters with ammonia*

Unlike the situation for iron clusters, ammonia dissociatively chemisorbs on nickel clusters. Furthermore, this ammonia decomposition is accompanied by loss of hydrogen molecules from the clusters. This is illustrated for Ni_7 in Fig. 3, which shows in the upper panel a portion of the TOF mass spectrum when ammonia at a pressure of 210 μ is allowed to interact with the clusters for 1.1 msec. As can be seen, the four principal species seen for Ni_7 at this pressure have four, five, six and seven ammonia molecules bound to them. However, the $Ni_7(NH_3)_7$ species also has mass peaks corresponding to loss of one, two and three H_2 molecules. The product that began with six ammonias loses one and two hydrogen molecules, that with five loses only one, and the four-ammonia species loses no hydrogens. We will return to these systematics below.

The initial reaction of the nickel clusters with ammonia appears to be facile, in that all clusters are very reactive. Uptake experiments suggest that at pressures in the 200 μ range the clusters are nearly saturated with ammonia and the reaction is at equilibrium. Yet there is still a kinetically controlled step in the overall process that leads to hydrogen desorption. This is illustrated in the lower spectrum in Fig. 3, which shows the result when the cluster – ammonia interaction time is reduced to 0.23 ms. In this case there is no hydrogen loss for any of the product species. (The smaller peaks in the spectrum are for the most part ammoniated species having nitrogen impurities.) It is also evident from this spectrum that the average number of "ammonias" on Ni_7 is approximately the same as in the upper spectrum, which satisfies our criterion for the chemisorption reaction being at equilibrium, i.e., independent of interaction time.

We can postulate the sequence of steps that lead to hydrogen loss by looking closely at the systematics shown in the top spectrum of Fig. 3. We assume that the final form of the nitrogen-containing species on the cluster is NH, so that each ammonia contributes two hydrogens to the cluster surface. The maximum hydrogen loss for $Ni_7(NH_3)_7$ is three H_2 molecules, leaving eight hydrogen atoms on the cluster. For $Ni_7(NH_3)_6$, two molecules are lost via desorption, again leaving eight. Similarly, for the other Ni_7 species, by this scenario the number of hydrogen atoms left on the cluster is eight. This is approximately the maximum number of hydrogens that a bare Ni_7 cluster can bind, as determined by H_2 uptake experiments, so eight represents a count of the number of sites on the cluster that can bind the hydrogens resulting from ammonia decomposition. Similar numerology is found for clusters in the Ni_4 to Ni_{22} size range.

IV. Conclusion

Transition metal clusters show a wide range of chemical behavior. In recent years we have been able to develop experimental techniques to probe the relatively simple nondissociative chemisorption reactions and the straightforward dissociative ones. Now we have examples of reactions in which adsorbate bonds are broken and new ones are formed on cluster surfaces, resulting in desorption of hydrogen. The new challenge we face is to develop similar techniques to probe these more complicated reactions. Since these reactions represent fundamental chemical processes that are promoted by the interactions between adsorbates and metal atoms, an understanding of their detailed sequence of steps will provide us with valuable information about the unique chemical properties of metal surfaces.

We thank Drs. B.H. Weiller and P.S. Bechthold for the collection and analysis of the Fe_n+D_2O data. and Dr. W.F. Hoffman III for the collection and analysis of the Ni_n+NH_3 data. We thank Dr. E.K. Parks for his work on the Fe_n+NH_3 system, and for constructive reading of this manuscript.

References

1. For a recent review, see Kaldor, A., Cox, D.M., Zakin, M.R.: In: Advances in chemical physics. Prigogine, I., Rice, S.A. (eds.) Vol. 70, Part 2, p. 111. New York: Wiley 1988
2. Parks, E.K., Liu, K., Richtsmeier, S.C., Pobo, L.G., Riley, S.J.: J. Chem. Phys. **82**, 5470 (1985)
3. Nieman, G.C., Parks, E.K., Richtsmeier, S.C., Liu, K., Pobo, L.G., Riley, S.J.: High Temp. Sci. **22**, 115 (1986)
4. Richtsmeier, S.C., Parks, E.K., Liu, K., Pobo, L.G., Riley, S.J.: J. Chem. Phys. **82**, 3659 (1985); Whetten, R.L., Cox, D.M., Trevor, D.J., Kaldor, A.: Phys. Rev. Lett. **54**, 1494 (1985)
5. Trevor, D.J., Whetten, R.L., Cox, D.M., Kaldor, A.: J. Am. Chem. Soc. **107**, 518 (1985)
6. St. Pierre, R.J., El-Sayed, M.A.: J. Phys. Chem. **91**, 763 (1987); St. Pierre, R.J., Chronister, E.L., El-Sayed, M.A.: J. Phys. Chem. **91**, 5228 (1987); Zakin, M.R., Cox, D.M., Kaldor, A.: J. Phys. Chem. **91**, 5224 (1987)
7. Weiller, B.H., Bechthold, P.S., Parks, E.K., Pobo, L.G., Riley, S.J.: (to be published)
8. a) Parks, E.K., Nieman, G.C., Pobo, L.G., Riley, S.J.: J. Chem. Phys. **86**, 1066 (1987); b) Parks, E. K., Weiller, B.H., Bechthold, P.S., Hoffman, W.F., Nieman, G.C., Pobo, L.G., Riley, S.J.: J. Chem. Phys. **88**, 1622 (1988)
9. Parks, E.K., Nieman, G.C., Pobo, L.G., Riley, S.J.: J. Chem. Phys. **88**, 6260 (1988)
10. Erley, W., Ibach, H.: J. Electron. Spec. Rel. Phenom. **31**, 161 (1983)
11. Bechthold, P.S.: unpublished results

Z. Phys. D – Atoms, Molecules and Clusters 12, 543–546 (1989)

Atoms, Molecules
Zeitschrift für Physik D and Clusters
© Springer-Verlag 1989

Cluster-model study of the interaction of halogen atoms with Ag clusters

G. Pacchioni*, P.S. Bagus, and M.R. Philpott
IBM Research, Almaden Research Center, 650 Harry Road, San Jose, CA 95120-6099, USA

Received 4 July 1988; final version 30 September 1988

Analysis of ab initio wave functions shows that the interaction between halogen atoms (F, Cl, and Br) and Ag clusters is ionic, and that the halogen ionicity is essentially −1. The interaction and the bonding arise, almost entirely, from the Coulomb attraction between the charged halogen and the metal and from polarization of the two subunits. Large shifts in equilibrium bond distances and vibrational energies are caused by an external electric field. These changes arise from a dominant Stark effect.

PACS: 31.20; 36.40

The nature of the bonding between halogen atoms and silver is a subject of considerable interest in surface science and electrochemistry [1–4], and the nature of the bonding interaction, if more ionic or more covalent, is still a matter of controversy.

We studied the interaction between F, Cl, and Br atoms and an Ag_4 cluster modeling the three-hollow site of the Ag (111) surface using ab initio Hartree–Fock wavefunctions. Attention has been focused on the character of the bonding and on the importance of the polarization effects for the dipole moment changes resulting from the bond formation. In addition we have considered the effect of an external electric field on the metal-adsorbate properties.

A major concern in adsorbate cluster-model studies is the dependence of the properties on the number of substrate atoms in the cluster. However, the analysis of the interaction of F atoms with Ag_6, Ag_7, and Ag_{10} clusters confirms the basic features of the bonding derived for Ag_4 [5]. Here we discuss only the results for Ag_4, but we believe that our conclusions about the character of the bonding will be valid for more general cluster geometries.

The molecular wave functions have been determined using flexible basis sets of contracted Gaussian type orbitals. In order to facilitate the calculations, the core electrons of Ag have been replaced by Effective Core Potentials [6] including explicitly in the valence only the $4s^2\ 4p^6\ 4d^{10}\ 5s^1$ electrons. The valence basis set for Ag is contracted to [4s4p3d]. The halogen atoms have been treated at all electron level and the contracted basis sets used are [5s4p1d], [5s4p2d], and [9s7p3d] for F, Cl, and Br, respectively [7]. The basis sets optimized for the neutral atoms [7] were extended with diffuse *s*, *p*, and *d* functions whose exponents have been determined in order to obtain a better approximation for the SCF electron affinity (EA), and for the dipole polarizability, α_D, of the respective anions (Table 1). The computed EAs are close to the near Hartree–Fock limit values (Table 1) and the differences with the experimental data are almost entirely due to the lack of correlation effects.

The halogen atoms to surface distance in Ag_n-X has been varied along the C_{3v} axis: the distance of *X* from the surface plane (*r*), the dissociation energy (D_e), the vibrational frequency (ω_e) and the dipole moment (μ) have been determined (Table 1).

For distances close to the equilibrium Ag-halogen distance, the ground state configuration is $(1a_1)^2\ (2a_1)^2\ (3a_1)^2\ (1e)^4\ (2e)^1\ (^2E)$; the halogen cores as well as the Ag $1s^2$ to $4d^{10}$ cores are not explicitly given. This state arises from the combination of the Ag_4 ground state $(a_1)^2\ (e)^2$ with the seven valence *ns* and *np* electrons of

* Present address: Dipartimento di Chimica Inorganica e Metallorganica, Università di Milano, 20133 Milano, Italy

Table 1. Properties of Ag_4-X clusters

		Ag_4-F	Ag_4-Cl	Ag_4-Br
EA (X)	eV SCF	1.35	2.08	2 52
	SCF[a]	1.37	2.58	2.58
	exp	3.40	3.62	3.36
α_D	Å^3 SCF	1.35	3.77	6.19
	HFL[b]	1.71	5.89	8 18
r (Ag-X)	au	3.310	4.347	4.721
ω_e	cm^{-1}	277	180	115
D_e (Ag_4+X) eV		2.72	2.68	2.76
$\mu(r)$ (Ag_4−X) au				
M_0		−1.09	−1.72	−2.00
M_1		−1.39	−1.60	−1.59
M_2		−0.09	−0.04	−0.01
M_3		0.01	0.03	−0.03
$\mu(r)$ (X) au				
M_0		−2.90	−3.31	−3.43
M_1		−1.00	−1.10	−1.11
M_2		0.10	−0.03	−0.04
M_3		−0.01	0.04	−0.01

[a] Ref. 14
[b] HFL = Hartree-Fock limit [15]

X in configuration $(a_1)^2$ $(a_1)^2$ $(e)^3$. The main bonding effects are in the outer e shells and can be either covalent or ionic. In this latter case the bonding can be considered as originating from the interaction of the ionized Ag_4^+ $(a_1)^2$ $(e)^1$ cluster and the closed shell anion X^- $(a_1)^2$ $(a_1)^2$ $(e)^4$. In the covalent view, a filled bonding $(1e)^4$ and a singly occupied antibonding $(2e)^1$ MOs are formed.

To determine the character of the Ag-halogen bonding as covalent or ionic we apply two measures. First, we make use of a projection operator [8] for the occupied orbitals of the free halogen anion X^-. The expectation value of the projection operator for a given spatial orbital of X^-, $\varphi_{nlm}\varphi^+_{nlm}$, is a measure of the extent to which φ_{nlm} is occupied in the total cluster wavefunction, $\Psi(Ag_nX)$. In the case that this projection gives a value close to 2, it is clear that φ_{nlm} of X^- is completely occupied in $\Psi(Ag_nX)$ [8]. For all X and for a wide range of distances about the minimum of the Ag_4-X energy curve, the projection of all φ_{nlm} is about 2. Summing over all the orbitals of X, we find 9.98, 17.95, and 35.92 electrons for F, Cl, and Br, respectively in Ag_4-X; these values are for the halogen near the minimum. They provide very strong evidence for an ionic interaction where the silver cluster is Ag_4^+ and the halogen is X^-.

A second measure of the bonding nature is based on the slope of the dipole moment curve as the distance of X from the cluster is varied [8]. For an ionic bond, the absolute magnitude of the slope should be −1. In fact, the non-zero component of the dipole moment of an ionic molecule represented by two point charges $+q$ and $-q$ will be $\mu = -q\,r$ and $d\mu/dr = -q$. Therefore, for an ideal fully ionic molecule where $q = 1$, $d\mu/dr = -1$ and the curve is a straight line. Generally, on the other hand, a small slope is characteristic of a covalent bond. Of course in a real system the polarization of the subunits, in our case the halogen and the cluster, can lead to a substantial change in the absolute magnitude of the dipole moment. In our case, this polarization must be quite important because we find a slope greater than 1, see Table 1; it is impossible that the ionicity of the halogen is greater than 1.

The dipole moment curve as function of the distance has been determined by a cubic polynomial fitting:

$$\mu(r) = M_0 + M_1(r-r_e) + M_2(r-r_e)^2 + M_3(r-r_e)^3$$

For each value of r it is possible to transform the Ag_4X cluster orbitals into those which are most like X, φ_1(loc, X), and the orthogonal complement of the SCF canonical space. This corresponding orbital transformation [9, 10] provides the transformation with maximally localized orbitals. The contribution to the total dipole moment of φ_1(loc, X) to the total cluster dipole is denoted by μ(X). The coefficients of the expansions of $\mu(Ag_4X)$ and $\mu(Ag_4)$ are reported in Table 1. The dipole moment curve, for a considerable distance about the equilibrium Ag-halogen bond, is a straight line (small M_2 and M_3) and the contribution from the chemisorbed X^- halogen to the slope of the curve is almost exactly −1 thus supporting the view of an essentially ionic bond.

In order to understand the origin of the Ag_n-X interaction, we applied a constrained space-orbital variation CSOV [11]. In this way it is possible to determine the individual contributions to the bonding, i.e. electrostatic effects, intraunit polarization, charge transfer, and covalent bonding. We start, CSOV step 0, with the superposed Ag_4^+ and X^- charge distributions; this gives the frozen-orbital (FO) interaction. Given the strong evidence that the bonding is ionic, we analyze the effect of replacing the X^- ion with a unit point charge, PC = −1, placed at the position of the halogen nucleus. In this case step 0 gives the pure electrostatic attraction with no repulsive contribution originating from non-bonding overlap of X^- and Ag_4^+ charge distributions (Pauli repulsion). Then we allow the Ag_4^+ orbitals to vary because of the presence of the X anion for both the PC and the extended representation of X^- (CSOV step 1); this allows the Ag_4^+ to polarize in the presence of X^-. The next step (step 2) is the Ag_4^+ orbital variation in the full space, indicating the amount of charge transfer from Ag_4^+ to X^-. In step

3 the Ag_4^+ orbitals are fixed and the X^- orbitals are varied in their own space. This permits the X^- unit to polarize as a result of the presence of Ag_4^+. In CSOV step 4 this variation occurs in the full space, so that both X^- to Ag_4^+ donation and X^- polarization are possible. In step 5 the covalent mixing between the valence e orbitals, $(1e)^4$ (X^-) and $(2e)^1$ (Ag_4^+), is allowed; all other orbitals are fixed as determined in step 4. The result of step 5 is compared to the full unconstrained SCF result and if they are nearly the same this means that all important bonding contributions have been considered. Of course, steps 2–6 have no meaning when X^- is replaced by a point charge. For each step, we consider the interaction energy E_{int}, with respect to separated Ag_4^+ and X^- fragments, and the dipole moment μ. In Table 2 the results of this analysis are reported for Ag_4-F at $r_e = 3.2$ near equilibrium bond distance. Clearly, the most important contribution comes from the electrostatic interaction. The difference between the PC and F^- values, 1.5 eV, represents the Pauli repulsion contribution. The electrostatic attraction is followed by the cluster polarization, about 2 eV, which reduces the repulsive contribution to the bonding. These two contributions alone account for about 90% of the whole binding energy. The other CSOV steps, X^- polarization and covalency, only slightly improve E_{int} (Table 2). Similar data have been obtained for Ag_4Cl and Ag_4Br [5]. The main difference is that, because of the larger equilibrium bond distances, the electrostatic interaction and the cluster polarization are smaller than in Ag_4F. On the other hand, the contribution from the Cl^- and Br^- polarization is higher, as expected on the basis of the larger dipole polarizabilities of these anions (Table 1).

Table 2. CSOV analysis for Ag_4-F and Ag_4-PC (PC = point charge)[a]

Step		E_{int}	ΔE_{int}	μ	$\Delta\mu$
0. Frozen orbital	PC	+4.82	. .	−4.49	. . .
	F^-	+3.37	. .	−4.21	. . .
1 Ag_4^+ polarization (Ag_4 basis)	PC	+6.92	+2.10	−1.52	+2.97
	F^-	+5.25	+1.88	−1.37	+2.84
2 Ag_4^+ polarization (full basis)	F^-	+5.35	+0.10	−1.43	−0.06
3. X^- polarization (X basis)	F^-	+5.59	+0.24	−1.15	+0.28
4. X^- polarization (full basis)	F^-	+5.84	+0.25	−1.05	+0.10
5 Covalent e sym interaction	F^-	+5.85	+0.01	−1.04	+0.01
6. Full SCF	F^-	+5.88	+0.03	−0.94	+0.10

[a] Interaction energies E_{int} in eV are computed with respect to separate ions; μ in au

The ionic nature of the silver-halogen interaction suggests that an external electric field F will produce large changes on the chemisorption properties. For a cluster in a uniform electric field F, the SCF first order perturbation theory energy, $E_p(F)$, is obtained as the difference between the $F = 0$ SCF energy and $\mu \times F$, where μ is the field-free dipole moment [12, 13]. This is a pure Stark effect and does not include chemical changes caused by the field. These electronic effects are explicitly considered in the SCF variational energy in the presence of the field, $E_{SCF}(F)$, obtained by addition of the term $(\Sigma r_i - \Sigma Q_N R_N) \times F$ to the field-free hamiltonian; i denotes electrons, N nuclei, and Q_N the nuclear charges [12, 13].

The fields considered are $+-0.01$ a.u. $= +-5.7 \times 10^7$ V/cm. The sign of F is such that $F < 0$ attracts electrons from the surface towards the ligand. Indeed, we observe that for $F = -0.01$ the X^- ions are pulled away from the surface giving a large increase in r; similarly, $F = +0.01$ pushes the ion toward the surface (Table 3). Consequently, the vibrational energies decrease for $F = -0.01$ and considerably increase for $F = +0.01$ (Table 4). This occurs because the adsorbed ligand is stretched against the wall represented by surface charge density. The changes caused by the external field are dominantly stark effects, while the chemical changes make only minor contributions (Tables 3 and 4).

In conclusion, we have shown that the bonding of F, Cl, and Br with Ag clusters is almost entirely ionic.

Table 3. Equilibrium bond distances in bohrs for halogens chemisorbed on Ag_4. When a field F = + −0.01 au is applied both SCF and Stark values are given. Δr are also given

F		Ag_4-F	Ag_4-Cl	Ag_4-Br
0.00	SCF	3.310/ .	4.347/ . . .	4.721/ . . .
+0.01	Stark	3.098/−0.21	4.055/−0.29	4.405/−0.32
	SCF	3.092/−0.22	4.058/−0.29	4.418/−0.30
−0.01	Stark	3.637/+0.33	4.893/+0.55	5.375/+0.65
	SCF	3.620/+0.31	4.904/+0.56	5.410/+0.69

Table 4. Vibrational frequencies in cm^{-1} for halogens chemisorbed on Ag_4. When a field F = + −0.01 au is applied, both SCF and Stark values are given. $\Delta\omega$ are also given

F		Ag_4-F	Ag_4-Cl	Ag_4-Br
0.00	SCF	277/ . .	180/ . . .	115/ . . .
+0.01	Stark	322/+45	222/+42	144/+29
	SCF	323/+46	222/+42	144/+29
−0.01	Stark	213/−64	118/−62	66/−49
	SCF	215/−62	116/−64	64/−51

We believe this is true also for surfaces in particular considering that the SCF wave function underestimates the electron affinities of the atoms and that a silver surface is more polarizable than a small cluster.

One of us (G.P.) is grateful to IBM Italy for partially supporting his stay at the IBM Almaden Research Center. This work has been supported, in part, by the Office of Naval Research.

References

1. Lamble, G.M., Brooks, R.S., Ferrer, S., King, D.A., Norman, D.: Phys. Rev. B **34**, 2975 (1986)
2. Goddard, P.J., Schwaha, K., Lambert, R.M.: Surf. Sci. **71**, 351 (1978)
3. Citrin, P.H., Eisenberger. P., Hervitt, R.C.: Phys. Rev. Lett. **41**, 309 (1978)
4. Wertheim, G.K., DiCenzo, S.B., Buchanan, D.N.E.: Phys. Rev. B **25**, 3020 (1982)
5. Bagus, P.S., Pacchioni, G., Philpott, M.R.: J. Chem. Phys. (1989) (in press)
6. Hay, P.J., Wadt, W.R.: J. Chem. Phys. **82**, 299 (1985)
7. van Duijeneveldt, F.B.: IBM Res. Rep. No RJ945 (1971) (unpublished)
8. Nelin, C.J., Bagus, P.S., Philpott, M.R.: J. Chem. Phys. **87**, 2170 (1987)
9. Seel, M., Bagus, P.S.: Phys. Rev. B **28**, 2023 (1983)
10. Agren, H., Bagus, P.S.: J. Am. Chem. Soc. **107**, 134 (1985)
11. Bagus, P.S., Hermann, K., Bauschlicher, C.W.: J. Chem. Phys **80**, 4378 (1984); **81**, 1966 (1984)
12. Bagus, P.S., Nelin, C.J., Muller, W., Philpott, M.R., Seki, H.: Phys. Rev. Lett. **58**, 59 (1987)
13. Bagus, P.S., Nelin, C.J., Hermann, K., Philpott, M.R.: Phys. Rev. B **36**, 8169 (1987)
14. Clementi, E., Roetti, C.: At. Data Nucl. Data Tables **14**, 177 (1974)
15. Fraga, S., Saxena, K.M., Lo, B.W.N.: At. Data Nucl. Data Tables **3**, 323 (1971)

Z. Phys. D – Atoms, Molecules and Clusters 12, 547–550 (1989)

Atoms, Molecules
Zeitschrift für Physik D and Clusters
© Springer-Verlag 1989

On the splitting of the CO 1π level in the chemisorption system CO/Ni(111): substrate or lateral interaction?

N. Rösch[1], P. Sandl[1], P. Knappe[1], A. Görling[1], and B.I. Dunlap[2]

[1] Lehrstuhl für Theoretische Chemie, Technische Universität München, D-8046 Garching, Federal Republic of Germany

[2] Theoretical Chemistry Section, Code 6119, Naval Research Laboratory, Washington, D.C. 20375, USA

Received 5 July 1988

LCGTO Xα model cluster calculations have been carried out to rationalize the shape of the CO 1π band of the chemisorption system CO/Ni(111) observed in angular resolved photoemission. The splitting induced by substrate interaction at twofold bridging sites (0.4 eV) is much smaller than the value deduced from experiment (1.2 eV). From calculations on $(CO)_n$ clusters lateral adsorbate interaction is estimated to cause a sizable broadening of the 1π band (1.2 eV, in satisfactory agreement with experiment), but essentially no splitting ($\lesssim$0.1 eV). Therefore, rehybridisation due to local interaction does not seem to suffice as an explanation for the observed shape of the CO 1π band.

PACS: 31.20.G

1. Introduction

Adsorption of carbon monoxide on various crystalline transition metal surfaces is probably the most extensively studied chemisorption problem, by experimentalists and theoreticians alike (for an extensive listing of relevant references see [1]). Nevertheless, ever increasing sophistication of experimental methods still furnish new information on this prototype system. Recently, investigations have focused on the coadsorption of carbon monoxide and alkali metals [2]. A better understanding of these systems is of great practical interest due to their relation to alkali promoted catalysts. Alkali coadsorption affects the surface bond as well as the CO bond, as indicated by changes of the adsorption site, of the adsorbate orientation, and of the photoelectron spectra and by a sometimes drastic lowering of the CO stretching frequency [2].

Several attempts have been made to provide a uniform description of the alkali metal-molecule coadsorbed complex [2] by alluding to a direct or substrate-mediated mechanism for alkali to CO charge transfer or to a direct ionic interaction between alkali metal and CO. Another approach is based on the idea of sp^2 rehybridisation [3] at the carbon and a concomitant reduction of the carbon oxygen binding, possibly after a change of adsorption site to a twofold bridge, as found in the system K + CO/Ru(001). In case of chemisorption at twofold bridging sites rehybridisation is expected to be correlated with the splitting of the CO 1π level due to local C_{2v} symmetry.

The chemisorption system $c(4\times2)$CO/Ni(111) features carbon monoxide adsorbed perpendicularly at twofold bridging sites and thus provides, without coadsorbate, an even simpler case for the study of rehybridisation [4]. Available information for this system includes rather extensive data from angular resolved photoemission experiments. Via a suitable choice of experimental geometry a rather broad band ($\sim$2 eV) below the nickel d band could be unambiguously identified as emission from the CO 1π level. This band has been interpreted as superposition of two CO 1π orbitals split by about 1.2 eV [4].

We decided to calculate this splitting using various cluster models in order to assess the amount of rehybridisation occurring in CO/Ni(111). The present study is part of our ongoing effort towards construction and application of chemisorption cluster models that are suitable for a realistic description of alkali CO coadsorption [5, 6].

2. Chemisorption models, computational details

In the present investigation the linear combination of Gaussian-type orbitals (LCGTO) Xα method, a variant of the density functional approach, has been used [7]. This approach to electronic structure problems has proven useful for a large variety of molecular [8], solid state [9] and surface problems [5, 6, 10].

Cluster models have been employed successfully to study quite a number of localized chemisorption phenomena [5, 6, 11]. Of course, such models must be chosen judiciously in order to avoid obvious shortcomings. We have used fractional occupation numbers [5, 6] which are compatible with the Xα model as a device to mimic the effect of level broadening associated with a proper embedding of the cluster into the remaining solid and to remove possible artifacts due to the high density of states around the Fermi level when geometry changes are investigated. A value of 0.3 eV was chosen for the Gaussian broadening parameter.

Here we report results on two clusters modeling CO chemisorption on twofold bridging sites of nickel surfaces: Ni_7CO for Ni(111) and $Ni_{14}CO$ for Ni(001) (see Fig. 1). Although the experimental data refer only to Ni(111) the latter cluster was included because it allows for an extension to a model of coadsorption, $Ni_{14}Na_2CO$, which is computationally feasible (due to its C_{2v} symmetry, see Fig. 1b). The distance between nearest neighbor substrate atoms was taken from bulk nickel. The adsorbate geometry in the (001) clusters was fixed at d(Ni–C) = 1.75 Å, d(C–O) = 1.20 Å. The two Na atoms augmenting the Ni(001) cluster were placed on fourfold sites at a distance corresponding to an effective Na radius of 1.60 Å as determined from a study of alkali chemisorption on various sites of nickel surfaces [5, 6].

Various clusters $(CO)_n$ modeling lateral interaction within an adsorbate layer were also calculated, among them a cluster $(CO)_7$ of C_{2v} symmetry representing a proper section of the $c(4\times2)$CO overlayer of Ni(111) and a series of C_{4v} clusters ($n=4$, 12, 16). The nearest neighbor distance within this overlayer is 3.29 Å.

In the following the arguments will be based on one-electron energies and not, as proper comparison would require, on total energy differences or equivalently on Slater's transition state energies. However, the relaxation of the adsorbate-derived levels is expected to be rather uniform according to previous experience with related systems [10]. The procedure should be acceptable for the present context where we concentrate on the splitting of orbitals of similar character.

The basis sets – for the expansion of the molecular orbitals and for the fitting of the electronic charge density and of the exchange potential – were taken from previous calculations of related chemisorption cluster models (Ni: no. 5 of Table I in [6]; Na: [5, 9]; C, O: [8, 10]). The exchange scaling parameter $\alpha = 0.7$ was used.

3. Results and discussion

The resulting values for the splitting and the width of the CO 1π band are collected in Table 1. The calculated values for the splitting of the CO 1π level are all

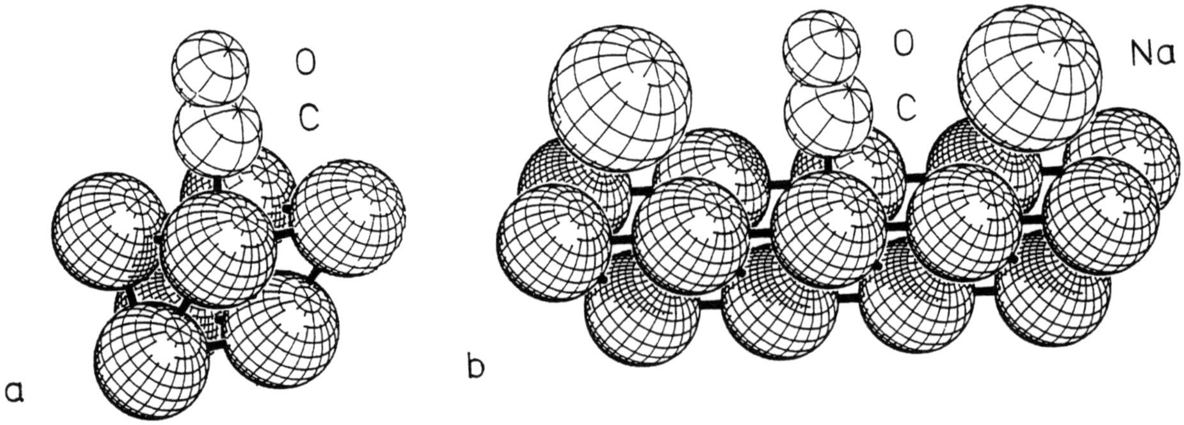

Fig. 1a and b. Model clusters for chemisorption of carbon monoxide on twofold bridging sites of nickel surfaces: **a** Ni_7(111)/CO; **b** Ni_{14}(001)/CO. In the latter case the cluster $Ni_{14}Na_2$(CO) is actually displayed as employed in a coadsorption study

substantially smaller than the value deduced from the analysis of the band observed in angular resolved photoemission. For the small model Ni_2CO [10], the groundstate splitting of 0.68 eV was larger than the computed 0.11 eV difference of ionization potentials. The calculated data of Table 1 should therefore be viewed as upper bounds for the splitting.

The accepted mechanism of the CO interaction with a transition metal surface involves backdonation into the CO 2π level, the donation to the substrate via the CO 5σ orbital being somewhat less important since it entails partial repulsive character [1, 12]. The CO 1π orbital is strongly localized at the oxygen and therefore not well suited for a strong overlap with substrate orbitals. A small splitting, as found in the present cluster calculations, is consistent with this orbital interaction scheme.

In Fig. 2 we present the cluster density of states for Ni_7(111)CO where the selfconsistently determined cluster Fermi energy is also shown. The characteristic $5\sigma - 1\pi$ inversion is clearly visible. Experiment shows the CO 5σ peak 1.7 eV below the center of the CO 1π band compared to only 0.7 eV in our calculation.

The local picture of the CO substrate interaction is well confirmed in a detailed analysis of the various cluster orbitals. Indeed, one may argue that the observed differences in the splitting is consistent with the asymmetry of the directions sampled by the two π orbitals in a C_{2v} environment. The threefold holes of a (111) surface are expected to be less pronounced than the fourfold holes of a (001) surface, in agreement with the data in Table 1. Adding two Na atoms at fourfold hollow sites of the (001) surface reduces the asymmetry of the potential felt by the electrons, consistent with the reduction of the 1π splitting. Furthermore, by the same token one anticipates and finds the largest splitting in the cluster Ni_2CO.

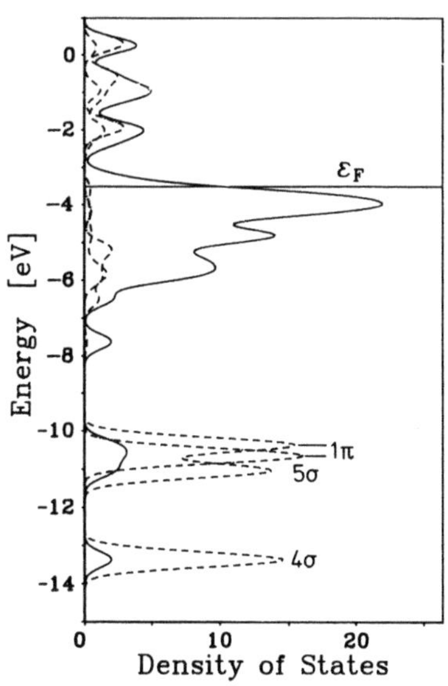

Fig. 2. Cluster density of states (in arbitrary units) generated by Gaussian broadening ($\sigma = 0.3$ eV) of the one-electron energy spectrum and cluster Fermi energy ε_F for the chemisorption model cluster Ni_7CO. Also shown are the localisations (× 9) on the carbon monoxide in levels of a′ (dashed line) and a″ (dot-dashed line) symmetry

A substantial splitting of the CO 1π band has been found for Ni(110)/CO [13]. In this system it was rationalized [13] by the asymmetry experienced by the two CO 1π orbitals which is essentially due to the tilting of the adsorbate molecules. For $c(4\times 2)$CO/Ni(111) the CO molecules stand essentially upright [4], but are exposed to a environment whose symmetry would allow for π splitting. The two sets of nearest neighbor CO–Ni–CO angles are 80° and 100°, quite close to C_{4v} symmetry where the two π orbitals are degenerate. Lateral interaction in the $(CO)_7$ cluster produces only a small splitting (<0.1 eV). By far the dominant effect of lateral interaction in the $c(4\times 2)$CO overlayer of Ni(111) is a dispersion giving rise to a rather broad 1π-derived band. Its width has been estimated from the corresponding part of the level spectrum of various $(CO)_n$ clusters by extrapolating to a cluster of infinite diameter leading to a value of 1.2 eV. This is larger than the value of 0.6 eV deduced from the experimental data. This suggests that the cluster results would not underestimated the later interaction within the CO overlayer.

Additional cluster investigations of the Ni/CO chemisorption system are currently under way. However, we do not expect the conclusions of the present analysis to be changed when based on results from the equilibrium geometry of the various model clusters and on ionization potentials. Further model studies aiming directly at the shape of the 1π band seem highly desirable in order to clear the disparity between experiment and theory.

Table 1. Comparison of theoretical and experimental data pertaining to the shape of the CO 1π band in the chemisorption system Ni/CO (all energies in eV)

System	Method	Ground state splitting	Bandwidth
Ni_7(111)/CO	LCGTO[a]	0.29	
Ni_{14}(100)/CO	LCGTO[a]	0.42	
Ni_{14}(100)Na_2/CO	LCGTO[a]	0.28	
Ni_2/CO	LCGTO[b]	0.68	
$(CO)_7$	LCGTO[a]	~0.1	
$(CO)_\infty$	LCGTO[a]		1.2
Ni(111)/CO	Exp.[c]	1.2 (±0.5)	0.6

[a] This work; [b] Ref. 10; [c] Ref. 4

We would like to thank Prof. D. Menzel and Dr. H.-P. Steinrück for stimulating discussions. This work was supported by the Deutsche Forschungsgemeinschaft through Sonderforschungsbereich 128, by the Fonds der Chemischen Industrie and by the Bund der Freunde der Technischen Universität München.

References

1. Sung S.-S., Hofmann, R.: J. Am. Chem. Soc. **107**, 578 (1985)
2. Bonzel, H.P.: Surf. Sci. Rep. **8**, 43 (1987)
3. Weimer, J.J., Umbach, E.: Phys. Rev. **B30**, 4863 (1984); Weimer, J.J., Umbach, E., Menzel, D.: Surf. Sci. **155**, 132 (1985)
4. Schneider, C., Steinrück, H.-P., Heimann, P., Pache, T., Glanz, M., Eberle, K., Umbach, E., Menzel, D.: Verh. Dtsch. Phys. Ges. **4**, 59 (1986)
5. Rösch, N., Sandl, P., Görling, A., Knappe, P.: Int. J. Quantum Chem. **S22**, 275 (1988)
6. Rösch, N., Knappe, P., Sandl, P., Görling, A., Dunlap, B.I.: In: Computational chemistry: the challenge of *d* and *f* electrons. Salahub, D.R., Zerner, M.C. (eds.) ACS Symposium Series (1989) (to be published)
7. Dunlap, B.I., Connolly, J.W.D., Sabin, J.R.: J. Chem. Phys. **71**, 3396, 4993 (1979)
8. Rösch, N., Kotzian, M., Jörg, H., Schröder, H., Rager, B., Metev, S.: J. Am. Chem. Soc. **108**, 4238 (1986); Rösch, N., Jörg, H., Kotzian, M.: J. Chem. Phys. **86**, 4038 (1987)
9. Rösch, N., Knappe, P., Dunlap, B.I., Bertel, E., Netzer, F.P.: J. Phys. C **21**, 3423 (1988)
10. Jörg, H., Rösch, N.: Surf. Sci. **163**, L627 (1985)
11. Kunz, B.: In: Smith, J.R. (ed.): Theory of chemisorption, p 115. Berlin, Heidelberg, New York: Springer 1980
12. Bauschlicher, C.W., Bagus, P.S.: J. Chem. Phys. **81**, 5889 (1984)
13. Kuhlenbeck, H., Neumann, M., Freund, H.-J.: Surf. Sci. **173**, 194 (1986)

Z. Phys. D – Atoms, Molecules and Clusters 12, 551–554 (1989)

Atoms, Molecules and Clusters
Zeitschrift für Physik D
© Springer-Verlag 1989

Chemisorption and chemical reactions on size selected clusters

M.F. Jarrold and J.E. Bower
AT&T Bell Laboratories, Murray Hill, NJ 07974, USA

Received 8 July 1988

Low energy ion beam techniques have been used to perform a detailed study of the reactions of Al_{25}^+ and Si_{25}^+ with a range of simple molecules (D_2, CH_4, O_2, C_2H_4, CO and N_2). The reactions were studied over a center of mass collision energy range from 0.2 eV up to 7 eV. Activation barriers for chemisorption onto the clusters were deduced from the experimental results. The activation barriers for chemisorption on Al_{25}^+ and Si_{25}^+ are generally similar and show a qualitative correlation with the electronic properties of the reactant molecule. However, the products of the chemical reactions of Al_{25}^+ and Si_{25}^+ which result from cluster fragmentation are quite different. Si_{25}^+ shows a tendency to undergo fission as observed in a number of recent studies of the dissociation of the bare clusters.

PACS: 36.40; 34.50; L

Introduction

How the chemical properties of atomic clusters change with cluster size and approach bulk behaviour is a topic of considerable fundamental interest. An important question that has arisen from the recent studies of cluster chemistry [1, 2] concerns the factors important in controlling reactivity, and in particular the relative importance of cluster geometry and electronic factors. In this paper we present a summary of some recent results we have obtained for the reactions of size selected Al_{25}^+ and Si_{25}^+ clusters with D_2, CH_4, O_2, C_2H_4, CO and N_2. The reactions were studied over a wide range of collision energies using low energy ion beam techniques. A unique feature of these experiments is the ability to directly observe a metastable adduct arising from chemisorption on the cluster. From the collision energy dependence of metastable adduct formation, the activation barrier for chemisorption on the cluster can be determined. Thus for systems which at room temperature would simply be unreactive it is possible to determine the size of the activation barrier.

Experimental methods

The experimental apparatus and techniques have been described in detail elsewhere [3] so only a brief description will be given here. Cluster ions are generated by pulsed laser vaporization in a continuous flow of helium buffer gas. After exiting the source the ions are focussed into a quadrupole mass spectrometer where a particular cluster size is selected. The size selected clusters are then focussed into a low energy ion beam and passed through the gas cell (2.5 cm long) where the neutral reagent is introduced to a pressure of 0.1–1.0 mtorr. After passing through the gas cell the products and unreacted ions are focussed into a second quadrupole mass spectrometer where they are analyzed. At the end of the quadrupole the ions are detected by an off-axis collision dynode and dual microchannel plates.

Results

Figure 1 shows a plot of the cross sections for metastable adduct formation against collision energy for

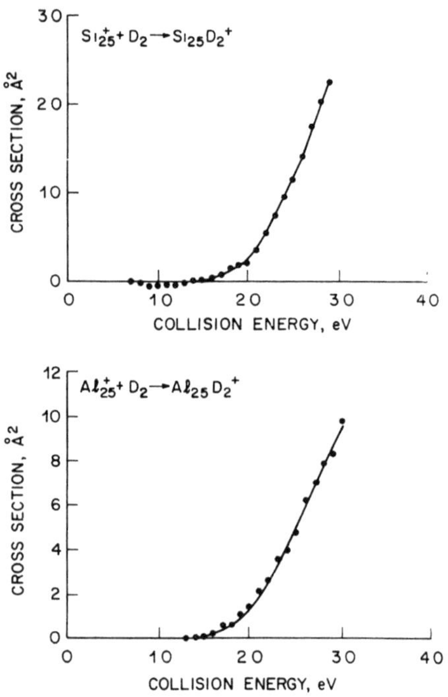

Fig. 1. Plots of the cross sections against collision energy for $Si_{25}D_2^+$ (upper) and $Al_{25}D_2^+$ (lower) formation in the reactions of Al_{25}^+ and Si_{25}^+ with D_2

the reactions between Al_{25}^+ and Si_{25}^+ with D_2. In these experiments the $Al_{25}D_2^+$ and $Si_{25}D_2^+$ adducts are not stabilized by further collisions in the gas cell (the pressure is too low) and so they contain enough energy to dissociate back to reactants. The adducts are metastable, however, and survive long enough to be detected because of the large number of internal degrees of freedom in the clusters. The points in Fig. 1 are the experimental data and the lines are the results of simulations to deduce the true collision energy thresholds. The simulations use an assumed cross section function and take in account the energy spread of the ion beam and the distribution of collision energies resulting from thermal motion of the D_2. It is apparent from Fig. 1 that the cross sections for $Al_{25}D_2^+$ and $Si_{25}D_2^+$ adduct formation show a similar collision energy dependence. The collision energy thresholds for adduct formation deduced from the simulation (shown in Table 1) are 2.0 eV for both Al_{25}^+ and Si_{25}^+. These thresholds can be related to the activation barriers for chemisorption on the clusters. Since the adducts must be quite strongly bound (> 1 eV) in order to survive long enough to be detected, the activation barriers determined from these measurements are for dissociative chemisorption rather than for molecular adsorption.

Methane also chemisorbs on both Al_{25}^+ and Si_{25}^+ to yield $Al_{25}CH_4^+$ and $Si_{25}CH_4^+$ adducts. The collision energy thresholds for adduct formation with CH_4 are shown in Table 1. The thresholds for Al_{25}^+ and Si_{25}^+ are similar and substantially larger than for chemisorption of D_2. The reactions with CH_4 were studied over a wider collision energy range than the reactions with D_2 (up to 6 eV). At the higher collision energies studied a number of other products were observed in addition to the adducts. These products probably arise mainly from dissociation of the adducts caused by the extra internal energy resulting from the higher collision energies. With Al_{25}^+ the main products were $Al_{24}H^+$ and $Al_{24}CH_3^+$, and with Si_{25}^+, $Si_{15}CH_4^+$ and Si_{15}^+ were observed. The Si_{15}^+ product probably arises from collision induced dissociation of Si_{25}^+, loss of Si_{10} is the dominant dissociation pathway of Si_{25}^+ in both photodissociation [4] and collision induced dissociation by argon [5]. Apparently, the $Si_{25}CH_4^+$ adduct also dissociates by loss of an Si_{10} unit to give $Si_{15}CH_4^+$.

With oxygen, adducts were not observed for either Si_{25}^+ or Al_{25}^+. For the clusters of both elements highly exothermic reactions follow dissociative chemisorption of oxygen, so the adduct does not survive long enough to be detected. With Al_{25}^+ the main product is Al_{21}^+ (arising from loss of two Al_2O molecules), and with Si_{25}^+ the main product is Si_{23}^+ (arising from loss of two SiO molecules or an Si_2O_2 species). The thresholds for the reactions (which can be related to activation barriers for chemisorption because the subsequent fragmentation of the adduct is so exothermic) are shown in Table 1. The activation barriers for chemisorption of O_2 on both clusters are quite small. Though the thresholds are similar, the reaction cross sections for Al_{25}^+ and Si_{25}^+ show different collision energy dependences. For Al_{25}^+ the cross sections rise sharply above threshold and level-off at a value ~50% of the geometric cross section. The cross sections for Si_{25}^+ increase more slowly with collision energy. A possible explanation for this behavior is steric effects (the orientation of the O_2 and where it strikes the cluster surface) are relatively unimportant for Al_{25}^+, but are important for Si_{25}^+.

Table 1. Collision energy thresholds for chemisorption on Al_{25}^+ and Si_{25}^+.

Reactant	Threshold, eV	
	Al_{25}^+	Si_{25}^+
D_2	2.0 ± 0 2	2.0 ± 0.3
CH_4	3.5 ± 0.3	2.9 ± 0.3
O_2	0.5 ± 0.3	0.3 ± 0.3
C_2H_4	< 0.2	< 0.2
CO	1.9 ± 0.3	3 1 ± 0.5
N_2	3.5 ± 0.3	5.0 ± 0.8

The reactions of C_2H_4 with Al_{25}^+ and Si_{25}^+ are complicated but remarkably similar. For both Al_{25}^+ and Si_{25}^+ there appear to be three different types of adduct formed over different collision energy ranges. C_2H_4 chemisorbs on Al_{25}^+ and Si_{25}^+ at the lowest collision energy studied (0.2 eV). As the collision energy is raised the cross sections for adduct formation fall and then at 0.5 eV start to rise as a different type of adduct becomes accessible. As the collision energy is raised further the cross sections peak and then fall. Above collision energies of 3.0–3.5 eV the cross sections for adduct formation start to rise again as yet another type of adduct becomes accessible. C_2H_4 is known to give a range of different products on bulk surfaces [6] so it is reasonable to assign the different adducts formed to different structures of the C_2H_4 molecule on the cluster. Unfortunately our experiments provide little clue as to what the structures might be, but it is surprising that Si_{25}^+ and Al_{25}^+ behave so similarly.

CO dissociatively chemisorbs on Al_{25}^+ to give an $Al_{25}CO^+$ adduct which is only observed over a relatively narrow collision energy range. The $Al_{25}CO^+$ adduct apparently undergoes a fairly facile dissociation, loosing an Al_2O molecule to give $Al_{23}C^+$. With Si_{25}^+ no adduct was observed. However, $Si_{24}C^+$ and Si_{19}^+ result from the reaction with CO. The $Si_{24}C^+$ product probably arises from loss of SiO from the adduct. This must be a facile process since the adduct is not directly observed. At the higher collision energies studied the Si_{19}^+ becomes important. This product probably arises from loss of an Si_5C species from $Si_{24}C^+$. Si_5C is an analog of Si_6 which is known to be a particularly stable cluster. Activation barriers for chemisorption of CO on Al_{25}^+ (from the threshold for adduct formation) and Si_{25}^+ (from the reaction threshold) are shown in Table 1. The activation barrier for Si_{25}^+ is significantly larger than for Al_{25}^+.

N_2 dissociatively chemisorbs on both Al_{25}^+ and Si_{25}^+ to give adducts. The collision energy thresholds for adduct formation are shown in Table 1. The activation barrier for chemisorption on Si_{25}^+ is substantially larger than for chemisorption on Al_{25}^+.

Discussion

A) Comparison of cluster and bulk surface chemistry

Of the molecules studied, bulk aluminum and silicon surfaces only react with O_2 and C_2H_4 at room temperature. As can be seen from Table 1 these molecules have the smallest activation barriers for chemisorption on the clusters, suggesting that there is a correlation between the reactivity observed for the clusters and the chemistry of the bulk surfaces. However, there appear to be a small activation barriers for the reactions of the clusters with oxygen, but there are no significant activation barriers for the bulk surfaces. Activation barriers for chemisorption of the other molecules we have studied have not been measured for the bulk surfaces.

B) What factors determine the size of the activation barriers

As can be seen from Table 1 the activation barriers for chemisorption on Al_{25}^+ and Si_{25}^+ are quite similar and show a large variation as the reactant molecule is changed. Recently, Whetten and coworkers [1] proposed that charge transfer from the metal cluster to the reactant molecule could be a factor in determining the rates of chemisorption reactions. Other workers have argued against this idea [2]. Figure 2 shows a comparison of the activation barriers measured for chemisorption on Al_{25}^+ and Si_{25}^+ with the cluster HOMO→ molecule LUMO promotion energies [7]. The data for C_2H_4 is separated from the other molecules studied in Figure 2 because with C_2H_4 several different types of adduct are formed over different collision energy ranges. Furthermore, C_2H_4 has a high lying HOMO and could act as a donor rather than an acceptor. There is a fairly good qualitative correlation between the size of the activation barriers and the cluster HOMO → molecule LUMO promotion energies. These results suggest that charge transfer stabilization of the transition state is a factor in determining the size of the activation barriers. However, this is certainly not the only factor that is important. The electronic

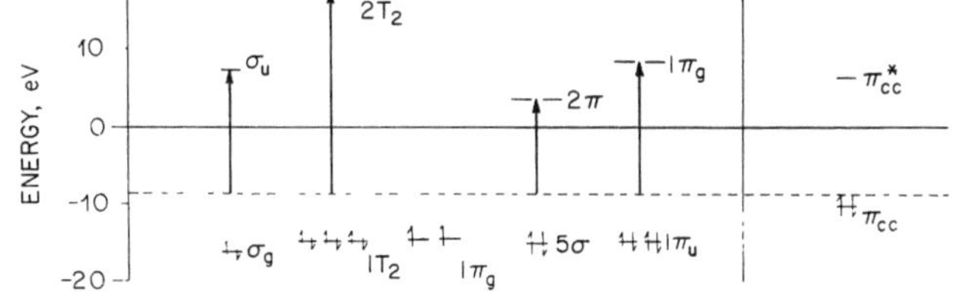

Fig. 2. Plot of the activation barriers for chemisorption on Al_{25}^+ and Si_{25}^+ compared with a plot of the HOMO and LUMO energies of the molecules studied (taken from [7])

structure of the metal cluster is important (transition metal clusters chemisorb hydrogen at room temperature [1, 2]) and probably the strength of the bonds being broken and formed is a factor (as suggested by Shustorovitch [8] for surfaces). Finally, geometric structure probably exerts a subtle influence as well.

References

1. Whetten, R.L., Cox, D.M., Trevor, D.J., Kaldor, A.: Phys. Rev. Lett. **54**, 1494 (1985)
2. Parks, E.K., Weiller, B.H., Bechthold, P.S., Hoffman, W.F., Nieman, G.C., Pobo, L.G., Riley, S.J.: J. Chem. Phys. **88**, 1622 (1988); Elkind, J.L., Weiss, F.D., Alford, J.M., Laaksonen, R.T., Smalley, R.E.: J. Chem. Phys. **88**, 5215 (1988)
3. Jarrold, M.F., Bower, J.E.: J. Am. Chem. Soc. **110**, 6706 (1988)
4. Zhang, Q.-L., Liu, Y., Curl, R.F., Tittel, F.K., Smalley, R.E.: J. Chem. Phys. **88**, 1670 (1988)
5. Jarrold, M.F., Bower, J.E.: J. Phys. Chem. **92**, 5702 (1988)
6. See, for example, Bertolini, J.C., Massardier, J.: The chemical physics of solid surfaces and heterogeneous catalysis. King, D.A., Woodruff, D.P. (eds.). Amsterdam: Elsevier 1984
7. Jorgensen, W.L., Salem, L.: The organic chemist's book of orbitals. New York: Academic 1973
8. Shustorovitch, V.: Surf. Sci. Rep. **6**,1 (1986)

Z. Phys. D – Atoms, Molecules and Clusters 12, 555–561 (1989)

Atoms, Molecules and Clusters
Zeitschrift für Physik D
© Springer-Verlag 1989

Sequential reactions of silicon clusters with SiD_4: constrained heterogeneous nucleation of deuterated silicon particles

M.L. Mandich and W.D. Reents, Jr.
AT&T Bell Laboratories, Murray Hill, NJ 07974, USA

Received 8 April 1988

The sequential clustering reactions of gas phase silicon cluster ions are ideally suited for studying heterogeneous nucleation of silicon particles. Stepwise reactions of silicon cluster ions with silane have been observed which lead to growth of larger deuterated silicon clusters. Extensive information is obtained about their exothermic gas phase reaction rates and product distributions as a function of cluster size and degree of hydrogenation. What is found is that each size of silicon cluster exhibits a unique pattern of growth. Furthermore, nearly all of the silicon clusters encounter chemical constraints to rapid nucleation. These constraints are a consequence of specific electronic and structural requirements in the chemical reactions. The nature of these requirements are deduced using experimental results in concert with ab initio electronic structure theoretical calculations of the energetics and structures of various species.

PACS: 81.20; 07.75

I. Introduction

Deleterious effects of gas phase particles as components of atmospheric pollution, dust contaminants of electronic devices, etc. have stimulated an increasing need to know how they are formed. Both experimental and theoretical studies of particle nucleation have highlighted the first stages of growth as critical factors in the rate of particle formation [1]. Here, the critical nucleus size is formed which can then undergo rapid spontaneous growth, limited only by factors such as diffusion and availability of monomers. Processes by which these critical nuclei form are poorly understood. The examples which have been the most studied involve the simple aggregation of monomeric species such as rare gases or stable molecules [2]. Formation of nuclei in this type of process is endoergic and proceeds relatively slowly. The rapid nucleation observed in most naturally occurring systems, however, requires that the condensation nucleus be formed efficiently and exothermically. One way that this can be accomplished in the gas phase is by a sequence of exothermic chemical reactions. Yet, such a sequence has never been clearly delineated.

The sequential clustering reactions of gas phase ions are particularly salient to the questions of silicon particle nucleation. Ions are currently believed to control the chemistry and kinetics of prenucleation of these aggregates under plasma conditions used for depositing silicon films in device fabrication [3]. We have undertaken a study of the sequential clustering reactions of positively charged silicon clusters with SiD_4. In these reactions, growth of larger silicon clusters is found to occur primarily by stepwise addition of $-SiD_2$ accompanied by loss of D_2. A silicon cluster species is sought which undergoes unlimited incremental growth since such a species could ultimately form the critical nucleus required for rapid growth of large deuterated silicon particles. What is found instead is that all of the bare silicon cluster ions encounter chemical constraints to rapid nucleation and growth. These constraints appear to arise because the chemical reactions of $Si_xD_y^+$ with SiD_4 have specific electronic and structural requirements. The nature of these requirements are deduced using experimental results in concert with ab initio electronic structure theoretical calculations of the energetics and structures of various species.

II. Experimental method

Si^+_{1-7} and Si^-_{1-6} clusters are created by laser evaporation of a bulk silicon target located outside the low pressure ($\sim 10^{-9}$ Torr) side of a differentially pumped dual ion cell of a modified Nicolet FT/MS-1000 Fourier transform mass spectrometer (FTMS). A single cluster size is selected from the resulting ion population using double resonance techniques and reacted with $1-10\times10^{-7}$ Torr 99.7 atom% deuterium-enriched SiD_4 for times ranging from 0.1–10 s. The progress of the reaction is interrogated by recording the mass spectrum of the ions in the cell at regular time intervals. Mass spectra are obtained with mass resolutions of at least 10,000 which allows for separation of mass doublets. Accurate mass determination within 20 ppm allows for straightforward assignment of reactant and product ion stoichiometries. Standard double resonance techniques are also used to confirm reaction pathways.

III. Results

Ground state Si^+_{1-3} and Si^+_5 react exothermically with SiD_4 in a sequential fashion to form larger hydrogenated silicon cluster ions. Si^+_4, Si^+_6 and Si^+_7 do not react with SiD_4 on the time scale of these experiments. No reactions are observed for the negatively charged silicon cluster ions, Si^-_{1-7}.

The clustering sequence and associated rates for each step of the sequence can be deduced from the time dependences of the evolving cluster ion populations. For example, the ion intensities as a function of time for Si^+_3 and its sequential reaction products with SiD_4 are shown in Fig. 1. Analysis of the exponential behavior of each ion intensity yields pseudo-first-order rate constants for each step of these reactions as well as those for the other starting cluster ions, Table 1. All of the products and rates are obtained under conditions which ensure that they pertain to the exothermic reactions of ground state species which are in thermal equilibrium with the cell wall temperature at 300 K.

IV. Discussion

The sequential clustering reactions of Si^+_{1-7} with SiD_4 are severely chemically constrained. In all but one case,

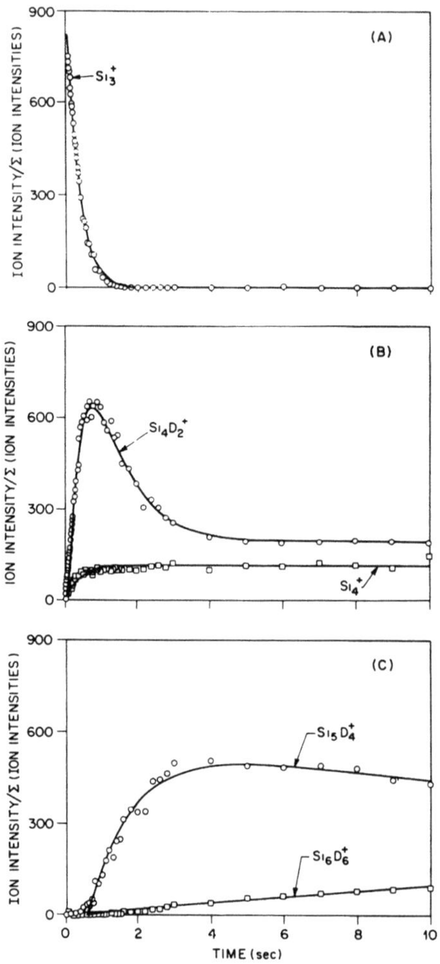

Fig. 1A–C. Time dependences of the evolving ion populations in the sequential clustering reactions of Si_3^+ with 1.7×10^{-7} Torr SiD_4. **A** The decay of the initial Si_3^+ ion population. **B** The rise and decay of the primary reaction products, $Si_4D_2^+$ and Si_4^+. Note that a portion of the $Si_4D_2^+$ plus the entire Si_4^+ populations do not react on this timescale. **C** The rise and decay of the secondary ion product, $Si_5D_4^+$ plus the rise of the tertiary $Si_6D_6^+$ product

each reaction gives remarkably similar products,

$$Si_n^+ + SiD_4 \rightarrow Si_{n+1}^+ + 2D_2 \quad (1)$$

$$\rightarrow Si_{n+1}D_2^+ + D_2 \quad (2)$$

Yet, all of the clustering sequences rapidly reach terminating structures, with the possible exception of Si_3^+. Once a terminating structure is reached, further growth is limited to slow bimolecular attachment of SiD_4 to the cluster ion. The underlying reasons for the constraints to infinite growth can be found in an examination of the structures and bonding of the various $Si_xD_y^+$ clusters involved in these reactions. Some of the possible cluster structures have been calculated by Raghavachari using ab initio electronic structure theory [4–5]. Overall, terminating cluster structures are found to be saturated in deuterium or

Table 1. Sequential clustering reactions of Si^{+}_{2-7} and $^{29}Si^{+}$ with SiD_4^{a}

Reaction[b]		Product fraction (%)	Rate constant (× 10^{10}) (cm^3 $molecule^{-1}$ s^{-1})	Reaction probability[c]
$*Si^{+} + SiD_4$	$\rightarrow *Si_2D_2^{+} + D_2$	85 ± 4	8.1 ± 0.4^{d}	0.82 ± 0.04
	$Si^{+} + *SiD_4$	15 ± 4		
$*Si_2D_2^{+} + SiD_4$	$\rightarrow *Si_3D_4^{+} + D_2$	88 ± 5	0.36 ± 0.04^{d}	0.036 ± 0.004
	$Si_2D_2^{+} + *SiD_4$	12 ± 5		
$*Si_3D_4^{+} + SiD_4$	$\rightarrow *Si_4D_6^{+} + D_2$	87 ± 5	2.0 ± 0.3^{d}	0.20 ± 0.03
	$Si_3D_4^{+} + *SiD_4$	13 ± 5		
$Si_4D_6^{+} + SiD_4$	$\rightarrow Si_5D_{10}^{+}$	100	0.0010 ± 0.0003	0.00010 ± 0.00003
$Si_2^{+} + SiD_4$	$\rightarrow Si_3D_2^{+} + D_2$	100	2.5 ± 0.4	0.25 ± 0.04
$Si_3D_2^{+} + SiD_4$	$\rightarrow Si_4D_6^{+}$	100	0.010 ± 0.003	0.0010 ± 0.0003
$Si_3^{+} + SiD_4$	$\rightarrow Si_4^{+} + 2D_2$	14 ± 2	4.8 ± 0.7	0.48 ± 0.07
	$Si_4D_2^{+} + D_2$	86 ± 2		
$Si_4D_2^{+} + SiD_4$	$\rightarrow Si_5D_4^{+} + D_2$	70 ± 3	1.8 ± 0.3	0.18 ± 0.03
	N.R.	30 ± 3	$\leqslant 0.001$	$\leqslant 0.0001$
$Si_5D_4^{+} + SiD_4$	$\rightarrow Si_6D_6^{+} + D_2$	100	0.036 ± 0.009	0.0036 ± 0.0009
$Si_4^{+} + SiD_4$	$\rightarrow$ N.R.	100	$\leqslant 0.006$	$\leqslant 0.0006$
$Si_5^{+} + SiD_4$	$\rightarrow Si_6D_2^{+} + D_2$	58 ± 5	0.8 ± 0.1	0.08 ± 0.01
	$Si_6D_3^{+} + D$	41 ± 5		
$Si_6D_2^{+} + SiD_4$	$\rightarrow$ N.R.	100	$\leqslant 0.01$	$\leqslant 0.001$
$Si_6D_3^{+} + SiD_4$	$\rightarrow$ N.R.	100	$\leqslant 0.01$	$\leqslant 0.001$
$Si_6^{+} + SiD_4$	$\rightarrow$ N.R.	100	$\leqslant 0.001$	$\leqslant 0.0001$
$Si_7^{+} + SiD_4$	$\rightarrow$ N.R.	100	$\leqslant 0.005$	$\leqslant 0.0005$

[a] "N.R." indicates that no reaction products have been observed for a particular reacting ion. Previously determined rates for Si^{+} reacting with SiD_4 can be found in Ref. 7

[b] Reactions here are all exothermic with ground state ions at 300 K. Neutral species are inferred from thermodynamics. Silicon-29 isotope labelling of Si^{+} and its sequentially formed cluster ions is designated with an asterisk. For example, $*Si_3D_4^{+} \equiv {}^{29}Si^{28}Si_2D_4^{+}$. No isotope labelling is included for Si^{+}_{2-7} and their sequentially formed clusters since isotope exchange with SiD_4 is not observed for these clusters

[c] The probability of reaction is calculated as the ratio of the measured reaction rate to the Langevin ion-molecule collision rate

[d] Rate constant includes the isotope exchange process

have other electronic constraints preventing further interaction with SiD_4. In some cases, there is sufficient theoretical and experimental information to quantitatively test the proposed silicon cluster structures using statistical phase space theory [6–7].

A. Clustering sequences of the reactive clusters, $Si^{+}_{1-3,5}$

1). $Si^{+} + SiD_4$. Si^{+} is observed to undergo three sequential additions of $-SiD_2$ to form $Si_4D_6^{+}$ [7]. The clustering reactions then stop and further growth of $Si_4D_6^{+}$ only occurs slowly by bimolecular attachment of SiD_4 to form $Si_5D_{10}^{+}$. If isotopically labelled Si^{+} is used, e.g. $^{29}Si^{+}$, another process is seen at each step of the clustering sequence where part of the isotopically labelled silicon is exchanged with $^{28}SiD_4$, Table 1. At each step, the exchange of label occurs on the same timescale with the formation of the $Si_{n+1}D_{2n}^{+}$ product. These observations indicate that the microscopic pathways of all of the clustering reactions of Si^{+} and its progeny have two significant energetic barriers. This allows for a branching between product formation in a forward process and isotopic exchange in a reverse process.

The mechanism and energetics for all of the reaction pathways have been derived by Raghavachari using ab initio electronic structure theory [4]. These calculations find that the two energetic barriers correspond to two important transition states in the reaction: a Si-D bond insertion transition state (TS1) and a D_2 elimination transition state (TS2). The structures of the growing hydrogenated silicon cluster ions are shown below.

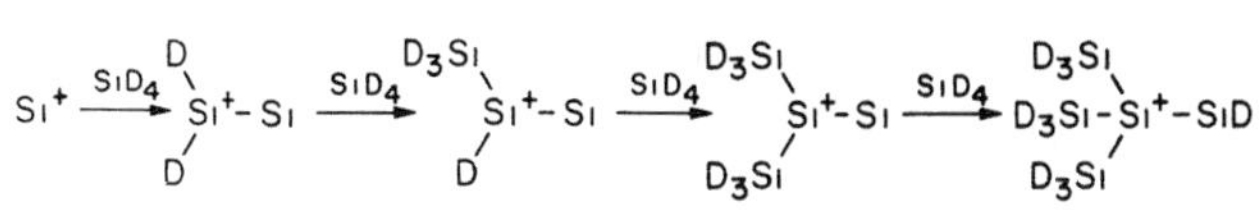

The reaction terminates with the highly branched $Si_5D_{10}^+$ structure. This structure is nearly saturated with deuterium which raises the energy of TS2, thus blocking loss of D_2. Using statistical phase space theory (PST) in conjunction with the experimentally derived forward and reverse reaction rates, we are able to test the accuracy of the calculated energetics of TS1, TS2, as well as the binding energy (CBE) of $Si_5D_{10}^+$ [6–7]. The results of these calculations are given in Table 2 [7]. Good quantitative agreement is found between the PST and ab initio results which confirms the validity of the calculated microscopic reaction pathways.

2). $Si_2^+ + SiD_4$. The first clustering reaction of Si_2^+ with SiD_4 occurs readily. By analogy to the reactions of Si^+, $Si_2D_2^+$, etc., the initial step of the Si_2^+ reaction with SiD_4 probably involves a Si-D insertion to form either $(D)(SiD_3)Si_2^+$ or $(SiD_3)Si_2D^+$. Both of these complexes have been investigated and are calculated to be energetically well below the starting reactants, $Si_2^+ + SiD_4$, at -1.4 and -0.87 eV, respectively [4, 8]. Elimination of D_2 follows, forming the $Si_3D_2^+$ product. The most energetically favored form for $Si_3D_2^+$ is the cyclic $^2A_1$ structure where the deuteriums reside on different silicon atoms [5]. If this is the final product, then the overall reaction is calculated to be exothermic by 0.78 eV. An alternative cyclic structure with the deuteriums on the same center is not favorable; its formation is calculated to be 0.52 eV endothermic. The linear $^2B_1$ product, $(D_2Si)Si_2^+$, is also estimated to be slightly endothermic, by ~ 0.1 eV. Further investigation of this structure, especially bending along the Si–Si–Si bond, is probably warranted since in Si_3 the bent form has been found to be more stable than the linear form by 0.4 eV [9].

The available theoretical data is insufficient to determine whether the $Si_3D_2^+$ product is cyclic or not. However, the next clustering step of the reaction to form $Si_4D_4^+$ does not occur and only the collisionally stabilized adduct, $Si_4D_6^+$ is seen. It is difficult to conceive of an energetic or structural reason why loss of D_2 would be blocked in any linear form of $Si_3D_2^+$, suggesting that the $Si_3D_2^+$ product has the stable cyclic form. To be consistent with this explanation, formation of the $Si_4D_6^+$ adduct would have to leave the $Si_3D_2^+$ cyclic form intact. From the experimental formation rate of the $Si_4D_6^+$ complex, we have performed a phase space calculation of its binding energy, showing that it lies between 1.2 and 1.4 eV below the reactants. This binding energy is consistent with either form of $Si_4D_6^+$ shown below. Additional theoretical calculations will prove invaluable for ascertaining why elimination of D_2 is endothermic from these structures.

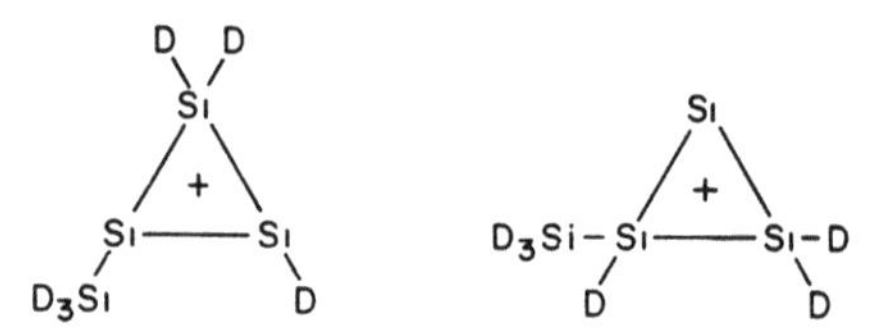

3). $Si_3^+ + SiD_4$. Si_3^+ exhibits the most extensive clustering with SiD_4 of all of the Si_{2-7}^+ clusters. On the timescale of our experiments, three additions of $-SiD_2$ occur. Nonetheless, the Si_3^+ clustering sequence does not follow a single course. The first reaction of Si_3^+ forms three products, two of which do not continue to react with SiD_4. Two of these products have the same stoichiometry, $Si_4D_2^+$, but are chemically distinct in further reactivity with SiD_4. This is a

Table 2. Comparison of PST and *ab initio* calculations of transition state (TS1-2) or intermediate complex binding (CBE) energies for each sequential clustering reaction of $^{29}Si^+$ with SiD_4 [4, 7]

Reaction	PST calculations TS1	PST calculations TS2	PST calculations CBE ($Si_5D_{10}^+$)	*ab initio* calculations TS1	*ab initio* calculations TS2	*ab initio* calculations CBE ($Si_5D_{10}^+$)
$*Si^+ + SiD_4 \rightarrow \begin{cases} *Si_2D_2^+ + D_2 \\ Si^+ + *SiD_4 \end{cases}$	-0.34 ± 0.05	-0.29 ± 0.04	—	-0.46	-0.01	—
$*Si_2D_2^+ + SiD_4 \rightarrow \begin{cases} *Si_3D_4^+ + D_2 \\ Si_2D_2^+ + *SiD_4 \end{cases}$	-0.16 ± 0.02	-0.14 ± 0.02	—	-0.21	-0.05	—
$*Si_3D_4^+ + SiD_4 \rightarrow \begin{cases} *Si_4D_6^+ + D_2 \\ Si_3D_4^+ + *SiD_4 \end{cases}$	-0.27 ± 0.02	-0.19 ± 0.02	—	-0.21	-0.06	—
$Si_4D_6^+ + SiD_4 \rightarrow Si_5D_{10}^+$	—	—	-0.80 ± 0.02	—	—	-0.98

clear example that pathways leading to multiple non-interconverting isomers are energetically and kinetically accessed starting with a given initial reactant Si_x^+ ion.

The first clustering reaction of Si_3^+ with SiD_4 forms both Si_4^+ and $Si_4D_2^+$. From the ab initio electronic structure determinations of the fragmentation energies for $Si_n^+ \rightarrow Si_{n-1}^+ + Si$, the formation of the lowest energy D_{2h} rhombus structure of Si_4^+ is calculated to be exothermic by 1.0 eV [8, 10]. Based on ab initio calculations of the relative energetics of alternative Si_4^+ structures, formation of any other geometric isomer of Si_4^+ would be endothermic by 0.1–2.4 eV. Additional confirmation that Si_4^+ formed from Si_3^+ has the lowest energy rhombus structure is that it is not observed to react further with SiD_4. This is exactly the same as for the Si_4^+ population which is formed initially from evaporation of the silicon target.

The other ionic product formed in the reaction of Si_3^+ with SiD_4 is $Si_4D_2^+$. As can be seen in its subsequent reaction with SiD_4, this product consists of two chemically distinct species (Table 1). As discussed below, Si_4^+ also does not react with SiD_4 because of its unusually stable rhombus structure. This immediately suggests that the nonreactive $Si_4D_2^+$ is structurally similar to D_{2h} Si_4^+ in such a way that the electronic configuration responsible for the stability of the Si_4^+ rhombus is not disrupted. Ab initio calculations find

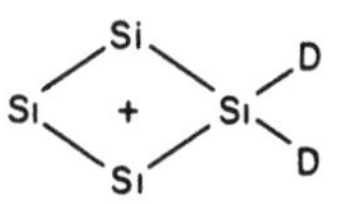

(I)

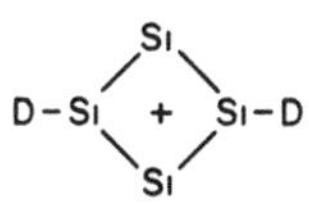

(II)

that structure (*I*) above is the only possibility which satisfies this criterion [5]. Theoretical investigations of other possible $Si_4D_2^+$ forms indicate that structure (*II*) is energetically well below other cyclic and non-cyclic forms [5]. In (*II*), however, the electronic configuration which results in the rhomboid shape in Si_4^+ has been sacrificed in order to bind the two deuterium atoms. The resulting structure (*II*) is nearly a D_{4h} square which is high enough in energy to react further with SiD_4. *A distinguishing feature of structures (I) and (II) is that, experimentally, they are found to be non-interconverting even via a single deuterium migration.* This is the first evidence that deuterium migration may be constrained in some of these highly unsaturated species.

The reactive portion of the $Si_4D_2^+$ population interacts further with SiD_4 exclusively by sequential addition of $-SiD_2$ to form $Si_5D_4^+$ and then $Si_6D_6^+$. Currently, there is insufficient experimental and theoretical information to ascertain the structures of $Si_5D_4^+$ and $Si_6D_6^+$. One possibility is that they are silyl ($-SiD_3$) substituted versions of structure (*II*). No $Si_7D_x^+$ products are observed during our experimental time scale, indicating that further growth must proceed at less than 1% of the collision rate.

3). $Si_5^+ + SiD_4$. Si_5^+ is only observed to undergo one clustering step with SiD_4, forming $Si_6D_2^+$ and $Si_6D_3^+$. Further clustering of $Si_6D_{2,3}^+$ does not occur within our experimental time limits (Table 1).

The reaction of Si_5^+ has a small rate of only about one-tenth of the Langevin rate. Nonetheless, the overall reaction energetics are sufficiently favorable that one of the products, $Si_6D_3^+$, results from the loss of a deuterium *atom*. This is the only reaction that is observed for the Si_n^+ clusters with SiD_4 where a radical product is formed. Based on the relative energetics of possible structures for $Si_6D_3^+$, loss of a deuterium atom can probably occur only if the addition of SiD_3 to Si_5^+ is accompanied by rearrangement and ring closure to form a deuterated Si_6^+ cluster similar to the bare Si_6^+ cluster itself. Formation of any functionalized Si_5^+ clusters, such as $(Si_5)(SiD_3)^+$, is estimated to be endothermic by 1.1–1.2 eV [5, 8, 11]. However, Si_6^+ is sufficiently stable relative to the Si_5^+ structure that the addition of a silicon atom to Si_5^+ to form the most stable edged capped trigonal bipyramid C_{2v} form of Si_6^+ is exothermic by 5.5 eV. This energy is comparable to that for addition of a silicon atom to Si_3^+. In the case of Si_3^+, the exothermicity appears to allow for elimination of *two* molecules of D_2 to form the stable D_{2h} rhombus structure. In contrast, for Si_6^+, three deuteriums remain on the cluster and loss of a deuterium atom occurs. This provides additional evidence that substantial ring closure and rearrangement can occur for some of these highly unsaturated silicon clusters as more silicon atoms are added to the framework.

B. The non-reacting clusters with SiD_4: Si_4^+, Si_6^+, and Si_7^+

Three clusters, Si_4^+, Si_6^+, and Si_7^+, are inert with respect to reaction with SiD_4. In addition, none of these clusters are observed to form collisionally stabilized bimolecular adducts with SiD_4. The nonreactivity of Si_6^+ and Si_7^+ is not unusual. To date, they have not been observed to react with any organic or organosilicon reagent, including CH_3SiH_3, CH_3C_2H, $(CH_2)_2C$, and C_2H_2, with the exception of CH_3OH [12]. The failure of Si_4^+ to react with SiD_4 as an isolated cluster size, however, is unprecedented for silicon cluster ions.

The non-reactivity of Si_4^+, Si_6^+ and Si_7^+ may well be related to their unusual stability. Photofragmentation and collisional fragmentation studies as well as ab initio electronic structure calculations have shown that these cluster sizes are energetically more stable than other small silicon cluster sizes [8, 10, 13, 14]. Furthermore, the calculated energies for the addition of a silicon atom to these clusters to form the ground state structure of the next larger cluster, $Si_n^+ + Si \rightarrow Si_{n+1}^+$, are the lowest of all of the silicon clusters containing between two and ten atoms [8, 10, 14]. As discussed above, addition of $-SiD_2$ to a bare Si_n^+ cluster appears to make a cluster which has a structure closely related to the lowest energy isomer of the Si_{n+1}^+ cluster. Thus, the energies calculated for the addition of a silicon atom can be used to provide a first approximation for the thermochemistry of the reaction involving addition of $-SiD_2$. Although the actual values should not be taken too seriously, this approximation shows that the addition of $-SiD_2$ to either Si_4^+ or Si_7^+ would be significantly less exothermic as compared to the other clusters, including Si_6^+. This suggests that the reaction of Si_4^+ and Si_7^+ with SiD_4 may well be endothermic or be so close to thermoneutral that any transition state energies would be endothermic. *The estimated reaction thermodynamics do not explain why Si_6^+ is unreactive.*

The most plausible reason for the inertness of Si_6^+ is that its dangling electrons have the wrong electronic character to insert into the Si-D bond of SiD_4. The net result is that the energy of the first transition state involving Si-D insertion rises well above the available reaction energy. Previously, we have postulated that a divalent (silylene) silicon center having a localized lone pair of non-bonding electrons is the chemically active site on the cluster for Si-D bond insertion [12]. This suggestion originates from the well-established solution phase organosilicon chemistry of divalent silicon centers [15]. Recently, extensive ab initio electronic structure calculations have been performed on the Si-D insertion transition states involved in the sequential clustering reactions of Si^+ with SiD_4. These calculations clearly indicate the essential role played by the silylene lone pair in both the attachment of the incoming D of SiD_4 and in the formation of the Si-D insertion transition state. The smaller silicon clusters, Si_{2-5}^+ contain divalent silicon centers. These divalent silicon centers are replaced by trivalent and higher valency centers in silicon clusters with six or more atoms. In this light, the non-reactiveness of Si_6^+ and Si_7^+ may well reflect not the overall reaction thermochemistry but rather the chemistry available to their valence electrons. Similarly, larger silicon clusters are predicted to be unreactive towards SiD_4.

V. Implications for prenucleation of hydrogenated silicon particles in silane plasmas and vapors

Chemical vapor deposition of epitaxial, polycrystalline and amorphous silicon films is widely used in the fabrication of VLSI circuits, solar cells and other silicon devices. Silane is the source gas of choice for deposition of polycrystalline and amorphous silicon films; it is also used to deposit epitaxial films when low deposition temperatures are required to minimize diffusion, autodoping, etc [6]. All of these silane processes share a serious problem: gas phase nucleation of silicon particles which ultimately contribute to poor film quality and device failure [16].

Despite numerous investigations, the formation mechanisms of these particles has yet to be proven [17]. Several chemical models have been developed to explain the formation of these particles [3]. At the heart of the models lies chemical condensation reactions of the form,

$$Si_xH_y^{(+)} + SiH_4 \rightarrow Si_{x+1}H_z^{(+)} + mH_2 + nH$$
$$y = 2m + n + z - 4 \qquad (3)$$

where the superscript "(+)" signifies that the species may be either neutral or positively charged. These models have become increasingly sophisticated in their treatments of the important chemical and physical conditions in high temperature vapors and plasma reactors. Nonetheless, they are severely limited by the lack of hard chemical data concerning the reaction rates and energetics for the quintessential condensation reactions embodied by (3). In the absence of this information, all of these models have been forced to make simplifying assumptions about the condensation reactions. One assumption is that a single reaction rate can be used to describe all of the condensation reactions embodied by (3), except for the association reaction rate ($n+m=0$) which involves three body collisions. As a corollary to this working hypothesis, cluster growth is characterized by the number of silicon atoms regardless of the number of hydrogens. Another assumption is that the structures of the evolving silane species do not significantly affect their chemistry. For example, it has been invoked that these may consist of linear silane polymer chains with an unsaturated "growing" end. Finally, it is assumed that the condensation reactions are limitless and that termination can only occur by another chemical or physical route, such as deposition on a surface.

The results of our studies of the sequential clustering reactions of silicon clusters cations with silane clearly indicate that simplifying assumptions about the chemistry of embryonic hydrogenated silicon particles should be postponed. We find that the rates and products of their reactions varies with the cluster size and number of deuteriums. Table 1 lists reaction rates that span three orders of magnitude for condensation reactions of the type in (3). Furthermore, higher $Si_xD_y^+$ species appear to have highly diverse structures involving branched chains, rings and more highly compact forms. These geometric structures have a profound influence on the electronic structures —and therefore the chemistry of these higher $Si_xD_y^+$ ions. Finally, all of the clustering sequences that we have monitored appear to be highly constrained. Most of them terminate and further cluster growth can only occur by inefficient bimolecular association reactions. In cases where the sequence termination is not directly observed, all but one of the clustering sequences proceeds to a reaction with a probability of less than 0.1% of the ion-molecule collision rate. All of these observations point to the need for further study of the individual reactions which lead to spontaneous growth of hydrogenated silicon particles. Key information on the prenucleation steps may eventually lead to rational design of intervention measures to prevent or control the occurrence of these particles.

VI. Conclusions

Bare silicon cluster ions undergo exothermic sequential clustering reactions with SiD_4. The overall reaction in all of the sequences is the addition of SiD_2 accompanied by loss of D_2. None of the clustering sequences exhibit unconstrained growth of increasingly larger $Si_xD_y^+$ species. Of the cluster ions studied, $Si_{2-7}^{+/-}$, only $Si_{2-3,5}^+$ react with SiD_4 in their ground state. No reactions are found for the negative silicon cluster ions.

Electronic structure theoretical calculations of the energetics of various possible forms are used to deduce the structures of the products ions in these clustering reactions. Overall, it appears that the structures of the growing $Si_xD_y^+$ ions are closely related to the lowest energy structures that have been found for the bare clusters when the degree of hydrogen saturation is low. This suggests that considerable rearrangement of the silicon framework in these clusters is possible during the reaction. Nonetheless, non-converting geometric isomers are also found which indicates that not all structural rearrangements are energetically accessible. The unreactiveness of the largest clusters appears to arise from the lack of the appropriate divalent silicon center which is necessary for the Si-D insertion transition state early along the reaction pathway. The nonavailability of the divalent silicon centers occurs because these larger silicon clusters have increased intramolecular silicon-silicon binding resulting in higher valencies.

The early bottlenecks found in the sequential clustering reactions of Si_x^+ with SiD_4 indicate that ground state, bare silicon cluster ions are not the important precursors in the formation of hydrogenated silicon dust in silane plasmas. Furthermore, the chemical constraints that are found for these clusters are undoubtedly ubiquitous in the reactivity of unsaturated $Si_xD_y^+$ species. A more complete understanding of the prenucleation of particulates awaits additional information about the structures and chemistry of these types of unusual species.

We thank R. Raghavachari for communicating numerous results to us prior to publication.

References

1. Hidy, G.M., Brock, J.R.: Dynamics of aerocolloidal systems. New York: Pergamon Press 1970
2. Nucleation phenomena. Zettlemoyer, A.C. (ed.) New York: Elsevier 1977
3. (a) Haller, I.: Appl. Phys. Lett. **37**, 282 (1980); (b) Chatham, H., Gallagher, A.: J. Appl. Phys. **58**, 159 (1985); (c) Weakliem, H.A., Estes, R.D., Longeway, P.A.: J. Vac. Sci. Technol. A **5**, 29 (1987); (d) Kushner, M.J.: J. Appl. Phys. **63**, 2532 (1988)
4. Raghavachari, K.: J. Chem. Phys. **88**, 1688 (1988)
5. Raghavachari, K.: unpublished results
6. Chesnavich, W.J., Bowers, M.T.: Progr. React. Kinet. **11**, 137 (1982)
7. Mandich, M.L., Reents, Jr., W.D., Jarrold, M.F.: J. Chem. Phys. **88**, 1703 (1988)
8. Raghavachari, K.: J. Chem. Phys. (to be published)
9. (a) Jones, R.O.: Phys. Rev. A **32**, 2589 (1985); (b) Grev, R.S., Schaefer, H.F.: Chem. Phys. Lett. **119**, 111 (1985); (c) Raghavachari, K.: J. Chem. Phys. **83**, 3520 (1985); (d) Balasubramanian, K.: Chem. Phys. Lett. **N125**, 400 (1986)
10. Raghavachari, K., Logovinsky, V.: Phys. Rev. Lett. **55**, 2853 (1985)
11. Berkowitz, J., Greene, J.P., Cho, H., Ruscic, B.: J. Chem. Phys. **86**, 1235 (1987)
12. (a) Mandich, M.L., Reents, Jr., W.D., Bondybey, V.E.: J. Phys. Chem. **90**, 2315 (1986); (b) Mandich, M.L., Reents, Jr., W.D., Bondybey, V.E.: In: Physics and chemistry of small clusters. Jena, P., Rao, B.K., Khanna, S.N. (eds.). New York: Plenum 1987; (c) Creasy, W.R., O'Keefe, A., McDonald, J.R.: J. Phys. Chem **91**, 2848 (1987)
13. (a) Bloomfield, L.A., Freeman, R.R., Brown, W.L.: Phys. Rev. Lett. **54**, 2246 (1985); (b) Liu, Y., Zhang, Q.-L., Tittel, F.K., Curl, R.F., Smalley, R.E.: J. Chem. Phys. **85**, 7434 (1986); (c) Begemann, W., Dreihofer, S., Gantefor, G., Siekmann, H.R., Meiwes-Broer, K.H., Lutz, H.O.: Elemental and molecular clusters. Martin, T.P. (ed.) Berlin, Heidelberg, New York: Springer (to be published)
14. Raghavachari, K., Rohlfing, C.M.: Chem. Phys. Lett. **143**, 428 (1988)
15. Gaspar, P.P.: In: Reactive intermediates, Vol. 3. Jones, Jr., M., Moss, R.A. (eds.). New York: Wiley 1985
16. VLSI Technology, Sze, S.M. New York: McGraw-Hill 1983
17. See, for example, (a) Haller, I.: J. Vac. Sci Technol. A **1**, 1376 (1983); (b) Roth, R.M., Spears, K.G., Stein, G.D., Wong, G.: Appl. Phys. Lett. **46**, 253 (1985); (c) Breiland, W.G., Ho, P., Coltrin, M.E.: J. Appl Phys. **60**, 1505 (1986)

Z. Phys. D – Atoms, Molecules and Clusters 12, 563–565 (1989)

Atoms, Molecules
Zeitschrift für Physik D and Clusters
© Springer-Verlag 1989

Monitoring adsorption on small particles

H. Burtscher, U. Müller and A. Schmidt-Ott
Laboratorium für Festkörperphysik, ETH-Hönggerberg, CH-8093 Zürich, Switzerland

Received 5 July 1988; final version 30 September 1988

A review on the application of photoelectron emission from small particles suspended in a gas to monitor the adsorption properties and to detect adsorbates on the particles is given. The relation between surface coverage with an adsorbate and the shift in photothreshold and photoelectric yield is demonstrated by two examples: The adsorption of O_2 on silver particles and of perylene on carbon particles. A third example concerns the detection of polyaromatic hydrocarbons (PAH) on particles arising from combustions.

PACS: 82.65.i; 79.60.i; 82.70.y

Introduction

For the interaction of particles with their surrounding, their adsorption properties are of great importance. A field where adsorption plays a crucial role is heterogenous catalysis. Further, ambient air aerosols often consist of a core coated with toxic species (e.g. polyaromatic hydrocarbons on combustion particles which, though neglegible in mass, are responsible for the mutagenicity of the particles).

In the following it will be demonstrated how photoelectron emission from particles, suspended in a carrier gas, can be used to detect adsorbates. Photoelectron emission is a process which is very sensitive to the particle surface. Small changes in the concentration of adsorbed material corresponding to fractions of monolayers can significantly alter photoemission, mainly the photothreshold Φ. Adsorbed atoms or molecules lead to the formation of electrical dipoles on the surface, which decrease Φ if the adsorbate is electropositive compared to the core and increase it if the adsorbate is electronegative. The shift $\Delta\Phi$ is related to the surface fraction Θ covered by the absorbate. Often it is proportional to Θ and can therefore be used directly as a measure for adsorption.

Φ can be determined by measuring the photoelectric yield Y as function of the photon energy $h\nu$. Y is the probability of electron emission per incident photon. Near threshold, Y and Φ are related by the Fowler–Nordheim law [1]

$$Y = c(h\nu - \Phi)^m, \tag{1}$$

where c is a constant, depending for example on light absorption, and m is 2 for metals [2].

Figure 1 shows a simplified diagram of an apparatus to measure photoemission. The measurement is done in situ, the gas containing the particles flowing through the system. In the photoemission chamber the particles are irradiated by UV-light of variable wavelength. With a probability given by (1), electrons are emitted and the particles remain with a positive charge. Due to their much higher mobility the electrons or small ions formed by them in the gas rapidly diffuse to the wall of the chamber, whereas the particles follow the gas flow. They flow through a filter, where their positive charge produces a current which is proportional to Y [3].

Experimental results

The first example concerns the adsorption of O_2 on silver particles of about 5 nm in diameter. The experiment is done in ulraclean Helium as carrier gas. Ag particles are produced by evaporation from a heated silver wire. After particle production, oxygen is added

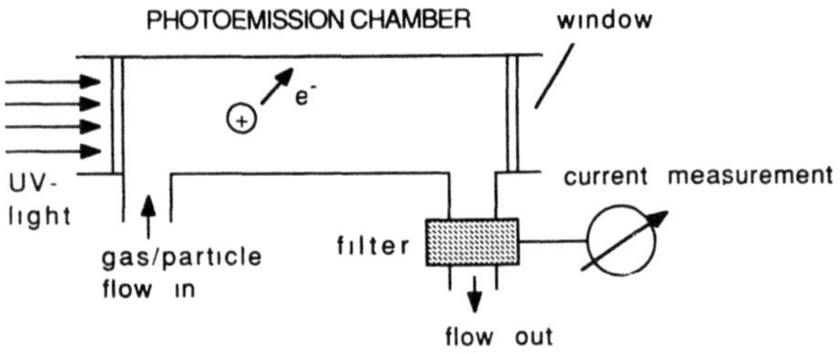

Fig. 1. Schematic diagram of a setup to measure photoelectron emission from particles in a gas flow

to the gas. A more detailed description of the experiment is given in [4]. Figure 2 shows the threshold shift as function of the O_2 concentration. For this system the surface coverage Θ can be written as [5]

$$\Theta = \Delta\Phi/\Delta\Phi_{max} \tag{2}$$

Calculating $\Theta(n_{O_2})$ by determining the flux of oxygen molecules to the particles leads to [4]

$$\Theta = \left(1 + \sqrt{\frac{2\pi m}{kT}} \frac{N_0}{\alpha_0 n_{O_2} t}\right)^{-1}, \tag{3}$$

where m is the molecular weight, n_{O_2} the oxygen concentration, α_0 the sticking coefficient at low surface coverage, N_0 the density of adsorption sites and t the exposure time.

The solid line in Fig. 2 is determined from (3) with α_0/N_0 fitted to match the experimental data. Whereas the shape of the theoretical result fits the experimental data very well, α_0/N_0 obtained by the fit is two orders of magnitude smaller than for flat surfaces. This difference cannot be explained so far, but together with other observations [4] it indicates basic differences in the process of adsorption on particles and flat surfaces. In this experiment Φ has been determined by measuring Y as function of $h\nu$. Another result of the experiment is that the constant c in (1) does not depend on the O_2-concentration. This means that once c is known it is no longer necessary to do the experiment with more than one light wavelength to determine Φ.

Another system investigated is perylene on C-particles. This is a simple model for particles arising from combustions. In this experiment the radius increase ΔR of the particles due to adsorption of the PAH was measured by a diffusion battery [6]. This method is sensitive enough to determine a ΔR due to fractions of monolayers. Comparing the radius increase with the photoemission signal again shows that photoemission yields information on the amount of PAH on the particles and on the adsorption sites available [7].

So far experiments with well defined systems have been presented, where the process of adsorption is monitored by photoemission. The example to follow deals with 'real' combustion aerosols. Photoemission from car exhaust particles is measured as a function of the temperature in a heated tube, preceeding the photoemission chamber. Figure 3 shows $\Phi(T)$ obtained from this experiment. With increasing temperature, material reducing Φ desorbs and Φ increases. At about 600°C the curve saturates at $\Phi = 4.7$ eV, which is the value expected for carbon particles. This indicates

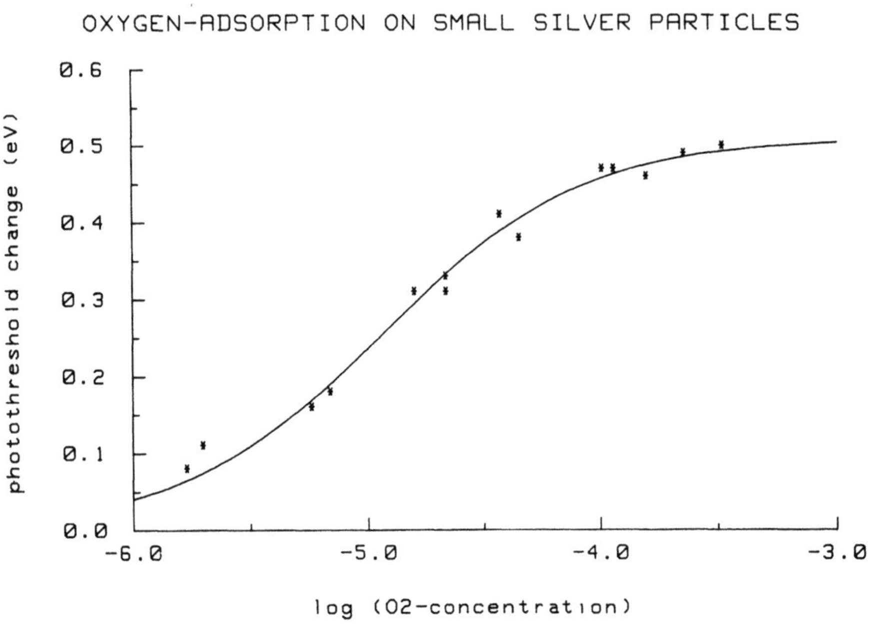

Fig. 2. Threshold shift versus O_2 concentration, *: exp. data, solid line: calculated from (3)

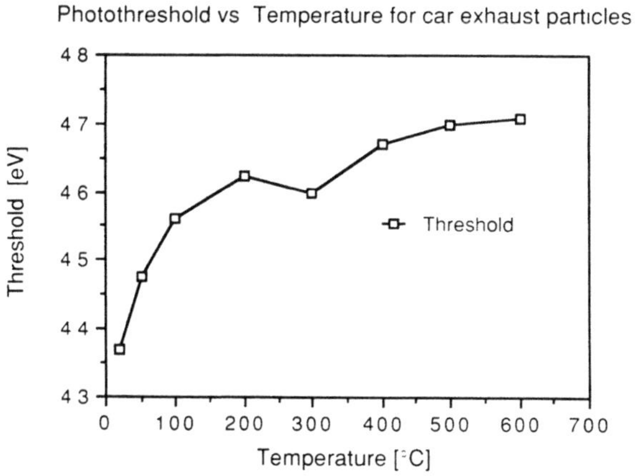

Fig. 3. Threshold shift versus temperature of car-exhaust particles

that at this temperature all adsorbates influencing Φ desorb.

Photoemission from combustion aerosols is dominated by the effect of PAH adsorption. They cause the strongest decrease in Φ. Comparing the total PAH concentration on the particles determined by gas chromatography with the photoemission signal shows a linear relation between these two quantities with a correlation coefficient of 0.98 [8].

Conclusions

Photoemission from small particles can be used to detect adsorbates on particles suspended in a gas and can be applied to study the adsorption process, which is important for example in heterogeneous catalysis. The method is in situ and very fast. A broad field of applications are combustion aerosols, where it allows a simple and sensitive detection of PAH's on the particles. The measurement can also be used to monitor and control combustion processes [9], since response times below one second can be achieved.

References

1. Fowler R.M.: Phys. Rev. **38**, 45 (1931)
2. Cardona M., Ley L.: Photoemission in solids I, pp. 23. Berlin, Heidelberg, New York: Springer 1978
3. Burtscher H., Scherrer L., Siegmann H.C., Schmidt-Ott A., Federer B.: J. Appl. Phys. **53**, 3787 (1982)
4. Müller U., Schmidt-Ott A., Burtscher H.: Phys. Rev. Lett. **58**, 1684 (1987)
5. Schmiedl E., Wissmann P.: Vak. Techn. **33**, 78 (1984)
6. Scheibel H.G., Porstendörfer J.. J. Aerosol. Sci. **15**, 673 (1984)
7. Burtscher H., Schmidt-Ott A.: J. Aerosol. Sci. **17**, 699 (1986)
8. McDow S., Giger W., Burtscher H., Schmidt-Ott A.. 3rd Int Conf. on Carb. Part., Berkely Oct. 5–8, Lawrence Berkely Lab. LBL-23997, p. 57 (1987)
9. Burtscher H., Schmidt-Ott A., Siegmann H.C.: Aerosol. Sci. Technol. **8**, 125 (1988)

Z. Phys. D – Atoms, Molecules and Clusters 12, 567–570 (1989)

Atoms, Molecules and Clusters
Zeitschrift für Physik D
© Springer-Verlag 1989

Reactions of small carbon cluster ions with HCN: a structural probe

D.C. Parent and S.W. McElvany
Chemistry Division/Code 6110, Naval Research Laboratory, Washington, DC 20375-5000, USA

Received 19 July 1988

The results of a detailed study of the primary and secondary reactions of carbon cluster ions, $C_n^+ (3 \leq n \leq 20)$, with HCN are presented and discussed. The experiments were performed in a Fourier transform (ICR) mass spectrometer, using direct laser vaporization of graphite to form the carbon cluster ions. Evidence for two structural forms of the $n=7$, 8 and 9 cluster ions is obtained from their differing reactivity with HCN. The C_7^+ ion is anomalous in its behavior in many respects, which is interpreted by an isomerization mechanism. The HCN reactions offer a contrast to the reactions with nonpolar neutrals studied previously. All HCN reactions produced ions of the type C_nX^+ (primary product) or C_nXY^+ (secondary product) where X, Y = H, CN or HCN. Fragmentation of the original carbon cluster was not observed, while radiative association is an important reaction channel. Low-energy collision-induced dissociation studies of the product ions support the mechanism of insertion into the H-CN bond and formation of covalent bonds at the carbene site for the primary reactions. In most secondary reactions however, the HCN associates weakly with the ion, rather than binding covalently.

PACS: 36.40; 82.30

I. Introduction

The emphasis of experimental and theoretical research on semiconductor and metal clusters in the last few years has been on the elucidation of their structures [1–3]. Using chemical reactivity as a probe of ion structure, recent work in our laboratory provided the first evidence for the existence of structural isomers of a semiconductor/metal cluster, C_7^+ [2]. An abrupt decrease in the reaction rate constant between C_9^+ and C_{10}^+ was also observed, which correlates with theoretical predictions of a change in structure from linear to monocyclic at $n=10$ [3]. The proposed structure for the small ($n<10$) clusters has reactive carbene sites (a pair of nonbonding electrons) at the end of the linear chains, which are absent in the $n \geq 10$ cyclic structures [4]. In the case of C_7^+ it was postulated that the isomer of high reactivity is linear while the other isomer, with a much lower reactivity, is cyclic.

For the present study, we chose HCN as the neutral reactant since it is highly polar in contrast to the nonpolar reagents used to date, and is very polarizable as well. Thus the behavior of the C_n^+/HCN system may differ significantly from those systems previously studied. Also, HCN is an abundant species in the interstellar medium (ISM) and its reactions with other ISM species such as carbon chains are of interest for chemical modeling [5]. We measured the rate constants and branching ratios for the primary and secondary reactions of $C_n^+ (n=3–20)$ with HCN, and used collision-induced dissociation (CID) to investigate the structure (bonding) of the product ions.

II. Experimental method

Experiments were performed in a Fourier transform (ICR) mass spectrometer (FTMS) [2]. The graphite sample is positioned so that it is flush with one of the trapping plates (90% transparent nickel mesh). The frequency-doubled Nd:YAG laser beam (532 nm, 1–5 mJ/pulse) was focused onto the graphite sample to produce the carbon clusters by direct laser vaporization (DLV) in a low pressure environment.

The resonant ejection capabilities of the FTMS allow the isolation of a particular mass ion for study. The reaction rate constant is determined by following the pseudo first-order decay of the reactant ion as a function of time at constant HCN pressure. The slope of this decay, divided by the HCN pressure (typically $1–4 \times 10^{-7}$ torr), yields the rate constant. Due to various sources of error, particularly in the determination of the absolute HCN pressure, the absolute rate constants reported could be in error by as much as a factor of 2. However, the differences in the relative rate constants should be significant, as they were measured under similar conditions.

III. Results

III.A. Primary reactions

The rate constants for the primary reactions, normalized by the C_4^+ reaction rate constant, are shown in Fig. 1. As observed in other reactions, there is a sharp drop in rate constant between C_9^+ and C_{10}^+, which is attributed to a change in structure from linear to cyclic [2, 6]. A new result from this work is that both C_8^+ and C_9^+, as well as C_7^+, have two components of differing reactivity with HCN. The parameters characterizing the two component decays are listed in Table 1.

For the "linear" isomers, Fig. 1 shows there is very little variation (factor of 2) in their reactivity with HCN. This is in contrast to the reactions with nonpolar neutrals, whose rate constants varied by a factor of 4 to 70 [2, 6]. Thus the large polarizability and

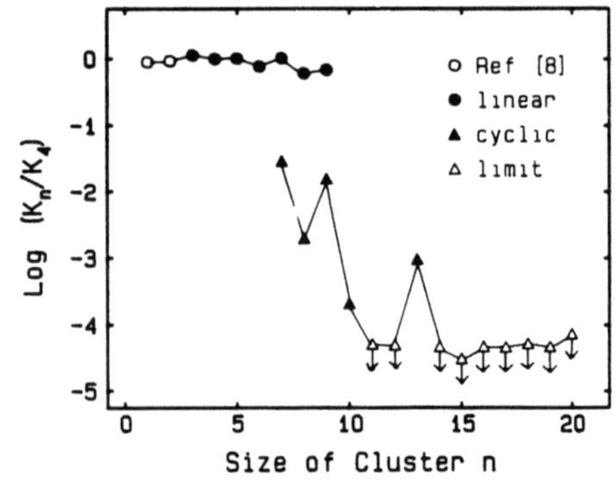

Fig. 1. The rate constants for the primary reactions of C_n^+ with HCN, normalized to $n=4$, are plotted versus cluster size for the linear and cyclic forms of the clusters, (circles and triangles respectively). The values for C^+ and C_2^+ [8], normalized by the C_4^+ reaction rate constant measured in the same work, are included for completeness. Arrows on the open triangles indicate these are upper limits

Table 1. Parameters for the biexponential decay curves

Size of cluster	Fraction of slow reactant	k_{fast}/k_{slow}
7	0.7	36
8	0.3	300
9	0.2	43

dipole moment of HCN appear to facilitate the reactions. The even/odd alternation of rate constants (odd clusters are more reactive) noted in previous work, is also observed for the HCN reactions, particularly for $n=5$ to 9. The "cyclic" $n=7$ to 10 and 13 cluster ions show quite pronounced variations in reactivity of over two orders of magnitude. Again, it is the odd clusters which are more reactive.

The products formed by the primary reactions are mostly addition of H, CN or HCN to the cluster ion, as shown in Fig. 2 for the "linear" isomers. (The "cyclic" isomers react only by radiative association). The branching ratios for the "linear" clusters show several superimposed trends. The first is that the smaller clusters stabilize the collision complex by eliminating part of the HCN, while the larger clusters can accommodate the excess energy long enough to radiatively stabilize. Secondly, H is preferentially eliminated, suggesting it has the weaker bonding to the cluster. Thirdly, the even clusters radiatively associate with HCN to a greater extent than the odd clusters.

No products resulting from breakup of the carbon cluster were observed, making the reactions with HCN unique. It is also interesting to note the facility with which the larger clusters undergo radiative association. This was not observed in any of the previous studies except for $n=7$ to 10 with C_2H_2 to a lesser extent [6]. The high degree of association and the lack of cluster breakup indicate that the C_n^+/HCN reaction is a low energy process.

III.B. Secondary reactions

The reactions of all the primary product ions, C_nH^+, C_nCN^+ and C_nHCN^+, were studied and the normalized reaction rate constants for each series of reactant ions are shown in Fig. 3. Note the different trends displayed by the three series of reactant ions. Most of the secondary reactions form only the association product. A small amount of proton transfer was observed with C_3H^+. C_4X^+ and C_6X^+ (X=H and CN) gave some HCN fragmentation products, suggesting that these reactions are more exothermic than those of the corresponding ions produced from the odd carbon clusters. No tertiary products were observed, which appears to agree with the previously

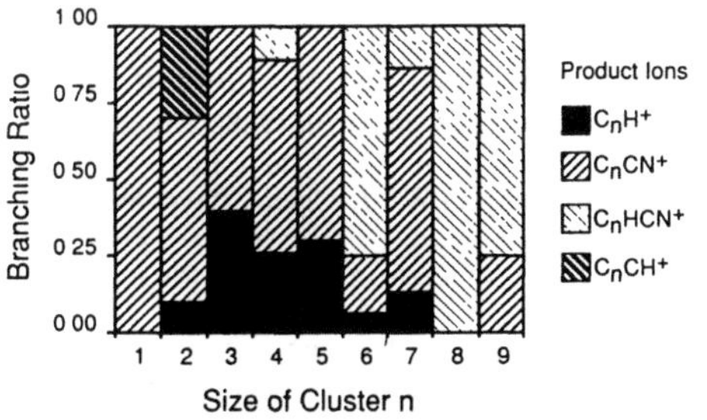

Fig. 2. The branching ratios for the primary reactions of the "linear" carbon cluster ions with HCN are shown. The values for C^+ and C_2^+ were taken from [8]

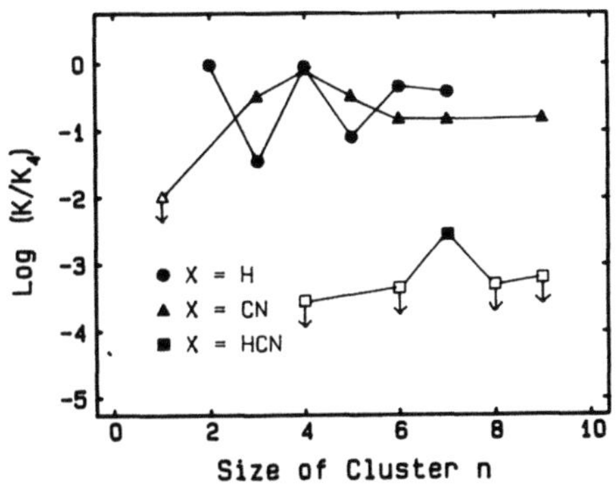

Fig. 3. The rate constants for the secondary reactions with HCN, normalized by the primary reaction rate constant for C_4^+, are plotted versus cluster size for the three series of reactant ions C_nX^+. The values for C_2H^+ and C_2N^+ were taken from [8]. Open symbols with arrows are upper limits

proposed mechanism, whereby reactions occur at the two carbene sites of the linear cluster ions [2].

III.C. Collision-induced dissociation

The primary products C_nH^+ and C_nCN^+ showed loss of H and CN respectively. Thus the weakest bond in the product is to the HCN fragment, in agreement with the observed lack of cluster breakup in the primary reactions. The "linear" C_nHCN^+ lost H and CN as well. These results are consistent with insertion into the H-CN bond and covalent bonding at the carbene site. Note that the ions produced in the low energy CID of the association product are the same as those formed by ion-molecule reactions, suggesting that C_nHCN^+ is the stabilized intermediate in these reactions. CID spectra of the cyclic C_nHCN^+ ($n = 7-9$) are quite different. For $n = 8$ and 9 a large loss of *only* HCN is observed at very low excitation energies, indicating that a weak bond is formed. The $n = 7$ ion loses H and CN at much higher excitation energy and more closely resembles the "linear" ions in its behavior.

The secondary product ions (C_nXHCN^+, X = H, CN or HCN) lose HCN at very low CID excitation energies. Coupled with the observation that most secondary reactions add HCN only, the CID results indicate that the second HCN does not react with the carbene end of the C_nX^+ to bond covalently. Instead, hydrogen bonding to the "X" end of the cluster ion is proposed to be the favored pathway, by analogy to van der Waals complexes of HCN with itself and unsaturated hydrocarbons.

IV. Discussion

Looking at the trends in reactivity with cluster size for all the reactions of C_n^+ and C_nX^+ studied to date, we can formulate the following empirical "rule". *If the ions in a series have an odd (even) number of electrons, the odd (even) n ions will be more reactive than the even (odd) n ions, irrespective of the neutral reactant.* By this rule, for C_n^+, C_nO^+ and C_nHCN^+ odd *n* ions will be more reactive whereas the even *n* ions will be more reactive for C_nH^+ (C_nD^+) and C_nCN^+. We do not know what the underlying physical explanation for this "rule" is, though it undoubtedly is based on structural and/or energetic considerations.

In a number of ways, C_7^+ exhibits rather unusual behavior. For example, the amount of "cyclic" form is high compared to C_8^+ and C_9^+. C_7HCN^+ is more reactive than the other C_nHCN^+. "Cyclic" C_7HCN^+ is more strongly bound than other "cyclic" C_nHCN^+ ions. The CID losses for "cyclic" C_7HCN^+ are the same as for "linear" C_7HCN^+, indicating covalent bonding, instead of associative as is seen for the other "cyclic" C_nHCN^+. And lastly, the shape of the decay curve for C_7^+ is not a simple biexponential as was found for C_8^+ and C_9^+. From these observations, we suggest that the cyclic C_7^+ isomerizes to the linear form during the reaction with HCN [7]. The relatively high abundance of the seven-carbon ring is surprising, as it appears to be less stable than the eight- or nine-carbon rings.

V. Conclusions

There are a number of interesting results which have come out of the study of HCN reactions with small positive carbon cluster ions. Besides C_7^+, C_8^+ and C_9^+ are also present as two different forms. The reactivity and CID studies suggest that the isomers are linear and cyclic in structure. The C_7^+ cluster behaves in an anomalous fashion, which we interpret in part as due to the isomerization of the cyclic form during reaction. The reactivity of carbon clusters and their derivatives follows a simple rule correlated to the radical/nonradical nature of the ion though we do not know the physical basis of this correlation. The first HCN seems to add covalently at the carbene site, but the second HCN preferentially associates with the ion, presumably by hydrogen bonding to the first HCN fragment. The efficient radiative association observed with HCN may have important implications for interstellar chemistry.

References

1. Some recent papers are: Raghavachari, K., Binkley, J.S.: J. Chem. Phys. **87**, 2191 (1987); Bréchignac, C., Cahuzac, Ph., Roux, J. Ph.: J. Chem. Phys. **87**, 229 (1987); Jarrold, M.F., Bower, J.E.: J Chem. Phys. **87**, 5728 (1987); Parks, E.K., Weiller, B.H., Bechthold, P.S., Hoffman, W.F., Nieman, G.C., Pobo, L.G., Riley, S.J.: J. Chem Phys. **88**, 1622 (1988); Zhang, Q.-L., Liu, Y., Curl, R.F., Tittel, F.K., Smalley, R.E.: J. Chem. Phys. **88**, 1670 (1988); Zakin, M.R., Brickman, R.O., Cox, D.M., Kaldor, A.. J. Chem. Phys. **88**, 3555 (1988) Ross, M.M., McElvany, S.W.: J. Chem. Phys. **89**, 4821 (1988)
2. McElvany, S.W., Dunlap, B.I., O'Keefe, A.: J. Chem. Phys. **86**, 715 (1987)
3. Bernholc, J., Phillips, J C.: J. Chem Phys. **85**, 3258 (1986)
4. Raksit, A.B., Bohme, D.K.: Int. J. Mass Spectrom. Ion Proc. **55**, 69 (1983)
5. Mann, A.P.C., Williams, D.A.: Nature (London) **283**, 721 (1980) Bohme, D.K., Wlodek, S., Raksit, A.B.: Can. J. Chem. **65**, 2057 (1987)
6. McElvany, S.W.: J. Chem. Phys. **89**, 2063 (1988)
7. Parent, D.C., McElvany, S.W.: J. Am. Chem. Soc. **111**, 2393 (1989)
8. Anicich, V.G., Huntress, W.T. Jr., McEwan, M.J.: J. Phys. Chem **90**, 2446 (1986)

Z. Phys. D – Atoms, Molecules and Clusters 12, 571–573 (1989)

Atoms, Molecules
Zeitschrift für Physik D and Clusters
© Springer-Verlag 1989

Reaction of Al_n clusters with oxygen and ammonia

K. Kaya, K. Fuke, S. Nonose, and N. Kikuchi
Department of Chemistry, Keio University, Yokohama, Japan

Received 5 July 1988; final version 30 September 1988

Aluminum clusters were generated by a laser vaporization method. Ionization potential of Al_2 was found to be lower than that of Al. In the reaction with oxygen at high concentration, the mixed cluster Al_9O_7 survives predominantly as a stable cluster. Ammonia was found to be adsorbed weakly on the Al_n surface.

PACS: 82.30; 36.40

Introduction

The structure and reactivity of metal clusters have been the subject of numerous studies which may bridge the gap between molecular and surface sciences. In these studies, much efffort has been paid to clarify the relationship between electronic and geometric structure of clusters and the reactivity toward molecules such as H_2 and O_2. In the past two years, the reactivity of aluminum cluster ions has been studied extensively by Jarrold and Bower [1], and also by Anderson and coworkers [2] under the selectivel collision energy condition. Cox et al. recently published the results of reactivity of neutral aluminum clusters toward several simple molecules [3].

In the present work, the reactions of aluminum clusters with oxygen and ammonia were studied under multiple collision condition using a fast flow reactor. Oxygen was found to react rapidly with Al_n and a stable mixed cluster Al_9O_7 survives predominantly after the multiple collisions of Al_n with oxygen. In contrast to oxygen, ammonia interacts weakly with Al_n in a physisorbed form.

Experimental

Aluminum clusters were generated by a Smalley type pulsed laser vaporization source [4] from an aluminum rod in 10 atm of He buffer gas. A YAG second harmonic was used as the vaporization laser. The fast flow reaction cylinder has a dimension 5 cm long by 5 mm diameter, into which is fired a second He pulse containing the reactant gas. Cluster reaction products were detected by a TOF mass spectrometer using ArF laser as an ionization source. LeCroy 9400 transient digitizer assisted by NEC 9800 microcomputer was used for the averaging of TOF mass spectra.

Results and discussion

(1) Ionization potential

We have tried to determine the IP values of Al clusters by photoionization using frequency doubled dye laser excited by a YAG laser in the wavelength region longer than 207 nm. Free Al atom was found to have IP value of 5.986 eV which is exactly the same as reported in the literature. Al_2 exhibits an ionization threshold 40 meV below that of free Al. The result is in contradiction to the predictions suggested by CID experiment and others. The explanation consistent both for our result and CID experiment is that the observed ionization at 5.94 eV comes from the photo-ionization from the lowest singlet excited state ($^1\Sigma_g$) instead of the ionization from the ground state.

Because of the fragmentation of ionized clusters, IP values of Al clusters are very hard to be determined. Following conclusions were drawn from our experiment.

(a) IPs of trimer to hexamer are higher than 6.42 eV.
(b) Al_{13} has the IP value very close to 6.42 eV.
(c) Al_n (*n* is larger than 14) has IP value less than 5.5 eV.
The conclusions *(a)* and *(b)* are the same as those drawn by Cox et al..

(2) Chemisorption of Al_n with O_2

Reaction of Al clusters with oxygen was studied using a fast flow reaction cell. Figure 1 displays TOF mass spectra of Al clusters reacted with O_2. In order to avoid saturation effect and multiphoton processes, the mass spectra were obtained using unfocused 20 μ Jcm^{-2} ArF excimer laser as an ionization photon source. In the figure, bottom trace (Fig. 1a) is the TOF mass spectrum of bare Al clusters where pure He was injected into the reaction cell. Figure 1b and c in the middle and upper trace represent the results of the reaction of Al_n with O_2 (10 and 20 Torr O_2 with 2 atm He for 1b and c). As seen in the figures, extensive reaction is observed for the higher partial pressure of oxygen, and the oxide products Al_5O_3, Al_7O_4 and Al_9O_7 dominate the mass spectra. In order to confirm the above product assignments, similar experiments were conducted with $^{18}O_2$. The results of these experiments supported the above assignments. The most significant result is that as the oxygen pressure is increased, the oxide Al_9O_7 survives as a stable cluster and no other peak is observed in the TOF spectrum. In order to clarify the mechanism of the observed enhancement, we mixed N_2 (50 torr) with a lower partial pressure of O_2. Then signal intensities of oxide peaks were found to grow up appreciably. Ionization laser fluence was found to have nothing to do with the enhancement of the oxide peaks. It was concluded that the enrichment of oxide peaks comes not from the fragmentation of cluster ions by the ionization laser but from collision induced fragmentation in the reaction tube. Another interesting aspect of this reaction is that stable oxides contains odd number of oxygen. This indicates that the reaction of oxygen on Al clusters is in a form of dissociative chemisorption where the adsorbed O_2 molecules decompose into atoms on the Al clusters. In order to understand the reactions more deeply, one needs to get structural information on the oxide clusters by spectroscopic or diffraction methods.

(3) Physisorption of Al_n with NH_3

Figure 2 shows the TOF mass spectra of Al_n mixed with NH_3 whose partial pressure is 8 torr with 2 atm He. The bottom trace is the mass spectrum of bare Al clusters. As is clear from the figures, the depletion of Al_n signals in the reaction with ammonia is not so prominent as those for oxygen even though the partial pressure of NH_3 is increased. Also, by increasing the

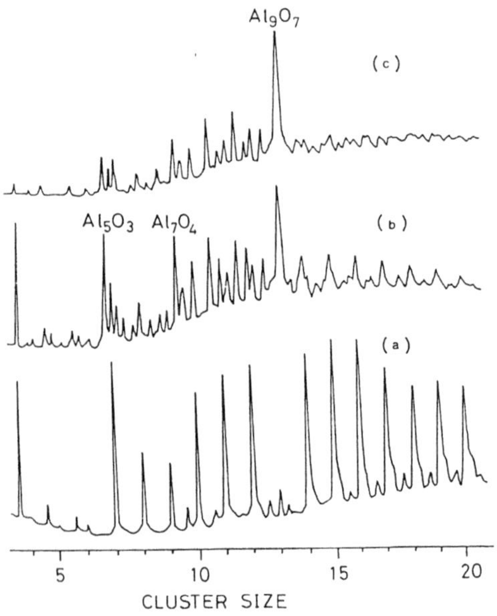

Fig. 1a–c. TOF mass spectra of Al_n reacted with oxygen. (**a**) is the spectrum of bare aluminum clusters. (**b**) and (**c**) represent the spectrum of Al_n mixed with 10 Torr (**b**) and 50 Torr (**c**) of oxygen in a fast flow reaction tube

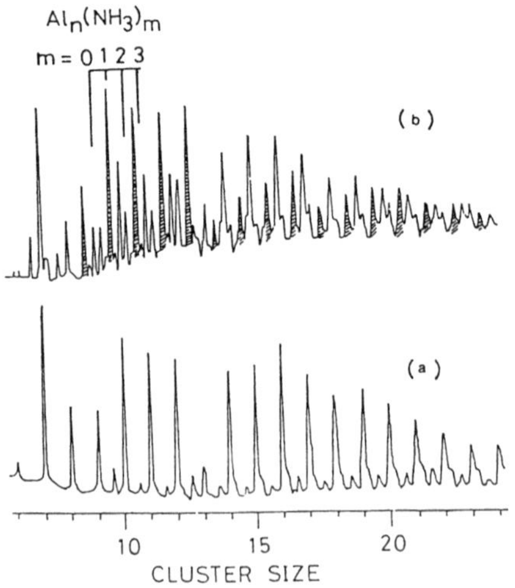

Fig. 2a and b. TOF mass spectra of Al_n reacted with ammonia. (**a**) stands for the bare aluminum clusters for comparison. (**b**) is the result of reaction of Al_n reacted with 8 Torr NH_3. Shaded peaks in the figure correspond to $Al_n(NH_3)_{m=1}$. Mass peaks corresponding to $m=2$ and 3 are also seen in the figure

fluence of the ionization laser, the adsorbed NH_3 molecules is easily desorbed and the mass spectrum pattern goes back to that of bare Al clusters. This seems to indicate that the ammonia reacts gently with Al clusters to form physisorbed products. We carefully examined the stability of Al_nNH_3 against the clustering number n. It was found that the stability increases from $n=7$ to 12 peaking at $n=9$, then drops abruptly at $n=13$, and then increases slowly as n is increased. This tendency seems to be explained by the n dependent variation of the polarizability of Al clusters in analogy with that of sodium and potassium clusters [5].

This work is supported by the Grant-in-Aid for Scientific Research on Priority Areas from Ministry of Education, Science and Culture.

References

1. Jarrold, M.F., Bower, J.E.: J. Chem. Phys. **87**, 5728 (1987)
2. Hanley, L., Ruatte, S.A., Anderson, S.L.: J. Chem. Phys. **87**, 260 (1987)
3. Cox, D.M., Trevor, D.M., Whetten, R.L., Kaldor, A.: J. Phys. Chem. **92**, 421 (1988)
4. Morse, M.D., Geusac, M.E., Heath, J.R., Smalley, R.E.: J. Chem. Phys. **83**, 2293 (1985)
5. Knight, W.D., Clemenger, K, de Haar, W.A., Saunders, W.A.: Phys. Rev. **B31**, 2539 (1985).

Z. Phys. D – Atoms, Molecules and Clusters 12, 575–578 (1989)

Atoms, Molecules
Zeitschrift für Physik D and Clusters
© Springer-Verlag 1989

Size effects in the CO-oxidation over supported Pd particles

N. Kruse* and E. Gillet

Laboratoire de Microscopie et Diffractions Electroniques, UA 797, Faculté des Sciences et Techniques de Saint-Jérôme, F-13397 Marseille Cedex, France

Received 6 July 1988, final version 30 September 1988

Low pressure CO oxidation was studied over small supported Pd particles by means of static secondary ion mass spectrometry (SSIMS). The SSIMS spectra of adsorbed layers on Pd particles with diameters below ~7 nm contained, besides other species, Pd_nX^+ ions (X=C, O, CO; n=1–3). In experiments with Pd films, Pd_nC^+ species were much less abundant or even absent from the mass spectra. We conclude that CO decomposed in appreciable amounts only on Pd particles. No significant influence of the support material (mica, MgO, α-Al_2O_3) was seen and posthumous TEM analysis revealed well defined particle morphologies with largely (111) and (100) facets. At total pressures below 10^{-4} Pa, the relative carbidic intensities Pd_nC^+/Pd_n^+, which were taken as proportional to the carbon surface concentrations, first increased with increasing temperatures and then decreased after they passed a maximum at T=400–440 K. The smaller the particle size, the more intense were the maxima. For T$\gtrsim$440 K, carbon was removed from the surface via reaction with oxygen. Strong hysteresis effects in the C_{ad} and CO_{ad} concentrations occurred during temperature variation measurements. The Pd_nC^+/Pd_n^+ ratios displayed maxima during heating but not during cooling. While the Pd_nCO^+/Pd_n^+ ratios were always high at 300 K and always low at T>550 K, they were more intense in the medium temperature range when subjecting the samples to heating. These hysteresis effects are discussed in terms of asymmetric adsorption and reaction behaviour of CO/O_2 mixtures, involving inhibition of CO_2 formation by adsorbed carbon and carbon monoxide.

PACS: 82.65; 82.30

1. Introduction

One of the most widely studied catalytic reactions is the oxidation of carbon monoxide over Pt group metals. This interest arises from the importance of this reaction in air pollution control as well as from the intricate kinetics associated with steady-state multiplicity and self-sustained oscillations (for recent reviews see [1, 2]). Despite this complexity, CO_2 production is generally believed to proceed according to a Langmuir-Hinshelwood mechanism between molecular CO and atomic oxygen. Recently, however, in studies with small supported Pd particles with diameters below ~7 nm and well defined morphology, CO_{ad} was found to partially decompose at temperatures both below the onset of and during the steady-state reaction [3]. Consequently, carbon formation and subsequent combustion with oxygen constitute an alternate pathway of CO_2 production over small Pd particles. These results may contribute to recent experimental and theoretical investigations, which indicate that cycles of catalyst deactivation by carbon and reactive regeneration by oxygen may be responsible for the CO_2 rate oscillations [4, 5].

* Current address: Technisch-Chemisches Labor, ETH-Zentrum, CH-8092 Zurich, Switzerland

In this paper, we report hysteresis effects in both adsorption of CO and catalytic carbon production during interaction of CO/O_2 stoichiometric gas mixtures with small supported Pd particles Static secondary ion mass spectrometry (SSIMS) was applied to monitor changes in the surface concentration of carbon, oxygen and carbon monoxide during temperature variation measurements. SSIMS was shown to be a valuable method capable of distinguishing between molecular and dissociative CO adsorption on metal surfaces [6].

2. Experimental

The SSIMS studies were performed in an ion pumped UHV chamber which was described in detail elsewhere [7]. During the course of this work a new 1–600 amu range quadrupole mass spectrometer was installed which permitted analysis of metal ion cluster formation up to Pd_5^+. In order to minimise ion damage effects of the Pd particles, static SIMS conditions with low Ar^+ primary ion current densities ($< 10^{-4}$ A/m^2 at 600–800 eV) were maintained throughout the experiments. Electrostatic charging of the insulator supports (mica, MgO, α-Al_2O_3) was compensated by an electron beam with energies up to 200 eV.

Pd particles were deposited on the support materials by in-situ evaporation from a Knudsen cell and stabilized before an experiment by annealing in mixtures of CO/O_2 at ~ 600 K. For Pd/mica samples this procedure produced epitaxial truncated cuboctahedral particles with mainly (111) and (100) facets, as concluded from detailed electron microscopy and diffraction studies [8]. Similar observations were made in studies with α-Al_2O_3 supports [9]. Figure 1 shows a transmission electron micrograph of a Pd/MgO sample previously subjected to CO oxidation studies and SSIMS analysis.

3. Results and discussion

The SSIMS results presented below were obtained under in-situ reaction conditions at steady total CO/O_2 pressures below 10^{-4} Pa. In studies with Pd particles, various ionic species, including those of the type Pd_nX^+ (X = C, O, CO; n = 1–3), were detected. We conclude that molecular as well as dissociative CO adsorption occurred on the Pd particle surfaces. The intensities of the carbide species, Pd_nC^+, were dependent on the particle size. Generally, the smaller the particles, the higher was the amount of Pd_nC^+ ions (or Pd_2C^+ ratios as in Fig. 2). For Pd films the carbidic species were less abundant or even absent from the mass spectra. These results demonstrate a particle size effect in the decomposition of CO_{ad} during reaction studies with CO/O_2 gas mixtures. Evidence for a disproportionation was also obtained by other authors in investigations of only CO adsorption on small Pd particles [10, 11].

The detection of Pd_nC^+ ions by SSIMS provides direct evidence that CO decomposition occurs on the particle surface. However, the ejection of any Pd_nX^+ species involves the rupture of metallic bonds and the creation of local defect sites. One might suspect that these sites are active in the CO decomposition, however, such an experimental artifact would not explain the occurrence of particle size dependent Pd_nC^+ intensities. In addition, Brown and Vickerman [12, 13] detected neither Pd_nC^+ nor Pd_nO^+ ions in SSIMS

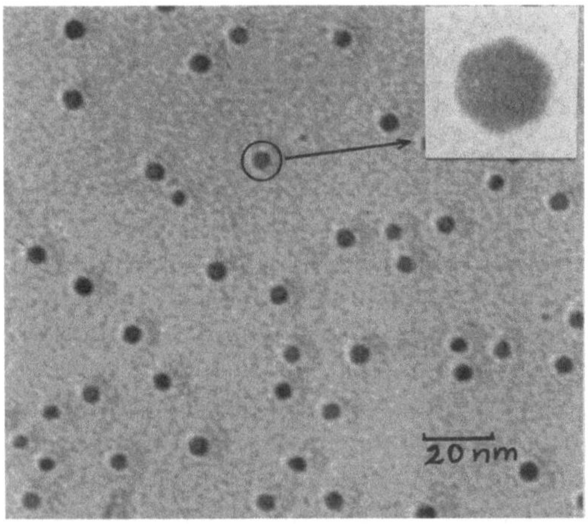

Fig. 1. TEM micrograph after SSIMS studies of CO/O_2 reaction on small Pd particles supported on MgO (100); mean particle diameter 4 nm, number density $2 \cdot 10^{11}$ cm^{-1}

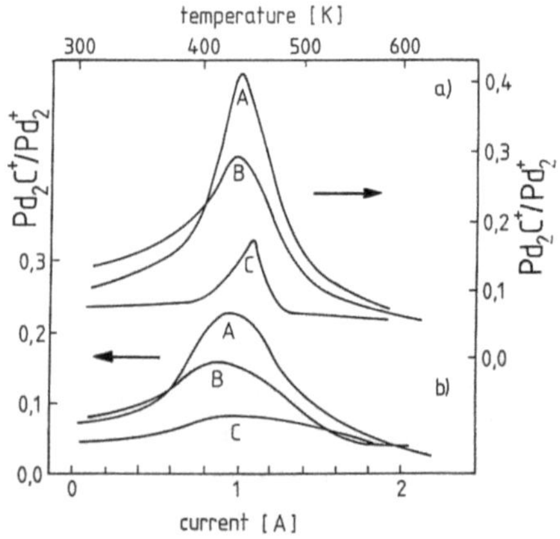

Fig. 2. SSIMS Pd_2C^+/Pd_2^+ intensity ratios versus heating current I (temperature) for different particle sizes (A: 3 nm, B: 10 nm, C: "continous" film) on mica supports; *a*) $P_{O_2} = 2.7 \cdot 10^{-5}$ Pa, $p_{CO} = 5.4 \cdot 10^{-5}$ Pa, *b*) $p_{O_2} = 2.7 \cdot 10^{-5}$ Pa, $p_{CO} = 1.3 \cdot 10^{-5}$ Pa

studies of CO adsorption on macroscopic (100) and (111) oriented Pd single crystal surfaces.

Absolute Pd_nX^+ secondary ion intensities do not directly represent surface concentrations of species X. Brown and Vickerman [12, 13] investigated CO adsorption by combined SSIMS and surface vibrational spectroscopy. The authors showed that the sum of the intensity ratios, Σ_n (Me_nCO^+/Me_n^+), is proportional to the CO surface coverage. Proportionality was also found during interaction and reaction of CO/O_2 gas mixtures with (111) oriented Pd films [14]. Similar relationships for carbon and oxygen coadsorbates were not clearly established so far. However, Barber et al. [6], in a study of CO adsorption on a number of different polycrystalline transition metals, reported that, if dissociation occurred, ion intensity ratios, MeC^+/Me^+ and MeO^+/Me^+, traced the deposited amounts of carbon and oxygen fairly well. In our recent paper [3] we showed that the dependence of Pd_2X^+/Pd_2^+ ratios on temperature, gas pressure and particle size consistently described the reaction behaviour of CO/O_2 mixtures on Pd particles. Accordingly, relative rather than absolute carbidic intensities are evaluated in Fig. 2. We mention, however, that the gross features of the particle size dependent CO decomposition remained unchanged when absolute ion intensities were plotted versus temperature.

As can be seen from Fig. 2, carbon formation is found most intense for small particles ($\phi = 3$ nm, sample A) in measurements with a stoichiometric gas mixture. For lower partial pressures of CO (lower set of data in Fig. 2) smaller amounts of carbon are observed. Generally, carbon concentration maxima occur in the medium temperature range between 400 and 440 K. At low temperatures, carbon formation seems to be inhibited by the presence of high CO_{ad} concentrations (see also Fig. 3). During heating, part of the CO_{ad} layer undergoes thermal desorption, thus liberating sites for the decomposition of remaining CO_{ad}. At high temperatures, i.e. beyond the peak maxima, carbon reacts readily with coadsorbed oxygen to form CO_2 which desorbs immediately and cannot be detected by SSIMS. Since the experiments were conducted under continuous gas supply, the high temperature Pd_2C^+/Pd_2^+ ratios reflect steady carbon concentrations maintained by formation from CO_{ad} and subsequent reaction with O_{ad}.

A small carbon peak is also apparent in experiments with Pd films (sample C) of 30 nm "thickness". This is not entirely surprising, since the film was not perfect, but rather had holes and an appreciable number of defect sites which could be active in CO decomposition.

The results shown in Fig. 2 were obtained by stepwise heating of the sample from 300 to 600 K.

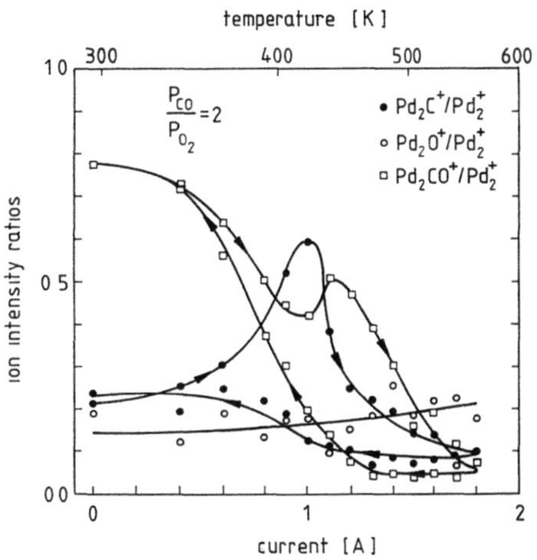

Fig. 3. SSIMS intensity ratios, Pd_2X^+/Pd_2^+, versus temperature during heating and cooling in CO/O_2 at $p_{CO} = 2.7 \cdot 10^{-7}$ Pa and $p_{O_2} = 1.3 \cdot 10^{-3}$ Pa; mean Pd particle size 4 nm on MgO substrate; number density: $5 \cdot 10^{11}$ cm^{-2}

Interestingly, carbon concentration maxima were not reproduced during stepwise cooling from high to low temperatures (Fig. 3). Moreover, strong hysteresis effects are also observed for CO_{ad}. While the Pd_2CO^+/Pd_2^+ ratios are always high at 300 K and always low above 550 K, they are more intense in the medium temperature range during heating. By contrast, the Pd_2O^+/Pd_2^+ ratios are quite low throughout the measurements. In experiments with non-stoichiometric gas mixtures and higher oxygen partial pressures, however, the Pd_2O^+/Pd_2^+ values increased considerably with increasing surface temperatures [3]. Note that the support material changed from mica in Fig. 2 to MgO (100) in Fig. 3. The hysteresis behaviour in the CO oxidation reaction as described by Fig. 3 was also seen when using the mica support [15].

According to Fig. 3, CO adsorption dominates at low temperatures. Under these conditions dissociative oxygen chemisorption is inhibited and CO_2 formation does not occur. This is in agreement with the well established adsorption and reaction behaviour of Pd surfaces [16, 17]. Increasing temperatures lead to gradual depletion of CO_{ad} due to its thermal desorption and reaction. At high temperatures the CO_{ad} concentration (in low pressure studies) is low and the CO_2 formation rate high. Interestingly, the Pd_2CO^+/Pd_2^+ ratios decrease non-monotonically during heating. In particular, the minimum in the CO_{ad} curve corresponds to the maximum in the carbon production, suggesting intermediate blocking of CO adsorption sites due to the presence of carbon, as described in detail elsewhere [3]. At high temperatures carbon concentrations are low and do not recover during stepwise cooling to room temperature. Consequently, site blocking effects are no longer observed.

Figure 3 establishes "clockwise" hysteresis in CO_{ad} and C_{ad} concentrations. By contrast, "anticlockwise" hysteresis was observed for CO_2 production in the gas phase. According to these results, which will be presented in detail elsewhere, the concentrations of CO_{ad} and C_{ad} intermediates are closely related to the CO_2 product formation. Generally, steady CO_2 gas pressures due to surface reaction were never observed before the carbon peak maximum occurred and/or the CO_{ad} species underwent partial thermal desorption to allow for oxygen chemisorption. The general features of the hysteresis loops were reproduced twice, in experiments one after the other, with ~40 min duration each.

Steady-state multiplicity can be explained if the kinetics of the elementary surface reactions vary nonlinearily with the concentrations of the reaction partners. This is the case for CO oxidation since it is well known that a saturated CO_{ad} layer inhibits oxygen chemisorption whereas the reverse is not the case [17]. Once oxygen has been incorporated into the adsorbed layer at higher temperatures, CO_2 formation can occur and persist even during cooling, provided a large activation energy prevents the reactive O–CO phase from attaining the low Gibbs energy of the only CO_{ad} layer. In this manner even Pd films can be expected to exhibit rate multiplicity, which, in fact, has been observed recently in our laboratory [15]. The hysteresis loops obtained in reaction studies over small Pd particles seem, however, more pronounced, probably due to the presence of carbon in the medium temperature range during heating procedures. Once the carbon surface concentrations are low at high temperatures, CO_2 formation occurs predominantly by direct reaction between CO_{ad} and O_{ad}. The lack of carbon concentration maxima during cooling must mean that CO decomposition in a reactive O–CO coadsorbate phase is strongly inhibited.

Recently, carbon segregation from the bulk to the surface and subsequent removal of the carbon by oxygen were proposed as the rate-controlling factors in long-period oscillations of the CO-oxidation over noble metals [4, 5]. According to the present results, surface carbon may be formed chemically by CO decomposition on small Pd particles. It is therefore tempting to incorporate in model calculations this reaction step as part of a reaction sequence exhibiting self-sustained rate oscillations.

References

1. Slin'ko, M.M., Slin'ko, M.G.: Kinet. Katal. **23**, 1421 (1982)
2. Razon, L.F., Schmitz, R.A.: Catal. Rev. Sci. Eng. **28**, (1) 89 (1986)
3. Matolin, V., Gillet, E., Kruse, N.: Sur. Sci. **186**, L541 (1987)
4. Burrows, V.A., Sundaresan, S., Chabal, Y.J., Christman, S.B.: Surf. Sci. **180**, 110 (1987)
5. Collins, N.A., Sundaresan, S., Chabal, Y.J.: Surf. Sci. **180**, 136 (1987)
6. Barber, M., Vickerman, J.C., Wolstenholme, J.: Surf. Sci. **68**, 130 (1977)
7. Gillet, E., Channakhone, S., Matolin, V.: J. Catal. **97**, 437 (1986)
8. Gillet, M.F., Channakhone, S.: J. Catal. **97**, 427 (1986)
9. Gillet, E., Legressus, C., Gillet, M.: J. Chim. Phys. **84**, 167 (1987)
10. Doering, D.L., Poppa, H., Dickinson, J.T.: J. Catal. **73**, 104 (1982)
11. Ichikawa, S., Poppa, H., Boudart, M.: J. Catal. **91**, 1 (1985)
12. Brown, A., Vickerman, J.C.: Surf. Sci. **124**, 267 (1983)
13. Brown, A., Vickerman, J.C.: Surf. Sci. **151**, 319 (1985)
14. Matolin, V., Channakhone, S., Gillet, E.: Surf. Sci. **164**, 209 (1985)
15. Matolin, V.: Thesis, University Aix-Marseille III (1987)
16. Engel, T., Ertl, G.: J. Chem. Phys. **69**, 1267 (1978)
17. Engel, T., Ertl, G.: Adv. Catal. **28**, 1 (1979)

Z. Phys. D – Atoms, Molecules and Clusters 12, 579–582 (1989)

Atoms, Molecules
Zeitschrift für Physik D and Clusters
© Springer-Verlag 1989

Influence of the precursors on the properties of alumina supported rhodium particles

B. Coq, F. Figueras, and T. Tazi
Laboratoire de Chimie Organique Physique et Cinétique Chimique Appliquées, U.A. 418 CNRS, ENSCM, 8, rue Ecole Normale, F-34075 Montpellier Cedex, France

Received 5 July 1988; final version 30 September 1988

The effect of chlorine on the properties of Rh particles supported on alumina was studied using the hydroconversion of alkanes as a molecular probe. Chloride ions on the alumina have little influence on the Rh particles. Chlorine adsorbed on the Rh particles decreases the chemisorption capacity and increases the metal-support interaction, which results in modifications of their catalytic properties. Depending on the alkane processed, specific activity can be depressed up to four orders of magnitude.

PACS: 36.40

Introduction

Catalyst preparation is of a great importance for determining its final properties. The nature of the precursor and support used, as well as activation procedure, have a big effect on the size, morphology, and electronic state of the supported metallic particles. A strong effect of chlorine on chemisorption and catalysis on supported ruthenium clusters was found recently [1–5]. Less attention was paid to rhodium catalysts so far, and the reported data appear more contradictory [5–8]. The aim of the present work was to investigate the properties of two series of supported Rh-alumina catalysts with different dispersions of Rh, prepared either from a chloride precursor or from a chlorine free one. Hydrogen chemisorption and alkane hydrogenolysis were taken as tests to study the changes in properties of Rh/alumina catalysts.

Experimental

All the chemicals and reactants were of a high purity grade. Rhodium was deposited either by ligand exchange with rhodium-acetylacetonate, $Rh(AcAc)_3$ in benzene solution, or by ion exchange with $RhCl_3$, $3H_2O$ or $(Rh(NH_3)_5Cl)Cl_2$ in acidic (HCl, pH $\simeq 2$) or basic (NH_4OH, pH $\simeq 12$) medium respectively. The supports used were γ-alumina (200 m^2/g) and silica (350 m^2/g). The procedures for the catalyst preparation have been published elsewhere [9]. For the sample (H/Rh = 1.5) a quantitative amount of Cl^- or NH_4^+ ions (Cl^-, $NH_4^+/Rh = 100$), in acidic or basic medium respectively, were added to the final reduced catalyst.

The chemisorption of hydrogen was carried out in a conventional volumetric apparatus at 298 K in the 0–20 kPa pressure range. The linear part of the isotherm was extrapolated to zero pressure value to determine the hydrogen uptake, which is proportional to the number of surface rhodium atoms. The size of the metallic particles was checked by Transmission Electron Microscopy (TEM).

The hydroconversions of *n*-hexane (nH), methylcyclopentane (MCP), 2,2,3,3-tetramethylbutane (2233TeMB) were carried out at atmospheric pressure in a microflow reactor in line with a gas chromatograph. The reaction temperature was 493 K and the molar ratios of the reaction mixture: $H_2/nH = 14.0$, $H_2/MCP = 15.5$, $H_2/2233TeMB = 44$.

Results and discussion

The data for chemisorption of hydrogen and the catalytic properties of supported rhodium clusters are summarized in Tables 1–3. In these Tables, the catalysts are organised from the left to the right side according to the decrease of the metallic particle size. Hence the ratio between surface and total number of Rh atoms, as well as the H/Rh ratio, increase from left to right. For some samples prepared from the $Rh(AcAc)_3$ precursor the H/Rh ratio is larger than one. This is characteristic of multiple adsorption of hydrogen on very small Rh clusters [10]. The activity is expressed as the number of reactant molecules converted per surface Rh atoms and per hour (Turnover Frequency, TOF). Alkane conversion at 493 K on Rh catalysts gives mainly hydrogenolysis, the different possible pathways are presented in Scheme 1.

Table 1. Turnover frequencies and selectivity fingerprints as a function of metallic dispersion of Rh/Al_2O_3 catalysts prepared from $RhCl_3$, in the hydrogenolysis of alkanes at 493 K

H/Rh		0.07	0.27	0.56	0.9	*a*
nH	TOF	185	110	75	38	24
	SFH	2.2	0.59	0.69	0.63	0.54
MCP	TOF	500	420	76	30	12
	SFM	27.3	14.5	7.5	6.0	4.0
2233TeMB	TOF	260	39	—	1.3	0.07
	SFT	68	6.9	—	1.67	0.88

[a] H/Rh not determined (uptake too slow); TEM shows particle size lower than 1 nm only

Table 2. Turnover frequencies and selectivity fingerprints as a function of metallic dispersion of Rh/Al_2O_3 catalysts prepared from $Rh(AcAc)_3$, in the hydrogenolysis of alkanes at 493 K

H/Rh		0.08	0.35	0.75	1.4	1.5
nH	TOF	65	70	70	130	185
	SFH	2.42	0.95	0.68	0.64	0.68
MCP	TOF	360	500	480	65	40
	SFM	14.3	17.4	23.4	4.4	4.2
2233TeMB	TOF	40	20	10	30	7
	SFT	65	23 4	11.8	0.42	0.1

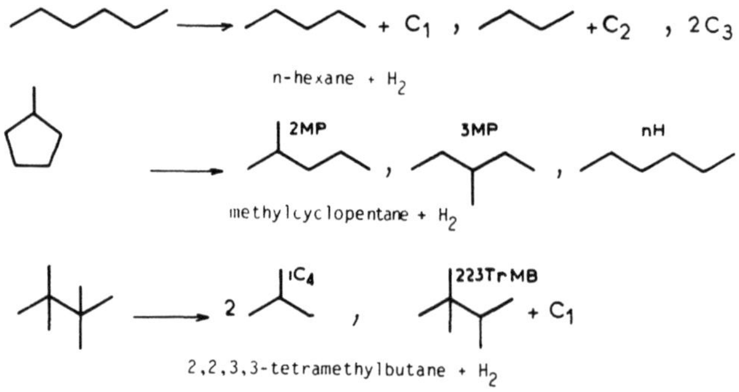

The changes of selectivity were characterized by a Selectivity Fingerprint (SF) specific of each molecule of reactant. Thus we defined for nH, $SFH=(C_1+C_5)/(C_2+C_4)$, for MCP, SFM = 2MP/nH, for 2233TeMB, $SFT=iC_4/223TrMB$. C_1, C_2, . . . 223TrMB represent the concentrations of reaction products in the effluents.

Whatever the nature of the precursor, when the particle size decreases the changes of selectivity, SFH, SFM, SFT, in the two series of studied catalysts follow the same trends (Tables 1, 2). A shift from terminal to internal C–C bond cleavage of nH, a more statistical ring opening of MCP, a pronounced demethylation of 2233TeMB were obtained on very small clusters. The selectivity changes come from the existence of different

Table 3. Turnover frequencies and selectivity fingerprints for supported Rh catalysts, in the conversion of alkanes at 493 K

		Rh/Al_2O_3 from $Rh(AcAc)_3$ (H/Rh = 1.5)		Rh/Al_2O_3	Rh/SiO_2
		$+100\ Cl^-/Rh$	$+100\ NH_4^+/Rh$	from $(Rh(NH_3)_5\ Cl)Cl_2$	
H/Rh		1.0	—	1.02	1.1
nH	TOF	45	105	160	180
	SFH	0.65	0.69	0.66	0.22
MCP	TOF	14	35	100	90
	SFM	4.5	4.7	3.3	4.4
2233TeMB	TOF	1.1	14.5	14	15
	SFT	0.46	0.22	0.4	0.66

kinds of surface complexes during the alkane hydrogenolysis, which depends on the coordination of surface atoms [11]. Thus the selectivity fingerprint can be taken as characteristic of a size of Rh particles. When the activity is taken into account the effect of particle size is different, according to the nature of the precursor. With catalysts prepared in organic medium, the size of Rh particles little influences the TOF which only decreases by a factor of 10 for MCP conversion. At variance Rh/Al_2O_3 prepared from chloride precursors behave differently, since a decrease of activity larger than three orders of magnitude sets up for 2233TeMB conversion. Obviously the smaller the particles are, the larger the effect of the precursor is. This different behaviour between the two series of catalysts probably derives from the presence of residual chlorine in samples prepared from $RhCl_3$ precursor. The chlorine removal from catalysts prepared in such a way is always difficult, and little amount of chlorine is released when both the Rh content is lower than 1% and the reduction temperature lower than 773 K [6]. The poisoning by chlorine may be interpreted as following:

(1)–chlorine adsorbed on the Rh clusters protects metallic sites against chemisorption and catalysis, as was shown for Ru particles supported on alumina [3, 5].
(2)–strong interaction between chlorine and Rh deeply modifies the local environment of Rh atoms, and results in changes of morphology of Rh particles.
(3)–chloride ions adsorbed on alumina influence the Rh particles through the support.

It is well known that chlorination modifies strongly the acid-base properties of Al_2O_3. However the alumina used for the two series of catalysts was treated with HCl at pH $\simeq 2$ previously. Hence Cl^- ions are always present on the support. Thereby the presence of Cl^- adsorbed on Al_2O_3 cannot explain the large difference observed between the two series of catalysts. Probably these differences are the result of direct interaction between chlorine and Rh particles.

To check this, chlorine was added to the catalyst with the smallest Rh clusters obtained from $Rh(AcAc)_3$ (H/Rh = 1.5). The solid obtained was then reduced at 673 K. The examination by TEM of this chlorinated sample does not show any change of the Rh particle size. Slight variations of selectivity, but a decrease of activity are observed (Table 3). This decrease cannot be correlated with the 30% change of chemisorption capacity. Hence the possible inhibiting effect of Cl^- on surface Rh atoms interferes with the decrease of activity for a lesser extent. It seems likely that strong modifications of particle morphology will be responsible for the observed effects (hypothesis 2). This conclusion would be supported by theoretical and experimental studies showing that adsorption of chlorine and CO results, on the one hand, in strong destabilization of the clusters and even disruption of the metallic bond, and on the other, in significant changes in the electronic states of Rh atoms [12, 13]. However these very small Rh particles remain in a reduced state, as it was shown by XPS measurements. Otherwise the addition of Cl to the final catalyst does not reproduce totally the properties of a sample prepared from $RhCl_3$ precursor (Tables 1, 3). The decrease of activity is much larger on the latter, which would indicate a specific effect of Cl during the nucleation process; this could be related to the strong interaction in acidic medium between $RhCl_3$ or Cl^- and alumina [6].

Supported rhodium clusters prepared from $Rh(AcAc)_3$, or with $(Rh(NH_3)_5Cl)Cl_2$ in basic medium, show similar behaviours, close to those of Rh clusters supported on silica (Table 3). In the Rh-SiO_2 system the metal-support interaction is low and small metal particles present tridimensional and even spherical shapes [7, 14, 15]. The breakdown of these shapes can happen when Cl is added to the catalyst, and the extent of this process will be larger if Cl is present from the beginning of the catalyst preparation procedure. The effect will be so much greater as the Rh particles will be smaller, "raft-like" aggregate may be then found [16].

In conclusion Cl little modifies the chemisorption behaviour, as well as hydrogenolysis of linear alkanes, of alumina supported Rh catalysts. However the effect of Cl is stronger for the hydrogenolysis of branched alkanes, the specific activity may be then decreased by four orders of magnitude. Probably chlorine strongly interacts with Rh particles, resulting in changes of metal-support interaction and morphology of the supported clusters.

References

1. Bond, G.C., Xu Yide J. Chem. Soc. **80**, 3103 (1984)
2. Burch, R., Bond, G.C., Rajaram, R.R.: J. Chem. Soc. **82**, 1985 (1986)
3. Narita, T., Miura, H., Sagiyama, K., Matsuda, T., Gonzalez, R.D. J. Catal. **103**, 492 (1987)
4. Lu, K., Tatarchuk, B.J.: J. Catal. **106**, 166 (1987)
5. Miura, H., Hondou, H., Sugiyama, S., Matsuda, T., Gonzalez, R.D.: In: Proceedings of the 9th International Congress on Catalysis, Vol. 3, p. 1307. Ottawa: The Chemical Institute of Canada 1988
6 Marquez da Cruz, G., Djega-Mariadassou, G., Bugli, G.: Appl. Catal **17**, 205 (1985)
7. Nunez, G.M., Patrignani, A.R., Rouco, A.J J Catal. **98**, 554 (1986)
8. Abderrahim, H., Duprez, D.: In: Proceedings 9th International Congress on Catalysis, Vol. 3, p 1246. Ottawa· The Chemical Institute of Canada 1988
9. Fuentes, S., Figueras, F. J. Catal. **61**, 443 (1980)

10. Kip, B.J., Duivenvoorden, F.B.M., Koningsberger, D.C., Prins, R.: J. Catal. **105**, 26 (1987)
11. Coq, B., Figueras, F.: J. Mol. Catal. **40**, 93 (1987)
12. Halachev, T., Ruckenstein, E.: J. Catal. **73**, 171 (1982)
13. van't Blik, H.F.J., van Zon, J.B.A.D., Huizinga, T., Vis, J.C., Koningsberger, D.C., Prins, R.: J. Am. Chem. Soc. **107**, 3139 (1985)
14. Yacaman, M.J., Romeu, D., Fuentes, S., Dominguez, J.M.: J. Chim. Phys. **78**, 861 (1981)
15. Vis, J.C.: Ph.D. thesis, Univ. of Tech. Eindhoven, The Netherlands, 1984
16. Yates, D.J.C., Murrell, L.L., Prestridge, E.B.: Growth and Properties of Metal Clusters, pp. 137–150. Amsterdam, Oxford, New York: Elsevier 1980

Z. Phys. D – Atoms, Molecules and Clusters 12, 583–586 (1989)

Atoms, Molecules and Clusters
Zeitschrift für Physik D
© Springer-Verlag 1989

Influence of dispersion on the energies of adsorption: H_2, CO, propylene and propyne on supported Pd or Pt

A. Guerrero[1], M. Reading[2], Y. Grillet[3], J. Rouquerol[3], J.P. Boitiaux[4], and J. Cosyns[4]

[1] Departamento de Quimica Inorganica, U.N.E.D, Cinidad Universitaria, Madrid, Spain
[2] ICI Paints Research Laboratory, Wescham Road, Slough, UK
[3] Centre de Thermodynamique et de Microcalorimétrie du C.N.R.S., 26 rue du 141 ème R.I.A., F-13003 Marseille, France
[4] Direction Cinétique et Catalyse, Institut Français du Pétrole, Ave. de Bois-Préau, F-92502 Rueil-Malmaison, France

Received 5 July 1988

The aim of this work is to contribute to an explanation of the *lowering of the specific activity* of Pd/Al_2O_3 catalysts, towards hydrogenation of acetylenic molecules, *as the dispersion increases.* For this purpose, the interaction energy between these catalysts and H_2, CO, propylene and propyne was directly measured by adsorption microcalorimetry at 323 K. In the case of propyne, the enthalpy of adsorption was found to be more than 50% higher for the catalysts with 97% dispersion than for the one with 26% dispersion, for coverages ranging from 0.2 to 0.7. *This high enthalpy of adsorption* (up to 290 $kJ.mol^{-1}$) is attributed to a different proportion of edges and crystalline planes in the smallest Pd crystallites and is indeed consistent with a *"blocking" chemisorption or with a hindered desorption of the products of a catalytic hydrogenation*, i.e. with a lowering of the specific activity.

PACS: 36.40; 82.65.Jv; 82.60.Cx

Introduction

This work follows a previous one by Boitiaux et al. [1] which showed, *at the highest dispersions of* Pd, *a decrease of specific activity* (i.e. per gram of Pd) of alumina-supported Pd catalysts towards the hydrogenation of the acetylenic bond. When using CO as a probe molecule to detect differences between the above catalysts, they could observe a shift, vs. dispersion, of the IR absorption bands for CO. This led us to the idea of checking, by direct microcalorimetric measurements, whether a change in dispersion could lead to a change in adsorption energy and then in catalytic activity. We therefore decided to follow the adsorption of CO but also of H_2, propylene and finally propyne since the phenomenon to be explained occurred for the hydrogenation of acetylenic molecules (like propyne) but not for the hydrogenation of ethylenic molecules (like propylene).

Experimentals

Samples

The two Pd/Al_2O_3 catalysts examined here have the same Pd content (0.5% in mass) but a dispersion of 26% and 97%, respectively. Details about their preparation and characterization may be found in [1] and [2]. For the sake of comparison, a Pt/Al_2O_3 catalyst (Pt content: 0.47% in mass, dispersion: 70%) and a Pt/SiO_2 catalyst (Pt content: 1.05%, dispersion: 3%) are also examined in the case of propyne adsorption.

Pre-treatment

For the Pd catalysts it is as follows: the *c.a.* 1g sample is successively heated (3C min^{-1}) under an air flow of 10 cm^{-3} min^{-1} up to 300°C where it is calcined during

90 min, then cooled down to 250°C under a nitrogen flow of 10 $cm^3 min^{-1}$ and kept 30 min in these conditions (to desorb oxygen and avoid any further strong reaction with hydrogen), reduced under a hydrogen flow of 20 $cm^3 min^{-1}$ during 2 h at 250°C and finally outgassed during 12 h, until a final vacuum of 10^{-6} mbar (with a liquid nitrogen trap to avoid any retro-diffusion of oil vapour).

The treatment of the Pt catalysts is similar, except, principally, the temperature and duration of the calcination (1 h at 400°C) and of the reduction step (also, 1 h at 400°C).

Adsorption microcalorimetry

The sample bulb is a glass cylinder of 16 mm diameter and 65 mm high, connected to a double concentrical tubing with two bellow-stopcocks used to provide a gas flow over the catalyst, when needed during the pretreatment. This bulb lies in the center of a Tian-Calvet microcalorimeter maintained at 50°C, and is also connected to an all-metallic equipment of adsorption volumetry with stainless bellow valves and kel-F seats, standing upon the microcalorimeter, as represented in Fig. 1. This equipment may be used either in the conventional way (with successive introductions of adsorptive) or following a quasi-equilibrium procedure, with continuous introduction of adsorptive allowing, when applicable, to record a continuous adsorption isotherm together with a continuous curve of derivative enthalpy of adsorption [3].

Results

The adsorption of H_2 and CO could be directly followed by the quasi-equilibrium procedure: for introduction rates ranging between 0.9 and 3 cm^3 STP per hour the conditions of equilibrium were checked to be satisfactory (*i.e.* after a sudden interruption of the introduction, the pressure signal was stable and the heat-flow signal dropped down to zero with the time constant of the microcalorimeter). The corresponding derivative enthalpies of adsorption (whose recording is continuous) are given in the curves of Fig. 2 (for H_2) and Fig. 3 (for CO). To easily compare the behaviours of the two Pd/Al_2O_3 catalysts, these curves were normalized with respect to the amount adsorbed under a pressure of 10 Torr[1] (for which we arbitrarily state $\theta = 1$).

The adsorption of propylene and propyne is a more complex phenomenon with slower kinetics which needed the use of the discontinuous method. The main part of the phenomenon was completed after 3 hours and this is the time after which we carried out our calorimetric and volumetric determinations. We nevertheless speak of "pseudo-adsorption isotherms" since a much weaker and slow gas uptake may then be observed over hours, at a relatively constant rate, even after 15 h. The enthalpy curves are reported in Figs. 4, 5 and 6, with a similar normalization as for Figs. 2 and 3, except that the reference pressure (for which $\theta = 1$, indicated in each figure caption) is lower than above.

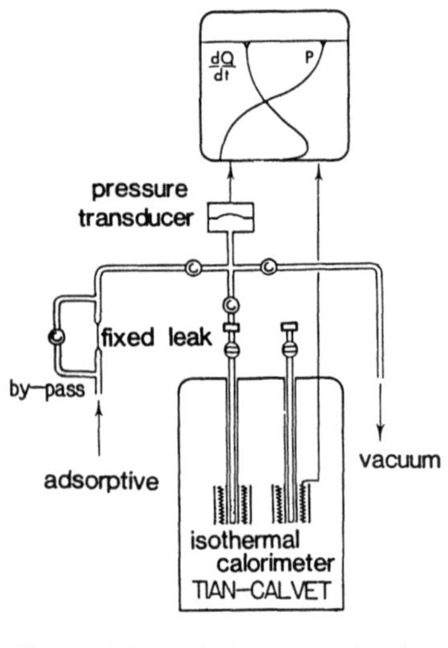

Fig. 1. Microcalorimetric and volumetric assembly

Discussion

The observation of Figs. 2 to 6 leads to the following remarks:

⇒ In all cases, except for CO, the derivative enthalpies of adsorption *are higher on the catalysts with the highest dispersion.*

⇒ The difference is specially large for the adsorption of propyne on Pd/Al_2O_3 catalysts (Fig. 6), *i.e.* with the systems modelling those raising a question in our introduction. For "coverages" ranging from 0.2 to 0.7, the derivative enthalpies of adsorption are more than 50% higher for the high dispersion catalyst.

⇒ Conversely, the dispersion increase of Pd has no measurable influence on the adsorption enthalpy of CO, showing that this often convenient probe molecule (specially in view of I.R. studies) does not bring us here the explanation required about the influence of dispersion towards other molecules.

⇒ A number of values of the derivative enthalpies of adsorption are given in Table 1. For H_2 and CO they correspond to the initial plateau. They are of the same order of magnitude as those reported: for adsorption

[1] 0.75 Torr = 100 Pa

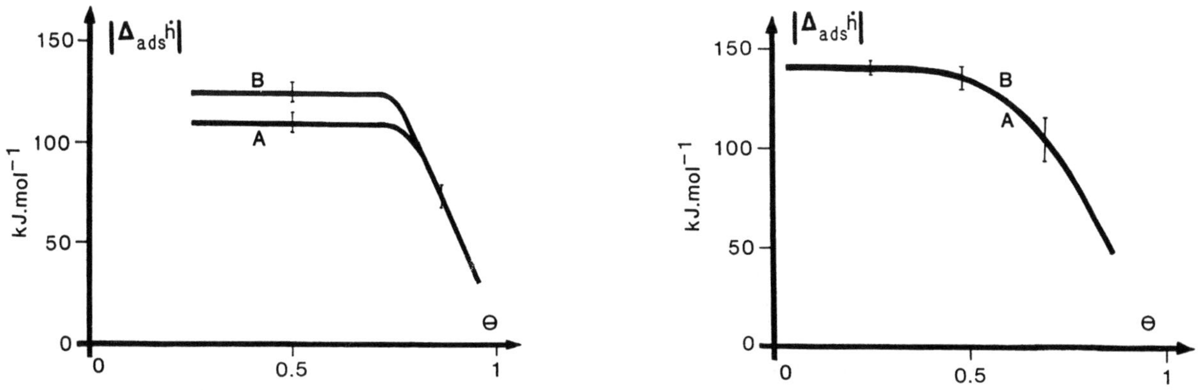

Fig. 2 and 3. Derivative enthalpies of adsorption, vs coverage, for the 26% dispersion (curves A) and 97% dispersion (curves B) Pt catalysts, Fig. 2: H_2 adsorption, $\theta = N_a/N_{a(10\,torr)}$, Fig. 3: Co adsorption, $\theta = N_a/N_{a(10\,torr)}$

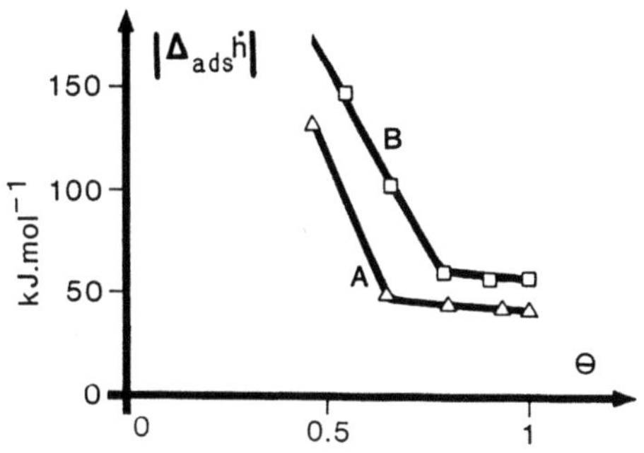

Fig. 4. Pseudo-derivative enthalpy of adsorption of propylene vs coverage for the 2 Pd catalysts. $\theta = N_a/N_{a(7.5\,torr)}$

Table 1. Derivatives enthalpies of adsorption/kJ.mol^{-1}

Adsorbate	H_2	CO	Propylene	Propyne
θ	0.25–0.75	0.05–0.25	0.5	0.1
Catalyst				
26% Pd	110±5	140±5	125	215
97% Pd	125±5	140±5	170	275
3% Pt				155[1]
70% Pt				210

[1] Extrapolated down to this coverage

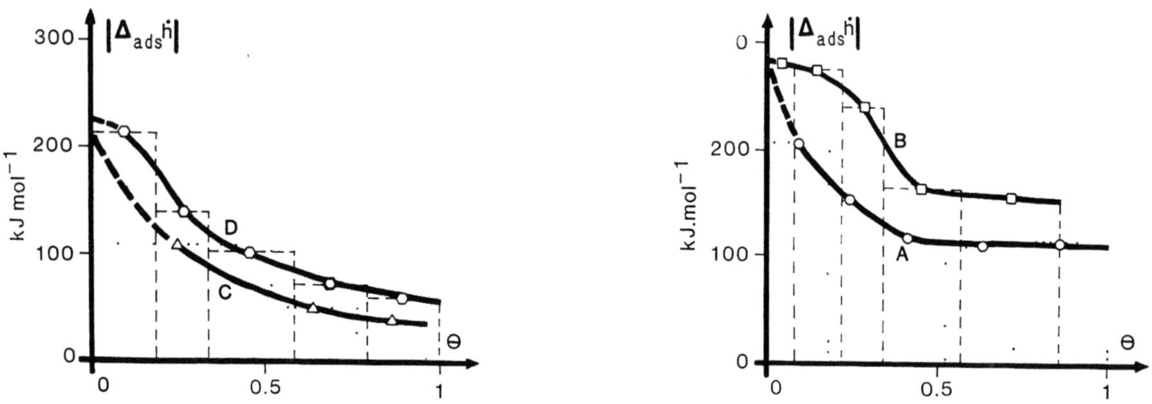

Fig. 5 and 6. Pseudo-derivative enthalpy of adsorption of propyne, vs coverage, for the Pt catalysts (3 and 70% dispersion for curves C and D, respectively) and the Pd catalysts (26 and 97% dispersion for curves A and B, respectively); Fig. 5: Pt catalysts, $\theta = N_a/N_{a(3\,torr)}$; Fig. 6: Pd catalysts, $\theta = N_a/N_{a(1.5\,torr)}$

on Pd films: 84 to 147 kJ.mol^{-1} for H_2 and 105 to 167 kJ.mol^{-1} for CO [4]. Nevertheless, these results are not consistent with those obtained by Takasu et al. [5, 6] who observe a lowering of the energy of adsorption of H_2 or CO when Pd dispersion increases. A support (graphite) different from ours may explain the difference, although Vannice et al. [7] seem to show, also from calorimetric measurements (but rapid and non isothermal, being of the DSC type) that the adsorption enthalpy of CO on Pd is practically the same whatever the support (SiO_2, Al_2O_3, TiO_2). . . . For propylene, the enthalpies given here compare well with the value of 176 kJ.mol^{-1} measured by Cerny et al. on a Pt film [8], whereas we have no point of comparison for the very high enthalpies of adsorption of propyne given in Table 1.

⇒ The values of over 250 kJ.mol^{-1} for coverages up to 0.35 (cf. Fig. 6) are enough to account for a kind of "blocking" chemisorption or for a hindered desorption of the products of a catalytic hydrogenation. The change in energy of adsorption observed with the change in cristallite size may well be explained by a different proportion of edges and crystalline planes of each type, following the observations of Muetterties et al. [9] about adsorption of ethylene or acetylene on different crystal faces of Pd.

References

1. Boitiaux, J.P., Cosyns, J., Vasudevan, S.: Preparation of catalysts III, pp. 123, Poncelet, G., Grange, P., Jacobs, P.A., (Eds.) Amsterdam: Elsevier 1983
2. Vasudevan, S.: Thèse Docteur-Ingénieur, Ecole Nationale du Pétrole et des Moteurs, Paris (1982)
3. Rouquerol, J., Rouquerol, F., Grillet, Y., Ward, R.J., Characterization of Porous solids, pp. 67–76, Unger, K.K., Rouquerol, T., Sing, K.S.W. (eds.) Amsterdam: Elsevier 1988
4. Toyoshima, I., Somorjai, G.A.: Catal. Rev. Sci. Eng. **19**, 105 (1979)
5. Takasu, Y., Akimaru, T., Kasahara, K., Matsuda, Y., Miura; H., Toyoshima, I.: J. Am. Chem. Soc. **104**, 5249 (1982)
6. Takasu, Y., Matsuda, Y., Toyoshima, I.: Chem. Phys. Lett. **108**, (n°4) 384 (1984)
7. Vannice, A., Chou, P.: J. Chem. Soc., Chem. Commun. 1590 (1984)
8. Palfi, S., Lisowski, W., Smutek, M., Cerny, S.: J. Catal. **88**, 300 (1984)
9. Gentle, T.M., Muetterties, E.L.: J. Phys. Chem. **87**, 2469 (1983)

Z Phys. D – Atoms, Molecules and Clusters 12, 587–590 (1989)

Atoms, Molecules
Zeitschrift für Physik D and Clusters
© Springer-Verlag 1989

Electronic effect induced by variation of size for Pd clusters in 1,3-butadiene hydrogenation

V. De Gouveia[1], B. Bellamy[1], Y. Hadj Romdhane[1], A. Masson[1], and M. Che[2]

[1] Laboratoire de Physico-Chimie des Surfaces associé au C. N. R. S., UA 425, Ecole Nationale Supérieure de Chimie de Paris, 11, rue Pierre et Marie Curie, Université P. et M. Curie, F-75005 Paris, France

[2] Laboratoire de Réactivité de Surface et Structure associé au C. N. R. S., UA 1106, Université P. et M. Curie, F-75532 Paris, France

Received 5 July 1988; final version 30 September 1988

The size of palladium aggregates is monitored with atomic beam technique condensed onto an SiO_2 amorphous substrate, by this way it is possible to define an average size that make explicit the intrinsic electronic size effect on a given reaction. The 1,3 hydrogenation of butadiene is followed and it is shown that intrinsic electronic properties observed by A.E.S. and U.P.S. spectroscopies can explain the variation of the turn over frequency.

PACS: 82.30; 79.20.F; 79.60

Introduction

As it is now possible to control the nucleation and growth processes of metallic aggregates in the gas phase or onto a substrate, size or mass selected distribution of clusters can be obtained. The variation of the size is believed to introduce specific electronic and crystallographic structures which lead to specific chemical reactivity. The size effect is one of the most important problems in heterogeneous catalysis [1] and we have attempted to model it by following the selective hydrogenation of 1,3-butadiene by size selected palladium clusters obtained by atomic beam techniques.

Experimental

The experimental apparatus consist of an UHV preparation chamber and of an UHV characterization chamber equipped with high resolution electron spectroscopy and a catalytic reactor designed to work at atmospheric pressure. The metallic phase (palladium) is prepared under high vacuum conditions (10^{-9} Torr) by vapour deposition on flat amorphous silica and carbon substrates (200 Å thick). The atomic beam source is a particular Knudsen cell [2]. The atomic flux was calibrated by Rutherford Backscattering measurements (R.B.S.). The electronic properties of the metallic clusters are followed by electron spectroscopies (A.E.S., U.P.S.). Finally, the sample is transferred under vacuum by means of a magnetic device in a catalytic reactor in which the catalytic reaction (1,3-butadiene hydrogenation) is performed. For other experimental details see [3].

Results and discussion

Growth mechanism of palladium film

As expected from thermodynamical arguments, the growth mechanism is of a Volmer-Weber type i.e characterized by a distribution of 3-D nuclei. The metallic cluster size and the nucleus density ($N_s = 3.2\,10^{12}$ nuclei/cm^2) at saturation have been obtained by T.E.M.. As for the small particles (< 12 Å) the Schmeisser's approach [4] allows us to calculate the diameter (d) of the particle knowing the impinging atoms flux (R_i), evaporation time (t) and substrate temperature (T) when N_s is constant, by the following formula:

$$d \propto (R_i \exp((E_a - E_d)/2\,kT)^{1/2}\, t^{1/2} \quad (1)$$

where E_a and E_d are respectively the adsorption energy and the activation energy for the diffusion of the atom on the surface. Other the things being equal the particle size is controlled only by the evaporation time (t). The validity of Schmeisser's approach has been experimentally checked for the Pd/SiO_2 and Pd/C systems (Fig. 1).

Electronic properties

U.P.S. The variation of the particle is believed to induce changes in the intrinsic electronic properties. On Fig. 2a, the valence band versus the palladium coverage (cluster size) is reported. On Fig. 2b we have represented the difference spectra obtained by substracting the silica valence band. We can observe a narrowing of the full width at half maximum (FWHM) of the valence band with decreasing coverage, i.e, with decreasing cluster size. These phenomena have been observed in all photoemission studies of supported metal clusters, by, for example, Liang et al. [5], Mason and Baetzold [6], Takasu et al. [7], Roulet et al. [8]. This is of course the result expected on the basis of decreasing coordination number, because the bandwidth in a tight-binding approximation is proportional to the square root of the number of nearest neighbors (Haydock and Kelly [9] Desjonqueres and Cyrot-Lackmann [10]) as expressed by:

$$W_d \propto W.(Z)^{1/2} \quad (2)$$

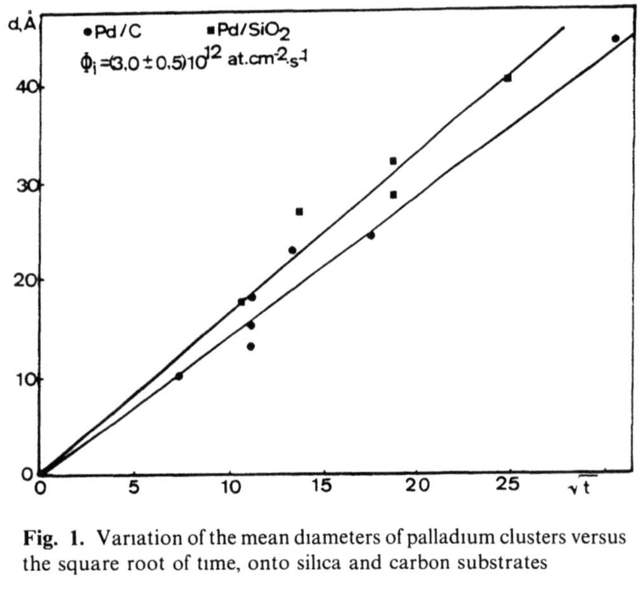

Fig. 1. Variation of the mean diameters of palladium clusters versus the square root of time, onto silica and carbon substrates

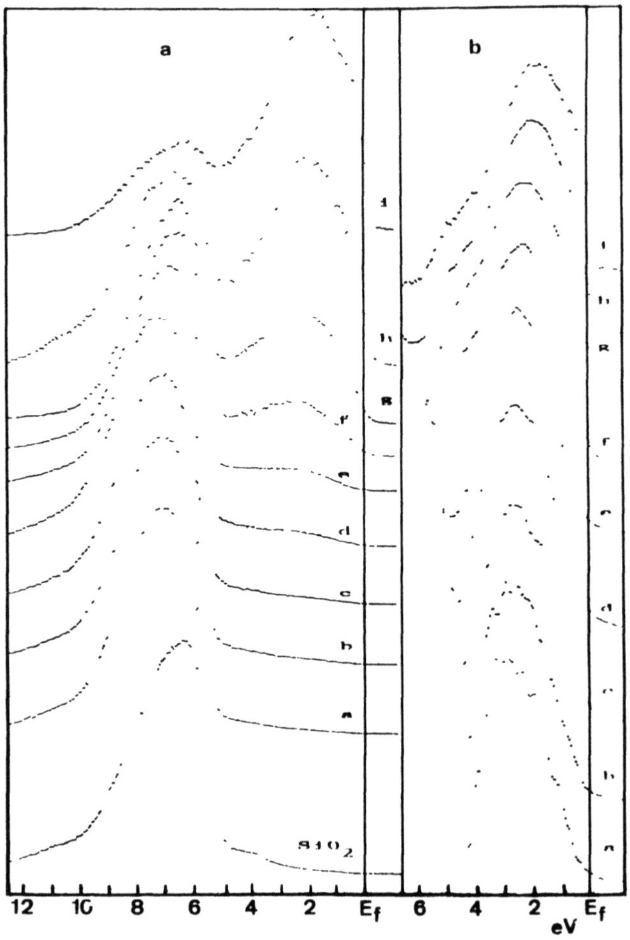

Fig. 2a and b. He I photoelectron spectra of the valence band of Pd particles for different coverages on a silica substrate. The coverage are respectively a) 0.14 b) 0.82 c) 1.6 d) 3.3 e) 8 f) 16 g) 33 h) 54 and i) 98×10^{14} atoms cm^{-2}. On the figure 2b the reponse of the silica substrate has been subtracted from the spectra of the clusters

where W is the exchange integral which represents the electron probability to jump from one site to another and Z is the coordination number. In Table 1, we can see that the variation of the calculated FWHM, is in good agreement with experimental observations (Fig. 2b). The shift of the Fermi level observed as the cluster size is varied may be interpreted in terms of the electronic properties differences induced by the change of the cluster size (size intrinsic effect: initial state and by the variation in the final state: relaxation processes.) [11–12].

A.E.S. The other way of probing the valence band is to follow a M-VV Auger transition as shown on Fig. 3. A sizeable shift of 0.75 eV toward low kinetic energies with decreasing coverage is observed. The energy shift is accompanied by a variation of FWHM with the cluster size. This energy shift may be due to three complementary effects:

Table 1

EXP		Model			
A	B	C	D	E	F
a = 0.13	5				
		13	5	5.5	32
b = 0.81	27	—	—	—	
c = 1.62	54	55	10	7.9	18
d = 3.24	108				
		147	15	9	13
e = 8.21	270				
		309	20	9.7	10
f = 16.2	540				
		561	25	10.1	8
		923	30	10.4	6.8
g = 32.4	1080				
		1415	35	10.6	5.8
h = 54	1800				
		2057	40	10.8	5.1
		2869	45	10.9	4.6
i = 97.5	3240				
		3871	50	11	4.1

EXP = experimentals datas
A = Palladium coverage (10^{14} atoms.cm^{-2})
B = atoms/particle
C = atoms/particle (models: icos or cubo)
D = diameter (Å)
E = average coordination number
F = variation of the FWHM (percentage-wise) compared with the bulk

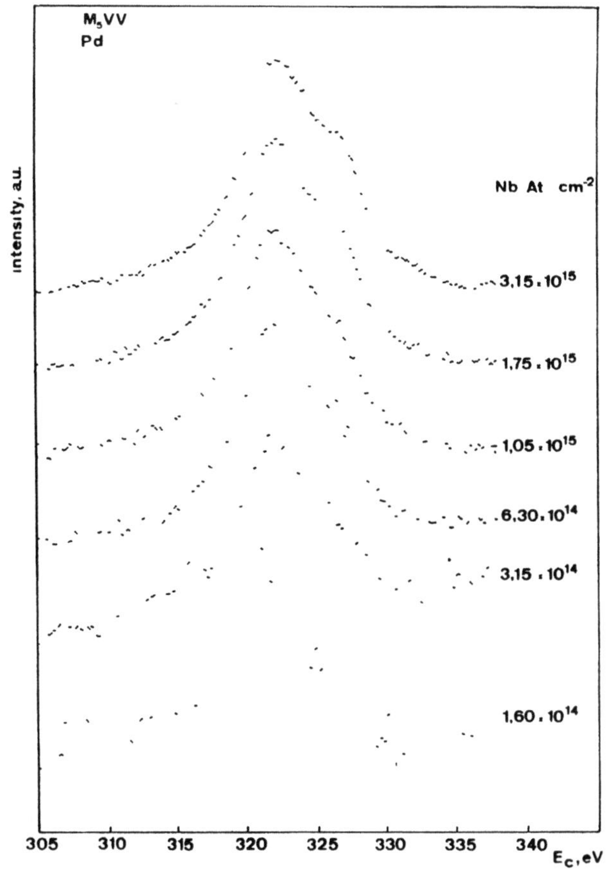

Fig. 3. M-VV Palladium transition as a function of palladium coverage for Pd/SiO_2 system. The energies on the abscissa are the kinetic energies

–Variation of the Fermi level [11]
–Modification of the core level binding energy [11]
–Change of the extra-atomic relaxation energy [11, 13].

The FWHM variation with the particle size can be qualitatively attributed to the variation of the coordination number Z which has been discussed above. To explicit more quantitatively this effect, we must take into account other parameters as well as correlation effects [13] and the broadening of the transition caused by the size distribution [14].

Reactivity

It is now possible to use our model system as an heterogeneous catalyst. In order to correlate the electronic properties of the clusters with their chemical reactivity, we have chosen the 1,3 butadiene hydrogenation. In Fig. 4 it is observed that the small clusters exhibit a weak activity which increases with the cluster size (between 10 Å and 30 Å).

We can explain this phenomenon by a progressive change of the electronic properties with the particle size, this implying a variation of the reactants adsorption energy with the average coordination number

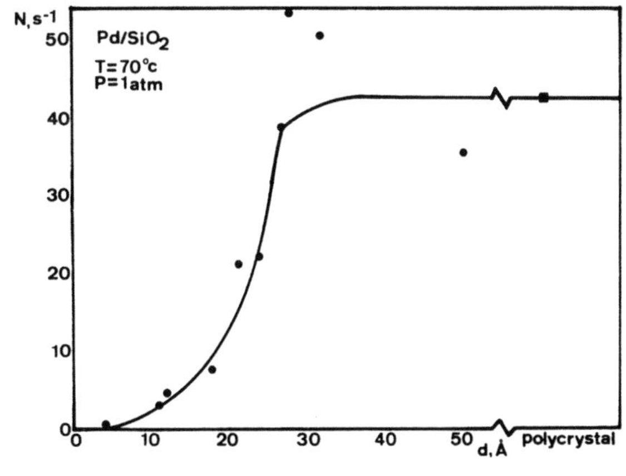

Fig. 4. Variation of the turnover number* as function of cluster size in the hydrogenation of 1.3 butadiene *the number of surface atoms has been calculated assuming that the equilibrium form of the clusters is an cubo-octahedra

[16]. The weak activity of the very small clusters (< 10 Å) would be due to an autoinhibition of the hydrocarbon induced by a too strong adsorption on low coordination sites [17, 18].

Conclusion

Because it is possible to monitor the size and the intrinsic electronic properties of metallic clusters, one can explain the behavior of the T.O.F. in the hydrogenation of 1,3 butadiene. Such an approach is being continued on bimetallic aggregates obtained by the same method [19].

References

1. Che, M., Bennett, C.O.: Adv. Catal. **36**, 55 (1989)
2. Bellamy, B., Colomer, G.: J. Vac. Sci. Technol. **A2**(4) 1604 (1984)
3. Hadj Romdhane, Y., Bellamy, B., De Gouveia, V., Masson, A., Che, M.: Appl. Surf. Sci. **31**, 383 (1988)
4. Schmeisser, H.: Thin Solid Films **22**, 83 (1974)
5. Liang, K.S., Salaneck, W.R., Aksay, I.A.: Solid State Commun. **19**, 329 (1979)
6. Mason, M.G., Baetzold, R.C.: J. Chem. Phys. **64**, 271 (1976)
7. Takasu, Y., Unwin, R., Tesche, B., Bradshaw, A.M., Grunze, M.: Surf. Sci. **77**, 219 (1978)
8. Roulet, H., Mariot, J.M., Dufour, G., Hague, C.F.: J. Phys. F. **10**, 1025 (1980)
9. Haydock, R., Kelly, M.J.: Surf. Sci. **38**, 139 (1973)
10. Desjonqueres, M.C., Cyrot-Lackmann, F.: Surf. Sci. **53**, 429 (1975)
11. Wertheim, G.K., DiCenzo, S.B., Youngquist, S.E.: Phys. Rev. Lett. **51**, 2310 (1983)
12. Mason, M.G.: Phys. Rev. B **27**, 748 (1983)
13. Kohiki S., Ikeda S.: Phys. Rev. B **34**, 3786 (1986)
14. Bellamy, B., Masson, A., De Gouveia, V., Desjonqueres, M.C., Spanjaard, D., Treglia, G.: J. Phys. F. (to be published)
15. De Crescenzi, M. Diociaiuti, M., Lozzi, L., Battistoni, C., Mattogno, G.: Surf. Sci. **178**, 282 (1986)
16. De Gouveia, V.: Thesis Paris **VI** (1988)
17. Boitiaux, J.P., Cosyns, J., Vasudevan, S.: Preparation of catalysts, III, pp.123. Amsterdam: Elsevier 1983
18. Ugo, R.: Catal. Rev. Sci. Eng. **11**, 225 (1975)
19. De Gouveia, V. Bellamy, B., Hadj Romdhame, Y., Masson, A., Che. M.: (to be published)

Z. Phys. D – Atoms, Molecules and Clusters 12, 591–595 (1989)

Atoms, Molecules and Clusters
Zeitschrift für Physik D
© Springer-Verlag 1989

Structure flexibility of iridium metal aggregates. Comparison with platinum and rhodium

G. Bergeret and P. Gallezot
Institut de Recherches sur la Catalyse-CNRS, 2, avenue Albert Einstein, F-69626 Villeurbanne Cédex, France

Received 5 July 1988; final version 30 September 1988

The atomic structure of iridium aggregates occluded in zeolite has been studied by radial electron distribution from X-ray scattering data taken under different atmospheres. The interatomic distances in 1–2 nm naked aggregates are contracted with respect to the bulk distances. The adsorption of hydrogen produces a relaxation but a residual contraction remains unlike in the case of platinum aggregates. The carbon monoxide adsorption produces little effect on the position of surface iridium atoms unlike on platinum and rhodium. The lack of structure flexibility of iridium aggregates is probably due to the higher cohesive energy of iridium.

PACS: 82.65; 82.70; 61.14.F

Introduction

The structure of naked metal clusters *Mm* with *m* smaller than hundred is still often viewed erroneously as rigid pieces of metal cut from the bulk under the form of regular polyhedra. However bulk atoms are outnumbered by surface atoms and there is a growing number of experimental evidences showing that the atomic positions are perturbed by the environment. Thus it has been shown by Radial Electron Distribution (RED) from X-ray scattering data [1, 2] and by EXAFS [3] that the interatomic distances in 1 nm platinum aggregates are contracted with respect to the normal f.c.c. packing of bulk platinum. The adsorption of hydrogen produces a relaxation to the normal distances [1, 2, 4] whereas the carbon monoxide chemisorption displaces the atoms from their equilibrium position. In the case of 1 nm rhodium aggregates, CO adsorption extracts progressively the surface atoms resulting in a complete disintegration of the aggregate into $Rh^{I}(CO)_2$ species [5]. The aim of the present paper is to study by RED, the structure of iridium aggregates supported on Y-type zeolite under different atmospheres. Comparison will be made with platinum and rhodium aggregates of similar size under the same environment.

Experimental

Materials and treatments

The IrY zeolite was prepared as described previously [6]. After activation at 520 K under oxygen, the zeolite was evacuated at 520 K and reduced under flowing hydrogen from 300 to 520 K (1 K/min). After evacuating, the zeolite was heated at 670 K under vacuum. The outgassed zeolite Ir-v(I) thus obtained was trans ferred from the quartz cell to an X-ray cell with a beryllium window which is used to record X-ray scattering data. The transfer was done in a glove box flushed with argon. All the adsorption and desorption treatments at room temperature have been carried out in the X-ray cell. The conditions of treatments of the various samples are given in Table 1.

Table 1. Treatments of iridium zeolites

Samples	Treatments
Ir-v(I)	$Ir(NH_3)_5Cl^{2+}$ exchanged *Y* zeolite; heated from 300 to 520 K under O_2; evacuated at 520 K; heated from 300 to 520 K in flowing H_2; evacuated at 520 K; heated at 670 K under vacuum, kept under Ar
Ir-H(I)	Ir-v(I) contacted at 300 K with 760 Torr of H_2; kept under H_2
Ir-H(II)	Treated as Ir-H(I) but reduced in static H_2
Ir-CO(II)	Ir-H(II) evacuated at 300 K; contacted at 300 K with 760 Torr of CO; kept under CO

RED Investigation

The X-ray diffraction patterns were recorded as described previously [6]. The radial electron distribution (RED) function $4\pi r^2\ \rho(r)$ was calculated, by Fourier analysis using the computer program RADIST [5]. The RED of a dehydrated NaHY zeolite was subtracted from the RED of the IrY zeolite. The resulting RED function shows the distribution of distances between Ir atoms and between Ir atoms and the atoms of the zeolite.

The distributions of model aggregates were computed with the program RADMOR [7] which calculates the interatomic distances from the atom positions generated by space-group symmetry transformations and unit-cell translations. The coordination numbers associated with each type of distance in the sample are determined for each coordination shell having a bin width of 0.01 Å. A differential RED function $4\pi r^2(\rho(r) - \rho_0)$ is obtained by applying a Gaussian broadening to each bin weighted by each atomic scattering factor. For a model fragment, i.e. a spherical particle of radius *R*, the RED function is modified by a spherical finite particle size function inversely proportional to *R* which corrects the relative frequency of occurrence of an interatomic distance in a structure of unbound dimensions to that which pertains within the spherical region of radius *R* being considered.

Results and discussion

Nuclearity of the aggregates

In the Pt-H sample studied previously [1, 2], the Pt aggregates were in the size range 0.9 + 0.3 nm as shown by high resolution Transmission Electron Microscopy (TEM) studies on ultramicrotome cuts of the zeolite grains. A more accurate estimation of the nuclearity was obtained comparing the experimental RED with a distribution calculated from model aggregates [8]. The best fit was obtained for a truncated tetrahedron

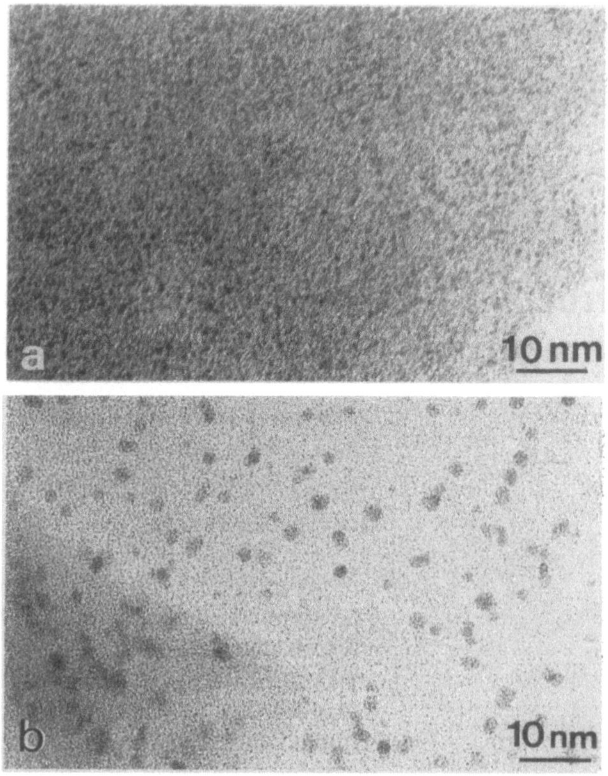

Fig. 1a and b. TEM image of two grains of a IrY zeolite from a same batch (ultramicrotome cut, magnification 10^6). Micrograph **a)** 1 nm or less Ir aggregates; micrograph **b)** 2 nm Ir aggregates

containing 40 atoms (36 surface atoms, 4 bulk atoms) which has the same symmetry as the zeolite. Rhodium aggregates in Y-zeolites were found even smaller (size < 1 nm, $m < 20$) [5].

The size distribution of aggregates in platinum and rhodium zeolites is unimodal throughout all the zeolite grains. In contrast, TEM views of the reduced iridium zeolites Ir-H(I) and Ir-H(II) show that some zeolite grains contain an homogeneous distribution of less than 1 nm Ir aggregates and the other an homogeneous distribution of 2 nm Ir aggregates. Figure 1a and b give TEM views taken on two different microcrystals of zeolite issued from the same preparation batch. The number of zeolite microcrystals with the 2 nm Ir aggregates is much smaller in Ir-H(I) than in Ir-H(II). This is corroborated by the comparison between the experimental RED of these samples and the RED calculated from models. Thus Fig. 2a gives the RED of Ir-H(II). In contrast with platinum aggregates [8] it was not possible to fit a distribution calculated from a single model, a combination of two distributions was required. A reasonable good fit is obtained in Fig. 2b by combining two distributions calculated for a 1.8 nm (141 atoms) and for a 1.0 nm (19 atoms) spherical f.c.c. models in the ratio 75:25. On the other hand the RED of Ir-H(I) fits with a distribution calculated for a mixture of 1.8 and 1.0 nm aggregates in the ratio 50:50.

Structure of naked and hydrogen-covered aggregates

Figure 3 gives the RED obtained from the X-ray scattering data taken on Ir-v(I) which has been outgassed at 670 K. Table 2 gives the interatomic distances in bulk iridium. Ir-H(I) and Ir-H(II). The distances in the naked and H_2-covered Pt aggregates found previously are given for comparison. The percentage of contraction with respect to bulk distance are given in parentheses.

It is noteworthy that the Ir-Ir distances in Ir-v(I) are contracted with respect to the normal distances of bulk iridium. The contraction is not even for the different coordination shells which means that there is a slight distortion with respect to the normal f.c.c. packing. This contraction is not unexpected for aggregates where most of the atoms are on the surface and are coordinatively unsaturated. Indeed a comparable contraction was found for the Pt-Pt distances in naked platinum aggregates (Table 2). However the contraction is slightly larger in iridium aggregates in spite of the fact that there are less surface atoms than in the case of platinum (the ratios of surface to bulk atoms are close to 9 and 2.5 respectively). The larger contraction of distances in iridium could be related with its

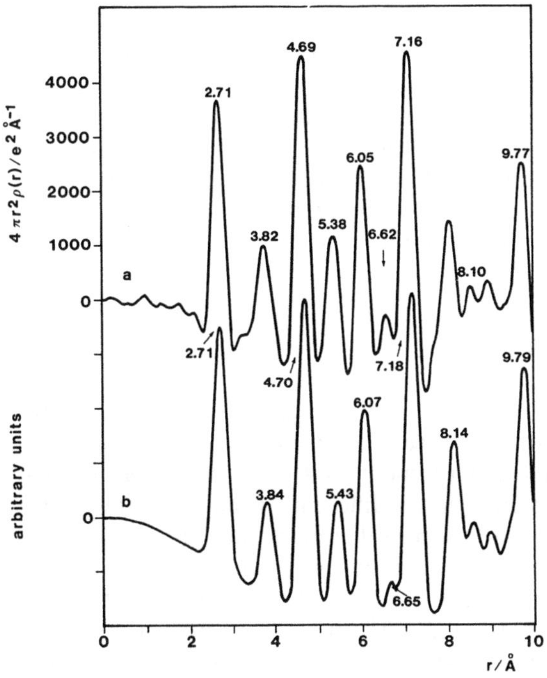

Fig. 2a. RED obtained from X-ray scattering of the IrY zeolite, sample Ir-H(II); see Table 1 for corresponding treatments; the RED of the zeolite has been subtracted *b.* RED obtained from model aggregates; mixture of 1.8 and 1.0 nm spherical f.c.c. particle in the ratio 75·25

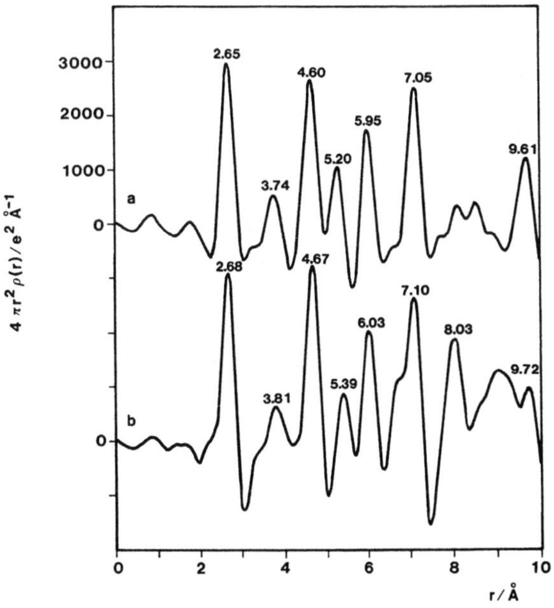

Fig. 3. RED of Ir-Y zeolite; *a)* Ir-v(I); *b)* Ir-v(I); see Table 1 for corresponding treatments

Table 2. Interatomic distances in naked and hydrogen-covered aggregates (in Angstrom Unit)

Bulk Ir	Ir-v(I)	Ir-H(I)	Ir-H(II)	Bulk Pt	Pt-v[b]	Pt-H[b]
2.715	2 65 (2.4)[a]	2.68 (1 3)	2 71 (n)	2.774	2.70 (2.7)	2.77 (n)
3.839	3 74 (2.6)	3.81 (0.8)	3 82 (0.5)	3.923	3.90 (0.6)	3.93 (n)
4 702	4.60 (2 2)	4 67 (0 7)	4.69 (0 3)	4.805	4 72 (1 8)	4.80 (n)
5 430	5 20 (4.2)	5.39 (0.7)	5.38 (0.9)	5.548	5.32 (4 1)	5.52 (0.5)
6.071	5.95 (2 0)	6 03 (0 7)	6.05 (0 3)	6.203	6.12 (1.3)	6 20 (n)
6 650	(sh)	(sh)	6.62 (0 4)	6.795	(sh)	(sh)
7 183	7.05 (1 9)	7.10 (1.1)	7 16 (0.3)	7.339	7.22 (1 6)	7.32 (0.2)

[a] Percentage of contraction with respect to bulk given in parentheses (n) means negligible, (sh) shoulder
[b] From Ref. 2

larger cohesive energy; the molecular enthalpy of atomization are 664 and 564 kJ/mole for Ir and Pt respectively.

Upon contacting the samples with hydrogen at 300 K, all the Ir-Ir distances increase (Ir-H(I), Table 2). This relaxation of the distance occurs because the coordination of the surface atoms is now partially fulfilled by the bonding of surface atoms with chemisorbed hydrogen atoms. This was also observed in the case of platinum aggregates (Table 2); however the Pt-Pt distances in the hydrogen-covered aggregates were almost similar to bulk distances whereas a substantial contraction remains in the case of iridium in spite of the larger enthalpy of H_2-adsorption on iridium than on platinum [9]. The residual contraction of the metal skeleton can be interpreted in terms of the larger Ir-Ir bond enthalpy (111 kJ/mole compared to 94 kJ/mole for Pt-Pt bonds) which make it less sensitive to perturbation induced by adsorbates bonded to the surface atoms.

Effects of CO – adsorption on iridium aggregates

The Ir-H(II) zeolite has been contacted with CO at 300 K. The RED corresponding to the zeolite covered with CO is given in Fig. 4b, the RED of Ir-H(II) is given for comparison in Fig. 4a. The two distributions are very similar which means that the replacement of hydrogen by carbon monoxide on the surface produces little change in the structure of the aggregates. This is in contrast with the CO adsorption on platinum and rhodium aggregates. The CO-adsorption on the 1 nm Pt-aggregate produced a 50% decrease of the intensity of the Pt-Pt peaks because the surface platinum atoms bonded to chemisorbed CO are displaced from the equilibrium positions of the f.c.c. lattice [2]. The structure of the CO covered aggregates can be compared to that of polynuclear carbonyl clusters where for a given coordination sphere the metal-metal bonds are distributed over a range of values with a mean distance close to that of bulk metal. The CO-adsorption on rhodium aggregates [5] produced an even more drastic effect since a complete disintegration of the Rh aggregates into monomeric $Rh^{I}(CO)_2$ species was observed. Even though there are only 70% of surface atoms in Ir-H(II) the RED would have been deeply modified if CO would have induced on the iridium surface a corrosive effect similar to that observed for platinum or rhodium. Again the weak mobility of surface atoms on iridium particles can be related to the high cohesive energy of the iridium lattice.

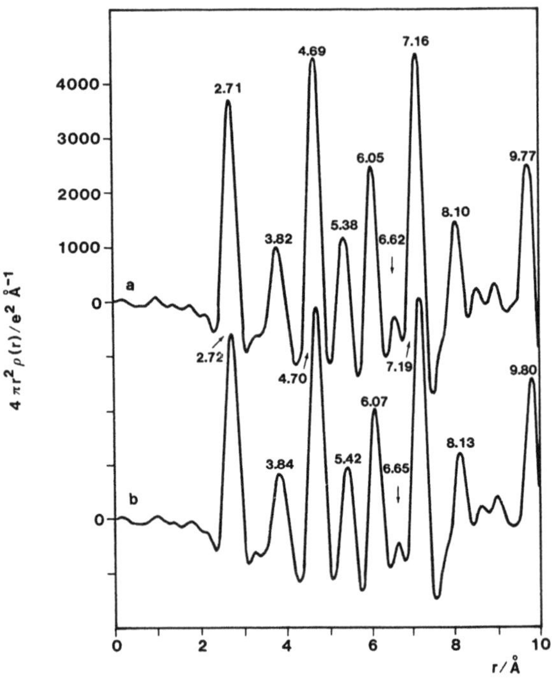

Fig. 4. RED of Ir-Y zeolites; *a)* Ir-H(II); *b)* Ir-CO(II); see Table 1 for corresponding treatments

Conclusion

The RED study of iridium aggregates occluded in the microporous lattice of Y-type zeolite shows that under the same environment and for comparable particle size, iridium behaves differently from platinum aggregates. The following differences can be interpreted in terms of the higher cohesive energy of the iridium lattice *(i)* the distances in the naked aggregates are more contracted *(ii)* the adsorption of hydrogen produces a relaxation of the distances but a residual contraction remains unlike on platinum *(iii)* the adsorption of CO produces almost no effect on the interatomic distances whereas platinum aggregates were deeply modified and rhodium totally disintegrated. This study shows that the structure flexibility of metal aggregates in response to environment effect depends not only upon the metal-adsorbate binding energy but also on the metal-metal bond enthalpy.

References

1. Gallezot, P., Bergeret, G.: J. Catal. **72**, 294 (1981)
2. Gallezot, P.: Zeolites **2**, 103 (1982)
3. Moraweck, B., Clugnet, G., Renouprez, A.: Surf. Sci. **81**, 1631 (1978)
4. Gallezot, P., Bienenstock, A., Boudart, M.: Nouv. J. Chim. **2**, 263 (1978)

5. Bergeret, G., Gallezot, P., Gelin, P., Ben Taarit, Y., Lefebvre, F., Naccache, C., Shannon, R.D.: J. Catal. **104**, 279 (1987)
6. Bergeret, G. Gallezot, P., Lefebvre, F.: New development in zeolite science technology. Proceedings of the 7th International Zeolite Conference Murakami, Y., Iijima, A., Ward, J. W. (eds), p. 401. Tokyo: Kodansha Ltd 1986
7. D'Antonio, P., Karle, J.: J. Appl. Crystallogr. **13**, 459 (1980)
8. Bergeret, G., Gallezot, P.: Proceedings of the 8th International Congress on Catalysis Berlin 1984, Vol. 5, p. 659. Weinheim: Verlag Chemie 1984
9. Toyoshima, I., Somorjai, G.A.: Catal. Rev. Sci. Eng. **19**, 355 (1979)

Z. Phys. D – Atoms, Molecules and Clusters 12, 597–600 (1989)

Atoms, Molecules and Clusters
Zeitschrift für Physik D
© Springer-Verlag 1989

Surface oxidation and reduction of small platinum particles observed by in-situ X-ray diffraction

V. Gnutzmann and W. Vogel
Fritz-Haber-Institut der Max-Planck-Gesellschaft, Faradayweg 4–6, D-1000 Berlin 33, Germany

Received 5 July 1988

A new method is presented, in which the constructive Bragg interference intensity of small metallic clusters during the formation or decay of a surface compound is measured by an in situ X-ray technique. First results for a series of platinum catalysts supported on silica gel show intensity losses on the 111-peak of up to 40% under oxygen following a standard pretreatment. These losses are restored reversibly under hydrogen. The average thickness of the disrupted surface layer ranges between 0.9 to 4.0 Å depending on the treatment temperature and platinum crystallite size.

PACS: 61.10

I. Introduction

Interest in the oxidation-reduction behavior of small platinum particles was raised in recent times by an EXAFS study by Joyner [1] and a temperature programmed reduction (TPR) study by Bond et at. [2], both performed on the standard catalyst EUROPT-1, i.e. 6.3% Pt on silica gel.

This paper focuses on the structural characterisation of small platinum crystallites under the action of oxygen and hydrogen by means of in situ X-ray diffraction [3].

The technique applied is to follow the changes of the integral 111 Bragg peak intensity as a function of temperature, pressure and time. We conclude, that those changes are induced by the oxidation and reduction of a certain platinum surface layer. This layer will not contribute to the constructive interferences of the metal core. In a first application, a series of four Pt/SiO_2 standard catalysts were investigated.

II. Experimental

The four Pt/SiO_2 catalysts listed in Table 1 are prepared and made available by Burwell and coworkers [4]. Detailed characterisation of these catalysts are given in a series of papers [4, 5, 6, 7].

Table 1. Codenames of catalysts as described in [4]. Pt loading and dispersion D_h obtained by hydrogen chemisorption are also taken from these authors. The surface mean radii R_S and the related dispersions D_X are obtained by Fourier line profile analysis

No.	Catalyst	Pt (wt%)	D_h (%)	R_S (Å)	D_X (%)
#88	7.1-SiO_2-PtCl-S	1.91	6.8	112.5	5.0
#69	21.5-SiO_2-IonX-L	1.48	21.5	31.4	17.8
#65	40-SiO_2-PtCl-L	1.10	39.8	14.3	39.2
#67	63.5-SiO_2-IonX-L	0.49	64.5	9.3	60 2

The X-ray measurements were performed by a commercial Guinier-counter diffractometer (HUBER), modified by a specially designed specimen cell which allows treatments at 77–900 K and 10^{-6}– 2 bar under different gas atmospheres. The samples were pressed with 8 tons to pellets 15*12*0.3 mm in size. All catalysts were given a standard pretreatment as described in [4].

The X-ray spectra were taken in 45°-transmission geometry using monochromatic Cu-$K_{\alpha 1}$ radiation. Primary beam intensity fluctuations were eliminated by synchronously monitoring the backscattered X-rays at the entrance slit. Monitoring appeared to be essential for the open slit measurements. In this mode the counter slit was set to the constant Bragg angle $\vartheta = 19.9°$ of the 111 platinum peak and opened to a width of $\pm 0.75°$ ϑ as indicated in Fig. 1. Alternatively the angular dependent X-ray patterns were measured with high resolution in the step scanning mode with a narrow receiving slit. A typical measurement of a full spectrum takes about 20 h and is performed under complete control of a LSI-11 computer.

III. Theory

In the case of a chemical surface reaction of a small platinum crystallite, a certain fraction of the surface atoms may either be "evaporated" to the gas phase or rearranged to form an epitaxial surface compound which in general will not be in registry with the underlying metallic core. In both cases a reduction in the metal phase Bragg intensity will occur which increases with decreasing crystallite size.

For clusters of less than $\sim 10^3$ atoms the scattering function can be calculated numerically by the Debye formula. This is shown in Fig. 1 for a cuboctahedron of 923 Pt atoms arranged in the perfect fcc symmetry (solid line). This ~ 30 Å sized crystallite exposes 282 atoms (31%) to the $\{100\}$ surface planes (total dispersion is 39%). After "evaporation" of these atoms the intensity changes as indicated by the dashed curve, while the dotted curve represents a random shift of these atoms by 0.5 and 0.1 near neighbour distances parallel and orthogonal to their individual $\{100\}$ netplanes. The latter case should be typical of an "amorphous" top layer.

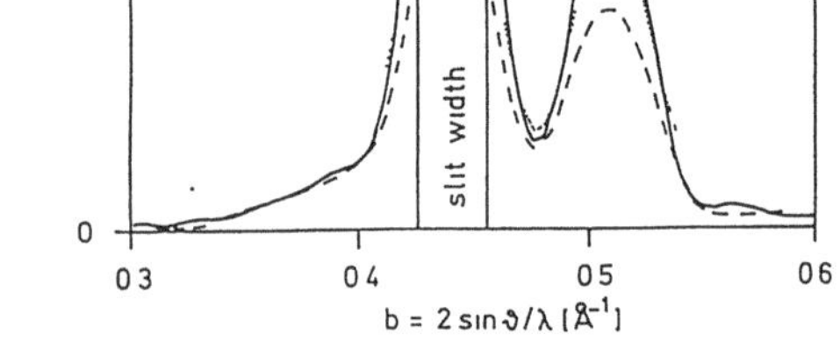

Fig. 1. Calculated interference function of a 923-atoms coboctahedron in perfect fcc symmetry (———),-with 282 $\{100\}$-surface atoms "evaporated" (– – –) and randomly shifted (.........)

The measuring slit (as indicated in Fig. 1 by vertical lines) receives an intensity deficiency $-\Delta I/I$ of 38% and 28% respectively, compared to the undisturbed crystallite. The cluster calculated above compares to the catalyst #65 with nearly the same surface average diameter $2R_S = 28.6$ Å, for which the measured oxygen-induced intensity loss ranges from 10–30% (see next section).

The situation is however more complicated, since for real catalysts the crystallite size is not uniform but in general follows a broad distribution function $H(R)$.

A rough estimate of a structural quantity, the average thickness ΔR of an "evaporated" surface layer may be obtained by a continuum theoretical approach assuming:

a) spherically shaped crystallites;

b) a surface layer of thickness ΔR is removed independently of the individual crystallite radius R.

If we call $H^0(R)$ the original fraction by number of crystallite radii, the new distribution function simply equals the old one, but now shifting to lower R-values by ΔR:

$$H(R) = H^0(R + \Delta R).$$

The n-th momentum J_n of this new distribution can then be expressed by a Taylor progression depending on the moments of the old distribution:

$$J_n = J_n^0 - n\Delta R J_{n-1}^0 + (1/2)n(n-1)\Delta R^2 J_{n-2}^0 - \ldots$$

Since the integral intensity I is proportional to the volume, i.e. the third momentum of $H(R)$, we get:

$$\Delta I/I = (J_3 - J_3^0)/J_3^0$$
$$= -3\Delta R(J_2^0/J_3^0 - \Delta R J_1^0/J_3^0 + \ldots)$$

We use the usual definitions for differently averaged radii:

Linear mean radius: $R_D = J_2^0/J_1^0$; surface mean radius: $R_S = J_3^0/J_2^0$; volume mean radius: $R_M = J_4^0/J_3^0$. In the second approximation we get:

$$\Delta I/I = -3\Delta R(1/R_S - \Delta R/(R_S R_D)) \qquad (1)$$

The quantities R_D, R_S were obtained by a modified Fourier line profile analysis. (This method will be described elsewhere). The mass mean radius R_M is directly related to the integral line width. It can be shown that this quantity is scarcely influenced (typically less than 2% at 15% intensity loss) by polydispersed systems.

IV. Results

Figure 2 shows the original X-ray spectra of catalyst #65 (dashed curve), taken at 1 bar, 300°C

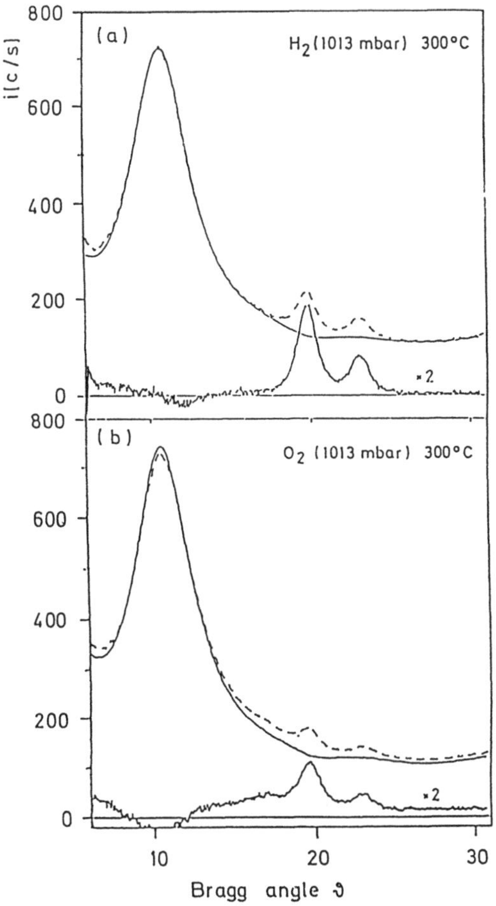

Fig. 2a and b. Measured X-ray spectra of cat. #65 **a)** at 300°C, 1 bar H_2 **b)** at 300°C, 1 bar O_2. The lower curves are difference spectra, the substrate scattering being subtracted

a) fully reduced. The difference spectrum (smooth curve: pure substrate scattering) at the bottom shows the 111 and 200 Bragg peaks as expected from a metallic fcc platinum phase.

b) The same sample after oxygen treatment for 1 h at 300°C.

The Bragg peak intensities are weakened and additional diffuse intensity is growing out especially on the left of the 111 peak. This indicates the development of a new, more or less amorphous phase. For the catalyst #88 with the lowest dispersion no such separate phase could be detected.

Figure 3 shows a typical time-dependent open-slit measurement of the 111 integral intensity of catalyst #69 at a pressure of 1 bar. The intensity changes $\Delta I/I$ were referenced to the maximal difference intensity indicated by the vertical bar in Fig. 3. This value was always reproduced under hydrogen. It drops rapidly (within less than 1 min) about 12% under oxygen and rises immediately to the old value under hydrogen. For the catalysts #65, #67 with the highest dispersions an

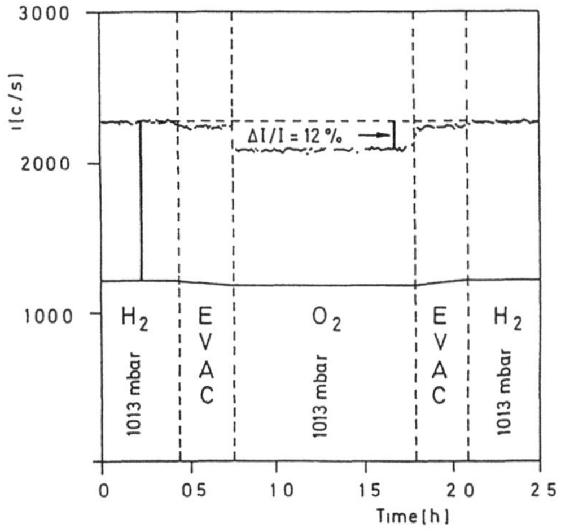

Fig. 3. Time dependent 111-open slit measurement of cat. #69. The vertical bars indicate the integral Pt-intensity under 1 bar H_2 and the loss of intensity under 1 bar oxygen at 20°C. (The smooth line at ~1200 c/s is the measured integral background due to SiO_2)

additional small time-dependent intensity loss was observed over a period of ~3 h.

Chemisorption of oxygen does not seem to disrupt the platinum surface seriously: The small intensity loss versus the hydrogen-level during the evacuation phase ($p \leqq 10^{-2}$ mbar) seen in Fig. 3 might be oxygen induced chemisorption but did not appear to be reproducible in all cases. The main jump in the Bragg intensity loss occurred at oxygen partial pressures above ~0.5 mbar.

Figure 4 shows the collected results of all catalysts plotted as the percentage of residual metallic platinum $1-\Delta I/I$ versus the reciprocal mean crystallite radius R_S. It is seen, that the relative amount of disrupted metal phase is maximal at ~300°C and increases to the smallest crystallites. From (1) a nearly linear decrease is to be expected, as long as the depth ΔR of "evaporated" Pt does not depend on the crystallite radius R_S. ΔR can be calculated by (1) leading to Fig. 5. In fact ΔR decreases with decreasing average crystallite size. This effect is even more pronounced at low temperature.

The stability of the proposed surface oxide to reduction of the oxygen partial pressure, i.e. to evacuation to $P_{O_2} < 10^{-2}$ mbar was found to increase with decreasing crystallite size and temperature of formation. At RT the oxygen coverage is completely removed for the largest crystallites (catal. #88).

V. Discussion

The ΔR-values are based on the unrealistic assumption of an evaporated surface layer. It was seen from

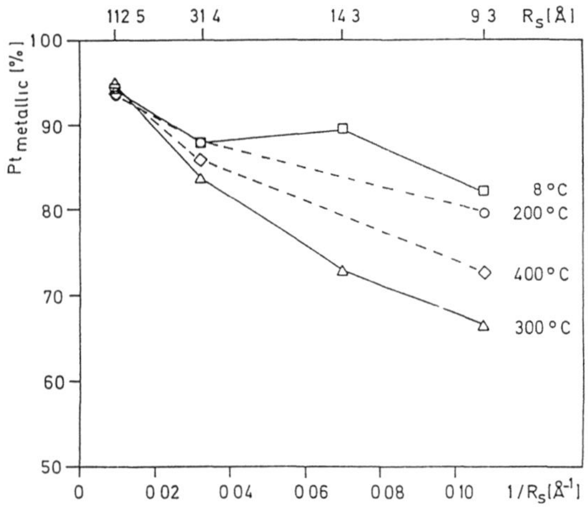

Fig. 4. Residual metallic platinum under oxygen for different surface mean radii R_S at different temperatures, applying cyclic H_2–O_2–H_2 atmospheres

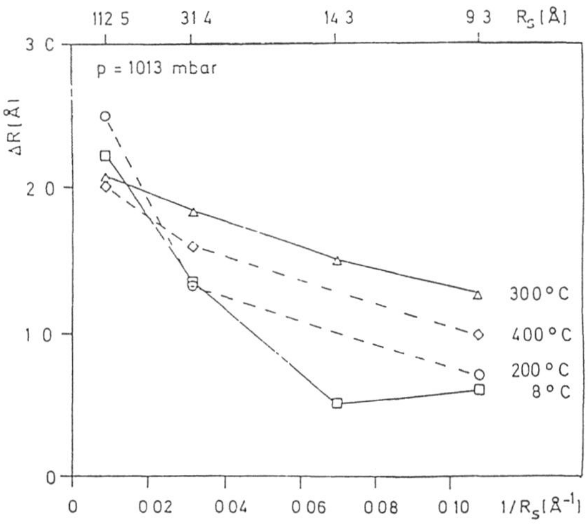

Fig. 5. Average "evaporated" surface layer thickness obtained from Fig. 4 by (1). ΔR is a lower limit for the depth of a disrupted platinum surface layer

the Debye-function calculation in Sect. III, that a disordered platinum surface layer may need to be thicker than an evaporated surface sheet in order to produce the same intensity drop. Assuming a similar law as that given in (1) for this quantity $\Delta R_{\text{dist.}}$, as a rough guide the factor (38/28) = 1.4 may be used to obtain the average disordered depth of a surface layer: $\Delta R_{\text{dist.}} = 1.4\,\Delta R$.

The disrupted surface thickness arrives at a maximum value of 3–4 Å for the largest crystallites as seen in Fig. 5. These are up to two (100)-monolayers of 1.96 Å spacing. The smallest average depth of disruption is 0.9 Å, i.e. about half of a monolayer at low temperatures and crystallite sizes.

These results are in good agreement with those of Uchijima et al. [4] on the same catalysts using hydrogen-oxygen titration: "Upon exposure to oxygen at 300 °C, faceting or penetration of oxide ions into the bulk appears to occur for larger particles of platinum since the oxygen content (of catalyst #88) considerably exceeds a monolayer."

All catalysts (except cat. #88) in the as-received state were partially oxidized as indicated by a positive 111-intensity jump under hydrogen. To the contrary, in the as- received Pt/SiO_2 catalyst EUROPT-1 a near 1:1 stoichiometry of Pt:O was observed [1, 2]. Further application of the technique described here will help to clarify the situation.

References

1. Joyner, R.W., Meehan, P.: Vacuum **33**, 691 (1983)
2. Bond, G.C., Gelsthorpe, M.R.: Appl. Catal. **35**, 169 (1987)
3. Gnutzmann, V., Vogel, W.: J. Catal. Lett. (submitted)
4. Uchijima, T., Herrmann, J.M., Yasunobu Inoue, Burwell, R., Jr., Butt, J.B., Cohen, J.B.: J. Catal. **50**, 464 (1977)
5. Sashital, R.S., Cohen, J.B., Burwell, R.L. Jr., Butt, J.B.: J. Catal. **50**, 479 (1977)
6. Otero-Schipper, P.H., Wachter, W.A., Butt, J.B., Burwell, R.L. Jr., Cohen, J.B.: J. Catal. **50**, 491 (1977)
7. Nandi, R.K., Molinaro, F., Tang, C., Cohen, J.B., Butt, J.B., Burwell, R.L. Jr.: J. Catal. **78**, 289 (1982)

Z. Phys. D – Atoms, Molecules and Clusters 12, 601–603 (1989)

Atoms, Molecules and Clusters
Zeitschrift für Physik D
© Springer-Verlag 1989

Optical and chemical studies of II–VI compound clusters

W.A. Saunders

Institut de Physique Expérimentale, Ecole Polytechnique Fédérale de Lausanne, PHB-Ecublens, CH-1015 Lausanne, Switzerland

Received 6 July 1988

Recent experimental results on the structural, chemical, and optical properties of II–VI compound clusters containing between 2 and 24 molecules are presented. Stability patterns in the mass spectra of MgO clusters correlate with stable structures predicted for ionic clusters. The time dependence of H_2O adsorption onto MgO clusters is measured and found to vary strongly as a function of the number of adsorbed molecules. A sharp peak in the photodissociation probability is observed for metal-excess SrO and CaO cluster ions near 2.0 eV.

PACS: 36.40.+d; 33.80.Gj; 31.70.Hq

Cluster properties and their relation to the corresponding properties of the bulk and molecular phases depend strongly on the material being investigated. Van der Waals [1], covalent [2], metallic [3], and ionic [4] cluster systems show a rich variety of unique physical and chemical properties. Bulk II–VI compounds [5] show both covalent and ionic character in their bonding and are widely employed as catalysts and in opto-electronics. We report here results of some recent investigations of II–VI compound clusters including measurements of wavelength dependent photodissociation and the time-dependence of H_2O adsorption.

Experiments are carried out in a triple quadrupole mass spectrometer [6] in which the first and third quadrupoles are used as mass analyzers and the middle (second) quadrupole serves as an ion guide for experimentation. Cluster are created by sputtering a pure metal target which is continuously flooded by a jet of oxygen, sulfur or selenium. Measurements indicate the clusters are sputtered from a thin layer on the metal surface. A mass spectrum of sputtered MgO cluster ions, which is typical of the ionic II–VI compounds, is shown in Fig. 1. Two distinct cluster species are observed: stoichiometric $(MgO)_n^+$ and metal-excess $Mg(MgO)_n^+$. The smooth overall decrease of the cluster intensities is modulated by a fine structure of strong drops following certain cluster sizes. These occur most noticeably after the clusters $(MgO)_{3n}^+$, with $n = 2-8$. Although no explicit calculation of MgO cluster structure are available, results of Coulomb-pair-potential models for NaCl [4] clusters suggest that the basis of the structure may be ring-like rather than cubic. This represents a significant departure from the bulk (rocksalt) structure. Collision induced fragmentation measurements confirm the special stability of the $(MgO)_3$ sub-unit [7]. The enhanced intensity of the $Mg(MgO)_{13}^+$ cluster correlates with the stable cubic cluster predicted for ionic clusters [4]. Following this cluster, an unexpected and interesting series of abundant clusters of the form $Mg(MgO)_{13+3n}^+$ for $n = 1-3$ are also observed. It is impossible to differentiate between conceivable structures [8] for these clusters without further investigation.

The study of the binding of simple molecules to MgO and other II–VI compound clusters offers the possibility to investigate some aspects of adsorption onto "imperfect" surfaces, which are known to be important in certain catalytic processes [9], and may be of interest with regards to interstellar grain surfaces [10]. In Fig. 2 we show the time dependence of the hydration of $Mg(MgO)_3^+$ measured by guiding a pulse

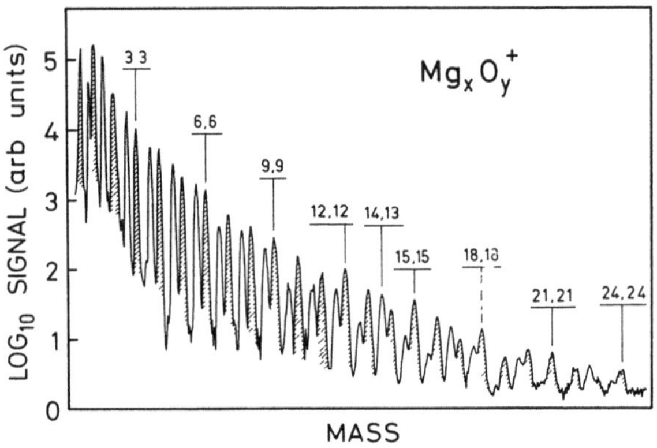

Fig. 1. A mass spectrum of sputtered MgO clusters. Two clusters species are observed: stoichiometric $(MgO)_n^+$ (shaded) and metal-excess $Mg(MgO)_n^+$. Pronounced features are observed for the series $(MgO)_{3n}^+$, $n = 2-8$. Enhanced abundance is also observed for $Mg(MgO)_{13}^+$ and $Mg(MgO)_{13+3n}^+$

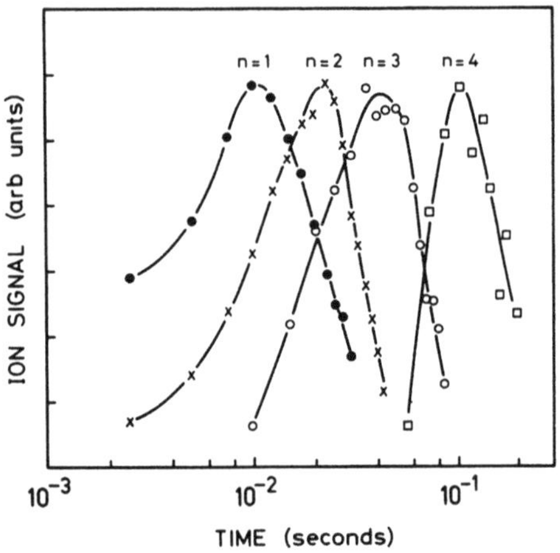

Fig. 2. The measured time dependence for the growth of $Mg(MgO)_3(H_2O)_n^+$ in approximately 10^{-3} mbar Ar with 0.05% H_2O. The first three water molecules are added quickly, while a substantially longer time is required to add the fourth

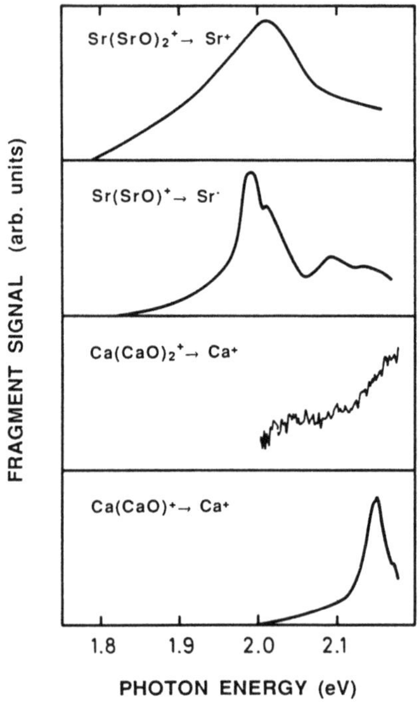

Fig. 3. The wavelength dependence of the photodissociation probabilities for $Sr(SrO)_n^+$ and $Ca(CaO)_n^+$ cluster ions. A sharp transition is observed for the $n = 1$ clusters, while the transition is broadened and blue-shifted for $n = 2$. The transition energies are close to the *s-p* transition energies of the respective free metal atoms

of parent clusters through 10^{-3} mbar of Ar mixed with approximately 0.05% H_2O. This technique is related to one used to investigate CO adsorption by Ni clusters [11]. The time dependence is measured using gated ion counting following a variable delay after a pulse of parent clusters. The cluster pulse is created by gating an electrode in the ion optics. We find the measured curves to depend strongly on the Ar density and only weakly on the H_2O density, suggesting that the pickup rate is limited by the removal of excess energy by collisions with the argon buffer rather than by the collision rate with H_2O molecules. Adhesion of the first and second molecules requires approximately 20 ms, while the third and fourth require an additional 20 ms and 60 ms respectively. Based on the above discussion, we suspect these variations are related to variations in the binding energies of successive molecules. For example, the fourth molecule may be weakly bound or the third may deposit a significant amount of energy into the system. Up to 7 water molecules are observed to stick to $Mg(MgO)_3^+$ under the present experimental conditions.

Photodissociation spectra (Fig. 3) of SrO and CaO clusters show a strong dissociative transition for clusters $Sr(SrO)_n^+$ and $Ca(CaO)_n^+$, $n \geq 1$ in the vicinity of 2.0 eV. Parent clusters are selected by the first quadrupole and the fragment intensity is monitored with the third as the wavelength of a *cw* dye laser is scanned. The data in Fig. 3 have been normalized for variation of the laser power with wavelength. For $Ca(CaO)^+$ a single peak in the dissociation probability of width 0.03 eV and centered at 2.15 eV is observed. A similar peak is observed for $Sr(SrO)^+$, though it is red-shifted by 0.15 eV and is split. A peak is also observed for the $n = 2$ and $n = 3$ clusters. With increasing cluster size it blue-shifts slightly and broadens.

The observed transition is directly related to the excess metal atom. The photofragmentation cross sections of the stoichiometric clusters are a factor of 10 to 100 smaller than those of the metal-excess clusters and show no pronounced wavelength dependence. Comparison to atomic data [12] shows that the observed peaks lie close to the atomic *s-p* transitions. The $^3P_0$ state of CaI lies at 1.879 eV, while the $^2P_{1/2}$ state of CaII lies at 3.124 eV. For Sr these energies are 1.775 eV and 2.941 eV, respectively. This correspondence suggests that the cluster transition may be closely related to the atomic *s-p* transition. The close proximity of the cluster transition to the neutral atomic transition may indicate a metal atom weakly bound to the cluster or a significant screening effect of an ion transition by the attached (polar) cluster. Comparison to atomic data shows that the splitting of the $Sr(SrO)^+$ peak is of the same order as the spin-orbit splitting in the free atom. The spin-orbit interaction in Ca is significantly smaller.

In summary, some recent experimental results on the structural, chemical and electronic properties of II–VI compound clusters have been presented. The mass spectra of MgO clusters indicate the stability of the $(MgO)_3$ sub-unit and show regular features extending to large cluster sizes. We have measured the time-dependence for water adsorption onto MgO cluster ions and have observed that the variations may be related to differing binding energies. Finally, measured photodissociation probabilities for SrO and CaO cluster have been presented. A sharp transition close in energy to the free metal-atom *s-p* transition has been observed.

References

1. Echt, O., Sattler, K., Recknagel, E.: Phys. Rev. Lett. **47**, 1121 (1981); Mark, T.D., Castleman, A.W.: Adv. At. Mol. Phys. **20**, 65 (1985)
2. Bloomfield, L.A., Freeman, R.R., Brown, W.L.: Phys. Rev. Lett. **54**, 2246 (1985); Kroto, H.W., Heath, J.R., O'Brian, S.C., Curl, R.F., Smalley, R.E.: Nature **318**, 162 (1985)
3. Knight, W.D., Clemenger, K., de Heer, W.A., Saunders, W.A., Chou, M.Y., Cohen, M.L.: Phys. Rev. Lett. **52**, 2141 (1984)
4. Martin, T.P.: Phys. Rep. **95**, 167 (1983)
5. Ashcroft, N.W., Mermin, N.D.: Solid state physics. New York: Holt, Rinehart, and Winston 1976
6. Fayet, P., Wöste, L.: Z. Phys. D–Atoms, Molecules and Clusters **3**, 177 (1986)
7. Saunders, W.A.: Phys. Rev. B**37**, 6583 (1988)
8. Conover, C.W.S., Yang, Y.A., Bloomfield, L.A.: Phys. Rev. B **38**, 3517 (1988)
9. See: Proceedings of the Symposium on Adsorption and Catalysis on Oxide Surfaces. Che, M., Bond, G.C. (eds.). Amsterdam: Elsevier 1985
10. Duley, W.W., Millar, T.J., Williams, D.A.: Mon. Not. R. Astr. Soc. **185**, 915 (1978); Aannestad, P.A., Purcell, E.M.: Ann. Rev. Astron. Astro. **11**, 309 (1973)
11. Fayet, P., McGlinchey, M.J., Wöste, L.: Am. Chem. Soc. **109**, 1733 (1987); Bréchignac, C., Bréchignac, Ph., Fayet, P., Saunders, W.A., Wöste, L.: J. Chem. Phys. **89**, 2419 (1988)
12. Moore, C.E.: Atomic energy levels. Washington D.C.: U.S. Government Printing Offices 1971

Z. Phys. D – Atoms, Molecules and Clusters 12, 605–607 (1989)

Atoms, Molecules
Zeitschrift für Physik D and Clusters
© Springer-Verlag 1989

Conclusions

J. Friedel
Physique des Solides*, Université Paris Sud, F-91405 Orsay, France

Looking back at the three previous such meetings (Lyon 1976, Lausanne 1980, Berlin 1984), it is clear that the field of clusters has gained in certainty, in depth and in width, but keeping generally along the initial lines already sketched more than 10 years ago. I see progresses on three fronts.

A. New and better techniques

This is clear for *sources*, especially vapour jets. But other sources, especially those using emission from liquids at the tip of a metallic needle, seem very promising for high fluxes and low effective temperatures, especially for fairly high melting elements.

Together with sources, we see the development of whole series of physical *instrumentation* – laser, synchrotron radiation, electron or ionic beams which together with mass charge and energy separation techniques, allow many physical properties and chemical reactions to be studied in flight.

Cluster *deposition* or growth from a solid (e.g. by irradiation or electric discharge) seems also very active, together with techniques based on electron microscopy (STEM, HREM) and tunnel microscopy which allow detailed local studies of shapes and chemistry of these clusters.

Finally although coming from widely different backgrounds, people involved in this field learn progressively to develop a common language. This is especially true for *computations*: ab initio computations involving a high degree of configuration interactions seem to converge for a wide number of small clusters. For larger ones, model computations with likely approximations seem often within reach and, at least for not too complex situations, to give results which explain reasonably experiments.

B. Basic boundings

Cluster of *'round' molecules* have properties which seem to be fairly well in hand. This includes not only rare gases but molecules such as O_2, CO, OH_2, NH_3 . . . The observation of geometric magic numbers up to nearly 1000 units, with icosahedral symmetry, confirms earlier analysis by J. Farges and others.

The beautiful work on surface and/or volume capture of electrons gives confidence on the understanding of detailed properties of such clusters.

In simple *metals* too, things seem well in hand, with a confirmation of Kubo's even/odd alternation and the occurrence of orbital magic numbers together with ellipsoïdal distortions. The analysis of plasmons in such clusters is still another analogy with nuclear physics; and further work should elucidate the mechanism which limits their life time, whether electron hole excitations or atomic evaporation. Analogy with nuclear physics also points to a possible analysis of fragmentation as a fission process, at least in liquid droplets.

Here again, the observation of large magic numbers is striking. It says that the atomic 'grain', both on the surface and in the volume, is compatible with the orbital symmetry involved. However a better grasp of the atomic structure of these large clusters and of their surface roughness, and how these vary with temperature, below and above melting are still missing.

The way kinetics of evaporation of single atoms or dimers play on the population distribution in a jet has been studied, I feel, in a very convincing way, even if the underlying thermodynamical model is somewhat schematic. This is a major advance in our comprehension of these populations and of a quantitative analysis of closed shells effects in free clusters.

The topic of metal/non metal appears in various ways. It was demonstrated as a closed *s* shell to *sd* hybridised band transition in Hg aggregates. This should be confirmed on other divalent metals. Similar

* Formation associeé au CNRS

$f \rightarrow d$ promotions have been observed in Ce. There might be a comparable $d \rightarrow s$ problem in some transition metal clusters, which has not been tackled. The more usual metal – insulator transition appears when the distance between successive delocalised states at the Fermi level becomes comparable with the thermal energy. This is a somewhat gradual cross over, except at very low temperatures. And I feel no convincing evidence can be drown from change in the screening of deep levels, which should be much more sensitive to changes in local initial potentials with the coordination numbers of the atoms concerned.

A third type of metal insulator transition can in principle come from increased correlations. In most metals, correlations can be fully accounted for in ab initio computations using fully blown configuration interactions techniques. But for large clusters, some short cut is necessary. A possible way is to use a simplified Hubbard model, where the effective 'Coulomb' interaction between electrons on the same atom, U, is renormalised to take into account intra-atomic correlations. This approximation, standard in solids, can be used either neglecting, with respect to the transfer integrals t, terms in U larger than the first (Hartree Fock), or keeping the U^2 terms, or treating U in an approximate variational way (Gutzwiller or Hubbard). This renormalisation explains why $|U/t|$ remain modest in most metals, despite the large (intra-atomic) correlations present in CI computations. An approximation of this kind would be especially useful to study transitional metal clusters, where $|U/t|$ is still small but sizeable enough to lead to observable correlation effects: magnetism in the $3d$ series or absence of even odd alternation for instance.

Not many studies of covalent structures have appeared in this meeting, although the situation for C and Si seems much clearer for small aggregates. It seems clear now that chains with π bondings dominate in the small carbon clusters and explain, as in metals, the alternations in stability and ionisation. The bending of linear chains into cyclic ones, perhaps already present for some short chains, but taking over for chains larger than 9 to 10 atoms seems also convincing. What remains to be studied in the same details are the intermediary configurations up till the famous C_{60} 'footballon', which can be looked at as a spherical $2d$ graphite like π bonding, and what occurs beyond. There is no doubt of a special stability of π bonding in Carbon, which subsists up to macroscopic crystals. In contrast, Si aggregates are $3d$ from very small sizes, with to start with compact configurations which do *not* fit either σ or π classical bonding. It is probably more useful to think of them as liquid like droplets, if one remember that Si (as Ge) is metallic in the liquid phase, with a number of first neighbours near to 6, showing a dominance of p bonding. It would be quite interesting to see at what critical size one observes the crossover to σ bonding. Preliminary studies had suggested that the icosahedral cluster with 20 atoms should be fairly stable. But it has an important number of broken bonds, so might be only stabilised by adsorption of say H.

Ionic clusters show, in their low temperature state, no such surprising variety of configurations. Indeed they look mostly as small chips from crystals. However, the thermal excitation of small ionic clusters into closed loops of ions of alternate size has been, I think, one of the great surprises in this field. As emphasised by Jortner, here is a high temperature energy state with no soft mode, no diffusion, no melting process!

C. More complex problems

It is clear that the finite size of the clusters is bound up with a number of more complex problems which are only very partly solved so far.

In situations that evolve rapidly in time (jets) or where clusters are perturbed by a substrate, it is not clear in what conditions the concepts of thermodynamics apply. Is one ever near equilibrium conditions? and in that case, what is the temperature of the cluster? are most often still open questions. Bound up with this is the fact that clusters have usually a large but finite number of isomers, so that one falls between the cases of small molecules with a small number of degrees of liberty and of macroscopic matter where statistical averages can be taken.

The fact that life with clusters is complex and usual concepts do not necessarily apply have appeared clearly in some instances studied in depth. One of the clearest is the study of isomerisation, surface melting and surface wetting of Ar on flat organic molecules. But the partial phase change near the surface of SF_6 clusters is another.

All problems where the size of the cluster is smaller than the 'coherence length' of the phenomenon studied are also leading necessarily to fluctuating regimes. Magnetism, superconduction and superfluidity are well known cases which, however, could be usefully revisited. But the case of internally adsorbed electrons in molecular clusters is another beautiful example.

Finally all chemical reactions on clusters pose complex problems which of course involve these complications.

D. Future tendancies?

It seems to me that *basic* studies on simple clusters must still go on. To take two examples, no systematics exist so far for transitional metals, or for III V compounds. Little is known of the kinetics of aggregation.

However it is also clear that our basic knowledge and our equipment are now ready to attack in a fundamental way some problems of *chemical reactions*, on size selected clusters, either in flight or supported, varying T, p in controlled ways, and studying the kinetics as well as the end products. Here, the first aim, I feel, is to deepen some typical cases, such as the ones shown in this meeting, more than to skim the surface of the many possible cases.

Two more fields should be mentioned. Despite their potential usefulness, I do not feel they are so urgent for our community. The first is the problem of *dendritic aggregation kinetics*, leading to fractal geometries of clusters. This topic, very fashionable with theoreticians using big computers, is only met in practice in somewhat extreme cases; and the end products are very sensitive to the details of the physics of aggregation.

The second is the various problems involved in *aggregation of clusters* into thin films or 3d solids. Sintering, recrystallisation are here standard problems, together with the taking into account of cooperative phenomena in physical properties of each cluster = optical, transport, magnetic, superconductive aspects have all been more or less treated in ways that again depend somewhat on the topology of aggregation. For very small clusters, these problems merge into those connected with amorphous structures, quasicrystals and with some complex crystalline structures based on clusters (Frank and Kasper phases, Chevrel phases . . .). Despite their intrinsic and practical interests, I feel that most of these various fields are too far removed from that of the present conference to be usefully discussed by the same people.

E. Future conferences

It is felt that the field of cluster is presently active enough to justify one meeting every two years, and that these meetings should not be necessarily restricted to Europe.

As a result, it has been decided to have the next meeting in Great Britain in 1990, in Brighton (Professor Stace), but in 1992 the meeting should be held in the States, possibly in Chicago. Japan should also be considered as one of the next possibilities.

Author Index

Indexed in *Current Contents*
Evaluated and abstracted
for PHYS on STN

MIX
Papier aus verantwortungsvollen Quellen
Paper from responsible sources
FSC® C105338

If you have any concerns about our products,
you can contact us on
ProductSafety@springernature.com

In case Publisher is established outside the EU,
the EU authorized representative is:
Springer Nature Customer Service Center GmbH
Europaplatz 3, 69115 Heidelberg, Germany

Printed by Libri Plureos GmbH
in Hamburg, Germany